T0263677

INSECT ENDOCRINOLOGY

INSECT ENDOCRINOLOGY

EDITED BY

LAWRENCE I. GILBERT

Department of Biology
University of North Carolina
Chapel Hill, NC

AMSTERDAM • BOSTON • HEIDELBERG • LONDON • NEW YORK • OXFORD
PARIS • SAN DIEGO • SAN FRANCISCO • SINGAPORE • SYDNEY • TOKYO
Academic Press is an imprint of Elsevier

Academic Press is an imprint of Elsevier
32 Jamestown Road, London NW1 7BY, UK
225 Wyman Street, Waltham, MA 02451, USA
525 B Street, Suite 1800, San Diego, CA 92101-4495, USA

First edition 2012

Copyright © 2012 Elsevier B.V. All Rights Reserved

No part of this publication may be reproduced, stored in a retrieval system
or transmitted in any form or by any means electronic, mechanical, photocopying,
recording or otherwise without the prior written permission of the publisher
Permissions may be sought directly from Elsevier's Science & Technology Rights
Department in Oxford, UK: phone (+ 44) (0) 1865 843830; fax (+44) (0) 1865 853333;
email: permissions@elsevier.com. Alternatively, visit the Science and Technology Books website at
www.elsevierdirect.com/rights for further information

Notice
No responsibility is assumed by the publisher for any injury and/or damage to persons
or property as a matter of products liability, negligence or otherwise, or from any use or
operation of any methods, products, instructions or ideas contained in the material herein.
Because of rapid advances in the medical sciences, in particular, independent verification of diagnoses
and drug dosages should be made

British Library Cataloguing-in-Publication Data
A catalogue record for this book is available from the British Library

Library of Congress Cataloging-in-Publication Data
A catalog record for this book is available from the Library of Congress

ISBN: 978-0-12-384749-2

For information on all Academic Press publications
visit our website at elsevierdirect.com

Typeset by TNQ Books and Journals Pvt Ltd.
www.tnq.co.in

CONTENTS

PREFACE

Some years have passed since the seven volume series *Comprehensive Molecular Insect Science* was published. One volume, *Insect Endocrinology*, was dedicated to my favorite field of research. The manuscripts for that volume were written in 2003 and 2004 and of the many references, the great majority were from 2003 and earlier. The series did very well and chapters were cited quite frequently, although, because of the price and the inability to purchase single volumes, the set was purchased mainly by libraries. In 2010 I was approached by Academic Press to think about bringing two major fields up to date with volumes that could be purchased singly, and would therefore be available to faculty members, scientists in industry and government, postdoctoral researchers, and interested graduate students. I chose *Insect Endocrinology* for one volume because of the remarkable advances that have been made in that field in the past half dozen years.

With the advice of several outside advisors, we chose the central chapters from the original series plus two additional chapters on insect peptide hormones. The volume includes articles on the juvenile hormones, circadian organization of the endocrine system, ecdysteroid chemistry and biochemistry, as well as new chapters on insulin-like peptides and the peptide hormone Bursicon. The original chapters have either been revised with additional material published in the last few years or for many, completely rewritten. All of the authors are among the leaders in their fields of interest and have worked hard to contribute the best chapters possible. I believe that this volume is among the very best I have edited in the past half century, and that it will be of great help to senior and beginning researchers in the fields covered. Further, it is competitively priced so that individuals and laboratories can have their own copies.

LAWRENCE I. GILBERT

CONTRIBUTORS

Michael Adams
Depts. of Entomology and Cell Biology &
Neuroscience, University of California, Riverside,
CA, USA

Yevgeniya Antonova
Department of Entomology, University of Arizona,
Tucson, AZ, USA

Anam J. Arik
Department of Entomology, University of Arizona,
Tucson, AZ, USA

Gary J. Blomquist
Department of Biochemistry, University of
Nevada, Reno, Reno, NV, USA

François Bonneton
Université de Lyon, Université Lyon 1, ENS de
Lyon, IGFL, CNRS, INRA, Lyon, France

Mark R. Brown
Department of Entomology, University of
Georgia, Athens, GA, USA

Geoffrey M. Coast
Birkbeck College, London, UK

Michel Cusson
Laurentian Forestry Centre, Canadian Forest
Service, Natural Resources Canada, Quebec
City, QC, Canada

C. Dauphin-Villemant
Université Pierre et Marie Curie, Paris, France

David L. Denlinger
Departments of Entomology and Evolution,
Ecology and Organismal Biology, Ohio State
University, Columbus, OH, USA

Douglas J. Emlen
Division of Biological Sciences, University of
Montana, Missoula, MT, USA

Walter Goodman
Department of Entomology,
University of Wisconsin-Madison, Madison,
WI, USA

Klaus Hartfelder
Departamento de Biologia Celular e Molecular e
Bioagentes Patogênicos, Faculdade de Medicina
de Ribeirão Preto, Universidade de São Paulo,
Ribeirão Preto, SP, Brazil

Vincent C. Henrich
Center for Biotechnology, Genomics, and
Health Research, University of North Carolina-
Greensboro, Greensboro, NC, USA

Frank M. Horodyski
Ohio University, Athens, OH, USA

Russell Jurenka
Department of Entomology, Iowa State
University, Ames, IA, USA

Rene Lafont
Université Pierre et Marie Curie, Paris, France

Vincent Laudet
Université de Lyon, Université Lyon 1, ENS de
Lyon, IGFL, CNRS, INRA, Lyon, France

Wendy Moore
Department of Entomology, University of Arizona,
Tucson, AZ, USA

H. Rees
University of Liverpool, Liverpool, UK

Michael A. Riehle
Department of Entomology, University of Arizona, Tucson, AZ, USA

Joseph P. Rinehart
Insect Genetics and Biochemistry Research Unit, USDA-ARS Red River Valley Agricultural Research Center, Fargo, ND, USA

Robert Rybczynski
Department of Psychiatry and Behavioral Sciences, Duke University Medical Center, Durham, NC, USA

Coby Schal
Department of Entomology, North Carolina State University, Raleigh, NC, USA

David A. Schooley
University of Nevada, Reno, NV, USA

Wendy Smith
Department of Biology, Northeastern University, Boston, MA, USA

Qisheng Song
Division of Plant Sciences-Entomology, University of Missouri, Columbia, MO, USA

Claus Tittiger
Department of Biochemistry, University of Nevada, Reno, Reno, NV, USA

J.T. Warren
University of North Carolina, Chapel Hill, NC, USA

George D. Yocum
Insect Genetics and Biochemistry Research Unit, USDA-ARS Red River Valley Agricultural Research Center, Fargo, ND, USA

Dusan Zitnan
Institute of Zoology, Slovak Academy of Sciences, Bratislava, Slovakia

1 Prothoracicotropic Hormone

Wendy Smith
Northeastern University, Boston, MA, USA
Robert Rybczynski
Duke University Medical Center, Durham, NC, USA

© 2012 Elsevier B.V. All Rights Reserved

Summary

This chapter focuses on prothoracicotropic hormone (PTTH), a brain neuropeptide hormone that stimulates the secretion of the molting hormone, ecdysone, from the prothoracic glands. Physiological, biochemical, and genetic studies on PTTH have shaped our current understanding of how insect molting and metamorphosis are controlled. This chapter reviews the discovery of PTTH followed by sections describing PTTH structure, regulation of PTTH release, and the cellular mechanisms by which PTTH stimulates ecdysteroid secretion. It ends with a brief discussion of developmental changes in the prothoracic glands and of roles played by factors aside from PTTH, which have modified our current picture of the regulation of ecdysteroid secretion in insects.

1.1. General Introduction and Historical Background

Many invertebrates, including insects, undergo striking and periodic post-embryonic changes in rates of development, morphology, and physiology. The physical growth of insects is restricted by hardened portions of the cuticle; insects solve this problem by periodic molts in which the old cuticle is partly resorbed and partly shed and a new cuticle is laid down, allowing for further growth. In addition, many insects undergo metamorphic molts in which significant changes in structure occur. A suite of hormones controls and coordinates these episodes of rapid developmental change. Ecdysteroids, polyhydroxylated steroids such as 20-hydroxyecdysone (20E) and ecdysone (E), are major hormonal regulators of molting

DOI: 10.1016/B978-0-12-384749-2.10001-9

and metamorphosis. For the actual pathway of ecdysteroid synthesis from cholesterol, which occurs in the prothoracic glands or homologous cells in the dipteran ring gland, see Chapter 4. This chapter focuses on PTTH, a brain neuropeptide hormone that plays a pivotal role in regulating the activity of the prothoracic glands.

The first section of this chapter reviews the field of PTTH research from the 1920s to the early 1980s. For further and more detailed pictures of this time period, see reviews by Ishizaki and Suzuki (1980), Gilbert *et al.* (1981), Granger and Bollenbacher (1981), Ishizaki and Suzuki (1984), and Bollenbacher and Granger (1985). Lepidopteran insects dominate this portion of the chapter, as it was from moths such as *Bombyx mori* (domestic silkmoth) and *Manduca sexta* (tobacco hornworm) that much of our knowledge of PTTH was derived. Subsequent sections of this chapter describe the structure of PTTH, regulation of PTTH release, and cellular mechanisms by which PTTH stimulates ecdysteroid secretion. The chapter ends with a discussion of developmental changes in the prothoracic glands and a brief review of their regulation by factors other than PTTH. In these sections, while Lepidoptera remain informative models, *Drosophila melanogaster* plays an increasingly important role in understanding peptide-regulated ecdysteroid secretion.

The first experiments demonstrating control of insect metamorphosis by the brain were conducted by Polish biologist Stefan Kopec (1922), using the gypsy moth, *Lymantria dispar*. Kopec employed simple techniques such as the ligature of larvae into anterior and posterior sections and brain extirpation. His data demonstrated that the larval–pupal molt of the gypsy moth was dependent upon the brain during a critical period in the last instar larval stage. After this stage, head ligation failed to prevent pupation. Kopec (1922) also found that this control was not dependent on intact nervous connections with regions posterior to the brain. The studies of Kopec represent the beginning of the field of neuroendocrinology, but their importance was not recognized at the time. Later workers, notably Wigglesworth (1934, 1940), took similar approaches with another hemipteran insect, *Rhodnius prolixus*, and demonstrated again that the brain was important for larval molting at certain critical periods and that the active principle appeared to be produced in a part of the brain containing large neurosecretory cells.

At the same time, other studies, again using simple techniques such as transplantation and transection, found that the brain-derived factor was not sufficient to stimulate molting and metamorphosis. Rather, a second component produced in the thorax was necessary (Burtt, 1938; Fukuda, 1941; Williams, 1947). This component originated in the prothoracic gland or ring gland and was later shown to be an ecdysteroid precursor to 20E; that is, 3-dehydroecdysone (3dE) in Lepidoptera (Kiriishi *et al.*,

1992), or E and makisterone A in *Drosophila* (Pak and Gilbert, 1987; see Chapter 4).

The chemical nature of PTTH was addressed in studies beginning in 1958, when Kobayashi and Kimura (1958) found that high doses of an ether extract of *B. mori* pupal brains induced adult development in a *Bombyx* dauer pupa assay. This result suggested that PTTH was a lipoidal moiety, such as a sterol. In contrast, Ichikawa and Ishizaki (1961) found that a simple aqueous extract of *Bombyx* brains could elicit development in a *Samia* pupal assay, an observation not supporting a lipoidal nature for PTTH, and in a later study, they provided direct evidence that *Bombyx* PTTH was a protein (Ichikawa and Ishizaki, 1963). Schneiderman and Gilbert (1964), using the *Samia* assay, explicitly tested the possibility that PTTH was cholesterol or another sterol and concluded that a sterol PTTH was highly unlikely. During the 1960s and 1970s, data accumulated from a number of studies that indicated *Bombyx* PTTH was proteinaceous in nature, but attempts to ascertain the molecular weight yielded at least two peaks of activity, one $\leq 5\,kDa$ and a second $\geq 20\,kDa$ (Kobayashi and Yamazaki, 1966).

In the 1970s and 1980s, a second lepidopteran, the tobacco hornworm *M. sexta*, gained favor as a model organism for the elucidation of the nature of PTTH. Initial experiments on PTTH with *Manduca* utilized a pupal assay (Gibbs and Riddiford, 1977), but a system measuring ecdysteroid synthesis by prothoracic glands *in vitro* soon gained prominence (Bollenbacher *et al.*, 1979). Using the pupal assay, Kingan (1981) estimated the molecular weight of *Manduca* PTTH to be $\approx 25,000$. A similar estimate (MW $\approx$ 29,000) resulted from assays employing the *in vitro* technique, but this approach also provided evidence for a second, smaller PTTH (MW = 6000–7000) (Bollenbacher *et al.*, 1984).

The site of PTTH synthesis within the brain was first investigated by Wigglesworth (1940) who, by implanting various brain subregions into host animals, identified a dorsal region of the *Rhodnius* brain containing large neurosecretory cells as the source of PTTH activity. Further studies, in species such as *Hyalophora cecropia*, *Samia cynthia*, and *M. sexta*, also provided evidence that PTTH was produced within regions of the dorsolateral or mediolateral brain that contained large neurosecretory cells. This methodology was refined and ultimately culminated in the identification of the source of PTTH as two pairs of neurosecretory cells (one pair per side) in the brain of *Manduca* (Agui *et al.*, 1979).

The corpus cardiacum (CC) of the retrocerebral complex was originally suggested to be the site of PTTH release based on structural characteristics of the organ and the observation that axons of some cerebral neurosecretory cells terminate in the CC. However, physiological studies involving organ transplants and extracts implicated the corpus allatum (CA), the other component of the retrocerebral complex, as the site of PTTH release. *In vitro* studies utilizing isolated CCs and CAs from *Manduca* revealed that the CAs contained considerably more PTTH than CCs (Agui *et al.*, 1979, 1980). This result was also obtained when high levels of K+ were used to stimulate PTTH release from isolated CCs and CAs; K+ caused a severalfold increase in PTTH release from both organs, but the increase from CCs barely matched the spontaneous release from CAs (Carrow *et al.*, 1981). PTTH activity in the CC could be explained by the fact that axons that terminate in the CA traverse the CC and isolation of the CC would carry along any PTTH caught in transit through this organ. Additional evidence that the CAs are the neurohemal organ for the PTTH-producing neurosecretory cells was the observation that CA PTTH content correlated positively with brain PTTH content (Agui *et al.*, 1980).

A variety of studies tried to determine when molt-stimulating PTTH peaks occurred and what factors influenced the release of PTTH into the hemolymph. Wigglesworth (1933) performed pioneering studies on the control of *Rhodnius* PTTH release. He found that distension of the intestinal tract was the most important determinant of blood-meal-dependent molting. Multiple small meals were not effective, but artificially increasing distension by blocking the anus with paraffin notably decreased the size of a blood meal required. A meal of saline can also trigger (but not sustain) a molt, indicating that for *Rhodnius*, nutritional signals are not critical (Beckel and Friend, 1964). Meal-related distension appears to be sensed by abdominal stretch or pressure receptors in insects such as *Rhodnius* (Wigglesworth, 1934). This type of regulation is not surprising in blood-sucking insects that rapidly change the fill state of their gut, but somewhat surprisingly, it appears also to occur in some insects that are continuous rather than episodic feeders, such as the milkweed bug *Oncopeltus fasciatus* (Nijhout, 1979). In this species, subcritical weight nymphs inflated with air or saline will commence molting.

In many other insects, the physical and physiological conditions that permit and control PTTH release are subtle and incompletely known. In *Manduca*, it appears that a critical weight must be achieved before a molt is triggered (Nijhout, 1981). However, this weight is not a fixed quantity; rather the critical weight is a function of the weight and/or size at the time of the last molt. Animals that are very small or very large at the fourth to fifth larval instar molt will exhibit critical weights smaller or larger than the typical individual. Further complications revolve around the element of time. Animals that remain below critical size for a long period may eventually molt to the next stage (*Manduca*: Nijhout, 1981), or undergo stationary larval molts without growth (*Galleria mellonella*: Allegret, 1964). PTTH release also appears to be a "gated" phenomenon, occurring only during a specific time of day and only in animals that meet critical criteria, such as size, weight, or allometric relationships between structures or physiological variables. Individuals that reach critical size within a diel cycle but after that day's gate period will not release PTTH until the next gate on the following day (Truman and Riddiford, 1974).

The concept of a critical size or similar threshold parameter that must be met to initiate PTTH release implies that there will be a single PTTH release per larval instar. This appears to be true through the penultimate larval instar, but careful assessment of PTTH activity in *Manduca* hemolymph revealed two periods of significant levels of circulating PTTH during the last larval instar (Bollenbacher and Gilbert, 1981). In this species, three small PTTH peaks were seen about four days after the molt to the fifth instar while a single larger peak occurred about two days later. The first period of PTTH release is gated, but the second appears to be neither gated nor affected by photoperiod (Bollenbacher, unpublished data cited in Bollenbacher and Granger, 1985). These two periods of higher circulating PTTH correspond to the two head critical periods (HCPs) that are found in the last larval instar of *Manduca*. HCPs represent periods of development when the brain is necessary for progression to the next molt, as determined by ligation studies. In the fifth (last) larval instar of *Manduca*, the two periods of higher hemolymph titers of PTTH activity immediately precede and partially overlap two peaks in circulating ecdysteroids (Bollenbacher and Gilbert, 1981). The first small ecdysteroid peak determines that the next molt will be a metamorphic one (the commitment peak: Riddiford, 1976) while the second large peak actually elicits the molt.

The existence of daily temporal gates for PTTH release suggested that a circadian clock plays a role in molting and metamorphosis (Vafopoulou and Steel, 2005). Such a clock does appear to function in *Manduca*, because shifting the photoperiod shifted the time of the gate (Truman, 1972). Furthermore, the entrainment of this circadian cycling was not dependent on retinal stimulation because larvae with cauterized ocelli still reacted to photoperiodic cues (Truman, 1972). A more overarching effect of the photoperiod can be seen in the production of diapause (see Chapter 10). For instance, in *Antheraea* and *Manduca*, which exhibit pupal diapause, short days experienced during the last larval instar result in the inhibition of PTTH release during early pupal–adult development (Williams, 1967; Bowen *et al.*, 1984a); the brain PTTH content of these diapausing pupae and that of non-diapausing pupae are similar, at least at the beginning of the pupal stage. Temperature often plays an important role in breaking diapause and an elevation of temperature has been interpreted to cause the release of PTTH (Bollenbacher and Granger, 1985), with a resultant rise in circulating ecdysteroids that triggers the resumption of development.

The interposition of hormonal signals between light-cycle sensing and the inhibition of PTTH release and/or synthesis can involve the juvenile hormones (JHs). In species exhibiting larval diapause, such as *Diatraea grandiosella*, the persistence of relatively high JH levels controls diapause (Yin and Chippendale, 1973). Ligation experiments indicated that the head was involved in JH-related larval diapause, resulting in the suggestion that JH inhibited the release of an ecdysiotropin, that is, PTTH (Chippendale and Yin, 1976).

In summary, by the 1980s, PTTH had progressed from being an undefined brain factor controlling molting (Kopec, 1922) to being identified as one or two brain neuropeptides that specifically regulated ecdysteroid synthesis by the prothoracic gland (Bollenbacher *et al.*, 1984) that was synthesized in a very limited set of brain neurosecretory cells (Agui *et al.*, 1979). In the remaining portion of this chapter the chemical characterization and purification of PTTH will be detailed, as will the intracellular signaling pathways by which PTTH activates the prothoracic gland, and there will be a brief discussion of other regulators of prothoracic gland steroidogenesis. This phase of PTTH research has involved a combination of genetic, biochemical, and physiological approaches, including more prevalent use of *Drosophila* as a model for the study of insect steroidogenesis (Nagata *et al.*, 2005; Mirth and Riddiford, 2007; Huang *et al.*, 2008; De Loof *et al.*, 2008; Marchal *et al.*, 2010).

1.2. Prothoracicotropic Hormone: Characterization, Purification, and Cloning

1.2.1. Characterization, Purification, and Cloning of Lepidopteran PTTH

As described previously, evidence suggested that for *B. mori* and *M. sexta*, PTTH activity resided in proteins exhibiting two distinct size ranges — one 4–7 kDa and the second 20–30 kDa. Early work with *Bombyx* brain extracts utilized a cross-species assay to determine PTTH activity. This assay, using debrained *Samia* pupae, suggested that the most effective *Bombyx* PTTH (4 K-PTTH) was the smaller protein. However, as efforts to purify and further characterize *Bombyx* PTTH progressed, a *Bombyx* pupal assay was developed (Ishizaki *et al.*, 1983a). This assay revealed that when *Bombyx* PTTH preparations were tested with debrained *Bombyx* pupae, *Bombyx* PTTH appeared to be the larger molecule whereas the small PTTH, identified with the *Samia* assay, was not effective at physiologically meaningful doses (Ishizaki *et al.*, 1983a,b). This result was confirmed when the small and large PTTHs were tested for their ecdysteroidogenic effect *in vitro* on *Bombyx* prothoracic glands (Kiriishi *et al.*, 1992).

The effort to isolate and sequence *Bombyx* PTTH shifted rapidly to the "large" PTTH form following the 1983 results; the smaller molecule, subsequently named bombyxin, is discussed briefly in Section 1.5.2., and at length in Chapter 2. Within a few years, the sequence of the amino terminus (13 amino acids) of *Bombyx* PTTH (called 22 K-PTTH) was obtained after a purification effort that involved processing 500,000 adult male heads (Kataoka *et al.*, 1987). The final steps of this purification involved HPLC separations of proteins, and this technique suggested that PTTH might exhibit heterogeneity in sequence and/or structure. However, the nature of that heterogeneity was unknown, since only one of four potential PTTH peaks was sequenced. In addition to this limited amount of amino acid sequence, this study is significant since it demonstrated that *Bombyx* PTTH was indeed a protein, and one that had no sequence homology to other proteins known at that time, including bombyxin. Also significant was the finding that PTTH was highly active when injected into brainless *Bombyx* pupae. A dose of only 0.11 ng was as effective as 100 to 400 ng of bombyxin, thus demonstrating again that bombyxin was not the *Bombyx* PTTH (Kataoka *et al.*, 1987). A second round of *Bombyx* PTTH purification was initiated, using 1.8 million heads (derived from 1.5 tons of moths), and this heroic effort resulted in the determination of nearly the entire PTTH amino acid sequence (104 amino acids), except for a small fragment (5 amino acids) at the carboxy terminus (Kataoka *et al.*, 1991). At the same time, the amino terminal sequence was used to produce a polyclonal mouse antibody against *Bombyx* PTTH, which was then successfully utilized in a cDNA expression library screening (Kawakami *et al.*, 1990). The empirically determined and deduced amino acid sequences of *Bombyx* PTTH agreed completely in the region of overlap. Furthermore, the amino acid sequence deduced from the cDNA sequence indicated that, like many neuropeptides, *Bombyx* PTTH appeared to be synthesized as a larger, precursor molecule (**Figure 1A**). Two slightly smaller sequences of PTTH were also obtained during PTTH purification (Kataoka *et al.*, 1991) and were missing six or seven amino acids from the N-terminal sequence obtained from the largest peptide. It is not known if this amino terminus heterogeneity was an artifact of sample storage or purification methods, or reflected a natural variation with possible biological effects on activity. A more detailed analysis of the amino acid sequence and structure of *Bombyx* PTTH and other PTTHs is provided in Section 1.2.2.

The large number of brains required to purify PTTH has precluded attempts to purify this protein from species other than *Bombyx*, with the exception of *M. sexta*. For *Manduca*, Bollenbacher and colleagues raised a monoclonal antibody against a putative purified big PTTH (O'Brien *et al.*, 1988) and used this antibody to construct an affinity column. This immunoaffinity column was then

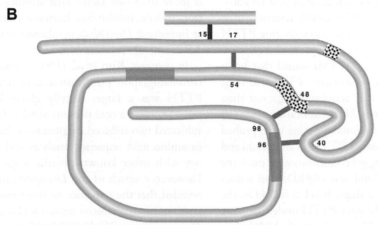

A

Signal Peptide   p2k   p6k

MITRPIILVILCYAILMIVQSFVPKAVAL KRKPDVGGFMVEDQRTHKSHNYMMK RARNDVLGDKE

NVRPNPYYTEPFDPDTSPEELSALIVDYANMIRNDVILLDNSVETRT RKRGNIQVENQAIPDPPC
                                                                 15

TCKYKKEIEDLGENSVPRFIETRNC NKT QQPTCRPPYICKESLYSITILKRRETKSQESLEIPNE
17   40   48   54

LKYRWVAESHPVSVACLCTRDYQLRYNNN
              96 98

B

Figure 1 (A) Amino acid sequence and main features of *Bombyx* PTTH. Dashed underlining: the putative signal peptide region; thin solid underlining: the mature, bioactive peptide; thick underlining: predicted peptide cleavage sites; light shading: putative 2 kDa peptide; dark shading: putative 6 kDa peptide; boxed amino acids: glycosylation site; *: cysteines involved in intra-monomeric bonds; and Δ: cysteine involved in dimer formation. Based on data from Kawakami *et al.* (1990) and Ishibashi *et al.* (1994). (B) The structure of mature PTTHs based on the *Bombyx* structure elucidated by Ishibashi *et al.* (1994). Shading indicates predicted hydrophobic regions in many (stippled) or all (dark) lepidopteran PTTH sequences (Rybczynski, unpublished analysis). Numbers indicate the positions of the cysteines involved in intra- and inter-monomeric bonds in the mature *Bombyx* PTTH sequence.

used to attempt a purification of *Manduca* PTTH from an extract made from about 10,000 brains (Muehleisen *et al.*, 1993). Partial amino acid sequence obtained from a protein purified via this affinity column showed no sequence homology with *Bombyx* PTTH; instead, the isolated protein appeared to be a member of the cellular retinoic acid binding protein family (Muehleisen *et al.*, 1993; Mansfield *et al.*, 1998). This surprising result was in direct contrast to the finding of Dai *et al.* (1994) that an anti-*Bombyx* PTTH antibody recognized only the cells previously identified as the *Manduca* prothoracicotropes by Agui *et al.* (1979), and the observation that *Manduca* PTTH could be immunoprecipitated by an anti-*Bombyx* PTTH antibody (Rybczynski *et al.*, 1996). In the latter case, the immunoprecipitated *Manduca* PTTH could be recovered in a still-active state (Rybczynski *et al.*, 1996). Subsequently, the isolation of a *Manduca* PTTH cDNA, utilizing a *Bombyx* probe, and expression of the derivative recombinant *Manduca* PTTH, confirmed that *Manduca* PTTH was indeed related to the *Bombyx* molecule (Gilbert *et al.*, 2000; Shionoya *et al.*, 2003). In an earlier molecular approach, Gray *et al.* (1994) used *Bombyx* PTTH-derived sequences to probe *Manduca* genomic DNA and a neuronal cDNA library. This search yielded an intron-less

sequence differing in only two amino acids from *Bombyx* PTTH but quite different from the *Manduca* PTTH later demonstrated to exhibit bioactivity (Gilbert *et al.*, 2000); note that the gene for *Bombyx* PTTH contains five introns (Adachi-Yamada *et al.*, 1994). This unexpected result, suggesting that *Manduca* may have a second large PTTH, has not been further pursued. Evidence suggesting that the *Manduca* brain might express a small PTTH in addition to the larger peptide is discussed in Section 1.5.2.

Preliminary determinations of the molecular weight of PTTH have been made for a few additional lepidopteran species such as the gypsy moth *L. dispar* (Kelly *et al.*, 1992; Fescemyer *et al.*, 1995), primarily using *in vitro* assays of ecdysteroidogenesis and variously treated brain extracts. These studies revealed two molecular weight ranges for PTTH, such as small and large. However, the molecular weight obtained for the large *Lymantria* PTTH (11,000–12,000) varied notably from those determined for the purified and cloned forms such as *Bombyx* PTTH (native weight ≈ 25,000–30,000). This difference is probably a function of the sample treatment (acidic organic extraction vs. aqueous extraction) and molecular weight analysis methods (HPLC vs. SDS-PAGE), since when essentially the same extraction and HPLC methodology were applied to *Manduca* brain extracts, a similarly low molecular weight determination for "big" PTTH was obtained (11,500; Kelly *et al.*, 1996).

1.2.2. Characterization, Purification, and Cloning of Non-Lepidopteran PTTH

Despite the importance of the hemipteran *Rhodnius* in early studies demonstrating a brain-derived prothoracicotropic factor (Wigglesworth, 1934, 1940), relatively little work has been done on molecularly characterizing PTTH activity from species other than Lepidoptera and Diptera. In *Rhodnius*, Vafopoulou *et al.* (1996) found that brain extracts contained a protease-sensitive PTTH activity. Later work suggested that this activity was greater than 10 kDa when immunoblots of *Rhodnius* brain proteins or medium from *in vitro* incubations of brains were probed with an antibody against *Bombyx* PTTH (Vafopoulou and Steel, 2002). If non-reducing gel conditions were used, the strongest immunoreactive band was ≈68 kDa, but reducing conditions yielded only a single band at ≈17 kDa, the same as that calculated for *Bombyx* PTTH under reducing electrophoresis conditions (Mizoguchi *et al.*, 1990). It is also very similar to those determined for *Manduca* PTTH, again employing the *Bombyx* antibody after reducing SDS-PAGE (≈16 kDa; Rybczynski *et al.*, 1996), and for the *Antheraea* PTTH, using an *Antheraea* PTTH antibody and strong reducing conditions (≈15 kDa; Sauman and Reppert, 1996a). Also, as demonstrated earlier for *Manduca* brain extract (Rybczynski *et al.*, 1996), an anti-*Bombyx* PTTH antibody removed PTTH activity from a *Rhodnius* brain extract (Vafopoulou and Steel, 2002).

The widespread use of *D. melanogaster* as a model organism for the study of development has only recently been accompanied by a concomitant increase in our knowledge of *Drosophila* PTTH. A gene for *Drosophila* PTTH was identified based on homology with lepidopteran PTTH (Rybczynski, 2005), and has since been cloned as discussed in the next section (McBrayer *et al.*, 2007). In the 1980s, the small size and rapid development of *Drosophila* made it difficult to isolate PTTH using a classic endocrinological approach, for example, measuring ecdysteroid titers, accumulating large quantities of brain proteins, or isolating the ring gland for *in vitro* studies of ecdysteroid synthesis. In addition, the composite nature of the ring gland, which contains multiple cell types in addition to the prothoracic gland cells, complicated the interpretation of experiments in a way that is not seen with the moth glands. Nevertheless, an *in vitro* assay for ring gland ecdysteroid synthesis was used to search for PTTH activity from the central nervous system of *Drosophila* (Redfern, 1983; Henrich *et al.*, 1987a,b; Henrich, 1995). While recent studies indicate that PTTH transcript is found only in the brain (McBrayer *et al.*, 2007), Henrich and colleagues found PTTH activity in both the brain and the ventral

ganglion. In addition, an antibody against *Bombyx* PTTH was seen to bind specific cells both in the brain and in ventral ganglionic regions of *Drosophila* (Zitnan *et al.*, 1993). Ultrafiltration through a 10 kDa filter resulted in PTTH activity in both the filtrate and retentate (Henrich, 1995). Pak *et al.* (1992), using column chromatography, found two peaks of PTTH activity: one of ≈4 kDa and the other larger, at ≈16 kDa. The possibility exists that there is more than one factor that stimulates ecdysone secretion, indeed, insulin-like hormones have been implicated as important *Drosophila* prothoracicotropins (see Section 1.5.2. and Chapter 2).

In contrast, Kim *et al.* (1997), using a more complex chromatographic purification, concluded that *Drosophila* PTTH was a large, heavily glycosylated molecule of ≈66 kDa, with a core protein of ≈45 kDa. This group also subjected two reduced fragments of their purified PTTH to amino acid sequence analysis and found no homology with other known protein sequences at that time. However, a search of the *Drosophila* molecular databases revealed that the sequence of these two fragments match portions of the deduced amino acid sequence encoded by *Drosophila* gene BG:*DS000180.7* (unpublished analysis). This protein is cysteine rich, with a MW of ≈45,000, and contains sequence motifs that place it in the EGF protein family, thus suggesting that it is a secreted protein. Based on its sequence, the *DS000180.7* gene product has been hypothesized to be involved in cell-cell adhesion (Hynes and Zhao, 2000). As a member of the EGF family of proteins, the *DS000180.7* gene product cannot be ruled out as an intercellular signaling molecule, that is, as a prothoracicotropic hormone, without a functional test. What is clear currently is that this protein shows no detectable amino acid similarity to the demonstrated PTTHs of other insects (unpublished analysis).

Assays of *Drosophila* PTTH have been clouded by the requirement for relatively large amounts of putative PTTH material and relatively poor ecdysteroidogenic activity. Thus, in the *Drosophila* studies, Henrich and co-workers (Henrich *et al.*, 1987a,b; Henrich, 1995) and Pak *et al.* (1992) needed to use approximately eight brain-ventral ganglion equivalents of extract to stimulate ecdysteroid synthesis to 2 to 3 times basal while Kim *et al.* (1997) employed 500 ng of purified protein to achieve a similar activation. In contrast, lepidopteran PTTHs are effective at much lower doses; for instance, in *Manduca*, 0.25 brain equivalents or 0.25 to 0.5 ng of pure PTTH consistently results in a 4- to 6-fold increase in ecdysteroidogenesis (Gilbert *et al.*, 2000) and activations of steroidogenesis of 8 to 10 times basal are not rare (personal observations). Further differences between putative *Drosophila* PTTH(s) and the lepidopteran PTTHs are discussed in the next section.

Mosquitoes, such as the African malaria mosquito *Anopheles gambiae* and the yellow fever mosquito *Aedes aegypti,* are important disease vectors and have been the subject of many studies incorporating basic and applied approaches to their biology and control. In adult females of these species, a blood meal stimulates both the release of brain-derived gonadotropins like ovary ecdysteroidogenic hormone and insulin-like molecules that stimulate ovarian ecdysteroid synthesis (see Chapter 2). Pre-adult development in mosquitoes appears to depend on ecdysteroid hormones, as in other insects; however, the morphological source of ecdysteroid production is not clear. Jenkins *et al.* (1992) provided evidence that the prothoracic gland of *A. aegypti*, which is part of a composite, ring-gland-like organ, is inactive and that unknown cell types in the thorax and abdomen produce ecdysteroids. Searches of the *A. gambiae* and *Culex quinquefasciatus* genome databases with lepidopteran and putative *Drosophila* PTTH amino acid sequences revealed candidate gene products, which are discussed in the next section.

Antibodies against PTTH afford an additional tool to search for PTTH-like proteins and have been used in a number of studies in addition to those previously discussed (*Drosophila*: Zitnan *et al.*, 1993; *Manduca*: Dai *et al.*, 1994; Rybczynski *et al.*, 1996; *Rhodnius*: Vafopoulou and Steel, 2002). Zavodska *et al.* (2003) used polyclonal anti-*Antheraea* PTTH antibodies in an immunohistochemical survey of the central nervous systems of 12 species of insects in 10 orders: Archaeognatha, Ephemerida, Odonata, Orthoptera, Plecoptera, Hemiptera, Coleoptera, Hymenoptera, Tricoptera, and Diptera. Positive signals were seen in all species surveyed except *Locusta migratoria* (Orthoptera), with immunoreactive cells found mainly in the protocerebrum and less commonly in the subesophageal ganglion such as found in the mayfly *Siphlonurus armatus* (Ephemerida) and damselfly *Ischnura elegans* (Odonata). The number of putative PTTH-containing cells ranged from two or three pairs (Archaeognatha, Hemiptera, and Hymenoptera) to five or more pairs (Plecoptera, Coleoptera, Tricoptera, and Diptera), and only in the stonefly *Perla burmeisteriana* (Plecoptera) was a signal detected in a potential neurohemal organ (the CC). These results are intriguing but must be interpreted with caution pending further information such as determination of the size of the recognized protein or immunoprecipitation of PTTH activity by the antibody. In this context, results obtained with an anti-*Manduca* PTTH antibody must serve as a cautionary tale. This monoclonal antibody recognized only a small number of cells in the CNS, including but not limited to the demonstrated prothoracicotropes (Westbrook *et al.*, 1993), and apparently also bound active PTTH (Muehleisen *et al.*, 1993). Yet the protein isolated by an immunoaffinity column constructed with this antibody proved to be an intracellular retinoid-binding protein (Mansfield *et al.*, 1998) and not a secreted neurohormone.

1.2.3. Sequence Analyses and Comparisons of PTTHs

The amino acid sequence and structure of *Bombyx* PTTH have been characterized in a number of studies, beginning with Kataoka *et al.* (1987). **Figure 1** summarizes the major features of this protein. *Bombyx* PTTH appears to be synthesized as a 224 amino acid prohormone that is cleaved to yield an active peptide of 109 amino acids (Kawakami *et al.*, 1990). The first 29 residues comprise a relatively hydrophobic region that is terminated by a trio of basic amino acids, fulfilling the criteria for a signal peptide. A second trio of basic residues immediately precedes the beginning of the mature PTTH hormone. Both of these basic areas were believed to define peptide cleavage sites. Between these two sites lies a pair of basic amino acids that has been hypothesized to be an additional cleavage site (Kawakami *et al.*, 1990). The presence of these three potential peptide processing sites has led to speculation that the region between the signal peptide and the mature PTTH sequence might be cleaved into two smaller peptides and that these small molecules serve a physiological function, such as modulating JH synthesis in the CA (Kawakami *et al.*, 1990; Ishizaki and Suzuki, 1992). However, there were, and remain, no experimental data to support this speculation.

A dimeric structure for *Bombyx* PTTH was long suspected, and this was confirmed as part of the massive PTTH purification that resulted in the nearly complete amino sequence (Kataoka *et al.*, 1991). This study also indicated that PTTH was a homodimer joined by one or more disulfide bonds, and that aspargine-linked glycosylation (see **Figure 1A**) was a likely cause for the disparity between the observed monomeric molecular weight of ≈17,000 and the predicted molecular weight of ≈12,700 (Kawakami *et al.*, 1990). Further elucidation of the structure of *Bombyx* PTTH was provided by the incisive work of Ishibashi *et al.* (1994). This group used enzyme digestions of partially reduced recombinant PTTH to determine the intra- and inter-dimeric disulfide bonds (see **Figure 1B**). This recombinant PTTH, expressed in bacteria (*Escherichia coli*) without glycosylation, had about 50% of the biological activity per nanogram of native PTTH. This result indicated that glycosylation is not necessary for biological activity, although it may be required for maximum activity. *Bombyx* and other lepidopteran PTTHs (see **Figure 2**) do not have any significant homologues among vertebrate proteins, based on amino acid sequence (Kawakami *et al.*, 1990 and unpublished BLAST analysis of vertebrate protein databases). However, Noguti *et al.* (1995) showed that *Bombyx* PTTH exhibits an arrangement of its intra-monomeric disulfide bonds that is very much like that seen in some members of the vertebrate growth factor superfamily (βNGF, TGF-β2, and PDGF-BB). These workers suggested that PTTH is a member of this superfamily and that PTTH and verte-brate growth factors share a common ancestor. Structural homologies between PTTH and the embryonic ligand Trunk were identified by Marchal and co-workers (2010) in a recent review article. That same year, the PTTH receptor in *Drosophila* was demonstrated to be the same as that which binds Trunk, that is, a tyrosine-kinase-linked receptor known as Torso (Rewitz *et al.*, 2009b and dis-cussed below in Section 1.4.1.).

Additional molecular studies using probes derived from the *Bombyx* PTTH sequence have yielded lepidopteran PTTH sequences from several species: *Samia cynthia ric-ini* (AAA29964.1; Ishizaki and Suzuki, 1994); *Antheraea pernyi* (AAB05259.1; Sauman and Reppert, 1996a), *H. cecropia* (AAG10517.1; Sehnal *et al.*, 2002), *Helicoverpa zea* (AAO18190.1; Xu *et al.*, 2003), *Heliothis virescens* (AAO18191.1; Xu and Denlinger, 2003), *H. armigera* (AAP41131.1; Wei *et al.*, 2005), *M. sexta* (AAG14368.1; Shionoya *et al.*, 2003), *Spodoptera exigua* (AAT64423.2; Xu *et al.*, 2007), and *Sesamia nonagrioides* (ACV41310.1; Perez-Hedo *et al.*, 2010b). PTTH homologue sequences can also be found in GenBank for two additional lepidopterans (*Antheraea yamamai*, AAR23822.1; *H. assulta*, AAV41397.1*)*, and a coleopteran (*Tr. castaneum* EEZ99381.1). Dipteran homologues of PTTH have been identified in *D. melanogaster* (NP_608537.2; McBrayer *et al.*, 2007) and 11 additional *Drosophila* species (GenBank: *D. simulans* XP_002077659.1; *D. erecta* XP_001968193.1; *D. sechellia* XP_002041604.1; *D. yakuba* XP_002087450.1; *D. pseudoobscura* XP_001356419.2; *D. persimilis* XP_002014561.1; *D. ananassae* XP_001961552.1; *D. mojav-ensis* XP_002003636.1; *D. grimshawi* XP_001996840.1; *D. virilis* XP_002052857.1; *D. willistoni* XP_002066708.1). PTTH homologues have also been identified for the mosquito *A. gambiae* (XP_555854; Rybczynski, 2005; Marchal *et al.*, 2010; and GenBank) and *C. quinquefasciatus* (XP_001844784.1; Marchal *et al.*, 2010).

Of these homologues, only a few have been expressed and PTTH functionality confirmed. These include *in vitro* ecdysteroidogenic activity for *Bombyx* (Kawakami *et al.*, 1990) and *Manduca* (Gilbert *et al.*, 2000) prothoracic glands, stimulation of adult development in debrained *Antheraea* pupae (Sauman and Reppert, 1996a), and dia-pause termination in *H. armigera* pupae (Wei *et al.*, 2005). The functionality of the non-lepidopteran sequences is less clear. The mosquito and *Tr.* PTTHs have not been examined for activity. In *Drosophila*, genetic abla-tion of the neurons expressing putative *Drosophila* PTTH delays metamorphosis, although the larvae do ultimately metamorphose into large pupae and adults (McBrayer *et al.*, 2007). PTTH ablation also leads to reductions in ecdysone steroidogenic gene transcription, decreased expression of ecdysone-sensitive genes, and is corrected by ecdysone feeding, that is, characteristics that are in keep-ing with a functional prothoracicotropin. Nonetheless,

Drosophila recombinant PTTH expressed by a *Drosophila* cell line did not consistently stimulate ecdysteroid secre-tion by isolated ring glands. The authors suggest that, given the direct innervation of *Drosophila* ecdysone-secreting cells by PTTH-expressing neurons, *Drosophila* prothoracic glands may respond poorly to PTTH sup-plied *in vitro*. By contrast, in Lepidoptera, *in vitro* assays more closely resemble the natural delivery of PTTH via the hemolymph from the corpora allata. Alternatively, in fruit flies, ecdysone-secreting cells may require additional factors for optimal response, for example insulin-like hor-mones, and hence are less affected by exposure to a single prothoracicotropin *in vitro* or loss of a single prothoraci-cotropin *in vivo*.

Figure 2 shows the amino acid sequences of several of the PTTHs discussed earlier, representing a sample of the major phyletic groups in which it has been identified. Since the putative receptor for PTTH is the same as that for the embryonic Trunk ligand (Rewitz *et al.*, 2009b), three Trunk sequences are also provided. The lepidopteran PTTHs appear to be synthesized as prohormones that begin with a hydrophobic signal peptide sequence. Following the signal peptide, these proteins exhibit only moderate sequence identity until an area preceding the mature PTTH sequence by about 35 amino acids. This cluster of greater amino acid conservation among the lepidopteran PTTHs might indicate that this region of the prohormone indeed has a function, as proposed by Ishizaki and Suzuki (1992).

Note that the position of basic amino acids between the signal peptide and the mature hormone (see **Figure 1A**) is not identical among the lepidopteran PTTHs, and thus the 2k and 6k peptide regions hypoth-esized for *Bombyx* (Kawakami *et al.*, 1990) do not have equivalents in other moths. A variety of pro-protein con-vertases exist that recognize monobasic, dibasic, and tri-basic amino acid patterns as well as specific sites lacking basic amino acids (see Loh *et al.*, 1984; Mains *et al.*, 1990; Seidah and Prat, 2002). Without experimental evidence, it is premature to predict where the preproPTTHs might be cleaved, in addition to the known site yielding the mature monomeric unit.

The cellular location for the processing of PTTH from the longer, initial translation product to the shorter mature peptide form is not definitively known. Immunoblots of *Bombyx* and *Manduca* brain proteins revealed only the mature form, indicating that the protein is probably cleaved totally within the confines of the brain (Mizoguchi *et al.*, 1990; Rybczynski *et al.*, 1996). These observations did not reveal the intracellular site of processing, that is, discerning whether the cleavage(s) occurred completely within the soma or if cleavage occurred within axonal secretory granules, before those axons left the brain (see Loh *et al.*, 1984). It is presumed that removal of the signal sequence involves co-translational processing and takes

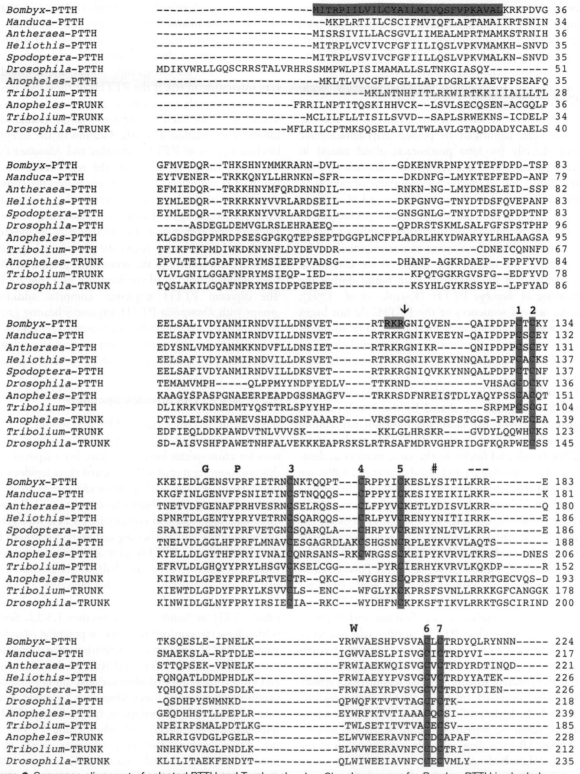

Figure 2 Sequence alignment of selected PTTH and Trunk molecules. Signal sequence for *Bombyx* PTTH is shaded green, three basic residues preceding mature PTTH are shaded blue, and the arrow indicates start of mature peptide for lepidopteran PTTH. Cysteines conserved in all PTTHs are numbered and shaded in yellow and number 1 represents cysteine participating in dimeric bonding. Amino acids conserved in all PTTH and Trunk sequences are indicated with letters; #, tyrosine conserved in all PTTH sequences; ---, tribasic sequence shared by lepidopteran PTTHs and Trunk. Accession numbers: *Anopheles gambiae* PTTH (XP_555854), *Antheraea pernyi* PTTH (AAB05259.1), *Bombyx mori* PTTH (NP_001037349.1), *Drosophila melanogaster* PTTH (NP_608537.2), *Heliothis virescens* PTTH (AAO18191.1), *Manduca sexta* PTTH (AAG14368.1), *Spodoptera exigua* PTTH (AAT64423.2), *Tr. castaneum* PTTH (EEZ99381.1), *An. gambiae* Trunk (XP_563293.3), *D. melanogaster* Trunk (NP_476767.2), *Tr. castaneum* Trunk (XP_971946.1).

place in the soma (see Loh *et al.*, 1984). The expected location of PTTH in neurosecretory particles has been confirmed by immunogold electron microscopy in both *Manduca* and *Bombyx*, using the same anti-*Bombyx* PTTH antibody (Dai *et al.*, 1994, 1995).

The mature lepidopteran PTTHs generally begin with glycine-asparagine (GN) or glycine-aspartic acid (GD) and the sequence from this glycine to the carboxy end has been confirmed to possess ecdysteroidogenic activity directly (*in vitro* prothoracic gland assays) in *Bombyx* (Kawakami *et al.*, 1990) and *Manduca* (Gilbert *et al.*, 2000), and indirectly (development of pupae) in *Antheraea* (Sauman and Reppert, 1996a) and *H. armigera* (Wei *et al.*, 2005). Proteins beginning with glycines can be modified by N-terminal myristylation, yielding a hydrophobic moiety that often serves as a membrane anchor. No evidence for this or any other N- or C-terminal post-translational modifications were found during the sequencing of *Bombyx* PTTH (Kataoka *et al.*, 1991), and the derived sequences of the PTTHs do not begin or end with the amino acids most likely to be so modified, for example, N-terminal alanines (methylation site) or serines (acetylation site) or C-terminal glycines (amidation substrate).

The mature PTTH peptides all contain seven cysteines, five of which are shared by the Trunk proteins (**Figure 2**). Given the close spacing-similarity among the sequences, it is likely that all PTTHs are linked by disulfide bonds and folded in the same manner as demonstrated for *Bombyx* PTTH (Ishibashi *et al.*, 1994; see **Figure 1B**). The lepidopteran PTTHs all contain consensus sites for N-linked glycosylation (Rybczynski, 2005). Given a consistent disparity of 4 to 5 kDa between the predicted size based on amino acid sequence and the larger size observed after SDS-PAGE, it seems likely that these proteins are indeed glycosylated (for *Bombyx*, compare Kawakami *et al.*, 1990 and Kataoka *et al.*, 1991; for *Manduca*, compare Rybczynski *et al.*, 1996 and Shionoya *et al.*, 2003; for *Antheraea*, see Sauman and Reppert, 1996a). The *Drosophila* PTTH amino acid sequence does not contain a consensus site for N-linked glycosylation but does possess one site for O-linked glycosylation within the putative mature peptide region (Rybczynski, 2005). The processed (mature) PTTH sequences are, on the whole, hydrophilic proteins, with three or four short hydrophobic regions, depending on the species. See **Figure 1B** for a diagrammatic representation of the location of these hydrophobic regions in the lepidopteran sequences. Whether or not these hydrophobic regions play significant roles in determining monomeric or dimeric structures, or participate in receptor interactions, are unanswered questions.

Similarities in the PTTH and Trunk sequences are summarized graphically in **Figure 3**, using a phylogenetic tree neighbor-joining program (ClustalW, http://www.ebi.ac.uk/tools/clustalw2/index.html; Larkin *et al.*, 2007) applied to the mature PTTH amino acid sequences. In this analysis, the Saturniid moths (*Antheraea*, *Hyalophora*, and *Samia*) form a clear-cut group. The *Hyalophora* and *Samia* sequences are intriguingly similar, and it would be very interesting to test if the PTTHs of these two species would significantly cross-activate prothoracic gland ecdysteroid synthesis. Such a comparison might enable some informed speculation as to the location of the receptor-binding regions of PTTHs. *Bombyx* and *Manduca* form their own subgroup in which the relationship of the two PTTH sequences do not closely match the conventional view of their taxonomic relationship. Yet another lepidopteran group is formed by *Spodoptera*, *Sesamia*, *Heliothis*, and *Helicoverpa*. The *Tr.* sequence comprises its own branch, which may reflect the phylogenetic position of Coleoptera relative to the other listed groups, or result from misidentification of the sequence as a true PTTH. The dipteran PTTH sequences comprise additional groups with *Drosophila* PTTH sequences bearing greater homology to each other than to mosquito sequences. As expected, the Trunk sequences form a separate branch with appropriate phylogenetic relationships within the group.

1.2.4. Are PTTHs Species Specific?

It is difficult to predict a priori if PTTHs should act across species at biologically meaningful concentrations. Actual tests for cross-species bioactivity have been equivocal on this question. Agui *et al.* (1983) tested three lepidopteran pupal brain extracts with prothoracic glands from the three species (*M. sexta*, *Mamestra brassicae*, and *B. mori*). Their data suggested that conspecific PTTHs were generally most efficacious at eliciting ecdysteroid synthesis but that heterospecific PTTHs were also effective at doses (brain equivalents) from one half to about four times that of the conspecific molecule. However, the brain extracts employed had not been size-selected to remove small PTTH or bombyxin (see Section 1.5.2.). Several studies have re-examined the issue of potential *Manduca*-*Bombyx* PTTH cross-species activation of ecdysteroidogenesis. Gray *et al.* (1994) and Rybczynski, Mizoguchi, and Gilbert (unpublished observations) did not find activation of *Manduca* ecdysteroidogenesis by crude, native (size-selected) or by recombinant (pure) *Bombyx* PTTH, respectively. Additionally, *Manduca* crude PTTH was not able to activate *Bombyx* prothoracic glands in an *in vivo* dauer assay (Ishizaki quoted in Gray *et al.*, 1994). Similarly, *Bombyx* crude PTTH did not stimulate development by debrained *Samia* pupae when separated from the much smaller bombyxin molecule originally thought to be PTTH (Ishizaki *et al.*, 1983b; see also Kiriishi *et al.*, 1992), and *Antheraea* recombinant PTTH did not activate *Manduca* prothoracic glands *in vitro* (Rybczynski and Gilbert, unpublished observations).

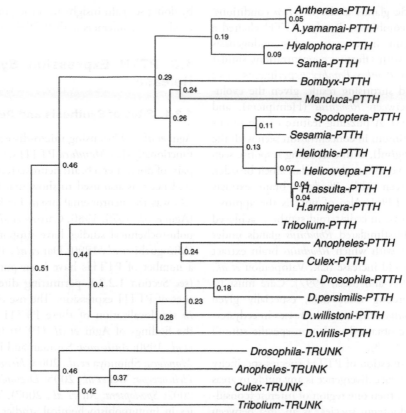

Figure 3 PTTH amino acid sequence relationships, using a phylogenetic tree neighbor-joining program (ClustalW; Larkin *et al.*, 2007, http://www.ebi.ac.uk/Tools/clustalw2/index.html) and illustrated with the Figtree graphic viewer (http://tree.bio. ed.ac.uk/software/figtree/). Accession numbers: *Anopheles gambiae* PTTH (XP_555854), *Antheraea pernyi* PTTH (AAB05259.1), *A. yamamai* PTTH (AAR23822.1), *Bombyx mori* PTTH (NP_001037349.1), *Drosophila melanogaster* PTTH (NP_608537.2), *D. persimilis* (XP_002014561.1), *D. virilis* (XP_002052857.1), *D. willistoni* (XP_002066708.1); *Culex quinquefasciatus* PTTH (XP_001844784.1), *Helicoverpa armigera* PTTH (AAP41131.1), *H. zea* PTTH (AAO18190.1), *H. assulta* PTTH (AAV41397.1), *Heliothis virescens* PTTH (AAO18191.1), *Hyalophora cecropia* PTTH (AAG10517.1), *Manduca sexta* PTTH (AAG14368.1), *Samia cynthia ricini* PTTH (AAA29964.1), *Sesamia nonagrioides* PTTH (ACV41310.1), *Spodoptera exigua* PTTH (AAT64423.2), *Tr. castaneum* PTTH (EEZ99381.1), *An. gambiae* Trunk (XP_563293.3), *Cu. quinquefasciatus* Trunk (XP_001844331.1), *D. melanogaster* Trunk (NP_476767.2), *Tr. castaneum* Trunk (XP_971946.1).

Yokoyama *et al.* (1996) tested brain extracts prepared from pupal day 0 brains of four species of swallowtail butterflies (*Papilio xuthus, P. machaon, P. bianor,* and *P. helenus*) and found cross-specific activation of prothoracic gland ecdysteroid synthesis *in vitro* with similar doses for half-maximal and maximal activation being relatively independent of the species of origin. These results are somewhat in contrast with the moth cross-species studies discussed previously and this difference may be a function of the evolutionary closeness of the *Papilio* species versus the evolutionary distance of the moth species.

The possibility of cross-species PTTH activity has also been examined using dipteran ring glands and brain extracts. Roberts and Gilbert (1986) found that ecdysteroid synthesis by ring glands from post-feeding larval *Sarcophaga bullata* was readily activated by extracts from *Sarcophaga* pre-pupal brains, and that these ring glands were not activated by *Manduca* brain extract. In contrast, *Manduca* prothoracic glands showed a stage-specific

response to *Sarcophaga* brain extract with larval but not pupal glands responding positively at relatively low doses of extract, that is, less than one brain equivalent. Henrich (1995) addressed the cross-species question with studies on the *Drosophila* ring gland. As discussed earlier, the activation of *Drosophila* ring gland ecdysteroidogenesis by *Drosophila* brain or ganglionic extracts is not robust. Nevertheless, in a small study, Henrich (1995) found that both *Manduca* small and large PTTH preparations were able to elicit increased steroidogenesis by the ring gland with the former possibly being more efficacious. In contrast, a preliminary study with pure recombinant *Manduca* PTTH was not able to detect stimulation of *Drosophila* ecdysteroid synthesis (Rybczynski, unpublished observations).

Both *Rhodnius* and *Bombyx* have served as important experimental organisms in research aimed at understanding PTTH. Vafopoulou and Steel (1997) explored the effect of *Bombyx* PTTH on ecdysteroid synthesis by

Rhodnius prothoracic glands under *in vitro* conditions. They found that recombinant *Bombyx* PTTH elicited a statistically significant increase in ecdysteroidogenesis relative to controls with effective concentrations similar to that observed in *Bombyx-Bombyx* experiments. This is an intriguing and surprising result, given the evolutionary distance between *Rhodnius* (Hemiptera), and *Bombyx* (Lepidoptera). A puzzling feature of the results is the very sharp optimum in concentration seen with the *Bombyx* PTTH (8 ng/ml), with decreasing response seen at 10 ng/ml and above, although a similar, albeit broader, optimum is also seen with *Rhodnius* brain extracts (Vafopoulou *et al.*, 1996). Also of note is the approximately twofold increase in ecdysteroid synthesis achieved with *Bombyx* PTTH-stimulated *Rhodnius* glands under the same conditions with which *Rhodnius* brain extract elicited nearly a fivefold increase (c.f., Vafopoulou *et al.*, 1996 and Vafopoulou and Steel, 1997). Care must be exercised in interpreting these data, especially given observations in *Manduca* that small increases in ecdysteroid synthesis can be obtained with non-specific stimuli (Bollenbacher *et al.*, 1983).

In summary, the question of PTTH species-specificity remains open. If sequence divergence is any guide to areas conferring specificity, then one region of interest (considering only the lepidopteran species) is an area between the fifth and sixth conserved cysteines, which contains a conserved triplet of basic amino acids located slightly past the middle of the mature peptide (see **Figure 2**, fourth of five sets of rows). This portion of the PTTH molecule lies in an area predicted to be within a hydrophilic area, extending out from a common fold (Noguti *et al.*, 1995). The most N-terminal region of the mature PTTH molecule does not appear to be involved in receptor binding since an antibody to this region failed to block PTTH-stimulated ecdysteroid synthesis by *Manduca* prothoracic glands (Rybczynski *et al.*, 1996). The cloning, sequencing, and expression of recombinant pure PTTHs offer more rigorous tools to investigate this topic and perhaps, by doing so, gain insight into the portion of PTTH that productively interacts with PTTH receptors.

1.3. PTTH: Expression, Synthesis, and Release

1.3.1. Sites of Synthesis and Release

Agui *et al.* (1979), using micro-dissections, demonstrated functionally that *Manduca* PTTH was present only in two pairs of dorsolateral brain neurosecretory cells. Functional evidence was also used to show that the CA and not the CC was the neurohemal organ for lepidopteran PTTH (Agui *et al.*, 1979, 1980; Carrow *et al.*, 1981) and immunohistochemical studies have supported this conclusion (Mizoguchi *et al.*, 1990; Dai *et al.*, 1994). Subsequently, a number of PTTHs have been purified and/or cloned (see Section 1.2.3.), permitting direct measurement of sites of PTTH expression. The use of cDNA probes for *in situ* localization of these PTTH mRNAs supported the findings of Agui *et al.* (1979) (*Bombyx*: Kawakami *et al.*, 1990; *Antheraea*: Sauman and Reppert, 1996a; and *Manduca*: Shionoya *et al.*, 2003; *Heliothis*: Xu *et al.*, 2003; *Helicoverpa*: Wei *et al.*, 2005; *Drosophila*: McBrayer *et al.*, 2007; *Spodoptera*: Xu *et al.*, 2007). The use of antibodies in immunohistochemical studies also supported the early *Manduca* work (*Bombyx*: Mizoguchi *et al.*, 1990; *Antheraea*: Sauman and Reppert, 1996a; and *Manduca*: Gilbert *et al.*, 2000, see **Figure 4**; *Helicoverpa*: Wei *et al.*, 2005). *Drosophila* also expresses PTTH in a pair of bilaterally symmetrical cells in the brain, but McBrayer *et al.* (2007) suggested that these neurons may be of a different derivation than those in Lepidoptera. Specifically, PTTH-expressing neurons in *Drosophila* terminate on the prothoracic gland rather than on the *Drosophila* equivalent of the CA.

In *H. armigera*, RT-PCR revealed expression of PTTH in tissues in addition to the brain, including ganglia, midgut, and fat body. PTTH protein was only observed in the

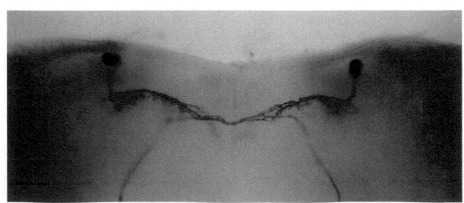

Figure 4 Immunocytochemical localization of PTTH in the pupal brain of *Manduca sexta* showing that PTTH is detectable only in the cell bodies and axons of two pairs of lateral neurosecretory cells. Reproduced with permission from *Gilbert et al.* (2000).

brain, however, as detected by Western blotting. A more striking difference in sites of PTTH expression was seen in *S. nonagrioides* (Perez-Hedo *et al.*, 2010a,b). In this moth, PTTH transcript and protein were seen in the gut as well as the brain, in keeping with the observation that headless larvae of this species can molt normally. Decapitation in this species actually increases PTTH expression in the gut (Perez-Hedo *et al.*, 2010b).

As discussed in Section 1.2.2., anti-PTTH antibodies have been used also in heterospecific immunohistochemical studies. Anti-*Bombyx* PTTH antibodies were used to probe *Locusta* (Goltzene *et al.*, 1992), *Drosophila* (Zitnan *et al.*, 1993), *Manduca* (Dai *et al.*, 1994), and *Samia* (Yagi *et al.*, 1995). In *Manduca* and *Samia* brains, the antibody decorated two pairs of lateral neurosecretory cells. These cells appeared to be the same as seen in *Bombyx* brain in the conspecific probing summarized earlier. Anti-*Helicoverpa* antibodies were used to probe the brain of *Spodoptera*, with similar results (Xu *et al.*, 2007). In *Drosophila*, Zitnan *et al.* (1993) found that anti-*Bombyx* PTTH antibodies labeled mediolateral cells, axons, and axon terminals, in keeping with PTTH expression later seen by McBrayer *et al.* (2007). However, in the Zitnan study, additional cells in the subesophageal, thoracic, and abdominal ganglia also bound anti-*Bombyx* PTTH antibodies, particularly in the adult, while adult PTTH expression was later seen by McBrayer *et al.* (2007) only in the brain. It is likely that the discrepancy resulted from cross-reactivity of the anti-*Bombyx* antibody to a protein besides PTTH. Questions regarding antibody specificity are a common concern in immunohistochemical studies; for example, PTTH immunoreactivity in *Locusta* is dependent on the antibody employed. Goltzene *et al.* (1992) found that a number of brain neurosecretory-type cells reacted positively with an anti-*Bombyx* PTTH antibody while the use of an anti-*Antheraea* PTTH antibody revealed no positive reactions (Zavodska *et al.*, 2003). This same anti-*Antheraea* PTTH antiserum was used to survey a wide variety of insects (Zavodska *et al.*, 2003; see Section 1.2.2.); until we know more about the nature of PTTH in these species, it is not possible to interpret the smorgasbord of immunohistochemical patterns seen among the surveyed taxa.

1.3.2. Timing and Control of PTTH Expression

In addition to locating sites of PTTH expression, sequence information has also been used to measure temporal changes in PTTH expression during development. In Lepidoptera, high levels of PTTH expression are seen at expected times, for example at pupation (*H. zea*: Xu *et al.*, 2003) and in early pupae (*Spodoptera exigua*: Xu *et al.*, 2007). PTTH is also expressed in pharate adults (Xu *et al.*, 2007) and at adult eclosion (Xu *et al.*, 2003) suggesting effects on tissues other than the prothoracic glands, which by

then have apoptosed (see Section 1.5.4.). Changes in PTTH expression are particularly apparent in those lepidopteran species undergoing pupal diapause (see Chapter 10). For example, in *A. pernyi*, PTTH is expressed at a relatively constant level from the embryo onward, with a notable decline only in diapausing pupae (Sauman and Reppert, 1996a). Xu and Denlinger (2003) found that PTTH mRNA levels in non-diapause *H. virescens* remained relatively constant through the fifth instar until about day 10 of pupal–adult development. However, in pre-diapause larvae, PTTH expression dropped at the onset of wandering and remained low in pupae until the termination of diapause. Similar expression patterns were seen in diapausing versus non-diapausing pupae of *H. armigera* (Wei *et al.*, 2005). In *Drosophila*, McBrayer *et al.* (2007) observed an 8 h periodicity of PTTH expression, with highest levels 12 h before pupariation. This is similar to the periodicity of 20E pulses seen in carefully staged third instar larvae (Warren *et al.*, 2006). Hence in *Drosophila*, PTTH delivery may be regulated at the level of expression. By contrast, in Lepidoptera, the most striking changes in PTTH expression are associated with longer term changes, such as the reductions during diapause described earlier. Because lepidopteran PTTH is not delivered directly to the target glands, it may be that alterations in PTTH release or prothoracic gland responsiveness to PTTH play a more critical role than changes in mRNA expression in acute regulation of ecdysteroid secretion in this order (see the following sections).

PTTH sequence information has also been used to investigate the regulation of PTTH transcription. A study by Shiomi *et al.* (2005) revealed that *Bombyx* myocyte-specific enhanced factor 2 (BmEF2), a multifunctional transcription factor involved in development (Potthoff and Olson, 2007), was expressed in PTTH-producing neurons and overexpression enhanced transcription of PTTH. Wei *et al.* (2010) recently analyzed cis-regulatory elements in the *Bombyx* PTTH gene promoter and found evidence for regulation by TATA box binding proteins, with predicted regulators of expression that included, among others, BmEF2. PTTH-containing neurons that innervate *Drosophila* were recently found to require the gap gene giant (gt) gene during embryonic development for later functionality (Ghosh *et al.*, 2010). Mutations in that gene led to developmental delays and large body size characteristic of the ablation of PTTH-producing neurons seen earlier by McBrayer *et al.* (2007).

In *Drosophila*, damaged imaginal discs delay pupariation, and in larvae with damaged discs, PTTH expression is also delayed (Halme *et al.*, 2010). The retinoid synthetic pathway was found to be required for this delay, that is, retinoids may exert an inhibitory effect on PTTH expression. The mechanism by which this occurs was not determined. While there may be no functional connection, the retinoid regulation of PTTH expression brings to mind earlier studies by Muehleisen *et al.* (1993) in which a peptide similar to a retinoid-binding protein (Mansfield *et al.*, 1998) was isolated using an antibody that also strongly cross-reacted with PTTH-containing neurons (see Section 1.2.1.).

1.3.3. PTTH Brain and Hemolymph Titer

The periodic increases in ecdysteroid synthesis and hemolymph titer that control molting and molt quality have long been interpreted to indicate that PTTH was released periodically into the hemolymph. However, testing this hypothesis has not been simple. Unlike E and 20E, a sensitive radioimmunoassay for PTTH is not widely available. Instead, the measurement of PTTH hemolymph titers has been done primarily via *in vitro* analysis of prothoracic gland ecdysteroidogenesis using hemolymph extracts.

Bollenbacher and Gilbert (1981) determined the brain and hemolymph titers of PTTH in *Manduca* during the fourth and fifth larval instars, using an *in vitro* bioassay system and discovered little concordance between measured brain- and hemolymph-derived PTTH activities. Brain titers of PTTH rose fairly steadily during this period, punctuated by minor decreases or plateaus. A small dip in brain PTTH was seen immediately after the molt to the fifth larval instar and a larger decrease was seen about one day prior to the pupal molt. In contrast, hemolymph PTTH activity was low throughout this period, with only a few, very sharp peaks. In the fourth instar, a single PTTH peak occurred late on day 1, also reported later by Bollenbacher *et al.* (1987), with a second small peak spanning the molt to the fifth larval instar. During the fifth larval instar proper, four peaks were noted. Three medium-sized peaks clustered on and about the fourth day, preceding and overlapping the commitment peak of ecdysteroid synthesis. A much larger PTTH hemolymph titer was measured on day six, at the rising edge of the major larval ecdysteroid titer that precedes the metamorphic molt to pupal–adult development and closely anticipating apolysis. The PTTH titer appeared to be rising again just before pupal ecdysis, but titers were not determined beyond this point. Later studies of *Manduca* brain PTTH content indicated that PTTH levels in the brain increased considerably during pupal–adult development (O'Brien *et al.*, 1986), but hemolymph titers have yet to be determined at this stage for *Manduca*. These studies suggested that PTTH titers in the brain are not closely correlated with circulating levels, unlike the levels in the CA (Agui *et al.*, 1980).

Shirai *et al.* (1993) first addressed the topic of PTTH hemolymph titers in *B. mori*. Again, an assay was used in which fifth larval instar hemolymph extracts were tested for PTTH ecdysteroidogenic activity with prothoracic glands *in vitro*. In addition, the secretion of PTTH activity by brain-endocrine complexes was measured with the

in vitro gland assay. Hemolymph ecdysteroid levels were also determined via radioimmunoassay. The results of this study suggested that five peaks of PTTH activity were present in the hemolymph of *Bombyx* fifth instar larvae. Two of these PTTH peaks occurred in the first half of the fifth instar, clearly before a small ecdysteroid peak on day seven or day eight to nine, depending on the particular *Bombyx* strain-cross utilized (*Shunrei x Shougetsu* and *Kinshu x Shouwa*, respectively). This small ecdysteroid peak was presumed to be the commitment peak, as seen in *Manduca*. A third PTTH peak occurred the day before the putative commitment peak, whereas the fourth peak encompassed the time span of the commitment peak and just overlapped the beginning of the large pre-molt ecdysteroid peak. The fifth hemolymph peak detected by Shirai *et al.* (1993) began almost immediately after the finish of the fourth peak and overlapped with approximately the first third of the large ecdysteroid peak. These workers also found four to five peaks of PTTH activity when brain-endocrine complexes were assayed *in vitro*. These peaks anticipated the hemolymph peaks by about one day.

The data obtained by Bollenbacher and Gilbert (1981) and Shirai *et al.* (1993) must be interpreted carefully. In both studies, hemolymph and incubation medium assays measured total prothoracicotropic activity, but did not distinguish activity stimulated by PTTH from that stimulated by other molecules such as small PTTH or bombyxin. Further, in the case of the Shirai data (1993), the peaks of PTTH activity obtained from hemolymph extracts and brain-endocrine incubations were surprisingly similar in height, suggesting that, at least in *Bombyx*, isolated brain-endocrine complexes might be released prematurely from negative controls that function *in vivo*. Nonetheless, results of both studies suggest that some pulses of PTTH are not directly ecdysteroidogenic, and may instead serve other functions. This possibility is discussed further in the following sections in the context of PTTH determined with immunologic techniques.

The development of a highly specific antibody against *Bombyx* PTTH (Mizoguchi *et al.*, 1990) afforded an opportunity to develop a more facile and sensitive method to determine hemolymph PTTH titers. Dai *et al.* (1995) and Mizoguchi *et al.* (2001, 2002) used this antibody in a time-resolved fluoroimmunoassay to study *Bombyx* PTTH hemolymph titers during larval–pupal and pupal–adult development. In addition, Dai *et al.* (1995) utilized this anti-PTTH antibody to assess the PTTH content of brain-retrocerebral complexes using immunogold electron microscopy.

In the first of these three studies, the hemolymph PTTH titer was moderately high at the beginning of the fourth instar and reached a minimum about 36 h later [scotophase of day 1, fourth instar (IV$_1$)] (Dai *et al.*, 1995). A larger peak ensued, with a slight dip in the photophase of IV$_2$, with a minimum again just before the molt to the fifth instar. The fifth instar PTTH titer exhibited a small but distinct peak during the first photophase, on V_0 with minima 12 h later (V_1) and just at the larval–pupal molt (V_9-P_0). In between the two minima, hemolymph levels climbed fairly steadily to a plateau from V_6 through V_8, at which point a rapid decline to the molt minimum began. The parallel determinations of hemolymph ecdysteroid titer and PTTH antibody reactivity in the CA (the site of PTTH release; see Section 1.3.1.) indicated that there was no obvious correlation between PTTH and ecdysteroid hemolymph titers, except during the fourth instar, when both of the PTTH peaks were associated with ecdysteroid peaks. The commitment and large pre-molt ecdysteroid increases of the fifth larval instar both occurred during the plateau period of high PTTH titer. In contrast, PTTH immunostaining in the CA was low during the early fifth instar peak (V_1) and during most of the later hemolymph plateau phase of PTTH titer; however, a striking transitory increase in PTTH immunostaining was seen on V_7 just before the rapid decline in hemolymph PTTH titer commenced.

Subsequent to these intriguing observations by Dai et al. (1995), it was discovered that the method used to extract the hemolymph PTTH for the fluoroimmunoassay included a hemolymph factor that contributed an unknown amount to the measured PTTH titer (Mizoguchi et al., 2001). Consequently, Mizoguchi et al. (2001) repeated the determination of *Bombyx* hemolymph and ecdysteroid titers, using a revised hemolymph treatment protocol. This study also extended through the period of pupal–adult development and included a parallel ecdysteroid titer determination, as well as an analysis of brain PTTH content, using the time-resolved fluoroimmunoassay. Mizoguchi et al. (2002) provided a third determination of fifth instar and pupal–adult *Bombyx* hemolymph PTTH and ecdysteroid titers, using the revised fluoroimmunoassay, in conjunction with a study directed at discovering potential rhythmicity in PTTH secretion and the influence of diel light cycles. These two more recent studies contrast in several ways with the data obtained by Dai et al. (1995) and Shirai et al. (1993). Like Dai et al. (1995), Mizoguchi et al. (2001) found two hemolymph PTTH peaks in the fourth instar, but only the second one was accompanied by a clear-cut rise in hemolymph ecdysteroids. During the fifth instar, Mizoguchi and colleagues detected three rises and falls of hemolymph PTTH, with the instar ending in a period of rapidly increasing PTTH levels that carried on into pupal–adult development. These studies involved three different crosses of *Bombyx* races and four separate determinations (in the following section, the name of the female race in a cross is given first, and the male second). The earliest of these increases took place on V_4 and rose just slightly above background PTTH levels. Of the two crosses for which complimentary ecdysteroid titers are available, the *Kinshu*

x *Showa* (Mizoguchi et al., 2001) exhibited ecdysteroid peaks corresponding to all three PTTH rises (albeit the first ecdysteroid peak is very small). In the *J106 x Daizo* hybrid (Mizoguchi et al., 2002), only the second and third PTTH peaks are correlated with measurable ecdysteroid peaks, that is, the small, presumptive commitment and the much larger pre-molt peaks. In contradistinction to the associated ecdysteroid peaks, the relative heights of the second and third PTTH peaks vary with the hybrid, for example, the V_5 peak is distinctly higher than the V_7 peak in the *Kinshu x Showa* hybrid (Mizoguchi et al., 2001) while in the *J106 x Daizo* hybrid the later peak (V_6) is much larger (Mizoguchi et al., 2002). The immunoassay survey of fifth instar brain PTTH by Mizoguchi et al. (2001) revealed a period of gradual increase, starting from the first day of the instar and peaking on the fifth and sixth days, followed by a rapid decline to a minimum on the ninth day. PTTH levels then rose slightly on the ninth day and remained static through pupation on the following day.

Pupal–adult PTTH hemolymph titers were analyzed four times by Mizoguchi and colleagues and indicated one or two high peaks of concentration extending over the first half of the stage followed by a period of much lower levels and ending with a variety of levels, low, intermediate, or high, depending on the cross (Mizoguchi et al., 2001, 2002). The two parallel ecdysteroid titers exhibited much less variation in temporal patterning. A large, fairly symmetric peak was centered about 48 h after pupal ecdysis and ecdysteroid levels declined rapidly and evenly to very low concentrations by the time of adult ecdysis. The major difference between the two pupal–adult ecdysteroid titers was that of concentration. The *Kinshu x Showa* racial hybrids reached levels of more than 7 μg equivalents per ml (Mizoguchi et al., 2001) while the *J106 x Daizo* hybrids peaked at only about 1.5 μg equivalents per ml (Mizoguchi et al., 2002).

The five immunologic determinations of *Bombyx* PTTH hemolymph titers, and the three of ecdysteroids, made by Dai et al. (1995) and Mizoguchi et al. (2001, 2002) paint a fairly similar picture of circulating PTTH levels in *Bombyx*, yet there are several important discrepancies among the results that need to be resolved. First, the small yet distinct early ecdysteroid peak seen in the fourth instar by Dai et al. (1995), occurring in concert with the first of the two PTTH rises, was not consistently found in other assays of fourth instar ecdysteroid titers; for example, possibly seen by Kiguchi and Agui (1981), but not by Gu and Chow (1996) or Mizoguchi et al. (2001). Verifying, or disproving, the existence of this ecdysteroid increase is necessary to understand the role(s) of PTTH at this time. Second, the status of hemolymph PTTH levels at the beginning of pupal–adult development is not clear. In two of four assays, a broad peak of hemolymph PTTH was found in early pupal–adult life centered on day 2

(*Kinshu x Showa*: Mizoguchi *et al.*, 2001; *Showa x Kinshu* raised at 25°C; Mizoguchi *et al.*, 2002). In a third profile (*Showa x Kinshu* raised at 25°C; Mizoguchi *et al.*, 2002), a peak on the second day after pupal ecdysis was followed by an approximately two-day plateau of intermediate values before finishing the last half of pupal–adult life with lower levels but with considerable variation, especially in the 36h before adult ecdysis. In the fourth instance (*J106 x Daizo*; Mizoguchi *et al.*, 2002), a high, narrow peak of hemolymph PTTH occurred on the first day after ecdysis, followed by a second, broader, peak about 36h later. A third topic of concern is the level of PTTH circulating at the end of pupal–adult development. Mizoguchi *et al.* (2001) reported an adult ecdysis peak (≈250pg rPTTH equivalents) that was as high as that seen at the beginning of pupal–adult development, when ecdysteroid levels are also very high. In contrast, PTTH hemolymph titers in the other three pupal–adult studies were only 25 to 50% of the earlier pupal–adult peaks (Mizoguchi *et al.*, 2002).

It is not possible currently to assess the reasons for the observed large temporal and concentration variations in PTTH titer. Multiple racial hybrids of *Bombyx* were used and some data came from animals raised on mulberry leaves (Shirai *et al.*, 1993), rather than on an artificial diet. The length of the fifth instar ranged from 8 to 12 days while pupal–adult development varied from 8 to 11 days. The possibility that further stage- or hybrid-specific factors are influencing the PTTH titer data cannot be ruled out. However, since hybrid-specific differences in ecdysteroid titers were also found, especially in the peak concentrations, and ecdysteroid assays are well standardized, it seems likely that the PTTH titer variations are mainly biologically based. Given this assumption, it is possible to present an "average" or summary picture of PTTH and ecdysteroid hemolymph titers in *Bombyx* (**Figure 5**).

Several conclusions can be drawn from these data. First, some correlation does exist between hemolymph PTTH

levels and ecdysteroid peaks in the fourth and fifth instar, particularly prior to the pre-ecdysial peak in the fourth instar, and the wandering and pre-pupal ecdysteroid peaks in the fifth (Mizoguchi *et al.*, 2001, 2002). However, it is clear that PTTH increases do not always result in a detectable increase in hemolymph ecdysteroid concentrations (e.g., in *Bombyx* on V_6 and P_0), even when glands are responsive *in vitro* (Okuda *et al.*, 1985). This discrepancy suggests that negative controls such as prothoracicostatic factors reduce glandular activity in intact larvae (see 1.5.2.3.). In addition, early in the fifth instar, PTTH is elevated at times when the glands are refractory to stimulation *in vivo* and *in vitro*. PTTH may have a priming effect on glandular activity at that time, insufficient to stimulate steroid secretion, but serving to augment overall steroidogenic capacity. Another obvious inference stemming from these observations is that PTTH might stimulate non-steroidogenic events, perhaps in tissues other than, or in addition to, the prothoracic glands. These data suggest that our understanding of factors involved in PTTH release and activity are still rudimentary. One missing factor in this equation may well be JH, whose values were not addressed in these studies, but other hormones and factors are also likely to be involved.

The variation found in PTTH and ecdysteroid titers and in developmental rates, originating from methodologically similar studies (Dai *et al.*, 1995; Mizoguchi *et al.*, 2001, 2002), clearly indicate that more data are desirable from *Bombyx* and other insects. Only then will we obtain a reliable picture of how this important piece of the insect endocrine system functions. The open questions are many. For instance, how much does the magnitude of a PTTH peak matter? Do very high levels have unique biological effects on the prothoracic glands, or do they simply delay processes that depend on low PTTH levels for their expression or regulation? What are the targets and consequences of the orphan peaks, that is, PTTH

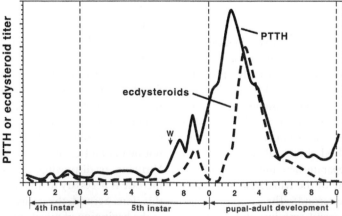

Figure 5 Typical PTTH and ecdysteroid titers in *Bombyx mori* hemolymph. Values are a composite of determinations. Reproduced with permission from *Dai et al.* (1995) and *Mizoguchi et al.* (2001, 2002). Any individual, original profile may vary notably from the constructed profile shown here.

increases that do not elicit ecdysteroidogenesis? What is the role of PTTH in late pupal–adult development and during adult life when prothoracic glands are generally absent due to programmed cell death?

1.3.4. Control of PTTH Release: Neurotransmitters

PTTH release is ultimately controlled by a number of factors, including photoperiod, size, and nutritional state of the animal. Proximately, a number of studies pointed out the likelihood that PTTH release is directed by cholinergic neurons synapsing with the prothoracicotropes. It must be emphasized that the data discussed in the following studies in this section do not definitively identify PTTH as the active molecule released from the brain, although this is most likely the case.

Carrow et al. (1981) showed that PTTH could be released by depolarizing *Manduca* (brain) neurons with a high K+ concentration in the presence of Ca^{2+}; a similar result was obtained when *Bombyx* brains were treated with high K^+ in the presence or absence of external Ca^{2+} (Shirai et al., 1995). Lester and Gilbert (1986, 1987) found that acetylcholine accumulation and α-bungarotoxin binding sites exhibited temporal peaks in the brain of fifth instar *Manduca* larvae, and that these peaks coincided with periods of PTTH release and increased ecdysteroid synthesis. It is unlikely that all of the changes in the cholinergic systems of the brain that occur at these stages directly feed into the PTTH-producing neurons, but these data are supportive for a role of cholinergic neurons in PTTH biology. Direct evidence that cholinergic stimulation can result in PTTH release comes from several *in vitro* studies of *Bombyx* and *Mamestra*. Agui (1989) showed that acetylcholine and cholinergic agonists caused cultured *Mamestra* brains to release PTTH into the medium while dopaminergic agents had little or no effect. Shirai et al. (1994) obtained similar results (using the acetylcholine agonist, carbachol) and further refined the analysis by using both muscarinic and nicotinic agents. They found that muscarine treatment caused PTTH release from *Bombyx* brains *in vitro*, and that carbachol-induced release was blocked by the muscarinic antagonist atropine. Nicotine had no effect on PTTH release, and the composite data indicated the involvement of a muscarinic acetylcholine receptor. Later immunohistochemical studies revealed that the PTTH-producing cells of *Bombyx* do indeed express a muscarinic acetylcholine receptor (Aizono et al., 1997). Shirai et al. (1994) further determined the effects of a calmodulin inhibitor, a phospholipase C inhibitor (PLC), and a protein kinase C (PKC) inhibitor on carbachol-induced PTTH release. All three drugs inhibited the effects of carbachol, suggesting roles for Ca^{2+}-calmodulin, PLC, and PKC in this system. A role

for Ca^{2+} was more directly demonstrated when it was found that Ca^{2+}-ionophore also caused PTTH release (Shirai et al., 1995). Finally, in regard to the cholinergic neurotransmission, Shirai et al. (1998) investigated the possible role for small G-proteins in carbachol-stimulated PTTH release. The poorly hydrolyzable GTP analog GTPγS was found to stimulate PTTH release in the presence of a PLC inhibitor; the use of the PLC inhibitor removed the positive effect that the GTPγS would have on the larger heterotrimeric G-protein directly associated with muscarinic receptors (see Hamilton et al., 1995). In this context, antibodies directed against PKC and the small GTPase Rab8 were found to co-localize with antibodies directed against PTTH in the brain of *Periplaneta americana* (Hiragaki et al., 2009). This finding supports the idea that PKC and small G-proteins regulate PTTH secretion, although further experimental evidence in this case was not provided.

The experiments of Agui (1989) indicated that dopamine might have a small effect on PTTH release *in vitro*. This possibility was tested with *Bombyx* brains and it was found that dopamine and serotonin increased PTTH release, whereas norepinephrine had no effect (Shirai et al., 1995). The effects of dopamine and serotonin were notably slower than that of carbachol, suggesting that the first two compounds might be acting at some distance from the PTTH-producing cells. It must be emphasized that there are no data that demonstrate that cholinergic agonists directly act upon the prothoracicotropes.

In contrast to observations in *Bombyx* (Shirai et al., 1995), serotonin had an inhibitory effect on PTTH release, as measured by prothoracic gland ecdysteroid synthesis, during short-term (3 h) co-cultures of cockroach (*P. americana*) brains and prothoracic glands (Richter et al., 2000). No effect of serotonin on isolated prothoracic gland ecdysteroid synthesis was found. Serotonin is the precursor to melatonin, the latter playing an important role as neuromodulator associated with darkness in diel light-dark cycles. In contrast to the results with serotonin, Richter et al. (2000) found that melatonin increased PTTH release in brains and prothoracic gland co-cultures during both short-term (photophase) and long-term (12 h; scotophase) incubations. A melatonin receptor antagonist, luzindole, blocked the melatonin effect. Again, neither melatonin nor luzindole had an effect on prothoracic gland ecdysteroid synthesis in the absence of the brain. These results raise the possibility that melatonin may play a role in modulating temporal patterns of PTTH release. However, as Richter et al. (2000) pointed out, there was a lack of daily rhythmicity in the molting of cockroaches in their colony, suggesting that there was no rhythmicity to PTTH release either. This indicated that an *in vivo* role for melatonin is problematic. See Section 1.3.6. for further information on cycles of PTTH release.

1.3.5. Control of PTTH Synthesis and Release: Juvenile Hormone

Insect molts, be they simple or metamorphic, are under the general control of 20-hydroxyecdysone (20E). This hormone does not act alone and there are additional hormones (i.e., ecdysis-triggering hormone and eclosion hormone) that act downstream from 20E, playing very proximate roles in controlling ecdysis behavior (see Chapter 7). However, JHs, which also participate in regulating molting and metamorphosis (see Chapter 8), modulate the action of 20E in a way that makes them partners with 20E rather than dependent factors. JHs are epoxidated sesquiterpenoids produced in, and released from, the CA, which is also the site of PTTH release. JH is an important hormone in some aspects of insect reproduction, but its relevant role regarding PTTH is the ability to direct a molt either into a simple or into a metamorphic pathway. Basically, the presence of circulating JH during an ecdysteroid peak results in the next molt being a simple one (larva to larva), but if JH levels are very low during an ecdysteroid peak, then the ensuing molt will be a metamorphic molt (larva to pupa). In the last larval instar of many or most insects a small release of PTTH appears to trigger a small ecdysteroid release from the prothoracic gland during a period of very low JH titer. This small increase in ecdysteroid titer is termed the commitment peak, because it commits the animal to a metamorphic molt (Riddiford, 1976).

How JHs modulate molt quality is not well understood. A number of studies have demonstrated that JH or JH analogs (JHAs) affect ecdysteroid titers when applied topically, injected, or fed to pre-adult insects (Chapter 8). In these studies involving intact larvae, pupae, debrained animals, or isolated abdomens, JHs appeared to modulate PTTH synthesis and/or release and perhaps affect the PTTH signaling pathway in the prothoracic gland or the gland's capacity for ecdysteroidogenesis. Two periods of JH influence have been identified in such studies during the last larval instar; the influence of JH during the penultimate instar is different and is treated separately in the next paragraph.

In Lepidoptera, JH hemolymph titers decline dramatically very early in the last larval instar (for *Manduca*, see Baker *et al.*, 1987; for *Bombyx*, see Nimi and Sakurai, 1997). The JH decline is followed by a release of PTTH and the resultant small commitment peak in ecdysteroids (see Section 1.3.3.). Application of juvenoids (JHs or JHAs) before the commitment peak results in a delay of development, for example, deferral of the cessation of feeding, and wandering (see Riddiford, 1994, 1996). In *Manduca*, Rountree and Bollenbacher (1986) found that prothoracic glands from early fifth instar larvae treated with JH-1 or ZR512 (a JHA) for 10 h prior to removal still responded to PTTH with increased ecdysteroid

synthesis, although the normal developmental increase in basal ecdysteroidogenesis was inhibited. However, when the brain-retrocerebral complexes from such animals were implanted into head-ligated larvae, they exhibited a delay in activating development relative to control complexes. This result suggested that JHs controlled the time of PTTH release, although these data do not rule out an effect on PTTH content/synthesis in the complexes as well. In a later study of *Manduca* glands, Watson and Bollenbacher (1988) did find a reduced response of the prothoracic glands to PTTH *in vitro*, but this study involved animals treated with a JHA for two days, starting at an earlier time point. Thus the conclusions reached may reflect the experimental paradigms as well as the animal's intrinsic biology.

In *Bombyx*, similar results have been obtained. Sakurai (1984), using the technique of brain removal, concluded that PTTH release occurred early in the fifth instar shortly after a precipitous decline in JH titer, as found for *Manduca*. Furthermore, allatectomy late in the fourth instar shortened the phagoperiod in the fifth instar, and brain removal performed on these larvae at 48 h into the fifth instar did not block metamorphosis. Sakurai also found that JH application early in the fifth instar prolonged the larval period. The combined data suggested again that JH in the early fifth instar blocked PTTH release. Sakurai *et al.* (1989) provided evidence that JH inhibits prothoracic gland activity in *Bombyx* by showing that allatectomy of early fifth instar larvae accelerated the period of PTTH sensitivity in glands, but that this acceleration was blocked by JH treatment.

The previously mentioned studies utilized either JHs or JHAs with JH-like chemical structures, such as terpenoids. However, the non-terpenoid chemical fenoxycarb, a carbamate, has also been shown to exhibit JH-like effects when applied to larvae. Leonardi *et al.* (1996) found that single, topically applied doses of fenoxycarb as low as 10 fg were sufficient to disrupt larval–pupal ecdysis in *Bombyx*. Also studying *Bombyx*, Monconduit and Mauchamp (1998) showed that very low doses of fenoxycarb, 1 ng and below, were sufficient to induce permanent larvae when administered in the food during the phagoperiod of the last larval instar. This treatment also considerably depressed the ecdysteroid synthesis of prothoracic glands when measured *in vitro*. The glands still exhibited a qualitative response to a (crude) PTTH preparation, but the quantitative response was greatly curtailed. Dedos and Fugo (1996) found that fenoxycarb significantly and directly inhibited prothoracic gland steroidogenesis under *in vitro* conditions; however, the effective doses required in these *in vitro* experiments were much higher (0.5 to 1 μg/gland) than employed by Leonardi *et al.* (1996) and Monconduit and Mauchamp (1998). Thus, the possibility that fenoxycarb at these doses had effects that were not attributable to JH-like activities cannot be ruled out, as suggested by Mulye and Gordon (1993) and Leonardi *et al.* (1996), among others.

When exogenous JH or JHAs are administered after the commitment peak period of last larval instars, quite different results are obtained — JHs accelerate development and shorten the period to pupation (see Sakurai and Gilbert, 1990; Smith, 1995). Brain removal or ligation after the commitment peak period does not affect this accelerative ability of JH, indicating that the brain and hence PTTH release are not the targets of JH at this time (Safranek *et al.*, 1980; Gruetzmacher *et al.*, 1984a). Intriguing in this regard are the observations of Sakurai and Williams (1989). They implanted prothoracic glands extirpated from feeding (day 2) fifth instar *Manduca* larvae into pupae lacking a cephalic complex. Treatment of these pupae with either 20E or the JH analog hydroprene resulted in an inhibition of steroidogenesis when the prothoracic glands were studied subsequently *in vitro*. However, if both 20E and hydroprene were administered, an increase in ecdysteroid synthesis was seen. These conditions simulate the pre-pupal period when JH (particularly JH acids) and 20E are both high. Thus, as Smith (1995) suggested, the pre-pupal ecdysteroid peak may be initiated by PTTH and maintained by JH, with the eventual decline of JH titer contributing to the decline of ecdysteroid levels that occurs shortly before larval–pupal ecdysis. An inhibitory factor, prothoracicostatic peptide (PTSP), is also likely to contribute to a fall in ecdysteroid levels (see Section 1.5.2.3.).

The relationship between JH and the brain-prothoracic gland axis observed in the last larval instar is not characteristic of every developmental stage. Stages prior to the last larval instar lack the commitment peak of ecdysteroids, along with the PTTH release that triggers it. In *Manduca*, the fourth (penultimate) larval instar is characterized by high to moderate JH levels throughout (see Riddiford, 1996 and Chapter 8). The fourth instar JH titer minimum is reached shortly before ecdysis to the fifth instar and ecdysteroid levels in the hemolymph begin rising while JH levels are relatively high, albeit falling. In *Bombyx*, the case is somewhat more complicated. Peaks in the JH titer occur during the early and middle-late portions of the penultimate instar with the minimum between the two peaks higher in concentration and shorter in duration than the minimum measured during the last (fifth) instar (Nimi and Sakurai, 1997).

Fain and Riddiford (1976) showed that head ligation of fourth instar *Manduca* larvae before a presumed HCP resulted in precocious pupation; this operation isolated the sources of both PTTH and JH from the prothoracic glands. Injection of 20E immediately after ligation resulted instead in a larval molt. Delaying the 20E treatment resulted in a progressive loss of the larval response, but JH injection maintained the 20E-induced larval molt. Lonard *et al.* (1996) found that application

of the JH analog methoprene to fourth instar *Manduca* larvae affected neither the ecdysteroid titer *in vivo* nor the ecdysteroidogenic activity of prothoracic glands measured *in vitro*; however, allatectomy abolished the ecdysteroid peak, which could be restored by the application of methoprene. Further methoprene experiments with debrained and neck-ligated animals indicated that the restorative effect of methoprene involved the brain and was interpreted to indicate the existence of JH-dependent PTTH release.

Sakurai (1983) observed results in *Bombyx* similar to those obtained by Fain and Riddiford (1976), such as head ligation early in the fourth instar resulted in precocious pupation and application of methoprene shifted the post-ligation molts back to larval or larval–pupal intermediates. Unlike the case of JH in *Manduca*, methoprene was effective at reversing precocious pupations only when administered about 24 h after ligation, with larvae treated directly after ligation simply showing no further development. This difference between the two species may reflect a fundamental difference in the ecdysteroidogenic competence of the prothoracic gland in the time immediately following larval ecdysis (see Section 1.5.1.). Sakurai (1983) also found that application of 20E to debrained *Bombyx*, still possessing CA, resulted in larval rather than pupal molts. These data, from *Manduca* and *Bombyx*, indicate that high levels of 20E in the relative absence of JHs will result in a pupal molt, even prior to the last larval instar. However, under ordinary circumstances, PTTH release and ecdysteroid synthesis take place against a significant JH titer prior to the last larval instar.

The relationship between JHs and the PTTH-prothoracic gland axis during pupal–adult development has not been studied extensively. In general, application of exogenous juvenoids to pupae shortens the time until adult ecdysis, and the juvenoid must be administered about the time of pupal ecdysis or termination of pupal diapause (see Sehnal, 1983). Using *Bombyx*, Dedos and Fugo (1999a) injected the potent juvenoid fenoxycarb at pupal ecdysis and found that ecdysteroids were considerably elevated above control during subsequent pupal–adult development. *In vitro* analysis of the prothoracic glands from treated animals showed elevated ecdysteroid secretion relative to controls; additionally, brain-retrocerebral complexes from fenoxycarb-treated animals secreted more PTTH than controls did after 3 h of *in vitro* incubation; however, there was no difference from controls if the incubation was only one hour. These data were interpreted to indicate that fenoxycarb stimulated ecdysteroid synthesis by promoting PTTH release. A later study indicated that PTTH levels in fenoxycarb-treated pupae were indeed elevated at the start of pupation, as assessed by time-resolved fluoroimmunoassay of hemolymph (Dedos *et al.*, 2002). However, PTTH did not remain elevated throughout the extended period of prothoracic gland activity; i.e., prothoracic gland activity was not stimulated entirely by the brain (Dedos *et al.*, 2002).

Although all the studies cited above, and many others not discussed, clearly point to the existence of cross-talk between the brain-prothoracic gland axis and JHs, considerable care must be exercised in interpreting these experiments. Ligations, extirpations, and tissue transplants have provided valuable insights into insect endocrinology over the years, but these methodologies are rarely clean. For instance, allatectomy removes not only the site of JH synthesis but the neurohemal organ for normal PTTH release as well. Similarly, brain removal eliminates a source of many neuropeptides as well as a biochemical clock involved in regulating many diel cycles (see Section 1.3.6.).

Experiments involving hormone additions to intact or non-intact animals also have to be interpreted carefully. Hormones are powerful drugs when given in large doses or at unusual times. For example, fenoxycarb-treated pupae that had elevated ecdysteroid titers also exhibited malformed rectums (Dedos and Fugo, 1999a), and whether or not this result tells us something informative about ecdysteroid titer regulation during normal adult development is open to debate. JH has often been suggested to have a negative effect on PTTH secretion in the last larval instar (see above), based on experiments in which larvae were treated with JH or JH analogs early in the instar. However, Mizoguchi (2001) measured the PTTH titer in JH-treated fourth and fifth instar *Bombyx* larvae using time-resolved fluoroimmunoassay. His data showed that JH treatment had just the opposite effect — JH treatment raised the PTTH hemolymph titer. Thus, experiments in which PTTH secretion was measured *in vitro* following JH treatment of the donor larvae may be misleading in our attempt to understand the relationship between JH and PTTH.

It must also be emphasized that direct effects of JHs on prothoracic glands are rare, indicating the probable participation of one or more additional cell types and/or hormones and/or neuronal connections in JH effects. In the rare cases where a direct effect has been demonstrated *in vitro*, the concentrations of JHs or JHAs employed are notably above physiological limits. In addition to the effect of 1 μg (≈65 μM) fenoxycarb on *Bombyx* prothoracic glands discussed earlier (Dedos and Fugo, 1996), Richard and Gilbert (1991) reported that JHIII-bisepoxide and JHIII both inhibited the secretion of ecdysteroids from *Drosophila* brain-ring-gland complexes; however, millimolar concentrations were needed to elicit the effect.

1.3.6. PTTH Rhythms and Cycles

The previous discussion of PTTH hemolymph titers concentrated on the issues of the timing and magnitude of stage-specific peaks. However, a number of studies yielded data showing that some PTTH peaks are "gated," that is, they occur only within a narrow window within a 24 hr light–dark cycle. Some evidence also suggests that a circadian variation in PTTH titer is overlaid upon the broader sweeps of changing PTTH levels. Review articles cover the subject of PTTH-related circadian rhythms in detail; only a brief treatment is given here (Vafopoulou and Steel, 2005; Steel and Vafopoulou, 2006).

Head critical period is a term used to refer to a developmental period before the presence of the head (brain) is necessary for later events to transpire. In the case of ecdysteroid-triggered events, the HCP has generally been used as a marker for PTTH release. Studies of the HCP for molting provided the first data indicating that an endocrine-dependent event in insect development involved circadian cycles (Truman, 1972). This work revealed that larval molting and the HCP for molt initiation in *Manduca* and *A. pernyi* were temporally gated. Since this HCP and ecdysis are ultimately controlled by PTTH via ecdysteroid levels, these observations implicated the involvement of a circadian clock in PTTH release. Further work in *Manduca* (Truman and Riddiford, 1974) demonstrated that the PTTH release responsible for inducing wandering in the last larval instar, as revealed through HCP timing, was also temporally gated — again indicating the participation of a circadian clock. In *S. cynthia ricini*, the HCP for gut purging was also gated (Fujishita and Ishizaki, 1982), but not every HCP is necessarily under the control of a photoperiodically controlled circadian clock. In *Samia*, in contrast to the HCP for gut purging, the HCP for pupal ecdysis, which follows wandering by two days, was unaffected by shifts in the photoperiod conditions and appeared to be a constant (≈96 h; (Fujishita and Ishizaki, 1982). Light-entrained circadian rhythms in endocrine-dependent events are not confined to the Lepidoptera, and have been studied in a number of other taxa, such as the Diptera. For instance, in *S. bullata* (Roberts, 1984) and *S. argyrostoma* (Richard *et al.*, 1986) larval wandering is under strong photoperiodic control. A similar result was seen with *D. melanogaster*, if they were raised under long-day (16:8) and low-density conditions (Roberts *et al.*, 1987).

Work on *R. prolixus* and the recent surveys of PTTH hemolymph titers have suggested that there might be diel variations in titer, and presumably PTTH release, which are superimposed upon the larger peaks and plateaus of circulating PTTH levels. In *Bombyx*, data from the fifth larval instar and from pupal-adult development (Sakurai *et al.*, 1998; Mizoguchi *et al.*, 2001) suggested that there may be such a daily cycling of PTTH hemolymph levels; a similar suggestion was seen in the *in vitro* secretory activity of prothoracic glands from the last four days of the fifth instar (Sakurai *et al.*, 1998). To examine this possibility more directly, Mizoguchi *et al.* (2001) phase-shifted two groups of initially synchronous *Bombyx* larvae on the first day of the fifth instar, such that their light-dark periods were displaced by 12 h from one another. After 5 days, both groups exhibited a peak in PTTH hemolymph titer that began at the scotophase-photophase transition and peaked during the photophase, supporting the hypothesis that light-dark cycles influence some periods of PTTH release. However, the careful temporal analysis of PTTH titers afforded by this study did not indicate that every apparent PTTH peak was so influenced. In both phase-shifted populations, a second PTTH peak was observed that varied between the two groups in time of appearance (30 h vs. 48 h), with one of the following peaks occurring during the scotophase and the other during the photophase. Thus, this study did not support the notion of small daily PTTH peaks. More recently, pulses of PTTH expression were seen in *Drosophila* at 8 h as opposed to daily intervals (McBrayer *et al.*, 2007). These were found to correspond with ecdysteroid pulses.

Strong, direct evidence for circadian rhythms in PTTH levels and consequent ecdysteroid hemolymph levels comes from studies of the bug *R. prolixus* and the cockroach *P. americana*. In *Rhodnius*, during the last two-thirds of the final larval instar, PTTH synthesis and release appears to be controlled by a circadian clock with peaks occurring during the scotophase (Vafopoulou and Steel, 1996; Steel and Vafopoulou, 2006). The PTTH-synthesizing cells are closely associated with the eight identified clock neurons in the *Rhodnius* brain, which also expresses clock proteins such as Period (PER) and Timeless (TIM), and pigment-dispersing factor (PDF: Vafopoulou *et al.*, 2007, 2010). The PTTH-producing neurons of *Drosophila* are also in close apposition to those containing PDF (McBrayer *et al.*, 2007). The implications of rhythmicity for ecdysteroid synthesis are not straightforward, as studies *in vitro* revealed that the *Rhodnius* prothoracic gland exhibited circadian cycling of ecdysteroid synthesis in the absence of PTTH with a phase shift to peaks during the photophase (Pelc and Steel, 1997). Steroid synthesis in animals that had been either decapitated or paralyzed with tetrodotoxin, to ablate PTTH release, exhibited a rhythmicity similar to that seen *in vitro*, and the composite results indicate that PTTH is an important entraining factor for the circadian clock regulating ecdysteroid synthesis. The endogenous clock of the prothoracic gland is light entrained, specifically by a "lights-off" cue (Vafopoulou and Steel, 1998, 2001). A lights-off cue also functions in controlling PTTH release, as shown in experiments where PTTH release was abolished under continuous light but promptly restored after transfer to dark conditions (Vafopoulou and Steel, 1998, 2001). Further complexity in the control of ecdysteroidogenesis by PTTH in *Rhodnius* was indicated by the discovery that prothoracic glands also express a daily rhythm in PTTH responsiveness, that is, PTTH receptivity is gated (Vafopoulou and Steel, 1999).

In *Periplaneta*, isolated prothoracic glands did not exhibit a circadian cycling of ecdysteroid synthesis *in vitro* when they had been extirpated from stages that exhibit daily cycles of ecdysteroids in the hemolymph, that is, the last several days of the last larval instar (Richter, 2001). However, if the *Periplaneta* glands were co-cultured with brains taken from such late instar animals, then a diel rhythm in ecdysteroidogenesis resulted with a peak in the scotophase matching the *in vivo* hemolymph pattern (Richter, 2001). As discussed earlier (Section 1.3.4.), melatonin dramatically increased the secretion of PTTH from brains *in vitro* during a scotophase, with lesser increases seen during a photophase (Richter, 2001). These data suggested that PTTH release might be, at least partially and at only some stages, regulated by melatoninergic neurons. This PTTH cycling would result in a circadian variation in ecdysteroid synthesis and hemolymph titer. Nonetheless, further support for this hypothesis has yet to be marshaled. A demonstration of PTTH cycling *in vivo*, preferably in the hemolymph titer, would be supportive because factors other than PTTH might be driving prothoracic gland cycles, for example, input from nerves that synapse with the prothoracic gland in *Periplaneta* (Richter, 1985).

The existence of light-dependent cyclic or rhythmic behavior by an organism, a tissue, or a single cell indicates the presence of a light-sensing system as well as an actual molecular-based clock. In cases where isolated brains or prothoracic glands cyclically release hormones under *in vitro* conditions, the sensing apparatus and clock(s) must reside within some or all of the cells under study. A light-sensing system implies the presence of a photoreceptive protein or a photopigment coupled to a transduction system. In the instance of PTTH and ecdysteroid cycling observed *in vitro* in *Rhodnius* tissues, such a sensing system must be extra-retinal. In *Drosophila*, one such protein is known, cryptochrome, which absorbs in the blue, and functions as an extra-ocular photoreceptor in the lateral neurons of the brain (see Stanewsky, 2002, 2003). In a series of clever experiments involving gland transplantation, selective illumination of larval regions and phase-shifting of gut purge, Mizoguchi and Ishizaki (1982) showed that the prothoracic gland of *S. cynthia* possesses both a photoreceptor and an endogenous clock; it seems likely that a cryptochrome is the photoreceptor in this system.

The actual cellular molecular clock has been studied extensively in *Drosophila* where the central oscillator is formed by a complex feedback loop involving the transcription, translation, and phosphorylation of multiple genes: *period, timeless, clock, cycle, doubletime, pdp1, clockwork orange, vrille,* and *shaggy,* among others (see Allada and Chung, 2010; Tomioka and Matsumoto, 2010). PER and TIM proteins have been immunologically detected in the *Rhodnius* prothoracic gland (Vafopoulou *et al.*, 2010) and in the prothoracic gland portion of the *Drosophila* ring gland (Myers *et al.*, 2003). At least some of the proteins involved in the *Drosophila* central oscillator shuttle between the nucleus and the cytoplasm, and cycles of abundance of both protein and mRNA are characteristic. For a more detailed explanation of these processes, see Allada and Chung (2010).

The expression of the PER protein in insect brains has been the subject of several immunohistochemical studies in insects other than *Drosophila*. In the *Rhodnius* brain, a small group of neurons adjacent to the prothoracicotropes (one per hemisphere) were immunopositive for the PER and TIM proteins (Steel and Vafopoulou, 2006; Vafopoulou *et al.*, 2010). Sauman and Reppert (1996a) found that two pairs of dorsolateral neurons per hemisphere were immunopositive for PER in the brain of the moth *A. pernyi*, and that the cell bodies were in contact with the two pairs of dorsolateral neurosecretory cells that were immunopositive for PTTH. Interestingly, the PER positive cells sent their axons to the CC, whereas PTTH-producing neurons course contralaterally to terminate in the CA. Thus, if the PTTH- and PER-producing cells communicate with one another it is likely to be at the level of the cell bodies, perhaps via tight junctions, or in the CC, where the PER cells could transmit information to boutons on the axons of the PTTH neurons. Interestingly, only cytoplasmic and axonal PER staining were observed in the *Antheraea* brain (Sauman and Reppert, 1996b); the expected nuclear presence was not seen, although temporal cycling of protein and mRNA were observed. Zavodska *et al.* (2003) surveyed the cephalic nervous systems of 14 species of insects, representing six orders, using immunohistochemical methods and antibodies against PER, PTTH, pigment-dispersing hormone (PDH, also known as pigment-dispersing factor), and eclosion hormone. PDH-expressing neurons are also important in circadian processes, although their role appears to be information transfer between cells rather than direct participation in the intracellular molecular clock oscillator (Stengl and Homburg, 1994; Helfrich-Förster *et al.*, 2000). A detailed recapitulation of the data obtained in this study is beyond the purposes of this review, but a few results are worth noting. First, PER was widely distributed with most species showing antibody reactivity in cells of the optic lobe, pars intercerebralis, and the dorsolateral protocerebrum. Second, PER staining was limited to the cytoplasm. Third, PDH exhibited the most consistent immunoreactivity, being found in the optic lobes of all 14 species. Finally, none of the cells immunopositive for any one of the four antigens was also immunopositive for any of the other three antigens. Thus, to the extent that these four proteins play a role in circadian cycles, these data suggest that such cycles are controlled, at least within the brain, by a nexus of cells rather than a discrete nucleus.

The studies reviewed here suggest that there are several ways in which the PTTH-prothoracic gland axis may entrain rhythms in steroid levels. It is not yet possible to say whether these varied circadian systems sort out along taxonomic, ecological, or other grounds. Furthermore, we know essentially nothing about the removal or degradation of PTTH from the insect circulation. Changes in the PTTH hemolymph titer must be regulated not only by synthesis and release, but also by processes of sequestration and/or proteolysis. Whether or not there are PTTH-specific uptake, elimination, or proteolytic pathways is totally unknown, but it seems unlikely that the prothoracic gland will be the only tissue found to be involved in these processes. More work on these problems is clearly needed, involving not only molecular genetics, but biochemistry and physiology as well.

1.3.7. PTTH and Diapause

The endocrine control of diapause is the subject of Chapter 10. Features particularly relevant to PTTH are briefly considered in this section. Diapause is a type of dormancy that involves perception of one or more predictable environmental cues that anticipate and predict regularly occurring, unfavorable conditions (see Chapter 10). Animals in diapause express physiological mechanisms that greatly increase their ability to resist environmental challenges that range from desiccation to freeze-damage resistance. The cues that trigger diapause are not necessarily the conditions that the behavior has evolved to escape or moderate, for example, organisms that enter a winter diapause state may be stimulated to do so by shorter day length rather than by falling temperatures. Diapause involves a genetically determined response controlled by the neuroendocrine system. Diapause is obligate or facultative and undergone by embryos, larvae, pupae, or adults, depending on the species.

The PTTH-ecdysteroid axis has been implicated in the control of both larval and pupal diapause. *Manduca* larvae molt into diapausing pupae after being raised under a short-day (e.g., 12 h) photoperiod. Bowen *et al.* (1984a) found that such animals exhibit a much reduced hemolymph ecdysteroid titer relative to developing pupae. Low ecdysteroid titers have been documented in other Lepidoptera and other taxa that undergo a pupal diapause, for example, the moths *H. cecropia* (McDaniel, 1979) and *H. virescens* (Loeb, 1982), and the flesh flies *S. crassipalpis* (Walker and Denlinger, 1980) and *S. argyrostoma* (Richard *et al.*, 1987). Bowen *et al.* (1984a) showed that two factors resulted in the low levels of circulating ecdysteroids during diapause. First, diapause prothoracic glands were found to be refractory to PTTH stimulation when tested *in vitro*. Smith *et al.* (1986a) found that the refractory state in diapausing pupae could be simulated in non-diapausing pupae by removal of the brain, indicating that the presence of a brain-derived factor is necessary to retain prothoracic gland responsiveness. PTTH is one candidate, but a role for other hormones cannot be ruled out (see Section 1.5.2.). Smith *et al.* (1986a) also determined that the refractory "blockade" in the *Manduca* prothoracic gland is downstream from the PTTH receptor, Ca^{2+}-influx and cAMP generation (see Section 1.4. for a discussion of the PTTH transduction cascade). Both calcium ionophore and a cell-permeable cAMP analog failed to stimulate ecdysteroid synthesis in the diapause glands, although these agents are as effective as PTTH when applied to non-diapause glands. However, PTTH stimulated cAMP accumulation within diapause glands at levels higher than seen under non-diapause conditions. This observation indicates that there is no block upstream from cAMP generation, such as fewer PTTH receptors or reduced activity of adenylyl cyclase. Similar data, indicating a signal transduction lesion downstream from cAMP generation, were obtained by Richard and Saunders (1987) studying diapause in the flesh fly *Calliphora vicina*. Basal protein kinase A activity in *Manduca* day 10 diapause glands was ≈20% lower than that seen in non-diapause day 0 pupal glands, but both pupal and diapause protein kinase activity approximately doubled in response to a cAMP analog, dibutyryl cAMP (dbcAMP; Smith *et al.*, 1987a). Thus, lowered PKA activity is likely contributory to the refractory nature of diapause prothoracic glands, but other factors are undoubtedly involved.

Bowen *et al.* (1984a) suggested that the second reason for low ecdysteroid levels in diapause pupae is a lack of PTTH release. These workers found that in *Manduca* during the first 20 days of diapause the content of PTTH in the brain or in the brain-retrocerebral complex did not differ significantly from that in non-diapausing, developing pupae. Bowen and colleagues construed these data to indicate that PTTH release, rather than content, was greatly curtailed during diapause, but the interpretation of this result is, unfortunately, not straightforward since it is possible that PTTH catabolism in the hemolymph was increased. Note that the hemolymph PTTH titer was not measured in this case. Bowen *et al.* (1986) studied the release of PTTH *in vitro* from *Manduca* brain-retrocerebral complexes and found no difference between complexes from non-diapause and diapause pupae. PTTH release *in vitro* from diapause complexes could have been an artifact stemming from dissection or incubation conditions, or it might indicate a release from an *in vivo* inhibitory factor that is lost or inhibited itself under the *in vitro* regimen. If PTTH release is indeed inhibited *in vivo* in diapausing pupae, then data from nematodes may provide a clue to one step in its control. Tissenbaum *et al.* (2000) found that the diapause-like dauer state of *Caenorhabditis elegans* and *Ancylostoma caninum* was broken rapidly by muscarinic agents. Furthermore, dauer recovery was inhibited by atropine, a muscarinic antagonist. This

finding is of note because both Agui (1989) and Shirai *et al.* (1994) showed that cholinergic agents caused the release *in vitro* of PTTH from lepidopteran larval complexes. A dopaminergic pathway may also play a role in diapause, but in contrast to the muscarinic system dopamine may be involved in inducing or maintaining diapause (see Chapter 10). Some insects that undergo pupal diapause may exhibit a more complex control of the PTTH-prothoracic gland axis. Williams (1967) was not able to detect PTTH activity in brains from diapause pupae of *H. cecropia* or *S. cynthia*. Similarly, PTTH mRNA levels were seen to be reduced in diapause pupae of *H. virescens* (Xu and Denlinger, 2003) and *H. armigera* (Wei *et al.*, 2005). This contrasts with little difference in *A. pernyi* between diapause versus non-diapause pupae in PTTH activity (Williams, 1967), PTTH protein, or PTTH mRNA (Sauman and Reppert, 1996a).

Data from *Manduca* indicate that the PTTH-producing neurosecretory cells undergo morphological and electrophysiological changes during diapause. Hartfelder *et al.* (1994) found no significant differences in cellular ultrastructure between developing pupae and early diapause pupae. However, as diapause progressed, a number of cellular structures exhibited changes: neurosecretory granules were concentrated into large clusters that were separated by well-organized rough endoplasmic reticulum, notably fewer Golgi complexes were seen, and interdigitations with glial cells were decreased. When diapause was terminated by exposure to warm conditions (26° vs. 4°C), the PTTH cells reverted to pre-diapause morphology within three days. Progressive differences in electrophysiological properties between diapausing and developing pupae were found by Tomioka *et al.* (1995), who noted that the diapausing PTTH cells gradually became less excitable. This relative refractoriness was caused by a rise in the threshold value for action potential generation and a decrease in the input resistance.

Reduced PTTH release during diapause suggests that the photoperiodic clock involved in the control of PTTH in short-day versus long-day developmental programs might reside in the brain where PTTH is synthesized. This hypothesis was elegantly addressed with experiments that involved *in vitro* reprogramming of larval brains with altered light cycles (Bowen *et al.*, 1984b). In brief, brains from short-day larvae were exposed to a long-day light treatment for three days *in vitro* and then implanted into short-day larvae destined to enter diapause. Approximately half of these host larvae did not enter diapause, that is, the implanted brains reversed the diapause program. If the transplanted brains received a short-day light treatment *in vitro* prior to implantation, then very few host larvae failed to enter diapause. The composite data showed that the three main regulators of pupal diapause in *Manduca* reside in the brain: photoreceptor, clock, and hormonal effector. The main hormonal actor in this developmental program is clearly PTTH, but other factors might well be involved. For instance, brains in animals destined to diapause might secrete an ecdysiostatic factor to downregulate prothoracic gland activity in addition to curtailing PTTH release after pupal ecdysis (see Section 1.5.2.3.).

The endocrine control of larval diapause appears to be more varied than pupal diapause. In the European corn borer, *Ostrinia nubilalis*, larval diapause is similar to *Manduca*. Hemolymph ecdysteroid titers are significantly lower in pre-diapause larvae than in developing animals (Bean and Beck, 1983), and brain PTTH levels were the same in diapause and non-diapause animals for at least the first four weeks of diapause (Gelman *et al.*, 1992). Gelman and colleagues also found that diapause prothoracic glands were highly refractory to PTTH stimulation, as seen in *Manduca*. It must be pointed out that although these results are very reminiscent of the *Manduca* findings, the only PTTH activity detected in diapause brains of *Ostrinia* had an apparent molecular weight of ≤5 kDa (Gelman *et al.*, 1992). In contrast, the southwestern corn borer *Diatraea grandiosella*, which undergoes up to three stationary molts during diapause, appears to maintain an active PTTH-prothoracic gland axis during larval diapause; its diapause is controlled by high JH levels (Chippendale and Yin, 1973; Yin and Chippendale, 1973; Chippendale, 1984; see Chapter 11). High JH levels are also responsible for provoking and sustaining larval diapause in the rice-stem borer *Chilo suppressalis* (Yagi and Fukaya, 1974); the exact state of the PTTH-prothoracic gland axis during larval life in this species appears to be unknown.

The literature on diapause in insects is very large (see Chapter 10) and only a very small fraction of it has been reviewed here. Given the long evolutionary history of insects, it is likely that there are additional variations in the ways in which environmental cues, receptors, PTTH, and the prothoracic glands interact to induce or modulate diapause.

1.4. PTTH: Effects on Prothoracic Gland Signaling and Steroidogenic Enzymes

This section considers the intracellular changes in the prothoracic gland elicited by PTTH and the relationship of these changes to ecdysteroidogenesis. The reader is referred to Chapter 4 for a discussion of the actual synthetic pathway leading from sterols to ecdysteroids. The present discussion has been organized to present the biochemical events of PTTH signal transduction beginning with the PTTH receptor and continuing through changes in protein synthesis that are likely to stimulate ecdysteroidogenesis. An overview of the PTTH signal transduction pathway is provided in **Figure 6**. Briefly, it appears that PTTH binds to a receptor tyrosine kinase, which leads to elevated intracellular Ca^{2+}, enhanced synthesis of cAMP,

and activation of downstream kinases including protein kinase A (PKA) and PKC. The mitogen-activated protein (MAP) kinase cascade is activated as is the target of rapamycin (TOR) signaling cascade with a resulting increase in protein synthesis. Ecdysteroid synthesis and secretion are increased by a translation-dependent step. Each of these aspects of PTTH action is discussed separately in the following sections. While they are presented sequentially, multiple PTTH-dependent phenomena occur in close temporal order; indeed, it would be a mistake to conceive of any signal transduction cascade as a linear, vectorial process, rather than as a set of interconnected pathways.

1.4.1. The PTTH Receptor and Tyrosine Kinase Activity

Identification of the PTTH receptor has been one of the most important recent developments in PTTH research (Rewitz *et al.*, 2009b). In the 1990s, PTTH was suggested to bind to a receptor with tyrosine kinase activity based on the predicted similarity of PTTH in three-dimensional conformation to mammalian growth factors (Noguti *et al.*, 1995; see Section 1.2.). A genomic search revealed sequence homologies between PTTH and Trunk, an embryonic ligand for a receptor tyrosine kinase known as Torso. Both PTTH and Trunk contain a cysteine-knot motif in the C-terminus and conserved cysteines responsible for intra-monomeric bonding (Marchal *et al.*, 2010; Rewitz *et al.*, 2009b). Trunk plays a key role in embryonic anterior-posterior patterning. Its receptor, Torso, became a prime candidate for the PTTH receptor when it was seen to be specifically expressed in larval *Drosophila* prothoracic glands (Rewitz *et al.*, 2009b). Trunk is not found in third instar larvae, suggesting that PTTH is the Torso ligand at this stage. RNAi-mediated inactivation of Torso in the prothoracic glands leads to delayed metamorphosis and increased size, mimicking knockdown of PTTH. These effects are reversed by feeding of 20E.

The ligand sensitivity of Torso was explored using the *Bombyx* Torso homologue, as active *Drosophila* PTTH is not available (Rewitz *et al.*, 2009b). *Bombyx* Torso, ectopically expressed in *Drosophila* S2 cells, is activated by 1 nM recombinant *Bombyx* PTTH, leading to ERK activation. In *Bombyx* prothoracic glands from fifth instar larvae, Torso expression is correlated with the acquisition of steroidogenic competence. Torso is found in *Bombyx* ovaries and testes, suggesting possible reproductive functions. This is in keeping with other studies suggesting a reproductive role for PTTH. For example, in *Drosophila*, viable adults that emerge following ablation of PTTH-producing neurons exhibit reduced fecundity (females) and aberrant courtship behavior (males; McBrayer *et al.*, 2007). In *Manduca*, PTTH is found in the adult brain and is capable of stimulating protein phosphorylation and protein synthesis in female and male reproductive tissues (Rybczynski *et al.*, 2009).

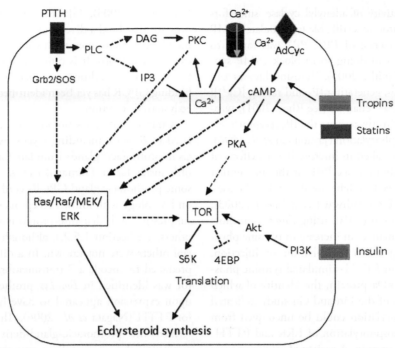

Figure 6 PTTH signal transduction cascade in prothoracic gland cells. Solid lines indicate demonstrated or highly likely interactions, and dashed lines indicate hypothetical relationships. Rectangles = receptors for PTTH (blue); tropins, i.e., other stimulatory factors (green); statins, i.e., inhibitory factors (red); insulin, including bombyxin (yellow); cylinder = plasma membrane calcium channel; diamond = calcium/calmodulin-sensitive adenylyl cyclase. Abbreviations: AdCyc, adenylyl cyclase; Akt, protein kinase B; cAMP, cyclic adenosine monophosphate; DAG, diacylglycerol; 4EBP, binding protein for ribosomal initiation factor 4E; ERK, extracellular signal-regulated kinase; Grb2/SOS, adapter and guanine-nucleotide-exchange factor for Ras; IP$_3$, inositol trisphosphate; MEK, MAPK/ERK kinase; PLC, phospholipase C; PKA, protein kinase A; PKC, protein kinase C; PI3K, phosphoinositide 3-kinase; Ras, Ras GTP-binding protein; Raf, Raf serine/threonine kinase; S6, ribosomal protein S6; S6K, 70 kDa S6 kinase; TOR, target of rapamycin.

As noted by Rewitz *et al.* (2009b), it is striking that a single tyrosine kinase was conserved for use in determining embryonic terminal cell fate as well as adult body size by terminating larval growth. Trunk is postulated to be cleaved by a protein, named torso-like, generating the active ligand at appropriate embryonic locations. This raises the possibility that cleavage of PTTH by specific prothoracic-gland-associated proteins may contribute to its activity at specific times in development.

Torso appears to act via the Ras-Raf-ERK pathway discussed in Section 1.4.3.3., as evidenced by similar phenotypes generated by Ras, Raf, or particularly ERK knockdown. Ras overexpression in the prothoracic gland rescues the Torso knockdown phenotype. A gap currently exists, however, in our understanding of the links between PTTH binding and ERK activation and between activation of a tyrosine kinase-linked receptor and increased levels of Ca2+ and cAMP.

In addition, it remains to be determined whether the tyrosine kinase activity previously associated with the stimulation of lepidopteran prothoracic glands by PTTH is mediated by Torso. In an early study of prothoracic gland tyrosine kinases, an antibody directed against the vertebrate insulin receptor was found to

immunoprecipitate a 178 kDa prothoracic gland protein in *Manduca* (Smith *et al.*, 1997). It is unlikely that the antibody, directed against the entire β subunit of the human insulin receptor, specifically immunoprecipitated the *Manduca* insulin receptor. Sequence information for the lepidopteran insulin receptor suggests a β subunit size closer to a conventional 90 kDa (Swevers and Iatrou, 2003; Koyama *et al.*, 2008). The Smith *et al.*, 1997 study was, however, the first demonstration of hormone-sensitive tyrosine kinase activity in the prothoracic glands, manifested as increased tyrosine phosphorylation of the 178 kDa protein in response to partially purified PTTH. In a later study, recombinant PTTH was shown to increase tyrosine phosphorylation of at least four prothoracic gland proteins seen on Western blots using antibody to phosphorylated tyrosine (Smith *et al.*, 2003). Phosphoproteins were not identified beyond migration on one-dimensional electrophoretic gels, although a smaller one appeared to be ERK based on size and sensitivity to MEK inhibition.

PP1, an inhibitor of the Src tyrosine kinase family, was seen to repress PTTH-stimulated cAMP and ecdysteroid synthesis in *Manduca* prothoracic glands (Smith *et al.*, 2003; Priester and Smith, 2005). PP1 did not appear to block the catalytic activity of adenylyl cyclase, since forskolin, a cyclase activator, was still able to stimulate cAMP production in the presence of PP1. PP1 was found to inhibit Ca^{2+} entry and in doing so to block cAMP synthesis (Priester and Smith, 2005). Tyrosine kinases can activate phospholipases, generating IP_3 and raising $[Ca^{2+}]i$ via the IP_3 receptor/ Ca^{2+} channel (see Rhee, 2001); tyrosine kinases can also directly activate the IP_3 receptor/Ca^{2+} channel through phosphorylation (Jayaraman et al., 1996). However, care must be taken in interpreting the effects of this single kinase inhibitor, since PP1, at the micromolar concentrations employed, might inhibit other kinases, including those in the ERK pathway (see Bain et al., 2003).

A study by Lin and Gu (2010) using Bombyx prothoracic glands also documented an increase in tyrosine phosphorylation following PTTH incubation or injection. The authors focused on PTTH-stimulated tyrosine phosphorylation of a 120 kDa protein, the identity of which is not known. Results of the Lin and Gu study indicated that tyrosine phosphorylation could be uncoupled from PTTH-stimulated phosphorylation of ERK and PTTH-stimulated ecdysone secretion. A variety of inhibitors were tested including the tyrosine kinase inhibitor genistein. Genistein blocked phosphorylation of the 120 kDa protein and ecdysone secretion, but did not block the phosphorylation of ERK. However, genistein is a plant-derived steroid that may block steroidogenesis by means other than kinase inhibition, including effects on the ecdysteroid receptor (Oberdorster et al., 2001). PP2, a Src-family kinase inhibitor similar to PP1, partially blocked tyrosine phosphorylation of the 120 Da protein without affecting ERK or steroidogenesis, in contrast to the effects of PP1 on Manduca prothoracic glands (Gu et al., 2010; Lin and Gu, 2010). An inhibitor known as HNMPA-(AM₃), directed against the insulin receptor, blocked both phosphorylation of ERK and steroidogenesis without affecting the 120 kDa protein (Gu et al., 2010; Lin and Gu, 2010). To reiterate, kinase inhibitors are not specific for their intended targets, hence some of these results are likely to be mediated by off-target enzymes. However, it is clear that not all tyrosine phosphorylation events that accompany PTTH stimulation are required for acute steroidogenesis. It remains to be determined which aspects of tyrosine phosphorylation are directly stimulated by PTTH-Torso interactions; of these, which are necessary for ecdysone secretion; and the specific role played by additional tyrosine kinases, such as those associated with insulin-like receptors, in prothoracic gland function (see Chapter 2).

1.4.2. G-protein-coupled Receptors and Second Messengers: Ca^{2+}, IP_3, and cAMP

1.4.2.1. G-protein-coupled receptors   For many years the PTTH receptor was thought to be a member of the G-protein-coupled receptor (GPCR) superfamily (see Rybczynski, 2005). GPCRs are a very large group of proteins that bind a diverse array of ligands, and span the membrane seven times, with connecting intracellular and extracellular loops. It is quite possible that PTTH binds to a GPCR in addition to Torso. However, no PTTH-binding GPCR has yet been identified that also stimulates ecdysteroid secretion.

A number of genes homologous to GPCRs were isolated in B. mori, including several expressed in high levels in prothoracic glands from late fourth instar and early fifth instar larvae (Yamanaka et al., 2008). Ligands for some prothoracic gland GPCRs could be predicted based on homology with known receptors (e.g., those for myosuppressin and diapause-promoting hormone, among others, see Section 1.5.2.), although the ligands for several others were not known. In a different paper, a gene predicted to encode a 7-transmembrane-spanning receptor was identified in Bombyx prothoracic glands, which upon expression appeared to have high binding affinity for PTTH (Nagata et al., 2006). This protein, currently termed the prothoracic-gland-derived-receptor (Pgdr), resembles the mammalian growth-hormone-inducible transmembrane protein (Ghitm) (Yoshida et al., 2006). The protein is not a conventional G-protein linked receptor, but rather belongs to a family of anti-apoptotic proteins with Bax-inhibitory domains (BI-1), found in C. elegans and Drosophila as well as mammals (Yoshida et al., 2006; Reimers et al., 2007). mRNA for Bombyx Pgdr was expressed in a variety of tissues, including brain, salivary glands, and testes, and was expressed in particularly high levels in post-wandering larvae. The homology of Pgdr and anti-apoptotic proteins raises the possibility that the Bombyx Pgdr may mediate anti-apoptotic actions of PTTH or other ligands. However, the function of Pgdr has not been characterized further in prothoracic glands.

1.4.2.2. Ca^{2+} and IP_3 Abundant evidence links the action of PTTH with second messengers traditionally linked to G-proteins, including Ca^{2+} and cAMP. The hormone-sensitive heterotrimeric G-proteins (α,β, and γ) are a diverse group, and they in turn regulate a variety of effector enzymes. For example, $G\alpha_s$ stimulates adenylyl cyclase, which converts ATP to cAMP, while $G\alpha_q$ can activate PLC (see Offermanns, 2003). PLC hydrolyzes phosphatidylinositol bisphosphate (PIP2) generating diacylglycerol and inositol trisphosphate (IP_3); IP_3 in turn, binds to calcium channels in the endoplasmic reticulum, which enhances release of Ca^{2+} into the cytosol (see Berridge et al., 2000).

Ca^{2+} in particular has been shown to play a critical, early role in the action of PTTH. The modes by which free Ca^{2+} enters the cytoplasm of prothoracic glands in response to PTTH stimulation have been addressed chiefly in Manduca, Galleria, and Bombyx. The composite data indicate that PTTH-stimulated increases in intracellular Ca^{2+} originate from two Ca^{2+} pools: the first an internal, IP_3-controlled endoplasmic reticulum store and the second from outside of the cell.

Smith *et al.* (1985) first noted a requirement for extracellular Ca^{2+} in PTTH-stimulated ecdysteroid synthesis by *Manduca* prothoracic glands. Ca^{2+}-influx appeared to be a very early event in the PTTH transduction pathway; cAMP analogs that stimulated ecdysteroid synthesis were still effective in the absence of extracellular Ca^{2+}, indicating that Ca^{2+}-influx preceded cAMP synthesis (see 1.4.2.3.). External Ca^{2+} was shown to be important for PTTH-stimulated ecdysteroidogenesis in at least two other insect species. PTTH-stimulated ecdysteroid synthesis by *Bombyx* prothoracic glands also required the presence of Ca^{2+} in the external medium (Gu *et al.*, 1998), and PTTH-containing neural extracts failed to stimulate ecdysteroid synthesis by *Drosophila* ring glands in the absence of external Ca^{2+} (Henrich, 1995). The importance of Ca^{2+} influx to PTTH-stimulated ecdysteroid synthesis was further demonstrated by the ability of the Ca^{2+} ionophore A23187 to stimulate ecdysteroidogenesis in *Manduca* (Smith and Gilbert, 1986) and *Bombyx* (Gu *et al.*, 1998) prothoracic glands.

The calcium requirement for basal steroidogenesis is less clear. Smith *et al.* (1985) reported that basal ecdysteroidogenesis was insensitive to the presence or absence of external Ca^{2+}, as also later noted by Gu *et al.* (1998) for *Bombyx* prothoracic glands, and Henrich (1995) for *Drosophila* ring glands. In contrast, Meller *et al.* (1990) found that basal ecdysteroidogenesis in *Manduca* glands was decreased, although not absent, in Ca^{2+}-free medium, and varied with stage of development. Dedos and Fugo (1999b) uncovered a similar, developmentally specific Ca^{2+}-dependency for basal ecdysteroid synthesis in *Bombyx* glands. Thus, whether or not basal ecdysteroidogenesis is responsive to short-term changes in external Ca^{2+} is not certain; differences in experimental technique might well determine a gland's sensitivity in this regard.

Identification of the precise types of Ca^{2+} channels that function in PTTH signaling has proven difficult. In *Manduca* and *Bombyx*, cationic lanthanum inhibited PTTH-stimulated ecdysteroid synthesis (Smith *et al.*, 1985; Birkenbeil and Dedos, 2002); lanthanum is a nonspecific blocker of many plasma membrane ion channels. The Ca^{2+} channel agonist Bay K8644, which activates voltage-gated Ca^{2+} channels, stimulated ecdysteroid synthesis in both *Manduca* and *Bombyx* prothoracic glands (Smith *et al.*, 1987b; Dedos and Fugo, 1999b), but it is not known how this compound interacts with PTTH-stimulated Ca^{2+} influx. In *Manduca*, the L-type channel blocker nitrendipine partially inhibited PTTH-stimulated ecdysteroid synthesis by prothoracic glands (Smith *et al.*, 1987b; Girgenrath and Smith, 1996). However, nitrendipine was not found to inhibit PTTH-stimulated Ca^{2+} influx by Birkenbeil (1996, 1998). To confuse the issue more,

an inhibitory effect of nifedipine, another L-type channel blocker, was demonstrated by Fellner *et al.* (2005) while Birkenbeil (1998) did not observe any effect of nifedipine. Birkenbeil did note an inhibitory effect of amiloride, a T-type Ca^{2+} channel blocker, on PTTH-stimulated Ca^{2+} influx, but, at the concentration used, amiloride affects other proteins in addition to the T-type Ca^{2+} channels. In *Bombyx*, similarly conflicting data have been obtained. The L-type channel inhibitors verapamil (Gu *et al.*, 1998) and nitrendipine (Dedos and Fugo, 2001) both inhibited PTTH-stimulated ecdysteroidogenesis in *Bombyx* prothoracic glands, but nitrendipine had no apparent effect on PTTH-stimulated cytoplasmic free Ca^{2+} ($[Ca^{2+}]_i$; Birkenbeil and Dedos, 2002). The contribution of T-type Ca^{2+} channels to *Bombyx* PTTH signaling is not clear; two inhibitors of this channel class, bepridil and mibefradil, had no effect on PTTH-stimulated $[Ca^{2+}]_i$, while the less specific amiloride inhibited a PTTH-stimulated increase in $[Ca^{2+}]_i$. These perplexing results may reflect methodological differences among the studies, as well as inherent problems associated with identification of insect channel types based on pharmacological reagents characterized in vertebrates (Girgenrath and Smith, 1996; Rybczynski and Gilbert, 2003).

Several studies using microfluorometric Ca^{2+} measurements explicitly verified the assumption that PTTH increases prothoracic gland $[Ca^{2+}]_i$ and that the external environment was the primary source of this Ca^{2+}. Birkenbeil (1996) showed that a PTTH preparation caused an increase in $[Ca^{2+}]_i$ in prothoracic glands of the wax moth *G. mellonella* while Birkenbeil and Dedos (2002) showed that PTTH caused Ca^{2+} influx in *Bombyx* glands. In *Manduca*, Birkenbeil (1998, 2000) showed that PTTH preparations stimulated Ca^{2+} influx, and this work has been extended with pure *Manduca* recombinant PTTH (Fellner *et al.*, 2005; Priester and Smith, 2005).

Fellner *et al.* (2005) investigated the kinetics of PTTH-stimulated increases in $[Ca^{2+}]_i$ in *Manduca* prothoracic glands and found that PTTH caused a rapid increase in $[Ca^{2+}]_i$ with levels plateauing within ≈ 30 sec to 1 min. Furthermore, when prothoracic glands were incubated in the absence of external Ca^{2+} a small, transient $[Ca^{2+}]_i$ peak was seen followed by a large peak if Ca^{2+} was added to the external medium (see **Figure 7A**). These results suggested that PTTH first causes a release of Ca^{2+} from internal stores followed by a larger influx of external Ca^{2+}, with the large influx of Ca^{2+} due substantially to the opening of store-operated channels. Store-operated channels are Ca^{2+} channels that respond to increases in internal Ca^{2+} (see Clementi and Meldolesi, 1996). The compound 2-aminoethoxydiphenyl borate (APB), an inhibitor of Ca^{2+} release from IP_3-sensitive stores in the endoplasmic reticulum, inhibited the PTTH-stimulated increase of $[Ca^{2+}]_i$, indicating that PTTH acts through the IP_3 receptor (**Figure 7B**). TMB-8, a second inhibitor of the release of Ca²⁺ from internal stores, also greatly diminished the PTTH-stimulated increase of [Ca²⁺]ᵢ, thus confirming the early participation of internal Ca²⁺ stores in the PTTH signal transduction pathway. It should be noted that Birkenbeil (1996, 1998) did not observe TMB-8 inhibition of PTTH-stimulated Ca²⁺ influx in *Galleria* or *Manduca* prothoracic glands. In the Fellner *et al.* (2005) study, the participation of an IP₃-controlled endoplasmic reticulum store in the early events of PTTH-stimulated Ca²⁺ mobilization was further confirmed by the ability of a PLC inhibitor (U73122) to block or interrupt [Ca²⁺]ᵢ increases (**Figure 7C**); PLC activity is required to generate IP₃ from its PIP₂ precursor.

The results with *Manduca* were largely congruent with those reached in studies utilizing *Bombyx* prothoracic glands (Birkenbeil and Dedos, 2002; Dedos *et al.*, 2005). A small [Ca²⁺]ᵢ peak at 1min was followed by a larger increase some minutes later. A difference was seen in that maximal levels of PTTH-stimulated [Ca²⁺]ᵢ were

reached more slowly in *Bombyx* (from ≈2 to >10 min) than in *Manduca* (≈30 sec to 1 min). Dedos *et al.* (2005) observed inhibitory effects of both TMB-8 and APB in the prothoracic glands of *Bombyx*, implicating a contribution from intracellular Ca²⁺ stores. The prolonged nature of the *Bombyx* [Ca²⁺]ᵢ response is surprising; when [Ca²⁺]ᵢ increases are a primary response to receptor activation, the kinetics of [Ca²⁺]ᵢ increases are generally rapid and temporally stereotyped. A later study revealed that phosphorylation of store-operated Ca²⁺ channels in *Bombyx* by PKA and PKC reduced the entry of Ca²⁺ (Dedos *et al.*, 2008). However, such an effect of kinase activation on Ca²⁺ channels would be expected to curtail, rather than to prolong, Ca²⁺ influx. Regardless of the differences between *Manduca* and *Bombyx*, the *Bombyx* data are consistent with the interpretation that PTTH causes release of Ca²⁺ from internal pools followed by the opening of store-operated channels and voltage-gated channels.

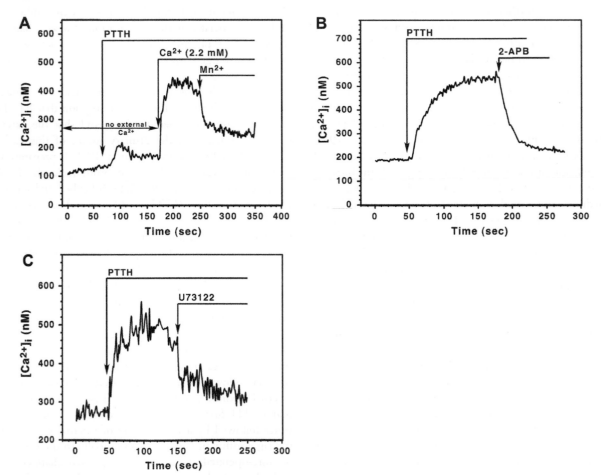

Figure 7 Free Ca²⁺ levels in prothoracic gland cells of *Manduca sexta* larvae, measured by fura-2 fluorescence. Data shown are representative plots from single cells. (A) The effect of PTTH on free cellular Ca²⁺ levels in the absence of external Ca²⁺, and following the addition of Ca²⁺ to the incubation medium. The drop in free cellular Ca²⁺ levels following the addition of Mn²⁺ to the external medium is due to the entry of Mn²⁺ through store-operated channels. Mn²⁺-dependent quenching of the Ca²⁺ fura-2 fluorescence indicates that the observed fluorescence results from Ca²⁺ binding to the Fura, and not the simple accumulation of the dye within the cell. (B) 2-APB, an inhibitor of Ca²⁺ release from IP₃ gated stores on free cellular Ca²⁺ levels, truncates the PTTH-stimulated rise in free cellular Ca²⁺. (C) U73122, an inhibitor of PLC, ablates the PTTH-stimulated rise in the concentration of free cellular Ca²⁺. Reproduced with permission from *Fellner et al.* (2005).

The existence of a PTTH-stimulated, small $[Ca^{2+}]_i$ peak in *Manduca*, and presumably in *Bombyx*, sensitive to APB and TMB-8 inhibition, suggested that inositol metabolism played a role in PTTH-stimulated signal transduction, that is, the generation of IP_3 by PLC and the subsequent release of IP_3-sensitive endoplasmic reticulum Ca^{2+} stores. Support for this hypothesis came from both *Bombyx* and *Manduca*. In the silkworm, the PLC inhibitor, 2-nitro-4-carboxyphenyl-*N*, *N*-diphenylcarbamate, blocked PTTH-stimulated ecdysteroidogenesis but had no effect on basal synthesis. In *Manduca*, the PLC inhibitor U73122, likewise blocked PTTH-stimulated ecdysteroidogenesis and was without effect on basal production (Rybczynski and Gilbert, 2006). Smith and Gilbert (1989) postulated the participation of PLC in PTTH-stimulated ecdysteroidogenesis, but attempts to detect PTTH-stimulated increases in PIP_2 metabolites like IP_3 have not been successful (Girgenrath and Smith, 1996; Dedos and Fugo, 2001). The failure to detect IP_3 might reflect a difficult-to-detect, small, and transient IP_3 peak, if the APB-sensitive change in $[Ca^{2+}]_i$ is any guide to IP_3 behavior in the prothoracic gland. Indirect evidence supporting the participation of the IP_3 pathway in ecdysteroidogenesis has also resulted from the study of *Drosophila* carrying mutations in the IP_3 receptor gene (Venkatesh and Hasan, 1997; Venkatesh *et al.*, 2001). Mutations in this gene caused molting delays and larval lethality that could be partially reversed by feeding 20E.

Intracellular compartmentalization of early-acting components of the signal transduction system might make small changes biologically meaningful and eliminate the need for large-scale intracellular increases in IP_3 or DAG. It is possible that a subtle increase in IP_3 is an effective stimulus for opening IP_3 receptors due to co-activation of adenylyl cyclase. In vertebrates, hormones such as parathyroid hormone have been found to sensitize IPc receptors by facilitating direct association of the IP_3 receptor with adenylyl cyclase (Tovey *et al.*, 2008); a similar effect may accompany prothoracic gland activation by PTTH.

Despite evidence for the involvement of PLC, and of IP_3 receptors, in the action of PTTH on Ca^{2+}, G-protein involvement is unlikely at this step. Pertussis toxin, an inhibitory toxin that uncouples G-proteins from receptors, had no effect on PTTH-stimulated ecdysteroid synthesis by *Manduca* prothoracic glands (Girgenrath and Smith, 1996), or on PTTH-stimulated Ca^{2+}influx (Birkenbeil, 2000). Several additional pharmacological agents that are known to affect GPCR–G-protein interactions have been used to further investigate the role of G-proteins in PTTH signaling. Mastoparan, a wasp venom that activates G-proteins, caused a dose-dependent increase in intracellular Ca^{2+} (Birkenbeil, 2000; Dedos *et al.*, 2007), but a dose-dependent decrease in ecdysteroid production in *Manduca* prothoracic gland cells (Birkenbeil, 2000). A later study revealed that the site

of action of mastoparan on Ca^{2+} influx differed between *Manduca* and *Bombyx* (Dedos *et al.*, 2007) with the Ca^{2+}-mobilizing effects of mastoparan in *Manduca* occurring through G-protein-regulated plasma membrane channels, distinct from plasma membrane channels regulated by PTTH. By contrast, the Ca^{2+}-mobilizing effects of mastoparan in *Bombyx* occurred from intracellular stores in a non-G-protein-dependent manner. The compound suramin, which uncouples G-proteins from GPCRs, blocked PTTH-stimulated Ca^{2+} influx in *Bombyx* (Birkenbeil, 2000) while in *Manduca*, suramin inhibited PTTH-stimulated ecdysteroid synthesis (Rybczynski and Gilbert, unpublished observations). Given the wide-ranging effects of suramin including, for example, acting as an antagonist of purinergic receptor binding (Lambrecht, 2000), results using this agent do not conclusively argue for involvement of G-proteins in prothoracic gland activity. Interestingly, store-operated channels in the plasma membrane of *Bombyx* prothoracic glands appeared to be opened by a functional IP_3 receptor based on pharmacological characteristics without G-proteins (Dedos *et al.*, 2005). Injection of GDP-β-S, an inhibitor of G-protein activation, failed to block PTTH-stimulated Ca^{2+} influx; PTTH-stimulated ecdysteroid synthesis was not measured in this context (Birkenbeil, 2000; Dedos *et al.*, 2005). The observed increase in intracellular Ca^{2+}, initially arising from release of an IP_3-sensitive intracellular store of Ca^{2+}, is best explained by a non-G-protein-mediated increase in the activity of PLC. This could be generated by an isoform of PLC that is activated by receptor tyrosine kinases, such as the vertebrate PLCγ (Poulin *et al.*, 2005; and see Harden *et al.*, 2009).

PTTH may also stimulate changes in glandular $[Ca^{2+}]_i$ indirectly. Electrophysiological studies using *Manduca* prothoracic glands *in vitro* revealed that Ca^{2+}-dependent action potentials could spread from cell to cell (Eusebio and Moody, 1986). PTTH and signals from nerves (see Section 1.5.2.3.4.) are candidate triggers for such action potentials, but whether or not this occurs *in vivo* is unknown. Gap junctions and intercellular bridges have been found in *Manduca* prothoracic glands and these structures may allow the spread of signaling molecules, such as Ca^{2+} and cAMP, among cells (Dai *et al.*, 1994). Thus, Ca^{2+}-dependent action potentials, gap junctions, and intercellular bridges all could serve to synchronize the behavior of the prothoracic gland cells in response to PTTH or other hormones. Measurements of resting $[Ca^{2+}]_i$ in *Manduca* cells indicated that cells of the prothoracic gland are indeed heterogeneous (Fellner, Rybczynski, and Gilbert, unpublished observations) but, as yet, the significance of cell-to-cell differences is unknown. If cells are heterogeneous *in vivo* in their sensitivity to PTTH or other signals, signal-spreading among cells could serve to make the whole prothoracic gland respond at signal levels set by the most responsive cells. Note that a synchronization of cellular responses could downregulate the prothoracic gland's response to PTTH just as readily as facilitate it.

Many, if not all, intracellular signaling systems exhibit downregulation after a time, with signaling intermediates returning to basal levels. During the continuous presence of PTTH, $[Ca^{2+}]_i$ levels persisted above basal for at least 25 min in *Bombyx* prothoracic glands, with a peak at about 15 min after initiating PTTH stimulation; after transfer of stimulated glands to PTTH-free conditions, $[Ca^{2+}]_i$ levels remained elevated for at least 20 min (Birkenbeil and Dedos, 2002). Similar $[Ca^{2+}]_i$ data for prolonged exposure to PTTH are not available for *Manduca* prothoracic glands. In the prothoracic glands of *Galleria* sixth and seventh (final) larval instars, Birkenbeil (1996) found developmental changes in $[Ca^{2+}]_i$, with peaks co-occurring coincident with peaks of *in vitro* ecdysteroid production. This observation suggested a relatively slow return to basal conditions and/or non-adaptation to continuous PTTH stimulation. In contrast, intracellular Ca^{2+} was essentially constant throughout the fifth instar in glands of *Manduca* larvae (Birkenbeil, 1998). The mechanisms by which prothoracic gland $[Ca^{2+}]_i$ is restored to the basal level has not been explicitly defined, but presumably involves buffering by Ca^{2+}-binding proteins, reloading of internal stores, and transport to the extracellular milieu.

1.4.2.3. Cyclic AMP   A role for cAMP in PTTH-stimulated ecdysteroid synthesis was first suggested by measurements of intracellular cAMP and of adenylyl cyclase in larval *Manduca* prothoracic glands (Vedeckis and Gilbert, 1973; Vedeckis *et al.*, 1974, 1976). An explicit demonstration of the role of cAMP in ecdysteroid synthesis was obtained a decade later when *Manduca* PTTH preparations were shown to stimulate cAMP synthesis in a Ca^{2+}-dependent manner, that is, extracellular Ca^{2+} was required for PTTH-stimulated cAMP synthesis (Smith *et al.*, 1984, 1985, 1986b). Cell-permeable, excitatory cAMP analogs stimulated ecdysteroid synthesis in *Manduca* (Smith *et al.*, 1984, 1985; Smith and Gilbert, 1986; Rybczynski and Gilbert, 1994) and *Bombyx* (Gu *et al.*, 1996, 1998; Dedos and Fugo, 1999c). In contrast to results with Lepidoptera, however, cAMP did not stimulate ecdysteroid production in *Drosophila* (see Huang *et al.*, 2008). A specific link between PTTH and the steroidogenic effects of cAMP came from studies using an antagonist form of cAMP (Rp-cAMP). The antagonist substantially, although not completely, inhibited PTTH-stimulated ecdysteroidogenesis in *Manduca* (Smith *et al.*, 1996). Further data supporting the upstream position of $[Ca^{2+}]_i$ relative to cAMP in the PTTH transduction cascade included the observation that calcium ionophore mimics PTTH in eliciting both steroid and cAMP synthesis (Smith *et al.*, 1985, 1986b). The cell-permeable cAMP analog dibutyryl cAMP (dbcAMP), as well as

agents that activated adenylyl cyclase and inhibited cAMP phosphodiesterase activity, stimulated ecdysteroid synthesis in prothoracic glands incubated in Ca^{2+}-free medium (Smith *et al.*, 1985, 1986b). Note that these *Manduca* studies utilized a partially purified PTTH, but identical results were later obtained with a recombinant PTTH (Gilbert *et al.*, 2000). Recombinant PTTH was also used to demonstrate that PTTH-stimulated Ca^{2+} influx was required for cAMP synthesis (Priester and Smith, 2005). Similar results have been obtained with *Bombyx* (Gu *et al.*, 1996, 1998; Dedos and Fugo, 1999b). Furthermore, *Bombyx* prothoracic glands stimulated with PTTH under Ca^{2+}-replete conditions continued to secrete ecdysteroids at elevated levels even after transfer to a Ca^{2+}-free medium (Birkenbeil and Dedos, 2002). Thus, if sufficient cAMP has been synthesized, extracellular Ca^{2+} is apparently no longer needed, at least under short-term *in vitro* conditions; any Ca^{2+}-dependent events that occur subsequent to cAMP generation must depend on $[Ca^{2+}]_i$ for time intervals of at least 2 h, a common *in vitro* incubation period.

Prothoracic gland adenylyl cyclase is stimulated by $G\alpha_s$, and by Ca^{2+} in conjunction with calmodulin. Meller and Gilbert (1990), using antibodies against mammalian $G\alpha$ and $G\beta$ proteins, found several types of G-proteins in *Manduca* prothoracic glands including $G\alpha_s$ associated with adenylyl cyclase. Both *Bombyx* and *Manduca* prothoracic glands responded to cholera toxin, an activator of $G\alpha_s$, with increased ecdysteroidogenesis (Girgenrath and Smith, 1996; Dedos and Fugo, 1999c). At present it appears likely that the prothoracic gland $G\alpha_s$-adenylyl cyclase complex is activated only indirectly by PTTH. The same G-protein-sensitive adenylyl cyclase may, however, be activated directly by other ligands such as diapause hormone for which a GPCR has been found in *Bombyx* prothoracic glands (Yamanaka *et al.*, 2008; see Section 1.5.2.2.).

Adenylyl cyclase, which synthesizes cAMP from ATP, is found in multiple forms in vertebrates. The activity of these forms may be stimulated or inhibited by or be insensitive to G-proteins or Ca^{2+}-calmodulin. In *Drosophila*, there appear to be nine isoforms of adenylyl cyclase, although five of these may be restricted to the male germ line (Cann *et al.*, 2000). Meller *et al.* (1988, 1990) found that the *Manduca* prothoracic gland expresses Ca^{2+}/calmodulin-dependent adenylyl cyclase activity that exhibits a developmental shift in some characteristics. Either of two calmodulin antagonists inhibited nearly all cyclase activity in membrane preparations (Meller *et al.*, 1988), suggesting that the vast majority, if not entirety, of gland adenylyl cyclase is calmodulin-dependent. GTP-γ-S, a poorly hydrolyzable GTP analog, stimulated cyclase activity in both larval and pupal membrane preparations; however, in larval glands, GTP-γ-S was not effective without the simultaneous presence of calmodulin. This developmental switch in the characteristics of glandular cyclase activity occurred about day 5 of the fifth larval instar, just prior to the beginning of the large pre-molt ecdysteroid peak. The switch was also characterized by changes in cyclase sensitivity to cholera toxin and NaF, two factors that interact with G-proteins (Meller *et al.*, 1990). It is not known if the change in cyclase characteristics represents a change in cyclase isoform, a post-translational modification, or a shift in the expression of proteins that interact with cyclase. In *Bombyx* prothoracic gland membrane preparations, calmodulin, Gpp(NH)p, a GTP analog, and NaF stimulated adenylyl cyclase activity while GDP-β-S, a GDP analog, inhibited activity (Chen *et al.*, 2001). NaF and calmodulin effects on cyclase activity were additive rather than synergistic, suggesting that both G-protein- and calmodulin-dependent adenylyl cyclases exist in the *Bombyx* prothoracic gland. It remains to be determined which of these putatively different adenylyl cyclases is coupled to the PTTH signal transduction cascade.

PTTH elicited a rapid generation of cAMP in *Manduca* prothoracic glands, readily detectable within two minutes of applying PTTH and peaking after 5 min (Smith *et al.*, 1984). Elevated cAMP levels persisted for at least 20 min, with continuous PTTH exposure. Data from *Bombyx* prothoracic glands were similar, with a rapid rise in cAMP that reached a maximum at 10 min and continued at about twice basal for at least an hour (Gu *et al.*, 1996). Analysis of cAMP levels in *Manduca* prothoracic glands during the fifth (final) larval instar revealed a large maximum occurring at about the time of the commitment peak in ecdysteroid production (Vedeckis *et al.*, 1976), suggesting, as in the case for $[Ca^{2+}]_i$, a slow return to basal levels following stimulation and/or non-adaptation to continuous PTTH stimulation.

Intracellular cAMP levels are a product of synthesis and breakdown. The low *in vivo* levels of cAMP found during the pre-molt ecdysteroid peak (Vedeckis *et al.*, 1976), as well as in *Manduca* glands stimulated *in vitro* with PTTH at this stage and later (Smith and Gilbert, 1986; Smith and Pasquarello, 1989), appear to result chiefly from developmental differences in phosphodiesterase activity. Analysis of cAMP phosphodiesterase activity revealed a nearly sixfold increase from a minimum on days 3 and 4 to a peak just prior to pupal ecdysis; at the time of the large, pre-molt ecdysteroid peak, when PTTH signaling is presumably high also, phosphodiesterase activity was ≈5 times the minimum (Smith and Pasquarello, 1989). This observation suggested that the importance of the cAMP pathway in PTTH signal transduction was greatly diminished in late larval and pupal glands (Vedeckis *et al.*, 1976). This proved clearly not to be the case, however, as demonstrated by the inhibitory action of a cAMP antagonist on PTTH-stimulated pupal ecdysteroid synthesis (Smith *et al.*, 1996).

1.4.3. Protein Kinases and Phosphatases

A very common, if not universal, response to increases in second messengers is the modulation of protein kinase and phosphatase activities. In this section, known and hypothesized kinase and phosphatase contributors to PTTH signaling are discussed. These enzymes are discussed separately, perhaps giving a linear view of their participation in the PTTH signal transduction cascade. This is clearly not the case and the enzyme-by-enzyme approach is followed merely to provide a structure to the discussion, not to imply simplicity of the pathway. A broad analysis of the effects of PTTH on kinase activity was undertaken by Rewitz *et al.* (2009a). These investigators measured changes in phosphorylated proteins stimulated by PTTH after 15- and 60-min incubations. Prothoracic glands were obtained from fifth instar *Manduca* larvae, prior to wandering (days 3 and 4), that is, at a time when glands are relatively large and PTTH-sensitive. Isolated phospho-peptides, obtained following digestion, iTRAQ labeling, and TiO2 column separation, were identified by nano-liquid chromatography tandem mass spectrometry and proteins identified by comparison with *Bombyx* protein databases. A number of changes were seen, including large increases in the phosphorylation of proteins in expected signaling pathways, including MEK, ERK, members of the cAMP and PKC pathways, and at least two protein phosphatases. Ribosomal protein S6 was phosphorylated, as was the steroidogenic enzyme Spook (Spo). Levels of non-phosphorylated Spo also increased, although the other steroidogenic enzymes (Phm, Dib, Sad) did not (see Chapter 4 for a discussion of these Halloween genes). The study suggests that a rate-limiting increase in Spo may be a primary means by which PTTH enhances ecdysone synthesis and provides an extremely useful catalog of other PTTH-sensitive proteins in the pre-metamorphic larval prothoracic gland.

1.4.3.1. Protein kinase A   Two isoforms of cAMP-dependent PKA (types I and II) have been found in the *Manduca* prothoracic gland, with the majority of the activity (≥95%) attributable to the Type II isoform (Smith *et al.*, 1986b). PTTH-dependent stimulation of PKA activity was found in both larval and pupal glands, although the addition of a phosphodiesterase inhibitor was necessary to measure PTTH-dependent PKA activity in the latter. The requirement of a phosphodiesterase inhibitor to measure PTTH-dependent PKA activity in pupal glands was not surprising, given the rapid metabolism of cAMP at this stage, as discussed earlier. PTTH stimulated PKA activity rapidly, with a patent increase above controls following 3 min of exposure, and with elevated levels still evident after 90 min (Smith *et al.*, 1986b). PKA is composed of two subunits: one catalytic and the other regulatory. In prothoracic glands of early fifth larval instars of *Manduca*, the catalytic subunit remains relatively constant per milligram of protein for the first four days while the regulatory subunit increases ≈ threefold (Smith *et al.*, 1993). This may insure that PKA is only activated when a fairly high threshold of cAMP has accumulated in response to PTTH (Smith *et al.*, 1993), but this hypothesis is tentative pending information not currently available on the relative abundances of the catalytic and regulatory subunits at other stages of development.

The rapid activation of PKA in response to PTTH stimulation suggests that PKA plays a proximate role in the control of PTTH-stimulated steroidogenesis. The effects of agonist and antagonist cAMP homologues is supportive but not conclusive evidence for this interpretation, since there are other cellular targets for cAMP besides PKA. For example, cAMP activates certain monomeric G-proteins in the Ras family by direct binding to guanine-nucleotide exchange factors (exchange protein directly activated by cAMP, EPAC) (Gloerich and Bos, 2010). However, the PKA inhibitor H89 is effective at blocking PTTH-stimulated ecdysteroid synthesis in *Manduca* glands, albeit at fairly high doses, again implicating PKA as a regulatory factor in this pathway (Smith *et al.*, 1996). The *in vivo* substrates for PTTH-activated PKA are poorly known. PTTH, dbcAMP, and Ca^{2+} ionophore all stimulate phosphorylation of a 34 kDa protein in *Manduca* prothoracic glands (Rountree *et al.*, 1987, 1992; Combest and Gilbert, 1992; Song and Gilbert, 1994) that has been identified as the S6 ribosomal protein (Song and Gilbert, 1995). It is unclear that S6 is directly phosphorylated by PKA, although when the (mammalian) catalytic subunit of PKA, or cAMP, was added to prothoracic gland homogenates, the phosphorylation of S6 was enhanced (Smith *et al.*, 1987a; Rountree *et al.*, 1992; Smith *et al.*, 1993; Combest and Gilbert, 1992), indicating that S6 was directly or indirectly a PKA target.

1.4.3.2. Protein kinase C

The possibility that PKC was involved in PTTH-stimulated ecdysteroid synthesis was first addressed, with negative results, by Smith *et al.* (1987a), using the potent PKC activator phorbol myristate acetate (PMA). A 90 min incubation with PMA had no effect on *Manduca* steroidogenesis, either alone or in conjunction with PTTH (see also Smith, 1993). However, in *Bombyx*, PMA boosted prothoracic gland ecdysteroid synthesis after 5 h of treatment but no effect was apparent after 2 h (Dedos and Fugo, 2001), suggesting that PKC did not play an acute role in steroidogenesis. Furthermore, studies on inositol metabolism in both *Manduca* and *Bombyx* failed to find PTTH-related generation of PIP_2 metabolites (see Section 1.4.2.2.). Diacylglycerol (DAG), one of these metabolites, is an activator of PKCs and is formed in a 1:1 ratio with IP_3. The apparent absence of IP_3 suggested that DAG was also absent, hence PKCs were not involved in PTTH action.

Three lines of evidence indicate that a PKC is part of the signal transduction network of PTTH, and that it occupies an early and important position in regulating prothoracic gland ecdysteroidogenesis. First, the PKC inhibitor chelerythrine inhibited PTTH-stimulated ecdysteroid synthesis by both *Bombyx* (Dedos and Fugo, 2001; Gu *et al.*, 2010) and *Manduca* prothoracic glands (Rybczynski and Gilbert, 2006). Second, PTTH-stimulated the phosphorylation of several (unidentified) prothoracic gland proteins, as recognized by an antibody against phosphorylated PKC substrate sequences, and chelerythrine inhibited these phosphorylations (Rybczynski and Gilbert, 2006). Third, while measurable changes in IP_3 or other PIP_2 metabolites have not been detected following PTTH stimulation, PLC inhibitors blocked PTTH-stimulated ecdysteroid synthesis in *Bombyx* (2-nitro-4-carboxyphenyl-*N*, *N*-diphenylcarbamate; Dedos and Fugo, 2001) and *Manduca* (U73122: Fellner *et al.*, 2005) prothoracic glands. U73122 also inhibited PTTH-stimulated ERK phosphorylation (Rybczynski and Gilbert, 2006) and the release of Ca^{2+} from IP_3-sensitive stores (see Section 1.4.2.2.). These data suggest that PTTH stimulates inositol metabolism and activates PKCs, but the inositol increases are too small or transient to have been detected. It does not appear that PLC is activated by a G-protein, as calcium influx is insensitive to pertussis toxin and non-hydrolyzable GTP analogs (see Section 1.4.2.2.). However, the PTTH receptor, being a tyrosine kinase, could directly activate subtypes of PLC such as PLCγ with SH2 domains (see Harden *et al.*, 2009).

The identity and function of proteins phosphorylated by PKC in response to PTTH stimulation of the prothoracic gland are currently unknown. PKCs can modulate the function of a wide variety of proteins including ion channels (see Shearman *et al.*, 1989), MAP kinases (Suarez *et al.*, 2003), plasma membrane receptors (Medler and Bruch, 1999), cytoskeletal elements (Gujdar *et al.*, 2003), and nuclear proteins (Martelli *et al.*, 2003). The observation that some putative PKC phosphorylations took place only after several hours of PTTH stimulation suggested that PKC might be involved in the long-term effects of PTTH, such as transcription, as well as in the acute regulation of steroid synthesis (Rybczynski and Gilbert, 2006). The position of PKC in the PTTH signaling hierarchy has not been determined but, given a probable requirement for DAG, it seems likely that PKC is among the first intracellular components to be called into action. Because inhibition of PKC activity blocks both MAP kinase phosphorylation and ecdysteroid secretion, it is likely to play a role early in the signaling pathway (Rybczynski and Gilbert, 2006).

1.4.3.3. Extracellular signal-regulated kinase The MAP kinases are a family of related proteins that are activated by a wide variety of extracellular signals, ranging from growth factors and peptide hormones to heat and osmotic stresses. MAP kinases, in turn, regulate a wide variety of intracellular processes from microtubule stability and protein synthesis to transcription. There are several families of MAP kinases, which are organized into three-level signaling modules: a MAP kinase, a MAP kinase kinase (MEK), and a MAP kinase kinase kinase (MEKK or MAP3K) (see Lewis *et al.*, 1998; Kyriakis, 2000). Perhaps the most studied of the MAP kinase modules is the extracellular signal-regulated kinase (ERK) cascade. This module typically starts with a protein called Raf and proceeds through a MEK to the ERK protein(s); in mammals there are two ERKs while *Drosophila* and *Manduca* appear to express a single ERK (see Lewis *et al.*, 1998; Biggs and Zipursky, 1992; Rybczynski *et al.*, 2001). A number of proteins lie upstream from Raf. The small G-protein Ras is a frequent but not ubiquitous activator of Raf, which may lie downstream of G-protein-coupled receptors, tyrosine kinase receptors, or cytokine receptors (see Lewis *et al.*, 1998). MAPKKKs and MEKs are activated by single phosphorylations whereas ERKs require a dual phosphorylation on threonine and tyrosine residues for activity.

PTTH stimulation of the ERK pathway in larval *Manduca* prothoracic glands was shown by Rybczynski *et al.* (2001) using activity assays, inhibitors of MEK (PD98059 or U0126) to block ERK phosphorylation, and antibodies that recognize the dually phosphorylated state. This study revealed that PTTH caused a rapid but transient increase in ERK phosphorylation and that inhibitors of ERK activation blocked both basal and PTTH-stimulated ecdysteroid synthesis. The data suggested that ERKs might be involved in regulating some aspects of the acute PTTH response in prothoracic glands. This finding was replicated in a study of *Bombyx* prothoracic glands in which MEK inhibitors were also found to block ERK phosphorylation and ecdysone secretion (Lin and Gu, 2007). *Bombyx* prothoracic glands are refractory to PTTH stimulation at the start of the fifth instar; glands removed at that time did not show a PTTH-stimulated increased in ERK phosphorylation, further linking this enzyme to functional PTTH response. However, a study by Gu *et al.* (2010), using a variety of inhibitors, suggested that ERK phosphorylation did not necessarily lead to an increase in ecdysteroidogenesis (Gu *et al.*, 2010). Some of the inhibitors used, however, may have blocked steroidogenesis downstream of ERK signaling (see Section 1.4.1.).

In *Manduca*, Rybczynski and Gilbert (2003) showed that PTTH-stimulated ERK phosphorylation was primarily a larval gland response. As *Manduca* larvae progressed through the fifth instar, PTTH-stimulated ERK phosphorylation in the prothoracic gland declined until it was undetectable in pupal glands (Rybczynski and Gilbert, 2003). Concurrently, total ERK levels declined in the glands, but the basal levels of phosphorylation increased in response to unknown factors. These two phenomena combined to cause the aforementioned decline in PTTH-stimulated ERK phosphorylation. In both *Manduca* and *Bombyx*, PTTH-stimulated phosphorylation of ERK was stimulated by the calcium ionophore A23187 (Rybczynski *et al.*, 2003; Gu *et al.*, 2010). ERK phosphorylation in *Bombyx* was also stimulated by the intracellular calcium-releaser thapsigargin, or the PKC activator, PMA (Gu *et al.*, 2010). In *Manduca*, ERK phosphorylation was blocked by removal of extracellular calcium, while in *Bombyx* it was only partially blocked by removal of extracellular calcium or by the calmodulin inhibitor calmidazolium A23187 (Rybczynski *et al.*, 2003; Gu *et al.*, 2010).

Evidence from *Drosophila* argues strongly for a central role of the ERK pathway in hormone-stimulated ecdysone synthesis. Overactivation of Ras and Raf in the larval *Drosophila* ring gland (prothoracic gland) leads to premature metamorphosis and small adults, strongly suggesting enhanced ecdysone production (Caldwell *et al.*, 2005). Conversely, dominant negative Ras and Raf in the prothoracic glands increase body size suggesting inhibition of ecdysone secretion (Caldwell *et al.*, 2005). The same effects are seen by manipulating PI3K activity, although the link to ERK in this case is not clear (see Section 1.5.2.1.). As mentioned earlier, prothoracic-gland-specific knockdown of the PTTH receptor (Torso) leads to the same delayed-metamorphosis phenotype as knockdown of Ras, Raf, or ERK, the latter being the most effective. Ras overexpression in the prothoracic gland rescues the Torso knockdown phenotype (Rewitz *et al.*, 2009b).

The mechanism by which PTTH activates the MAP kinase cascade is not yet known. This link speaks directly to the primary action of ligand-activated PTTH receptor. The most direct pathway for activation of MAP kinases, based upon the PTTH receptor being a tyrosine kinase, is the recruitment of an SH2-domain-containing adapter such as Grb2, which in turn would recruit the guanine-nucleotide exchange factor SOS to the vicinity of Ras in the plasma membrane, and initiate a canonical MAP kinase cascade. However, this "direct" route does not explain the stimulatory effects of calcium on prothoracic gland ERK or the requisite role for calcium early in the PTTH signaling pathway.

Some PKC isotypes require Ca^{2+} for activation and thus are candidates for regulation of the ERK cascade (see Harden *et al.*, 2006). PKC inhibitors can block PTTH-stimulated ERK phosphorylation in *Manduca* (Rybczynski and Gilbert, 2006), but not in *Bombyx* (Gu *et al.*, 2010), which suggests species differences in PKC regulation of the MAP kinase pathway. In *Manduca*, PKC may be able to activate enzymes such as Raf, isoforms of which have been shown to be PKC targets in other systems (Liebman, 2001). The protein kinase Pyk2, which can be activated by increases in [Ca^{2+}] and by PKC (Lev *et al.*, 1995), is also a reasonable candidate to be an upstream regulator of the ERK cascade.

Rybczynski and Gilbert (2003) found that PTTH-stimulated ERK phosphorylation in *Manduca* was not cAMP-dependent; that is, in early fifth instar larval glands, cAMP analogs did not stimulate significant ERK phosphorylation. Nonetheless, the ERK inhibitor U0126 blocks the ability of cAMP analogs to stimulate ecdysone secretion (Smith, unpublished observations). It may be that PKA activates a subset of MAP kinases in the prothoracic glands (e.g., associated with mitochondria) that are not observed to change in whole glandular homogenates. Activation of a PKA-sensitive, mitochondria-specific MEK has previously been suggested to underlie PKA-stimulated steroidogenesis in vertebrates (Poderoso *et al.*, 2008).

The list of potential targets of ERK phosphorylation is long and encompasses transcription factors, steroid and growth factor receptors, various membrane-associated proteins, cytoskeletal proteins, kinases, and proteins involved in regulating translation (see Roux and Blenis, 2004). In most cell types, some fraction of dually phosphorylated ERKs migrate into the nucleus where they modulate transcription; no conclusive information about the prothoracic gland on this topic is presently available. In vertebrate steroidogenic cells, ERK may regulate the activity of steroidogenic acute regulatory (StAR) protein (see Section 1.4.4.) and hence steroid synthesis at the level of mitochondria (Poderoso *et al.*, 2008). It should be noted that PTTH also stimulates ERK phosphorylation and protein synthesis in the reproductive system of *Manduca* adult males and females (Rybczynski *et al.*, 2009). The function of PTTH-stimulated ERK activation at this stage of development is not known.

1.4.3.4. p70 S6 Kinase   The S6 protein is a prominent ribosomal phosphoprotein that can be phosphorylated by PKA and a 70 kDa S6 kinase (p70S6K; Palen and Traugh, 1987). S6 phosphorylation is a frequent and presumably important event in various intercellular signaling pathways (see Erickson, 1991; Volarevic and Thomas, 2001). Phosphorylation of S6 by p70S6K, but apparently not by PKA, can increase the rate of translation of some mRNAs (Palen and Traugh, 1987). PTTH has long been known to induce the phosphorylation of a $\approx$34 kDa protein in prothoracic glands of *Manduca*, which was hypothesized to be S6 (Gilbert *et al.*, 1988); this identification proved to be correct (see Section 1.4.3.5.).

Song and Gilbert (1994) showed that p70S6K was likely to be responsible for PTTH-stimulated S6 phosphorylation in *Manduca* prothoracic glands. They found that the macrolide immunosuppressant rapamycin blocks p70S6K phosphorylation-dependent kinase activity (see

Dufner and Thomas, 1999). Specifically, they found that PTTH-stimulated phosphorylation of an S6 peptide substrate was inhibited by rapamycin, as tested using an S6K1-enriched protein fraction prepared from glands treated with both PTTH and rapamycin. Furthermore, rapamycin inhibited PTTH-stimulated ecdysteroid synthesis at low (nM) concentrations. These observations implicate TOR (the *Drosophila* gene is called dTOR) and FRAP (FKBP12-rapamycin-associated protein) in PTTH action. TOR is a serine/threonine kinase; rapamycin forms an inhibitory complex with FRAP that blocks the activity of TOR (see Fumagalli and Thomas, 2000; Soulard *et al.*, 2009). TOR has a stimulatory effect on S6K1, and also reduces the activity of the translation inhibitor 4EPB (initiation factor 4E binding protein). A typical TOR activation cascade begins with the activation of a receptor tyrosine kinase, followed by activation of PI3K and of protein kinase B/Akt, phosphorylation and inactivation of inhibitory proteins comprising the tuberous sclerosis complex (TSC), and hence activation of the small G-protein Rheb which in turn activates TOR. Akt also phosphorylates an inhibitor of TOR known as PRAS40, which dissociates from TOR enhancing its activation. One would predict that a PTTH-receptor-tyrosine kinase would enhance the activity of TOR through PI3K, and yet inhibitors of PI3K such as wortmannin and LY294002 do not inhibit PTTH-stimulated ecdysteroid secretion (Smith, unpublished observations). Hence the mechanism by which PTTH activates TOR is a currently unresolved question. A study by Blancquaert *et al.* (2010) suggests one possibility, based on the discovery that PRAS40 is also a substrate of PKA. This finding explains the ability of thyroid-stimulating hormone, which acts through cAMP, to activate TOR in the absence of increased activity of PI3K or Akt.

1.4.3.5. Ribosomal protein S6   Ribosomal protein S6 is a small ($\approx$34 kDa), basic protein present at one copy per 40S ribosome near the tRNA/mRNA binding site (see Fumagalli and Thomas, 2000; Meyuhas, 2008). Phosphorylation of S6 is associated with mitogenic stimulation of non-dividing cells and occurs in vertebrates in response to tropic factors as follicle-stimulating hormone and thyroid-stimulating hormone (Das *et al.*, 1996; Suh *et al.*, 2003). S6 phosphorylation at six serine residues near the carboxy-terminal of the protein has been correlated with the selective increase of certain mRNA, especially those containing polypyrimidine tracts near the transcription initiation site (see Erickson, 1991; Fumagalli and Thomas, 2000; Meyuhas and Hornstein, 2000; Volarevic and Thomas, 2001).

Rountree *et al.* (1987) and others discovered that PTTH stimulates the phosphorylation of a $\approx$34 kDa protein in *Manduca* prothoracic glands that was hypothesized to be the S6 ribosomal protein (see Section 1.4.3.1.). Song and Gilbert (1994, 1995, 1997) confirmed this identification using the TOR inhibitor rapamycin and, more rigorously, by two-dimensional analysis of ribosomal proteins. As discussed earlier, they found that rapamycin blocked S6 phosphorylation and ecdysteroid synthesis; furthermore, rapamycin inhibited the accumulation of specific, newly synthesized proteins, including a heat shock 70 kDa cognate protein (Hsc70; Song and Gilbert, 1995) that had been hypothesized to be necessary for PTTH-stimulated ecdysteroid synthesis (Rybczynski and Gilbert, 1994, 1995a; see 1.4.4.).

cDNA cloning of the *Manduca* S6 mRNA revealed that the sequence coded for a polypeptide of ≈29,000 MW with an estimated isoelectric point of 11.5 (Song and Gilbert, 1997). Like the mammalian S6 proteins, the *Manduca* sequence contained a cluster of 6 serines in the final (carboxy-terminal) 20 amino acids. Phosphoamino acid analysis indicated that, following PTTH stimulation, the *Manduca* S6 protein was phosphorylated solely on serines. The two-dimensional analysis of ribosomal proteins showed that PTTH stimulation resulted in phosphorylation of up to five sites; non-stimulated glands yielded an S6 with an apparent single phosphorylation (Song and Gilbert, 1995, 1997). The location of the PTTH-stimulated serine phosphorylations is not known since, in addition to the six terminal serines, the *Manduca* protein was deduced to contain an additional ten residues. The combination of Northern blotting, using the S6 cDNA as a probe, and immunoblotting, using polyclonal antibodies against the S6 protein terminus, showed that S6 RNA and protein levels in the prothoracic gland were not correlated (Song and Gilbert, 1997). S6 RNA per unit total RNA was highest in the first three days of the fifth instar and decreased linearly to a much lower level by day 7. This low level persisted through day 4 of pupal life, interrupted only by a minor increase on the first day following pupal ecdysis. In contrast, S6 proteins showed two peaks of abundance. The first occurred on days 6 and 7 of the fifth instar, when the prothoracic gland is beginning the large surge of ecdysteroid synthesis that precedes the larval–pupal molt. The second peak fell on days 3 and 4 of pupal–adult development, the beginning of the large pupal–adult period of ecdysteroid synthesis. As a control, S6 protein levels in the fat body were assayed. These showed no relationship to S6 in the prothoracic gland or to hemolymph ecdysteroid titers. Rather, S6 protein in fat body showed a steady decline from the beginning to the end of the fifth instar and then fairly constant low levels through day 4 of pupal–adult development. These data suggested that S6 levels in the prothoracic glands might support high levels of ecdysteroid synthesis.

Although these composite data are in keeping with the notion that the TOR–p70S6K–S6 axis is essential for PTTH-stimulated ecdysteroid synthesis to occur, there are also data that do not readily fit this scheme. Both Rountree *et al.* (1987) and Smith *et al.* (1987a) discovered

that, like PTTH, a cAMP analog, dbcAMP, stimulated S6 phosphorylation, and Rountree *et al.* (1987) found that JH and a JHA inhibited S6 phosphorylation in glands studied *in vitro* (Rountree *et al.*, 1987). However, under the same *in vitro* conditions, JH had no effect on PTTH-stimulated cAMP synthesis and ecdysteroid synthesis (Rountree and Bollenbacher, 1986). A second puzzling observation, revolving again around cAMP, stems from the study of the effects of rapamycin on steroidogenesis. As described previously, rapamycin inhibits both PTTH-stimulated S6 phosphorylation and ecdysteroidogenesis. However, dbcAMP was able to override the rapamycin block and stimulate ecdysteroid synthesis to the usual levels above basal, even though rapamycin blocked the dbcAMP-dependent S6 phosphorylation (Song and Gilbert, 1994). These two sets of data suggest that a vigorous, six-site S6 phosphorylation response may not be required to support PTTH-stimulated ecdysteroid synthesis. In this respect, it would be interesting to use the more sensitive two-dimensional gel analysis of ribosomal proteins to determine if JH and rapamycin blocked all phosphorylations above the basal level; it might be that the single-dimension gel analysis used was not sensitive enough to detect partial S6 phosphorylation in these experiments. It is also possible that there are alternate pathways by which PTTH stimulates ecdysteroid synthesis, and that these pathways can substitute for the TOR-dependent one. For instance, ERK, through the activation of 90 kDa ribosomal S6 kinase (RSK), can also upregulate translation (see Anjum and Blenis, 2008; Meyuhas, 2008). The signaling pathway by which PTTH enhances ecdysone production in *Drosophila* is likely to differ from the Lepidoptera, given an apparent lack of effect of cAMP on fruit fly prothoracic glands (see Huang *et al.*, 2008). PTTH in this species may, like the insulins, enhance PI3K activity and TOR. Alternatively, the MAP kinase pathway, which mimics PTTH (Caldwell *et al.*, 2005; Rewitz *et al.*, 2009b), may mediate PTTH-stimulated ecdysteroidogenesis in this insect, independently of TOR.

1.4.3.6. Protein phosphatases The stimulation of ecdysteroidogenesis *in vivo* by PTTH is transient, and with varying biosynthetic rates, depending on the developmental stage. Given that PTTH stimulates the phosphorylation of a number of kinases and an unknown number of target proteins, it is reasonable to assume that downregulation of PTTH-stimulated ecdysteroid synthesis involves the activity of protein phosphatases. This topic has received scant attention in regard to the action of PTTH. Song and Gilbert (1996) used okadaic acid and calyculin, highly effective inhibitors of the serine/threonine-specific protein phosphatases PP1 and PP2A (not to be confused with Src kinase inhibitors discussed in Section 1.4.1.), to address this topic. They found that treatment of *Manduca* prothoracic glands with phosphatase inhibitors enhanced the basal phosphorylation of S6, and of several unidentified proteins, but that PTTH-stimulated phosphorylation of S6 was not augmented. However, although okadaic acid and calyculin increased S6 phosphorylation, these drugs inhibited both basal and PTTH-stimulated ecdysteroid and protein synthesis; included in the latter effect was the apparent inhibition of the syntheses of β tubulin and Hsc70. Finally, Song and Gilbert (1996) found that PTTH-stimulated phosphatase activity was capable of dephosphorylating S6, and perhaps other proteins, and that this activity was inhibited by okadaic acid, implicating the phosphatases PP1 and/or PP2A in the downregulation of PTTH effects. The combined data indicated that the multiple phosphorylation of S6, by itself, was not sufficient to drive increased ecdysteroidogenesis. They also suggested that the non-phosphorylated state of one or more proteins was necessary for normal ecdysteroid synthesis, assuming that neither okadaic acid nor calyculin had effects on proteins other than phosphatases. That PTTH both stimulated S6 phosphorylation and dephosphorylation processes was expected given the likelihood that negative feedback loops exist in the prothoracic gland to restore its ecdysteroidogenic activity to basal levels.

While serine/threonine-specific phosphatases such as PP1 and PP2A in some manner positively affect steroidogenesis, overactivation of tyrosine kinase activity can block steroidogenesis. In *Manduca* larvae infected with parasitoid wasps, associated bracoviruses induced overexpression of tyrosine phosphatases in fat body and prothoracic glands (Falabella *et al.*, 2006). While S6 kinase activity and general protein synthesis were unimpaired, ecdysteroid secretion was blocked, suggesting a site of action separate from that regulating translation. In keeping with an inhibitory role of tyrosine phosphatases on steroid secretion, sodium orthovanadate, a tyrosine phosphatase inhibitor, stimulated ecdysteroidogenesis by *Bombyx* prothoracic glands (Lin and Gu, 2010).

The MAP kinases, like ERK (see Section 1.4.3.), can be dephosphorylated by a number of phosphatases (see Camps *et al.*, 2000). Dephosphorylation of ERKs at either the threonine or tyrosine sites results in loss of ERK kinase activity; PP2A and several tyrosine phosphatases (e.g., PTP-SL, STEP, and HePTP) have all been shown to inactivate vertebrate ERKs (see Camps *et al.*, 2000). In addition, ERKs and other MAP kinases are dephosphorylated by unique dual-specificity phosphatases (DSPs) that remove phosphates from both threonine and tyrosine residues. These DSPs are probably the major agents of vertebrate ERK inactivation (see Camps *et al.*, 2000). Mammals express at least nine DSP genes and several homologs are found in *Drosophila* including puckered, MKP, MKP3, and MKP4 (Lee *et al.*, 2000; Kim *et al.*, 2004; Sun *et al.*, 2008). Currently, the contributions of various protein phosphatases to the regulation of prothoracic gland ERK

phosphorylation are unknown. Serine/threonine phosphatases are likely regulators of proteins, like MEK, that are upstream from ERK and their role can be assessed in part by using well-characterized inhibitors like okadaic acid. Understanding the role of DSPs is more difficult and will require the availability of appropriately targeted inhibitors (Vogt *et al.*, 2003).

1.4.4. Translation and Transcription

In vertebrate steroidogenic tissues, the final signal transduction step in peptide-regulated steroid hormone production is the rapid translation of one or more proteins. In vertebrates, two proteins, StAR and DBI (diazepam-binding inhibitor), facilitate the delivery of cholesterol across the mitochondrial membrane to the side-chain cleavage P_{450} enzyme, which converts cholesterol to pregnenolone (see Manna *et al.*, 2009; Papadopoulos *et al.*, 1997; Batarseh and Papadopoulos, 2010; and Chapter 4). This step is rate-limiting and is alleviated by rapid translation of StAR and/or DBI that is stimulated by a peptide hormone such as adrenocorticotropic hormone (ACTH). A *Manduca* DBI homologue has been cloned and evidence suggests a role for DBI in basal ecdysteroid synthesis (Snyder and Feyereisen, 1993; Snyder and Van Antwerpen, 1998), but experiments to explicitly test its role in acute PTTH-stimulated ecdysteroidogenesis have not been performed. A DBI homologue has also been cloned in *H. armigera*, and its expression in the prothoracic glands increases in response to the steroidogenic-activating diapause hormone (Liu *et al.*, 2005). In *D. melanogaster*, a gene for a StAR-related lipid transfer domain (START) protein has been found that resembles the vertebrate cholesterol transport protein metastatic lymph node 64 (MLN64; Roth *et al.*, 2004). These proteins share a START domain with StAR but also possess a transmembrane spanning domain. START homologues were also found in *D. pseudoobscura*, *A. gambiae*, and *Bombyx*. *D. melanogaster* START is expressed in the larval brain/ring gland, but also imaginal discs, salivary glands, and adult tissue including the ovaries and testes. It is not seen in ecdysoneless (ecd-1) mutants, suggesting regulation of START expression by ecdysteroid positive feedback (Roth *et al.*, 2004). In *Bombyx*, a START homologue was expressed in the prothoracic glands in conjunction with elevated hemolymph ecdysteroids (Sakudoh *et al.*, 2005).

Translation inhibitors like cycloheximide and puromycin block PTTH-stimulated ecdysteroid synthesis, but have no effect on basal ecdysteroid synthesis under standard *in vitro* protocols (Keightley *et al.*, 1990; Kulesza *et al.*, 1994; Rybczynski and Gilbert, 1995b). In *Drosophila*, the movement of an ecdysteroid precursor(s) between intracellular compartments (endoplasmic reticulum and mitochondrion) involves a carrier protein, although the necessity of translation in this species has not been demonstrated (Warren *et al.*, 1996; Warren and Gilbert, 1996).

Analysis of protein synthesis in *Manduca* prothoracic glands, using ^{35}S-methionine to label newly synthesized proteins, revealed that PTTH stimulated overall translation at most developmental stages, and that some proteins, perhaps as many as ten, were differentially translated above this general increase (see **Figure 8**) (Keightley *et al.*, 1990; Rybczynski and Gilbert, 1994; Gilbert *et al.*, 1997). Two of these proteins have been identified and cloned: a β tubulin (Rybczynski and Gilbert, 1995a, 1998) and a 70 kDa heat shock cognate protein (Hsc70) (Rybczynski and Gilbert, 1995b, 2000). PTTH-stimulated synthesis of Hsc70 rose slowly and was prolonged, suggesting that Hsc70 might function as a chaperone in supporting translation, and in other long-term effects of PTTH not directly connected to acute changes in ecdysteroidogenesis, for example, an Hsc70 protein appears to be necessary for proper function of the ecdysone receptor (Arbeitman and Hogness, 2000). This has been confirmed in *H. armigera* in which Hsc70 was shown to bind Ultraspiracle (USP), the heterodimeric binding partner of the ecdysone receptor (EcR), and in which RNAi knockdown of Hsc70 decreased nuclear content of EcRB1 and USP (Zheng *et al.*, 2010). In *Manduca*, PTTH stimulated Hsc70 synthesis above basal at all stages examined during the fifth instar and pupal–adult development; peaks occurred on the third day of the fifth instar (V_3) and the first day of pupal–adult development (P_1) (Rybczynski and Gilbert, 1995b). In contrast,

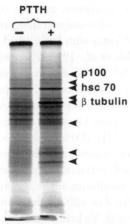

Figure 8 PTTH stimulates the synthesis of specific proteins in the prothoracic gland of *Manduca sexta*. Glands were incubated with ^{35}S-methionine ± PTTH for 1 h and then subject to SDS-PAGE and autoradiography. The arrowheads indicate newly synthesized proteins whose accumulation in the prothoracic gland is upregulated by PTTH. Hsc70, β tubulin, and p100 were identified in the first studies on this system. Reproduced with permission from *Rybczynski and Gilbert* (1994, 1995a,b). The additional PTTH-regulated translation products were only detected after further refinement of the SDS-PAGE conditions (Rybczynski and Gilbert, unpublished data).

PTTH-stimulated β tubulin synthesis occurred rapidly and was transitory, suggesting a more direct role in ecdysteroidogenesis for this cytoskeletal protein (Rybczynski and Gilbert, 1995a). Peaks in PTTH-stimulated β tubulin synthesis occurred on V_3 and P_1 (Rybczynski and Gilbert, 1995a), as observed for Hsc70. Indeed, treatment of pupal prothoracic glands with microtubule-disrupting drugs inhibited PTTH-stimulated ecdysteroid synthesis (Watson *et al.*, 1996), but this effect was not seen in larval glands (Rybczynski and Gilbert, unpublished data). Furthermore, the pupal inhibition effect was associated with translation inhibition that was overridden by treatment of glands with a cAMP analog, suggesting that tubulin may be required for physical coupling between the PTTH receptor and a protein acting early in the signaling cascade (Rybczynski and Gilbert, unpublished data). The reason for the larval–pupal difference in microtubule effects on ecdysteroid synthesis is unknown. Note that neither study (Watson *et al.*, 1996; Rybczynski and Gilbert, unpublished data) explicitly monitored the effects of the microtubule disrupting drugs on cytoskeletal structures or cell shape. Note also that the role of the microtubule cytoskeleton in vertebrate steroidogenic cells is not clear. Basal steroidogenesis seems to be generally insensitive to microtubule-acting drugs, while peptide hormone-stimulated steroidogenesis is variably inhibited or not, depending on the cell type and drug concentration (see Feuilloley and Vaudry, 1996). In systems where peptide hormone-stimulated steroidogenesis is inhibited, the data suggest that the pharmacological disruption of microtubules disturbs early events in the signal transduction cascade, for example, receptor-G-protein-cyclase interactions (see Feuilloley and Vaudry, 1996), which is a hypothesis consistent with the previously discussed ability of a cAMP analog to override such a disruption.

PTTH-stimulated protein synthesis is not limited to specific proteins but can also be detected in overall rates of translation (Keightley *et al.*, 1990; Rybczynski and Gilbert, 1994; Kulesza *et al.*, 1994). The amount by which PTTH stimulates overall translation above basal appears to vary with the stage, with V_3 glands responding at about six times the rate of V_1 glands (Rybczynski and Gilbert, 1994). The incubation of non-stimulated prothoracic glands with ^{35}S-methionine revealed that the pattern of translated proteins varied among several stages in the fifth larval instar (Lee *et al.*, 1995); how these changes relate to changes in the ecdysteroidogenic capability of the prothoracic gland or to PTTH signaling is unknown. A more detailed developmental analysis of protein synthesis, both basal and PTTH-stimulated, by larval and pupal prothoracic glands remains to be done. As mentioned previously, phosphorylation of ribosomal protein S6 preferentially increases the translation of mRNAs containing a 5'-polypyrimidine tract. The identification of these RNAs is undoubtedly incomplete, but recent evidence reveals that in vertebrate cells, many such RNAs code for ribosomal proteins (see Pearson and Thomas, 1995), suggesting that some of the unidentified PTTH-dependent translation products are also components of the ribosome.

PTTH stimulates transcription in prothoracic glands based on the incorporation of radiolabeled uridine into acid-precipitable RNA in response to dbcAMP (Keightley et al., 1990). Inhibition of transcription using actinomycin D resulted in a partial inhibition of both PTTH- and dbcAMP-stimulated ecdysteroid synthesis (Keightley et al., 1990; Rybczynski and Gilbert, 1995b). These data suggested that transcription may be necessary for a maximal ecdysteroidogenic response to PTTH. However, actinomycin D had a larger effect on translation ($\approx$50% inhibition in a 45 min incubation) than might be expected unless an unprecedented fraction of prothoracic gland proteins were translated from mRNAs with very short half-lives (Rybczynski and Gilbert, 1995b). In other words, the possibility that actinomycin D interfered with translation, as well as with transcription, cannot be ruled out, thus making it difficult to determine the role of transcription in acute increases of ecdysteroidogenesis.

cDNA clones for *Manduca* β1 tubulin and hsc70 were obtained with the hope that the expression pattern of these two genes might shed some light on their function in PTTH signaling. In the prothoracic gland, β1 tubulin mRNA exhibited a peak in the fifth instar just before wandering (Rybczynski and Gilbert, 1998). β1 tubulin protein levels paralleled the mRNA profile during the fifth instar, but in contrast to the mRNA profile, β1 tubulin protein abundance exhibited a second peak on day 2 (P_2) of pupal–adult development. These data might indicate that β1 tubulin protein levels are important to glands that are preparing to synthesize large amounts of ecdysteroids but do little to explain the very rapid synthesis that occurs upon PTTH stimulation (Rybczynski and Gilbert, 1995b). Hsc70 mRNA levels also showed a peak in the fifth instar, just before wandering (V_4), with lesser peaks on V_7 and P_2 (Rybczynski and Gilbert, 2000). Hsc70 protein levels showed peaks on V_3, V_7, and P_0, yielding a pattern of expression that does not clearly relate to either prothoracic gland growth or ecdysteroid synthesis. Although the profiles of β1 tubulin and hsc70 mRNAs and proteins did not lend themselves to understanding the roles these gene products play in PTTH-stimulated ecdysteroid synthesis, analyses of brain and fat body indicated that the expression patterns of these products were tissue-specific (Rybczynski and Gilbert, 1998, 2000). Complicating the question of determining what special roles β1 tubulin and Hsc70 might play in PTTH signaling in the prothoracic gland is the fact that both proteins are abundant and participate in various housekeeping duties in the cell. Discerning a PTTH-specific function against this background is difficult. Clearly, the topic of

transcription in PTTH-stimulated ecdysteroid synthesis has not been exhausted and is worthy of further scrutiny.

1.5. The Prothoracic Gland

In this section, an overview is presented of developmental changes in the prothoracic glands and factors, aside from conventional PTTH, that affect prothoracic gland function. As will be seen from the brief descriptions in the following sections, the field of peptide regulators of prothoracic gland function is rapidly expanding beyond what can be covered in a single chapter. Several chapters in this book including chapters 2, 4, and 10, address important aspects of ecdysteroid secretion. A recent comprehensive review providing a broad perspective on prothoracic gland function can also be found in Marchal et al. (2010).

1.5.1. Developmental Changes in the Prothoracic Gland

A number of components in the PTTH transduction cascade change as the fifth larval instar progresses, as discussed in the previous section. These include an increase in cAMP phosphodiesterase activity, a decrease in the ERK phosphorylation response, an increase in the amount of the regulatory subunit of PKA, and a change in calmodulin sensitivity, rendering pupal adenylyl cyclase less responsive than the larval enzyme. The prothoracic glands undergo many other developmental changes. First, prothoracic gland cell size changes dynamically and rapidly accompanied by a number of structural changes. In *Manduca* and *Spodoptera*, cell size (i.e., diameter) is roughly correlated with ecdysteroidogenic capacity during the last larval instar with large-celled glands the most productive (Sedlak et al., 1983; Zimowska et al., 1985; Hanton et al., 1993). While the prothoracic gland cells do not divide, DNA synthesis occurs, and prothoracic gland cells have been identified as polyploid (*Samia*, Oberlander et al., 1965; *Oncopeltus*, Dorn and Romer, 1976; *Bombyx*, Gu and Chow, 2001, 2005a). As expected, as cell size increases, the protein content of the gland also increases. In *Manduca*, total extractable and particulate protein increases from a minimum at the beginning of the fifth instar of $\approx$3 μg/gland to $\approx$20 μg/gland on V_4; protein levels stay high through V_7 before reaching a low of 10–12 μg/gland on P_0 (Smith and Pasquarello, 1989; Meller et al., 1990; Lee et al., 1995; Rybczynski and Gilbert, 1995b). Protein content rises again through at least P_4 (Rybczynski and Gilbert, 1995b). Although size matters, the ecdysteroidogenic potential of prothoracic glands is not a simple function of cell size or protein content. An analysis of protein content and ecdysteroidogenesis in V_1 to V_3 *Manduca* fifth instar prothoracic glands revealed that basal and PTTH-stimulated synthesis per microgram protein increased during this period; that is, ecdysteroid synthesis per gland increased faster than did the gland's total protein content (Rybczynski and Gilbert, 1994). A re-analysis of data gathered throughout the fifth instar of *Manduca* (provided by Smith and Pasquarello, 1989) confirms this observation, and revealed that a peak of basal ecdysteroid synthesis per unit protein occurred on V_4. Levels after V_4 declined but never reached the low seen on V_1. PTTH-stimulated synthesis per microgram protein exhibited a different pattern. A peak was again seen on V_4, but levels were high and relatively similar from V_2 through V_7, followed by a dip on V_8 and V_9, and then a sharp rise to near V_4 levels on P_0. Note that PTTH-stimulated synthesis per microgram protein was essentially the same on V_2 and V_3 using the data from Smith and Pasquarello (1989), while Rybczynski and Gilbert (1994) found V_2 levels to be considerably lower than V_3 levels.

An additional developmental change in prothoracic glands has been described for *Bombyx*. *Bombyx* prothoracic glands from the early fifth instar do not respond to PTTH stimulation with either increased ecdysteroid or cAMP synthesis, and are also distinguished by an exceedingly low basal level of steroid synthesis *in vitro* (Sakurai, 1983; Gu *et al.*, 1996; Gu and Chow, 2005b). However, these glands apparently respond with increased ecdysteroid synthesis to dbcAMP, suggesting that there is an upstream developmental "lesion" in one or more components of the PTTH signaling pathway (Gu *et al.*, 1996, 1997; Gu and Chow, 2005b). In contrast, Takaki and Sakurai (2003) found that dbcAMP did not elicit ecdysteroid synthesis in early fifth instar glands. The reason for the discrepancy in results is not known but might lie in methodological or strain-specific differences among the studies. Further work indicated that early fifth instar glands from *Bombyx* possess an adenylyl cyclase that is much less responsive to Ca^{2+}/calmodulin stimulation than that characterized later in the instar when the gland is responsive to PTTH (Chen *et al.*, 2001). It is not known if this cyclase difference is sufficient to explain fully the refractory nature of early fifth instar glands, nor is it known what underlies the cyclase behavior.

Regardless of the site of refractoriness, *Bombyx* glands must regain steroidogenic capacity and ability to respond to PTTH. It appears that an absence of JH and ecdysteroids leads to this change, with no requisite involvement of a trophic factor beyond the glands themselves (Sakurai, 1983). Direct evidence for this was obtained by Mizoguchi and Kataoka (2005) using long-term cultures of *Bombyx* prothoracic glands. In this study, glands from fourth instar larvae exposed to 20E for 24 h in culture gradually lost ecdysone secretory activity, while those not exposed to 20E remained active. Newly ecdysed prothoracic glands were inactive for 24 h in culture and then began to spontaneously secrete ecdysone. Glands co-cultured with a source of JH, in the form of corpora allata, remained inactive for 8 days. Yamanaka *et al.* (2007)

also saw increased steroidogenic cytochrome P450 gene expression in long-term prothoracic gland cultures. Gene expression was inhibited by co-culture with CA, similar to the acquisition of steroidogenic capacity and the sensitivity to JH seen by Mizoguchi and Kataoka (2005).

The prothoracic glands are sensitive to changes in nutritional state (see Section 1.5.2.1.). Based on abundant evidence from *Drosophila*, the insect insulin-like hormones have taken center stage as systemic nutrient signals, regulating growth of the glands as well as other tissues (see Chapter 2 and Section 1.5.2.1.). The nutritional regulation of prothoracic gland growth in Lepidoptera is also well documented, although the factors responsible are not well defined. Chen and Gu (2006, 2008) found that starvation inhibits DNA synthesis in the prothoracic glands of *Bombyx* until day 3 of the fifth instar. Starvation also blocks protein synthesis (Chen and Gu, 2006) and ecdysone secretion (Gu and Chow, 2005b). After day 3, ecdysteroidogenesis occurs in glands from starved larvae, despite low protein levels, and is actually accelerated by about a day relative to controls. This is not accompanied by an increase in PTTH release, assessed by bioassay. Responsiveness to PTTH and exogenously added cAMP analogs are not affected. *Drosophila* similarly accelerates pupation upon starvation after acquiring an appropriate body weight (Mirth *et al.*, 2005) as does the beetle *Onthophagus* (Shafiei *et al.*, 2001). The underlying regulatory mechanisms have not been identified.

As an activator of receptor tyrosine kinase activity, PTTH is well suited to promote glandular growth. Functionally homologous vertebrate peptide hormones, acting alone or in concert with other factors, often have a differentiation and mitogenic effect on their target tissues in addition to their steroidogenic activity (see Adashi, 1994; Richards, 1994; Richards *et al.*, 2002). PTTH was shown to stimulate two categories of protein synthesis in the *Manduca* prothoracic glands: one class of proteins comprised a small group of molecules whose translation appears to be linked to PTTH-stimulated ecdysteroidogenesis (see Section 1.4.4.), while the second class of translation products was less well defined and thought to include housekeeping proteins in the prothoracic gland (Rybczynski and Gilbert, 1994). Synthesis of the second class of products was enhanced by PTTH, but not by Ca^{2+} and/or cAMP. Given that the PTTH preparation used in these experiments was not purified, it is likely that factors other than PTTH contributed to the observed response (Rybczynski and Gilbert, 1994).

Vertebrate steroidogenic hormones enhance the transcription of steroidogenic enzymes and associated proteins (see Simpson and Waterman, 1988; Orme-Johnson, 1990), and the same has proven true for PTTH. Many of the enzymes responsible for conversion of cholesterol to 20E have now been identified (see Chapter 4 and Huang *et al.*, 2008). Briefly, cholesterol is converted to

7-dehydrocholesterol by the action of a non-rate-limiting dehydrogenase. An enzyme encoded by a gene termed *neverland (nvd)* has been implicated in this step. Then 7-dehydrocholesterol is converted to a diketol by reactions termed the "black box." The enzymes encoded by the cytochrome P450 Halloween genes, *spook* and *spookier (spo)*, have been implicated here. The diketol is converted to ecdysone by cytochrome P450 enzymes (C25-, C22-, C2-, and C20-hydroxyls, respectively) encoded by *phantom (phm)*, *disembodied (dib)*, and *shadow (sad)*. Ecdysone is released from the prothoracic glands and converted to 20E by the enzyme encoded by an additional halloween gene, *shade (shd)*, found in peripheral tissues. With specific regard to regulation of these enzymes by PTTH, the expression of dib in *Bombyx* prothoracic glands is stimulated within 2 h of PTTH exposure. Neither phm nor sad showed a similar response (Niwa *et al.*, 2005). Ablation of PTTH-containing neurons in *Drosophila* led to the repression of a number of ecdysteroidogenic genes including dib, phm, nvd, and spo (but not sad; McBrayer *et al.*, 2005). Ecdysteroid secretion, however, did eventually occur (McBrayer *et al.*, 2005). In *Manduca*, PTTH increased both the amount and phosphorylation of Spo within an hour of exposure, suggesting control by PTTH at the translational and post-translational levels (Rewitz *et al.*, 2009a).

Cholesterol trafficking is also an important aspect of steroid hormone synthesis. Because dietary cholesterol is used in insects for ecdysone synthesis, the uptake of cholesterol into the prothoracic glands is a critical component of ecdysteroid secretion (Huang *et al.*, 2008). Genes responsible for human neurodegenerative Neimann-Pick type C disease (*npc1* and *npc2*) regulate endosomal movement of cholesterol, and loss of npc causes a fatal disorder in which unesterified cholesterol accumulates (Carstea *et al.*, 1997). *Drosophila* npc1 mutants undergo larval arrest similar to other ecdysteroid-deficient mutants (Huang *et al.*, 2005, 2007). A conserved ring-gland-specific, cis-regulatory element in the promoter for this gene was used to construct a reporter for *npc1* expression in the ring glands, and expression was found to be positively regulated by the broad gene (*br;* previously called the Broad-Complex or BR-C) (Xiang *et al.*, 2010). *br* is an ecdysone-inducible transcription factor that stimulates pupal commitment (Zhou and Riddiford, 2002). It is expressed during the final instar of holometabolous insects, for example, in the third instar in *Drosophila*, in the ring glands, and in other tissues. *br* was recently shown to strongly regulate prothoracic gland utilization of cholesterol (Xiang *et al.*, 2010). When an isoform of the gene, *br-Z3*, is overexpressed in the ring glands during the second instar, larval molting is impaired, and the prothoracic glands degenerate prematurely (Zhou *et al.*, 2004). Mutants of *br*, or ring-gland-specific knockdown of *br*, show reduced *npc1* expression (Xiang *et al.*, 2010). Similar knockdowns lead to third instar larval arrest animals in which Torso and insulin receptor are also greatly reduced. The results implicate *br* in enhanced responsiveness to both PTTH and insulin, and increased steroidogenic capacity required for metamorphosis (Xiang *et al.*, 2010), but subsequently enhancing apoptosis (Zhou *et al.*, 2004).

1.5.2. Regulatory Peptides Other than Conventional PTTH

1.5.2.1. Bombyxins and other insulin-like hormones
The role of insulin-like hormones in insect development is discussed in detail in Chapter 2, including the important role of insect insulins in stimulating ecdysone secretion by the mosquito ovary (Graf *et al.*, 1997; Riehle and Brown, 1999). A brief discussion of insect insulin-like hormones is warranted here in the specific context of PTTH and prothoracic gland activities.

The history of discovery of the small insulin-like neuropeptide bombyxin was introduced in Section 1.2.1. Bombyxin was initially characterized as a *Bombyx* PTTH, using a *Samia* pupal development assay (Ichikawa and Ishizaki, 1961, 1963). Bombyxin was later found to have little prothoracicotropic activity in *Bombyx* pupae (Ishizaki *et al.*, 1983a,b). The purification and amino acid sequencing of bombyxin revealed that there was actually a family of bombyxin proteins, with sequence homology to the A and B chains of human insulin (see Ishizaki and Suzuki, 1994). Jhoti *et al.* (1987) modeled the three-dimensional structure of a bombyxin and found that it could assume an insulin-like tertiary structure. Nearly 40 bombyxin genes in *Bombyx* have been characterized (Kondo *et al.*, 1996; see Nagata *et al.*, 2005). An insulin-like growth factor, distinct from bombyxin, was also recently identified (Okamoto *et al.*, 2009). The ability of *Bombyx* bombyxin, effective at 30 pM, to stimulate adult development in debrained *Samia* pupae is still not understood (see Ishizaki and Suzuki, 1994). Nagata *et al.* (1999) used the deduced amino acid sequence of two *Samia* bombyxin homologues to synthesize the corresponding peptides. These peptides were then used in the *Samia* debrained pupae, where they proved to be effective at stimulating adult development at estimated hemolymph concentrations of 1 to 5 nM, a concentration at least 30 times higher than the *Bombyx* bombyxin. Given the plethora of bombyxin genes in *Bombyx*, and presumably other Lepidoptera, the possibility that one of these genes might code for a true prothoracicotropic protein cannot be ruled out. An additional factor must be kept in mind: the ability of bombyxins to activate adult development does not prove that bombyxins directly stimulate prothoracic gland ecdysteroidogenesis. In addition to the study by Kiriishi *et al.* (1992), demonstrating the relative ineffectiveness of bombyxin in stimulating ecdysteroid synthesis by isolated prothoracic glands (see Section 1.2.1.), Vafopoulou and Steel (1997) found that *Bombyx* bombyxin was much less effective than *Bombyx* PTTH at stimulating ecdysteroid synthesis by *Rhodnius* glands *in vitro*. Similarly, the prothoracicotropic effect of *Samia* bombyxin-related peptides was less than *Bombyx* bombyxin and far less than *Samia* PTTH. Other tests *in vitro* of small ecdysteroidogenic molecules have used large amounts of size-fractionated brain extracts (e.g., 50 brain equivalents per prothoracic gland), making it difficult to determine the physiological relevance of the stimulatory molecules, presumed to be bombyxins (Endo *et al.*, 1990; Fujimoto *et al.*, 1991).

Long-term (8 h) incubation with mammalian insulin does directly enhance ecdysteroid secretion in *Bombyx* larvae. The effect is seen in glands removed from fourth instar larvae, and also in glands removed later than day 3 of the fifth instar (Gu *et al.*, 2009). By contrast, mammalian insulin has no steroidogenic effect on *Manduca* prothoracic glands removed from day 2 or day 5 fifth instar larvae, even following 8 h exposure (Walsh and Smith, 2011). Unlike the early, steady growth seen in *Manduca* prothoracic glands during the fifth instar, the majority of the growth in the prothoracic glands of *Bombyx* occurs just before, and following, the cessation of feeding (Mizoguchi *et al.*, 2001). It thus appears that insulin has a stimulatory effect on ecdysone secretion when growth and substantial steroidogenesis occur concurrently as opposed to discrete processes as seen during the last instar in *Manduca*.

Concurrent, insulin-stimulated effects on glandular growth and steroid secretion also occur in *Drosophila*. The *Drosophila* genome contains seven genes that, like the bombyxins, code for proteins containing characteristic insulin motifs (Brogiolo *et al.*, 2001; see Gronke *et al.*, 2010 and Chapter 2). These seven genes (*Drosophila* insulin-like proteins; Dilp 1–7) do not show significant homology to bombyxins outside of the insulin motifs and, unlike the bombyxins, are not limited in expression to neuronal cells. Increasing evidence from this species indicates that insulin-like hormones play a key role in regulating the timing of ecdysone secretion. In *Drosophila*, there seems to be a direct link between insulin, prothoracic gland size, and the determination of final body size. For example, ring-gland-specific overexpression of PTEN, a phosphatase in the insulin pathway that is responsible for suppressing growth, resulted in smaller prothoracic glands and larger adults (Mirth *et al.*, 2005). The overexpression of PTEN only impacted overall growth of the larva when expressed in prothoracic glands and corpora allata, not when it was expressed in corpora allata alone, indicating that the prothoracic glands were responsible for body size assessment (Mirth *et al.*, 2005). Similarly, expression of a dominant-negative (dn) PI3 K in prothoracic glands of flies resulted in smaller glands and larger flies. By contrast, when PI3 K was overexpressed, the resulting glands were larger and the adults were smaller (Mirth *et al.*, 2005). These changes in overall body size are not dependent upon prothoracic gland growth per se. Mutations that caused the upregulation of other growth-enhancing proteins, dMyc and cyclin D/cdk4, were able to increase the size of the prothoracic gland, but did not impact final body size (Colombani *et al.*, 2005). Conversely, mutations in the Ras signaling pathway that did not cause growth led to smaller adult body size (Caldwell *et al.*, 2005).

Based on these findings, it follows that changes in the insulin pathway impact the ability of *Drosophila* glands to produce and secrete ecdysone. While enhancement of the insulin pathway did not detectably increase ecdysteroid titers (Colombani *et al.*, 2005; Mirth *et al.*, 2005), it did lead to the transcriptional activation of the ecdysone target gene, E74B and enhanced transcription of two of the steroidogenic Halloween genes, *phm* and *dib* (Colombani *et al.*, 2005; Caldwell *et al.*, 2005). Premature release of Dilp2, stimulated by genetic manipulation of Dilp2-secreting cells, also led to precocious metamorphosis (Walkiewicz and Stern, 2009). In this case, because systemic insulin levels were elevated, adult body size was not altered. This treatment also increased transcription of dib, and induced transcription of the ecdysone-inducible reporter gene E74B (Walkiewicz and Stern, 2009). Taken together these data demonstrate that insulin signaling in the prothoracic glands is able to directly stimulate ecdysone secretion, which leads to the cessation of feeding and the determination of final body size (see Shingleton, 2005). However, the exact mechanism by which insulin stimulates ecdysone secretion remains largely unknown.

Downregulation of TOR activity by overexpression of TSC or TOR RNAi led to a reduction in prothoracic gland size, delayed molting, and overgrowth similar to that seen following downregulation of the insulin pathway (Layalle *et al.*, 2008). However, the role of TOR may not be relevant to normally fed animals. Overactivation of TOR in the prothoracic glands through overexpression of dRheb, an upstream activator of TORC1, or of an inducible RNAi against TSC2, reduced the developmental delay induced by starvation. However, in fed larvae, no effect was seen, suggesting that TOR activation is normally not limiting, but rather provides a sensing mechanism when nutrition falls below a threshold value (Layalle *et al.*, 2008). Starvation did not affect levels of PTTH expression, and while inhibition of TOR in the PTTH-producing neurons reduced the size of those neurons, it had no effect on larval development time or on the size of adults. Thus the PTTH-secreting neurons do not appear to be using TOR-mediated changes to time larval development, but rather, such a mechanism appears to strongly reside in the prothoracic glands.

1.5.2.2. Other stimulatory factors

The prothoracic glands are potentially regulated by any number of the hormones found in the hemolymph, in addition to PTTH and insulin-like hormones. The influence of the

JHs has been discussed (see Section 1.3.3. and Chapter 8), but it bears repeating that the evidence is weak for direct regulation of prothoracic gland physiology by JHs. Aside from JHs and the ecdysteroids, most, if not all, of the other regulators of the prothoracic glands are peptides or small proteins. Some of these are stimulatory (discussed in this section while others are inhibitory (see Section 1.5.2.3.).

Small PTTH: As discussed in Section 1.2.2., size-fractionated brain extracts frequently exhibit two peaks of ecdysteroidogenic activity, with one peak at 20–30 kDa and a second, smaller peak below 10 kDa , giving rise to characterized "big" and "small" PTTHs. In most cases, there are simply no further data beyond approximate molecular weight and dose (brain equivalents). The observations indicate a much larger dose requirement for the so-called small PTTHs. Additionally, a smaller maximum response (Yokoyama *et al.*, 1996, Endo *et al.*, 1990, 1997) suggests, but does not prove, that these so-called small PTTHs may be bombyxin homologues. As noted for *Bombyx* prothoracic glands tested *in vitro* (Kiriishi *et al.*, 1992), PTTH stimulated ecdysteroidogenesis much more effective than does bombyxin. However, there are some data from *Manduca* that suggest that small PTTH in this species may be a molecule other than a bombyxin. Bollenbacher *et al.* (1984) described a small PTTH (6–7 kDa) that has subsequently been partially characterized in regard to its developmental expression and the signal transduction events it elicits in prothoracic glands. *Manduca* small PTTH appears to be much more abundant in pupal than in larval brains, while "big" PTTH shows only a slight increase in the pupal versus larval brain (O'Brien *et al.*, 1986). Bollenbacher *et al.* (1984) reported that small PTTH was notably less effective at stimulating ecdysteroid synthesis in pupal prothoracic glands than in larval glands; however, a study by O'Brien *et al.* (1986) showed no such difference. Like big PTTH, *Manduca* small PTTH appears to stimulate ecdysteroid synthesis via Ca^{2+} influx and Ca^{2+}-dependent cAMP accumulation (Watson *et al.*, 1993; Hayes *et al.*, 1995). These observations concerning intracellular signaling suggest that *Manduca* small PTTH might not be a bombyxin homologue because, as members of the insulin peptide family, bombyxin would not be expected to stimulate either Ca^{2+} influx or Ca^{2+}-dependent cAMP (see Combettes-Souverain and Issad, 1998). It should also be pointed out that for small PTTH to have the same effects as PTTH on Ca^{2+} influx and Ca^{2+}-dependent cAMP generation is surprising and confusing, especially if small PTTH does not preferentially activate larval glands (c.f,. conflicting data in Bollenbacher *et al.*, 1984 and O'Brien *et al.*, 1986). It has been hypothesized that small PTTH might be an active proteolytic fragment of PTTH (Gilbert *et al.*, 2002); if so, it is unclear whether small PTTH is generated as an artifact during PTTH extractions or if it is a naturally occurring processed form.

FXPRL-amide peptides: Stimulatory factors in the FXPRL-amide family bind to GPCRs and include hormones such as PBAN (regulating sex pheromone synthesis, see Chapter 12) and diapause hormone (stimulating the induction of embryonic diapause, see Chapter 10). *Bombyx* diapause hormone receptor (BmDHR) was found to be highly expressed in the prothoracic glands (Yamanaka *et al.*, 2006; Watanabe *et al.*, 2007). While the BmDHR mediates embryonic diapause (Homma *et al.*, 2006), at later stages it appears to have the opposite effect. The diapause hormone of *H. armigera* was found to bind to pupal prothoracic glands and stimulate ecdysteroid secretion, and hence terminate diapause (Zhang *et al.*, 2004). In *Bombyx*, diapause hormone was found to stimulate ecdysteroid secretion via increases in intracellular calcium and cAMP, particularly in the late fifth instar where it may play an accessory role with PTTH in stimulating pupation (Watanabe *et al.*, 2007).

Autocrine factors: A stimulatory factor produced by the prothoracic glands was reported for *Bombyx* (Gu, 2006, 2007), as well as *L. migratoria* and *Schistocerca gregaria* (Vandermissen *et al.*, 2007). These autocrine factors are produced, paradoxically, under conditions of cellular stress, and stimulate ecdysone secretion as well as DNA synthesis. They were originally characterized as a non-ecdysteroid factor secreted by single glands incubated in very small volumes of incubation medium (5 μl), or with 5 glands in 50 μl of medium (Gu, 2006, 2007). The chemical nature of autocrine factor(s) has not yet been determined, beyond a relatively small molecular weight of 1–3 kDa based on membrane filtration and susceptibility to proteases, suggesting a peptide (Gu, 2007).

Gut peptides: Other non-cerebral ecdysiotropic activities have been described, e.g., factors extracted from the proctodaea of *Manduca*, the European corn borer *O. nubilalis*, and the gypsy moth *L. dispar* (Gelman *et al.*, 1991; Gelman and Beckage, 1995). These data are not completely surprising, given the precedent of the vertebrate gut as an endocrine organ. However, a definitive role for such factors *in vivo* has not been established.

1.5.2.3. Inhibitory factors   PTTH and other stimulatory factors are only one side of the coin with regard to the regulation of ecdysteroid secretion. Inhibitory factors also appear to play a part in determining both the onset and termination of prothoracic gland activity during the molting cycle. Evidence for the importance of such prothoracicostatic factors has been growing in recent years from the point of view of the factors themselves, that is, chemistry and release, as well as their actions on identified G-protein-linked receptors in developing prothoracic glands.

1.5.2.3.1. Prothoracicostatic peptide   A nonapeptide termed PTSP has been purified from *Bombyx* brain (Hua *et al.*, 1999). PTSPs are members of the W(X)6Wamide peptide family, which share a common C-terminal sequence motif. They are also known as allatostatin B (Lorenz *et al.*, 1995) and myoinhibitory peptide (Blackburn *et al.*, 1995) according to earlier characterized effects. *Bombyx* PTSP, identical in sequence to a myoinhibitory peptide first isolated from *Manduca* ventral nerve cord (Mas-MIP I; Blackburn *et al.*, 1995), was shown to inhibit basal (Hua *et al.*, 1999; Dedos *et al.*, 2001) and PTTH-stimulated (Hua *et al.*, 1999) ecdysteroidogenesis *in vitro* in a dose-dependent manner. A receptor for PTSP was identified by gene expression; surprisingly, it turned out to be a homologue of the *Drosophila* sex peptide receptor (Yamanaka *et al.*, 2010). Mammalian cells ectopically expressing the *Bombyx* homologue of the sex peptide receptor responded to *Bombyx* PTSPs at relatively low concentrations (0.1 μM), but not other neuropeptides (PTTH or bommo-myosuppressins, discussed next).

PTSP blocked "basal" Ca^{2+} influx via dihydropyridine-sensitive channels in *Bombyx* prothoracic glands (Dedos *et al.*, 2001; Dedos and Birkenbeil, 2003), but PTSP did not block PTTH-stimulated Ca^{2+} influx (Dedos and Birkenbeil, 2003). It is difficult to reconcile the latter observation with the finding that PTSP blocked PTTH-stimulated ecdysteroid synthesis (Hua *et al.*, 1999), given that PTTH-stimulated ecdysteroid synthesis requires Ca^{2+} influx. Furthermore, rather large doses were required (≥100 nM) to achieve an ecdysteroidogenic inhibition relative to the dose needed for myoinhibition (1 nM; Blackburn *et al.*, 1995). More recent studies indicate that PTSP blocked basal and ecdysone-stimulated synthesis of cAMP, although at doses about 100-fold higher than those for a different prothoracicostatin, bommo-myosuppressin (BMS; discussed next) (Yamanaka *et al.*, 2006). PTSP was effective when injected *in vivo*, and it was reported to inhibit ribosomal S6 protein phosphorylation (Liu *et al.*, 2004).

The PTSP receptor was expressed in high levels in the prothoracic glands just before each ecdysis. Receptors were also found in the brain and Malpighian tubules (Yamanaka *et al.*, 2010). The receptor was induced by 20-hydroxyecdysone *in vitro*, particularly when administered as a 6h pulse rather than continuously. Davis *et al.* (2003), using immunohistochemical techniques, have found an apparent release of MIP I (PTSP) from *Manduca* epitracheal glands that coincides with the rapid drop in ecdysteroid hemolymph titer that occurs just prior to larval ecdysis. PTSP appears to be released from the epitracheal gland at ecdysis (Davis *et al.*, 2003). This suggests that PTSP ensures the suppression of ecdysteroidogenesis at ecdysis, and is likely to have additional effects beyond the prothoracic glands.

1.5.2.3.2. Myosuppressins

A decapeptide myoinhibitory hormone, BMS, also has prothoracicostatic activity. Myosuppressins are found in other insects, including flies

and cockroaches, and have been shown to decrease spontaneous contractions of gut muscle and also to increase contractions in flight muscle and stimulate digestive enzyme release (Nichols, 2003). The myosuppressins are FMRFamide related peptides (FaRPs) that each contain a FMRFamide C-terminus with variable N-termini and multiple functions. BMS is identical in structure to *Manduca* FLRFamide1 (an FMR-amide-related peptide; Lu *et al.*, 2002). BMS is expressed in the brain, with a pattern suggesting that it suppresses ecdysone secretion as larvae feed during the first half of the fifth instar. This pattern is the reverse of that seen for PTSP, which acts at the end of the instar. The receptor for BMS is a GPCR, found in the prothoracic glands but also in the gut and Malpighian tubules (Yamanaka *et al.*, 2005). It blocks basal and PTTH-stimulated ecdysteroid secretion by fifth instar larval prothoracic glands at doses between 0.1 and 1 nM. Doses of BMS higher than 1 nM do not block ecdysone secretion. This is in contrast to the effects of BMS on PTTH-stimulated cAMP synthesis, which is strongly blocked at doses of 0.1 nM and higher.

The odd lack of effect of BMS at high doses (>1nM) may be due to its differential effects on cAMP synthesis and Ca^{2+} influx. Doses of BMS >1nM enhance Ca^{2+} influx in mammalian cells ectopically expressing the BMS receptor. Thus BMS at high doses may activate pathways that simultaneously promote (Ca^{2+} influx) and block (cAMP inhibition) PTTH-stimulated ecdysone secretion. Four additional *Bombyx* FMRFamide genes (BRFas) were found in the thoracic ganglia (Yamanaka *et al.*, 2006). They were shown to act through the same receptor as BMS, and to block PTTH-stimulated ecdysone and cAMP synthesis, although at higher doses in keeping with direct release from nerves to the prothoracic glands. BMS and BRFas block PTTH-stimulated expression of the steroidogenic enzyme, *phm*, although the signaling pathway by which they do so is not known (Yamanaka *et al.*, 2007).

Neurons supplying BRFas to the prothoracic gland surface were more active during times of low ecdysteroid secretion, that is, the BRFa peptides are likely to be released at the same time as BMS, with a similar function of suppressing ecdysone secretion, as larvae feed. BRFas were postulated to provide localized control to the prothoracic glands, while BMS may have broader effects body wide (Yamanaka *et al.*, 2006; Truman, 2006).

1.5.2.3.3. Trypsin-modulating oostatic factor

An ovarian-derived dipteran oostatic factor (trypsin modulating oostatic factor, TMOF) appears to exert dose-dependent stimulatory and inhibitory effects on ring gland and prothoracic gland basal and PTTH-stimulated ecdysteroid production (Hua and Koolman, 1995; Gelman and Borovsky, 2000). Two types of TMOF have been identified. One is a decapeptide from the ovaries of *A. aegypti* (Gelman and Borovsky, 2000), the other a completely different hexapeptide from the ovaries of the flesh fly *Neobellieria bullata* (Byelmans *et al.*, 1995). The name arises from an inhibitory effect of both hormones on trypsin biosynthesis in the gut, which impairs digestion, and thus inhibits vitellogenin synthesis and oogenesis. *Aedes* TMOF has only been shown to be ecdysiostatic in prothoracic glands from larval *L. dispar*, while that of *Neobellieria* inhibits its own larval ring glands as well as those of *C. vicina*. The ecdysiostatic effect of TMOF, a hexapeptide, on *C. vicina* ring glands was attributed to a TMOF-dependent rise in cAMP, with the latter inhibiting basal ecdysteroidogenesis (Hua and Koolman, 1995). However, an earlier study showed that dbcAMP stimulated *Calliphora* ring gland ecdysteroidogenesis (Richard and Saunders, 1987), and it is not currently possible to reconcile these conflicting observations about cAMP. How TMOF exerts a prothoracicotropic effect rather than an ecdysiostatic effect under some conditions is also not known. Thus, these observations are intriguing, but the *in vivo* significance of TMOF in regard to prothoracic gland function remains to be rigorously determined.

1.5.2.3.4. Neural control As discussed in Section 1.2.3., PTTH appears to be delivered to *Drosophila* prothoracic glands directly by innervating neurons, providing a means for neural stimulation of prothoracic gland activity (McBrayer *et al.*, 2007). The BRFas in *Bombyx* also appear to be neurally delivered factors, in this case inhibitory factors that ensure low steroidogenic activity while fifth instar larvae feed (Yamanaka *et al.*, 2006). For the most part, however, the role of innervation in chronic or acute regulation of ecdysteroid synthesis is not well understood. In *P. americana*, the activity of a nerve originating from the prothoracic ganglion was correlated positively with the secretion of ecdysteroids from the prothoracic gland (Richter and Gersch, 1983), but experimental discharge of this nerve did not have a significant effect on ecdysteroidogenesis (Richter, 1985). Sectioning the prothoracic gland nerve of last instar larvae prevented the normal ecdysteroid peak seen between days 20 and 24, but the subsequent larger peak seen after day 26 was unaffected, suggesting a positive role for neuronal input during the first, but not the second, ecdysteroid peak. In the moth *Mamestra*, the neurons innervating the prothoracic gland are inhibitory in regard to basal ecdysteroid synthesis, but their effect on PTTH-stimulated synthesis is not known (Okajima and Watanabe, 1989; Okajima and Kumagai, 1989; Okajima *et al.*, 1989). Alexander (1970) proposed an inhibitory role in *Galleria* for the prothoracic gland nerve originating in the subesophageal ganglion while Mala and Sehnal (1978) argued for an excitatory (steroidogenic) influence. Conflicting data on the role of innervation complicates any attempt to generalize at this time.

1.5.3. Ecdysteroid Feedback on the Prothoracic Gland

Relatively little data has been gathered to test the hypothesis that the prothoracic glands are subject to both rapid and delayed feedback from ecdysteroids. Effects could be stimulatory or inhibitory, depending upon the timing of their occurrence within a molting cycle. Sakurai and Williams (1989) showed that incubating larval *Manduca* prothoracic glands for 24 h with low concentrations of 20E (e.g., 0.20 μM) resulted in an inhibition of 20 to 60% of basal ecdysteroid synthesis. By contrast, 20E generally stimulated pupal glands. This was true of basal secretion by prothoracic glands from both non-diapausing and diapausing pupae; PTTH-stimulated secretion was not tested. A study of the effect of shorter 20E incubations on basal and PTTH-stimulated *Manduca* glands revealed that, when a high dose (10 μM) of 20E was employed, significant inhibition of basal ecdysteroidogenesis (≈75%) was seen after one hour of incubation (Song and Gilbert, 1998). Note that in these experiments, prothoracic glands were pre-incubated with exogenous 20E, but their ecdysteroid production was measured subsequently in medium lacking the supplemental 20E. Somewhat surprisingly, this regime of 20E pre-treatment had no effect on PTTH-stimulated ecdysteroid synthesis, relative to controls not incubated with 10 μM 20E, unless the pre-incubation was for more than 3 h. These observations suggest that the repression of basal ecdysteroid synthesis after short incubations was due to a reversible block in the steroidogenic pathway that could be overcome by PTTH stimulation. What constitutes this block is open to speculation.

Incubation of *Manduca* prothoracic glands with 10 μM 20E altered the expression pattern of USP proteins within 6 h (Song and Gilbert, 1998). USP is an obligate partner to the EcR protein in forming the functional ecdysteroid receptor (Chapter 5), and this change might be a sign of an altered transcription pattern in the gland. Perhaps the rapid 20E effect on basal steroidogenesis was due to direct, competitive inhibition by 20E of some step in ecdysteroid synthesis. This possibility could be tested by using lower concentrations of 20E and non-steroidal compounds that bind to the ecdysone receptor, but they would not be expected to interact with the biosynthetic enzymes involved in ecdysteroidogenesis. Jiang and Koolman (1999) used the non-steroidal ecdysone agonist RH-5849 in such a test, using *C. vicina* ring glands. Their results revealed that RH-5849 inhibited basal ecdysone secretion rapidly, with more than 50% inhibition seen within 30 min of incubation. These data suggest that ecdysteroid feedback on the prothoracic gland involves regulation of transcription or non-genomic effects, perhaps mediated through second messenger signaling initiated via a plasma membrane ecdysteroid receptor (see Chapters 4 and 5). Longer incubations (>3 h), however, resulted in the downregulation of steroidogenesis, which was not fully reversible with PTTH (brain extract) treatment. More recently, Gu *et al.* (2008) tested the effects of the ecdysone agonist RH-5992 on *Bombyx* fourth instar larval prothoracic glands. The agonist had an inhibitory effect on basal steroidogenesis without blocking PTTH-stimulated changes in cAMP, ERK phosphorylation, or steroidogenesis. How the ecdysone receptor might interact with the synthesis pathway is unknown. A direct interaction of native ecdysone with the prothoracic glands through ecdysone receptors was demonstrated in *Rhodnius* (Vafopoulou and Steel, 2006; Vafopoulou, 2009). The receptor was localized to microtubules and mitochondria, in addition to the nuclei, suggesting novel sites of action for the ecdysone receptor in the prothoracic glands.

Non-lepidopteran model insects are providing new insights into the regulation of ecdysone secretion by ecdysone-sensitive genes. For example, in *Drosophila*, an ecdysteroid responsive transcription factor, DHR4, is expressed in high levels at the onset of the third instar, particularly in the ring gland where it represses other ecdysone-inducible genes (King-Jones *et al.*, 2005). Mutation of DHR4 accelerates the initiation of metamorphosis, leading to small adults in a manner similar to insulin-stimulated growth of the prothoracic glands (King-Jones *et al.*, 2005). One gene repressed by DHR4 is βFTZ-F1 (fushi-tarazu factor 1, beta form), the insect homologue of vertebrate steroidogenic factor 1 (SF1), a regulator of vertebrate steroidogenic genes (Val *et al.*, 2003). In *Drosophila*, βFTZ-F1 has been found to increase the expression of phm and dib in the ring glands and has been associated with increased steroidogenic capacity (Parvy *et al.*, 2005). On the other hand, in the hemimetabolous insect *Blattella*, βFTZ-F1 expression in the prothoracic glands is negatively correlated with hemolymph ecdysteroids (i.e., present early in the last instar) declining at the time of the ecdysteroid peak, and rising again prior to ecdysis (Cruz *et al.*, 2008). Apoptosis of the glands normally occurs just after adult ecdysis. Knockdown of βFTZ-F1 in *Blatella* accelerates the pre-metamorphic ecdysteroid peak, but developmental abnormalities result and ecdysis itself is blocked. In addition, the prothoracic glands fail to degenerate. These results suggest that in *Blattella*, early expression of βFTZ-F1 delays steroidogenesis, while later it enhances apoptosis (Cruz *et al.*, 2008).

The effects of 20E feedback on prothoracic gland activity may be indirect, reducing sensitivity to growth factors, or enhancing the secretion of inhibitory hormones. For example, in *Drosophila*, ecdysteroids reduce sensitivity of the prothoracic glands and other tissues to insulin, resulting in reduced glandular growth (Colombani *et al.*, 2005). The effects of ecdysteroids are mediated by the fat body, which in some manner reduces systemic responsiveness to insulin-like hormones. In *Bombyx*, 20E may facilitate the release of inhibitory hormones. Injection of 20E reduces basal ecdysteroid secretion, but less so when the brain and corpora allata are removed from larvae prior to 20E injection (Takaki and Sakurai, 2003). These results suggest that 20E might downregulate prothoracic gland activity by stimulating the release of an inhibitory brain or CA factor (see Section 1.5.2.). Ecdysone does enhance glandular content of receptors for PTSP, supporting an ecdysteroid-mediated increase in sensitivity to inhibitory factors (Yamanaka *et al.*, 2010).

In *Bombyx*, genes induced by 20E were isolated from the brain of fifth instar larvae (Hossain *et al.*, 2006, 2008). All were found in lateral neurosecretory cells known to produce PTTH, providing a potential mechanism for ecdysteroid feedback (see Section 1.5.3.). Knockdown of one such gene, bombe.1-2, led to defective larval–pupal molting, and adult wing and leg malformations. In larvae, the ecdysteroid receptors EcRA and EcRB1 were seen exclusively in PTTH-containing neurons, while in pupae they were seen in other neuronal cell types. Hossain's work suggests that because feedback regulation by ecdysteroids in the larval brain are seen exclusively in PTTH-expressing neurons, these neurons may be responsible for regulating other ecdysteroid-sensitive neurohemal responses.

1.5.4. Apoptosis of the Prothoracic Gland

Programmed cell death or apoptosis is a normal component of development in multicellular organisms, including insects, and involves a complex, multistep cascade of intracellular events (see Schwartz, 2008; Kourtis and Tavernakis, 2009). The only demonstrated function of the prothoracic glands is to produce ecdysteroids and thus to control and coordinate the molting process. It is not surprising, therefore, that prothoracic glands undergo apoptosis during pupal–adult development or early in adult life, once sufficient ecdysteroids have been produced to accomplish this last molt. Programmed cell death of the prothoracic glands occurs both in insects that possess a structurally distinct gland, like Lepidoptera, as well as in species possessing a ring gland, where the prothoracic gland is part of a composite, multi-tissue organ.

In *Drosophila*, the degeneration of the prothoracic gland portion of the ring gland during pupal–adult development takes place at a time when whole animal ecdysteroid titers are high and production of ecdysteroids *in vitro* is low (Dai and Gilbert, 1991). This observation led to the suggestion that the source of ecdysteroids at this time was either another tissue (e.g. oenocytes) or the conversion of previously produced hormone from an inactive conjugate to an active form, with the conjugate not being recognized by the radioimmunoassay employed to assess steroid hormone levels (Dai and Gilbert, 1991). Degeneration of the prothoracic gland cells appears to be stimulated by expression of products of the ecdysteroid-responsive broad gene complex (br) (Zhou *et al.*, 2004; see Section 1.5.1.).

The structural and hormonal correlates of programmed cell death of the *Manduca* prothoracic gland, a "free-standing" tissue, has been addressed by Dai and Gilbert, (1997, 1998, 1999). In this species, apoptosis of the gland was initiated on day 6 (P_6) of pupal–adult development (Dai and Gilbert, 1997). P_6 is approximately two days after the pupal–adult peak of ecdysteroidogenesis and is a time when fat body and hemocytes begin to envelop the gland (Dai and Gilbert, 1997). Ecdysteroid synthesis fell rapidly after P_6, as the glands continued to degenerate, and by P_{14}, only cellular debris was present. The external factors controlling apoptosis of the prothoracic gland have been partially determined. The absence of JHs at the beginning of pupal–adult development was early believed to be an enabling condition (Wigglesworth, 1955; Gilbert, 1962) and subsequent data supported this proposal, that is, injection of JH into lepidopteran pupae at this stage prevented gland death (Gilbert, 1962; Dai and Gilbert, 1997). The absence of JH was not sufficient to initiate apoptosis, because prothoracic glands extirpated from early *Manduca* pupae and placed in JH-free culture conditions did not undergo apoptosis (Dai and Gilbert, 1999). Nevertheless, a hormonal basis for gland apoptosis was indicated because the addition of 20E to these cultures efficiently promoted programmed cell death if the steroid exposure was ≥24 h. The 20E doses found to be effective *in vitro* were well within the physiological range seen in the days preceding normal, *in vivo* cell death. These cultured-gland experiments also revealed an extra level of complexity in the control of prothoracic gland apoptosis, namely, that JH did not protect glands *in vitro* against 20E-induced cell death (Dai and Gilbert, 1999). A similar regulation of prothoracic gland apoptosis was seen in a hemimetabolous insect, *O. fasciatus*, by Smith and Nijhout (1982, 1983). Apoptosis was stimulated by 20E *in vitro*, and inhibited by JH *in vivo*, but not *in vitro*. Thus, once again, the role of JHs in modulating prothoracic gland physiology was indirect, possibly involving another tissue or hormone.

In *Blattella germanica*, apoptosis of the prothoracic glands is also stimulated by 20E acting in the absence of JH in the last nymphal stage (Mane-Padros *et al.*, 2010). The mechanism by which 20E stimulates apoptosis appears to involve βFTZ-F1 (Mane-Padros *et al.*, 2010). 20E triggers nuclear receptor transcription in the prothoracic glands of *Blattella*, including members of the early gene family E75 (designated BgE75 for this species), required for ecdysteroid levels to increase, and ending with a marked increase in βFTZ-F1. Without βFTZ-F1 the glands do not degenerate, and ectopic expression of βFTZ-F1 leads to premature degeneration. Apoptosis is inhibited in *Blattella* by an inhibitor of apoptosis protein (IAP1), which is present in the glands until the time of degeneration just after nymphal-adult ecdysis. Downregulation of IAP1 only occurs in adult prothoracic glands, and is correlated with high levels of βFTZ-F1. However, the link between IAP1 and βFTZ-F1 was not found.

A clear connection exists in a variety of species between ecdysteroids and apoptosis of the prothoracic glands. One might also conclude that 20E-induced apoptosis of the prothoracic gland represents the ultimate negative feedback loop for this steroid hormone producing gland.

1.5.5. Parasitoids and Ecdysteroidogenesis

Insect parasitoids lay their eggs on or in host insects. Following hatching, the parasitoid develops until finally emerging, with the host generally dying as a result of the infection. Parasitoids manipulate the physiology of the host organism to their advantage and this often includes disrupting the normal function of the host's endocrine system (see Beckage, 1985, 1997, 2002; Beckage and Gelman, 2004). It must also be pointed out that parasitoid infections are biologically complex, involving not only the parasitoid, but also ovarian proteins, symbiotic viruses, and venoms. These additional factors can have major effects on the host's physiology, even in the absence of the parasitoids.

Parasitoids utilize a wide variety of tactics to increase the suitability of their hosts as growth chambers and even a modest review of the subject is beyond the constraints of this chapter (see Beckage, 1985, 1997, 2002; Beckage and Gelman, 2004). However, several endocrine effects are frequent enough to deserve mention in the context of PTTH. One of the most common effects of parasitoid infestation appears to be elevated JH levels. These high titers result in developmental arrest of the host and stem from multiple factors, including increased secretion by the host, decreased degradation, and JH secretion by the parasitoid and perhaps by parasitoid-derived teratocytes (Cole *et al.*, 2002). Elevated levels of JH may act partly through blocking PTTH release since increased levels of PTTH in brains of parasitized *Manduca* have been observed (Zitnan *et al.*, 1995). In some Lepidoptera-parasitoid interactions, the prothoracic gland disintegrates, e.g., *H. virescens* parasitized by *Campoletis sonorensis*, an ichneumonid wasp (Dover and Vinson, 1990; Dover *et al.*, 1995). In other species, the gland persists. In prothoracic glands from *Manduca* parasitized by *Cotesia congregata*, a braconid wasp, both basal and brain-extract-stimulated ecdysteroid synthesis *in vitro* was lower than synthesis from glands of non-parasitized larvae (Kelly *et al.*, 1998). This is strongly reminiscent of the effect of experimentally augmented JH levels on prothoracic glands discussed in Section 1.3.5., and is consistent with a view that this parasitism interaction, which blocks development in the fifth instar, is essentially a juvenilizing event. Evidence from *in vitro* studies suggested that a portion of the elevated JH in this example was secreted by the parasitoid and did not simply result from decreased JH catabolism (Cole *et al.*, 2002).

In vivo ecdysteroid levels at this time were not strongly suppressed relative to control early fifth instar larvae, but the high pre-molt peak seen normally in the second half of the fifth instar was absent. Ligation experiments revealed that it was likely that some ecdysteroids were secreted by the parasitoid *Cotesia* (Gelman *et al.*, 1998). A stronger suppression of ecdysteroid hemolymph titer has been described for precocious prepupae of *Trichoplusia ni* parasitized by the wasp *Chelonus* near *curvimaculatus* (Jones *et al.*, 1992); in this interaction, an ecdysteroid peak observed shortly before the emergence of the parasitoid larvae was also believed to originate with the parasitoid.

Evidence that parasitism has a direct effect on the functioning of the prothoracic gland comes primarily from work on *H. virescens* parasitized by the braconid wasp *Cardiochiles nigriceps*. In this interaction, basal ecdysteroid synthesis by the glands was considerably depressed when measured *in vitro* (Pennacchio *et al.*, 1997, 1998a); JH levels were elevated, and JH metabolism depressed, in parasitized larvae during the first four days of the final larval instar (Li *et al.*, 2003). Such glands showed no activation when challenged with a crude PTTH extract or forskolin (an adenylyl cyclase activator) or dbcAMP; but all three reagents stimulated ecdysteroid synthesis in control glands. Further investigation revealed several biochemical lesions downstream from cAMP generation (Pennacchio *et al.*, 1997, 1998a), including depressed basal protein synthesis and phosphorylation and a failure to increase protein phosphorylation and synthesis upon PTTH stimulation. No difference in RNA synthesis was found between control and parasite-derived glands (Pennacchio *et al.*, 1998a). Surprisingly, the effects of *Cardiochiles* parasitism on the prothoracic gland proved to be largely the result of co-infection with the *C. nigriceps* polydnavirus that is injected into the host along with the *Cardiochiles* eggs and venom (Pennacchio *et al.*, 1998b). Venoms also play a role in parasitoid manipulation of host prothoracic glands. The venom of the ectoparasitic wasp *Eulophus pennicornis*, when injected into larvae of the tomato moth (*Lacanobia oleraceae*), resulted in depressed basal ecdysteroid synthesis by prothoracic glands when assayed 48 h later, and furthermore, such glands failed to respond to forskolin with increased steroidogenesis (Marris *et al.*, 2001). However, incubation of prothoracic glands with venom extract for 3 h had no effect. This observation suggests that either venom requires a longer time to disrupt the PTTH transduction cascade, or that an intermediate tissue is involved in the venom effect. An inhibitory effect on the prothoracic glands of parasitized larvae, as well as on the prothoracicotropic activity of the brain, was demonstrated to be caused by the bracovirus associated with wasp parasitoid *Chelonius inanitus* in the host insect *Spodoptera littoralis* (Pfister-Wilhem and Lanzrein, 2009). The resulting low levels of ecdysteroids led to precocious onset of metamorphosis, in this case stimulated by ecdysteroids provided by the parasitoid.

These and similar observations suggest that the roles of viruses and venoms in manipulating host endocrine systems can be great. How parasitoids and their co-evolved, symbiotic viruses influence the PTTH-prothoracic gland axes of their hosts is far from resolved and might well involve a number of tactics that include altering hormone synthesis and prohormone conversion by their hosts, hormone release, and feedback systems.

1.6. PTTH: The Future

Recent studies have greatly enhanced our understanding of the nature of the PTTH receptor, key aspects of PTTH signaling, and the complexity of prothoracic gland growth and activity. Of course, these findings and others raise new questions, and a number of old questions still remain. Future work is likely to be shaped by two trends apparent from this chapter and other recent reviews (Mirth and Riddiford, 2007; Huang et al., 2008; Marchal et al., 2010). One trend is the use of *Drosophila* to study prothoracic gland development and ecdysteroid synthesis. Manipulation of individual genes in specific target tissues such as the prothoracic glands or PTTH-producing cells is a powerful approach, particularly in concert with studies on other insects in which physiological changes such as ecdysteroid titers are more easily measured. A second trend is a shift in viewpoint from the prothoracic glands as minions of PTTH, to the prothoracic glands as integrators of multiple signals (both stimulatory and inhibitory) that determine the timing of molting and metamorphosis and adult size. From additional tropins, such as the insulin-like hormones and diapause hormone, to statins such as PTSP and BRFas, the regulation of ecdysone secretion, like that of vertebrate steroids, is multifaceted.

Each of these trends presents promises and challenges. Genetically targeting specific proteins in the brain and prothoracic glands allows clearer identification of the roles of such proteins than, for example, invasive surgeries or relatively non-specific chemical inhibitors. However, *Drosophila* is an evolutionarily and ecologically distinct insect from *Bombyx, Manduca,* or other models such as *Tr.* and *Blattella,* leading to differences in the control of pivotal metamorphic events, such as timing of the commitment ecdysteroid peak. For example, the role of PTTH in *Drosophila* development is different than that characterized for moths such as *Bombyx* and *Manduca,* both with regard to its means of delivery to the prothoracic glands (neural vs. hormonal), and its relative importance to metamorphosis (with Diptera relatively adept at metamorphosing, albeit slowly, even in the absence of PTTH). Such differences call for requisite caution in applying data from one group to understanding the biology of another, and argue for continued cross-phyletic studies.

Gaps exist in several areas among all studied insects. In particular, despite abundant PTTH sequence information, we still know very little about the structural characteristics of PTTH responsible for its activity and species-specificity. Identification of a receptor provides a critical, long-missing component to investigate structure-function relationships between ligand and receptor. The relatively high species-specificity of PTTH, as compared to ecdysteroids, provides an opening for PTTH-directed control strategies in which specific insect groups can be selectively targeted. With regard to signaling pathways, a clear gap exists for all studied groups regarding the specific links between Torso, the MAP kinase pathway, and second messengers such as calcium and cyclic AMP. And further downstream in the signaling pathways, it is still not clear how PTTH accelerates ecdysteroid secretion, that is, specific links between kinases such as ERK, PKA, PKC, and the synthesis and phosphorylation of steroidogenic enzymes such as Spook. This work will be facilitated by the remarkable progress that has been made in identifying enzymes responsible for many steps in ecdysone synthesis. The necessity for new protein synthesis in PTTH-stimulated ecdysteroidogenesis is well established, but the role of such protein(s) has not been determined. If one of these proteins facilitates steroid precursor movement, as in vertebrate steroidogenesis, what do the other proteins do? And responses of the prothoracic glands to ecdysteroids brings into the limelight ecdysteroid-responsive transcriptional regulators such as βFTZ-F1 and Broad, well-studied in insect development, yet only recently implicated as regulating specific aspects of ecdysteroid secretion.

With regard to the complex regulation of prothoracic gland function, the fact that multiple factors control ecdysteroid secretion is hardly surprising, and must be taken into account in predicting the activity of the prothoracic glands at a given time in development. Further, we know relatively little about the factors that coordinate the release of PTTH or other regulatory peptides that control ecdysteroid secretion and glandular secretory capacity. The importance of insulin-like hormones is clear, but among the many putative neuropeptide receptors that are present in the brain (see, e.g., Yamanaka *et al.*, 2008), one can envision a control network that resembles, and likely provided a prototype for, the network of hormonal cues that regulate vertebrate post-developmental changes such as puberty. Further, given that PTTH is present in adults and influences reproductive function, the role of PTTH beyond metamorphosis is also an area ripe for further pursuit.

Our understanding of ecdysteroidogenesis, and developmental changes that result from ecdysone secretion, will in future years be unlikely to be as PTTH-centric as seen in the past. It is clear, however, that 90 years of PTTH research have been integral to our understanding of molting and metamorphosis and have shaped the framework from which future models of insect development will emerge.

Acknowledgments

Many thanks to the members of the Smith laboratory who provided help and advice with the manuscript, and to Smith's husband and family for their patience and good humor. This chapter is a revision of an earlier work by Robert Rybczynski.

References

Adachi-Yamada, T., Iwami, M., Kataoka, H., Suzuki, A., & Ishizaki, H. (1994). Structure and expression of the gene for the prothoracicotropic hormone of the silkmoth *Bombyx mori*. *Eur. J. Biochem.*, *220*, 633–643.

Adashi, E. Y. (1994). Endocrinology of the ovary. *Hum. Reprod.*, *9*, 815–827.

Agui, N. (1989). *In vitro* release of prothoracicotropic hormone (PTTH) from the cultured brain of *Mamestra brassicae* L. effects of neurotransmitters on PTTH release. In J. Mitsuhashi (Ed.), *Invertebrate Cell System Applications Vol. 1.* (pp. 111–119). Boca Raton, FL: CRC Press.

Agui, N., Bollenbacher, W. E., & Gilbert, L. I. (1983). *In vitro* analysis of prothoracicotropic hormone specificity and prothoracic gland sensitivity in Lepidoptera. *Experientia*, *39*, 984–988.

Agui, N., Bollenbacher, W. E., Granger, N. A., & Gilbert, L. I. (1980). Corpus allatum is release site for insect prothoracicotropic hormone. *Nature*, *285*, 669–670.

Agui, N., Granger, N. A., Bollenbacher, W. E., & Gilbert, L. I. (1979). Cellular localization of the insect prothoracicotropic hormone: *In vitro* assay of a single neurosecretory cell. *Proc. Natl. Acad. Sci. USA*, *76*, 5694–5698.

Aizono, Y., Endo, Y., Sattelle, D. B., & Shirai, Y. (1997). Prothoracicotropic hormone-producing neurosecretory cells in the silkworm, *Bombyx mori*, express a muscarinic acetylcholine receptor. *Brain Res.*, *763*, 131–136.

Alexander, N. J. (1970). A regulatory mechanism of ecdysone release in *Galleria mellonella*. *J. Insect Physiol.*, *16*, 271–276.

Allada, R., & Chung, B. Y. (2010). Circadian organization of behavior and physiology in *Drosophila*. *Annu. Rev. Physiol.*, *72*, 605–624.

Allegret, P. (1964). Interrelationship of larval development, metamorphosis and age in a pyralid lepidopteran, *Galleria mellonella* (L.), under the influence of dietetic factors. *Exp. Gerontol.*, *1*, 49–66.

Anjum, R., & Blenis, J. (2008). The RSK family of kinases: emerging roles in cellular signalling. *Nature Rev. Mol. Cell Biol.*, *9*, 747–758.

Arbeitman, M. N., & Hogness, D. S. (2000). Molecular chaperones activate the *Drosophila* ecdysone receptor, an RXR heterodimer. *Cell*, *101*, 67–77.

Bain, J., McLauchlan, H., Elliott, M., & Cohen, P. (2003). The specificities of protein kinase inhibitors: an update. *Biochem. J.*, *371*, 199–204.

Baker, F. C., Tsai, L. W., Reuter, C. C., & Schooley, D. A. (1987). *In vivo* fluctuation of JH, JH acid, and ecdysteroid titer, and JH esterase activity during development of fifth stadium *Manduca sexta*. *Insect Biochem.*, *17*, 989–996.

Batarseh, A., & Papadopoulos, V. (2010). Regulation of translocator protein 18 kDa (TSPO) expression in health and disease states. *Mol. Cell. Endocrinol.*, *327*, 1–12.

Bean, D. W., & Beck, S. D. (1983). Haemolymph ecdysteroid titres in diapause and nondiapause larvae of the European corn borer, *Ostrinia nubilalis*. *J. Insect Physiol.*, *29*, 687–693.

Beckage, N. E. (1985). Endocrine interactions between endoparasitic insects and their hosts. *Annu. Rev. Entomol.*, *30*, 371–413.

Beckage, N. E. (1997). New insights: how parasites and pathogens alter the endocrine physiology and development of insect hosts. In N. E. Beckage (Ed.), *Parasites and Pathogens: Effects on Host Hormones and Behavior* (pp. 3–36). London: Chapman & Hall.

Beckage, N. E. (2002). Parasite- and pathogen-mediated manipulation of host hormones and behavior. In D. Pfaff, A. Arnold, A. Etgen, S. Fahrbach, & R. Rubin (Eds.), *Hormones, Brain, and Behavior Vol. 3.* (pp. 281–315). New York: Academic Press.

Beckage, N. E., & Gelman, D. B. (2004). Wasp parasitoid disruption of host development: Implications for new biologically based strategies for insect control. *Annu. Rev. Entomol.*, *49*, 299–330.

Beckel, W. E., & Friend, W. (1964). The relation of abdominal distension and nutrition to molting in *Rhodnius prolixus* (Stahl) (Hemiptera). *Can. J. Zool.*, *42*, 71–78.

Berridge, M. J., Lipp, P., & Bootman, M. D. (2000). The versatility and universality of calcium signalling. *Nat. Rev. Mol. Cell Biol.*, *1*, 11–21.

Biggs, W. H., III, & Zipursky, S. L. (1992). Primary structure, expression, and signal-dependent tyrosine phosphorylation of a *Drosophila* homologue of extracellular signal-regulated kinase. *Proc. Natl. Acad. Sci. USA*, *89*, 6295–6299.

Birkenbeil, H. (1996). Involvement of calcium in prothoracicotropic stimulation of ecdysone synthesis in *Galleria mellonella*. *Arch. Insect Biochem. Physiol.*, *33*, 39–52.

Birkenbeil, H. (1998). Intracellular calcium in PTTH-stimulated prothoracic glands of *Manduca sexta* (Lepidoptera: Sphingidae). *Eur. J. Entomol.*, *96*, 295–298.

Birkenbeil, H. (2000). Pharmacological study of signal transduction during stimulation of prothoracic glands from *Manduca sexta*. *J. Insect Physiol.*, *46*, 1409–1414.

Birkenbeil, H., & Dedos, S. G. (2002). Ca²⁺ as second messenger in PTTH-stimulated prothoracic glands of the silkworm, *Bombyx mori*. *Insect Biochem. Mol. Biol.*, *32*, 1625–1634.

Blackburn, M. B., Wagner, R. M., Kochansky, J. P., Harrison, D. J., Thomas-Lemont, P., & Raina, A. K. (1995). The identification of two myoinhibitory peptides, with sequence similarities to the galanins, isolated from the ventral nerve cord of *Manduca sexta*. *Regul. Pept.*, *57*, 213–219.

Blancquaert, S., Wang, L., Paternot, S., Coulonval, K., Dumont, J. E., Harris, T. E., & Roger, P. P. (2010). cAMP-dependent activation of mammalian target of rapamycin (mTOR) in thyroid cells. Implication in mitogenesis and activation of CDK4. *Mol. Endocrinol.*, *24*, 1453–1468.

Bollenbacher, W. E., Agui, N., Granger, N. A., & Gilbert, L. I. (1979). *In vitro* activation of insect prothoracic glands by the prothoracicotropic hormone. *Proc. Natl. Acad. Sci. USA*, *76*, 5148–5152.

Bollenbacher, W. E., & Gilbert, L. I. (1981). Neuroendocrine control of postembryonic development in insects. In D. S. Farmer, & K. Lederi (Eds.), *Neurosecretion* (pp. 361–370). New York: Plenum Press.

Bollenbacher, W. E., & Granger, N. A. (1985). Endocrinology of the prothoracicotropic hormone. In G. A. Kerkut, & L. I. Gilbert (Eds.), *Comprehensive Insect Physiology, Biochemistry and Pharmacology Vol. 7.* (pp. 109–151). New York: Pergamon Press.

Bollenbacher, W. E., Granger, N. A., Katahira, E. J., & O'Brien, M. A. (1987). Developmental endocrinology of larval moulting in the tobacco hornworm, *Manduca sexta*. *J. Exp. Biol.*, *128*, 175–192.

Bollenbacher, W. E., Katahira, E. J., O'Brien, M. A., Gilbert, L. I., Thomas, M. K., Agui, N., & Baumhover, A. H. (1984). Insect prothoracicotropic hormone: Evidence for two molecular forms. *Science*, *224*, 1243–1245.

Bollenbacher, W. E., O'Brien, M. A., Katahira, E. J., & Gilbert, L. I. (1983). A kinetic analysis of the action of the insect prothoracicotropic hormone. *Mol. Cell. Endocrinol.*, *32*, 27–46.

Bowen, M. F., Bollenbacher, W. E., & Gilbert, L. I. (1984a). *In vitro* studies on the role of the brain and prothoracic glands in the pupal diapause of *Manduca sexta*. *J. Exp. Biol.*, *108*, 9–24.

Bowen, M. F., Gilbert, L. I., & Bollenbacher, W. E. (1986). Endocrine control of insect diapause: An *in vitro* analysis. *In vitro Invert. Horm. Genes*, *C210*, 1–14.

Bowen, M. F., Saunders, D. S., Bollenbacher, W. E., & Gilbert, L. I. (1984b). *In vitro* reprogramming of the photoperiodic clock in an insect brain-retrocerebral complex. *Proc. Natl. Acad. Sci. USA*, *81*, 5881–5884.

Brogiolo, W., Stocker, H., Ikeya, T., Rintelen, F., Fernandez, R., & Hafen, E. (2001). An evolutionarily conserved function of the *Drosophila* insulin receptor and insulin-like peptides in growth control. *Curr. Biol.*, *11*, 213–221.

Burtt, E. T. (1938). On the corpora allata of dipterous insects II. *Proc. Roy. Soc. Lond. B*, *126*, 210–223.

Bylemans, D., Hua, Y. J., Chiou, S. J., Koolman, J., Borovsky, D., & De Loof, A. (1995). Pleiotropic effects of trypsin modulating oostatic factor (Neb-TMOF) of the fleshfly *Neobellieria bullata* (Diptera: Calliphoridae). *Eur. J. Entomol.*, *92*, 143–143.

Caldwell, P. E., Walkiewicz, M., & Stern, M. (2005). Ras activity in the *Drosophila* prothoracic gland regulates body size and developmental rate via ecdysone release. *Curr. Biol.*, *15*, 1785–1795.

Camps, M., Nichols, A., & Arkinstall, S. (2000). Dual specificity phosphatases: A gene family for control of MAP kinase function. *FASEB J.*, *14*, 6–16.

Cann, M. J., Chung, E., & Levin, L. R. (2000). A new family of adenylyl cyclase genes in the male germline of *Drosophila melanogaster*. *Dev. Genes Evol.*, *210*, 200–206.

Carrow, G., Calabrese, R. L., & Williams, C. M. (1981). Spontaneous and evoked release of prothoracicotropin from multiple neurohemal organs of the tobacco hornworm. *Proc. Natl. Acad. Sci. USA*, *78*, 5866–5870.

Carstea, E. D., Morris, J. A., Coleman, K. G., Loftus, S. K., Zhang, D., Cummings, C., Gu, J., Rosenfeld, M. A., Pavan, W. J., Krizman, D. B., Nagle, J., Polymeropoulos, M. H., Sturley, S. L., Ioannou, Y. A., Higgins, M. E., Comly, M., Cooney, A., Brown, A., Kaneski, C. R., Blanchette-Mackie, E. J., Dwyer, N. K., Neufeld, E. B., Chang, T. Y., Liscum, L., Strauss, J. F.R., Ohno, K., Zeigler, M., Carmi, R., Sokol, J., Markie, D., O'Neill, R. R., van Diggelen, O. P., Elleder, M., Patterson, M. C., Brady, R. O., Vanier, M. T., Pentchev, P. G., & Tagle, D. A. (1997). Niemann-Pick C1 disease gene: Homology to mediators of cholesterol homeostasis. *Science, 277*, 228–231.

Chen, C. H., & Gu, S. H. (2006). Stage-dependent effects of starvation on the growth, metamorphosis, and ecdysteroidogenesis by the prothoracic glands during the last larval instar of the silkworm, *Bombyx mori. J. Insect Physiol., 52*, 968–974.

Chen, C. H., & Gu, S. H. (2008). Inhibitory effects of starvation on prothoracic gland cell DNA synthesis during the last larval instar of the silkworm, *Bombyx mori. J. Exp. Zool. A Ecol. Genet. Physiol., 309*, 399–406.

Chen, C. H., Gu, S. H., & Chow, Y. S. (2001). Adenylyl cyclase in prothoracic glands during the last larval instar of the silkworm, *Bombyx mori. Insect Biochem. Mol. Biol., 31*, 659–664.

Chippendale, G.M. (1984). Environmental signals, the neuroendocrine system, and the regulation of larval diapause in the southwestern corn borer, *Diatraea grandiosella*. In R. Porter (Ed.), Ciba Foundation Symposium, Vol. 104, 259–276.

Chippendale, G. M., & Yin, C. M. (1973). Endocrine activity retained in diapause insect larvae. *Nature, 246*, 511–513.

Chippendale, G. M., & Yin, C. M. (1976). Diapause of the southwestern corn borer, *Diatraea grandiosella* Dyar (Lepidoptera: Pyralidae): Effects of a juvenile hormone mimic. *Bull. Entomol. Res., 66*, 75–79.

Clementi, E., & Meldolesi, J. (1996). Pharmacological and functional properties of voltage-independent Ca^{2+} channels. *Cell Calcium, 19*, 269–279.

Cole, T. J., Beckage, N. E., Tan, F. F., Srinivasan, A., & Ramaswamy, S. B. (2002). Parasitoid-host endocrine relations: self-reliance or co-optation? *Insect Biochem. Mol. Biol., 32*, 1673–1679.

Colombani, J., Bianchini, L., Layalle, S., Pondeville, E., Dauphin-Villemant, C., Antoniewski, C., Carre, C., Noselli, S., & Leopold, P. (2005). Antagonistic actions of ecdysone and insulins determine final size in *Drosophila. Science, 310*, 667–670.

Combest, W. L., & Gilbert, L. I. (1992). Polyamines modulate multiple protein phosphorylation pathways in the insect prothoracic gland. *Mol. Cell. Endocrinol., 83*, 11–19.

Combettes-Souverain, M., & Issad, T. (1998). Molecular basis of insulin action. *Diabetes Metab., 24*, 477–489.

Cruz, J., Nieva, C., Mane-Padros, D., Martin, D., & Belles, X. (2008). Nuclear receptor BgFTZ-F1 regulates molting and the timing of ecdysteroid production during nymphal development in the hemimetabolous insect *Blattella germanica. Dev. Dynam., 237*, 3179–3191.

Dai, J. -D., & Gilbert, L. I. (1991). Metamorphosis of the corpus allatum and degeneration of the prothoracic gland during larval-pupal-adult transformation of *Drosophila melanogaster*: A cytophysiological analysis of the ring gland. *Dev. Biol., 144*, 309–326.

Dai, J. -D., & Gilbert, L. I. (1997). Programmed cell death of the prothoracic glands of *Manduca sexta* during pupal–adult metamorphosis. *Insect Biochem. Mol. Biol., 27*, 69–78.

Dai, J. -D., & Gilbert, L. I. (1998). Juvenile hormone prevents the onset of programmed cell death in the prothoracic glands of *Manduca sexta. Gen. Comp. Endocrinol., 109*, 155–165.

Dai, J. -D., & Gilbert, L. I. (1999). An in vitro analysis of ecdysteroid-elicited cell death in the prothoracic gland of *Manduca sexta. Cell Tissue Res., 297*, 319–327.

Dai, J. -D., Mizoguchi, A., & Gilbert, L. I. (1994). Immunoreactivity of neurosecretory granules in the brain-retrocerebral complex of *Manduca sexta* to heterologous antibodies against *Bombyx* prothoracicotropic hormone and bombyxin. *Invert. Reprod. Dev., 26*, 187–196.

Dai, J. -D., Mizoguch, A., Satake, S., Ishizaki, H., & Gilbert, L. I. (1995). Developmental changes in the prothoracicotropic hormone content of the *Bombyx mori* brain-retrocerebral complex and hemolymph: analysis by immunogold electron microscopy, quantitative image analysis, and time-resolved fluoroimmunoassay. *Dev. Biol., 171*, 212–223.

Das, S., Maizels, E. T., DeManno, D., St Clair, E., Adam, S. A., & Hunzicker-Dunn, M. (1996). A stimulatory role of cyclic adenosine 3',5'-monophosphate in follicle-stimulating hormone-activated mitogen-activated protein kinase signaling pathway in rat ovarian granulosa cells. *Endocrinology, 137*, 967–974.

Davis, N. T., Blackburn, M. B., Golubeva, E. G., & Hildebrand, J. G. (2003). Localization of myoinhibitory peptide immunoreactivity in *Manduca sexta* and *Bombyx mori*, with indications that the peptide has a role in molting and ecdysis. *J. Exp. Biol., 206*, 1449–1460.

Dedos, S. G., & Birkenbeil, H. (2003). Inhibition of cAMP signalling cascade-mediated Ca^{2+} influx by a prothoracicostatic peptide (Mas-MIP I) via dihydropyridine-sensitive Ca^{2+} channels in the prothoracic glands of the silkworm, *Bombyx mori. Insect Biochem. Mol. Biol., 33*, 219–228.

Dedos, S. G., & Fugo, H. (1996). Effects of fenoxycarb on the secretory activity of the prothoracic glands in the fifth instar of the silkworm, *Bombyx mori. Gen. Comp. Endocrinol., 104*, 213–224.

Dedos, S. G., & Fugo, H. (1999a). Disturbance of adult eclosion by fenoxycarb in the silkworm, *Bombyx mori. J. Insect Physiol., 45*, 257–264.

Dedos, S. G., & Fugo, H. (1999b). Interactions between Ca^{2+} and cAMP in ecdysteroid secretion from the prothoracic glands of *Bombyx mori. Mol. Cell. Endocrinol., 154*, 63–70.

Dedos, S. G., & Fugo, H. (1999c). Downregulation of the cAMP signal transduction cascade in the prothoracic glands is responsible for the fenoxycarb-mediated induction of permanent 5th instar larvae in *Bombyx mori. Insect Biochem. Mol. Biol., 29*, 723–729.

Dedos, S. G., & Fugo, H. (2001). Involvement of calcium, inositol-1,4, 5 trisphosphate and diacylglycerol in the prothoracicotropic hormone-stimulated ecdysteroid synthesis and secretion in the prothoracic glands of *Bombyx mori. Zool. Sci., 18*, 1245–1251.

Dedos, S. G., Kaltofen, S., & Birkenbeil, H. (2008). Protein kinase A and C are "Gatekeepers" of capacitative Ca^{2+} entry in the prothoracic gland cells of the silkworm, *Bombyx mori. J. Insect Physiol., 54*, 878–882.

Dedos, S. G., Nagata, S., Ito, J., & Takamiya, M. (2001). Action kinetics of a prothoracicostatic peptide from *Bombyx mori* and its possible signaling pathway. *Gen. Comp. Endocrinol.*, *122*, 98–108.

Dedos, S. G., Szurdoki, F., Szekacs, A., Mizoguchi, A., & Fugo, H. (2002). Induction of dauer pupae by fenoxycarb in the silkworm, *Bombyx mori*. *J. Insect Physiol.*, *48*, 857–865.

Dedos, S. G., Wicher, D., Fugo, H., & Birkenbeil, H. (2005). Regulation of capacitative Ca²⁺ entry by prothoracicotropic hormone in the prothoracic glands of the silkworm, *Bombyx mori*. *J. Exp. Zool. Comp. Exp. Biol.*, *303*, 101–112.

Dedos, S. G., Wicher, D., Kaltofen, S., & Birkenbeil, H. (2007). Different Ca²⁺ signalling cascades manifested by mastoparan in the prothoracic glands of the tobacco hornworm, *Manduca sexta*, and the silkworm, *Bombyx mori*. *Arch. Insect Biochem. Physiol.*, *65*, 52–64.

De Loof, A. (2008). Ecdysteroids, juvenile hormone and insect neuropeptides: Recent successes and remaining major challenges. *Gen. Comp. Endocrinol.*, *155*, 3–13.

Dorn, A., & Romer, F. (1976). Structure and function of prothoracic glands and oenocytes in embryos and last larval instars of *Oncopeltus fasciatus* Dallas (Insecta, Heteroptera). *Cell Tissue Res.*, *171*, 331–350.

Dover, B. A., Tanaka, T., & Vinson, S. B. (1995). Stadium-specific degeneration of host prothoracic glands by *Campoletis sonorensis* calyx fluid and its association with host ecdysteroid titers. *J. Insect Physiol.*, *41*, 947–955.

Dover, B. A., & Vinson, S. B. (1990). Stage-specific effects of *Campoletis sonorensis* parasitism on *Heliothis virescens* development and prothoracic glands. *Physiol. Entomol.*, *15*, 405–414.

Dufner, A., & Thomas, G. (1999). Ribosomal S6 kinase signaling and the control of translation. *Exp. Cell Res.*, *1253*, 100–109.

Endo, K., Fujimoto, Y., Kondo, M., Yamanaka, A., Watanabe, M., Weihua, K., & Kumagai, K. (1997). Stage-dependent changes of the prothoracicotropic hormone (PTTH) activity of brain extracts and of the PTTH sensitivity of the prothoracic glands in the cabbage armyworm, *Mamestra brassicae*, before and during winter and aestival pupal diapause. *Zool. Sci.*, *14*, 127–133.

Endo, K., Fujimoto, Y., Masaki, T., & Kumagai, K. (1990). Stage-dependent changes in the activity of the prothoracicotropic hormone (PTTH) in the brain of Asian comma butterfly, *Polygonia c-aureum* L. *Zool. Sci.*, *7*, 695–702.

Erikson, R. L. (1991). Structure, expression, and regulation of protein kinases involved in the phosphorylation of ribosomal protein S6. *J. Biol. Chem.*, *266*, 6007–6010.

Eusebio, E. J., & Moody, W. J. (1986). Calcium-dependent action potentials in the prothoracic gland of *Manduca sexta*. *J. Exp. Biol.*, *126*, 531–536.

Fain, M. J., & Riddiford, L. M. (1976). Reassessment of the critical periods for prothoracicotropic hormone and juvenile hormone secretion in the larval molt of the tobacco hornworm *Manduca sexta*. *Gen. Comp. Endocrinol.*, *30*, 131–141.

Falabella, P., Caccialupi, P., Varricchio, P., Malva, C., & Pennacchio, F. (2006). Protein tyrosine phosphatases of *Toxoneuron nigriceps* bracovirus as potential disrupters of host prothoracic gland function. *Arch. Insect Biochem. Physiol.*, *61*, 157–169.

Fellner, S. K., Rybczynski, R., & Gilbert, L. I. (2005). Ca2+ signaling in prothoracicotropic hormone-stimulated prothoracic gland cells of *Manduca sexta*: Evidence for mobilization and entry mechanisms. *Insect Biochem. Mol. Biol.*, *35*, 263–275.

Fescemyer, H. W., Masler, E. P., Kelly, T. J., & Lusby, W. R. (1995). Influence of development and prothoracicotropic hormone on the ecdysteroids produced *in vitro* by the prothoracic glands of female gypsy moth (*Lymantria dispar*) pupae and pharate adults. *J. Insect Physiol.*, *41*, 489–500.

Feuilloley, M., & Vaudry, H. (1996). Role of the cytoskeleton in adrenocortical cells. *Endocr. Rev.*, *17*, 269–288.

Fujimoto, Y., Endo, K., Watanabe, M., & Kumagai, K. (1991). Species-specificity in the action of big and small prothoracicotropic hormones (PTTHs) of four species of lepidopteran insects, *Mamestra brassicae*, *Bombyx mori*, *Papilio xuthus* and *Polygonia c-aureum*. *Zool. Sci.*, *8*, 351–358.

Fujishita, M., & Ishizaki, H. (1982). Temporal organization of endocrine events in relation to the circadian clock during larval-pupal development in *Samia cynthia ricini*. *J. Insect Physiol.*, *28*, 77–84.

Fukuda, S. (1941). Role of the prothoracic gland in differentiation of the imaginal characters in the silkworm pupa. *Annot. Zool. Japon.*, *20*, 9–13.

Fumagalli, S., & Thomas, G. (2000). S6 phosphorylation and signal transduction. In N. Sonenberg, J. W.B. Hershey, & M. B. Mathews (Eds.), *Translational Control of Gene Expression* (2nd ed.). (pp. 695–717). New York: Cold Spring Harbor Laboratory Press.

Gelman, D. B., & Beckage, N. E. (1995). Low molecular weight ecdysiotropins in proctodaea of fifth instars of the tobacco hornworm, *Manduca sexta* (Lepidoptera: Sphingidae), and host parasitized by the braconid wasp *Cotesia congregata* (Hymenoptera: Braconidae). *Eur. J. Entomol.*, *92*, 123–129.

Gelman, D. B., & Borovsky, D. (2000). *Aedes aegypti* TMOF modulates ecdysteroid production by prothoracic glands of the gypsy moth, *Lymantria dispar*. *Arch. Insect Biochem. Physiol.*, *45*, 60–68.

Gelman, D. B., Reed, D. A., & Beckage, N. E. (1998). Manipulation of fifth-instar host (*Manduca sexta*) ecdysteroid levels by the parasitoid wasp *Cotesia congregata*. *J. Insect Physiol.*, *44*, 833–843.

Gelman, D. B., Thyagaraja, B. S., Kelly, T. J., Masler, E. P., Bell, R. A., & Borkovec, A. B. (1991). The insect gut: a new source of ecdysiotropic peptides. *Cell. Mol. Life Sci.*, *47*, 77–80.

Gelman, D. B., Thyagaraja, B. S., Kelly, T. J., Masler, E. P., Bell, R. A., & Borkovec, A. B. (1992). Prothoracicotropic hormone levels in brains of the European corn borer, *Ostrinia nubilalis*: Diapause vs. the non-diapause state. *J. Insect Physiol.*, *38*, 383–395.

Ghosh, A., McBrayer, Z., & O'Connor, M. B. (2010). The *Drosophila* gap gene giant regulates ecdysone production through specification of the PTTH-producing neurons. *Dev. Biol.*, *347*, 271–278.

Gibbs, D., & Riddiford, L. M. (1977). Prothoracicotropic hormone in *Manduca sexta*. Localization by a larval assay. *J. Exp. Biol.*, *66*, 255–266.

Gilbert, L. I. (1962). Maintenance of the prothoracic gland by the juvenile hormone in insects. *Nature*, *193*, 1205–1207.

Gilbert, L. I., Bollenbacher, W. E., Agui, N., Granger, N. A., Sedlak, B. J., Gibbs, D., & Buys, C. M. (1981). The prothoracicotropes: Source of the prothoracicotropic hormone. *Am. Zool., 21,* 641–653.

Gilbert, L. I., Combest, W. L., Smith, W. A., Meller, V. H., & Rountree, D. B. (1988). Neuropeptides, second messengers and insect molting. *BioEssays, 8,* 153–157.

Gilbert, L. I., Rybczynski, R., Song, Q., Mizoguchi, A., Morreale, R., Smith, W. A., Matubayashi, H., Shionoya, M., Nagata, S., & Kataoka, H. (2000). Dynamic regulation of prothoracic gland ecdysteroidogenesis: *Manduca sexta* recombinant prothoracicotropic hormone and brain extracts have identical effects. *Insect Biochem. Mol. Biol., 30,* 1079–1089.

Gilbert, L. I., Rybczynski, R., & Warren, J. T. (2002). Control and biochemical nature of the ecdysteroidogenic pathway. *Annu. Rev. Entomol., 47,* 883–916.

Gilbert, L. I., Song, Q., & Rybczynski, R. (1997). Control of ecdysteroidogenesis: activation and inhibition of prothoracic gland activity. *Invert. Neurosci., 3,* 205–216.

Girgenrath, S., & Smith, W. A. (1996). Investigation of presumptive mobilization pathways for calcium in the steroidogenic action of big prothoracicotropic hormone. *Insect Biochem. Mol. Biol., 26,* 455–463.

Gloerich, M., & Bos, J. L. (2010). Epac: Defining a new mechanism for cAMP action. *Annu. Rev. Pharmacol. Toxicol., 50,* 355–375.

Goltzene, F., Holder, F., Charlet, M., Meister, M., & Oka, T. (1992). Immunocytochemical localization of *Bombyx*-PTTH-like molecules in neurosecretory cells of the brain of the migratory locust, *Locusta migratoria*. A comparison with neuroparsin and insulin-related peptide. *Cell Tissue Res., 269,* 133–140.

Graf, R., Neuenschwander, S., Brown, M. R., & Ackermann, U. (1997). Insulin-mediated secretion of ecdysteroids from mosquito ovaries and molecular cloning of the insulin receptor homologue from ovaries of bloodfed *Aedes aegypti*. *Insect Mol. Biol., 6,* 151–163.

Granger, N. A., & Bollenbacher, W. E. (1981). Hormonal control of insect metamorphosis. In L. I. Gilbert, & E. Frieden (Eds.), *Metamorphosis* (2nd ed.). (pp. 105–137). New York: Plenum Press.

Gray, R. S., Muehleisen, D. P., Katahira, E. J., & Bollenbacher, W. E. (1994). The prothoracicotropic hormone (PTTH) of the commercial silkmoth, *Bombyx mori*, in the CNS of the tobacco hornworm, *Manduca sexta. Peptides, 15,* 777–782.

Gronke, S., Clarke, D. F., Broughton, S., Andrews, T. D., & Partridge, L. (2010). Molecular evolution and functional characterization of *Drosophila* insulin-like peptides. *PLoS Genet., 6,* e1000857.

Gruetzmacher, M. C., Gilbert, L. I., Granger, N. A., Goodman, W., & Bollenbacher, W. E. (1984a). The effect of juvenile hormone on prothoracic gland function during larval–pupal development of the tobacco hornworm, *Manduca sexta. J. Insect Physiol., 30,* 331–340.

Gu, S. H. (2006). Autocrine activation of DNA synthesis in prothoracic gland cells of the silkworm, *Bombyx mori*. *J. Insect Physiol., 52,* 136–145.

Gu, S. H. (2007). Autocrine activation of ecdysteroidogenesis in the prothoracic glands of the silkworm, *Bombyx mori*. *J. Insect Physiol., 53,* 538–549.

Gu, S. H., & Chow, Y. S. (2001). Induction of DNA synthesis by 20-hydroxyecdysone in the prothoracic gland cells of the silkworm *Bombyx mori* during the last larval instar. *Gen. Comp. Endocrinol., 124,* 269–276.

Gu, S. H., & Chow, Y. S. (2005a). Temporal changes in DNA synthesis of prothoracic gland cells during larval development and their correlation with ecdysteroidogenic activity in the silkworm, *Bombyx mori. J. Exp. Zool., 303A,* 249–258.

Gu, S. H., & Chow, Y. S. (2005b). Analysis of ecdysteroidogenic activity of the prothoracic glands during the last larval instar of the silkworm, *Bombyx mori. Arch. Insect Biochem. Physiol., 58,* 17–26.

Gu, S. H., Lin, J. L., & Lin, P. L. (2010). PTTH-stimulated ERK phosphorylation in prothoracic glands of the silkworm, *Bombyx mori*: role of Ca(2+)/calmodulin and receptor tyrosine kinase. *J. Insect Physiol., 56,* 93–101.

Gu, S. H., Lin, J. L., Lin, P. L., & Chen, C. H. (2009). Insulin stimulates ecdysteroidogenesis by prothoracic glands in the silkworm, *Bombyx mori. Insect Biochem. Mol. Biol., 39,* 171–179.

Gu, S. H., Lin, J. L., Lin, P. L., Kou, R., & Smagghe, G. (2008). Effects of RH-5992 on ecdysteroidogenesis of the prothoracic glands during the fourth larval instar of the silkworm, *Bombyx mori. Arch. Insect Biochem. Physiol., 68,* 197–205.

Gu, S. -H., & Chow, Y. -S. (1996). Regulation of juvenile hormone biosynthesis by ecdysteroid levels during the early stages of the last two larval instars of *Bombyx mori. J. Insect Physiol., 42,* 625–632.

Gu, S. -H., Chow, Y. -S., Lin, F. -J., Wu, J. -L., & Ho, R. -J. (1996). A deficiency in prothoracicotropic hormone transduction pathway during the early last larval instar of *Bombyx mori. Mol. Cell. Endocrinol., 120,* 99–105.

Gu, S. -H., Chow, Y. -S., & O'Reilly, D. R. (1998). Role of calcium in the stimulation of ecdysteroidogenesis by recombinant prothoracicotropic hormone in the prothoracic glands of the silkworm, *Bombyx mori. Insect Biochem. Mol. Biol., 28,* 861–867.

Gu, S. -H., Chow, Y. -S., & Yin, C. -M. (1997). Involvement of juvenile hormone in regulation of prothoracicotropic hormone transduction during the early last larval instar of *Bombyx mori. Mol. Cell. Endocrinol., 127,* 109–116.

Gujdar, A., Sipeki, S., Bander, E., Buday, L., & Farago, A. (2003). Phorbol ester-induced migration of HepG2 cells is accompanied by intensive stress fibre formation, enhanced integrin expression and transient down-regulation of p21-activated kinase 1. *Cell. Signal., 15,* 307–318.

Halme, A., Cheng, M., & Hariharan, I. K. (2010). Retinoids regulate a developmental checkpoint for tissue regeneration in *Drosophila. Curr. Biol., 20,* 458–463.

Hamilton, S. E., McKinnon, L. A., Jackson, D. A., Goldman, P. S., Migeon, J. C., Habecker, B. A., Thomas, S. L., & Nathanson, N. M. (1995). Molecular analysis of the regulation of muscarinic receptor expression and function. *Life Sci., 56,* 939–943.

Hanton, W. K., Watson, R. D., & Bollenbacher, W. E. (1993). Ultrastructure of prothoracic glands during larval–pupal development of the tobacco hornworm, *Manduca sexta*: A reappraisal. *J. Morphol., 216,* 95–112.

Harden, T. K., Hicks, S. N., & Sondek, J. (2009). Phospholipase C isozymes as effectors of Ras superfamily GTPases. *J. Lipid Res., 50*(Suppl.), S243–S248.

Hartfelder, K., Hanton, W. K., & Bollenbacher, W. E. (1994). Diapause-dependent changes in prothoracicotropic hormone-producing neurons of the tobacco hornworm, *Manduca sexta*. *Cell Tissue Res.*, *277*, 69–78.

Hayes, G. C., Muehleisen, D. P., Bollenbacher, W. E., & Watson, R. D. (1995). Stimulation of ecdysteroidogenesis by small prothoracicotropic hormone: role of calcium. *Mol. Cell. Endocrinol.*, *115*, 105–112.

Helfrich-Förster, C., Tauber, M., Park, J. H., Muhlig-Versen, M., Schneuwly, S., & Hofbauer, A. (2000). Ectopic expression of the neuropeptide pigment-dispersing factor alters behavioral rhythms in *Drosophila melanogaster*. *J. Neurosci.*, *20*, 3339–3353.

Henrich, V. C. (1995). Comparison of ecdysteroid production in *Drosophila* and *Manduca*: pharmacology and cross-species neural reactivity. *Arch. Insect Biochem. Physiol.*, *30*, 239–254.

Henrich, V. C., Pak, M. D., & Gilbert, L. I. (1987a). Neural factors that stimulate ecdysteroid synthesis by the larval ring gland of *Drosophila melanogaster*. *J. Comp. Physiol. B*, *157*, 543–549.

Henrich, V. C., Tucker, R. L., Maroni, G., & Gilbert, L. I. (1987b). The ecdysoneless (ecd1ts) mutation disrupts ecdysteroid synthesis autonomously in the ring gland of *Drosophila melanogaster*. *Dev. Biol.*, *120*, 50–55.

Hiragaki, S., Uno, T., & Takeda, M. (2009). Putative regulatory mechanism of prothoracicotropic hormone (PTTH) secretion in the American cockroach, *Periplaneta americana* as inferred from co-localization of Rab8, PTTH, and protein kinase C in neurosecretory cells. *Cell Tissue Res.*, *335*, 607–615.

Hossain, M., Shimizu, S., Fujiwara, H., Sakurai, S., & Iwami, M. (2006). EcR expression in the prothoracicotropic hormone-producing neurosecretory cells of the *Bombyx mori* brain. *FEBS J.*, *273*, 3861–3868.

Hossain, M., Shimizu, S., Matsuki, M., Imamura, M., Sakurai, S., & Iwami, M. (2008). Expression of 20-hydroxyecdysone-induced genes in the silkworm brain and their functional analysis in post-embryonic development. *Insect Biochem. Mol. Biol.*, *38*, 1001–1007.

Hua, Y. -J., & Koolman, J. (1995). An ecdysiostatin from flies. *Regulat. Pept.*, *57*, 263–271.

Hua, Y. -J., Tanaka, Y., Nakamura, K., Sakakibara, M., Nagata, S., & Kataoka, H. (1999). Identification of a prothoracicostatic peptide in the larval brain of the silkworm, *Bombyx mori*. *J. Biol. Chem.*, *274*, 31169–31173.

Huang, X., Suyama, K., Buchanan, J. A., Zhu, A. J., & Scott, M. P. (2005). A *Drosophila* model of the Niemann-Pick type C lysosome storage disease: dnpc1a is required for molting and sterol homeostasis. *Development*, *132*, 5115–5124.

Huang, X., Warren, J. T., Buchanan, J. A., Gilbert, L. I., & Scott, M. P. (2007). *Drosophila* Niemann-Pick Type C-2 genes control sterol homeostasis and steroid biosynthesis: a model of human neurodegenerative disease. *Development*, *134*, 3733–3742.

Huang, X., Warren, J. T., & Gilbert, L. I. (2008). New players in the regulation of ecdysone biosynthesis. *J. Genet. Genomics*, *35*, 1–10.

Hynes, R. O., & Zhao, Q. (2000). The evolution of cell adhesion. *J. Cell Biol.*, *150*, F89–F95.

Ichikawa, M., & Ishizaki, H. (1961). Brain hormone of the silkworm, *Bombyx mori*. *Nature*, *191*, 933–934.

Ichikawa, M., & Ishizaki, H. (1963). Protein nature of the brain hormone of insects. *Nature*, *198*, 308–309.

Ishibashi, J., Kataoka, H., Isogai, A., Kawakami, A., Saegusa, H., Yagi, Y., Mizoguchi, A., Shizaki, H., & Suzuki, A. (1994). Assignment of disulfide bond location in prothoracicotropic hormone of the silkworm, *Bombyx mori*: A homodimeric protein. *Biochemistry*, *33*, 5912–5919.

Ishizaki, H., Mizoguchi, A., Fujishita, M., Suzuki, A., Moriya, I., O'oka, H., Kataoka, H., Isogai, A., Nagasawa, H., Tamura, S., & Suzuki, A. (1983b). Species specificity of the insect prothoracicotropic hormone (PTTH): the presence of *Bombyx*- and *Samia*-specific PTTHs in the brain of *Bombyx mori*. *Dev. Growth Differ.*, *25*, 593–600.

Ishizaki, H., & Suzuki, A. (1980). Prothoracicotropic hormone. In T. A. Miller (Ed.), *Neurohormonal Techniques in Insects* (pp. 244–276). New York: Springer Verlag.

Ishizaki, H., & Suzuki, A. (1984). Prothoracicotropic hormone of *Bombyx mori*. In J. Hoffman, & M. Prochet (Eds.), *Biosynthesis, Metabolism and Mode of Action of Invertebrate Hormones* (pp. 63–77). Berlin: Springer Verlag.

Ishizaki, H., & Suzuki, A. (1992). Brain secretory peptides of the silkmoth *Bombyx mori*: Prothoracicotropic hormone and bombyxin. In J. Joose, R. M. Buijs, & J. H. Tilder (Eds.), *Progress in Brain Research* (Vol. 92, pp. 1–14). Amsterdam: Elsevier.

Ishizaki, H., & Suzuki, A. (1994). The brain secretory peptides that control moulting and metamorphosis of the silkmoth, *Bombyx mori*. *Int. J. Dev. Biol.*, *38*, 301–310.

Ishizaki, H., Suzuki, A., Moriya, I., Mizoguchi, A., Fujishita, M., O'oka, H., Kataoka, H., Isogai, A., Nagasawa, H., & Suzuki, A. (1983a). Prothoracicotropic hormone bioassay: Pupal–adult *Bombyx* assay. *Dev. Growth Differ.*, *25*, 585–592.

Jayaraman, T., Ondrias, K., Ondriasova, E., & Marks, A. R. (1996). Regulation of the inositol 1,4,5-trisphosphate receptor by tyrosine phosphorylation. *Science*, *272*, 1492–1494.

Jenkins, S. P., Brown, M. R., & Lea, A. O. (1992). Inactive prothoracic glands in larvae and pupae of *Aedes aegypti*: Ecdysteroid release by tissues in the thorax and abdomen. *Insect Biochem. Mol. Biol.*, *22*, 553–559.

Jhoti, H., McLeod, A. N., Blundell, T. L., Ishizaki, H., Nagasawa, H., & Suzuki, A. (1987). Prothoracicotropic hormone has an insulin-like tertiary structure. *FEBS Lett.*, *219*, 419–425.

Jiang, R. -J., & Koolman, J. (1999). Feedback inhibition of ecdysteroids: Evidence for a short feedback loop repressing steroidogenesis. *Arch. Insect Biochem. Physiol.*, *41*, 54–59.

Jones, D., Gelman, D., & Loeb, M. (1992). Hemolymph concentrations of host ecdysteroids are strongly suppressed in precocious prepupae of *Trichoplusia ni* parasitized and pseudoparasitized by *Chelonus* near *curvimaculatus*. *Arch. Insect Biochem. Physiol.*, *21*, 155–165.

Kataoka, H., Nagasawa, H., Isogai, A., Ishizaki, H., & Suzuki, A. (1991). Prothoracicotropic hormone of the silkworm, *Bombyx mori*: Amino acid sequence and dimeric structure. *Agric. Biol. Chem.*, *55*, 73–86.

Kataoka, H., Nagasawa, H., Isogai, A., Tamura, S., Mizoguchi, A., Fujiwara, Y., Suzuki, C., Ishizaki, H., & Suzuki, A. (1987). Isolation and partial characterization of prothoracicotropic hormone of the silkworm, *Bombyx mori*. *Agric. Biol. Chem.*, *51*, 1067–1076.

Kawakami, A., Kataoka, H., Oka, T., Mizoguchi, A., Kimura-Kawakami, M., Adachi, T., Iwami, M., Nagasawa, H., Suzuki, A., & Ishizaki, H. (1990). Molecular cloning of the *Bombyx mori* prothoracicotropic hormone. *Science, 247,* 1333–1335.

Keightley, D. A., Lou, K. J., & Smith, W. A. (1990). Involvement of translation and transcription in insect steroidogenesis. *Mol. Cell. Endocrinol., 74,* 229–237.

Kelly, T. J., Gelman, D. B., Reed, D. A., & Beckage, N. E. (1998). Effects of parasitization by *Cotesia congregata* on the brain-prothoracic gland axis of its host, *Manduca sexta. J. Insect Physiol., 44,* 323–332.

Kelly, T. J., Kingan, T. G., Masler, C. A., & Robinson, C. H. (1996). Analysis of the ecdysiotropic activity in larval brains of the tobacco hornworm, *Manduca sexta. J. Insect Physiol., 42,* 873–880.

Kelly, T. J., Masler, E. P., Thyagaraja, B. S., Bell, R. A., & Imberski, R. B. (1992). Development of an *in vitro* assay for prothoracicotropic hormone of the gypsy moth, *Lymantria dispar* (L.) following studies on identification, titers and synthesis of ecdysteroids in last-instar females. *J. Comp. Physiol. B, 162,* 81–587.

Kiguchi, K., & Agui, N. (1981). Ecdysteroid levels and developmental events during larval moulting in the silkworm, *Bombyx mori. J. Insect Physiol., 27,* 805–812.

Kim, A. -J., Cha, G. -H., Kim, K., Gilbert, L. I., & Lee, C. C. (1997). Purification and characterization of the prothoracicotropic hormone of *Drosophila melanogaster. Proc. Natl. Acad. Sci. USA, 94,* 1130–1135.

Kim, M., Cha, G. H., Kim, S., Lee, J. H., Park, J., Koh, H., Choi, K. Y., & Chung, J. (2004). MKP-3 has essential roles as a negative regulator of the Ras/mitogen-activated protein kinase pathway during *Drosophila development. Mol. Cell. Biol., 24,* 573–583.

King-Jones, K., Charles, J. P., Lam, G., & Thummel, C. S. (2005). The ecdysone-induced DHR4 orphan nuclear receptor coordinates growth and maturation in *Drosophila. Cell, 121,* 773–784.

Kingan, T. G. (1981). Purification of the prothoracicotropic hormone from the tobacco hornworm *Manduca sexta. Life Sci., 28,* 2585–2594.

Kiriishi, S., Nagasawa, H., Kataoka, H., Suzuki, A., & Sakurai, S. (1992). Comparison of the *in vivo* and *in vitro* effects of bombyxin and prothoracicotropic hormone on prothoracic glands of the silkworm, *Bombyx mori. Zool. Sci., 9,* 149–155.

Kobayashi, M., & Kimura, J. (1958). The "brain" hormone in the silkworm, *Bombyx mori* L. *Nature, 181,* 1217.

Kobayashi, M., & Yamazaki, M. (1966). The proteinic brain hormone in an insect, *Bombyx mori* L. (Lepidoptera: Bombycidae). *Appl. Entomol. Zool., 12,* 53–60.

Kondo, H., Ino, M., Suzuki, A., Ishizaki, H., & Iwami, M. (1996). Multiple gene copies for bombyxin, an insulin-related peptide of the silkmoth *Bombyx mori:* structural signs for gene rearrangement and duplication responsible for generation of multiple molecular forms of bombyxin. *J. Mol. Evol., 259,* 926–937.

Kopec, S. (1922). Studies on the necessity of the brain for the inception of insect metamorphosis. *Biol. Bull., 42,* 323–342.

Kourtis, N., & Tavernarakis, N. (2009). Autophagy and cell death in model organisms. *Cell Death Differ., 16,* 21–30.

Kulesza, P., Lee, C. Y., & Watson, R. D. (1994). Protein synthesis and ecdysteroidogenesis in prothoracic glands of the tobacco hornworm (*Manduca sexta*): Stimulation by big prothoracicotropic hormone. *Gen. Comp. Endocrinol., 93,* 448–458.

Kyriakis, J. M. (2000). Mammalian MAP kinase pathways. In J. Woodgett (Ed.), *Protein Kinase Functions* (pp. 40–156). Oxford: Oxford University Press.

Lambrecht, G. (2000). Agonists and antagonists acting at P2X receptors: selectivity profiles and functional implications. *Naunyn-Schmiedebergs Arch. Pharmacol., 362,* 340–350.

Larkin, M. A., Blackshields, G., Brown, N. P., Chenna, R., McGettigan, P. A., McWilliam, H., Valentin, F., Wallace, I. M., Wilm, A., & Lopez, R. (2007). Clustal W and Clustal X version 2.0. *Bioinformatics, 23,* 2947–2948.

Layalle, S., Arquier, N., & Leopold, P. (2008). The TOR pathway couples nutrition and developmental timing in *Drosophila. Dev. Cell., 15,* 568–577.

Lee, C. Y., Lee, K. J., Chumley, P. H., Watson, C. J., Abdur-Rahman, A., & Watson, R. D. (1995). Capacity of insect (*Manduca sexta*) prothoracic glands to secrete ecdysteroids: relation to glandular growth. *Gen. Comp. Endocrinol., 100,* 404–412.

Lee, W. J., Kim, S. H., Kim, Y. S., Han, S. J., Park, K. S., Ryu, J. H., Hur, M. W., & Choi, K. Y. (2000). Inhibition of mitogen-activated protein kinase by a *Drosophila* dual-specific phosphatase. *Biochem. J., 349,* 821–828.

Leonardi, M. G., Cappellozza, S., Ianne, P., Cappellozza, L., Parenti, P., & Giordana, B. (1996). Effects of topical application of an insect growth regulator (fenoxycarb) on some physiological parameters on the fifth instar larvae of the silkworm *Bombyx mori. Comp. Biochem. Physiol., 113B,* 361–365.

Lester, D. S., & Gilbert, L. I. (1986). Developmental changes in choline uptake and acetylcholine metabolism in the larval brain of the tobacco hornworm, *Manduca sexta. Brain Res., 391,* 201–209.

Lester, D. S., & Gilbert, L. I. (1987). Characterization of acetylcholinesterase activity in the larval brain of *Manduca sexta. Insect Biochem., 17,* 99–109.

Lev, S., Moreno, H., Martinez, R., Canoll, P., Peles, E., Musacchio, J. M., Plowman, G. D., Rudy, B., & Schlessinger, J. (1995). Protein tyrosine kinase PYK2 involved in Ca^{2+}-induced regulation of ion channel and MAP kinase functions. *Nature, 376,* 737–745.

Lewis, T. S., Shapiro, P. S., & Ahn, N. G. (1998). Signal transduction through MAP kinase cascades. *Adv. Cancer Res., 74,* 49–139.

Li, S., Falabella, P., Kuriachan, I., Vinson, S. B., Borst, D. W., Malva, C., & Pennacchio, F. (2003). Juvenile hormone synthesis, metabolism, and resulting haemolymph titre in Heliothis virescens larvae parasitized by *Toxoneuron nigriceps. J. Insect Physiol., 49,* 1021–1030.

Liebmann, C. (2001). Regulation of MAP kinase activity by peptide signalling pathway: paradigms of multiplicity. *Cell Signal, 13,* 777–785.

Lin, J. L., & Gu, S. H. (2007). *In vitro* and *in vivo* stimulation of extracellular signal-regulated kinase (ERK) by the prothoracicotropic hormone in prothoracic gland cells and its developmental regulation in the silkworm, *Bombyx mori. J. Insect Physiol., 53,* 622–631.

Lin, J. L., & Gu, S. H. (2010). Prothoracicotropic hormone induces tyrosine phosphorylation in prothoracic glands of the silkworm, *Bombyx mori*. *Arch. Insect Biochem. Physiol.*, [epub ahead of print].

Liu, M., Zhang, T. Y., & Xu, W. H. (2005). A cDNA encoding diazepam-binding inhibitor/acyl-CoA-binding protein in *Helicoverpa armigera*: molecular characterization and expression analysis associated with pupal diapause. *Comp. Biochem. Physiol. C Toxicol. Pharmacol.*, *141*, 168–176.

Liu, X., Tanaka, Y., Song, Q., Xu, B., & Hua, Y. (2004). *Bombyx mori* prothoracicostatic peptide inhibits ecdysteroidogenesis in vivo. *Arch. Insect Biochem. Physiol.*, *56*, 155–161.

Loeb, M. J. (1982). Diapause and development in the tobacco budworm, *Heliothis virescens*: A comparison of haemolymph ecdysteroid titres. *J. Insect Physiol.*, *28*, 667–673.

Loh, Y. P., Brownstein, M. J., & Gainer, H. (1984). Proteolysis in neuropeptide processing and other neural functions. *Annu. Rev. Neurosci.*, *7*, 189–222.

Lonard, D. M., Bhaskaran, G., & Dahm, K. H. (1996). Control of prothoracic gland activity by juvenile hormone in fourth instar *Manduca sexta* larvae. *J. Insect Physiol.*, *42*, 205–213.

Lorenz, M. W., Kellner, R., & Hoffmann, K. H. (1995). A family of neuropeptides that inhibit juvenile hormone biosynthesis in the cricket, *Gryllus bimaculatus*. *J. Biol. Chem.*, *270*, 21103–21108.

Lu, D., Lee, K. Y., Horodyski, F. M., & Witten, J. L. (2002). Molecular characterization and cell-specific expression of a *Manduca sexta* FLRFamide gene. *J. Comp. Neurol.*, *446*, 377–396.

Mains, R. E., Dickerson, I. M., May, V., Stoffers, D. A., Perkins, S. N., Ouafik, L., Husten, E. J., & Eipper, B. A. (1990). Cellular and molecular aspects of peptide hormone biosynthesis. *Frontiers Neuroendocrinol.*, *11*, 52–89.

Mala, J., & Sehnal, F. (1978). Role of the nerve cord in the control of prothoracic glands in *Galleria mellonella*. *Experientia*, *34*, 1233–1235.

Mane-Padros, D., Cruz, J., Vilaplana, L., Nieva, C., Urena, E., Belles, X., & Martin, D. (2010). The hormonal pathway controlling cell death during metamorphosis in a hemimetabolous insect. *Dev. Biol.*, *346*, 150–160.

Manna, P. R., Dyson, M. T., & Stocco, D. M. (2009). Regulation of the steroidogenic acute regulatory protein gene expression: present and future perspectives. *Mol. Hum. Reprod.*, *15*, 321–333.

Mansfield, S. G., Cammer, S., Alexander, S. C., Muehleisen, D. P., Gray, R. S., et al. (1998). Molecular cloning and characterization of an invertebrate cellular retinoic acid binding protein. *Proc. Natl. Acad. Sci. USA*, *95*, 6825–6830.

Marchal, E., Vandersmissen, H. P., Badisco, L., Van de Velde, S., Verlinden, H., Iga, M., Van Wielendaele, P., Huybrechts, R., Simonet, G., Smagghe, G., & Vanden Broeck, J. (2010). Control of ecdysteroidogenesis in prothoracic glands of insects: a review. *Peptides*, *31*, 506–519.

Marris, G. C., Weaver, R. J., Bell, J., & Edwards, J. P. (2001). Venom from the ectoparasitoid wasp *Eulophus pennicornis* disrupts host ecdysteroid production by regulating host prothoracic gland activity. *Physiol. Entomol.*, *26*, 229–238.

Martelli, A. M., Faenza, I., Billi, A. M., Fala, F., Cocco, L., & Manzoli., L. (2003). Nuclear protein kinase C isoforms: key players in multiple cell functions? *Histol. Histopathol.*, *18*, 1301–1312.

McBrayer, Z., Ono, H., Shimell, M., Parvy, J. P., Beckstead, R. B., Warren, J. T., Thummel, C. S., Dauphin-Villemant, C., Gilbert, L. I., & O'Connor, M. B. (2007). Prothoracicotropic hormone regulates developmental timing and body size in *Drosophila Dev. Cell*, *13*, 857–871.

McDaniel, C. N. (1979). Hemolymph ecdysone concentrations in *Hyalophora cecropia* pupae, dauer pupae, and adults. *J. Insect Physiol.*, *25*, 143–145.

Medler, K. F., & Bruch, R. C. (1999). Protein kinase Cβ and δ selectively phosphorylate odorant and metabotropic glutamate receptors. *Chemical Senses*, *24*, 295–299.

Meller, V. H., Combest, W. L., Smith, W. A., & Gilbert, L. I. (1988). A calmodulin-sensitive adenylate cyclase in the prothoracic glands of the tobacco hornworm, *Manduca sexta*. *Mol. Cell. Endocrinol.*, *59*, 67–76.

Meller, V. H., & Gilbert, L. I. (1990). Occurrence, quaternary structure and function of G protein subunits in an insect endocrine gland. *Mol. Cell. Endocrinol.*, *74*, 133–141.

Meller, V. H., Sakurai, S., & Gilbert, L. I. (1990). Developmental regulation of calmodulin-dependent adenylate cyclase in an insect endocrine gland. *Cell Regul.*, *1*, 771–780.

Meyuhas, O. (2008). Physiological roles of ribosomal protein S6: one of its kind. *Int. Rev. Cell Mol. Biol.*, *268*, 1–37.

Meyuhas, O., & Hornstein, E. (2000). Translational control of TOP mRNAs. In N. Sonenberg, J. W.B. Hershey, & M. B. Mathews (Eds.), *Translational Control of Gene Expression* (2nd ed.). (pp. 671–693). New York: Cold Spring Harbor Laboratory Press.

Mirth, C., & Riddiford, L. M. (2007). Size assessment and growth control: how adult size is determined in insects. *Bioessays*, *29*, 344–355.

Mirth, C., Truman, J. W., & Riddiford, L. M. (2005). The role of the prothoracic gland in determining critical weight for metamorphosis in *Drosophila melanogaster*. *Curr. Biol.*, *15*, 1796–1807.

Mizoguchi, A. (2001). Effects of juvenile hormone on the secretion of prothoracicotropic hormone in the last- and penultimate-instar larvae of the silkworm *Bombyx mori*. *J. Insect Physiol.*, *47*, 767–775.

Mizoguchi, A., Dedos, S. G., Fugo, H., & Kataoka, H. (2002). Basic pattern of fluctuation in hemolymph PTTH titers during larval-pupal and pupal-adult development of the silkworm, *Bombyx mori*. *Gen. Comp. Endocrinol.*, *127*, 181–189.

Mizoguchi, A., & Ishizaki, H. (1982). Prothoracic glands of the saturniid moth *Samia cynthia* ricini possess a circadian clock controlling gut purge timing. *Proc. Natl. Acad. Sci. USA*, *79*, 2726–2730.

Mizoguchi, A., & Kataoka, H. (2005). An in vitro study on regulation of prothoracic gland activity in the early last-larval instar of the silkworm *Bombyx mori*. *J. Insect Physiol.*, *51*, 871–879.

Mizoguchi, A., Ohashi, Y., Hosoda, K., Ishibashi, J., & Kataoka, H. (2001). Developmental profile of the changes in the prothoracicotropic hormone titer in hemolymph of the silkworm *Bombyx mori*: correlation with ecdysteroid secretion. *Insect Biochem. Mol. Biol.*, *31*, 349–358.

Mizoguchi, A., Oka, T., Kataoka, H., Nagasawa, H., Suzuki, A., & Ishizaki, H. (1990). Immunohistochemical localization of prothoracicotropic hormone-producing cells in the brain of *Bombyx mori*. *Dev. Growth Differ., 32*, 591–598.

Monconduit, H., & Mauchamp, B. (1998). Effects of ultralow doses of fenoxycarb on juvenile hormone-regulated physiological parameters in the silkworm, *Bombyx mori* L. *Arch. Insect Biochem. Physiol., 37*, 178–189.

Muehleisen, D. P., Gray, R. S., Katahira, E. J., Thomas, M. K., & Bollenbacher, W. E. (1993). Immunoaffinity purification of the neuropeptide prothoracicotropic hormone from *Manduca sexta*. *Peptides, 14*, 531–541.

Mulye, H., & Gordon, R. (1993). Effects of fenoxycarb, a juvenile hormone analog, on lipid metabolism of the eastern spruce budworm, *Choristoneura fumiferana*. *J. Insect Physiol., 39*, 721–727.

Myers, E. M., Yu, J., & Sehgal, A. (2003). Circadian control of eclosion: Interaction between a central and peripheral clock in *Drosophila melanogaster*. *Curr. Biol., 13*, 526–533.

Nagata, S., Kataoka, H., & Suzuki, A. I. (2005). Silk moth neuropeptide hormones: Prothoracicotropic hormone and others. *Ann. N.Y. Acad. Sci., 1040*, 38–52.

Nagata, K., Maruyama, K., Kojima, K., Yamamoto, M., Tanaka, M., Kataoka, H., Nagasawa, H., Isogai, A., Ishizaki, H., & Suzuki, A. (1999). Prothoracicotropic activity of SBRPs, the insulin-like peptides of the saturniid silkworm *Samia cynthia ricini*. *Biochem Biophys. Res. Commun., 266*, 575–578.

Nagata, S., Namiki, T., Ko, R., Kataoka, H., & Suzuki, A. (2006). A novel type of receptor cDNA from the prothoracic glands of the silkworm, *Bombyx mori*. *Biosci. Biotechnol. Biochem., 70*, 554–558.

Nichols, R. (2003). Signaling pathways and physiological functions of *Drosophila melanogaster* FMRFamide-related peptides. *Annu. Rev. Entomol., 48*, 485–503.

Nijhout, H. F. (1979). Stretch-induced molting in *Oncopeltus fasciatus*. *J. Insect Physiol., 25*, 277–281.

Nijhout, H. F. (1981). Physiological control of molting in insects. *Am. Zool., 21*, 631–640.

Nimi, S., & Sakurai, S. (1997). Development changes in juvenile hormone and juvenile hormone acid titers in the hemolymph and in-vitro juvenile hormone synthesis by corpora allata of the silkworm, *Bombyx mori*. *J. Insect Physiol., 43*, 875–884.

Niwa, R., Sakudoh, T., Namiki, T., Saida, K., Fujimoto, Y., & Kataoka, H. (2005). The ecdysteroidogenic P450 Cyp302a1/disembodied from the silkworm, *Bombyx mori*, is transcriptionally regulated by prothoracicotropic hormone. *Insect Mol. Biol., 14*, 563–571.

Noguti, T., Adachi-Yamada, T., Katagiri, T., Kawakami, A., Iwami, M., et al. (1995). Insect prothoracicotropic hormone: a new member of the vertebrate growth factor superfamily. *FEBS Lett., 376*, 251–256.

Oberdorster, E., Clay, M. A., Cottam, D. M., Wilmot, F. A., McLachlan, J. A., & Milner, M. J. (2001). Common phytochemicals are ecdysteroid agonists and antagonists: a possible evolutionary link between vertebrate and invertebrate steroid hormones. *J. Steroid Biochem. Mol. Biol., 77*, 229–238.

Oberlander, H., Berry, S. J., Krishnakumaran, A., & Schneiderman, H. A. (1965). RNA and DNA synthesis during activation and secretion of the prothoracic glands of saturniid moths. *J. Exp. Zool., 159*, 15–31.

O'Brien, M. A., Granger, N. A., Agui, N., Gilbert, L. I., & Bollenbacher, W. E. (1986). Prothoracicotropic hormone in the developing brain of the tobacco hornworm *Manduca sexta*: Relative amounts of two molecular forms. *J. Insect Physiol., 32*, 719–725.

O'Brien, M. A., Katahira, E. J., Flanagan, T. R., Arnold, L. W., Haughton, G., & Bollenbacher, W. E. (1988). A monoclonal antibody to the insect prothoracicotropic hormone. *J. Neurosci., 8*, 3247–3257.

Offermanns, S. (2003). G-proteins as transducers in transmembrane signalling. *Prog. Biophys. Mol. Biol., 83*, 101–130.

Okajima, A., & Kumagai, K. (1989). The inhibitory control of prothoracic gland activity by the neurosecretory neurones in a moth, *Mamestra brassicae*. *Zool. Sci., 6*, 851–858.

Okajima, A., Kumagai, K., & Watanabe, M. (1989). The involvement of afferent chemoreceptive activity in the nervous regulation of the prothoracic gland in a moth, *Mamestra brassicae*. *Zool. Sci., 6*, 859–866.

Okajima, A., & Watanabe, M. (1989). Electrophysiological identification of neuronal pathway to the prothoracic gland and the change in electrical activities of the prothoracic gland innervating neurons during larval development of a moth, *Mamestra brassicae*. *Zool. Sci., 6*, 459–468.

Okamoto, N., Yamanaka, N., Satake, H., Saegusa, H., Kataoka, H., & Mizoguchi, A. (2009). An ecdysteroid-inducible insulin-like growth factor-like peptide regulates adult development of the silkmoth *Bombyx mori*. *FEBS J., 276*, 1221–1232.

Okuda, M., Sakurai, S., & Ohtaki, T. (1985). Activity of the prothoracic gland and its sensitivity to prothoracicotropic hormone in the penultimate and last-larval instar of *Bombyx mori*. *J. Insect Physiol., 31*, 455–461.

Orme-Johnson, N. R. (1990). Distinctive properties of adrenal cortex mitochondria. *Biochim. Biophys. Acta, 1020*, 213–231.

Pak, J. -W., Chung, K. W., Lee, C. C., Kim, K., Namkoong, Y., & Koolman, J. (1992). Evidence for multiple forms of the prothoracicotropic hormone in *Drosophila melanogaster* and indication of a new function. *J. Insect Physiol., 38*, 167–176.

Pak, M. D., & Gilbert, L. I. (1987). A developmental analysis of ecdysteroids during the metamorphosis of *Drosophila melanogaster*. *J. Liq. Chromatogr., 10*, 2591–2611.

Palen, E., & Traugh, J. A. (1987). Phosphorylation of ribosomal protein S6 by cAMP-dependent protein kinase and mitogen-stimulated S6 kinase differentially alters translation of globin mRNA. *J. Biol. Chem., 262*, 3518–3523.

Papadopoulos, V., Amri, H., Boujrad, N., Cascio, C., Culty, M., et al. (1997). Peripheral benzodiazepine receptor in cholesterol transport and steroidogenesis. *Steroids, 62*, 21–28.

Parvy, J. P., Blais, C., Bernard, F., Warren, J. T., Petryk, A., Gilbert, L. I., O'Connor, M. B., Dauphin-Villemant, C., (2005). A role for betaFTZ–F1 in regulating ecdysteroid titers during post-emboyonic development in. *Drosophila melanogaster*. *Dev. Biol., 282*, 84–94.

Pearson, R. B., & Thomas, G. (1995). Regulation of p70s6 k/p85s6 k and its role in the cell cycle. *Prog. Cell Cycle Res., 1*, 21–32.

Pelc, D., & Steel, C. G.H. (1997). Rhythmic steroidogenesis by the prothoracic glands of the insect *Rhodnius prolixus* in the absence of rhythmic neuropeptide input: Implications for the role of prothoracicotropic hormone. *Gen. Comp. Endocrinol., 108*, 358–365.

Pennacchio, F., Falabella, P., Sordetti, R., Varricchio, P., Malva, C., & Vinson, B. S. (1998a). Prothoracic gland inactivation in *Heliothis virescens* (F.) (Lepidoptera: Noctuidae) larvae parasitized by *Cardiochiles nigriceps* Viereck (Hymenoptera: Braconidae). *J. Insect Physiol., 44*, 845–857.

Pennacchio, F., Falabella, P., & Vinson, S. B. (1998b). Regulation of *Heliothis virescens* prothoracic glands by *Cardiochiles nigriceps* polydnavirus. *Arch. Insect Biochem. Physiol., 38*, 1–10.

Pennacchio, F., Sordetti, R., Falabella, P., & Vinson, S. B. (1997). Biochemical and ultrastructural alterations in prothoracic glands of *Heliothis virescens* (F.) (Lepidoptera, Noctuidae) last instar larvae parasitized by *Cardiochiles nigriceps* Viereck (Hymenoptera, Braconidae). *Insect Biochem. Mol. Biol., 27*, 439–450.

Perez-Hedo, M., Eizaguirre, M., & Sehnal, F. (2010a). Brain-independent development in the moth *Sesamia nonagrioides*. *J. Insect Physiol., 56*, 594–602.

Perez-Hedo, M., Pena, R. N., Sehnal, F., & Eizaguirre, M. (2010b). Gene encoding the prothoracicotropic hormone of a moth is expressed in the brain and gut. *Gen. Comp. Endocrinol.*, [epub ahead of print].

Pfister-Wilhelm, R., & Lanzrein, B. (2009). Stage dependent influences of polydnaviruses and the parasitoid larva on host ecdysteroids. *J. Insect Physiol., 55*, 707–715.

Poderoso, C., Converso, D. P., Maloberti, P., Duarte, A., Neuman, I., Galli, S., Maciel, F. C., Paz, C., Carreras, M. C., Poderoso, J. J., & Podesta, E. J. (2008). A mitochondrial kinase complex is essential to mediate an ERK1/2-dependent phosphorylation of a key regulatory protein in steroid biosynthesis. *PLoS ONE, 3*, e1443.

Potthoff, M. J., & Olson, E. N. (2007). MEF2: a central regulator of diverse developmental programs. *Development, 134*, 4131–4140.

Poulin, B., Sekiya, F., & Rhee, S. G. (2005). Intramolecular interaction between phosphorylated tyrosine-783 and the C-terminal Src homology 2 domain activates phospholipase C-gamma1. *Proc. Natl. Acad. Sci. USA, 102*, 4276–4281.

Priester, J., & Smith, W. A. (2005). Inhibition of tyrosine phosphorylation blocks hormone-stimulated calcium influx in an insect steroidogenic gland. *Mol. Cell. Endocrinol., 229*, 185–192.

Redfern, C. P. F. (1983). Ecdysteroid synthesis by the ring gland of *Drosophila melanogaster* during late-larval, pre-pupal and pupal development. *J. Insect Physiol., 29*, 65–71.

Reimers, K., Choi, C. Y., Bucan, V., & Vogt, P. M. (2007). The growth-hormone inducible transmembrane protein (Ghitm) belongs to the Bax inhibitory protein-like family. *Int. J. Biol. Sci., 3*, 471–476.

Rewitz, K. F., Larsen, M. R., Lobner-Olesen, A., Rybczynski, R., O'Connor, M. B., & Gilbert, L. I. (2009a). A phosphoproteomics approach to elucidate neuropeptide signal transduction controlling insect metamorphosis. *Insect Biochem. Mol. Biol., 39*, 475–483.

Rewitz, K. F., Yamanaka, N., Gilbert, L. I., & O'Connor, M. B. (2009b). The insect neuropeptide PTTH activates receptor tyrosine kinase torso to initiate metamorphosis. *Science, 326*, 1403–1405.

Rhee, S. G. (2001). Regulation of phosphoinositide-specific phospholipase C. *Annu. Rev. Biochem., 70*, 281–312.

Richard, D. S., & Gilbert, L. I. (1991). Reversible juvenile hormone inhibition of ecdysteroid and juvenile hormone synthesis by the ring gland of *Drosophila melanogaster*. *Experientia, 47*, 1063–1066.

Richard, D. S., & Saunders, D. S. (1987). Prothoracic gland function in diapause and non-diapause destined *Sarcophaga argyrostoma* and *Calliphora vicina*. *J. Insect Physiol., 33*, 385–392.

Richard, D. S., Saunders, D. S., Egan, V. M., & Thomson, R. C.K. (1986). The timing of larval wandering and puparium formation in the flesh-fly *Sarcophaga argyrostoma*. *Physiol. Entomol., 11*, 53–60.

Richard, D. S., Warren, J. T., Saunders, D. S., & Gilbert, L. I. (1987). Haemolymph ecdysteroid titres in diapause and non-diapause destined larvae and pupae of *Sarcophaga argyrostoma*. *J. Insect Physiol., 33*, 115–122.

Richards, J. S. (1994). Hormonal control of gene expression in the ovary. *Endocr. Rev., 15*, 725–751.

Richards, J. S., Russell, D. L., Ochsner, S., Hsieh, M., Doyle, K. H., Falender, A. E., Lo, Y. K., & Sharma, S. C. (2002). Novel signaling pathways that control ovarian follicular development, ovulation, and luteinization. *Recent Progr. Horm. Res., 57*, 195–220.

Richter, K. (1985). Physiological investigations on the role of the innervation in the regulation of the prothoracic gland in *Periplaneta americana*. *Arch. Insect Biochem. Physiol., 2*, 319–329.

Richter, K. (2001). Daily changes in neuroendocrine control of moulting hormone secretion in the prothoracic gland of the cockroach *Periplaneta americana* (L.). *J. Insect Physiol., 47*, 333–338.

Richter, K., & Gersch, M. (1983). Electrophysiological evidence of nervous involvement in the control of the prothoracic gland in *Periplaneta americana*. *Experientia, 39*, 917–918.

Richter, K., Peschke, E., & Peschke, D. (2000). A neuroendocrine releasing effect of melatonin in the brain of an insect, *Periplaneta americana* (L.). *J. Pineal Res., 28*, 129–135.

Riddiford, L. M. (1976). Hormonal control of insect epidermal cell commitment *in vitro*. *Nature, 259*, 115–117.

Riddiford, L. M. (1994). Cellular and molecular actions of juvenile hormone. I. General considerations and premetamorphic actions. *Adv. Insect Physiol., 24*, 213–274.

Riddiford, L. M. (1996). Juvenile hormone: the status of its "status quo" action. *Arch. Insect Biochem. Physiol., 32*, 271–286.

Riehle, M. A., & Brown, M. R. (1999). Insulin stimulates ecdysteroid production through a conserved signaling cascade in the mosquito *Aedes aegypti*. *Insect Biochem. Mol. Biol., 29*, 855–860.

Roberts, B. (1984). Photoperiodic regulation of prothoracicotropic hormone release in late larval, prepupal and pupa; stages of *Sarcophaga bullata*. In: R. Porter and G. M. Collins (Eds.), *Ciba Foundation Symposium Vol. 104* (pp. 170–188). Chicester UK. John Wiley and Sons.

Roberts, B., & Gilbert, L. I. (1986). Ring gland and prothoracic gland sensitivity to interspecific prothoracicotropic hormone extracts. *J. Comp. Physiol. B*, *156*, 767–771.

Roberts, B., Henrich, V., & Gilbert, L. I. (1987). Effects of photoperiod on the timing of larval wandering in *Drosophila melanogaster*. *Physiol. Entomol.*, *12*, 175–180.

Roth, G. E., Gierl, M. S., Vollborn, L., Meise, M., Lintermann, R., & Korge, G. (2004). The *Drosophila* gene *Start*1: A putative cholesterol transporter and key regulator of ecdysteroid synthesis. *Proc. Natl. Acad. Sci. USA*, *101*, 1601–1606.

Rountree, D. B., & Bollenbacher, W. E. (1986). The release of the prothoracicotropic hormone in the tobacco hornworm, *Manduca sexta*, is controlled intrinsically by juvenile hormone. *J. Exp. Biol.*, *120*, 41–58.

Rountree, D. B., Combest, W. L., & Gilbert, L. I. (1987). Protein phosphorylation in the prothoracic glands as a cellular model for juvenile hormone-prothoracicotropic hormone interactions. *Insect Biochem.*, *17*, 943–948.

Rountree, D. B., Combest, W. L., & Gilbert, L. I. (1992). Prothoracicotropic hormone regulates the phosphorylation of a specific protein in the prothoracic glands of the tobacco hornworm, *Manduca sexta*. *Insect Biochem. Mol. Biol.*, *22*, 353–362.

Roux, P. P., & Blenis, J. (2004). ERK and p38 MAPK-activated protein kinases: a family of protein kinases with diverse biological functions. *Microbiol. Mol. Biol. Rev.*, *68*, 320–344.

Rybczynski, R. (2005). Prothoracicotropic hormone. In L. I. Gilbert, K. Iatrou, & S. S. Gill (Eds.), *Comprehensive Molecular Insect Science Vol. 3*, (pp. 61–123). Amsterdam: Elsevier.

Rybczynski, R., Bell, S. C., & Gilbert, L. I. (2001). Activation of an extracellular signal-regulated kinase (ERK) by the insect prothoracicotropic hormone. *Mol. Cell. Endocrinol.*, *184*, 1–11.

Rybczynski, R., & Gilbert, L. I. (1994). Changes in general and specific protein synthesis that accompany ecdysteroid synthesis in stimulated prothoracic glands of *Manduca sexta*. *Insect Biochem. Mol. Biol.*, *24*, 175–189.

Rybczynski, R., & Gilbert, L. I. (1995a). Prothoracicotropic hormone elicits a rapid, developmentally specific synthesis of β tubulin in an insect endocrine gland. *Dev. Biol.*, *169*, 15–28.

Rybczynski, R., & Gilbert, L. I. (1995b). Prothoracicotropic hormone-regulated expression of a hsp 70 cognate protein in the insect prothoracic gland. *Mol. Cell. Endocrinol.*, *115*, 73–85.

Rybczynski, R., & Gilbert, L. I. (1998). Cloning of a β 1 tubulin cDNA from an insect endocrine gland: Developmental and hormone-induced changes in mRNA expression. *Mol. Cell. Endocrinol.*, *141*, 141–151.

Rybczynski, R., & Gilbert, L. I. (2000). cDNA cloning and expression of a hormone-regulated heat shock protein (hsc 70) from the prothoracic gland of *Manduca sexta*. *Insect Biochem. Mol. Biol.*, *30*, 579–589.

Rybczynski, R., & Gilbert, L. I. (2003). Prothoracicotropic hormone stimulated extracellular signal-regulated kinase (ERK) activity: the changing roles of Ca^{2+}- and cAMP-dependent mechanisms in the insect prothoracic glands during metamorphosis. *Mol. Cell. Endocrinol.*, *205*, 159–168.

Rybczynski, R., & Gilbert, L. I. (2006). Protein kinase C modulates ecdysteroidogenesis in the prothoracic gland of the tobacco hornworm, *Manduca sexta*. *Mol. Cell. Endocrinol.*, *251*, 78–87.

Rybczynski, R., Mizoguchi, A., & Gilbert, L. I. (1996). *Bombyx* and *Manduca* prothoracicotropic hormones: an immunologic test for relatedness. *Gen. Comp. Endocrinol.*, *102*, 247–254.

Rybczynski, R., Snyder, C. A., Hartmann, J., Gilbert, L. I., & Sakurai, S. (2009). *Manduca sexta* prothoracicotropic hormone: evidence for a role beyond steroidogenesis. *Arch. Insect Biochem. Physiol.*, *70*, 217–229.

Safranek, L., Cymborowski, B., & Williams, C. M. (1980). Effects of juvenile hormone on ecdysone-dependent development in the tobacco hornworm, *Manduca sexta*. *Biol. Bull.*, *158*, 248–256.

Sakudoh, T., Tsuchida, K., & Kataoka, H. (2005). BmStart1, a novel carotenoid-binding protein isoform from *Bombyx mori*, is orthologous to MLN64, a mammalian cholesterol transporter. *Biochem. Biophys. Res. Commun.*, *336*, 1125–1135.

Sakurai, S. (1983). Temporal organization of endocrine events underlying larval-larval ecdysis in the silkworm, *Bombyx mori*. *J. Insect Physiol.*, *29*, 919–932.

Sakurai, S. (1984). Temporal organization of endocrine events underlying larval–pupal metamorphosis in the silkworm, *Bombyx mori*. *J. Insect Physiol.*, *30*, 657–664.

Sakurai, S., & Gilbert, L. I. (1990). Biosynthesis and secretion of ecdysteroids by the prothoracic glands. In E. Ohnishi, & H. Ishizaki (Eds.), *Molting and Metamorphosis* (pp. 83–106). New York: Springer-Verlag.

Sakurai, S., Kaya, M., & Satake, S. (1998). Hemolymph ecdysteroid titer and ecdysteroid-dependent developmental events in the last-larval stadium of the silkworm, *Bombyx mori*: Role of low ecdysteroid titer in larval-pupal metamorphosis and a reappraisal of the head critical period. *J. Insect Physiol.*, *44*, 867–881.

Sakurai, S., Okuda, M., & Ohtaki, T. (1989). Juvenile hormone inhibits ecdysone secretion and responsiveness to prothoracicotropic hormone in prothoracic glands of *Bombyx mori*. *Gen. Comp. Endocrinol.*, *75*, 222–230.

Sakurai, S., & Williams, C. M. (1989). Short-loop negative and positive feedback on ecdysone secretion by prothoracic gland in the tobacco hornworm, *Manduca sexta*. *Gen. Comp. Endocrinol.*, *75*, 204–216.

Sauman, I., & Reppert, S. M. (1996a). Molecular characterization of prothoracicotropic hormone (PTTH) from the giant silkmoth *Antheraea pernyi*: developmental appearance of PTTH-expressing cells and relationship to circadian clock cells in central brain. *Dev. Biol.*, *178*, 418–429.

Sauman, I., & Reppert, S. M. (1996b). Circadian clock neurons in the silkmoth *Antheraea pernyi*: novel mechanisms of *Period* protein regulation. *Neuron*, *17*, 889–900.

Schneiderman, H. A., & Gilbert, L. I. (1964). Control of growth and development in insects. *Science*, *143*, 325–333.

Schwartz, L. M. (2008). Atrophy and programmed cell death of skeletal muscle. *Cell Death Differ.*, *15*, 1163–1169.

Sedlak, B. J., Marchione, L., Devorkin, B., & Davino, R. (1983). Correlations between endocrine gland ultrastructure and hormone titers in the fifth larval instar of *Manduca sexta*. *Gen. Comp. Endocrinol.*, *52*, 291–310.

Sehnal, F. (1983). Juvenile hormone analogues. In R. Downder, & H. Laufer (Eds.), *Insect Endocrinology* (pp. 657–672). New York: Alan R. Liss.

Sehnal, F., Hansen, I., & Scheller, K. (2002). The cDNA-structure of the prothoracicotropic hormone (PTTH) of the silkmoth *Hyalophora cecropia. Insect Biochem. Mol. Biol., 32,* 233–237.

Seidah, N. G., & Prat, A. (2002). Precursor convertases in the secretory pathway, cytosol and extracellular milieu. *Essays Biochem., 38,* 79–94.

Shafiei, M., Moczek, A. P., & Nijhout, H. F. (2001). Food availability controls the onset of metamorphosis in the dung beetle *Onthophagus taurus* (Coleoptera: Scarabaeidae). *Physiol. Entomol., 26,* 173–180.

Shearman, M. S., Sekiguchi, K., & Nishizuka, Y. (1989). Modulation of ion channel activity: A key function of the protein kinase C enzyme family. *Pharmacol. Rev., 41,* 211–237.

Shingleton, A. W. (2005). Body-size regulation: combining genetics and physiology. *Curr. Biol., 15,* R825–R827.

Shiomi, K., Fujiwara, Y., Atsumi, T., Kajiura, Z., Nakagaki, M., Tanaka, Y., Mizoguchi, A., Yaginuma, T., & Yamashita, O. (2005). Myocyte enhancer factor 2 (MEF2) is a key modulator of the expression of the prothoracicotropic hormone gene in the silkworm, *Bombyx mori. FEBS. J., 272,* 3853–3862.

Shionoya, M., Matsubayashi, H., Asahina, M., Kuniyoshi, H., Nagata, S., Riddiford, L. M., & Kataoka, H. (2003). Molecular cloning of the prothoracicotropic hormone from the tobacco hornworm, *Manduca sexta. Insect Biochem. Mol. Biol., 33,* 795–801.

Shirai, Y., Aizono, Y., Iwasaki, T., Yanagida, A., Mori, H., Sumida, M., & Matsubara, F. (1993). Prothoracicotropic hormone is released five times in the 5th-larval instar of the silkworm *Bombyx mori. J. Insect Physiol., 39,* 83–88.

Shirai, Y., Iwasaki, T., Matsubara, F., & Aizono, Y. (1994). The carbachol-induced release of prothoracicotropic hormone from brain-corpus cardiacum-corpus allatum complex of the silkworm, *Bombyx mori. J. Insect Physiol., 40,* 469–473.

Shirai, Y., Shimazaki, K., Iwasaki, T., Matsubara, F., & Aizono, Y. (1995). The in vitro release of prothoracicotropic hormone (PTTH) from the brain-corpus cardiacum-corpus allatum complex of silkworm, *Bombyx mori. Comp. Biochem. Physiol. C, 110,* 143–148.

Shirai, Y., Uno, T., & Aizono, Y. (1998). Small GTP-binding proteins in the brain-corpus cardiacum-corpus allatum complex of the silkworm, *Bombyx mori:* involvement in the secretion of prothoracicotropic hormone. *Arch. Insect Biochem. Physiol., 38,* 177–184.

Simpson, E. R., & Waterman, M. R. (1988). Regulation of the synthesis of steroidogenic enzymes in adrenal cortical cells by ACTH. *Annu. Rev. Physiol., 50,* 427–440.

Smith, W. A. (1993). Second messengers and the action of prothoracicotropic hormone in *Manduca sexta. Am. Zool., 33,* 330–339.

Smith, W. A. (1995). Regulation and consequences of cellular changes in the prothoracic glands of *Manduca sexta* during the last larval instar: a review. *Arch. Insect Biochem. Physiol., 30,* 271–293.

Smith, W. A., Bowen, M. F., Bollenbacher, W. E., & Gilbert, L. I. (1986a). Cellular changes in the prothoracic glands of *Manduca sexta* during pupal diapause. *J. Exp. Biol., 120,* 131–142.

Smith, W. A., Combest, W. L., & Gilbert, L. I. (1986b). Involvement of cyclic AMP-dependent protein kinase in prothoracicotropic hormone-stimulated ecdysone synthesis. *Mol. Cell. Endocrinol., 47,* 25–33.

Smith, W. A., Combest, W. L., Rountree, D. B., & Gilbert, L. I. (1987a). Neuropeptide control of ecdysone biosynthesis. In J. H. Law (Ed.), *Molecular Entomology* (pp. 129–139). New York: Alan R. Liss.

Smith, W. A., Gilbert, L. I., & Bollenbacher, W. E. (1984). The role of cyclic AMP in the regulation of ecdysone synthesis. *Mol. Cell. Endocrinol., 37,* 285–294.

Smith, W. A., Gilbert, L. I., & Bollenbacher, W. E. (1985). Calcium-cyclic AMP interactions in prothoracicotropic hormone stimulation of ecdysone synthesis. *Mol. Cell. Endocrinol., 39,* 71–78.

Smith, W. A., & Gilbert, L. I. (1986). Cellular regulation of ecdysone synthesis by the prothoracic glands of *Manduca sexta. Insect Biochem., 16,* 143–147.

Smith, W. A., & Gilbert, L. I. (1989). Early events in peptide-stimulated ecdysteroid secretion by the prothoracic glands of *Manduca sexta. J. Exp. Zool., 252,* 262–270.

Smith, W. A., Koundinya, M., McAllister, T., & Brown, A. (1997). Insulin receptor-like tyrosine kinase in the tobacco hornworm, *Manduca sexta. Arch. Insect Biochem. Physiol., 35,* 99–110.

Smith, W. A., & Nijhout, H. F. (1982). Synchrony of juvenile hormone-sensitive periods for internal and external development in last-instar larvae of *Oncopeltus fasciatus. J. Insect Physiol., 28,* 797–799.

Smith, W. A., & Nijhout, H. F. (1983). In vitro stimulation of cell death in the moulting glands of *Oncopeltus fasciatus* by 20-hydroxyecdysone. *J. Insect Physiol., 29,* 169–171.

Smith, W. A., & Pasquarello, T. J. (1989). Developmental changes in phosphodiesterase activity and hormonal response in the prothoracic glands of *Manduca sexta. Mol. Cell. Endocrinol., 63,* 239–246.

Smith, W., Priester, J., & Morais, J. (2003). PTTH-stimulated ecdysone secretion is dependent upon tyrosine phosphorylation in the prothoracic glands of *Manduca sexta. Insect Biochem. Mol. Biol., 33,* 1317–1325.

Smith, W. A., Rountree, D. B., Bollenbacher, W. E., & Gilbert, L. I. (1987b). Dissociation of the prothoracic glands of *Manduca sexta* into hormone-responsive cells. In A. Borkovec, & D. Gelman (Eds.), *Progress in Insect Neurochemistry and Neurophysiology* (pp. 319–322). Clifton, New Jersey: Humana Press.

Smith, W. A., Varghese, A. H., & Lou, K. J. (1993). Developmental changes in cyclic AMP-dependent protein kinase associated with increased secretory capacity of *Manduca sexta* prothoracic glands. *Mol. Cell. Endocrinol., 90,* 187–195.

Smith, W. A., Varghese, A. H., Healy, M. S., & Lou, K. J. (1996). Cyclic AMP is a requisite messenger in the action of big PTTH in the prothoracic glands of pupal *Manduca sexta. Insect Biochem. Mol. Biol., 26,* 161–170.

Snyder, M. J., & Feyereisen, R. (1993). A diazepam binding inhibitor (DBI) homologue from the tobacco hornworm, *Manduca sexta. Mol. Cell. Endocrinol., 94,* R1–R4.

Snyder, M. J., & Van Antwerpen, R. (1998). Evidence for a diazepam-binding inhibitor (DBI) benzodiazepine receptor-like mechanism in ecdysteroidogenesis by the insect prothoracic gland. *Cell Tissue Res., 294,* 161–168.

Song, Q., & Gilbert, L. I. (1994). S6 phosphorylation results from prothoracicotropic hormone stimulation of insect prothoracic glands: A role for S6 kinase. *Dev. Genet., 15,* 332–338.

Song, Q., & Gilbert, L. I. (1995). Multiple phosphorylation of ribosomal protein S6 and specific protein synthesis are required for prothoracicotropic hormone-stimulated ecdysteroid biosynthesis in the prothoracic glands of *Manduca sexta. Insect Biochem. Mol. Biol., 25,* 591–602.

Song, Q., & Gilbert, L. I. (1996). Protein phosphatase activity is required for prothoracicotropic hormone-stimulated ecdysteroidogenesis in the prothoracic glands of the tobacco hornworm, *Manduca sexta. Arch. Insect Biochem. Physiol., 31,* 465–480.

Song, Q., & Gilbert, L. I. (1997). Molecular cloning, developmental expression, and phosphorylation of ribosomal protein S6 in the endocrine gland responsible for insect molting. *J. Biol. Chem., 272,* 4429–4435.

Song, Q., & Gilbert, L. I. (1998). Alterations in ultraspiracle (USP) content and phosphorylation state accompany feedback regulation of ecdysone synthesis in the insect prothoracic gland. *Insect Biochem. Mol. Biol., 28,* 849–860.

Soulard, A., Cohen, A., & Hall, M. N. (2009). TOR signaling in invertebrates. *Curr. Opin. Cell Biol., 21,* 825–836.

Stanewsky, R. (2002). Clock mechanisms in *Drosophila. Cell Tissue Res., 309,* 11–26.

Stanewsky, R. (2003). Genetic analysis of the circadian system in *Drosophila melanogaster* and mammals. *J. Neurobiol., 54,* 111–147.

Steel, C. G., & Vafopoulou, X. (2006). Circadian orchestration of developmental hormones in the insect, *Rhodnius prolixus. Comp. Biochem. Physiol. A Mol. Integr. Physiol., 144,* 351–364.

Stengl, M., & Homberg, U. (1994). Pigment-dispersing hormone-immunoreactive neurons in the cockroach *Leucophaea maderae* share properties with circadian pacemaker neurons. *J. Comp. Physiol. A, 175,* 203–213.

Suarez, C., Diaz-Torga, G., Gonzalez-Iglesias, A., Vela, J., Mladovan, A., et al. (2003). Angiotensin II phosphorylation of extracellular signal-regulated kinases in rat anterior pituitary cells. *Am. J. Physiol. Endocrinol. Metab., 85,* E645–E653.

Suh, J. M., Song, J. H., Kim, D. W., Kim, H., Chung, H. K., et al. (2003). Regulation of the phosphatidylinositol 3-kinase, Akt/protein kinase B, FRAP/mammalian target of rapamycin, and ribosomal S6 kinase 1 signaling pathways by thyroid-stimulating hormone (TSH) and stimulating type TSH receptor antibodies in the thyroid gland. *J. Biol. Chem., 278,* 21960–21971.

Sun, L., Yu, M. C., Kong, L., Zhuang, Z. H., Hu, J. H., & Ge, B. X. (2008). Molecular identification and functional characterization of a *Drosophila* dual-specificity phosphatase DMKP-4 which is involved in PGN-induced activation of the JNK pathway. *Cell. Signal., 20,* 1329–1337.

Takaki, K., & Sakurai, S. (2003). Regulation of prothoracic gland ecdysteroidogenic activity leading to pupal metamorphosis. *Insect Biochem. Mol. Biol., 33,* 1189–1199.

Tissenbaum, H. A., Hawdon, J., Perregaux, M., Hotez, P., Guarente, L., & Ruvkun, G. (2000). A common muscarinic pathway for diapause recovery in the distantly related nematode species *Caenorhabditis elegans* and *Ancylostoma caninum. Proc. Natl. Acad. Sci. USA, 97,* 460–465.

Tomioka, K., Agui, N., & Bollenbacher, W. E. (1995). Electrical properties of the prothoracicotropic hormone cells in diapausing and non-diapausing pupae of the tobacco hornworm, *Manduca sexta. Zool. Sci., 12,* 165–173.

Tomioka, K., & Matsumoto, A. (2010). A comparative view of insect circadian clock systems. *Cell. Mol. Life Sci., 67,* 1397–1406.

Tovey, S. C., Dedos, S. G., Taylor, E. J., Church, J. E., & Taylor, C. W. (2008). Selective coupling of type 6 adenylyl cyclase with type 2 IP3 receptors mediates direct sensitization of IP3 receptors by cAMP. *J. Cell Biol., 183,* 297–311.

Truman, J. W. (1972). Physiology of insect rhythms. II. The silkmoth brain as the location of the biological clock controlling eclosion. *J. Comp. Physiol., 81,* 99–114.

Truman, J. W. (2006). Steroid hormone secretion in insects comes of age. *Proc. Natl. Acad. Sci. USA, 103,* 8909–8910.

Truman, J. W., & Riddiford, L. M. (1974). Physiology of insect rhythms. 3. The temporal organization of the endocrine events underlying pupation of the tobacco hornworm. *J. Exp. Biol., 60,* 371–382.

Vafopoulou, X. (2009). Ecdysteroid receptor (EcR) is associated with microtubules and with mitochondria in the cytoplasm of prothoracic gland cells of *Rhodnius prolixus* (Hemiptera). *Arch. Insect Biochem. Physiol., 72,* 249–262.

Vafopoulou, X., Sim, C. -H., & Steel, C. G. (1996). Prothoracicotropic hormone in *Rhodnius prolixus*: in vitro analysis and changes in amounts in the brain and retrocerebral complex during larval–adult development. *J. Insect Physiol., 42,* 407–415.

Vafopoulou, X., & Steel, C. G.H. (1996). The insect neuropeptide prothoracicotropic hormone is released with a daily rhythm: Re-evaluation of its role in development. *Proc. Natl. Acad. Sci. USA, 93,* 3368–3372.

Vafopoulou, X., & Steel, C. G. (1997). Ecdysteroidogenic action of *Bombyx* prothoracicotropic hormone and bombyxin on the prothoracic glands of Rhodnius prolixus *in vitro. J. Insect Physiol., 43,* 651–656.

Vafopoulou, X., & Steel, C. G. (1998). A photosensitive circadian oscillator in an insect endocrine gland: Photic induction of rhythmic steroidogenesis in vitro. *J. Comp. Physiol. A, 182,* 343–349.

Vafopoulou, X., & Steel, C. G. (1999). Daily rhythm of responsiveness to prothoracicotropic hormone in prothoracic glands of *Rhodnius prolixus. Arch. Insect Biochem. Physiol., 41,* 117–123.

Vafopoulou, X., & Steel, C. G.H. (2001). Induction of rhythmicity in prothoracicotropic hormone and ecdysteroids in *Rhodnius prolixus*: roles of photic and neuroendocrine Zeitgebers. *J. Insect Physiol., 47,* 935–941.

Vafopoulou, X., & Steel, C. G.H. (2002). Prothoracicotropic hormone of *Rhodnius prolixus*: Partial characterization and rhythmic release of neuropeptides related to *Bombyx* PTTH and bombyxin. *Invert. Reprod. Dev., 42,* 111–120.

Vafopoulou, X., & Steel, C. G.H. (2005). Circadian organization of the endocrine system. In L. I. Gilbert, K. Iatrou, & S. S. Gill (Eds.), *Comprehensive Molecular Insect Science* (pp. 551–614). Amsterdam: Elsevier.

Vafopoulou, X., & Steel, C. G. (2006). Ecdysteroid hormone nuclear receptor (EcR) exhibits circadian cycling in certain tissues, but not others, during development in *Rhodnius prolixus* (Hemiptera). *Cell Tissue Res.*, *323*, 443–455.

Vafopoulou, X., Steel, C. G., & Terry, K. L. (2007). Neuroanatomical relations of prothoracicotropic hormone neurons with the circadian timekeeping system in the brain of larval and adult *Rhodnius prolixus* (Hemiptera). *J. Comp. Neurol.*, *503*, 511–524.

Vafopoulou, X., Terry, K. L., & Steel, C. G. (2010). The circadian timing system in the brain of the fifth larval instar of *Rhodnius prolixus* (Hemiptera). *J. Comp. Neurol.*, *518*, 1264–1282.

Val, P., Lefrancois-Martinez, A. M., Veyssiere, G., & Martinez, A. (2003). SF-1 a key player in the development and differentiation of steroidogenic tissues. *Nucl. Recept.*, *1*, 8.

Vandersmissen, T., De Loof, A., & Gu, S. H. (2007). Both prothoracicotropic hormone and an autocrine factor are involved in control of prothoracic gland ecdysteroidogenesis in *Locusta migratoria* and *Schistocerca gregaria*. *Peptides*, *28*, 44–50.

Vedeckis, W. V., Bollenbacher, W. E., & Gilbert, L. I. (1974). Cyclic AMP as a possible mediator of prothoracic gland activation. *Zool. Jahrb. Physiol.*, *78*, 440–448.

Vedeckis, W. V., Bollenbacher, W. E., & Gilbert, L. I. (1976). Insect prothoracic glands: a role for cyclic AMP in the stimulation of a ecdysone secretion. *Mol. Cell. Endocrinol.*, *5*, 81–88.

Vedeckis, W. V., & Gilbert, L. I. (1973). Production of cyclic AMP and adenosine by the brain and prothoracic glands of *Manduca sexta*. *J. Insect Physiol.*, *19*, 2445–2457.

Venkatesh, K., & Hasan, G. (1997). Disruption of the IP3 receptor gene of *Drosophila* affects larval metamorphosis and ecdysone release. *Curr. Biol.*, *7*, 500–509.

Venkatesh, K., Siddhartha, G., Joshi, R., Patel, S., & Hasan, G. (2001). Interactions between the inositol 1,4,5-trisphosphate and cyclic AMP signaling pathways regulate larval molting in *Drosophila*. *Genetics*, *158*, 309–318.

Vogt, A., Cooley, K. A., Brisson, M., Tarpley, M. G., Wipf, P., & Lazo, J. S. (2003). Cell-active dual specificity phosphatase inhibitors identified by high-content screening. *Chem. Biol.*, *10*, 733–742.

Volarevic, S., & Thomas, G. (2001). Role of S6 phosphorylation and S6 kinase in cell growth. *Prog. Nucl. Acid Res. Mol. Biol.*, *65*, 101–127.

Walker, G. P., & Denlinger, D. L. (1980). Juvenile hormone and moulting hormone titres in diapause- and non-diapause destined flesh flies. *J. Insect Physiol.*, *26*, 661–664.

Walkiewicz, M. A., & Stern, M. (2009). Increased insulin/insulin growth factor signaling advances the onset of metamorphosis in *Drosophila*. *PLoS ONE*, *4*, e5072.

Walsh, A. L., & Smith, W. A. (2011). Nutritional sensitivity of fifth instar prothoracic glands in the tobacco hornworm, *Manduca sexta*. *J. Insect Physiol.* [E pub ahead of print]

Warren, J. T., Bachman, J. S., Dai, J. -D., & Gilbert, L. I. (1996). Differential incorporation of cholesterol and cholesterol derivatives by the larval ring glands and adult ovaries of *Drosophila melanogaster*: A putative explanation for the (*l(3)ecd*[l]) mutation. *Insect Biochem. Mol. Biol.*, *26*, 931–943.

Warren, J. T., & Gilbert, L. I. (1996). Metabolism *in vitro* of cholesterol and 25-hydroxycholesterol by the larval prothoracic glands of *Manduca sexta*. *Insect Biochem. Mol. Biol.*, *26*, 917–929.

Warren, J. T., Yerushalmi, Y., Shimell, M. J., O'Connor, M. B., Restifo, L. L., & Gilbert, L. I. (2006). Discrete pulses of molting hormone, 20-hydroxyecdysone, during late larval development of *Drosophila melanogaster*: Correlations with changes in gene activity. *Dev. Dynam.*, *235*, 315–326.

Watanabe, K., Hull, J. J., Nimi, T., Imai, K., Matsumoto, S., Yaginuma, T., & Kataoka, H. (2007). FXPRL-amide peptides induce ecdysteroidogenesis through a G-protein coupled receptor expressed in the prothoracic gland of *Bombyx mori*. *Mol. Cell. Endocrinol.*, *273*, 51–58.

Watson, R. D., Ackerman-Morris, S., Smith, W. A., Watson, C. J., & Bollenbacher, W. E. (1996). Involvement of microtubules in prothoracicotropic hormone-stimulated ecdysteroidogenesis by insect (*Manduca sexta*) prothoracic glands. *J. Exp. Biol.*, *276*, 63–69.

Watson, R. D., & Bollenbacher, W. E. (1988). Juvenile hormone regulates the steroidogenic competence of *Manduca sexta* prothoracic glands. *Mol. Cell. Endocrinol.*, *57*, 251–259.

Watson, R. D., Yeh, W. E., Muehleisen, D. P., Watson, C. J., & Bollenbacher, W. E. (1993). Stimulation of ecdysteroidogenesis by small prothoracicotropic hormone: Role of cyclic AMP. *Mol. Cell. Endocrinol.*, *92*, 221–228.

Wei, Z. J., Yu, M., Tang, S. M., Yi, Y. Z., Hong, G. Y., & Jiang, S. T. (2010). Transcriptional regulation of the gene for prothoracicotropic hormone in the silkworm, *Bombyx mori*. *Mol. Biol. Rep.*, [E pub ahead of print].

Wei, Z. J., Zhang, Q. R., Kang, L., Xu, W. H., & Denlinger, D. L. (2005). Molecular characterization and expression of prothoracicotropic hormone during development and pupal diapause in the cotton bollworm, *Helicoverpa armigera*. *J. Insect. Physiol.*, *51*, 691–700.

Westbrook, A. L., Regan, S. A., & Bollenbacher, W. E. (1993). Developmental expression of the prothoracicotropic hormone in the CNS of the tobacco hornworm, *Manduca sexta*. *J. Comp. Neurol.*, *327*, 1–16.

Wigglesworth, V. B. (1933). The physiology of the cuticle and of ecdysis in *Rhodnius prolixus* (Triatomidae, Hemiptera); with special reference to the function of oenocytes and of the dermal glands. *Q. J. Microsc. Sci.*, *76*, 296–318.

Wigglesworth, V. B. (1934). The physiology of ecdysis in *Rhodnius prolixus* (Hemiptera). II. Factors controlling moulting and "metamorphosis." *Q. J. Microsc. Sci.*, *77*, 191–222.

Wigglesworth, V. B. (1940). The determination of characters at metamorphosis in *Rhodnius prolixus*. *J. Exp. Biol.*, *17*, 201–222.

Wigglesworth, V. B. (1955). The breakdown of the thoracic gland in the adult insect, *Rhodnius prolixus*. *J. Exp. Biol.*, *32*, 485–491.

Williams, C. M. (1947). Physiology of insect diapause II. Interaction between the pupal brain and prothoracic glands in the metamorphosis of the giant silkworm, *Platysamia cecropia*. *Biol. Bull.*, *93*, 89–98.

Williams, C. M. (1967). The present status of the brain hormone. In J. W. L. Beamont, & J. E. Treherne (Eds.), *Insects and Physiology* (pp. 133–139). Edinburgh: Oliver and Boyd.

Xiang, Y., Liu, Z., & Huang, X. (2010). br regulates the expression of the ecdysone biosynthesis gene npc1. *Dev. Biol.*, *344*, 800–808.

Xu, J., Su, J., Shen, J., & Xu, W. (2007). Molecular characterization and developmental expression of the gene encoding the prothoracicotropic hormone in the beet armyworm, *Spodoptera exigua*. *Sci. China C Life Sci.*, *50*, 466–472.

Xu, W. H., & Denlinger, D. L. (2003). Molecular characterization of prothoracicotropic hormone and diapause hormone in *Heliothis virescens* during diapause, and a new role for diapause hormone. *Insect Mol. Biol.*, *12*, 509–516.

Xu, W. H., Rinehart, J. P., & Denlinger, D. L. (2003). Structural characterization and expression analysis of prothoracicotropic hormone in the corn earworm, *Helicoverpa zea*. *Peptides*, *24*, 1319–1325.

Yagi, S., & Fukaya, M. (1974). Juvenile hormone as a key factor regulating larval diapause of the rice stem borer, *Chilo suppressalis*. *Appl. Entomol. Zool.*, *9*, 247–255.

Yagi, Y., Ishibashi, J., Nagata, K., Kataoka, H., Suzuki, A., Mizoguchi, A., & Ishizaki, H. (1995). The brain neurosecretory cells of the moth *Samia cynthia ricini*: Immunohistochemical localization and developmental changes of the *Samia* homologues of the *Bombyx* prothoracicotropic hormone and bombyxin. *Dev. Growth Differ.*, *37*, 505–516.

Yamanaka, N., Honda, N., Osato, N., Niwa, R., Mizoguchi, A., & Kataoka, H. (2007). Differential regulation of ecdysteroidogenic P450 gene expression in the silkworm, *Bombyx mori*. *Biosci. Biotechnol. Biochem.*, *71*, 2808–2814.

Yamanaka, N., Hua, Y. J., Mizoguchi, A., Watanabe, K., Niwa, R., Tanaka, Y., & Kataoka, H. (2005). Identification of a novel prothoracicostatic hormone and its receptor in the silkworm *Bombyx mori*. *J. Biol. Chem.*, *280*, 14684–14690.

Yamanaka, N., Hua, Y. J., Roller, L., Spalovska-Valachova, I., Mizoguchi, A., Kataoka, H., & Tanaka, Y. (2010). Bombyx prothoracicostatic peptides activate the sex peptide receptor to regulate ecdysteroid biosynthesis. *Proc. Natl. Acad. Sci. USA*, *107*, 2060–2065.

Yamanaka, N., Yamamoto, S., Zitnan, D., Watanabe, K., Kawada, T., Satake, H., Kaneko, Y., Hiruma, K., Tanaka, Y., Shinoda, T., & Kataoka, H. (2008). Neuropeptide receptor transcriptome reveals unidentified neuroendocrine pathways. *PLoS ONE*, *3*, e3048.

Yamanaka, N., Zitnan, D., Kim, Y. J., Adams, M. E., Hua, Y. J., Suzuki, Y., Suzuki, M., Suzuki, A., Satake, H., Mizoguchi, A., Asaoka, K., Tanaka, Y., & Kataoka, H. (2006). Regulation of insect steroid hormone biosynthesis by innervating peptidergic neurons. *Proc. Natl. Acad. Sci. USA*, *103*, 8622–8627.

Yin, C. M., & Chippendale, G. M. (1973). Juvenile hormone regulation of the larval diapause of the southwestern corn borer, *Diatraea grandiosella*. *J. Insect Physiol.*, *19*, 22403–22420.

Yokoyama, I., Endo, K., Yamanaka, A., & Kumagai, K. (1996). Species-specificity in the action of big and small prothoracicotropic hormones (PTTHs) of the swallowtail butterflies, *Papilio xuthus*, *P. machaon*, *P. bianor* and *P. helenus*. *Zool. Sci.*, *13*, 449–454.

Yoshida, T., Nagata, S., & Kataoka, H. (2006). Ghitm is an orthologue of the *Bombyx mori* prothoracic gland-derived receptor (Pgdr) that is ubiquitously expressed in mammalian cells and requires an N-terminal signal sequence for expression. *Biochem. Biophys. Res. Commun.*, *341*, 13–18.

Zavodska, R., Sauman, I., & Sehnal, F. (2003). Distribution of PER protein, pigment-dispersing hormone, prothoracicotropic hormone, and eclosion hormone in the cephalic nervous system of insects. *J. Biol. Rhythm*, *18*, 106–122.

Zhang, T. Y., Sun, J. S., Zhang, Q. R., Xu, J., Jiang, R. J., & Xu, W. H. (2004). The diapause hormone-pheromone biosynthesis activating neuropeptide gene of *Helicoverpa armigera* encodes multiple peptides that break, rather than induce, diapause. *J. Insect Physiol.*, *50*, 547–554.

Zheng, W. W., Yang, D. T., Wang, J. X., Song, Q. S., Gilbert, L. I., & Zhao, X. F. (2010). Hsc70 binds to ultraspiracle resulting in the upregulation of 20-hydroxyecdsone-responsive genes in *Helicoverpa armigera*. *Mol. Cell. Endocrinol.*, *315*, 282–291.

Zhou, X., & Riddiford, L. M. (2002). Broad specifies pupal development and mediates the "status quo" action of juvenile hormone on the pupal-adult transformation in *Drosophila* and *Manduca*. *Development*, *129*, 2259–2269.

Zhou, X., Zhou, B., Truman, J. W., & Riddiford, L. M. (2004). Overexpression of broad: A new insight into its role in the *Drosophila* prothoracic gland cells. *J. Exp. Biol.*, *207*, 1151–1161.

Zimowska, G., Handler, A. M., & Cymborowski, B. (1985). Cellular events in the prothoracic glands and ecdysteroid titres during the last larval instar of *Spodoptera littoralis*. *J. Insect Physiol.*, *31*, 331–340.

Zitnan, D., Kingan, T. G., Kramer, S. J., & Beckage, N. E. (1995). Accumulation of neuropeptides in the cerebral neurosecretory system of Manduca sexta larvae parasitized by the braconid wasp Cotesia congregata. *J. Comp. Neurol.*, *356*, 83–100.

Zitnan, D., Sehnal, F., & Bryant, P. J. (1993). Neurons producing specific neuropeptides in the central nervous system of normal and pupariation-delayed *Drosophila*. *Dev. Biol.*, *156*, 117–135.

2 Insulin-Like Peptides: Structure, Signaling, and Function

Yevgeniya Antonova, Anam J. Arik, Wendy Moore
and Michael A. Riehle
University of Arizona, Tucson, AZ, USA
Mark R. Brown
University of Georgia, Athens, GA, USA

© 2012 Elsevier B.V. All Rights Reserved

Summary

Insulin-like peptides are encoded by multiple, distinct genes within each insect species examined to date. Upon secretion, they serve as hormones, neurotransmitters, and growth factors in the post-embryonic life stages of insects and activate the insulin receptor and a series of intracellular proteins, comprising the insulin signaling pathway. This pathway is responsive to nutrient intake and signaling through the "target of rapamycin" pathway, and together regulate growth, development, and reproduction in insects, as best characterized for *Drosophila melanogaster*. For other insects, more is known about the structural conservation of insulin-like peptides and proteins in both pathways than the functional conservation. With the application of genetic techniques and improvements in biochemical and bioassay approaches, our understanding of this fundamental regulatory pathway should greatly advance, and may reveal, intricacies unique to insects.

2.1. Introduction

Insulin-like peptides (ILPs) are paragons for the conservation of hormone structure and function among invertebrates and higher animals. They are encoded by multiple, distinct genes within each species and, upon secretion, serve as hormones, neurotransmitters, and growth factors during post-embryonic life stages. These diverse messages are transduced in target cells through an insulin receptor and a signaling network of activated proteins that directly affect biochemical pathways and gene expression. Of the diverse invertebrate groups, our accumulated knowledge of ILP endocrinology is the deepest and broadest for insects. This knowledge has revealed that essentially all of the proteins that make up the mechanisms for ILP processing, secretion, and signaling are remarkably similar to those of vertebrates, in effect conserving the pleiotropic effects of ILPs between insects and vertebrates. But what still needs to be resolved is the extent to which these mechanisms and functions are conserved in the hemimetabolous/endopterygote and apterygote insects as well as other arthropod and invertebrate groups, beyond the model nematode *Caenorhabditis elegans*, for which ILP signaling is conserved but not considered herein.

DOI: 10.1016/B978-0-12-384749-2.10002-0

In vertebrates, the insulin family includes insulin, insulin-like growth factors (IGFs), and relaxins/ILPs — ten such peptides are known for humans (Shabanpoor *et al.*, 2009). Insulin is a gut hormone, and IGFs and relaxins/ILPs are secreted in the nervous system and other tissues. The multifunctionality of vertebrate ILPs is due to their activation of different receptor types and signaling pathways. At least three different forms of receptor tyrosine kinase are known to form hybrid dimer receptors with different affinities for insulin and IGFs and activate the canonical insulin signaling pathway (ISP; insulin receptor substrate (IRS)/phosphoinositol 3-kinase (PI3 K)/Akt), the mitogen-activated protein kinase (MAPK; extracellular signal-regulated kinase, ERK) pathway, or the target of rapamycin (TOR) pathway (Taniguchi *et al.*, 2006; Belfiore *et al.*, 2009). Relaxins/ILPs bind to two different G-protein-coupled receptors (GPCRs) that activate or inhibit cAMP signaling (Kong *et al.*, 2010). At least one homologue of relaxin/ILP GPCRs exists in the insect species examined, but it is a functional orphan (Van Loy *et al.*, 2008).

In this chapter, we offer a summary of the literature since 2005 related to ILPs and the ISP in diverse insects. Several reviews of ILP endocrinology provide an extensive collation of earlier studies and their significance for insects and other invertebrates (Smit *et al.*, 1998; Claeys *et al.*, 2002; Wu and Brown, 2006; Gronke *et al.*, 2010). Molecular genetic studies of the model insect, *Drosophila melanogaster* (hereafter referred to as *Drosophila*), continue to elucidate the remarkable complexities of the insulin pathway and its interaction with other signaling pathways through its key roles in metabolism, growth, reproduction, and aging. Studies of other insects provide additional evidence for the structural conservation of ILPs and the ISP, as well as correlated changes in the expression of ILP and ISP genes with physiological or developmental processes. Still, little is known about the secretion of ILPs, receptor interactions, and their direct effects on specific processes in the target tissues of insects.

We also review the structural phylogeny of ILPs and the insulin receptor (IR) in insects and other invertebrates. This phylogeny is set in an evolutionary context with other arthropods and mollusks. Multiple ILP genes exist in each of the insect and invertebrate species examined to date, but we need to discern whether they are conserved as structural and even functional subfamilies. The existence of a single IR in most insects examined leads to questions as to whether individual ILPs bind to this receptor or isoforms, activate one or all of the conserved signal pathways, and exhibit overall functional redundancy or specificity. Answers for these questions are worthy goals for future studies of ILP endocrinology in insects that may provide insights into controlling notorious insect pests and vectors of disease pathogens.

2.2. Discovery of Invertebrate Insulin-Like Peptides

In recent years, identification of ILPs in insects and other invertebrates has rapidly advanced due to genomic or EST sequencing of selected species and the subsequent bioinformatic mining of these data. This strategy was first used to uncover genes encoding seven ILPs in *Drosophila* (DILP1-7; Brogiolo *et al.*, 2001), which presumably are ligands for the *Drosophila* IR, first characterized in 1985 (Petruzzelli *et al.*, 1985). Homologues for the DILPs were subsequently identified in the other 11 *Drosophila* species sequenced and are included in our phylogenetic analysis (**Figure 1**) (*Drosophila* 12 Genomes Consortium, 2007; Stark *et al.*, 2007).

Because mosquitoes are important vectors of disease pathogens, the development of genomic databases for representative species was a priority. Five ILPs were first identified in the African malaria mosquito *Anopheles gambiae* (Riehle *et al.*, 2002; Krieger *et al.*, 2004), then, eight ILPs for *Aedes aegypti* (Riehle *et al.*, 2006) and five ILPs for *Culex pipiens* (Sim and Denlinger, 2008, 2009a). With this accumulated sequence information, five ILPs were identified in *An. stephensi* without a genomic database using molecular methods (unpublished data, included in **Figure 1**). These studies also characterized the expression of mosquito ILPs in different life stages and tissues. Recently, this continued effort to sequence the genomes of arthropod vector species enabled identification of three ILPs in the kissing bug *Rhodnius prolixus*, one ILP in the human body louse *Pediculus humanus humanus*, three partial ILP sequences in the deer tick *Ixodes scapularis*, and one partial ILP in the American dog tick *Dermacentor variabilis* (Figure 1).

As the technology for genomics becomes less expensive, databases for more insect and invertebrate species are being made available, based on their evolutionary position or importance as an agricultural pest. The beetle *Tr. castaneum* has four ILP genes (*Tr.* Genome Sequencing Consortium, 2008). The pea aphid *Acyrthosiphon pisum* has 10 ILPs (Huybrechts *et al.*, 2010). In the order Hymenoptera, the honey bee *Apis mellifera* has two ILPs (Wheeler *et al.*, 2006), and the parasitic wasps *Nansonia longicornis* and *Nansonia giraulti* have one ILP identified so far, as does the leaf cutter ant *Atta cephalotes*. Pre-genomic efforts had characterized a significant number of ILP genes in other insect and invertebrate groups. The first confirmation of an ILP in an invertebrate was for the silkworm *Bombyx mori*. At least 38 ILPs have been described in the silkworm (which has some 500 genetic strains) and more are expected. Later, ILPs were identified in other lepidopteran species: *Samia cynthia* with five ILPs (Kimura-Kawakami *et al.*, 1992) and *Agruis convolvuli* with two ILPs (Iwami *et al.*, 1996). More recently, two ILPs were identified in *Spodoptera littoralis* (Van de Velde *et al.*, 2007) and one in *Lonomia oblique*. Single ILPs are known for two orthopteran species, *Locusta migratoria* (Lagueux *et al.*, 1990) and *Schistocerca gregaria* (Badisco *et al.*, 2008). Multiple ILPs are also encoded in species of another group of the superphylum Protostomia: Mollusca: sea hare, *Aplysia californica*; pond snail, *Lymnaea stagnalis*; Pacific oyster, *Crassostrea gigas*; and limpet, *Lottia gigantea* (Smit *et al.*, 1996; Floyd *et al.*, 1999; Hamano *et al.*, 2005; Veenstra, 2010). Multiple ILPs were also encoded in species representing early groups of the superphylum Deuterostomia: sea squirt, *Ciona intestinalis* (Olinski *et al.*, 2006; Kawada *et al.*, 2010) and starfish, *Asterina pectinifera* (Mita *et al.*, 2009). In total, 210 ILPs have been identified from insects, other invertebrates, and the early deuterostomes (**Figure 1**).

Interestingly, several insects with completed genomes seem to lack an IGF orthologue, which is characterized by a short C-peptide with proteolytic processing sites and a C-terminal extension. The first putative IGF-like peptide in insects was described in the American cockroach *Periplaneta americana* where it was shown to be expressed in neuronal cells and axonal fibers of the central nervous system and in the midgut epithelium (Verhaert *et al.*, 1989). Recently, an IGF-like peptide was purified from the hemolymph of *B. mori* larvae (Okamoto *et al.*, 2009a). In Diptera, IGF-like peptides have been identified in the mosquito *Ae. aegypti* and *Drosophila* (Riehle *et al.*, 2006; Okamoto *et al.*, 2009b; Slaidina *et al.*, 2009).

2.3. Phylogenetic Analysis of Invertebrate Insulin-Like Peptides and Insulin Receptors

Although several studies have compared the phylogeny of discrete groups of ILPs, a comprehensive phylogenetic analysis of all known invertebrate ILPs has not been undertaken. Therefore, we analyzed the 210 invertebrate ILPs (minus the *C. elegans* ILPs) that have either been previously published or were identified from recently completed genomic sequences. The neighbor-joining tree recovered from analysis of the ILPs is presented in **Figure 1**. Due to the high degree of sequence variably and the short length of the ILPs, most well-supported clades contain ILPs isolated from close relatives. There are very few well-supported clades in this analysis that contain ILPs isolated from species classified in more than one genus, and all of them include two or more genera of mosquitoes. One such clade, marked with a vertical line in **Figure 1**, contains mosquitoes (ILP5), *Drosophila* (ILP7), and two molluscs, *L. gigantea* (ILP4) and *A. californica* (ILP1), indicating that an ancestral copy of this ILP was present in the common protostomian ancestor of Arthropoda and Mollusca. Interestingly, all of the ILPs in this group possess an additional amino acid between the second and third conserved cysteine residues and this cluster represents the oldest ILP variant. One member of this group, DILP7, is a putative relaxin homologue (see the next paragraph).

In contrast to the short and highly variable ILP sequences, the 48 invertebrate IRs currently identified are well conserved. The maximum likelihood tree for the IR sequences is shown in **Figure 2** and largely agrees with expected phylogenies based on other genetic and morphological studies. That the IR is prone to undergo duplication events is evinced by the fact that there are four copies known from the water flea *Daphnia pulex* and two copies each from the trematode, *Shistosoma japonicum*; flour beetle, *Tr. castaneum*; and honeybee *Ap. mellifera*. Based on our phylogenetic analysis it appears that the IR may have also duplicated early in the evolution of insects (see arrow in **Figure 2**) and that today different copies are present in different species within the Diptera and Hymenoptera, and both copies are present within the only coleopteran, *Tr. castaneum*, included in our analysis.

One surprising result is the seemingly spurious placement of *Aplysia* (Mollusca) within a clade of closely related *Drosophila* sequences. The placement is so remarkable that one might suspect contamination. However, while the protein sequence of *Aplysia* is similar to the *Drosophila* sequences, it is also unique. Therefore, if it is a contaminant it represents a *Drosophila* species that has yet to have an IR identified. If it is not a contaminant, then the striking similarity between the *Drosophila* and *Aplysia* sequences suggests that this copy of the IR has changed very little since the most recent common ancestor of arthropods and mollusks. However, this scenario is unlikely given that our phylogenetic analysis recovers the tree topology within Diptera that one would predict based upon other data, including both morphology- and molecular-based phylogenetic studies. Also unusual is the placement of one of the honeybee IRs within the non-insect outgroup taxa. However, this second honeybee IR is extremely divergent from all other known IRs, and it represents the longest branch in the tree. Therefore its placement should be treated with some skepticism.

2.4. Expression, Processing, and Secretion of ILPs

Approximately 40 genes encoding a great variety of peptide hormones, including the ILPs, are expressed in specific cells in both the nervous system and midgut of insects (Veenstra *et al.*, 2008; Veenstra, 2009). It is interesting to note that the ILPs are the only peptide hormones to be encoded as multiple genes, whereas all others are proteolytically cleaved and processed from a single propeptide into different forms. Past and current studies of ILP expression in insects have focused on the nervous system because ILPs were first isolated from brains or

were localized by immunocytochemistry in specific clusters of brain neurosecretory cells (Wu and Brown, 2006). Another likely source is the midgut because it contains hundreds to thousands of isolated endocrine cells, yet only a few earlier studies detected ILP expression in this tissue (Wu and Brown, 2006). In the nematode *C. elegans*, specific cells in the intestine/gut are an ILP source (INS-7), thus making this important metabolic tissue a center for insulin signaling (Murphy *et al.*, 2007). As peptide hormones, ILPs can be stored or secreted from a cellular source in a bioactive form, but there is no evidence for insect ILPs that transcript levels in cells correlate with release or storage (Geminard *et al.*, 2009), although this is generally assumed. Release of ILPs into the hemolymph for transport to target cells or tissues likely rises and falls in response to internal and external cues, and ILPs are likely degraded by proteases in the hemolymph or after binding to the IR and being internalized.

Many studies have focused on the expression of ILPs in the nervous system of *Drosophila* to confirm the effects of targeted genetic manipulations, but other tissues or cells are likely ILP sources. Two recent studies provided a comprehensive survey of the expression of *DILP* and other peptide hormone genes, as well as their known receptor genes, in the adult nervous system and all regions of both the larval and adult gut in *Drosophila* (Veenstra *et al.*, 2008; Veenstra, 2009). Insulin receptor transcription was evident in all tissues sampled and showed little variation, in contrast to the more specific expression of peptide GPCRs. In general, *DILP2*, *DILP3*, and *DILP5* expression was highest in the adult brain and *DILP6* and *DILP7* in the thoracic/abdominal ganglia. Only *DILP3* expression was detected in the larval and adult gut, and its expression was localized to muscle bands in the midgut of *DILP3* promoter-gal4/UAS-LacZ transgenic flies, suggesting a novel source for an ILP. Previous studies detected *DILP2*, *DILP4*, *DILP5*, and *DILP6* transcripts in the embryonic or larval gut (Brogiolo *et al.*, 2001) and *DILP5* transcripts in the follicle cells of ovaries (Ikeya *et al.*, 2002).

Two similar studies provided transcript expression profiles (whole body) for all seven *DILP* genes that spanned embryonic to pupal development and included data for males and females (Okamoto *et al.*, 2009b; Slaidina *et al.*, 2009). Further characterization of the *DILP* genes exhibiting the highest expression in the pupal stage resulted in the localization of *DILP6* expression to the fat body, suggesting yet another ILP source.

Most studies of ILP expression and function in *Drosophila* have focused on *DILP2*, because it is highly expressed in bilaterally paired clusters of neurosecretory cells (~14) in the medial region of the brain (MNC, melian neurosecretory cells) from the larval to adult stage. In all insects, the MNC are identically located in the pars intercerebralis region of the brain (de Velasco *et al.*, 2007) and are presumed to be the primary source of circulating ILPs. Bundled axons from these cells extend out of the brain to neurohemal release sites associated with the aorta and corpora cardiaca (CC) and along the gut (Wu and Brown, 2006; Geminard *et al.*, 2009). *DILP1, 3, 4,* and *5* are co-expressed with *DILP2* in the MNC, likely reflecting their clustering within a 26 kb section of chromosome III. Expression of the individual *DILP* genes varies in the MNC during development and is compensatory in response to different cues and *DILP* gene deletions, suggesting different promoter/enhancer elements (Gronke *et al.*, 2010; Broughton *et al.*, 2008; Zhang *et al.*, 2009). In a series of elegant *in vivo* and *in vitro* studies Geminard *et al.* (2009) showed that DILP2 and DILP5 accumulate in the MNC of nutrient-deprived or starved larvae, and that secretion of the DILPs was enhanced by feeding yeast or specific amino acids (L-leucine and L-isoleucine), targeted depolarization, and a fat-body derived specific factor. In such conditions, DILP secretion from the MNC was mostly complete after 30 min, as quantified by immunofluorescence microscopy, and notably, there was no increase in *DILP* expression. Other studies have implicated serotonin (Kaplan *et al.*, 2008) and short neuropeptide F (Lee *et al.*, 2008; Lee *et al.*, 2009) in the regulation

Figure 1 Phylogenitic analysis of invertebrate insulin-like peptides. 210 insulin-like peptides were aligned using Opal (Wheeler and Kececioglu 2007) as implemented in Mesquite 2.74 (Maddison and Maddison 2010). Insect ILPs are color coded according to order while other invertebrate ILPs are in black. The DILP7/mosquito ILP5 cluster is indicated by a solid line on the right. Clades containing similar ILPs from various *Drosophila* species were collapsed into a single branch. Numbers following the combined *Drosophila* clades indicate the represented species as follows: 1. *D. ananassae*, 2. *D. erecta*, 3. *D. grimshawi*, 4. *D. melanogaster*, 5. *D. mojavensis*, 6. *D. persimilis*, 7. *D. pseudoobscura*, 8. *D. sechellia*, 9. *D. simulans*, 10. *D. virilis*, 11. *D. willistoni*, and 12. *D. yakuba*. Alignments of the ILP sequences were subsequently modified by hand using knowledge of conserved cysteine patterns within the A-chain and B-chain. Extreme variation in sequence length and high rates of evolution made alignment outside of the A-chain and B-chain impossible. Therefore, these regions were excluded from our phylogenetic analysis. Models of protein evolution were selected with the aid of ProtTest (Drummond & Strimmer 2001; Guindon and Gascuel 2003; Abascal et al 2005). The WAG + I+ Γ model of protein evolution best fit the IR alignment by both the Akaike Information Criterion (AIC) and the Bayesian Information Criterion (BIC). Searches for trees of highest likelihood were performed on the phylocluster at the California Academy of Sciences (San Francisco, CA) using RAxML v 7.0.0 (Stamatakis 2006, Stamatakis et al. 2008) and the WAG +I+Γ model of evolution with 500 rapid bootstrap analyses. Insulin-like peptide sequences were submitted to neighbor-joining analyses in PAUP*4.0 (Swofford, 2002).

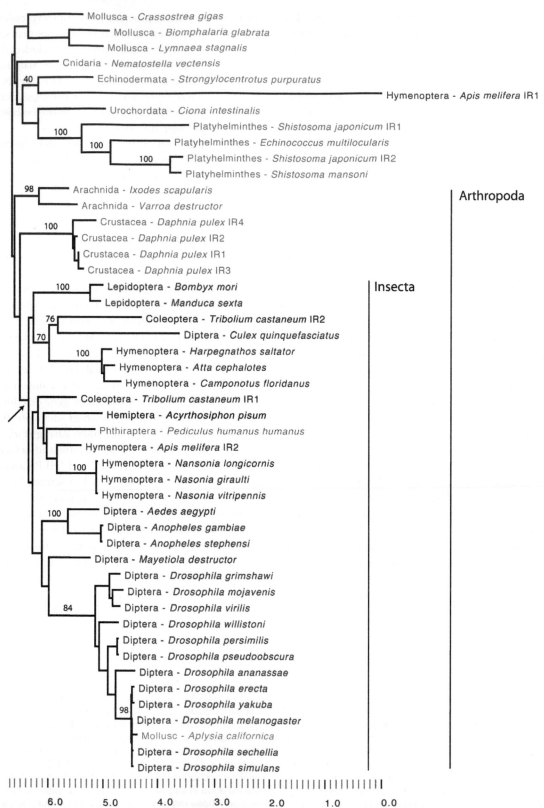

Figure 2 Phylogenetic analysis of invertebrate insulin receptors. Forty-eight insulin receptor (IR) sequences were aligned using Opal (*Wheeler and Kececioglu*, 2007) as implemented in Mesquite 2.74 (*Maddison and Maddison*, 2010). Insect IRs are color coded according to order while other invertebrate IRs are in gray. Unalignable sites were excluded from phylogenetic analysis. The model of protein evolution that best fit the data was selected with the aid of ProtTest. (*Drummond and Strimmer* 2001, *Guindon and Gascuel* 2003 and *Abascal et al.* 2005). The WAG + I+ Γ model of protein evolution best fit the alignment of IRs as selected by the Akaike Information Criterion (AIC). Searches for the tree of highest likelihood were performed using RAxML v 7.2.6. Reproduced with permission from *Stamatakis et al.* (2008). Rapid bootstrap analysis was also performed in RAxML using 500 random starting trees.

of DILP secretion from the MNC based on the proximity of stained axons/neurons and genetic manipulations.

The MNC originate from a single pair of neuroblasts that divide and differentiate into *DILP*-expressing cells at the end of embryogenesis (Wang *et al.*, 2007). The secretory capacity of these cells is maintained by the expression of *dimmed*, which encodes a basic helix-loop-helix protein required for neuroendocrine cell differentiation. It also regulates expression of another gene encoding a peptidylglycine α-hydroxylating monooxygenase, suggesting amidated peptides may also be secreted by the MNC (Hewes *et al.*, 2006). The site of MNC origin is adjacent to other neuroblasts that separate and differentiate into the CC cells that are part of the ring gland and produce adipokinetic hormone (Wang *et al.*, 2007). Both sets of neuroblasts originate in a domain of the neuroectoderm of *Drosophila* embryos that shares homology with the hypothalamic-pituitary axis in the vertebrate brain, based on marker gene expression (de Velasco *et al.*, 2007; Wang *et al.*, 2007). Studies have confirmed that insulin is not produced in the vertebrate brain; however, IGF and relaxins are expressed and functional in the brain (Callander and Bathgate, 2010; Torres-Aleman, 2010). In vertebrates, insulin production is shifted to endocrine cells that differentiate in the gut. These cells form islets in the mammalian pancreas with other peptide-secreting cells. The organization and function of these cells is regulated by the transcription factor PAX6. Similarly, expression of its homologue gene, *eyeless*, in *Drosophila* is required for DILP MNC differentiation (Clements *et al.*, 2008). This provides a striking example of functional conservation across animals and their tissues.

DILP6, the IGF homologue, and DILP7 are structurally different than the other DILPs, as is their tissue expression pattern. *DILP6* expression is predominantly localized to the fat body (Okamoto *et al.*, 2009b; Slaidina *et al.*, 2009). This tissue is the functional equivalent of the vertebrate liver and fat tissue adipocytes, which secrete IGF. *DILP6* expression surges late in the last larval instar and falls prior to adult emergence. DILP7 is a putative relaxin-like peptide that is specifically localized to eight pairs of cells in the thoracic/abdominal ganglia (Miguel-Aliaga *et al.*, 2008; Yang *et al.*, 2008). Axons from these cells project to the *DILP* MNC and out of the ganglia to the midgut, hindgut, and female reproductive tract. During embryogenesis, these neurons first set their axonal connections and then begin to express *DILP7* in response to bone morphogenetic protein signaling as coded through four transcription factors: *Abdominal-B*, *Hb9*, *Fork Head*, and *dimmed* (Miguel-Aliaga *et al.*, 2008). Based on its neuronal distribution, DILP7 may function more as a neurotransmitter than as a peptide hormone.

Other recent studies have characterized ILP expression in the nervous system and other tissues of insects. In mosquitoes, seven of the eight *AeaILP* genes were expressed in the brain (head) across all life stages of *Ae. aegypti* (Riehle *et al.*, 2006), as were four of the five *AngILP* genes in *An. gambiae* (Krieger *et al.*, 2004; Arsic and Guerin, 2008). *AeaILP5* and *AngILP5* encoding DILP7 orthologues were not detected in brains, but their transcripts, along with those of *AeaILP2*, *AngILP2*, and *AeaILP6* (IGF homologue), were detected in other body regions including the fat body and thoracic/abdominal ganglia, and in specific female tissues. The midgut was one such tissue, and endocrine cells are a possible source, given that a subset was immunostained with an ILP antibody. A subsequent study showed that AeaILP3 was localized in the MNC of females and not in other cells in the brain or midgut (Brown *et al.*, 2008). Brain MNC are the likely sources of two ILPs structurally characterized for the cotton leafworm, *Sp. littoralis* (Van de Velde *et al.*, 2007) and an ILP isolated and characterized for the desert locust, *Sc. gregaria* (Badisco *et al.*, 2008). Transcript expression of *ScgILP* was highest in the brain and fat body of male and female locusts. The two ILP genes in *Ap. mellifera*, are transcribed in the brains (heads) of different castes (Corona *et al.*, 2007; Ament *et al.*, 2008), but it is not known whether other tissues were examined.

The most significant recent advance is the characterization of an IGF-like peptide in *B. mori* (*Bommo*-IGFLP, BIGFLP), which was first detected in the hemolymph during the pupal stage (Okamoto *et al.*, 2009a). The fat body is the source of this 8 kDa peptide that has B, C, and A domains, and although the C domain has apparent protease processing sites, it is secreted as a single unprocessed chain similar to IGF. Its hemolymph titer rises early in the pupal stage to 800 nM in females and 200 nM in males. BIGFLP is structurally similar to AeaILP6 and DILP6, which is likely secreted by the fat body given the expression pattern of the *DILP6* gene (see above).

Expression of ILP genes in insects generally appears to be cell- and tissue-specific, but the dynamics of ILP processing, storage, and release remain uncharacterized with few exceptions. It is also possible that processed and unprocessed ILPs may be secreted by undifferentiated cells or tissues during embryogenesis and metamorphosis, as reported for proinsulin during vertebrate development (Hernandez-Sanchez *et al.*, 2006). Surprisingly, proteomic studies of insect brains or nervous system have not identified ILP fragments, whereas many other peptide hormones were found. Whether or not this is due to low quantities of ILPs or insufficient extraction from tissues is not known. The recent structural characterization of *Sc. gregaria* ILP and BIGFLP isolated from the brain or hemolymph provides insight into ILP processing and the nature of bioactive forms, as had earlier characterizations of locust and lepidopteran ILPs (Wu and Brown, 2006).

The bioactivity of chemically synthesized ILPs has been characterized for two lepidopteran species, *B. mori* and *Sa. cynthia*, as well as the mosquito *Ae. aegypti*. This achievement required complex synthesis of predicted bioactive forms, purification to homogeneity, and demonstration of dose-responsive bioactivity and receptor binding for a *B. mori* ILP (bombyxin II), two *Sa. cynthia* ILPs (ILP-A1, ILP-B1), and two *Ae. aegypti* ILPs (ILP3 and ILP4) (Brown *et al.*, 2008; Fullbright and Bullesbach, 2000; Nagata, 2010; Wen *et al.*, 2010). Recently a recombinant form of DILP5 was shown to competitively displace radiolabeled insulin bound to human and *Drosophila* cells (Sajid *et al.*, 2011). Chemical synthesis of ILPs also provides sufficient pure peptide for other types of structural studies (Nagata, 2010).

Because ILPs are destined for secretion, they are presumably packed in secretory granules with their processing enzymes, as in islet β-cells in the vertebrate pancreas (Suckale and Solimena, 2010). The C peptide of vertebrate proinsulin is cleaved by two prohormone convertases and a carboxypeptidase, after disulfide linkage of B and A peptides (Steiner *et al.*, 2009). These same enzymes process pro-forms of numerous neuropeptides in vertebrates, and their homologues likely have similar functions in insects. In *Drosophila*, the gene encoding a prohormone convertase 2 (PC2) homologue (*amontillado*) is expressed in the nervous system and gut (Rayburn *et al.*, 2009), and co-expression of a helper protein, 7B2, is likely required for activity (Hwang *et al.*, 2000). Genetic manipulations show that *PC2* expression is required for normal *Drosophila* development, in particular for adipokinetic hormone (AKH) function and processing in the CC (Rhea *et al.*, 2010). This same study claimed that artificial expression of the AKH gene in the fat body of *Drosophila* with blocked *PC2* expression in the CC restored AKH function. Other studies reported on the additive or restorative effects of ubiquitous *DILP* and even insulin gene expression in *Drosophila* of different backgrounds (Tsuda *et al.*, 2010). No evidence, however, is ever offered for *PC2* expression or proper processing of AKH or ILP in these tissues. Nevertheless, such results suggest that enzymatic processing of secreted, bioactive peptides, such as ILPs, may be a common feature of insect tissues.

The C peptide and extended B peptides cleaved from ILPs are released along with the processed ILP and are known to be functional in locusts (Bermudez *et al.*, 1991; Clynen *et al.*, 2003), as is the C peptide in mammals (Hills and Brunskill, 2009). With the exception of the IGF homologues, all insect ILPs have C peptides and extended B chains that typically are longer than vertebrate ILPs, suggesting an even greater range of ILP functionality. C-peptide-specific immunoassays are a well-accepted method for determining insulin plasma concentration. Such assays may be suitable for characterizing ILP hemolymph titer in insects, as only profiled for *B. mori* during post-embryonic development (Saegusa *et al.*, 1992).

After secretion into the hemolymph, ILPs may be transported to target cells by binding proteins (BP) that facilitate ILP action, as known for vertebrate IGF. In vertebrates, IGFBPs protect and store circulating IGFs and both directly and indirectly activate IGF receptor signaling and other intracellular pathways (Mohan and Baylink, 2002). One group of IGFBPs binds IGFs with high affinity, and a group of related proteins (IGFBP-rPs) with limited sequence similarity binds IGFs with low affinity. Putative IGFBP-rPs have been characterized for the fall armyworm *Spodotera frugiperda*, *Drosophila*, and the tick *Amblyomma americanum*. The first insect IGFBP-rP to be characterized was a 27 kDa protein secreted by *Sp. fugiperda* cells (Sf9 cells) that bound human insulin and IGFs in culture (Andersen *et al.*, 2000). The expression and functional significance of this protein in Lepidoptera is not known, but studies of the *Drosophila* homologue, imaginal morphogenesis protein-late 2 (Imp-L2), suggest a role in ILP signaling. Imp-L2 is expressed in brain DILP MNCs and the CC, and it binds unprocessed, tagged DILP2 in solution (Honegger *et al.*, 2008) or a recombinant form of DILP5 (Sajid *et al.*, 2011). Loss-of-function mutations showed that *Imp-L2* expression is not required for normal development, but it is essential for larval survival during starvation (Honegger *et al.*, 2008). Imp-L2 protein also binds to an orthologue of the vertebrate acid labile subunit (ALS) as well as DILP2 in extracts of transfected *Drosophila* S2 cells (Arquier *et al.*, 2008). In vertebrates, secreted ALS stabilizes IGFBP/IGF complex in circulation. *Drosophila* ALS is expressed in the brain DILP NSCs and in the fat body. Overexpression of tagged ALS and DILP2 in larval fat body enabled detection of a putative complex in hemolymph, whereas reduced *ALS* expression limits carbohydrate and lipid mobilization in fed larvae (Arquier *et al.*, 2008). In *Am. americanum*, expression of *IGFBP-rP* genes in different tissues was responsive to blood feeding (Mulenga and Khumthong, 2010). Silencing the genes by RNA interference (RNAi) resulted in ticks taking smaller blood meals and increased mortality during feeding. Other genes encoding putative IGFBP-rPs were identified in a mosquito baculovirus (Afonso *et al.*, 2001) or found to be expressed in the venom gland of the wasp *Eumenes pomiformis* (Baek and Lee, 2010), thus suggesting other levels of ILP interaction in insects.

Neuroparsin was first identified as a neuropeptide in locusts, and is now regarded as a putative IGFBP-rP based on limited sequence similarity and its binding of the *Sc. gregaria* ILP in solution (Badisco *et al.*, 2007, 2008). Neuroparsin homologues have been identified in all insect and arthropod groups examined, but not in the *melanogaster* subgroup of *Drosophila* (Veenstra, 2010). Many functions have been reported for neuroparsin in locusts, and the expression of alternatively spliced transcripts in different tissues (though no peptides are characterized) complicates assignment of a functional role (Veenstra, 2010; Badisco *et al.*, 2007). Neuroparsin and ILPs are secreted by different, but neighboring, brain neurosecretory cells in locusts, other *Drosophila spp.*, and other insects (Veenstra, 2010). Ovary ecdysteroidogenic hormone (OEH) is the neuroparsin homologue in the mosquito *Ae. aegypti* (Brown *et al.*, 1998). This neurohormone is secreted by brain NSCs, and has the same but independent gonadotropic activity, both *in vivo* and *in vitro*, as endogenous ILPs (Wen *et al.*, 2010; Brown *et al.*, 1998). This seemingly functional redundancy has yet to be explained and awaits characterization of the neuroparsin/OEH signaling pathway or other mechanisms that may intersect with ILP signaling.

Circulating ILPs likely are degraded by proteases, but once the ILPs are bound by the IR, this complex may be internalized in target cells where the ILP is degraded and the IR returned to the surface. An insulin degrading enzyme (IDE; a zinc metalloendopeptidase) is highly conserved from yeast to mammals, and a *Drosophila* homologue (110 kDa) is expressed in all life stages and cell lines (Stoppelli *et al.*, 1988). *Drosophila* IDE binds and cleaves mammalian insulin under the same conditions as human IDE (Garcia *et al.*, 1988). Interestingly, homozygous *IDE* knockout flies were viable and fertile with increased trehalose levels (Tsuda *et al.*, 2010), which suggests that DILP regulation of metabolism was specifically affected. Degradation of native ILPs has yet to be demonstrated for an insect IDE, and the enzyme may favor other peptide hormones or growth factors, as reported for human IDE (Fernandez-Gamba *et al.*, 2009). Internalization of ILPs may facilitate their interaction with cytoplasmic and nuclear proteins and activation of signaling pathways, as characterized for insulin, IGFs, and IGFBPs in mammalian cells (Mohan and Baylink, 2002; Harada *et al.*, 1999).

2.5. Insulin Signaling Pathway and Interactions with Other Pathways

The existence of multiple ILPs in an insect or invertebrate species is now well established, but their binding to a single encoded IR has yet to be demonstrated, with the exception of *Ae. aegypti* ILP3 (Brown *et al.*, 1998, 2008; Wen *et al.*, 2010). Insect ILPs are known to activate only the ISP and TOR pathway. Their activation of the MAPK pathway has yet to be established, although the existence of IGF homologues in insects does suggest this is a possibility. There is also no evidence for insect ILP binding to relaxin GPCR homologues in insects. Thus, these two pathways will not be covered in depth in this chapter.

Mining the genomes of particular insects has revealed all components of the mammalian ISP, even though it is more complex due to the variety of protein isoforms (Wu and Brown, 2006; Taniguchi *et al.*, 2006; Okada *et al.*, 2010). The general scenario in mammals and invertebrates unfolds as follows (**Figure 3**). Signaling is initiated by ILP binding to the α subunit of the IR that causes a conformational change leading to autophosphorylation of tyrosine residues on the cytosolic side of the β subunit.

The activated IR phosphorylates an adaptor scaffold protein, IRS, providing docking sites for downstream signaling proteins possessing src-homology-2 (SH2) domains (Taniguchi *et al.*, 2006). Two key SH2-domain-containing proteins, PI3 K and the adaptor protein Shc, interact with IRS and represent a major branching point in the ISP cascade leading to PI3 K/Akt or MAPK signaling (Taniguchi *et al.*, 2006). The *Drosophila* IR possesses 200–400 amino acid extensions in the N- and C-termini, including 3 YXXM motifs in the C-terminus extension, suggesting a functional IRS-like domain (Teleman, 2009). Other IRs in *Drosophila spp.* also possess this C-terminal extension. The presence of these motifs allows the *Drosophila* IR to bind directly to PI3 K and bypass IRS (Ruan *et al.*, 1995; Yenush *et al.*, 1996). Two proteins have been shown to serve as IRS in *Drosophila* and bind to IR: Chico, an orthologue of mammalian IRS1, and Lnk, an orthologue of the SH2B adaptor protein 1 (Bohni *et al.*, 1999; Werz *et al.*, 2009). Both *chico* and *Lnk* mutants have reduced body size and exhibit elevated lipids levels in hemolymph. Flies with a double mutation in both *chico* and *Lnk* are not viable, suggesting that these two proteins are key components of the ISP in *Drosophila* (Teleman, 2009; Werz *et al.*, 2009).

2.5.1. Canonical Insulin Signaling Pathway

Signaling through PI3 K/Akt branch is initiated when the 85 kDa regulatory subunit of PI3 K (p85) (also known as p60 in *Drosophila*) binds to tyrosine phosphorylated IRS through its SH2 domain. Upon binding, the 110 kDa catalytic subunit of PI3 K (p110) becomes activated and phosphorylates membrane-anchored phosphatidylinositol bisphosphate (PIP_2) converting it into phosphatidylinositol trisphosphate (PIP_3) (Taniguchi *et al.*, 2006; Leevers *et al.*, 1996). Proteins containing pleckstrin-homology (PH) domains can then bind to PIP_3 resulting in membrane localization and interactions with other membrane associated proteins. Two major components of ISP, 3-phosphoinositide-dependent protein kinase-1 (PDK-1) and Akt (also known as protein kinase B), are among PH-domain-containing proteins binding to PIP_3. As a result of this binding PDK-1 phosphorylates and activates Akt at a threonine residue in the activation center loop (Thr308 in humans and Thr342 in *Drosophila*; Taniguchi *et al.*, 2006). Full activation of Akt also requires phosphorylation of a serine residue in the hydrophobic motif of the C-terminus (Ser473 in humans and Ser505 in *Drosophila*). In both mammals and *Drosophila*, rapamycin-insensitive companion of TOR kinase (Rictor) in the TOR-complex 2 (TOR-C2) phosphorylates Akt at this residue (Hietakangas and Cohen, 2007) and can be used as a read-out of TOR-C2 activity (Yang *et al.*, 2006). In addition to Rictor, over 90 kinases can potentially phosphorylate Akt at Ser473 (Chua *et al.*, 2009), including DNA-protein kinase, some protein kinase C molecules,

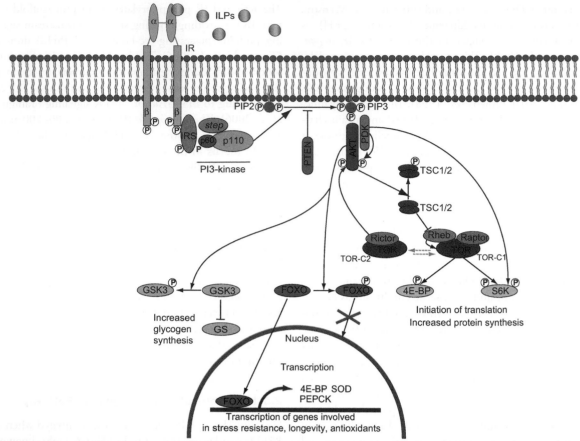

Figure 3 Insulin signaling pathway and its interaction with the TOR pathway. Generalized interactions among the intracellular components of the insulin signaling pathway in insects. Abbreviations: ILPs, insulin-like peptides; IR, IR; α subunit, β subunit; IRS, insulin receptor substrate; step, steppke guanine nucleotide exchange factor; PI3 K, phosphatidylinositol 3-kinase with p85 adaptor subunit and p110 catalytic subunit; PIP2, phosphatidylinositol-4,5- biphosphate; PIP3, phosphatidylinositol-3,4,5-triphosphate; Akt, Akt or protein kinase B; PDK1, phosphatidylinositol-dependent kinase 1; PTEN, phosphatase and tensin homologue; TSC, tuberous sclerosis complex; Rictor, rapamycin-insensitive companion of TOR kinase; Rheb, Ras homologue enriched in brain; Raptor, rapamycin-sensitive companion of TOR kinase; FOXO, Forkhead box-containing protein O subfamily; S6 K, p70 ribosomal S6 kinase; 4E-BP, translational factor 4e binding protein; SOD, superoxide dismutase; PEPCK, phosphoenolpyruvate carboxykinase; GSK3, glycogen synthase kinase 3; and GS, glycogen synthase.

and choline kinase (Chua *et al.*, 2009; Sarbassov *et al.*, 2004, 2005). In addition to phosphorylating Akt, PDK1 also phosphorylates p70 S6 ribosomal kinase (S6K) in the activation loop at Thr229 in mammals (Dennis *et al.*, 1996) and at Thr238 in *Drosophila* (Stewart *et al.*, 1996; Rintelen *et al.*, 2001; Barcelo and Stewart, 2002). In *Drosophila*, PI3 K/PDK1/Akt signaling is required for cell growth and survival. Overexpression of these ISP components leads to tissue overgrowth, whereas loss-of-function mutations lead to reduced tissue growth and lethality (Leevers *et al.*, 1996; Rintelen *et al.*, 2001; Verdu *et al.*, 1999; Weinkove *et al.*, 1999). In total, the structural conservation of Akt and other proteins in the ISP reflects their functional interactions through scaffolding and phosphorylation for insects, and likely other invertebrates.

Another signaling protein, steppke (step) was recently identified in *Drosophila*. Step encodes a member of cytohesin family of guanine nucleotide exchange factors (GEFs),

which mediate the conversion of GDP to GTP (Fuss *et al.*, 2006). The Step family consists of four proteins in mammals and one protein in *Drosophila* and mosquitoes (Fuss *et al.*, 2006). Step has two motifs characteristic of cytohesins: a Sec7 domain responsible for GTPase activity and a PH domain. Step, which is upstream of PI3 K, is required for proper activation of Akt and is necessary for insulin signaling. However, the exact mechanism of step activity requires further investigation (Fuss *et al.*, 2006).

Akt is a nexus for multiple signaling pathways involved in protein synthesis, cell survival and proliferation, glucose metabolism, neuroendocrine signaling, and stress response (Taniguchi *et al.*, 2006). It phosphorylates multiple downstream signaling molecules, such as glycogen synthase kinase-3 (GSK-3) and the *Drosophila* orthologue shaggy (Sgg), which regulates glycogen synthesis (Frame *et al.*, 2001), and As160, a Rab-GTPase-activating protein (Rab-GAP) involved in cytoskeletal reorganization and translocation of the glucose transporter GLUT4 to plasma membrane (Sano *et al.*, 2003). An additional target of Akt is tuberous sclerosis complex (TSC1/2), which is a negative regulator of the TOR kinase pathway involved in the cell growth, protein metabolism, and reproduction (Harris and Lawrence, 2003). Phosphorylation and inactivation of TSC1/2 by Akt leads to increased TOR signaling, translation, and protein synthesis (Taniguchi *et al.*, 2006). However, the best studied targets of Akt are the FOX transcription factors. Mammals and insects have multiple FOX proteins with more than 43 FOX genes present in the human genome (Katoh and Katoh, 2004), 16 in *Drosophila* (Junger *et al.*, 2003), and 18 in the mosquito *Ae. aegypti* (Hansen *et al.*, 2007). FOX factors were categorized into 19 groups from A to S (FOXA to FOXS) based on the amino acid sequence of their conserved DNA-binding domain (Kaestner *et al.*, 2000; Greer and Brunet, 2005). Akt phosphorylates and inhibits FOXO and FOXA1 by preventing their entry into the nucleus.

FOX factors of group O, FOXO, regulate cell fate, cell survival, cell differentiation, detoxification, and metabolism through the transcription of numerous genes (Lam *et al.*, 2006). In *Drosophila* cells, up to 2000 genes are regulated by FOXO (Teleman *et al.*, 2008). The best studied gene targets of FOXO are involved in cell cycle regulation (cyclins), reactive oxygen species (ROS) detoxification, apoptosis (Fas ligand, Bcl-6; Bcl-2), and metabolism (glucose-6-phosphotase (G6Pase) and phosphoenolpyruvate carboxykinase (PEPCK), as reviewed in Glauser and Schlegel (2007). In insects, FOXO factors play a role in crucial physiological processes, such as longevity and reproduction (Hwangbo *et al.*, 2004). In contrast to humans, the *Drosophila* genome contains only a single FOXO gene, an orthologue of FOXO3a (Junger *et al.*, 2003). *Drosophila* FOXO controls gene expression in the brain and fat body, and regulates both fertility and life span (Hwangbo *et al.*, 2004). *Drosophila* FOXO also regulates cAMP signaling by directly inducing expression of an adenylate cyclase gene in the corpora allata (CA), thereby increasing starvation resistance and limiting organismal growth. Thus, FOXO integrates cAMP signaling and the ISP to adapt organismal growth to the existing nutritional conditions (Mattila *et al.*, 2009). In the mosquitoes *Ae. aegypti* and *An. gambiae*, forkhead transcription factors are important regulators of reproduction in the fat body (Hansen *et al.*, 2007). In addition to FOXO, an orthologue of human FOXO3a, six additional forkhead transcription factors (FOXF, K1, K2, L, N1, and N2) are expressed within the abdominal fat body of *Ae. aegypti* mosquitoes where yolk synthesis occurs. Four of the seven FOX factors (FOX L, N1, N2, and O) regulate vitellogenin (VG) expression and yolk deposition (Hansen *et al.*, 2007).

Since the ISP is critical for so many physiological processes, it is tightly regulated. Negative regulation of the ISP is achieved by a number of mechanisms that lead to the inactivation of the IR, IRS, PI3 K, and Akt (reviewed in Taniguchi *et al.*, 2006). The IR is negatively regulated through direct dephosphorylation of tyrosine residues on the β subunit by protein tyrosine phosphatases (PTP) and PTPB1, which directly dephosphorylates tyrosine resides on the IR (Elchebly *et al.*, 1999). Other mechanisms negatively regulating the IR are serine phosphorylation and ligand-induced downregulation (Taniguchi *et al.*, 2006). Proteins involved in the negative regulation of the IR include suppressors of cytokine signaling-1 and -3, growth-factor-receptor bound protein 10 (Grb10), and plasma-cell-membrane glycoprotein-1 (PC1) (Ueki *et al.*, 2004). These proteins sterically block the interaction between the IR with IRS, modifying the kinase activity of the IR (Ueki *et al.*, 2004). Finally, the IR protein is also downregulated by ligand-mediated internalization and degradation (Taniguchi *et al.*, 2006). In both *Drosophila* and mammals the activity of the IR can also be regulated at the transcriptional level by FOXO (Puig and Tjian, 2005). IRS is negatively regulated by serine phosphorylation by multiple kinases activated by ISP including: MAPK (ERK) (Bouzakri *et al.*, 2003), c-Jun N-terminal kinases (JNK; Miller *et al.*, 1996), and p70 S6 ribosomal kinase (S6 K) (Harrington *et al.*, 2004). It is possible that phosphorylation of IRS by S6 kinase is conserved in *Drosophila*, since RNAi knockdown of S6 K in *Drosophila* S2 cells results in increased phosphorylation of Akt (Yang *et al.*, 2006). In addition to serine phosphorylation, IRS activity is regulated by decreased expression via the PI3 K/Akt pathway (Hirashima *et al.*, 2003) and increased ubiquitin-mediated degradation (Rui *et al.*, 2002). In both mammals and insects the activity of PI3 K is negatively regulated by phosphatase and tensin homologue (PTEN) and SH2-containing inositol 5′-phosphatase-2 (SHIP2), where PTEN dephosphorylates and inactivates PIP_3 at the 3′ position and SHIP at the 5′ position (Maehama and Dixon, 1998; Goberdhan *et al.*, 1999; Vazquez *et al.*, 2001). Akt is inhibited by protein-phosphatase-2A (PP2A) in *Drosophila* through its subunit, Wdb (widerborst) (Vereshchagina *et al.*, 2008), and PH-domain leucine-rich repeat protein phosphatase (PHLPP) that dephosphorylate Akt at Thr308 (PP2A) and Ser473 (both PP2A and PHLPP). Tribbles-3 (TRB3) is also capable of binding to unphosphorylated Akt, preventing its phosphorylation and activation (Du *et al.*, 2003).

2.5.2. MAPK Signaling

Signaling through the MAPK branch is activated primarily by IGF binding to IGF receptors, and the pathway is considered to be a branch of the ISP in vertebrates. This pathway regulates crucial cellular processes in mammals, such as cell proliferation, differentiation, and cell survival (Gollob *et al.*, 2006). Insects and other invertebrates as well possess this cascade (Chang and Karin, 2001), and exogenous mammalian insulin appears to activate MAPK (ERK) in

Drosophila Schneider cells (Kim *et al.*, 2004) and in the midgut of *An. stephensi* females when fed via an artificial blood meal (Kang *et al.*, 2008). The primary components of the MAPK cascade include RAS, Raf, MEK kinases and the downstream signaling kinases MAPK (ERK) 1/2, JNK, and p38 proteins. Activation of the MAPK cascade leads to activation of multiple transcriptional factors including p53; cMyc; and cJun (Whittaker *et al.*, 2010). Cross-talk between MAPK and PI3 K has been demonstrated in *Drosophila* where Ras is required for maximal PI3 K signaling (Orme *et al.*, 2006). Furthermore, Ras mutations result in smaller flies with dramatically reduced egg production suggesting interplay between MAPK and PI3 K signaling (Teleman, 2009; Orme *et al.*, 2006).

2.5.3. Target of Rapamycin Pathway

The target of rapamycin (TOR) pathway is activated by nutrient sensing through amino acid transporters or by upstream elements of the ISP (**Figure 3**). It is involved in the regulation of metabolism through the initiation of translation, biogenesis of the ribosome, storage of nutrients, and endocytosis. In both mammals and *Drosophila* cationic amino acid transporters (CATs) such as slimfast (Colombani *et al.*, 2003), proton assisted amino acid transporters (PATs; Goberdhan *et al.*, 2005), and heterodimeric amino acid transporter CD98 (Reynolds *et al.*, 2007) sense the presence of amino acids, initiate their uptake, and interact with the TOR pathway to modulate its activity (Goberdhan *et al.*, 2005; Colombani *et al.*, 2005). TOR kinase is highly conserved among eukaryotes. With the exception of yeast, which has two TOR genes, other eukaryotes possess a single TOR kinase gene, which is negatively regulated by TSC. TSC consists of two subunits: TSC1 (hamartin) and TSC2 (tuberin). TSC2 functions as a GTPase-activating protein (GAP) for the small GTPase Rheb (Ras homologue enriched in brain) to which TSC2 binds and inactivates. Phosphorylation of TSC2 results in the release of Rheb, which can subsequently activate TOR kinase. TOR participates in two distinct molecular complexes called TOR complex 1 (TOR-C1) and 2 (TOR-C2), which have both shared and distinct components (reviewed in Teleman, 2009; Yang *et al.*, 2006; Wullschleger *et al.*, 2006). One of the major differences between the complexes is the regulatory associated proteins of TOR, where rapamycin-sensitive companion of TOR (Raptor) is associated with TOR-C1 and rapamycin-insensitive companion of TOR (Rictor) is specific for TOR-C2 (Wullschleger *et al.*, 2006). TOR-C1 is capable of phosphorylating downstream targets such as S6 kinase (S6 K) and 4E-BP (*Drosophila* Thor), an inhibitor of translation. Activation of S6 K leads to the phosphorylation and activation of the 40S ribosomal protein S6. Together, with phosphorylation and inactivation of the translation inhibitor 4E-PB, TOR signaling can mediate

the initiation of translation. Mammalian TOR-C1 can also be activated autonomously by sensing free amino acids, ATP:ADP ratio (indicator of cellular activity levels), and oxygen levels (Colombani *et al.*, 2003; Wullschleger *et al.*, 2006; Hansen *et al.*, 2004). It is likely that a similar mechanism of autonomous TOR-C1 activation exists in insects. TOR-C2 is less well studied, but Rictor, an exclusive component of TOR-C2, phosphorylates Akt at the hydrophobic serine motif, suggesting a role in modulating the ISP (Yang *et al.*, 2006; Sarbassov *et al.*, 2005).

The ISP and TOR pathway are thought to interact on several levels. TOR-C1 kinase can be activated by the ISP molecule Akt, since Akt can phosphorylate and inactivate TSC2 (Inoki *et al.*, 2002; Manning *et al.*, 2002; Potter *et al.*, 2002). Also PDK1 acts in concert with TOR-C1 to phosphorylate S6 K at two sites, one in the T-loop of the kinase domain by PDK1 and the other in the C-terminus hydrophobic motif by TOR, leading to its full activation (Wullschleger *et al.*, 2006). In addition, several negative feedback loops exist between the TOR pathway and the ISP. Mammalian studies demonstrate that excess amino acid availability inhibits signaling through the IR-PDK1-Akt pathway (Wullschleger *et al.*, 2006; Tremblay *et al.*, 2005), and a similar negative feedback loop in insects is likely. Activated S6 K phosphorylates and inactivates IRS1 leading to reduced PI3 K signaling, and can also inhibit IRS1 at the translational level (Wullschleger *et al.*, 2006; Um *et al.*, 2004). Moreover, in *Drosophila* cells, when Rheb is knocked-down and TOR-C1 activity is decreased, TOR-C2 dependent phosphorylation of Akt is increased, suggesting a reciprocal relationship between TOR-C1 and TOR-C2 activity and existence of a feedback loop between TOR and Akt signaling (Yang *et al.*, 2006).

2.6. Functions of Insulin-Like Peptides and the Insulin Signaling Pathway

The great range of physiological functions regulated by ILPs through the ISP has largely been elucidated from studies of *Drosophila*. The genetic tractability of this insect allowed investigators to create an astonishing array of specific genetic mutations that reveal informative phenotypes. Recent reviews cover many of these ILP/ISP functions (Gronke *et al.*, 2010; Nässel and Winther, 2010). However, it must be noted that no endogenous DILP has been shown to activate the ISP or directly affect any physiological process in *Drosophila*. Mammalian insulin is used as a substitute and does appear to have some bioactivity, as it does in other insects. In contrast, there are a few recent advances in establishing or confirming a functional role for ILPs in other insects. These advances, however, are often based on a demonstration of the direct bioactivity of the ILP, as expected for peptide hormones. Overviews of the various roles of ILPs are provided in Section 2.6 for broad functional areas.

2.6.1. Growth and Development

Growth and post-embryonic development of insects are regulated through diverse signaling pathways activated by juvenile hormones (JH), ecdysteroid hormones (ECD), and neuropeptides, including ILPs. The graded acquisition and apportionment of nutrients determines the completion of development for individual insects and ultimately reproductive fitness. Over the past decade, a general consensus, based almost entirely on studies of *Drosophila*, has emerged that coordination of growth and development are regulated jointly by the ISP via ILP activation and by the TOR pathway through nutrient sensing. As well, TOR signaling can activate the ISP intracellularly, as discussed earlier. Although not investigated, expression of some *DILP* genes suggests the ISP may play an important role during embryonic development in *Drosophila* (Brogiolo *et al.*, 2001).

In-depth summaries of the *Drosophila* studies that support the previous consensus are available (Geminard *et al.*, 2009; Garofalo, 2002; Goberdhan and Wilson, 2003; Flatt *et al.*, 2005; Edgar, 2006; Mirth and Riddiford, 2007; Grewal, 2009; Walkiewicz and Stern, 2009; Shingleton, 2010), and in the following paragraphs is a brief overview. Ablation of the DILP MNC or knockout of genes encoding ILPs or proteins in the ISP delays development or produces smaller larvae and adults. In contrast, overexpression of these genes restores growth. Knockout or overexpression of genes for putative amino acid transporters or proteins in the TOR pathway has the same cognate effects on overall growth, and the effects of these genetic manipulations are mimicked by starvation or nutrient restriction and feeding, respectively. As well, this pathway can activate downstream components of the ISP, thus complicating interpretation of the results. The critical contributions of TOR signaling are its activation by ingested nutrients and its additive signaling through S6 kinase and crossactivation of the ISP. Both pathways induce and maintain expression of genes that promote organismal growth during the larval stages and organ-specific growth and remodeling in the pupal or adult stages. Because growth of insects is limited by the exoskeleton, this process must be limited and even stopped to allow molting into a larger or different exoskeleton. Attainment of a critical weight is sensed in the brain and results in the episodic release of a neurohormone, prothoracicotropic hormone (PTTH), which stimulates the prothoracic glands (PG) to produce ECD (see Chapter 1). Ecdysteroid signaling antagonizes the action of the ISP, thus preventing further growth, and activates expression of genes that enable formation of the next exoskeleton, organ remodeling, and finally molting/ecdysis, so that growth can resume in the larval stages or reproduction completed in the adult stage (see Chapter 5). For most insects, larger larvae give rise to larger females, and larger females are more successful at mating and can produce more eggs (Yuval *et al.*, 1993; Okanda *et al.*, 2002). Thus, promotion of growth through the ISP and TOR pathway ultimately impacts reproduction.

Growth promotion was demonstrated for an endogenous ILP first discovered in the hemolymph of *B. mori* larvae (Okamoto *et al.*, 2009a). The isolated peptide has an IGF-like structure (BIGFLP), is secreted by fat body, and stimulates the growth of genital imaginal discs *in vitro*, along with BrdU incorporation. Secretion of this peptide by the fat body occurs only in early pupae and is regulated by a surge of ECD that occurs at this time. The apparent structural homologue in *Drosophila*, DILP6, also promotes growth, based on the negative effects of *DILP6* mutations, and its expression in pupae is regulated by ECD (Okamoto *et al.*, 2009b; Slaidina *et al.*, 2009). The receptor and signaling pathway activated by insect IGFs have yet to be determined.

The existence of seven ILPs and only a single IR in *Drosophila* was taken as a priori evidence for their functional redundancy. Two recent studies investigated this hypothesis by knocking out the expression of individual and combinations of *DILPs* and determining the effects on parameters associated with growth, metabolism, reproduction, and life span (Gronke *et al.*, 2010; Zhang *et al.*, 2009). Mutants for a single *DILP* were all viable and generally exhibited normal growth/adult body weight, with *DILP6* mutants being the smallest. This strongly suggests compensatory or redundant function through the release of the other DILPs. Combined *DILP* knockouts revealed that one lacking all but *DILP6* developed into adults, whereas combinations with the *DILP6* knockout were lethal, suggesting that the fat body expressed DILP6 and the MNC expressed DILPs act redundantly (Gronke *et al.*, 2010). Mutation combinations of *DILP1-5*, all expressed in the MNC, were viable, but showed reduced growth/body weight, again supporting the redundancy of DILP action in regulating growth and development.

During the development of insects, most organs or tissues, such as the fat body, grow as body size increases, but some, such as the imaginal discs that give rise to adult structures, remain in an arrested state during the larval stages and subsequently grow during adult development. The same hormones that coordinate insect development and growth also regulate organ-specific processes. Several studies demonstrate that ILP treatment directly stimulates the growth or proliferation of cells in different organs. Imaginal discs of lepidopteran species are model organs for *in vitro* studies of the hormonal regulation of growth. Wing discs from last instar *Precis coenia* or *Manduca sexta* are synergistically stimulated to grow *in vitro* by 20-hydroxyecdysone (20-ECD) and an ILP (bombyxin-II or partially purified brain extract) (Nijhout and Grunert, 2002; Nijhout *et al.*, 2007). Insulin signaling is activated in these discs by ILP treatment and not by 20-ECD (Nijhout *et al.*, 2007). *In vivo* studies showed that commitment to pupal development in wing discs was stimulated by the injection of mammalian insulin or purified *M. sexta* ILP into newly molted, starved last instar *M. sexta* larvae, and further indicated that these treatments suppress JH action later in the instar (Koyama *et al.*, 2008). Hematopoietic organs produce hemocytes in Lepidoptera, and hemocyte discharge by these organs *in vitro* is stimulated in a dose-responsive manner by synthetic ILP (bombyxin-II; 100 nM maximum effect) and bovine insulin (1 μM maximum effect) (Nakahara *et al.*, 2006). The addition of 20-ECD enhanced the effect of low ILP and insulin concentrations. The midgut of insects is assaulted by pathogens and ingested molecules, and damaged/dead cells are replaced by the stem cell division. This process is reduced in the midgut of adult *Drosophila* with ablated MNC or negative IR and IRS mutations, but is restored in mutants overexpressing the IR or *DILP*2 (Amcheslavsky *et al.*, 2009). Cultured midgut stem cells obtained from lepidopteran larvae also proliferate after treatment with synthetic ILP (bombyxin-II) and insulin, and this effect is enhanced by the presence of fat body extract (Goto *et al.*, 2005), suggesting the presence of an IGF. As shown by these studies, insect tissues and cells are amendable to culture and experiments to investigate the direct effects of ILPs on the ISP and their specific functions.

2.6.1.1. Regulation of ecdysteroid hormone production

Ecdysteroid hormones are primary activators of insect development, and their roles in the regulation of molting and metamorphosis are well characterized (Chapters 5–7; Spindler *et al.*, 2009). In the larval/nymphal stages, ECD are synthesized from dietary cholesterol in the PG and released into the hemolymph for circulation to target tissues, where processing into other active forms, signaling, or degradation occurs (Chapters 4 and 5). A surprising number of neuropeptides are known to stimulate or inhibit ECD production by the PGs of insects (Marchal *et al.*, 2010). The first to be structural and functionally characterized was PTTH (Chapter 1), and its receptor, Torso — a receptor tyrosine kinase — was only recently identified for *Drosophila* and *B. mori* (Rewitz *et al.*, 2009) . Purification of PTTH from *B. mori* also yielded the first ILP (bomybxin; first called "small PTTH") to be characterized for an invertebrate, and this ILP was isolated based on its stimulation of pupal molting and ECD production by the PG of another silkworm, *Sa. cynthia* (reviewed in Marchal *et al.*, 2010). Similarly, two synthetic *Sa. cynthia* ILPs (A1 and B1) at doses of 1 to 5 nM stimulated pupal–adult molt in *Sa. cynthia*, but the doses were 20- to 100-fold higher than required for synthetic *B. mori* ILP (bombyxin-II; Nagata *et al.*, 1999). This same synthetic ILP exhibits cross-species activity by stimulating ECD production by the PG of the hemipteran *R. prolixus* (Steel and Vafopoulou, 1997).

A later study showed that bovine-insulin-stimulated ECD production by PG from last instar *B. mori* in the range of 1.7 to 17 μM (Gu *et al.*, 2009), but only after day 4 when the PG are responsive to PTTH. Insulin activated the ISP in the PG but not MAPK signaling, given the effects of different inhibitors and phosphorylation of IR and Akt. Insulin treatment also stimulated PG activity *in vivo*, but, as with the *in vitro* effects, required 8 to 24 h to be evident. This recent study provides more evidence for ILP regulation of ECD production by the PG of lepidopteran species, perhaps by signaling nutritional state and acting redundantly with PTTH. Also, there is the possibility that another of the multiple ILPs in *B. mori* and other species may be as potent as PTTH in stimulating the PG, and this is best demonstrated with species-specific bioassays and native ILPs.

Because growth and development in *Drosophila* is closely regulated by nutrition and the ISP, ECD production by the larval ring gland is presumed to be in part regulated by PTTH and one or more DILPs (Mirth *et al.*, 2009). Ectopic expression of *PI3K* in the ring gland increased gland size, with the expectation that the ISP and ECD production would be activated, but instead, larval growth rate was reduced (Colombani *et al.*, 2005; Mirth *et al.*, 2005). Although ECD levels in *PI3K* mutants were not significantly different than that of controls, the presumption was made that larger PGs made more ECD based on increased transcript abundance of genes encoding two cytochrome P450 enzymes involved in ECD biosynthesis. Based on this presumption, ECD and DILPs are thought to be antagonistic in their regulation of growth (Colombani *et al.*, 2005). A direct comparison of PTTH and DILP activity on the *Drosophila* ring gland would provide insight into whether their co-regulation of ECD production is conserved in immature insects or unique to Lepidoptera.

2.6.1.2. Regulation of juvenile hormone production

Juvenile hormones are the primary "morphostatic" modulators of ECD signaling and action during development (Riddiford, 2008; Chapter 8). Although other tissues can produce JH, the most important source is the CA, which is innervated by axon tracts from the brain, separate from the CC, and near the aorta and foregut (Chapter 8). Synthesis of different JH forms by the CA is activated by the neuropeptide, allatotropin, and inhibited by neuropeptides termed allatostatins. Although most insects examined possess these neuropeptides (*D. melanogaster* has no allatotropin), their activity on the CA is not universal (Weaver and Audsley, 2009; Chapter 8). The ISP has also been implicated in the regulation of JH synthesis in *Drosophila*, because the IR is present in the CA of larvae and adults (Belgacem and Martin, 2006). However the role is unresolved. *In vitro* synthesis of JH is reduced in CA from adult flies with deleterious mutations in the IR and *chico* (Tatar *et al.*, 2001; Tu *et al.*, 2005), but another study clearly demonstrated greater JH synthesis by the CA in IRS mutants, given the smaller body size (Richard *et al.*, 2005). A later study found that the targeted reduction of IR transcripts in the CA by RNAi disrupted sexual dimorphic locomotor activity in the same way as CA targeted RNAi for a gene encoding an enzyme involved in JH biosynthesis pathway, 3-hydroxy-3-methylglutaryl CoA reductase (HMGCR; Belgacem and Martin, 2007). This enzyme is also expressed in other tissues (Belgacem and Martin, 2007; Belles *et al.*, 2005), but the CA specific effects suggested that the ISP specifically regulates HMGCR expression and sexual dimorphic behaviors. Other mechanisms regulating JH titer include the secretion of JH esterases that degrade circulating JH and the uptake of JH by tissues (Chapter 8). The clearance of JH is a key requirement for the induction of adult development. Further investigations may yet reveal a direct role for ILPs and the ISP in the regulation of JH secretion and action or interactions with other neuropeptide regulators (Weaver and Audsley, 2009).

2.6.2. Metabolism

Given the small size of insects and the high energy demands of locomotion, especially flight, regulation of metabolism must be highly sensitive to internal and external cues and responsive to the ingestion of food, so that nutrients are allocated between immediate energy needs and storage to facilitate growth, development, and reproduction. When insects are starved or not feeding during the late molt period, diapause, or the pupal and even adult stages, regulation of the reallocation or mobilization of nutrient reserves is necessary to prolong survival. Two types of neuropeptides and their signaling pathways balance the energetic needs of insects between storage and mobilization of nutrients. Of the multiple ILPs encoded in insects, one or more are likely functional homologues of vertebrate insulin, as first shown for mosquitoes (Wen *et al.*, 2010). Insulin induces cells to take up glucose and convert it to glycogen, to inhibit glycogen breakdown and gluconeogenesis, and generally to shift from catabolic to anabolic lipid and protein metabolism. AKH is the structural and function homologue of glucagon in vertebrates in many different insects (Gade, 2004; Chapter 9). Acting through a GPCR and cAMP/inositol triphosphate signaling, AKH activates the mobilization of carbohydrate, proline, and lipid stores.

The fat body is the primary site of intermediary metabolism and nutrient storage, and thus fully responsive to ILP/TOR and AKH signaling. These processes also occur in the midgut and in oenocytes distributed throughout the body and govern the fat body through unknown mechanisms (Gutierrez *et al.*, 2007). The particulars of ILP and AKH signaling and interaction in the fat body and during larval and adult stages are best understood for *Drosophila* and are the subject of recent comprehensive reviews (Teleman, 2009; Baker and Thummel, 2007; Leopold and Perrimon, 2007; Taguchi and White, 2008; Hietakangas and Cohen, 2009). Recent studies suggest that DILP1-5 either alone or redundantly regulates carbohydrate and lipid metabolism in an insulin-like fashion, based on single and combinatorial *DILP* deletions (Gronke *et al.*, 2010; Zhang *et al.*, 2009). Reliance on genetic mutations takes time to reveal informative phenotypes, but does not advance our understanding of the direct endocrine actions of DILPs. The first attempt to establish such an action showed that a single dose (0.2 pmol) of recombinant DILP5 transiently (20 min) reduced trehalose levels *in vivo* in females in the absence of control treatments (Sajid *et al.*, 2011). Many studies have used purified mammalian insulins to characterize the functional conservation of ILP metabolic action in *Drosophila* and other insects (Wu and Brown, 2006) and the shrimp *Penaeus vannamei* (Gutierrez *et al.*, 2007). These insulins, however, differ greatly in primary sequence from insect ILPs and likely may not bind to insect IRs, as demonstrated for the *Ae. aegypti* IR (Brown *et al.*, 2008) or are poor competitors, as determined for the *Drosophila* IR (Sajid *et al.*, 2011). Mammalian insulin bioactivity and activation of the ISP in insects requires pharmacological concentrations, suggesting other mechanisms or aberrant interactions with the insect IR, yet to be explained.

Only two studies have reported that synthetic ILPs directly affect metabolic processes in insects. A silkmoth ILP (bombyxin-II), when injected into neck-ligated larvae, lowered trehalose and glycogen levels, apparently promoting consumption of carbohydrate reserves and not the accumulation of reserves, as for insulin in mammals (Satake *et al.*, 1997, 1999). To resolve whether different ILPs redundantly act as insulins, synthetic *Ae. aegypti* ILP3 and ILP4 were injected into decapitated, sugar-fed female *Ae. aegypti*. Only ILP3 stimulated trehalose depletion and glycogen and lipid storage in a dose-responsive manner (Brown *et al.*, 2008; Wen *et al.*, 2010). This result shows that ILPs differ in their ability to activate anabolic processes in the fat body, but as discussed previously, the two ILPs activated ovarian ECD production *in vitro* at similar doses. This differential action likely reflects the binding of ILP3 to the IR and ILP4 to another membrane protein, as determined for ovary membranes, but needs to be confirmed for fat body. Previous studies of insulin-like factors in insects relied on the classic endocrine method of decapitation to remove the neuropeptide source for comparison of nutrient levels or stores to that in intact or extract injected ones (Wu and Brown, 2006). Recent studies have turned to quantitative expression analyses of genes associated with the ISP to determine the effect of feeding or starvation (Liu *et al.*, 2010) and to identify other genes affected by nutrient intake (Gershman *et al.*,

2007). Notably in the Gershman *et al.* study, none of the DILPs showed any significant change in expression for up to 13 h after starved *Drosophila* females were fed, nor did Akt, FOXO, or TOR. However, expression of IR, IRS, and PDK1 were greatly decreased within 2 h after feeding and for the whole period. These trends suggest that the regulation of DILP and ISP gene expression is not uniformly responsive to nutrient intake or state.

2.6.3. Reproduction

As with growth and development, JH, ECD, and neuropeptides, including the ILPs, are implicated in the regulation of insect reproduction (Raikhel, 2004). For all insect groups with the exception of a few dipteran species, JH is considered to be the primary gonadotropic hormone in females, because it activates the fat body and ovarian cells to secrete vitellogenin (VG) and other yolk proteins (YP) for uptake by the maturing oocytes (Tufail and Takeda, 2008). Ovaries are the primary source of circulating ECD in females, because the PG degenerate during the transition to the adult stage in most insects (Rees *et al.*, 2010; Chapter 4), and follicle cells surrounding the oocytes are the site of ECD biosynthesis. These steroids are either stored in the oocytes for use during embryonic development or released into the hemolymph. In those insects for which JH functions as a gonadotropin, ECDs appear to be involved in the termination of VG production (Hatle *et al.*, 2003).

In female dipterans, however, JH does not regulate VG/YP secretion, but it is required for the development and differentiation of the fat body and ovaries (Raikhel, 2004). It plays a similar role in the termination of reproductive diapause in female *Cu. pipiens* mosquitoes induced by RNAi knockdown of ILP-1, as demonstrated by the application of a JH mimic (Sim and Denlinger, 2009a,b). Many dipteran females require a protein meal to initiate and maintain egg production, and ingestion of a meal stimulates ECD production by the ovaries. Ecdysteroid signaling in turn stimulates VG/YP secretion and uptake by developing oocytes. The interaction between nutrition, JH, ECD, and ILP/ISP is best characterized for *Drosophila* (Gruntenko and Rauschenbach, 2008) and mosquito females (Attardo *et al.*, 2005), but there is an important difference in their reproductive biology. *Drosophila* females continuously feed and produce eggs, whereas female mosquitoes take discrete blood meals from hosts and produce eggs in batches. These different feeding strategies affect their reproductive endocrinology, as discussed in the following paragraph.

For *Drosophila*, many studies of ILP and ISP mutations include reproduction parameters in their examination of effects on growth, metabolism, and longevity. A recent study comparing the effects of knocking out single and multiple DILPs found that flies lacking only DILP6,

the IGF homologue, had the greatest reduction (46%) in lifetime fecundity compared to wild-type flies, and other single DILP mutations had less of an effect. For example, DILP2 mutation reduced egg production by 25% (Gronke *et al.*, 2010). A similar range resulted from the combined DILP deletions: together DILP2, DILP3, and DILP5 reduced fecundity by 69% whereas DILP1-4 only resulted in a 14% reduction. Deletion of DILP7 did not reduce egg production, although another study suggested that localization of DILP7 in thoracic and abdominal neurons with axons to the female reproductive tract indicated a role in decision making during oviposition (Yang *et al.*, 2008). *Drosophila* females with hyperpolarized DILP7-expressing neurons do not display oviposition behavior and fail to lay eggs. These different outcomes may be due to incomplete knockdown of DILP7 in the mutants.

The mechanisms by which DILPs regulate egg maturation in *Drosophila* females are not defined, but a functional ISP is required for the initiation of oogenesis and YP synthesis and uptake. Oogenesis is continuous in fed *Drosophila* females, as germline and follicle stem cells in the germarium divide and form egg chambers around the oocytes. This process is dependent on DILP activation of the ISP in the stem cells and their progeny in females (LaFever and Drummond-Barbosa, 2005; Hsu *et al.*, 2008) and males (Ueishi *et al.*, 2009). The TOR pathway differentially affects the maintenance, division, and proliferation of the two types of stem cells in females via ISP dependent and independent mechanisms (LaFever *et al.*, 2010). Yolk protein synthesis is dependent on ISP and ECD signaling (Gruntenko and Rauschenbach, 2008; LaFever and Drummond-Barbosa, 2005). JH and the ISP are implicated in the activation of ovarian ECD production, given that ECD production is impaired in the ovaries of some *Drosophila* IR mutants (Tu *et al.*, 2002). Both the fat body and ovarian follicle cells produce YP in this species, and these YP are structurally more related to lipases than VG (Tufail and Takeda, 2008). Females with a homozygous mutation in IRS (chico) are small and sterile but maintain a normal range of JH and ECD production given their size, thus suggesting a primary role for DILP activation of the ISP (Richard *et al.*, 2005). Ovaries from these females, when implanted into normal females, did not complete YP uptake, whereas the reciprocal transplants did. These results demonstrate that the IRS mutation was ovary specific and disrupted YP uptake. Mutations in the ISP reveal the pivotal role of nutritional state in tipping toward reproduction or longevity, and even affects mating and sexually dimorphic behavior (Belgacem and Martin, 2006; Belgacem and Martin, 2007; Toivonen and Partridge, 2009; Wigby *et al.*, 2011).

The mechanisms by which ILPs regulate ovarian ECD production and VG in the fat body are well defined for female *Ae. aegypti* that require a blood meal to initiate egg maturation (Wen *et al.*, 2010; Attardo *et al.*, 2005).

Following the blood meal, neuropeptides, including ILPs and OEH, are released from brain MNC and stimulate the ovaries to produce ECD (Wen et al., 2010; Brown et al., 1998), which in turn activate the fat body to secrete VG/YP into the hemolymph for uptake by developing oocytes. Induction and maintenance of VG/YP synthesis and secretion depend on amino acids digested from the ingested blood, ECD, and insulin signaling (Roy et al., 2007). These processes end by 36 h post blood ingestion, and after another 36 h egg development and oviposition is complete. The first evidence for the role of ILPs in mosquito reproduction came from the demonstration that mammalian insulins stimulated ECD production in vitro by ovaries of Ae. aegypti, and that an IR homologue was expressed in ovaries (Graf et al., 1997). Insulin treatment activated the ISP in ovaries as revealed by inhibitors and activators (Riehle and Brown, 1999). Similarly, this process and the ISP are activated in the ovaries of the blowfly Phormia regina (Maniere et al., 2004, 2009). The question of whether endogenous ILPs directly regulate ovarian ECD production in Ae. aegypti was answered in later studies by chemically synthesizing ILPs and testing their activity in bioassays. Both synthetic ILP3 and ILP4 directly stimulated ECD production by ovaries in a dose-responsive manner, and the activating concentrations were 100-fold less than that required for mammalian insulin (Wen et al., 2010; Brown et al., 1998). Their effect was not additive in vitro, and both stimulated the ISP in ovaries. Competitive ligand binding and crosslinking assays established that ILP3 directly bound to the IR in ovary membranes and ILP4 bound to an as yet unidentified 55 kDa protein, suggesting that the ovarian IR has a signaling partner (Wen et al., 2010). Notably, this is the first confirmation of binding between an endogenous ILP and IR for invertebrates. Nevertheless, ILP4 activity on ovaries required IR expression as demonstrated by IR RNAi. The direct effect of ILP3 and ILP4 on ovary ECD production in part explains their action in vivo. Again both ILPs, when injected individually into decapitated blood-fed females, stimulated VG/YP secretion and uptake by oocytes, but controls showed no uptake (Wen et al., 2010; Brown et al., 1998).

The fat body of Ae. aegypti females is the site where the ISP, ECD signaling, and nutrient sensing/TOR signaling converge to activate VG/YP secretion. Expression of VG transcripts in fat body is synergistically stimulated during in vitro tissue culture by the combination of bovine insulin and 20-ECD, and this activation requires IR and Akt expression as determined by IR and Akt RNAi (Roy et al., 2007), thus establishing a key role for the ISP in vitellogenesis. The TOR pathway plays a direct role in VG production by the fat body, which is dependent on amino acid sensing, by phosphorylation of S6 K and 4EBP and the activation of GATA transcriptional factors that are crucial for VG transcription (Hansen et al., 2004, 2005;

Attardo et al., 2005; Raikhel et al., 2002). Two cationic amino acid transporters (slimfast and iCAT2) allow amino acid uptake by the fat body and stimulate TOR signaling (Attardo et al., 2005). The combination of amino acids and 20-ECD greatly enhance VG transcription and translation in the fat body in vitro (Hansen et al., 2004; Attardo et al., 2006). Inhibition of TOR signaling in the midgut of mosquitoes by RNAi significantly reduced trypsin activity and thus amino acid levels, resulting in significantly smaller eggs (Brandon et al., 2008). As digestion of the blood meal and ovarian ECD production ends, negative regulators of the ISP play a role in the termination of VG production by the fat body. Further evidence of the key role of ISP in mosquito VG is demonstrated by RNAi knockdown of FOXL, FOXN1, FOXN2, or FOXO, which halts VG expression by fat body in vitro and significantly decreases the number of eggs laid (Hansen et al., 2007). Knockdown of PTEN in the fat body and ovaries resulted in significantly higher egg production without any effects on their viability (Arik et al., 2009). Sugar and blood feeding also affected the expression of genes for ILPs, IR, and TOR pathway components in the mosquito An. gambiae, thus supporting the role of these pathways in the regulation of mosquito reproduction (Arsic and Guerin, 2008).

Sex determination and sexual differentiation in insects are regulated by specific genes, chromosome differences, and environmental factors (Sanchez, 2008), and hormones are thought to be secondarily involved (De Loof, 2008). This is not so for another arthropod group, the Crustacea, in which the ILP-like androgenic gland hormone (AGH) directly determines primary and secondary male characteristics and masculinizes females (Chang and Sagi, 2008). The AGH was extracted from or identified in the androgenic glands attached to the ejaculatory region of the vas deferens of isopods, crayfish, shrimp, and crabs (Ventura et al., 2009; Sroyraya et al., 2010). In vivo silencing of AGH expression in the prawn Macrobachium rosenbergii temporally arrested male sex differentiation, growth, and spermatogenesis (Ventura et al., 2009), thus offering support for the more classical endocrine manipulations that led to the characterization of this ILP. The structural and functional conservation of this ILP within the phylum Arthropoda has yet to be resolved.

2.6.4. Diapause

Diapause is a period of endocrine-mediated metabolic and developmental arrest induced by changes in abiotic cues that indicate the onset of adverse environmental conditions, such as winter (Chapter 10). This arrest occurs in a species-specific life stage, and in adult diapause, reproduction halts. Induction of this arrest typically stimulates nutrient/metabolite storage, so that a low level of metabolic activity can be maintained and sufficient reserves are left to continue development or reproduction. Two of the first reports on the effects of mammalian insulin in insects showed that insulin injection terminated diapause and promoted development in pupae of two lepidopterans, *Sa. cynthia ricini* (Wang *et al.*, 1986) and *Pieris brassicae* pupae (Arpagaus, 1987). Interestingly, it also stimulated ECD production in the *P. brassicae* pupae. More recent studies show that alterations in the ISP affect adult diapause in dipterans (Allen, 2007). In *Drosophila*, deletion of the gene encoding the catalytic subunit of PI3K (Dp110) resulted in a higher proportion of flies entering diapause, whereas overexpression of Dp110 in the nervous system decreased the proportion of individuals in diapause (Williams and Sokolowski, 1993, 2009; Williams *et al.*, 2006). Female *Cu. pipiens* mosquitoes enter diapause in the autumn with large fat reserves sufficient to survive through the winter, and this state is characterized by arrested ovarian development. Knockdown of IR expression by RNAi in non-diapausing mosquitoes arrests primary oocytes development, as in diapausing females (Sim and Denlinger, 2008). In contrast, knockdown of FOXO expression in females pre-programmed for diapause greatly decreased lipid storage and life span. Moreover, whole body expression of ILP1 and ILP5 transcripts is lower in females programmed for diapause, and RNAi knockdown of ILP1 halts ovarian development in non-diapausing females (Sim and Denlinger, 2009a). Considered altogether, results reported for both the lepidopteran and dipteran species suggest that inactivation of the ISP facilitates diapause induction and its activation terminates or prevents diapause. This concept seems to contradict the anabolic or nutrient storage role established for the ISP in non-diapausing insects, but it is in accord with the role of the ISP in the nematode, *C. elegans* (Fielenbach and Antebi, 2008).

2.6.5. Behavior

Signaling pathways driven by nutritional and hormonal cues are linked to the complex mechanisms in the brain that regulate insect behavior. Different ILPs are expressed in the central nervous system of insects and likely activate the ISP in specific neuronal pathways that govern behaviors associated with nutrient acquisition, as demonstrated for *Drosophila* and the honeybee. Downregulation of DILP expression and the ISP in fed *Drosophila* larvae increased food ingestion and acceptance of food containing noxious chemicals, also characteristic of starved larvae; whereas overexpression reduced these behaviors in starved larvae (Wu *et al.*, 2005). This pathway also negatively interacts with neuropeptide F signaling that promotes foraging over a range of aversive conditions (Lingo *et al.*, 2007). Mutations in ISP genes increase ethanol sensitivity in *Drosophila* (Corl *et al.*, 2005), thus providing evidence that DILPs and the ISP are involved in addictive behaviors associated with foods.

Two decades after the first report of ILPs in the honey bee *Ap. mellifera* (O'Connor and Baxter, 1985), a potential role for the ISP in caste determination and behavioral differentiation was suggested by differences in the expression of key ILP and ISP genes. Expression levels of ILP-1 and the IR were higher in queen larvae than in worker larvae during certain larval instars (Wheeler *et al.*, 2006). In contrast, ILP-2 is expressed at greater levels in workers than in queens (de Azevedo and Hartfelder, 2008). Genes encoding PI3K, IRS, and PDK1 are located in the "pollen" quantitative trait loci (QTL) associated with the genetic variation of honeybee foraging behavior, thus suggesting that the ISP is an upstream mediator of foraging behavior and division of labor (Hunt *et al.*, 2007). In a bee colony, duties shift from young bees nursing the larvae to older bees foraging for nectar and pollen (see Chapter 11). Foragers were shown to have significantly higher transcript levels of ILP-1 in heads and of InR1 and InR2 in abdomens than found in nurse bees (Ament *et al.*, 2008). Evidence suggests that the ISP via the TOR pathway also influences the timing of behavioral maturation of honeybees in a seasonally dependent manner (Ament *et al.*, 2008). One of the primary food-related behaviors of forager bees is to decide between nectar (a carbohydrate source) and pollen (a protein source). Expression of IRS in peripheral tissues of bees appears to regulate mechanisms that help foragers to determine food preference (Wang *et al.*, 2010). Strains of bees that are artificially selected for high pollen hoarding behavior have higher IRS expression than bees selected for low pollen hoarding behavior. Bees with reduced IRS expression due to RNAi displayed a potential bias toward foraging for pollen rather than nectar (Wang *et al.*, 2010).

2.6.6. Immunity

To defend against bacterial and fungal pathogens, insects rely on an innate immune system that coordinates pathogen recognition with induction of defenses to block infection. The presence of pathogens is detected through pathogen-recognition receptors that activate either the Toll or "immune deficiency" (IMD) signaling pathway depending on the pathogen type (Lemaitre and Hoffmann, 2007). These pathways in turn activate nuclear factor-κB transcriptional factors (NF-κB) that induce expression of numerous immune effector genes, including antimicrobial peptides (AMPs), such as defensin and cecropin (Lemaitre and Hoffmann, 2007). The connection between the ISP and immunity in invertebrates was first reported for the nematode *C. elegans* in which mutants expressing disabled IR *(daf-2)* or PI3K *(age-1)* are not only long lived, but also resistant to infection by gram negative and gram positive bacteria (Garsin *et al.*, 2003). Later studies showed that the pathogen resistance displayed by ISP mutants was associated with reduced bacterial colonization, enhanced bacterial clearance, and increased expression of multiple innate immune genes (Evans *et al.*, 2008).

Studies of *Drosophila* also have shown that pathogen infection suppresses the ISP and alters metabolism and the immune response. The first example came from a study of *Drosophila* infected with *Mycobacterium marinum*, intracellular bacteria closely related to *M. tuberculosis* (Dionne *et al.*, 2006). Infection reduced Akt and glycogen synthase kinase 3 (GSK-3) phosphorylation and expression of key anabolic genes, resulting in hyperglycemic flies that lost fat and glycogen stores and died sooner than infected FOXO mutants, which exhibited less wasting. Reduced Akt signaling was not due to a defect in the path between the IR and Akt, because injected mammalian insulin restored Akt phosphorylation. Mutations in the IRS gene (*chico*), which suppress the ISP, also increased pathogen resistance to gram positive and gram negative bacteria, as well as increased life span (Libert *et al.*, 2008). These studies support the concept that suppression of the ISP in the infected flies allowed energy stores to be mobilized for an enhanced immune response. A later study established a more direct link between the ISP and innate immunity. *Drosophila* mutants with a constitutively active Toll receptor expressed in fat body, the primary immune and metabolite storage tissue, exhibit induced AMP expression but decreased Akt phosphorylation and triglyceride storage (Diangelo *et al.*, 2009). Infection with gram positive bacteria *Micrococcus luteus* or the fungus *Beauvaria bassiana*, which activate the Toll pathway, had the same effects in wild-type flies, but not in flies with a constitutively active IMD pathway. Over expression of the Toll receptor also suppressed larval and adult growth, thus suggesting that mounting an immune response through this pathway can suppress the ISP throughout the organism.

Starvation also attenuates the ISP in *Drosophila*, and in such animals, the immune response is activated without pathogen challenge, as shown by increased expression of AMP genes (Becker *et al.*, 2010). A comprehensive investigation into this effect found that mutations of the IRS (*chico*) and GEF gene (*steppke*), which also blocks the ISP, resulted in higher AMP expression in non-infected larvae. Feeding wild-type adult flies with SecinH3, a GEF inhibitor that blocks PI3 K signaling, as well as starvation of larvae, which induces nuclear localization of FOXO, also increased AMP expression. In contrast, AMP expression is reduced in FOXO mutant flies starved or treated with SecinH3, suggesting a direct link between FOXO signaling and activation of AMP expression. This FOXO-dependent regulation of AMPs appears to be independent of Toll and IMD signaling, because FOXO activity can increase AMP gene expression in Toll and IMD mutants that fail to respond to immune challenges. FOXO binding sites are present in the regulatory regions of AMP genes, as are other forkhead binding sites and insulin response elements. In total, this study provides a new mechanism

for the activation of innate immunity that is dependent on nutrient state. Reduced insulin signaling in response to starvation or other stressors results in FOXO activation of AMP protection in advance of infection, whereas in normal, fed, or unstressed insects, innate immunity via AMPs is activated when pathogens invade and are recognized through the IMD or Toll pathways, which also limits the ISP as discussed earlier. In both scenarios, immunity depends on the mobilization of energy stores.

Interactions between the ISP and a different immune pathway have been characterized in the midgut of mosquitoes that take pathogen-infected blood meals. Feeding high concentrations of human insulin to *An. stephensi* and *An. gambiae* mosquitoes infected with the malaria pathogen, *Plasmodium falciparum*, results in a twofold increase in the number of oocysts on their midguts (Beier *et al.*, 1994). Whether this exogenous insulin alters mosquito digestion, metabolism, or the immune response to facilitate *Plasmodium* development is not known, but this study did provide the first evidence that exogenous insulin activates the ISP in mosquitoes. *Plasmodium* derived glycosylphosphatidylinositol (GPI), an important signaling molecule of parasite infection and mammalian insulin mimetic (Lim *et al.*, 2005), when fed to anopheline females induces Akt phosphorylation and nitrogen oxide synthase (NOS) expression in the midgut epithelium. Females limit *P. falciparum* infection by the induction of NOS, which produces inflammatory levels of toxic reactive NO (Luckhart *et al.*, 1998, 2003; Peterson *et al.*, 2007) that may stimulate parasite apoptosis (Hurd and Carter, 2004; Luckhart and Riehle, 2007). Human insulin induces NO production in cultured *An. stephensi* cells and in the female midgut (Lim *et al.*, 2005). The effect of insulin on NO production is also known for mammalian systems, thus suggesting conservation of this immune mechanism (Luckhart and Riehle, 2007). Human insulin also induces synthesis of hydrogen peroxide in the cultured mosquito cells and in the midgut of *An. stephensi*, thus reducing super oxide dismutase (SOD) and manganese SOD activity resulting in increased oxidative stress in the mosquito midgut (Kang *et al.*, 2008). A recent study confirms the importance of the midgut ISP in regulating mosquito immunity. Overexpression of an active, myristoylated Akt in the midgut of female *An. stephensi* leads to a dramatic decrease in *P. falciparum* infection (Corby-Harris *et al.*, 2010). Because Akt is a critical signaling node in eukaryotic cells (Manning and Cantley, 2007), the observed phenotype could be due to multiple mechanisms, including increased oxidative stress, molecular and structural changes, or upregulation of anti-*Plasmodium* genes in the midgut. One other study points to a link between infection in the insect gut and regulation by the ISP. Male dragonflies *Libellula pulchella* infected with related gut parasites (*Apicomplexa*, *Hoplorhynchus*) exhibit poor mating success due to poor flight performance and abnormally high trehalose levels in hemolymph and lipid stores in the thorax (Schilder and Marden, 2006). Injection of human insulin (18.5 pmol) reduced trehalose levels in healthy males and increased levels in infected ones. Interestingly, uninfected males given water containing only excreted parasite products also had high trehalose levels, suggesting that the midgut is also a key signaling center for the regulation of metabolism in insects. Together, these studies suggest that insects chronically infected with pathogens also display the "metabolic syndrome" associated with chronic infection, obesity, and insulin resistance of type 2 diabetes in humans.

2.6.7. Aging

Dietary, genetic, and molecular interventions that directly or indirectly reduce insulin signaling are known to increase life span in many organisms including *Drosophila* and mosquitoes. This active research area is extensively reviewed (Toivonen and Partridge, 2009; Piper and Partridge, 2007; Paaby and Schmidt, 2009; Fontana *et al.*, 2010). Exploring the role of the ISP in the nervous system, fat body, and midgut has expanded our understanding of the dynamics of aging in *Drosophila*. Reduced expression of DILP2 in the nervous system increases life span (Gronke *et al.*, 2010), as do mutations in the IR (Ikeya *et al.*, 2009), IRS (*chico*) (Clancy *et al.*, 2001), and the SH2B protein (*Lnk*) (Slack *et al.*, 2010) expressed in other tissues. Similarly, overexpression of the ISP antagonists FOXO and PTEN in the nervous system and fat body also extends life span (Hwangbo *et al.*, 2004; Giannakou *et al.*, 2004; Wessells *et al.*, 2004). In mosquitoes, the midgut potentially acts as a "signaling center" for the regulation of aging (Kang *et al.*, 2008). Overexpression of an active Akt specifically in the midgut of transgenic mosquitoes significantly reduced adult life span (Corby-Harris *et al.*, 2010). Similarly, the intestine of the nematode *C. elegans* plays a prominent role in life span regulation (Murphy *et al.*, 2007; Libina *et al.*, 2003).

Dietary or calorie restriction increases the life span of insects and other animals (Fontana *et al.*, 2010; Mair and Dillin, 2008; Partridge *et al.*, 2010). Nutrient sensing through the ISP or TOR pathway is directly affected by this external stressor. Earlier work showed that mutation of a gene encoding a transporter of Krebs cycle intermediates (*Indy*) extends the life span of *Drosophila* with no reduction in fecundity or other activities, and that this gene is expressed in fat body and midgut (Rogina *et al.*, 2000). Both calorie restriction and mutant *Indy* phenotypes exhibit reduced lipid stores, weight, and starvation resistance (Wang *et al.*, 2009). DILP expression is decreased in the heads of both phenotypes, indicating a silencing of the ISP, as shown by increased nuclear localization of FOXO. Calorie restriction reduces *Indy* expression in normal flies, and *Indy* mutants show no decrease

in food intake. Because the ISP is downregulated, nutrient storage would be minimal, thus suggesting activated nutrient mobilization supports the increased physical activity, also noted for the Indy mutant and starved flies. Aging in organisms is also modulated through the TOR pathway, both through its nutrient sensing role and interaction with the ISP (Partridge *et al.*, 2010; Kaeberlein and Kennedy, 2008; Parrella and Longo, 2010). Reduced expression of genes in the TOR pathway (*Tsc1*, *Tsc2*, *TOR*, and *S6K*) extends life span (Kapahi *et al.*, 2004). The potent inhibitor of TOR-C1, rapamycin, when fed to *Drosophila* at concentrations as high as 400 μM, directly blocked steps in the TOR pathway and significantly increased longevity but reduced fecundity (Bjedov *et al.*, 2010). However, the ISP was not affected, indicating that the TOR pathway can independently affect aging. Results of another study implied an integration of these mechanisms. Expression of the tumor suppressor p53, an inhibitor of the ISP in *Drosophila*, and dietary restriction extend life span but requires expression of TOR pathway elements (Bauer *et al.*, 2010). Together these results show that *Drosophila* and likely all insects are highly adapted to calorie restricted conditions commonly encountered in nature, through a number of mechanisms that include down regulation of the TOR and insulin pathways and a reapportionment of nutrients to enhance longevity in anticipation of finding food. Life span extension due to dietary restriction was also reported for *Ae. aegypti*, and this extension was exhibited regardless of whether nutrients were limited in the larval or adult stage (Joy *et al.*, 2010). These advances in our understanding of the ISP and TOR pathway show how tissue specificity and the synergistic action of signaling pathways underlies the complex molecular and cellular mechanisms regulating aging. Further studies are needed to understand how disruption of the ISP leads to life span extension; for example, how an inhibition of a transcription factor, FOXO, and the genes it transcribes leads to such a profound phenotype.

Given these advances for *Drosophila*, which appear to be relevant to mosquito biology, it is surprising that there are so few comparable investigations of other insects. The work on *Drosophila* is based on the hope that insight may be gained into age-related diseases in humans (Partridge *et al.*, 2010). For mosquitoes, the hope is to engineer short-lived mosquitoes incapable of transmitting disease pathogens, such as *Plasmodium spp.* (Corby-Harris *et al.*, 2010). It may be that other mechanisms affect aging in other insects, as suggested by the report that the dietary restriction of feeding only sucrose to male houseflies did not result in increased life span (Cooper *et al.*, 2004). Differences in the life span of queen and worker honeybees are thought to be modulated through the ISP, but differently than in *Drosophila*. Enriched nutrition in queens and nurses is associated with low expression of genes encoding the ILPs and IRs (Corona *et al.*, 2007).

Certainly, studies that identify mechanisms that decrease the longevity of pest insects would be welcome and may provide additional evidence for the conservation of these mechanisms and new control measures.

2.7. Conclusions

The great diversity and success of insects is due to the central regulation exerted by multiple ILPs through the ISP and associated pathways that coordinate growth, development, and reproduction, in response to external and internal cues regarding food ingestion, nutritional state, and pathogen infection. This concept is drawn from an ever-increasing number of studies based on the genetic manipulations and analyses of gene/protein expression in one highly derived fly species, *D. melanogaster*. Over the last five years, studies of other insects have confirmed the structural conservation of ILPs and proteins in the ISP and other pathways and indirectly substantiated some of the regulatory properties ascribed to ILP/ISP regulation, largely through changes in gene expression in response to nutritional or physiological manipulations. Still, few studies have confirmed direct and dose-responsive effects of ILPs and their binding to the IR or other receptors in insects or other invertebrates. The endocrinology of ILPs in female *Ae. aegypti* is the best characterized and provides important leads for future studies of other model insects. One important but unanswered set of questions is what cues regulate the release of ILPs, what are their circulating titers, and how does titer affect binding to the IR and other proteins? A closely related issue is how do multiple ILPs interact with a single IR and coordinate diverse physiological and behavioral processes in different tissues?

Our understanding of the intricacies of insulin signaling in *Drosophila* has also uncovered the complexity of its interactions with other signaling pathways. There is substantial evidence that ILPs acting through the ISP may be the primary regulators of ECD and JH secretion in different life stages, and thus direct or coordinate JH and ECD signaling at key points in the development and reproduction of insects. Sensing of nutrients acquired through ingestion or mobilized/stored internally through different mechanisms affects both the ISP and TOR pathway in an apparent positive feedback loop, and key proteins, such as Akt, can directly modulate expression of many genes or any number of different biochemical processes in tissues. Studies of insects may best reveal how the ISP independently or through its interactions with these other pathways activates particular genes or processes in tissues. Another question of particular interest is how does pathogen infection alter ISP regulation of metabolic homeostasis and modulate the Toll and IMD pathways that direct innate immune responses. With the genomes of insects and invertebrates being characterized at an unprecedented pace, the expansion of powerful genetic techniques into non-model organisms, and improvements in ILP synthesis and biochemical approaches, our understanding of this fundamental signaling cascade should significantly advance in the next few years.

Acknowledgments

The work of M. R. Brown was supported by National Institutes of Health Grant AI031108 and the Georgia Agricultural Experiment Station. The work of M. A. Riehle is supported by National Institutes of Health Grant AI073745 and the Bill and Melinda Gates Foundation.

References

Abascal, F., Zardoya, R., & Posada, D. (2005). ProtTest: selection of best-fit models of protein evolution. *Bioinformatics*, *21*, 2104–2105.

Afonso, C. L., Tulman, E. R., Lu, Z., Balinsky, C. A., Moser, B. A., et al. (2001). Genome sequence of a baculovirus pathogenic for *Culex nigripalpus*. *J.Virol.*, *75*, 11157–11165.

Allen, M. J. (2007). What makes a fly enter diapause? *Fly (Austin)*, *1*, 307–310.

Amcheslavsky, A., Jiang, J., & Ip, Y. T. (2009). Tissue damage-induced intestinal stem cell division in *Drosophila*. *Cell Stem Cell*, *4*, 49–61.

Ament, S. A., Corona, M., Pollock, H. S., & Robinson, G. E. (2008). Insulin signaling is involved in the regulation of worker division of labor in honey bee colonies. *Proc. Natl. Acad. Sci.USA*, *105*, 4226–4231.

Andersen, L., Jorgensen, P. N., Jensen, L. B., & Walsh, D. (2000). A new insulin immunoassay specific for the rapid-acting insulin analog, insulin aspart, suitable for bioavailability, bioequivalence, and pharmacokinetic studies. *Clin. Biochem.*, *33*, 627–633.

Arik, A. J., Rasgon, J. L., Quicke, K. M., & Riehle, M. A. (2009). Manipulating insulin signaling to enhance mosquito reproduction. *BMC Physiol.*, *9*, 15.

Arpagaus, M. (1987). Vertebrate insulin induces diapause termination in Pieris brassicae pupae. *Roux's Arch. Dev. Biol.*, *196*, 527–530.

Arquier, N., Geminard, C., Bourouis, M., Jarretou, G., Honegger, B., et al. (2008). *Drosophila* ALS regulates growth and metabolism through functional interaction with insulin-like peptides. *Cell Metab.*, *7*, 333–338.

Arsic, D., & Guerin, P. M. (2008). Nutrient content of diet affects the signaling activity of the insulin/target of rapamycin/p70 S6 kinase pathway in the African malaria mosquito *Anopheles gambiae*. *J. Insect Physiol.*, *54*, 1226–1235.

Attardo, G. M., Hansen, I. A., & Raikhel, A. S. (2005). Nutritional regulation of vitellogenesis in mosquitoes: implications for anautogeny. *Insect Biochem. Mol. Biol.*, *35*, 661–675.

Attardo, G. M., Hansen, I. A., Shiao, S. H., & Raikhel, A. S. (2006). Identification of two cationic amino acid transporters required for nutritional signaling during mosquito reproduction. *J. Exp. Biol.*, *209*, 3071–3078.

Badisco, L., Claeys, I., Van Hiel, M., Franssens, V., et al., (2007). Neuroparsins, a family of conserved arthropod neuropeptides. *Gen. Comp. Endocrinol. 153*, 64–71.

Badisco, L., Claeys, I., Van Hiel, M., Clynen, E., Huybrechts, J., et al. (2008). Purification and characterization of an insulin-related peptide in the desert locust, Schistocerca gregaria: Immunolocalization, cDNA cloning, transcript profiling and interaction with neuroparsin. *J. Mol. Endocrinol., 40*, 137–150.

Baek, J. H., & Lee, S. H. (2010). Differential gene expression profiles in the venom gland/sac of Eumenes pomiformis (Hymenoptera: Eumenidae). *Toxicon, 55*, 1147–1156.

Baker, K. D., & Thummel, C. S. (2007). Diabetic larvae and obese flies-emerging studies of metabolism in *Drosophila. Cell Metab., 6*, 257–266.

Barcelo, H., & Stewart, M. J. (2002). Altering *Drosophila* S6 kinase activity is consistent with a role for S6 kinase in growth. *Genesis, 34*, 83–85.

Bauer, D. C., Buske, F. A., & Bailey, T. L. (2010). Dual-functioning transcription factors in the developmental gene network of *Drosophila melanogaster. BMC Bioinformatics, 11*, 366.

Becker, T., Loch, G., Beyer, M., Zinke, I., Aschenbrenner, A. C., et al. (2010). FOXO-dependent regulation of innate immune homeostasis. *Nature, 463*, 369–373.

Beier, M. S., Pumpuni, C. B., Beier, J. C., & Davis, J. R. (1994). Effects of para-aminobenzoic acid, insulin, and gentamicin on Plasmodium falciparum development in anopheline mosquitoes (Diptera: Culicidae). *J. Med. Entomol., 31*, 561–565.

Belfiore, A., Frasca, F., Pandini, G., Sciacca, L., & Vigneri, R. (2009). Insulin receptor isoforms and insulin receptor/insulin-like growth factor receptor hybrids in physiology and disease. *Endocr. Rev., 30*, 586–623.

Belgacem, Y. H., & Martin, J. R. (2006). Disruption of insulin pathways alters trehalose level and abolishes sexual dimorphism in locomotor activity in *Drosophila. J. Neurobiol., 66*, 19–32.

Belgacem, Y. H., & Martin, J. R. (2007). Hmgcr in the corpus allatum controls sexual dimorphism of locomotor activity and body size via the insulin pathway in *Drosophila. PLoS One, 2*, e187.

Belles, X., Martin, D., & Piulachs, M. D. (2005). The mevalonate pathway and the synthesis of juvenile hormone in insects. *Annu. Rev. Entomol., 50*, 181–199.

Bermudez, I., Beadle, D. J., Trifilieff, E., Luu, B., & Hietter, H. (1991). Electrophysiological activity of the C-peptide of the Locusta insulin-related peptide. Effect on the membrane conductance of Locusta neurones *in vitro. FEBS Lett., 293*, 137–141.

Bjedov, I., Toivonen, J. M., Kerr, F., Slack, C., Jacobson, J., et al. (2010). Mechanisms of life span extension by rapamycin in the fruit fly *Drosophila melanogaster. Cell Metab., 11*, 35–46.

Bohni, R., Riesgo-Escovar, J., Oldham, S., Brogiolo, W., Stocker, H., et al. (1999). Autonomous control of cell and organ size by CHICO, a *Drosophila* homologue of vertebrate IRS1-4. *Cell, 97*, 865–875.

Bouzakri, K., Roques, M., Gual, P., Espinosa, S., Guebre-Egziabher, F., et al. (2003). Reduced activation of phosphatidylinositol-3 kinase and increased serine 636 phosphorylation of insulin receptor substrate-1 in primary culture of skeletal muscle cells from patients with type 2 diabetes. *Diabetes, 52*, 1319–1325.

Brandon, M. C., Pennington, J. E., Isoe, J., Zamora, J., Schillinger, A. S., et al. (2008). TOR signaling is required for amino acid stimulation of early trypsin protein synthesis in the midgut of *Aedes aegypti* mosquitoes. *Insect Biochem. Mol. Biol., 38*, 916–922.

Brogiolo, W., Stocker, H., Ikeya, T., Rintelen, F., Fernandez, R., et al. (2001). An evolutionarily conserved function of the *Drosophila* insulin receptor and insulin-like peptides in growth control. *Curr. Biol., 11*, 213–221.

Broughton, S., Alic, N., Slack, C., Bass, T., Ikeya, T., et al. (2008). Reduction of DILP2 in *Drosophila* triages a metabolic phenotype from lifespan revealing redundancy and compensation among DILPs. *PLoS One, 3*, e3721.

Brown, M. R., Clark, K. D., Gulia, M., Zhao, Z., Garczynski, S. F., et al. (2008). An insulin-like peptide regulates egg maturation and metabolism in the mosquito Aedes aegypti. *Proc. Natl. Acad. Sci. USA, 105*, 5716–5721.

Brown, M. R., Graf, R., Swiderek, K. M., Fendley, D., Stracker, T. H., et al. (1998). Identification of a steroidogenic neurohormone in female mosquitoes. *J. Biol. Chem., 273*, 3967–3971.

Callander, G. E., & Bathgate, R. A. (2010). Relaxin family peptide systems and the central nervous system. *Cell Mol. Life Sci., 67*, 2327–2341.

Chang, L., & Karin, M. (2001). Mammalian MAP kinase signalling cascades. *Nature, 410*, 37–40.

Chang, E. S., & Sagi, A. (2008). Male reproductive hormones. In E. Mente (Ed.), *Reproductive Biology of Crustaceans* (pp. 299–318). Enfield, New Hampshire: Science Publishers.

Chua, B. T., Gallego-Ortega, D., Ramirez de Molina, A., Ullrich, A., Lacal, J. C., et al. (2009). Regulation of Akt(ser473) phosphorylation by choline kinase in breast carcinoma cells. *Mol. Cancer, 8*, 131.

Claeys, I., Simonet, G., Poels, J., Van Loy, T., Vercammen, L., et al. (2002). Insulin-related peptides and their conserved signal transduction pathway. *Peptides, 23*, 807–816.

Clancy, D. J., Gems, D., Harshman, L. G., Oldham, S., Stocker, H., et al. (2001). Extension of life-span by loss of CHICO, a *Drosophila* insulin receptor substrate protein. *Science, 292*, 104–106.

Clements, J., Hens, K., Francis, C., Schellens, A., & Callaerts, P. (2008). Conserved role for the *Drosophila* Pax6 homologue Eyeless in differentiation and function of insulin-producing neurons. *Proc. Natl. Acad. Sci. USA, 105*, 16183–16188.

Clynen, E., Huybrechts, J., Baggerman, G., Van Doorn, J., Van Der Horst, D., et al. (2003). Identification of a glycogenolysis-inhibiting peptide from the corpora cardiaca of locusts. *Endocrinology, 144*, 3441–3448.

Colombani, J., Bianchini, L., Layalle, S., Pondeville, E., Dauphin-Villemant, C., et al. (2005). Antagonistic actions of ecdysone and insulins determine final size in *Drosophila. Science, 310*, 667–670.

Colombani, J., Raisin, S., Pantalacci, S., Radimerski, T., Montagne, J., et al. (2003). A nutrient sensor mechanism controls *Drosophila* growth. *Cell, 114*, 739–749.

Cooper, T. M., Mockett, R. J., Sohal, B. H., Sohal, R. S., & Orr, W. C. (2004). Effect of caloric restriction on life span of the housefly, *Musca domestica. FASEB J., 18*, 1591–1593.

Corby-Harris, V., Drexler, A., Watkins de Jong, L., Antonova, Y., Pakpour, N., et al. (2010). Activation of Akt signaling reduces the prevalence and intensity of malaria parasite infection and lifespan in *Anopheles stephensi* mosquitoes. *PLoS Pathog., 6,* e1001003.

Corl, A. B., Rodan, A. R., & Heberlein, U. (2005). Insulin signaling in the nervous system regulates ethanol intoxication in *Drosophila melanogaster. Nat. Neurosci., 8,* 18–19.

Corona, M., Velarde, R. A., Remolina, S., Moran-Lauter, A., Wang, Y., et al. (2007). Vitellogenin, juvenile hormone, insulin signaling, and queen honey bee longevity. *Proc. Natl. Acad. Sci. USA, 104,* 7128–7133.

de Azevedo, S. V., & Hartfelder, K. (2008). The insulin signaling pathway in honey bee (*Apis mellifera*) caste development — differential expression of insulin-like peptides and insulin receptors in queen and worker larvae. *J. Insect Physiol., 54,* 1064–1071.

De Loof, A. (2008). Ecdysteroids, juvenile hormone and insect neuropeptides: Recent successes and remaining major challenges. *Gen. Comp. Endocrinol., 155,* 3–13.

de Velasco, B., Erclik, T., Shy, D., Sclafani, J., Lipshitz, H., et al. (2007). Specification and development of the pars intercerebralis and pars lateralis, neuroendocrine command centers in the *Drosophila* brain. *Dev. Biol., 302,* 309–323.

Dennis, P. B., Pullen, N., Kozma, S. C., & Thomas, G. (1996). The principal rapamycin-sensitive p70(s6k) phosphorylation sites, T-229 and T-389, are differentially regulated by rapamycin-insensitive kinase kinases. *Mol. Cell Biol., 16,* 6242–6251.

Diangelo, J. R., Bland, M. L., Bambina, S., Cherry, S., & Birnbaum, M. J. (2009). The immune response attenuates growth and nutrient storage in *Drosophila* by reducing insulin signaling. *Proc. Natl. Acad. Sci. USA, 106,* 20853–20858.

Dionne, M. S., Pham, L. N., Shirasu-Hiza, M., & Schneider, D. S. (2006). Akt and FOXO dysregulation contribute to infection-induced wasting in *Drosophila. Curr. Biol., 16,* 1977–1985.

Drosophila 12 Genomes Consortium Clark, A. G., Eisen, M. B., Smith, D. R., & Bergman, C. M., et al. (2007). Evolution of genes and genomes on the *Drosophila* phylogeny. *Nature, 450,* 203–218.

Drummond, A., & Strimmer, K. (2001). PAL: an object-oriented programming library for molecular evolution and phylogenetics. *Bioinformatics, 17,* 662–663.

Du, K., Herzig, S., Kulkarni, R. N., & Montminy, M. (2003). TRB3: A tribbles homologue that inhibits Akt/PKB activation by insulin in liver. *Science, 300,* 1574–1577.

Edgar, B. A. (2006). How flies get their size: Genetics meets physiology. *Nat. Rev. Genet., 7,* 907–916.

Elchebly, M., Payette, P., Michaliszyn, E., Cromlish, W., Collins, S., et al. (1999). Increased insulin sensitivity and obesity resistance in mice lacking the protein tyrosine phosphatase-1B gene. *Science, 283,* 1544–1548.

Evans, E. A., Chen, W. C., & Tan, M. W. (2008). The DAF-2 insulin-like signaling pathway independently regulates aging and immunity in *C. elegans. Aging Cell, 7,* 879–893.

Fernandez-Gamba, A., Leal, M. C., Morelli, L., & Castano, E. M. (2009). Insulin-degrading enzyme: Structure-function relationship and its possible roles in health and disease. *Curr. Pharm. Des., 15,* 3644–3655.

Fielenbach, N., & Antebi, A. (2008). C. elegans dauer formation and the molecular basis of plasticity. *Genes Dev., 22,* 2149–2165.

Flatt, T., Tu, M. P., & Tatar, M. (2005). Hormonal pleiotropy and the juvenile hormone regulation of *Drosophila* development and life history. *Bioessays, 27,* 999–1010.

Floyd, P. D., Li, L., Rubakhin, S. S., Sweedler, J. V., Horn, C. C., et al. (1999). Insulin prohormone processing, distribution, and relation to metabolism in *Aplysia californica. J. Neurosci., 19,* 7732–7741.

Fontana, L., Partridge, L., & Longo, V. D. (2010). Extending healthy life span — from yeast to humans. *Science, 328,* 321–326.

Frame, S., Cohen, P., & Biondi, R. M. (2001). A common phosphate binding site explains the unique substrate specificity of GSK3 and its inactivation by phosphorylation. *Mol. Cell, 7,* 1321–1327.

Fullbright, G., & Bullesbach, E. E. (2000). The receptor binding conformation of bombyxin is induced by alanine(B15). *Biochemistry, 39,* 9718–9724.

Fuss, B., Becker, T., Zinke, I., & Hoch, M. (2006). The cytohesin Steppke is essential for insulin signalling in *Drosophila. Nature, 444,* 945–948.

Gade, G. (2004). Regulation of intermediary metabolism and water balance of insects by neuropeptides. *Annu. Rev. Entomol., 49,* 93–113.

Garcia, J. V., Fenton, B. W., & Rosner, M. R. (1988). Isolation and characterization of an insulin-degrading enzyme from *Drosophila melanogaster. Biochemistry, 27,* 4237–4244.

Garofalo, R. S. (2002). Genetic analysis of insulin signaling in *Drosophila. Trends Endocrinol. Metab., 13,* 156–162.

Garsin, D. A., Villanueva, J. M., Begun, J., Kim, D. H., Sifri, C. D., et al. (2003). Long-lived *C. elegans* daf-2 mutants are resistant to bacterial pathogens. *Science, 300,* 1921.

Geminard, C., Rulifson, E. J., & Leopold, P. (2009). Remote control of insulin secretion by fat cells in *Drosophila. Cell Metab., 10,* 199–207.

Gershman, B., Puig, O., Hang, L., Peitzsch, R. M., Tatar, M., et al. (2007). High-resolution dynamics of the transcriptional response to nutrition in *Drosophila:* a key role for dFOXO. *Physiol. Genomics, 29,* 24–34.

Giannakou, M. E., Goss, M., Junger, M. A., Hafen, E., Leevers, S. J., et al. (2004). Long-lived *Drosophila* with overexpressed dFOXO in adult fat body. *Science, 305,* 361.

Glauser, D. A., & Schlegel, W. (2007). The emerging role of FOXO transcription factors in pancreatic beta cells. *J. Endocrinol., 193,* 195–207.

Goberdhan, D. C., Meredith, D., Boyd, C. A., & Wilson, C. (2005). PAT-related amino acid transporters regulate growth via a novel mechanism that does not require bulk transport of amino acids. *Development, 132,* 2365–2375.

Goberdhan, D. C., Paricio, N., Goodman, E. C., Mlodzik, M., & Wilson, C. (1999). *Drosophila* tumor suppressor PTEN controls cell size and number by antagonizing the Chico/PI3-kinase signaling pathway. *Genes Dev., 13,* 3244–3258.

Goberdhan, D. C., & Wilson, C. (2003). The functions of insulin signaling: size isn't everything, even in *Drosophila. Differentiation, 71,* 375–397.

Gollob, J. A., Wilhelm, S., Carter, C., & Kelley, S. L. (2006). Role of Raf kinase in cancer: therapeutic potential of targeting the Raf/MEK/ERK signal transduction pathway. *Semin. Oncol.*, *33*, 392–406.

Goto, S., Loeb, M. J., & Takeda, M. (2005). Bombyxin stimulates proliferation of cultured stem cells derived from *Heliothis virescens* and *Mamestra brassicae* larvae. *In Vitro Cell Dev. Biol. Anim.*, *41*, 38–42.

Graf, R., Neuenschwander, S., Brown, M. R., & Ackermann, U. (1997). Insulin-mediated secretion of ecdysteroids from mosquito ovaries and molecular cloning of the insulin receptor homologue from ovaries of bloodfed *Aedes aegypti*. *Insect Mol. Biol.*, *6*, 151–163.

Greer, E. L., & Brunet, A. (2005). FOXO transcription factors at the interface between longevity and tumor suppression. *Oncogene*, *24*, 7410–7425.

Grewal, S. S. (2009). Insulin/TOR signaling in growth and homeostasis: a view from the fly world. *Int. J. Biochem. Cell Biol.*, *41*, 1006–1010.

Gronke, S., Clarke, D. F., Broughton, S., Andrews, T. D., & Partridge, L. (2010). Molecular evolution and functional characterization of *Drosophila* insulin-like peptides. *PLoS Genet.*, *6*, e1000857.

Gruntenko, N. E., & Rauschenbach, I. Y. (2008). Interplay of JH, 20E and biogenic amines under normal and stress conditions and its effect on reproduction. *J. Insect Physiol.*, *54*, 902–908.

Gu, S. H., Lin, J. L., Lin, P. L., & Chen, C. H. (2009). Insulin stimulates ecdysteroidogenesis by prothoracic glands in the silkworm, *Bombyx mori*. *Insect Biochem. Mol. Biol.*, *39*, 171–179.

Guindon, S., & Gascuel, O. (2003). A simple, fast, and accurate algorithm to estimate large phylogenies by maximum likelihood. *Syst. Biol.*, *52*, 696–704.

Gutierrez, A., Nieto, J., Pozo, F., Stern, S., & Schoofs, L. (2007). Effect of insulin/IGF-I like peptides on glucose metabolism in the white shrimp *Penaeus vannamei*. *Gen. Comp. Endocrinol.*, *153*, 170–175.

Hamano, K., Awaji, M., & Usuki, H. (2005). cDNA structure of an insulin-related peptide in the Pacific oyster and seasonal changes in the gene expression. *J. Endocrinol.*, *187*, 55–67.

Hansen, I. A., Attardo, G. M., Park, J. H., Peng, Q., & Raikhel, A. S. (2004). Target of rapamycin-mediated amino acid signaling in mosquito anautogeny. *Proc. Natl. Acad. Sci. USA*, *101*, 10626–10631.

Hansen, I. A., Attardo, G. M., Roy, S. G., & Raikhel, A. S. (2005). Target of rapamycin-dependent activation of S6 kinase is a central step in the transduction of nutritional signals during egg development in a mosquito. *J. Biol. Chem.*, *280*, 20565–20572.

Hansen, I. A., Sieglaff, D. H., Munro, J. B., Shiao, S. H., Cruz, J., et al. (2007). Forkhead transcription factors regulate mosquito reproduction. *Insect Biochem. Mol. Biol.*, *37*, 985–997.

Harada, S., Smith, R. M., & Jarett, L. (1999). Mechanisms of nuclear translocation of insulin. *Cell Biochem. Biophys.*, *31*, 307–319.

Harrington, L. S., Findlay, G. M., Gray, A., Tolkacheva, T., Wigfield, S., et al. (2004). The TSC1-2 tumor suppressor controls insulin-PI3K signaling via regulation of IRS proteins. *J. Cell Biol.*, *166*, 213–223.

Harris, T. E., & Lawrence, J. C., Jr. (2003). TOR signaling. *Sci.STKE*, re15.

Hatle, J. D., Juliano, S. A., & Borst, D. W. (2003). Hemolymph ecdysteroids do not affect vitellogenesis in the lubber grasshopper. *Arch. Insect Biochem. Physiol.*, *52*, 45–57.

Hernandez-Sanchez, C., Mansilla, A., de la Rosa, E. J., & de Pablo, F. (2006). Proinsulin in development: New roles for an ancient prohormone. *Diabetologia*, *49*, 1142–1150.

Hewes, R. S., Gu, T., Brewster, J. A., Qu, C., & Zhao, T. (2006). Regulation of secretory protein expression in mature cells by DIMM, a basic helix-loop-helix neuroendocrine differentiation factor. *J. Neurosci.*, *26*, 7860–7869.

Hietakangas, V., & Cohen, S. M. (2007). Re-evaluating AKT regulation: role of TOR complex 2 in tissue growth. *Genes Dev.*, *21*, 632–637.

Hietakangas, V., & Cohen, S. M. (2009). Regulation of tissue growth through nutrient sensing. *Annu. Rev. Genet.*, *43*, 389–410.

Hills, C. E., & Brunskill, N. J. (2009). Cellular and physiological effects of C-peptide. *Clin. Sci. (Lond.)*, *116*, 565–574.

Hirashima, Y., Tsuruzoe, K., Kodama, S., Igata, M., Toyonaga, T., et al. (2003). Insulin down-regulates insulin receptor substrate-2 expression through the phosphatidylinositol 3-kinase/Akt pathway. *J. Endocrinol.*, *179*, 253–266.

Honegger, B., Galic, M., Kohler, K., Wittwer, F., Brogiolo, W., et al. (2008). Imp-L2, a putative homologue of vertebrate IGF-binding protein 7, counteracts insulin signaling in *Drosophila* and is essential for starvation resistance. *J. Biol.*, *7*, 10.

Hsu, H. J., LaFever, L., & Drummond-Barbosa, D. (2008). Diet controls normal and tumorous germline stem cells via insulin-dependent and -independent mechanisms in *Drosophila*. *Dev. Biol.*, *313*, 700–712.

Hunt, G. J., Amdam, G. V., Schlipalius, D., Emore, C., Sardesai, N., et al. (2007). Behavioral genomics of honeybee foraging and nest defense. *Naturwissenschaften*, *94*, 247–267.

Hurd, H., & Carter, V. (2004). The role of programmed cell death in Plasmodium-mosquito interactions. *Int. J. Parasitol.*, *34*, 1459–1472.

Huybrechts, J., Bonhomme, J., Minoli, S., Prunier-Leterme, N., Dombrovsky, A., et al. (2010). Neuropeptide and neurohormone precursors in the pea aphid, Acyrthosiphon pisum. *Insect Mol. Biol.*, *19*(Suppl. 2), 87–95.

Hwang, J. R., Siekhaus, D. E., Fuller, R. S., Taghert, P. H., & Lindberg, I. (2000). Interaction of *Drosophila melanogaster* prohormone convertase 2 and 7B2. Insect cell-specific processing and secretion. *J. Biol. Chem.*, *275*, 17886–17893.

Hwangbo, D. S., Gershman, B., Tu, M. P., Palmer, M., & Tatar, M. (2004). *Drosophila* dFOXO controls lifespan and regulates insulin signalling in brain and fat body. *Nature*, *429*, 562–566.

Ikeya, T., Broughton, S., Alic, N., Grandison, R., & Partridge, L. (2009). The endosymbiont Wolbachia increases insulin/IGF-like signalling in *Drosophila*. *Proc. Biol. Sci.*, *276*, 3799–3807.

Ikeya, T., Galic, M., Belawat, P., Nairz, K., & Hafen, E. (2002). Nutrient-dependent expression of insulin-like peptides from neuroendocrine cells in the CNS contributes to growth regulation in *Drosophila*. *Curr. Biol.*, *12*, 1293–1300.

Inoki, K., Li, Y., Zhu, T., Wu, J., & Guan, K. L. (2002). TSC2 is phosphorylated and inhibited by Akt and suppresses mTOR signalling. *Nat. Cell Biol.*, *4*, 648–657.

Iwami, M., Furuya, I., & Kataoka, H. (1996). Bombyxin-related peptides: cDNA structure and expression in the brain of the hornworm Agrius convolvuli. *Insect Biochem. Mol. Biol.*, *26*, 25–32.

Joy, T. K., Arik, A. J., Corby-Harris, V., Johnson, A. A., & Riehle, M. A. (2010). The impact of larval and adult dietary restriction on lifespan, reproduction and growth in the mosquito *Aedes aegypti*. *Exp. Gerontol.*, *45*, 685–690.

Junger, M. A., Rintelen, F., Stocker, H., Wasserman, J. D., Vegh, M., et al. (2003). The *Drosophila* forkhead transcription factor FOXO mediates the reduction in cell number associated with reduced insulin signaling. *J. Biol.*, *2*, 20.

Kaeberlein, M., & Kennedy, B. K. (2008). Protein translation, 2008. *Aging Cell, 7*, 777–782.

Kaestner, K. H., Knochel, W., & Martinez, D. E. (2000). Unified nomenclature for the winged helix/forkhead transcription factors. *Genes Dev.*, *14*, 142–146.

Kang, M. A., Mott, T. M., Tapley, E. C., Lewis, E. E., & Luckhart, S. (2008). Insulin regulates aging and oxidative stress in Anopheles stephensi. *J. Exp. Biol.*, *211*, 741–748.

Kapahi, P., Zid, B. M., Harper, T., Koslover, D., Sapin, V., et al. (2004). Regulation of lifespan in *Drosophila* by modulation of genes in the TOR signaling pathway. *Curr. Biol.*, *14*, 885–890.

Kaplan, D. D., Zimmermann, G., Suyama, K., Meyer, T., & Scott, M. P. (2008). A nucleostemin family GTPase, NS3, acts in serotonergic neurons to regulate insulin signaling and control body size. *Genes Dev.*, *22*, 1877–1893.

Katoh, M., & Katoh, M. (2004). Human FOX gene family (Review). *Int. J. Oncol.*, *25*, 1495–1500.

Kawada, T., Sekiguchi, T., Sakai, T., Aoyama, M., & Satake, H. (2010). Neuropeptides, hormone peptides, and their receptors in Ciona intestinalis: An update. *Zool. Sci.*, *27*, 134–153.

Kim, S. E., Cho, J. Y., Kim, K. S., Lee, S. J., Lee, K. H., et al. (2004). *Drosophila* PI3 kinase and Akt involved in insulin-stimulated proliferation and ERK pathway activation in Schneider cells. *Cell Signal.*, *16*, 1309–1317.

Kimura-Kawakami, M., Iwami, M., Kawakami, A., Nagasawa, H., Suzuki, A., et al. (1992). Structure and expression of bombyxin-related peptide genes of the moth *Samia cynthia ricini*. *Gen. Comp. Endocrinol.*, *86*, 257–268.

Kong, R. C., Shilling, P. J., Lobb, D. K., Gooley, P. R., & Bathgate, R. A. (2010). Membrane receptors: Structure and function of the relaxin family peptide receptors. *Mol. Cell Endocrinol.*, *320*, 1–15.

Koyama, T., Syropyatova, M. O., & Riddiford, L. M. (2008). Insulin/IGF signaling regulates the change in commitment in imaginal discs and primordia by overriding the effect of juvenile hormone. *Dev. Biol.*, *324*, 258–265.

Krieger, M. J., Jahan, N., Riehle, M. A., Cao, C., & Brown, M. R. (2004). Molecular characterization of insulin-like peptide genes and their expression in the African malaria mosquito, *Anopheles gambiae*. *Insect Mol. Biol.*, *13*, 305–315.

LaFever, L., & Drummond-Barbosa, D. (2005). Direct control of germline stem cell division and cyst growth by neural insulin in *Drosophila*. *Science*, *309*, 1071–1073.

LaFever, L., Feoktistov, A., Hsu, H. J., & Drummond-Barbosa, D. (2010). Specific roles of Target of rapamycin in the control of stem cells and their progeny in the *Drosophila* ovary. *Development*, *137*, 2117–2126.

Lagueux, M., Lwoff, L., Meister, M., Goltzene, F., & Hoffmann, J. A. (1990). cDNAs from neurosecretory cells of brains of *Locusta migratoria* (Insecta, Orthoptera) encoding a novel member of the superfamily of insulins. *Eur. J. Biochem.*, *187*, 249–254.

Lam, E. W., Francis, R. E., & Petkovic, M. (2006). FOXO transcription factors: key regulators of cell fate. *Biochem. Soc. Trans.*, *34*, 722–726.

Lee, K. S., Hong, S. H., Kim, A. K., Ju, S. K., Kwon, O. Y., et al. (2009). Processed short neuropeptide F peptides regulate growth through the ERK-insulin pathway in *Drosophila* melanogaster. *FEBS Lett.*, *583*, 2573–2577.

Lee, K. S., Kwon, O. Y., Lee, J. H., Kwon, K., Min, K. J., et al. (2008). *Drosophila* short neuropeptide F signalling regulates growth by ERK-mediated insulin signalling. *Nat. Cell Biol.*, *10*, 468–475.

Leevers, S. J., Weinkove, D., MacDougall, L. K., Hafen, E., & Waterfield, M. D. (1996). The *Drosophila* phosphoinositide 3-kinase Dp110 promotes cell growth. *EMBO J.*, *15*, 6584–6594.

Lemaitre, B., & Hoffmann, J. (2007). The host defense of *Drosophila* melanogaster. *Annu. Rev. Immunol.*, *25*, 697–743.

Leopold, P., & Perrimon, N. (2007). *Drosophila* and the genetics of the internal milieu. *Nature*, *450*, 186–188.

Libert, S., Chao, Y., Zwiener, J., & Pletcher, S. D. (2008). Realized immune response is enhanced in long-lived puc and chico mutants but is unaffected by dietary restriction. *Mol. Immunol.*, *45*, 810–817.

Libina, N., Berman, J. R., & Kenyon, C. (2003). Tissue-specific activities of *C. elegans* DAF-16 in the regulation of lifespan. *Cell*, *115*, 489–502.

Lim, J., Gowda, D. C., Krishnegowda, G., & Luckhart, S. (2005). Induction of nitric oxide synthase in *Anopheles stephensi* by Plasmodium falciparum: mechanism of signaling and the role of parasite glycosylphosphatidylinositols. *Infect. Immun.*, *73*, 2778–2789.

Lingo, P. R., Zhao, Z., & Shen, P. (2007). Co-regulation of cold-resistant food acquisition by insulin- and neuropeptide Y-like systems in *Drosophila* melanogaster. *Neuroscience*, *148*, 371–374.

Liu, Y., Zhou, S., Ma, L., Tian, L., Wang, S., et al. (2010). Transcriptional regulation of the insulin signaling pathway genes by starvation and 20-hydroxyecdysone in the *Bombyx* fat body. *J. Insect Physiol.*, *56*, 1436–1444.

Luckhart, S., Li, K., Dunton, R., Lewis, E. E., Crampton, A. L., et al. (2003). *Anopheles gambiae* immune gene variants associated with natural Plasmodium infection. *Mol. Biochem. Parasitol.*, *128*, 83–86.

Luckhart, S., & Riehle, M. A. (2007). The insulin signaling cascade from nematodes to mammals: Insights into innate immunity of Anopheles mosquitoes to malaria parasite infection. *Dev. Comp. Immunol.*, *31*, 647–656.

Luckhart, S., Vodovotz, Y., Cui, L., & Rosenberg, R. (1998). The mosquito *Anopheles stephensi* limits malaria parasite development with inducible synthesis of nitric oxide. *Proc. Natl. Acad. Sci. USA*, *95*, 5700–5705.

Maddison, W. P., & Maddison, D. R. (*Mesquite: a modular system for evolutionary analysis*. http://mesquiteproject.org2010, Version 2.74.

Maehama, T., & Dixon, J. E. (1998). The tumor suppressor, PTEN/MMAC1, dephosphorylates the lipid second messenger, phosphatidylinositol 3,4,5-trisphosphate. *J. Biol. Chem., 273*, 13375–13378.

Mair, W., & Dillin, A. (2008). Aging and survival: the genetics of life span extension by dietary restriction. *Annu. Rev. Biochem., 77*, 727–754.

Maniere, G., Rondot, I., Bullesbach, E. E., Gautron, F., Vanhems, E., et al. (2004). Control of ovarian steroidogenesis by insulin-like peptides in the blowfly (*Phormia regina*). *J. Endocrinol., 181*, 147–156.

Maniere, G., Vanhems, E., Rondot, I., & Delbecque, J. P. (2009). Control of ovarian steroidogenesis in insects: A locust neurohormone is active *in vitro* on blowfly ovaries. *Gen. Comp. Endocrinol., 163*, 292–297.

Manning, B. D., & Cantley, L. C. (2007). AKT/PKB signaling: navigating downstream. *Cell, 129*, 1261–1274.

Manning, B. D., Tee, A. R., Logsdon, M. N., Blenis, J., & Cantley, L. C. (2002). Identification of the tuberous sclerosis complex-2 tumor suppressor gene product tuberin as a target of the phosphoinositide 3-kinase/akt pathway. *Mol. Cell, 10*, 151–162.

Marchal, E., Vandersmissen, H. P., Badisco, L., Van de Velde, S., Verlinden, H., et al. (2010). Control of ecdysteroidogenesis in prothoracic glands of insects: a review. *Peptides, 31*, 506–519.

Mattila, J., Bremer, A., Ahonen, L., Kostiainen, R., & Puig, O. (2009). *Drosophila* FoxO regulates organism size and stress resistance through an adenylate cyclase. *Mol. Cell Biol., 29*, 5357–5365.

Miguel-Aliaga, I., Thor, S., & Gould, A. P. (2008). Postmitotic specification of *Drosophila* insulinergic neurons from pioneer neurons. *PLoS Biol., 6*, e58.

Miller, B. S., Shankavaram, U. T., Horney, M. J., Gore, A. C., Kurtz, D. T., et al. (1996). Activation of cJun NH2-terminal kinase/stress-activated protein kinase by insulin. *Biochemistry, 35*, 8769–8775.

Mirth, C., Truman, J. W., & Riddiford, L. M. (2005). The role of the prothoracic gland in determining critical weight for metamorphosis in *Drosophila melanogaster*. *Curr. Biol., 15*, 1796–1807.

Mirth, C. K., & Riddiford, L. M. (2007). Size assessment and growth control: how adult size is determined in insects. *Bioessays, 29*, 344–355.

Mirth, C. K., Truman, J. W., & Riddiford, L. M. (2009). The ecdysone receptor controls the post-critical weight switch to nutrition-independent differentiation in *Drosophila* wing imaginal discs. *Development, 136*, 2345–2353.

Mita, M., Yoshikuni, M., Ohno, K., Shibata, Y., Paul-Prasanth, B., et al. (2009). A relaxin-like peptide purified from radial nerves induces oocyte maturation and ovulation in the starfish, *Asterina pectinifera*. *Proc. Natl. Acad. Sci. USA, 106*, 9507–9512.

Mohan, S., & Baylink, D. J. (2002). IGF-binding proteins are multifunctional and act via IGF-dependent and -independent mechanisms. *J. Endocrinol., 175*, 19–31.

Mulenga, A., & Khumthong, R. (2010). Disrupting the Amblyomma americanum (L.) CD147 receptor homologue prevents ticks from feeding to repletion and blocks spontaneous detachment of ticks from their host. *Insect Biochem. Mol. Biol., 40*, 524–532.

Murphy, C. T., Lee, S. J., & Kenyon, C. (2007). Tissue entrainment by feedback regulation of insulin gene expression in the endoderm of *Caenorhabditis elegans*. *Proc. Natl. Acad. Sci. USA, 104*, 19046–19050.

Nagata, K. (2010). Studies of the structure–activity relationships of peptides and proteins involved in growth and development based on their three-dimensional structures. *Biosci. Biotechnol. Biochem., 74*, 462–470.

Nagata, K., Maruyama, K., Kojima, K., Yamamoto, M., Tanaka, M., et al. (1999). Prothoracicotropic activity of SBRPs, the insulin-like peptides of the saturniid silkworm *Samia cynthia ricini*. *Biochem. Biophys. Res. Commun., 266*, 575–578.

Nakahara, Y., Matsumoto, H., Kanamori, Y., Kataoka, H., Mizoguchi, A., et al. (2006). Insulin signaling is involved in hematopoietic regulation in an insect hematopoietic organ. *J. Insect Physiol., 52*, 105–111.

Nässel, D. R., & Winther, A. M. (2010). *Drosophila* neuropeptides in regulation of physiology and behavior. *Prog. Neurobiol., 92*, 42–104.

Nijhout, H. F., & Grunert, L. W. (2002). Bombyxin is a growth factor for wing imaginal disks in Lepidoptera. *Proc. Natl. Acad. Sci. USA, 99*, 15446–15450.

Nijhout, H. F., Smith, W. A., Schachar, I., Subramanian, S., Tobler, A., et al. (2007). The control of growth and differentiation of the wing imaginal disks of *Manduca sexta*. *Dev. Biol., 302*, 569–576.

O'Connor, K. J., & Baxter, D. (1985). The demonstration of insulin-like material in the honey bee, *Apis mellifera*. *Comp. Biochem. Physiol., 81B*, 755–760.

Okada, Y., Miyazaki, S., Miyakawa, H., Ishikawa, A., Tsuji, K., et al. (2010). Ovarian development and insulin-signaling pathways during reproductive differentiation in the queenless ponerine ant Diacamma sp. *J. Insect Physiol., 56*, 288–295.

Okamoto, N., Yamanaka, N., Satake, H., Saegusa, H., Kataoka, H., et al. (2009a). An ecdysteroid-inducible insulin-like growth factor-like peptide regulates adult development of the silkmoth *Bombyx mori*. *FEBS J., 276*, 1221–1232.

Okamoto, N., Yamanaka, N., Yagi, Y., Nishida, Y., Kataoka, H., et al. (2009b). A fat body-derived IGF-like peptide regulates postfeeding growth in *Drosophila*. *Dev. Cell, 17*, 885–891.

Okanda, F. M., Dao, A., Njiru, B. N., Arija, J., Akelo, H. A., et al. (2002). Behavioural determinants of gene flow in malaria vector populations: Anopheles gambiae males select large females as mates. *Malar. J., 1*, 10.

Olinski, R. P., Dahlberg, C., Thorndyke, M., & Hallbook, F. (2006). Three insulin-relaxin-like genes in *Ciona intestinalis*. *Peptides, 27*, 2535–2546.

Orme, M. H., Alrubaie, S., Bradley, G. L., Walker, C. D., & Leevers, S. J. (2006). Input from Ras is required for maximal PI(3)K signalling in *Drosophila*. *Nat. Cell Biol., 8*, 1298–1302.

Paaby, A. B., & Schmidt, P. S. (2009). Dissecting the genetics of longevity in *Drosophila melanogaster*. *Fly (Austin), 3*, 29–38.

Parrella, E., & Longo, V. D. (2010). Insulin/IGF-I and related signaling pathways regulate aging in nondividing cells: From yeast to the mammalian brain. *Sci. World J., 10*, 161–177.

Partridge, L., Alic, N., Bjedov, I., & Piper, M. D. (2011). Ageing in *Drosophila*: The role of the insulin/Igf and TOR signalling network. *Exp. Gerontol.*, *46*, 376–381.

Peterson, T. M., Gow, A. J., & Luckhart, S. (2007). Nitric oxide metabolites induced in *Anopheles stephensi* control malaria parasite infection. *Free Radic. Biol. Med.*, *42*, 132–142.

Petruzzelli, L., Herrera, R., Garcia-Arenas, R., & Rosen, O. M. (1985). Acquisition of insulin-dependent protein tyrosine kinase activity during *Drosophila* embryogenesis. *J. Biol. Chem.*, *260*, 16072–16075.

Piper, M. D., & Partridge, L. (2007). Dietary restriction in *Drosophila*: Delayed aging or experimental artefact? *PLoS Genet.*, *3*, e57.

Potter, C. J., Pedraza, L. G., & Xu, T. (2002). Akt regulates growth by directly phosphorylating Tsc2. *Nat. Cell Biol.*, *4*, 658–665.

Puig, O., & Tjian, R. (2005). Transcriptional feedback control of insulin receptor by dFOXO/FOXO1. *Genes Dev.*, *19*, 2435–2446.

Raikhel, A. S. (2004). Vitellogenesis of Disease Vectors, from Cell Biology to Genes. In B. J. Beaty, & W. Marquardt (Eds.), *Biology of Vectors* (pp. 329–346). New York: Elsevier-Academic Press.

Raikhel, A. S., Kokoza, V. A., Zhu, J., Martin, D., Wang, S. F., et al. (2002). Molecular biology of mosquito vitellogenesis: from basic studies to genetic engineering of antipathogen immunity. *Insect Biochem. Mol. Biol.*, *32*, 1275–1286.

Rayburn, L. Y., Rhea, J., Jocoy, S. R., & Bender, M. (2009). The proprotein convertase amontillado (amon) is required during *Drosophila* pupal development. *Dev. Biol.*, *333*, 48–56.

Rees, D. A., Giles, P., Lewis, M. D., & Ham, J. (2010). Adenosine regulates thrombomodulin and endothelial protein C receptor expression in folliculostellate cells of the pituitary gland. *Purinergic Signal.*, *6*, 19–29.

Rewitz, K. F., Yamanaka, N., Gilbert, L. I., & O'Connor, M. B. (2009). The insect neuropeptide PTTH activates receptor tyrosine kinase torso to initiate metamorphosis. *Science*, *326*, 1403–1405.

Reynolds, B., Laynes, R., Ogmundsdottir, M. H., Boyd, C. A., & Goberdhan, D. C. (2007). Amino acid transporters and nutrient-sensing mechanisms: new targets for treating insulin-linked disorders? *Biochem. Soc. Trans.*, *35*, 1215–1217.

Rhea, J. M., Wegener, C., & Bender, M. (2010). The proprotein convertase encoded by amontillado (amon) is required in *Drosophila* corpora cardiaca endocrine cells producing the glucose regulatory hormone AKH. *PLoS Genet.*, *6*, e1000967.

Richard, D. S., Rybczynski, R., Wilson, T. G., Wang, Y., Wayne, M. L., et al. (2005). Insulin signaling is necessary for vitellogenesis in *Drosophila* melanogaster independent of the roles of juvenile hormone and ecdysteroids: Female sterility of the chico1 insulin signaling mutation is autonomous to the ovary. *J. Insect Physiol.*, *51*, 455–464.

Riddiford, L. M. (2008). Juvenile hormone action: a 2007 perspective. *J. Insect Physiol.*, *54*, 895–901.

Riehle, M. A., & Brown, M. R. (1999). Insulin stimulates ecdysteroid production through a conserved signaling cascade in the mosquito *Aedes aegypti*. *Insect Biochem. Mol. Biol.*, *29*, 855–860.

Riehle, M. A., Fan, Y., Cao, C., & Brown, M. R. (2006). Molecular characterization of insulin-like peptides in the yellow fever mosquito, *Aedes aegypti*: Expression, cellular localization, and phylogeny. *Peptides.*, *27*, 2547–2560.

Riehle, M. A., Garczynski, S. F., Crim, J. W., Hill, C. A., & Brown, M. R. (2002). Neuropeptides and peptide hormones in *Anopheles gambiae*. *Science*, *298*, 172–175.

Rintelen, F., Stocker, H., Thomas, G., & Hafen, E. (2001). PDK1 regulates growth through Akt and S6K in *Drosophila*. *Proc. Natl. Acad. Sci. USA*, *98*, 15020–15025.

Rogina, B., Reenan, R. A., Nilsen, S. P., & Helfand, S. L. (2000). Extended life-span conferred by cotransporter gene mutations in *Drosophila*. *Science*, *290*, 2137–2140.

Roy, S. G., Hansen, I. A., & Raikhel, A. S. (2007). Effect of insulin and 20-hydroxyecdysone in the fat body of the yellow fever mosquito, *Aedes aegypti*. *Insect Biochem. Mol. Biol.*, *37*, 1317–1326.

Ruan, Y., Chen, C., Cao, Y., & Garofalo, R. S. (1995). The *Drosophila* insulin receptor contains a novel carboxyl-terminal extension likely to play an important role in signal transduction. *J. Biol. Chem.*, *270*, 4236–4243.

Rui, L., Yuan, M., Frantz, D., Shoelson, S., & White, M. F. (2002). SOCS-1 and SOCS-3 block insulin signaling by ubiquitin-mediated degradation of IRS1 and IRS2. *J. Biol. Chem.*, *277*, 42394–42398.

Saegusa, H., Mizoguchi, A., Kitahora, H., Nagasawa, H., Suzuki, A., et al. (1992). Changes in the titer of bombyxin-immunoreactive material in hemolymph during the postembryonic development of the silkmoth *Bombyx mori*. *Dev. Growth Differ.*, *34*, 595–605.

Sajid, W., Kulahin, N., Schluckebier, G., Ribel, U., Henderson, H. R., et al. (2011). Structural and biological properties of the *Drosophila* insulin-like peptide 5 show evolutionary conservation. *J. Biol. Chem.*, *286*, 661–673.

Sanchez, L. (2008). Sex-determining mechanisms in insects. *Int. J. Dev. Biol.*, *52*, 837–856.

Sano, H., Kane, S., Sano, E., Miinea, C. P., Asara, J. M., et al. (2003). Insulin-stimulated phosphorylation of a Rab GTPase-activating protein regulates GLUT4 translocation. *J. Biol. Chem.*, *278*, 14599–14602.

Sarbassov, D. D., Ali, S. M., Kim, D. H., Guertin, D. A., Latek, R. R., et al. (2004). Rictor, a novel binding partner of mTOR, defines a rapamycin-insensitive and raptor-independent pathway that regulates the cytoskeleton. *Curr. Biol.*, *14*, 1296–1302.

Sarbassov, D. D., Guertin, D. A., Ali, S. M., & Sabatini, D. M. (2005). Phosphorylation and regulation of Akt/PKB by the rictor-mTOR complex. *Science*, *307*, 1098–1101.

Satake, S., Masumura, M., Ishizaki, H., Nagata, K., Kataoka, H., et al. (1997). Bombyxin, an insulin-related peptide of insects, reduces the major storage carbohydrates in the silkworm *Bombyx mori*. *Comp. Biochem. Physiol. B. Biochem. Mol. Biol.*, *118*, 349–357.

Satake, S., Nagata, K., Kataoka, H., & Mizoguchi, A. (1999). Bombyxin secretion in the adult silkmoth *Bombyx mori*: Sex-specificity and its correlation with metabolism. *J. Insect Physiol.*, *45*, 939–945.

Schilder, R. J., & Marden, J. H. (2006). Metabolic syndrome and obesity in an insect. *Proc. Natl. Acad. Sci. USA*, *103*, 18805–18809.

Shabanpoor, F., Separovic, F., & Wade, J. D. (2009). The human insulin superfamily of polypeptide hormones. *Vitam. Horm.*, *80*, 1–31.

Shingleton, A. W. (2010). The regulation of organ size in *Drosophila*: Physiology, plasticity, patterning and physical force. *Organogenesis*, *6*, 76–87.

Sim, C., & Denlinger, D. L. (2008). Insulin signaling and FOXO regulate the overwintering diapause of the mosquito *Culex pipiens*. *Proc. Natl. Acad. Sci. USA*, *105*, 6777–6781.

Sim, C., & Denlinger, D. L. (2009a). A shut-down in expression of an insulin-like peptide, ILP-1, halts ovarian maturation during the overwintering diapause of the mosquito *Culex pipiens*. *Insect Mol. Biol.*, *18*, 325–332.

Sim, C., & Denlinger, D. L. (2009b). Transcription profiling and regulation of fat metabolism genes in diapausing adults of the mosquito *Culex pipiens*. *Physiol. Genomics*, *39*, 202–209.

Slack, C., Werz, C., Wieser, D., Alic, N., Foley, A., et al. (2010). Regulation of lifespan, metabolism, and stress responses by the *Drosophila* SH2B protein, Lnk. *PLoS Genet.*, *6*, e1000881.

Slaidina, M., Delanoue, R., Gronke, S., Partridge, L., & Leopold, P. (2009). A *Drosophila* insulin-like peptide promotes growth during nonfeeding states. *Dev. Cell.*, *17*, 874–884.

Smit, A. B., Spijker, S., Van Minnen, J., Burke, J. F., De Winter, F., et al. (1996). Expression and characterization of molluscan insulin-related peptide VII from the mollusc *Lymnaea stagnalis*. *Neuroscience*, *70*, 589–596.

Smit, A. B., van Kesteren, R. E., Li, K. W., Van Minnen, J., Spijker, S., et al. (1998). Towards understanding the role of insulin in the brain: Lessons from insulin-related signaling systems in the invertebrate brain. *Prog. Neurobiol.*, *54*, 35–54.

Spindler, K. D., Honl, C., Tremmel, C., Braun, S., Ruff, H., et al. (2009). Ecdysteroid hormone action. *Cell Mol. Life Sci.*, *66*, 3837–3850.

Sroyraya, M., Chotwiwatthanakun, C., Stewart, M. J., Soonklang, N., Kornthong, N., et al. (2010). Bilateral eyestalk ablation of the blue swimmer crab, Portunus pelagicus, produces hypertrophy of the androgenic gland and an increase of cells producing insulin-like androgenic gland hormone. *Tissue Cell*, *42*, 293–300.

Stamatakis, A., Hoover, P., & Rougemont, J. (2008). A rapid bootstrap algorithm for the RAxML Web servers. *Syst. Biol.*, *57*, 758–771.

Stark, A., Lin, M. F., Kheradpour, P., Pedersen, J. S., Parts, L., et al. (2007). Discovery of functional elements in 12 *Drosophila* genomes using evolutionary signatures. *Nature*, *450*, 219–232.

Steel, C. G., & Vafopoulou, X. (1997). Ecdysteroidogenic action of Bombyx prothoracicotropic hormone and bombyxin on the prothoracic glands of Rhodnius prolixus *in vitro*. *J. Insect Physiol.*, *43*, 651–656.

Steiner, D. F., Park, S. Y., Stoy, J., Philipson, L. H., & Bell, G. I. (2009). A brief perspective on insulin production. *Diabetes Obes. Metab.*, *11*(Suppl. 4), 189–196.

Stewart, M. J., Berry, C. O., Zilberman, F., Thomas, G., & Kozma, S. C. (1996). The *Drosophila* p70s6k homologue exhibits conserved regulatory elements and rapamycin sensitivity. *Proc. Natl. Acad. Sci. USA*, *93*, 10791–10796.

Stoppelli, M. P., Garcia, J. V., Decker, S. J., & Rosner, M. R. (1988). Developmental regulation of an insulin-degrading enzyme from *Drosophila melanogaster*. *Proc. Natl. Acad. Sci. USA*, *85*, 3469–3473.

Suckale, J., & Solimena, M. (2010). The insulin secretory granule as a signaling hub. *Trends Endocrinol. Metab.*, *21*, 599–609.

Swofford, D. L. (2002). PAUP*. Phylogenetic Analysis Using Parsimony (*and Other Methods). *Sinauer Associates*, Version 4.0b10.

Taguchi, A., & White, M. F. (2008). Insulin-like signaling, nutrient homeostasis, and life span. *Annu. Rev. Physiol.*, *70*, 191–212.

Taniguchi, C. M., Emanuelli, B., & Kahn, C. R. (2006). Critical nodes in signalling pathways: Insights into insulin action. *Nat. Rev. Mol. Cell Biol.*, *7*, 85–96.

Tatar, M., Kopelman, A., Epstein, D., Tu, M. P., Yin, C. M., et al. (2001). A mutant *Drosophila* insulin receptor homologue that extends life-span and impairs neuroendocrine function. *Science*, *292*, 107–110.

Teleman, A. A. (2009). Molecular mechanisms of metabolic regulation by insulin in *Drosophila*. *Biochem. J.*, *425*, 13–26.

Teleman, A. A., Hietakangas, V., Sayadian, A. C., & Cohen, S. M. (2008). Nutritional control of protein biosynthetic capacity by insulin via Myc in *Drosophila*. *Cell Metab.*, *7*, 21–32.

Toivonen, J. M., & Partridge, L. (2009). Endocrine regulation of aging and reproduction in *Drosophila*. *Mol. Cell Endocrinol.*, *299*, 39–50.

Torres-Aleman, I. (2010). Toward a comprehensive neurobiology of IGF-I. *Dev. Neurobiol.*, *70*, 384–396.

Tremblay, F., Gagnon, A., Veilleux, A., Sorisky, A., & Marette, A. (2005). Activation of the mammalian target of rapamycin pathway acutely inhibits insulin signaling to Akt and glucose transport in 3T3-L1 and human adipocytes. *Endocrinology*, *146*, 1328–1337.

Tr. Genome Sequencing Consortium Richards, S., Gibbs, R. A., Weinstock, G. M., & Brown, S. J. (2008). The genome of the model beetle and pest Tr. castaneum. *Nature*, *452*, 949–955.

Tsuda, M., Kobayashi, T., Matsuo, T., & Aigaki, T. (2010). Insulin-degrading enzyme antagonizes insulin-dependent tissue growth and Abeta-induced neurotoxicity in *Drosophila*. *FEBS Lett.*, *584*, 2916–2920.

Tu, M. P., Yin, C. M., & Tatar, M. (2002). Impaired ovarian ecdysone synthesis of *Drosophila melanogaster* insulin receptor mutants. *Aging Cell*, *1*, 158–160.

Tu, M. P., Yin, C. M., & Tatar, M. (2005). Mutations in insulin signaling pathway alter juvenile hormone synthesis in *Drosophila melanogaster*. *Gen. Comp. Endocrinol.*, *142*, 347–356.

Tufail, M., & Takeda, M. (2008). Molecular characteristics of insect vitellogenins. *J. Insect Physiol.*, *54*, 1447–1458.

Ueishi, S., Shimizu, H., & H Inoue, Y. (2009). Male germline stem cell division and spermatocyte growth require insulin signaling in *Drosophila*. *Cell Struct. Funct.*, *34*, 61–69.

Ueki, K., Kondo, T., & Kahn, C. R. (2004). Suppressor of cytokine signaling 1 (SOCS-1) and SOCS-3 cause insulin resistance through inhibition of tyrosine phosphorylation of insulin receptor substrate proteins by discrete mechanisms. *Mol. Cell Biol.*, *24*, 5434–5446.

Um, S. H., Frigerio, F., Watanabe, M., Picard, F., Joaquin, M., et al. (2004). Absence of S6K1 protects against age- and diet-induced obesity while enhancing insulin sensitivity. *Nature, 431*, 200–205.

Van de Velde, S., Badisco, L., Claeys, I., Verleyen, P., Chen, X., et al. (2007). Insulin-like peptides in Spodoptera littoralis (Lepidoptera): Detection, localization and identification. *Gen. Comp. Endocrinol., 153*, 72–79.

Van Loy, T., Vandersmissen, H. P., Van Hiel, M. B., Poels, J., Verlinden, H., et al. (2008). Comparative genomics of leucine-rich repeats containing G protein-coupled receptors and their ligands. *Gen. Comp. Endocrinol., 155*, 14–21.

Vazquez, F., Grossman, S. R., Takahashi, Y., Rokas, M. V., Nakamura, N., et al. (2001). Phosphorylation of the PTEN tail acts as an inhibitory switch by preventing its recruitment into a protein complex. *J. Biol. Chem., 276*, 48627–48630.

Veenstra, J. A. (2009). Peptidergic paracrine and endocrine cells in the midgut of the fruit fly maggot. *Cell Tissue Res., 336*, 309–323.

Veenstra, J. A. (2010). Neurohormones and neuropeptides encoded by the genome of *Lottia gigantea*, with reference to other mollusks and insects. *Gen. Comp. Endocrinol., 167*, 86–103.

Veenstra, J. A., Agricola, H. J., & Sellami, A. (2008). Regulatory peptides in fruit fly midgut. *Cell Tissue Res., 334*, 499–516.

Ventura, T., Manor, R., Aflalo, E. D., Weil, S., Raviv, S., et al. (2009). Temporal silencing of an androgenic gland-specific insulin-like gene affecting phenotypical gender differences and spermatogenesis. *Endocrinology, 150*, 1278–1286.

Verdu, J., Buratovich, M. A., Wilder, E. L., & Birnbaum, M. J. (1999). Cell-autonomous regulation of cell and organ growth in *Drosophila* by Akt/PKB. *Nat. Cell Biol., 1*, 500–506.

Vereshchagina, N., Ramel, M. C., Bitoun, E., & Wilson, C. (2008). The protein phosphatase PP2A-B' subunit Widerborst is a negative regulator of cytoplasmic activated Akt and lipid metabolism in *Drosophila*. *J. Cell. Sci., 121*, 3383–3392.

Verhaert, P. D., Downer, R. G., Huybrechts, R., & De Loof, A. (1989). A substance resembling somatomedin C in the American cockroach. *Regul. Pept., 25*, 99–110.

Walkiewicz, M. A., & Stern, M. (2009). Increased insulin/insulin growth factor signaling advances the onset of metamorphosis in *Drosophila*. *PLoS One, 4*, e5072.

Wang, P. Y., Neretti, N., Whitaker, R., Hosier, S., Chang, C., et al. (2009). Long-lived Indy and calorie restriction interact to extend life span. *Proc. Natl. Acad. Sci. USA, 106*, 9262–9267.

Wang, S., Tulina, N., Carlin, D. L., & Rulifson, E. J. (2007). The origin of islet-like cells in *Drosophila* identifies parallels to the vertebrate endocrine axis. *Proc. Natl. Acad. Sci. USA, 104*, 19873–19878.

Wang, Y., Mutti, N. S., Ihle, K. E., Siegel, A., Dolezal, A. G., et al. (2010). Down-regulation of honey bee IRS gene biases behavior toward food rich in protein. *PLoS Genet., 6*, e1000896.

Wang, Z. S., Zheng, W. H., & Guo, F. (1986). Effect of Bombyx 4K-PTTH and bovine insulin on testis development of indian silkworm, Philosamia cynthia ricini. In A. B. Borkovec, & D. B. Gelman (Eds.), *Insect Neurochemistry and Neurophysiology* (pp. 335–338). Clifton, New Jersey: Humana Press.

Weaver, R. J., & Audsley, N. (2009). Neuropeptide regulators of juvenile hormone synthesis: Structures, functions, distribution, and unanswered questions. *Ann. N.Y. Acad. Sci., 1163*, 316–329.

Weinkove, D., Neufeld, T. P., Twardzik, T., Waterfield, M. D., & Leevers, S. J. (1999). Regulation of imaginal disc cell size, cell number and organ size by *Drosophila* class I(A) phosphoinositide 3-kinase and its adaptor. *Curr. Biol., 9*, 1019–1029.

Wen, Z., Gulia, M., Clark, K. D., Dhara, A., Crim, J. W., et al. (2010). Two insulin-like peptide family members from the mosquito Aedes aegypti exhibit differential biological and receptor binding activities. *Mol. Cell Endocrinol., 328*, 47–55.

Werz, C., Kohler, K., Hafen, E., & Stocker, H. (2009). The *Drosophila* SH2B family adaptor Lnk acts in parallel to chico in the insulin signaling pathway. *PLoS Genet., 5*, e1000596.

Wessells, R. J., Fitzgerald, E., Cypser, J. R., Tatar, M., & Bodmer, R. (2004). Insulin regulation of heart function in aging fruit flies. *Nat. Genet., 36*, 1275–1281.

Wheeler, D. E., Buck, N., & Evans, J. D. (2006). Expression of insulin pathway genes during the period of caste determination in the honey bee, *Apis mellifera*. *Insect Mol. Biol., 15*, 597–602.

Wheeler, T. J., & Kececioglu, J. D. (2007). Multiple alignment by aligning alignments. *Bioinformatics, 23*, i559–i568.

Whittaker, S., Marais, R., & Zhu, A. X. (2010). The role of signaling pathways in the development and treatment of hepatocellular carcinoma. *Oncogene, 29*, 4989–5005.

Wigby, S., Slack, C., Gronke, S., Martinez, P., Calboli, F. C., et al. (2011). Insulin signalling regulates remating in female *Drosophila*. *Proc. Biol. Sci., 278*, 424–431.

Williams, K. D., Busto, M., Suster, M. L., So, A. K., Ben-Shahar, Y., et al. (2006). Natural variation in *Drosophila melanogaster* diapause due to the insulin-regulated PI3-kinase. *Proc. Natl. Acad. Sci. USA, 103*, 15911–15915.

Williams, K. D., & Sokolowski, M. B. (1993). Diapause in *Drosophila melanogaster* females: A genetic analysis. *Heredity, 71*(Pt 3), 312–317.

Williams, K. D., & Sokolowski, M. B. (2009). Evolution: how fruit flies adapt to seasonal stresses. *Curr. Biol., 19*, R63–R64.

Wu, Q., & Brown, M. R. (2006). Signaling and function of insulin-like peptides in insects. *Annu. Rev. Entomol., 51*, 1–24.

Wu, Q., Zhang, Y., Xu, J., & Shen, P. (2005). Regulation of hunger-driven behaviors by neural ribosomal S6 kinase in *Drosophila*. *Proc. Natl. Acad. Sci. USA, 102*, 13289–13294.

Wullschleger, S., Loewith, R., & Hall, M. N. (2006). TOR signaling in growth and metabolism. *Cell, 124*, 471–484.

Yang, C. H., Belawat, P., Hafen, E., Jan, L. Y., & Jan, Y. N. (2008). *Drosophila* egg-laying site selection as a system to study simple decision-making processes. *Science, 319*, 1679–1683.

Yang, Q., Inoki, K., Kim, E., & Guan, K. L. (2006). TSC1/TSC2 and Rheb have different effects on TORC1 and TORC2 activity. *Proc. Natl. Acad. Sci. USA, 103*, 6811–6816.

Yenush, L., Fernandez, R., Myers, M. G., Jr., Grammer, T. C., Sun, X. J., et al. (1996). The *Drosophila* insulin receptor activates multiple signaling pathways but requires insulin receptor substrate proteins for DNA synthesis. *Mol. Cell Biol., 16*, 2509–2517.

Yuval, B., Wekesa, J. W., & Washino, R. K. (1993). Effect of body size on swarming behavior and mating success of male *Anopheles freeborni* (Diptera: Culicidae). *J. Insect Behav., 6*, 333–342.

Zhang, H., Liu, J., Li, C. R., Momen, B., Kohanski, R. A., et al. (2009). Deletion of *Drosophila* insulin-like peptides causes growth defects and metabolic abnormalities. *Proc. Natl. Acad. Sci. USA, 106*, 19617–19622.

3 Bursicon, a Neuropeptide Hormone that Controls Cuticle Tanning and Wing Expansion

Qisheng Song
University of Missouri, Columbia, MO, USA

© 2012 Elsevier B.V. All Rights Reserved

3.1. Introduction

Bursicon is a neuropeptide responsible for cuticle tanning (melanization and sclerotization) and wing expansion in insects. It was first described over four decades ago as a bioactive cuticle tanning factor present in the central nervous system (CNS) and hemolymph of newly emerged blowflies *Calliphora erythrocephala* in a neck-ligated blowfly assay (Fraenkel and Hsiao, 1962, 1965; Cottrell, 1962a). However, the molecular identity of bursicon was not characterized until four decades later. It is now clear that functional bursicon is a heterodimer consisting of two cystine knot subunits referred to as bursicon (burs) or bursicon α (burs α) and partner of bursicon (pburs) or bursicon β (burs β) (Luo *et al.*, 2005; Mendive *et al.*, 2005). Burs and pburs are encoded by two individual genes: *CG13419* for burs and *CG15284* for pburs. Bursicon acts via a specific *Drosophila* G-protein-coupled receptor (GPCR) DLGR2, encoded by the *rickets* gene (Baker and Truman, 2002). Once activated, DLGR2 stimulates production of cAMP (Luo *et al.*, 2005; Mendive *et al.*, 2005) and activation of cAMP-dependent protein kinase (PKA), resulting in phosphorylation of tyrosine hydroxylase (Davis *et al.*, 2007), which in turn regulates the conversion of tyrosine to 3,4-dihydroxyphenylanaline (DOPA) in the metabolic pathway leading to cuticle tanning. Several excellent reviews summarize early physiological studies of bursicon action (Seligman, 1980; Reynolds, 1983) and recent advances in identification and functional characterization of bursicon (Ewer and Reynolds, 2002; Honegger *et al.*, 2008). This chapter is focused on bursicon, which has already been demonstrated to bind DLGR2 and activate cAMP/PKA signaling, and provides insights into novel bursicon functions beyond cuticle tanning and wing expansion.

3.2. Identification of Bursicon

3.2.1. Discovery of Bursicon

Insects are encased in a semi-rigid exoskeleton (cuticle) that provides insect protection and attachment sites for muscles and internal organs, which facilitates locomotion. On the other hand, it also limits insect growth. Insects must shed their old exoskeleton periodically (molt) to allow growth and metamorphosis. The newly formed exoskeleton after each molt is usually soft, flexible, and lacks physical strength for protection. Tanning of the cuticle must occur in a relatively short period of time after each molt for the new exoskeleton to be functional for protection. Insects could not survive without properly hardened cuticle.

Hormonal control of cuticle tanning in insects was first reported in the blowfly, *C. erythocephala*, by Fraenkel and Hsiao (1962) and Cottrell (1962a). When the blowfly was neck-ligated immediately after emergence, this

DOI: 10.1016/B978-0-12-384749-2.10003-2

neck-ligation prevented the thoracic and abdominal cuticle from being tanned. To demonstrate the probable hormonal nature of the tanning factor, hemolymph collected from a fly shortly after emergence was injected into the posterior portion of a neck-ligated fly with untanned cuticle. The cuticle of the latter animal quickly darkened as a result of the hemolymph transfusion. Fraenkel and Hsiao (1965) concluded that a blood-borne factor must be responsible for triggering the cuticular tanning process of newly emerged flies and named the factor "bursicon" (from the Greek *bursikos*, pertaining to tanning). The newly emerged, neck-ligated blowflies with untanned thoracic and abdominal regions were adopted as standard bioassay animals for detecting and localizing bursicon.

3.2.2. Molecular Nature of Bursicon

The molecular nature of bursicon was first reported by Cottrell (1962b). Evidence from his studies indicated that the active factor from the blowfly *Calliphora* was non-dialyzable and relatively insoluble in organic solvents. It was also inactivated by alcohol and by the bacterial protease subtilisin, suggesting its protein or polypeptide nature. Fraenkel and Hsiao (1965) also found that bursicon in *Calliphora* could be precipitated like a protein with loss of activity using reagents such as trichloroacetic acid, alcohol, and acetone. Full bursicon activity was retained after ammonium sulfate precipitation, but could be destroyed after treatment with trypsin or pronase in *Manduca* (Taghert and Truman, 1982a). Thus the protein nature of bursicon was clearly established.

After the protein nature of bursicon was established, purification and molecular identification of bursicon became a major focus for the next four decades. Based on the functional bioassay of partially purified samples, bursicon was initially thought to be a single protein with an estimated molecular size of around 30 to 60 kDa, depending on the gel filtration methods used and insect species analyzed (Fraenkel *et al.*, 1966; Seligman and Doy, 1973; Reynolds 1976, 1977; Mills and Lake, 1966; Mills and Nielsen, 1967; Taghert and Truman, 1982a). Using more precise two-dimensional (2D) SDS-PAGE as the key approach, Kaltenhauser *et al.* (1995) characterized the *Tr. molitor* bursicon with a molecular weight of 30 kDa. Similarly, Kostron *et al.* (1995) also attributed bursicon activity to proteins around 30 kDa in species such as *C. erythocephala*, *Glycymeris bimaculata*, *Periplaneta americana*, and *Locusta migratoria*. Surprisingly, bursicon activity could be retained even after the non-denatured SDS-PAGE separation (Kaltenhauser *et al.*, 1995; Kostron *et al.*, 1995), making the estimation of its molecular size much more accurate and easier.

Using partially purified bursicon from the American cockroach, *P. americana*, Honegger *et al.* (2002, 2004) obtained several partial amino acid sequences of the *P. americana* bursicon. By comparing the partial sequences with the available *Drosophila* genomic sequence (Adams *et al.*, 2000), Dewey *et al.* (2004) identified the *Drosophila melanogaster* gene *CG13419* as a candidate bursicon gene. A further study revealed that mutations in *CG13419* showed defects in cuticle tanning and wing expansion in *Drosophila* adults (Dewey *et al.*, 2004), indicating that *CG13419* should be a bursicon gene. However, the molecular weight of *CG13419* product is about 15 kDa, which is only one-half the molecular weight of the detected bursicon from gel filtration and 2D gel (non-denatured) electrophoresis. This indicated that bursicon might be a homo- or heterodimer (Honegger *et al.*, 2002). However, recombinant protein from the *CG13419* gene alone shows no cuticle tanning activity in the neck-ligated fly assay, suggesting that the functional bursicon is unlikely to be a homodimer. Therefore, identification of a heterodimer partner of functional bursicon was extended.

The breakthrough studies came simultaneously from two independent groups showing that CG15284, the only other cystine knot protein (CKP) present in the *Drosophila* genome, was identified as the heterodimer partner of functional bursicon using the available *CG13419* sequence (Luo *et al.*, 2005; Mendive *et al.*, 2005). These two groups reported that only the conditioned media of cells co-transfected with *CG13419* and *CG15284* initiated tanning of the neck-ligated flies while no bursicon activity was detected in the conditioned medium transfected with *CG13419* or *CG15284* alone. Thus, after more than four decades, the functional bursicon has finally been identified as a heterodimer consisting of two subunits named burs (CG13419) and pburs (CG15284) with a molecular weight of 15 kDa for each (Dewey *et al.*, 2004; Luo *et al.*, 2005; Mendive *et al.*, 2005). These results are consistent with the early estimate of bursicon size at about 30 kDa from gel filtration and SDS-PAGE analysis of different insect species.

3.2.3. Cystine Knot Protein

Both burs and pburs belong to a class of vertebrate signal proteins containing a consensus framework of cystine knots produced by disulfide bonds. The vertebrate CKP family contains six subgroups: (1) transforming growth factor-beta (TGF-β) like subgroup, (2) glycoprotein hormone subgroup, (3) platelet-derived growth factor (PDGF) like subgroup, (4) mucin like subgroup, (5) Slit-like subgroup, and (6) jagged-like subgroup (Vitt *et al.*, 2001). The TGF-β-like subgroup and PDGF-like subgroup are growth factors, containing nerve growth factor (NGF), TGF-β, and PDGF (Hearn and Gomme, 2000). These growth factors form hetero- or homodimers with different receptor-binding specificities (Hart and Bowen-Pope, 1990). The glycoprotein hormone subgroup includes the anterior pituitary hormones follitropin (FSH), lutropin (LH), thyrotropin (TSH), and the placental chorionic gonadotropin (CG). These hormones are heterodimers formed by an α subunit and a β subunit, which establish the structural and functional identity of the individual heterodimer (Butler *et al.*, 1999). However, a homodimer of the β subunit of CG has also been proposed to be a marker for many epithelial tumors (Butler *et al.*, 1999). In addition to the cystine knot motif, the mucin-like subgroup displays a conserved CXXCX{13} C signature in all members of this group, including the von Willebrand factor (vWF), Norrie disease (pseudoglioma; NDP), and the bone morphogenetic protein (BMP) antagonists. Both Slit- and jagged-like subgroups contain EGF signatures and are cysteine rich proteins. Only limited members from each of these two subgroups are found in human.

CKPs have also been identified in invertebrates. For example, the *daf-7* gene, which controls larval development in *Caenorhabditis elegans*, is predicted to contain a cystine motif of the TGF-β superfamily (Ren *et al.*, 1996). The *Drosophila* Slit protein, which is involved in the development of midline glial and commissural axon pathways, has a CKP motif as well (Rothberg *et al.*, 1990). Burs and pburs are the most recently identified members of CKPs in insects and are most closely related to the BMP antagonists in the mucin-like subgroup (Luo *et al.*, 2005). Bursicon was the first member of the invertebrate CKP family to have a defined hormonal function. They contain 11 cysteine residues in conserved positions. The core region between the 11 cysteines is highly conserved in arthropods starting at position C1, indicating that it plays a pivotal role in the structural integrity of these ligands. Based on the cystine knot model presented by Vitt *et al.* (2001), Luo *et al.* (2005) presented a hypothetical three-dimensional structure of burs and pburs (**Figure 1**). Cysteine C6 is thought to be essential for covalent dimer formation and C5 may strengthen the bond.

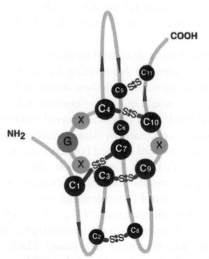

Figure 1 Predicted cystine knot structure of burs and pburs. Adapted from *Luo et al.* (2005), Figure 1.

3.2.4. Localization of Bursicon

To identify the site of bursicon synthesis, extracts from variety parts of the CNS and hemolymph samples at different time periods of test insects were injected into neckligated blowflies. Bursicon activity was demonstrated in almost all parts of the CNS including brain, corpora cardiaca (CC) and corpora allata (CA) complex, and thoracic and abdominal ganglia, depending upon the insect species analyzed (see reviews by Seligman, 1980; Reynolds, 1983; Ewer and Reynolds, 2002; Honegger *et al.*, 2008).

In dipterans such as *Calliphora*, *Sarcophaga*, and *Lucilia*, the fused thoracic–abdominal ganglion was found to have greater bursicon activity than the brain. More important, disappearance of bursicon activity from the thoracic and abdominal region was concurrent with its increase in the hemolymph (Fraenkel and Hsiao, 1963, 1965). In *M. sexta*, bursicon was presumably released via the abdominal ganglion because a clear loss of bursicon activity from the abdominal perivisceral organs was coincident with the appearance of bursicon in the hemolymph (Truman, 1973; Taghert and Truman, 1982a,b). In *P. americana*, release of bursicon was ascribed to the abdominal ganglion because when a ligation was placed between the thoracic and abdominal segments, cuticle darkening was only observed in the posterior portion (Mills, 1965; Mills *et al.*, 1965). In *Leucophaea madera*, the abdomen isolated before ecdysis did not tan while the similarly isolated thorax continued to darken (Srivastava and Hopkins, 1975). Using similar approaches, Vincent (1971, 1972) and Padgham (1976) showed the abdominal and thoracic ganglion to be release sites for bursicon in *L. migratoria* and *Schistocerca gregaria*, respectively. In the coleopteran *T. molitor*, transections of the ventral nerve cord followed by microscopic observations showed that bursicon was released from the perisympathetic organs of the anterior abdominal ganglion (Delachambre, 1971; Grillot *et al.*, 1976).

Subcellular localization of bursicon in the CNS was not definitive until specific antibodies and DNA probes to burs and pburs were developed. Both burs and pburs have been shown to co-localize to a set of neurons in the subesophageal, thoracic, and abdominal ganglia in *Drosophila* and several other insect species (for details see the review by Honegger *et al.*, 2008). **Figure 2** shows the burs and pburs mRNA expression patterns in the subesophageal, thoracic, and abdominal ganglia of *D. melanogaster* and *Musca domestica* (Wang *et al.*, 2008). Immunochemical staining revealed that both burs and pburs subunits are co-localized in a set of crustacean cardioactive peptide (CCAP) containing neurons as well (Peabody *et al.*, 2008; Loveall and Deitcher, 2010). However, as in some insects (Honegger *et al.*, 2002), not all CCAP neurons express bursicon. Bursicon transcripts were not detected in two pairs of CCAP-immunoreactive neurons in the brain or

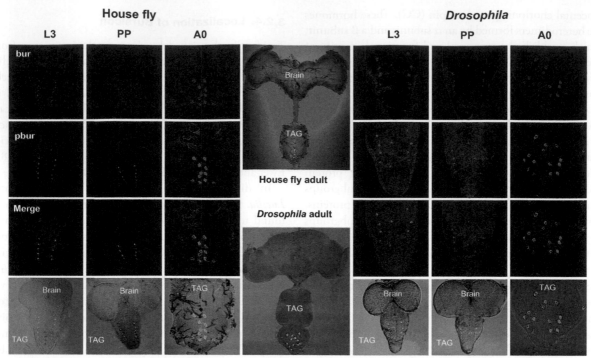

Figure 2 Fluorescence *in situ* hybridization of the cuticle tanning neuropeptide burs (red) and pburs (green) transcripts in the CNS of third instar larvae, pre-pupae and newly emerged adults of *D. melanogaster* (right panel) and *M. domestica* (left panel). The CNS samples were visualized under a confocal microscope and photographed. Yellow color in the merged pictures indicates the neurons with the co-localized burs and pburs signals. Transparent pictures show the images of the CNS. Adapted from Wang *et al.* (2008).

in two pairs in the first thoracic neuromeres. When the CCAP neurons were disrupted by ectopic expression of the fly cell death gene *reaper*, the resulting flies displayed two classes of distinguished phenotypes: one exhibited cuticular deformities and unexpanded wings remarkably similar to *burs* and *rk* mutants and the other was incapable of performing head eversion, a crucial event at pupal ecdysis, resembling the typical phenotype of CCAP deficiency (Park *et al.*, 2003).

3.2.5. Control of Bursicon Release

The release pattern of bursicon from CNS to hemolymph was described as a brief surge in almost all tested species. When hemolymph samples, collected periodically from individual *Manduca* moths via a cannula inserted through the metanotum of the thorax, were monitored for bursicon activity using a *Manduca* wing assay, Reynolds *et al.* (1979) demonstrated that bursicon activity was found to appear in hemolymph within 2 min after adult emergence. By 4–12 min, bursicon activity reached the maximum followed by a decline steadily over next 2 h.

The signal that triggers bursicon release varies in different insect species. Fraenkel and Hsiao (1962) and Cottrell (1962a) demonstrated that when *Calliphora* adults were kept from surfacing through sand or sawdust

at the moment of emergence, cuticle tanning was delayed. Similarly in *M. sexta*, bursicon release could be delayed for at least 24 h by forcing the newly emerged moth to dig (Truman, 1973). This suggested that environmental cues triggered bursicon release in the blowflies and the tobacco hornworm. In locust and cockroach, the movement of exuviae over the new cuticle's sense organs might be the stimulus for bursicon secretion because the timing for bursicon release was closely associated with ecdysis (Cottrell, 1962b; Fraenkel and Hsiao, 1965; Mills *et al.*, 1965).

In terms of intrinsic control mechanisms, Fraenkel and Hsiao (1962) reported that severing the ventral nerve cord and recurrent nerves in *Calliphora* eliminated the tanning response. This would imply that bursicon release is under nervous control. Mills (1967) supported this notion using the American cockroach by demonstrating that severing the ventral nerve cord stopped bursicon secretion. Recent evidence revealed that there are two distinct groups of CCAP-expression neurons: one for bursicon release and the other for control of bursicon release. The first group of CCAP-expressing neurons (N_{CCAP}-c929) is responsible for release of bursicon into the hemolymph. Suppression of CCAP activity within this group blocks bursicon release into the hemolymph together with tanning and wing expansion. The second group (N_{CCAP}-R) consists of N$_{CCAP}$ neurons outside the N$_{CCAP}$-c929 pattern. Suppression of synaptic transmission and PKA activity throughout N$_{CCAP}$, but not in N$_{CCAP}$-c929, also blocks tanning and wing expansion, indicating that neurotransmission and PKA are required in N$_{CCAP}$-R to regulate bursicon secretion from N$_{CCAP}$-c929. Enhancement of electrical activity in N$_{CCAP}$-R by expression of the bacterial sodium channel NaChBac also blocks tanning and wing expansion and leads to depletion of bursicon from central processes (Luan, 2006).

Chip is a ubiquitous, multifunctional protein cofactor interacting with transcription factor(s) of the LIM-HD family and has been shown to play a critical role in a set of neurons in *Drosophila* that control the well-described post-eclosion behavior. When the function of Chip was disrupted by genetic manipulation, cuticle tanning and wing expansion did not occur (Hari *et al.*, 2008) and the nature of the deficit was due to the absence of bursicon in the hemolymph of newly eclosed flies, whereas the response to bursicon in these flies remained normal. This suggests that Chip regulates bursicon release directly or indirectly. Thus, these findings provide the first evidence for the involvement of transcriptional mechanisms in the development of the neuronal circuit that regulates post-eclosion behavior in *Drosophila*. Despite all of these studies, we cannot rule out the possibility that other unidentified factors are involved in the regulation of bursicon release.

3.2.6. Evolutionary Conservation of Bursicon

Over the past four decades, bursicon activity, and most recently bursicon sequences, has been identified in many arthropods including insects, crustaceans, and arachnids, suggesting conservation of the two bursicon monomers burs and pburs among arthropods (see reviews by Seligman, 1980; Reynolds, 1983; Ewer and Reynolds, 2002; Honegger *et al.*, 2008). Surprisingly, burs and pburs sequences have also been identified in Echinoidea, a non-arthropod class in the phylum echinodermata (Van Loy *et al.*, 2007). All burs and pburs sequences identified in arthropod and non-arthropod species contain 11 cysteine residues in conserved positions, indicating a pivotal role in the structural integrity of these ligands (see review by Honegger *et al.*, 2008). Multiple bursicon sequence alignment of different species also show high sequence similarity among arthropods, indicating bursicon may have cross-species activities.

The extracts from the CNS or hemolymph of several dipteran species including, but not limited to, *Sarcophaga bullata* (Baker and Truman, 2002; Cottrell, 1962b; Fogal and Fraenkel, 1969), *Phormia regina* (Fraenkel and Hsiao, 1965), and Lucilia spp. (Cottrell, 1962b; Seligman and Doy, 1972) have also been shown to have cross-species activities in the neck-ligated fly assay.

However, it is not clear whether the cross-species activities from these assays are truly due to direct bursicon action or to the existing tanning agents already induced by bursicon in the donor hemolymph. The bioactive cuticle tanning factor present in the CNS and hemolymph of newly emerged adults has now been identified through molecular cloning and confirmed in the neck-ligated fly assay using the r-bursicon heterodimer (Luo *et al.*, 2005; Mendive *et al.*, 2005). The cross-species activity of bursicon has also been demonstrated using the r-bursicon heterodimer (An *et al.*, 2008). For example, *Drosophila* r-bursicon heterodimer expressed in a mammalian cell line exhibited strong tanning activity in the house fly, *M. domestica* and vice versa (An *et al.*, 2008; **Figure 3**). This is not surprising since *Drosophila* and *Musca* burs and pburs share 79% identity in their amino acid sequences. If the signal peptide at the N-terminal is excluded from the sequences, the mature *Drosophila* and *Musca* burs and pburs share much higher sequence identity with each other, up to 92 and 93%. What is surprising is that nervous system homogenates of the lobster *Homarus americanus* were reported to have bursicon activity in the ligated fly bioassay (Kostron *et al.*, 1995), and this observation was supported by the presence of a pburs-like transcript in the cDNA database of the lobster *H. americanus* (Van Loy *et al.*, 2007), illustrating the remarkable conservation of bursicon in invertebrate species. However, this cross-class bursicon activity needs to be verified using r-bursicon or highly purified bursicon of *H. americanus* in the ligated fly assay.

3.3. Mode of Action of Bursicon

3.3.1. Bursicon Receptor

In recent years, a subfamily of leucine-rich GPCRs (LGRs) was identified that had large ecto-domain-containing leucine-rich repeats. Genes in this family are conserved in invertebrates and vertebrates (Nishi *et al.*, 2000). LGRs can be divided into three groups: group A, vertebrate glycoprotein hormone receptors; group B, *D. melanogaster* LGR2 (DLGR2) (Eriksen *et al.*, 2000) and vertebrate orphan receptors LGR4, 5, and 6 (Hsu, 2003); and group C, mammalian relaxin_INSL3 receptors (Hsu *et al.*, 2002; Kumagai *et al.*, 2002).

Genetic analyses in *D. melanogaster* reveal that bursicon mediates the cuticle tanning and wing expansion process via a specific GPCR called DLGR2, encoded by the *rickets* gene. Mutation of *rickets* or *burs* causes defects in cuticle tanning and wing expansion (Baker and Truman, 2002; Dewey *et al.*, 2004). It has been demonstrated recently that the recombinant bursicon (r-bursicon) heterodimer binds to and activates DLGR2, which in turn leads to dose-dependent intracellular increase in adenyl cyclase activity and cAMP production in mammalian 293 T cells

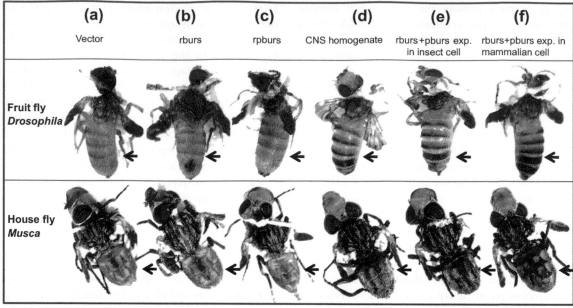

Figure 3 Functional assay of the r-bursicon heterodimer in neck-ligated flies. Newly emerged flies were ligated between the head and thorax at emergence and injected with 0.5 μl of cell culture transfected with blank pcDNA 3.One vector as a sham control (a) or with the purified r-burs, (b) r-pburs, (c) r-bursicon heterodimer expressed in insect High Five™ cells, (e) and in mammalian HEK293 cells (f). A CNS homogenate (0.5 CNS equivalent/fly) from newly emerged flies was used as a positive control (d). The arrow indicates the area with the unsclerotized cuticle (light color) in control (a–c) and the sclerotized cuticle (darkened) in the flies injected with *Drosophila* r-bursicon heterodimer. A strong cuticle tanning was observed when the neck-ligated house flies were injected with *Drosophila* r-bursicon or *Drosophila* CNS homogenate (bottom panel). Adapted from An *et al.* (2008).

and COS-7 cells that overexpress DLGR2 (Luo *et al.*, 2005; Mendive *et al.*, 2005). Thus, DLGR2 has been well recognized as the bursicon receptor.

3.3.2. Bursicon Signaling Pathway

Downstream of the bursicon receptor DLGR2, cyclic AMP (cAMP) has been shown to be involved in the bursicon-mediated cuticle tanning process. cAMP is an important second messenger for many signal transduction pathways. Earlier studies on blowflies have implicated an increase in cAMP upon the action of bursicon (extract or hemolymph; Reynolds, 1980; Seligman and Doy, 1972, 1973; Von Knorre *et al.*, 1972). Recent studies using an *in vitro* binding assay also indicated that r-bursicon stimulated cAMP production in a dose-dependent manner (Luo *et al.*, 2005; Mendive *et al.*, 2005). Injection of the *rickets* mutated flies with a cAMP analog resulted in melanization of the abdominal tergites. These results strongly indicate that DLGR2 is the receptor of bursicon and that DLGR2 mediates the bursicon signaling pathway through cAMP.

Upon activation by bursicon, DLGR2 has been shown to then activate the cAMP/PKA signaling pathway, eventually leading to cuticle sclerotization (Kimura *et al.*, 2004; Luo *et al.*, 2005; Mendive *et al.*, 2005). Davis *et al.* (2007) recently investigated the role of CCAP

neurons, bursicon, and rk in the tanning pathway. They showed that the levels of epidermal dopa decarboxylase and of epidermal tyrosine hydroxylase (encoded by the gene pale, *ple*) transcripts did not change at eclosion. By contrast, the levels of the epidermal tyrosine hydroxylase fell prior to eclosion then increased sharply immediately following eclosion. Flies mutants for *burs* (Dewey *et al.*, 2004) and *rk* (Baker and Truman, 2002) showed defects in tanning and were defective in tyrosine hydroxylase activation (Davis *et al.*, 2007). Whereas these mutant flies showed relatively normal increases in epidermal tyrosine hydroxylase expression after eclosion, they failed to phosphorylate tyrosine hydroxylase. When cAMP, the second messenger for bursicon, was injected, tyrosine hydroxylase was highly phosphorylated. However, tyrosine hydroxylase was not phosphorylated in the presence of a PKA inhibitor. These results demonstrate that bursicon binds to the DLGR2 receptor, stimulates cAMP production, and activates PKA to phosphorylate tyrosine hydroxylase. The activated tyrosine hydroxylase then converts tyrosine to DOPA, the precursor of tanning agents. This is the first report to provide convincing data that bursicon plays a role in cuticle tanning by acting at a strategic point in the biochemistry, and at a strategic location, in the insect. Since the cuticle tanning process is a complex biochemical process, it is hypothesized to be mediated by multiple components in the bursicon signaling pathway. In this regard, our knowledge of signaling components involved in the bursicon-stimulated cuticle tanning and wing expansion processes remains preliminary.

3.4. Bursicon Function

3.4.1. Cuticle Tanning

Cuticle tanning includes two separate processes, sclerotization and melanization. Before eclosion, the newly formed insect cuticle consists of a thin layer of hydrophobic, waxy, chitin-free epicuticle and a thick layer of protein- and chitin-rich procuticle (Locke, 2001). At eclosion, the outer portion of procuticle is quickly cross-linked by tanning agents N-acetyldopamine (NADA) and N-β-alanyldopamine (NBAD) to form a layer of hardened exocuticle and the inner portion of the procuticle remains uncross-linked to form endocuticle.

NADA and NBAD are synthesized from the precursor amino acid tyrosine, which is first hydroxylated to DOPA by tyrosine hydroxylase and then decarboxylated to dopamine by dopa decarboxylase (Andersen, 2005). For sclerotization, dopamine is N-acylated to NADA and NBAD by dopamine N-acetyltransferase. Both derivatives are secreted from epidermal cells into the cuticle where they are then enzymatically oxidized and cross-linked with different nucleophilic protein residues and chitin. As a result, the cuticle becomes hardened and more hydrophobic. In general, cuticles sclerotized exclusively by NADA are colorless or light colored, and cuticles sclerotized with the increased contribution of NBAD produce the darker colored cuticle. Thus, cuticle darkening can occur without the participation of melanin.

For cuticle melanization, dopamine is also converted to insoluble melanin via 5,6-dihydroxyindole. Melanin can be linked to granular proteins or may be distributed throughout the cuticular matrix, giving the cuticle a dark color. Melanin probably also forms bonds with cuticular proteins, contributing to cuticular strength (Andersen, 2005). Thus, dopamine is the central molecule for both sclerotization and melanization. Seligman et al. (1969) hypothesized that the conversion of tyrosine to DOPA is regulated by bursicon and the conversion process is initially thought to be in hemocytes (Mills and Whitehead, 1970). However, later evidence clearly indicates that the conversion of tyrosine to DOPA takes place in epidermal cells (Seligman, 1980; Reynolds, 1983). It is most likely that bursicon may act on LGR2 receptors in the epidermal walls to stimulate the hydroxylation of tyrosine to DOPA since the epidermis is the main tissue involved in either the production of the metabolites necessary for tanning or as the cell layer through which the dopamine metabolites must cross to reach the extracellular space.

The cuticle tanning process uses several enzymes to mediate the metabolic process. These include, but are not limit to, diphenoloxidases, laccases, peroxidase, tyrosine hydroxylase, dopa decarboxylase, and N-acetyltransferase. The activities of these enzymes could be regulated at the transcriptional, translational, and post-translational levels. In the red flour beetle, Tr. castaneum, laccase 2 is the only phenoloxidase whose activity is necessary for cuticle tanning. RNAi of the laccase 2 gene causes defects in cuticle tanning (Arakane et al., 2005), suggesting that it is regulated at the transcriptional level. However, not all genes involved in the tanning process are regulated at the transcriptional level. For example, the mRNA level of tyrosine hydroxylase, a key enzyme that mediates the conversion of tyrosine to DOPA in the tanning agent biosynthesis process, remains unchanged during the cuticle tanning period before and after adult emergence (Davis et al., 2007). This indicates that the activity of tyrosine hydroxylase during the sclerotization process (0 to 3 h after eclosion) is not regulated at the transcriptional level. Tyrosine hydroxylase is transiently activated during cuticle tanning by a post-translational mechanism, that is, phosphorylation by PKA (Davis et al., 2007). Another example is dopa decarboxylase, which catalyzes the conversion of DOPA to dopamine, a compound of central importance to sclerotization and melanization. The dopa decarboxylase mRNA level peaks at 24 h before eclosion and decreases thereafter. It is almost unchanged or decreases minutely from 0 to 3 h after eclosion. Dopa decarboxylase is transcribed and translated before eclosion to ensure that enzyme activity is present when substrate becomes available (Davis et al., 2007).

All evidence indicates that bursicon is the neuropeptide responsible for cuticle tanning and wing expansion in newly eclosed adults. However, recent data indicate that bursicon may not be necessary for cuticle sclerotization in all adult insects. Exposure of leg fragments from freshly ecdysed adult cockroach Blaberus craniifer to 20 μM 20E for 3 days induced apolysis of the newly synthesized cuticle (Weber, 1995), which is a classic function of 20E. What was surprising is that a new exocuticle was deposited and sclerotized. Since bursicon is not present under these culture conditions, the new cuticle must have been sclerotized in the absence of bursicon. In T. castaneum, injection of dsRNA for the genes encoding the subunits of bursicon (burs and pburs) and its receptor (Tcrk) into pharate pupae did not lead to visible defects in adult sclerotization or pigmentation, but it did cause defects in wing expansion (Arakane et al., 2008). However, when dsRNA of these genes was injected into early stage last instar larvae, cuticle tanning and wing expansion were indeed affected (Bai and Palli, 2010), indicating that the timing of RNA interference of the target genes is critical for analyzing gene function.

Although bursicon has been demonstrated to mediate cuticle tanning and wing expansion in newly eclosed adult insects, it does not imply that the hormone exhibits any physiological action leading to cuticle tanning in larval and pupal stages. Different blood-borne tanning factors have been found at different developmental stages in the same insect species. Sclerotization of the puparium in higher Diptera is controlled by a bursicon-like molecule called the puparion tanning factor (PTF; Zdarek and Fraekel, 1969; Zdarek, 1985). Studies on *S. bullata* led to the identification of two additional hormones, the anterior retraction factor (ARF) and the puparium immobilization factor (PIF), which are involved in the control of puparium formation and the actual tanning of the old larval cuticle to form the hardened puparium (Sivasubramaniun *et al.*, 1974; Zdarek, 1985). PTF of the larval blowfly did not exhibit tanning activity in the adult blowfly bursicon assay, and its molecular weights and distribution pattern in the CNS differed from that of bursicon.

3.4.2. Wing Expansion

In addition to eliciting the insect cuticle tanning process, bursicon is also involved in insect wing expansion. The insect wing expansion process is another physiological event occurring after ecdysis and accompanies cuticle tanning. These processes occur at the same time, but the mechanisms under which they are controlled and regulated are still not clear. Recent research showed that mutation in the bursicon genes causes the failure of initiation of the behavioral program for wing expansion and results in defects in wing expansion in *D. melanogaster* (Dewey *et al.*, 2004). Similarly, injection of RNAi of the burs gene also results in defects in wing expansion in the silkworm *Bombyx mori* (Huang *et al.*, 2007). When dsRNA of bursicon genes are injected into the early stage last instar larvae of *T. castaneum*, cuticle tanning and wing expansion are also affected (Bai and Palli, 2010). These studies indicated that bursicon is the hormone required to initiate both cuticle tanning and wing expansion processes in these, and perhaps all, insects. Another line of support for this observation is from genetic studies of the bursicon receptor DLGR2. It has been shown that mutation of the *rickets* gene, which encodes the bursicon receptor DLGR2, inhibits wing expansion in *Drosophila* (Kimura *et al.*, 2004) and *T. castaneum* (Bai and Palli, 2010).

Wing development and expansion are strongly associated with programmed cell death and removal of cell debris from the wing tissue (Johnson and Milner, 1987; Kiger *et al.*, 2007; Kimura *et al.*, 2004; Natzle *et al.*, 2008; Seligman *et al.*, 1972, 1975). Extensive studies have shown that both cell death and the removal process are controlled by changes in the levels of the steroid hormone 20-hydroxyecdysone (20E) through regulation of a series of early response and late response genes, including the expression of the cell death activator genes, *rpr* and *hid*, which then repress the *Drosophila* inhibitor of apoptosis proteins (IAPs) *Diap2* (Jiang *et al.*, 1997). Though 20E-dependent regulation is the possible pathway controlling cell death, it does not exclude the possibility of the involvement of other hormones in wing expansion and programmed cell death. It is very likely that bursicon co-regulates programmed cell death along with other factors during the post-eclosion wing maturation process. It has been shown that stimulation with components downstream of bursicon, such as a membrane permeate analog of cAMP, or ectopic expression of constitutively active forms of G proteins or PKA, induce precocious cell death; and conversely, cell death was inhibited in wing clones lacking G protein or PKA function. Bursicon has already been demonstrated to bind to DLGR2 and activate cAMP/PKA signaling; therefore, bursicon activation of the cAMP/PKA signaling pathway is likely required for transduction of the hormonal signal that induces wing epidermal cell death after eclosion (Kimura *et al.*, 2004).

3.5. Evidence for Bursicon's Roles Beyond Cuticle Tanning and Wing Expansion

3.5.1. Integumentary Structure Development and Expansion

Recent evidence indicated that bursicon also plays important roles other than the classic functions attributed to it, such as cuticle tanning and wing expansion. Injection of the bursicon receptor Tcrk dsRNA into early stage last instar *T. castaneum* larvae inhibits development and expansion of integumentary structures and adult eclosion, in addition to cuticle tanning (Bai and Palli, 2010). These results indicate that bursicon could also be released before eclosion to regulate development and expansion of integumentary structures. PCR analysis of burs and pburs transcripts in *Drosophila* and *Musca* clearly shows that the burs and pburs transcripts are highly expressed in larval and pupal stages (Luo *et al.*, 2005; Mendive *et al.*, 2005; An *et al.*, 2008; Wang *et al.*, 2008), suggesting that the products of *burs* and *pburs* may be released into hemolymph during larval and pupal development. Loveall and Deitcher (2010) reported that the release of bursicon in *Drosophila* surprisingly occurs in two waves: preceding and following ecdysis, indicating that the release preceding ecdysis may play a role in development and expansion of integumentary structures as shown in the red flour beetle (Bai and Palli, 2010) and the release following ecdysis is responsible for cuticle tanning and wing expansion.

3.5.2. Immune Response

It has been clearly demonstrated that bursicon mediates cuticle tanning and wing expansion via a specific DLGR2, encoded by the *rickets* gene (Baker and Truman, 2002; Dewey *et al.*, 2004; Luo *et al.*, 2005; Mendive *et al.*, 2005), leading to activation of the cAMP/PKA signaling pathway and tyrosine hydroxylase in the tanning pathway (Davis *et al.*, 2007). To identify the signaling pathway downstream of the bursicon receptor DLGR2 and adenylate cyclase, as well as the genes regulated by bursicon, An *et al.* (2008) performed a DNA microarray analysis in the neck-ligated flies using r-bursicon as a probe (**Figure 3**) and identified a set of 87 genes whose expression were up- or downregulated by r-bursicon in *D. melanogaster*. Analysis of these genes by inference from the fly database (http://flybase.bio.indiana.edu) revealed that only 57 genes show significant similarities to known proteins or functional domains in the *Drosophila* database and 30 did not. The 57 known genes encode proteins with diverse functions, including cell signaling, gene transcription, DNA/RNA binding, ion trafficking, proteolysis-peptidolysis, metabolism, cytoskeleton formation, immune response, and cell-adhesion. Most of these known genes could be linked to cuticle sclerotization and wing expansion, directly or indirectly.

However, not all identified genes could be related to the cuticle sclerotization and wing expansion processes, suggesting that bursicon might have unidentified functions. From the microarray identified genes, seven immune-related genes were upregulated by r-bursicon. These genes include three from the turandot gene family (*turandot* X, *turandot* F, and *turandot* C), two from the attacin gene family and one each from the cecropin and immunoglobulin families (An *et al.*, 2008). Although no direct association between bursicon and antibacterial peptides is obvious, the newly ecdysed insect is soft (before cuticle sclerotization and melanization) and more susceptible to injury and infection. Perhaps a more vigorous antibacterial defense is necessary at this time when there may be perforations in the soft cuticle due to predators or simple contact injuries.

Since bursicon is a heterodimer, it is quite possible that in addition to its classic roles in cuticle tanning and wing expansion, each subunit may have its own physiological function. Indeed, our preliminary data (Song lab, unpublished information) reveal that when neck-ligated flies were injected with r-bursicon, r-burs, and r-pburs separately, each could induce a set of immune response genes in the final molting cycle of *D. melanogaster*. Such induction was both time- and dose-dependent. When the homogenates from the flies injected with r-burs or r-pburs were incubated with the gram negative bacterium *Escherichia coli*, the homogenate from each fly equivalent was able to completely clear bacterial growth while the control that received the cell medium transfected with blank vector could not.

In *D. melanogaster*, two distinct signaling pathways, the Toll and IMD pathways, are responsible for the recognition of pathogens and the activation of the antimicrobial immune response (Iwanga and Lee, 2005). The Toll pathway is thought to be triggered in response to infections by fungi and gram positive bacteria, whereas the IMD pathway is thought to be triggered by gram negative bacteria (Hoffman, 2003). Activation of the Toll pathway by fungi and gram positive bacteria and the IMD pathway by gram negative bacteria will ultimately lead to the activation of *D. melanogaster* members of the nuclear factor-κB (NF-κB) family of inducible transactivators: dorsal-related immunity factor (DIF) and Relish, respectively. Our preliminary data indicate that bursicon mediates the immune response via the IMD pathway by activation of the transcription factor relish (unpublished data). Thus, a novel function of bursicon has been identified. It is not clear whether burs or pburs forms a homodimer or heterdimerizes with a yet unidentified factor to mediate the immune response. It is also not clear if burs or burs bind to DLGR2 or to an as yet unidentified receptor leading to the regulation of antibacterial peptide expression.

Although bursicon's roles in larval and pupal development have not been explored, each of the bursicon subunits (burs and pburs) is expressed throughout the larval and pupal stages suggesting potential roles in larval and pupal development. **Figure 4** summarizes the bursicon signaling pathways in cuticle tanning/wing expansion/cuticle structure development and the insect immune response.

3.6. Prospect

Bursicon has now been studied for over four decades. Physiological and molecular characterizations of bursicon have been well studied, but many questions still remain unresolved: What are the functions of one-third of the bursicon-regulated unknown genes identified in the DNA microarray? What are the signal transduction components involved in bursicon-stimulated cuticle sclerotization and wing expansion in addition to PKA and tyrosine hydroxylase? Does bursicon have a morphogenetic role in larval and pupal development? Does each bursicon subunit have its own physiological function? Does each bursicon subunit bind to DLGR2, the receptor for the heterodimer, to regulate the immune response? With the recombinant bursicon protein, well-established bursicon bioassay systems, and available genetic and molecular approaches the answers to some of these questions may occur in the near future.

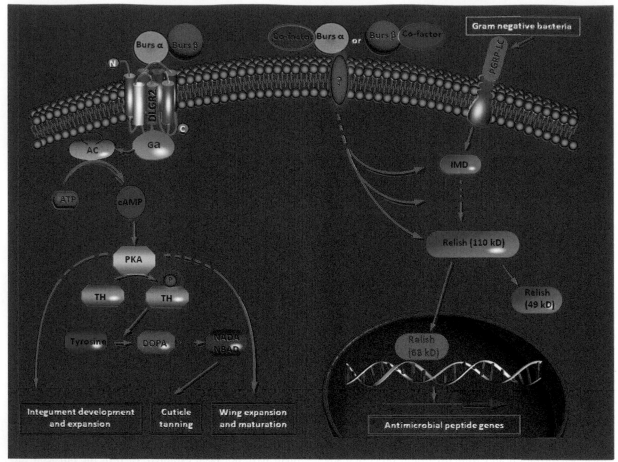

Figure 4 Map of bursicon signaling pathways.

References

Adams, M. D., Celniker, S. E., Holt, R. A., Evans, C. A., Gocayne, J. D., Amanatides, P. G., Scherer, S. E., Li, P. W., Hoskins, R. A., Galle, R. F., et al. (2000). The genome sequence of *Drosophila melanogaster*. *Science*, *287*, 2185–2195.

An, S., Wang, S., Gilbert, L., Beerntsen, B., Ellersieck, M., & Song, Q. (2008). Global identification of bursicon-regulated genes in *Drosophila melanogaster*. *BMC Genomics*, *9*, 424.

Andersen, S. O. (2005). Cuticular sclerotization and tanning. In L. I. Gilbert, K. Iatrou, & S. S. Gill (Eds.), *Comprehensive Molecular Insect Science Vol. 4* (pp. 145–170). Amsterdam: Elsevier Pergamon.

Arakane, Y., Li, B., Muthukrishnan, S., Beeman, R. W., Kramer, K. J., & Park, Y. (2008). Functional analysis of four neuropeptides, EH, ETH, CCAP and bursicon, and their receptors in adult ecdysis behavior of the red flour beetle. *Tr. castaneum. Mech. Dev.*, *125*, 984–995.

Arakane, Y., Muthukrishnan, S., Beeman, R. W., Kanost, M. R., & Kramer, K. J. (2005). Laccase 2 is the phenoloxidase gene required for beetle cuticle tanning. *Proc. Natl. Acad. Sci. USA*, *102*, 11337–11342.

Bai, H., & Palli, S. R. (2010). Functional characterization of bursicon receptor and genome-wide analysis for identification of genes affected by bursicon receptor RNAi. *Dev. Biol.*, *344*, 248–258.

Baker, J., & Truman, J. W. (2002). Mutations in the *Drosophila* glycoprotein hormone receptor, rickets, eliminate neuropeptide-induced tanning and selectively block a stereotyped behavioral program. *J. Exp. Biol.*, *205*, 2555–2565.

Butler, S. A., Laidler, P., Porter, J. R., Kicman, A. T., Chard, T., Cowan, D. A., & Iles, R. K. (1999). The subunit of human chorionic gonadotrophin exists as a homodimer. *J. Mol. Endocrinol.*, *22*, 185–192.

Cottrell, C. B. (1962a). The imaginal ecdysis of blowflies. The control of cuticular hardening and darkening. *J. Exp. Biol.*, *39*, 395–411.

Cottrell, C. B. (1962b). The imaginal ecdysis of blowflies. Detection of the blood-borne darkening factor and determination of some of its properties. *J. Exp. Biol.*, *39*, 413–430.

Davis, M. M., O'Keefe, S. L., Primrose, D. A., & Hodgetts, R. B. (2007). A neuropeptide hormone cascade controls the precise onset of post-eclosion cuticular tanning in *Drosophila melanogaster*. *Development*, *134*, 4395–4404.

Delachambre, J. (1971). Le tannage de la cuticle adulte de Tr. molitor: mise en evidence dune action hormonale induite par la region cephalique. *J. Insect Physiol.*, *17*, 2481–2490.

Dewey, E. M., McNabb, S. L., Ewer, J., Kuo, G. R., Takanishi, C. L., Truman, J. W., & Honegger, H. W. (2004). Identification of the gene encoding bursicon, an insect neuropeptide responsible for cuticle sclerotization and wing spreading. *Curr. Biol., 14*, 1208–1213.

Eriksen, K. K., Hauser, F., Schiott, M., Pedersen, K. M., Sondergaard, L., & Grimmelikhuijzen, C. J. (2000). Molecular cloning, genomic organization, developmental regulation, and a knock-out mutant of a novel leu-rich repeats-containing G protein-coupled receptor (DLGR-2) from *Drosophila melanogaster. Genome Res, 10*, 924–938.

Ewer, J., & Reynolds, S. (2002). Neuropeptide control of molting in insects. In D. W. Pfaff, A. P. Arnold, S. E. Fahrbach, A. M. Etgen, & R. T. Rubin (Eds.), *Hormones, Brain and Behavior* (pp. 1–92). San Diego: Academic Press.

Fogal, W., & Fraenkel, G. (1969). The role of bursicon in melanization and endocuticle formation in the adult fleshfly, *Sarcophaga bullata. J. Insect Physiol., 15*, 1235–1247.

Fraenkel, G., & Hsiao, C. (1962). Hormonal and nervous control of tanning in the fly. *Science, 138*, 27–29.

Fraenkel, G., & Hsiao, C. (1963). Tanning in the adult fly: a new function of neurosecretion in the brain. *Science, 141*, 1050–1058.

Fraenkel, G., & Hsiao, C. (1965). Bursicon, a hormone which mediates tanning of the cuticle in the adult fly and other insects. *J. Insect Physiol., 11*, 513–556.

Fraenkel, G., Hsiao, C., & Seligman, M. (1966). Properties of bursicon: An insect protein hormone that controls cuticular tanning. *Science, 151*, 91–93.

Grillot, J. P., Delachambre, J., & Provansal, A. (1976). Role des organs peristmpathiques et dynamique de la secretion de la bursicon chez Tr. molitor. *J. Insect Physiol., 22*, 763–780.

Hari, P., Deshpande, M., Sharma, N., Rajadhyaksha, N., Ramkumar, N., Kimura, K. -I., Rodrigues, V., & Tole, S. (2008). Chip is required for posteclosion behavior in *Drosophila. J. Neurosci., 28*, 9145–9150.

Hart, C. E., & Bowen-Pope, D. F. (1990). Platelet-derived growth factor receptor: Current views of the two-subunit model. *J. Invest. Dermatol., 94*, 53–57.

Hearn, M. T., & Gomme, P. T. (2000). Molecular architecture and biorecognition processes of the cystine knot protein superfamily: Part I. The glycoprotein hormones. *J. Mol. Recognit., 13*, 223–278.

Hoffman, J. A. (2003). The immune response of *Drosophila. Nature, 426*, 33–38.

Honegger, H. W., Market, D., Pierce, L. A., Dewey, E. W., Kostron, B., Wilson, M., Choi, D., Klukas, K. A., & Mesce, K. A. (2002). Cellular localization of bursicon using antisera against partial peptide sequences of this insect cuticle-sclerotizing neurohormone. *J. Comp. Neurol., 452*, 163–177.

Honegger, H. W., Dewey, E. M., & Kostron, B. (2004). From bioassays to *Drosophila* genetics: strategies for characterizing an essential insect neurohormone, bursicon. *Symp. Biol. Hung., 55*, 91–102.

Honegger, H. W., Dewey, E. M., & Ewer, J. (2008). Bursicon, the tanning hormone of insects: Recent advantages following the discovery of its molecular identity. *J. Comp. Physiol. A, 194*, 989–1005.

Hsu, S. Y. (2003). New insights into the evolution of the relaxin-LGR signaling system. *Trends Endocrinol. Metab., 7*, 303–309.

Hsu, S. Y., Nakabayashi, K., Nishi, S., Kumagai, J., Kudo, M., Sherwood, O. D., & Hsueh, A. J. (2002). Activation of orphan receptors by the hormone relaxin. *Science, 295*, 671–674.

Huang, J., Zhang, Y., Li, M., Wang, S., Liu, W., Couble, P., Zhao, G., & Huang, Y. (2007). RNA interference-mediated silencing of the bursicon gene induces defects in wing expansion of silkworm. *FEBS Lett., 581*, 697–701.

Iwanga, S., & Lee, B. L. (2005). Recent advances in the innate immunity of invertebrate animals. *J. Bio. Mol. Biol., 38*, 128–150.

Jiang, C., Baehrecke, E. H., & Thummel, C. S. (1997). Steroid regulated programmed cell death during *Drosophila* metamorphosis. *Development, 124*, 4673–4683.

Johnson, S. A., & Milner, M. J. (1987). The final stages of wing development in *Drosophila melanogaster. Tissue Cell, 19*, 505–513.

Kaltenhauser, U., Kellermann, J., Andersson, K., Lottspeich, F., & Honegger, H. W. (1995). Purification and partial characterization of bursicon, a cuticle sclerotizing neuropeptide in insects, from *Tr. molitor. Insect Biochem. Mol. Biol., 25*, 525–533.

Kiger, J. A., Natzle, J. E., Kimbrell, D. A., Paddy, M. R., Kleinhesselink, K., & Green, M. M. (2007). Tissue remodeling during maturation of the *Drosophila* wing. *Dev. Biol., 301*, 178–191.

Kimura, K., Kodama, A., Hayasaka, Y., & Ohta, T. (2004). Activation of the cAMP/PKA signaling pathway is required for post-ecdysial cell death in wing epidermal cells of *Drosophila melanogaster. Development, 131*, 1597–1606.

Kostron, B., Marquardt, K., Kaltenhauser, U., & Honegger, H. (1995). Bursicon, the cuticle sclerotizing hormone-comparison of its molecular mass in different insects. *J. Insect Physiol., 41*, 1045–1053.

Kumagai, J., Hsu, S. Y., Matsumi, H., Roh, J. S., Fu, P., Wade, J. D., Bathgate, R. A., & Hsueh, A. J. (2002). INSL3/Leydig insulin-like peptide activates the LGR8 receptor important in testis descent. *J. Biol. Chem., 277*, 31283–31286.

Locke, M. (2001). The Wigglesworth Lecture: insects for studying fundamental problems in biology. *J. Insect Physiol., 47*, 495–507.

Loveal, B. J., & Deitcher, D. L. (2010). The essential role of bursicon during *Drosophila* development. *BMC Dev. Biol., 10*, 92.

Luan, H., Lemon, W. C., Peabody, N. C., Pohl, J. B., Zelensky, P. K., Wang, D., Nitabach, M. N., Holmes, T. C., & White, B. H. (2006). Functional dissection of a neuronal network required for cuticle tanning and wing expansion in *Drosophila. J. Neurosci., 26*, 573–584.

Luo, C. W., Dewey, E. M., Sudo, S., Ewer, J., Hsu, S. Y., Honegger, H. W., & Hsueh, A. J. (2005). Bursicon, the insect cuticle-hardening hormone, is a heterodimeric cystine knot protein that activates G protein-coupled receptor LGR2. *Proc. Natl. Acad. Sci., 102*, 2820–2825.

Mendive, F. M., Van Loy, T., Claeysen, S., Poels, J., Williamson, M., Hauser, F., Grimmelikhuijzen, C. J. P., Vassart, G., & Vanden-Broeck, J. (2005). *Drosophila* molting neurohormone bursicon is a heterodimer and the natural agonist of the orphan receptor DLGR2. *FEBS Lett., 579*, 2171–2176.

Mills, R. R. (1965). Hormonal control of tanning in the American cockroach-II. Assay for the hormone and the effect of wound healing. *J. Insect Physiol., 11,* 1269–1275.

Mills, R. R. (1967). Control of cuticular tanning in the cockroach: bursicon release by nervous stimulation. *J. Insect Physiol., 13,* 815–820.

Mills, R. R., & Lake, C. R. (1966). Hormonal control of tanning in the American cockroach. IV. Preliminary purification of the hormone. *J. Insect Physiol., 12,* 1395–1401.

Mills, R. R., Mathur, R. B., & Guerra, A. A. (1965). Studies on the hormonal control of tanning in the American cockroach-I. release of an activation factor from the terminal abdominal ganglion. *J. Insect Physiol., 11,* 1047–1053.

Mills, R. R., & Nielsen, D. J. (1967). Changes in the diuretic and antidiuretic properties of the haemolymph during the six-day vitellogenic cycle in the American cockroach. *Gen. Comp. Endocrinol., 9,* 380–382.

Mills, R. R., & Whitehead, D. L. (1970). Hormonal control of tanning in the American cockroach: Changes in blood cell permeability during ecolysis. *J. Insect Physiol. 16,* 331–340.

Natzle, J. E., Kiger, J. A., & Green, M. M. (2008). Bursicon signaling mutations separate the epithelial-mesenchymal transition from programmed cell death during *Drosophila melanogaster* wing maturation. *Genetics, 180,* 885–893.

Padgham, D. E. (1976). Bursicon-mediated control of tanning in melanizing and non-melanizing first instar larvae of *Schistocerca gregaria. J. Insect Physiol., 22,* 1447–1452.

Park, J. H., Schroeder, A. J., Helfrich-Förster, C., Jackson, F. R., & Ewer, J. (2003). Targeted ablation of CCAP neuropeptide-containing neurons of *Drosophila* causes specific defects in execution and circadian timing of ecdysis behavior. *Development, 130,* 2645–2656.

Peabody, N. C., Diao, F., Luan, H., Wang, H., Dewey, E. M., Honegger, H. W., & White, B. H. (2008). Bursicon functions within the *Drosophila* central nervous system to modulate wing expansion behavior, hormone secretion, and cell death. *J. Neurosci., 28,* 14379–14391.

Ren, P., Lim, C.-S., Johnsen, R., Albert, P. S., Pilgrim, D., & Riddle, D. L. (1996). Control of *C. elegans* larval development by neuronal expression of a TGF-homologue. *Science, 274,* 1389–1391.

Reynolds, S. E. (1976). Hormonal regulation of cuticle extensibility in newly emerged adult blowflies. *J. Insect Physiol., 22,* 529–534.

Reynolds, S. E. (1977). Control of cuticle extensibility in the wings of adult *Manduca* at the time of eclosion: Effects of eclosion hormone and bursicon. *J. Exp. Biol., 70,* 27–39.

Reynolds, S. E. (1980). Integration of behavior and physiology in ecdysis. *Adv. J. Insect Physiol., 15,* 475–595.

Reynolds, S. E. (1983). Bursicon. In R. G.H. Downer, & H. Laufer (Eds.), *Endocrinology of Insects* (pp. 235–348). New York: Alan R. Liss.

Reynolds, S. E., Taghert, P., & Truman, J. W. (1979). Eclosion hormone and bursicon titres and the onset of hormonal responsiveness during the last day of adult development in *Manduca sexta* (L). *J. Exp. Biol., 78,* 77–86.

Rothberg, J. M., Jacobs, J. R., Goodman, C. S., & Artavanis-Tsakonas, S. (1990). Slit: An extracellular protein necessary for development of midline glia and commissural axon pathways contains both EGF and LRR domains. *Genes Dev., 12A,* 2169–2187.

Seligman, I. M., Friedman, S., & Fraenkel, G. (1969). Bursicon mediation of tyrosine hydroxylation during tanning in their adult cuticle in the fly *Sarcophaga bullata. J. Insect Physiol., 15,* 553–561.

Seligman, L. M., Doy, E. A., & Crossley, A. C. (1975). Hormonal control of morphogenetic cell death of the wing hypodermis in *Lucilia cuprina. Tissue Cell, 7,* 281–296.

Seligman, M., & Doy, F. A. (1972). Studies on cyclic AMP mediation of hormonally induced cytolysis of the alary hypodermal cells and of hormonally controlled dopa synthesis in *Lucilia cuprina. Israel J. Entomol., 7,* 129–142.

Seligman, M., & Doy, F. A. (1973). Hormonal regulation of disaggregation of cellular fragments in the haemolymph of *Lucilia cuprina. J. Insect Physiol., 19,* 125–135.

Seligman, I. M. (1980). Bursicon. In T. A. Miller (Ed.), *Neurohormonal Techniques in Insects* (pp. 137–153). Berlin: Springer.

Sivasubramanian, P., Friedman, S., & Fraenkel, G. (1974). Nature and role of proteinaceous hormonal factors acting during puparium formation in flies. *Biol. Bull., 147,* 163–185.

Srivastava, B. B., & Hopkins, T. L. (1975). Bursicon release and activity in haemolymph during metamorphosis of the cockroach, *Leucophaea maderae. J. Insect Physiol., 21,* 1985–1993.

Taghert, P. H., & Truman, J. W. (1982a). The distribution and molecular characteristics of the tanning hormone, bursicon, in the tobacco hornworm *Manduca sexta. J. Exp. Biol., 98,* 373–383.

Taghert, P. H., & Truman, J. W. (1982b). Identification of the bursicon-containing neurones in abdominal ganglia of the tobacco hornworm, *Manduca sexta. J. Exp. Biol., 98,* 385–401.

Truman, J. W. (1992). The eclosion hormone system of insects. *Prog. Brain, 92,* 361–374.

Truman, J. W. (1973). Physiology of insect ecdysis — III. Relationship between the hormonal control of eclosion and of tanning in the tobacco hornworm, *Manduca sexta. J. Exp. Biol., 58,* 821–829.

Van Loy, T., Van Hiel, M. B., Vandersmissen, H. P., Poels, J., Mendive, F., Vassart, G., & Vanden Broeck, J. (2007). Evolutionary conservation of bursicon in the animal kingdom. *Gen. Comp. Endocrinol., 153,* 59–63.

Vincent, J. F.V. (1971). Effects of bursicon on cuticular properties in *Locustia migratoria migratorioides. J. Insect Physiol., 17,* 625–636.

Vincent, J. F.V. (1972). The dynamics of release and the possible identity of bursicon in *Locusta migratoria migratorioides. J. Insect Physiol., 18,* 757–780.

Vitt, U. A., Hsu, S. Y., & Hsueh, A. J. (2001). Evolution and classification of cysteine knot-containing hormones and related extracellular signal molecules. *Mol. Endocrinol., 15,* 681–694.

Von Knorre, D., Gersch, M., & Kusch, T. (1972). Zur Frage der Beeinflussung des "tanning" phanomens durch zyklisches-3', 5' AMP. *Zool. Jahrb. (Physiol.)*, *76*, 434–440.

Wang, S. J., An, S., & Song, Q. (2008). Transcriptional expression of bursicon and novel bursicon-regulated genes in the house fly *Musca domestica*. *Arch. Insect Biochem. Physiol.*, *68*, 100–112.

Weber, F. (1995). Cyclic layer deposition in the cockroach (*Blaberus craniifer*) endocuticle: A circadian rhythm in leg pieces cultures *in vitro*. *J. Insect Physiol.*, *41*, 153–161.

Ždćárek, J. (1985). Regulation of pupariation in flies. In G. A. Kerkut, & L. I. Gilbert (Eds.), *Comparisons in Insect Physiology, Biochemistry and Pharmacology*: *Vol. 8* (pp. 301–333). Oxford: Pergamon Press.

Ždćárek, J., & Fraenkel, G. (1969). Correlated effects of ecdysone and neurosecretion in puparium formation (pupariation) of flies. *Proc. Natl. Acad. Sci.*, *64*, 565–572.

4   Ecdysteroid Chemistry and Biochemistry

Rene Lafont and C. Dauphin-Villemant
Université Pierre et Marie Curie, Paris, France
J.T. Warren
University of North Carolina, Chapel Hill, NC, USA
H. Rees
University of Liverpool, Liverpool, UK

© 2012 Elsevier B.V. All Rights Reserved

4.1. Introduction

It is not the aim here, owing to the limitations of space, to review all the literature concerning the chemistry and biochemistry of ecdysteroids, and the reader should consult older articles in several excellent reviews by Denis Horn (Horn, 1971; Horn and Bergamasco, 1985) and the books edited by Jan Koolman (1989) and Guy Smagghe (2009). After a short historical introduction, we will focus on recent data concerning (1) the techniques used for ecdysteroid analysis and (2) the biosynthesis and metabolism of these hormones. However, we will not give details about chemical synthesis procedures, and we will limit this topic to a few chemical reactions that can be performed by (trained) biologists and do not require an experienced chemist.

The endocrine control of insect molting was established in part by the pioneering work of Stefan Kopec (1922), who demonstrated the need for a diffusible factor originating from the brain (see Chapter 1) for metamorphosis to take place in the gypsy moth, *Lymantria dispar*. This role of the brain was further substantiated by Wigglesworth (1934) with experiments using the blood-sucking bug

Rhodnius prolixus to define critical periods for the necessity of the brain to produce a molting factor. The presence in the hemolymph of a molting/metamorphosis hormone was also established by other approaches: Frew (1928) cultivated imaginal discs *in vitro* in hemolymph from either larvae or pupae, and he observed their evagination in the second medium only and Koller (1929) and Von Buddenbrock (1931) observed that the injection of hemolymph of molting animals into younger ones was able to accelerate molting of the latter. Fraenkel (1935) ligated larvae of *Calliphora erythrocephala* and observed pupation in the anterior part only; this provided a basis for the establishment of a bioassay for molting hormone activity, which was later used for the isolation of ecdysone (E) by Butenandt and Karlson (1954). The use of double ligation experiments with *Bombyx mori* larvae allowed Fukuda (1940) to demonstrate that a factor originating in the brain (i.e., the "brain hormone"; see Chapter 1) was relayed to an endocrine gland situated in the prothorax that produced the true "molting hormone". This prothoracic gland was first described by Ke (1930). It is termed "ventral gland" in locusts, and in higher Diptera it

DOI: 10.1016/B978-0-12-384749-2.10004-4

is part of a complex endocrine structure, the "ring gland." The direct demonstration that prothoracic glands produce E came much later (Chino *et al.*, 1974; King *et al.*, 1974; Borst and Engelmann, 1974) with the analysis of the secretory products of prothoracic glands cultivated *in vitro*. One year later, Hagedorn *et al.* (1975) demonstrated that the ovary of adult mosquitoes was another source of E, and this was the basis for establishing the role of ecdysteroids in the control of reproduction.

The first attempts to purify the molting hormone were performed by Becker and Plagge (1939) using the *Calliphora* bioassay derived from Fraenkel's work. α-Ecdysone was isolated in 1954 by Butenandt and Karlson after 10 years of attempting to do so. These authors used *B. mori* pupae because this material was available in large amounts. Ten more years of work were necessary to elucidate E's structure, which was finally achieved using X-ray crystallography analysis (Huber and Hoppe, 1965). A detailed description of this story can be found in Karlson and Sekeris (1966), and it is of great interest to follow how E was progressively characterized and to compare this work with what can be accomplished with presently available tools.

A second molecule, 20-hydroxyecdysone (20E) (initially termed β-ecdysone) was described soon thereafter, and this molecule was also isolated from a crayfish (Hampshire and Horn, 1966). It was rapidly established that 20E was the major molting hormone of all arthropods. Amazingly, E, which was the first of these compounds to be isolated, is at best a minor compound in most insect species when detectable, with the exception of young lepidopteran pupae, that is, the material used by Butenandt and Karlson. Other closely related molecules were then isolated and the generic term of "ecdysteroids" was proposed for this new steroid family (Goodwin *et al.*, 1978). Meanwhile, ecdysteroids had also been isolated from plants, hence the terms "zooecdysteroids" and "phytoecdysteroids."

4.2. Definition, Occurrence, and Diversity

Ecdysteroids could be defined either by their biological activity (molting hormones) or their chemical structure. They represent a specific family of sterol derivatives comprising more than 300 members that bear common structural features: a *cis* (5β-H) junction of rings A and B, a 7-ene-6-one chromophore, and a trans (14α-OH or H) junction of rings C and D (**Figure 1**). Most of them also bear the 3β-OH already present in their sterol precursor, and a set of other hydroxyl groups located both on the steroid nucleus and the side chain (**Figure 2**) that renders them water soluble. The major insect ecdysteroid is 20E, a 27-carbon (C_{27}) molecule derived from cholesterol (C), but some insect species contain its C_{28} or C_{29} homologues, makisterone A, and makisterone C, respectively (**Figure 3**).

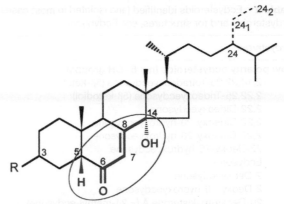

Figure 1 Common structural features of ecdysteroids (major characteristics are indicated in red/blue; R = 3β-OH in most cases).

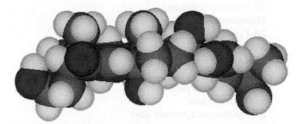

Figure 2 Space-filling model of the 20-hydroxyecdysone molecule. Blue: carbon atoms; red: oxygen atoms; white: hydrogen atoms. (Courtesy of Dr J.-P. Girault).

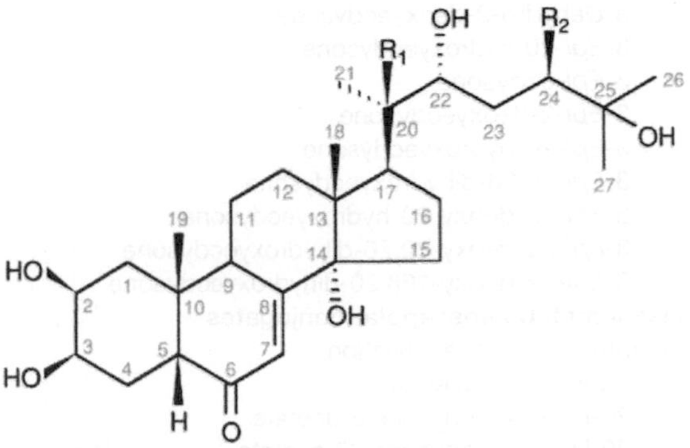

Figure 3 Structures of the major insect ecdysteroids. $R_1 = R_2 = H$: Ecdysone (E); $R_1 = OH$, $R_2 = H$: 20-Hydroxyecdysone (20E); $R_1 = OH$, $R_2 = Methyl$: Makisterone A; $R_1 = OH$, $R_2 = Ethyl$: Makisterone C.

The huge chemical diversity of ecdysteroids results from variations in the number and position of OH groups, which can be either free or conjugated to various polar or apolar moieties, giving rise to the 70 different molecules presently described in insects listed in **Table 1** (more information about these molecules can be found at www.ecdybase.org). The greatest diversity has been found in the eggs/embryos of Orthoptera and Lepidoptera, probably because they contain very large amounts of these molecules (up to 40 μg.g^{-1} fresh weight, i.e., 10 to 100 times the concentrations found in larvae or pupae).

Table 1 Ecdysteroids identified (and isolated in most cases) from insects (see also Rees, 1989; for a complete list of ecdysteroids and for structures, see Ecdybase)

Ecdysteroid	Species	Reference
Low polarity ecdysteroids (n ≤ 5 -OH groups)		
2,14,22,25-Tetradeoxyecdysone (5β-ketol)	*Locusta migratoria* ovaries	Hetru *et al.* (1978)
2,22,25-Trideoxyecdysone (5β-ketodiol)	*Locusta migratoria* ovaries	Hetru *et al.* (1978)
2,22-Dideoxyecdysone	*Locusta migratoria* ovaries	Hetru *et al.* (1978)
2,22-Dideoxy-20-hydroxyecdysone	*Bombyx mori* ovaries	Ikekawa *et al.* (1980)
2,22-Dideoxy-23-hydroxyecdysone	*Bombyx mori*	Kamba *et al.* (2000a)
22-Deoxy-20-hydroxyecdysone	*Bombyx mori*	Kamba *et al.* (1994)
Ecdysone	*Bombyx mori* pupae	Butenandt and Karlson (1954)
2-Deoxyecdysone	*Bombyx mori* ovaries/eggs	Ohnishi *et al.* (1977)
2-Deoxy-20-hydroxyecdysone	*Schistocerca gregaria*	Isaac *et al.* (1981a)
20-Deoxy-makisterone A (= 24-methyl-ecdysone)	*Drosophila* ring glands	Redfern (1984)
24-*Epi*-20-deoxy-makisterone A	*Drosophila* ring glands	Blais *et al.* (2010)
14-Deoxy-20-hydroxyecdysone	*Gryllus bimaculatus*	Hoffmann *et al.* (1990)
Phase I metabolites		
+ *20-OH*		
20-Hydroxyecdysone	*Bombyx mori* pupae	Hocks and Wiechert (1966)
Makisterone A	*Oncopeltus fasciatus*	Kaplanis *et al.* (1975)
24-Epi-makisterone A	*Acromyrmex octospinosus*	Maurer *et al.* (1993)
Makisterone C	*Dysdercus fasciatus*	Feldlaufer *et al.* (1991)
+ *26-OH (and the subsequent oxidation products)*		
20,26-Dihydroxyecdysone	*Manduca sexta* pupae	Thompson *et al.* (1967)
26-Hydroxyecdysone	*Manduca sexta* eggs	Kaplanis *et al.* (1973)
Ecdysonoic acid	*Schistocerca gregaria* eggs	Isaac *et al.* (1983)
20-Hydroxyecdysonoic acid	*Schistocerca gregaria* eggs;	Isaac *et al.* (1983);
	Pieris brassicae pupae	Lafont *et al.* (1983)
20-Hydroxyecdysone 26-aldehyde (2 hemiacetals)	*Chironomus tentans* cells	Kayser *et al.* (2002)
+ *3-oxo/3α-*		
3-Dehydroecdysone	*Calliphora erythrocephala*	Karlson *et al.* (1972)
3-Dehydro-20-hydroxyecdysone	*Calliphora vicina*	Koolman and Spindler (1977)
3-Dehydro-2-deoxyecdysone	*Locusta migratoria*	Tsoupras *et al.* (1983a)
3-*Epi*-20-hydroxyecdysone	*Manduca sexta*	Thompson *et al.* (1974)
3-*Epi*-ecdysone	*Manduca sexta*	Kaplanis *et al.* (1979)
3-*Epi*-2-deoxyecdysone	*Schistocerca gregaria*	Isaac *et al.* (1981a)
3-*Epi*-26-hydroxyecdysone	*Manduca sexta*	Kaplanis *et al.* (1980)
3-*Epi*-20,26-dihydroxyecdysone	*Manduca sexta*	Kaplanis *et al.* (1979)
3-*Epi*-22-deoxy-20-hydroxyecdysone	*Bombyx mori*	Mamiya *et al.* (1995)
3-*Epi*-22-deoxy-20,26-dihydroxyecdysone	*Bombyx mori*	Kamba *et al.* (2000b)
3-*Epi*-22-deoxy-16β,20-dihydroxyecdysone	*Bombyx mori*	Kamba *et al.* (2000b)
Phase II metabolites: apolar conjugates		
Acetates: 3- or 22-acetylation		
Ecdysone 3-acetate	*Schistocerca gregaria*	Isaac *et al.* (1981b)
20-Hydroxyecdysone 3-acetate	*Locusta migratoria*	Modde *et al.* (1984)
20-Hydroxyecdysone 22-acetate	*Drosophila melanogaster*	Maroy *et al.* (1988)
Acyl esters: 22-acylation		
Ecdysone 22-palmitate	*Gryllus bimaculatus*	Hoffmann *et al.* (1985)
Ecdysone 22-palmitoleate	*Gryllus bimaculatus*	Hoffmann *et al.* (1985)
Ecdysone 22-oleate	*Gryllus bimaculatus*	Hoffmann *et al.* (1985)
Ecdysone 22-linoleate	*Gryllus bimaculatus*	Hoffmann *et al.* (1985)
Ecdysone 22-stearate	*Gryllus bimaculatus*	Hoffmann *et al.* (1985)
20-Hydroxyecdysone 22-palmitate	*Heliothis virescens*	Kubo *et al.* (1987)
20-Hydroxyecdysone 22-oleate	*Heliothis virescens*	Kubo *et al.* (1987)
20-Hydroxyecdysone 22-linoleate	*Heliothis virescens*	Kubo *et al.* (1987)
20-Hydroxyecdysone 22-stearate	*Heliothis virescens*	Kubo *et al.* (1987)
Phase II metabolites: polar conjugates		
Glucosides: 22-glucosylation		
26-Hydroxyecdysone 22-glucoside	*Manduca sexta*	Thompson *et al.* (1987a)
Phosphates		
2-Phosphates		
Ecdysone 2-phosphate	*Schistocerca gregaria*	Isaac *et al.* (1984)
26-Hydroxyecdysone 2-phosphate	*Manduca sexta*	Thompson *et al.* (1987b)

Table 1   Ecdysteroids identified (and isolated in most cases) from insects (see also Rees, 1989; for a complete list of ecdysteroids and for structures, see Ecdybase)—Cont'd

Ecdysteroid	Species	Reference
3-*Epi*-22-deoxy-20-hydroxyecdysone 2-phosphate	*Bombyx mori*	Mamiya *et al.* (1995)
3-*Epi*-22-deoxy-20,26-dihydroxyecdysone 2-phosphate	*Bombyx mori*	Kamba *et al.* (2000b)
3-*Epi*-22-deoxy-16β,20-dihydroxyecdysone 2-phosphate	*Bombyx mori*	Kamba *et al.* (2000b)
3α-Phosphates		
3-*Epi*-2-deoxyecdysone 3-phosphate	*Locusta migratoria*	Tsoupras *et al.* (1982b)
3-*Epi*-20-hydroxyecdysone 3-phosphate	*Pieris brassicae*	Beydon *et al.* (1987)
3β-Phosphates		
Ecdysone 3-phosphate	*Locusta migratoria*	Lagueux *et al.* (1984)
22-Deoxy-20-hydroxyecdysone 3-phosphate	*Bombyx mori*	Kamba *et al.* (1994)
2,22-Dideoxy-20-hydroxyecdysone 3-phosphate	*Bombyx mori*	Hiramoto *et al.* (1988)
2,22-Dideoxy-23-hydroxyecdysone 3-phosphate	*Bombyx mori*	Kamba *et al.* (2000a)
22-Phosphates		
2-Deoxyecdysone 22-phosphate	*Schistocerca gregaria*	Isaac *et al.* (1982)
2-Deoxy-20-hydroxyecdysone 22-phosphate	*Locusta migratoria*	Tsoupras *et al.* (1982a)
Ecdysone 22-phosphate	*Schistocerca gregaria*	Isaac *et al.* (1982)
20-Hydroxyecdysone 22-phosphate	*Locusta migratoria*	Tsoupras *et al.* (1982a)
3-Epi-2-deoxyecdysone 22-phosphate	*Bombyx mori*	Kamba *et al.* (1995)
3-Epi-ecdysone 22-phosphate	*Bombyx mori*	Kamba *et al.* (1995)
26-Phosphates		
26-Hydroxyecdysone 26-phosphate	*Manduca sexta*	Thompson *et al.* (1985b)
Phase II metabolites/complex conjugates		
Acetyl-phosphates		
3-Acetylecdysone 2-phosphate	*Schistocerca gregaria*	Isaac *et al.* (1984)
3/2-Acetylecdysone 22-phosphate	*Schistocerca gregaria*	Isaac and Rees (1984)
3-Acetyl-20-hydroxyecdysone 2-phosphate	*Locusta migratoria;*	Modde *et al.* (1984);
	Schistocerca gregaria	Isaac and Rees (1984)
3/2-Acetyl-20-hydroxyecdysone 22-phosphate	*Locusta migratoria*	Tsoupras *et al.* (1983b)
Ecdysone 2,3-diacetate 22-phosphate	*Locusta migratoria*	Lagueux *et al.* (1984)
Nucleotides		
2-Deoxyecdysone 22-AMP	*Locusta migratoria*	Tsoupras *et al.* (1983a)
Ecdysone 22-AMP	*Locusta migratoria*	Hétru *et al.* (1984)
Ecdysone 22-isopentenyl-AMP	*Locusta migratoria*	Tsoupras *et al.* (1983c)
Ecdysteroid-related steroids		
Bombycosterol	*Bombyx mori*	Fujimoto *et al.* (1985a)
Bombycosterol 3-phosphate	*Bombyx mori*	Hiramoto *et al.* (1988)

The nomenclature of insect ecdysteroids is generally derived from the ecdysone name, but this is not always the case, and the trivial name of many ecdysteroids (especially phytoecdysteroids) is related to the name of the species from which they have been isolated; standardized abbreviations have been proposed for insect ecdysteroids (Lafont *et al.*, 1993a) and are used throughout this chapter.

4.3. Methods of Analysis

4.3.1. Methods of Identification

4.3.1.1. Physicochemical Properties

4.3.1.1.1. UV spectroscopy   The 7-en-6-one group of ecdysteroids has a relatively strong absorption at $\lambda_{max} \sim 243$ nm with an ε of ~10–16,000: E has a λ_{max} at 242 nm and an ε of 12,400 in EtOH. This maximum of absorbance is shifted to 247 nm in water.

4.3.1.1.2. IR spectroscopy   The presence of a number of hydroxyl groups on most ecdysteroids ensures a strong absorption in the infrared (IR) spectrum in the region of 3340–3500 cm^{-1}. The α,β-unsaturated ketone results in a characteristic cyclohexenone absorption at 1640–1670 cm^{-1}, with a weaker alkene stretch at approximately 1612 cm^{-1}. An additional carbonyl absorption is seen with 3-dehydroecdysteroids: thus 3-dehydroecdysone (3DE) and 3-dehydro-20-hydroxyecdysone (3D20E) show a carbonyl absorption at 1712 cm^{-1} and 1700 cm^{-1}, respectively, in addition to the usual 7-ene-6-one signal. Among conjugates, the absorption of phosphates ranges from 1100 to 1000 cm^{-1}. Acetates and acyl esters can be characterized by additional bands corresponding to the carbonyl and to the ester group. Ecdysteroids are not soluble in the usual infrared solvents (CS$_2$, CCl$_4$, CHCl$_3$), so IR spectra are done on KBr discs or mineral oil (Nujol) mulls. The development of Fourier transform spectrometers (FT-IR) allows the solubility problem to be overcome, as dilute solutions can be analyzed. Thus, IR spectra can even be recorded in-line during high-performance liquid chromatography (HPLC) analysis (Louden *et al.*, 2001, 2002).

4.3.1.1.3. Fluorescence In the presence of sulfuric acid or aqueous ammonia ecdysteroids fluoresce (Horn, 1971). Typical values for excitation and emission wavelengths are in the region of 380 and 430 nm, respectively. Some variations in excitation and emission wavelength and fluorescence intensity are observed with ecdysteroids (Koolman, 1980).

4.3.1.2. Mass spectrometry

Mass spectrometry (MS) has been invaluable in the elucidation of ecdysteroid structures and has been particularly useful for insect samples containing low levels of ecdysteroids. Most of the early work on ecdysteroids has been performed using relatively unsophisticated direct introduction techniques (electron impact, EI), but new ionization techniques are now increasingly employed.

4.3.1.2.1. Modes of ionization Electron impact (EI-MS) — In EI-MS, an electron beam is used for the ionization and the collision will break the molecule into several characteristic fragments. With nonvolatile substances like ecdysteroids, the use of a 70 eV electron beam results in extensive fragmentation. The abundant fragments correspond to the steroid nucleus or to the side chain and are highly informative for structural elucidation.

Chemical ionization (CI-MS) — This method uses a reagent gas that is ionized by the electron beam and gives rise to a set of ions that in turn can react with ecdysteroids (Lafont *et al.*, 1981). The ionization reaction proceeds as follows:

$$M + HR^+ \rightarrow MH^+ \text{ (pseudo-molecular ion)} + R$$
$$M + HR^+ \rightarrow MHR^+ \text{ (adduction)}$$

where R = the reagent gas such as ammonia. The relative intensity of these ions is rather high. Among the various reagent gases, ammonia is most widely used, giving both MH^+ and $(M+NH_4)^+$ ions of high intensity. This method is of interest for the determination of the molecular mass (M) of ecdysteroids. Both positive and negative ionization can be used. The chemical ionization/desorption (CI/D) techniques were analyzed in depth by Lusby *et al.* (1987) who considered the role of parameters such as source pressure and source temperature on fragmentation.

Field desorption (FD-MS) — With FD-MS, ionization proceeds in a high electrostatic field; such conditions allow ecdysteroid molecules to lose one electron ($\rightarrow M^+$) without excessive heating. In such cases, the organic molecules have not received any large amount of energy and, therefore, almost no fragmentation is observed (Koolman and Spindler, 1977).

Fast-atom bombardment (FAB-MS) — This method uses a beam of neutral atoms (argon or xenon) bearing a high kinetic energy for the ionization of involatile ionic or non-ionic compounds present either in the solid state or as a solution in a glycerol matrix. FAB-MS is particularly suitable for polar ecdysteroids such as phosphate conjugates (see Isaac *et al.*, 1982; Modde *et al.*, 1984) and ecdysonoic acids (Isaac *et al.*, 1984), but it also works with non-ionic ecdysteroids (Rees and Isaac, 1985). Sodium or potassium salts of conjugates and ecdysonoic acids give, respectively, $[M+Na]^+$ and $[M+K]^+$ ions. FAB$^+$ has been used to characterize fatty acyl derivatives of ecdysteroids (Dinan, 1988), but FAB$^-$ can also be used (Wilson *et al.*, 1988a). When using a deuterated matrix, extensive deuterium exchange takes place and it allows the number of hydroxyl groups present in the molecule to be determined (Pís and Vaisar, 1997).

Electrospray (ES-MS) — This method has been introduced more recently (Hellou *et al.*, 1994). It allows a high intensity of molecular ions to be obtained and is recommended for fragile molecules. Moreover, it can be conveniently coupled with HPLC for in-line analysis, and it has become the new standard for most analyses.

Low/high-resolution mass spectrometry (HR-MS) — All of the above methods provide high- and low-resolution mass spectra. Such data result from an approximation, as the exact molecular mass of atoms (with the exception of the ^{12}C carbon atom, which by definition is equal to 12) is not a whole number (e.g., the mass of ^1H is 1.00783 and that of ^{16}O is 15.99491). By using high-resolution instruments, the mass resolution is much increased and the masses of fragments can be obtained with a great accuracy, allowing the spectroscopist to unambiguously determine the elementary composition of ions and of their parent molecules. This technique has fully replaced the former elementary analysis.

Tandem mass spectrometry (MS-MS) — The determination of fragment structure and filiation is by no means easy. The MS-MS techniques correspond to the sequential operation of several chambers (quadrupoles) where fragment separation takes place. Thus, the first quadrupole is used to select one ion, and following the passage of that ion through a "fragmentation cell," the second one is used to analyze the further fragmentation of this selected ion. An extensive analysis of this type allows the establishment of genetic relationships between all of the observed fragments and is of considerable help in understanding fragmentation mechanisms. The same method can also be used to search for the presence of a specific compound in a crude mixture (Mauchamp *et al.*, 1993).

4.3.1.2.2. General rules for mass spectrometry The fragmentation patterns (with EI-MS and CI-MS) of ecdysteroids are dominated by ions resulting from the ecdysteroid nucleus and those due to the side chain (Nakanishi,

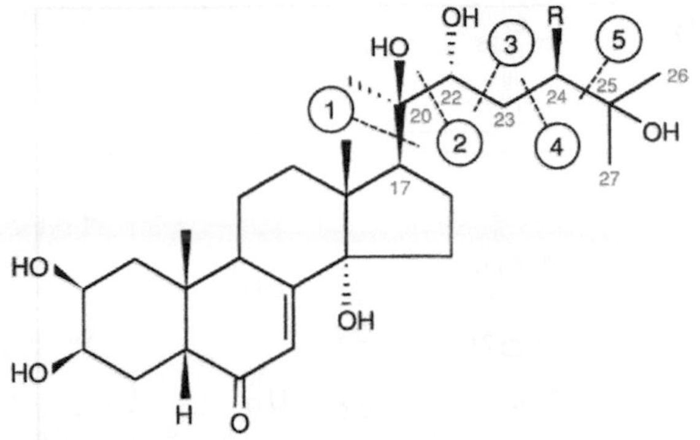

Figure 4 Major and minor fragmentations of ecdysteroid side chain. 1: C-17–C-20 cleavage; chain 2: C-20–C-22 cleavage (major if 20,22-diol); chain 3: C-22–C-23 cleavage (minor); chain 4: C-23–C-24 cleavage (if branching at C-24); chain 5: C-24–C-25 cleavage (for TMS ethers).

1971; **Figure 4**). Because of the ready elimination of water from these polyhydroxy compounds, molecular ions, if present, are usually weak (<1%) when using EI-MS, although [M-18]+ ions are often readily apparent. The high mass region of the spectra of ecdysteroids tends to be characterized by ions 18 mass units apart resulting from the sequential losses of water (Horn, 1971). The presence of a 20,22-vicinal diol allows the cleavage of the side chain (**2**), giving rise to a prominent fragment at m/z 99 for 20E (otherwise dependent on the structure of the side chain), frequently forming the base peak of the EI spectrum. An ion at m/z 81 is also frequently observed, which probably results from the elimination of water from the side chain fragment (Horn, 1971; Nakanishi, 1971). In the absence of a 20,22-diol, fragmentation (**1**) between C-17 and C-20 is observed. This cleavage is less dominant than the 20,22-scission, and the result is a more complex pattern. Where the side chain has been modified to include methyl or ethyl substituents at C-24 (e.g., the makisterones A and C), the fragments at m/z 99 and 81 are replaced by ions 14 and 28 mass units higher (Nakanishi, 1971). With ES-MS various fragmentations of the side chain take place including additional cleavages between C-22 and C-23 as well as C-23 and C-24 (Blais *et al.*, 2010).

4.3.1.3. Nuclear magnetic resonance Nuclear magnetic resonance (NMR) spectroscopy is the method of choice for determining the structure of ecdysteroids. Proton ^{1}H-NMR and carbon ^{13}C-NMR are of general use, whereas the use of ^{31}P-NMR is restricted to phosphate conjugates (Hétru *et al.*, 1985; Rees and Isaac, 1985) and will not be discussed here.

Obtaining a high-quality ^{1}H-NMR spectrum using modern equipment requires about 50 μg of a pure compound, and this is generally sufficient to determine elementary changes like epimerization, appearance of an extra-OH group, or the position involved in a conjugation process. More important changes require ^{13}C-NMR

studies, which until recently required amounts in the milligrams. Complete NMR spectra of E and 20E have been published (Kubo *et al.*, 1985b; Girault and Lafont, 1988; Lee and Nakanishi, 1989), and general tables are given in several reviews (Hoffmann and Hétru, 1983; Horn and Bergamasco, 1985; Rees and Isaac, 1985; see also www.ecdybase.org). Most NMR data have been obtained using C_5D_5N as solvent, except for very apolar compounds (apolar conjugates or acetate derivatives analyzed in $CDCl_3$) and polar conjugates (analyzed in D_2O). These solvents may be of interest for use with specific compounds in connection with either solubility problems or the presence of signals, which would be masked by signals due to the classical solvents. For ^{1}H-NMR studies using modern machines, the water solubility of many common ecdysteroids is high enough to obtain excellent spectra (Girault and Lafont, 1988).

4.3.1.3.1. Proton NMR There have been spectacular improvements in proton NMR with the introduction of high-field machines and sophisticated software, and this has resulted in the possibility of using smaller amounts of ecdysteroids and obtaining more information from the spectra. Classical proton NMR spectroscopy (in C_5D_5N) allows the easy determination of methyl groups (usually C-18, C-19, C-21, C-26, and C-27, δ 0.8–1.4 ppm) and of a few additional signals corresponding to 7-H (vinylic proton, δ ca. 6.2 ppm) and to primary or secondary alcoholic functions (2-H, 3-H, and 22-H δ 4.0–4.2 ppm). Other signals (CH or CH_2 groups) overlap in the same area of the spectrum (δ 1.5–2.3 ppm), with the noticeable exception of 9-H (δ ca. 3.5 ppm) and 5-H (δ ca. 3.0 ppm) (Horn, 1971). Furthermore, two-dimensional techniques such as 2D-COSY and relayed correlated spectroscopy (COSY) have allowed the complete assignment of all protons of the ecdysone molecule (Kubo *et al.*, 1985a; Girault and Lafont, 1988). The proton NMR spectrum strongly depends upon the solvent and, to a lesser extent, on the temperature used. In most cases the solvent used was C_5D_5N or $CDCl_3$ (for acetate derivatives essentially), but for some compounds such data are lacking and the available spectra were obtained in D_2O, CD_3OD, …, so they are more difficult to compare. Coupling constants also provide useful information concerning the stereochemistry of the ecdysteroid molecule; for example, they allow 3α- and 3β-OH compounds to be distinguished.

4.3.1.3.2. Carbon NMR The introduction of ^{13}C NMR (Lukacs and Bennett, 1972) represented a major advance in the determination of ecdysteroid structures. The signals of the carbon atoms are spread over 200 ppm. Thus, ^{13}C NMR spectra are better resolved than ^{1}H spectra. Due to the low natural abundance of ^{13}C, relatively large samples (1–10 mg) are required. Off-resonance decoupled spectra show the coupling of carbon atoms to protons, allowing

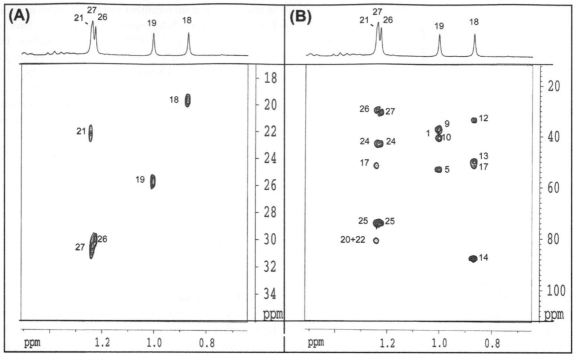

Figure 5 Inverse correlation NMR spectra of 20-hydroxyecdysone in D_2O. (A) HMQC allows the observation of the direct $^1J^{13}C$-H couplings (e.g., between the 18-Me protons and C-18) and (B) HMBC allows the observation of $^2J^{13}C$-H (e.g., between 18-Me protons and C-13) and $^3J^{13}C$-H (e.g., between 18-Me protons and C-12, C-14 and C-17) long-range couplings. (Courtesy of Dr J.-P. Girault).

the carbons to be distinguished from the number (0–3) of protons they bear and allowing an easier interpretation of spectra.

New methods have recently become available for collecting ^{13}C information thanks to inverse correlations with proton signals: the heteronuclear multiple-quantum correlation (HMQC) and the heteronuclear multiple-bond correlation (HMBC), which make use of the direct ($^1J^{13}C$-H) and long range ($^2J^{13}C$-H and $^3J^{13}C$-H) couplings, respectively (**Figure 5**; Girault, 1998). They allow ^{13}C data to be obtained with about 100 μg of ecdysteroid with the most recent machines.

4.3.2. Methods of Purification

Ecdysteroids form a group of polar compounds and, as a consequence, their extraction is usually performed using a polar solvent such as MeOH. The next step usually involves one or more solvent partitions with the aim of removing the bulk of polar and non-polar contaminants prior to chromatography (**Figure 6**).

4.3.2.1. Extraction procedures   Usually, a polar solvent (MeOH, at least 10x the sample weight v/w) is used. Two or three repeats are necessary to extract ecdysteroids quantitatively. Alternative solvents include the less toxic EtOH, or Me_2CO, MeCN, $MeOH$-H_2O or H_2O. Extraction can be performed at room temperature, but slight

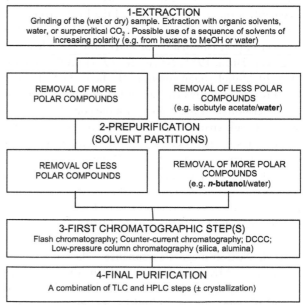

Figure 6 Example of an extraction and purification chart for the large-scale isolation of ecdysteroids. Reproduced with permission from *Lafont et al.* (1994b).

heating (<60 °C) or even refluxing conditions (Soxhlet apparatus) may significantly improve its efficiency.

4.3.2.2. Partition techniques   The crude extract is concentrated under reduced pressure to give an oily residue. The rather peculiar polarity of ecdysteroids allows their purification by solvent partitioning. This applies to compounds having a polarity in the same range as E and 20E. In the cases where polar or apolar ecdysteroids are present (e.g., phosphate or acyl esters), it is clear that those derivatives will partition in a very different way according to their respective polarity, and they may be lost.

4.3.2.2.1. Solvent partitioning
In early studies (Horn, 1971), the initial partition was made between an aqueous concentrate and Bu-1-OH. The Bu-1-OH residue, into which the ecdysteroids were extracted, was then partitioned between aqueous MeOH and hexane to remove non-polar materials, such as lipids. However, reversing this order of operations was found to be beneficial and led to reduced emulsion formation during the partition step and less problems with frothing during evaporation. Other suitable solvents for removing lipids include hexane-MeOH (7:3, v/v), light petroleum (b.p. 40–60°C), or Pr-1-OH-hexane (3:1, v/v), the ecdysteroids remaining in the aqueous phase. The addition of $(NH_4)_2SO_4$ improves phase separation. The separation of polar impurities from the ecdysteroids can be achieved by partition between H_2O and Bu-1-OH (ecdysteroids partition into the organic phase) and H_2O-EtOAc (ecdysteroids remain in the aqueous phase). The major factors governing the choice of solvent partition system are the type of contaminants to be removed (i.e., mainly lipids or mainly polar compounds, etc.) and the nature of the ecdysteroids to be isolated. Thus, the addition or removal of one -OH group, or conjugation to polar (e.g., sulfate) or non-polar (e.g., acetate or fatty acyl) groups can significantly affect partition ratios. A combination of two successive partitions allows elimination of both polar and apolar contaminants. The use of two systems having one solvent in common avoids the need for an evaporation step between the two partitions. It is thus possible to combine first $CHCl_3/H_2O$ followed by H_2O/Bu-1-OH. $CHCl_3$ can be replaced by a more polar organic solvent, such as isobutyl acetate, which allows ecdysteroids to remain in the water phase (Matsumoto and Kubo, 1989).

4.3.2.2.2. Countercurrent distribution (CCD)
Countercurrent distribution between Bu-1-OH and H_2O is effective for removing polar contaminants and several other systems have been successfully used in the past (Horn and Bergamasco, 1985). For instance, Butenandt and Karlson (1954) used this method to purify ecdysone from *Bombyx* pupae by means of Craig CCD with butanol-cyclohexane-water (6:4:10, v/v/v). This method allowed an efficient separation of two different fractions containing E and 20E, respectively. The same two ecdysteroids were also isolated from an extract of 15 kg *Calliphora* pupae (Karlson, 1956).

4.3.2.2.3. Droplet countercurrent chromatography (DCCC)
Droplet countercurrent chromatography provides an efficient means for purifying samples up to the gram range. It belongs to the family of liquid–liquid partition chromatographic methods. In favorable cases, DCCC can enable the preparation of pure compounds (Kubo *et al.*, 1985a). Generally, a subsequent HPLC step is required to get pure ecdysteroids. In a related procedure, high-speed countercurrent chromatography (HSCCC), the column is a multilayer coil and efficient mixing is achieved through planetary rotation (Ito, 1986). The centrifugal forces allow separations to be achieved within hours instead of days with DCCC.

4.3.2.3. Column chromatography (low pressure)
4.3.2.3.1. Preparative columns
This technique can be easily scaled-up according to sample size; it just suffices to keep a low sample to sorbent ratio. It can be used with various stationary phases (NP: silica, alumina; RP: hydrophobic resins, polyamide, or C18-bonded silica; ion-exchange phases, etc.). Elution usually proceeds with a step-gradient of the mobile phase with increasing eluting power; fractions are collected and assayed, for example, by TLC or HPLC.

4.3.2.3.2. Small columns and/or disposable cartridges
Normal phase — Low-pressure chromatography on a small column was used in very early methods for the fractionation of crude extracts. Silica or alumina columns (normal phase systems) were generally eluted with binary mixtures, such as a step-gradient of alcohol in chloroform or benzene.

Reversed phase — The availability of hydrophobic phases (resins like Amberlite® or hydrocarbon-bonded silica) has led to a complete renewal of the early procedures. Small cartridges or syringes containing 0.2–1 g of HPLC phase are ideally suited for a rapid clean up of small samples (Lafont *et al.*, 1982; Watson and Spaziani, 1982; Pimprikar *et al.*, 1984; Rees and Isaac, 1985; Lozano *et al.*, 1988b; Wilson *et al.*, 1990b). They can also be used for desalting purposes (e.g., direct adsorption of ecdysteroids from culture media) from buffers used for enzymatic hydrolysis of conjugates or from reversed-phase HPLC fractions when an involatile buffer is used.

4.3.3. Methods of Separation

The general characteristics of separation methods are summarized in **Table 2**.

4.3.3.1. Thin-layer chromatography (HPTLC, OPLC)
4.3.3.1.1. Chromatographic procedures
Normal-phase (absorption) chromatography on silica gel has been used extensively for the isolation of ecdysteroids and for

Table 2 Methods of analysis and of characterization

Method	Subtypes	Detection methods	Comments	Selected references
TLC	1D/2D, AMD	UV/fluorescence quenching		Mayer and Svoboda (1978), Wilson (1985), Wilson and Lafont (1986)
	NP/RP	Color reactions		Wilson and Lewis (1987)
	HPTLC	Radioactivity/ autoradiography		Wilson (1988), Wilson et al. (1988b)
				Báthori et al. (2000),
	OPLC	Off-line MS		Read et al. (1990), Wilson et al. (1990c)
		Immuno- or bioassay		Lafont et al. (1993b)
LC	Low-pressure/high-pressure (HPLC)	UV	Analytical or preparative	Lafont and Wilson (1990)
	NP, RP	Fluorescent derivatives MS (+IR, NMR)		Kubo and Komatsu (1986, 1987) Marco et al. (1993), Evershed et al. (1993),Wainwright et al. (1997), Louden et al. (2002)
		Immuno- or bioassay		Warren et al. (1986)
	Ion-exchange, affinity	Radioactivity		Lozano et al. (1988a,b)
	Nano-LC	Online MS/MS	Analytical	Blais et al. (2010)
SFC		UV and MS	Very fast and efficient	Raynor et al. (1988, 1989), Morgan et al. (1988), Morgan and Huang (1990)
DCCC	Ascending, descending	UV	Preparative	Kubo et al. (1985a), Báthori (1998)
CE, MCE		UV	Analytical only	Large et al. (1992), Davis et al. (1993), Yasuda et al. (1993)
GLC	Filled or capillary columns	FID ECD MS	Requires derivatization	Poole et al. (1975), Bielby et al. (1986), Evershed et al. (1987)

metabolic studies (Horn, 1971; Morgan and Poole, 1976; Morgan and Wilson, 1989).

Normal-phase and reversed-phase systems — A general normal-phase system consists of 95% EtOH-CHCl3 (1:4, v/v). For reproducible results, plates should be heated for 1 h at 120 °C, and then deactivated to constant activity over saturated saline, and thin-layer chromatography (TLC) tanks should be saturated with the vapor of the solvent used for the chromatography before the plates are developed. Reversed-phase TLC on either bonded or paraffin-coated plates has been widely used (Wilson et al., 1981, 1982b; Wilson, 1985). Bonded silicas with various alkyl chain lengths of C_2 to C_{18} or, alternatively, various degrees of coating, are available and all seem suitable for ecdysteroids. $MeOH-H_2O$, $Pr-2-OH-H_2O$, $EtOH-H_2O$, $MeCN-H_2O$, and Me_2CO-H_2O solvent systems have been used for chromatography; $MeOH-H_2O$ mixtures are most commonly used (Wilson et al., 1981).

Two-dimensional TLC — A given biological sample can contain a very complex mixture of ecdysteroids. In such cases, two-dimensional techniques have proved to be very efficient; for example, silica plates developed with (1) toluene-Me_2CO-EtOH-25% aqueous ammonia (100:140:32:9 v/v), then (2) $CHCl_3$-MeOH-benzene (25:5:3, v/v) (Báthori et al., 2000).

Visualization techniques — Detection of ecdysteroids on the TLC plate can be accomplished using a variety of techniques of varying specificities. Non-specific techniques include iodine vapors or heating in the presence of

ammonium carbonate, which produces fluorescent spots or fluorescence quenching when a fluor (ZnS) is incorporated into the silica. More specific reagents, such as the vanil-linsulfuric acid spray, can be used to give spots of characteristic color (Horn, 1971). A slightly more specific fluorescence reaction than that obtained using ammonia detection can be achieved by spraying the plate with sulfuric acid.

Scanners — Plate analysis can benefit from the wide array of available detectors (scanners) such as UV, UV-Vis, fluorescence, FT-IR, radioactivity, or mass spectrometry (Wilson et al., 1988a, 1990a,c).

4.3.3.1.2. General rules for TLC
Ecdysteroid behavior on TLC is directly related to the number and position of free hydroxyl groups and the presence of various substituents. The migration of a given compound is given by its R_f (= reference front), defined as migration of the substance/migration of the solvent front. From the R_f value, another parameter can be calculated, the R_m, defined as $R_m = \log (1/R_f - 1)$. A single modification of an ecdysteroid results in a change in its R_m, expressed as ΔR_m, which is to some extent characteristic of the modification (Koolman et al., 1979).

4.3.3.1.3. New TLC techniques
Several methods providing improved resolution when compared with conventional TLC are available.

Automated or programmed multiple development (AMD and PMD) — In automated or programmed multiple development (AMD and PMD), the plate is developed repeatedly with the same solvent, which is allowed to migrate further and further with each development. This method allows a reconcentration of ecdysteroids at each run, in particular by suppressing tailing, and this finally results in sharper bands and improved resolution (Wilson and Lewis, 1987).

Overpressure TLC (OPLC) — In this method, the plate is held under an inert membrane under hydrostatic pressure (ca. 25 bar) and the solvent is forced through the layer by an HPLC pump. Therefore, the flow of the solvent is controlled and is no longer due to capillarity, and the plate can be developed within minutes. To be efficient, this technique requires the use of HPTLC plates. OPLC has been used with ecdysteroids in only a few instances (Read et al., 1990).

4.3.3.2. High-performance liquid chromatography

HPLC is the most popular technique for ecdysteroid separations, both for analytical and preparative purposes. It offers a wide choice of techniques that can be optimized for polar or apolar metabolites. The identification of any ecdysteroid by comigration with a reference compound must rely on the simultaneous use of several (at least two) different HPLC systems, generally one normal-phase (NP) and one reversed-phase (RP) system (Lafont et al., 1980, 1981; Touchstone, 1986; Lafont, 1988; Thompson et al., 1989; Lafont and Wilson, 1990).

4.3.3.2.1. Chromatographic procedures NP systems —

Silica columns are usually run using mixtures of a chlorinated solvent (e.g., CH_2Cl_2, dichloroethane) and an alcohol (e.g., MeOH, EtOH, Pr-2-OH). Water added to just below saturation partially deactivates silica and this results in more symmetrical peaks; a classical mixture is CH_2Cl_2–Pr-2-OH–H_2O (Lafont et al., 1979). The respective proportions of the three components can be adapted to suit the sample polarity. Thus, specific mixtures can be prepared for non-polar compounds, for example, acetates or E precursors (125:15:1 v/v/v), for medium-polarity compounds (125:25:2 or 125:30:2 for E and 20E), or for more polar ecdysteroids (125:40:3 for 26-hydroxyecdysteroids) including glycoside conjugates (100:40:3). Polar-bonded columns (-diol or -polyol,-APS aminopropylsilane) can also be used instead of silica (Dinan et al., 1981). Diol-bonded columns have been used with the phasmid *Carausius morosus* for the separation of a wide array of metabolites (Fournier and Radallah, 1988), whereas the APS phase proved particularly efficient for the separation of mixtures of 3α-OH, 3β-OH, and 3-oxo ecdysteroids (Dinan et al., 1981). In addition, these columns allow the use of gradients without the problems linked with the long re-equilibration times encountered with silica. CH_2Cl_2-based solvents, although very efficient for chromatographic separations, suffer from a high UV-cutoff and the quenching properties of this compound, which preclude diode-array detection or efficient in-line radioactivity monitoring (see the next section). This problem may be overcome with cyclohexane based mixtures (Lafont et al., 1994a), which also display a different selectivity.

RP systems — C_{18}-bonded silicas are the most widely used phases, with elution performed with MeOH-H_2O. MeCN-H_2O or MeCN-buffer mixtures, and are more efficient, especially when polar conjugates and/or ecdysonoic acids are present. The use of ion suppression (e.g., using an acidic buffer) is absolutely essential in the case of polar conjugates or 26-acids. Otherwise, the ionized molecules do not partition well, resulting in variable retention times. Systems have been designed for polar or apolar metabolites, both of which may exist within the same animal. The use of various pHs (Pís et al., 1995a) can selectively modify the retention time of polar (ionizable) metabolites and give access to the pK value of ionizable groups, which can help to characterize conjugates.

Detection procedures — Several types of detectors can be used for the monitoring of ecdysteroids in the eluates.

UV detector. Ecdysteroids possess a chromophore that is strongly absorbing at 245 nm, which allows the easy detection of 10 pmol amounts.

Fluorescence detector. There have been several attempts to prepare fluorescent derivatives of ecdysteroids. The potential interest is great, as this would enhance the sensitivity of detection by at least 100 times and possibly also the selectivity as compared to UV. Pre-column or post-column derivatization can be used. There are several prerequisites for pre-column reactions: (1) they must be simple and give a single derivative for each ecdysteroid; (2) they must be quantitative, even at very low concentrations of ecdysteroids present in a crude sample; (3) they must be specific; and (4) the excess reagent (if fluorescent itself) must be removed prior to HPLC analysis. Two approaches have been described: the first used phenanthrene boronic acid (Poole et al., 1978), which is specific for diols and would therefore react also with sugars; the second used 1-anthroyl nitrile (Kubo and Komatsu, 1986), which reacts with alcohols, in this case the 2-OH of ecdysteroids. Interest in the second case would essentially be increased sensitivity, but it is not a specific reaction in that the authors use the same reagent for the determination of prostaglandins. These methods have not received further applications, and we must emphasize that the use of pre-column derivatization would negate one major advantage of HPLC over GLC. Post-column derivatization could make use of the sulfuric acid-induced fluorescence (see 4.3.1.1.3.), but this has not been investigated.

Diode-array detectors. Such detectors represent valuable tools since they provide the absorbance spectrum of all eluted peaks. Thus, in the case of ecdysteroids, whether or not a compound that co-migrates with a reference ecdysteroid has a UV absorbance with a maximum at 242 nm can be directly checked. This is an additional criterion for assessing the identity of peaks, and it works equally well with very small amounts of ecdysteroids (less than 100 ng).

Radioactivity monitors. On-line radioactivity measurement in ecdysteroid metabolic or biosynthesis studies is of considerable interest because: (1) it provides an immediate result; (2) it saves time and avoids the need for collecting many fractions, filling scintillation vials, and waiting for their measurement; (3) it provides direct comparison with the signal of another detector, such as a UV monitor, and allows an easy check for the coelution of radioactive peaks with reference compounds; and (4) it does not significantly decrease the resolution when compared with the UV signal. This means that at a flow rate of 1 ml.min^{-1} for the column effluent and 3 ml.min^{-1} for the scintillation cocktail, with a detector cell size of 0.5 ml, the result is equivalent or even better than collecting 0.2 ml fractions (i.e., 200 tubes for a 40 min analysis).

Mass spectrometry. Online HPLC-MS provides useful information on the structure of ecdysteroids eluting from the column, and has become a routine technique thanks to the thermospray and electrospray techniques. Two groups (Evershed *et al.*, 1993; Marco *et al.*, 1993) reported the successful use of HPLC-MS with ecdysteroids. These techniques work in the nanogram range with single ion monitoring (SIM) detection (Evershed *et al.*, 1993). Their use with acetonide derivatives results in a stabilization of the molecules and the production of abundant molecular ions (Marco *et al.*, 1993). A further improvement was obtained with atmospheric pressure chemical ionization (APCIMS) resulting in a much lower limit of detection — 10 pg with E and 20E (Wainwright *et al.*, 1996, 1997).

NMR. Using super heated D$_2$O as eluent, it has been possible to get online proton NMR spectra, but this requires high (100–400 µg) amounts of ecdysteroids to be injected on the column (Louden *et al.*, 2002).

4.3.3.2.2. Different aims of HPLC

Analytical and preparative HPLC — Columns of different sizes exist with the same packings, so it is very easy to scale up any chromatographic method. Analytical (inside diameter 4.6 mm), semi-preparative (inside diameter 9.4 mm), and preparative (inside diameter ca. 20 mm or more) columns are available. By increasing the flow rate in proportion to the cross-section, similar separations with increased maximal load are obtained.

Microbore HPLC — This method has been designed essentially for coupling with mass spectrometry. It uses columns of small internal diameter (e.g., 1 mm) run with a reduced solvent flow rate (ca. 50 µl.min^{-1}), which is compatible with a direct coupling with a mass spectrometer. Even narrower columns are presently available (inside

diameter 75 µm) and they are used for nanoLC with a flow rate of 240 nL.min^{-1} (Blais *et al.*, 2010).

Quantitative analyses — HPLC has been used for the direct quantification of individual ecdysteroids in biological samples. This requires high sensitivity because of the low concentrations encountered and adequate sample clean up. Quantification is best obtained if an internal standard is added to the sample either before HPLC analysis, or better, before sample purification (Lafont *et al.*, 1982). Many phytoecdysteroids can be used as internal standards (Wilson *et al.*, 1982a). The fundamental place of HPLC in ecdysteroid analysis will be exemplified by a detailed analytical protocol applied to locust eggs (**Figure 7**). A combination of HPLC at various pHs and of enzymatic hydrolyses may allow a rational diagnosis of ecdysteroid types present in a given biological extract (**Table 3**).

4.3.3.3. Gas-liquid chromatography

This method was first introduced for ecdysteroids by Katz and Lensky (1970), and developed during the early 1970s (Morgan and Woodbridge, 1971; Ikekawa *et al.*, 1972; Miyazaki *et al.*, 1973). Despite the advantages possessed by GLC in terms of sensitivity and specificity as compared to other techniques, its use for ecdysteroids has been confined to only a few groups (Borst and O'Connor, 1974; Morgan and Poole, 1976; Koreeda and Teicher, 1977; Lafont *et al.*, 1980; Webster, 1985; Bielby *et al.*, 1986; Evershed *et al.*, 1987, 1990). As all but a few biosynthetic intermediates of ecdysteroids are non-volatile, it is necessary to convert them into volatile trimethylsilyl (TMS) ether derivatives. Detection can be carried out using either a flame ionization detector (FID; poor sensitivity), or better still, an electron capture detector (ECD) or an MS. From 1970 to 1980, GC-MS represented a reference method for analyzing ecdysteroid mixtures (Lafont *et al.*, 1980).

With filled columns (1–3 m long), chromatography is performed on non-polar stationary phases such as OV-1 and OV-101 coated on Gas Chrom Q and Gas Chrom P. Typical operation temperatures for the columns are in the range of 280 °C, with carrier gas flow rates of 50 mL· min^{-1} (Bielby *et al.*, 1986). Retention behavior on GLC is the result of both molecular mass and polarity, depending on the degree of silylation. Capillary GC using flexible fused silica columns provides improved resolution and sensitivity. Thus, 10 pg (20E) or 100 pg (E) could be detected by GC coupled with mass spectrometric (SIM) detection (Evershed *et al.*, 1987).

4.3.3.4. Supercritical fluid chromatography

Supercritical fluid chromatography (SFC) uses a supercritical fluid as mobile phase, which behaves like a compressible fluid of very low viscosity, where mass transfer operates very rapidly, allowing the use of high linear velocities and, therefore, very short analysis times without any loss of efficiency. SFC may use either conventional packed

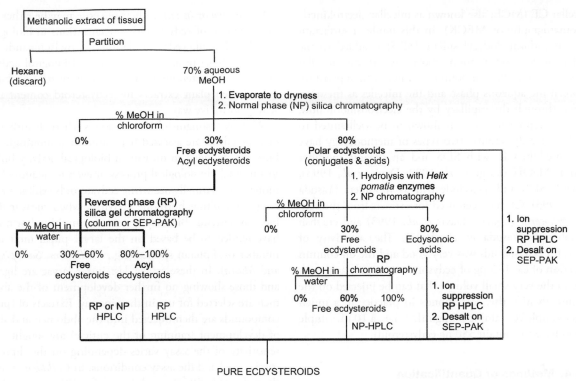

Figure 7 A general procedure for the purification of ecdysteroids combining several chromatographic steps. Reproduced with permission from *Russell et al.* (1989).

Table 3 Strategy for a rationale diagnosis of ecdysteroids

Polarity	Electrically charged (Yes/No)	pKa (if applicable)	Susceptibility to hydrolysis (Yes/No)	Conclusion
High	Y	6.5–7.0	Y	Phosphate esters
	Y	4.5	N	26-Oic acids
	Y	2.0	Y	Sulfate esters[a]
	N	n.a.	Y	Glucosides
Medium	N	n.a.	N	26-OH derivatives
	N	n.a.	N	E, 20E and related
Low	N	n.a.	N	Precursors
	N	n.a.	Y	Acyl esters

[a]Sulfate esters can only be observed during *in vitro* metabolic studies with insect tissue preparations.

HPLC columns or capillary tubes, and the mobile phase is generally supercritical CO_2 containing a small amount of an organic modifier, such as MeOH (Morgan *et al.*, 1988; Raynor *et al.*, 1988, 1989). Chromatographic parameters that can be monitored are the temperature, the pressure, and the percentage of the organic modifier. Supercritical CO_2 is a non-polar fluid and, therefore, SFC is more or less equivalent to NP HPLC; the eluting power is increased by adding MeOH. Various types of packed columns can be used, including silica or bonded silica; alternatively, fused silica capillary columns can be used (Raynor *et al.*, 1988). A major advantage of SFC over HPLC is the shorter retention times. Compounds elute as sharp peaks and the sensitivity of detection is accordingly increased. A second advantage concerns the high transparency of CO_2 in the

IR, which allows the use of FT-IR detectors. Furthermore, SFC is compatible with both FID (the universal detection used with GLC) and MS. No derivatization is required, and this represents an important advantage over GLC.

4.3.3.5. Capillary electrophoresis (CE) This is a recently introduced technique that allows extremely efficient separation of peptides, oligonucleotides, and also steroid hormones. The migration of substances is due to a combination of electroendosmotic flow of buffer toward the cathode and the (slower) migration of negatively charged solutes toward the anode. CE gives very efficient separations for a wide range of ionized and unionized compounds. As most of the ecdysteroids are not ionizable, CE has to be performed using some form of micellar CE (MCE; also known as micellar electrokinetic chromatography or MECK). In this mode, a surfactant such as sodium dodecyl sulfate (SDS) is added to the buffer at a concentration above its critical micellar concentration. The compounds of interest then partition between the aqueous phase and the micelles as these are drawn through the capillary by the electroosmotic flow. Such an approach has been shown to be well suited to the CE of ecdysteroids. Two types of methodology have been used in CE: with SDS and an organic modifier such as MeOH (Large *et al.*, 1992; Davis *et al.*, 1993); or with SDS and γ-cyclodextrin as modifiers (Yasuda *et al.*, 1993). CE has been used for extracts from the eggs of *Schistocerca gregaria* (Davis *et al.*, 1993) and crayfish hemolymph (Yasuda *et al.*, 1993). The sensitivity of MCE for ecdysteroids was very good and the on-column detection of ca. 175 pg of ecdysone was readily achieved. Due to the very small volume that can be injected on the column (5 nL), this corresponds in practice to a much larger sample containing 35 μg.mL^{-1} (i.e., a 10 μL sample would have to contain 350 ng ecdysteroid).

4.3.4. Methods of Quantification

One can use the physicochemical techniques as previously discussed (see **Table 4**) or, alternatively, bioassays and immunoassays (IAs; **Table 5**) performed either on crude extracts or on fractions collected after a chromatographic separation (TLC, HPLC).

4.3.4.1. Bioassays Various kinds of bioassays are used for testing ecdysteroids from animal or plant extracts,

either *in vivo* or *in vitro*. Their purpose may be either the measurement of ecdysteroid concentrations in biological samples (they allowed the isolation of E by Butenandt and Karlson, 1954) to analyze the relative biological activity of various ecdysteroids (structure–activity relationships), or to screen plant extracts for ecdysteroid content in a semiquantitative way.

We may distinguish between assays where disturbances or toxic effects are assessed (e.g., abnormal molting), and bioassays designed to measure a biological activity linked to a normal physiological process or even to quantify hormones. The naturally occurring ecdysteroids exhibit a wide range of molting hormone activities when measured in bioassay systems. Bioassay systems for molting hormones have tended to be based on the larval–pupal molt of a number of dipteran species (e.g., *Calliphora*, *Sarcophaga*, and *Musca*). In these assays, last instar larvae are ligated and those showing no further development of the abdomen are selected for use in the bioassay. Extracts of (pure) compounds are then injected into the abdomen and signs of development (tanning of the cuticle) are sought. The sensitivity of the assay varies depending on the dipteran species used and the assay conditions: in *Calliphora* it is of the order of 5–50 ng per abdomen for 20E versus 5–6 ng for *Musca*. A variation on the specialized larval–pupal molt used in the dipteran bioassays is the locust abdomens of fourth instar *S. gregaria* and is based on a larval-to-larval molt. Another useful bioassay, which does not require the injection of the test material, is based on the use of ligated larvae of *Chilo suppressalis* (Sato *et al.*, 1968). These are simply dipped into methanolic solutions of the test compound of extract. An *in vitro* variation of this assay uses

Table 4 Physicochemical methods of quantification: characteristics and limits of detection

Method	Limits of detection	Specificity	References
Direct or after a chromatographic separation			
UV absorbance (ca. 242 nm),	1 mg	Poor	
Chugaev's color reagent (380 nm)	<10 mg	Limited	Kholodova, (1977), (1981); Bondar *et al.*, (1993)
Sulfuric acid-induced fluorescence	10 pg	Limited	Gilgan and Zinck, (1972); Koolman, (1980)
Bioassay	ca. 5 ng	Good	Cymborowski, (1989)
Immunoassay	5–10 pg	(Very) good	Porcheron *et al.*, (1989)
TLC			
UV absorbance (scanner)	500 ng	Low	Wilson *et al.*, (1990c); Báthori and Kalász, (2001)
Fluorescence quenching (fluorescent TLC plates)	500 ng	Low	Mayer and Svoboda, (1978)
Mass spectrometer		High	Lafont *et al.*, (1993b)
HPLC			
UV absorbance	<10 ng	Fair	Lafont *et al.*, (1982); Wright and Thomas, (1983)
Fluorescence (requires derivatization)	<10 pg	?	Poole *et al.*, (1978); Kubo and Komatsu, (1986), (1987)
Mass spectrometer	50–100 pg	Very good	Evershed *et al.*, (1993); Le Bizec *et al.*, (2002)
nanoLC-MS/MS	<5 pg	Very high	Blais *et al.*, (2010)
GLC			
Flame ionization detector	50 ng	Poor	Katz and Lensky, (1970)
Electron capture detector	1–10 pg	Good	Poole *et al.*, (1975), (1980); Bielby *et al.*, (1986)
Mass spectrometer	10–100 pg	Very good	Evershed *et al.*, (1987)

pieces of integument from ligated larvae of *Chilo* and has a reported sensitivity of About 15 ng per test (Agui, 1973). In all *in vivo* test systems, it is difficult to assess whether the observed biological activity derives from the injected compound or an active metabolite. An obvious example of this is the conversion of 2dE and E into the more active metabolite, 20E (as shown by comparison with their activity in *in vitro* bioassays). The same considerations may apply to 3-oxo (3-dehydro) ecdysteroids.

In vitro systems, where it is possible to determine the biological activity of a compound in the quasi-absence of its metabolism, are therefore very useful for the determination of structure–activity relationships. Among them, a microplate bioassay using a *Drosophila* cell line (B II) was designed by Clément and Dinan (1991) and Clément *et al.* (1993) with good sensitivity and accuracy. This assay relies on a turbidimetric measurement of cell density in a system where ecdysteroids inhibit cell proliferation. More recently, new methods have been proposed, which use either *Drosophila* Kc cells transfected with a reporter plasmid containing the 5′ region of the *hsp27* promoter and the firefly luciferase gene (Mikitani, 1995) or SL2 cells with the addition of an EcR expression plasmid (Baker *et al.*, 2000). Cells were cultured for 24 h in the presence of ecdysteroids, and then cell lysates were used to measure luciferase activity.

A number of *in vitro* bioassay systems exist, including puffing of polytene chromosomes (Bergamasco and Horn, 1980; Cymborowski, 1989) and morphogenesis in imaginal discs. As Bergamasco and Horn (1980) pointed out, in bioassays where quantitative data are sought, the purity of the sample, especially the absence of other active ecdysteroids, is of paramount importance.

4.3.4.2. Immunoassays They use antibodies obtained by immunization with ecdysteroids. As ecdysteroids are not directly immunogenic, they must be coupled to a

Table 5 Bioassays and immunoassays

Method	Subtypes	Detection modes	Comments	References
Bioassays			Coupling with a separation method enhances the informative power	Review: Cymborovski, (1989)
	In vivo	Induction of molting or development in hormone-deprived systems	Fundamental role for ecdysone isolation; still used for some SAR* studies	*Calliphora* test: Adelung and Karlson, (1969); Karlson and Shaaya, (1964); Thomson *et al.*, (1970); Fraenkel and Zdarek, (1970) *Musca* test: Kaplanis *et al.*, (1966) *Sarcophaga* test: Ohtaki *et al.*, (1967) Diapausing *Samia cynthia* test: Williams, (1968) *Chilo* dipping test: Sato *et al.*, (1968); Koreeda *et al.*, (1970) *Limulus* bioassay: Jegla and Costlow, (1979) Blowfly abdominal bags: Agrawal and Scheller, (1985) *Galleria* test: Sláma *et al.*, (1993) *Dermestes* test: Sláma *et al.*, (1993)
	In vitro	Organ development		Integument pieces: Agui, (1973) Imaginal discs: Kuroda, (1968); Oberlander, (1974)
		Cell proliferation/ differentiation (s.l.)	SAR studies	Kc cells: Cherbas *et al.*, (1980) BII cells: Clément and Dinan (1991); Clément *et al.*, (1993)
		Chromosome puffing	SAR studies	Richards, (1978)
		Reporter gene (e.g. luciferase or GFP) activation	SAR studies	Mikitani, (1995); Baker *et al.*, (2000); Swevers *et al.*, (2004)
		Radio Receptor Assay	Uses *Drosophila* Kc cells cytosol as receptor preparation	Mikitani, (1996)
Immunoassays			Coupling with a separation method enhances the informative power	Reviews: Warren and Gilbert, (1988); Reum and Koolman, (1989)
	RIA	Radioactivity	Radioactive tracer required	Borst and O'Connor, (1972)
	CLA	Chemiluminescence	Chemiluminescent tracer required	Reum *et al.*, (1984)
	EIA	Colorimetry	Enzymatic tracer required	Porcheron *et al.*, (1989)
	ELISA	Colorimetry	No tracer required	Kingan, (1989)

*SAR: structure-activity relationship.

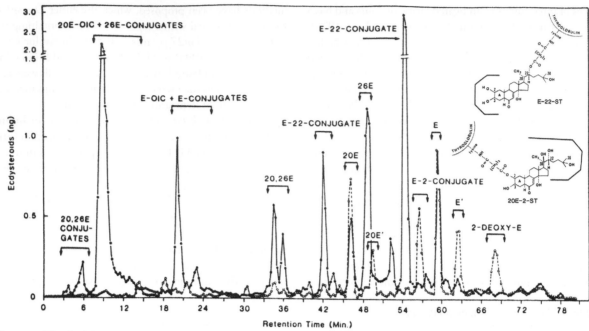

Figure 8 RP-HPLC/RIA analysis of ecdysteroids from the gut of females during day 14 of pupal–adult development in *M. sexta*. The chromatograms show the elution from a C18-silica column by a neutral buffered water/acetonitrile system of a female gut sample. Analysis is by differential RIA employing two complimentary antibodies. One antibody was elicited by an ecdysone-22-succinylthyroglobulin immunogen conjugate (E-22-ST) where the ecdysteroid hapten was conjugated via its C22-hydroxyl group to the immunogenic carrier protein (solid line). The other antibody was elicited by a 20-hydroxyecdysone-2-succinylthyroglobulin immunogen conjugate (20E-2-ST) where the ecdysteroid was conjugated via the C2 hydroxyl group (dashed line). In general, substrate discrimination by antibodies is greatest for molecules that differ in their architecture at a place far from the point of hapten attachment to the immunogenic protein carrier molecule. E′: 3α-epiE, 20E′, 3α-epi20E; 26E, 26-hydroxyE; 20,26E, 20,26-dihydroxyE; E-oic, ecdysonoic acid; 20E-oic, 20-hydroxyecdysonoic acid; E-22- (or E-2)-conjugate, E conjugated at the 22 (or 2) position. Modified from *Warren and Gilbert* (1986).

suitable protein before immunization. Generally, they are derivatized to create a carboxylic function that can react with the amino groups of protein carriers (BSA, HSA or thyroglobulin). The protein–ecdysone complex is strongly immunogenic and high titers of antibodies are generally elicited after the first injection.

The specificity of antibodies is rarely absolute. They generally will recognize, with different affinities, both E and 20E, as well as a series of closely related ecdysteroids. The degree of hapten binding by antibodies is primarily determined by the structure of the immunogen, especially regarding the position of the linkage between the ecdysteroid and the protein. For instance, protein esterification via the C-2 hydroxyl on the A-ring or the C-22 hydroxyl on the side chain will generally result in antisera specificity against ecdysteroid ligands exhibiting molecular changes in either the side chain or the steroid nucleus, respectively (figure 8, see also Warren and Gilbert, 1988).

Immunoassays need tracer molecules. In radioimmunoassay (RIA), tracers are radiolabeled with a high specific activity. Two kinds of analogs are used: commercially available [³H]-ecdysone (50–80 Ci/ mmol) and laboratory-made iodinated analogs (up to 2000 Ci/ mmol), which provide a higher sensitivity (Hirn and Delaage, 1980). The RIA has several drawbacks due to

the utilization of a radioisotope, which requires special precautions, and for that reason other immunoassays have been developed: a chemilumino-immunoassay (CIA), which uses ecdysone-6-carboxy-methoxime aminobutylethylisoluminol (ABEI) to form a chemoluminescent tracer (Reum *et al.*, 1984); and enzyme immunoassays (EIA) with an ecdysone derivative coupled to an enzyme such as acetylcholinesterase (Porcheron *et al.*, 1989; Royer *et al.*, 1993 1995) or peroxidase (Delbecque *et al.*, 1995; Von Gliscynski *et al.*, 1995).

After incubation of antibodies, tracer, and biological extract (or calibration references), the separation of free and bound fractions is performed in various ways (Reum and Koolman, 1989). The lower limit of sensitivity was estimated, in the best cases, as 4.10^{-10} mol (1.6 pg; Hirn and Delaage, 1980), but the sensitivity currently achieved by most ecdysteroid IAs is in the range of 20–200 fmol (10–100 pg). An example of RIA performed after an HPLC separation is given in **Figure 8**.

4.3.5. Sources of Ecdysteroids

Only a few ecdysteroids are commercially available. All of them have been isolated from plants and they include only a few insect ecdysteroids (2d20E, E, 20E, and

Table 6 Just a little bit of chemistry: Total or partial synthesis (side-chain modifications) and simple modifications (e.g. oxidation at C-3, removal of -OH groups (e.g. 25-OH), conjugation to acetic acid, fatty acids, glucose, sulfate)

Starting material	Derivative	References
20E	3D20E	Koolman and Spindler (1977); Spindler *et al.* (1977); Dinan and Rees (1978); Girault *et al.* (1989)
	3-*epi*-20E	Dinan and Rees (1978)
	Ponasterone A	Dinan (1985)
	Inokosterone	Yingyongnarongkul and Suksamrarn (2000)
	Acetates	Horn (1971)
	Glucosides	Pís *et al.* (1994)
	Poststerone	Hikino *et al.* (1969); Petersen *et al.* (1993)
	Rubrosterone	Hikino *et al.* (1969)
	2d20E	Suksamrarn and Yingyongnarongkul (1996)
	14d20E	Harmatha *et al.* (2002); Zhu *et al.* (2002)
	Integristerone A	Kumpun *et al.* (2007)
E	3DE	Spindler *et al.* (1977); Dinan and Rees (1978); Girault *et al.* (1989)
	3-*epi*-E	Dinan and Rees (1978)
	25dE	Pís *et al.* (1995b)
	22-Acyl esters	Dinan (1988)
	Sulfates	Pís *et al.* (1995a)
2dE	2,25dE	Pís *et al.* (1995b)
2,22,25dE	2,22dE	Niwa *et al.* (2005)

makisterone A) and some phytoecdysteroids (e.g., ponasterone A, muristerone A). Because of its use as an anabolic substance by many bodybuilders (Lafont and Dinan, 2003), 20E can now be purchased in large quantities at a low price. When an ecdysteroid other than those noted earlier is required, there are three ways in which it can be obtained: (1) it can be requested from someone who has already isolated (or synthesized) this compound, (2) the molecule can be isolated from an adequate plant source, or (3) it can be synthesized from an available precursor (usually 20E). This does not always require complicated chemistry; **Table 6** lists some syntheses that are not highly complex and can be performed in the absence of an experienced chemist.

4.4. Ecdysteroid Biosynthesis

4.4.1. Sterol Precursors and Dealkylation Processes

Owing to the lack of a squalene synthase (and/or other enzymes), insects are unable to synthesize sterols *de novo* (Clayton, 1964; Svoboda and Thompson, 1985; Santos and Lehmann, 2004). Consequently, they rely on dietary sterols for their supply. Zoosterols are essentially cholesterol; thus, insects feeding on animal-derived food get cholesterol directly. On the other hand, those feeding on plants obtain mainly phytosterols, which differ from cholesterol by the presence of various alkyl substituents (ethyl, methyl, methylene) on carbon 24. However, it should be noted that most plants also contain small but detectable amounts of cholesterol. Nevertheless, many,

but not all, phytophagous insect species possess the ability to first dealkylate phytosterols to cholesterol, which is subsequently converted in endocrine tissues to the active C_{27} molting hormone 20E. By contrast, some species, mainly *Hemiptera* and *Hymenoptera*, are unable to dealkylate phytosterols and consequently produce 24-alkylated ecdysteroids (i.e., C_{28} makisterone A and/or C_{29} makisterone C; Feldlaufer *et al.*, 1986; Feldlaufer, 1989; Mauchamp *et al.*, 1993).

In some cases, due to a combination of specific mechanisms including (1) selective concentration of the dietary cholesterol (Thompson *et al.*, 1963; Svoboda *et al.*, 1980) and/or (2) its selective delivery to steroidogenic organs, the insect can produce mainly C_{27} ecdysteroids even if cholesterol is a very minor sterol component of its diet. The ecdysteroid pool usually represents only 0.1–0.2% of total sterols, as illustrated by the low conversion rate of radioactive cholesterol when injected *in vivo* for long-term labeling studies (Beydon *et al.*, 1981, 1987).

Finally, some insects have become independent of a dietary sterol supply thanks to the production of sterols by yeast-like or fungal symbionts; this situation has been described in several insect species, particularly in aphids and beetles (Svoboda and Thompson, 1985). Yeast-like symbionts provide mainly 24-alkylated sterols (e.g., ergosta-5,7,24(28)-trienol). Thus, those insects still require a dealkylation process in order to produce cholesterol (Weltzel *et al.*, 1992; Nasir and Noda, 2003). According to the nature of their dietary sterols and their ability to dealkylate phytosterols, insects produce different ecdysteroids (**Table 7**).

4.4.1.1. General scheme of cholesterol formation from phytosterols A given species usually contains a complex sterol cocktail, with a few major compounds and many minor ones. Sterol diversity concerns (1) the number of carbon atoms (27 to 29) as discussed earlier and (2) the number and position of unsaturations, classically termed Δ, and mainly Δ^5 and Δ^7 on the steroid nucleus and Δ^{22} and Δ^{24} on the side chain (**Figure 9**). In addition, sterols can be in free or conjugated form, such as sterol acyl esters with various fatty acids or sterol acylglucosides, where a sugar moiety is linked both to the 3β-OH of the sterol and to a fatty acid (Eichenberger, 1977; Grunwald, 1980).

Ingested sterols are subjected to the action of several types of enzymes: (1) hydrolytic enzymes (esterases or glycosidases) from the insect gut (see Section 4.5.), which will release free sterols from their conjugates; (2) reductases and/or oxidases; and (3) enzymes involved in the

so-called "dealkylation process." Reactions (2) and (3) have been investigated extensively from the late 1960s by a combination of several strategies: feeding experiments with diets containing one particular sterol, *in vivo* metabolic studies with [3H] or [14C] labeled sterols allowing the isolation of labeled intermediates, enzymatic studies using isolated tissues or cell-free preparations, and observing the effects of various pharmacological inhibitors of reactions known in mammals/humans (i.e., hypocholesterolemic agents). All of the biochemical pathways have been fully elucidated, at least regarding the metabolism of the most common Δ^5 plant sterols (Fujimoto *et al.*, 1985b; Svoboda and Thompson, 1985; Svoboda and Feldlaufer, 1991; Svoboda and Weirich, 1995; Svoboda, 1999), and are summarized in **Figure 10**.

4.4.1.2. Dealkylating enzymes The first evidence for dealkylation of ergosterol into Δ^{22}-cholesterol was obtained with the cockroach *Blatella germanica* (Clark and Bloch, 1959). Further work, performed in particular with *Manduca sexta* and *B. mori* larvae, allowed the identification of all intermediates. Four to five enzymes are necessary to dealkylate phytosterols. The first three reactions involve the formation of an epoxide intermediate, and the removal of the side chain is accompanied by the formation of a Δ^{24} bond, thus forming desmosterol or its $\Delta^{22,24}$ analog (**Figure 10**). The second part (one or two steps) concerns the reduction of the side chain. For example, when

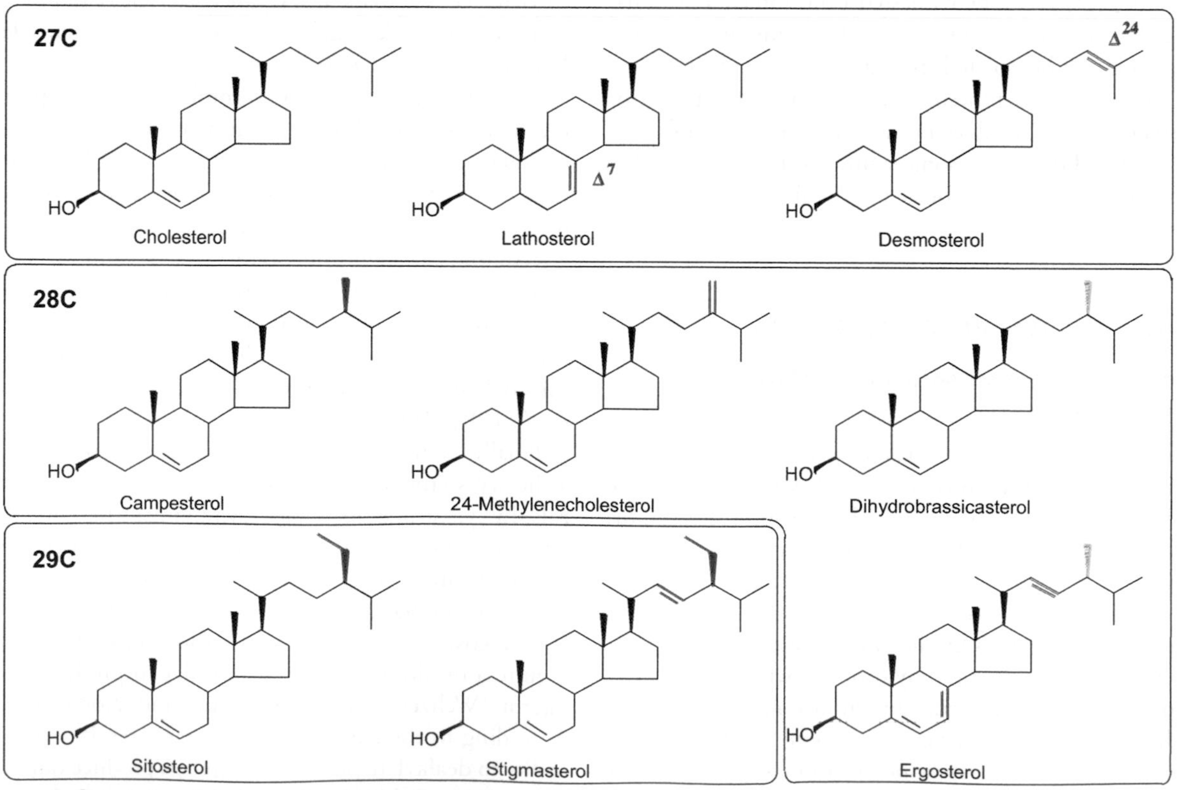

Table 7 Relationship between dietary sterols and ecdysteroid structures

Dietary sterols	Dealkylation	Major ecdysteroid
27C	Indifferent	20-Hydroxyecdysone (or Ponasterone A)
28C/29C	+	20-Hydroxyecdysone
28C	−	Makisterone A (or 24-*epi*-Makisterone A)
29C	−	Makisterone C

Figure 9 Structures of cholesterol and of representative phytosterols.

starting from sitosterol or campesterol, the dealkylation product (desmosterol) bears a single (Δ^{24}) double bond, and can be reduced to cholesterol by most insect species that have been analyzed. With insects fed on Δ^{22} sterol (e.g., stigmasterol or ergosterol), the dealkylation product bears a 22,24-diene and two reductases are required. It has been shown that the reduction at C-22 must precede that at C-24 and cannot be performed by all insect species (Svoboda and Thompson, 1985).

It was expected that dealkylation takes place in the insect gut. Cell-free extracts prepared from larval silkworm guts

when incubated with adequate intermediates resulted in the formation of desmosterol and cholesterol (Fujimoto *et al.*, 1985b). Midgut microsomes are the most active source of enzymes (e.g., *Spodoptera littoralis*; Clarke *et al.*, 1985), but midgut is not the only active tissue in this respect (Awata *et al.*, 1975). Unfortunately, most of these early experiments were discontinued, and as yet none of these enzymes has been fully characterized.

The desmosterol 24-reductase is present both in insects and vertebrates, catalyzing the last step in cholesterol biosynthesis. There seems to be a significant structural

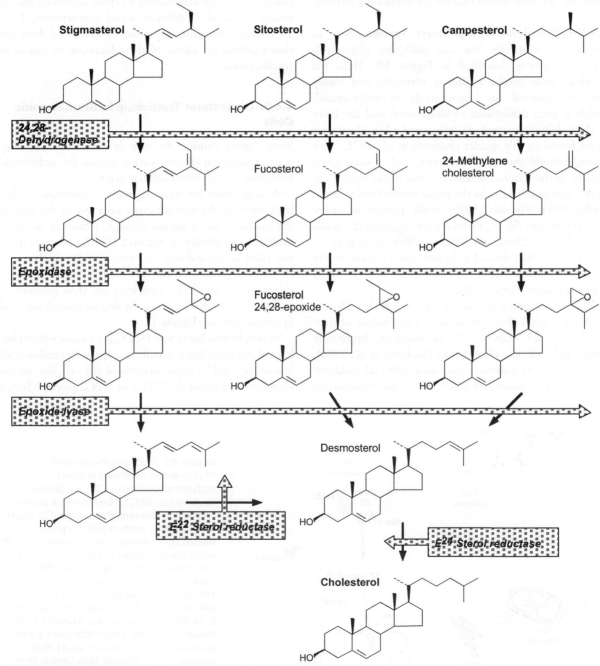

Figure 10 From phytosterols to cholesterol: the dealkylation processes.

conservation, as the insect enzyme can be inhibited by many hypocholesterolemic agents used in humans to reduce endogenous cholesterol biosynthesis. Azasteroids and various amines are also active on the insect enzyme and will consequently inhibit insect development (Svoboda and Weirich, 1995), but they cannot be used as insecticides due to their phytotoxicity. There is significant sequence conservation between the human and nematode proteins, and even human and plant sequences share 30% identity (Choe *et al.*, 1999; Waterham *et al.*, 2001). Unfortunately, available data on insect genomes concern species that are believed to be unable to dealkylate phytosterols, and thus would lack the corresponding enzyme.

4.4.1.3. Variations among insect species

Not all phytophagous species that can dealkylate phytoterols use the pathway described in **Figure 10**. Thus, the Mexican bean beetle, *Epilachna varivestis*, first totally reduces sitosterol (or stigmasterol) to stigmastanol, which is then dealkylated to cholestanol and the latter is then dehydrogenated to lathosterol (Δ^7), the major sterol found in this species (Svoboda *et al.*, 1975). This example highlights the importance of unsaturations present in the steroid nucleus. In some insect species, Δ^7 or $\Delta^{5,7}$ compounds can be the major sterols (Svoboda and Lusby, 1994; Svoboda, 1999), while *Drosophila pachea* has evolved specific requirements for ingesting Δ^7 sterols to produce E (Heed and Kircher, 1965; Warren *et al.*, 2001; see Section 4.4.2.1.). In summary, insects' ability to modify sterol nucleus unsaturations differs greatly between species (Rees, 1985).

In some instances, species unable to dealkylate phytosterols may adapt their physiology to the nature of their sterol diet, for example, *D. melanogaster*. Apparently unable to dealkylate phytosterols (Svoboda *et al.*, 1989), *D. melanogaster* produces a mixture of 20E and makisterone A, the proportions of which reflect the composition of its sterol diet (Redfern, 1986; Feldlaufer *et al.*, 1995; Pak and Gilbert, 1987). This strategy can operate because in this species E receptors have a very similar affinity for both ecdysteroids (Dinan *et al.*, 1999; Baker *et al.*, 2000). Bees (Hymenoptera) seem to follow a similar scheme (Svoboda *et al.*, 1983; Feldlaufer *et al.*, 1986). It was recently shown that *Drosophila* also produces 24-*epi*-makisterone A, which is likely produced from yeast sterols (Blais *et al.*, 2010). The same ecdysteroid is also found in leafcutter ants that feed on fungal sterols, which differ from plant sterols in their stereochemistry at C-24 (Maurer *et al.*, 1993). More generally, different enzymatic machinery and/or ecdysteroid receptor specificity among insects will result in different sterol requirements. These are revealed by feeding experiments on defined diets and thus represent an aspect of the adaptation of insects to specific foods.

4.4.2. Cholesterol Trafficking in Steroidogenic Cells

Since insects cannot *de novo* synthesize sterols, they depend solely on dietary sterols uptake for ecdysteroid biosynthesis as for other cellular needs.

A prerequisite for steroidogenesis is therefore cholesterol entry in the steroidogenic cells. As in the case of mammalian cells, it occurs through a classical receptor-mediated low-density lipoprotein (LDL) endocytic pathway (Rodenburg and van der Horst, 2005), which targets cholesterol to the endosomes. Conserved mechanisms of cholesterol intracellular trafficking are then involved to promote cholesterol access to the first enzyme(s) involved in steroidogenesis (**Figure 11**).

In vertebrates, Niemann-Pick C is a disease where cholesterol and other lipids accumulate in the late endosomal/lysosomal (LE/LY) compartment of the cell, due to the dysfunction of either the NPC1 or NPC2 protein (Storch

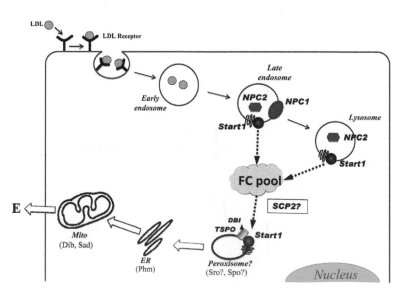

Figure 11 Proposed involvement of cholesterol trafficking in insect ecdysteroidogenesis. LDL, low density lipoproteins; NPC1, Niemann-Pick protein C1; NPC2, Niemann-Pick protein C2; Start1, steroidogenic acute regulatory protein related lipid transfer domain protein 1; SCP2, sterol carrier protein 2; TSPO, mitochondrial translocator protein (also named PBR, peripheral benzodiazepine receptor); DBI, diazepam binding inhibitor; Mito, mitochondrion; ER, endoplasmic reticulum; E, ecdysone. Enzymes are referred by their names as encoded from halloween genes in *Drosophila* (see Section 4.4.3.): Nvd, neverland; Sro, Shroud; Spo, Spook; Phm, Phantom; Dib, Disembodied; Sad, Shadow.

and Xu, 2009; Peake and Vance, 2010). NPC1, a multi-spanning membrane protein of LE/LY, and NPC2, a soluble luminal protein of LE/LY, play complementary roles to facilitate cholesterol absorption and subcellular transport from LE/LY and make it accessible for other subcellular organelles. Similarly, in *Drosophila*, inactivation of the *npc1* homologue gene *npc1a* (Huang *et al.*, 2005; Fluegel *et al.*, 2006) results in early larval lethality, which can be rescued by feeding the mutants excess 20E or its basic precursor cholesterol. Moreover, while a single *npc2* gene is present in yeast, worm, mouse, and human genomes, a family of eight *npc2* genes (*npc2a-h*) exists in *Drosophila* (Huang *et al.*, 2007). The double mutants of *npc2a; npc2b* are not viable, but they can also be rescued by 20E or cholesterol. The ecdysteroid deficiency in *npc* mutants is due to their inability to use cholesterol, although this compound massively accumulates as observed by filipin staining. The simplest explanation is that the accumulated sterol is stored in multilamellar and multivesicular compartments and is not available for ecdysteroid synthesis. These results highlight the importance of early steps of sterol transport in insect steroidogenic cells.

During ecdysteroid biosynthesis, intracellular movements of all sterol intermediates (and not just C) must occur, as the various enzymes involved are localized into different subcellular compartments. In vertebrates, several proteins (in addition to the enzymatic machinery) are expressed more or less specifically in the steroidogenic tissues, where they promote cholesterol movements inside the cells. These proteins play a particularly important role during the acute regulation of steroidogenesis. Three major steps are thought to be involved, although their mechanisms are still not fully understood. First, the sterol carrier protein 2 (SCP-2) mediates the transfer of C, whether dissolved in the cell membrane, the endoplasmic reticulum (ER), or within intracellular lipid droplets, to the outer mitochondrial membrane of steroidogenic cells. SCP-2 is encoded by alternative splicing from the *scp-x/scp-2* gene (Gallegos *et al.*, 2001). Then, the activation of the mitochondrial 18 kDa peripheral-type benzodiazepine receptor (PBR), also named mitochondrial translocator protein (mitoTSPO) by its endogenous ligand diazepam binding inhibitor (DBI) or endozepine (Papadopoulos *et al.*, 1998; Papadopoulos *et al.*, 2006), is involved with associated proteins (Lacapere and Papadopoulos, 2003) in the formation of a pore complex, creating contact sites between the outer and inner membranes of the mitochondria. Finally (at least in the adrenals), the steroidogenic acute regulatory protein (StAR), by means of its StAR-related lipid transfer domain (START), enables a rapid increase in the transport of cholesterol from the outer to the inner mitochondrial membrane (Stocco, 2001) where the first enzyme of steroidogenesis, the cholesterol side chain cleavage cytochrome P450 (CYP11A), resides (Miller, 1988; Stocco and Clark, 1996; Black *et al.*, 1994;

Miller, 2007). A similar function has been attributed to the vertebrate cholesterol transporter MLN64 (metastatic lymph node 64) in the placenta (Moog-Lutz *et al.*, 1997). MLN64 shares the START domain with the StAR protein, but it contains an additional transmembrane domain that targets the protein to the late endosomal compartment (Alpy and Tomasetto, 2006).

Homologues of the previously mentioned proteins are found across the animal kingdom, including insects. Recent studies dealing with their role in insects are consistent with their putative implication in the regulation of ecdysteroid biosynthesis (**Figure 11**). The *scp-x/scp-2* gene structure appears to be conserved in several insect species, such as *B. mori* (Gong *et al.*, 2006) and *M. sexta* (Kim and Lan, 2010). In *Drosophila*, an *scp-x* gene has been reported (Kitamura *et al.*, 1996), but only a 1.6 kb transcript was mentioned with high expression levels in the midgut of embryos. On the other hand, four genes encoding single SCP-2 domain proteins have been identified in *Aedes aegypti* (Krebs and Lan, 2003; Vyazunova *et al.*, 2007; Dyer *et al.*, 2009). Mosquito SCP-2 (AeSCP-2) and SCP-2-like proteins have very similar temporal and spatial expression, but vary in their affinity to different lipids (Dyer *et al.*, 2009). Only the *Aedes* SCP-2 protein presents a three-dimensional structure similar to mammalian SCP-2/SCP-X proteins (Stolowich *et al.*, 2002; Dyer *et al.*, 2003), and has a higher affinity toward cholesterol, suggesting a conserved role as a cholesterol transporter (Krebs and Lan, 2003; Dyer *et al.*, 2008). As in other insects species that have been investigated, it presents a broad tissue distribution, but further studies are needed to analyze the expression of a SCP-2-like protein in arthropod steroidogenic tissues.

In various vertebrate steroidogenic cell models, the activation of the mitochondrial 18 kDa PBR (also named mitoTSPO) by its ligand DBI or endozepine, confers them the ability to take up and release cholesterol (Papadopoulos *et al.*, 1998; Papadopoulos *et al.*, 2006). DBI is a 10 kDa protein, which has been identified in *M. sexta* (Snyder and Feyereisen, 1993) and *D. melanogaster* (Kolmer *et al.*, 1993, 1994). Homologous sequences are also found in other dipteran and lepidopteran species. In *M. sexta* prothoracic glands, DBI expression varies concomitantly with ecdysteroid production, and *in vitro* ecdysteroid production is stimulated by a diazepam analog (Snyder and Antwerpen, 1998). Changes in prothoracic glands of both mRNA and DBI protein levels suggest a tissue-specific regulation of this gene, which is consistent with a role in the regulation of ecdysteroid biosynthesis.

Finally, in vertebrates, there is a predominant role of the labile proteins StAR and its close analog MLN64 (Stocco, 2000, 2001) in the acute regulation of steroidogenesis. Similarly, the stimulation of insect steroidogenesis is rapidly inhibited by cycloheximide, an inhibitor of protein synthesis, indicating the requirement of a labile protein (Keightley *et al.*, 1990; Dauphin-Villemant *et al.*, 1995). Analysis of genome databases allowed the identification of proteins with a START domain in insects. In *Drosophila*, Start1, is homologous not with StAR, but with MLN64. This may fit because in insects the first enzymatic step of steroidogenesis is not mitochondrial. Another feature of Start1, specific to insects, is that the protein contains, in addition to the four putative transmembrane domains (helices) of MLN64, a stretch of 122 amino acids with a yet unknown function. The spatial and temporal pattern of both Start1 mRNA and Start1 protein expression in *Drosophila* is consistent with its importance in ecdysteroidogenesis (Roth *et al.*, 2004). Start1 homologues, expressed in steroidogenic tissues, have also been identified in mosquitoes (Sieglaff *et al.*, 2005) and in *Bombyx* (Sakudoh *et al.*, 2005). In *B. mori*, a shorter splicing isoform of Start1 has also been identified. This carotenoid binding protein (CBP) shares the START domain with Start1 but lacks the transmembrane domains. Although both isoforms appear to be expressed in steroidogenic tissues (prothoracic gland and gonads), it appears that CBP would bind carotenoids rather than cholesterol (Tabunoki *et al.*, 2002; Sakudoh *et al.*, 2005). Similarly, in *Drosophila*, shorter splicing isoforms of Start1 have been detected (C. Dauphin-Villemant, unpublished data). These shorter isoforms remind us of the vertebrate StAR. Therefore, it is probable that, as in vertebrates, several StAR-related proteins may exist in arthropods. Clearly, further studies are needed to precisely define the involvement of these proteins in the regulation of ecdysteroidogenesis.

4.4.3. The Biosynthetic Pathway of Ecdysteroids from Cholesterol

Although it has received considerable attention (Rees, 1985, 1995; Warren and Hétru, 1990; Sakurai and Gilbert, 1990; Grieneisen, 1994; Dauphin-Villemant *et al.*, 1998; Gilbert *et al.*, 2002; Gilbert and Warren, 2005; Lafont *et al.*, 2005; Gilbert and Rewitz, 2009), the biosynthetic pathway of ecdysteroids is still not fully understood. Biochemical approaches focusing on the nature of ecdysteroid synthetic intermediates have long formed the bulk of the results. However, unlike vertebrate steroidogenesis, where all the metabolites between cholesterol and the various active steroid hormones have been isolated and identified, no intermediate between the initial dehydrogenation product of cholesterol (7-dehydrocholesterol, 7dC) and the first recognizable ecdysteroid-like product, 5β-ketodiol (2,22,25-trideoxyecdysone, 2,22,25dE) has ever been observed. These reactions are still a "black box" as defined by Dennis Horn (**Figure 12**). During invertebrate steroidogenesis, biosynthetic intermediates do not accumulate, either because the first step after 7dC is the rate-limiting one, or because

these compounds are very unstable, or both. A great step toward elucidating the biosynthetic mechanism in the insect was made when it was demonstrated that the actual secretory product of lepidopteran prothoracic glands (Warren *et al.*, 1988a,b; Sakurai *et al.*, 1989; Kiriishi *et al.*, 1990) and of most other insect prothoracic glands and crustacean Y-organs (Spaziani *et al.*, 1989; Böcking *et al.*, 1994) was 3-dehydroecdysone (3DE) and not E. This finding greatly strengthened previous proposals of the early intermediacy of a 3-oxo-Δ^4-sterol in arthropod ecdysteroidogenesis (Davies *et al.*, 1981; Rees, 1985; Grieneisen *et al.*, 1991), as it is classical in mammalian steroidogenesis (Brown, 1998) and bile acid biosynthesis (Björkhem, 1969). However, the precise step at which the 3β-hydroxyl group of sterol is oxidized to a ketone remains to be determined. Although many of the early oxidation steps in ecdysteroid biosynthesis are still incompletely understood, the biosynthetic pathway appears to involve a cascade of sequential hydroxylations catalyzed by several cytochrome P450 enzymes.

4.4.3.1. Cholesterol 7,8-dehydrogenation Early feeding studies showed that 7dC can support insect growth in the absence of C or phytosterols, suggesting that this compound is an early intermediate in ecdysteroid biosynthesis (Horn and Bergamasco, 1985). Subsequent kinetic studies revealed that 7dC (**Figure 12**) was the first intermediate in the formation of either E or 3DE from C, both in insects and crustaceans (Milner *et al.*, 1986; Warren *et al.*, 1988a; Grieneisen *et al.*, 1991, 1993; Rudolph *et al.*, 1992; Böcking *et al.*, 1993, 1994; Warren and Gilbert, 1996). Based on labeling experiments with radioactive precursors, the mechanism of cholesterol 7,8-dehydrogenation is known to involve the stereospecific removal of the 7β- and 8β-hydrogens from the more sterically hindered "top" (β-surface) of the planar C molecule (Rees, 1985).

Although its activity may become substrate limited, the cholesterol 7,8-dehydrogenase does not appear to be the rate-limiting enzyme in the biosynthesis of ecdysteroids. In many insect species, including *Manduca* (Warren *et al.*, 1988a; Grieneisen *et al.*, 1991) and *Bombyx* (Sakurai *et al.*, 1986), the prothoracic glands always contain considerable amounts of both C and 7dC, even though both sterols undergo wild contralateral fluctuations during E biosynthesis. In a few insect species, such as *Tr. confusum*, *Atta cephalotes*, and *Acromyrmex octospinosus*, there is a clear preponderance of $\Delta^{5,7}$ sterols in the whole body (Grieneisen, 1994). In crabs, a ratio of 7dC:C ranging from 1:20 to 1:100 in the hemolymph was observed, while in the molting glands, the ratio was closer to 1:1 (Lachaise *et al.*, 1989; Rudolph *et al.*, 1992).

In the desert fly, *D. pachea*, the 7,8-dehydrogenase reaction is completely absent (Warren *et al.*, 2001) and so this insect depends on an alternative source of 7dC. It lives on a single species of cactus that supplies it with the Δ^7-sterol lathosterol, which it alternatively dehydrogenates to 7dC (Goodnight and Kircher, 1971) by a reaction similar to the synthesis of 7dC in vertebrates (Brown, 1998). Like *D. pachea*, the "low E" *Drosophila woc* mutant, also shows an impairment of 7,8-dehydrogenase activity (Warren *et al.*, 2001). However, in *woc* mutants, while late third instar larval ecdysteroid production and subsequent larval–pupal metamorphosis can be rescued by feeding 7dC, continued normal development is not observed (as in *D. pachea*), and the animals die as early pupae. This may be explained by the fact that *woc* encodes a transcription factor (Wismar *et al.*, 2000) with pleiotropic functions. In insect prothoracic glands, *woc* gene seems necessary for tubulin detyrosination, which may be required for sterol transport and utilization (Jin *et al.*, 2005).

The 7,8-dehydrogenation of sterols does not have strict substrate requirements in arthropods. Cholesterol, along with the major phytosterols campesterol, sitosterol, and stigmasterol (Sakurai *et al.*, 1986), as well as the more polar synthetic substrate, 25-hydroxycholesterol (25C) (Böcking *et al.*, 1994; Warren and Gilbert, 1996; Warren *et al.*, 1996, 1999, 2001), are all efficiently converted into their respective 7-dehydro derivatives, both *in vivo* and *in vitro*. This is also the case for the unusual oxysterols α-5,6-epoxycholesterol (αepoC) and α-5,6-iminocholesterol, but not for their respective β-isomers, which are competitive inhibitors of this reaction (Warren *et al.*, 1995). Only the former substrates were efficiently converted into either α-5,6-epoxy-7dC, an often-mentioned prospective "black box" intermediate (Rees, 1985) and a potentially potent alkylating agent (Nashed *et al.*, 1986), or α-5,6-imino7dC, respectively. A 46 kDa protein became increasingly radiolabeled following the incubation of ^{3}H-αepoC with *Manduca* prothoracic gland microsomes. Unfortunately, the resultant tagged product–enzyme adduct has so far proved too unstable for a classical purification and sequencing of the enzyme. Nevertheless, the substrate specificity of the enzyme for the α-isomer suggests a mechanism involving an attack on the β-face of the molecule.

In several arthropod species the reaction is basically irreversible (Grieneisen, 1994) and supposed to be catalyzed by a microsomal cytochrome P450 enzyme (CYP) based on its biochemical characteristics, such as enzyme subcellular localization, requirement for NADPH, or inhibition by CO and fenarimol (Grieneisen *et al.*, 1991, 1993; Warren and Gilbert, 1996). However, it was recently demonstrated that in *Caenorhabditis elegans*, a Rieske-like oxygenase (DAF-36), is responsible for the 7,8-dehydrogenation of cholesterol in the biosynthetic pathway leading to particular steroids (dafachronic acids) regulating the worm's development (Rottiers *et al.*, 2006; Gerisch *et al.*, 2007). This gene (*neverland*) is conserved in *B. mori* and *D. melanogaster*. In these species, it is specifically expressed in the steroidogenic tissues. In *Drosophila*, the loss of *nvd* function in the ring gland causes growth arrest during larval stages that is due to a reduced ecdysteroid titer. Mutant phenotype is rescued by feeding 7dC but not C, suggesting that the neverland protein is involved in 7,8-dehydrogenation (Yoshiyama *et al.*, 2006). When it is expressed in a *Drosophila* S2 cell system, nematode or insect Neverland enzyme actually catalyzes the conversion of C to 7dC (Niwa *et al.*, 2010). Consistently with this function as a 7,8-dehydrogenase, it was also shown that the *neverland* gene presents several mutations in *D. pachea*, which make the corresponding protein unable to convert C to 7dC (Lang *et al.*, submitted).

4.4.3.2. Early steps after the formation of 7dC The black box

Conversion of 7dC to the 5β-ketodiol or 5β-diketol, via the long-hypothesized Δ^4-diketol (14α-hydroxy-cholest-4,7-diene-3,6-dione) intermediate, constitutes the so-called "black box" (**Figures 12** and **13**). The name is most apt, as both the mechanism of this apparent multistep oxidation, as well as the subcellular location of all these reactions, is still a matter of debate. Several modifications of 7dC appear to take place early and seemingly simultaneously, such as oxidation of the 3β-alcohol to the ketone (with stereospecific loss of the 3α-hydrogen), C-6 oxidation with additional loss of the 4β and 6-hydrogens and formation of the 6-keto group, and finally 14α-hydroxylation. These observations are consistent with the intermediacy of Δ^4-diketol (Davies *et al.*, 1981; Rees, 1985; Grieneisen *et al.*, 1991; Blais *et al.*, 1996; Dauphin-Villemant *et al.*, 1998). Analogous to the rate-limiting step in mammalian steroidogenesis, it has long been thought that most of (or all) these reactions take place in the mitochondria and may be catalyzed by one or more P450 enzymes. That is, 7dC (or 7-dehydro-25C) could not be converted into ecdysteroids by the crude microsomal fraction, but only by the crude mitochondrial fraction of *Manduca* prothoracic glands. However, as no intermediates could be detected, no reaction mechanism was apparent (Grieneisen *et al.*, 1993; Warren and Gilbert, 1996).

Nevertheless, some "negative" evidence has been obtained from metabolic studies concerning these early biochemical modifications. Concerning 14α-hydroxylation, experiments with *Manduca* prothoracic glands (Bollenbacher *et al.*, 1977a) or *Locusta migratoria* prothoracic glands or ovaries (Haag *et al.*, 1987) suggested that this reaction must take place before (or in concert with) the introduction of the 6-keto group. Thus 3β-hydroxy-5β-cholest-7-en-6-one (ketol) is efficiently converted to 14-deoxy-ecdysone, but not to E.

Several hypotheses have also been tested concerning the introduction of the 6-keto group. Even though 6-hydroxy-sterol derivatives have been isolated in some biological models, ecdysteroid analogs bearing a 6-hydroxy group

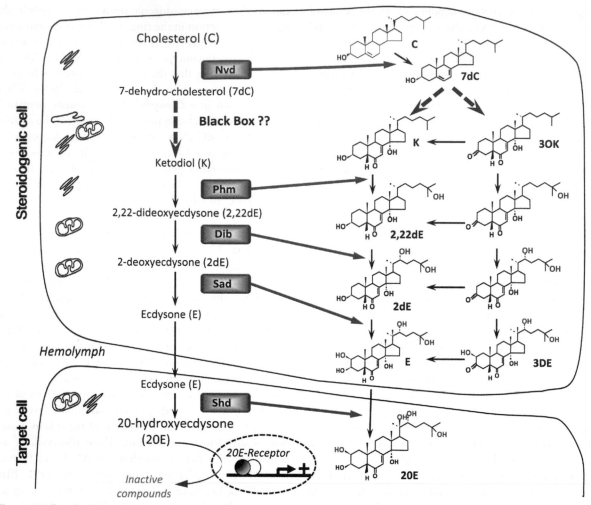

Figure 12 Putative biosynthetic pathway of ecdysteroids. Enzymes are referred by their names in *Drosophila* (see Section 4.4.3.): Nvd, neverland; Phm, Phantom; Dib, Disembodied; Sad, Shadow; Shd, Shade. Molecules described in the biosynthetic pathway: C, cholesterol; 7dC, 7-dehydrocholesterol; 3OK, 3-dehydro- 2,22,25-trideoxyecdysone (diketol); K, 2,22,25-trideoxyecdysone (ketodiol); 2,22dE, 2,22-dideoxyecdysone (ketotriol); 2dE, 2-deoxyecdysone; 3DE, 3-dehydroecdysone; E, ecdysone; 20E, 20-hydroxyecdysone. The subcellular localization of the different enzymes is indicated as follows: ⟿ cytosol ⚡ microsomes ∞ mitochondria.

could not be oxidized to true ecdysteroids with a 6-ketone group (Schwab and Hétru, 1991; Grieneisen, 1994; Gilbert *et al.*, 2002). The possibility that either a 3-dehydro-α-5,6-epoxy-7dC molecule (Grieneisen *et al.*, 1991; Warren *et al.*, 1995) or a 3-dehydro-5α,8α-epidioxide (J. Warren, C. Dauphin-Villemant, and R. Lafont, unpublished data) could be converted to a 3-dehydro-Δ⁴-sterol intermediate has also been investigated. However, when incubated with insect prothoracic glands or crustacean Y-organs, these compounds were never converted into known ecdysteroids (Gilbert *et al.*, 2002).

Finally, if Δ⁴-diketol (**Figure 13**) were an intermediate in ecdysteroid biosynthesis, its formation would also require an early oxidation step at C-3. However, every attempt to understand when and how this step takes place has failed. In studies with either prothoracic glands or Y-organs, the early oxidation of C (or 25C) to

3-dehydro-C (or 3-dehydro-25C) has never been directly observed. The facile conversion of these oxidized sterols, by either acid or base catalysis, to their distinctive isomerization products, 3-dehydro-Δ⁴-C or 3-dehydro-Δ⁴-25C, has also never been detected *in vivo* or *in vitro* (J. Warren and R. Lafont, unpublished data). Due to greater electron delocalization, 3-dehydro-7dC is very unstable, instantly converting into its more stable 3-dehydro-Δ⁴,⁷ sterol isomer when placed in an aqueous environment (Antonucci *et al.*, 1951; Lakeman *et al.*, 1967). However, the formation of either of these compounds (or their 25-hydroxylated analogs) has never been observed during ecdysteroid biosynthesis studies with radiolabeled precursors. To overcome the unstability problem, Warren *et al.* (2009) designed an original protocol using a stable ketal derivative of 3-dehydro-7dC, the 3-o-nitrophenylethyleneglycol ketal of 3-oxo-7dC. This compound was taken

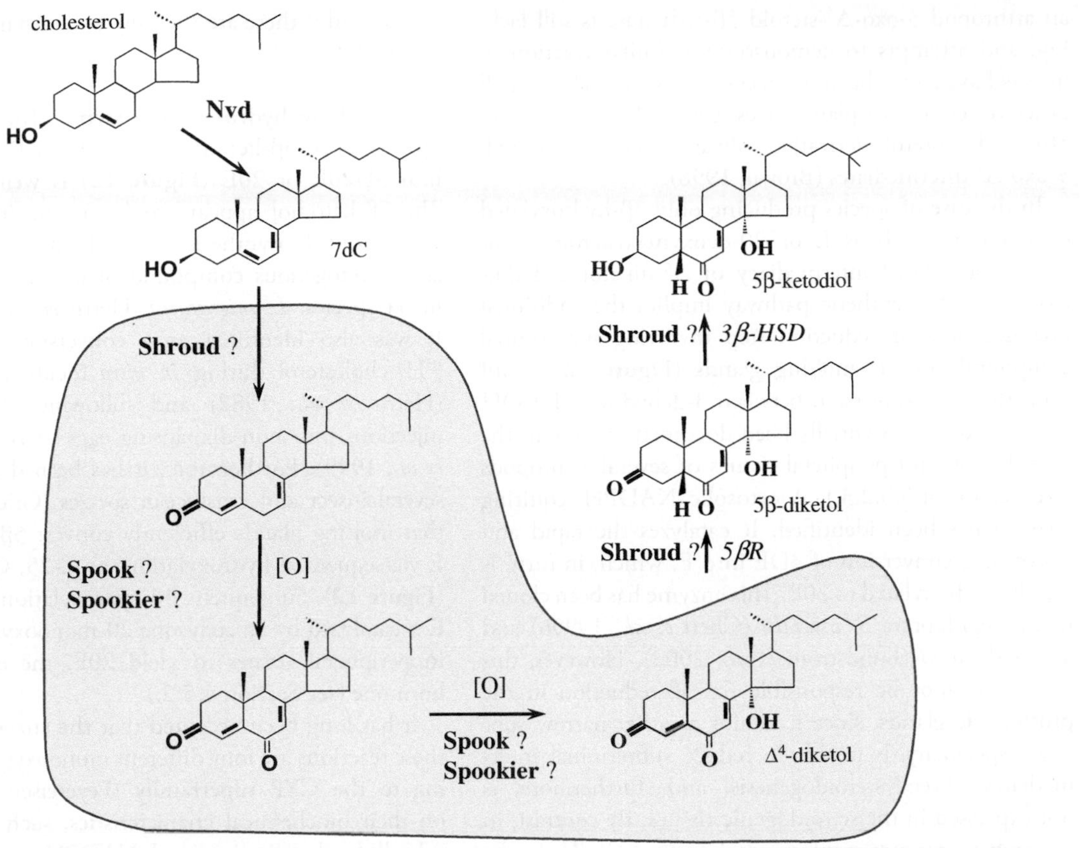

Figure 13 A possible sequence for the black box reactions. [O] means an as yet undefined oxidation/dehydrogenation process performed by a CYP or an oxidase-type enzyme. See text for details.

up by molting glands of *Manduca* and only thereafter cleaved to 3-dehydro-7dC by irradiation with long-wave UV light (365 nm), thus within steroidogenic cells, and in this case it was efficiently converted into ecdysteroids. This result strongly supports an oxidation of 7dC as a very early step of the black box.

Oxidation at C-3 must occur in concert with subsequent oxidations of very unstable intermediates. It has been proposed that all these modifications could be performed by a single enzyme, perhaps a CYP enzyme localized in the mitochondria (Gilbert *et al.*, 2002). However, this hypothesis is not supported for *Drosophila*. All six families of putative mitochondrial CYPs have been tentatively identified on the basis of their conserved motifs associated with the specific cofactor (adrenodoxin) binding to the cytochrome (Tijet *et al.*, 2001). According to tissue expression studies in *Drosophila* (C. Dauphin-Villemant and M. O'Connor, unpublished data; Chung *et al.*, 2009), no mitochondrial CYP other than those already identified as catalyzing the final 2- and 22-hydroxylation steps has been localized to the ring glands. Alternatively, it remains a possibility that these reactions take place in another subcellular compartment, such as the ER or in peroxisomes, which are involved in the oxidation of numerous substrates (including some mammalian sterols; see Section 4.6.2.). A current hypothesis is that the initial

dehydrogenation of 7dC to 3-dehydro-7dC is followed by further extensive rearrangements and oxidations, maybe catalyzed by the very atypical microsomal cytochrome P450 enzyme Spook and its paralogues Spookier and Spookiest (Namiki *et al.*, 2005; Ono *et al.*, 2006; Gilbert and Rewitz, 2009), ultimately leading to Δ^4-diketol. Of special interest is the finding that prothoracicotropic hormone (PTTH) stimulation specifically induced phosphorylation and translation of Spook, making it a possible rate-limiting enzyme in the ecdysteroid biosynthetic pathway (Rewitz *et al.*, 2009).

The 5β-reduction of a proposed 3-oxo-Δ^4 ecdysteroid intermediate might constitute the last step of the black box. Even though there is no direct evidence for the intermediacy of the Δ^4-diketol in the biosynthesis of ecdysteroids, it still remains a good candidate. Δ^4-diketol is efficiently converted first to 5β-diketol (and subsequently to 3DE and E) by a 5β-reductase present primarily in Y-organs of crustaceans, such as the crab *Carcinus maenas* (Blais *et al.*, 1996) and the crayfish *Orconectes limosus* (C. Blais, C. Dauphin-Villemant, and R. Lafont, unpublished data). The reaction proceeds only with 3-oxo-Δ^4 substrates, and no reduction of 3β-OH-Δ^4-sterol intermediate(s) was observed either in crab Y-organs (Blais *et al.*, 1996) or in insect prothoracic glands (Bollenbacher *et al.*, 1979). However, the complete characterization of an arthropod 3-oxo-Δ^4-steroid 5β-reductase is still lacking, and attempts to demonstrate a similar reaction in insects have so far been unsuccessful. Nevertheless, in all other vertebrates or plants investigated, the formation of 5β[H]-cholesterol derivatives always proceeds through 3-oxo-Δ^4 intermediates (Brown, 1998).

In the case of species producing only 3β-hydroxylated ecdysteroids, such as E or 20-deoxymakisterone A or C, the postulated intermediacy of Δ^4-diketol and diketol in the biosynthetic pathway implies the additional presence of a 3β-reduction step involving subterminal compounds in the molting glands (**Figure 12**). Until now, the interconversion between 3-dehydro and 3-OH compounds has essentially been documented only in the hemolymph and peripheral tissues of several arthropods (see Section 4.5.3.2.3.). A cytosolic NADPH-requiring enzyme has been identified. It catalyzes the rapid and irreversible conversion of 3DE into E, which, in turn, is rapidly hydroxylated to 20E. This enzyme has been cloned in the lepidoptera *S. littoralis* (Chen *et al.*, 1999b) and *Trichoplusia ni* (Lundström *et al.*, 2002). However, this enzyme cannot be responsible for 3β-reduction in the prothoracic glands, since it retains a rather narrow substrate specificity, is unable to reduce subterminal intermediates of ecdysteroidogenesis, and, furthermore, is not expressed in the steroidogenic tissues. By contrast, in the crab *C. maenas*, a microsomal enzyme similar to the 3β-hydroxysteroid dehydrogenase (3βHSD) of vertebrates has been detected specifically in the Y-organs (Dauphin-Villemant *et al.*, 1997). In the presence of NADH, it catalyzes the 3β-reduction of 5β-diketol, 3-dehydro-2, 22-dideoxyecdysone and 3-dehydro-2-deoxyecdysone, but not that of 3DE. Interestingly, in the presence of NAD$^+$, the same subcellular fraction can re-oxidize 2-deoxyecdysone and 2,22-deoxyecdysone back to the original 3-dehydro substrates, suggesting a reversible reaction consistent with a true 3βHSD activity.

Besides *spook*, another gene coding for a putative steroidogenic enzyme has recently been identified in the genetic models *D. melanogaster* and *B. mori*. *Shroud (sro)* belongs to the halloween genes in *Drosophila*. While it was first considered that *shroud* mutation affects the *Drosophila Fos* gene (Giesen *et al.*, 2003), it has now been demonstrated that *sro* (and its orthologue in *Bombyx nonmolting glossy nm-g*) encodes an oxidoreductase belonging to the short chain reductase family, which also includes several steroid dehydrogenases (Niwa *et al.*, 2010). One possibility is that Nm-g/Sro might act as the ecdysteroid 5β-reductase converting Δ^4-diketol to diketol or as the 3β-reductase, which are the putative last steps in the black box. Alternatively, Nm-g/Sro might catalyze the 3β-dehydrogenation of 7dC to 3-oxo-7dC (cholesta-5, 7-diene-3-one), the putative initial step in the black box. Further studies are needed to identify the exact steps catalyzed by spook and Nmg/Sro in ecdysteroid pathway and

understand if these are the only two enzymes involved in the black box.

4.4.3.3. Late hydroxylation steps
The conversion of 5β-diketol or 5β-ketodiol into 3DE or E and ultimately into 3D20E or 20E (**Figure 12**) is well documented. The 5β-ketodiol appears to be a true intermediate in ecdysteroid biosynthesis, since it has been identified as an endogenous compound of ovaries at least in one insect species, *L. migratoria* (Hétru *et al.*, 1978, 1982). It was also identified as a conversion product from [^{3}H]-cholesterol during *in vitro* incubations of ovaries (Hétru *et al.*, 1982) and following [^{3}H]-cholesterol injections into non-diapausing eggs of *B. mori* (Sonobe *et al.*, 1999). Furthermore, it has been demonstrated in several insect and crustacean species (Grieneisen, 1994) that molting glands efficiently convert 5β-ketodiol into E via sequential hydroxylations at C-25, C-22, and C-2 (**Figure 12**). Subsequent 20-hydroxylation of circulating E is catalyzed by an ecdysone 20-monooxygenase present in peripheral tissues to yield 20E, the active molting hormone (see Section 4.5.2.).

It has long been accepted that the enzymes catalyzing these reactions are four different monooxygenases belonging to the CYP superfamily (Feyereisen, 2005) based on their biochemical characteristics, such as membrane subcellular localization and NADPH and O$_2$ requirement and inhibition by CO or by classical CYP inhibitors like fenarimol, metyrapone, or piperonyl butoxide (Grieneisen, 1994). In this respect, the 2-hydroxylase presents some peculiar features, since it is insensitive to CO and apparently can function with some Krebs cycle intermediates (succinate or isocitrate) in addition to NADPH (Kappler *et al.*, 1986, 1988). A combination of molecular genetics and biochemistry was first used in *Drosophila* to identify the genes corresponding to these enzymes. One phenotype predicted for mutations that disrupt the production of ecdysteroids would be the inability to produce a cuticle at any stage, the earliest being the synthesis of embryonic cuticle during the second half of embryogenesis. Among the mutants screened by Nüsslein-Volhard and her colleagues in their landmark characterization of embryonic patterning (Jürgens *et al.*, 1984; Nüsslein-Volhard *et al.*, 1984; Wieschaus *et al.*, 1984), several were later grouped under the name of the halloween mutants (Chávez *et al.*, 2000). They were selected for their similar abnormal cuticular patterning and low ecdysteroid titers, suggestive of a defect in the ecdysteroid pathway. In all of them, several developmental and physiological defects (i.e., absence of head involution, dorsal closure, and gut development) were associated with a very poor differentiation of embryonic cuticle, which led to lethality before the end of embryonic development. The first halloween gene, *disembodied* (*dib*) (*Cyp302a1*) was located by classical molecular techniques (Chávez *et al.*, 2000), whereas the other halloween genes were identified by searching the *Drosophila* genome for the presence of cytochrome P450 enzyme-type sequences near their known cytogenetic map location. At least four other halloween genes — *phantom* (*phm*), *spook* (*spo*), *shadow* (*sad*), and *shade* (*shd*) — were shown to encode CYP enzymes, such as CYP 306A1, CYP307A1, CYP315A1, and CYP314A1, respectively (Warren *et al.*, 2002, 2004; Petryk *et al.*, 2003), as they retained the conserved domains characteristic of this superfamily (particularly the heme-binding domain at the C-terminal end). Owing to a high divergence in their overall sequence as compared to other already known sequences, these enzymes are members of new CYPs families.

The initial hypothesis that CYPs enzymes encoded by *Phm, Dib, Sad*, and *Shd* might catalyze steps in the ecdysteroid biosynthetic pathway has been confirmed by biochemical characterization. The cDNA coding sequences of these genes have been transiently transfected into *Drosophila* S2 cells as expression plasmids under the control of the actin 5C promoter. When the transfected cells are incubated with specific ecdysteroid precursor substrates, they efficiently and selectively convert them to product (**Figure 12**), indicating the 25-, 22-, 2-, and 20-hydroxylase functions for PHM, DIB, SAD, and SHD, respectively.

Cytochrome P450 enzymes, as membrane-bound enzymes, have been found in the ER or in the mitochondria of eukaryotes. Their amino acid sequences at the N-terminus present conserved structural features that direct their subcellular localization (Omura and Ito, 1991; von Wachenfeldt and Johnson, 1995; van den Broek *et al.*, 1996; Feyereisen, 2005). Biochemical approaches showed that the 25-hydroxylase is a microsomal enzyme, whereas the 2- and 22-hydroxylases are strictly mitochondrial. The N-terminal sequences deduced from *dib* and *sad* cDNAs are consistent with their mitochondrial localization (Feyereisen, 2005) as previously shown for the C_2- and C_{22}-hydroxylases from subcellular fractionation studies in *Locusta* (Kappler *et al.*, 1986, 1988) and *Manduca* (Grieneisen *et al.*, 1993). Both contain numerous positively charged residues, in addition to many polar groups within the first 20 to 30 amino acids. The mitochondrial localization of the *Dib* and *Sad* cytochromes has been confirmed by *in situ* localization of the corresponding C-terminal, epitope-tagged (His or HA) proteins expressed in S2 cells (Warren *et al.*, 2002; Petryk *et al.*, 2003). The predicted N-terminal sequence of *Drosophila Phm* (Warren *et al.*, 2004) could present an anomaly, as it shows no typical hydrophobic stretch in the first 20 amino acids. However, if one considers that the real start of the protein is the second methionine at position 14, then the next 20 amino acids are hydrophobic and could represent an ER import sequence. Identification of *Phm* homologues in various insect and crustacean species confirmed these

characteristics (Warren *et al.*, 2004; Rewitz *et al.*, 2006a; Asazuma *et al.*, 2009; Iga and Smagghe, 2010; Mykles, 2011). The subcellular localization of *Phm* protein has further been directly demonstrated in *Drosophila* S2 cells, as the tagged enzyme was localized in the ER of transfected S2 cells (Warren *et al.*, 2004), a finding consistent with previous biochemical evidence (Kappler *et al.*, 1988).

In contrast, the 20-hydroxylase has been shown to be microsomal and/or mitochondrial (Bollenbacher *et al.*, 1977b; Feyereisen and Durst, 1978; Smith *et al.*, 1979; Smith, 1985; Mitchell and Smith, 1986; Smith and Mitchell, 1986; see Section 4.5.2.). The N-terminus of SHD (the 20-monooxygenase) consists of a string of about 20 or more hydrophobic amino acids devoid of charged residues, suggesting that this CYP enzyme is targeted to the ER. However, when a similarly tagged *shd* construct was transfected into S2 cells, it was unexpectedly found that, like *dib* and *sad*, it localized in the mitochondria, and not in the ER (Petryk *et al.*, 2003).

When phylogenetic trees with various CYPs are constructed, SHD always clusters (along with DIB and SAD) with the mitochondrial CYPs, primarily because of shared sequences associated with binding to its redox partner adrenodoxin (Tijet *et al.*, 2001). However, a closer examination of the SHD N-terminus shows a possible mitochondrial import sequence containing numerous charged residues is present immediately downstream from the apparent microsomal targeting sequence. Protease-mediated cleavage of the N-terminus of SHD, eliminating the initial string of hydrophobic residues, could result in the re-targeting of this enzyme to the mitochondria. Such differential post-translational modification of SHD could explain the distibution of ecdysone 20-monoxygenase activity into either the mitochondria or ER or both, depending on insect, tissue, or stage. A similar situation has been reported for mammalian CYP1A1, which has a chimeric N-terminal signal that facilitates the targeting of the protein to the ER and, following protease action, to the mitochondria (Addya *et al.*, 1997). Not clear, however, is how the P450 enzyme manages to interact with apparently separate reductase cofactors when it is present in these two different membrane environments.

By analogy with steroid hormone biosynthesis in vertebrates, the identified hydroxylases are expected to be expressed specifically in the endocrine tissues (or peripheral tissues for the 20-monooxygenase) and perhaps act through a preferential sequence of hydroxylations. Yet, early metabolic studies (Rees, 1985; Kappler *et al.*, 1988; Grieneisen *et al.*, 1993) reported 2-, 22-, and 25-hydroxylase activities in a number of tissues, not only larval prothoracic glands and adult ovaries, but also in fat body, Malpighian tubules, gut, and epidermis. However, the specific activity of the observed enzymatic activities was seldom mentioned, and a closer examination always revealed a substrate conversion that was much more efficient in true steroidogenic organs (prothoracic glands, Y-organs, ovaries) than in other tissues, at least as concerns the 22- and 25-hydroxylations (Meister *et al.*, 1985; Rees, 1985, 1995; Grieneisen, 1994). While the evidence for a specific 2-hydroxylase activity outside the ring gland remains strong, the low 22- and 25-hydroxylations observed with crude preparations might result from unspecific reactions by CYPs involved in detoxification processes. In *Drosophila*, *Phm*, *Dib*, and *Sad* mRNAs are initially expressed in the epidermis during early embryogenesis and then later in classical steroidogenic tissues, such as in ring glands from late embryogenesis to metamorphosis and in the nurse and/or follicle cells of adult female ovaries (Warren *et al.*, 2002, 2004; Niwa *et al.*, 2004). RT-PCR and/or qPCR analysis of tissue expression has now been performed in various insects and confirm the preferential, if not exclusive, expression of these enzymes in steroidogenic tissues (Warren *et al.*, 2004; Niwa *et al.*, 2004; Rewitz *et al.*, 2006b; Iga and Smagghe, 2010). In contrast, while *shd* is also expressed in the early embryonic epidermis, it is not expressed in late embryonic or larval ring glands, but is prominent in peripheral larval tissues and adult ovaries (Petryk *et al.*, 2003; see Section 4.5.2.).

Another characteristic of the enzymes involved in vertebrate steroidogenesis is that the reactions generally occur in a preferred sequence, both due to their precise substrate specificities and their subcellular compartmentalization (Brown, 1998). This applies also to ecdysteroid biosynthesis, even though additional enzymological studies are still necessary to make this conclusion. The use of exogenously applied substrates (e.g., [^{3}H]5β-ketodiol) in insect and crustacean systems has resulted in the accumulation of many intermediates that have never been identified as endogenous intermediates (Rees, 1985; Grieneisen, 1994). Most probably they represent dead-end products and their formation results from an abnormal trafficking. More precisely, it has been shown both in insects (Dollé *et al.*, 1991; Rees, 1995) and crustaceans (Lachaise *et al.*, 1989; Pís *et al.*, 1995b) that premature 2-hydroxylation of either 5β-ketodiol (2,22,25dE) or 5β-ketotriol (2,22dE) prevents further hydroxylations, since both 22,25-dideoxyecdysone and 22-deoxyecdysone are poor substrates of the 22-hydroxylase (Warren *et al.*, 2002). Taken together, these results suggest that the last steps in E biosynthesis within the endocrine gland also follow a preferential sequence of (25-, 22-, then 2-) hydroxylations. Thus, after hydroxylation at C-25 in the ER, the resulting 5β-ketotriol (or its 3-oxo form) translocates to mitochondria, where 22- and the 2-hydroxylations take place, followed by the secretion of the product from the cell as either E (and/or 3DE).

Additional information on the substrate specificities of these enzymes has also been obtained using transfection of halloween genes in S2 cells. For instance,

dib-transfected cells 22-hydroxylate 2,22dE to 2dE, but cannot convert 22dE to E, confirming the strict substrate requirement (i.e., the absence of a 2-OH) of this enzyme (Warren *et al.*, 2002). On the other hand, cells transfected with *sad* can 2-hydroxylate not only 2dE and 2,22dE (Warren *et al.*, 2002), but also 2,22-dideoxy-20-hydroxyecdysone (C. Blais, C. Dauphin-Villemant, and R. Lafont, unpublished data), suggesting that side chain modifications do not necessarily interfere with the active site of this enzyme. Similarly, cells transfected with *phm* convert 2,22,25dE to 2,22dE, but also they 25-hydroxylate both the 3β,5α-ketodiol and the 3α,5α-ketodiol to their respective isomeric ketotriol products (Warren *et al.*, 2004). Apparently, the stereochemistry A/B-ring junction and the 3-OH are not important determinants for PHM activity. On the other hand, cells transfected with PHM did not metabolize 25dE to E or ponasterone A to 20E, indicating its inability (as was observed with Dib) to use substrates prematurely hydroxylated at C-2. Extensive enzymatic studies are now possible using the stable, heterologous expression in non-steroidogenic cell lines of single or multiple steroidogenic P450 enzymes targeted to mitochondria or/and ER.

4.4.4. Evolution of Ecdysteroid Biosynthesis

4.4.4.1. Various secretion products: Single or multiple metabolic pathways?
Ecdysteroid biosynthesis has been studied in various model systems like steroidogenic tissues originating from animals producing different and sometimes exclusive ecdysteroids. There is a general implicit postulate that production of the same ecdysteroid proceeds through a common pathway in all investigated arthropods. This means that orthologous enzymes are expected to be identified in the various species. The increasing amount of data from species where the genome has been sequenced confirms this assumption in Insects and Crustaceans (see 4.4.4.2.). On the other hand, the occurrence of different secretory products in different species raises the question of whether they are the result of profound changes in ecdysteroid biosynthesis or just the effect of minor differences in pathways. The reason for the different oxidation state at C-3 in the two major ecdysteroid products identified in many arthropod species (E or 3DE), might involve just variability in the activity or expression of a terminal reductase in or near the steroidogenic tissue. Such a hypothesis is substantiated by the fact that, generally, in 3DE-producing species, the conversion of the 5β-diketol leads preferentially to 3DE, while the conversion of the 5β-ketodiol is also efficient but leads to E (Böcking *et al.*, 1993; Grieneisen, 1994). Similarly, the preferential production of 25-deoxy ecdysteroids by crab Y-organs (Lachaise *et al.*, 1989; Rudolph and Spaziani, 1992) may be due simply to the low expression of the 25-hydroxylase in these crustaceans.

However, the mechanisms for these differences may be more complex and could involve different substrate specificities of the corresponding enzymes.

4.4.4.2. Conservation and/or evolution of steroidogenic enzymes?

Different approaches have been engaged over the past decades to gain structural information about the genes and proteins involved in ecdysteroid biosynthesis. Regarding the biosynthetic enzymes, a first approach has taken advantage of expected sequence homologies and looked for CYP enzymes expressed specifically in prothoracic glands (Snyder *et al.*, 1996). Using this strategy, CYP enzymes specific to *M. sexta* prothoracic glands (Feyereisen, 2005) and crayfish (*O. limosus*) Y-organs (Dauphin-Villemant *et al.*, 1997, 1999) were identified. Curiously, both belong to the CYP4 family, but their precise function has not yet been elucidated.

In *Drosophila*, the availability of mutants represented the first powerful tool to identify genes involved in ecdysteroid biosynthesis. Several mutations, including *ecdysoneless* (Garen *et al.*, 1977, Warren *et al.*, 1996, Gaziova *et al.*, 2004), *DTS3*, *dre4* (Sliter and Gilbert, 1992), *giant* (Schwartz *et al.*, 1984), *molting defective* (Neubueser *et al.*, 2005), and *woc* (Wismar *et al.*, 2000) affect E titers and consequently development and metamorphosis. However, most of the corresponding genes appear to be more linked to the upstream regulation machinery of steroidogenesis rather than to encode biosynthetic enzymes themselves. For instance, *giant* encodes a βZIP (basic-leucine zipper) transcription factor (Capovilla *et al.*, 1992) while *woc* encodes a transcription factor with eight zinc fingers (Wismar *et al.*, 2000) that appears (at least) to regulate the expression of the cholesterol 7,8-dehydrogenase (Warren *et al.*, 2001). An enhancer-trap approach has also been used (Harvie *et al.*, 1998) to characterize the function of *Drosophila* ring glands during larval development. Several genes specifically expressed in the lateral cells of the ring gland (i.e., the prothoracic gland cells) have been identified, but there is presently no evidence that they are involved in ecdysteroid biosynthesis. More integrated approaches using microarray technology are presently available and should constitute new tools to identify larger sets of genes specifically expressed in steroidogenic cells and thus potentially linked to ecdysteroidogenesis.

Recently the increasing number of genome sequencing projects, first in *Drosophila* and then in several insects or crustaceans, facilitated a rapid search of steroidogenic enzymes orthologues once they were identified in any species (Rewitz and Gilbert, 2008; Christiaens *et al.*, 2010; Iga and Smagghe, 2010). Comparative studies and rapid progress in insect genomics have allowed confirmation that these orthologous genes actually catalyze the same reactions. Following similar heterologous expression in *Drosophila* S2 cells, it was determined that a *Bombyx* gene analogous to *phm* also codes for a P450 enzyme

that catalyzes specific ecdysteroid 25-hydroxylation (Warren *et al.*, 2004). A similar approach was used to demonstrate the functional conservation of *Dib*, *Sad*, or *Shd* from *Anopheles gambiae* (Pondeville *et al.*, 2008). Furthermore, as was shown for the *Drosophila* ring-gland-specific halloween genes (Parvy *et al.*, 2005), *Bombyx phm* expression in the prothoracic gland cells during the last larval–larval molt undergoes a dramatic downregulation by the beginning of the last larval instar (Warren *et al.*, 2004), perhaps as a result of feedback inhibition by the product, 20E, operating via its nuclear receptor (see Chapter 5). In general, a single orthologue of CYPs encoding insect steroidogenic enzymes is found in any arthropod species investigated so far, reflecting the high pressure of selection on hormone synthesis. There are two important exceptions that concern *Cyp307* and *Cyp18* genes. In *D. melanogaster*, two paralogues, *Cyp307a1* (*spook-spo*) and *Cyp307a2* (*spookier-spok*), arise from a complex evolutionary scenario (Sztal *et al.*, 2007) and both play essential roles in ecdysone synthesis. They display different expression pattern with *spok* being expressed in the ring gland, whereas *spo* is expressed in the ovary and early embryo, but may possess biochemically redundant functions (in the black box). Two similar *spo/spok*-like genes were also recently detected in the pea aphid (Christiaens *et al.*, 2010). In other insects and in *Daphnia*, a single *Cyp307a-spo* gene is found, which appears to resume the biological functions of *Drosophila spo* and *spok*, according to genome microsyntheny (conserved arrangement of adjacent genes) and expression pattern in steroidogenic tissues (Namiki *et al.*, 2005; Rewitz and Gilbert, 2008). In addition to *spo-Cyp307a1* and *spok-Cyp307a2*, a third paralogue, *spot (spookiest)-Cyp307b1*, was found in species from several insect orders (but not in *Drosophila*; Ono *et al.*, 2006). Phylogenetic analyses showed a distinct separation of the *spo-Cyp307a1/spok-Cyp307a2* clade from the *spot-Cyp307b1* genes, indicating that they originate from an early duplication probably following the split between insects and crustaceans and a secondary loss in some insects (Rewitz *et al.*, 2007; Rewitz and Gilbert, 2008). The function of *spookiest-Cyp307b1* has not yet been investigated. A similar situation is observed concerning the *Cyp18a1* gene. This gene encodes the 26-hydroxylase involved in ecdysteroid catabolism (Guittard *et al.*, 2010; see Section 4.5.) and orthologues are present in most insects and in crustaceans (Guittard *et al.*, 2011; Rewitz and Gilbert, 2008). In addition, the microsyntheny of *Cyp18a1* and *phm* present in *Drosophila* species, *Apis mellifera*, *B. mori* and *Daphnia pulex* (Claudianos *et al.*, 2006; Rewitz and Gilbert, 2008), also supports orthology of *Cyp18a1* genes. A striking exception is *An. gambiae*, where no clear orthologue of *Cyp18a1* could be detected (Feyereisen, 2006), suggesting a gene loss. On the contrary, an additional *Cyp18b1* paralogue has been detected in the *Bombyx* genome, but the corresponding enzyme does not display ecdysteroid 26-hydroxylase activity (T. Kozaki, C. Dauphin-Villemant, and T. Shinoda, unpublished data).

At a more general level, phylogenetic analyses of all CYPs involved in vertebrate or ecdysozoan steroid biosynthesis suggest that steroidogenesis was independently elaborated in those phyla (Markov *et al.*, 2009). It is postulated that all cytochrome P450 enzymes are derived from a few ancient CYPs (Nelson, 1998; Feyereisen, 2006), and the metazoan CYP family is accordingly divided into clans. As concerns the mito clan that clusters mitochondrial proteins in vertebrates and insects, all mitochondrial CYPs identified to date in vertebrates are involved in metabolism of endogenous compounds (e.g., CYP27A1 for bile acids) or hormone biosynthesis (CYP11A and CYP11B for steroid hormones; Omura, 2006). By contrast, arthropod mitochondrial CYPs include several xenobiotic-metabolizing proteins (e.g., CYP12) and enzymes catalyzing ecdysteroid biosynthesis (CYP302, CYP314, and CYP315; review in Feyereisen, 2006). Markov *et al.* (2009) observed that vertebrate (CYP11A and CYP11B) and arthropod (CYP302, CYP314, and CYP315) steroidogenic enzymes do not form a monophyletic clade, and are rather dispersed at various places in the tree, often linked to non-steroidogenic proteins. The most parsimonious scenario is that the different steroidogenic activities arose independently in arthropods and vertebrates. Similar conclusions can be drawn for the microsomal CYPs. As concerns the oxidoreductase activities, such as that expected to be catalyzed by *sro*, the picture is even more complex since, in vertebrates, convergent acquisition of the same biochemical activity can be seen by enzymes from different protein families (Markov *et al.*, 2009).

In conclusion, most enzymes known to be involved in steroidogenesis in arthropods, nematodes, or vertebrates seem conserved within each phylum but have no clear orthologues outside their respective metazoan phyla. This indicates that the steroidogenic enzymes have evolved independently within each phylum, through lineage-specific duplications, and subsequent neofunctionalization.

4.4.4.3. Electron transport systems involved in cytochrome P450 reactions

Monooxygenations catalyzed by cytochrome P450 enzymes require molecular oxygen and two electrons during their catalytic cycle. In vertebrate steroidogenesis, the electron transfer systems associated with the mitochondrial and microsomal CYP enzymes are different. In the mitochondrial system, reduction occurs with FAD as the flavoprotein (adrenodoxin reductase), which transfers electrons to an iron-sulfur protein (adrenodoxin) and finally to the CYP. In contrast, in the microsomal system, cytochrome P450 reductase, a flavoprotein containing both Flavin Mono Nucleotide (FMN) and Flavin-Adenin Dinucleotide (FAD), passes electrons directly from NADPH to the CYP. The second electron may alternatively arise from NADPH

via cytochrome b5, and a number of CYPs show an obligatory requirement for cytochrome b5 (Miller, 1988; Schenkman and Jansson, 2003).

There is some evidence in arthropods that ecdysteroid hydroxylations catalyzed by CYPs involve electron transfer systems analogous to those used in vertebrates. The obligatory function of cytochrome P450 reductase and the adrenodoxin/adrenodoxin reductase system in the function of microsomal and mitochondrial CYPs, respectively, has been indicated by immunoinhibition studies. Anti-adrenodoxin and anti-adrenodoxin reductase antibodies effectively inhibited E 20-monooxygenase activity in *Spodoptera* fat body mitochondria (Chen *et al.*, 1994a). Similarly, antibodies raised against cytochrome P450 reductase inhibited microsomal E 20-monooxygenase activity in *B. mori* embryos (Horike and Sonobe, 1999). Electron Paramagnetic Resonance (EPR) spectroscopic studies have corroborated the presence of ferredoxin (adrenodoxin) and cytochrome P450 enzyme in the foregoing mitochondria (Shergill *et al.*, 1995) and in *Manduca* fat body mitochondria (Smith *et al.*, 1980).

The electron transfer systems appear to be structurally and functionally conserved. The cytochrome P450 reductase has been cloned in several insect species (Koener *et al.*, 1993; Hovemann *et al.*, 1997; Horike *et al.*, 2000). Similarly, the adrenodoxin reductase gene (DARE) has been identified in *Drosophila* (Freeman *et al.*, 1999). Taken together, these data underline the similarity of cytochrome P450 enzyme function in both vertebrates and invertebrates. As there are numerous cytochrome P450 enzymes involved in various metabolic processes, the expression patterns of the electron transfer proteins are not specific to steroidogenic tissues. However, specific transcriptional or post-transcriptional regulation may occur in steroidogenic tissues, as has been shown in vertebrates. Molecular studies are still needed to substantiate such hypotheses.

4.4.5. The Ecdysteroidogenic Tissues

In arthropods as in other organisms, several criteria have been classically used to demonstrate that a given tissue or cell is steroidogenic. A straight biochemical approach is to show that a tissue is able to metabolize cholesterol (which may be radiolabeled) into the final compound, here classically ecdysone, since the activation to 20-hydroxyecdysone proceeds in many target tissues of ecdysteroids (see Section 4.5.). Another way to identify "true" steroidogenic tissues has been gained more recently with the identification of molecular players in the biosynthetic pathway. A cell (or a tissue) may be considered as steroidogenic if it co-expresses all enzymes needed for the synthesis of ecdysteroids from cholesterol. Even if a black box remains in this pathway, major steps, both early and late, are now identified and it is possible to detect the corresponding transcripts and proteins. However, as is the case in vertebrates, the steroidogenic capacity of some tissues such as ovaries, may require cellular cooperation between different cell populations.

4.4.5.1. Prothoracic glands Classically, prothoracic glands are considered as the major source of ecdysteroids during post-embryonic development, but they usually degenerate prior to the early adult stage. The development of analytical techniques such as RIA (Borst and O'Connor, 1972) allowed direct evidence to be obtained for ecdysone secretion by prothoracic glands. Using *in vitro* incubations of these organs, several authors could demonstrate that they actually secrete measurable amounts of ecdysone, and that the variations of this secretory activity are consistent with the fluctuations of hemolymph ecdysteroid levels (Chino *et al.*, 1974; King *et al.*, 1974; Borst and Engelmann, 1974). However, this was not a full demonstration for *de novo* synthesis. Only the *in vitro* conversion of radiolabeled cholesterol (or 25-hydroxycholesterol) by prothoracic glands and Y-organs provided the final argument for the *de novo* synthesis of ecdysteroids by these endocrine organs (Hoffmann *et al.*, 1977; Warren *et al.*, 1988a,b; Böcking *et al.*, 1994). Recent molecular analyses have now demonstrated, at least in classical models such as *Drosophila*, *Bombyx*, or *Manduca*, that all steroidogenic enzymes and related proteins are concomitantly expressed in the prothoracic gland cells (the lateral part of the ring gland in *Drosophila*) from the end of embryogenesis until metamorphosis (Warren *et al.*, 2002, 2004; Petryk *et al.*, 2003; Niwa *et al.*, 2004, 2010; Ono *et al.*, 2006; Parvy *et al.*, 2005; Yoshiyama *et al.*, 2006). Genetic approaches are also useful in *Drosophila*. For example, using RNAi corresponding to genes involved in the ecdysteroid biosynthetic pathway and the UAS-GAL4 system allowed the demonstration of specific silencing of the selected gene in the ring gland was sufficient to impair ecdysteroid production and to alter development (Yoshiyama *et al.*, 2006; Niwa *et al.*, 2010).

4.4.5.2. Gonads Adult insects also contain ecdysteroids, as was found in *Bombyx* (Karlson and Stamm-Menéndez, 1956). The same methods used for prothoracic glands established that ovaries are a temporary ecdysteroid source, usually in adults and/or pharate adults, depending on the stage when the oocytes develop (Hagedorn *et al.*, 1975; Hagedorn, 1985). When oogenesis proceeds synchronously, as in mosquitoes (see Chapter 9), ecdysteroids peak to high levels during vitellogenesis. In *Locusta*, careful dissections allowed demonstration that the follicle cells surrounding terminal oocytes are able to produce these compounds *de novo* from cholesterol (Goltzené *et al.*, 1978). In *Drosophila*, where oogenesis proceeds asynchronously, no hormonal peak can be easily detected, but *in situ* detection of enzymes was mainly observed in vitellogenic follicles (stage 10; Brown *et al.*,

2009), suggesting local (paracrine) effects of ecdysteroids, as expected from the inhibitory effect of ecdysteroids on selective stages of oocyte maturation (Soller *et al.*, 1999). These maternal ecdysteroids can also be stored as conjugates by developing oocytes (see Section 4.5.) and/or are released into the hemolymph for distribution within the adult. There, they may retain specific endocrine functions such as the control of vitellogenesis, of sexual pheromone biosynthesis, or of ovarian cyclic activity (see Chapter 9). Recent studies, at least in *Drosophila*, are puzzling. Depending on the steroidogenic enzymes, expression is detected either in the nurse cells or in the follicular cells. Further studies are needed to find the identity of the ovarian steroidogenic cells and to understand if a cellular cooperation is needed between different cell populations to produce ecdysteroids.

Insect testes also contain ecdysteroids and it has been proposed that the interstitial tissue of the testis plays a role similar to the Leydig cells of vertebrates. When cultured *in vitro*, testes release small amounts of a complex mixture of immunoreactive compounds (Loeb *et al.*, 1982), and this release can be stimulated by an ecdysiotropic peptide originating in the brain (Loeb *et al.*, 1988). They efficiently convert 2,22,25-trideoxyecdysone into E and 20E (Jarvis *et al.*, 1994b). However, the direct evidence for a complete *de novo* biosynthetic pathway in testes is still lacking, and thus a partial synthesis starting from some downstream biosynthetic intermediate or ecdysteroid conjugate cannot be excluded. The amounts of ecdysteroids involved are always low as compared to those produced by ovaries, and the absence of a prominent component within the ecdysteroids released *in vitro* does not strongly argue for an endocrine function. Consistent with this hypothesis is the lack of *dib* and *phm* expression in testes, as noticed in the RT-PCR study of various *Drosophila* tissues, although further studies are needed to confirm this result. By contrast, *A. gambiae* adult males appear unique for their high production and storage of 20E, in the male accessory glands (MAGs). Moreover, remarkable amounts of 20E are transferred to females during mating, strongly suggesting that 20E may act as an allohormone modulating post-mating effects, rather than as a male sex steroid, even if we cannot presently exclude endogenous effects of ecdysteroids at low concentrations (Pondeville *et al.*, 2008).

4.4.5.3. Other steroidogenic tissues? The idea of alternative sites of ecdysteroid production has arisen from experimental data that do not fit with the classical scheme (Redfern, 1989; Delbecque *et al.*, 1990). For instance, the prothoracic glands of *Te. molitor* (Coleoptera) degenerate during the pharate pupal stage, so the pharate pupal and pupal ecdysteroid peaks must arise from another source (Glitho *et al.*, 1979). *In vitro* cultivated epidermal fragments of *Te.* pupae release significant amounts of ecdysteroids, and they convert 2,22,25-trideoxyecdysone (but not C) into E. This release is connected with the epidermis cell cycle, as it can be abolished by treatment with mitomycin (Delbecque *et al.*, 1990 and references therein). The prothoracic glands of larvae and pupae of the mosquito, *Ae. aegypti*, do not secrete any ecdysteroid when cultured *in vitro*, whereas fragments of thorax and abdomen do secrete ecdysteroids (Jenkins *et al.*, 1992). In agreement with these results, expression of steroidogenic enzymes is detected by RT-PCR in both thorax and abdomens of *An. gambiae* larvae (E. Pondeville and C. Dauphin-Villemant, unpublished data). In the same way, epidermal fragments from last instar nymphs of locusts secrete ecdysteroids *in vitro* (Cassier *et al.*, 1980), and cell lines derived from imaginal discs display a rhythmic ecdysteroid production (Mesnier *et al.*, 2000). It should be emphasized that prothoracic glands have an ectodermal origin. Therefore, the general epidermis is expected to be the source of molting hormones (Bückmann, 1984; Lachaise *et al.*, 1993). Interestingly, it was recently demonstrated that skin is also a primary steroidogenic tissue in humans (Thiboutot *et al.*, 2003).

Other approaches have raised similar questions. Insects deprived of molting glands can still molt (e.g., *Periplaneta americana*, Gersch, 1979; *M. sexta*, Sakurai *et al.*, 1991); ovariectomized adult insects still contain/produce ecdysteroids (Delbecque *et al.*, 1990). Thus, there are data that cannot be explained, unless alternative ecdysteroid sources are postulated. At this level, we should clearly distinguish between primary and secondary sources (Delbecque *et al.*, 1990). Primary sources can be defined as tissues that are capable of the *de novo* synthesis of ecdysteroids starting from C or related phytosterols. Secondary sources release ecdysteroids following the hydroxylation of late intermediates and/or the hydrolysis of conjugates, therefore they do not contain the whole set of enzymes required for *de novo* synthesis. With the characterization of various steroidogenic enzymes, molecular tools have become available for various insect species (e.g., *Aedes*, *Te.*), and they will allow a more precise identification of the tissues that express steroidogenic enzymes in insects.

Finally, the importance and mechanism of ecdysteroid biosynthesis during embryonic development is still not well understood. Early metabolic studies performed in *Locusta* (Sall *et al.*, 1983) or *Drosophila* (Bownes *et al.*, 1988; Grau and Gutzeit, 1990; Kozlova and Thummel, 2003) suggested that free ecdysteroids were provided to the developing embryo by the hydrolysis of maternally derived ecdysteroid conjugates. However, the embryonic phenotype of halloween mutants and the results of *in situ* hybridizations using *halloween* genes probes argue for a more complex picture. During early embryogenesis in *Drosophila* (Chávez *et al.*, 2000; Warren *et al.*, 2002, 2004; Niwa *et al.*, 2004) and in *Bombyx* embryos (Sonobe *et al.*, 1999), the enzymes catalyzing the terminal steps of ecdysteroid biosynthesis first appear as characteristic stripes in

ectodermal domains, and their expression remains localized in segmented epidermal cells by the time of germ-band extension. By contrast, *spook* (an unknown step of the black box) is mainly expressed in the amnioserosa and yolk nuclei (Namiki *et al.*, 2005; Ono *et al.*, 2006). Nvd, which catalyzes the first step of ecdysteroid biosynthesis, is first expressed in the primordium of the ring gland (Yoshiyama *et al.*, 2006), where expression of all steroidogenic enzymes is restricted at the end of embryogenesis.

4.5. Ecdysteroid Metabolism

4.5.1. Introduction

Early reviews are available on ecdysteroid metabolism (Rees and Isaac, 1984, 1985; Lagueux *et al.*, 1984; Lafont and Koolman, 1984), including two on synthetic (Smith, 1985) and degradative (Koolman and Karlson, 1985) aspects of regulation of ecdysteroid titer in the first edition of this series, together with several articles in the book edited by Koolman (1989) and a more recent one (Rees, 1995).

In view of the diverse nature of the insect class, there is always a danger of attempting to unify their metabolic processes into common schemes. Certainly, there is considerable species variation in ecdysteroid metabolic pathways. The pathways have been primarily deduced from examination of the fate of $[^3H]$-ecdysteroids in various species of whole insects at various stages of development or in various tissues incubated *in vitro*. In many cases the metabolic products have been incompletely characterized, and such studies have been complemented by the isolation and physical characterization of metabolites.

Defined changes in the ecdysteroid titer are mandatory for correct functioning of the hormone system during development. Regulation of the ecdysteroid titer involves the rates of ecdysone biosynthesis, conversion into the generally more active 20E, inactivation, and excretion. The presence of ecdysteroid carrier proteins, which may protect the hormone to a certain extent from inactivating enzymes, may influence the fate of circulating ecdysteroids. From the limited information available, there is an apparent species-dependent heterogeneity in ecdysteroid transport with some species lacking binding proteins, while others differ in relation to the specificity of protein binding (Karlson, 1983; Whitehead, 1989; Koolman, 1990).

One difficulty is ensuring that some ecdysteroids that have been generally regarded as hormonally inactive metabolites do not possess biological activity in particular systems. For example, although 20E is regarded as the principal active ecdysteroid in most species, there is evidence that E, 3-dehydro-, and 26-hydroxy-ecdysteroids may have hormonal activity in certain systems (see Lafont and Connat, 1989; Dinan, 1989). The tissues that are most active in metabolizing ecdysteroids are the fat body, Malpighian tubules, and the midgut. In larvae, ecdysteroids are excreted via the gut or Malpighian tubules and occur in feces as unchanged hormones and various metabolites including conjugates, ecdysteroid acids, and 3-epiecdysteroids (Lafont and Koolman, 1984; Koolman and Karlson, 1985).

Over the years, there have been a plethora of studies on the metabolism of E in various species and the major modifications that have been detected are summarized in **Figure 14**. This concentrates on insects and does not include some more specialized transformations such as found in vertebrates. Since most of the metabolic studies have utilized [^{3}H]-ecdysone labeled in the side chain, the importance of side chain cleavage may well have been underestimated, since the labeled side chain fragment may not have been detected under the conditions used. This section will concentrate on work published since 1985. Studies *in vitro* on E metabolism in different tissues of various species (Koolman and Karlson, 1985), together with the enzymes catalyzing the reactions (Weirich, 1989), have been tabulated and reviewed.

4.5.2. Ecdysone 20-Monooxygenase

E, either produced directly by the prothoracic glands or in many Lepidoptera via reduction of 3dE by hemolymph 3-dehydroecdysone 3β-reductase, undergoes ecdysone 20-monooxygenase-catalyzed hydroxylation to yield 20E (**Figure 12**). The 20-hydroxylation reaction is important in the production of 20E, which is generally far more active than E. Earlier work on ecdysone 20-monooxygenase (ecdysone 20-hydroxylase; EC 1.14.99.22; E20MO) is covered in several reviews (Weirich *et al.*, 1984; Lafont and Koolman, 1984; Smith, 1985; Weirich, 1989; Grieneisen, 1994; Rees, 1995).

4.5.2.1. Properties Species and tissues — Early studies on E20MO in certain tissues of various insect species, together with the biochemical properties of some of these enzymes, have been reviewed (Weirich *et al.*, 1984; Smith, 1985). Recently, the gene coding for E20MO activity in *Drosophila* (*shd, Cyp314a1*) has been identified and characterized (Petryk *et al.*, 2003; see Section 4.4.3.1.). The enzyme is expressed in several peripheral tissues, primarily in the fat body, Malpighian tubules, and midgut. Although lower activity may occur in certain other tissues including the ovaries and integument, it is absent from prothoracic glands.

Subcellular distribution — Depending on species and tissue, E20MO may be either mitochondrial or microsomal or occur in both subcellular fractions. For example, a thorough study has established the occurrence of both high mitochondrial and microsomal activities in midgut from *M. sexta* (Weirich *et al.*, 1985). Although the mitochondrial enzyme has the higher V_{max}, since it also has the higher apparent K_M (E) (1.63 10^{-5} M vs. 3.67 10^{-7} M for the microsomal one), at physiological E concentrations (10^{-7} - 10^{-8} M), it is only one-eighth to one-tenth as active as the microsomal enzyme. Thus, the microsomal E20MO is the primary enzyme activity in *Manduca* midgut (Weirich *et al.*, 1996). Similarly, in larvae of *Drosophila* and fed adult females of *Ae. aegypti*, dual localization of enzymatic activity has been shown (Smith and Mitchell, 1986; Mitchell and Smith, 1986; see Section 4.4.3.1.). Furthermore, in larvae of the housefly, *Musca domestica*, although appreciable E20MO activity occurs in both mitochondrial and microsomal fractions, pre-treatment of larvae with E results in an increase in V_{max} and a decrease in K_M values in mitochondria, but not in microsomes. This suggested that the mitochondrial E20MO is under regulatory control by E in the larval stage and that only such activity has a physiological role during development in *M. domestica* (Agosin *et al.*, 1988).

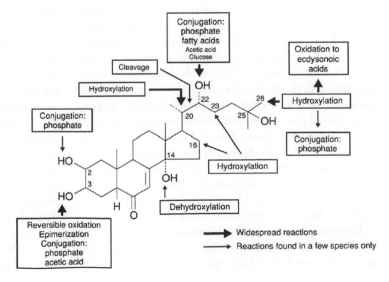

Figure 14 Major reactions of ecdysone metabolism. Modified from *Lafont and Connat* (1989).

Since one of the mitochondrial cytochrome P450 species fractionated from E-induced mitochondria catalyzed the reaction at significantly higher rates than the other five species in a reconstituted system, it was postulated that the 20-monooxygenase is a mitochondrial process that requires the induction of a low K_M P450 species by ecdysone (Agosin and Srivatsan, 1991). In adults of both sexes of the cricket, *Gryllus bimaculatus*, E20MO activity is located primarily in the microsomal fraction of the midgut, with peak activity at day 4 following ecdysis; no detectable activity was observed in fat body, ovaries, Malpighian tubules, and carcass tissues (Liebrich *et al.*, 1991; Liebrich and Hoffmann, 1991).

Cofactor requirements and cytochrome P450 properties — E20MO requires NADPH as a source of reducing equivalents. With vertebrates the mitochondrial system can use other secondary sources of reducing equivalents such as NADH or Krebs cycle intermediates, presumably by intramitochondrial generation of NADPH via transhydrogenation of $NADP^+$ (Greenwood and Rees, 1984; Smith, 1985). Similarly, a requirement of the 20-monooxygenase for oxygen has been firmly established. Evidence for involvement of cytochrome P450 enzyme has been obtained by inhibition of activity with carbon monoxide, its maximal reversal with light at 450 nm as revealed by a photochemical action spectrum, and use of a range of specific inhibitors of cytochrome P450 enzymes together with demonstration of the presence of the protein by CO difference spectroscopy (see Smith, 1985).

Substrate specificity — Although reconstituted pure vertebrate steroid hydroxylating cytochrome P450 enzymes exhibit substrate specificity, it is not absolute and may hydroxylate more than one position on the steroid substrate albeit at different rates (Sato *et al.*, 1978). Since pure E20MO has not been available for such reconstitution experiments (see Section 4.4.3.1.), only preliminary data are available using mitochondrial or microsomal preparations where multiple P450 enzymes may occur. *Locusta migratoria* microsomes do not react with 2β-acetoxy-, 3-dehydroecdysone, or with the 5α-epimer of E, and they will not hydroxylate 20E or 3D20E at any other position (Feyereisen and Durst, 1978). With *Manduca* fat body mitochondria the results of a competition assay, which is presumed to reflect enzyme–substrate binding, surprisingly showed that the most effective competitor was not E, but 2,25-dideoxyecdysone, followed in order of decreasing competition by E, 22-deoxyecdysone, 2-deoxyecdysone, 22,25-dideoxyecdysone, and 2,22,25-trideoxyecdysone (Smith *et al.*, 1980). Interestingly, 20-hydroxylation of 26-hydroxyecdysone, ecdysonoic acid, and 3-epiecdysone has been observed in *Pieris brassicae* (Lafont *et al.*, 1980).

Kinetics and product inhibition — The K_M and V_{max} values determined for the various mitochondrial and microsomal E20MO systems show substantial differences (Weirich *et al.*, 1984; Smith, 1985). However, the extent to which these differences are real or reflect different detailed methodologies is uncertain (Smith, 1985). However, several studies have demonstrated competitive inhibition of the enzyme system by its product, 20E.

4.5.2.2. Control during development Early studies on regulation of E20MO activity have been comprehensively reviewed (Smith, 1985).

4.5.2.2.1. Developmental changes in ecdysone 20-monooxygenase activity Although many early studies on changes in the metabolism of E *in vivo* suggested that E20MO activities vary during development, definitive demonstration of this required *in vitro* studies, since several factors could conceivably complicate the former (Smith, 1985).

In the migratory locust, *L. migratoria*, peaks in E20MO activity, microsomal cytochrome P450 levels, and NADPH cytochrome c reductase activity in both fat body and Malpighian tubules coincided with the peak of 20E in the hemolymph (Feyereisen and Durst, 1980). This suggested that the increases in E20MO activities were due to increased amounts of the monooxygenase components. However, in several species, the peak of E20MO activity preceded that of the hormone titer during the last larval instar (*S. gregaria* Malpighian tubules: Johnson and Rees, 1977; Gande *et al.*, 1979; *Calliphora vicina*: Young, 1976). Similarly, in the fifth larval instar of *Manduca*, fat body and midgut monooxygenase activities exhibited 10- and 60-fold fluctuations, respectively, that were not temporally coincident with one another, or with the major hemolymph ecdysteroid titer peak (Smith *et al.*, 1983). These peak activities for fat body and midgut monooxygenases were temporally coincident and succedent, respectively, with the small hemolymph peak of ecdysteroid titer on day 4 responsible for the change in commitment during larval–pupal reprogramming; both activities were at basal levels during the major hemolymph ecdysteroid titer peak on day 7–8.

The exact significance of these E20MO developmental profiles remains obscure. Since the apparent K_M values for the fat body and midgut systems were fairly constant during the instar, whereas the apparent V_{max} values in each tissue fluctuated in a manner that was temporally and quantitatively coincident with the fluctuations in monooxygenase activity, it suggested that the changes in these activities were due to changes in amounts of the enzymes. More recently, the peak in *Manduca* midgut E20MO activity at the onset of wandering on day 5 has been further examined (Weirich, 1997). During the day preceding the peak, the microsomal activity increased 60-fold (total activity) or 115-fold (specific activity) and decreased gradually within 2 days after the peak. In contrast, the mitochondrial activity increased only 1.3- to 2.4-fold (total and specific activities, respectively) before the peak, but declined more rapidly than the microsomal activity after the peak. This indicates that mitochondrial and microsomal E20MO activities are controlled independently and that changes in the physiological rate of E 20-hydroxylation in the midgut are affected primarily by changes in the microsomal E20MO activities. Further studies on *M. sexta* showed that basal levels of E20MO activity occurred during the later stages of embryogenesis and that a peak of monooxygenase activity occurred late in the fourth instar in both fat body and midgut at the time of spiracle apolysis (Mitchell *et al.*, 1999). Both the fat body and midgut hydroxylase activities were basal during the pupal stage and only rose late in pharate–adult development, just prior to adult eclosion.

Following the characterization of the halloween genes, *shade* (*shd*; *Cyp314a1*), that encode E20MO, in *Drosophila* (Petryk *et al.*, 2003), the orthologous gene (Ms*Shd*) has been characterized in *M. sexta* (Rewitz *et al.*, 2006b). Ms*Shd* shows high homology to *Shd* from *Bombyx* (a species from the same order), but less so to *Shd* from dipterans, which are highly diverged from the Lepidoptera. That Ms*Shd* protein catalyzes the conversion of E into 20E was confirmed by heterologous expression in a *Drosophila* S2 cell system (Rewitz *et al.*, 2006b). Expression of Ms*shd* was primarily in the midgut, Malpighian tubules, and fat body, consistent with the enzymatic assays summarized earlier. During last larval instar and early pupal–adult development, there is an almost perfect temporal correlation in the fat body and midgut between changes in the abundance of the Ms*shd* transcript and E20MO activity, indicating that Ms*shd* encodes the sole enzyme catalyzing E 20-hydroxylation. At the time of the commitment peak, the fat body is likely the primary site for formation of 20E from E, since there is no significant Ms*shd* expression and E20MO activity in other tissues at this developmental time. In the midgut, both Ms*shd* expression and E 20-monooxygenase activity increase considerably within 1–2 days following the commitment peak, but decline appreciably by the time of the major peak in hemolymph 20E concentration on day 7, indicating that the midgut monooxygenase activity may not be responsible for that peak in 20E concentration. It has been suggested that this enzymatic activity in midgut at that time may elicit morphogenetic changes in that organ (Rewitz *et al.*, 2006b). Preliminary results suggest that maximum expression of Ms*shd* in Malpighian tubules correlates temporally with the major hemolymph ecdysteroid titer peak, suggesting that the tissue may be involved in maintaining the hemolymph peak in 20E titer that elicits the molt to the pupa. As expected, Ms*shd* expression and E20MO activity are low in fat body, midgut, and Malpighian tubules during the increase in pupal hemolymph ecdysteroid titer during the first six days of pupal–adult development, when the hormone is principally ecdysone. However, during days

7–8 of pupal development, Ms*shd* expression was high in epidermis, indicating that the tissue may contribute to ecdysone 20-hydroxylation at that time.

In another lepidopteran, the gypsy moth *L. dispar*, preliminary studies with homogenates of various tissues taken at different times in the fifth instar surprisingly showed that E 20-monooxygenase activity in fat body, midgut, and Malpighian tubules exhibited peak activity coinciding with the peak in the hemolymph ecdysteroid titer on the penultimate day of the instar (Weirich and Bell, 1997). In *P. brassicae*, E20MO activity occurs in microsomes of pupal wings, with an apparent K_M of 58 nM for E. The activity varied during pupal–adult development with a maximum on day 4, a time when ecdysteroid titers are high (Blais and Lafont, 1986). Although the activity is low, the peak activity is sufficient to hydroxylate 25% of endogenous E in pupae. In eggs of the silkworm *B. mori*, during embryonic development, E 20-monooxygenase activity was primarily microsomal and remained low in diapause eggs, whereas activity in non-diapause eggs increased from the gastrula stage (Horike and Sonobe, 1999). This increase in E20MO activity was prevented by actinomycin D and α-amanitin, suggesting the requirement for gene transcription. Interestingly, a cDNA encoding E20MO from developing *B. mori* eggs has been cloned. According to the authors, the corresponding protein sequence has characteristics of microsomal P450 enzyme (Maeda *et al.*, 2008), although this interpretation may be discussed as it clearly clusters with other SHD proteins in a phylogenetic tree (see Section 4.3.3.3.). Furthermore, the developmental profile of expression of the *B. mori* E20MO transcript (reaching a maximum in early organogenesis) correlated with changes in E20MO activity (maximum in late organogenesis) in non-diapause eggs. In the last larval stage of *Diploptera punctata*, maximal E20MO activity of midgut homogenates, which was primarily microsomal, occurred on days 5–9, whereas the hemolymph ecdysteroid titer exhibited a small peak on day 10 with a major one on day 17 (Halliday *et al.*, 1986).

In *Drosophila*, E20MO activity has been assayed in homogenates throughout the life cycle (Mitchell and Smith, 1988; Petryk *et al.*, 2003; see Section 4.4.3.1.). There was a small peak of enzymatic activity in the egg that was temporally coincident with the major peak in the egg ecdysteroid titer. This was followed by several larger fluctuations in activity during the first, second, and early third larval stadia, with a large peak during the wandering stage that remained high during early pupariation before dropping to basal levels by pupation and remaining at these low levels through to adults. The major peak in E20MO activity during the wandering stage and early pupariation is temporally coincident with the major ecdysteroid titer peak that occurs at puparium formation. Similarly, in the Diptera *Neobellieria bullata* and *Parasarcophaga argyrostoma* fat body microsomal E20MO activity exhibits a peak at the beginning of wandering behavior in the third instar (Darvas *et al.*, 1993). Clearly, the exact significance of the developmental profiles in E20MO activities during development and, particularly, in relation to the ecdysteroid titer in various species is far from clear.

4.5.2.2.2. Factors affecting ecdysone 20-monooxygenase activity

See Section 4.5.2.2.1. for consideration of possible regulation of transcription of E20MO. Little is known about the physiological significance of the developmental fluctuations in E20MO during development, and the factors and mechanisms responsible for such changes are ill understood. As alluded to earlier, the product of the reaction, 20E, is a competitive inhibitor, although on the basis of K_i values and hormone titers, there is doubt as to whether 20E could affect the monooxygenase activity under physiological conditions in all species examined (see Smith, 1985).

In certain species, there is evidence that E or 20E may induce E20MO systems. For example, in last instar larvae of *L. migratoria*, where the peak in ecdysteroid titer and E20MO activity are coincident, injection of E or 20E at an earlier time led to selective induction of cytochrome P450 systems, including E20MO, lauric acid ω-hydroxylase, NADPH-cytochrome c reductase, and cytochrome P450 enzyme (Feyereisen and Durst, 1980). This induction is apparently a specific effect of active molting hormones since 22-isoecdysone and 22,25-dideoxyecdysone were inactive. Such induction apparently requires protein synthesis, since it is abolished after simultaneous injection of E and actinomycin D, puromycin, and cycloheximide. Furthermore, involvement of the brain in this induction by E is ruled out, since it was observed in isolated abdomens.

Studies on housefly larvae corroborate regulation of E20MO activity by E (Srivatsan *et al.*, 1987). Interestingly, although both mitochondria and microsomes catalyze the reaction, only the former activity appears to be regulated by E.

In the midgut of final larval instar tobacco hornworm, *Manduca*, the 50-fold increase in E20MO activity at the onset of the wandering stage is prevented by actinomycin D and cycloheximide, indicating a requirement for transcription and protein synthesis (Keogh *et al.*, 1989). This increase in E20MO activity could also be elicited in head (but not thoracic) ligated animals by a brain retrocerebral complex factor(s) released at the time of PTTH release (see Chapter 1). It was reported that E or 20E could elicit the increase in monooxygenase activity in both head- and thoracic-ligated animals, suggesting the operation of a neuroendocrine–endocrine axis (PTTH → prothoracic glands → E → midgut) in the induction. Furthermore, it was reported that the ecdysteroid receptor agonist, RH 5849, could also elicit the increase in midgut E20MO activity in head or thoracic ligated

larvae (Keogh and Smith, 1991). However, the finding that 20E and RH5849 induced ecdysteroid 26-hydroxylase activity instead of E 20-monooxygenase activity in the cotton leafworm, *S. littoralis* (Chen *et al.*, 1994b; see Section 4.5.3.2.2.), prompted the reinvestigation of the situation in *Manduca* midgut. In this study, it was shown that although 20E and RH5849 could induce E20MO in subcellular fractions of midgut of *Manduca*, there was a much stronger induction of ecdysteroid 26-hydroxylase activity (Williams *et al.*, 1997). In the original studies (Keogh *et al.*, 1989; Keogh and Smith, 1991), 26-hydroxylation was not detected, presumably because the TLC system used for the assay may not have resolved 20E and 26-hydroxyecdysone.

In fat body of sixth larval instar *S. littoralis*, E20MO activity is predominantly mitochondrial with much less activity in the microsomes (Hoggard and Rees, 1988). The former activity exhibits a peak of activity during the instar at 72 h, when the larvae stop feeding (Chen *et al.*, 1994a). Various antibodies raised against components of vertebrate mitochondrial steroidogenic enzyme systems have been used as potential probes for the corresponding insect proteins to discover the molecular mechanism underlying this increased activity. Since correlation was observed between developmental changes in mitochondrial E20MO activity and the abundance of polypeptides recognized by cytochrome $P450_{11\beta}$ antibody and a polypeptide recognized by the adrenodoxin reductase antibody, it suggests that developmental changes in the abundance of components of the monooxygenase system may be important in developmental regulation of the enzyme expression (Chen *et al.*, 1994a).

There is evidence that E20MO activity may also be influenced by dietary factors, including plant flavonoids and other allelochemicals. For example, an assay of *in vitro* activities of E20MO from certain species has demonstrated that a range of plant flavonoids and other allelochemicals, including azadirachtin, significantly inhibit the activity in a dose-dependent manner, whereas certain flavonols and other allelochemicals stimulate activity (Smith and Mitchell, 1988; Mitchell *et al.*, 1993). However, none of the compounds tested elicited effects at very low concentrations. Furthermore, studies on the influence of dietary allelochemicals on midgut microsomal E20MO activity in the fall armyworm, *S. frugiperda*, have shown that various indoles, flavonoids, monoterpenes, sesquiterpenes, coumarins, methylenedioxyphenyl compounds, and ketohydrocarbons all caused significant stimulation of activity (Yu, 1995, 2000). Significantly, enzyme inducibility was different between E20MO and xenobiotic-metabolizing cytochrome P450 monooxygenases.

In addition, there have been numerous studies on the effects of a plethora of synthetic inhibitors on E20MO activity in various species (Kulcsar *et al.*, 1991; Darvas *et al.*, 1992; Jarvis *et al.*, 1994a).

4.5.2.3. Potential modulation by covalent modification The initial indication that hormonal factors might regulate ecdysteroid metabolism was furnished by the demonstration that a forskolin-induced increase in intracellular cyclic AMP in *Calliphora* fat body *in vitro* led to decreased rates of conversion of E into 20E and of the latter to other metabolites (Lehmann and Koolman, 1986, 1989). Thus, it was suggested that since forskolin mimics the action of peptide hormones by activating the adenylate cyclase system, a peptide hormone, possibly a neuropeptide, might be involved in the regulation of E metabolism. Such activation of adenylate cyclase leads to an increase in intracellular cAMP, which enhances protein (enzyme) phosphorylation via cAMP-dependent protein kinases. Such a process as part of a reversible phosphorylation–dephosphorylation of enzymes provides a mechanism for rapid modulation of enzymatic activities. In the case of vertebrate mitochondria, such a control mechanism has been reported only for a few enzyme complexes. In such systems, relevant endogenous mitochondrial protein kinase and phosphoprotein phosphatase activities have been demonstrated (see Hoggard and Rees, 1988). Experimental support for such modulation of E20MO activity by phosphorylation–dephosphorylation has been obtained. Both the microsomal and broken mitochondrial fractions from the fat body of *S. littoralis*, experiments involving treatments with phosphatase inhibitors, exogenous phosphatase, protein kinase, or the adenylate cyclase activator, forskolin, provide indirect evidence that the enzyme system may exist in an active phosphorylated state and inactive dephosphorylated state (Hoggard and Rees, 1988; Hoggard *et al.*, 1989). However, confirmation of this notion requires more definitive direct evidence, including a direct demonstration of the incorporation of labeled phosphate into components of the E20MO, together with demonstration of the occurrence of an appropriate mitochondrial protein kinase and protein phosphatase. In this respect, the recently described phosphoproteomic approach seems very promising (Rewitz *et al.*, 2009).

There are many reports of the phosphorylation of cytochrome P450 enzymes (Koch and Waxman, 1991), including steroidogenic ones as CYP11B from bovine adrenal mitochondria (Defaye *et al.*, 1982), with limited evidence for the physiological significance of the phenomenon. The activity of the microsomal cytochrome-P450-dependent cholesterol 7α-hydroxylase may also be modulated by phosphorylation–dephosphorylation (Goodwin *et al.*, 1982).

Since the pharmacologically inactive forskolin derivative, 1,9-dideoxy-forskolin, as well as forskolin, induced a dose-dependent inhibition of E20MO, the effect must be direct rather than via activation of adenylate cyclase (Keogh *et al.*, 1992). Thus, care must be exercised in the design of such experiments to avoid possible misinterpretation of data.

4.5.3. Ecdysteroid Inactivation and Storage

Insect development is absolutely dependent on defined changes in ecdysteroid titer. We have already considered the major hormone activation step, the E20MO-catalyzed reaction, and will now consider the major pathways contributing to hormone inactivation. As alluded to in Section 4.3.1., there are many reviews covering ecdysteroid metabolism and inactivation. It is difficult to establish unequivocally that a particular compound is an inactivation product, since an ecdysteroid that is essentially hormonally inactive in one assay may have activity in other systems (Dinan, 1989; Lafont and Connat, 1989).

Early studies on [^{3}H]-ecdysteroid metabolism in insects and other arthropods frequently reported the formation of polar products that were hydrolyzable with various hydrolytic enzymes, especially the so-called "*Helix pomatia* sulfatase," releasing free ecdysteroids. However, unless the enzyme is absolutely pure, no conclusion can be drawn regarding the nature of the conjugate.

It is convenient to consider metabolism and inactivation of ecdysteroids in relation to (1) potential reutilization of the products from adult females during embryogenesis or (2) removal of hormonal activity and excretion.

4.5.3.1. Ecdysteroid storage, utilization, and inactivation Arthropod gonadal ecdysteroidogenesis and ecdysteroid metabolism have been recently reviewed (Brown *et al.*, 2009).

4.5.3.1.1. Phosphate conjugates Ecdysteroid "storage" has been reviewed (Isaac and Slinger, 1989). Initially, a relatively high concentration of polar conjugated ecdysteroids was reported in *Bombyx* eggs (Mizuno and Ohnishi, 1975). However, the demonstration that ecdysteroids in locust ovaries/eggs occurred almost exclusively as polar conjugates (Dinan and Rees, 1981a) and their identification as the 22-phosphate derivatives in *S. gregaria* (Rees and Isaac, 1984) and *L. migratoria* (Lagueux *et al.*, 1984) opened the way for more definitive investigations in various species. Significantly, these maternal conjugates are phosphorylated at C-22 and are presumably inactive, since a free $22R$ hydroxyl group is important for high hormonal activity. The ovarian ecdysiosynthetic tissue has been shown to be the follicle cells (Lagueux *et al.*, 1984) and the cytosolic phosphotransferase from *S. gregaria* follicle cells has been characterized (Kabbouh and Rees, 1991a). The enzyme is dependent on ATP/Mg^{2+} for high activity, and activity varies during ovarian development, reaching a peak at the end of oogenesis in agreement with the titer of ecdysteroid 22-phosphates in the oocytes. Furthermore, enzymatic activity can be induced in terminal oocytes by E or 20E treatment, which suggests a physiological mechanism of increasing conjugation as biosynthesis of ecdysteroids in follicle cells proceeds (Kabbouh and Rees, unpublished data). Surprisingly, in *S. littoralis*, the newly laid eggs contain primarily 2-deoxyecdysone 22-phosphate (Rees, 1995).

The majority of the ovarian ecdysteroid conjugates, together with some free hormone, was passed into the oocytes (Dinan and Rees, 1981a; Rees and Isaac, 1984; Hoffmann and Lagueux, 1985). During embryogenesis, locusts undergo several molts and cycles of cuticulogenesis coincident with peak titers of free ecdysteroids (Sall *et al.*, 1983). Support for the hypothesis that the ecdysteroid 22-phosphates, following enzymatic hydrolysis, could at least provide one source of hormone in embryogenesis, particularly before differentiation of the prothoracic glands, was provided by demonstration of enzymatic hydrolysis of conjugates in a cell-free system from embryos (Isaac *et al.*, 1983). Significantly, activity of the enzyme increases around the time of greater utilization of the conjugates, which also corresponds to an increase in free ecdysteroid titer (Scalia *et al.*, 1987). Moreover, an ecdysteroid-phosphate phosphatase was recently purified and cloned from *Bombyx* embryos and shown to be active only in non-diapausing eggs (Yamada and Sonobe, 2003). In several species, there is no doubt that embryonic synthesis of ecdysteroids occurs in the later stages. Reviews on the role of ecdysteroids in reproduction and embryonic development are available (Hagedorn, 1985; Hoffmann and Lagueux, 1985).

The egg is a closed system and, therefore, decreases in the embryonic hormone titer occur largely by metabolic inactivation with accumulation of products prior to emergence (Dinan and Rees, 1981b). This accumulation facilitated the identification of a number of new ecdysteroid derivatives presumed to be inactivation products, including ecdysteroid 2-phosphate 3-acetate derivatives, 3-epi-2-deoxyecdysone 3-phosphate, and ecdysteroid 26-oic acids (see Rees and Isaac, 1984; Hoffmann and Lagueux, 1985; Rees and Isaac, 1985). **Figure 15** shows probable metabolic pathways of the major ecdysteroid 22-phosphates in developing eggs of *S. gregaria* (Rees and Isaac, 1984).

In *S. gregaria*, at the beginning of embryonic development, the ecdysteroids (conjugates and free) occur only in the yolk, whereas after blastokinesis, they occur in the embryo (Scalia *et al.*, 1987). *L. migratoria* produce diapause eggs under short-day (SD) conditions and non-diapause eggs under long-day (LD) conditions. The demonstration that the total detected ecdysteroids (primarily phosphate conjugates) in newly laid eggs were more than three times higher in non-diapause eggs than in diapause eggs suggests that ecdysteroids may be involved in the control of embryonic diapause in this species (Tawfik *et al.*, 2002).

Surprisingly, in *Manduca*, the major conjugate in newly laid eggs is 26-hydroxyecdysone 26-phosphate (Thompson *et al.*, 1985b). This compound undergoes hydrolysis during embryogenesis, with 20-hydroxylation of the released hormone occurring, yielding 20,26-dihydroxyecdysone, followed by reconjugation of 26-hydroxyecdysone to produce the 22-glucoside derivative; formation of 3α-epimers and 26-oic acids also occurs during embryogenesis (Warren *et al.*, 1986; Thompson *et al.*, 1987b, 1988). The proposed pathways of metabolism of 26-hydroxyecdysone 26-phosphate in the developing embryos of *Manduca* are given in **Figure 16**.

Early in embryogenesis, much of the 26-hydroxyecdysone 26-phosphate is hydrolyzed, resulting in a peak of 26-hydroxyecdysone just prior to the appearance of the first serosal cuticle (Dorn *et al.*, 1987). The 20,26-dihydroxyecdysone titer reaches a peak at about the time of deposition of the first larval cuticle late in embryogenesis, raising the possibility that this ecdysteroid may have a hormonal role at this stage of development (Warren *et al.*, 1986). Significantly, the concentrations of E and 20E are minimal throughout embryogenesis. That the foregoing metabolic changes are related to embryogenesis is indicated by their virtual absence in incubated unfertilized eggs of *Manduca* (Feldlaufer *et al.*, 1988).

A large amount of work on ecdysteroids, phospho-conjugation, and dephosphorylation reactions during ovarian development and embryogenesis has been undertaken in *B. mori*; these studies have been reviewed (Sonobe and Ito, 2009; Sonobe and Yamada, 2004). A plethora of free ecdysteroids and phosphate conjugates have been isolated from ovaries of the silkworm, *B. mori*, and include the following: E, 20E, 2-deoxyecdysone, 2-deoxy-20-hydroxyecdysone, 2,22-dideoxy-20-hydroxyecdysone and their 22-phosphates, bombycosterol and bombycosterol 3-phosphate (Ohnishi, 1986; Ohnishi *et al.*, 1989), 22-deoxy-20-hydroxyecdysone and its 3-phosphate (Kamba *et al.*, 1994), 3-epi-22-deoxy-20-hydroxyecdysone and its 2-phosphate (Mamiya *et al.*, 1995), 3-epi-22-deoxy-20,26-dihydroxyecdysone, plus 3-epi-22-deoxy-16β,20-dihydroxyecdysone and their 2-phosphates (Kamba *et al.*, 2000b), as well as 2,22-dideoxy-23-hydroxyecdysone and its 3-phosphate (Kamba *et al.*, 2000a). Although the metabolic relationships of many of the foregoing ecdysteroids can be surmised, they await experimental verification. The occurrence of 23- and 16β-hydroxy ecdysteroids is unusual. An ATP ecdysteroid phosphotransferase that is active on a number of ecdysteroid substrates has been demonstrated in *B. mori* ovaries (Takahashi *et al.*, 1992).

The properties of the ecdysteroid 22-kinase (EcKinase) and ecdysteroid phosphate phosphatase (EPPase) from *Bombyx* have been reviewed (Sonobe and Ito, 2009). In *Bombyx*, the ecdysone 22-kinase activity in ovarian homogenates during ovarian development increases in

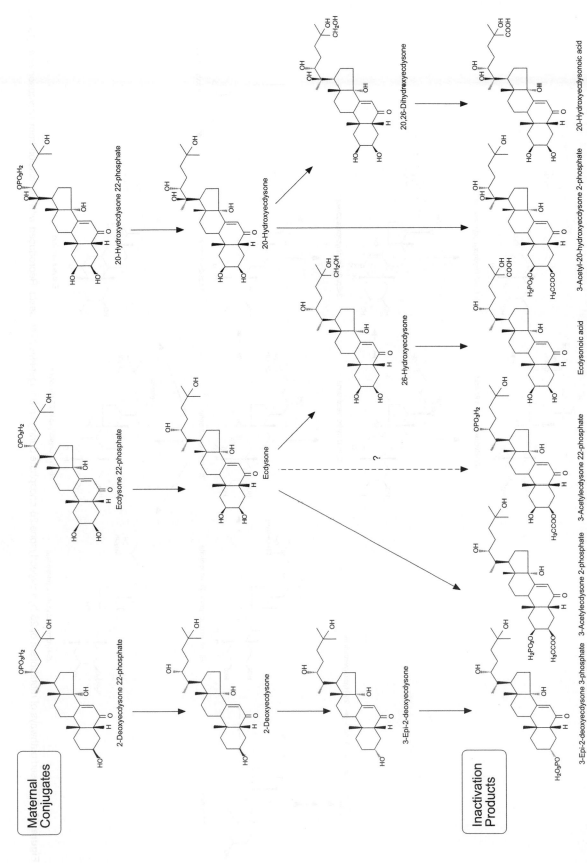

Figure 15 Probable metabolic pathways of the major ecdysteroid 22-phosphates in developing eggs of *Schistocerca gregaria*. Modified from *Rees and Isaac* (1984).

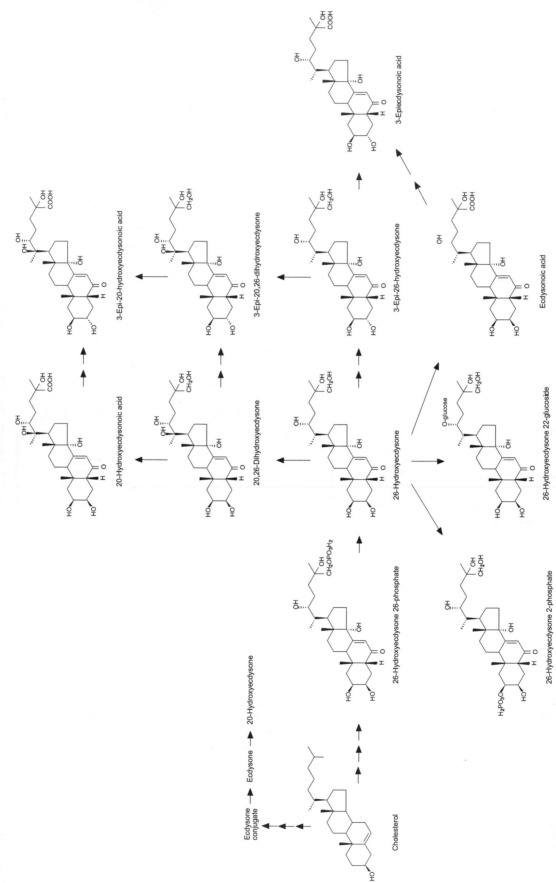

Figure 16 Proposed pathways of metabolism of 26-hydroxyecdysone 26-phosphate during embryogenesis of *M. sexta*. Reproduced with permission from *Thompson et al.* (1988).

parallel with an increase in quantities of ecdysteroid phosphates in the ovaries of both non-diapause egg-producing pupae and diapause egg-producing pupae (Sonobe *et al.*, 2006). Enzymatic activity was detected predominantly in the supernatant fraction (cytosol), with little activity in the yolk granule fraction. The *Bombyx* EcKinase shares numerous properties with a 2-deoxyecdysone 22-kinase from follicle cells of *S. gregaria* (Kabbouh and Rees, 1991a), including cytosolic localization, dependence on ATP and Mg^{2+}, inhibition by Ca^{2+}, and similar pH and temperature optima. EcKinase has been purified from *Bombyx* ovarian cytosol, partially sequenced and the cDNA cloned.

The full-length cDNA open reading frame encoded a protein of 386 amino acid residues (molecular mass, 44 kDa). Interestingly, a highly conserved cluster, Brenner's motif, $HXDhX_3Nh_3...D$ (where h represents a large hydrophobic amino acid, and X is any amino acid), which occurs in many phosphotransferases, was observed in EcKinase. Furthermore, the ATP binding sites, proposed in choline kinase, are also conserved in EcKinase (Sonobe *et al.*, 2006). The protein belongs to class 2.7.1, which is composed of enzymes that phosphorylate hydroxyl groups from ATP, but the amino acid sequence of EcKinase had less than 30% identity with any other known sequences, suggesting that it does not belong to a known kinase family. The predicted three-dimensional structure of EcKinase is quite similar to that of choline kinase, despite no significant similarity in amino acid sequences between the proteins. Besides containing the conserved Brenner's phosphotransferase motif alluded to earlier, EcKinase also contains three highly conserved residues from the ATP binding sites proposed in choline kinase. Furthermore, according to the NCBI conserved domain database, EcKinase shares ~45% identity with "a domain of unknown function" (DUF227) that occurs in *D. melanogaster* and *C. elegans* (Sonobe *et al.*, 2006).

Investigation of the sites of synthesis and phosphorylation of free ecdysteroids in the *B. mori* ovary revealed that the mRNAs of the two P450s involved in ecdysteroidogenesis, CYP306A1 (25-hydroxylase) and CYP314A1 (20-hydroxylase), are expressed mainly in follicle cells, that EcKinase mRNA localizes in the oocyte and nurse cells and that EcKinase immunoreactivity localizes mainly in the external region of the oocyte, not in nurse cells or follicle cells (Ito *et al.*, 2008; Ito and Sonobe, 2009). These results suggest that ecdysteroids in *B. mori* ovary are synthesized in follicle cells and transferred into the oocyte, where they are phosphorylated primarily in the external region by EcKinase, whose mRNA originates from nurse cells and the oocyte itself.

The ecdysteroid profiles and transformations, including ecdysteroid dephosphorylation, occurring in early *Bombyx* embryogenesis have been reviewed (Sonobe and Yamada, 2004; Sonobe and Ito, 2009). As alluded to earlier, an ecdysteroid phosphate phosphatase (EPPase) has

been purified and cloned from *Bombyx* embryos (Yamada and Sonobe, 2003). EPPase has a cytosolic subcellular localization and differs from non-specific lysosomal acid phosphatases and alkaline phosphatase in various properties (Yamada and Sonobe, 2003). Ecdysteroid phosphates in *B. mori* eggs are exclusively hydrolyzed by EPPase. The enzyme has a greater specificity for ecdysteroid phosphates having a phosphate group at the C-22 position, rather than at C-3. The full-length cDNA of EPPase has an open reading frame encoding a protein of 331 amino acid residues. EPPase mRNA was expressed predominantly during gastrulation and organogenesis in non-diapause eggs, but was not detectable in diapause eggs, which were developmentally arrested at the late gastrula stage. The developmental changes in expression of EPPase mRNA corresponded closely to changes in the enzymatic activity (suggesting transcriptional control) and in free ecdysteroids in eggs (Yamada and Sonobe, 2003).

In *Bombyx*, approximately 80% of ecdysteroids in the mature ovary is converted into ecdysteroid 22-phosphates that are mostly bound to vitellin (Vn) and stored in yolk granules in the mature oocytes (Sonobe and Ito, 2009). Yolk granules from the cellular blastoderm stage to the early organogenesis stage contained two populations, light yolk granules (LYGs) and dense yolk granules (DYGs). Wheras LYGs remained neutral, DYGs became acidic by action of V-ATPase from the cellulart blastoderm stage to the early organogenesis stage. This acidification of DYGs causes dissociation of ecdysteroid phosphates from the Vn–ecdysteroid phosphate complex into the cytosol of yolk cells (Yamada *et al.*, 2005). EPPase synthesized de novo in yolk cells then dephosphorylates ecdysteroid phosphates. However, 20-hydroxyecdysone is produced by dephosphorylation of the 22-phosphate as well as by hydroxylation of precursors of 20-hydroxyecdysone (Sonobe *et al.*, 1999; Sonobe and Ito, 2009). It has been shown that [^{3}H]2,22,25-trideoxyecdysone (5β–ketodiol) is formed from cholesterol and that [^{3}H]5β–ketodiol is converted into 20-hydroxyecdysone following injection of these labeled precursors into *Bombyx* eggs (Sonobe *et al.*, 1999). Furthermore, by incubation of dissected yolk cells *in vitro* with [^{3}H]5β–ketodiol and [^{3}H]ecdysone, conversion into 20-hydroxyecdysone was observed. By inference, it is assumed that transformation of cholesterol into 5β–ketodiol is also located in yolk cells of the oocytes (Sonobe and Yamada, 2004). The schemes for 20E production in yolk cells from Vn–ecdysteroid phosphate complexes stored in yolk granules of mature oocytes and from de novo synthesis, are depicted in **Figure 17**.

Although the *B. mori* EPP was originally concluded to belong to a novel phosphatase family (Yamada and Sonobe, 2003), it has subsequently been shown that it resides in the large histidine phosphatase superfamily related to cofactor-dependent phosphoglycerate mutase (Davies *et al.*, 2007). Interestingly, analysis of genomic

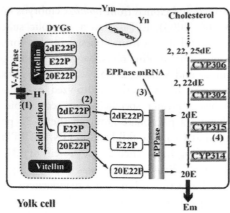

Figure 17 Scheme for 20-hydroxyecdysone production in yolk cells from Vn-ecdysteroid phosphate complexes stored in yolk granules of mature oocytes and from de novo synthesis. Abbreviations: DYGs, dense yolk granules; Ym, yolk cell membranes; Yn, yolk cell nucleus; Em, embryo; EPPase, ecdysteroid phosphate phosphatase. Reproduced with permission from *Sonobe and Ito* (2009).

and transcript data for various insect species showed that EPP may exist in both the single domain form characterized from *Bombyx* (Yamada and Sonobe, 2003) and in a longer multidomain form. Surprisingly, the latter form has quite an unexpected relationship in sequence and domain architecture to certain vertebrate proteins, including Sts-1, which is a key regulator of t-cell activity. Particularly interesting is that long-form putative *D. melanogaster* Epp, human Sts-1, and a related protein from *C.elegans* have been cloned and produced in recombinant form and that all catalyze the hydrolysis of ecdysteroid and steroid phosphates (Davies *et al.*, 2007). Obviously, the full physiological implications of these findings await elucidation.

The crystal structure of the *B. mori* EPPase C-terminal phosphoglycerate mutase (PGM) homology domain (amino acid residues 69-330) has been determined (Chen *et al.*, 2008). Interestingly, it is more related to the mammalian Sts-1$_{PGM}$ protein than to any other family member. Furthermore, *Bombyx* EPPase$_{PGM}$ has been shown to possess a new protein tyrosine phosphatase (PTP) activity *in vitro* (Chen *et al.*, 2008); whether this activity is physiologically relevant remains to be investigated.

The activities of both *de novo* ecdysteroid biosynthesis and ecdysteroid phosphate dephosphorylation in diapause eggs of *Bombyx* remains very low in comparison with that in non-diapause eggs (Makka and Sonobe, 2000).

Evidence has been furnished suggesting that 20E, acting via the receptor, is responsible for the developmental difference between diapause and non-diapause in *B. mori* embryos and that a continuous supply of 20E may be required to induce embryonic development (Makka *et al.*, 2002). In agreement with this is evidence suggesting that 20-hydroxylation of E may be a rate-limiting step in diapause silkworm eggs (Sonobe *et al.*, 1999; Maeda

et al., 2008). Furthermore, dephosphorylation of ecdysone 22-phosphate and its subsequent hydroxylation at C_{20} and C_{26} are prominent in non-diapause eggs, whereas phosphorylation of E is a major process in diapause eggs (Makka and Sonobe, 2000). EPPase mRNA and EPPase activity were hardly detected in diapause eggs (Yamada and Sonobe, 2003). Significantly, the immunoreactivity to anti-V-ATPase antiserum in diapause eggs was far weaker than in non-diapause eggs, consistent with the fact that the internal pH of DYGs remained neutral and ecdysteroid phosphates were hardly released from DYGs in diapause eggs (Yamada *et al.*, 2005).

In both diapause and non-diapause eggs, the developmental profiles of expression of *Bombyx* 20-hydroxylase mRNA correlated with changes in the hydroxylase activity (Maeda *et al.*, 2008). Furthermore, the expression profile of the 20-hydroxylase mRNA correlated significantly with that of the EPPase mRNA. Since ecdysone hydroxylation in early embryonic development also requires cytochrome P450 reductase expression, the results suggest that genes encoding ecdysone 20-monooxygenase, cytochrome P450 reductase, and EPPase, all involved in increased 20E in non-diapause eggs, may well be regulated by a common transcription factor (Maeda *et al.*, 2008). The fushi tarazu factor 1 (FTZ-F1) transcription factor is required for transcription of *dib, phm*, and probably *sad* in *Drosophila* (Parvy *et al.*, 2005). Significantly, preliminary evidence indicates in *Bombyx* that the upstream regions of the 20-monooxygenase, cytochrome P450 reductase, and EPPase genes contain candidate (FTZ)-F1 binding sites, suggesting that all three genes might be regulated by that transcription factor (Maeda *et al.*, 2008). 20E has been shown to promote development of diapause embryos. Thus, of relevance is the demonstration that EPPase is induced by incubation of diapause eggs at 5°C, that extracellular signal-regulated kinase (ERK) is similarly activated under the same conditions and regulates transcription of EPPase, which leads to increase in hormone levels (Fujiwara *et al.*, 2006).

4.5.3.1.2. Fatty acyl conjugates Quite different types of conjugates, such as apolar ecdysteroid 22-fatty acyl esters, originally found in tick ovaries and eggs (Wigglesworth *et al.*, 1985; Crosby *et al.*, 1986a) also occur in such tissues in certain Orthoptera. In the tick, *Boophilus microplus*, these conjugates appear to function as inactive, storage forms of maternal hormone utilized in early embryogenesis and undergoing subsequent inactivation by 20E-oic acid formation in conjunction with re-conjugation (Crosby *et al.*, 1986b). However, the situation is complicated, since eggs of all tick species do not contain significant ecdysteroid 22-fatty acyl esters with some containing appreciable free hormones.

Analogous apolar ecdysone 22-fatty acyl esters, produced by the ovaries, have been identified in newly laid eggs of the house cricket, *Acheta domesticus* (Whiting and Dinan, 1989). The rate of fatty acylation by ovaries is developmentally regulated, increasing as the ovaries increase in size (Whiting *et al.*, 1997). Evidence suggests that the ecdysone 22-fatty acyl esters undergo hydrolysis during embryogenesis with the resulting E converted to 20E, which is then inactivated by double conjugation, first with an uncharacterized polar moiety followed by 22-fatty acylation (Whiting *et al.*, 1993; Dinan, 1997).

The apolar esters in eggs of another orthopteran, the cockroach, *P. americana*, differ slightly in chromatographic properties from ecdysteroid 22-fatty acyl esters, although free ecdysteroids are released by hydrolysis with *H. pomatia* enzymes or esterases (Slinger *et al.*, 1986; Slinger and Isaac, 1988a). A storage role is suggested for these apolar conjugates, since they undergo hydrolysis during the first third of embryogenesis when there is also an increase in free and polar conjugated ecdysteroids (Slinger and Isaac, 1988b). Synthesis of such apolar esters by an ovary microsomal fraction from *P. americana* requires fatty acyl-CoA as cosubstrate, indicating that the enzyme is an acyl-CoA ecdysone acyltransferase and not a hydrolase (Slinger and Isaac, 1988c).

Interestingly, significant levels of apolar or polar ecdysteroid conjugates do not occur in newly laid eggs of another cricket species, *G. bimaculatus* (Espig *et al.*, 1989). Similarly, it is noteworthy that significant amounts of detectable maternal ecdysteroid conjugates have not been reported in newly laid eggs of many insect species (Hoffmann and Lagueux, 1985; Isaac and Slinger, 1989).

Comparison of the sequence similarity of *Drosophila* vitellins and part of the porcine triacylglycerol lipase led to the intriguing hypothesis that ecdysteroid fatty acyl ester may bind to vitellin in the eggs as storage forms with the conjugates made available for enzymic hydrolysis as the vitellin is degraded during embryogenesis (Bownes *et al.*, 1988; Bownes, 1992). Such candidate ecdysteroid 22-fatty acyl esters are produced in adult female *Drosophila*, together with ecdysone 22-phosphate (Grau and Lafont, 1994b; Grau *et al.*, 1995). However, the significance of binding of ecdysone 22-phosphate to an approximately 50 kDa protein in *Drosophila* ovaries is unclear at present (Grau *et al.*, 1995).

4.5.3.2. Inactivation and excretion in immature stages Ecdysteroids such as E and 20E may be excreted per se or as metabolites via the gut or Malpighian tubules. Ecdysteroid metabolizing enzymes, particularly in the fat body, midgut, and Malpighian tubules, may facilitate excretion in addition to inactivating the hormone. Variation in the rate of excretion during development appears to be an important factor in determining the hormone titer. Furthermore, hormone inactivation by the gut may be important in protection of the insect from any ingested phytoecdysteroids (see the next section).

In closed stages of insects, where excretion is impossible (egg and pupae), irreversible inactivation of ecdysteroids is important with metabolites accumulating in the gut (Koolman and Karlson, 1985; Isaac and Slinger, 1989; Lafont and Connat, 1989).

A comparative study of the metabolism of E and 20E in representatives of arthropods, mollusks, annelids, and mammals has led to the conclusion that the metabolic pathways differ strongly between species and that each class has evolved specific reactions (Lafont *et al.*, 1986).

4.5.3.2.1. Conjugates and overall metabolism Aspects of conjugate formation will be considered initially, since such compounds feature prominently in the metabolism and inactivation of ecdysteroids in immature stages of most insect species. Early studies on enzymatic aspects of ecdysteroid conjugation/hydrolysis have been reviewed (Weirich, 1989). Originally, the identification of various conjugated ecdysteroid inactivation products (e.g., 3-acetylecdysone 2-phosphate) together with the ecdysteroid 26-oic acids from developed eggs of locusts before hatching facilitated elucidation of inactivation pathways in other systems (Lagueux *et al.*, 1984; Rees and Isaac, 1984). In larval stages of locusts, the inactivation pathways (Gibson *et al.*, 1984; Modde *et al.*, 1984) proved to be very similar to the embryonic routes (see **Figure 15**), with pathways in other species having much in common (Lafont and Connat, 1989). In *Manduca*, when labeled C was injected into fifth instar larvae and the metabolites isolated from day 8 pupae, the predominant ecdysteroid metabolites were 20E, 20,26-dihydroxyecdysone, 20-hydroxyecdysonoic acid, and 3-epi-20-hydroxyecdysonoic acid. Smaller amounts occurred as conjugates (phosphates) of 26-hydroxyecdysone, 3-epi-20-hydroxyecdysone, 20,26-dihydroxyecdysone, and its 3α-epimer (Lozano *et al.*, 1988a,b). At various stages throughout pupal–adult development, the predominant metabolites were 3-epi-20-hydroxyecdysonoic acid and the phosphate conjugates of 3-epi-20-dihydroxyecdysone (Lozano *et al.*, 1989).

In feeding the last larval stage of another lepidopteran, *P. brassicae*, injected 20E was converted primarily into 20-hydroxyecdysonoic acid, 3-dehydro-20-hydroxyecdysone, and 3-epi-20-hydroxyecdysone 3-phosphate. Use of E as a substrate produced the same metabolites together with ones lacking the 20-hydroxyl group (Beydon *et al.*, 1987). Similarly, in *P. brassicae* eggs, [³H]-ecdysone metabolism is similar to that in larvae and pupae, namely hydroxylation at C-20 and C-26, with further oxidation to ecdysteroid 26-oic acids. Injection of [³H]-cholesterol into female pharate adults indicated that the egg ecdysteroids/metabolites consisted of 20-hydroxyecdysonoic acid, ecdysonoic acid, 20E, and E (Beydon *et al.*, 1989). In another lepidopteran, the tobacco budworm, *Heliothis virescens*, incubation of testes with labeled E yielded a range of products including 20,26-dihydroxyecdysone, 3-epi-20-hydroxyecdysone, 20E, and highly polar metabolites (Loeb and Woods, 1989).

In female adults of the cricket, *G. bimaculatus*, injected [³H]-ecdysone was converted into at least 24 metabolites, the highest number from the gut and feces, which included 26-hydroxyecdysone and 14-deoxyecdysone among the free ecdysteroids. In addition, polar (phosphate) and apolar (fatty acyl) conjugates were also observed. Following the feeding of [³H] ecdysone, most of the label remained within the gut and was excreted within 48 h, with 14-deoxyecdysone as a major metabolite (Thiry and Hoffmann, 1992). Both apolar and polar ecdysteroid conjugates are widely distributed in tissues of males of this species (Hoffmann and Wagemann, 1994). In vitellogenic females of *Labidura riparia* (Dermaptera), [³H] 20E was metabolized primarily through conjugation, in particular acetylation at C-3 and phosphate ester formation at C-2 (Sayah and Blais, 1995).

Incubation of midgut cytosol of *Manduca*, where prominent 3-epimerization of ecdysteroid occurs, with [³H]-ecdysone and Mg^{2+}/ATP results in detection of four E phosphoconjugates and two 3-epiecdysone phosphoconjugates (Weirich *et al.*, 1986). ATP- and Mg^{2+}-dependent ecdysteroid phosphotransferase activities have been demonstrated in cytosolic extracts of ovaries and fat body from mature female pupae of *B. mori* (Takahashi *et al.*, 1992). The midgut cytosol of the cotton leafworm, *S. littoralis*, contains Mg^{2+}/ATP-dependent ecdysteroid 2- and 22-phosphotransferase activities, with a predominance of the former (Webb *et al.*, 1995). The phosphotransferase activities were high early in the final larval instar and then declined. The relatively high K_M value (21 μM) of the E 2-phosphotransferase (Webb *et al.*, 1996), together with its expression during the feeding period when the hemolymph ecdysteroid titer is low, may suggest that these phosphotransferase enzymes are involved in inactivation of any dietary ecdysteroids.

The enzymes catalyzing formation of E 3-acetate and its phosphorylation in production of 3-acetylecdysone 2-phosphate have been characterized in larval tissues of *Schistocerca*. 3-Acetylation of E and 20E was characterized in gastric ceca from final larval instar *S. gregaria*, since high activity occurred in that tissue. The microsomal enzyme had a somewhat stronger affinity for 20E (K_M 53.5 μM) than E (71 μM) and utilized acetyl coenzyme A, but not acetic acid, as cosubstrate. This indicated that the enzyme is the acetyl coenzyme A (CoA) ecdysteroid acetyltransferase and not a hydrolase. Furthermore, esterification of E was not observed when long chain fatty acyl-CoA derivatives were substituted as cosubstrates (Kabbouh and Rees, 1991b).

In final instar *S. gregaria* larvae, ATP E 3-acetate 2-phosphotransferase activity was low in gastric ceca and gut, but was somewhat higher in fat body than in Malpighian

tubules. Thus, the soluble fat body enzyme (ca. 45 kDa) was characterized. The Mg^{2+}/ATP-dependent phosphotransferase utilized E 3-acetate as the preferred substrate (K_M 10.5 μM), with no significant phosphorylation of E and E 2-acetate. The physiological significance of the increase in phosphotransferase activity during the second half of the instar, well before the rise in ecdysteroid titer, is unclear (Kabbouh and Rees, 1993).

In some lepidopteran species, for example, *H. armigera* (cotton bollworm, Robinson *et al.*, 1987) and *H. virescens* (cotton budworm, Kubo *et al.*, 1987), a major fate of ingested ecdysteroids is formation of 22-fatty acyl esters in the gut for fecal excretion. Characterization of the ecdysteroid-22-O-acyltransferase activity in crude homogenates of midgut from *H. virescens* has shown that the enzyme can use fatty acyl-CoA as cosubstrate, but not phosphatidylcholine or free fatty acid, indicating that the enzyme differs appreciably from the cholesterol acyltransferase, where all such cosubstrates can be utilized (Zhang and Kubo *et al.*, 1992a). Furthermore, unlike the situation for the latter enzyme, divalent cations inhibit the ecdysteroid acyltransferase. Subsequently, it has been demonstrated that the ecdysteroid 22-O-acyltransferase is located in the plasma membrane of the gut epithelial cells (gut brush border membrane; Kubo *et al.*, 1994). The acyltransferase was active only during feeding stages; its activity decreased as the larvae became committed to pupation and was not enhanced by feeding ecdysteroids to the fifth instar larvae (Zhang and Kubo *et al.*, 1992b). It has been suggested that dietary phytoecdysteroid detoxification is the major function of the acyltransferase (Zhang and Kubo, 1992b, 1994). In both *H. virescens* and *B. mori* injected E was converted into 20E, 26-hydroxyecdysone, and 20,26-dihydroxyecdysone, which was oxidized to 20E 26-oic acid for excretion. These transformations also occurred in the case of ingested E in these species, but as already seen in *H. virescens*, an additional prominent route involved E-22-O-acyl ester formation, which was largely excreted. In contrast, in *B. mori*, the latter pathway was replaced by formation of 3-epi-E and a polar conjugate tentatively identified as the sulfate, which was excreted, in the case of ingested E (Zhang and Kubo, 1993). In the tomato moth, *Lacanobia oleracea*, and the cotton leafworm, *S. littoralis*, the situation is similar to that in *H. virescens*, where injected/endogenous ecdysteroids undergo conversion into 20-hydroxyecdysonoic acid, with ingested ecdysteroids undergoing detoxification yielding apolar 22-fatty acyl esters (Blackford and Dinan, 1997a; Blackford *et al.*, 1997). Surprisingly, the death's head hawkmoth, *Aclerontia atropos*, is unaffected by dietary 20E, which it excretes in an unmetabolized form (Blackford and Dinan, 1997b). Interestingly, in four other Lepidoptera (*Aglais urticae*, *Inachis io*, *Cynthia cardini*, and *Tyria jacobaeae*) ingested and injected 20E followed the same fate, with prominent polar compounds and no detectable formation of 22-fatty acyl esters (Blackford and Dinan, 1997c). Ingested ecdysteroids were eliminated in the lepidopteran, *Plodia interpunctella,* both in the free form and as fatty acyl conjugates, whereas *Ostrinia nubilalis* only produces fatty acyl conjugates with the aphid, *Myzus persicae,* producing mainly a 22-glucoside conjugate (Rharrabe *et al.,* 2007). Clearly, phytophagous insects employ a diversity of detoxification mechanisms to overcome potential harmful effects of ecdysteroids in the food.

In adult *D. melanogaster,* following injection of [³H]-E, E-22-fatty acyl esters were the major metabolites followed by 3DE, 26-hydroxyecdysone, ecdysonoic acid, 20-hydroxyecdysone, and E 22-phosphate (Grau and Lafont, 1994a; Grau *et al.,* 1995). 20E was metabolized in a similar manner (Grau and Lafont, 1994a). The E-22-phosphate was produced in the ovaries, whereas E was efficiently converted into 3DE and ecdysone 22-fatty acyl esters by gut/Malpighian tubule complexes (Grau and Lafont, 1994b). In third instar *D. melanogaster* larvae, although E metabolism proceeds through different pathways including C-20 hydroxylation, C-26 hydroxylation, and oxidation to 26-oic acids, C-3 oxidation and epimerization followed by conjugation. 3-Dehydroecdysteroids are the major metabolites with C-3 oxidation occurring in various tissues (Sommé-Martin *et al.,* 1988).

4.5.3.2.2. 26-Hydroxylation and ecdysonoic acid formation
Ecdysonoic acid and 20-hydroxyecdysonoic acid were originally identified from developing eggs of *Schistocerca,* pupae of *Spodoptera* (Isaac *et al.,* 1983), and *Pieris,* where their formation has been shown to occur in several tissues (Lafont *et al.,* 1983). Since that time, it has become apparent that 26-oic acid formation is a widespread and prominent pathway of ecdysteroid inactivation, as alluded to in several studies considered earlier.

It has been shown that administration of E, 20E, or an ecdysteroid agonist, RH-5849, to last instar larvae of the cotton leafworm, *S. littoralis,* leads to induction of ecdysteroid 26-hydroxylase activity in fat body mitochondria (Chen *et al.,* 1994b). This induction occurred in both early last instar larvae and in older larvae that had been head-ligated to prevent the normal developmental increase in E 20-monooxygenase activity. The induction of hydroxylase activity requires both RNA and protein synthesis. E-26-oic acid and a compound tentatively identified as the 26-aldehyde derivative of E were also formed from E in the RH 5949-induced systems. Direct demonstration in a cell-free system of the conversion of 26-hydroxyecdysone into the aldehyde and the corresponding 26-oic acid (ecdysonoic acid), established the following inactivation pathway: ecdysteroid → 26-hydroxyecdysteroid → ecdysteroid 26-aldehyde → ecdysteroid 26-oic acid.

Induction of the 26-hydroxylase activity might be expected to result from enhanced cytochrome P450 gene transcription and protein synthesis (Chen *et al.,* 1994b).

Similarly, both 20E and RH-5849 caused RNA- and protein-synthesis-dependent induction of ecdysteroid 26-hydroxylase activity in midgut mitochondria and microsomes in *Manduca* (Williams *et al.,* 1997). This induction of an ecdysteroid inactivation pathway (and others, see the following section) by 20E and RH-5849 is reminiscent of the regulation of vitamin D inactivation in vertebrates where 1α,25-dihydroxyvitamin D₃ stimulates its own degradation by induction of a vitamin D response element to initiate degradation to excretory metabolites (see Williams *et al.,* 1997 for references). However, the possibility cannot be discounted that at least some of the observed induction events may be indirect. Evidence has been furnished that the 26-hydroxylase activity is a typical NADPH-requiring cytochrome-P450-dependent monooxygenase that is sensitive to inhibition by CO and imidazole/triazole-based fungicides (Kayser *et al.,* 1997; Williams *et al.,* 2000). Furthermore, indirect evidence suggested that the *Manduca* mitochondrial and microsomal 26-hydroxylases could exist in a less active dephosphorylated state or more active phosphorylated state (Williams *et al.,* 2000). Interestingly, long-term culture of a *Chironomus tentans* (Diptera) cell line in the presence of 20E resulted in selection of subclones that are resistant to the steroid, but respond normally to the agonist, tebufenozide (RH5992). Evidence was obtained that ecdysteroid resistance of the cell clones is due to high metabolic inactivation of the hormone by 26-hydroxylation (Kayser *et al.,* 1997). In this cell system, 20,26-dihydroxyecdysone is further metabolized to two compounds that are slowly interconvertible geometrical isomers, which presumably arise from hemiacetal formation between the 26-aldehyde and the 22R-hydroxyl groups to yield a tetrahydropyrane ring in the side chain (Kayser *et al.,* 2002). Induction of ecdysteroid 26-hydroxylase by the agonists RH-5849, RH-5992, and RH-0345 appears universal in lepidopteran species that show susceptibility to the agonists (Williams *et al.,* 2002). It appears that the more potent ecdysteroid agonists in Lepidoptera, RH-5992 and RH-0345, show a greater induction of 26-hydroxylase than RH-5849. Feeding RH-5849 to the dipteran *M. domestica* results in induction of an ecdysteroid phosphotransferase. The low toxicity of the ecdysteroid agonists in orthopteran and coleopteran orders also correlates with a lack of induction of ecdysteroid 26-hydroxylase activity. It has been proposed that in species where ecdysteroid agonists are effective in stimulating an untimely premature molt, a response to a state of apparent hyperecdysonism elicited by the agonists is induction of enzymes of ecdysteroid inactivation (Williams *et al.,* 2002).

The irreversible inactivation of numerous plant and insect hormones occurs by 26-hydroxylation, which is apparently an example of an evolutionarily conserved mechanism (Meaney, 2005). However, all the P450 enzymes involved may not show a close phylogenetic relationship (Guittard *et al.*, 2011). The cytochrome P450 enzyme, CYP18A1, had been long suspected of catalyzing ecdysteroid 26-hydroxylation and has been cloned and characterized in the Lepidoptera, *S. littoralis*, and *M. sexta* (Davies *et al.*, 2006). In *S. littoralis*, the mRNA transcript was expressed at times of increasing ecdysteroid titer in final instar larvae and was induced in midgut and fat body by the ecdysteroid receptor agonist, RH-5992 (tebufenozide). Surprisingly, in both *S. littoralis* and *M. sexta*, transcript expression was also detected in the prothoracic glands, a major ecdysteroid biosynthetic tissue, at a time of increasing ecdysteroid titer. In *B. mori* as well, the expression of *Cyp18a1* during larval development closely coincided with the ecdysteroid titer in the hemolymph (Ai *et al.*, 2008). However, recently demonstration that CYP18A1 catalyzes conversion of ecdysteroids into their 26-hydroxy derivatives and of these into the 26-oic acids has been furnished in *Drosophila* (Guittard *et al.*, 2011). When *Drosophila Cyp18a1* was transfected in *Drosophila* S2 cells, extensive conversion of 20E into 20E-oic acid was observed, with detection of much less 20,26E, a transformation that is also obtained in short times by membrane preparations of the transfected S2 cells. Since transfected cells also converted 20,26E into 20Eoic acid, the *Cyp18a1* encodes a 26-hydroxylase enzyme that also catalyzes further oxidation of the 26-hydroxy group to the carboxylic acid. This is analogous to the situation during oxidation of the terminal primary hydroxyl group in various vertebrate steroids, for example, in the case of rabbit liver cytochrome P450 enzyme CYP27, which catalyzes the complete conversion of 3α,7α,12α-trihydroxy-5β-cholestane into 3α,7α,12α-trihydroxy-5β-cholestanoic acid (Beltsholtz and Wikvall, 1995). The fact that 20E conversion to 20E-oic acid by CYP18A1 was efficient, with little accumulation of 20,26E and that exogenous 20,26E was poorly converted suggests that, in *D. melanogaster*, the multistep conversion proceeds with a limited release of intermediates. *Cyp18a1* transcript expression is higher in early third larval stage (when ecdysteroids are low) than at the end of that instar (when ecdysteroid titers peak). Thus, *Cyp18a1* transcript levels fluctuate in an opposite manner to the hormone titers and expression of the genes encoding the biosynthetic enzymes. At the time of high *Cyp18a1* expression early in the third larval instar, its strong expression was demonstrated in many ecdysteroid target tissues, including the epidermis, fat body, salivary glands, eye-antenna imaginal discs, and was also observed in gastric cecae and midgut, but not in hindgut and Malpighian tubules (Guittard *et al.*, 2011). Interestingly, *Cyp18a1* expression was also detected in the prothoracic cells in the ring gland, but only early in the third instar. As alluded to earlier, CYP18A1 activity in Lepidoptera is known to be induced by 20E and agonists (Davies *et al.*, 2006). In *D. melanogaster*, *Cyp18a1* expression peaks at the beginning of the third larval instar

during basal ecdysteroid levels, several hours after the peak of 20E in the second larval instar (Guittard *et al.*, 2011). A similar delayed induction was also observed following injection of ecdysteroid agonists in Lepidoptera (Chen *et al.*, 1994b; Williams *et al.*, 1997, 2000) and may be due to involvement of another factor in *in vivo* regulation of CYP18A1.

As indicated earlier, there are several ecdysteroid inactivation systems besides formation of ecdysteroid-oic acids, for example, oxidation and epimerization at C-3 and/or conjugation. Despite this apparent redundancy, *Cyp18a1* is required for *Drosophila* metamorphosis. After ubiquitous inactivation of *Cyp18a1* by RNAi, or in *Cyp18a1* null mutants, massive pupal lethality is observed that coincides with an extended pupal ecdysteroid titer peak, which indicates that CYP18A1 is required for catabolism of 20E. These data suggest that CYP18A1 is not essential for early development and that it primarily contributes to the regulation of ecdysteroid titer during the third larval instar and the onset of metamorphosis. Although *Cyp18a1* is expressed in late embryos and first and second instar larvae, lack of an early stage observable phenotype could be explained by the occurrence of alternative catabolic reactions or direct excretion of 20E. Furthermore, whereas 20E could be excreted by open larval systems, it might be toxic when retained in the closed pupal system. That overexpression of CYP18A1 is lethal to the embryo with rescue of the phenotype by 20E, which is presumably explained simply by removal of essential 20E for embryogenesis (Guittard *et al.*, 2011). Surprisingly, there is transient expression of *Cyp18a1* in the ring gland of *Drosophila*, analogous to that reported in the prothoracic glands of two lepidopteran species (Davies *et al.*, 2006). Specific knockdown of *Cyp18a1* in *Drosophila* ring gland induced lethality at the third larval instar and pupal stages, suggesting that CYP18A1 is involved in an aspect of regulation of ecdysteroid homeostasis directly in the ring gland (Guittard *et al.*, 2011). Putative *Cyp18a1* orthologues occur in many insect and crustacean species, indicating that the gene is well conserved in these groups.

4.5.3.2.3. 3-epimerization   An oxygen and NAD(P)H-dependent "epimerase" activity was originally reported in *Manduca* midgut cytosol, which irreversibly converted E into its 3α-isomer (3-epiecdysone; Nigg *et al.*, 1974; Mayer *et al.*, 1979). Subsequently, this transformation has been shown to occur via the oxygen-dependent E oxidase catalyzed formation of 3DE, which is then irreversibly reduced to 3-epiecdysone, catalyzed by NAD(P)H-dependent 3-dehydroecdysteroid 3α-reductase (**Figure 18**; *Pieris*: Blais and Lafont, 1984; *Spodoptera*: Milner and Rees, 1985; *Manduca*: Weirich, 1989).

The 3-epiecdysteroid product may also undergo phosphoconjugation. However, many tissues also contain an NAD(P)H-dependent 3-dehydroecdysteroid

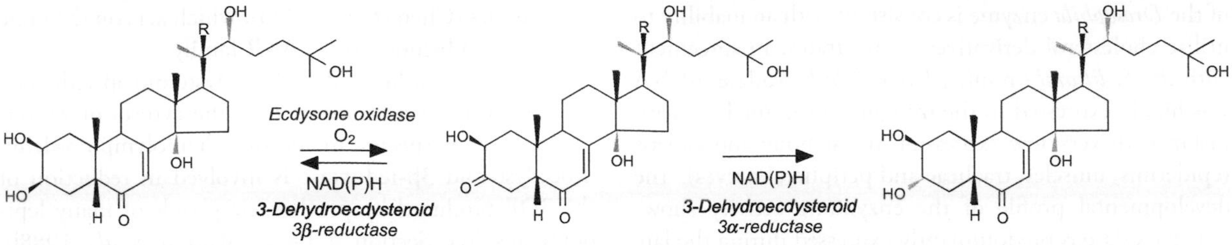

Figure 18 Enzymatic interconversion of 3β-ecdysteroids, 3-dehydroecdysteroids, and 3α-ecdysteroids. R = H: Ecdysone (metabolites); R = OH: 20-hydroxyecdysone (metabolites). Modified from *Rees* (1995).

3β-reductase, which reduces the 3DE back to E (see Lafont and Koolman, 1984); the exact significance of this apparent futile cycle is unclear. The foregoing pathways occur with E and 20E and have been reviewed (Lafont and Koolman, 1984; Koolman and Karlson, 1985; Thompson *et al.*, 1985a; Weirich, 1989; Rees, 1995). It is apparent from the foregoing sections of this chapter that 3-epiecdysteroid formation is a widespread ecdysteroid metabolic pathway among species, with the 3-dehydro-ecdysteroid 3α-reductase apparently restricted to gut tissues. 3-Epiecdysteroid formation is certainly emphasized in Lepidoptera.

The kinetic properties and developmental expression of E oxidase, 3-dehydroecdysteroid 3α-reductase, and 3β-reductase enzymes from midgut cytosol of *Manduca* (Weirich *et al.*, 1989, 1991, 1993) and *Spodoptera* have been investigated (Webb *et al.*, 1995, 1996). In the former species, the larval midgut was the only organ exhibiting substantial specific activities of E oxidase and 3α-reductase, with both activities increasing up to the seventh day after ecdysis (Weirich *et al.*, 1989). Hemolymph and fat body had only moderate to high 3β-reductase and low E oxidase activities, whereas the epidermis did not contain significant activities of any of the enzymes. Thus, apparently, only the larval midgut has a role in the inactivation of ecdysteroids by 3-epimeriza-tion in *Manduca*.

It has been shown that administration of the hormone agonist, RH5849, but not of 20E, to *Manduca* results in RNA- and protein-dependent induction of the midgut cytosol E oxidase and 3-epiecdysone phosphotransferase activities (Williams *et al.*, 1997). Clearly, elucidation of the significance of these results requires further work.

Ecdysone oxidase from *Calliphora* was among the first enzymes catalyzing ecdysteroid transformations to be investigated extensively (Koolman, 1978; Koolman and Karlson, 1978; Koolman, 1985) and has been purified considerably (Koolman and Karlson, 1978). In *C. vicina*, injected 3DE is rapidly reduced *in vivo*, but relative forma-tion of E and 3-epiecdysone cannot be ascertained since the TLC fractionation system employed lacks resolution (Karlson and Koolman, 1973). The reaction catalyzed by E oxidase (EC 1.1.3.16) may be written as:

$$E + O_2 = 3\text{-dehydroecdysone} + H_2O_2$$

The enzyme would be expected to possess a prosthetic group, but none has been identified so far. E oxidase efficiently oxidizes a number of ecdysteroids, but not 22,25-dideoxyecdysone nor cholesterol (Koolman and Karlson, 1978; Koolman, 1985). E oxidase represents the only oxidase in eukaryotic animals known to catalyze oxy-gen-dependent oxidation of steroids; in contrast, oxida-tion of steroids in vertebrates occurs via NAD(P)+-linked dehydrogenases.

The native enzyme from *Calliphora* has an apparent molecular mass of approximately 240 kDa (Koolman, 1985). However, native E oxidase purified from midgut cytosol of *Spodoptera* might be a trimer (approximately 190 kDa), since the apparent Mr of the oxidase sub-unit was 64 kDa by SDS-PAGE (Chen *et al.*, 1999a). The latter was corroborated by cloning the cDNA (2.8 kb) encoding a 65 kDa protein (Takeuchi *et al.*, 2001). Northern blotting demonstrated that the mRNA tran-script was expressed in midgut during the pre-pupal stage of the last larval instar at a time corresponding to an ecdysteroid titer peak. Conceptual translation of the cDNA followed by database searching indicated that the enzyme is an FAD flavoprotein that belongs to the glucose-methanol-choline oxidoreductase superfamily. Injection of the agonist, RH-5992, induced the tran-scription of E oxidase, suggesting that it is an ecdysteroid responsive gene. The gene encoding the enzyme consists of five exons and sequences similar to the binding motifs for Broad-Complex and FTZ-F1 that occur in the 5′ flanking region, consistent with ecdysteroid regulation. Southern blotting indicated that E oxidase is encoded by a single-copy gene. The E oxidase gene from *D. melano-gaster* has also been cloned and characterized (Takeuchi *et al.*, 2005). Despite availability of the *Spodoptera* sequence, species hopping to *D. melanogaster* was non-trivial. A BLAST search of the *D. melanogaster* genome with the *S. littoralis* sequence produced no fewer than 15 candidates, but using complementary bioinformatics approaches, phylogenomics and model structure analy-sis, they were reduced to one favored candidate that was shown to encode E oxidase by expression in COS7 cells. Although *Drosophila* E oxidase shares only 27% amino acid sequence identity with the *Spodoptera* protein, key substrate-binding residues are all well conserved. A model of the *Drosophila* enzyme is consistent with an inability to utilize cholesterol derivatives as substrates. In agreement with the *S. littoralis* results, *Drosophila* E oxidase mRNA was highly expressed in the midgut during the last larval instar, with very low expression in fat body and carcass (epidermis, muscles, tracheae and peripheral nerves). The developmental profile of the enzymatic activity shows that E oxidase is predominantly expressed during the late stage of each larval instar with the expression profile of the mRNA corresponding closely, suggesting that the *Drosophila* enzyme is regulated at the transcriptional level. The developmental profile of *S. littoralis* E oxidase activity is also similar to that for *D. melanogaster*. There is high expression of E oxidase in the late stage of larval instars, when the ecdysteroid titer is high, suggesting that the enzyme may play a role in modification of endogenous ecdysteroids. mRNA for *Drosophila* E oxidase is induced by the ecdysteroid agonist, RH-0345, analogous to induction of enzyme activity by agonists in lepidopteran species. It is likely that expression of the *Drosophila* E oxidase gene may be induced by endogenous ecdysteroids, since its expression during metamorphosis was significantly suppressed in EcR (ecdysone receptor) mutant larvae (Li and White, 2003). The provocative possibility has been considered that *Drosophila* E-oxidase could be involved in ligand activation in relation to evidence suggesting that 3-dehydro- and 3-epi-ecdysteroids may be functionally active as ligands in a novel, atypical ecdysteroid signaling pathway involving the *Drosophila* orphan nuclear receptor, DHR 38, in addition to being merely hormone inactivation products (Takeuchi *et al.*, 2005).

One major 3-dehydroecdysteroid 3α-reductase (active with both NADPH and NADH) together with three major 3-dehydroecdysteroid 3β-reductases (active only with NADPH as cosubstrate) have been fractionated from *Manduca* midgut cytosol (Weirich and Svoboda, 1992). In the case of *Spodoptera*, two forms of 3-dehydroecdysone 3α-reductase have been observed during purification: a 26 kDa form that may be a trimer with an apparent Mr of approximately 76 kDa and a second 51 kDa form that appears to be a monomer by gel filtration (Chen *et al.*, 1999a). The cDNA (1.2 kb) encoding the 26 kDa protein has been cloned and Northern blotting showed that the mRNA transcript was expressed in Malpighian tubules during the early stage of the last larval instar (Takeuchi *et al.*, 2000). Conceptual translation of the cDNA followed by database searching revealed that the enzyme belongs to the short chain dehydrogenases/reductases (SDR) superfamily. Furthermore, the enzyme is a novel eukaryotic 3-dehydrosteroid 3α-reductase member of that family, whereas vertebrate 3-dehydrosteroid 3α-reductases belong to the aldo-keto reductase (AKR) superfamily. Surprisingly, no similarity was observed between the 3-dehydroecdysone 3α-reductase, and a reported 3-dehydroecdysone 3β-reductase from the

same species (Chen *et al.*, 1999b), which acts on the same substrate and belongs to the AKR family.

As alluded to earlier, 3-dehydroecdysteroid 3β-reductase activity has been demonstrated in the cytosol of various tissues of many species. In addition, hemolymph 3-dehydroecdysteroid 3β-reductase is involved in reduction of the 3DE produced by prothoracic glands of many lepidopterans (see Section 4.3.4.2.; Warren *et al.*, 1988b; Sakurai *et al.*, 1989; Kiriishi *et al.*, 1990).

Hemolymph NAD(P)H-dependent 3-dehydroecdysone 3β-reductase activity has been shown to undergo developmental fluctuation in several species, exhibiting high activity at the time of the peak in ecdysteroid titer near the end of the last larval instar (*Manduca*: Sakurai *et al.*, 1989; *Spodoptera*: Chen *et al.*, 1996; *P. americana*: Richter *et al.*, 1999). In *P. americana*, some enzyme is located on the outer surface of the prothoracic gland cells as well. The exact source of the hemolymph 3-dehydroecdysone 3β-reductase and the mechanism regulating its expression during development have not been elucidated. The 3β-reductase has a wide tissue expression (*Pieris*: Blais and Lafont, 1984; European corn borer, *O. nubilalis*: Gelman *et al.*, 1991; *Bombyx*: Nomura *et al.*, 1996), including embryonated eggs of the gypsy moth, *Lymantria* (Kelly *et al.*, 1990). Of significance to ecdysteroid biosynthesis is the demonstration that various 3-dehydroecdysteroid precursors may be reduced to the corresponding 3β-hydroxy compounds in prothoracic glands and follicle cells of *Locusta* (Dollé *et al.*, 1991; Rees, 1995).

The 3-dehydroecdysteroid 3β-reductases have been purified from hemolymph of *Bombyx* (Nomura *et al.*, 1996) and *Spodoptera* (Chen *et al.*, 1996), demonstrating that the native enzyme is a monomer of Mr of approximately 42 kDa and 36 kDa, respectively. In both enzymes, a lower K_M was observed with NADPH than with NADH as cosubstrate (*B. mori*, 0.90 μM vs. 5.4 μM; *S. littoralis*, 0.94 μM vs. 22.8 μM). Kinetic analysis of the *S. littoralis* hemolymph 3β-reductase, using either NADPH or NADH as cofactor, revealed that the enzyme exhibits maximal activity at low 3DE substrate concentrations with a drastic inhibition of activity at higher concentrations (>5 μM). Similar substrate inhibition was observed for the cytosolic NADPH-dependent 3-dehydroecdysone 3α-reductase of *Manduca* midgut, which showed apparent 3DE substrate inhibition above 10 μM (Weirich *et al.*, 1989). Although the K_M for the *S. littoralis* hemolymph 3-dehydroecdysone 3β-reductase has not been determined due to the substrate inhibition, it is conceivable that the K_M value could be less than 5 μM. The results suggest that the 3β-reductase has a high-affinity (low K_M) binding site for the 3-dehydroecdysteroid substrate, together with a lower affinity inhibition site (Chen *et al.*, 1996). Cloning of the full-length 3-dehydroecdysone 3β-reductase cDNA from *S. littoralis* and conceptual translation yield a protein with a predicted Mr of 39 591 of which the first 17 amino acids appear to constitute a signal peptide to yield a predicted Mr for the mature protein of 37 689 (Chen *et al.*, 1999b), which agrees with the approx. Mr (36 kDa) of the purified protein on SDS-PAGE. Similarly, database searching suggested that the enzyme is a new member of the third (AKR) superfamily of oxidoreductases.

Northern blot analysis revealed that highest expression is detected in Malpighian tubules, followed by midgut and fat body with low-level expression in hemocytes and none detectable in central nervous system. Surprisingly, preliminary results of Western blotting experiments revealed 3β-reductase detection in hemolymph and hemocytes with little in midgut and none in Malpighian tubules in agreement with the results for *B. mori* (Nomura *et al.*, 1996). This suggests that selective translation of mRNA may contribute to regulation of expression of 3-dehydroecdysteroid 3β-reductase (Chen *et al.*, 1999b). The developmental profile of the mRNA revealed that the gene is only transcribed in the second half of the sixth instar essentially at the time of increasing enzymatic activity and ecdysteroid titer. Thus, it appears that the change in 3β-reductase activity is, at least in part, regulated by gene expression at the transcriptional level, although post-transcriptional regulation may also be important in certain tissues. Southern analysis indicates that the 3-dehydroecdysone 3β-reductase is encoded by a single gene, which probably contains at least one intron (Chen *et al.*, 1999b). Recently, a gene encoding a protein that shows 42.5% identity to *S. littoralis* 3-dehydroecdysteroid 3β-reductase has been cloned from the cabbage looper, *T. ni* and the protein has been shown to possess low 3β-reductase activity (Lundström *et al.*, 2002). Northern blotting has indicated that highest expression of the gene is detected in fat body; the expression is highly induced after bacterial challenge. However, by immunohistochemistry, the protein was localized exclusively in the epidermis and the cuticle, but its significance is unclear.

4.5.3.2.4. 22-glycosylation

The formation of 26-hydroxyecdysone 22-glucoside in late embryos of *Manduca* has already been alluded to (Thompson *et al.*, 1987a), and the aphid, *M. persicae*, excretes a 22-glucoside conjugate of ingested [³H]20E (Rharrabe *et al.*, 2007). In other cases, the formation of glucosides is performed by viruses.

In the baculovirus, *Autographa californica* nuclear polyhedrosis virus, expression of the egt gene produces UDP-glycosyl transferase, which apparently inactivates the host ecdysteroids by catalyzing production of ecdysteroid 22-glycoside. Thus, egt gene expression allows the virus to interfere with insect development by blocking molting in infected larvae of the fall armyworm, *S. frugiperda* (O'Reilly and Miller, 1989; O'Reilly *et al.*, 1991). Although kinetic analysis indicates that the ecdysteroid UDP-glucosyltransferase (EGT) has broadly similar specificities for UDP-galactose and UDP-glucose (Evans and O'Reilly, 1998), there is evidence that *in vivo*, ecdysteroids

are conjugated with galactose (O'Reilly *et al.*, 1992; O'Reilly, 1995). EGT appears to be an oligomer of 3 to 5 subunits (Mr approx 56 kDa) and, surprisingly, E seems to be the optimal substrate, whereas 3DE is seven times less favorable, with 20E conjugated very poorly, more than 34 times less readily than 3DE (Evans and O'Reilly, 1998). The physiological significance of this finding is uncertain. Analogous EGT enzymes have been reported in several species (Park *et al.*, 1993; Burand *et al.*, 1996; Caradoc-Davies *et al.*, 2001; Clarke *et al.*, 1996; Manzan *et al.*, 2002; Khan *et al.*, 2003; Pinedo *et al.*, 2003). Interestingly, while an entomopoxvirus (AdhoEPV) and a granulovirus (AdhoGV) both prevent pupation of *Adoxophyes honmai* larvae, they appear to employ different mechanisms, although these are unclear at present (Nakai *et al.*, 2004). There appeared to be differences in profiles of both juvenile hormone (JH) esterase and ecdysteroid titers between the two species. The EGT gene promoter from *A. californica* multicapsid nucleopolyhedrovirus has been investigated and the transcriptional activity shown to require transactivation of viral factor(s) (Shen *et al.*, 2004).

Formation of inactive glycosides is not the only way used by parasites to interfere with ecdysteroid titers. Interestingly, the entomogenous fungus, *Nomuraea rileyi*, inhibits molting of its host, *B. mori*, by production of an enzyme that oxidizes the C-22 hydroxyl group of hemolymph ecdysteroids (Kiuchi *et al.*, 2003). Whether this enzyme is an oxidase or an NAD(P)⁺-dependent dehydrogenase has not been established.

4.5.3.2.5. Neglected pathways?

Given the huge diversity of insects, there is no reason to consider that metabolic pathways are restricted to those described earlier. A few examples will illustrate that the story is not finished.

Side chain cleavage between C-20 and C-22 resulting in the formation of poststerone was described long ago in *Bombyx* by Hikino *et al.* (1975); moreover, other experiments have demonstrated the ability of certain insects to produce C_{21} steroids from cholesterol in exocrine glands of aquatic beetles (Schildknecht, 1970) and also ovaries of *Manduca* (Thompson *et al.*, 1985c), providing the evidence for a desmolase activity in some insect species. Available [³H]-ecdysteroids are labeled on the side chain, which has precluded further analysis of this problem as the resulting poststerone would not be labeled and the radioactivity would become "volatile." This problem awaits re-examination when nuclear labeled ecdysteroids are available.

Minor ecdysteroids have been isolated from *Bombyx* that bear an –OH group in positions 16β (Kamba *et al.*, 2000b) or 23 (Kamba *et al.*, 2000a), which indicates that during some developmental stages at least (here in adults) additional CYPs must be active.

14-Deoxy metabolites of E and 20E have been identified in *G. bimaculatus* feces, and their formation is probably due to bacterial metabolism in the insect gut (Hoffmann *et al.*, 1990).

Labeling studies with *Drosophila* adults have shown very complex patterns that include some as yet unidentified metabolites (Grau and Lafont, 1994a,b). Their chromatographic behavior is consistent with the presence of one additional –OH group in position 11β(?), but positions 16β and 23 as observed in *Bombyx* (see previous section)are also plausible.

Finally, it should be noted that sulfate conjugates, although never isolated as endogenous metabolites, can be formed *in vitro* by insect/tissue homogenates incubated with ecdysteroids and a sulfate donor (*Prodenia eridania* gut: Yang and Wilkinson, 1972; *Ae. togoi* whole larvae/pupae/adult: Shampengtong and Wong, 1989; *Sarcophaga peregrina* fat body: Matsumoto *et al.*, 2003). In *Sarcophaga*, a 43 kDa protein responsible for this reaction was purified by affinity chromatography on a 3′-phosphoadenosine 5′-phosphate (PAP)-agarose column (Matsumoto *et al.*, 2003). Similarly, one of the four cytosolic sulfotransferase isoenzymes in *D. melanogaster*, produced in recombinant form, showed a low, but not negligible, activity toward 20E (Hattori *et al.*, 2008). However, the physiological significance of such activities *in vivo* in ecdysteroid metabolism remains to be firmly established, since it is conceivable that the activity observed might be ascribed to another primary activity with low specificity for ecdysteroids.

4.6. Some Prospects for the Future

4.6.1. Evolution and Appearance of Ecdysteroids

Zooecdysteroids are not restricted to arthropods. Trace amounts of E, 20E, and other ecdysteroids have been detected by HPLC-immunoassays or GC-MS in annelids, nematodes, helminths, and mollusks (Spindler, 1988; Franke and Käuser, 1989; Barker *et al.*, 1990). More recently, ecdysteroids were also found in nemerteans (Okazaki *et al.*, 1997). It must be emphasized that at the moment there is no evidence that these animals are able to produce ecdysteroids themselves, as they seem to lack some key biosynthetic enzymes, a conclusion based on metabolic studies performed with biosynthetic intermediates efficiently converted in arthropods (**Table 8**, Lafont *et al.*, 1995; Lafont, 1997). An evolutionary scenario has been proposed for the progressive appearance of enzymes, which resulted in an increasing number of hydroxyl groups on the sterol molecule (the last ones were the three mitochondrial enzymes, i.e., the 22-, 2-, and the 20-hydroxylases). A similar phylogenetic emergence seems to have operated in the case of vertebrate-type steroid hormones (Bolander, 1994).

Several findings, however, do not fit with this scheme. In particular, large amounts of ecdysteroids have been isolated from Zoanthids (e.g., Sturaro *et al.*, 1982; Suksamrarn *et al.*, 2002) and even from various sponges (Diop *et al.*, 1996; Cafieri *et al.*, 1998; Costantino *et al.*, 2000). The large amounts seem to preclude a hormonal role and fit with an allelochemical (protective) role, perhaps against predation, as in the case of pycnogonids (Tomaschko, 1995). With Zoanthids it is conceivable that ecdysteroids arise from their food (e.g., copepods), but this does not seem evident in the case of sponges, which are expected to feed on bacteria, protozoa, or dissolved matter, none of which are known to contain ecdysteroids. This awaits additional experiments to establish whether or not sponges are able to produce ecdysteroids. It should be noted that sponges also contain an incredible number of steroids (Aiello *et al.*, 1999), including steroidal $\Delta^{4,7}$-3, 6-diketones (Malorni *et al.*, 1978) or 4,6,8-triene-3-ones (Kobayashi *et al.*, 1992), which resemble putative ecdysteroid precursors (Grieneisen *et al.*, 1991; Grieneisen, 1994; Gilbert *et al.*, 2002). When looking for oxysterol diversity, we are faced with multiple ways for functionalization of the A/B rings (Lafont and Koolman, 2009). We expect that molecular approaches will allow a search for homologues of insect biosynthetic enzymes that will provide an adequate answer to this fundamental problem in comparative endocrinology. In the same way, a search for ecdysone nuclear receptor homologues was recently performed on mollusks and annelids (Laguerre and Veenstra, 2010), and may help to understand the possible co-evolution of steroidogenic enzymes and steroid receptors, as was suggested for vertebrates (Baker, 2011).

4.6.2. Elucidating the Black Box

Recently, all enzymes involved in already identified reactions (i.e., 7-8-dehydrogenation or hydroxylations at positions 25, 2, 22, and 20) have been identified, thanks to the conjunction of genetic/molecular approaches (identification of *Drosophila* halloween mutants) and to the expression of recombinant enzymes and the determination of their catalytic activity. Whether this approach will allow the identification of the remaining unknown reactions (the so-called black box) is debatable. There are already two candidate genes (*sro* and *spo/spok*) that probably encode enzymes from the black box. Subsequent to the conversion of C to 7dC by the 7,8-dehydrogenase present in the ER, oxidized early intermediates do not accumulate under normal conditions. In theory we may expect to increase their concentrations by using one (or a combination) of the following strategies: (1) the use of mutants defective in one or several terminal enzymes; (2) the use of specific inhibitors of these enzymes (Burger *et al.*, 1988, 1989; Mauvais *et al.*, 1994); or (3) the use of subcellular fractions instead of whole cells/tissues to have only a few biosynthetic enzymes present in the preparation, or alternatively the use of recombinant

Table 8 Presence/absence of some key hydroxylases involved in ecdysteroid biosynthesis and/or metabolism in invertebrates

Animals	Hydroxylations at positions				
	C2	C20	C22	C25	Other
ARTHROPODS					
Insects	+	+	+	+	16β, 23, 26
Crustaceans	+	+	+	+	26
Ticks	+	+	+	+	26
Myriapods	-	+	+	+	?
NON-ARTHROPODS					
Molluscs (Gastropods)	-	+	-	+	16β
Annelids					
Achaetes	-	+	-	+	16β
Polychaetes	-	+	-	+	26
Oligochaetes	-	-	-	+	18
Nematodes	-	?	?	?	?

? : does not mean that the reaction is absent, but only that there is at the moment no direct demonstration for it.
After Lafont et al., 1995.

enzymes. An additional technical problem to overcome will be the instability of such intermediates. This problem has already been encountered when using a tritiated $\Delta^{4,7}$-3,6-diketone (Blais et al., 1996) or α-5,6-epoxy7dC (Warren et al., 1995). Furthermore, recent experiments (see Section 4.4.3.2.) suggest that the long-hypothesized, rate-limiting black box oxidations of 7dC, leading to the Δ^4-diketol, cannot be catalyzed by a mitochondrial P450 enzyme and may even occur outside mitochondria. Studies implicating the mitochondria in these novel transformations employed crude membrane fractions from *Manduca* prothoracic glands (Warren and Gilbert, 1996) and there was clear evidence for the co-sedimentation of a large proportion of microsomes with the crude mitochondrial pellet, as has been observed with other insect prothoracic gland tissues (Kappler et al., 1988). Moreover, insect peroxisomes also normally sediment with the mitochondrial pellet (Kurisu et al., 2003) and mammalian peroxisomes are known to participate in the oxidation of both steroids and fatty acids (Breitling et al., 2001). Thus, it is possible that the oxidation of 7dC (or 3-oxo7dC) may instead occur in this membrane-bound subcellular organelle. In addition, the peroxisome has been shown to be a target of sterol carrier protein 2 (Schroeder et al., 2000). Also, at least in *S. frugiperda* cells, this organelle has been shown to be devoid of catalase activity, a normal marker enzyme in mammalian peroxisomes that acts to destroy hydrogen peroxide (Kurisu et al., 2003). If a normal synthesis of hydrogen peroxide is also observed in insect peroxisomes, then the result could be an ideal source of oxygen for the hypothesized black box oxidations (Gilbert et al., 2002).

Until now, the black box has been investigated through trial-and-error experiments analyzing the metabolic fate of radiolabeled putative intermediates, but this has mainly led to the elimination of several hypotheses (Grieneisen, 1994). In an alternative strategy, the so-called "isotope trap," steroidogenic organs would be incubated with a labeled early precursor in the presence of an excess of an unlabeled putative downstream intermediate. If the latter is indeed an intermediate (and is stable enough to survive the incubation), it is expected to trap radioactivity due to isotopic dilution and saturation of the subsequent steps. In addition, the original protocol using a stable ketal derivative of 3-dehydro-7dC, the 3-o-nitrophenylethyleneglycol ketal of 3-oxo-7dC may give a strong opportunity to overcome the instability problem (Warren et al., 2009). It could also be possible to use systems capable of producing larger amounts of ecdysteroids (other than what can be achieved through the tedious dissection of thousands of prothoracic glands or ring glands); for example, embryos or suitable plant cell or organ cell cultures, which would allow the use of precursors labeled with stable isotopes (^{2}H, ^{13}C) and subsequent NMR analysis of terminal ecdysteroid products (Fujimoto et al., 1997, 2000).

Thus, the problem of the early steps in ecdysteroid biosynthesis remains an open question, even if important progress has been made. In addition, it may well be that different biosynthetic pathways have been developed between insects and plants (Fujimoto et al., 1989, 1997; Reixach et al., 1999), and also within the diverse array of insects and other arthropods. Nevertheless, the black box reactions appear to be quite specific to arthropods (and maybe a few plants). As such, they make a very appealing target for the development of novel biochemical strategies for the control of insect pests.

In addition to the expression of enzymes, the function of steroidogenic cells requires multiple layers of regulation. For instance, although the precise targets remain to be identified, it has recently been demonstrated that in *Drosophila* alteration of the sumoylation pathway in the ring gland alters the capacity to produce ecdysteroids, leading to lethality at metamorphosis (Talamillo et al., 2008). Finally, another yet underinvestigated question is to understand if the secretion of ecdysteroids is passive (as is usually assumed for steroids) or if it may require any specific mechanism of exocytosis. The latter hypothesis was raised because inactivating the expression of *Sec10*, a member of exocytosis machinery specifically in the ring gland, leads to lethality at metamorphosis presumably due to altered ecdysteroid production (Andrews et al., 2002). Taken together, these studies show that besides the elucidation of all enzymes involved in ecdysteroid biosynthesis, the understanding of the function of steroidogenic cells in Arthropods obviously requires a better understanding of a number of other cellular mechanisms.

4.6.3. Additional Functions for Ecdysteroids?

We have discussed that tissues other than prothoracic glands and gonads may produce ecdysteroids, possibly as autocrine factors. Whatever the biosynthetic site, this corresponds to an involvement of ecdysteroids in the control of development and reproduction processes. The recent discovery of large ecdysteroid concentrations in the defensive secretions of chrysomelid beetles (Laurent *et al.*, 2003) suggests a possible allelochemical role of ecdysteroids in certain cases. The situation described in chrysomelid beetles is similar to that of the pycnogonids, where ecdysteroids have been demonstrated conclusively to represent an efficient protection against predation by crabs (Tomaschko, 1995). Another example is the phytoecdysteroid accumulation by many plant species, seen as a protection against phytophagous insects (Dinan, 2001). However, it should be emphasized that previous analyses of defensive secretions of chrysomelid beetles allowed the isolation of 2,14,22-trideoxy-ecdysone 3-sophorose and other ecdysteroid-related compounds lacking the Δ^7 bond (Pasteels *et al.*, 1994). Thus, the story of ecdysteroid biosynthesis could turn out to be considerably more complex than it already appears to be.

Acknowledgments

René Lafont wishes to thank Dr Jean-Pierre Girault for providing the materials of **Figures 2** and **5**. Huw Rees is grateful to the Biotechnology and Biological Sciences Research Council and The Leverhulme Trust for financial support of work in his laboratory, Dr Hajime Takeuchi for preparing **Figures 14–16** and **18** for this review, and Dr Lyndsay Davies for help with the manuscript. This work was supported in part by National Science Foundation Grant IBN0130825.

References

Addya, S., Anandatheerthavarada, H., Biswas, G., Bhagwat, S., Mullick, J., et al. (1997). Targeting of NH2-terminal-processed microsomal protein to mitochondria: A novel pathway for the biogenesis of hepatic mitochondrial P450MT2. *J. Cell Biol.*, *139*, 589–599.

Adelung, D., & Karlson, P. (1969). Eine verbesserte, sehr empfindliche Methode zur biologischen Auswertung des Insektenhormones Ecdyson. *J. Insect Physiol.*, *15*, 1301–1307.

Agosin, M., & Srivatsan, J. (1991). Role of microsomal cytochrome P-450 in the formation of ecdysterone in larval house fly. *Comp. Biochem. Physiol.*, *99B*, 271–274.

Agosin, M., Srivatsan, J., & Weirich, M. (1988). On the intracellular localization of the ecdysone 20-monooxygenase in Musca domestica. *Arch. Insect Biochem. Physiol.*, *9*, 107–117.

Agrawal, O. P., & Scheller, K. (1985). An *in vivo* system has the advantages of an *in vitro* system as well: Blowfly abdominal bags. *Naturwissenschaften*, *72*, 436–437.

Agui, N. (1973). Quantitative bioassay for moulting hormone *in vitro*. *Appl. Entomol. Zool.*, *8*, 236–239.

Ai, J., Wang, G., Li, Y., Yu, Q., et al. (2008). Molecular cloning, sequence analysis and transcriptional activity determination of cytochrome P450 gene CYP18A1 in the silkworm, *Bombyx mori*. *Acta Entomol. Sinica*, *51*, 237–245.

Aiello, A., Fattorusso, E., & Menna, M. (1999). Steroids from sponges: Recent reports. *Steroids*, *64*, 687–714.

Alpy, F., & Tomasetto, C. (2006). MLN64 and MENTHO, two mediators of endosomal cholesterol transport. *Biochem. Soc. Trans.*, *34*, 343–345.

Andrews, H. K., Zhang, Y. Q., Trotta, N., & Broadie, K. (2002). *Drosophila* sec10 is required for hormone secretion but not general exocytosis or neurotransmission. *Traffic*, *3*, 906–921.

Antonucci, R., Bernstein, S., Giancola, D., & Sax, K. (1951). $\Delta 5,7$ steroids. IX. The preparation of $\Delta 4,7$ and $\Delta 4,7,9$ 3-ketosteroid hormones. *J. Org. Chem.*, *16*, 1453–1457.

Asazuma, H., Nagata, S., & Nagasawa, H. (2009). Inhibitory effect of molt-inhibiting hormone on phantom expression in the Y-organ of the kuruma prawn, *Marsupenaeus japonicus*. *Arch. Insect Biochem. Physiol.*, *72*, 220–233.

Awata, N., Morisaki, M., & Ikekawa, N. (1975). Carbon–carbon cleavage of fucosterol-24,28-oxide by cell-free extracts of silkworm *Bombyx mori*. *Biochem. Biophys. Res. Commun*, *64*, 157–161.

Baker, K. D., Warren, J. T., Thummel, C. S., Gilbert, L. I., & Mangelsdorf, D. J. (2000). Transcriptional activation of the *Drosophila* ecdysone receptor by insect and plant ecdysteroids. *Insect Biochem. Mol. Biol.*, *30*, 1037–1043.

Baker, M. E. (2011). Origin and diversification of steroids: Co-evolution of enzymes and nuclear receptors. *Mol. Cell Endocrinol. 334*, 14–20.

Barker, G. C., Chitwood, D. J., & Rees, H. H. (1990). Ecdysteroids in helminths and annelids. *Invert. Reprod. Dev.*, *18*, 1–11.

Báthori, M. (1998). Purification and characterization of plant ecdysteroids of *Silene* species. *Trends Anal. Chem.*, *17*, 372–383.

Báthori, M., Blunden, G., & Kalász, H. (2000). Two-dimensional thin-layer chromatography of plant ecdysteroids. *Chromatographia*, *52*, 815–817.

Báthori, M., & Kalász, H. (2001). Separation methods for phytoecdysteroids. *LC-GC Europe*, *14*, 626–633.

Becker, E., & Plagge, E. (1939). Über das Pupariumbildung auslösende Hormon der Fliegen. *Biol. Zbl.*, *59*, 326.

Bergamasco, R., & Horn, D. H.S. (1980). The biological activities of ecdysteroids and ecdysteroid analogues. In J. A. Hoffmann (Ed.), *Progress in Ecdysone Research* (p. 299). Amsterdam: Elsevier/North-Holland.

Beydon, P., Claret, J., Porcheron, P., & Lafont, R. (1981). Biosynthesis and inactivation of ecdysone during the pupal–adult development of the cabbage butterfly, *Pieris brassicae*. *Steroids*, *38*, 633–650.

Beydon, P., Girault, J. P., & Lafont, R. (1987). Ecdysone metabolism in *Pieris brassicae* during the feeding last larval instar. *Arch. Insect Biochem. Physiol*, *4*, 139–149.

Beydon, P., Permana, A. D., Colardeau, J., Morinière, M., & Lafont, R. (1989). Ecdysteroids from developing eggs of *Pieris brassicae*. *Arch. Insect Biochem. Physiol.*, *11*, 1–11.

Bielby, C. R., Morgan, E. D., & Wilson, I. D. (1986). Gas chromatography of ecdysteroids as their trimethylsilyl ethers. *J. Chromatogr.*, *351*, 57–64.

Björkhem, I. (1969). Stereochemistry of the enzymatic conversion of Δ^4-3-oxosteroid into a 3-oxo-5β-steroid. Bile acids and Steroids 208. *Eur. J. Biochem.*, *7*, 413–417.

Black, S. M., Harikrishna, J. A., Szklarz, G. D., & Miller, W. L. (1994). The mitochondrial environment is required for activity of the cholesterol side-chain cleavage enzyme, cytochrome P450scc. *Proc. Natl. Acad. Sci. USA*, *91*, 7247–7251.

Blackford, M. J.P., Clarke, B. S., & Dinan, L. (1997). Distribution and metabolism of exogenous ecdysteroids in the Egyptian Cotton leafworm *Spodoptera littoralis* (Lepidoptera: Nocturidae). *Arch. Insect Biochem. Physiol.*, *34*, 329–346.

Blackford, M., & Dinan, L. (1997a). The tomato *moth Lacanobia oleracea* (Lepidoptera: Noctuidae) detoxifies ingested 20-hydroxyecdysone, but is susceptible to the ecdysteroid agonists RH-5849 and RH-5992. *Insect Biochem. Mol. Biol.*, *27*, 167–177.

Blackford, M., & Dinan, L. (1997b). The effects of ingested ecdysteroid agonists (20-hydroxyecdysone, RH5849 and RH5992) and an ecdysteroid antagonist (cucurbitacin B) on larval development of two polyphagous lepidopterans *Acherontia atropus* and *Lacanobia oleracea*. *Entomol. Exp. Appl.*, *83*, 263–276.

Blackford, M. J.P., & Dinan, L. (1997c). The effects of ingested 20-hydroxyecdysone on the larvae of *Aglais urticae*, *Inachis io*, *Cynthia cardui* (Lepidoptera: Nymphalidae) and *Tyria jacobeae* (Lepidoptera: Arctiidae). *J. Insect Physiol.*, *43*, 315–327.

Blais, C., Blasco, T., Maria, A., Dauphin-Villemant, C., & Lafont, R. (2010). Characterization of ecdysteroids in *Drosophila melanogaster* by enzyme immunoassay and nano-liquid chromatography-tandem mass spectrometry. *J. Chromatogr. B Analyt. Technol. Biomed. Life Sci.*, *878*, 925–932.

Blais, C., Dauphin-Villemant, C., Kovganko, N., Girault, J.-P., Descoins, C., Jr., et al. (1996). Evidence for the involvement of 3-oxo-Δ^4 intermediates in ecdysteroid biosynthesis. *Biochem. J.*, *320*, 413–419.

Blais, C., & Lafont, R. (1984). Ecdysteroid metabolism by soluble enzymes from an insect: Metabolic relationship between 3β-hydroxy-, 3α-hydroxy-, and 3-oxo-ecdysteroids. *H.-S.Z. Physiol. Chem.*, *365*, 809–817.

Blais, C., & Lafont, R. (1986). Ecdysone 20-hydroxylation in imaginal wing discs of *Pieris brassicae* (Lepidoptera): Correlations with ecdysone and 20-hydroxyecdysone titers in pupae. *Arch. Insect Biochem. Physiol.*, *3*, 501–512.

Böcking, D., Dauphin-Villemant, C., Sedlmeier, D., Blais, C., & Lafont, R. (1993). Ecdysteroid biosynthesis in molting glands of the crayfish *Orconectes limosus*: Evidence for the synthesis of 3-dehydroecdysone by *in vitro* synthesis and conversion studies. *Insect Biochem. Mol. Biol.*, *23*, 57–63.

Böcking, D., Dauphin-Villemant, C., Toullec, J.-Y., Blais, C., & Lafont, R. (1994). Ecdysteroid formation from 25hydroxycholesterol by arthropod molting glands *in vitro*. *C.R. Acad. Sci. Paris*, *317*, 891–898.

Bolander, F. F. (Ed.), (1994). *Molecular Endocrinology* (2nd ed) (p. 501). New York: Academic Press.

Bollenbacher, W., Faux, A. F., Galbraith, N., Gilbert, L. I., Horn, D. H.S., et al. (1979). *In vitro* metabolism of possible ecdysone precursors by prothoracic glands of the tobacco hornworm, *Manduca sexta*. *Steroids*, *34*, 509–526.

Bollenbacher, W., Galbraith, N., Horn, D. H.S., & Gilbert, L. I. (1977a). *In vitro* metabolism of 3β-hydroxy-, and 3β-, 14α-hydroxy-[3α-3H]-5β-cholest-7-en-6-one by the prothoracic glands of *Manduca sexta*. *Steroids*, *29*, 47–63.

Bollenbacher, W., Smith, S., Wielgus, J., & Gilbert, L. I. (1977b). Evidence for an α-ecdysone cytochrome P-450 mixed function oxidase in insect fat body mitochondria. *Nature*, *268*, 660–663.

Bondar, O. P., Saad, L. M., Shatursky, Ya. P., & Kholodova, Yu. D. (1993). Comparative characteristics of color reactions of ecdysteroids. *Ukr. Biokhim. Zh.*, *65*(4), 83–87.

Borst, D. W., & Engelmann, F. (1974). *In vitro* secretion of α-ecdysone by prothoracic glands of a hemimetabolous insect, *Leucophaea maderae* (Blattaria). *J. Exp. Zool.*, *189*, 413–419.

Borst, D. W., & O'Connor, J. D. (1972). Arthropod molting hormone: Radioimmune assay. *Science*, *178*, 418–419.

Borst, D. W., & O'Connor, J. D. (1974). Trace analysis of ecdysones by gas-liquid chromatography, radioimmunoassay and bioassay. *Steroids*, *24*, 637–656.

Bownes, M. (1992). Why is there sequence similarity between insect yolk proteins and vertebrate lipases? *J. Lipid Res.*, *33*, 677–690.

Bownes, M., Shirras, A., Blair, M., Collins, J., & Coulson, A. (1988). Evidence that insect embryogenesis is regulated by ecdysteroids released from yolk proteins. *Proc. Natl. Acad. Sci. USA*, *85*, 1554–1557.

Breitling, R., Marijanovic, Z., Perovic, D., & Adamski, J. (2001). Evolution of 17 beta-HSD type 4, a multifunctional protein of beta-oxidation. *Mol. Cell. Endocrinol.*, *171*, 205–210.

Brown, G. D. (1998). The biosynthesis of steroids and triterpenoids. *Nat. Prod. Rep.*, *15*, 653–696.

Brown, M. R., Sieglaff, D. H., & Rees, H. H. (2009). Gonadal ecdysteroidogenesis in Arthropoda: Occurrence and regulation. *Annu. Rev. Entomol.*, *54*, 105–125.

Bückmann, D. (1984). The phylogeny of hormones and hormonal systems. *Nova Acta Leopoldina NF*, *56*, 437–452.

Burand, J. P., Park, E. J., & Kelly, T. J. (1996). Dependence of ecdysteroid metabolism and development in host larvae on the time of baculovirus infection and the activity of the UDP-glucosyltransferase gene. *Insect Biochem. Mol. Biol.*, *26*, 845–852.

Burger, A., Colobert, F., Hétru, C., & Luu, B. (1988). Acetylenic cholesteryl derivatives as irreversible inhibitors of ecdysone biosynthesis. *Tetrahedron*, *44*, 1141–1152.

Burger, A., Roussel, J.-P., Hétru, C., Hoffmann, J. A., & Luu, B. (1989). Allenic cholesteryl derivatives as inhibitors of ecdysone biosynthesis. *Tetrahedron*, *45*, 155–164.

Butenandt, A., & Karlson, P. (1954). Über die Isolierung eines Metamorphosehormons der Insekten in kristallisierter Form. *Z. Naturforsch.*, *9b*, 389–391.

Cafieri, F., Fattorusso, E., & Taglialatela-Scafati, O. (1998). Novel bromopyrrole alkaloids from the sponge Agelas dispar. *J. Nat. Prod.*, *61*, 122–125.

Capovilla, M., Eldon, E. D., & Pirrotta, V. (1992). The giant gene of *Drosophila* encodes a b-ZIP DNA-binding protein that regulates the expression of other segmentation gap genes. *Development, 114*, 99–112.

Caradoc-Davies, K., Graves, S., O'Reilly, D. R., Evans, O. P., & Ward, V. K. (2001). Identification and *in vivo* characterization of the *Epiphyas postvittana* nucleopolyhedrovirus ecdysteroid UDP-glucosyltransferase. *Virus Genes, 22*, 255–264.

Cassier, P., Baehr, J.-C., Caruelle, J.-P., Porcheron, P., & Claret, J. (1980). The integument and ecdysteroids: *in vivo* and *in vitro* studies. In J. A. Hoffmann (Ed.), *Progress in Ecdysone Research* (pp. 235–246). Amsterdam: Elsevier/ North-Holland Biomedical Press.

Chávez, V. M., Marqués, G., Delbecque, J.-P., Kobayashi, K., Hollingsworth, M., et al. (2000). The *Drosophila* disembodied gene controls late embryonic morphogenesis and codes for a cytochrome P450 enzyme that regulates embryonic ecdysone levels. *Development, 127*, 4115–4126.

Chen, J.-H., Hara, T., Fisher, M. J., & Rees, H. H. (1994a). Immunological analysis of developmental changes in ecdysone 20-monooxygenase expression in the cotton leafworm, *Spodoptera littoralis. Biochem. J., 299*, 711–717.

Chen, J.-H., Kabbouh, M., Fisher, M. J., & Rees, H. H. (1994b). Induction of an inactivation pathway for ecdysteroids in larvae of the cotton leafworm, *Spodoptera littoralis. Biochem. J., 301*, 89–95.

Chen, J.-H., Powls, R., Rees, H. H., & Wilkinson, M. C. (1999a). Purification of ecdysone oxidase and 3-dehydroecdysone 3β-reductase from the cotton leafworm, *Spodoptera littoralis. Insect Biochem. Mol. Biol., 29*, 899–908.

Chen, J.-H., Turner, P. C., & Rees, H. H. (1999b). Molecular cloning and characterization of hemolymph 3-dehydroecdysone 3β-reductase from the cotton leafworm, *Spodoptera littoralis* —a new member of the third superfamily of oxidoreductases. *J. Biol. Chem., 274*, 10551–10556.

Chen, J.-H., Webb, T. J., Powls, R., & Rees, H. H. (1996). Purification and characterization of hemolymph 3-dehydroecdysone 3β-reductase in relation to ecdysteroid biosynthesis in the cotton leafworm *Spodoptera littoralis. Eur. J. Biochem., 242*, 394–401.

Chen, Y., Jakoncic, J., Wang, J., Zheng, X., Carpino, N., & Nassar, N. (2008). Structural and functional characterization of the C-terminal domain of the ecdysteroid phosphate phosphatase from *Bombyx mori* reveals a new enzymatic activity. *Biochemistry, 47*, 12135–12145.

Cherbas, L., Yonge, C. D., Cherbas, P., & Williams, C. M. (1980). The morphological response of Kc-H cells to ecdysteroids: Hormonal specificity. *Roux Arch. Dev. Biol., 189*, 1–15.

Chino, H., Sakurai, S., Ohtaki, T., Ikekawa, N., Miyazaki, H., et al. (1974). Biosynthesis of α-ecdysone by prothoracic glands *in vitro. Science, 183*, 529–530.

Choe, S., Dilkes, B. P., Gregory, B. D., Ross, A. S., Yuan, H., et al. (1999). The *Arabidopsis* dwarf1 mutant is defective in the conversion of 24-methylenecholesterol to campesterol in brassinosteroid biosynthesis. *Plant Physiol., 119*, 897–907.

Christiaens, O., Iga, M., Velarde, R. A., Rougé, P., & Smagghe, G. (2010). Halloween genes and nuclear receptors in ecdysteroid biosynthesis and signalling in the pea aphid. *Insect Mol. Biol., 19*, 187–200.

Chung, H., Sztal, T., Pasricha, S., Sridhar, M., Batterham, P., & Daborn, P. J. (2009). Characterization of *Drosophila melanogaster* cytochrome P450 genes. *Proc. Natl. Acad. Sci. USA, 106*, 5731–5736.

Clark, A. J., & Bloch, K. (1959). Conversion of ergosterol to 22-dehydrocholesterol in *Blatella germanica. J. Biol. Chem., 234*, 2589–2593.

Clarke, E. E., Tristem, M., Cory, J. S., & O'Reilly, D. R. (1996). Characterization of the ecdysteroid UDP-glucosyltransferase gene from *Mamestra brassicae* nucleopolyhedrovirus. *J. Gen. Virol., 77*, 2865–2871.

Clarke, G. S., Baldwin, B. C., & Rees, H. H. (1985). Inhibition of the fucosterol-24(28)-epoxide cleavage enzyme of sitosterol dealkylation in *Spodoptera littoralis* larvae. *Pestici. Biochem. Physiol., 24*, 220–230.

Claudianos, C., Ranson, H., Johnson, R. M., et al. (2006). A deficit of detoxification enzymes: pesticide sensitivity and environmental response in the honeybee. *Insect Mol. Biol., 15*, 615–636.

Clayton, R. B. (1964). The utilization of sterols by insects. *J. Lipid Res., 5*, 3–19.

Clément, C., Bradbrook, D. A., Lafont, R., & Dinan, L. N. (1993). Assessment of a microplate-based bioassay for the detection of ecdysteroid-like or antiecdysteroid activities. *Insect Biochem. Mol. Biol., 23*, 187–193.

Clément, C., & Dinan, L. N. (1991). Development of an assay for ecdysteroid-like and anti-ecdysteroid activities in plants. In I. Hardy (Ed.), *Insect Chemical Ecology* (pp. 221–226). The Hague: Academia Prague and SBP Academic Publishers.

Costantino, V., Dell'Aversano, C., Fattorusso, E., & Mangoni, A. (2000). Ecdysteroids from the Caribbean sponge Iotrochota birotulata. *Steroids, 65*, 138–142.

Crosby, T., Evershed, R. P., Lewis, D., Wigglesworth, K. P., & Rees, H. H. (1986a). Identification of ecdysone 22-longchain fatty acyl esters in newly laid eggs of the cattle tick *Boophilus microplus. Biochem. J., 240*, 131–138.

Crosby, T., Wigglesworth, K. P., Lewis, D., & Rees, H. H. (1986b). Moulting hormones in the development of the cattle tick, Boophilus microplus. In D. Borovsky, & A. Spielman (Eds.), *Proceedings of the Vero Beach Symposium; Host Regulated Developmental Mechanisms in Vector Arthropods* (pp. 37–45). Florida: University of Florida.

Cymborowski, B. (1989). Bioassay of ecdysteroids. In J. Koolman (Ed.), *Ecdysone, From Chemistry to Mode of Action* (pp. 144–149). Stuttgart: Georg Thieme Verlag.

Darvas, B., Rees, H. H., & Hoggard, N. (1993). Ecdysone 20-monooxygenase systems in flesh-flies (Diptera: Sarcophagidae), *Neobellieria bullata and Parasarcophaga argyrostoma. Comp. Biochem. Physiol., 105B*, 765–773.

Darvas, B., Rees, H. H., Hoggard, N., El-Din, M. H.T., Kuwano, E., et al. (1992). Cytochrome P-450 inducers and inhibitors interfering with ecdysone 20-monooxygenases and their activities during postembryonic development of *Neobellieria bullata* Parker. *Pestici. Sci., 36*, 135–142.

Dauphin-Villemant, C., Blais, C., & Lafont, R. (1998). Towards the elucidation of the ecdysteroid biosynthetic pathway. *Ann. NY Acad. Sci., 839*, 306–310.

Dauphin-Villemant, C., Böcking, D., Blais, C., Toullec, J.-Y., & Lafont, R. (1997). Involvement of a 3β-hydroxysteroid dehydrogenase in ecdysteroid biosynthesis. *Mol. Cell Endocrinol.*, *128*, 139–149.

Dauphin-Villemant, C., Böcking, D., & Sedlmeier, D. (1995). Regulation of ecdysteroidogenesis in crayfish molting glands: Involvement of protein synthesis. *Mol. Cell Endocrinol.*, *109*, 97–103.

Dauphin-Villemant, C., Böcking, D., Tom, M., Maïbèche, M., & Lafont, R. (1999). Cloning of a novel cytochrome P-450 (CYP4C15) differentially expressed in the steroidogenic glands of an arthropod. *Biochem. Biophy. Res. Commun.*, *264*, 413–418.

Davies, L., Williams, D. R., Turner, P. C., & Rees, H. H. (2006). Characterization in relation to development of an ecdysteroid agonist-responsive cytochrome P450, CYP18A1, in Lepidoptera. *Arch. Biochem. Biophys.*, *453*, 2–10.

Davies, L., Anderson, I. P., Turner, P. C., et al. (2007). An unsuspected ecdysteroid/steroid phosphatase activity in the key T-cell regulator, Sts-1: Surprising relationship to insect ecdysteroid phosphate phosphatase. *Proteins: Structure, Function and Bioinformatics*, *67*, 720–731.

Davies, T. G., Dinan, L. N., Lockley, W. J.S., Rees, H. H., & Goodwin, T. W. (1981). Formation of the A/B cis ring junction of ecdysteroids in the locust, *Schistocerca gregaria*. *Biochem. J.*, *194*, 53–62.

Davis, P., Lafont, R., Large, T., Morgan, E. D., & Wilson, I. D. (1993). Micellar capillary electrophoresis of the ecdysteroids. *Chromatographia*, *37*, 37–42.

Defaye, G., Mornier, N., Guidicelli, C., & Chambaz, E. M. (1982). Phosphorylation of purified mitochondrial cytochromes P-450 (cholesterol desmolase and 11β-hydroxylase) from bovine adrenal cortex. *Mol. Cell Endocrinol.*, *27*, 157–168.

Delbecque, J. P., Weidner, K., & Hoffmann, K.-H. (1990). Alternative sites for ecdysteroid production in insects. *Invert. Reprod. Dev.*, *18*, 29–42.

Delbecque, J.-P., Beydon, P., Chapuis, L., & Corio-Costet, M.-F. (1995). *in vitro* incorporation of radiolabeled cholesterol and mevalonic acid into ecdysteroid by hairy root cultures of a plant, *Serratula tinctonia*. *Eur. J. Entomol.*, *92*, 301–307.

Dinan, L. (1985). Ecdysteroid receptors in a tumorous blood cell line of *Drosophila melanogaster*. *Arch. Insect Biochem. Physiol.*, *2*, 295–317.

Dinan, L. (1988). The chemical synthesis of ecdysone 22-long-chain fatty acyl esters in high yield. *J. Steroid Biochem.*, *31*, 237–245.

Dinan, L. (1989). Ecdysteroid structure and hormonal activity. In J. Koolman (Ed.), *Ecdysone, From Chemistry to Mode of Action* (pp. 345–354). Stuttgart: Georg Thieme Verlag.

Dinan, L. (1997). Ecdysteroids in adults and eggs of the house cricket, *Acheta domesticus* (Orthoptera: Gryllidae). *Comp. Biochem. Physiol.*, *116B*, 129–135.

Dinan, L. (2001). Phytoecdysteroids: Biological aspects. *Phytochemistry*, *57*, 325–339.

Dinan, L., Hormann, R. E., & Fujimoto, T. (1999). An extensive ecdysteroid CoMFA. *J. Comp. Aid. Mol. Design*, *13*, 185–207.

Dinan, L. N., Donnahey, P. L., Rees, H. H., & Goodwin, T. W. (1981). High-performance liquid chromatography of ecdysteroids and their 3-epi, 3-dehydro and 26-hydroxy derivatives. *J. Chromatogr.*, *205*, 139–145.

Dinan, L. N., & Rees, H. H. (1978). Preparation of 3-epi-ecdysone and 3-epi-20-hydroxyecdysone. *Steroids*, *32*, 629–638.

Dinan, L. N., & Rees, H. H. (1981a). The identification and titers of conjugated and free ecdysteroids in developing ovaries and newly-laid eggs of Schistocerca gregaria. *J. Insect Physiol.*, *27*, 51–58.

Dinan, L. N., & Rees, H. H. (1981b). Incorporation *in vivo* of [4-^{14}C]-cholesterol into the conjugated ecdysteroids in ovaries and eggs of Schistocerca gregaria. *Insect Biochem.*, *11*, 255–265.

Diop, M., Samb, A., Costantino, V., Fattorusso, E., & Mangoni, A. (1996). A new iodinated metabolite and a new alkyl sulfate from the Senegalese sponge Ptilocaulis spiculifer. *J. Nat. Prod.*, *59*, 271–272.

Dollé, F., Hétru, C., Roussel, J. P., Rousseau, B., Sobrio, F., et al. (1991). Synthesis of a tritiated 3-dehydroecdysteroid putative precursor of ecdysteroid biosynthesis in *Locusta migratoria*. *Tetrahedron*, *47*, 7067–7080.

Dorn, A., Bishoff, S. T., & Gilbert, L. I. (1987). An incremental analysis of the embryonic development of the tobacco hornworm, *Manduca sexta*. *Int. J. Invert. Reprod. Dev.*, *11*, 137–158.

Dyer, D. H., Lovell, S., Thoden, J. B., Holden, H. M., Rayment, I., et al. (2003). The structural determination of an insect sterol carrier protein-2 with a ligand bound C16 fatty acid at 1.35 Å resolution. *J. Biol. Chem.*, *278*, 39085–39091.

Dyer, D. H., Vyazunova, I., Lorch, J. M., Forest, K. T., & Lan, Q. (2009). Characterization of the yellow fever mosquito sterol carrier protein-2 like 3 gene and ligand-bound protein structure. *Mol. Cell Biochem.*, *326*, 67–77.

Dyer, D. H., Wessely, V., Forest, K. T., & Lan, Q. (2008). Three-dimensional structure/function analysis of SCP-2-like2 reveals differences among SCP-2 family members. *J. Lipid Res.*, *49*, 644–653.

Eichenberger, W. (1977). Steryl glycosides and acylated steryl glycosides. In M. Tevini, & H. K. Lichtenthaler (Eds.), *Lipids and Lipid Polymers in Higher Plants* (pp. 169–182). Berlin: Springer-Verlag.

Espig, W., Thiry, E., & Hoffman, K. (1989). Ecdysteroids during ovarian development and embryogenesis in the cricket, *Gryllus bimaculatus* de Greer. *Invert. Reprod. Dev.*, *15*, 143–154.

Evans, O. P., & O'Reilly, D. R. (1998). Purification and kinetic analysis of a baculovirus ecdysteroid UDP-glucosyl-transferase. *Biochem. J.*, *330*, 1265–1270.

Evershed, R. P., Kabbouh, M., Prescott, M. C., Maggs, J. L., & Rees, H. H. (1990). Current status and recent advances in the chromatography and mass spectrometry of ecdysteroids. In A. R. McCaffery, & I. D. Wilson (Eds.), *Chromatography and Isolation of Insect Hormones and Pheromones* (p. 103). London: Plenum Press.

Evershed, R. P., Mercer, J. G., & Rees, H. H. (1987). Capillary gas chromatography — mass spectrometry of ecdysteroids. *J. Chromatogr.*, *390*, 357–369.

Evershed, R. P., Prescott, M. C., Kabbouh, M., & Rees, H. H. (1993). High-performance liquid chromatography/mass spectrometry with thermospray ionization of free ecdysteroids. *Rapid Commun. Mass Spectrom., 7*, 477–481.

Feldlaufer, M. F. (1989). Diversity of molting hormones in insects. In J. Koolman (Ed.), *Ecdysone, From Chemistry to Mode of Action* (pp. 308–312). Stuttgart: Georg Thieme Verlag.

Feldlaufer, M. F., Herbert, E. W., Jr., Svoboda, J. A., & Thompson, M. J. (1986). Biosynthesis of makisterone A and 20-hydroxyecdysone from labeled sterols in the honey bee, *Apis mellifera. Arch. Insect Biochem. Physiol., 3*, 415–421.

Feldlaufer, M. F., Svoboda, J. A., Thompson, M. J., & Wilzer, K. R. (1988). Fate of maternally-acquired ecdysteroids in unfertilized eggs of *Manduca sexta. Insect Biochem., 18*, 219–221.

Feldlaufer, M. F., Weirich, G. F., Imberski, R. B., & Svoboda, J. A. (1995). Ecdysteroid production in *Drosophila* reared on defined diets. *Insect Biochem. Mol. Biol., 25*, 709–712.

Feldlaufer, M. F., Weirich, G. F., Lusby, W. R., & Svoboda, J. A. (1991). Makisterone C: A 29-carbon ecdysteroid from developing embryos of the cotton stainer bug, *Dysdercus fasciatus. Arch. Insect Biochem. Physiol., 18*, 71–79.

Feyereisen, R. (2005). Insect cytochrome P450. In L. I. Gilbert, K. Iatrou, & S. S. Gill (Eds.), *Comprehensive Molecular Insect Science* (Vol. 4), (pp. 1–77). Amsterdam: Elsevier.

Feyereisen, R. (2006). Evolution of insect P450. *Biochem. Soc. Trans., 34*, 1252–1255.

Feyereisen, R., & Durst, F. (1978). Ecdysterone biosynthesis: A microsomal cytochrome P-450 linked ecdysone 20-monooxygenase from tissues of the African migratory locust. *Eur. J. Biochem., 88*, 37–47.

Feyereisen, R., & Durst, F. (1980). Development of microsomal cytochrome P-450 monooxygenases during the last larval instar of the locust, *Locusta migratoria*: Correlation with the hemolymph 20-hydroxyecdysone titer. *Mol. Cell Endocrinol., 20*, 157–169.

Fluegel, M. L., Parker, T. J., & Pallanck, L. J. (2006). Mutations of a *Drosophila* NPC1 gene confer sterol and ecdysone metabolic defects. *Genetics, 172*, 185–196.

Fournier, B., & Radallah, D. (1988). Ecdysteroids in *Carausius morosus* eggs during embryonic development. *Arch. Insect Biochem. Physiol., 7*, 211–224.

Fraenkel, G. (1935). A hormone causing pupation in the blowfly, *Calliphora erythrocephala. Proc. R. Soc. Lond. B, 118*, 1–12.

Fraenkel, G., & Zdarek, J. (1970). The evaluation of the "*Calliphora* test" as an assay for ecdysone. *Biol. Bull., 139*, 138–150.

Franke, S., & Käuser, G. (1989). Occurrence and hormonal role of ecdysteroids in non-Arthropods. In J. Koolman (Ed.), *Ecdysone, From Chemistry to Mode of Action* (pp. 296–307). Stuttgart: Georg Thieme Verlag.

Freeman, M. R., Dobritsa, A., Gaines, P., Segraves, W. A., & Carlson, J. R. (1999). The dare gene: Steroid hormone production, olfactory behaviour, and neural degeneration in *Drosophila. Development, 126*, 4591–4602.

Frew, J. G. (1928). A technique for the cultivation of insect tissues. *Br. J. Exp. Biol., 6*, 1–11.

Fujimoto, Y., Hiramoto, M., Kakinuma, K., & Ikekawa, N. (1989). Elimination of C6-hydrogen during the formation of ecdysteroids from cholesterol in *Locusta migratoria* ovaries. *Steroids, 53*, 477–485.

Fujimoto, Y., Kushiro, T., & Nakamura, K. (1997). Biosynthesis of 20-hydroxyecdysone in *Ajuga* hairy roots: Hydrogen migration from C-6 to C-5 during cis-A/B ring formation. *Tetrahedron Lett., 38*, 2697–2700.

Fujimoto, Y., Miyasaka, S., Ikeda, T., Ikekawa, N., Ohnishi, E., et al. (1985a). An unusual ecdysteroid, (20S)-cholesta-7,14-diene-3β,5α,6α,20,25-pentaol (bombycosterol) from the ovaries of the silkworm, *Bombyx mori. J. Chem. Soc., Chem. Commun.*, 10–12.

Fujimoto, Y., Morisaki, M., & Ikekawa, M. (1985b). Enzymatic dealkylation of phytosterols in insects. *Method. Enzymol., 111*, 346–352.

Fujimoto, Y., Ohyama, K., Nomura, K., Hyodo, R., Takahashi, K., et al. (2000). Biosynthesis of sterols and ecdysteroids in *Ajuga* hairy roots. *Lipids, 35*, 279–288.

Fujiwara, Y., Tanaka, Y., Iwata, K., Rubio, R. O., Yaginuma, T., et al. (2006). ERK/MAPK regulates ecdysteroid and sorbitol metabolism for embryonic diapause termination in the silkworm, *Bombyx mori. J. Insect Physiol., 52*, 569–575.

Fukuda, S. (1940). Induction of pupation in silkworm by transplanting the prothoracic gland. *Proc. Imperial Acad. (Tokyo), 16*, 417–420.

Gallegos, A. M., Atshaves, B. P., Storey, S. M., Starodub, O., Petrescu, A. D., et al. (2001). Gene structure, intracellular localization, and functional roles of the sterol carrier protein-2. *Prog. Lipid Res., 40*, 498–563.

Gande, A. R., Morgan, E. D., & Wilson, I. D. (1979). Ecdysteroid levels throughout the life cycle of the desert locust, *Schistocerca gregaria. J. Insect Physiol., 25*, 669–675.

Garen, A., Kaurar, L., & Lepesant, J. -A. (1977). Roles of ecdysone in *Drosophila* development. *Proc. Natl. Acad. Sci. USA, 74*, 5099–5103.

Gaziova, I., Bonnette, P. C., Henrich, V. C., & Jindra, M. (2004). Cell-autonomous roles of the ecdysoneless gene in *Drosophila* development and oogenesis. *Development, 131*, 2715–2725.

Gelman, D. B., De Milo, A. B., Thyagaraja, B. S., Kelly, T. J., Masler, E. P., et al. (1991). 3-Oxoecdysteroid 3β-reductase in various organs of the European corn borer, *Ostrinia nubilalis* (Hubner). *Arch. Insect Biochem. Physiol., 17*, 93–106.

Gerisch, B., Rottiers, V., Li, D., Motola, D. L., Cummins, C. L., et al. (2007). A bile acid-like steroid modulates *Caenorhabditis elegans* lifespan through nuclear receptor signaling. *Proc. Natl. Acad. Sci. USA, 104*, 5014–5019.

Gersch, M. (1979). Molting of insects without molting gland: Results with larvae of Periplaneta americana. *Experientia, 33*, 228–230.

Gibson, J. M., Isaac, R. E., Dinan, L. N., & Rees, H. H. (1984). Metabolism of [³H] ecdysone in *Schistocerca gregaria*; formation of ecdysteroid acids together with free and phosphorylated ecdysteroid acetates. *Arch. Insect Biochem. Physiol., 1*, 385–407.

Giesen, K., Lammel, U., Langehans, D., Krukkert, K., Bunse, I., et al. (2003). Regulation of glial cell number and differentiation by ecdysone and Fos signaling. *Mech. Dev., 120*, 401–413.

Gilbert, L. I., & Rewitz, K. F. (2009). The function and evolution of the haloween genes. In G. Smagghe (Ed.), *Ecdysone: Structures and Functions* (pp. 231–269). Berlin: Springer Science.

Gilbert, L. I., Rybczynski, R., & Warren, J. T. (2002). Control and biochemical nature of the ecdysteroidogenic pathway. *Annu. Rev. Entomol., 47,* 883–916.

Gilbert, L. I., & Warren, J. (2005). A molecular genetic approach to the biosynthesis of the insect steroid molting hormone. *Vitam. Horm., 73,* 31–57.

Gilgan, M. W., & Zinck, M. E. (1972). Estimation of ecdysterone from sulphuric acid-induced fluorescence. *Steroids, 20,* 95–104.

Girault, J.-P., Blais, C., Beydon, P., Rolando, C., & Lafont, R. (1989). Synthesis and N.M.R. study of 3-dehydroec-dysteroids. *Arch. Insect Biochem. Physiol., 10,* 199–213.

Girault, J.-P. (1998). Determination of ecdysteroid structure by 1D and 2D NMR. *Russ. J. Plant Physiol, 45,* 306–309.

Girault, J.-P., & Lafont, R. (1988). The complete ^{1}H-NMR assignment of ecdysone and 20-hydroxyecdysone. *J. Insect Physiol., 34,* 701–706.

Glitho, I., Delbecque, J. -P., & Delachambre, J. (1979). Prothoracic gland involution related to moulting hormone levels during the metamorphosis of *Te. molitor* L. *J. Insect Physiol., 25,* 187–191.

Goltzené, F., Lagueux, M., Charlet, M., & Hoffmann, J. A. (1978). The follicle cell epithelium of maturing ovaries of, *Locusta migratoria*: A new biosynthetic tissue for ecdysone. *H.-S.Z. Physiol. Chem., 359,* 1427–1434.

Gong, J., Hou, Y., Zha, X. F., Lu, C., et al. (2006). Molecular cloning and characterization of *Bombyx mori* sterol carrier protein x/sterol carrier protein 2 (SCPx/SCP2) gene. *Mitochondr. DNA, 17,* 326–333.

Goodnight, K. C., & Kircher, H. W. (1971). Metabolism of lathosterol by *Drosophila melanogaster*. *Lipids, 6,* 166–169.

Goodwin, C. D., Cooper, B. W., & Margolis, S. (1982). Rat liver cholesterol 7α-hydroxylase. Modulation of enzyme activity by changes in phosphorylation state. *J. Biol. Chem., 257,* 4469–4472.

Goodwin, T. W., Horn, D. H.S., Karlson, P., Koolman, J., Nakanishi, K., et al. (1978). Ecdysteroids: A new generic term. *Nature, 272,* 111.

Grau, V., & Gutzeit, H. (1990). Asymmetrically distributed ecdysteroid-related antigens in follicles and young embryos of *Drosophila melanogaster*. *Roux. Arch. Dev. Biol, 198,* 295–302.

Grau, V., & Lafont, R. (1994a). Metabolism of ecdysone and 20-hydroxyecdysone in adult *Drosophila melanogaster*. *Insect Biochem. Mol. Biol, 24,* 49–58.

Grau, V., & Lafont, R. (1994b). The distribution of ecdysone metabolites within the body of adult *Drosophila* melanogaster females and their sites of production. *J. Insect Physiol., 40,* 87–96.

Grau, V., Pis, J., & Lafont, R. (1995). Ovary-specific interaction of ecdysone 22-phosphate with proteins in adult *Drosophila melanogaster* (Diptera: Drosophilidae). *Eur. J. Entomol., 92,* 189–196.

Greenwood, D. R., & Rees, H. H. (1984). Ecdysone 20-monooxygenase in the desert locust, *Schistocerca gregaria*. *Biochem. J., 223,* 837–847.

Grieneisen, M., Warren, J. T., & Gilbert, L. I. (1993). Early steps in ecdysteroid biosynthesis: Evidence for the involvement of cytochrome P-450 enzymes. *Insect Biochem., 23,* 13–23.

Grieneisen, M., Warren, J. T., Sakurai, S., & Gilbert, L. I. (1991). A putative route to ecdysteroids: Metabolism of cholesterol *in vitro* by mildly disrupted prothoracic glands of *Manduca sexta*. *Insect Biochem., 21,* 41–51.

Grieneisen, M. L. (1994). Recent advances in our knowledge of ecdysteroid biosynthesis in insects and crustaceans. *Insect Biochem. Mol. Biol., 24,* 115–132.

Grunwald, C. (1980). Steroids. In E. A. Bell, & B. V. Charlwood (Eds.), *Encyclopedia of Plant Physiology, New Series, Vol. 8. Secondary Plant Products* (p. 221). Berlin: Springer-Verlag.

Guittard, E., Blais, C., Maria, A., Parvy, J. P., et al. (2011). *Drosophila melanogaster* CYP18A1: A key enzyme of molting hormone (20-hydroxyecdysone) inactivation essential for metamorphosis. *Dev. Biol., 349,* 35–45.

Haag, T., Meister, M.-F., Hétru, C., Kappler, C., Nakatani, Y., et al. (1987). Synthesis of a labelled putative precursor of ecdysone. II. [^{3}H4]3β-hydroxy-5β-cholest-7-ene-6-one: Critical re-evaluation of its role in *Locusta migratoria*. *Insect Biochem., 17,* 291–301.

Hagedorn, H. H. (1985). The role of ecdysteroids in reproduction. In G. A. Kerkut, & L. I. Gilbert (Eds.), *Comprehensive Insect Physiology, Biochemistry, and Pharmacology* (Vol. 8), (pp. 205–262). Oxford: Pergamon Press.

Hagedorn, H. H., O'Connor, J. D., Fuchs, M. S., Sage, B., Schlaeger, D. A., et al. (1975). The ovary as a source of α-ecdysone in an adult mosquito. *Proc. Natl. Acad. Sci. USA, 72,* 3255–3259.

Halliday, W. R., Farnsworth, D. E., & Feyereisen, R. (1986). Hemolymph ecdysteroid titer and midgut ecdysone 20-monooxygenase activity during the last larval stage of *Diploptera punctata*. *Insect Biochem., 16,* 627–634.

Hampshire, F., & Horn, D. H.S. (1966). Structure of crustecdysone, a crustacean moulting hormone. *J. Chem. Soc. Chem. Commun, 37–38.*

Harmatha, J., Budesinsky, M., & Vokac, K. (2002). Photochemical transformation of 20-hydroxyecdysone: production of monomeric and dimeric ecdysteroid analogues. *Steroids, 67,* 127–135.

Harvie, P. D., Filippova, M., & Bryant, P. J. (1998). Genes expressed in the ring gland, the major endocrine organ of *Drosophila melanogaster*. *Genetics, 149,* 217–231.

Hattori, K., Motohashi, N., Kobayashi, I., Tohya, T., et al. (2008). Cloning, expression and characterization of cytosolic sulfotransferase isozymes from *Drosophila melanogaster*. *Biosci. Biotechnol. Biochem., 72,* 540–547.

Heed, W. B., & Kircher, H. W. (1965). Unique sterol in the ecology and nutrition of *Drosophila pachea*. *Science, 149,* 758–761.

Hellou, J., Banoub, J., Gentil, E., Taylor, D. M., & O'Keefe, P. G. (1994). Electrospray tandem mass spectrometry of ecdysteroid moulting hormones. *Spectroscopy, 12,* 43–53.

Hétru, C., Fraisse, D., Tabet, J.-C., & Luu, B. (1984). Confirmation, par spectrométrie de masse en mode FAB, de l'identification des esters d'AMP d'ecdystéroïdes dans les oeufs du Criquet migrateur. *C.R. Acad. Sci. Paris (Sér. II), 299,* 429–432.

Hétru, C., Kappler, C., Hoffmann, J. A., Nearn, R., Luu, B., et al. (1982). The biosynthetic pathway of ecdysone: Studies with vitellogenic ovaries of *Locusta migratoria* (Orthoptera). *Mol. Cell Endocrinol.*, *26*, 51–81.

Hétru, C., Lagueux, M., Luu, B., & Hoffmann, J. A. (1978). Adult ovaries of *Locusta migratoria* contain the sequence of biosynthetic intermediates for ecdysone. *Life Sci.*, *22*, 2141–2154.

Hétru, C., Luu, B., & Hoffmann, J. A. (1985). Ecdysone conjugates: Isolation and identification. *Method Enzymol.*, *111*, 411–419.

Hikino, H., Hikino, Y., & Takemoto, T. (1969). Rubrosterone, a metabolite of insect metamorphosing substance from *Achyranthes rubrofusca*: Synthesis. *Tetrahedron*, *25*, 3389–3394.

Hikino, H., Oizumi, Y., & Takemoto, T. (1975). Steroid metabolism in *Bombyx mori*. I. Catabolism of ponasterone A and ecdysterone in *Bombyx mori*. *H.-S.Z. Physiol. Chem*, *356*, 309–314.

Hiramoto, M., Fujimoto, Y., Kakinuma, K., Ikekawa, N., & Ohnishi, E. (1988). Ecdysteroid conjugates in the ovaries of the silkworm, *Bombyx mori*: 3-phosphates of 2,22-dideoxy-20-hydroxyecdysone and bombycosterol. *Experientia*, *44*, 823–825.

Hirn, M. L., & Delaage, M. A. (1980). Radioimmunological approaches to the quantification of ecdysteroids. In J. A. Hoffmann (Ed.), *Progress in Ecdysone Research* (pp. 69–82). Amsterdam: Elsevier/North-Holland Biomedical Press.

Hocks, P., & Wiechert, R. (1966). 20-Hydroxyecdyson, isoliert aus Insekten. *Tetrahedron Lett.*, *26*, 2989–2993.

Hoffmann, J. A., & Hétru, C. (1983). Ecdysone. In R. G.H. Downer, & H. Laufer (Eds.), *Invertebrate Endocrinology, Vol 1: Endocrinology of Insects* (pp. 65–88). New York: Alan R. Liss.

Hoffmann, J. A., Koolman, J., & Beyler, C. (1977). Rôle des glandes prothoraciques dans la production d'ecdysone au cours du dernier stade larvaire de *Locusta migratoria* L. *C.R. Acad. Sci. Paris, (Sér. D.)*, *280*, 733–737.

Hoffmann, J. A., & Lagueux, M. (1985). Endocrine aspects of embryonic development in insects. In G. A. Kerkut, & L. I. Gilbert (Eds.), *Comprehensive Insect Physiology, Biochemistry and Pharmacology* (Vol. 1), (pp. 435–460). Oxford: Pergamon Press.

Hoffmann, K. H., Bulenda, D., Thiry, E., & Schmid, E. (1985). Apolar ecdysteroid esters in adult female crickets, *Gryllus bimaculatus*. *Life Sci.*, *37*, 185–192.

Hoffmann, K. H., Thiry, E., & Lafont, R. (1990). 14-Deoxyecdysteroids in an insect (*Gryllus bimaculatus*). *Z. Naturforsch.*, *45C*, 703–708.

Hoffmann, K. H., & Wagemann, M. (1994). Age dependency and tissue distribution of ecdysteroids in adult male crickets, *Gryllus bimaculatus* De Geer (Ensifera, Gryllidae). *Comp. Biochem. Physiol. A*, *109*, 293–302.

Hoggard, N., Fisher, M. J., & Rees, H. H. (1989). Possible role for covalent modification in the reversible activation of ecdysone 20-monooxygenase activity. *Arch. Insect Biochem. Physiol.*, *10*, 241–253.

Hoggard, N., & Rees, H. H. (1988). Reversible activation–inactivation of mitochondrial ecdysone 20-mono-oxygenase: A possible role for phosphorylation–dephosphorylation. *J. Insect Physiol.*, *34*, 647–653.

Horike, N., & Sonobe, H. (1999). Ecdysone 20-monooxygenase in eggs of the silkworm, *Bombyx mori*: Enzymatic properties and developmental changes. *Arch. Insect Biochem. Physiol.*, *41*, 9–17.

Horike, N., Takemori, H., Nonaka, Y., Sonobe, H., & Okamoto, M. (2000). Molecular cloning of NADPH-cytochrome P450 oxidoreductase from silkworm eggs. Its involvement in 20-hydroxyecdysone biosynthesis during embryonic development. *Eur. J. Biochem.*, *267*, 6914–6920.

Horn, D. H.S. (1971). The ecdysones. In M. Jacobson, & D. G. Crosby (Eds.), *Naturally Occurring Insecticides* (pp. 333–459). New York: Marcel Dekker.

Horn, D. H.S., & Bergamasco, R. (1985). Chemistry of ecdysteroids. In G. A. Kerkut, & L. I. Gilbert (Eds.), *Comprehensive Insect Physiology, Biochemistry, and Pharmacology* (Vol. 7), (pp. 185–248). Oxford: Pergamon Press.

Hovemann, B. T., Sehlmeyer, F., & Malz, J. (1997). *Drosophila melanogaster* NADPH-cytochrome P450 oxidoreductase: Pronounced expression in antennae may be related to odorant clearance. *Gene*, *189*, 213–219.

Huang, X., Suyama, K., Buchanan, A., Zhu, A. J., & Scott, M. P. (2005). A *Drosophila* model of the Niemann-Pick type C lysosome storage disease: *dnpc 1* is required for molting and sterol homeostasis. *Development*, *132*, 5115–5124.

Huang, X., Warren, J. T., Buchanan, J., Gilbert, L. I., & Scott, M. P. (2007). *Drosophila Niemann-Pick type C-2* genes control sterol homeostasis and steroid biosynthesis: A model of human neurodegenerative disease. *Development*, *134*, 3733–4372.

Huber, R., & Hoppe, W. (1965). Zur Chemie des Ecdysons. VII. Die Kristall- und Molekülstruktur- analyse des Insektenverpuppungshormons Ecdyson mit des automatisierten Faltmolekülmethode. *Chem. Ber.*, *98*, 2403–2424.

Iga, M., & Smagghe, G. (2010). Identification and expression profile of Halloween genes involved in ecdysteroid biosynthesis in *Spodoptera littoralis*. *Peptides*, *31*, 456–467.

Ikekawa, N., Hattori, F., Rubio-Lightbourn, J., Miyazaki, H., Ishibashi, M., et al. (1972). Gas chromatographic separation of phytoecdysteroids. *J. Chromatogr. Sci.*, *10*, 233–242.

Ikekawa, N., Ikeda, T., Mizuno, E., Ohnishi, E., & Sakurai, S. (1980). Isolation of a new ecdysteroid, 2,22-dideoxy-20-hydroxyecdysone from the ovaries of the silkworm, *Bombyx mori*. *J. Chem. Soc. Chem. Commun.*, 448–449.

Isaac, R. E., Desmond, H. P., & Rees, H. H. (1984). Isolation and identification of 3-acetyl-ecdysone 2-phosphate, a metabolite of ecdysone, from developing eggs of *Schistocerca gregaria*. *Biochem. J.*, *217*, 239–243.

Isaac, R. E., Milner, N. P., & Rees, H. H. (1983). Identification of ecdysonoic acid and 20-hydroxyecdysonoic acid isolated from developing eggs of Schistocerca gregaria and pupae of *Spodoptera littoralis*. *Biochem. J.*, *213*, 261–265.

Isaac, R. E., & Rees, H. S. (1984). Isolation and identification of ecdysteroid phosphates and acetylecdysteroid phosphates from developing eggs of the locust, *Schistocerca gregaria*. *Biochem. J.*, *231*, 459–464.

Isaac, R. E., Rees, H. H., & Goodwin, T. W. (1981a). Isolation of 2-deoxy-20-hydroxyecdysone and 3-epi-2-deoxyecdysone from eggs of the desert locust, *Schistocerca gregaria* during embryogenesis. *J. Chem. Soc. Chem. Commun.*, 418–420.

Isaac, R. E., Rees, H. H., & Goodwin, T. W. (1981b). Isolation of ecdysone 3-acetate as a major ecdysteroid from the developing eggs of the desert locust, *Schistocerca gregaria*. *J. Chem. Soc. Chem. Commun.*, 594–595.

Isaac, R. E., Rose, M. E., Rees, H. H., & Goodwin, T. W. (1982). Identification of ecdysone-22-phosphate and 2-deoxyecdysone-22-phosphate in eggs of the desert locust, *Schistocerca gregaria*, by fast atom bombardment mass spectrometry and N.M.R. spectroscopy. *J. Chem. Soc. Chem. Commun.*, 249–251.

Isaac, R. E., & Slinger, A. J. (1989). Storage and excretion of ecdysteroids. In J. Koolman (Ed.), *Ecdysone, From Chemistry to Mode of Action* (pp. 250–253). Stuttgart: George Thieme Verlag.

Isaac, R. E., Sweeney, F. P., & Rees, H. H. (1983). Enzymic hydrolysis of ecdysteroid phosphate during embryogenesis in the desert locust (*Schistocerca gregaria*). *Biochem. Soc. Trans.*, *11*, 379–380.

Ito, Y. (1986). Trends in Countercurrent chromatography. *Trends Analyt. Chem.*, *5*, 142–147.

Ito, Y., & Sonobe, H. (2009). The role of ecdysteroid 22-kinase in the accumulation of ecdysteroids in ovary of silkworm *Bombyx mori*. *Ann. N.Y. Acad. Sci.*, *1163*, 421–424.

Ito, Y., Yasuda, A., & Sonobe, H. (2008). Synthesis and phosphorylation of ecdysteroids during development in the silkworm, *Bombyx mori*. *Zool Sci.*, *25*, 721–727.

Jarvis, T. D., Earley, F. G.P., & Rees, H. H. (1994a). Inhibition of the ecdysteroid biosynthetic pathway in ovarian follicle cells of *Locusta migratoria*. *Pestici. Biochem. Physiol.*, *48*, 153–162.

Jarvis, T. D., Earley, F. G.P., & Rees, H. H. (1994b). Ecdysteroid biosynthesis in larval testes of *Spodoptera littoralis*. *Insect Biochem. Mol. Biol.*, *24*, 531–537.

Jegla, T. C., & Costlow, J. D. (1979). The *Limulus* bioassay for ecdysteroids. *Biol. Bull.*, *156*, 103–114.

Jenkins, S. P., Brown, M. R., & Lea, A. O. (1992). Inactive prothoracic glands in larvae and pupae of *Aedes aegypti*: Ecdysteroid release by tissues in the thorax and abdomen. *Insect Biochem. Mol. Biol.*, *22*, 553–559.

Jin, X., Sun, X., & Song, Q. (2005). Woc gene mutation causes 20E-dependent alpha-tubulin detyrosination in *Drosophila melanogaster*. *Arch. Insect Biochem. Physiol.*, *60*, 116–129.

Johnson, P., & Rees, H. H. (1977). The mechanism of C-20 hydroxylation of α-ecdysone in the desert locust, *Schistocerca gregaria*. *Biochem. J.*, *168*, 513–520.

Jürgens, G., Wieschaus, E., Nüsslein-Volhard, C., & Kluding, H. (1984). Mutations affecting the pattern of the larval cuticle in *Drosophila melanogaster*. II Zygotic loci on the third chromosome. *Roux Arch. Dev. Biol.*, *193*, 283–295.

Kabbouh, M., & Rees, H. H. (1991a). Characterization of the ATP: 2-Deoxyecdysone 22-phosphotransferase (2-deoxyecdysone 22-kinase) in the follicle cells of *Schistocerca gregaria*. *Insect Biochem.*, *21*, 57–64.

Kabbouh, M., & Rees, H. H. (1991b). Characterization of acetyl-CoA: Ecdysone 3-acetyltransferase in *Schistocerca gregaria* larvae. *Insect Biochem.*, *21*, 607–613.

Kabbouh, M., & Rees, H. H. (1993). Characterization of ATP: Ecdysone 3-acetate 2-phosphotransferase (ecdysone 3-acetate 2-kinase) in *Schistocerca gregaria* larvae. *Insect Biochem. Mol. Biol.*, *23*, 73–79.

Kamba, M., Mamiya, Y., Sonobe, H., & Fujimoto, Y. (1994). 22-Deoxy-20-hydroxy-ecdysone and its phosphoric acid ester from ovaries of the silkworm, *Bombyx mori*. *Insect Biochem. Mol. Biol.*, *24*, 395–402.

Kamba, M., Sonobe, H., Mamiya, Y., Hara, N., & Fujimoto, Y. (1995). Isolation and identification of 3-epiecdyster-oids from diapause eggs of the silkworm, *Bombyx mori*. *J. Sericult. Sci. Japan*, *64*, 333–343.

Kamba, M., Sonobe, H., Mamiya, Y., Hara, N., & Fujimoto, Y. (2000a). 2,22-Dideoxy-23-hydroxyecdysone and its 3-phosphate from ovaries of the silkworm, *Bombyx mori*. *Nat. Prod. Lett.*, *14*, 349–356.

Kamba, M., Sonobe, H., Mamiya, Y., Hara, N., & Fujimoto, Y. (2000b). 3-Epi-22-deoxy-20,26-dihydroxyecdysone and 3-epi-22-deoxy-16β,20-dihydroxyecdysone and their 2-phosphate from eggs of the silkworm, *Bombyx mori*. *Nat. Prod. Lett.*, *14*, 469–476.

Kaplanis, J. N., Dutky, S. R., Robbins, W. E., Thompson, M. J., & Lindquist, E. L. (1975). Makisterone A: A new 28-carbon hexahydroxy moulting hormone from the embryo of the milkweed bug. *Science*, *190*, 681–682.

Kaplanis, J. N., Robbins, W. E., Thompson, M. J., & Dutky, S. R. (1973). 26-Hydroxyecdysone: New insect molting hormone from the egg of the tobacco hornworm. *Science*, *180*, 307–308.

Kaplanis, J. N., Tabor, L. A., Thompson, M. J., Robbins, W. E., & Shortino, T. J. (1966). Assay for ecdysone (moulting hormone) activity using the housefly, *Musca domestica* L. *Steroids*, *8*, 625–631.

Kaplanis, J. N., Thompson, M. J., Dutky, S. R., & Robbins, W. E. (1979). The ecdysteroids from the tobacco hornworm during pupal–adult development five days after peak titer of molting hormone activity. *Steroids*, *34*, 333–345.

Kaplanis, J. N., Thompson, M. J., Dutky, S. R., & Robbins, W. E. (1980). The ecdysteroids from young embryonated eggs of the tobacco hornworm. *Steroids*, *36*, 321–336.

Kappler, C., Kabbouh, M., Durst, F., & Hoffmann, J. A. (1986). Studies on the C-2 hydroxylation of 2-deoxyecdysone in *Locusta migratoria*. *Insect Biochem.*, *16*, 25–32.

Kappler, C., Kabbouh, M., Hétru, C., Durst, F., & Hoffmann, J. A. (1988). Characterization of the three hydroxylases involved in the final steps of biosynthesis of the steroid hormone ecdysone in *Locusta migratoria* (Insecta, Orthoptera). *J. Steroid Biochem.*, *31*, 891–898.

Karlson, P. (1956). Biochemical studies on insect hormones. *Vitam. Horm.*, *14*, 227–266.

Karlson, P. (1983). Eighth Adolf Butenandt Lecture. Why are so many hormones steroids? H.-S.Z. *Physiol. Chem. (S)*, *364*, 1067–1087.

Karlson, P., Bugany, H., Döpp, H., & Hoyer, G. A. (1972). 3-Dehydroecdyson, ein Stoffwechselprodukt des Ecdysons bei der Schmeissfliege *Calliphora erythrocephala* Meigen. *H.-S.Z. Physiol. Chem.*, *358*, 1610–1614.

Karlson, P., & Koolman, J. (1973). On the metabolic fate of ecdysone and 3-dehydroecdysone in *Calliphora vicina*. *Insect Biochem.*, *3*, 409–417.

Karlson, P., & Sekeris, C. E. (1966). Ecdysone, an insect steroid hormone, and its mode of action. *Recent Prog. Horm. Res.*, *22*, 473–502.

Karlson, P., & Shaaya, E. (1964). Der Ecdysontiter während der Insektenentwicklung. I. Eine Methode zur Bestimmung des Ecdysongehalts. *J. Insect Physiol., 10*, 797–804.

Karlson, P., & Stamm-Menéndez, M. D. (1956). Notiz über den Nachweis von Metamorphose-Hormon in den Imagines von *Bombyx mori. H.-S.Z. Physiol. Chem., 306*, 109–111.

Katz, M., & Lensky, Y. (1970). Gas chromatographic analysis of ecdysone. *Experientia, 26*, 1043.

Kayser, H., Ertl, P., Eilinger, P., Spindler-Barth, M., & Winkler, T. (2002). Diastereomeric ecdysteroids with a cyclic hemiacetal in the side chain produced by cytochrome P450 in hormonally resistant cells. *Arch. Biochem. Biophys., 400*, 180–187.

Kayser, H., Winkler, T., & Spindler-Barth, M. (1997). 26-Hydroxylation of ecdysteroids is catalyzed by a typical cytochrome P-450-dependent oxidase and related to ecdysteroid resistance in an insect cell line. *Eur. J. Biochem., 248*, 707–716.

Ke, O. (1930). Morphological variation of the prothoracic gland in the domestic and the wild silkworm. *Bull. Sci. Fac. Terkult Ksuju Imp. Uni., 4*, 12–21, (cited in Koolman, 1989).

Keightley, D. A., Lou, K. J., & Smith, W. A. (1990). Involvement of translation and transcription in insect steroidogenesis. *Mol. Cell Endocrinol., 74*, 229–237.

Kelly, T. J., Thyagaraja, B. S., Bell, R. A., Masler, E. P., Gelman, D. B., et al. (1990). Conversion of 3-dehydroecdysone by a ketoreductase in post-diapause, pre-hatch eggs of the Gypsy Moth *Lymantria dispar. Arch. Insect Biochem. Physiol., 14*, 37–46.

Keogh, D. P., Johnson, R. F., & Smith, S. L. (1989). Regulation of cytochrome P-450 dependent steroid hydroxylase activity in *Manduca sexta*: Evidence for the involvement of a neuroendocrine-endocrine axis during larval-pupal development. *Biochem. Biophys. Res. Commun., 165*, 442–448.

Keogh, D. P., Mitchell, M. J., Crooks, J. R., & Smith, S. L. (1992). Effects of the adenylate cyclase activator forskolin and its inactive derivative 1,9-dideoxy-forskolin on insect cytochrome P-450 dependent steroid hydroxylase activity. *Experientia, 48*, 39–41.

Keogh, D. P., & Smith, S. L. (1991). Regulation of cytochrome P-450 dependent steroid hydroxylase activity in *Manduca sexta*: Effects of the ecdysone agonist RH 5849 on ecdysone 20-monooxygenase activity. *Biochem. Biophys. Res. Commun., 176*, 522–527.

Khan, S., Sneddon, K., Firlding, B., Ward, V., et al. (2003). Functional characterization of the ecdysteroid UDP-glucosyl transferase gene of *Helicoverpa armigera* single-enveloped nucleopolyhedrovirus isolated in South Africa. *Virus Genes 27*, 17–27.

Kholodova, Yu., D. (1977). Use of the Chugaev reaction for the quantitative determination of ecdysones. *Khim. Prir. Soedin,* 227–230.

Kholodova, Yu., D. (1981). Chugaev's reaction for the analysis of steroids. *Proc. Symp. Analysis Steroids, Eger, Hungary,* 519.

Kim, M. S., & Lan, Q. (2010). Sterol carrier protein-x gene and effects of sterol carrier protein-2 inhibitors on lipid uptake in *Manduca sexta. BMC Physiol., 10*, 9.

King, D. S., Bollenbacher, W. E., Borst, D. W., Vedeckis, W. V., O'Connor, J. D., et al. (1974). The secretion of α-ecdysone by the prothoracic glands of *Manduca sexta in vitro. Proc. Natl Acad. Sci. USA, 71*, 793–796.

Kingan, T. G. (1989). A competitive enzyme-linked immunosorbent assay: Application in the assay of peptides, steroids and cyclic nucleotides. *Anal. Biochem., 183*, 283–289.

Kiriishi, S., Rountree, D. B., Sakurai, S., & Gilbert, L. I. (1990). Prothoracic gland synthesis of 3-dehydroecdysone and its hemolymph 3β-reductase mediated conversion to ecdysone in representative insects. *Experientia, 46*, 716–721.

Kitamura, T., Kobayashi, S., & Okada, M. (1996). Regional expression of the transcript encoding sterol carrier protein x-related thiolase and its regulation by homeotic genes in the midgut of *Drosophila* embryos. *Dev. Growth Differ., 38*, 373–381.

Kiuchi, M., Yasui, H., Hayasaka, S., & Kamimura, M. (2003). Entomogenous fungus *Nomuraea rileyi* inhibits host insect molting by C22-oxidizing inactivation of hemolymph ecdysteroids. *Arch. Insect Biochem. Physiol., 52*, 35–44.

Kobayashi, M., Krishna, M. M., Ishida, K., & Anjaneyulu, V. (1992). Marine sterols. XXII. Occurrence of 3-oxo-4,6,8(14)-triunsaturated steroids in the sponge *Dysidea herbacea. Chem. Pharm. Bull, 40*, 72–74.

Koch, J. A., & Waxman, D. J. (1991). P450 phosphorylation in isolated hepatocytes and *in vivo. Method. Enzymol., 206*, 305–315.

Koener, J. F., Carino, F. A., & Feyereisen, R. (1993). The cDNA and deduced protein sequence of house fly NADPH-cytochrome P450 reductase. *Insect Biochem. Mol. Biol., 23*, 439–447.

Koller, G. (1929). Die innere Sekretion bei wirbellosen Tiere. *Biol. Rev., 4*, 269–306.

Kolmer, M., Alho, H., Costa, E., & Pani, L. (1993). Cloning and tissue-specific functional characterization of the promoter of the rat diazepam binding inhibitor, a peptide with multiple biological actions. *Proc. Natl Acad. Sci. USA, 90*, 8439–8443.

Kolmer, M., Roos, C., Tirronen, M., Myohanen, S., & Alho, H. (1994). Tissue-specific expression of the diazepam-binding inhibitor in *Drosophila melanogaster*: Cloning, structure, and localization of the gene. *Mol. Cell Biol., 14*, 6983–6995.

Koolman, J. (1978). Ecdysone oxidase in insects. *H.-S.Z. Physiol. Chem., 359*, 1315–1321.

Koolman, J. (1980). Analysis of ecdysteroids by fluorometry. *Insect Biochem., 10*, 381–386.

Koolman, J. (1985). Ecdysone oxidase. *Method Enzymol., 111*, 419–429.

Koolman, J. (Ed.), (1989). *Ecdysone, From Chemistry to Mode of Action*. Stuttgart: Georg Thieme Verlag.

Koolman, J. (1990). Ecdysteroids. *Zool. Sci., 7*, 563–580.

Koolman, J., & Karlson, P. (1978). Ecdysone oxidase: Reaction and specificity. *Eur. J. Biochem., 89*, 453–460.

Koolman, J., & Karlson, P. (1985). Regulation of ecdysteroid titer: Degradation. In G. A. Kerkut, & L. I. Gilbert (Eds.), *Comprehensive Insect Physiology, Biochemistry, and Pharmacology* (Vol. 7), (pp. 343–361). Oxford: Pergamon Press.

Koolman, J., Reum, L., & Karlson, P. (1979). 26-Hydroxy-ecdysone, 20,26-dihydroxyecdysone and inokosterone detected as metabolites of ecdysone in the blowfly, *Calliphora vicina* by radiotracer experiments. *H.-S.Z. Physiol. Chem.*, *360*, 1351–1355.

Koolman, J., & Spindler, K. -D. (1977). Enzymatic and chemical synthesis of 3-dehydro-ecdysterone, a metabolite of the moulting hormone of insects. *H.-S.Z. Physiol. Chem.*, *358*, 1339–1344.

Kopec, S. (1922). Studies on the necessity of the brain for the inception of insect metamorphosis. *Biol. Bull.*, *42*, 323–342.

Koreeda, M., Nakanishi, K., & Goto, M. (1970). Ajugalactone, an insect moulting hormone inhibitor as tested by the Chilo dipping method. *J. Am. Chem. Soc.*, *92*, 7512–7513.

Koreeda, M., & Teicher, B. A. (1977). Chemical analysis of insect molting hormones. In R. B. Turner (Ed.), *Analytical Biochemistry of Insects* (pp. 207–240). Amsterdam: Elsevier Scientific Publishing Company.

Kozlova, T., & Thummel, C. S. (2003). Essential roles for ecdysone signalling during *Drosophila* mid-embryonic development. *Science*, *301*, 1911–1914.

Krebs, K. C., & Lan, Q. (2003). Isolation and expression of a sterol carrier protein-2 gene from the yellow fever mosquito, *Aedes aegypti. Insect Mol. Biol.*, *12*, 51–60.

Kubo, I., & Komatsu, S. (1986). Micro analysis of prostaglandins and ecdysteroids in insects by high-performance liquid chromatography and fluorescence detection. *J. Chromatogr.*, *362*, 61–70.

Kubo, I., & Komatsu, S. (1987). Simultaneous measurement of prostaglandins and ecdysteroids in insects by high-performance liquid chromatography and fluorescence labeling. *Agr. Biol. Chem.*, *51*, 1305–1309.

Kubo, I., Komatsu, S., Asaka, Y., & De Boer, G. (1987). Isolation and identification of apolar metabolites of ingested 20-hydroxyecdysone in frass of *Heliothis virescens* larvae. *J. Chem. Ecol.*, *13*, 785–794.

Kubo, I., Matsumoto, A., & Asano, S. (1985a). Efficient isolation of ecdysteroids from the silkworm, *Bombyx mori* by droplet counter-current chromatography. *Insect Biochem.*, *15*, 45–47.

Kubo, I., Matsumoto, A., & Hanke, F. J. (1985b). The ^{1}H NMR assignment of 20-hydroxyecdysone. *Agr. Biol. Chem.*, *49*, 243–244.

Kubo, I., Zhang, M., de Boer, G., & Uchima, K. (1994). Location of ecdysteroid 22-O-acyltransferase in the larvae of *Heliothis virescens. Entomol. Exp. Appl.*, *70*, 263–272.

Kulcsar, P., Darvas, B., Brandtner, H., Koolman, J., & Rees, H. H. (1991). Effects of the pyridine-containing P-450 inhibitor, fenarimol, on the formation of 20-OH ecdysone in flies. *Experientia*, *47*, 261–263.

Kumpun, S., Yingyongnarongkul, B.-E., Lafont, R., Girault, J.-P., & Suksamrarn, A. (2007). Stereoselective synthesis and moulting hormone activity of integristerone A and analogues. *Tetrahedron*, *63*, 1093–1099.

Kurisu, M., Morita, M., Kashiwayama, Y., Yokata, S., Hayashi, H., et al. (2003). Existence of catalase-less peroxisomes in Sf21 insect cells. *Biochem. Biophys. Res. Commun.*, *306*, 169–176.

Kuroda, Y. (1968). Effects of ecdysone analogues on differentiation of eye-antennal discs of *Drosophila melanogaster* in culture. *Annu. Rep. Natl. Inst. Genet. Japan*, *19*, 22–23.

Lacapere, J. J., & Papadopoulos, V. (2003). Peripheral-type benzodiazepine receptor: Structure and function of a cholesterol-binding protein in steroid and bile acid biosynthesis. *Steroids*, *68*, 569–585.

Lachaise, F., Carpentier, G., Sommé, G., Colardeau, J., & Beydon, P. (1989). Ecdysteroid synthesis by crab Y-organs. *J. Exp. Zool.*, *252*, 283–292.

Lachaise, F., Le Roux, A., Hubert, M., & Lafont, R. (1993). The molting glands of crustaceans: Localization, activity and endocrine control (a review). *J. Crust. Biol.*, *13*, 198–234.

Lafont, R. (1997). Ecdysteroids and related molecules in animals and plants. *Arch. Insect Biochem. Physiol.*, *35*, 3–20.

Lafont, R. (1988). HPLC analysis of ecdysteroids in plants and animals. In H. Kalász, & L. S. Ettre (Eds.), *Chromatography '87* (pp. 1–15). Budapest: Akadémiai Kiadó.

Lafont, R., Beydon, P., Blais, C., Garcia, M., Lachaise, F., et al. (1986). Ecdysteroid metabolism — a comparative study. *Insect Biochem.*, *16*, 11–16.

Lafont, R., Beydon, P., Mauchamp, B., Sommé-Martin, G., Andrianjafintrimo, M., et al. (1981). Recent progress in ecdysteroid analytical methods. In F. Sehnal, A. Zabza, & B. Cymborowski (Eds.), *Regulation of Insect Development and Behaviour* (pp. 125–144). Wroclaw, Poland: Technical University of Wroclaw Press.

Lafont, R., Beydon, P., Sommé-Martin, G., & Blais, C. (1980). High-performance liquid chromatography of ecdysone metabolites applied to the cabbage butterfly, *Pieris brassicae. Steroids*, *36*, 185–207.

Lafont, R., Blais, C., Beydon, P., Modde, J.-F., Enderle, U., et al. (1983). Conversion of ecdysone and 20-hydroxyecdysone into 26-oic derivatives is a major pathway in larvae and pupae of species from three insect orders. *Arch. Insect Biochem. Physiol.*, *1*, 41–58.

Lafont, R., Blais, C., Harmatha, J., & Wilson, I. D. (2000). Ecdysteroids: Chromatography. In I. D. Wilson, E. R. Adlard, M. Cooke, & C. F. Poole (Eds.), *Encyclopedia of Separation Science* (pp. 2631–2640). London: Academic Press.

Lafont, R., & Connat, J.-L. (1989). Pathways of ecdysone metabolism. In J. Koolman (Ed.), *Ecdysone, From Chemistry to Mode of Action* (p. 167). Stuttgart: Georg Thieme Verlag.

Lafont, R., Connat, J.-L., Delbecque, J.-P., Porcheron, P., Dauphin-Villemant, C., et al. (1995). Comparative studies on ecdysteroids. In E. Ohnishi, S. Y. Takahashi, & H. Sonobe (Eds.), *Recent Advances in Insect Biochemistry and Molecular Biology* (pp. 45–91). Nagoya, Japan: University of Nagoya Press.

Lafont, R., Dauphin-Villemant, C., Warren, J., & Rees, H. H. (2005). Ecdysteroid Chemistry and Biochemistry. In L. I. Gilbert, K. Iatrou, & S. S. Gill (Eds.), *Comprehensive Molecular Insect Science* (Vol. 3), (p. 125–196). Amsterdam: Elsevier.

Lafont, R., & Dinan, L. N. (2003). Practical uses for ecdysteroids in mammals and humans: An update. *J. Insect Sci.*, *3.7* http://www.insectscience.org/3.7/.

Lafont, R., Kaouadji, N., Morgan, E. D., & Wilson, I. D. (1994a). Selectivity in the HPLC analysis of ecdysteroids. *J. Chromatogr., 658,* 55–67.

Lafont, R., & Koolman, J. (1984). Ecdysone metabolism. In J. A. Hoffmann, & M. Porchet (Eds.), *Biosynthesis, Metabolism and Mode of Action of Invertebrate Hormones* (pp. 196–226). Berlin: Springer-Verlag.

Lafont, R., & Koolman, J. (2009). Diversity of ecdysteroids in animal species. In G. Smagghe (Ed.), *Ecdysone: Structures and Functions* (pp. 47–71). Berlin: Springer Science.

Lafont, R., Koolman, J., & Rees, H. H. (1993a). Standardized abbreviations for common ecdysteroids. *Insect Biochem., 23,* 207–209.

Lafont, R., Morgan, E. D., & Wilson, I. D. (1994b). Chromatographic procedures for phytoecdysteroids. *J. Chromatogr., 658,* 31–53.

Lafont, R., Pennetier, J.-L., Andrianjafintrimo, M., Claret, J., Modde, J.-F., et al. (1982). Sample processing for high-performance liquid chromatography of ecdysteroids. *J. Chromatogr., 236,* 137–149.

Lafont, R., Porter, C. J., Williams, E., Read, H., Morgan, E. D., et al. (1993b). The application of off-line HPTLC-MS-MS to the identification of ecdysteroids in plant and arthropod samples. *J. Planar Chromatogr., 6,* 421–424.

Lafont, R., Sommé-Martin, C., & Chambet, J.-C. (1979). Separation of ecdysteroids by using high pressure liquid chromatography on microparticulate supports. *J. Chromatogr., 170,* 185–194.

Lafont, R., & Wilson, I. D. (1990). Advances in ecdysteroid HPLC. In A. R. McCaffery, & I. D. Wilson (Eds.), *Chromatography and Isolation of Insect Hormones and Pheromones* (pp. 79–94). London: Plenum Press.

Laguerre, M., & Veenstra, J. A. (2010). Ecdysone receptor homologues from mollusks, leeches and a polychaete. *FEBS Lett., 584,* 4458–4462.

Lagueux, M., Hoffmann, J. A., Goltzené, F., Kappler, C., Tsoupras, G., et al. (1984). Ecdysteroids in ovaries and embryos of *Locusta migratoria.* In J. A. Hoffmann, & M. Porchet (Eds.), *Biosynthesis, Metabolism and Mode of Action of Invertebrate Hormones* (pp. 168–180). Berlin: Springer-Verlag.

Lakeman, J., Speckamp, W. W., & Huisman, H. O. (1967). Addition to steroid polyenes IV. *Tetrahedron Lett., 38,* 3699–3703.

Lang, M., Murat, S., Gouppil, G., et al. (2011). Mutations in *neverland* have turned *Drosophila pachea* into an obligate specialist species (submitted for publication).

Large, T., Lafont, R., Morgan, E. D., & Wilson, I. D. (1992). Micellar capillary electrophoresis of ecdysteroids. *Anal. Proc., 29,* 386–388.

Laurent, P., Braekman, J.-C., Daloze, D., & Pasteels, J. M. (2003). An ecdysteroid (22-acetyl-20-hydroxyecdysone) from the defense gland secretions of an insect: *Chrysolina carnifex* (Coleoptera: Chrysomelidae). *Chemoecology, 13,* 109–111.

Le Bizec, B., Antignac, J.-P., Monteau, F., & André, F. (2002). Ecdysteroids: One potential new anabolic family in breeding animals. *Anal. Chim. Acta, 473,* 89–97.

Lee, S. S., & Nakanishi, K. (1989). ¹³C-NMR assignments of some insect molting hormones. *Kaohsiung J. Med. Sci., 5,* 564–568.

Lehmann, M., & Koolman, J. (1986). The influence of forskolin on the metabolism of ecdysone and 20-hydroxyecdysone in isolated fat body of the blowfly, *Calliphora vicina. H.-S.Z. Physiol. Chem, 367,* 387–393.

Lehmann, M., & Koolman, J. (1989). Regulation of ecdysone metabolism. In J. Koolman (Ed.), *Ecdysone, From Chemistry to Mode of Action* (pp. 217–220). Stuttgart: Georg Thieme Verlag.

Li, T. R., & White, K. P. (2003). Tissue-specific gene expression and ecdysone-regulated genomic networks in *Drosophila. Dev. Cell, 5,* 59–72.

Liebrich, W., Dumberger, B. B., & Hoffmann, K. H. (1991). Ecdysone 20-monooxygenase in a cricket (*Gryllus bimaculatus,* Ensifera, Gryllidae): Activity throughout adult lifecycle. *Comp. Biochem. Physiol. A Physiol., 99,* 597–602.

Liebrich, W., & Hoffmann, K. H. (1991). Ecdysone 20-monooxygenase in a cricket, *Gryllus bimaculatus* (Ensifera, Gryllidae): Characterization of the microsomal midgut steroid hydroxylase in adult females. *J. Comp. Physiol. B, 161,* 93–99.

Loeb, M. J., Brandt, E. P., Woods, C. W., & Bell, R. A. (1988). Secretion of ecdysteroid by sheets of testes of the gipsy moth, *Lymantria dispar,* and its regulation by testis ecdysiotropin. *J. Exp. Zool., 248,* 94–100.

Loeb, M. J., & Woods, C. W. (1989). Metabolism of ecdysteroid by testes of the tobacco budworm, *Heliothis virescens. Arch. Insect Biochem. Physiol., 10,* 83–92.

Loeb, M. J., Woods, C. W., Brandt, E. P., & Borkovec, A. B. (1982). Larval testes of the tobacco budworm. A new source of insect ecdysteroids. *Science, 218,* 896–898.

Louden, D., Handley, A., Lafont, R., Taylor, S., Sinclair, I., et al. (2002). HPLC analysis of ecdysteroids in plant extracts using superheated deuterium oxide with multiple on-line spectroscopic analysis (UV, IR, ¹H NMR, and MS. *Anal. Chem., 74,* 288–294.

Louden, D., Handley, A., Taylor, S., Lenz, E., Miller, S., et al. (2001). Spectroscopic characterization and identification of ecdysteroids using high performance liquid chromatography combined with on-line diode-array, FT-infrared, ¹H-Nuclear Magnetic Resonance and time-of-flight mass spectrometry. *J. Chromatogr. A, 910,* 237–246.

Lozano, R., Thompson, M. J., Lusby, W. R., Svoboda, J. A., & Rees, H. H. (1988a). Metabolism of [¹⁴C]-cholesterol in *Manduca sexta* pupae: Isolation and identification of sterol sulfates, free ecdysteroids and ecdysteroid acids. *Arch. Insect Biochem. Physiol., 7,* 249–266.

Lozano, R., Thompson, M. J., Svoboda, J. A., & Lusby, W. R. (1988b). Isolation of acidic and conjugated ecdysteroid fractions from *Manduca sexta* pupae. *Insect Biochem., 18,* 163–168.

Lozano, R., Thompson, M. J., Svoboda, J. A., & Lusby, W. R. (1989). Profiles of free and conjugated ecdysteroids and ecdysteroid acids during pupal-adult development of *Manduca sexta. Arch. Insect Biochem. Physiol., 12,* 63–74.

Lukacs, G., & Bennett, C. R. (1972). Résonance magnétique du ¹³C de produits naturels et apparentés. VII (18. L'α-ecdysone). *Bull. Soc. Chim. Fr., 10,* 3996–4000.

Lundström, A., Kang, D., Liu, G., Fernandez, C., Warren, J. T., et al. (2002). A protein from the cabbage looper, Trichoplusia ni, regulated by a bacterial infection is homologous to 3-dehydroecdysone 3β-reductase. *Insect Biochem. Mol. Biol., 32,* 829–837.

Lusby, W. R., Oliver, J. E., & Thompson, M. J. (1987). Application of desorption chemical ionization techniques for analysis of biologically active compounds isolated from insects. In J. D. Rosen (Ed.), *Applications of New Mass Spectrometry Techniques in Pesticide Chemistry* (pp. 99–115). John Wiley.

Maeda, S., Nakashima, A., Yamada, R., Hara, N., et al. (2008). Molecular cloning of ecdysone 20-hydroxylase and expression pattern of the enzyme during embryonic development of silkworm *Bombyx mori*. *Comp. Biochem. Physiol. B, 149*, 507–516.

Makka, T., Seino, A., Tomita, S., Fujiwara, H., & Sonobe, H. (2002). A possible role of 20-hydroxyecdysone in embryonic development of the silkworm *Bombyx mori*. *Arch. Insect Biochem. Physiol., 51*, 111–120.

Makka, T., & Sonobe, H. (2000). Ecdysone metabolism in diapause eggs and non-diapause eggs of the silkworm, *Bombyx mori*. *Zool. Sci., 17*, 89–95.

Malorni, A., Minale, L., & Riccio, K. (1978). Steroids from sponges Occurence of steroidal $\Delta^{4,7}$-3,6-diketones in the marine sponge *Raphidostila incisa*. *Nouv. I. Chimie, 2*, 351–354.

Mamiya, Y., Sonobe, H., Yoshida, K., Hara, N., & Fujimoto, Y. (1995). Occurrence of 3-epi-22-deoxy-20-hydroxyec-dysone and its phosphoric ester in diapause eggs of the silkworm, *Bombyx mori*. *Experientia, 51*, 363–367.

Manzan, M. A., Lozano, M. E., Sciocco-Cap, A., Ghiringhelli, P. D., et al. (2002). Identification and characterizatiion of the ecdysteroid UDP-Glycosyltransferase gene of *Epinotia aporema* granulovirus. *Vitus Genes, 24*, 119–130.

Marco, M.-P., Sánchez-Baeza, F. J., Camps, F., & Coll, J. (1993). Phytoecdysteroid analysis by high-performance liquid chromatography-thermospray mass spectrometry. *J. Chromatogr., 641*, 81–87.

Markov, G. V., Tavares, R., Dauphin-Villemant, C., Demeneix, B. A., Baker, M. E., & Laudet, V. (2009). Independent elaboration of steroid hormone signaling pathways in metazoans. *Proc. Natl. Acad. Sci. USA, 106*, 11913–11918.

Maroy, P., Kauffmann, G., & Dübendorfer, A. (1988). Embryonic ecdysteroids of *Drosophila melanogaster*. *J. Insect Physiol., 34*, 633–637.

Matsumoto, T., & Kubo, I. (1989). Ecdysone from Vitex strickeri as insect growth inhibitor. Application JP 87–295802 (cited in *Chem. Abstr., 111*, 169367).

Matsumoto, E., Matsui, M., & Tamura, H. (2003). Identification and purification of sulfotransferase for 20-hydroxyec-dysteroid from the larval fat body of a fleshfly, *Sarcophaga peregrina*. *Biosci. Biotechnol. Biochem., 67*, 1780–1785.

Mauchamp, B., Royer, C., Kerhoas, L., & Einhorn, J. (1993). MS/MS analyses of ecdysteroids in developing *Dysdercus fasciatus*. *Insect Biochem. Mol. Biol., 23*, 199–205.

Maurer, P., Girault, J.-P., Larchevêque, M., & Lafont, R. (1993). 24-Epi-makisterone A (not makisterone A) is the major ecdysteroid in the leaf-cutting ant *Acromyrmex octospinosus* (Reich) (Hymenoptera, Formicidae: Attini). *Arch. Insect Biochem. Physiol., 23*, 29–35.

Mauvais, A., Burger, A., Rousset, J.-P., Hétru, C., & Luu, B. (1994). Acetylenic inhibitors of C-22 hydroxylase of ecdysone biosynthesis. *Bioorg. Chem., 22*, 36–50.

Mayer, R. T., Durrant, J. L., Holman, G. M., Weirich, G. F., & Svoboda, J. A. (1979). Ecdysone 3-epimerase from the midgut of *Manduca sexta* (L.). *Steroids, 34*, 555–562.

Mayer, R. T., & Svoboda, J. A. (1978). Thin-layer chromatographic in situ analysis of insect ecdysones via fluorescence-quenching. *Steroids, 31*, 139–150.

Meaney, S. (2005). Is C-26 hydroxylation an evolutionary conserved steroid inactivation process? *FASEB J., 19*, 1220–1224.

Meister, M. F., Dimarcq, J.-L., Kappler, C., Hétru, C., Lagueux, M., et al. (1985). Conversion of a labelled ecdysone precursor, 2,22,25-trideoxyecdysone, by embryonic and larval tissues of *Locusta migratoria*. *Mol. Cell Endocrinol., 41*, 27–44.

Mesnier, M., Partiaoglou, N., Oberlander, H., & Porcheron, P. (2000). Rhythmic autocrine activity in cultured insect epidermal cells. *Arch. Insect Biochem. Physiol., 44*, 7–16.

Mikitani, K. (1995). Sensitive, rapid and simple method for evaluation of ecdysteroid agonist activity based on the mode of action of the hormone. *J. Sericult. Sci. Japan, 64*, 534–539.

Mikitani, K. (1996). An automated ecdysteroid receptor binding assay using a 96-well microplate. *J. Sericult. Sci. Japan, 65*, 141–144.

Miller, W. L. (1988). Molecular biology of steroid hormone synthesis. *Endocr. Rev., 9*, 295–318.

Miller, W. L. (2007). Steroidogenic acute regulatory protein (StAR), a novel mitochondrial cholesterol transporter. *Biochim. Biophys. Acta, 1771*, 663–676.

Milner, N. P., Nali, M., Gibson, J. M., & Rees, H. H. (1986). Early stages of ecdysteroid biosynthesis: The role of 7-dehydrocholesterol. *Insect Biochem., 16*, 17–23.

Milner, N. P., & Rees, H. H. (1985). Involvement of 3-dehydroecdysone in the 3-epimerization of ecdysone. *Biochem. J., 231*, 369–374.

Mitchell, M. J., Crooks, J. R., Keogh, D. P., & Smith, S. L. (1999). Ecdysone 20-monooxygenase activity during larval-pupal-adult development of the tobacco hornworm, *Manduca sexta*. *Arch. Insect Biochem. Physiol., 41*, 24–32.

Mitchell, M. J., Keogh, D. P., Crooks, J. R., & Smith, S. L. (1993). Effects of plant flavonoids and other allelochemicals on insect cytochrome P-450 dependent steroid hydroxylase activity. *Insect Biochem. Mol. Biol., 23*, 65–71.

Mitchell, M. J., & Smith, S. L. (1986). Characterization of ecdysone 20-monooxygenase activity in wandering stage larvae of *Drosophila melanogaster*: Evidence for mitochondrial and microsomal cytochrome P-450 dependent systems. *Insect Biochem., 16*, 525–537.

Mitchell, M. J., & Smith, S. L. (1988). Ecdysone 20-monooxygenase activity throughout the life cycle of *Drosophila melanogaster*. *Gen. Comp. Endocrinol., 72*, 467–470.

Miyazaki, H., Ishibashi, M., Mori, C., & Ikekawa, N. (1973). Gas phase microanalysis of zooecdysones. *Anal. Chem., 45*, 1164–1168.

Mizuno, T., & Ohnishi, E. (1975). Conjugated ecdysone in the eggs of the silkworm, *Bombyx mori*. *Dev. Growth Differ., 17*, 219–225.

Modde, J.-F., Lafont, R., & Hoffmann, J. A. (1984). Ecdysone metabolism in *Locusta migratoria* larvae and adults. *Int. J. Invert. Reprod. Dev., 7*, 161–183.

Moog-Lutz, C., Tomasetto, C., Regnier, C. H., Wendling, C., Lutz, Y., et al. (1997). MLN64 exhibits homology with the steroidogenic acute regulatory protein (STAR) and is overexpressed in human breast carcinomas. *Int. J. Cancer, 71*, 183–191.

Morgan, E. D., & Huang, H.-P. (1990). Analysis of ecdysteroids by supercritical fluid chromatography. In A. R. McCaffery, & I. D. Wilson (Eds.), *Chromatography and Isolation of Insect Hormones and Pheromones* (pp. 95–102). London: Plenum Press.

Morgan, E. D., Murphy, S. J., Games, D. E., & Mylchreest, I. (1988). Analysis of ecdysteroids by supercritical-fluid chromatography. *J. Chromatogr., 441,* 165–169.

Morgan, E. D., & Poole, C. F. (1976). The extraction and determination of ecdysones in arthropods. *Adv. Insect Physiol., 12,* 17–62.

Morgan, E. D., & Wilson, I. D. (1989). Methods for separation and physico-chemical quantification of ecdysteroids. In J. Koolman (Ed.), *Ecdysone, From Chemistry to Mode of Action* (pp. 114–130). Stuttgart: Georg Thieme Verlag.

Morgan, E. D., & Woodbridge, A. P. (1971). Insect moulting hormones (ecdysones). Identification as derivatives by gas chromatography. *J. Chem. Soc. Chem. Commun.,* 475–476.

Mykles, D. L. (2011). Ecdysteroid metabolism in crustaceans. *J. Steroid Biochem. Mol. Biol.,* in the press. doi:10.1016/j.jsbmb.2010.09.001.

Nakai, M., Shiotsuki, T., & Kunimi, Y. (2004). An entomopoxvirus and a granulovirus use different mechanisms to prevent pupation of Adoxophyes honmai. *Virus Res., 101,* 185–191.

Nakanishi, K. (1971). The ecdysones. *Pure Appl. Chem., 25,* 167–195.

Namiki, T., Niwa, R., Sakudoh, T., Shirai, K., et al. (2005). Cytochrome P450 CYP307A1/Spook: A regulator for ecdysone synthesis in insects. *Biochem. Biophys. Res. Commun., 337,* 367–374.

Nashed, N. T., Michaud, D. P., Levin, W., & Jerina, D. M. (1986). 7-Dehydrocholesterol 5,6β-oxide as a mechanism-based inhibitor of microsomal cholesterol oxide hydrolase. *J. Biol. Chem., 261,* 2510–2513.

Nasir, H., & Noda, H. (2003). Yeast-like symbiotes as a sterol source in Anobiid beetles (Coleoptera, Anobiidae): Possible metabolic pathways from fungal sterols to 7-dehydrocholesterol. *Arch. Insect Biochem. Physiol., 52,* 175–182.

Nelson, D. R. (1998). Metazoan cytochrome P450 evolution. *Comp. Biochem. Physiol. C Pharmacol. Toxicol. Endocrinol., 121,* 175–182.

Neubueser, D., Warren, J. T., Gilbert, L. I., & Cohen, S. M. (2005). *Molting defective* is required for ecdysone biosynthesis. *Dev. Biol., 280,* 362–372.

Nigg, H. A., Svoboda, J. A., Thompson, M. J., Kaplanis, J. N., Dutky, S. R., et al. (1974). Ecdysone metabolism: Ecdysone dehydrogenase-isomerase. *Lipids, 9,* 971–974.

Niwa, R., Matsuda, T., Yoshiyama, T., et al. (2004). CYP306A1, a cytochrome P450 enzyme, is essential for ecdysteroid biosynthesis in the prothoracic glands of *Bombyx* and *Drosophila. J. Biol. Chem., 279,* 35942–35949.

Niwa, R., Sakudoh, T., Namiki, T., et al. (2005). The ecdysteroidogenic P450 *Cyp302a1/disembodied* from the silkworm, *Bombyx mori,* is transcriptionally regulated by prothoracicotropic hormone. *Insect Mol. Biol., 14,* 563–571.

Niwa, R., Namiki, T., Ito, K., Shimada-Niwa, Y., Kiuchi, M., et al. (2010). *Non-molting glossy/shroud* encodes a short-chain dehydrogenase/reductase that functions in the "Black Box" of the ecdysteroid biosynthesis pathway. *Development, 137,* 1991–1999.

Nomura, Y., Komatsuzaki, M., Iwami, M., & Sakurai, S. (1996). Purification and characterization of hemolymph 3-dehydroecdysteroid 3β-reductase of the silkworm, *Bombyx mori. Insect Biochem. Mol. Biol., 26,* 249–257.

Nüsslein-Volhard, C., Wieschaus, E., & Kluding, H. (1984). Mutations affecting the pattern of the larval cuticle in *Drosophila melanogaster.* I. Zygotic loci on the second chromosome. *Roux Arch. Dev. Biol., 183,* 267–282.

Oberlander, H. (1974). Biological activity of insect ecdysones and analogues *in vitro. Experientia, 30,* 1409–1410.

Ohnishi, E. (1986). Ovarian ecdysteroids of *Bombyx mori:* Retrospect and prospect. *Zool. Sci., 3,* 401–407.

Ohnishi, E., Hiramoto, M., Fujimoto, Y., Kakinuma, K., & Ikekawa, N. (1989). Isolation and identification of major ecdysteroid conjugates from the ovaries of *Bombyx mori. Insect Biochem., 19,* 95–101.

Ohnishi, E., Mizuno, T., Chatani, F., Ikekawa, N., & Sakurai, S. (1977). 2-Deoxy-α-ecdysone from ovaries and eggs of the silkworm, *Bombyx mori. Science, 197,* 66–67.

Ohtaki, T., Milkman, R. D., & Williams, C. M. (1967). Ecdysone and ecdysone analogues: Their assay on the flesh fly *Sarcophaga peregrina. Proc. Natl. Acad. Sci. USA, 85,* 981–984.

Okazaki, R. K., Snyder, M. J., Grimm, C. C., & Chang, E. S. (1997). Ecdysteroids in nemerteans: Further characterization and identification. *Hydrobiologia, 365,* 281–285.

Omura, T. (2006). Mitochondrial P450s. *Chem. Biol. Interact., 163,* 86–93.

Omura, T., & Ito, A. (1991). Biosynthesis and intracellular sorting of mitochondrial forms of cytochrome P450. *Method. Enzymol., 206,* 75–81.

Ono, H., Rewitz, K. F., Shinoda, T., Itoyama, K., et al. (2006). *Spook* and *Spookier* code for stage-specific components of the ecdysone biosynthetic pathway in Diptera. *Deve. Biol., 298,* 555–570.

O'Reilly, D. R. (1995). Baculovirus encoded ecdysteroid UDP-glucosyl-transferases. *Insect Biochem. Mol. Biol., 25,* 541–550.

O'Reilly, D. R., Brown, M. R., & Miller, L. K. (1992). Alteration of ecdysteroid metabolism due to baculovirus infection of the fall armyworm *Spodoptera frugiperda:* Host ecdysteroids are conjugated with galactose. *Insect Biochem. Mol. Biol., 22,* 313–320.

O'Reilly, D. R., Howarth, O. W., Rees, H. H., & Miller, L. K. (1991). Structure of the ecdysone glucoside formed by a baculovirus ecdysteroid UDP-glucosyltransferase. *Insect Biochem., 21,* 795–801.

O'Reilly, D. R., & Miller, L. K. (1989). A baculovirus blocks insect molting by producing ecdysteroid UDP-glucosyl-transferase. *Science, 245,* 1110–1112.

Pak, M. D., & Gilbert, L. I. (1987). A developmental analysis of ecdysteroids during the metamorphosis of *Drosophila melanogaster. J. Liq. Chromatgr., 10,* 2591–2611.

Papadopoulos, V., Baraldi, M., Guilarte, T. R., Knudsen, T. B., Lacapere, J. J., et al. (2006). Translocator protein (18kDa): New nomenclature for the peripheral-type benzodiazepine receptor based on its structure and molecular function. *Trends Pharmacol. Sci., 27,* 402–409.

Papadopoulos, V., Widmaier, E. P., Amri, H., Zilz, A., Li, H., et al. (1998). *In vivo* studies on the role of the peripheral benzodiazepine receptor (PBR) in steroidogenesis. *Endocr. Res., 24,* 479–487.

Park, E. J., Burand, J. P., & Yin, C.-M. (1993). The effect of baculovirus infection on ecdysteroid titer in Gypsy Moth larvae (*Lymantria dispar*). *J. Insect Physiol.*, *39*, 791–796.

Parvy, J. P., Blais, C., Bernard, F., Warren, J. T., Petryk, A., et al. (2005). A role for βFTZ-F1 in regulating ecdysteroid titers during post-embryonic development in *Drosophila melanogaster*. *Dev. Biol.*, *282*, 84–94.

Pasteels, J. M., Rowell-Rahier, M., Braekman, J. C., & Daloze, D. (1994). Chemical defense of adult leaf beetles updated. In P. H. Jolivet, M. L. Cox, & E. Petitpierre (Eds.), *Novel Aspects of the Biology of Chrysomelidae* (pp. 289–301). Dordrecht: Kluwer.

Peake, K. B., & Vance, J. E. (2010). Defective cholesterol trafficking in Niemann-Pick C-deficient cells. *FEBS Lett.*, *584*, 2731–2739.

Petersen, Q. R., Cambie, R. C., & Russel, G. B. (1993). Jones oxidation of 20-hydroxyecdysone (crustecdysone). *Aust. J. Chem*, *46*, 1961–1964.

Petryk, A., Warren, J. T., Marqués, G., Jarcho, M. P., Gilbert, L. I., et al. (2003). Shade: The *Drosophila* P450 enzyme that mediates the hydroxylation of ecdysone to the steroid insect molting hormone 20-hydroxyecdysone. *Proc. Natl. Acad. Sci. USA*, *100*, 13773–13778.

Pimprikar, G. D., Coign, M. J., Sakurai, H., & Heits, J. R. (1984). High-performance liquid chromatographic determination of ecdysteroid titers in the house fly. *J. Chromatogr.*, *317*, 413–419.

Pinedo, F. J.R., Moscardi., F., Luque, T., Olszewski, J. A., et al. (2003). Inactivation of the ecdysteroid UDP-glucosyltransferase (*egt*) gene of *Anticarsia gemmaltais* nucleopolyhedrovirus (AgMNPV) improves its virulence towards its insect host. *Biol. Control*, *27*, 336–344.

Pís, J., Girault, J.-P., Grau, V., & Lafont, R. (1995a). Analysis of ecdysteroid conjugates: Chromatographic characterization of glucosides, phosphates and sulfates. *Eur. J. Entomol.*, *92*, 41–50.

Pís, J., Girault, J.-P., Larchevêque, M., Dauphin-Villemant, C., & Lafont, R. (1995b). A convenient synthesis of 25-deoxyecdysone, a major secretory product of Crustacean Y-organs and of 2,25-dideoxyecdysone, its putative immediate precursor. *Steroids*, *60*, 188–194.

Pís, J., Hykl, J., Budesinski, M., & Harmatha, J. (1994). Regioselective synthesis of 20-hydroxyecdysone glycosides. *Tetrahedron*, *50*, 9679–9690.

Pís, J., & Vaisar, T. (1997). H/D isotopic exchange in the Fast Atom Bombardment of ecdysteroids. *J. Mass Spectrom.*, *32*, 1050–1056.

Pondeville, E., Maria, A., Jacques, J. C., Bourgoin, C., & Dauphin-Villemant, C. (2008). *Anopheles gambiae* males produce and transfer the vitellogenic steroid hormone 20-hydroxyecdysone to females during mating. *Proc. Natl. Acad. Sci. USA*, *105*, 19631–19636.

Poole, C. F., Morgan, E. D., & Bebbington, P. M. (1975). Analysis of ecdysone by gas chromatography using electron capture detection. *J. Chromatogr.*, *104*, 172–175.

Poole, C. F., Singhawangcha, S., Zlatkis, A., & Morgan, E. D. (1978). Determination of bifunctional compounds. Part III. Polynuclear aromatic boronic acids as selective fluorescent reagents for HPTLC and HPLC. *J. High Resol. Chromatogr. Chromatogr. Commun.*, *2*, 96–97.

Poole, C. F., Zlatkis, A., Sye, W.-F., Singhawangcha, S., & Morgan, E. D. (1980). The determination of steroids with and without natural electrophores by gas chromatography and electron-capture detection. *Lipids*, *15*, 734–744.

Porcheron, P., Morinière, M., Grassi, J., & Pradelles, P. (1989). Development of an enzyme immunoassay for ecdysteroids using acetylcholinesterase as label. *Insect Biochem.*, *19*, 117–122.

Raynor, W. M., Kithinji, J. P., Barker, I. K., Bartle, K. D., & Wilson, I. D. (1988). Supercritical fluid chromatography of ecdysteroids. *J. Chromatogr.*, *436*, 497–502.

Raynor, W. M., Kithinji, J. P., Bartle, K. D., Games, D. E., MylchReest, I. C., et al. (1989). Packed column supercritical fluid chromatography and linked supercritical fluid chromatography — mass spectrometry for the analysis of phytoecdysteroids from Silene nutans and Silene otites. *J. Chromatogr.*, *467*, 292–298.

Read, H., Wilson, I. D., & Lafont, R. (1990). Overpressure thin-layer chromatography of ecdysteroids. In A. R. McCaffery, & I. D. Wilson (Eds.), *Chromatography and Isolation of Insect Hormones and Pheromones* (p. 127). London: Plenum Press.

Redfern, C. P.F. (1984). Evidence for the presence of makisterone A in *Drosophila* larvae and the secretion of 20deoxymakisterone A by the ring gland. *Proc. Natl. Acad. Sci. USA*, *81*, 5643–5647.

Redfern, C. P.F. (1986). Changes in patterns of ecdysteroid secretion by the ring gland of *Drosophila* in relation to the sterol composition of the diet. *Experientia*, *42*, 307–309.

Redfern, C. P.F. (1989). Ecdysiosynthetic tissues. In J. Koolman (Ed.), *Ecdysone, From Chemistry to Mode of Action* (pp. 182–187). Stuttgart: Georg Thieme Verlag.

Rees, H. H. (1985). Biosynthesis of ecdysone. In G. A. Kerkut, & L. I. Gilbert (Eds.), *Comprehensive Insect Physiology, Biochemistry, and Pharmacology* (Vol. 7), (pp. 249–293). Oxford: Pergamon Press.

Rees, H. H. (1989). Zooecdysteroids structures and occurrence. In J. Koolman (Ed.), *Ecdysone, From Chemistry to Mode of Action* (pp. 28–38). Stuttgart: Georg Thieme Verlag.

Rees, H. H. (1995). Ecdysteroid biosynthesis and inactivation in relation to function. *Eur. J. Entomol.*, *92*, 9–39.

Rees, H. H., & Isaac, R. E. (1984). Biosynthesis of ovarian ecdysteroid phosphates and their metabolic fate during embryogenesis in *Schistocerca gregaria*. In J. A. Hoffmann, & M. Porchet (Eds.), *Biosynthesis, Metabolism and Mode of Action of Invertebrate Hormones* (pp. 181–195). Berlin: Springer-Verlag.

Rees, H. H., & Isaac, R. E. (1985). Biosynthesis and metabolism of ecdysteroids and methods of isolation and identification of the free and conjugated compounds. *Method. Enzymol.*, *111*, 377–410.

Reixach, N., Lafont, R., Camps, F., & Casas, J. (1999). Biotransformation of putative phytoecdysteroid biosynthetic precursors in tissue cultures of Polypodium vulgare. *Eur. J. Biochem.*, *266*, 608–615.

Reum, L., Klinger, W., & Koolman, J. (1984). A new immunoassay for ecdysteroids based on chemiluminescence. In L. J. Kricka, P. E. Stanley, G. H.G. Thorpe, & T. C. Whitehead (Eds.), *Analytical Applications of Bioluminescence and Chemiluminescence* (pp. 249–252). New York: Academic Press.

Reum, L., & Koolman, J. (1989). Radioimmunoassay of ecdysteroids. In J. Koolman (Ed.), *Ecdysone, From Chemistry to Mode of Action* (p. 131). Stuttgart: Georg Thieme Verlag.

Rewitz, K. F., & Gilbert, L. I. (2008). *Daphnia* Halloween genes that encode P450s mediating the synthesis of the arthropod molting hormone: Evolutionary implications. *BMC Evol. Biol., 8,* 60.

Rewitz, K. F., Larsen, M. R., Lobner-Olesen, A., Rybczynski, R., et al. (2009). A phosphoproteomics approach to elucidate neuropeptide signal transduction controlling insect metamorphosis. *Insect Biochem. Mol. Biol., 39,* 475–483.

Rewitz, K. F., O'Connor, M. B., & Gilbert, L. I. (2007). Molecular evolution of the insect Halloween family of cytochrome P450s: Phylogeny, gene organization and functional conservation. *Insect Biochem. Mol. Biol., 37,* 741–753.

Rewitz, K. F., Rybczynski, R., Warren, J., & Gilbert, L. I. (2006a). Identification, characterization and developmental expression of Halloween genes encoding P450 enzymes mediating ecdysone biosynthesis in the tobacco hornworm, *Manduca sexta. Insect Biochem. Mol. Biol., 36,* 188–199.

Rewitz, K. F., Rybczynski, R., Warren, J., & Gilbert, L. I. (2006b). Developmental expression of *Manduca shade*, the P450 mediating the final step in molting hormone synthesis. *Mol. Cell Endocr., 247,* 166–174.

Rharrabe, K., Alla, S., Maria, A., Sayah, F., & Lafont, R. (2007). Diversity of detoxification pathways of ingested ecdysteroids among phytophagous insects. *Arch. Insect Biochem. Physiol., 65,* 65–73.

Richards, G. (1978). The relative biological activity of α- and β-ecdysone and their 3-dehydro derivatives in the chromosome puffing assay. *J. Insect Physiol., 24,* 329–335.

Richter, K., Böhm, G.-A., & Leubert, F. (1999). 3-Dehydroecdysone secretion by the molting gland of the cockroach, *Periplaneta americana. Arch. Insect Biochem. Physiol., 41,* 107–116.

Robinson, P. D., Morgan, E. D., Wilson, I. D., & Lafont, R. (1987). The metabolism of ingested and injected [3H]ecdysone by final instar larvae of *Heliothis armigera. Physiol. Entomol., 12,* 321–330.

Rodenburg, K. W., & Van der Horst, D. J. (2005). Lipoprotein-mediated lipid transport in insects: Analogy to the mammalian lipid carrier system and novel concepts for the functioning of LDL receptor family members. *Biochim. Biophys. Acta, 1736,* 10–29.

Roth, G. E., Gierl, M. S., Vollborn, L., Meise, M., Lintermann, R., et al. (2004). The *Drosophila* gene start 1: A putative cholesterol transporter and key regulator of ecdysteroid synthesis. *Proc. Natl. Acad. Sci. USA, 101,* 1601–1606.

Rottiers, V., Motola, D. L., Gerisch, B., Cummins, C. L., Nishiwaki, K., et al. (2006). Hormonal control of *C. elegans* dauer formation and life span by a Rieske-like oxygenase. *Dev. Cell, 10,* 473–482.

Royer, C., Porcheron, P., Pradelles, P., Kerhoas, L., Einhorn, J., et al. (1995). Rapid and specific enzyme immunoassay for makisterone A. *Insect Biochem. Mol. Biol., 25,* 235–240.

Royer, C., Porcheron, P., Pradelles, P., & Mauchamp, B. (1993). Development and use of an enzymatic tracer for an enzyme immunoassay of makisterone A. *Insect Biochem. Mol. Biol., 23,* 193–197.

Rudolph, P. H., & Spaziani, E. (1992). Formation of ecdysteroids by Y-organs of the crab, Menippe mercenaria. II. Incorporation of cholesterol into 7-dehydrocholesterol and secretion products *in vitro. Gen. Comp. Endocrinol, 88,* 235–242.

Rudolph, P. H., Spaziani, E., & Wang, W. L. (1992). Formation of ecdysteroids by Y-organs of the crab, *Menippe mercenaria.* I. Biosynthesis of 7-dehydrocholesterol *in vivo. Gen. Comp. Endocrinol, 88,* 224–234.

Russell, G. B., & Greenwood, D. R. (1989). Methods of isolation of ecdysteroids. In J. Koolman (Ed.), *Ecdysone, From Chemistry to Mode of Action* (p. 97). Stuttgart: Georg Thieme Verlag.

Sakudoh, T., Tsuchida, K., & Kataoka, H. (2005). BmStart1, a novel carotenoid-binding protein isoform from *Bombyx mori,* is orthologous to MLN64, a mammalian cholesterol transporter. *Biochem. Biophys. Res. Commun., 336,* 1125–1135.

Sakurai, S., & Gilbert, L. I. (1990). Biosynthesis and secretion of ecdysteroids by the prothoracic glands. In E. Ohnishi, & H. Ishizaki (Eds.), *Molting and Metamorphosis* (pp. 83–106). Tokyo: Japan Scientific Society Press.

Sakurai, S., Warren, J. T., & Gilbert, L. I. (1989). Mediation of ecdysone synthesis in *Manduca sexta* by a hemolymph enzyme. *Arch. Insect Biochem. Physiol., 10,* 179–197.

Sakurai, S., Warren, J. T., & Gilbert, L. I. (1991). Ecdysteroid synthesis and molting by the tobacco hornworm, *Manduca sexta,* in the absence of prothoracic glands. *Arch. Insect Biochem. Physiol., 18,* 13–36.

Sakurai, S., Yonemura, N., Fujimoto, Y., Hata, F., & Ikekawa, N. (1986). 7-Dehydrosterols in prothoracic glands of the silkworm, *Bombyx mori. Experientia, 42,* 1034–1036.

Sall, C., Tsoupras, G., Kappler, C., Lagueux, M., Zachary, D., et al. (1983). Fate of maternal conjugated ecdysteroids during embryonic development in *Locusta migratoria. J. Insect Physiol., 29,* 491–507.

Santos, A. C., & Lehmann, R. (2004). Isoprenoids control germ cell migration downstream of HMGCoA reductase. *Dev. Cell, 6,* 283–293.

Sato, H., Ashida, N., Suhara, K., Itagaki, E., Takemori, S., et al. (1978). Properties of an adrenal cytochrome P-450 (P-450 11b) for the hydroxylations of corticosteroids. *Arch. Biochem. Biophys., 190,* 307–314.

Sato, Y., Sakai, M., Imai, S., & Fujioka, S. (1968). Ecdysone activity of plant-originated molting hormones applied on the body surface of lepidopterous larvae. *Appl. Entomol. Zool., 3,* 49–51.

Sayah, F., & Blais, C. (1995). Ecdysteroid metabolism in *Labidura riparia* females (Insecta, Dermaptera). *Neth. J. Zool., 45,* 89–92.

Scalia, S., Sbrenna-Miacciarelli, A., Sbrenna, G., & Morgan, E. D. (1987). Ecdysteroid titers and location in developing eggs of *Schistocerca gregaria. Insect Biochem., 17,* 227–236.

Schenkman, J. B., & Jansson, I. (2003). The many roles of cytochrome b5. *Pharmacol. Ther., 97,* 139–152.

Schildknecht, H. (1970). The defensive chemistry of land and water beetles. *Angew. Chem., Int. Ed., 9,* 1–9.

Schroeder, F., Frolov, A., Starodub, O., Atshaves, B. B., Russell, W., et al. (2000). Pro-sterol carrier protein-2: Role of the N-terminal presequence in structure, function and peroxisomal targeting. *J. Biol. Chem., 275,* 25547–25555.

Schwab, C., & Hétru, C. (1991). Synthesis and conversion study of a radiolabeled putative ecdysone precursor, 5β-cholest-7-ene-3β,6α,14α-triol in *Locusta migratoria* prothoracic glands. *Steroids, 56,* 316–319.

Schwartz, M. B., Imberski, R. B., & Kelly, T. J. (1984). Analysis of metamorphosis in *Drosophila melanogaster*: Characterization of giant, an ecdysteroid-deficient mutant. *Dev. Biol., 103,* 85–95.

Shampengtong, L., & Wong, K. P. (1989). An *in vitro* assay of 20-hydroxyecdysone sulfotransferase in the mosquito, *Aedes togoi. Insect Biochem., 19,* 191–196.

Shen, X.-J., Yi, Y.-Z., Tang, S.-M., Zhang, Z.-F., Li, Y.-R., & He, J.-L. (2004). The ecdysteroid UDP-glucosyltransferase gene promoter from *Autographa californica* multicapsid nucleopolyhedrovirus. *Z. Naturforsch., 59C,* 749–754.

Shergill, J. K., Cammack, R., Chen, J.-H., Fisher, M. J., Madden, S., et al. (1995). EPR spectroscopic characterization of the iron-sulphur proteins and cytochrome P-450 in the mitochondria from the insect *Spodoptera littoralis* (cotton leafworm). *Biochemical J., 307,* 719–728.

Sieglaff, D. H., Duncan, K. A., & Brown, M. R. (2005). Expression of genes encoding proteins involved in ecdysteroidogenesis in the female mosquito, *Aedes aegypti. Insect Biochem. Mol. Biol., 35,* 471–490.

Sláma, K., Abubakirov, N. K., Gorovits, M. B., Baltaev, U. A., & Saatov, Z. (1993). Hormonal activity of ecdysteroids from certain Asiatic plants. *Insect Biochem. Mol. Biol., 23,* 181–185.

Slinger, A. J., Dinan, L. N., & Isaac, R. E. (1986). Isolation of apolar ecdysteroid conjugates from newly-laid oothecae of *Periplaneta americana. Insect Biochem., 16,* 115–119.

Slinger, A. J., & Isaac, R. E. (1988a). Synthesis of apolar ecdysteroid conjugates by ovarian tissue from *Periplaneta americana. Gen. Comp. Endocrinol., 70,* 74–82.

Slinger, A. J., & Isaac, R. E. (1988b). Ecdysteroid titers during embryogenesis of the cockroach, *Periplaneta americana. J. Insect Physiol., 34,* 1119–1125.

Slinger, A. J., & Isaac, R. E. (1988c). Acyl-CoA: Ecdysone acyltransferase activity from the ovary of *P. americana. Insect Biochem., 18,* 779–784.

Sliter, T., & Gilbert, L. I. (1992). Developmental arrest and ecdysteroid deficiency resulting from mutations at the dre4 locus of *Drosophila* encodes a product with DNA binding activity. *Genetics, 130,* 555–568.

Smagghe, G. (Ed.), (2009). *Ecdysone: Structures and Functions* Berlin: Springer Science 583 pp.

Smith, S. L. (1985). Regulation of ecdysteroid titer: Synthesis. In G. A. Kerkut, & L. I. Gilbert (Eds.), *Comprehensive Insect Physiology, Biochemistry, and Pharmacology* (Vol. 7), (p. 295). Oxford: Pergamon Press.

Smith, S. L., Bollenbacher, W., Cooper, D., Schleyer, H., Wielgus, J., et al. (1979). Ecdysone 20-monooxygenase: Characterization of an insect cytochrome P-450 dependent steroid hydroxylase. *Mol. Cell Endocrinol., 15,* 111–133.

Smith, S. L., Bollenbacher, W. E., & Gilbert, L. I. (1980). Studies on the biosynthesis of ecdysone and 20-hydroxyecdysone in the tobacco hornworm, Manduca sexta. In J. A. Hoffmann (Ed.), *Progress in Ecdysone Research* (p. 139). Amsterdam: Elsevier/North-Holland Biomedical Press.

Smith, S. L., Bollenbacher, W. E., & Gilbert, L. I. (1983). Ecdysone 20-monooxygenase activity during larval–pupal development of *Manduca sexta. Mol. Cell Endocrinol., 31,* 227–251.

Smith, S. L., & Mitchell, M. J. (1986). Ecdysone 20-monooxygenase systems in a larval and an adult dipteran. *Insect Biochem., 16,* 49–55.

Smith, S. L., & Mitchell, M. J. (1988). Effects of azadirachtin on insect cytochrome P-450 dependent ecdysone 20-monooxygenase activity. *Biochem. Biophys. Res. Commun., 154,* 559–563.

Snyder, M. J., & Antwerpen, R. (1998). Evidence for a diazepam-binding inhibitor (DBI) benzodiazepine receptor-like mechanism in ecdysteroidogenesis by the insect prothoracic gland. *Cell Tissue Res., 294,* 161–168.

Snyder, M. J., & Feyereisen, R. (1993). A diazepam binding inhibitor (DBI) homologue from the tobacco hornworm, *Manduca sexta. Mol. Cell Endocrinol., 94,* R1–4.

Snyder, M. J., Scott, J. A., Andersen, J. F., & Feyereisen, R. (1996). Sampling P450 diversity by cloning polymerase chain reaction products obtained with degenerate primers. *Method. Enzymol., 272,* 304–312.

Soller, M., Bownes, M., & Kubli, E. (1999). Control of oocyte maturation in sexually mature *Drosophila* females. *Dev. Biol., 208,* 337–351.

Sommé-Martin, G., Colardeau, J., & Lafont, R. (1988). Conversion of ecdysone and 20-hydroxyecdysone into 3-dehydroecdysteroids is a major pathway in third instar *Drosophila melanogaster* larvae. *Insect Biochem., 18,* 729–734.

Sonobe, H., & Ito, Y. (2009). Phosphoconjugation and dephosphorylation reactions of steroid hormone in insects. *Mol. Cell Endocr., 307,* 25–35.

Sonobe, H., Ohira, T., Ieki, K., Maeda, S., et al. (2006). Purification, kinetic characterization, and molecular cloning of a novel enzyme, ecdysteroid 22-kinase. *J. Biol. Chem., 281,* 29513–29524.

Sonobe, H., Tokushige, H., Makka, T., Tsutsumi, H., Hara, N., et al. (1999). Comparative studies of ecdysteroid metabolism between diapause eggs and non-diapause eggs of the silkworm, *Bombyx mori. Zool. Sci., 16,* 935–943.

Sonobe, H., & Yamada, R. (2004). Ecdysteroids during early embryonic development in silkworm *Bombyx mori*: Metabolism and functions. *Zool. Sci., 21,* 503–516.

Spaziani, E., Rees, H. H., Wang, W. L., & Watson, R. D. (1989). Evidence that Y-organs of the crab *Cancer antennarius* secrete 3-dehydroecdysone. *Mol. Cell Endocrinol., 66,* 17–25.

Spindler, K.-D. (1988). Parasites and hormones. In H. Mehlhorn (Ed.), *Parasites* (pp. 465–476). New York: Springer.

Spindler, K.-D., Koolman, J., Mosora, F., & Emmerich, H. (1977). Catalytical oxidation of ecdysteroids to 3-dehydro products and their biological activities. *J. Insect Physiol., 23,* 441–444.

Srivatsan, J., Kuwahara, T., & Agosin, M. (1987). The effect of alpha-ecdysone and phenobarbital on the alpha-ecdysone 20-monooxygenase of house fly larva. *Biochem. Biophys. Res. Commun., 148,* 1075–1080.

Stocco, D. M. (2000). Intramitochondrial cholesterol transfer. *Biochim. Biophys. Acta, 1486,* 184–197.

Stocco, D. M. (2001). StAR protein and the regulation of steroid hormone biosynthesis. *Annu. Rev. Physiol., 63,* 193–213.

Stocco, D. M., & Clark, B. J. (1996). Regulation of the acute production of steroids in steroidogenic cells. *Endocr. Rev.*, *17*, 221–244.

Stolowich, N., Petrescu, A. D., Huang, H., Martin, G. G., Scott, A. I., et al. (2002). Sterol carrier protein-2: Structure reveals function. *Cell Mol. Life Sci.*, *59*, 193–212.

Storch, J., & Xu, Z. (2009). Niemann-Pick C2 (NPC2) and intracellular cholesterol trafficking. *Biochim. Biophys. Acta*, *1791*, 671–678.

Sturaro, A., Guerriero, A., De Clauser, R., & Pietra, F. (1982). A new unexpected marine source of a molting hormone. Isolation of ecdysterone in large amounts from the zoanthid *Gerardia savaglia*. *Experientia*, *38*, 1184–1185.

Suksamrarn, A., Jankam, A., Tarnchompoo, B., & Putchakarn, S. (2002). Ecdysteroids from a *Zoanthus* sp. *J. Nat. Prod.*, *65*, 1194–1197.

Suksamrarn, A., & Yingyongnarongkul, B. (1996). Synthesis and biological activity of 2-deoxy-20-hydroxyecdysone and derivatives. *Tetrahedron*, *52*, 12623–12630.

Svoboda, J. A. (1999). Variability of metabolism and function of sterols in insects. *Crit. Rev. Biochem. Mol. Biol.*, *34*, 49–57.

Svoboda, J. A., Agarwal, H., Robbins, W. E., & Nair, A. M.G. (1980). Lack of conversion of C29 phytosterols to cholesterol in the Khapra beetle, *Trogoderma granarium*. *Experientia*, *36*, 1029–1030.

Svoboda, J. A., & Feldlaufer, M. F. (1991). Neutral sterol metabolism in insects. *Lipids*, *26*, 614–618.

Svoboda, J. A., Herbert, E. W., Jr., & Thompson, M. J. (1983). Definitive evidence for lack of phytosterol dealkylation in honey bees. *Experientia*, *39*, 1120–1121.

Svoboda, J. A., Imberski, R. B., & Lusby, W. R. (1989). *Drosophila* melanogaster does not dealkylate [^{14}C]sitosterol. *Experientia*, *45*, 983–985.

Svoboda, J. A., & Lusby, W. R. (1994). Variability of sterol utilization in stored-product insects. *Experientia*, *50*, 72–74.

Svoboda, J. A., & Thompson, M. J. (1985). Steroids. In G. A. Kerkut, & L. I. Gilbert (Eds.), *Comprehensive Insect Physiology, Biochemistry, Pharmacology* (Vol. 10) (pp. 137–175). Oxford: Pergamon Press.

Svoboda, J. A., Thompson, M. J., Robbins, W. E., & Elden, T. C. (1975). Unique pathways of sterol metabolism in the Mexican bean beetle, a plant-feeding insect. *Lipids*, *10*, 524–527.

Svoboda, J. A., & Weirich, G. F. (1995). Sterol metabolism in the tobacco hornworm, *Manduca sexta* — a review. *Lipids*, *30*, 263–267.

Swevers, L., Kravariti, L., Ciolfi, S., Xenou-Kokoletsi, M., Ragousis, N., et al. (2004). A cell-based high-throughput screening system for detecting ecdysteroid agonists and antagonists in plant extracts and libraries of synthetic compounds. *FASEB J.*, *18*, 134–136.

Sztal, T., Chung, H., Gramzow, L., Daborn, P. J., et al. (2007). Two independent duplications forming the Cyp307a genes in *Drosophila*. *Insect Biochem. Mol. Biol.*, *37*, 1044–1053.

Tabunoki, H., Sugiyama, H., Tanaka, Y., Fujii, H., Banno, Y., et al. (2002). Isolation, characterization and cDNA sequence of a carotenoid binding protein from the silk gland of *Bombyx mori* larvae. *J. Biol. Chem.*, *277*, 32133–32140.

Takahashi, S. Y., Okamoto, K., Sonobe, H., Kamba, M., & Ohnishi, E. (1992). *In vitro* synthesis of ecdysteroid conjugates by tissue extracts of the silkworm, *Bombyx mori. Zool. Sci.*, *9*, 169–174.

Takeuchi, H., Chen, J.-H., O'Reilly, D. R., Rees, H. H., & Turner, P. C. (2000). Regulation of ecdysteroid signaling: Molecular cloning, characterization and expression of 3-dehydroecdysone 3α-reductase, a novel eukaryotic member of the short-chain dehydrogenases/reductases superfamily from the cotton leafworm, *Spodoptera littoralis. Biochem. J.*, *349*, 239–245.

Takeuchi, H., Chen, J.-H., O'Reilly, D. R., Turner, P. C., & Rees, H. H. (2001). Regulation of ecdysteroid signaling: Cloning and characterization of ecdysone oxidase — a novel steroid oxidase from the cotton leafworm, *Spodoptera littoralis. J. Biol. Chem.*, *276*, 26819–26828.

Takeuchi, H., Rigden, D. J., Ebrahimi, B., Turner, P. C., & Rees, H. H. (2005). Regulation of ecdysteroid signalling during *Drosophila* development: Identification, characterization and modelling of ecdysone oxidase, an enzyme involved in control of ligand concentration. *Biochem. J.*, *389*, 637–645.

Talamillo, A., Sánchez, J., Cantera, R., Pérez, C., et al. (2008). Smt3 is required for *Drosophila melanogaster* metamorphosis. *Development*, *135*, 1659–1668.

Tawfik, A. I., Tanaka, Y., & Tanaka, S. (2002). Possible involvement of ecdysteroids in embryonic diapause of *Locusta migratoria. J. Insect Physiol.*, *48*, 743–749.

Thiboutot, D., Jabara, S., McAllister, J. M., Sivarajah, A., Gilligand, K., et al. (2003). Human skin is a steroidogenic tissue: Steroidogenic enzymes and cofactors are expressed in epidermis, normal sebocytes, and an immortalized sebocyte cell line (SEB-A). *J. Invest. Dermatol.*, *120*, 905–914.

Thiry, Y., & Hoffmann, K. H. (1992). Dynamics of ecdysone and 20-hydroxyecdysone metabolism after injection and ingestion in *Gryllus bimaculatus. Zool. Jahrb. Allg. Zool. 96*, 17–38.

Thompson, M. J., Feldlaufer, M. F., Lozano, R., Rees, H. H., Lusby, W. R., et al. (1987a). Metabolism of [^{14}C] 26-hydroxyecdysone 26-phosphate in the tobacco hornworm, *Manduca sexta* (L.), to a new ecdysteroid conjugate: [^{14}C] 26-hydroxyecdysone 22-glucoside. Arch. *Insect Biochem. Physiol.*, *4*, 1–15.

Thompson, M. J., Kaplanis, J. N., Robbins, W. E., Dutky, S. R., & Nigg, H. N. (1974). 3- *Epi*-20-hydroxyecdysone from meconium of the tobbacco hornworm. *Steroids*, *24*, 359–366.

Thompson, M. J., Kaplanis, J. N., Robbins, W. E., & Yamamoto, R. T. (1967). 20,26-Dihydroxyecdysone, a new steroid with moulting hormone activity from the tobacco hornworm, *Manduca sexta* (Johannson). *J. Chem. Soc. Chem. Commun.*, 650–653.

Thompson, M. J., Louloudes, S. J., Robbins, W. E., Waters, J. A., Steele, J. A., et al. (1963). The identity of the major sterol from house flies reared by the CSMA procedure. *J. Insect Physiol.*, *9*, 615–622.

Thompson, M. J., Svoboda, J. A., & Feldlaufer, M. F. (1989). Analysis of free and conjugated ecdysteroids and polar metabolites of Insects. In W. D. Nes, & E. J. Parish (Eds.), *Analysis of Sterols and Other Biologically Significant Steroids* (pp. 81–105). San Diego: Academic Press.

Thompson, M. J., Svoboda, J. A., Lozano, R., & Wilzer, K. R. (1988). Profile of free and conjugated ecdysteroids and ecdysteroid acids during embryonic development of *Manduca sexta* (L.) following maternal incorporation of [^{14}C] cholesterol. *Arch. Insect Biochem. Physiol.*, *7*, 157–172.

Thompson, M. J., Svoboda, J. A., Lusby, W. R., Rees, H. H., Oliver, J. E., et al. (1985c). Biosynthesis of a C21 steroid conjugate in an insect. The conversion of [^{14}C]-cholesterol to 5- [^{14}C]-pregnen-3β,20β-diol glucoside in the tobacco hornworm, *Manduca sexta*. *J. Biol. Chem.*, *260*, 15410–15412.

Thompson, M. J., Svoboda, J. A., Rees, H. H., & Wilzer, K. R. (1987b). Isolation and identification of 26-hydroxyecdysone 2-phosphate: An ecdysteroid conjugate of eggs and ovaries of the tobacco hornworm, *Manduca sexta*. *Arch. Insect Biochem. Physiol.*, *4*, 183–190.

Thompson, M. J., Weirich, G. F., Rees, H. H., Svoboda, J. A., Feldlaufer, M. F., et al. (1985b). New ecdysteroid conjugate: Isolation and identification of 26-hydroxyec-dysone 26-phosphate from eggs of the tobacco hornworm, *Manduca sexta* (L.). *Arch. Insect Biochem. Physiol.*, *2*, 227–236.

Thompson, M. J., Weirich, G. F., & Svoboda, J. A. (1985a). Ecdysone 3-epimerase. *Method. Enzymol.*, *111*, 437–442.

Thomson, J. A., Imray, F. P., & Horn, D. H.S. (1970). An improved *Calliphora* bioassay for insect moulting hormones. *Aust. J. Exp. Biol.*, *48*, 321–328.

Tijet, N., Helvig, C., & Feyereisen, R. (2001). The cytochrome P450 gene superfamily in *Drosophila melanogaster*: Annotation, intron-exon organization and phylogeny. *Gene*, *262*, 189–198.

Tomaschko, K.-H. (1995). Autoradiographic and morphological investigations of the defensive ecdysteroid glands in adult *Pycnogonum litorale* (Arthropoda: Pantopoda). *Eur. J. Entomol.*, *92*, 105–112.

Touchstone, J. C. (1986). Ecdysteroids. In *CRC Handbook. Chromatography. Steroids* (pp. 119–135). Boca Raton, FL: CRC Press.

Tsoupras, G., Hétru, C., Luu, B., Constantin, E., Lagueux, M., et al. (1983a). Identification and metabolic fate of ovarian 22-adenosine monophosphoric ester of 2-deoxyecdysone in ovaries and eggs of an insect, *Locusta migratoria*. *Tetrahedron*, *39*, 1789–1796.

Tsoupras, G., Hétru, C., Luu, B., Lagueux, M., Constantin, E., et al. (1982b). The major conjugates of ecdysteroids in young eggs and in embryos of *Locusta migratoria*. *Tetrahedron Lett.*, *23*, 2045–2048.

Tsoupras, G., Luu, B., Hétru, C., Müller, J. F., & Hoffmann, J. A. (1983b). Conversion *in vitro* de 20-hydroxyecdysone en métabolites phosphorylés et acétylés par des complexes tube digestif-tubes de Malpighi de larves de *Locusta migratoria*. *C.R. Acad. Sci. Paris*, *(Sér. III)*, *296*, 77–80.

Tsoupras, G., Luu, B., & Hoffmann, J. A. (1982a). Isolation and identification of three ecdysteroid conjugates with a C-20 hydroxyl group in eggs of *Locusta migratoria*. *Steroids*, *40*, 551–560.

Van den Broek, P. J. A., Barroso, M., & Lechner, M. C. (1996). Critical amino-terminal segments in insertion of rat liver cytochrome P450 3A1 into the endoplasmic reticulum membrane. *Experientia*, *52*, 851–855.

Von Buddenbrock, W. (1931). Untersuchungen über die Häutungshormone des Schmetterlinge. *Z. Vergleich. Physiol.*, *14*, 415–428.

Von Gliscynski, U., Delbecque, J.-P., Böcking, D., Sedlmeier, D., Dircksen, H., et al. (1995). Three new antisera with high sensitivity to ecdysone, 3-dehydroecdysone and other A-ring derivatives: Production and characterization. *Eur. J. Entomol.*, *92*, 75–79.

Von Wachenfeldt, C., & Johnson, E. F. (1995). Structures of eukaryotic cytochrome P450 enzymes. In P. R. Ortiz de Montellano (Ed.), *Cytochromes P450, Structure, Mechanism, and Biochemistry* (pp. 183–223). New York: Plenum Press.

Vyazunova, I., Wessley, V., Kim, M., & Lan, Q. (2007). Identification of two sterol carrier protein-2 like genes in the yellow fever mosquito, *Aedes aegypti*. *Insect Mol. Biol.*, *16*, 305–314.

Wainwright, G., Prescott, M. C., Lomas, L. O., Webster, S. G., & Rees, H. H. (1996). Trace analysis of arthropod hormones by liquid chromatography-atmospheric pressure chemical ionization/mass spectrometry. *Biochem. Soc. Trans.*, *24*, 476S.

Wainwright, G., Prescott, M. C., Lomas, L. O., Webster, S. G., & Rees, H. H. (1997). Development of a new high-performance liquid chromatography-mass-spectrometric method for the analysis of ecdysteroids in biological extracts. *Arch. Insect Biochem. Physiol.*, *35*, 21–31.

Warren, J. T., Bachman, J. S., Dai, J. D., & Gilbert, L. I. (1996). Differential incorporation of cholesterol and cholesterol derivatives into ecdysteroids by the larval ring glands and adult ovaries of *Drosophila melanogaster*: A putative explanation for the l(3)ecd^1 mutation. *Insect Biochem. Mol. Biol.*, *26*, 931–943.

Warren, J. T., Dai, J., & Gilbert, L. I. (1999). Can the insect nervous system synthesize ecdysteroids? *Insect Biochem. Mol. Biol.*, *29*, 571–579.

Warren, J. T., & Gilbert, L. I. (1986). Ecdysone metabolism and distribution during the pupal–adult development of *Manduca sexta*. *Insect Biochem.*, *16*, 65–82.

Warren, J. T., & Gilbert, L. I. (1988). Radioimmunoassay of ecdysteroids. In L. I. Gilbert, & T. A. Miller (Eds.), *Immunological Techniques in Insect Biology* (p. 181). Berlin: Springer-Verlag.

Warren, J. T., & Gilbert, L. I. (1996). Metabolism *in vitro* of cholesterol and 25-hydroxycholesterol by the larval prothoracic glands of *Manduca sexta*. *Insect Biochem. Mol. Biol.*, *26*, 917–929.

Warren, J. T., & Hétru, C. (1990). Ecdysone biosynthesis: Pathways, enzymes and the early steps problem. *Invert. Reprod. Dev.*, *18*, 91–99.

Warren, J. T., O'Connor, M. B., & Gilbert, L. I. (2009). Studies on the black box: Incorporation of 3-oxo-7-dehydrocholesterol into ecdysteroids by *Drosophila melanogaster* and *Manduca sexta*. *Insect Biochem. Mol. Biol. 39*, 677–687.

Warren, J. T., Petryk, A., Marqués, G., Jarcho, M., Parvy, J.-P., et al. (2002). Molecular and biochemical characterization of two P450 enzymes in the ecdysteroidogenic pathway of *Drosophila melanogaster*. *Proc. Natl. Acad. Sci., USA*, *99*, 11043–11048.

Warren, J. T., Petryk, A., Marqués, G., Parvy, J.-P., Shinoda, T., et al. (2004). Phantom encodes the 25-hydroxylase of *Drosophila melanogaster* and *Bombyx mori*: A P450 enzyme critical in ecdysone biosynthesis. *Insect Biochem. Mol. Biol.*, *34*, 991–1010.

Warren, J. T., Rybczynski, R., & Gilbert, L. I. (1995). Stereospecific, mechanism-based, suicide inhibition of a cytochrome P450 involved in ecdysteroid biosynthesis in the prothoracic glands of *Manduca sexta*. *Insect Biochem. Mol. Biol., 29*, 679–695.

Warren, J. T., Sakurai, S., Rountree, D. R., & Gilbert, L. I. (1988a). Synthesis and secretion of ecdysteroids by the prothoracic glands of *Manduca sexta*. *J. Insect Physiol., 34*, 571–576.

Warren, J. T., Sakurai, S., Rountree, D. R., Gilbert, L. I., Lee, S.-S., et al. (1988b). Regulation of the ecdysteroid titer of *Manduca sexta*: Reappraisal of the role of the prothoracic glands. *Proc. Natl. Acad. Sci. USA, 85*, 958–962.

Warren, J. T., Steiner, B., Dorn, A., Pak, M., & Gilbert, L. I. (1986). Metabolism of ecdysteroids during the embryogenesis of *Manduca sexta*. *J. Liq. Chromatogr., 9*, 1759–1782.

Warren, J. T., Wismar, J., Subrahmanyam, B., & Gilbert, L. I. (2001). Woc (without children) gene control of ecdysone biosynthesis in *Drosophila melanogaster*. *Mol. Cell Endocrinol, 181*, 1–14.

Waterham, H. R., Koster, J., Romeijn, G. J., Hennekam, R. C.M., Vreken, P., et al. (2001). Mutations in the 3β-hydroxysterol Δ^{24}-reductase gene cause desmosterolosis, an autosomal recessive disorder of cholesterol biosynthesis. *Am. J. Hum. Genet., 69*, 685–694.

Watson, R. D., & Spaziani, E. (1982). Rapid isolation of ecdysteroids from crustacean tissues and culture media using Sep-Pak C18 cartridges. *J. Liq. Chromatogr., 5*, 525–535.

Webb, T. J., Powls, R., & Rees, H. H. (1995). Enzymes of ecdysteroid transformation and inactivation in the midgut of the cotton leafworm, *Spodoptera littoralis*: Properties and developmental profiles. *Biochem. J., 312*, 561–568.

Webb, T. J., Powls, R., & Rees, H. H. (1996). Characterization, fractionation and kinetic properties of the enzymes of ecdysteroid 3-epimerization and phosphorylation isolated from the midgut cytosol of the cotton leaf-worm, *Spodoptera littoralis*. *Insect Biochem. Mol. Biol., 26*, 809–816.

Webster, S. G. (1985). Catalyzed derivatization of trimethylsilyl ethers of ecdysterone. A preliminary study. *J. Chromatogr., 333*, 186–190.

Weirich, G. F. (1989). Enzymes involved in ecdysone metabolism. In J. Koolman (Ed.), *Ecdysone, From Chemistry to Mode of Action* (pp. 174–180). Stuttgart: Georg Thieme Verlag.

Weirich, G. (1997). Ecdysone 20-hydroxylation in *Manduca sexta*. (Lepidoptera: Sphingidae) midgut: Development-related changes of mitochondrial and microsomal ecdysone 20-monooxygenase activities in the fifth larval instar. *Eur. J. Entomol., 94*, 57–65.

Weirich, G. F., & Bell, R. A. (1997). Ecdysone 20-hydroxylation and 3-epimerization in larvae of the Gypsy Moth, *Lymantria dispar* L: Tissue distribution and developmental changes. *J. Insect Physiol., 43*, 643–649.

Weirich, G. F., Feldlaufer, M. F., & Svoboda, J. A. (1993). Ecdysone oxidase and 3-oxoecdysteroid reductases in *Manduca sexta*: Developmental changes and tissue distribution. *Arch. Insect Biochem. Physiol., 23*, 199–211.

Weirich, G. F., & Svoboda, J. A. (1992). 3-Oxoecdysteroid reductases in *Manduca sexta* midgut. *Arch. Insect Biochem. Physiol., 21*, 91–102.

Weirich, G. F., Svoboda, J. A., & Thompson, M. J. (1984). Ecdysone 20-monooxygenases. In J. A. Hoffmann, & M. Porchet (Eds.), *Biosynthesis, Metabolism and Mode of Action of Invertebrate Hormones* (pp. 227). Berlin: Springer-Verlag.

Weirich, G. F., Svoboda, J. A., & Thompson, M. J. (1985). Ecdysone 20-monooxygenase in mitochondria and microsomes of *Manduca sexta* L. midgut: Is the dual localization real?. *Arch. Insect Biochem. Physiol., 2*, 385–396.

Weirich, G. F., Thompson, M. J., & Svoboda, J. A. (1986). *In vitro* ecdysteroid conjugation by enzymes of *Manduca sexta* midgut cytosol. *Arch. Insect Biochem. Physiol., 3*, 109–126.

Weirich, G. F., Thompson, M. J., & Svoboda, J. A. (1989). Ecdysone oxidase and 3-oxoecdysteroid reductases in *Manduca sexta* midgut: Kinetic parameters. *Arch. Insect Biochem. Physiol., 12*, 201–218.

Weirich, G. F., Thompson, M. J., & Svoboda, J. A. (1991). Enzymes of ecdysteroid 3-epimerization in midgut cytosol of *Manduca sexta*: pH optima, consubstrate kinetics, and sodium chloride effect. *Insect Biochem., 21*, 65–71.

Weirich, G. F., Williams, V. P., & Feldlaufer, M. F. (1996). Ecdysone 20-hydroxylation in *Manduca sexta* midgut: Kinetic parameters of mitochondrial and microsomal ecdysone 20-monooxygenases. *Arch. Insect Biochem. Physiol., 31*, 305–312.

Weltzel, J. M., Ohnishi, M., Fujita, T., Nakanishi, K., Naya, Y., et al. (1992). Diversity in steroidogenesis of symbiotic microorganisms from planthoppers. *J. Chem. Ecol., 18*, 2083–2094.

Whitehead, D. L. (1989). Ecdysteroid carrier proteins. In J. Koolman (Ed.), *Ecdysone, From Chemistry to Mode of Action* (p. 232). Stuttgart: George Thieme Verlag.

Whiting, P., & Dinan, L. (1989). Identification of the endogenous apolar ecdysteroid conjugates present in newly-laid eggs of the house cricket (*Acheta domesticus*) as 22-long-chain fatty acyl esters of ecdysone. *Insect Biochem., 19*, 759–765.

Whiting, P., Sparks, S., & Dinan, L. (1993). Ecdysteroids during embryogenesis of the house cricket, *Acheta domesticus*: Occurrence of novel ecdysteroid conjugates in developing eggs. *Insect Biochem. Mol. Biol., 23*, 319–329.

Whiting, P., Sparks, S., & Dinan, L. (1997). Endogenous ecdysteroid levels and rates of ecdysone acylation by intact ovaries *in vitro* in relation to ovarian development in adult female house crickets, *Acheta domesticus*. *Arch. Insect Biochem. Physiol., 35*, 279–299.

Wieschaus, E., Nüsslein-Volhard, C., & Jürgens, G. (1984). Mutations affecting the pattern of the larval cuticle in *Drosophila melanogaster*. III. Zygotic loci on the X-chromosome and fourth chromosome. *Roux Arch. Dev. Biol., 193*, 296–307.

Wigglesworth, K. P. W., Lewis, D., & Rees, H. H. (1985). Ecdysteroid titer and metabolism to novel apolar derivatives in adult female *Boophilus microplus*. *Arch. Insect Biochem. Physiol., 2*, 39–54.

Wigglesworth, V. B. (1934). Factors controlling moulting and "metamorphosis" in an insect. *Nature, 133,* 725–726.

Williams, C. M. (1968). Ecdysone and ecdysone-analogues: Their assay and action on diapausing pupae of the *Cynthia* silkworm. *Biol. Bull., 134,* 344–355.

Williams, D. R., Chen, J.-H., Fisher, M. J., & Rees, H. H. (1997). Induction of enzymes involved in molting hormone (ecdysteroid) inactivation by ecdysteroids and an agonist, 1,2- dibenzoyl-1-tert-butylhydrazine (RH -5849). *J. Biol. Chem., 272,* 8427–8432.

Williams, D. R., Fisher, M. J., & Rees, H. H. (2000). Characterization of ecdysteroid 26-hydroxylase: An enzyme involved in molting hormone inactivation. *Arch. Biochem. Biophys., 376,* 389–398.

Williams, D. R., Fisher, M. J., Smagghe, G., & Rees, H. H. (2002). Species specificity of changes in ecdysteroid metabolism in response to ecdysteroid agonists. *Pest. Biochem. Physiol., 72,* 91–99.

Wilson, I. D. (1985). Thin-layer chromatography of ecdysteroids. *J. Chromatogr., 318,* 373–377.

Wilson, I. D. (1988). Reversed phase thin-layer chromatography of the ecdysteroids. *J. Planar Chromatogr., 1,* 116–122.

Wilson, I. D., Bielby, C. R., & Morgan, E. D. (1982a). Evaluation of some phytoecdysteroids as internal standards for the chromatographic analysis of ecdysone and 20-hydroxyecdysone in arthropods. *J. Chromatogr., 236,* 224–229.

Wilson, I. D., Bielby, C. R., & Morgan, E. D. (1982b). Studies on the reversed-phase thin-layer chromatography of ecdysteroids on C-12 bonded and paraffin-coated silica. *J. Chromatogr., 242,* 202–206.

Wilson, I. D., & Lafont, R. (1986). Thin-layer chromatography and high-performance thin-layer chromatography of [³H] metabolites of 20-hydroxyecdysone. *Insect Biochem., 16,* 33–40.

Wilson, I. D., Lafont, R., Kingston, R. G., & Porter, C. J. (1990a). Thin-layer chromatography-tandem mass spectrometry directly from the adsorbent: Application to phytoecdysteroids from *Silene otites. J. Planar Chromatogr., 3,* 359–361.

Wilson, I. D., Lafont, R., Porter, C. J., Kingston, R. G., Longden, K., et al. (1990c). Thin-layer chromatography of ecdysteroids: Detection and identification. In A. R. McCaffery, & I. D. Wilson (Eds.), *Chromatography and Isolation of Insect Hormones and Pheromones* (pp. 117–126). London: Plenum Press.

Wilson, I. D., Lafont, R., & Wall, P. (1988a). TLC of ecdysteroids with off-line identification by fast atom bombardment mass spectrometry directly from the adsorbent. *J. Planar Chromatogr., 1,* 357–359.

Wilson, I. D., & Lewis, S. (1987). Separation of ecdysteroids by high-performance thin-layer chromatography using automated multiple development. *J. Chromatogr., 408,* 445–448.

Wilson, I. D., Morgan, E. D., & Murphy, S. J. (1990b). Sample preparation for the chromatographic analysis of ecdysteroids using solid phase extraction methods. *Anal. Chim. Acta, 236,* 145–155.

Wilson, I. D., Morgan, E. D., Robinson, P., Lafont, R., & Blais, C. (1988b). A comparison of radio-thin-layer and radio-high-performance liquid chromatography for ecdysteroid metabolism studies. *J. Insect Physiol, 34,* 707–711.

Wilson, I. D., Scalia, S., & Morgan, E. D. (1981). Reversed-phase thin-layer chromatography for the separation and analysis of ecdysteroids. *J. Chromatogr., 212,* 211–219.

Wismar, J., Habtemichael, N., Warren, J. T., Dai, J. D., Gilbert, L. I., et al. (2000). The mutation without children (rgl) causes ecdysteroid deficiency in third-instar larvae of *Drosophila melanogaster. Dev. Biol., 226,* 1–17.

Wright, J. E., & Thomas, B. R. (1983). Boll weevil: Determination of ecdysteroids and juvenile hormones with high pressure liquid chromatography. *J. Liq. Chromatogr., 6,* 2055–2066.

Yamada, R., & Sonobe, H. (2003). Purification, kinetic characterization, and molecular cloning of a novel enzyme ecdysteroid-phosphate phosphatase. *J. Biol. Chem., 278,* 26365–26373.

Yamada, R., Yamahama, Y., & Sonobe, H. (2005). Release of ecdysteroid-phosphates from egg yolk granules and their déphosphorylation during early embryonic development in silkworm, *Bombyx mori. Zool. Sci., 22,* 187–198.

Yang, R. S., & Wilkinson, C. F. (1972). Enzymatic sulphation of p-nitrophenol and steroids by larval gut tissues of the southern armyworm (*Prodenia eridania* Cramer). *Biochem. J., 130,* 487–493.

Yasuda, A., Ikeda, M., & Naya, Y. (1993). Analysis of hemolymph ecdysteroids in the crayfish *Procambarus clarkii* by high-performance capillary electrophoresis. *Nippon Suisan Gakk, 59,* 1793–1799.

Yingyongnarongkul, B., & Suksamrarn, A. (2000). 25-Deoxyecdysteroids: Synthesis and moulting hormone activity of two C-25 epimers of inokosterone. *Sci. Asia, 26,* 15–20.

Yoshiyama, T., Namiki, T., Mita, K., Kataoka, H., & Niwa, R. (2006). Neverland is an evolutionarily conserved Rieske-domain protein that is essential for ecdysone synthesis and insect growth. *Development, 133,* 2565–2574.

Young, N. L. (1976). The metabolism of ³H-molting hormone in *Calliphora erythrocephala* during larval development. *J. Insect Physiol., 22,* 153–155.

Yu, S. J. (1995). Allelochemical stimulation of ecdysone 20-monooxygenase in fall armyworm larvae. *Arch. Insect Biochem. Physiol., 28,* 365–375.

Yu, S. J. (2000). Allelochemical induction of hormone-metabolizing microsomal monooxygenases in the fall armyworm. *Zool. Stud., 39,* 243–249.

Zhang, M., & Kubo, I. (1992a). Characterization of ecdysteroid 22-O-acyltransferase from tobacco budworm, *Heliothis virescens. Insect Biochem. Mol. Biol., 22,* 599–603.

Zhang, M., & Kubo, I. (1992b). Possible function of ecdysteroid 22-O-acyltransferase in the larval gut of tobacco budworm, *Heliothis virescens. J. Chem. Ecol., 18,* 1139–1149.

Zhang, M., & Kubo, I. (1993). Metabolic fate of ecdysteroids in larval *Bombyx mori* and *Heliothis virescens. Insect Biochem. Mol. Biol., 23,* 831–843.

Zhang, M., & Kubo, I. (1994). Mechanism of *Heliothis virescens* resistance to exogenous ecdysteroids. *Bioreg. Crop Protect. Pest Contr.*, *557*, 182–201.

Zhu, W. M., Zhu, H. J., Tian, W. S., et al. (2002). The selective dehydroxylation of 20-hydroxyecdysone by Zn powder and anhydrous acetic acid. *Synthetic Commun.*, *32*, 1385–1391.

Relevant Web Site

http://www.ecdybase.org — This Web site contains datasheets on more than 400 different ecdysteroids and a large amount of bibliographic data.

5 The Ecdysteroid Receptor

Vincent C. Henrich
University of North Carolina-Greensboro,
Greensboro, NC, USA

© 2012 Elsevier B.V. All Rights Reserved

5.1. Overview

5.1.1. Molecular Identity of the Insect Ecdysteroid Receptor

The morphogenetic events associated with insect development are largely triggered by the action of a single class of steroid hormones, the ecdysteroids.[1] The canonical mode of ecdysteroid action, as originally described in *Drosophila melanogaster*, has more or less been verified in species across all orders in the class Insecta. Generally, the mechanism by which the ecdysteroid-induced orchestration of transcriptional responses that underlie cellular changes leading to molting, metamorphosis, and reproductive maturation are mediated by a heterodimer comprised of the ecdysone receptor (EcR; Koelle *et al.*, 1991) and ultraspiracle (USP; Oro *et al.*, 1990; Shea

---

[1]Ecdysteroids refers to the family of ecdysteroids including natural and artificial steroids. Individual forms, including the classic active "molting" hormone, 20-hydroxyecdysone (20E, formerly B-ecdysone), will be specified throughout the chapter, as will its precursor, ecdysone (formerly a-ecdysone). Ecdysteroid agonists refer to molecules that are not steroidal, but are capable of inducing one or more insect EcR genes in a given experimental regime.

et al., 1990; Henrich *et al.*, 1990). The dimer is stabilized by the molting hormone, 20-hydroxyecdysone (20E), and recognizes specific promoter elements in the insect genome to regulate transcription (Yao *et al.*, 1992, 1993; Thomas *et al.*, 1993). Both proteins belong to a superfamily of nuclear receptors that mediates transcriptional responses to steroids and other lipophilic molecules (Evans, 2005; King-Jones and Thummel, 2005, and references therein). A second ecdysteroid-signaling pathway involving USP and another nuclear receptor, DHR38, has been described (Baker *et al.*, 2003). EcR mediated activity has been described in *Drosophila* that does not require USP (Costantino *et al.*, 2008), suggesting that EcR may be able to dimerize with another partner, form homodimers, and/or form protein complexes that have yet to be described. In scorpions, EcR apparently does not require USP as a heterodimeric partner (Nakagawa *et al.*, 2007) perhaps revealing an evolutionary aspect of ecdysteroid action.

The insect EcR is a distant relative of the vertebrate farnesol X receptor (FXR; Forman *et al.*, 1995) and LXR (Willy *et al.*, 1995) and was first characterized in the fruit fly, *D. melanogaster*. **Table 1** provides an updated listing

DOI: 10.1016/B978-0-12-384749-2.10005-6

Table 1   EcR and USP/RXR genes and GenBank Accession numbers

Phylum/Class/Order	Species	EcR, USP (or RXR)	GenBank Accession Nos.
Phylum Arthropoda			
Class Insecta			
Diptera	Aedes aegypti	EcR	U02021
		EcRA	AY345989
		USPa	AF305213
		USPb	AF305214
		RXR (partial)	XM_001656028
	Ae. albopicus	EcR	AF210733
		USP	AF210734
	Bradysia hygida	EcR (partial)	AF121910
	Calliphora vicina	EcR	AF325360
	Ceratitis capitata	"EcR"	AJ224341
		EcR-A (partial)	AJ438575
	Drosophila melanogaster	EcR-A	NM_165464
		EcR-B1	NM_165465
		EcR-B2	NM_165462
		USP	NM_057433
	D. pseudoobscura	EcR	XM_001361914
	Lucilia cuprina	EcR	U75355
		USP	AY007213
	Lucilia sericata	EcR (partial)	AB118974
		USP (partial)	AB118975
	Sarcophaga crassipalpis	EcR (partial)	AF239825
		USP (partial)	AF239826
	S. similis	EcR (partial)	AB196921
Lepidoptera	Bicyclus anynana	EcR (partial)	AJ251810
	Bombyx mori	EcR-A	NM_001173377
		EcR-B1	NM_001173375
		EcR-B2	NM_001043866
		USP (CF1)	NM_001044005 (U06073)
	Chilo suppressalis	EcR-A	AB067811
		EcR-B1	AB067812
		USP	AB081840
	Chironus tentans	EcRH	S60739
		USP-1	AF045891
	Choristoneura fumiferana	EcR-A	AF092030
		EcR-B	U29531
		USP	AF016368
	Helicoverpa armigera	EcR (2 partial)	EU526831/EF174331
		USP-1	EU526832
		USP-2	EF174333
	Heliothis virescens	EcR-B1	Y09009
		USP	BD183648
	Junonia coenia	EcR (partial)	AJ251809
	Manduca sexta	EcR	U19812
		EcRA (partial)	U49246
		USP-1	U44837
		USP-2 (partial)	U57921
	Orgyia recens	EcR-A	AB086421
		EcR-B1	AB86422
	Omphisa fuscidentalis	EcR-A	EF667890
		EcR-B1	EF667891
	Plodia interpunctella	EcR	AY489269
		USP	AY619987
	Spodoptera litura	EcR	EU180021
		USP	EU180022
	Spodoptera frugiperda	EcR (partial)	AF411254
		USP (partial)	AF411255
	Trichoplusia ni	EcR (partial)	AF411256
		USP (partial)	AF411257

Table 1   EcR and USP/RXR genes and GenBank Accession numbers—Cont'd

Phylum/Class/Order	Species	EcR, USP (or RXR)	GenBank Accession Nos.
Hymenoptera	Apis mellifera	EcR-A	NM_001098215
			AB490017
			AB267886
		EcR-B1	AB490050
			NM_001159355
		USP	NM_001011634
			AF263459
			AY273778
	Copidosoma floridanum	EcR (putative, partial)	S64705
	Camonotus japonicas	EcR-A	AB296080
		EcR α	AB296081
	Lepopilina heterotoma	EcR (partial)	AY157932
		USP (partial)	AY157931
	Nasonia vitripennis	EcR-A	NM_001159356
		EcR-B1	NM_001159357
	Pheidole megacephala	EcR-A	AB194765
		EcR-B (partial)	AB194766
	Polistes dominulus	EcR (partial)	DQ083517
	Sarcophaga depilis	USP	DQ190542
Coleoptera	Anthonomus grandis	EcR	EU935856
	Leptinotarsa decemlineata	EcR-A	AB211191
		EcR-B1	AB211192
		USP	AB211193
	Tr. castaneum	EcR-A	NM_001114178
		EcR-B	NM_001141918
		USP	NM_001114294
	Te. molitor	EcR	Y1153
		USP	AJ251542
	Harmonia axyridis	EcR-A	AB506666
		EcR-B1	AB506665
		USP-1	AB506667
		USP-2	AB506668
	Epilachna vigintioctopunctata	EcR-A	AB506670
		EcR-B1	AB506669
		USP-1	AB506671
		USP-2	AB506672
Orthoptera	Blattella germanica	EcR-A	AM039690
		RXR-S	AJ854489
		RXR-L	AJ854490
	Locusta migratoria	EcR	AF049136
		RXR	AF136372
		RXR-I	AY348873
Hemiptera	Bemica tabaci	EcR (partial)	EF174329
		USP (partial)	EF174330
	Myzus persicae	EcR	EF174334
		USP (partial)	EF174335
	Acrythosiphon pisum	EcR-A	NM_001159359
		EcR-B1	NM_001159360
		USP	NM_001161668
Dictyoptera	Periplanta americana	USP (RXR) (partial)	AY157928
Collembola	Folsomia candida	USP (RXR) (partial)	AY157930
Myriapoda	Lithobius forficatus	USP (RXR) (partial)	AY157929
Urochordata	Polyandrocarpa miaskiensis	USP (RXR) (partial)	AB030318
Trichoptera	Hydropsyche incognita	EcR (partial)	DQ083516
Mecoptera	Panorpa germanica	EcR (partial)	DQ083515
		USP (RXR) (partial)	DQ083512
Strepsiptera	Xenos vesparum	EcR (partial)	DQ083518
Class Crustacea			
	Americamysis bahia	EcR	DD410574
	Carcinus maenas	EcR (partial)	AY496927
		RXR-I	EU683888

(Continued)

Table 1   EcR and USP/RXR genes and GenBank Accession numbers—Cont'd

Phylum/Class/Order	Species	EcR, USP (or RXR)	GenBank Accession Nos.
	Daphnia magna	EcR-A1	AB274820
			AB274821
		EcR-A2	AB274822
			AB274823
		EcR-B	AB274824
		USP	AB274819
	Gecarcinus lateralis	RXR	DQ067280
	Marsupenaeus japonicas	EcR	AB295492
		RXR	AB295493
	Celuca pugilator	EcR	AF034086
		RXR	AF032983
	Tigriopus japonicus	EcR	GQ351503
	Amblyomma americanum	EcR-A1	AF020187
		EcR-A2	AF020188
		EcR-A3	AF020186
		RXR-1	AF035577
		RXR-2	AF035578
	Ornithodoros moubata	EcR-A	AB191193
		RXR	AB353290
Class Arachnida			
	Liocheres australasiae	EcR	AB297929
		USP	AB297930
Phylum Nematoda			
	Dirofilaria immitis	EcR	GQ200815
		RXR-1	AF438229
	Pristionchus pacificus	EcR*	
		RXR*	
	Haemondus contortus	EcR	GU250809
	Brugia malayi	EcR-A	EF362469
		EcR-C	EF262470
Phylum Platyhelminthes Class Trematoda			
	Schistosoma mansoni	RXR	AF158102
Phylum Mollusca			
	Lottia gigantea	EcR*	
	Euprymna scolopes	EcR*	
	Aplysia californica	EcR	
Phylum Annelida Class Polychaeta			
	Capitella teleta	EcR*	
Class Clitellata			
	Hirudo medicinalis	EcR*	
	Helobdella robusta	EcR*	

*Indicates sequences that have been reported but have not yet been assigned an Accession Number.

of the species and orders for which EcR and USP have been identified, which now includes other classes within the Arthropoda. Recently, a functional EcR orthologue has been reported outside the arthropod phylum (Shea *et al.*, 2010).

The USP protein is an orthologue of the retinoid X receptor (RXR), which in turn, is also a heterodimeric partner for several other nuclear receptors (Mangelsdorf *et al.*, 1990; Oro *et al.*, 1990). Both EcR and USP have diverged evolutionarily among the insect orders (see Chapter 6), indicating that the functional properties of EcR and USP have also diverged among them (Iwema *et al.*, 2009). In particular, the dimerization interface surfaces of EcR and USP have co-evolved so that the contact points, and thus the shape and binding properties of the ligand-binding pocket have changed. This diversity will be discussed in more depth and later.

5.1.2. Historical Perspective

The pursuit of the insect ecdysteroid receptor and its activities has led to a convergence of several basic and applied research disciplines. The ability of a single steroid hormone to induce widespread and coordinated changes in gene transcription has attracted basic scientists interested in the mechanistic processes associated with its variable gene and cellular action. For insect biologists, the diversity of developmental responses triggered by a single class of steroid hormones among the insect orders reveals the essential importance of this process for understanding the evolutionary diversity and life cycle adaptation seen among individual insect species. In turn, the variety of response among the insect orders has led to attempts to disrupt ecdysteroid receptor action with species-specific agonists and antagonists such as the bisacylhydrazines (Wing, 1988) for insecticidal purposes (Palli *et al.*, 2005; Nakagawa and Henrich, 2009). Finally, because ecdysteroids exert no harmful effects in humans and other organisms, the ecdysteroid receptor is now viewed as a potential inducer of beneficial transcriptional responses in plant and animal cells. Much of the work on ecdysteroid action that was reported prior to 1985 will not be discussed here (Riddiford, 1985; O'Connor, 1985; Yund and Osterbur, 1985) except for those reports that continue to have a direct bearing on recent investigations.

The effect of ecdysteroids upon gene transcriptional activity became apparent through studies of the effects evoked by the "molting hormone" 20E, along the polytene chromosomes of insects such as *D. melanogaster* and *Chironomus tentans* (midge). Polytene chromosomes result from several rounds of DNA replication unaccompanied by cytokinesis, so that structural features such as euchromatic regions appear as bands along the chromosome copies lying side by side (see **Figure 1**). Puffing, indicative of unraveling of a region within the chromosome to allow transcription, was first reported in 1933 in *D. melanogaster* by E. Heitz, and in the early 1950s Wolfgang Beerman performed seminal work on puff changes along the polytene chromosomes in the larval salivary glands of *C. tentans*. Later, Ulrich Clever demonstrated that the insect molting hormone induced puffing changes and thus reported the first evidence that steroid hormones act directly upon the transcriptional activity of specific target genes, now regarded as a central feature of steroid hormone action in all animals. A variety of biochemical studies also established the presence of a protein in cellular extracts that binds to ecdysteroids, further indicating that an intracellular receptor mediates the transcriptional response (O'Connor, 1985; Yund and Osterbur, 1985, and older references therein).

Conceptually, the polytene chromosomes can be viewed as an *in situ* expression microarray, since the puff sites disclose the transcription of specific genes (by chromosomal location rather than by sequence). This natural array has the added and unique benefit of showing temporal changes in puff size that roughly reflect continuous changes in transcriptional rate. Hans Becker noted in 1959 that "early" puffs appear when incubated with the ecdysteroidogenic ring gland and regress as other "late puffs" appear later (Becker, 1959). Michael Ashburner

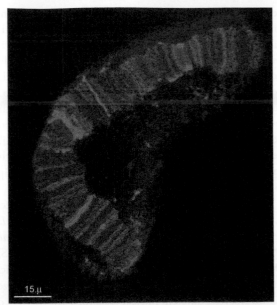

Figure 1 Pre-pupal chromosome III from *C. tentans* labeled with antibodies against EcR (green) and RNA polymerase II (red). Yellow signals indicate co-localization of the two antibodies. Green signal is a fixation artifact. Reproduced with permission from *Wegmann et al.* (1995) for methods.

later showed that blocking the protein translation of early puff RNAs with cycloheximide treatment prevents the regression of some early puffs, and simultaneously prevents the appearance of many late puffs (Ashburner *et al.*, 1974) From these findings, Ashburner postulated that an intracellular ecdysteroid receptor directs a transcriptional response at early puff sites in the presence of 20E. Further, the early puff gene products regulate the appearance of later puffs and also feedback to repress their own expression. This original model for ecdysteroid action has proven remarkably durable over the years, although it is often overlooked that several puffs described by Ashburner have not been placed within the regulatory hierarchy and that some features of the model have still not been pursued at the molecular level.

The role of a receptor was also implicated by the identification of an ecdysone response element (EcRE) in the ecdysteroid-inducible promoter of the 27 kDa heat shock protein (*hsp27*) of *D. melanogaster* (Riddihough and Pelham, 1987). The palindromic inverted nucleotide repeat sequence of this ecdysteroid receptor target resembles the motifs that at the time of its discovery were associated with a growing class of nuclear receptors for several vertebrate hormones, including the glucocorticoids and estrogen, raising anticipation that the ecdysteroid receptor is evolutionarily conserved, just as the process of steroid action on transcriptional activity is mechanistically conserved (Evans, 2005, and references therein).

Much of the progress made on the molecular genetic basis of ecdysteroid action can be traced to the molecular identity of the "puff" genes targeted by ecdysteroid action[2]. One of the early puff genes, E75, encodes a member of the nuclear hormone receptor superfamily (as defined by its two cysteine–cysteine zinc fingers that are responsible for the receptor protein's recognition of specific promoter elements). E75 has proven to be a receptor for heme, whose redox state is tied to its function as a ligand (Reinking et al., 2005). Several other ecdysteroid-responsive gene products in Drosophila, and numerous other superfamily members, are "orphan receptors," that is, they carry the zinc finger configuration but do not mediate transcriptional activity through any known ligand.

The E75 coding region was used as a probe to screen a cDNA library, which resulted in the isolation of the D. melanogaster ecdysone receptor gene (DmEcR; Koelle et al., 1991). EcR dimerizes with a second nuclear receptor, USP (DmUSP; Yao et al., 1993; Thomas et al., 1993), which structurally resembles the vertebrate RXR that dimerizes itself with several different nuclear receptors. The recovery of EcR, USP, and orphans such as E75 inspired a more exhaustive search for nuclear receptors in D. melanogaster and its annotated genome includes 21 nuclear receptors; orthologues for many of these receptors have been found in other insects and in some cases, other arthropods (see King-Jones and Thummel, 2005, and references therein). Methyl farnesoate (MF), the immediate precursor of juvenile hormone (JH), has been implicated as a ligand for USP and will be discussed later (Jones et al., 2006, 2010; see Chapter 8).

The recovery and characterization of EcR, USP, and other orphan receptors, along with various ecdysteroid-inducible targets has led to a proliferation of DNA probes, antibodies, and double-stranded RNAs (dsRNAs) with which to study developmental processes. Several insect and arthropod researchers have identified orthologues for EcR and USP as well as the "early puff" gene targets mentioned earlier (see **Table 1**). The abundance of information accumulated in the Drosophila system has heavily influenced mechanistic interpretations of ecdysteroid action.

5.1.3. Chapter Organization

The straightforward thesis of the 20E-triggered hierarchy based on the chromosomal puff response seems to contrast with the complexity found in many in vivo experiments. Ecdysteroid responsiveness of target genes varies widely by tissue or cell type, gene promoter organization, species, incubation conditions, developmental time,

activating ligand, ligand concentration, and treatment duration, presumably because the functional capabilities of the ecdysteroid receptor are modified by its cellular and environmental context. Nevertheless, several common and conserved themes have emerged from this work that are valuable not only for insect biologists, but also for other geneticists, cell biologists, and endocrinologists.

The organization of this chapter is based largely on the criteria by which the DmEcR was originally defined by Koelle et al. (1991): (1) the deduced EcR amino acid sequence includes the cysteine-cysteine zinc fingers and the twelve α-helices of the ligand-binding domain (LBD) that typify nuclear receptors, (2) extracts from cells expressing EcR bind to radiolabeled ponA (see Chapter 4) and binding to this phytoecdysteroid disappears with the addition of anti-EcR antibodies, (3) cellular extracts contain a protein that binds to an hsp27 EcRE which does not display such binding when the extract is pretreated with an anti-EcR antibody, (4) ecdysteroid-inducible transcription is restored to an ecdysone-insensitive Drosophila cell line by transfecting the cells with EcR cDNA, and (5) the pattern of spatial and temporal expression of EcR during pre-metamorphic development is consistent with its role as a mediator of ecdysteroid response (Koelle et al., 1991).

This chapter will examine the information gathered about the ecdysteroid receptor at three levels and provide brief descriptions and references for the tools used to undertake those experiments. First, the receptor will be viewed as a structural entity with particular emphasis on the sequence characteristics of EcR and USP as well as their biophysical and biochemical properties. These studies build upon the first three properties noted for EcR in its original characterization. Secondly, the transcriptional function of the ecdysteroid receptor will be examined with particular regard to its interactions with ligand, other protein factors, and promoter elements. This section will also describe a few of the ecdysteroid-inducible systems that have been developed for various applications. Third, the ecdysteroid receptor will be viewed from a cellular and developmental perspective in vivo with special emphasis on the spatial and temporal diversity of ecdysteroid-mediated action, along with the approaches used to address these questions. Finally, a prognosis will be offered concerning unresolved and arising issues surrounding ecdysteroid action via its receptors along with possible experimental technologies and strategies that might clarify them.

5.2. Ecdysteroid Receptor Structure, Biophysics, and Biochemistry

5.2.1. Domain Organization and Amino Acid Alignment

Both EcR and USP belong to the superfamily of nuclear receptors first described for several steroid hormones and vitamins among the vertebrates. Like their evolutionary

---

[2]The puff sites in Drosophila melanogaster were originally identified by their cytological location among the 102 subintervals arbitrarily delineated across the sex chromosome (X or chromosome 1, subintervals 1-20) and three autosomes (chromosome 2, subintervals 21-60; chromosome 3, subintervals 61-100, chromosome 4, subintervals 101-102) that comprise the D. melanogaster genome. The puffs are further designated as E (early or early-late) or L (late) based on their temporal appearance after 20E incubation.

counterparts, EcR and USP are structurally modular, that is, they are composed of distinct domains responsible for specific molecular functions (**Figure 2**). Individual domains are at least partially autonomous in their function, since domains of different nuclear receptors can be swapped to create structural and functional chimera. Chimerae derived from the EcR and USP of different insect species have been used to compare and differentiate their functional capabilities (Suhr *et al.*, 1998; Wang *et al.*, 2000; Henrich *et al.*, 2000). Fusion constructs designed to test the responsiveness of the ligand-binding domain to a variety of test compounds have been developed for both *in vitro* and *in vivo* investigations (Palanker *et al.*, 2006; LaPenna *et al.*, 2008; Beck *et al.*, 2009, and references therein).

The basic organization of nuclear receptors has been described extensively in other reviews, and will be examined here exclusively in terms of the functional features found in insects and associated with each of the domains: (1) the N-terminal (A/B) domain through which nuclear receptors interact with other transcriptional factors; (2) the DNA binding (DBD) or C domain, which is responsible for the receptor's recognition of specific DNA response elements in the genome and is comprised of two cysteine-cysteine zinc fingers that define proteins as members of the nuclear receptor superfamily; (3) the hinge region (D domain), which has been implicated in ligand-dependent heterodimerization, nuclear localization, and DNA recognition; (4) the E- or ligand-binding domain (LBD), which is typically comprised of 12 α-helices that form a ligand-binding pocket; and (5) the F domain, which is

found as a nonconserved sequence in all of the insect EcRs but not in any known USP or RXR sequence.

5.2.1.1. The A/B domain This domain is sometimes referred to as the transactivation domain because this portion of the nuclear receptor interacts with the cell's transcriptional machinery and is responsible for a ligand-independent transcriptional activation function (AF1). The A/B domain tends to be variable among nuclear receptors, although there is modest similarity in this portion of EcR and USP among insect species, as revealed by PCR cloning of the EcR N-terminal region for 51 insect species spanning all the insect orders (Watanabe *et al.*, 2010). The number of A/B isoforms of EcR and/or USP varies among species and arises from the activity of alternative promoters and/or alternative pre-mRNA splicing. The occurrence of multiple isoforms for EcR and USP does not follow a rigorously predictable pattern, although most species encode both an A and B isoform, each of which carries distinct motifs that have not been extensively studied. There are also functional motifs within both the A and B isoforms that are restricted evolutionarily to specific orders and have been classified into specific subsets (Watanabe *et al.*, 2010; also see Chapter 6). The *Dm*EcR gene encodes three isoforms (A, B1, and B2; Talbot *et al.*, 1993) that arise from two alternative promoters (A and B). The B1 and B2 isoforms share some motifs in the N-terminal domain, but B1 also includes a large N-terminal region as a result of alternative splicing. Other biochemical and functional properties of the insect EcR isoforms and

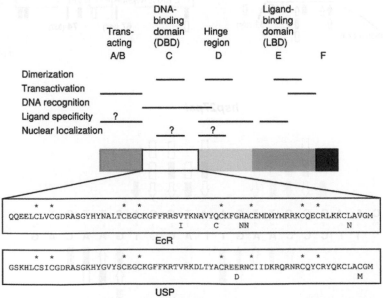

Figure 2 Domain organization of nuclear receptor superfamily members. Bars denote regions of the receptor associated with specific subfunctions or suspected subfunctions (as indicated with a ?). Boxed regions indicate consensus amino acid sequences for the DNA-binding domain of insect EcR and USP. Asterisks (*) indicate positions of two sets of four cysteines that ligand to zinc ion to form each of the two zinc fingers.

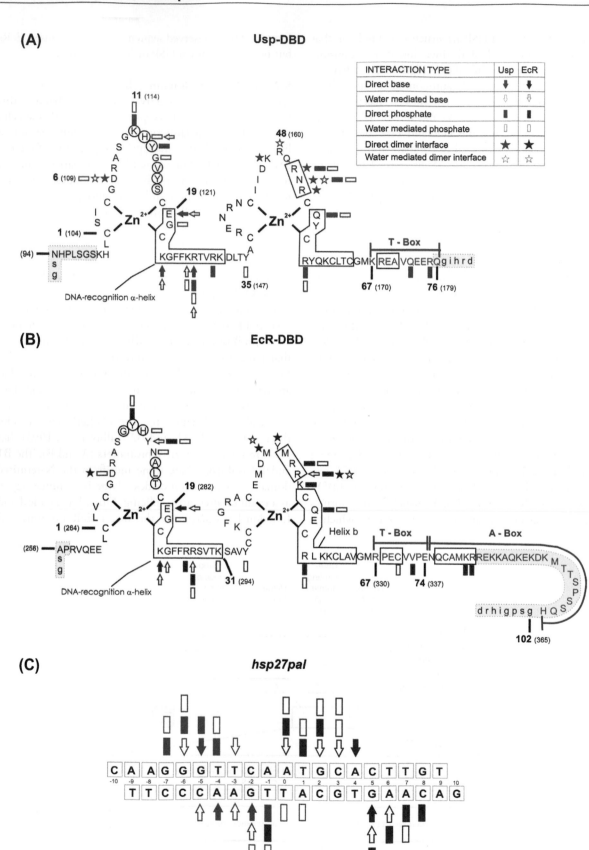

their distinct developmental roles will be discussed in Sections 5.3.3. and 5.4.5.

By contrast to the extensive sequence characterization of the EcR isoforms, relatively little work has been done to compare the A/B domain of insect USP/RXR sequences. In at least one species, *C. tentans*, multiple USP isoforms differing in their A/B domain occur along with a single EcR isoform; interestingly, the two isoforms differ markedly in both their developmental expression and in their transcriptional capabilities (Vogtli *et al.*, 1999). The migratory locust (*Locusta migratoria; Lm*) also expresses multiple USP isoforms, although the USP variants do not arise from differential splicing in the A/B domain (Hayward *et al.*, 2003). Along with two EcR isoforms, two USP isoforms exist in *Aedes aegypti* (Kapitskaya *et al.*, 1996), *Manduca sexta* (Jindra *et al.*, 1997), and *Bombyx mori* (Tzertzinis *et al.*, 1994). Two RXR/USP isoforms also exist in *Amblyomma americanum* along with its three EcR isoforms (Guo *et al.*, 1998). **Table 1** also designates the species and sequence identity of RXR/USPs that have been characterized so far.

5.2.1.2. The C domain or DBD The members of the nuclear receptor superfamily are defined by a 66-68 amino acid region known as the cysteine-cysteine zinc finger (**Figures 2** and **3**). The EcR DBD closely resembles the vertebrate FXR, and both EcR/USP and FXR/RXR recognize a palindromic inverted repeat sequence separated by a single nucleotide (IR1); the *hsp27* EcRE follows this arrangement, but the contact points for several amino acid residues in the EcR-DBD are different for the IR1 compared to the *hsp27* EcRE (**Figure 4**). EcR also carries a conserved C-terminal extension (CTE) of the DBD consisting of several amino acids, which is necessary for high-affinity recognition of the *hsp27* EcRE and is unique among nuclear receptors (**Figure 3**; Jakob *et al.*, 2007). The plasticity of the EcR DBD is reflected in the relatively variable consensus EcRE sequence based on the investigation of several functional IR1 elements is 5'-PuG(G/T)T(C/G) A(N)TG(C/A(C/A(C/t)Py (Antoniewski *et al.*, 1993). The EcR/USP complex is also capable of recognizing direct repeat elements (D'Avino *et al.*, 1995; Antoniewski *et al.*, 1996).

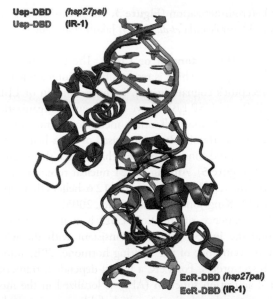

Figure 4 A superposition of the DmEcR-DBD/DmUSP-DBD/*hsp27* EcRE DNA complex and the DmEcR/DmUSP/IR1 DNA complex. Note the bending of the *hsp27* palindromic EcRE imposed by the EcR/USP interaction. Adapted from *Jakob et al.* (2007).

The amino acid sequence is highly conserved among the Diptera and Lepidoptera, but a few residues are substituted in the EcR DBD among insects; **Figure 2**).

5.2.1.3. The D domain or hinge region The D domain includes the amino acids that lie between the zinc fingers and the LBD. Portions of this region are highly conserved among all the insect EcRs, including a T-box motif that plays a role in DNA recognition (Devarakonda *et al.*, 2003), an A-box, and a helical structure essential for high-affinity recognition of a DNA response element (Niedziela-Majka *et al.*, 2000) and comprises the CTE mentioned earlier (Jakob *et al.*, 2007; **Figure 3**). The D region of EcR is also essential for ligand-dependent heterodimerization with USP (Suhr *et al.*, 1998; Perera *et al.*, 1999). The USP hinge region includes a highly conserved T-box motif, which plays a role in site recognition of DNA response elements but is not essential

Figure 3 The protein constructs and DNA response element constructs used for crystallization study, designating amino acid contact sites with noted DNA interactions. (A) Sequence of *D. melanogaster* EcR and USP DBD with liganded zinc fingers; T-box (green) and A-box (red). (B) Sequence of *D. melanogaster* USP DBD with liganded zinc fingers and T-box. Numbering of amino acid residues is relative to the first conserved cysteine, with the authentic numbers appearing in parentheses. In gray boxes, the N- and C-terminal residues not visible in electron density maps are listed; lower case letters are cloning artifacts. The α-helices are boxed and the residues from the β-sheet are circled. (C) Schematic diagram showing amino acid residue interactions with the *hsp27* EcRE and the EcR-USP complex. Red indicates USP interaction sites and blue indicates EcR interaction sites. Solid arrows indicate a direct base interaction and hollow arrows indicate a water-mediated base interaction. Solid bars indicate a direct phosphate interaction, and hollow bars indicate a water-mediated phosphate interaction. Adapted from *Jakob et al.* (2007).

for heterodimerization (**Figure 3**; Niedziela-Majka *et al.*, 2000; Devarakonda *et al.*, 2003; Jakob *et al.*, 2007).

5.2.1.4. The E domain or LBD The domain of the nuclear receptor, which is responsible for interacting with the receptor's cognate ligand, is the E domain or LBD. The EcR and USP LBDs have the canonical organization observed for almost all other members of the nuclear receptor superfamily. It is comprised of 12 α-helices that form a ligand-binding pocket that holds the cognate ligand (**Figure 5**). Among the 21 nuclear receptors in *D. melanogaster*, all but two carry the 12 α-helices that typify the LBD (King-Jones and Thummel, 2005).

The sequence of the EcR LBD is highly conserved among the insects, which is consistent with the widespread occurrence of the molting hormone, 20E, among the insect orders. For EcR, a ligand-dependent transcriptional activation function (AF2) is localized in the most carboxy-terminal helix 12, which folds over the pocket to hold the ligand molecule inside (**Figure 5**). This folding creates an interactive surface with other proteins that ultimately modulates the receptor's transcriptional activity. For nuclear receptors generally, a dimerization interface lies along helices 9 and 10 (Perlmann *et al.*, 1996) and ligand-independent transcriptional functions (AF1) are also been associated with the EcR LBD (Hu *et al.*, 2003). Later the functional features of the LBD based on biochemical, cellular, and *in vivo* experiments will be discussed as

well as the effects of LBD modification and species-based sequence differences upon EcR and USP capabilities.

A possible ligand for USP has been a subject of intense investigation over the past decade. Using a fluorescence assay, *Drosophila* USP shows a nanomolar affinity for MF and a mutational ablation of putative binding sites for MF significantly reduces this affinity (Jones *et al.*, 2006). Using a reporter gene assay system, Wang and LeBlanc (2009) showed that the crustacean RXR is responsive to MF, but only when dimerized with EcR.

The USP LBD has diverged considerably over evolutionary time and the EcR/USP dimerization interface has co-evolved with it. Iwema *et al.* (2009) phylogenetically subdivided the USP LBD into three categories: (1) the dipteran and lepidopteran USP LBD, which appears to have no pocket and has undergone rapid evolutionary divergence; (2) the LBD of Coleoptera and Hymenoptera, which resembles mammalian RXR but also does not bind to RXR ligands; and (3) the USP LBD of lower insects, which has been reported to bind to the RXR ligand, 9-cis retinoic acid (Nowickyj *et al.*, 2008). The amino acid residues responsible for dimerization between these USP subclasses and EcR are schematically shown in **Figure 6** Iwema *et al.*, 2009 and Hemiptera EcR/USP dimers is shown in **Figure 5** (Carmichael *et al.*, 2005).

5.2.1.5. F domain Only a few members of the nuclear receptor superfamily possess the F domain; the *D. melanogaster* EcR carries up to 226 amino acids (in *D. melanogaster*) beyond the helix 12 region, although these are not found among other insect EcR sequences. Apparently, the F domain of *D. melanogaster* EcR is totally dispensable, since its cleavage produces a receptor that has the same transcriptional capabilities as the full-length EcR (Hu *et al.*, 2003).

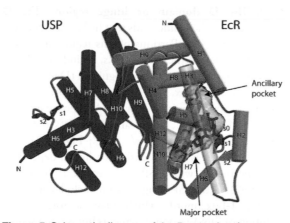

Figure 5 Schematic diagram of the *B. tabaci* ecdysone receptor LBD heterodimer showing the ecdysteroid-binding pocket. Bt-EcR-LBD is shown in yellow and BtUSP-LBD is shown in cyan. Individual helices are shown as cylinders, and individual β-strands are shown as arrows. The observed termini of each LBD are labeled (N-terminal, N, and C-terminal, C). Ponasterone A is shown in green with oxygen atoms in red. Helices H3 and H12 of BtEcR-LBD are rendered transparent to enable viewing of the ponA moiety. The surface of the binding pocket is shown in transparent gray. Ancillary pocket denotes a space within the ligand pocket that is formed by several amino acid residues conserved among insect EcR sequences. Figure produced using MOLSCRIPT, CONSCRIPT, and RASTER-3D software. Reproduced with permission from *Carmichael et al.* (2005).

5.2.2. Ecdysteroid Action and the Role of Other Nuclear Receptors

The initial identification of EcR using a genomic probe from the early responsive gene, E75, in *Drosophila* gave an early indication that other members of the nuclear receptor superfamily are involved in the course of ecdysteroid-regulated transcription events.

Another "early late" puff gene product (early-induced puff at 78C; Eip78) is a member of the nuclear receptor superfamily (DHR78; Stone and Thummel, 1993) as is DHR3, which is also a heterodimeric partner of EcR (White *et al.*, 1997). Still another nuclear receptor, FTZ-F1, a transcriptional regulator of the embryonic segmentation gene, *fushi tarazu*, encodes a β-isoform that is the product of the stage-specific 75CD chromosomal puff that occurs during the mid-prepupal ecdysteroid peak (King-Jones and Thummel, 2005).

More systematic screens have yielded orphans, many of which are also ecdysteroid-regulated (Sullivan and Thummel, 2003). A yeast two-hybrid assay employing USP as bait led to the isolation of an interacting orphan receptor, DHR38, which is orthologous to the vertebrate nerve growth factor 1B (NGF1B; Sutherland *et al.*, 1995). The *seven-up* (*svp*) gene, originally defined by a mutation that disrupts photoreceptor function, encodes the *Drosophila* orthologue of another vertebrate orphan, the chicken ovalbumin upstream promoter-transcription factor (COUP-TF; Mlodzik *et al.*, 1990). COUP-TF heterodimerizes with RXR, and analogously, SVP forms a functional heterodimer with USP as shown by its ability to compete for dimerization with EcR in both *D. melanogaster* (Zelhof *et al.*, 1995a) and *A. aegypti* (Zhu *et al.*, 2003a). Only those nuclear receptors directly associated with EcR and USP action, either as partners or as downstream co-regulators, will be discussed further.

Based on the sequence information provided by the nuclear receptors originally cloned and described in *D. melanogaster* with classic probe hybridization methods, the use of PCR primers based on conserved amino acid regions within many of these orphan receptors have been used to identify receptor orthologues in other arthropod and some non-arthropod species (Jindra *et al.*, 1994a,b; Mouillet *et al.*, 1997, 1999). The pace of nuclear receptor

gene discovery has been further facilitated through screening of insect genome databases (Christiaens *et al.*, 2010). In summary, the modular domain organization of the orphan nuclear receptors not only resemble EcR and USP structurally, some also play a role in the orchestration of ecdysteroid response, both as targets of EcR and USP-mediated transcriptional activity and as alternative heterodimeric partners for the two functional components of the ecdysteroid receptor.

5.2.3. Structural analysis of EcR and USP

The need to understand the interaction of EcR with its natural ligand and other ecdysteroid agonists has prompted the effort to elucidate the crystal structure of the EcR LBD. Homology models based on a comparison of the EcR LBD with the known crystal structures of the vertebrate retinoic acid receptor (RAR) and vitamin D receptor (VDR) had been employed to formulate predictions about the three-dimensional structure of the EcR LBD (Wurtz *et al.*, 2000). The shape and size of the pocket predicted by this model also indicate that the more compact bisacylhydrazines that behave as non-steroidal agonists of EcR fill up only a portion of the pocket.

The crystal structure of *Hv*EcR reveals the EcR LBD as highly flexible, with a ligand-binding pocket whose

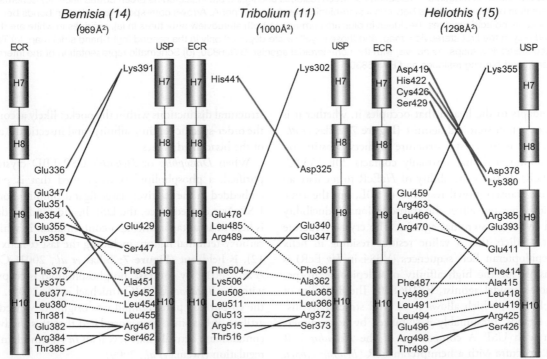

Figure 6 Schematic diagram showing residue interactions between EcR and USP in three species representative of three orders (Hemiptera, *Bemisia*; Coleoptera, *Tr.*; Lepidoptera, *Heliothis*) possessing each of the three USP subtypes (see text). The interactions show the co-evolved interface between EcR and USP across orders. The hydrophobic core is presented as dashed lines. Conserved electrostatic and polar interactions are presented as black lines. The electrostatic and polar interactions specific to one species are in orange, whereas *Heliothis/Bemisia*-specific interactions are in cyan. Reproduced with permission from *Iwema et al.* (2009).

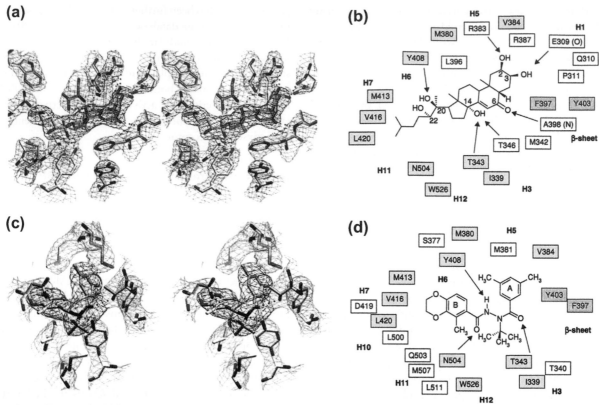

Figure 7 Two ligand-binding modes for *Hv*EcR LBD as described in Billas et al. (2003). (a) Two weighted omit stereoview maps of the electron density for ponA-bound HvEcR-LBD at 2.9 Angstoom resolution. Ligand shown in blue and selected amino acid residues are shown in magenta. Interactions between residues and ligand are indicated as green dotted lines. (b) Schematic representation of the interactions between ligand-binding residues and ponA. Arrows correspond to hydrogen bonds between ligand and amino acid residues. Residues in blue are common to both stereoview structures in (a), residues in white are those depicted only in the first stereoview map, and those in yellow are depicted only in the second (right-hand side) map. (c) Two weighted stereoview maps, as above, with the non-steroidal agonist, BYI06830. (d) Schematic representation of the interactions between ligand-binding residues and BYI06830.

shape adapts to the ligand that occupies it, whether it is a 20E or a non-steroidal agonist (**Figure 7**; Billas *et al.*, 2003). Just as the crystal structure predicts, substitution of an amino acid that normally contacts 20E (A398P in *Hv*EcR) destroys the ability of *Hv*EcR to mediate an inducible transcriptional response to 20E, but the same mutation does not affect the transcriptional inducibility caused by a non-steroidal agonist. The crystal structure further predicts that a valine residue residing in helix 5 of lepidopteran EcR sequences (V384 in *Hv* EcR) is responsible for the high affinity of a lepidopteran-specific non-steroidal agonist, BY106830. The flexibility of the EcR LBD, which allows it to bind structurally disparate ligands, presumably is stabilized by heterodimerization with USP. A comparison of the *Heliothis* EcR crystal structure with a hemipteran EcR (*Bemisia tabaci*, sweet potato whitefly) shows high similarity in terms of the residues within the ligand-binding pocket that are contacted by 20E (Carmichael *et al.*, 2005). However, there are considerable differences in the packing of residues within the pocket that are not bound by 20E. The

structural distinctions within the pocket likely account for the order-specific binding affinity (and insecticidal action) of the bisacylhydrazines.

When *Drosophila* or *Heliothis* USP LBD crystals are purified, a phospholipid is co-purified that is partially embedded in the relatively large ligand-binding pocket of USP. As a consequence, the USP ligand-binding domain is held in an antagonistic, apo-conformation in which the carboxy-terminal region, including the last α-helix (helix 12), is held out (**Figure 7**; Billas *et al.*, 2001; Clayton *et al.*, 2001). By contrast, no phospholipid is co-purified with the *B. tabaci* USP (Carmichael *et al.*, 2005). While the co-purified phospholipid was originally viewed as an artifact, it has since been postulated that this inadvertent contaminant actually may reveal an *in vivo* mechanism of regulation (Iwema *et al.*, 2009).

The crystal structure of the EcR and USP DBD at the perfect IR1palindrome compared to the natural *hsp27* EcRE DNA-binding site reveals the nature and location of DNA-binding residues for the two DBD regions (**Figure 3**). The binding to the hsp27 EcRE imposes a bend in the DNA structure that is different than IR1 binding. The points of protein-protein and protein-DNA interaction with the *hsp27* EcRE are schematically shown (**Figure 3c**). While the mammalian RXR forms a functional dimer with EcR (Christopherson *et al.*, 1992; Yao *et al.*, 1993), the physical interaction between USP/RXR and EcR is also different than it is for RXR with several of its natural partners (Devarakonda *et al.*, 2003).

The ability of the DHR38/USP heterodimer to induce transcription in response to some ecdysteroids, even though the complex does not physically interact with its ligand, prompted a crystal structure analysis of the DHR 38 LBD. DHR38 lacks a true ligand-binding pocket — the space is filled with four phenylalanine side chains. The dimerization interface apparently lies along helix 10, but DHR38 lacks helix 12, and its AF2 function may reside in a unique sequence lying between helix 9 and 10 in an activated (agonistic) conformation (Baker *et al.*, 2003).

5.2.4. Biochemical analysis of EcR and USP

Numerous biochemical experiments have been performed to assess the properties associated with EcR and USP function in insects, including: (1) the affinity of EcR for ecdysteroids and non-steroidal agonists, (2) the ability of EcR and USP to dimerize and interact with DNA target sequences, (3) the physical interaction of EcR and USP with other orphan receptors and proteins, (4) the detection and demonstration of post-translational and other modified forms of EcR and USP, and (5) the investigation of intramolecular interactions in EcR and USP.

The necessity for heterodimerization between EcR and USP to produce a functional ecdysteroid receptor was demonstrated by showing that the *in vitro* translated EcR and USP products bind to a radiolabeled *hsp27* EcRE, but that neither product alone is capable of binding to the same element, based on the results of an electrophoretic mobility shift assay (EMSA). Moreover, this effect is enhanced by the simultaneous presence of murA or 20E, indicating that the hormone stabilizes the heterodimer at an EcRE site (Yao *et al.*, 1993). Further, the affinity of radiolabeled ^{125}I-iodoponA, an ecdysteroid with high specific activity and affinity for EcR (Cherbas *et al.*, 1988), is substantially higher for EcR/USP dimers than for EcR alone (~K_d = 1.1 nM, Yao *et al.*, 1993). A similar level of ligand affinity is obtained with extracts taken from cultured *Drosophila* Kc and S2 cells, which contain endogenously expressed USP (Koelle *et al.*, 1991). EcR also heterodimerizes with the vertebrate RXR and this heterologous dimer is responsive to muristerone A (murA), but not 20E, suggesting that USP plays some role in determining the ligand specificity of the complex (Christopherson *et al.*, 1992; Yao *et al.*, 1993). There has been no explicit evidence to demonstrate that the USP from a given species substantially alters the ligand specificity of EcR, although heterologous pairing of EcR and USP from different species can substantially alter levels of transcriptional activity (Beatty *et al.*, 2009).

The EcR/USP heterodimers from other insects show an affinity for [^{125}I]-iodoponA and [^{3}H]-ponA (ponA), which approximates the DmEcR/USP heterodimer (e.g., K_d ≈1.1 nM; Nakagawa and Henrich, 2009, and references therein). For *A. aegypti* EcR and USP, the detection of EcR/USP heterodimers in a mammalian cell culture extract increases approximately 25-fold with the addition of 5 μM 20E. This accompanies a decrease in the quantity of AaUSP/AaSVP heterodimers, thus illustrating the competitive dynamics of these two protein-protein interactions involving USP (Zhu *et al.*, 2003). In a yeast two-hybrid fusion protein, the association rate of a DmEcR LBD fusion protein with ponA is modestly elevated by the presence of USP, and the dissociation rate is reduced by about 20-fold, so that the dissociation constant is reduced from about 40 nM in EcR alone to about 0.5 nM in the presence of USP. At least one mutation of the *Drosophila* EcR LBD (N626K) increases the rate of dissociation without affecting association rate, and perhaps ponA enters and exits the EcR ligand-binding pocket by different routes, although this effect has not been tested in the intact EcR or in *vivo*. Purified DmUSP has the potential to form homodimers and other oligomers, even in the absence of a DNA-binding site, but this capability is reduced by removal of the A/B domain (Rymarczyk *et al.*, 2003).

The ability of non-steroidal agonists, such as the diacylhydrazines, RH5849 (1,2-dibenzoyl-1-tert-butylhydrazine) and RH5992 (tebufenozide) to displace radiolabeled ponA has been used to assess EcR specificity for these compounds. When extracts from a Lepidopteran (Sf9) and Dipteran insect cell line (Kc) are compared for the ability of the diacylhydrazines to displace [^{3}H]- ponA, the affinity of RH compounds is considerably lower in the Kc cells (Nakagawa and Henrich, 2009, and references therein). Also, several RH compounds displaced [^{3}H]-ponA in extracts containing the migratory locust's EcR and USP, indicating that the Orthopteran EcR, like those of other ancient insect orders, possesses substantially different binding properties than those of the Lepidoptera (Hayward *et al.*, 2003). This has also been shown for other primitive orders; RH5992 competes away from labeled ponA when tested with *Helicoverpa armigera* EcR, but no such competition is detected when *B. tabaci* EcR is tested (Carmichael *et al.*, 2005). The receptor of *Leptinotarsa decemlineata* (Colorado potato beetle) shows relatively high affinity for RH0345, RH5992, and methoxyfenozide (RH2485), but toxicity to halofenozide (RH0345) is much higher than the others (Carton *et al.*, 2003). The potency of RH0345 on EcR transcriptional assays is about the same as it is for RH5992 and methoxyfenozide (RH2485) indicating that pharmacokinetic differences among these compounds may be an important aspect of toxicity in such compounds (Beatty *et al.*, 2009). Ligand binding requires the entire LBD, and deletion of a portion of helix 12 at the carboxy-terminal end of EcR is sufficient to eliminate ponA binding (Perera *et al.*, 1999; Hu *et al.*, 2003; Grebe *et al.*, 2003).

Along with its ligand-binding properties, the ability to recognize a DNA element, usually the consensus *hsp27* EcRE by an EcR/USP or EcR/RXR dimer, is a standard for identifying a functional ecdysteroid receptor. The *hsp27* EcRE was originally localized by DNAse I footprinting and its ability to confer ecdysteroid-inducibility on an otherwise non-inducible promoter in S1 *Drosophila* cultured cells. The sequence is 23 bp long and arranged in an inverted palindrome separated by a single nucleotide (Riddihough and Pelham, 1987). The EcR and USP zinc finger interactions with an idealized palindromic (IR1) and the *hsp27* EcRE are shown graphically (**Figure 3**; Jakob *et al.*, 2007), and the nature of these amino acid-nucleotide interactions is depicted in **Figure 3C** (Jakob *et al.*, 2007). The EcR/USP heterodimer also recognizes half sites arranged into direct repeats (DR) separated by 0–5 nucleotide spacers (DR0-DR5) with the DR4 element showing the highest affinity. Various direct repeat elements were tested with the ecdysteroid-inducible and fat-body-specific *fbp1* promoter. When these direct repeat elements are connected to the minimal *fbp1* promoter, they are unable to induce higher transcription in the presence of ecdysteroids. However, when the DR0 and DR3 elements are substituted for the natural *fbp1* EcRE in its normal promoter context, both elements are ecdysteroid-inducible and fat-body-specific (Antoniewski *et al.*, 1996). Context is also important for the intermolt, *sgs4* gene promoter, which requires that an EcRE is surrounded by serum element binding protein (SEBP) binding sites to mediate a receptor-inducible response (Lehman and Korge, 1995). Promoter context surrounding an *hsp27* EcRE also affects reporter gene activity in cell culture experiments (Braun *et al.*, 2009).

A perfect palindrome carrying the half-site sequence, GAGGTCA, separated by a single A/T nucleotide (PAL1) shows the highest affinity for the DmEcR/DmUSP complex in extracts taken from *Drosophila* S2 cells. By testing the ability of other elements to compete for this element, the order of affinity has been determined: PAL1>DR4>DR5>PAL0>DR2>DR1>*hsp27*,DR3>DR0.

In all cases, affinity was elevated by the presence of murA. In cell culture experiments testing the inducibility of these elements when attached to a minimal promoter, inducibility correlates with affinity, except that the *hsp27* EcRE is a stronger inducing element than DR1, DR2, or PAL0. When DR4 is placed in a reverse orientation relative to the promoter, it maintains its ability to mediate a transcriptional response (Vogtli *et al.*, 1998). A similar correlation between DNA affinity and inducibility has been noted for the AaEcR/AaUSP complex and its response elements (Wang *et al.*, 1998).

A natural direct repeat EcRE has been identified in the ecdysteroid-regulated intermolt 3C puff of *D. melanogaster* separated by 12 nucleotides (DR12) capable of conferring ecdysteroid-responsiveness to a heterologous promoter in cell cultures (D'Avino *et al.*, 1995). The affinity of the EcR/USP complex for the DR12 element is offset by the presence of other orphan receptors (DHR38, DHR39, βFTZF1), and a weak inductive response to 20E is also offset by the presence of these orphans (Crispi *et al.*, 1998). A similar competition scenario exists between DmEcR and DmSVP as heterodimeric partners with DmUSP. The SVP/USP dimer preferentially binds to a DR1 element, whereas the EcR/USP dimer displays a heightened affinity for an EcRE associated with a gene induced in Kc cells (Zelhof *et al.*, 1995a).

The recognition of these canonical elements also depends upon specific features of the ecdysteroid receptor. In at least one instance, DNA recognition depends upon isoform-specific EcR and USP combinations. The EcR heterodimers formed with each of the two USP/RXR isoforms in the ixodid tick, *A. americanum*, display substantially different affinities for DNA response element sequences, with an RXR1/EcR dimer showing strong affinity for both palindromic and direct repeat sequences, whereas an RXR2/EcR dimer shows only weak affinity for a PAL1 element (Palmer *et al.*, 2002).

Ligand-binding specificity also plays a role in determining the affinity of EcR/USP for a response element. AaEcR/DmUSP recognition of an *hsp27* EcRE on EMSAs is normally enhanced by the presence of the 20E precursor, ecdysone, but only 20E enhances affinity of the DmEcR on this assay. Domain swapping between the AaEcR and the DmEcR localizes a region within the LBD responsible for the differential effect of ecdysone in the two EcRs. Within the swapped region, a single amino acid conversion in the DmEcR (Y611F) to the corresponding AaEcR sequence results in a mutated receptor that shows the ability to dimerize in the presence of both 20E and ecdysone (Wang *et al.*, 2000). All insect EcR sequences encode either a tyrosine or phenylalanine at this position in helix 10, which lies along a dimerization interface. The subtle yet functionally significant difference in sequence, when considered in the framework of the numerous residue differences that exist among the insect EcR LBDs, illustrates the dimensions of potential functional diversity among them.

Generally, the interpretation of EMSA results must be handled circumspectively because purified EcR/USP proteins do not always exhibit the same DNA-binding properties as cell extracts, which include the receptor components (Lan *et al.*, 1999; Palmer *et al.*, 2002). This is exemplified by the observation that affinity column purified DmEcR and DmUSP, when mixed together, fail to bind to the *hsp27* EcRE unless other chaperone proteins are added. Whereas no single chaperone is essential for DNA binding, the elimination of any one of them proportionally reduces EcRE recognition. The formation of this chaperone/receptor complex also requires ATP, although none of these chaperones is necessary for ligand-binding. Only two of the six chaperones described, Hsp90 and Hsc70, are *Drosophila* proteins; the others used in the study were human chaperone proteins (Arbeitman and Hogness, 2000).

By expressing GST fusion proteins in *Escherichia coli* and using affinity chromatography to purify them, preparations of *Chironomus* EcR and USP have been recovered that retain essential DNA-binding and ligand-binding characteristics as long as detergents are absent from the purification process. Scatchard plots of bacterially expressed *Ct*EcR show two high-affinity ecdysteroid binding sites, as do extracts from a *Chironomus* epithelial cell line (Grebe *et al.*, 2000, 2002, and references therein). Bacterial GST-fusion proteins expressing the *Drosophila* EcR and USP DNA-binding domains have also been isolated for the purpose of testing their affinity to DNA elements (Jakob *et al.*, 2007, and references therein). The recovery of a fusion protein encoding the DmEcR LBD is enhanced by cleaving the C-terminal F domain found in *D. melanogaster*. Purification methods have been developed for the isolation of EcR ligand-binding domains for the purpose of assessing their ligand-binding properties (Halling *et al.*, 1999, Graham *et al.*, 2007). As EcR LBDs from numerous pest species have been crystalized and ligand interactions modeled by quantitative structure-activity relations (QSARs), the opportunity to develop and test new insecticidal candidates intended to be safe for public health and highly toxic to targeted pests has expanded dramatically in recent years (Fujita and Nakagawa, 2007; Arai *et al.*, 2008; Harada *et al.*, 2009; Holmwood and Schindler, 2009; Birru *et al.*, 2010). Since EcR and USP function as part of a protein complex, orphan receptors and cofactors described throughout this chapter typically require the discovery and/or demonstration of a physical interaction. These demonstrations include the use of yeast two-hybrid assays (Sutherland *et al.*, 1995; Beckstead *et al.*, 2001, Bergman *et al.*, 2004), the interaction of *in vitro* translated EcR and USP with a protein or protein domain (Bai *et al.*, 2000), and the identification of interacting proteins that coprecipitate with EcR and/or USP (Sedkov *et al.*, 2003; Badenhorst *et al.*, 2005; van der Knaap *et al.*, 2010; Bitra and Palli, 2009). A member of another class of proteins, the immunophilins, which are known to interact with vertebrate steroid receptor complexes, co-precipitates as part of an EcR/USP complex taken from *M. sexta* prothoracic (ecdysteroidogenic) glands. Ligand affinity of the complex falls within an expected range, although it is unknown whether the interaction between the immunophilin FKBP46 and EcR/USP is direct. It is also unknown whether this interaction affects transcriptional activity (Song *et al.*, 1997).

5.3. Functional Characterization of the Ecdysteroid Receptor

5.3.1. Cell Culture Studies: Rationale

The study of EcR and USP *in vivo* is complicated because ecdysteroid responses vary both spatially and temporally. The effects of ecdysteroids can be obscured by the heterogeneity of response at the promoter of a single gene (Andres and Cherbas, 1992, 1994; Huet *et al.*, 1993). Therefore, it is often difficult to unravel *in vivo* transcriptional and cellular responses mechanistically, even in a single cell type, since the response to ecdysteroids typically triggers ongoing changes in the target cell, that in turn, affect later ecdysteroid responses.

Cell cultures provide the benefit of working with a stable, relatively homogeneous cell type that does not require the rigorous staging and rearing conditions necessary for meaningful studies *in vivo*. Thus, cultured cells can provide a variety of important and useful insights about EcR and USP by establishing a foundation for subsequent *in vivo* hypothesis testing. Because they are easy to grow, it is also relatively easy to recover cellular extracts for biochemical testing. Nevertheless, stably cultured cells probably do not duplicate any actual cell exactly, so observations reflect what a cell can do, but not necessarily what a cell does. Therefore, subsequent *in vivo* verification is essential for insights garnered through cell culture experiments.

Numerous experiments involving cultured cell lines have been conducted over the years to test the ability of the ecdysteroid receptor to regulate transcriptional activity induced by ecdysteroid treatment. Early experiments focused on endogenous changes in cell morphology and gene expression in insect cells challenged with ecdysteroids such as 20E, which express EcR and USP endogenously. For instance, a *Drosophila* B II cell line, whose cells respond to 20E exposure by forming clumps that can be measured as a change in turbidity, has been used to compare relative agonist activity of several natural ecdysteroids (Harmantha and Dinan, 1997).

As transfection technology has developed, increasingly sophisticated cell systems have been developed to analyze and dissect receptor functions. In recent years these systems have been developed to screen for novel agonists and devise "new and improved" ecdysteroid induction systems for agricultural, biomedical, and commercial application.

Cell culture studies have generated numerous insights concerning (1) the responsiveness and activity of target DNA sequences in promoters; (2) the effect of various ecdysteroids, agonists, and antagonists upon receptor activity; (3) the properties of specific domains on receptor function and the effect of structural modifications upon transcriptional activity; (4) the effect of heterodimerization and other cofactors, including orphan receptors, upon transcriptional activity; or (5) genome-wide responses to ecdysteroids, as measured by transcriptomic changes.

5.3.2. Cell Cultures: Basic Features and Use

The most prominent cell line for early studies of ecdysteroid action was the *D. melanogaster* Kc cell line, and derivatives of it are still employed today (Gauhar *et al.*, 2009). In response to 20E at 10^{-7} to 10^{-8} M, Kc cells start to develop extensions within hours, produce acetylcholinesterase, and undergo a proliferative arrest a few days later. These cells are even more responsive to the phytoecdysteroids, ponA and murA, than to 20E (Cherbas *et al.*, 1991, and references therein). With prolonged exposure to ecdysteroids, Kc cells become insensitive to 20E and EcR levels become depressed (Koelle *et al.*, 1991). The non-steroidal agonist RH5849 not only mimics the effects of 20E when tested on sensitive Kc cells, it also fails to evoke a response from insensitive Kc cells, thus arguing that the RH5849 acts via a common mechanism with ecdysteroids (Wing, 1988). Other cell lines, Schneider 2 and 3 (S2 and S3), also become insensitive to ecdysteroid action by reducing their titer of EcR, and have been used because they are easy to transfect transiently with fusion and reporter genes. As noted earlier, the identity of the DmEcR gene was partly demonstrated by introducing the DmEcR cDNA into ecdysteroid-insensitive S2 cells and thus restoring their responsiveness to 20E (Koelle *et al.*, 1991). More recently, Kc cell lines have been improved by developing protocols for stable integration of transgenes by P-element transposition into the genome (Segal *et al.*, 1996) and by using parahomologous gene targeting to "knock out" an endogenous gene target (Cherbas and Cherbas, 1997). In this way, a cell line, L57-3-11, containing no endogenous EcR has been produced that allows the introduction of modified versions of EcR for subsequent experimentation, without the complications normally posed by the presence of endogenous EcR (Hu *et al.*, 2003).

The Kc cell line has been used to identify ecdysteroid-regulated transcriptional targets. By comparing hormone-treated and control Kc cells, the synthesis of three ecdysteroid-inducible peptides (EIPs) named by their molecular weight in kiloDaltons — EIP28, EIP29, and EIP40 — is elevated by tenfold at 10^{-8} M 20E. Juvenile hormone supplementation of the cell culture medium inhibits acetylcholinesterase induction but not EIP induction indicating that the modulatory influence of JH III on 20E involves specific genes rather than a generalized effect (Cherbas *et al.*, 1989, and references therein). The implications of this early finding on recent work will be discussed in Section 5.4.7.

The use of ecdysteroid-responsive lines for other insects is expanding continuously, largely to understand the causes of insecticidal resistance to ecdysteroid agonists and to identify novel insecticidal candidates. A *C. tentans* epithelial cell line was used to isolate several stable and ecdysteroid-resistant clones that metabolized 20E relatively quickly (Spindler-Barth and Spindler, 1998). Clones from the same line were later selected for their resistance to the inductive effects of tebufenozide (RH5992). Neither differential metabolism of RH5992 nor the loss of EcR was the cause for the resistance. A variety of differences in EcR ligand-binding characteristics were noted among the selected 20E-resistant sublines tested that presumably reflect a variety of roles played by other factors (Grebe *et al.*, 2000). Another widely used Lepidopteran cell line is Sf9, derived from ovarian cells of the fall armyworm, *Spodoptera frugiperda*. The affinity of ponA for Sf9 and Kc cells is very similar, but the non-steroidal agonist, ANS-118 (chromafenozide), displays a much higher affinity for extracts derived from Sf9 cells than from Kc cells (Toya *et al.*, 2002). Similarly, cell lines derived from *C. fumiferana* show a much greater responsiveness to RH5992 than *Drosophila* cultured cells, which do not respond to the compound and excrete it at an elevated rate via an ABC transporter system (Retnakaran *et al.*, 2001). *Choristoneura* cell lines, with prolonged exposure to RH5992, become irreversibly insensitive to ecdysteroids, although the mechanism is unknown (Hu *et al.*, 2004). Numerous lines have been developed and used for studying ecdysteroid action in economically important insect species including (but not only) the European corn borer (*Ostrinia nubilalis*: Trisyono *et al.*, 2000; Belloncik *et al.*, 2007), the tent caterpillar (*Malacosoma disstria*: Palli *et al.*, 1995), the silkmoth (*B. mori*: Swevers *et al.*, 2004, 2008), the fall armyworm (*S. frugiperda*: Chen *et al.*, 2002a), the beet armyworm (*S. exigua*: Swevers *et al.*, 2008), the cotton boll weevil (*Anthonomus grandis*: Dhadialla and Tzertzinis, 1997), the Colorado potato beetle (*L. decemlineata*: Charpentier *et al.*, 2002), the Asian tiger mosquito (*A. albopictus*: Fallon and Gerenday, 2010, and references therein), and the Indian mealmoth (*Plodia interpunctella*: Lalouette *et al.*, 2010, and references therein). The ecdysteroid responsiveness has been established for many of these lines by showing that orthologues for evolutionarily conserved and ecdysteroid responsive *D. melanogaster* genes, such as the protooncogenes E74 and E75, are activated transcriptionally when the cells are challenged with ecdysteroids such as 20E or a non-steroidal agonist.

A major benefit of using insect cell lines is that they provide the cofactors and machinery necessary for transcription to occur. On the other hand, other factors may modify or obscure an aspect of ecdysteroid response in a given experimental regimen as silent and unidentified participants. As is evident from *in vivo* studies, a given experiment will generate substantially different cellular responses depending upon the milieu of proteins that contribute to ecdysteroid responsiveness in a given cell type. For this reason, mammalian cell lines containing no endogenous ecdysteroid responsiveness have been used to reconstruct an ecdysteroid responsive transcriptional system introducing the genes encoding EcR, USP, cofactors, and reporter genes to analyze their individual effects on ecdysteroid action (Christopherson *et al.*, 1992; Yao *et al.*, 1993; Palli *et al.*, 2003; LaPenna *et al.*, 2008; Henrich *et al.*, 2009). Heterologous cells and cell lines pose their own special difficulties, since it is conceivable that endogenous proteins provide novel functions or act as surrogates for similar insect proteins. Conversely, these cells might render false negative results in some experiments because they are missing one or more crucial cofactors.

5.3.3. Cell Cultures: Functional Receptor Studies

5.3.3.1. Promoter studies The *hsp27* EcRE connected to a weak constitutive promoter is capable of inducing the transcription of a reporter gene in a *Drosophila* cell line by over 100-fold when challenged with 20E (Riddihough and Pelham, 1987). When *hsp27* EcREs are organized in a tandem repeat and attached to a weak promoter, the gene is not only responsive to 20E, but it is also repressed in the absence of hormone, suggesting a repressive role for the ecdysteroid receptor via this element (Dobens *et al.*, 1991).

A plethora of cell culture transcriptional tests and their accompanying EMSAs have been reported that followed from this early work and they are important for determining the functionality of an EcR/USP heterodimer, as it was in the original report (Koelle *et al.*, 1991).

The most fundamental task concerning promoter activity is to define the functional promoter elements to which the ecdysteroid receptor complex binds and exerts its effects on transcription. This was accomplished for the genes encoding the ecdysone-induced proteins EIP28 and EIP29, which are translated from alternatively spliced mRNAs of the same gene in Kc cell lines (Cherbas *et al.*, 1989, 1991, and references therein). Three specific ecdysone response elements are ecdysteroid-responsive *in vivo* (Andres *et al.*, 1992; Andres and Cherbas, 1994). One of these Eip28/29 EcREs lies in the promoter region, although this is not required for ecdysteroid inducibility in Kc cells, possibly because this element is not used in the embryonic hemocytes from which the cells are derived. The other two elements lie downstream from the polyadenylation site, that is, on the 3′ side of the gene. The EIP40 gene is also ecdysteroid-inducible *in vivo* (Andres and Cherbas, 1992), and the genomic region that includes the EIP40 gene has been analyzed (Rebers, 1999).

Studies of the three *Drosophila* EcR isoforms in mammalian cells have shown that each has varying capabilities for DNA-binding, recognition of specific EcREs, and ligand-dependent transcriptional activity (Braun *et al.*, 2009), and that the A or B1 N-terminal domains of *Drosophila* EcR physically interact with helix 12 in the

LBD in the absence of 20E as evidenced by Förster resonance energy transfer (FRET) analysis. This configuration results in a repressive state that is stabilized by K497, a conserved amino acid residue that lies in helix 5; in other nuclear receptors this residue is a cofactor binding site that stabilizes the closure of the ligand-binding pocket by the folding of helix 12 over the ligand-binding pocket (via K497). The A/B domain of EcR-B2 is relatively short and lacks this interactive capability (Tremmel *et al.*, 2010); it also lacks repressive functions seen with the A and B1 isoforms in cell culture experiments (Mouillet *et al.*, 2001; Hu *et al.*, 2003). In *Drosophila* Kc167 cell lines that differentiate in response to treatment by 20E, EcR and USP are co-localized to over 500 genomic DNA sites. EcR recognizes another 91 sites without USP, and USP is found at about 950 other genomic sites without EcR. This is likely because USP heterodimerizes with several other nuclear receptors. The studies, which tested only the EcRB1 isoform, also revealed that many of the EcR/USP co-localized sites did not contain a recognizable EcRE, and that only some of these binding sites were associated with 20E-inducible transcription in the Kc167 cells, although many of the sites had been previously associated with 20E-induction in other tissues. Finally, a subset of genes was downregulated by 20E over 1 to 48 h and EcR/USP binding was fairly constant over that time range, but EcR/USP binding of upregulated genes occurred primarily in the first few hours of exposure and then receded (Gauhar *et al.*, 2009). One of the early gene promoter regions (E75A), which is responsive to both 20E and JH (Dubrovskaya *et al.*, 2004), is involved in the JH-mediated suppression of 20E-mediated induction of early genes, including the *broad* gene (Dubrovsky *et al.*, 2004). It has been subjected to computational, biochemical, and functional analysis in S2 cells. Seven EcR binding sites were identified in the region, and each one displayed unique binding and activation properties (Bernardo *et al.*, 2009). Needless to say, a fuller understanding of EcR and USP binding sites will depend upon the analysis of specific elements in 20E-responsive promoters.

Cell culture experiments have been used to identify the functional ecdysteroid response elements within a promoter region as well as to examine the possibility that different factors compete for these sites to modulate transcriptional activity. The relationship of *M. sexta* EcRB1 and two *Ms*USP isoforms has been explored in *M. sexta* GV1 cell cultures by observing the ecdysteroid-inducible MHR3 promoter (Lan *et al.*, 1997). Both EcR/USP complexes display about the same level of ligand affinity, but the EcRB1/USP1 complex induces much higher levels of expression in response to 20E from an intact promoter and also represses basal expression in the absence of 20E. Further, the EcRB1/USP1 complex binds to a canonical EcRE in the promoter and activates transcription, but the addition of USP2 prevents transcription and blocks binding to the EcRE. *In vitro* translated EcRB1/USP2 can bind to this same EcRE, indicating that other cellular cofactors are responsible for the differential action of the two *Ms*USP isoforms (Lan *et al.*, 1999). The experiments described here allude to the fact that target gene regulation in an insect cell is dependent upon a milieu of factors, including the relative abundance of specific EcR and USP isoforms. While the complexity of these interactions may be difficult to assess directly, these may be important for crucial cellular events during the course of development.

The two isoforms of *C. tentans* USP lead to different ligand-binding capabilities of CtEcR, and one of the forms recognizes a DR1 element that is not recognized by the other on EMSAs. Both show about the same ability to induce transcription with DmEcR in human HeLa cells, although the CtEcR does not possess the ability to induce transcription in cell cultures (Vogtli *et al.*, 1999).

5.3.3.2. Transactivation studies   The appearance of multiple EcR isoforms in *D. melanogaster* has prompted comparisons of their transcriptional responsiveness in cell cultures and in yeast. The B1 and B2 N-terminal domains from DmEcR, when combined with a GAL4 DBD and transfected into yeast, are capable of mediating transcription via a GAL4-responsive universal activation site (UAS) in the promoter. This AF1 (ligand-independent) transcriptional function is further elevated by removing an inhibitory region within the B1 domain (Mouillet *et al.*, 2001). In human HeLa cells, DmEcRB1 and DmEcRB2 showed about the same level of AF1 function and both also possessed AF2 (ligand-dependent) transcriptional activity. By contrast, EcRA reduced basal transcriptional activity below the level obtained from an EcR with no A/B domain at all. Similar results have been obtained in a mammalian Chinese hamster ovary (CHO) line, except that the B1 isoform displays a highly elevated AF1 function compared to the A and B2 isoforms (Beatty *et al.*, 2006). In heterologous lines, a deletion of the USP DBD has no effect on the inducibility of the EcRB1 isoform, which may explain why issues that are mutant for lethal *usp* point mutations of conserved USP DBD residues still maintain the ability to mediate 20E-inducible responses *in vitro* and *in vivo* (Schubiger *et al.*, 2000; Ghbeish *et al.*, 2001, 2002). Nevertheless, EcRA and EcRB2 fail to mediate induction when paired with a USP construct that lacks a DBD (Beatty *et al.*, 2006).

In the EcR-deficient L57-3-11 line, all three DmEcR isoforms are responsive to ecdysteroids, but the A domain and the DmUSP A/B both contribute no AF1 function. The B1 domain contains several regions, that when deleted, impose a modest impact on AF1 functions. By contrast, a specific point mutation (E9 K) in the B2 domain dramatically reduces AF1 functions (Hu *et al.*, 2003) and alters a residue that is the site of an interaction with the bZip transcription factor, Cryptocephalic (CRC).

Lezzi *et al.* (2002) devised a yeast two-hybrid assay that fused the EcR and USP LBD to the yeast GAL4 activation domain and DNA-binding domain, respectively. A variety of site-directed mutations of the LBDs were used to test their effects on functional dimer formation and both AF1 and AF2 (ligand-dependent) transcriptional activity as measured by GAL4/UAS-mediated lacZ activity. The wild-type LBDs dimerize and induce activity at a low rate even in the absence of murA, and this rate increases more than tenfold in the presence of murA (but not other ecdysteroids such as 20E). As predicted, mutations of critical residues in EcR's helix 10 all but eliminate both dimerization and transcriptional activity, and AF2 function is eliminated by mutations in helix 12 and by mutations of ligand-binding residues in the ligand-binding pocket (Grebe *et al.*, 2003; Bergman *et al.*, 2004). The previously noted K497E mutation in EcR-B2 alters a consensus cofactor binding site resulting in the elevation of AF1 in both yeast and mammalian cells, which may indicate the elimination of a repressive function. Virtually all point mutations in the USP LBD eliminate transcriptional activity, although many retain the ability to dimerize with EcR, suggesting that USP is also required for normal AF1 activity. Deletion of the carboxy-terminal region of DmUSP, including helix 12, does not disrupt homodimerization detected by gel filtration, but it does eliminate ligand-binding. Substitutions of specific residues in helix 12 modestly reduce basal transcriptional activity in yeast. However, while ligand-binding for the mutant USP is only slightly reduced, the ligand-dependent transcription of the EcR/USP complex is virtually eliminated (Przibilla *et al.*, 2004). This result is counter-intuitive because the crystallized DmUSP is locked into an antagonistic conformation that implies AF2 activity is impossible. It also contrasts with the normal activity seen in several helix 12 mutations of USP that were tested in the *Drosophila* L57-3-11 cell line, which, nonetheless, may express enough DmUSP to mask the mutational effects (Hu *et al.*, 2003).

5.3.3.3. Ligand effects   Transcriptional activity of EcR varies by species and isoform. Nevertheless, cell culture studies have shown that ligand-dependent transcriptional activity in response to a given ligand is predictable for a given species' EcR. At least three issues pertaining to ligand responsiveness of the ecdysteroid receptor have been addressed through the use of cell cultures. One has focused on EcR responsiveness to ecdysteroids, the other agonists, and novel compounds as measured by ligand potency (the agonist concentration necessary to evoke a half-maximal response) and ligand efficacy (the maximal level of change evoked by the tested compound). Secondly, modified receptors have been tested for their capacity to respond to a given ligand or to assess amino acid residue as ligand-interacting sites. Finally, other studies have investigated the possibility that other orphan receptors are responsive to ecdysteroids and/or other ligands.

By testing ecdysteroid-induced transcriptional activity in cell culture, many of the several hundred phytoecdysteroids (www.ecdybase.org) and other synthetic agonists have been tested and characterized (Mikitani, 1996; Arai *et al.*, 2008; Birru *et al.*, 2010; also see Nakagawa and Henrich, 2009, and references therein). The potency of several ecdysteroids has been compared systematically in mammalian cells expressing RXR and transfected with EcR-encoding vectors for a variety of insect species, revealing distinctive, species-specific profiles of ligand responsiveness. These are useful not only for the purposes of identifying order-specific features of the ligand-binding pocket, but also for the development of "gene switches" discussed further in Section 5.3.4. (**Figure 8**; LaPenna *et al.*, 2008). Antagonists for EcR have also been identified, notably cucurbitacin B (Oberdorster *et al.*, 2001).

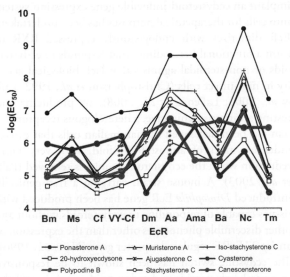

Ponasterone A — Muristerone A — Iso-stachysterone C — 20-hydroxyecdysone — Ajugasterone C — Cyasterone — Polypodine B — Stachysterone C — Canescensterone

Figure 8 Potency level of selected ecdysteroids–log(EC$_{50}$) 91 M as a function of EcR arranged in phylogenic order. Lepidopteran EcRs appear on the left, and non-lepidopteran EcRs on the right. Each horizontal line represents a different ligand. Species identities (left to right) are Bm (silkworm, *B. mori*); Ms (tobacco hornworm, *M. sexta*); Cf (spruce budworm, *C. fumiferana*), and VY-Cf (*C. fumiferana* with E274V/V390E/Y410E mutations); Dm (fruit fly, *D. melanogaster*); Aa (yellow fever mosquito, *A. aegypti*); Ama (ixodid tick, *A. americanum*); Ba (silverleaf whitefly, *Bemisia argentifolia*); Nc (leaf hopper, *Nephotettix cinciteps*); and Tm (yellow mealworm, *Te. molitor*). Dotted lines indicate an inversion of potency (an orthogonality) with respect to the two ligands and EcRs on either side of the crossover (e.g., high response of VY-Cf EcR and low response of BaEcR to cyasterone; high response of BaEcR and low response of VY-CfEcR to canescensterone). Dotted lines indicate cyasterone-VYCf EcR/canescensterone-BaEcR (red) and cyasterone-VY-CfEcR/polypodine B-AaEcR (green) orthogonalities. Reproduced with permission from *LaPenna et al.* (2008).

Environmental pollutants, including polyaromatic hydrocarbons (PAH) and polychlorinated biphenyls (PCBs), affect ecdysteroid-responsive genes in cell cultures, suggesting an EcR-mediated basis for developmental defects commonly found among aquatic species residing in polluted water (Oberdorster *et al.*, 1999). The use of cell culture assays has also been utilized to analyze the effects of JH and its analogs upon transcriptional activity and cell responses (Fang *et al.*, 2005; Beatty *et al.*, 2006, 2009; Soin *et al.*, 2008). The topic of JH and its effects upon EcR and USP will be discussed in Section 5.4.7.

Based on lines of evidence gleaned from experimental studies over the years, the possibility that endogenous ecdysteroids other than 20E display biological activity has been speculated about and even suspected for years (Baker *et al.*, 2000, and references therein; Beckstead *et al.*, 2007). In S2 cells, the 20-hydroxylated derivates of the major and natural ecdysteroid products of the *D. melanogaster* larval ring gland, 20E, and makisterone A are the most efficacious in activating a reporter gene whose promoter carried three tandem copies of the *hsp27* EcRE. By contrast, precursor and metabolite ecdysteroids, as well as nonsteroidal agonists, evoke considerably less activity in cell culture assays. As noted in other studies, ponA displays the highest potency, and two other phytoecdysteroids, murA and cyasterone, also are more inducible than 20E in these cells (Baker *et al.*, 2000). Nevertheless, the transcriptional responsiveness of *Drosophila* EcR is not identical with those observed in other species. In mammalian CHO cells, DmEcR and USP display ligand-reponsive properties that resemble those seen in *Drosophila* cell lines, but *L. decemlineata* EcR/USP responds very weakly to murA compared to 20E and responds relatively strongly to RH0345, RH2485, and RH5992 (Beatty *et al.*, 2009). In other words, the receptor, rather than the cellular milieu, is the primary determinant of ligand-induced transcriptional activity.

Receptor modifications have also been utilized to delineate the basis for differences in ligand specificity, as noted earlier for the effects of the A398P mutation to test predictions based on the *Hv*EcR crystal structure (Billas *et al.*, 2003). Similarly, a substitution in the *Cf*EcR (A393P) eliminates both its ligand-binding affinity for ponA and its transcriptional inducibility in cell cultures, possibly by impairing the receptor's interaction with a co-activator (GRIP1), although the mutation exerts no debilitative effect on responsiveness to a non-steroidal agonist (Kumar *et al.*, 2002). A chimeric *Drosophila/Bombyx* EcR carrying only the *Bombyx* D domain is responsive to RH5992 in mammalian CV-1 cells. The effect is largely attributable to the relatively high level of RH5992-dependent dimerization between the D domain of the chimera and endogenous RXR or co-transfected USP (Suhr *et al.*, 1998). Based on a deletion analysis of EcR, both its D and E domain are required for ligand-dependent dimerization with USP (Perera *et al.*, 1999).

As already noted, insect cells may provide endogenous components that influence ecdysteroid activity. EcR masks a second ecdysteroid responsive activity in *Drosophila* S2 cells. A fusion protein carrying the DHR38 LBD was tested for its ability to dimerize with *Dm*USP in response to several ecdysteroids using S2 cells from which endogenous EcR activity was eliminated by RNA interference (RNAi). A response to 20E occurs and is further elevated by adding an RXR activator, indicating that the DHR38/RXR heterodimer is responsible for the effect. Similarly, co-transfection with a chimeric and constitutively active form of USP also confers the ability of DHR38 to respond to 20E (Baker *et al.*, 2003). The DHR38/USP dimer also responds to at least six other ecdysteroids, including 3-dehydro-20E and 3-dehydro-makisterone A, although these are not found during late larval development in whole body titers (Warren *et al.*, 2006) and do not elicit a response from selected tissues *in vivo* (Beck *et al.*, 2009). Cell culture experiments have also been used to demonstrate the effect of orphan receptors and receptor cofactors upon ecdysteroid response, including *A. aegypti* AHR38 (Zhu *et al.*, 2000), AaSVP (Zhu *et al.*, 2003a), DmSVP (Zelhof *et al.*, 1995a), and DHR78 (Zelhof *et al.*, 1995b).

5.3.4. Ecdysteroid-inducible Cell Systems

The use of ecdysteroid-inducible cellular systems for medical and agricultural applications has emerged rapidly in recent years because ecdysteroids and non-steroidal agonists are inexpensive and harmless to humans (see Palli *et al.*, 2005; Tavva *et al.*, 2007).

Yeast assays have been developed to screen for ecdysteroid agonists of potential commercial value. When EcR is transformed into yeast cells, it is capable of inducing a high level of transcription that is ligand-independent and not appreciably increased by the presence of USP or RXR (De la Cruz and Mak, 1997). Through modifications of the A/B domain, an inducible system was successfully produced (De la Cruz *et al.*, 2000).

A second ecdysteroid-inducible yeast system was produced by making at least two modifications. First, by removing the AaEcR A/B domain, the receptor's AF1 function was reduced when transformed into yeast. Second, the addition of the mammalian receptor coactivator GRIP1 confers yeast cells with the ability to be induced more strongly by a range of ecdysteroid agonists. Conversely, the addition of the receptor corepressor, SMRT, represses transcription (Tran *et al.*, 2001a). A yeast-inducible assay utilizing the CfEcR and CfUSP also requires the removal of the A/B domain from both EcR and USP, along with the addition of the GRIP1 co-activator (Tran *et al.*, 2001b). The USP A/B domain

often displays repressive properties or is inactive in heterologous cells, thus necessitating its deletion and/or replacement with a constitutively active N-terminal domain, such as VP16 (Tran *et al.*, 2001b; Hu *et al.*, 2003; Henrich *et al.*, 2003).

The field use of the bisacylhydrazines as insecticides has also inspired the development of ecdysteroid-inducible systems into plants that could be used to promote the expression of beneficial genes. Non-steroidal, agonist-inducible expression has been obtained from *Zea mays* (corn) protoplasts (Martinez *et al.*, 1999), tobacco, and *Arabidopsis* (Padidam *et al.*, 2003) via EcR fusion proteins. The induction in *Arabidopsis* has been successfully linked to the expression of a coat protein gene from the tobacco mosaic virus, which confers resistance to TMV infection (Koo *et al.*, 2004) and also induces the expression of a factor in maize that restores fertility to a male-sterile strain (Unger *et al.*, 2002).

As noted, ecdysteroids evoke no discernible responses upon mammalian cells, and therefore, attempts to implant an ecdysteroid-inducible gene expression system into cells for therapeutical purposes has been undertaken. EcR dimerizes with endogenously expressed RXR to form a functional heterodimer that responds to ecdysteroids or non-steroidal agonists that lack biological activity in non-insect cells (Christopherson *et al.*, 1992; Palli *et al.*, 2003; LaPenna *et al.*, 2008). By producing and testing a variety of chimerae, a heterologous receptor system has been developed in mammalian cells that evokes induction rates of almost 9000-fold, with very rapid reduction when the ecdysteroid agonist is removed (Palli *et al.*, 2003). A mouse strain carrying a transgenically introduced *Drosophila* EcR gene has been produced with tissues capable of responding to ecdysteroids without any other discernible phenotypes other than the expression of an *hsp27* EcRE-regulated reporter gene (No *et al.*, 1996). The ecdysone switch has been shown to be responsive to several ecdysteroids and ecdysteroid agonists, and the maximal response can be further elevated by the addition of RXR activators, although RXR superinduction requires that EcR be previously bound by its cognate ligand (Saez *et al.*, 2000). Ecdysteroid-inducible expression of medically important genes has been accomplished in human cell lines (Choi *et al.*, 2000), as well as the introduction of an ecdysteroid-inducible gene into mice via an adenovirus vector used for somatic gene therapy (Hoppe *et al.*, 2000; Karzenowski *et al.*, 2005). The use of EcR as a ligand-dependent transcription factor for regulating gene expression has continued to develop in vertebrates such as the zebra fish, because the transcriptional response evoked by non-steroidal agonists is rapid, relatively transient, and robust (Esengil *et al.*, 2007). The RXR partner for EcR is typically expressed at high levels and broadly among vertebrate cell types. Insect RXRs, that is, those classes of USP belonging to primitive orders such as Orthoptera (e.g., *L. migratoria*) actually confer a higher sensitivity to ecdysteroid ligands than human RXR, and have been further optimized by site-directed mutagenesis (Tavva *et al.*, 2006; Singh *et al.*, 2010). Systematic testing of several phytoecdysteroids has shown that the EcR of a given insect species shows differences in responsiveness to these agonists, as measured by differences in potency on transcriptional assays that exceeds two orders of magnitude (**Figure 8**; LaPenna *et al.*, 2008). Therefore, it should be possible to develop orthogonal gene switches that can independently trigger different subsets of genes in the same cell type.

5.4. Cellular, Developmental, and Genetic Analysis

5.4.1. The Salivary Gland Hierarchy: A Model for Steroid Action

Many of the insights concerning steroid regulation of transcriptional activity were first recognized in the larval salivary gland of several insect species. By observing the location, timing, and size of these chromosomal puffs, the activity of an ecdysteroid-inducible gene can be inferred. The progression of events and their relation to developmental changes in the salivary gland and the whole animal are described here for *D. melanogaster* (see Thummel, 2002, and references therein). During the period of the third instar that precedes the wandering stage, "intermolt puffs" appear at several specific chromosomal sites. In response to the late larval ecdysteroid peak, these puffs regress and another set of "early" ecdysteroid-responsive puffs appear. The timing and duration of these early puffs varies (some are referred to as "early-late" puffs) and these eventually regress as another set of "late puffs" appear at other chromosomal sites. The puffing pattern accompanies the production and secretion of glue protein from the salivary gland, which is extruded as the larva becomes immobile and tanning of the larval cuticle begins. Several hours later, a second smaller pulse of ecdysteroid induces a second wave of early puff activity, which is followed by a second wave of "late" puff activity. The two peaks evoke similar, but not identical, transcriptional changes, and these subtle differences underlie important developmental consequences. The second wave culminates in the expression of βFTZ-F1, a specific isoform of an orphan receptor encoded by the stage-specific (i.e., the pre-pupal-specific) gene puff located at the chromosomal interval, 75C. βFTZ-F1, in turn, sets off a cascade that ultimately includes the expression of another stage-specific puff, E93, involved in the histolysis of the gland (Lee *et al.*, 2002).

As expected, 20E shows the greatest potency in terms of puff induction, followed by 3-dehydro-20E, which is an endogenously produced ecdysteroid in the *Drosophila* third instar (see Chapter 4). Significantly, however, while the potency of individual ecdysteroids varies, no individual ecdysteroid evokes a unique puffing site, or alternatively, fails to evoke puffing at a site induced by other ecdysteroids, although microarray analysis may yet provide a level of resolution that allows for the detection of specific changes that are not evident from puffing studies.

5.4.2. Molecular and Genetic Characterization of the Puff Hierarchy

Numerous genetic and molecular studies have verified the basic tenets of the Ashburner model. Ashburner proposed that the ecdysteroid receptor directly induces the early puffs while simultaneously repressing late puff genes, and that the early puff gene products not only increase late puff transcription, but also feed back to downregulate their own expression (Ashburner et al., 1974 and references therein).

Another line of inquiry has focused on demonstrating that EcR and USP interact directly with puff site regions. Immunostaining has shown specific instances in which EcR and USP co-localize to early puff sites (Yao *et al.*, 1993), and that EcR and RNA polymerase II antibodies co-localize on *C. tentans* polytene chromosomes (**Figure 1**). EcR has also been associated with late puff sites (Talbot *et al.*, 1993). Further, the products of the early puff gene, E75, have been associated by immunostaining with early and late puff sites, as predicted by the Ashburner model (Hill *et al.*, 1993). Chemically modified forms of EcR and USP have more recently been used to demonstrate that EcR and USP co-localize at early puff sites, as well as several hundred other sites (Gauhar *et al.*, 2009).

The puff hierarchy actually begins before the onset of pupariation in *Drosophila* with the appearance of several intermolt puffs (see Lehman, 1995; Thummel, 2002, and references therein). The *sgs4* gene encodes a salivary glue protein and is regulated by the synergistic interplay of an ecdysone receptor element and a secretion enhancer binding protein (SEBP3) site. A second SEBP binding site has been associated with products of the early puff product, *broad*, indicating that the proper regulatory interplay between the ecdysone receptor and downstream transcriptional factors depends upon the surrounding context (von Kalm *et al.*, 1994; Lehman and Korge, 1995). As ecdysteroid titers increase, the intermolt puffs regress.

Three early puff genes have been the subject of intensive ongoing study in *D. melanogaster* and other insects. All three genes encode transcription factors, consistent with a role in regulating late puff expression: the *Broad-Complex* (*Br-C*) encodes 4 or 5 zinc finger isoforms whose isoform-specific domains have been evolutionarily conserved in various species (Spokony and Restifo, 2007, and references therein). E74 is an *ets* protooncogene encoding two isoforms via alternative promoters: one activated at 100 nM 20E, the other activated at 1μM 20E. As noted, E75 encodes a heme receptor and produces three isoforms, one of which includes only the second zinc finger in the DBD. All three genes are complex, utilizing alternative promoters and splicing not evident from the cytological studies alone (Thummel, 2002, and references therein). The expression of the *broad* isoforms varies temporally and spatially (Huet *et al.*, 1993; Mugat *et al.*, 2000; Brennan *et al.*, 2001; Riddiford *et al.*, 2003, 2010). The *broad* Z1 isoform is not only a target of 20E activity, but is ectopically induced by the exogenous application of JH and JH analogs (Zhou and Riddiford, 2002). Further, the *broad* gene has been identified as a downstream target of the bHLH-PAS transcription factor, methoprene-tolerant (MET; Konopova and Jindra, 2008). There is a growing body of evidence to suggest that the *broad* gene lies at an important intersection of hormonal activity, just before and during insect metamorphosis and during adult reproductive maturation. The switchover of expression among combinations of BrC isoforms has been associated with the modulatory regulation of many ecdysteroid-inducible genes including the transcription factor gene *hedgehog* (*hh*, Brennan *et al.*, 1998), micro-RNA (Sempere *et al.*, 2003), and *fbp1* (Mugat *et al.*, 2000), vitellogenin synthesis (Zhu *et al.*, 2007), and the course of metamorphic development (Suzuki *et al.*, 2008; Riddiford *et al.*, 2010).

The relationship between ecdysteroid receptor activity and chromosome remodeling has become an intense focus of interest. For instance, the role of a chromatin remodeling factor, cohesin, has been investigated in larval salivary glands of *D. melanogaster*. This factor has been associated previously with sister chromatid cohesion prior to mitosis, and also plays a role in the transcription of some ecdysteroid-inducible genes, including E74 (Pauli *et al.*, 2010). A histone chaperone, DEK, is also essential for proper 20E-induced puffing in salivary glands (Sawatsubashi *et al.*, 2010). Other factors involved in chromatin remodeling, notably the nucleosome reorganization factor (NURF; Badenhorst *et al.*, 2005) and trithorax-related (TRR; Sedkov *et al.*, 2003), have also been implicated in the regulation of EcR/USP-mediated activity, and these will be discussed in Section 5.4.6.

The potential effects of USP phosphorylation have also been investigated in *Drosophila* larval salivary glands by examining the protein profiles evoked by 20E exposure in the presence and absence of a protein kinase inhibitor. The profile of response, including early genes, is substantially altered when phosphorylation is pharmacologically inhibited, as is the subcellular localization of USP (Song *et al.*, 2003; Sun *et al.*, 2008).

5.4.3. Developmental Regulation of Ecdysteroid Response in *D. Melanogaster*

Transcript levels of EcR, the intermolt genes, and the individual isoforms of early puff genes such as BrC, E74, and E75, vary temporally among individual tissues during the late larval/pre-pupal period in *D. melanogaster* (Huet *et al.*, 1993). It is apparent that the relative abundance of the early puff isoforms changes continuously as ecdysteroid titers surge and decline, that the profile of early puff expression varies among tissues at any given time, and that the pattern changes are specific for a given tissue. For instance, EcR levels are very low throughout the larval/pre-pupal period in imaginal discs, whereas they remain elevated in the gut throughout this period. The relative abundance of EcR and its early puff products plays some role in the timing of subsequent gene expression, as exemplified by the *Ddc* (dopa decarboxylase) gene. It behaves as an early-late gene and is induced by the combined action of EcR and BrC, which is offset, in turn, by a repressive promoter element in epidermis and imaginal discs (Chen *et al.*, 2002b). Temporally and spatially diverse responses have been seen for many ecdysteroid-inducible genes. For instance, the fat body protein-1 (*fbp1*) gene is ecdysteroid-inducible and expressed only in the fat body during the onset of metamorphosis (Maschat *et al.*, 1991). The aforementioned Eip28/29 gene is ecdysteroid-responsive but displays several different patterns of expression in individual tissues at the onset of metamorphosis (Andres and Cherbas, 1992, 1994). The expression of the E93 gene sets off the salivary gland histolysis, although its expression in other tissues is not responsive to 20E (Lee *et al.*, 2002, and references therein).

The complexity of *in vivo* response to ecdysteroids is also revealed by comparing the puffing response seen along the polytene chromosomes of the larval fat body during the same time that the salivary gland puff response occurs (Richards, 1982). While many aspects of the fat body response resemble those seen in the salivary gland, individual fat body puff sites are responsive to both ecdysone and 20E in ranges below 10^{-6} M, whereas salivary gland puffing is much more responsive to 20E. While one obvious explanation lies with differences in metabolism, the fat body response to ecdysone does not lag the response to 20E in the fat body, as would be expected if conversion to 20E preceded receptor-mediated response. Fat body chromosomes also evoke different puffing patterns than salivary gland chromosomes, and these presumably reflect functional differences between the two compared cell types, and generally in larval tissues hormonal challenge with ecdysone evokes different transcriptional changes than 20E (Beckstead *et al.*, 2007).

The *D. melanogaster* imaginal discs and ring gland are notable because they apparently fail to respond to ecdysteroids during the larval/pre-pupal period, as measured with a GAL4-EcR/GAL4-USP *in vivo* system in which lacZ staining reports the level of ligand-dependent heterodimerization between the two fusion proteins (Kozlova and Thummel, 2002). The lack of responsiveness may reflect the effect of cofactors (or their absence) and/or mechanisms by which 20E is either removed from the cells via a transporter mechanism (Hock et al., 2000) or metabolized.

The same GAL4 in vivo reporter system indicates that EcR and USP heterodimerization occurs in response to the ecdysteroid peak during D. melanogaster mid-embryogenesis and that the receptor plays a role in germ band retraction and head involution (Kozlova and Thummel, 2003). Mutations of the early-late gene, DHR3, cause embryonic lethality (Carney et al., 1997), further implicating not only EcR but the entire ecdysteroid hierarchy as a conserved and essential gene network over the course of development.

The existence of functionally distinct EcR isoforms provides an important basis for differential ecdysteroid responses. During the late larval stage, the EcRA isoform of D. melanogaster is generally predominant in imaginal discs, whereas the B isoforms are associated primarily with larval tissues that undergo histolysis during metamorphosis (Talbot et al., 1993). Issues surrounding the diversity of isoform function will be discussed further in Section 5.4.5.

5.4.4. Conservation of Transcriptional Hierarchy Among Insects

The isolation of EcR and USP from other insects and the assembly of their developmental profiles allows for direct analysis of ecdysteroid action in other organisms. So far, these studies show general consistency with those obtained in D. melanogaster, specifically, EcR expression varies over time and peak levels of expression coincide with ecdysteroid peaks, and different isoforms predominate among different tissues. For instance, the expression patterns of the two EcR isoforms in M. sexta vary among cell types (Jindra et al., 1996) and expression levels of the two USP isoforms switch in conjunction with molts (Jindra et al., 1997) providing a regulatory framework for the early-late response of MHR3 discussed earlier (Lan et al., 1999).

The recovery of ecdysteroid-inducible target genes such as E74, E75, FTZF1, and HR3 orthologues also allows a comparison of the ecdysteroid-inducible cascade in other organisms, and the resulting proteins play biological roles in a variety of regulatory processes that is too extensive to discuss here. The comparative analysis over pre-adult development is most complete in M. sexta, and the essential features of the ecdysteroid-inducible hierarchy are maintained (Riddiford et al., 2003, and references therein). Similarly, the players and timing of early ecdysteroid-responsive genes during B. mori choriogenesis are conserved (Swevers and Iatrou, 2003), as they are in A. aegypti vitellogenesis (Zhu et al., 2003). The hierarchical ecdysteroid response in the Aedes vitellogenin gene reveals interplay between specific Broad isoforms, E74B, FTZ-F1, and EcR/USP for proper regulation (Sun et al., 2005; Zhu et al., 2006, 2007). As already noted, USP

also interacts with other orphan nuclear receptors, such as DHR38 and SVP, which are involved in specific developmental processes. In flies, SVP is involved in eye differentiation (Mlodzik et al., 1990) and DHR38 is involved in cuticle formation (Fisk and Thummel, 1998) at the onset of metamorphosis. Both HR38 and SVP are involved in mosquito vitellogenesis (Zhu et al., 2003a). A dimer between A. aegypti AHR38 and USP has been noted prior to vitellogenesis. The nonresponsive heterodimer is displaced by the EcR/USP heterodimer in response to a 20E titer peak following a blood meal (Zhu et al., 2000). As the 20E peak that stimulates A. aegypti vitellogenin expression ensues, a similar competition between EcR and SVP for USP as a heterodimeric partner leads to a downregulation of A. aegypti vitellogenin gene transcription after egg laying (Zhu et al., 2003a), illustrating a mechanism by which an ecdysteroid response can be terminated.

Ectopic expression of the svp+ gene in flies causes lethality, but this effect can be rescued by simultaneous ectopic expression of usp+ (Zelhof et al., 1995a). This interaction apparently is a relevant aspect of photoreceptor differentiation, since both SVP and USP function are essential for this process to occur normally (Mlodzik et al., 1990; Zelhof et al., 1997). The potential relevance of the SVP/USP interaction has also been established by demonstrating that SVP competes with EcR for USP's partnership, reducing ecdysteroid-induced transcription in cell cultures. Further, the functional SVP/USP dimer preferentially interacts with DR1 DNA elements, whereas the EcR/USP dimer interacts with the Eip28/29 element. These mechanisms illustrate the possibility that combinations of the EcR/SVP/DHR38/USP orphan receptor group (and perhaps other orphan receptors) might be incorporated into various ecdysteroid-regulated processes for the purpose of regulating a linear series of events such as mosquito vitellogenesis and fly eye differentiation.

Historically, numerous studies of puffing on polytene chromosomes preceded and have continued along with the work on D. melanogaster. The salivary gland response of Sciara coprophila (fungal fly) provides an interesting comparison to D. melanogaster, and illustrates the importance of investigating the features of each system without pretense. The S. coprophila II/9A puff appearing in response to ecdysteroid treatment reflects the initiation of both bidirectional DNA replication (Liang et al., 1993) and the elevated transcription of two similar mRNA species that bear no resemblance to ecdysteroid-responsive genes found in Drosophila, even though the promoter remains active when the gene is transgenically introduced into flies.

In yet another sciarid fly, Bradysia hygida, 20E induces DNA amplification and transcription of a temporally late puff gene (BhC4-1). When this gene is transgenically introduced into D. melanogaster (Bienz-Tadmor et al., 1991), it retains its late puff characteristics in terms of salivary gland timing, but transcription of the gene is not blocked by repression of protein translation with cycloheximide, suggesting that the late induction is actually a form of derepression in the *D. melanogaster* salivary gland (Basso *et al.*, 2002). EcR is immunologically detected on *B. hygida* polytene chromosomes, but its role seems to be restricted to the regulation of transcription, with no role in gene amplification (Candido-Silva *et al.*, 2008). As evidenced by the cases presented in this chapter, beyond the superficial similarity, there are numerous unsolved mysteries about the integration of ecdysteroid response into biological processes, especially when one considers that the most complete hierarchical comparisons so far involve only Lepidopteran and Dipteran species. Several ecdysteroid-inducible orphan receptor genes have been identified in other insects (Jindra *et al.*, 1995, and references therein; Mouillet *et al.*, 1999; Christiaens *et al.*, 2010).

5.4.5. *In vivo* Molecular and Genetic Analysis of EcR and USP

The pattern of EcR gene transcription during pre-metamorphic development fluctuates dramatically, with high levels accompanying each of the embryonic and larval molts, followed by peaks of expression during the prepupal and pupal–adult stages of metamorphosis (Koelle *et al.*, 1991). Moreover, the EcR gene encodes three different isoforms with the A isoform's transcription governed by a promoter and two other isoforms (B1 and B2) generated by alternative splicing via a second promoter (Talbot *et al.*, 1993). All three isoforms share a common sequence that includes a small portion of the A/B domain and other domains, but the amino-terminal side of the A/B region for each isoform is unique. The A isoform predominates in fly tissues that undergo morphogenetic changes during metamorphosis, including the imaginal discs; the ring gland; and the imaginal rings of the foregut, hindgut, and salivary gland (Talbot *et al.*, 1993); and in a heterogeneous subset of neurons that degenerate after adult emergence (Truman *et al.*, 1994) whereas the B1 isoform is found in larval tissues including the salivary gland, the fat body, larval muscle (Talbot *et al.*, 1993), and proliferative neurons (Schubiger *et al.*, 2005, and references therein).

Functional studies of EcR based on the isolation of mutations within the *D. melanogaster* gene have led to further insights about EcR's developmental role, and essentially confirmed its ecdysteroid-mediating function. As might be expected, null mutations that disrupt a common region of EcR disrupt all three isoforms resulting in embryonic lethality, whereas mutations that retain a reduced ability to mediate ecdysteroid response allow mutant survival through some or all of the larval stages. Some of these weaker mutations affect nonconserved residues, but at least one of these prepupal lethal mutations involves a highly conserved phenylalanine in the DBD; conversely,

there are nonconserved residues that cause embryonic lethality when substituted. One of the lethal mutations (A483T) replaces the amino acid residue that interacts with the SMRTER corepressor and is a conditional third instar lethal mutation. Presumably, the lability of the mutant protein disrupts the normal interaction between EcR and SMRTER at higher temperatures (29°C). The A483T mutation (along with other EcR mutations) also disrupts adult female fecundity at its restrictive temperature, indicating that the corepressor interaction is essential for at least late larval development and oogenesis (Bender *et al.*, 1997; Carney and Bender, 2000).

Considerable experimental evidence has been reported in recent years that further distinguishes the functional roles of the three *Drosophila* isoforms, as noted by the effects of isoform-specific mutations (Davis *et al.*, 2005) and isoform-specific overexpression (Schubiger *et al.*, 2003). B-specific mutations cause early larval lethality (Schubiger *et al.*, 1998); an A-specific EcR mutation reduces but does not eliminate mutant survival to the adult stage and disrupts the normal expression of EcRB1 (D'Avino and Thummel, 2000; Schubiger *et al.*, 2003). Other A-specific mutations cause pupal lethality (Davis *et al.*, 2005). Over developmental time, the B2 isoform, when expressed under the control of a heat shock promoter, rescues larval development in EcR-null mutants, although heat shocks are required in each instar to accomplish it. The A and B1 isoforms rescue development through the first instar, but fail thereafter, suggesting that a common function is required during the first instar and more differentiated EcR functions are essential later. The rescued B2 transformants become sluggish during the wandering stage of the late third instar, become immobile, and eventually die (Li and Bender, 2000).

Isoform-specific mutations in conjunction with transformation rescue have further delineated EcR-based functions in fly development. As expected, an EcRB1-specific mutation disrupts developmental activities in those cells where it is expressed. For instance, the salivary gland's ecdysteroid-induced puffing is disrupted in B1 mutants (but not eliminated), and only transformation with EcRB1 restores normal puffing of various early and early-late genes, although B2 exerts a partial rescue. Similarly, B1-expressing abdominal histoblasts and midgut cells develop abnormally (Bender *et al.*, 1997). Neuronal remodeling during metamorphosis is disrupted in genetic mosaics that do not express either of the two B isoforms in proliferating neurons (Schubiger *et al.*, 1998), but remodeling is rescued by the expression of either B isoform within these cells (Schubiger *et al.*, 2005). Remodeling is not disrupted in mutations of the early puff genes indicating that this aspect of the cascade is not specifically tied to the failure of B isoform mutant effects (Lee *et al.*, 2000).

The advent of RNAi as a tool for reducing (if not eliminating) gene function should allow for expanded investigation of both general and isoform-specific functions of EcR and USP in other insects. The utility of the approach has been demonstrated in the red flour beetle, *Tr. castaneum*. In this species, RNAi-mediated knockdown of EcR-A expression exerts more severe effects upon development, although loss of EcR-B function also leads to developmental defects (**Figure 9**). Interestingly, isoform-specific loss of USP function caused no defects; only a common region USP RNAi led to abnormalities, possibly indicating a redundancy of function (Tan and Palli, 2008). The development of a GAL4/UAS system in *Tr.* that allows for tissue-specific expression of RNAi and mutant versions of EcR and USP will undoubtedly be a catalyst for even more sophisticated future studies (Schinko *et al.*, 2010).

A related strategy for discriminating EcR functions involves introducing an isoform or fusion protein transgenically into flies, expressing the transgene ectopically, and then observing the dominant negative phenotypic consequences of such expression. At least three variations have been reported: (1) expression of a wild-type isoform ectopically under the control of a UAS promoter regulated by the yeast GAL4 transcription factor (Schubiger *et al.*, 2003), (2) the expression of a GAL4-EcR LBD fusion protein that forms an inactive dimer with cellular USP in a non-isoform-specific manner (Kozlova and Thummel, 2002, 2003), and (3) UAS-controlled expression of a dominant negative EcR isoform (F645A in *D. melanogaster*) resulting in an inactive dimer with intracellular USP, which is then tested for rescuability by the concomitant expression of a specific isoform (Cherbas *et al.*, 2003). In the case of wild-type overexpression, each isoform generates a unique pattern of phenotypes. Overexpression of EcR-A suppresses posterior puparial tanning and affects ecdysteroid-inducible gene expression in posterior compartments of the wing disc, but does not affect viability. By contrast, overexpression of EcRB1 and B2 during puparial formation greatly reduces viability (Schubiger *et al.*, 2003). Ectopic expression of a GAL4-EcR during the late third instar causes a failure of puparial contraction

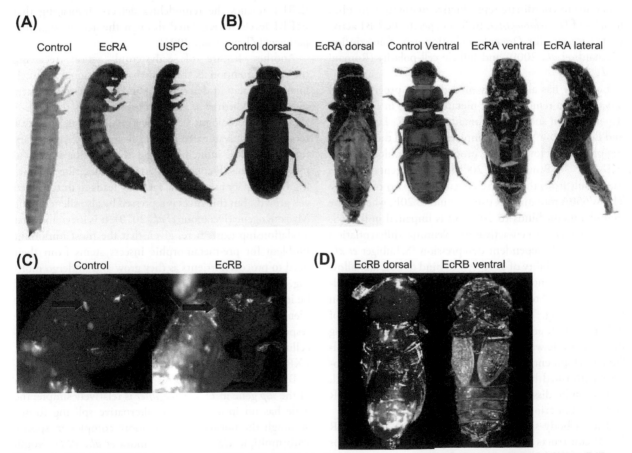

Figure 9 Effect of dsRNA injection on metamorphic development in *T. castaneum*. (A) Developmental effects of EcR-A RNAi and USPC (common region) RNAi during quiescent stage. Desiccation occurred within 24 h after injections. Injections of dsRNA for each of the two specific *T. castaneum* USP isoforms evoked no developmental effects. (B) Developmental effects of low concentration injections (0.04 µg) of EcRA dsRNA during quiescent stage resulting in pupal–adult intermediate with defects in wing development. Abnormal wing did not cover the dorsal side and animals died of dehydration. (C) Abnormal compound eye development in larva injected with EcR-B dsRNA. (D) EcR-B RNAi animals showing both pupal and adult structures. Reproduced with permission from *Tan and Palli* (2008).

and cuticular tanning that resembles the traits displayed by mutant EcR larvae (Kozlova and Thummel, 2002). The isoform-specific rescue of the EcR[F645A] mutants, however, reveals a poor correlation between tissue-specific effects and intracellular titers of each isoform, although the accumulation of disrupted responses in several individual tissues leads to a stage-specific developmental arrest in the third instar (Cherbas et al., 2003).

The EcR-Dominant negative (DN) system, expressing an EcR that carries either an F645A or W650A mutation, has been successfully employed to assess EcR roles in developmental processes. Expression of the EcR-DN causes a disruption of follicle cell migration in the adult female ovary, and produced eggshell phenotypes previously associated with genetically evoked failures in the Ras oncogene signaling pathway (Hackney et al., 2007). In ovaries, all three EcR isoforms are expressed at various times during ovarian development, and another EcR-dependent process, border cell migration (Jang et al., 2009), has not been associated with a specific EcR isoform. It is possible that the isoforms carry out specific subfunctions to coordinate reproductive maturation in adult females of D. melanogaster. Isoform-specific EcR-B1 activity is necessary for Drosophila chorion gene transcription (Bernardi et al., 2009) and follicle cell polarity (Romani et al., 2009).

EcR-DN has already revealed that the progression of larval neuron remodeling at metamorphosis in Drosophila depends upon specific subfunctions of EcR. EcR-A[W650A] and EcR-B1[W650A] almost completely blocked the remodeling processes. However, the process was not blocked with EcR-B1[F645A] (Brown et al., 2005). The different outcomes may result from the diverse effects of the two mutations. The W650A mutant EcR cannot bind to 20E, whereas the F645A mutant binds to 20E and is impaired in its ability to interact with co-activators. Neuronal differentiation involves a 20E-dependent derepression (Schubiger et al., 2005), so the ability of EcR[F645A] to bind to 20E may allow it to perform a normal derepressive role, whereas the failure of EcR[W650A] to bind to 20E prevents this derepression from occurring. The findings illustrate that the effect of individual EcR and USP mutations may vary, depending upon their role within a developmental process. By corollary, developmental processes may sometimes require subroles performed by individual isoforms in the same cell, which can be dissected through the judicious use of EcR and USP mutations.

The fat-body-specific expression of EcR[F645A] and EcR RNAi constructs have shown that 20E is responsible for the EcR-mediated suppression of the Myc oncogene. The loss of this suppression (i.e., the activation) of Myc leads to ribosomal biogenesis and larval growth (Delanoue et al., 2010). The same general strategy has been used to show that fat body histolysis, which is normally triggered by caspase induction by 20E, although it is offset by JH (Liu et al., 2009), as will be discussed further in Section 5.4.7.

Further insights about the functional differences associated with insect EcR isoforms will likely emerge by understanding the basis for isoform-specific mRNA transcription via the aforementioned A and B promoters. Two promoters in Drosophila, one associated with the A isoform and the other associated with the two B isoforms, regulate the appearance of these transcripts. Promoter segments responsible for high levels of EcR-A expression during metamorphosis have been identified (Sung and Robinow, 2000). While the level of EcR-A detected is homogeneous among those neurons expressing EcR-A during metamorphosis, the underlying promoter regulation is surprisingly heterogeneous among them, indicating that the observed pattern of expression obscures an underlying regulatory complexity. A TGF-β/activin signaling pathway has been implicated in the regulation of EcRB1 transcription. In activin mutants, EcRB1 transcription is reduced and neurons in the mushroom bodies of the brain fail to undergo remodeling during metamorphosis. The expression of EcRB1 rescues the remodeling defects, indicating that EcRB1 levels are regulated through the activin signaling pathway (Zheng et al., 2003).

Another feature of ecdysteroid action receiving increased attention is the role that nutritional and cellular states play in determining and modifying ecdysteroid action (Mirth et al., 2009, and references therein). A Drosophila homologue of the mammalian Dor, a nuclear receptor co-activator tied to insulin signaling, is necessary for normal 20E inducibility and required for successful metamorphosis (Francis et al., 2010). As described, silencing EcR activity in the larval fat body leads to unregulated cell growth, but this effect is repressed by also silencing the Myc oncogene (Delanoue et al., 2010). It is inevitable that a relationship exists here, given that the most important problem for pre-metamorphic insects stems from their need to process nutrients as they grow and undergo molting. The role of orphan receptors and their ligands could be important to explore this possibility, since most of the identified orphans among vertebrates have proven to be responsive to a variety of dietary compounds and intracellular metabolites, including EcR's vertebrate relatives, FXR and LXR (King-Jones and Thummel, 2005).

By comparison with EcR's complexity, the regulation of the usp gene in D. melanogaster is relatively simple; the gene has no introns and no alternative splicing forms, although the normal profile is more complex in species with multiple USP isoforms (Jindra et al., 1997; Vogtli et al., 1999). In flies, the expression of USP through development is relatively stable, but it is unclear whether EcR or USP is the rate-limiting partner in developing tissues. The usp gene is defined by several recessive early larval lethal mutations in D. melanogaster: three missense substitutions (usp[3], usp[4], and usp[5]) that mutate amino acid residues in the USP DBD and directly contact phosphate residues in the DNA backbone, along with a non-sense mutation, usp^2, which truncates the DBD (Oro *et al.*, 1990; Henrich *et al.*, 1994; Lee *et al.*, 2000). The null-usp^2 allele evokes a different effect on gene expression than the other *usp* alleles. Whereas all the mutations disrupt the normal repression of the BrC-Z1 isoform, only USP2 is incapable of activating BrC-Z1 transcription. USP3 and USP4 are also able to mediate ecdysteroid-induced gene transcription through an *hsp27* EcRE-regulated promoter (Ghbeish *et al.*, 2001).

This dual capability has been analyzed *in vivo* during ommatidial assembly and differentiation in the eye disc. Briefly, the differentiation of the eight retinula cells in each of the 700 or so ommatidia that become the compound eye occurs through the recruitment of undifferentiated cells as a wave moves from the posterior to anterior end of the eye disc. Cells along this progressing wave, the morphogenetic furrow, undergo a host of transcriptional changes and some aspects of the process are ecdysteroid-dependent *in vitro*. Furrow advancement accelerates in mutant *usp* patches on the eye disc (Zelhof *et al.*, 1997), whereas an ecdysteroid deficit retards its advancement (Brennan *et al.*, 1998). The apparently paradoxical action results from the failure of a repressive *usp* function as evidenced by the abnormal appearance of Br-C Z1 expression in mutant patches lying along the front of the moving furrow. The usp^2 allele causes an absence of Z1 behind the furrow, where it is normally present, further demonstrating the dual roles. USP has been similarly implicated in both wing margin bristle differentiation (Ghbeish and McKeown, 2002) and the repression of premature neuronal differentiation in the wing imaginal discs (Schubiger and Truman, 2000). The maintenance of USP3 and USP4 activation functions likely explains the normal appearance of mutant *usp* imaginal clones (Oro *et al.*, 1992).

Maternal contribution of normal *usp* transcript is also essential for the completion of embryogenesis (Oro *et al.*, 1992). Mutant *usp* larvae are rescued through the third instar with a USP connected to a heat shock promoter. The lethal phenotype of these partially rescued larvae is reminiscent of the effects seen with larval EcR mutations, and internal morphology is similar at the time of larval arrest. Only the *usp* mutants, however, develop a supernumerary larval cuticle and fail to wander off the food (Hall and Thummel, 1998; Li and Bender, 2000). DHR38 mutants also undergo abnormal cuticle apolysis (Kozlova *et al.*, 1998), suggesting that the DHR38/USP dimer may be essential for this aspect of pre-metamorphic development, rather than the EcR/USP dimer.

Possibly distinct roles of USP during larval and metamorphic development are evidenced *in vivo* by the fact that a chimeric USP transgene (in which the *Drosophila* USP LBD is replaced by the equivalent domain from *C. tentans*) completely rescues larval development in *usp*

mutants that otherwise die in the first instar. Such genetically rescued larvae, however, suddenly die as the onset of the pre-pupal stage approaches (Henrich *et al.*, 2000).

With eye differentiation in flies, morphogenetic furrow movement during the late third instar is dependent on 20E, but not on normal EcR gene activity (Brennan *et al.*, 2001). Instead it depends upon the repressive action of USP (Ghbeish and McKeown, 2002), illustrating how the partners' interplay is essential for proper development. USP-mediated repression is offset by the positive regulation of *Hedgehog* (*Hh*) transcription that involves EcR (Sedkov *et al.*, 2003). *Hh*, in turn, is a transcription factor that facilitates a switchover from the Z2 to the Z1 isoform at the ecdysteroid-responsive Br-C gene locus. The temporal coordination of hormonal signaling with the regulatory activities along the furrow, therefore, provide a balance of signals that both stimulates morphogenetic furrow movement (in the form of EcR-mediated Hh expression) and regulates its rate of progression (in the form of a repressive USP function). The mechanism of furrow movement is not fully elucidated, and it is conceivable that a second hormonal signaling pathway is involved in the process.

The ecdysteroid response is highly heterogeneous among cell types because of differences in the quantitative levels of EcR and USP isoforms, rates of 20E conversion from ecdysone, and 20E metabolism and cellular exclusion. Yet it is evident that the circulation of ecdysteroids in the insect hemolymph provides the organism with a means to coordinate its individual developmental programs. Therefore, it is expected that ecdysteroids set off general responses and specific responses that involve either EcR and USP or targets that are regulated by them.

For organismal investigations, the dramatic changes associated with the late larval ecdysteroid peak in *D. melanogaster* give rise to substantial changes in transcript levels that can be detected using genomic approaches. As already noted, a combination of standard technology and a genomic outlook has motivated the assembly of a detailed profile of orphan receptor transcript levels through pre-adult development (Sullivan and Thummel, 2003). A limitation of the whole body approach is evident in the case of DHR38, which is widely expressed but at such low levels that the transcript is barely evident on Northern blots. An earlier study using subtractive hybridization led to the characterization of several ecdysteroid-inducible genes, most of which were unknown and not directly associated with ecdysteroid-inducible events (Hurban and Thummel, 1993).

Microarray analysis has been used to investigate the changes in transcription on a genome-wide basis, using the onset of the white pre-pupal (puparial) stage as a reference point. The early puff genes change in a predictable, ecdysteroid-regulated manner, and genes involved in processes such as myogenesis, apoptosis, and imaginal disc differentiation are upregulated at appropriate times during the period. The analysis has shown that as ecdysteroid levels peak, transcription of most genes encoding glycolytic enzymes are substantially reduced, as are genes whose enzymatic products regulate the citric acid and fatty acid cycles, oxidative phosphorylation, and amino acid metabolism (White et al., 1999; Zinke et al., 2002). This could be notable since adult flies heterozygous for a lethal EcR mutation survive longer than wild-type flies and show greater resistance to oxidative stress, heat, and dry conditions, although activity levels are normal and a possible connection to metabolic function has not been made (Simon et al., 2003). Mechanistic activities at the cellular level, however, will continue to require tissue-specific analysis. Microarray analysis has also been employed not only to investigate developmental changes in gene transcription at the onset of metamorphosis, but also to examine the effects of ecdysone, 20E, and/or JH treatment on larval tissues as the onset of metamorphosis approaches (Beckstead et al., 2005, 2007). This approach has also been used to compare expression in Kc167 and SL2 cell lines (Neal et al., 2003) and has to follow expression changes throughout larval and metamorphic development. The results suggest that while the larval stage is relatively quiescent, there are unique patterns of expression associated with this early developmental time, which in some cases may reflect repressive functions (Arbeitman et al., 2002). In this respect, microarrays also may prove important for gaining a better grasp of the mode of action of JHs during larval stages (Beckstead et al., 2007). Inevitably, the genomic approach will unravel entire gene networks that interact with and operate within the ecdysteroid-regulated hierarchy (Stathopoulos and Levine, 2002).

5.4.6. Ecdysteroid Receptor Cofactors

As noted earlier, the ecdysteroid receptor is part of a complex of proteins that affects both its inductive and repressive transcriptional functions. In the case of Hsc70, a chaperone that physically interacts with DmEcR, this functional importance is demonstrated by the genetic interaction between EcR and Hsc70 mutations (hsc4 gene) in D. melanogaster. Trans-heterozygotes for these two mutations develop severely blistered wings and bent legs typically associated with the mutational impairment of EcR function (Bender et al., 1997; Arbeitman and Hogness, 2000). In H. armigera, RNAi analysis of Hsc70 showed that the loss of its expression prevented 20E inducibility. Further, the protein is found predominantly in the cytoplasm in the absence of 20E, whereas it is located in the nucleus in its presence. Immunoprecipitation revealed that Hsc70 binds to USP (Zheng et al., 2009).

Based on findings from experiments with vertebrate nuclear receptors, transcriptional cofactors and their roles have begun to be recognized and explored. For example,

the process of eye ommatidial differentiation during the third instar of D. melanogaster that was described earlier has become the focus of efforts to understand the role of cofactors in development. The dominant negative EcR^{F645A} mutation severely disrupts eye development (Cherbas et al., 2003; Sedkov et al., 2003), and USP function is also required for normal furrow progression (Zelhof et al., 1997). One cofactor implicated in the process is the product of the trithorax-related (trr) gene, a histone methyltransferase that plays a role in remodeling the chromatin in promoter regions, just as vertebrate factors do to facilitate receptor-mediated transcription (Stallcup, 2001). The role of TRR as an ecdysone-dependent coactivator is evidenced by the fact that it is recovered as part of a complex that includes EcR and a trimethylated form of the histone-3 protein (modified by TRR) from the ecdysteroid-inducible promoter of the Drosophila hedgehog (hh) gene in extracts derived from 20E challenged S2 cells. The methylation is reduced in complexes taken from trr mutant embryos. Further, the trans-heterozygotic combination of a trr mutation and the EcR^{F645A} mutation causes almost complete lethality, revealing an essential in vivo interaction. The aforementioned Hedgehog (Hh) protein regulates the progression of eye cell differentiation, and trr-mutant somatic cell clones express Hh at reduced levels. It follows that TRR is normally a coactivator associated with elevated ecdysteroid-inducible activity leading to higher Hh levels. EcR retains its repressive capabilities in the absence of its inducible activity, and another cofactor, the corepressor SMRTER, has also been implicated in the regulation of Hh expression and eye differentiation. The lethal and mutant phenotypic effects of the dominant negative EcR^{F645A} mutation are reversed by a hypomorphic mutation of the Smr corepressor (i.e., a partial loss of repression), just as EcR mutant effects are overcome by increasing TRR co-activator activity (Sedkov et al., 2003). Mutations in D. melanogaster of the large subunit of NURF disrupt normal 20E induction of several known EcR/USP gene targets and disrupt metamorphosis. NURF normally regulates the movement of nucleosomes and thereby determines the availability of transcription start sites. It also interacts directly with EcR-A and EcR-B2 (Badenhorst et al., 2005). NURF is also required for 20E-mediated regulation of germ stem cell proliferation (Godt and Tepass, 2009; Ables et al., 2010). Other chromatin remodeling enzymes, GMP synthetase and ubiquitin-specific protease 7, are also necessary for 20E induction (van der Knaap et al., 2010).

Both TRR and SMRTER carry the LXXLL amino acid motifs (L refers to leucine and X refers to any amino acid) associated with nuclear receptor interactions, and physical interaction sites with EcR have been mapped in both cases (Tsai et al., 1999; Sedkov et al., 2003). Moreover, the SMRTER interactive site in DmEcR has been mapped to an amino acid residue in helix 5 that when mutated

(A483T) results in conditional larval lethality (Bender et al., 1997; Tsai et al., 1999).

Another *Drosophila* co-activator, Taiman (TAI), was first identified in a screen for mutations that disrupt oogenesis. TAI resembles a human orthologue (AIB1) belonging to the p160 steroid receptor co-activator family (SRC), whose members are typified by a basic helix-loop-helix (bHLH) domain, two PAS domains and several LXXLL motifs associated with nuclear receptor interaction, and several glutamine-rich stretches. These proteins form a bridge between hormone receptors, chromatin-modifying enzymes such as the histone acetyl transferases, and the transcriptional machinery. TAI expression co-localizes with EcR and USP in the border cells of the ovary and also co-localizes with USP on the polytene chromosomes of *Drosophila* larval salivary glands. TAI also elevates ecdysteroid-inducible transcription in a cell culture and co-precipitates with EcR, but not USP (Bai et al., 2000). A second cofactor, Abrupt, offsets the TAI-mediated potentiation of 20E induction, co-regulating the movement of border cells in the adult female ovary (Jang et al., 2009). Still another EcR-interacting protein containing the LXXLL motif, *rigor mortis* (*rig*), is required for ecdysteroid signaling during larval development. Mutant *rig* larvae fail to survive beyond the advent of metamorphosis while displaying phenotypes resemblant of other mutations defective in ecdysteroid synthesis or response. RIG is required as a co-activator for induction of the E74A isoform, which normally appears as ecdysteroid titers increase, but is not required for E75A, EcR, or USP transcription. The effect on transcription is likely to be indirect since RIG contains no DNA-binding motifs. The protein interacts physically with EcR and USP and also with the orphan receptors DHR3, βFTZ-F1, and SVP. Even when helix 12 is deleted from βFTZ-F1, its interaction with Rig is detectable, suggesting that the relationship between Rig and nuclear receptors is ligand-independent (Gates et al., 2004). A *D. melanogaster* corepressor, *Alien*, which is highly conserved phylogenetically, also interacts with several receptors, including EcR, SVP, and βFTZ-F1 (but not DHR3, DHR38, DHR78, or DHR96; Dressel et al., 1999). No mutations of *Alien* have been reported so far.

The pattern of receptor interactions seen with TAI, Rig, and Alien suggests that a level of regulation remains to be elucidated in connection with the ecdysteroid hierarchy. This is further highlighted by the effects of another cofactor, Bonus, which belongs to a class of proteins (TIF1) that do not bind to DNA directly; instead they repress transcriptional activity. Homozygous *bon* mutants display many of the developmental phenotypes shown by βFTZ-F1 phenotypes, and Bon physically interacts with helix 12 of βFTZ-F1 via an LXXLL motif. In some *bon* mutants, transcript levels of EcR B1, E74A and B, and Br-C are reduced, but DHR3 transcript levels are elevated (Beckstead et al., 2001). Non-DNA binding cofactors,

MBF1 and MBF2, associated with *Bombyx* βFTZ-F1 activity, have also been identified that interact with the TATA binding protein to induce transcription (Li et al., 1994).

Finally, it is notable that the *Methoprene-tolerant* (Met) mutation in *D. melanogaster* defines a gene that specifies another bHLH-PAS transcription factor. Cellular extracts from flies homozygous for *Met* mutations show little binding to the juvenile hormone analog, methoprene (Wilson et al., 2006, and references therein; see also Chapter 8). The implications of this identity and its significance for explaining the modulatory effects of JH on ecdysteroid-inducible transcriptional activity will be addressed later.

5.4.7. Juvenile Hormone Effects Upon Ecdysteroid Action

A central tenet of insect biology is that ecdysteroids mediate larval–larval molts in the presence of the sesquiterpenoid, juvenile hormone. In the absence of JH, by contrast, ecdysteroid triggers a larval–pupal transition. Three non-exclusive views about JH action and its relationship with EcR/USP have emerged from experimentation throughout the years: JH via its receptor competes with the EcR/USP dimer at target genes (Dubrovsky et al., 2004), USP is a JH receptor (Jones et al., 2001), or JH modifies the activity of the ecdysteroid receptor (Cherbas et al., 1989; Henrich et al., 2003; Fang et al., 2005; Beck et al., 2009). Nevertheless, no uniform explanation for the action of JH, alone or in conjunction with ecdysteroids, has been forthcoming. The exact nature of the interaction between JH and ecdysteroid action remains elusive, despite the clear recognition that there must be cross-talk between the two hormones. As noted by early cell culture studies, JH modulates the activity of specific 20E-inducible genes (Cherbas et al., 1989). A transcriptomic analysis of 20E and/or JH-mediated effects on larval organ tissues reveals that the expression of many genes inducible by 20E, including E74 and E75, are modulated by the presence of JH. Other genes are also induced by ecdysone and JH (Beckstead et al., 2007).

The exploration of this potential interaction has been bolstered by the demonstration that RNAi-mediated knockout of the bHLH-PAS protein, MET, in *T. castaneum*, eliminates the abnormalities normally seen when JH analogs are applied exogenously to developing larval cuticles (Konopova and Jindra, 2007). The *Drosophila* MET was originally identified as a mutation that conferred resistance to the toxic effects of the JH analog, methoprene. Further, methoprene binding to cytosolic extracts is greatly reduced in MET mutant flies and has since been shown to bind to JH directly (Ashok et al., 1998); MET has been shown to bind to JH (Miura et al., 2005).

The *Tr.* MET is also necessary for the regulation of the 20E-inducible *broad* gene, further suggesting a direct interplay between bHLH-PAS proteins and nuclear receptors (Konopova and Jindra, 2008). Among vertebrates, this structural interaction has been well established, and a physical interaction between EcR, USP, and MET has also been shown in *Tr.* (Bitra and Palli, 2009).

There are indications, however, that the interaction of bHLH-PAS proteins with nuclear receptors and with each other will prove to be combinatorial, resulting in a complex regulatory interaction. The *Drosophila* MET is actually a duplicated gene derived from another bHLH-PAS-encoding gene, *germ-cell expressed (gce)*. Godlewski *et al.* (2006) showed that the MET and GCE PAS domains dimerize in the absence of the JH analog, methoprene, and that the addition of methoprene causes the dimer to dissociate. It is unknown whether MET and GCE dimerize at a specific promoter, or whether the release of one or both proteins allows them to interact with other factors, including EcR and USP. Perversely, it is the *Drosophila* GCE that more closely resembles the *Tr.* MET (as well as the *A. aegypti* MET), and so far, only *Drosophila* carries two MET paralogues in its genome (Wang *et al.*, 2007). There are some indications, however, that the issue will prove even more complex. The steroid co-activator, p160, is also a bHLH-PAS factor recruited by the orphan receptor, βFTZ-F1, and potentiates the ecdysteroid response of the vitellogenin gene in *A. aegypti*. As noted earlier, the p160 is a bHLH-PAS protein (Zhu *et al.*, 2006) that closely resembles the *Drosophila* TAI. Interestingly, the conformation of EcR's ligand-binding pocket, which is dependent on the presence of specific ligands, also affects its ability to interact with TAI (Kamar *et al.*, 2004) and conceivably affects its interaction with other cofactors as well. What, if any, role exists for JH via TAI is a matter of speculation. Nevertheless, the observation highlights the possibility that the interaction of the ecdysteroid receptor with TAI, MET, GCE, and other cofactors may underlie the difficulty with finding a true receptor for JH.

Along a different line of reasoning, experiments have been undertaken based on the sequence similarity between the EcR LBD and the vertebrate FXR LBD, which surprisingly, is responsive to JH III in the 10–50 μM range with an RXR dimer partner in mammalian cell cultures (Forman *et al.*, 1995). When tested in this regimen, JH III potentiates the ability of the EcR/USP complex to induce a weak transcriptional induction via an *hsp27* EcRE in the presence of submaximal murA levels. The potentiation apparently requires prior binding of EcR to a natural ecdysteroid as JH exerts no effects by itself or when a nonsteroidal agonist is bound to EcR (Beatty *et al.*, 2009). When tested with 20E, only EcRB2 among the three *Drosophila* isoforms is additionally potentiated with JH III (Henrich *et al.*, 2003), and as noted, the B2 isoform is the only one capable of rescuing larval development

in EcR mutants (Bender *et al.*, 1997). The potentiation is not analogous to the superinduction described earlier with RXR activators and the EcR/RXR dimer (Saez *et al.*, 2000). The low ecdysteroid levels that accompany JH titers during larval development perhaps reveal the possible biological relevance for regulating 20E-dependent developmental processes during larval stages. In both *Drosophila* and *Manduca*, one of the isoforms of Broad (Z1) is ectopically induced in cuticle by the application of JH analogs, further supporting the possibility that JH III acts directly upon the ecdysteroid receptor's transcriptional capability (Zhou and Riddiford, 2002), since the BR-C gene is a direct target of the receptor in both *Drosophila* and *Manduca*.

JH III has been proposed as a ligand for USP based on fluorescence binding assays (Jones *et al.*, 2001, and references therein). More recently, several precursors of JH in the mevalonate pathway, including MF, which is secreted from the *Drosophila* corpus allatum (CA), shows affinity for USP using the fluorescence assay (Jones *et al.*, 2006). The affinity of USP for MF ($K_d \approx 40$ nM) is reduced drastically when two amino acids that normally protrude into the ligand-binding pocket are substituted. RNAi-mediated, CA-specific ablation of HMG coreductase, an enzyme that lies in an early portion of the pathway leading to JH synthesis, leads to numerous developmental defects at the onset of metamorphosis (Jones *et al.*, 2010).

A direct comparison of *Drosophila* USP with the more primitive *Tr.* USP and the vertebrate RXR has been made by transgenically introducing them into *Drosophila* and subjecting tissues (which express the *Drosophila* EcR) to a variety of hormonal challenges. Using this paradigm, it is apparent that RXR is responsive to RXR activators such as 9-cis retinoic acid and methoprene acid, which is known to be an RXR activator (Harmon *et al.*, 1995). Neither JH III nor MF had any direct effect on activation of a target reporter gene *in vivo*. However, the preincubation of tissues with JH III prior to exposure to 20E prevented the normal response in all cases. These findings led to the conclusion that the effects of JH and its analogs involve other cofactors that interact with USP or RXR in *Drosophila* tissues (Beck *et al.*, 2009). Using GAL4-LBD constructs of EcR and USP, the effects of JH and JH analogs were also tested *in vivo* (Palanker *et al.*, 2006). In this assay the phytocompound angelicin and the insecticide fenoxycarb evoked some activity from USP (although EcR was not required); the effect of the latter was attributed to xenobiotic activity involving an unknown cofactor.

Generally, these studies with EcR and USP lead back to the suggestion that JH interacts with cofactors, possibly the bHLH-PAS proteins, to modulate the activity of the ecdysteroid receptor. As noted already, some transcriptional activities of MET have been shown to be

JH-dependent. Developmental evidence in *D. melanogaster*, based on the study of *Met* mutants, RNAi for *gce*, and genetic ablation of the CA (the site of JH synthesis), has further implicated the role of MET and/or GCE for mediating JH-dependent processes associated with programmed cell death in the larval fat body (Liu *et al.*, 2009; Baumann *et al.*, 2010) and the timing of EcRB1 expression in differentiating ommatidial cells during the pre-pupal period (Riddiford *et al.*, 2010). The criteria that were outlined earlier in the chapter to identify the ecdysone receptor must now be applied to the bHLH-PAS proteins, notably MET and GCE, to determine whether these are components of an insect JH receptor.

5.5. Prognosis

A growing body of evidence has demonstrated that the ecdysteroid receptor is structurally complex and composed of various EcR and/or USP isoforms with the capacity to form a variety of complexes with other nuclear receptors and cofactors. Among the insect orders, the functional diversity of the ecdysteroid receptor, as illustrated by differences in ligand specificity and transcriptional activity, brings another level of complexity to the interpretation of 20E action in biological processes. These complexities must be considered even before considering that the ecdysteroid receptor operates at the cellular level in a context of cofactors and promoters of diverse cell types and developmental times. Ultimately, the anlagen of the process is evident as it requires a coordinated response of cells to ecdysone, 20E, and/or other ecdysteroids at critical times during organismal development.

While it is naïve to make general statements about ecdysteroid receptor function when considering the layers of complexity involved, the powerful tools of modern molecular biology, genetics, and the "omics" make it possible to separate individual components of the process in order to assess their roles. Over the past five years, considerable progress has been made in several specific areas which together suggest that the ecdysteroid receptor will continue to be an absorbing and informative focus of future research: (1) the emergence of genomic, transcriptomic, and proteomic profiles for investigating the effects of ecdysteroids *in toto*; (2) the adaptation of transgenic and genomic methods that have been useful in *D. melanogaster* into receptor studies of other insects; (3) the *in silico* modeling and crystal structure analysis of insect receptor complexes along with tests of these models through both *in vitro* and *in vivo* experimentation; and the (4) recognition that ecdysteroids, and the ecdysteroid receptor, are active players not only for regulating transcriptional responses, but also in the regulation of molecular-based processes such as insulin signaling, cell migration and growth, programmed cell death, and chromatin remodeling. Ironically, the increasing network of processes tied to ecdysteroid action highlight the need, more than ever, to examine and understand the structure and function of EcR and USP, along with its associated partners.

The emergence of clever methodologies, such as the *in vivo* lacZ reporter system reported by Kozlova and Thummel (2003), EcR-DN system (Cherbas *et al.*, 2003), and UAS-GAL4 systems should provide *in vivo* tools for further *Drosophila* work, and hopefully, similar transgenic systems will also be developed for other insects, as has begun to occur with *T. castaneum*. The assembly of databases and the design of new molecular constructs have also produced a variety of opportunities which remain to be further examined.

The ability to assess the effects of EcR, USP, and genes targeted by the ecdysteroid receptor in other insects will be greatly enhanced by continued development of transformation procedures. The ability to produce "null" mutations via RNA interference introduced with transgenic constructs will be particularly important for assessing functional processes in other insects. Uhlirova *et al.* (2003) illustrated this possibility by showing that Sindbis virus-induced transformation with an RNAi eliminates Br-C expression in *B. mori*. This interference exacerbates the same developmental defects in *Bombyx* as previously noted for the effects of Br-C null mutations in flies.

Related to the discovery and characterization of ecdysteroid responses in insect processes will be the continued examination of the insect receptors. The use of chimerae and mutational analysis, in conjunction with the predictive powers provided by improved modeling and crystal structures, will lead to further insights concerning the capabilities of receptor function and its response to ligands, which may include a broad spectrum based on the flexibility of its ligand-binding pocket (Billas *et al.*, 2003). Many mutations affect receptor function non-specifically (Bender *et al.*, 1997; Bergman *et al.*, 2004), and site-directed mutagenesis provides a particularly useful approach for finding mutations that specifically disrupt EcR and USP subfunctions essential for interpreting receptor function in cellular and *in vivo* systems (e.g., Cherbas *et al.*, 2003).

Needless to say, the advent and development of direct molecular detection (nanomolecular) methods could dramatically alter the outlook for insect endocrine research by eliminating the need to interpret the results of "ensemble averaging" and interpreting data from batches of tissues; this need is particularly evident for ecdysteroids and juvenoids, which often cannot be ascertained for individual tissues or organisms.

From a biological perspective, ecdysteroid action has been associated with a variety of diverse functions, including long-term courtship memory (Ishimoto *et al.*, 2009), longevity (Simon *et al.*, 2003), innate immunity (Flatt *et al.*, 2008), embryonic morphogenesis (Chavoshi *et al.*,

2010), and sleep cycles (Ishimoto and Kitamoto, 2010). The regulation of EcR and USP gene expression has been undertaken and could also lead to important insights about endocrine regulation generally, and the interaction of ecdysteroid regulation with biological processes and with environmental toxins such as endocrine disruptors (Sung and Robinow, 2000; Zheng *et al.*, 2003; Planello *et al.*, 2008, 2010). EcR undergoes fairly dramatic fluctuations in transcription, although little is known about the relative titers of EcR and USP intracellularly (Koelle *et al.*, 1991) partly because of the limited capability for quantifying proteins within individual cells.

When ecdysteroid action is viewed as a challenge to cellular equilibrium, it is seen as a continuous and equilibrating process rather than a directed one, even though the consequences of 20E action bring about a permanent change in cellular state. Interestingly, there are still unanswered questions about ecdysteroid regulation within the *Drosophila* salivary gland hierarchy, which could be particularly useful for examining ecdysteroid response and processing, particularly with the judicious use of transgenic constructs.

Even as work continues to understand the systems biology associated with ecdysteroid receptor activity and function is embedded within insect processes, the collection of tools and insights seems to be a remarkable opportunity to examine nontraditional roles for ecdysteroids (e.g., Srivastava *et al.*, 2005). One particular area of future interest will likely center on the EcR N-terminal domain, which includes phylogenetically conserved motifs, others confined to specific orders, and still others that are apparently unique. Another aspect of EcR and USP function (as well as other cofactors and nuclear receptors) at the molecular level will focus on the interaction of the domains within and between nuclear receptors, as well as how they might affect biological function. Processes such as subcellular localization and intramolecular regulation, for instance, have not yet been explored extensively (Dutko-Gwodz *et al.*, 2008; Krusinski *et al.*, 2010). Relatively little work has focused on post-translational modifications of EcR and USP, such as phosphorylation, even though these have been described in the literature. Finally, based on the study of other nuclear receptors, it seems almost inevitable that the ecdysteroid receptor possesses activities that have been ignored, largely because its canonical role as a ligand-activated nuclear receptor has been such a fruitful line of investigation. Nevertheless, the possibility of other nuclear receptor functions lurks and must be appreciated as it may be encountered.

This basic information about structure-function of the ecdysteroid receptor will likely prove to be useful and important for a variety of biomedical and industrial applications in the future. The flexibility of the EcR ligand-binding pocket (and the diverse ligand specificity found among species), its ability to interact with other partners, and its apparent diversity in regulating transcriptional activity, point to its potential utility as a transcription factor that can be employed and modified to regulate complex cell networks.

In summary, the information reported here represents only a fraction of the progress on ecdysteroid receptor action that has taken place over the last 20 years, particularly in recent years as new technologies have been employed to address molecular questions. More important, the convergence of evolutionary, molecular genetic, biochemical, and computational approaches discussed here has created a synergy among them that will continue to yield important new insights about the regulation of gene activity in developmental processes, endocrine action and its effects on development, new possibilities for insecticidal discovery, and new applications for the use of ecdysteroid-regulated transcriptional cell systems.

Acknowledgments

The author wishes to thank Drs. Dino Moras, Markus Lezzi, Subbha Palli, Robert Hormann, Anrezej Okzyhar, and Ronald Hill for contributing figures for this chapter. The author also acknowledges Drs. Margarethe Spindler-Barth and Klaus Spindler, Yoshiaki Nakagawa, and Tom Wilson for useful discussions related to the material presented in this manuscript, along with Dr. Larry Gilbert for his advice on the writing and his friendship. The author is responsible for any inaccuracies in the content and apologizes for the exclusion of many excellent contributions because of space constraints.

References

Ables, E. T., & Drummond-Barbosa, D. (2010). The steroid hormone ecdysone functions with instrinsic chromatin remodeling factors to control female germline stem cells in *Drosophila*. *Cell. Stem Cell*, 7, 581–592.

Andres, A. J., & Cherbas, P. (1992). Tissue-specific ecdysone responses: Regulation of the *Drosophila* genes Eip28/29 and Eip40 during larval development. *Development*, 116, 865–876.

Andres, A. J., & Cherbas, P. (1994). Tissue-specific regulation by ecdysone: distinct patterns of Eip28/29 expression are controlled by different ecdysone response elements. *Dev. Genet.*, 15, 320–331.

Antoniewski, C., Laval, M., & Lepesant, J. A. (1993). Structural features critical to the activity of an ecdysone receptor binding site. *Insect Biochem. Mol. Biol.*, 23, 105–114.

Antoniewski, C., Mugat, B., Delbac, F., & Lepesant, J.-A. (1996). Direct repeats bind the EcR/USP receptor and mediate ecdysteroid responses in *Drosophila melanogaster*. *Mol. Cell Biol.*, 16, 2977–2986.

Arai, H., Watanabe, B., Nakagawa, Y., & Miyagawa, H. (2008). Synthesis of ponasterone A derivatives with various steroid skeleton moieties and evaluation of their binding to the ecdysone receptor of Kc cells. *Steroids*, 73, 1452–1464.

Arbeitman, M. N., Furlong, E. E., Imam, F., Johnson, E., Null, B. H., Baker, B. S., Krasnow, M. A., Scott, M. P., Davis, R. W., & White, K. P. (2002). Gene expression during the life cycle of *Drosophila melanogaster. Science, 297,* 2270–2275.

Arbeitman, M. N., & Hogness, D. S. (2000). Molecular chaperones activate the *Drosophila* ecdysone receptor, an RXR heterodimer. *Cell, 101,* 67–77.

Ashburner, M., Chihara, C., Meltzer, P., & Richards, G. (1974). Temporal control of puffing activity in polytene chromosones. *Cold Spring Horbon Symp. Quant. Biol., 38,* 655–662.

Ashok, M., Turner, C., & Wilson, T. G. (1998). Insect juvenile hormone resistance gene homology with the bHLH-PAS family of transcriptional regulators. *Proc. Natl. Acad. Sci. USA, 95,* 2761–2766.

Badenhorst, P., Xiao, H., Cherbas, L., Kwon, S. Y., Voas, M., Rebay, I., Cherbas, P., & Wu, C. (2005). The *Drosophila* nucleosome remodeling factor NURF is required for ecdysteroid signaling and metamorphosis. *Genes Dev., 19,* 2540–2545.

Bai, J., Uehara, Y., & Montell, D. J. (2000). Regulation of invasive cell behavior by Taiman, a *Drosophila* protein related to AIB1, a steroid receptor coactivator amplified in breast cancer. *Cell, 103,* 1047–1058.

Baker, K. D., Shewchuk, L. M., Kozlova, T., Makishima, M., Hassell, A., Wisely, B., Caravella, J. A., Lambert, M. H., Reinking, J. L., Krause, H., Thummel, C. S., Wilson, T. M., & Mangelsdorf, D. J. (2003). The *Drosophila* orphan nuclear receptor DHR38 mediates an atypical ecdysteroid signaling pathway. *Cell, 113,* 731–742.

Baker, K. D., Warren, J. T., Thummel, C. S., Gilbert, L. I., & Mangelsdorf, D. J. (2000). Transcriptional activation of the *Drosophila* ecdysone receptor by insect and plant ecdysteroids. *Insect Biochem. Mol. Biol., 30,* 1037–1043.

Basso, L. R., Vasconcelos, C., Fontes, A. M., Hartfelder, K., Silva, J. A., Coelho, P. S.R., Monesi, N., & Paco-Larson, M. L. (2002). The induction of DNA puff BhC4-1 is a late response to the increase in 20-hydroxyecdysone titers in last instar dipteran larvae. *Mech. Dev., 110,* 15–26.

Baumann, A., Barry, J., Wang, S., Fujiwara, Y., & Wilson, T. G. (2010). Paralogous genes involved in juvenile hormone action in *Drosophila melanogaster. Genetics, 185,* 1327–1336.

Beatty, J., Fauth, T., Callender, J. L., Spindler-Barth, M., & Henrich, V. C. (2006). Analysis of transcriptional activity mediated by the *Drosophila melanogaster* ecdysone receptor isoforms in a heterologous cell culture system. *Insect Mol. Biol., 15,* 785–795.

Beatty, J. M., Smagghe, G., Ogura, T., Nakagawa, Y., Spindler-Barth, M., & Henrich, V. C. (2009). Properties of ecdysteroid receptors from diverse insect species: A basis for identifying novel insecticides. *FEBS J., 276,* 3087–3098.

Beck, Y., Delaporte, C., Moras, D., Richards, G., & Billas, I. M.L. (2009). The ligand-binding domains of the three RXR-USP nuclear receptor types support distinct tissue and ligand specific hormonal responses in transgenic *Drosophila. Dev. Biol., 330,* 1–11.

Becker, H. J., (1959). Die puffs der speicheldrü senchromosomen von *Drosophila melanogaster. Chromosoma, 10,* 654–678.

Beckstead, R., Ortiz, J. A., Sanchez, C., Prokopenko, S. N., Chambon, P., Losso, R., & Bellen, H. J. (2001). Bonus, a *Drosophila* homologue of TIF1 proteins, interacts with nuclear receptors and can inhibit βFTZ-F1-dependent transcription. *Mol. Cell, 7,* 753–765.

Beckstead, R. B., Lam, G., & Thummel, C. S. (2005). The genomic response to 20-hydroxyecdysone at the onset of *Drosophila* metamorphosis. *Genome Biol., 6,* R99.

Beckstead, R. B., Lam, G., & Thummel, C. S. (2007). Specific transcriptional responses to juvenile hormone and ecdysone in *Drosophila. Insect Biochem. Mol. Biol., 37,* 570–578.

Belloncik, S., Petcharawan, O., Couillard, M., Charpentier, G., Larue, B., Gaurdado, H., Chareonsak, S., & Imanishi, S. (2007). Development and characterization of a continuous cell line, AFKM-On-H, from hemocytes of the European corn borer *Ostrinia nubilalis* (Hubner) (Lepidoptera, Pyralidae). *In vitro Cell Dev. Biol. Anim, 43,* 245–254.

Bender, M., Imam, F. B., Talbot, W. S., Ganetzky, B., & Hogness, D. S. (1997). *Drosophila* ecdysone receptor mutations reveal functional differences among receptor isoforms. *Cell, 91,* 777–788.

Bergman, T., Henrich, V. C., Schlattner, U., & Lezzi, M. (2004). Ligand control of interaction in vivo between ecdysteroid receptor and ultraspiracle ligand-binding domain. *Biochem. J., 378,* 779–784.

Bernardi, F., Romani, P., Tzertzinis, G., Gargiulo, G., & Cavaliere, V. (2009). EcR-B1 and USP nuclear hormone receptors regulate expression of the VM32E eggshell gene during *Drosophila* oogenesis. *Dev. Biol., 328,* 541–551.

Bernardo, T. J., Dubrovskaya, V. A., Jannat, H., Maughan, B., & Dubrovsky, E. B. (2009). Hormonal regulation of the E75 gene in *Drosophila*: identifying functional regulatory elements through computational and biological analysis. *J. Mol. Biol., 387,* 794–808.

Bienz-Tadmor, B., Smith, H. S., & Gerbi, S. A. (1991). The promoter of DNA puff gene II/9-1 *Sciara coprophila* is inducible by ecdysone in late prepupal salivary glands of *Drosophila melanogaster. Cell Regul., 2,* 875–888.

Billas, I. M., Iwema, T., Garnier, J.-M., Mitschier, A., Rochel, N., & Moras, D. (2003). Structural adaptability in the ligand-binding pocket of the ecdysone hormone receptor. *Nature, 426,* 91–96.

Billas, I. M., Moulinier, I., Rochel, N., & Moras, D. (2001). Crystal structure of the ligand-binding domain of the ultraspiracle protein USP, the orthologue of retinoid X receptors in insects. *J. Biol. Chem., 276,* 7465–7474.

Birru, W., Fernley, R. T., Graham, L. D., Grusovin, J., Hill, R. J., Hofmann, A., Howell, L., James, P. J., Jarvis, K. E., Johnsone, W. M., Jones, D. A., Leitner, C., Liepa, A. J., Lovrecz, G. O., Lu, L., Nearn, R. H., O'Driscoll, B. J., Phan, T., Pollard, M., Turner, K. A., & Winkler, D. A. (2010). Synthesis, binding, and bioactivity of gamma-methylene gamma-lactam ecdysone receptor ligands: Advantages of QSAR models for flexible receptors. *Bioorg. Med. Chem., 18,* 5647–5660.

Bitra, K., & Palli, S. R. (2009). Interaction of proteins involved in ecdysone and juvenile hormone signal transduction. *Arch. Insect Biochem. Physiol., 70,* 90–105.

Braun, S., Azopoitei, A., & Spindler-Barth, M. (2009). DNA-binding properties of *Drosophila* ecdysone receptor isoforms and their modification by the heterodimerization partner, ultraspiracle. *Arch. Insect Biochem. Physiol., 72*, 172–191.

Brennan, C. A., Ashburner, M., & Moses, K. (1998). Ecdysone pathway is required for furrow progression in the developing *Drosophila* eye. *Development, 125*, 2653–2664.

Brennan, C. A., Ashburner, M., & Moses, K. (2001). Broad-Complex, but not ecdysone receptor, is required for progression of the morphogenetic furrow in the *Drosophila* eye. *Development, 128*, 1–11.

Brown, H. L.D., Cherbas, L., Cherbas, P., & Truman, J. W. (2005). Use of time-lapse imaging and dominant negative receptors to dissect the steroid receptor control of neuronal remodeling in *Drosophila*. *Development, 133*, 275–285.

Candido-Silva, J. A., Carvalho, J. A., Coelho, G. R., & de Almeida, J. C. (2008). Indirect immune detection of ecdysone receptor (EcR) during the formation of DNA puffs in *Bradysia hygida* (Diptera, Sciaridae). *Chromosome Res., 16*, 609–622.

Carmichael, J. A., Lawrence, M. C., Graham, L. D., Pilling, P. A., Chandana Epa, V., Noyce, L., Lovrecz, G., Winkler, D. A., Pawlak-Skrezecz, A., Eaton, R. E., Hannan, G. N., & Hill, R. J. (2005). The X-ray structure of a hemipteran ecdysone receptor ligand-binding domain: Comparison with a Lepidopteran ecdysone receptor ligand-binding domain and implications for insecticide design. *J. Biol. Chem., 280*, 22258–22269.

Carney, G. E., & Bender, M. (2000). The *Drosophila* ecdysone receptor (EcR) gene is required maternally for normal oogenesis. *Genetics, 154*, 1203–1211.

Carney, G. E., Wade, A. A., Sapra, R., Goldstein, E. S., & Bender, M. (1997). DHR3, an ecdysone-inducible early-late gene encoding a *Drosophila* nuclear receptor, is required for embryogenesis. *Proc. Natl. Acad. Sci. USA, 94*, 12024–12029.

Carton, B., Smagghe, G., & Tirry, L. (2003). Toxicity of two ecdysone agonists, halofenozide and methoxyfenozide, against the multicoloured Asian lady beetle, *Harmonia axyridis* (Coleoptera; Coccinellidae). *J. Appl. Entomol., 127*, 240–242.

Chavoshi, T. M., Moussian, B., & Uv, A. (2010). Tissue-autonomous EcR functions are required for concurrent organ morphogenesis in the *Drosophila* embryo. *Mech. Dev., 127*, 308–319.

Charpentier, G., Tian, L., Cossette, J., Lery, X., & Belloncik, S. (2002). Characterization of cell lines developed from the Colorado potato beetle, *Leptinotarsa decemlineata* Say (Coleoptera: Chrysomelidae). *In vitro Cell Dev. Biol. Anim., 38*, 73–78.

Chen, J. H., Turner, P. C., & Roes, H. H. (2002a). Molecular cloning and induction of nuclear receptors from insect cell lines. *Insect Biochem. Mol. Biol., 32*, 657–667.

Chen, L., O'Keefe, S. L., & Hodgetts, R. B. (2002b). Control of Dopa decarboxylase gene expression by the Broad-Complex during metamorphosis in *Drosophila*. *Mech. Devel., 119*, 145–156.

Cherbas, L., & Cherbas, P. (1997). Parahomologous gene targeting in *Drosophila* cells: An efficient, homology-dependent pathway of illegitimate recombination near a target site. *Genetics, 145*, 349–358.

Cherbas, L., Hu, X., Zhimulev, I., Belyaeva, E., & Cherbas, P. (2003). EcR isoforms in *Drosophila*: Testing tissue-specific requirements by targeted blockade and rescue. *Development, 130*, 271–284.

Cherbas, L., Koehler, M. M.D., & Cherbas, P. (1989). Effects of juvenile hormone on the ecdysone response in *Drosophila* Kc cells. *Dev. Genet., 10*, 177–188.

Cherbas, L., Lee, K., & Cherbas, P. (1991). Identification of ecdysone response elements by analysis of the *Drosophila* Eip28/29 gene. *Genes Dev., 5*, 120–131.

Cherbas, P., Cherbas, L., Lee, S.-S., & Nakanishi, K. (1988). 26-[125I]Iodoponasterone A is a potent ecdysone and a sensitive radioligand for ecdysone receptors. *Proc. Natl. Acad. Sci. USA, 85*, 2096–2100.

Choi, D. S., Wang, D., Tolbert, L., & Sadee, W. (2000). Basal signaling activity of human dopamine D2L receptor demonstrated with an ecdysone-inducible mammalian expression system. *J. Neurosci. Meth., 94*, 217–225.

Christiaens, O., Iga, M., Vetarde, R. A., Rouge, P., & Smagghe, G. (2010). Halloween genes and nuclear receptors in ecdysteroid biosynthesis and signalling in the pea aphid. *Insect Mol. Biol., 19*, 187–200.

Christopherson, K. S., Mark, M. R., Bajaj, V., & Godowski, P. J. (1992). Ecdysteroid-dependent regulation of genes in mammalian cells by a *Drosophila* ecdysone receptor and chimeric transactivators. *Proc. Natl. Acad. Sci. USA, 89*, 6314–6318.

Clayton, G. M., Peak-Chew, S. Y., Evans, R. M., & Schwabe, J. W.R. (2001). The structure of the ultraspiracle ligand-binding domain reveals a nuclear receptor locked in an inactive conformation. *Proc. Natl. Acad. Sci. USA, 98*, 1549–1554.

Costantino, F. B., Bricker, D., Alexandre, K., Shen, K., Merriam, J., Callender, J., Henrich, V., Presente, A., & Andres, A. J. (2008). A novel ecdysone (20E) receptor mediates steroid-regulated developmental events prior to the onset of metamorphosis in *Drosophila*. *PLoS Genetics*, online.

Crispi, S., Giordano, E., D'Avino, P. P., & Furia, M. (1998). Cross-talking among *Drosophila* nuclear receptors at the promiscuous response element of the ng-1 and ng-2 intermolt genes. *J. Mol. Biol., 275*, 561–574.

D'Avino, P. P., Drispi, S., Cherbas, L., Cherbas, P., & Furia, M. (1995). The moulting hormone ecdysone is able to recognize target elements composed of direct repeats. *Mol. Cell Endocrinol., 113*, 1–9.

D'Avino, P. P., & Thummel, C. S. (2000). The ecdysone regulatory pathway controls wing morphogenesis and integrin expression during *Drosophila* metamorphosis. *Dev. Biol., 220*, 211–224.

Davis, M. B., Carney, G. E., Robertsone, A. E., & Bender, M. (2005). Phenotypic analysis of EcR-A mutants suggests that EcR isoforms have unique functions during *Drosophila* development. *Dev. Biol., 282*, 385–396.

De la Cruz, F., & Mak, P. (1997). *Drosophila* ecdysone receptor functions as a constitutive activator in yeast. *J. Steroid Biochem. Mol. Biol., 62*, 353–369.

De la Cruz, F. E., Kirsch, D. R., & Heinrich, J. N. (2000). Transcriptional activity of *Drosophila* melanogaster ecdysone receptor isoforms and ultraspiracle in *Saccharomyces cerevisiae*. *J. Mol. Endocrinol., 24*, 183–191.

Delanoue, R., Slaidina, M., & Leopold, P. (2010). The steroid hormone ecdysone controls systemic growth by repressing dMyc function in *Drosophila* fat cells. *Dev. Cell, 18*, 1012–1021.

Devarakonda, S., Harp, J. M., Kim, Y., Ozyhar, A., & Rastinejad, F. (2003). Structure of the heterodimeric ecdysone receptor DNA-binding complex. *EMBO J., 22*, 5827–5840.

Dhadialla, T. S., & Tzertzinis, G. (1997). Characterization and partial cloning of ecdysteroid receptor from a cotton boll weevil embryonic cell line. *Arch. Insect Biochem. Physiol., 35*, 45–57.

Dobens, L., Rudolph, K., & Berger, E. M. (1991). Ecdysterone regulatory elements function as both transcriptional activators and repressors. *Mol. Cell Biol., 11*, 1846–1853.

Dressel, U., Thormeyer, D., Altincicek, B., Paululat, A., Eggert, M., Schneider, S., Tenbaum, S. P., Renkawitz, R., & Banaihmad, A. (1999). Alien, a highly conserved protein with characteristics of a corepressor for members of the nuclear hormone receptor superfamily. *Mol. Cell Biol., 19*, 3383–3394.

Dubrovskaya, V. A., Berger, E. M., & Dubrovsky, E. B. (2004). Juvenile hormone regulation of the E75 nuclear receptor is conserved in Diptera and Lepidoptera. *Gene., 13*, 171–177.

Dubrovsky, E. B., Dubrovskaya, V. A., & Berger, E. M. (2004). Hormonal regulation and functional role of *Drosophila* E75A orphan nuclear receptor in the juvenile hormone signaling pathway. *Dev. Biol., 268*, 258–270.

Dutko-Gwozdz, J., Gwozdz, T., Orlowski, M., Greb-Markiewicz, B., Dus, D., Dobrucki, J., & Ozyhar, A. (2008). The variety of complexes formed by EcR and USP nuclear receptors in the nuclei of living cells. *Mol. Cell Endocrinol., 294*, 45–51.

Esengil, H., Chang, V., Mich, J. K., & Chen, J. K. (2007). Small-molecular regulation of zebra fish gene expression. *Nat. Chem. Biol., 3*, 154–155.

Evans, R. M. (2005). The nuclear receptor superfamily: a Rosetta stone for physiology. *Mol. Endocrinol., 19*, 1429–1438.

Fallon, A. M., & Gerenday, A. (2010). Ecdysone and the cell cycle: investigations in a mosquito cell line. *J. Insect Physiol., 56*, 1396–1401.

Fang, F., Xu, Y., Jones, D., & Jones, G. (2005). Interactions of ultraspiracle with ecdysone receptor in the transduction of ecdysone and juvenile hormone signaling. *FEBS J., 272*, 1577–1589.

Flatt, T., Heland, A., Rus, F., Porpiglia, E., Sherlock, C., Yamamoto, R., Garbuzov, A., Palli, S. R., Tatar, M., & Silverman, M. (2008). Hormonal regulation of the humoral innate immune response in *Drosophila melanogaster. J. Exp. Biol., 211*, 2712–2724.

Forman, B. M., Goode, E., Chen, J., Oro, A. E., Bradley, D. J., Perlmann, T., Noonan, D. J., Burka, L. T., McMorris, T., & Lamph, W. W. (1995). Identification of a nuclear receptor that is activated by farnesol metabolites. *Cell, 81*, 687–693.

Francis, V. A., Zorzano, A., & Teleman, A. A. (2010). dDOR is an EcR coactivator that forms a feed-forward loop connecting insulin and ecdysone signaling. *Curr. Biol., 20*, 1799–1808.

Fujita, T., & Nakagawa, Y. (2007). QSAR and mode of action studies of insectidal ecdysone agonists. *SAR QSAR Environ. Res., 18*, 77–88.

Gates, J., Lam, G., Ortiz, J. A., Losson, R., & Thummel, C. S. (2004). rigor mortis encodes a novel nuclear receptor interacting protein required for ecdysone signaling during *Drosophila* larval development. *Development, 131*, 25–36.

Gauhar, Z., Sun, I. V., Hua, S., Mason, C. E., Fuchs, F., Li, T. R., Boutros, M., & White, K. P. (2009). Genomic mapping of binding regions for the ecdysone receptor protein complex. *Genome Res., 19*, 1006–1013.

Ghbeish, N., & McKeown, M. (2002). Analyzing the repressive function of ultraspiracle, the *Drosophila* RXR, in *Drosophila* eye development. *Mech. Dev., 111*, 89–98.

Ghbeish, N., Tsai, C. -C., Schubiger, M., Zhou, J. Y., Evans, R. M., & McKeown, M. (2001). The dual role of ultraspiracle, the *Drosophila* retinoid X receptor, in the ecdysone response. *Proc. Natl. Acad. Sci. USA, 98*, 3867–3872.

Godlewski, J., Wang, S., & Wilson, T. G. (2006). Interaction of bHLH-PAS proteins involved in juvenile hormone reception in *Drosophila. Biochem. Biophys. Res. Commun., 342*, 1305–1311.

Godt, D., & Tepass, U. (2009). Breaking a temporal barrier: signaling crosstalk regulates the initiation of border cell migration. *Nat. Cell Biol., 11*, 536–538.

Graham, L. D., Pilling, P. A., Eaton, R. E., Gorman, J. J., Braybrook, C., Hannan, G. N., Pawlak-Skrzec, A., Noyce, L., Lovrecz, G. O., Lu, L., & Hill, R. J. (2007). Purification and characterization of recombinant ligand-binding domains from the ecdysone receptors of four pest insects. *Protein Expr. Purif., 53*, 309–324.

Grebe, M., Przibilla, S., Henrich, V. C., & Spindler-Barth, M. (2003). Characterization of the ligand-binding domain of the ecdysteroid receptor from *Drosophila melanogaster. Biol. Chem., 384*, 93–104.

Grebe, M., Rauch, P., & Spindler-Barth, M. (2000). Characterization of subclones of the epithelial cell line from *Chironomus tentans* resistant to the insecticide RH5992, a nonsteroidal moulting hormone agonist. *Insect Biochem. Mol. Biol., 30*, 591–600.

Grebe, M., & Spindler-Barth, M. (2002). Expression of ecdysteroid receptor and ultraspiracle from *Chironomus tentans* (Insecta, Diptera) in *E. coli* and purification in a functional state. *Insect Biochem. Mol. Biol., 32*, 167–174.

Guo, X., Xu, Q., Harmon, M., Jin, X., Laudet, V., Mangelsdorf, D. F., & Palmer, M. J. (1998). Isolation of two functional retinoid X receptor subtypes from the ixodid tick Amblyomma americanum (L.). *Insect Biochem. Mol. Biol., 27*, 945–962.

Hackney, J. F., Pucci, C., Naes, N., & Dobens, L. (2007). Ras signaling modulates activity of the ecdysone receptor EcR during cell migration in the *Drosophila* ovary. *Dev. Dynam., 236*, 1213–1226.

Hall, B. L., & Thummel, C. S. (1998). The RXR homologue Ultraspiracle is an essential component of the *Drosophila* ecdysone receptor. *Development, 125*, 4709–4717.

Halling, B. P., Yuhas, D. A., Eldridge, R. R., Gilbery, S. N., Deutsch, V. A., & Herron, J. D. (1999). Expression and purification of the hormone binding domain of the *Drosophila* ecdysone and ultraspiracle receptors. *Protein Expr. Purif., 17*, 373–386.

Harada, T., Nakagawa, Y., Akamatsu, M., & Miyagawa, H. (2009). Evaluation of hydrogen bonds of ecdysteroids in the ligand-receptor interactions using a protein modeling system. *Bioorg. Med. Chem., 17,* 5868–5873.

Harmatha, J., & Dinan, L. (1997). Biological activity of natural and synthetic ecdysteroids in the B$_{II}$ bioassay. *Arch. Insect Biochem. Physiol., 35,* 219–225.

Harmon, M. A., Boehm, M. F., Heyman, R. A., & Mangelsdorf, D. J. (1995). Activation of mammalian retinoid X receptors by the insect growth regulator methoprene. *Proc. Natl. Acad. Sci. USA, 92,* 6157–6160.

Hayward, D. C., Dhadialla, T. S., Zhou, S., Kuiper, M. J., Ball, E. E., Wyatt, G. R., & Walker, V. K. (2003). Ligand specificity and developmental expression of RXR and ecdysone receptor in the migratory locust. *J. Insect Physiol., 49,* 1135–1144.

Henrich, V. C., Burns, E., Yelverton, D. P., Christensen, E., & Weinberger, C. (2003). Juvenile hormone potentiates ecdysone receptor-dependent transcription in a mammalian cell culture system. *Insect Biochem. Mol. Biol., 33,* 1239–1247.

Henrich, V. C., Beatty, J. M., Ruff, H., Callender, J., Grebe, M., & Spindler-Barth, M. (2009). The multidimensional partnership of EcR and USP. In G. Smagghe (Ed.), *Ecdysone: Structures and Functions* (pp. 361–375). New York: Springer Science.

Henrich, V. C., Szekely, A. A., Kim, S. J., Brown, N., Antoniewski, C., Hayden, M. A., Lepesant, J.-A., & Gilbert, L. I. (1994). Expression and function of the ultraspiracle (usp) gene locus during development in *Drosophila melanogaster. Dev. Biol., 165,* 38–52.

Henrich, V. C., Vogtli, M., Grebe, M., Przibilla, S., Spindler-Barth, M., & Lezzi, M. (2000). Developmental effects of chimeric ultraspiracle gene derived from *Drosophila* and *Chironomus. Genesis, 28,* 125–133.

Hill, R. J., Segraves, W. A., Choi, D., Underwood, P. A., & Macavoy, E. (1993). The reaction with polytene chromosomes of antibodies raised against *Drosophila*E75A protein. *Insect Biochem. Mol. Biol., 23,* 99–104.

Hock, T., Cottrill, T., Keegan, J., & Garza, D. (2000). The E23 early gene of *Drosophila* encodes an ecdysone-inducible ATP-binding cassette transporter capable of repressing ecdysone-mediated gene activation. *Proc. Natl. Acad. Sci. USA, 15,* 9519–9524.

Holmwood, G., & Schindler, M. (2009). Protein structure based rational design of ecdysone agonists. *Bioorg. Med. Chem., 17,* 4064–4070.

Hoppe, U. C., Marban, E., & Johns, D. C. (2000). Adenovirus-mediated inducible gene expression in vivo by a hybrid ecdysone receptor. *Mol. Ther., 1,* 159–164.

Hu, W., Cook, B. J., Ampasala, D. R., Zheng, S., Caputo, G., Krell, P. J., Retnakaran, A., Arif, B. M., & Feng, Q. (2004). Morpholoigcal and molecular effects of 20-hydroxyecdysone and its agonist tebufenozide on CF-203, a midgut derived cell line from the spruce budworm, *Choristoneura fumiferana. Arch. Insect Biochem. Physiol., 55,* 68–78.

Hu, X., Cherbas, L., & Cherbas, P. (2003). Transcription activation by the ecdysone receptor (EcR/USP): Identification of activation functions. *Mol. Endocrinol., 17,* 716–731.

Huet, F., Ruiz, C., & Richards, G. (1993). Puffs and PCR: the in vivo dynamics of early gene expression during ecdysone responses in *Drosophila. Development, 118,* 613–627.

Hurban, P., & Thummel, C. S. (1993). Isolation and characterization of fifteen ecdysone-inducible *Drosophila* genes reveal unexpected complexities in ecdysone regulation. *Mol. Cell Biol., 13,* 7101–7111.

Ishimoto, H., & Kitamoto, T. (2010). The steroid molting hormone ecdysone regulates sleep in adult *Drosophila melanogaster. Genetics, 185,* 269–281.

Ishimoto, H., Sakai, T., & Kitamoto, T. (2009). Ecdysone signaling regulates the formation of long-term courtship memory in adult *Drosophila melanogaster. Proc. Natl. Acad. Sci. USA, 106,* 6381–6386.

Iwema, T., Chaumot, A., Studer, R. A., Robinson-Rechavi, M., Billas, I. M.L., Moras, D., Laudet, V., & Bonneton, F. (2009). Structural and evolutionary innovation of the heterodimerization interface between USP and the ecdysone receptor EcR in insects. *Mol. Biol. Evol., 26,* 753–768.

Jakob, M., Kolodziejczyk, R., Orlowski, M., Krzyda, S., Kowalska, A., Dutko-Gwozdz, J., Gwozdz, T., Kochman, M., Jaskolski, M., & Ozyhar, A. (2007). Novel DNA-binding element within the C-terminal extension of the nuclear receptor DNA-binding domain. *Nucleic Acids Res., 35,* 2705–2718.

Jang, A. C., Chang, Y. C., Bai, J., & Montell, D. (2009). Border-cell migration requires integration of spatial and temporal signals by the BTB protein Abrupt. *Nat. Cell Biol., 11,* 569–579.

Jindra, M., Huang, J.-Y., Malone, F., Asahina, M., & Riddiford, L. M. (1997). Identification and developmental profiles of two ultraspiracle isoforms in the epidermis and wings of *Manduca sexta. Insect Mol. Biol., 6,* 41–53.

Jindra, M., Malone, F., Hiruma, K., & Riddiford, L. M. (1996). Developmental profiles and ecdysteroid regulation of the mRNAs for two ecdysone receptor isoforms in the epidermis and wings of the tobacco horroworm. *Manduca Sexta. Devel. Biol., 180,* 258–272.

Jindra, M., Sehnal, F., & Riddiford, L. M. (1994a). Isolation, characterization, and developmental expression of the ecdysteroid-induced E75 gene of the wax moth *Galleria mellonella. Eur. J. Biochem., 221,* 665–675.

Jindra, M., Sehnal, F., & Riddiford, L. M. (1994b). Isolation and developmental expression of the ecdysteroid-induced GHR3 gene of the wax moth *Galleria mellonella. Insect Biochem. Mol. Biol., 24,* 763–773.

Jindra, M., Sehnal, F., & Riddiford, L. M. (1995). Ecdysteroid-induced expression of the GmE75 and GHR3 orphan receptor genes in Galleria mellonella (Lepidoptera: Pyralidae) larvae and cultured silk glands. *Eur. J. Entomol., 92,* 235–236.

Jones, D., Jones, G., Teal, P., Hammac, C., Messmer, L., Osborne, K., Belgacem, Y. H., & Martin, J. R. (2010). Suppressed production of methyl farnesoid hormones yields developmental defects and lethality in *Drosophila* larvae. *Gen. Comp. Endocrinol., 165,* 244–254.

Jones, G., Jones, D., Teal, P., Sapa, A., & Wozniak, M. (2006). The retinoid-X receptor orthologue, ultraspiracle, binds with nanomolar affinity to an endogenous morphogenetic ligand. *FEBS J., 273,* 4983–4996.

Jones, G., Wozniak, M., Chu, Y., Dhar, S., & Jones, D. (2001). Juvenile hormone III-dependent conformational changes of the nuclear receptor ultraspiracle. *Insect Biochem. Mol. Biol., 32,* 33–49.

Kamar, M. B., Potter, D. W., Hormann, R. E., Edwards, A., Tice, C. M., Smith, H. C., Dipietro, M. A., Polley, M., Lawless, M., Wolohan, P. R.N., Kethidi, D. R., & Palli, S. R. (2004). Highly flexible ligand binding pocket of ecdysone receptor: A single amino acid change leads to discrimination between two groups of nonsteroidal ecdysone agonists. *J. Biol. Chem., 279*, 27211–27222.

Kapitskaya, M., Wang, S., Cress, D. E., Dhadialla, T. S., & Raikhel, A. S. (1996). The mosquito ultraspiracle homologue, a partner of ecdysteroid receptor heterodimer: Cloning and characterization of isoforms expressed during vitellogenesis. *Mol. Cell Endocrinol., 12*, 119–132.

Karzenowski, D., Potter, D. W., & Padidam, M. (2005). Inducible control of transgene expression with ecdysone receptor: Gene switches with high sensitivity, robust expression, and reduced size. *BioTechniques, 39*, 191–197.

King-Jones, K., & Thummel, C. S. (2005). Nuclear receptors — a perspective from *Drosophila*. *Nat. Rev. Genet., 6*, 311–323.

Koelle, M. R., Talbot, W. S., Segraves, W. A., Bender, M. T., Cherbas, P., & Hogness, D. S. (1991). The *Drosophila* EcR gene encodes an ecdysone receptor, a new member of the steroid receptor superfamily. *Cell, 4*, 59–77.

Konopova, B., & Jindra, M. (2007). Juvenile hormone resistance gene Methoprene-tolerant controls entry into metamorphosis in the beetle *Tr. castaneum*. *Proc. Natl. Acad. Sci. USA, 104*, 10488–10493.

Konopova, B., & Jindra, M. (2008). Broad-Complex acts downstream of Met in juvenile hormone signaling to coordinate primitive holometabolan metamorphosis. *Development, 135*, 559–568.

Koo, J. C., Asurmendi, S., Bick, J., Woodford-Thomas, T., & Beachy, R. N. (2004). Ecdysone agonist-inducible expression of a coat protein gene from tobacco mosaic virus confers viral resistance in transgenic Arapidopsis. *Plant J., 37*, 439–448.

Kozlova, T., Pokholkova, G. V., Tzertzinis, G., Sutherland, J. D., Zhimulev, I. F., & Kafatos, F. C. (1998). *Drosophila* hormone receptor 38 functions in metamorphosis: A role in adult cuticle formation. *Genetics, 149*, 1465–1475.

Kozlova, T., & Thummel, C. S. (2002). Spatial patterns of ecdysteroid receptor activation during the onset of *Drosophila* metamorphosis. *Development, 129*, 1739–1750.

Kozlova, T., & Thummel, C. S. (2003). Essential roles for ecdysone signaling during *Drosophila* mid-embryonic development. *Science, 301*, 1911–1914.

Krusinski, T., Ozyhar, A., & Dobryszycki, P. (2010). Dual FRET assay for detecting receptor protein interaction with DNA. *Nucleic Acids Res., 38*, e108, (online).

Kumar, M. B., Fujimoto, T., Potter, D. W., Deng, Q., & Palli, S. R. (2002). A single point mutation in ecdysone receptor leads to increased specificity: Implications for gene switch applications. *Proc. Natl. Acad. Sci. USA, 99*, 14710–14715.

Lalouette, L., Renault, D., Ravaux, J., & Siaussat, D. (2010). Effects of cold-exposure and subsequent recovery on cellular proliferation with influence of 20-hydroxyecdysone in a lepidopteran cell line (IAL-PID2). *Comp. Biochem. Physiol. A Mol. Integr. Physiol., 155*, 407–414.

Lan, Q., Hiruma, K., Hu, X., Jindra, M., & Riddiford, L. M. (1999). Activation of a delayed early gene encoding MHR3 by the ecdysone receptor heterodimer EcR-B1-USP-1 but not by EcR-B1-USP-2. *Mol. Cell Biol., 19*, 4897–4906.

Lan, Q., Wu, Z. N., & Riddiford, L. M. (1997). Regulation of the ecdysone receptor, USP, E75, and MHR3 genes by 20-hydroxyecdysone in the GV1 cell line of the tobacco hornworm, *Manduca sexta*. *Insect. Mol. Biol., 6*, 3–10.

LaPenna, S., Friz, J., Barlow, A., Palli, S. R., Dinan, L., & Hormann, R. E. (2008). Ecdysteroid ligand-receptor selectivity-exploring trends to design orthogonal gene switches. *FEBS J., 275*, 5785–5809.

Lee, C. Y., Simon, C. R., Woodard, C. T., & Baehrecke, E. H. (2002). Genetic mechanism for the stage and tissue-specific regulation of steroid triggered programmed cell death in *Drosophila*. *Dev. Biol., 252*, 138–148.

Lee, T., Marticke, S., Sung, C., Robinow, S., & Luo, L. (2000). Cell-autonomous requirement of the USP/EcR-B ecdysone receptor for mushroom body neuronal remodeling in *Drosophila*. *Neuron, 28*, 807–818.

Lehman, M. (1995). *Drosophila* Sgs genes: stage and tissue specificity of hormone responsiveness. *BioEssays, 18*, 47–54.

Lehmann, M., & Korge, G. (1995). Ecdysone regulation of the *Drosophila* Sgs-4 gene is mediated by the synergistic action of ecdysone receptor and SEBP 3. *EMBO J., 14*, 716–726.

Lezzi, M., Bergman, T., Henrich, V. C., Vogtli, M., Fromei, C., Grebe, M., Przibilla, S., & Spindler-Barth, M. (2002). Ligand-induced heterodimerization between ligand binding domains of *Drosophila* ecdysteroid receptor (EcR) and Ultraspiracle (USP). *Eur. J. Biochem., 269*, 3237–3246.

Li, F.-Q., Ueda, H., & Hirose, S. (1994). Mediators of activation of *fushi tarazu* gene transcription by BmFTZ-F1. *Mol. Cell Biol., 14*, 3013–3021.

Li, T.-R., & Bender, M. (2000). A conditional rescue system reveals essentials functions for the ecdysone receptor (EcR) gene during molting and metamorphosis in *Drosophila*. *Development, 127*, 2897–2905.

Liang, C., Spitzer, J. D., Smith, H. S., & Gerbi, S. A. (1993). Replication initiates at a confined region during DNA amplification in *Sciara* DNA puff II/9A. *Genes Dev., 7*, 1072–1084.

Liu, Y., Sheng, Z., Liu, H., Wen, D., He, Q., Wang, S., Shao, W., Jiang, R. J., An, S., Sun, Y., Bendena, W. G., Wang, J., Gilbert, L. I., Wilson, T. G., Song, Q., & Li, S. (2009). Juvenile hormone counteracts the bHLH-PAS transcription factors MET and GCE to prevent caspase-dependent programmed cell death in *Drosophila*. *Development, 136*, 2015–2025.

Mangelsdorf, D. J., Ong, E. S., Dyck, J. A., & Evans, R. M. (1990). Nuclear receptor that identifies a novel retinoic acid response pathway. *Nature, 345*, 224–229.

Martinez, A., Sparks, C., Drayton, P., Thompson, J., Greenland, A., & Jepson, I. (1999). Creation of ecdysone receptor chimeras in plants for controlled regulation of gene expression. *Mol. Gen. Genet., 261*, 546–552.

Maschat, F., Dubertret, M. L., & Lepesant, J. A. (1991). Transformation mapping of the regulatory elements of the ecdysone-inducible P1 gene of *Drosophila melanogaster*. *Mol. Cell. Biol., 11*, 2913–2917.

Mikitani, K. (1996). A new nonsteroidal chemical class of ligand for the ecdysteroid receptor 3, 5-di-tert-butyl-4-hydroxy-N-isobutyl-benzamid shows apparent insect molting hormone activities at molecular and cellular levels. *Biochem Biophys. Res. Commun., 227*, 427–432.

Mirth, C. K., Truman, J. W., & Riddiford, L. M. (2009). The ecdysone receptor controls the post-critical weight switch to nutrition-independent differentiation in *Drosophila* wing imaginal discs. *Development, 136,* 2345–2353.

Miura, K., Oda, M., Makita, S., & Chinzei, Y. (2005). Characterization of the *Drosophila Methoprene-tolerant* gene product: juvenile hormone binding and ligand-dependent gene regulations. *FEBS J., 272,* 1169–1172.

Mlodzik, M., Hiromi, Y., Weber, U., Goodman, C. S., & Rubin, G. M. (1990). The *Drosophila* seven-up gene, a member of the steroid receptor gene superfamily, controls photoreceptor cell fates. *Cell, 60,* 211–224.

Mouillet, J. F., Bousquet, F., Sedano, N., Alabouvette, J., Nicolai, M., Zelus, D., Laudet, V., & Delachambre, J. (1999). Cloning and characterization of new orphan nuclear receptors and their developmental profiles during *Te.* metamorphosis. *Eur. J. Biochem., 265,* 972–981.

Mouillet, J. F., Delbecque, J. P., Quennedey, B., & Delachambre, J. (1997). Cloning of two putative ecdysteroid receptor isoforms from Te. molitor and their developmental expression in the epidermis *during metamorphosis. Eur. J. Biochem., 248,* 856–863.

Mouillet, J.-F., Henrich, V. C., Lezzi, M., & Vogtli, M. (2001). Differential control of gene activity by isoforms A, B1, and B2 of the *Drosophila* ecdysone receptor. *Eur. J. Biochem., 268,* 1811–1819.

Mugat, B., Brodu, V., Kejzloarova-Lepesant, J., Antoniewski, C., Bayer, C. A., Fristrom, J. W., & Lepesant, J. A. (2000). Dynamic expression of broad-complex isoforms mediates temporal control of an ecdysteroid target gene at the onset of *Drosophila* metamorphosis. *Dev. Biol., 227,* 104–117.

Nakagawa, Y., & Henrich, V. C. (2009). Arthropod nuclear receptors and their role in molting. *FEBS J., 276,* 6128–6157.

Nakagawa, Y., Sakai, A., Magata, F., Ogura, T., Miyashita, M., & Miyagawa, H. (2007). Molecular cloning of the ecdysone receptor and the retinoid X receptor from the scorpion, *Liocheles australasiae. FEBS J., 274,* 6191–6203.

Neal, S. J., Gibson, M. L., & Westwood, J. T. (2003). Construction of a cDNA-based microarray for *Drosophila melanogaster*: a comparison of gene transcription profiles from SL2 and Kc167 cells. *Genome, 46,* 879–892.

Niedziela-Majka, A., Kochman, M., & Ozyhar, A. (2000). Polarity of the ecdysone receptor complex interaction with the palindromic response element from the hsp27 gene promoter. *Eur. J. Biochem., 267,* 507–519.

No, D., Yao, T. P., & Evans, R. M. (1996). Ecdysone-inducible gene expression in mammalian cells and transgenic mice. *Proc. Natl. Acad. Sci. USA, 93,* 3346–3351.

Nowickyj, S. M., Chithalen, J. V., Cameron, D., Tyshenko, M. G., Petkovich, M., Wyatt, G. R., Jones, G., & Walker, V. K. (2008). Locust retinoid X receptors: 9-cis-retinoic in embryos from a primitive insect. *Proc. Natl. Acad. Sci. USA, 105,* 9540–9545.

Oberdorster, E., Clay, M. A., Cottam, D. M., Wilmot, F. A., McLachlan, J. A., & Milner, M. J. (2001). Common phytochemicals are ecdysteroid agonists and antagonists: A possible evolutionary link between vertebrate and invertebrate steroid hormones. *J. Steroid Biochem. Mol Biol., 77,* 229–238.

Oberdorster, E., Cottam, D. M., Wilmot, F. A., Milner, M. J., & McLachian, J. A. (1999). Interaction of PAHs and PCBs with ecdysone-dependent gene expression and cell proliferation. *Toxicol. Appl. Pharmacol., 160,* 101–108.

O'Connor, J. D. (1985). Ecdysteroid action at the molecular level. In *Comprehensive Insect Physiology, Biochemistry, and Pharmacology, Vol. 8, Endocrinology II.* (pp. 85–98). Oxford, UK: Pergamon Press.

Oro, A. E., McKeown, M., & Evans, R. M. (1990). Relationship between the product of the *Drosophila ultraspiracle* locus and the vertebrate retinoid X receptor. *Nature, 347,* 298–301.

Oro, A. E., McKeown, M., & Evans, R. M. (1992). The *Drosophila* retinoid X receptor homologue ultraspiracle functions in both female reproduction and eye morphogenesis. *Development, 115,* 449–462.

Padidam, M., Gore, M., Lu, D. L., & Smirnova, O. (2003). Chemical inducible, ecdysone receptor based gene expression system for plants. *Transgenic Res., 12,* 101–109.

Palanker, L., Necakov, A. S., Sampson, H. M., Ni, R., Hu, C., Thummel, C. S., & Krause, H. M. (2006). Dynamic regulation of *Drosophila* nuclear receptor activity in vivo. *Development, 133,* 3549–3562.

Palli, S. R., Hormann, R. E., Schlattner, U., & Lezzi, M. (2005). Ecdysteroid receptors and their applications in agriculture and medicine. *Vitam. Horm., 73,* 59–100.

Palli, S. R., Kaptiskaya, M. Z., Kumar, M. B., & Cress, D. E. (2003). Improved ecdysone receptor based inducible gene regulation system. *Eur. J. Biochem., 270,* 1308–1315.

Palmer, M. J., Warren, J. T., Jin, X., Guo, X., & Gilbert, L. I. (2002). Developmental profiles of ecdysteroids, ecdysteroid receptor mRNAs and DNA-binding properties of ecdysteroid receptors in the ixodid tick *Amblyomma americanum* (L.). *Insect Biochem. Mol. Biol., 32,* 465–476.

Pauli, A., van Bemmel, J. G., Oliveira, R. A., Itoh, T., Shirahige, K., van Steensel, B., & Nasmyth, K. (2010). A direct role for cohesin in gene regulation and ecdysone response in *Drosophila* salivary glands. *Curr. Biol., 20,* 1787–1798.

Perera, S. C., Sundaram, M., Krell, P. J., Retnakaran, A., Dhadialla, T. S., & Palli, S. R. (1999). An analysis of ecdysone receptor domains required for heterodimerization with Ultraspiracle. *Arch. Insect Biochem. Physiol., 41,* 61–70.

Perlmann, T., Umesono, K., Rangarajan, P. N., Froman, B. M., & Evans, R. M. (1996). Two distinct dimerization interfaces differentially modulate target gene specificity of nuclear hormone receptors. *Mol. Endocrinol., 10,* 958–966.

Planello, R., Martinez-Guitarte, J. L., & Morcillo, G. (2008). The endocrine disruptor bisphenol A increases the expression of HSP70 and ecdysone receptor genes in the aquatic larvae of *Chironomus riparius. Chemosphere, 71,* 1870–1876.

Planello, R., Martinez-Guitarte, J. L., & Morcillo, G. (2010). Effect of acute exposure to cadmium on the expression of heat-shock and hormone-nuclear receptor genes in the aquatic midge *Chironomus riparius. Sci. Total Environ, 408,* 1598–1603.

Przibilla, S., Hitchcock, W. W., Szecsi, M., Grebe, M., Beatty, J., Henrich, V. C., & Spindler-Barth, M. (2004). Functional studies on the ligand-binding domain of Ultraspiracle from *Drosophila melanogaster. Biol. Chem., 385,* 21–30.

Rebers, J. E. (1999). Overlapping antiparallel transcripts induced by ecdysone in a *Drosophila* cell line. *Insect Biochem. Mol. Biol.*, *29*, 293–302.

Reinking, J., Lam, M. M., Pardee, K., Sampson, H. M., Liu, S., Yang, P., Williams, S., White, W., Lajoie, G., Edwards, A., & Krause, H. M. (2005). The *Drosophila* nuclear receptor E75 contains heme and is gas responsive. *Cell*, *122*, 195–207.

Retnakaran, A., Gelbic, I., Sundaram, M., Tomkins, W., Ladd, T., Primavera, M., Feng, Q., Arif, B., Palli, R., & Krell, P. (2001). Mode of action of the ecdysone agonist tebufenozide (RH-5992), and an exclusion mechanism to explain resistance to it. *Pest. Manag. Sci.*, *57*, 951–957.

Richards, G. (1982). Sequential gene activation by ecdysteroids in polytene chromosomes of *Drosophila melanogaster* VII. Tissue specific puffing. *Roux Arch.*, *191*, 103–111.

Riddiford, L. (1985). Hormone Action at the Cellular Level. In *Comprehensive Insect Physiology, Biochemistry, and Pharmacology, Vol. 8, Endocrinology II* (pp. 37–84). Oxford, UK: Pergamon Press.

Riddiford, L. M., Hiruma, K., Zhou, X., & Nelson, C. A. (2003). Insights into the molecular basis of the hormonal control of molting and metamorphosis from *Manduca sexta* and *Drosophila melanogaster*. *Insect Biochem. Mol. Biol.*, *33*, 1327–1338.

Riddiford, L. M., Truman, J. W., Mirth, C. K., & Shen, Y. C. (2010). A role for juvenile hormone in the prepupal development of *Drosophila melanogaster*. *Development*, *137*, 1117–1126.

Riddihough, G., & Pelham, H. R.B. (1987). An ecdysone response element in the *Drosophila* hsp27 promoter. *EMBO J.*, *6*, 3729–3734.

Romani, P., Bernardi, F., Hackney, J., Dobens, L., Gargiulo, G., & Cavaliere, V. (2009). Cell survival and polarity of *Drosophila* follicle cells require the activity of ecdysone receptor B1 isoform. *Genetics*, *181*, 165–175.

Rymarczyk, G., Grad, I., Rusek, A., Oswiecimska-Rusin, K., Niedziela-Majli, A., Kochman, M., & Ozyhar, A. (2003). Purification of *Drosophila melanogaster* ultraspiracle protein and analysis of it's a/B region-dependent dimerization in vitro. *Biol. Chem.*, *384*, 59–69.

Saez, E., Nelson, M. C., Eshelman, B., Banayo, E., Koder, A., Cho, G. J., & Evans, R. M. (2000). Identification of ligands and coligands for the ecdysone-regulated gene switch. *Proc. Natl. Acad. Sci. USA*, *97*, 14512–14517.

Sawatsubashi, S., Murata, T., Lim, J., Fujiki, R., Ito, S., Suzuki, E., Tanabe, M., Zhao, Y., Kimura, S., Fujiyama, S., Ueda, T., Umetsu, D., Ito, T., Takeyama, K., & Kato, S. (2010). A histone chaperone, DEK, transcriptively coactivates a nuclear receptor. *Genes Dev.*, *24*, 159–170.

Schinko, J. B., Weber, M., Viktorinova, I., Kiupakis, A., Averof, M., Klingler, M., Wimmer, E. A., & Bucher, G. (2010). Functionality of the GAL4/UAS system in *Tr.* requires the use of endogenous core promoters. *BMC Dev. Biol.*, *10*, 53, 10.1186/1471-213X-10-53.

Schubiger, M., Carre, C., Antoniewski, C., & Truman, J. W. (2005). Ligand-dependent de-repression via EcR/USP acts as a gate to coordinate the differentiation of sensory neurons in the *Drosophila* wing. *Development*, *132*, 5239–5248.

Schubiger, M., & Truman, J. W. (2000). The RXR orthologue USP suppresses early metamorphic processes in *Drosophila* in the absence of ecdysteroids. *Development*, *127*, 1151–1159.

Schubiger, M., Truman, J. W., Wade, A. A., Carney, G. E., Truman, J. W., & Bender, M. (1998). *Drosophila* EcR-B ecdysone receptor isoforms are required for larval molting and for neuron remodeling during metamorphosis. *Development*, *125*, 2053–2062.

Sedkov, Y., Cho, E., Petruk, S., Cherbas, L., Smith, S. T., Jones, R. S., Cherbas, P., Canaani, E., Jaynes, J. B., & Mazo, A. (2003). Methylation at lysine 4 of histone H3 in ecdysone-dependent development of *Drosophila*. *Nature*, *426*, 78–83.

Segal, D., Cherbas, L., & Cherbas, P. (1996). Genetic transformation of *Drosophila* cells in culture by P-element mediated transposition. *Somat. Cell Mol. Genet.*, *22*, 159–165.

Sempere, L. F., Sokol, N. S., Dubrovsky, E. B., Berger, E. M., & Ambros, V. (2003). Temporal regulation of microRNA expression in *Drosophila melanogaster* mediated by hormonal signals and broad-Complex gene activity. *Dev. Biol.*, *259*, 9–18.

Shea, C., Richer, J., Tzertzinis, G., & Maina, C. V. (2010). An EcR homologue from the filarial parasite, *Dirofilaria immitis* requires a ligand-activated partner for transactivation. *Mol. Biochem. Parasitol.*, *171*, 55–63.

Shea, M. J., King, D. L., Conboy, M. J., Mariani, B. D., & Kafatos, F. C. (1990). Proteins that bind to *Drosophila* chorion cis-regulatory elements: A new C2H2 zinc finger protein and a C2C2 steroid receptor-like component. *Genes Dev.*, *4*, 1128–1140.

Schubiger, M., Tomita, S., Sung., C., Robinow, S., & Troman, J. W. (2003). Isoform specific control of gene activity *invivo* by the *Drosophila* ecdysone receptor. *Mech. Devel.*, *120*, 909–918.

Simon, A. F., Shih, C., Mack, A., & Benzer, S. (2003). Steroid control of longevity in *Drosophila* melanogaster. *Science*, *299*, 1407–1410.

Singh, A. K., Tavva, V. S., Collins, G. B., & Palli, S. R. (2010). Improvement of ecdysone receptor gene switch for applications in plants: *Locusta migratoria* retinoid X receptor (LmRXR) mutagenesis and optimization of translation start site. *FEBS J*, *277*, 4640–4650.

Soin, T., Swevers, L., Mosallanejad, H., Efrose, R., Labropoulou, V., Iatrou, K., & Smagghe, G. (2008). Juvenile hormone analogs do not affect directly the activity of the ecdysteroid receptor complex in insect culture cell lines. *J. Insect Physiol.*, *54*, 429–438.

Song, Q., Alnemri, E. S., Litwack, G., & Gilbert, L. I. (1997). An immunophilin is a component of the insect ecdysone receptor (EcR) complex. *Insect Biochem. Mol. Biol.*, *216*, 973–982.

Song, Q., Sun, X., & Jin, X. Y. (2003). 20E regulated USP expression and phosphorylation in *Drosophila melanogaster*. *Insect Biochem. Mol. Biol.*, *33*, 1211–1218.

Spindler-Barth, M., & Spindler, K.-D. (1998). Ecdysteroid resistant subclones of the epithelial cell line from Chironomus tentans (Insecta, Diptera. I.) Selection and characterization of resistant clones. *In vitro Cell Dev. Biol. Anim.*, *36*, 116–122.

Spokony, R. F., & Restifo, L. L. (2007). Anciently duplicated Broad Complex exons have distinct temporal functions during tissue morphogenesis. *Dev. Genes. Evol., 217,* 499–513.

Srivastava, D. P., Yu, E. J., Kennedy, K., Chatwin, H., Reeale, V., Hamon, M., Smith, T., & Evans, P. D. (2005). Rapid, nongenomic responses to ecdysteroids and catecholamines mediated by a novel *Drosophila* G-protein coupled receptor. *J. Neurosci., 25,* 6145–6155.

Stallcup, M. R. (2001). Role of protein methylation in chromatin remodeling and transcriptional regulation. *Oncogene, 20,* 3014–3020.

Stathopoulos, A., & Levine, M. (2002). Whole-genome expression profiles identify gene batteries in *Drosophila*. *Dev. Cell, 3,* 464–465.

Stone, B. L., & Thummel, C. S. (1993). The *Drosophila* 78C early late puff contains E78, an ecdysone-inducible gene that encodes a novel member of the nuclear hormone receptor superfamily. *Cell, 75,* 307–320.

Suhr, S. T., Gil, E. B., Senut, M.-C., & Gage, F. H. (1998). High level transactivation by a modified *Bombyx* ecdysone receptor in mammalian cells without exogenous retinoid X receptor. *Proc. Natl. Acad. Sci. USA,* 7999–8004.

Sullivan, A. A., & Thummel, C. S. (2003). Temporal profiles of nuclear receptor gene expression reveal coordinate transcriptional responses during *Drosophila* development. *Mol. Endocrinol., 17,* 2125–2137.

Sun, Y., An, S., Henrich, V. C., & Song, Q. (2008). Proteomic identification of PKC-mediated expression of 20E-induced protein in *Drosophila melanogaster*. *J. Proteomic Res., 6,* 4478–4488.

Sun, G., Zhu, J., Chen, L., & Raikhel, A. S. (2005). Synergistic action of E74B and ecdysteroid receptor in activating a 20-hydroxyecdysone effector gene. *Proc. Natl. Acad. Sci. USA, 102,* 15506–15511.

Sung, C., & Robinow, S. (2000). Characterization of the regulatory elements controlling neuronal expression of the A-isoform of the ecdysone receptor gene of *Drosophila melanogaster*. *Mech. Dev., 91,* 237–248.

Sutherland, J. D., Kozlova, T., Tzertzinis, G., & Kafatos, F. C. (1995). *Drosophila* hormone receptor 38: a second partner for *Drosophila* Vsp suggests an unexpected role for nuclear receptors of the nerve growth factor-induced protein B type. *Proc. Natl. Acad. Sci. USA, 92,* 7966–7970.

Suzuki, Y., Truman, J. W., & Riddiford, L. M. (2008). The role of *Broad* in the development of *Tr. castaneum*: Implications for the evolution of the holometabolouos insect pupa. *Development, 135,* 569–577.

Swevers, L., & Iatrou, K. (2003). The ecdysone regulatory cascade and ovarian development in Lepidopteran insects: Insights from the silkmoth paradigm. *Insect Biochem. Mol. Biol., 33,* 1285–1297.

Swevers, L., Kravariti, L., Ciolfi, S., Xenov-Kokoletsi, M., Ragoussis, N., Smagghe, G., Nakagawa, Y., Mazomenos, B., & Iatrou, K. (2004). A cell-based high-throughput screening system for detecting ecdysteroid agonists and antagonists in plant extracts and libraries of synthetic compounds. *FASEB J., 18,* 134–136.

Swevers, L., Soin, T., Mosallanejad, H., Iatrou, K., & Smagghe, G. (2008). Ecdysteroid signaling in ecdysteroid-resistant cell lines from the polyphagous noctuid pest *Spodoptera exigua*. *Insect Biochem. Mol. Biol., 38,* 825–833.

Talbot, W. S., Swywryd, E. A., & Hogness, D. S. (1993). *Drosophila* tissues with different metamorphic responses to ecdysone express different ecdysone receptor isoforms. *Cell, 73,* 1323–1337.

Tan, A., & Palli, S. R. (2008). Ecdysone receptor isoforms play distinct roles in controlling molting and metamorphosis in the red flour beetle, *Tr. castaneum. Mol. Cell Endocrinol., 291,* 42–49.

Tavva, V. S., Dinkins, R. D., Palli, S. R., & Collins, G. B. (2006). Development of a methoxyfenozide-responsive gene switch for applications in plants. *Plant J., 45,* 457–469.

Tavva, V. S., Palli, S. R., Dinkins, R. D., & Collins, G. B. (2007). Applications of EcR gene switch technology in functional genomics. *Arch. Insect Biochem. Physiol., 65,* 164–179.

Thomas, H. E., Stuttenberg, H. G., & Stewart, A. F. (1993). Heterodimerization of the *Drosophila* ecdysone receptor with retinoid X receptors and ultraspiracle. *Nature, 362,* 471–475.

Thummel, C. S. (2002). Ecdysone-regulated puff genes 2000. *Insect Biochem. Mol. Biol., 32,* 113–120.

Toya, T., Fukasawa, H., Masui, A., & Endo, Y. (2002). Potent and selective partial ecdysone agonist activity of chromafenozide in Sf9 cells. *Biochem Biophys. Res. Commun., 292,* 1087–1091.

Tran, H. T., Askari, H. B., Shaaban, S., Price, L., Palli, S. R., Dhadialla, T. S., Carlson, G. R., & Butt, T. R. (2001a). Reconstruction of ligand-dependent transactivation of *Choristoneura fumiferana* ecdysone receptor in yeast. *Mol. Endocrinol., 15,* 1140–1153.

Tran, H. T., Shaaban, S., Askari, H. B., Walfish, P. G., Raikhel, A. S., & Butt, T. R. (2001b). Requirement of co-factors for the ligand-mediated activity of the insect ecdysteroid receptor in yeast. *J. Mol. Endocrinol., 27,* 191–209.

Tremmel, C., Schaefer, M., Azoitei, A., Ruff, H., & Spindler-Barth, M. (2010). Interaction of the N-terminus of ecdysone receptor isoforms with the ligand binding domain. *Mol. Cell. Endocrinol, 332,* 293–300.

Trisyono, A., Goodman, C. L., Grasela, J. J., McIntosh, A. H., & Chippendale, G. M. (2000). Establishment and characterization of an *Ostrinia nubilalis* cell line and its response to ecdysone agonists. *in vitro Cell Dev. Biol., 36,* 300–404.

Tsai, C. C., Kao, H. Y., Yao, T. P., Mckeown, M., & Evans, R. M. (1999). SMRTER, a *Drosophila* nuclear receptor coregulator, reveals that EcR mediated repression is critical for development. *Mol. Cell, 4,* 175–186.

Tzertzinis, G., Malecki, A., & Kafatos, F. C. (1994). BmCF1, a *Bombyx mori* RXR-type receptor related to the *Drosophila* ultraspiracle. *J. Mol. Biol., 238,* 479–486.

Uhlirova, M., Foy, B. D., Beaty, B. J., Olson, K. E., Riddiford, L. M., & Jindra, M. (2003). Use of Sindbis virus-mediated RNA interference to demonstrate a conserved role of Broad-Complex in insect metamorphosis. *Proc. Natl. Acad. Sci. USA, 23,* 15607–15612.

Unger, E., Cigan, A. M., Trimnell, M., Xu, R. J., Kendall, T., Roth, B., & Albertson, M. (2002). A chimeric ecdysone receptor facilitates methoxfenozide-dependent restoration of male fertility in ms45 maize. *Transgenic Res., 11*, 455–465.

van der Knaap, J. A., Kozhevnikova, E., Langenberg, K., Moshkin, Y. M., & Verrijzer, C. P. (2010). Biosynthetic enzyme GMP synthetase cooperates with ubiquitin-specific protease 7 in transcriptional regulation of ecdysteroid target genes. *Mol. Cell Biol., 30*, 736–744.

Vogtli, M., Elke, C., Imhof, M. O., & Lezzi, M. (1998). High level transactivation by the ecdysone receptor complex at the core recognition motif. *Nucleic Acids Res., 10*, 2407–2414.

Vogtli, M., Imhof, M. O., Brown, N. E., Rauch, P., Spindler-Barth, M., Lezzi, M., & Henrich, V. C. (1999). Functional characterization of two Ultraspiracle forms (CtUSP-1 and CtUSP-2) from Chironomus tentans. *Insect Biochem. Mol. Biol., 29*, 931–942.

von Kalm, L., Crossgrove, K., Von Seggern, D., Guild, G. M., & Beckendorf, S. K. (1994). The Broad-Complex directly controls a tissue-specific response to the steroid hormone ecdysone at the onset of *Drosophila* metamorphosis. *EMBO J., 13*, 3505–3516.

Wang, S., Baumann, A., & Wilson, T. G. (2007). *Drosophila melanogaster* Methoprene-tolerant (Met) gene homologues from three mosquito species: members of PAS transcriptional factor family. *J. Insect Physiol., 53*, 246–253.

Wang, S.-F., Ayer, S., Segraves, W. A., Williams, D. R., & Raikhel, A. S. (2000). Molecular determinants of differential ligand sensitivities of insect ecdysteroid receptors. *Mol. Cell Biol., 20*, 3870–3879.

Wang, S. F., Li, C., Sun, G., Zhu, J., & Raikhel, A. S. (2002). Differential expression and regulation by 20-hydroxyecdysone of mosquito ecdysteroid receptor isoforms A and B. *Mol. Cell Endocrinol., 196*, 29–42.

Wang, S.-F., Miura, K., Miksicek, R. J., Segraves, W. A., & Raikhel, A. S. (1998). DNA-binding and transactivation characteristics of the mosquito ecdysone receptor-Ultraspiracle complex. *J. Biol. Chem., 273*, 27531–27540.

Wang, Y. H., & LeBlanc, G. A. (2009). Interactions of methyl farnesoate and related compounds with a crustacean retinoid X receptor. *Mol. Cell Endocrinol., 309*, 109–116.

Warren, J. T., Yerushalmi, Y., Shimell, J. M., O'Connor, M. B., Restifo, L., & Gilbert, L. I. (2006). Discrete pulses of molting hormone, 20-hydroxyecdysone, during late larval development of *Drosophila melanogaster*: Correlations with changes in gene activity. *Dev. Dynam, 235*, 315–326.

Watanabe, T., Takeuchi, H., & Kubo, T. (2010). Structural diversity and evolution of the N-terminal isoform specific region of ecdysone receptor A and B1 isoforms in insects. *BMC Evol. Biol., 10*, 40.

Wegmann, I. S., Quack, S., Spindler, K. D., Dorsch-Hasler, K., Vogtli, M., & Lezzi, M. (1995). Immunological studies on the developmental and chromosomal distribution of ecdysteroid receptor proteins in Chironomus tentans. *Arch. Insect Biochem. Physiol., 30*, 95–114.

White, K. P., Hurban, P., Watanabe, T., & Hogness, D. S. (1997). Coordination of *Drosophila* metamorphosis by two ecdysone-induced nuclear receptors. *Science, 276*, 114–117.

White, K. P., Rifkin, S. A., Hurban, P., & Hogness, D. S. (1999). Microarray analysis of *Drosophila* development during metamorphosis. *Science, 286*, 2179–2184.

Willy, P. J., Umesono, K., Ong, E. S., Evans, R. M., Heyman, R. A., & Mangelsdorf, D. J. (1995). LXR, a nuclear receptor that defines a distinct retinoid response pathway. *Genes Dev., 9*, 1033–1045.

Wilson, T. W., Wang, S., Beno, M., & Farkas, R. (2006). Wide mutational spectrum of a gene involved in hormone action and insecticide resistance in *Drosophila melanogaster*. *Mol. Gen. Genomics, 276*, 294–303.

Wing, K. D. (1988). RH5849, a nonsteroidal ecdysone agonist: effects on a *Drosophila* cell line. *Science, 241*, 467–469.

Wurtz, J.-M., Guilot, B., Fagart, J., Moras, D., Tietjen, K., & Schindler, M. (2000). A new model for 20-hydroxyecdysone and dibenzoylhydrazine binding: A homology modeling and docking approach. *Protein Sci., 9*, 1073–1084.

Yao, T., Forman, B. M., Jiang, Z., Cherbas, L., Chen, J. D., McKeown, M., Cherbas, P., & Evans, R. M. (1995). *Drosophila* ultraspiracle modulates ecdysone receptor function via heterodimer formation. *Cell, 71*, 63–72.

Yao, T. P., Forman, B. M., Jiang, Z., Cherbas, L., Chen, J. D., McKeown, M., Cherbas, P., & Evans, R. M. (1993). Functional ecdysone receptor is the product of EcR and ultraspiracle genes. *Nature, 366*, 476–479.

Yund, M. A., & Osterbur, D. L. (1985). Ecdysteroid Receptors and Binding Proteins. In *Comprehensive Insect Physiology, Biochemistry, and Pharmacology, Vol. 7, Endocrinology I* (pp. 473–490). Oxford, UK: Pergamon Press.

Zelhof, A. C., Ghbeish, N., Tsai, C., Evans, R. M., & McKeown, M. (1997). A role for Ultraspiracle, the *Drosophila* RXR, in morphogenetic furrow movement and photoreceptor cluster formation. *Development, 124*, 2499–2505.

Zelhof, A. C., Yao, T. P., Chen, J. D., Evans, R. M., & McKeown, M. (1995a). Seven-up inhibits ultraspiracle-based signaling pathways in vitro and in vivo. *Mol. Cell Biol., 15*, 6736–6745.

Zelhof, A. C., Yao, T., Evans, R. M., & McKeown, M. (1995b). Identification and characterization of a *Drosophila* nuclear receptor with the ability to inhibit the ecdysone response. *Proc. Natl. Acad. Sci. USA, 92*, 10477–10481.

Zheng, W. W., Yang, D. T., Wang, J. X., Song, Q. S., Gilbert, L. I., & Zhao, X. F. (2009). Hsc70 binds to ultraspiracle resulting in the upregulation of 20-hydroxyecdysone responsive genes in *Helicoverpa armigera*. *Mol. Cell Endocrinol., 315*, 282–291.

Zheng, X., Wang, J., Haerry, T. E., Wu, A. Y.-H., Martin, J., O'Connor, M. B., Lee, C.-H.J., & Lee, T. (2003). TGF-B signaling activates steroid hormone receptor expression during neuronal remodeling in the *Drosophila* brain. *Cell, 112*, 303–315.

Zhou, Z., & Riddiford, L. (2002). Broad specifies pupal development and mediates the "status quo" action of juvenile hormone on the pupal-adult transformation in *Drosophila* and *Manduca*. *Development, 129*, 2259–2269.

Zhu, J., Chen, L., & Raikhel, A. S. (2007). Distinct roles of Broad isoforms in regulation of the 20-hydroxyecdsyone effector gene, Vitellogenin, in the mosquito *Aedes aegypti*. *Mol. Cell Endocrinol., 267*, 97–105.

Zhu, J., Chen, L., Sun, G., & Raikhel, A. S. (2006). The competence factor beta Ftz-F1 potentiates ecdysone receptor activity via recruiting a p160/SRC coactivator. *Mol. Cell Biol.*, *26*, 9402–9412.

Zhu, J., Miura, K., Chen, L., & Raikhel, A. S. (2000). AHR38, a homologue of NGF1-B, inhibits formation of the functional ecdysteroid receptor in the mosquito *Aedes aegypti*. *EMBO J.*, *19*, 253–262.

Zhu, J., Miura, K., Chen, L., & Raikhel, A. S. (2003). Cyclicity of mosquito vitellogenic ecdysteroid-mediated signaling is modulated by alternative dimerization of the RXR homologue, Ultraspiracle. *Proc. Natl. Acad. Sci. USA*, *100*, 544–549.

Zinke, I., Schutz, C. S., Katzenberger, J. D., Bauer, M., & Pankratz, M. J. (2002). Nutrient control of gene expression in *Drosophila*: microarray analysis of starvation and sugar-dependent response. *EMBO J.*, *21*, 6162–6173.

6 Evolution of Nuclear Receptors in Insects

François Bonneton and Vincent Laudet
Université de Lyon, Lyon, France

© 2012 Elsevier B.V. All Rights Reserved

6.1. Nuclear Receptors in Animals

6.1.1. Ligand Activated Transcription Factors

Nuclear receptors are metazoan transcription factors activated by small lipophilic ligands, such as steroid hormones, thyroid hormones, retinoids, vitamin D, or fatty acids. The availability of the ligand controls the transcriptional activity of nuclear receptors. However, this general and traditional definition has to be put into perspective, since natural ligands are still missing for several receptors in mammals and for most of them in insects. Furthermore, it appears that some nuclear receptors are true orphans. However, most members of this family are expected to be regulated by one or several ligands. Testing this hypothesis has proven to be an arduous challenge, especially when endocrinological knowledge is non-existent for a given receptor.

Nuclear receptors are involved in a considerable diversity of developmental and physiological processes (Laudet and Gronemeyer, 2002). In insects, their role has been well characterized in embryo segmentation, development of the nervous system, molting, and metamorphosis (King-Jones and Thummel, 2005; see Chapter 5). Nuclear receptors are also promising targets for the control of insect pests (Palli *et al.*, 2005). The family is divided into seven subfamilies (Nuclear Receptors Nomenclature Committee, 1999). In animals, proteins of the bHLH-PAS family are the only other known transcription factors activated by small lipophilic ligands (Furness *et al.*,

2007). The dual nature of these two classes of proteins, being both ligand receptors and transcription factors, provides the organism with specific tools to respond, at the gene expression level, to environmental cues. Thus it is not surprising that interactions between nuclear receptors and bHLH-PAS proteins seem to be an important physiological link in animals.

6.1.2. Modularity in Structure and Function

Nuclear receptors are modular proteins with five to six regions, designated A to F, from the N-terminal to the C-terminal end (**Figure 1**). The regions A/B, D, and F, not all present in all receptors, are usually poorly conserved and their structure is unknown. By contrast, the DNA-binding domain (DBD, region C) and the ligand-binding domain (LBD, region E) are highly conserved and their structures have been determined for several receptors (Li *et al.*, 2003).

The DBD is made up of two non-equivalent zinc-finger structures (C4-zinc fingers) with each zinc atom necessary to retain stable domain structure and function (Khorasanizadeh and Rastinejad, 2001; **Figure 2A**). Nuclear receptors bind to the regulatory regions of target genes as homodimers or heterodimers, but more rarely as monomers. Recent genomic studies have shown that the number of nuclear receptor binding sites range from several hundreds (Gauhar *et al.*, 2009) to several thousands (Cheung and Kraus, 2010). The canonical hormone

DOI: 10.1016/B978-0-12-384749-2.10006-8

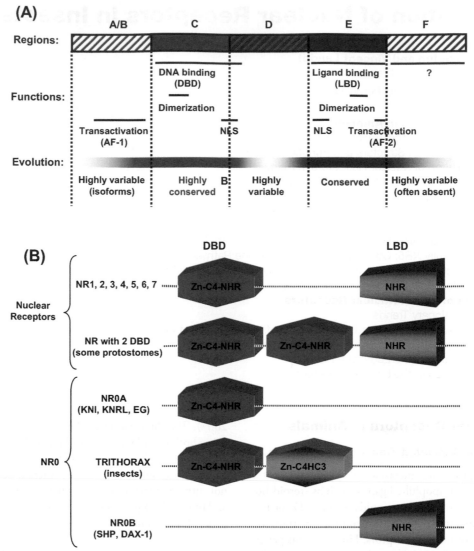

Figure 1 Modularity of nuclear receptors. (A) Schematic illustration of the six regions, designated A to F, from the N-terminal to the C-terminal end. The functions of the regions are depicted below the scheme, and most of these are derived from analyses of vertebrate receptors. Within the transcription activation functions, AF-1 and AF-2, autonomous transactivation domains have been defined in most receptors. The activation domain of the E region has been mapped within the C-terminal helix H12. Many receptor genes encode different isoforms harboring different A/B regions, and thus different AF-1 and activation domains. The evolutionary conserved regions C and E are indicated as red and blue boxes, respectively. Region F may be absent in some receptors. (B) Organization of the DBD and LBD domains among the different types of nuclear receptors. The DBD is shown in red as a zinc finger with 4 cysteines (Zn-C4-NHR). The LBD is shown in blue as a Nuclear Hormone Receptor (NHR). *Trithorax* has been originally classified within the NR0, because it contains a nuclear receptor DBD. However, this very large protein with multiple domains is actually related to the MLL family of transcription factors. Consequently, it will not be discussed further in this chapter.

response element (HRE) has the core sequence RGGTCA (Laudet and Gronemeyer, 2002). Mutations, extensions, duplications, and distinct relative orientations of repeats of this motif generate response elements selective for a given class of receptor. Some receptors, such as FTZ-F1, HR3, or E75, can bind to DNA as monomers through a single core sequence. In this case, an A/T-rich region 5′ to the core element governs the binding specificity (Laudet and Adelmant, 1995). All members of subfamily 2 are able to form homodimers on direct repeat sequence.

Steroid receptors (NR3) bind as homodimers to HREs containing two core motifs separated by 1–3 nucleotides and organized as palindromes. Most of the receptors of subfamilies 1 and 4 heterodimerize with RXR (NR2B, homologous to the insect USP) and bind to direct repeats (DRs) of the core motif (Brelivet *et al.*, 2004). The spacing between the two halves of the DR dictates the type of heterodimer that will bind. For example, a DR separated by five nucleotides (DR5) will usually be recognized by RXR:RAR and a DR4 by RXR:TR.

A

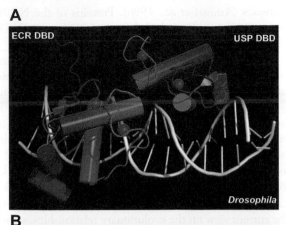

ECR DBD   USP DBD

Drosophila

B

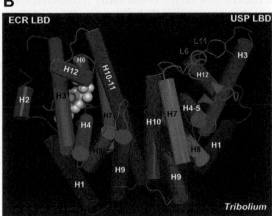

ECR LBD   USP LBD

Tribolium

Figure 2 Structure of the heterodimer ECR-USP. (A) Crystal structure of the DBD of the *D. melanogaster* (Diptera) heterodimer ECR-USP bound to the natural pseudo-palindromic response element hsp27 (PDB: 2HAN; *Jakob et al.*, 2007). Loops and helices are colored from blue (N-terminal) to red (C-terminal). The DNA is shown in white. Zinc atoms of the zinc fingers are mauve spheres. Part of the C-terminal extension of the EcR DBD folds into a helix located in the minor groove, a location never observed for other nuclear receptors. (B) Crystal structure of the LBD of the *T. castaneum* (Coleoptera) heterodimer ECR-USP (PDB: 2NXX; *Iwema et al.*, 2007). Helices are colored from blue (H1) to red (H12). The ECR ligand, ponasterone (25-deoxy-20-hydroxyecdysone), is represented with white spheres. In *Tr.* USP, helices H6 and H11 are transformed into loops L6 (green) and L11 (orange) that fill the site of the ligand-binding pocket. Thus, on this figure, we can see the structure of a hormone receptor (ECR) and of a real orphan (USP).

The LBD is composed of 10 to 12 α-helices that form 3 antiparallel helical layers that combine to make an α-helical sandwich (**Figure 2B**). The LBD fulfills three main functions: ligand binding, dimerization, and recruitment of co-regulators. The ligand-binding pocket (LBP) of the receptor is located in the interior of the structure and is formed by a subset of the surrounding helices. The core of the dimerization interface is mainly constituted by helices H9 and H10 (more than 75% of the total surface), together with other residues from helices H7 and H11 as well as from loops L8-9 and

L9-10 (Bourguet *et al,*. 2000; Gampe *et al.*, 2000). The heterodimeric arrangement closely resembles that of a homodimer, except that the heterodimer interfaces are asymmetric (Folkertsma *et al.*, 2005). The LBD harbors a ligand-dependent activation function, called AF-2, which is responsible for the recruitment of transcriptional co-regulators. Many unliganded (apo) nuclear receptors are transcriptional silencers as a result of interaction with corepressors. The LBD domain undergoes a conformational change upon ligand binding (holo), allowing the interaction with co-activators and the transactivation of target genes. Importantly, this conformational change can be induced not exclusively by ligand but also by other mechanisms, such as phosphorylation (Rochette-Egly, 2003) and protein-protein interactions (Moore *et al.*, 2010). Real orphan receptors, like *Drosophila* HR38 (NR4A4) or *Tr.* USP, have no pocket but are nonetheless active without any ligand (Baker *et al.*, 2003; Iwema *et al.*, 2007; **Figure 2B**). In other cases, the irreversible binding of a ligand inside the pocket stabilizes an active conformation in an unconventional manner when compared to the hormonal receptors. Such structural ligands, like fatty acids, have been observed in the USP LBD of *Heliothis* and *Drosophila* (Billas *et al.*, 2001; Clayton *et al.*, 2001). Even more surprising was the discovery that the *Drosophila* E75 is regulated by gas (NO and CO), which controls the redox status of an iron heme molecule permanently bound to its ligand binding pocket (Reinking *et al.*, 2005; de Rosny *et al.*, 2006). All of these results have considerably modified our view of nuclear receptor ligands (Sladek, 2011). In conclusion, the LBD could be viewed as a domain responsible for an allosteric modification of the receptor allowing the exchange between corepressor and co-activator complexes.

6.1.3. Evolution of Nuclear Receptors in Animals

Nuclear receptors are present in all metazoans, and only in metazoans. They are absent from the genome of the unicellular *Monosiga brevicollis*, which belongs to choanoflagellates, the sister group of animals (King *et al.*, 2008). Their origin is still unknown. The DBD and the LBD of nuclear receptors share no significant similarity with other domains outside animals. It should be noted here that some fungi have ligand-regulated transcription factors that share many characteristics with nuclear receptors (Näär and Thakur, 2009). Indeed, in the budding yeast, the heterodimeric transcription factors Oaf1/Pip2 are bound and regulated by fatty acids (oleate) through a mechanism that is very similar to PPAR/RXR (Phelps *et al.*, 2006). Similarly, the yeast transcription factors Pdr1p/Pdr3p are regulated by xenobiotics, like the PXR nuclear receptor (Thakur *et al.*, 2008). These proteins might contain an LBD with a putative nuclear receptor folding. However, given the sequence divergence, these resemblances could be due to either homology or homoplasic evolution.

To clarify the history of nuclear receptors, phylogenetic trees have been produced using the conserved DBD and LBD (Laudet *et al.*, 1992; Laudet, 1997; Bertrand *et al.*, 2004; Bridgham *et al.*, 2010). These phylogenies provided a framework to establish a nomenclature for the family (Nuclear Receptors Nomenclature Committee, 1999). According to this nomenclature, the family is divided into seven monophyletic groups called subfamilies (**Figure 3**). Note that the NR7 subfamily, which has been lost independently in vertebrates and in ecdysozoans, was discovered only very recently (Bertrand *et al.*, 2011). Another recent discovery is the identification of unusual nuclear receptors with two DBD in tandem with one LBD in the genome of three flatworm species, a mollusc, and *Daphnia* (Wu *et al.*, 2006, 2007). They are related to the subfamily NR1 and still await a proper group name. In addition, the official nomenclature artificially gathers proteins that have only one of the two conserved domains (DBD or LBD) into a paraphyletic subfamily called NR0. Without a LBD, proteins of the group NR0A act as classical ligand-independent transcription factors. A good example is the gap segmentation gene *knirps*, a transcriptional repressor

in insects (Arnosti *et al.*, 1996). Proteins of the NR0B group (DAX-1, SHP), which have an LBD but no DBD, are not found in insects.

Since there is no outgroup to root the tree of the family, the relationships between the subfamilies are still not fully resolved. However, it is likely that the whole family originated from an ancestral NR2 gene. Indeed, it is the only subfamily found in all of the major groups of animals. Furthermore, Porifera (sponges), which is the most anciently branching metazoan phylum, have only a couple of nuclear receptors that are both related to NR2 (Bridgham *et al.*, 2010). It is now clear that subfamilies 1 and 4 are closely related, as well as subfamilies 5 and 6. The current view on the evolutionary relationships within the nuclear receptor family is shown on **Figure 3**. Several rounds of duplications led to the ancestral set of bilaterian nuclear receptors, which was probably around 25 genes, already distributed into the seven subfamilies (Bertrand *et al.*, 2004; Bridgham *et al.*, 2010). After complex lineage-specific events of domain shuffling, gene losses, gene duplications, and even genome duplications (in vertebrates), the size of the extant families ranges from 20 members in insects to 17 to 70 in chordates (Bertrand *et al.*, 2010), with a puzzling amplification in some nematodes

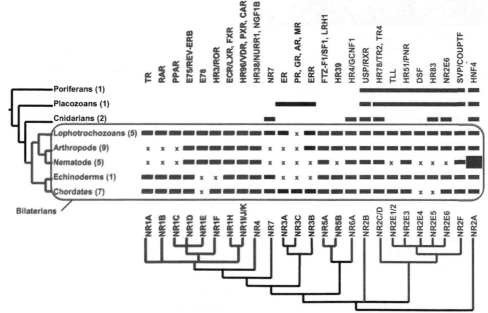

Figure 3 Simplified phylogenetic distribution of nuclear receptors in animals. Rectangular boxes colored according to the subfamilies indicate the presence of a nuclear receptor in a taxonomic group. Red cross: absence of a nuclear receptor in a bilaterian phylum. On the left, a tree indicates the evolutionary relationships between major groups of metazoans. Poriferans: sponges; Placozoans: *Trichoplax adhaerens;* Lophotrochozoans: Annelids, Molluscs, Platyhelminthes; Arthropods: Insects, *Daphnia pulex*; Echinoderms: sea urchin; Chordates: Cephalochordates (amphioxus), Urochordates (sea squirt), and Vertebrates. The number of genomes where nuclear receptors have been annotated is indicated in brackets for each taxon. On the bottom, a tree indicates the evolutionary relationships between the 25 groups of bilaterian nuclear receptors. Trivial names of insect and vertebrate nuclear receptors are given above the diagram. For the sake of simplicity, lineage specific amplifications are not illustrated, except for the group NR2A in nematodes (a larger box), which contains 10 to 250 genes, depending on the species. We have also omitted a few groups of nuclear receptors specific for a given taxon and not yet fully described in the official nomenclature. Data were obtained from different sources and collected mainly by Bertrand *et al.* (2004), Bridgham *et al.* (2010), and G. Markov (personal communication).

(30–280) (Stein *et al.*, 2003; Robinson-Rechavi *et al.*, 2005; Abad *et al.*, 2008).

Regarding the functional evolution of the family, very little is known about target genes and co-regulators. By contrast, much attention has been given to the evolution of ligand-binding across the phylum of animals (Escriva *et al.*, 2000). The emerging theory is that the ancestral nuclear receptor would have been a sensor able to bind with low affinity to a wide range of lipophilic nutriments and xenobiotics brought inside the body by nutrition. The best candidates for such ligands are fatty acids and ubiquitous metabolic intermediate and signaling lipids, which bind to several nuclear receptors most notably from the two basal groups NR2A (HNF-4) and NR2B (USP-RXR). By interacting with these metabolites, the ancestral receptor was able to act as a homeostatic rheostat by adapting the transcriptional regulation of target genes to environmental conditions. This allowed a fine tuning of processes that require an adapted metabolism, such as physiology and development (Markov *et al.*, 2010). Nuclear receptors with high-affinity ligand binding (hormone receptors) or with ligand-independent activation (orphans) would have been acquired several times independently by evolutionary tinkering of the LBD conformation (Bridgham *et al.*, 2010).

6.2. An Overview of Insect Nuclear Receptors

The sequenced genomes of insects contains between 19 and 21 nuclear receptor genes, distributed into 22 groups (**Figure 4**). This set of genes is very similar in number and identity between holometabolous and heterometabolous insects. Small variations are found among the subfamilies NR0 (1 to 3 genes) and NR2 (8 or 9 genes). In contrast to crustaceans, nematodes, and chordates, it seems that the family of nuclear receptors did not experience any lineage-specific expansions in insects. However, it would be premature to draw conclusions at the level of the phylum, since we know the complete genome sequence of at least one species for only six out of approximately 30 orders of insects. Indeed, we lack exhaustive data for the nuclear receptors of several taxa, such as the holometabolous superorder of Neuropterida (lacewings, alderfly, snakefly, etc.), as well as large groups like Polyneoptera (Dictyoptera, Orthoptera, etc.), Paleoptera (mayfly, dragonfly, etc.), and Apterygota hexapods (**Figure 4**). Recently, functional analysis in the Coleoptera *Tr. castaneum* and in a hemimetabolous insect, the cockroach *Blattella germanica* (Dictyoptera), has provided valuable results to complete our understanding of the evolution of nuclear receptors in insects (Martin, 2010). In this chapter, we provide an overview of the evolution and function of all insect nuclear receptors excluding ECR (NR1H1; CG1765) and USP (NR2B4; CG4380), which will be considered separately (see Chapter 5).

6.2.1. Subfamily NR1: The Core of the Ecdysone Pathway

In insects, subfamily NR1 contains five receptors, four of which are directly involved in the control of the ecdysone pathway. One group includes E75 (NR1D3), E78 (NR1E1), and HR3 (NR1F4). A second group contains the ecdysone receptors ECR (NR1H1) and HR96 (NR1J1). The vertebrate homologues of the first group are unable to form heterodimers with RXR, whereas homologues of the second group can form heterodimers with RXR (Brelivet *et al.*, 2004). This feature may not be fully conserved in insects, since USP (RXR homologue) heterodimerizes with ECR, but probably not with HR96. Three famous groups of subfamily 1 were present in the ancestor of Bilateria and lost in ecdysozoans: thyroid hormone receptors (TRs, NR1A), retinoic acid receptors (RARs, NR1B), and peroxisome proliferator-activated receptors (PPAR, NR1C; **Figure 3**).

6.2.1.1. E75 The *E75* gene (Eip75B, NR1D3; CG8127) has been studied in *Drosophila melanogaster* (Feigl *et al.*, 1989; Segraves and Hogness, 1990), *Aedes aegypti* (Pierceall *et al.*, 1999), several species of Lepidoptera (Segraves and Woldin, 1993; Jindra *et al.*, 1994a; Palli *et al.*, 1995, 1997b; Zhou *et al.*, 1998; Swevers *et al.*, 2002a; Dubrovskaya *et al.*, 2004), *T. castaneum* (Tan and Palli, 2008a; Xu *et al.*, 2010) and *B. germanica* (Mané-Padrós *et al.*, 2008). It was also analyzed in crustaceans (Chan, 1998; Kim *et al.*, 2005; Priya *et al.*, 2010). These proteins are homologous to the vertebrate receptors REVERB-α and REVERB-β, which have been shown to be constitutive transcriptional repressors involved in the pacemaker controlling the circadian clock in vertebrates (Preitner *et al.*, 2002). In insects, E75 acts as a repressor of HR3, probably through direct interaction, both in *Drosophila* (White *et al.*, 1997) and in *Bombyx* (Swevers *et al.*, 2002b). One of the most amazing discoveries in the field of nuclear receptor that occurred over the past six years (2005–2010) is that *Drosophila* E75 contains a heme prosthetic group within its LBD (Reinking *et al.*, 2005; de Rosny *et al.*, 2006). The oxidation state of the heme iron and the binding of gases (NO or CO) modulate the activity of E75 and its interaction with HR3. The vertebrate homologues REVERB-α and -β also act as redox and gas sensors (Yin *et al.*, 2007; Marvin *et al.*, 2009). Thus, not only are nuclear receptors of the NR1D group no longer orphans, but they also broaden our notion of ligand activation. The possible implication of E75 in circadian rhythm is still not known. In *Drosophila*, inactivation of all E75 functions causes first instar larval lethality, but isoform specific null mutations reveal different functions for each of the three isoforms (Bialecki *et al.*, 2002). In *Tr.*, E75 is important for successful embryogenesis and for both larval–pupal as

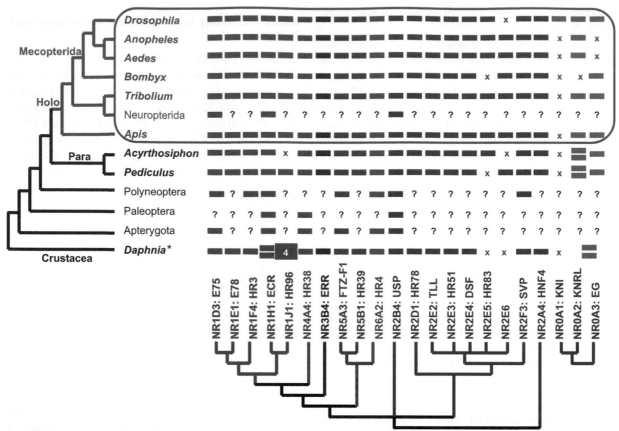

Figure 4 Phylogenetic distribution of nuclear receptors in insects. On the left, a tree indicates the evolutionary relationships between the groups of insects (Kristensen, 1981; Whiting *et al.*, 1997; Grimaldi and Engel, 2005; Wiegmann *et al.*, 2009). Holo: Holometabolous, highlighted in pink; Para: Paraneoptera. On the bottom, a tree indicates the evolutionary relationships between the 22 groups of insect nuclear receptors. For each nuclear receptor, a colored box indicates its presence in a taxonomic group. Questions marks: no sequence data. Red cross: absence of a nuclear receptor in a sequenced genome. *Drosophila* species, Diptera (King-Jones and Thummel, 2005); *An. gambiae*, Diptera (Cruz *et al.*, 2009); yellow fever mosquito *Ae. aegypti*, Diptera (Cruz *et al.*, 2009); silkworm *B. mori*, Lepidoptera (Cheng *et al.*, 2008); red flour beetle *T. castaneum*, Coleoptera (Bonneton *et al.*, 2008; Tan and Palli, 2008a); honey bee *A. mellifera*, Hymenoptera (Velarde *et al.*, 2006); pea aphid *Acyrthosiphon pisum*, Hemiptera (Christiaens *et al.*, 2010; Shigenobu *et al.*, 2010); human body louse *P. humanus* (Kirkness *et al.*, 2010); *D. pulex*, Crustacea (Thomson *et al.*, 2009). *: note that, for the sake of simplicity, the groups NR1M and NR1N, that were newly defined in *Daphnia*, are not shown on this figure. For the same reason, the three *Daphnia* genes HR97a, b, and g (group NR1L) are shown here as duplicates of NHR96. Finally, the two NR0 genes of *Daphnia* are so divergent from their insect homologues it is impossible to classify them as NR0A2 or NR0A3. Some nuclear receptors were cloned individually from different species belonging to Neuropterida and various paraphyletic groups: Polyneoptera (including *B. germanica*), Paleoptera, and Apterygota (see text for details).

well as pupal–adult metamorphosis (Tan and Palli, 2008a; Xu *et al.*, 2010). The complex role of this gene is not fully understood, but expression and hormonal induction data suggest that its involvement in early ecdysone responses during molting and metamorphosis may be shared among arthropods (Jindra *et al.*, 1994a; Palli *et al.*, 1997b; Chan, 1998; Zhou *et al.*, 1998; Mané-Padrós *et al.*, 2008; Priya *et al.*, 2010). It should be noted that the term ecdysone in this review is used in a general sense and in most cases refers to the principal molting hormone 20-hydroxyecdysone. Furthermore, E75 seems to act as a central link in the interplay between ecdysone and juvenile hormone (JH) pathways (Dubrovsky *et al.*, 2004; Keshan *et al.*, 2006). It also plays a role during oogenesis and vitellogenesis in

Drosophila (Bryant *et al.*, 1999; Terashima and Bownes, 2006), *Aedes* (Pierceall *et al.*, 1999; Raikhel *et al.*, 2002), *Bombyx* (Swevers *et al.*, 2002a), and *Tr.* (Xu *et al.*, 2010).

6.2.1.2. E78 The *E78* gene (*Eip78C*, NR1E1; CG18023) is a paralogue of E75 that experienced a rapid evolutionary rate. This duplication is probably very ancient and may have occurred at the origin of bilaterians. Indeed, *Cainorhabditis elegans* has an orthologue of E75 (*nhr-85*) together with an orthologue of E78 (*SEX-1*) (Kostrouch *et al.*, 1995; Carmi *et al.*, 1998; Unnasch *et al.*, 1999; Sluder and Maina, 2001; Crossgrove *et al.*, 2002). Furthermore, the identification of an orthologue of E78 in the parasitic trematod *Schistosoma mansoni* (Wu *et al.*,

2008) and a genomic survey revealed that E78 is actually present in all protostomes, and only in protostomes. Thus, NR1E is the only group of bilaterian nuclear receptor that is absent in deuterostomes (**Figure 3**). Mutant phenotypes indicate that the NR1E gene probably plays different roles in insects and in nematodes. Indeed, in *Drosophila*, E78 homozygous mutants are viable and fertile with subtle defects in regulation of some puffs and in formation of dorsal chorionic appendages (Russell *et al.*, 1996; Bryant *et al.*, 1999). In *Tr.*, the knockdown in the expression of E78 partially blocked embryogenesis but did not affect metamorphosis or reproduction (Tan and Palli, 2008a; Xu *et al.*, 2010). By contrast, the E78 orthologue (*SEX-1*) of *Caenorhabditis* is required for sex determination and viability of hermaphrodites (Carmi *et al.*, 1998).

6.2.1.3. HR3 Orthologues of HR3 (*Hr46*, NR1F4; CG33183) have been studied in Diptera (Koelle *et al.*, 1992; Kapitskaya *et al.*, 2000; Bertrand *et al.*, 2004), Lepidoptera (Palli *et al.*, 1992, 1995, 1996; Jindra *et al.*, 1994b; Matsuoka and Fujiwara, 2000; Debernard *et al.*, 2001; Zhao *et al.*, 2004), *T. castaneum* (Tan and Palli, 2008a; Xu *et al.*, 2010) and *B. germanica* (Cruz *et al.*, 2007). This receptor is conserved in animals, with three homologues in vertebrates: ROR α, β, and γ. These are transcriptional activators probably regulated by sterols (Jetten *et al.*, 2009). RORs and REVERB bind to the same response element and RORs are competitors for REVERB. Together, they play an important role in circadian rhythm (Jetten *et al.*, 2009). HR3 is an ecdysone inducible early–late gene that plays a key role during *Drosophila* metamorphosis by repressing early genes and directly inducing the pre-pupal regulator FTZ-F1 (Lam *et al.*, 1997; White *et al.*, 1997). This function in molting and metamorphosis is probably conserved throughout insects, as suggested by studies of expression and hormonal induction in *Manduca* (Palli *et al.*, 1992; Lan *et al.*, 1997, 1999; Langelan *et al.*, 2000), *Choristoneura fumiferana* (Palli *et al.*, 1996, 1997a), *Te.* (Mouillet *et al.*, 1999), and *B. germanica* (Cruz *et al.*, 2007; Martin, 2010). It may even be conserved in ecdysozoans, since inhibition of *nhr-23* expression showed that this HR3 homologue is also required for *Caenorhabditis* molting (Kostrouchova *et al.*, 1998, 2001). As for E75, HR3 is a functionally complex gene involved in many aspects of the ecdysone pathway. For example, *Drosophila* HR3 regulates cell-autonomous growth by modulating the S6 kinases that integrate nutrient and insulin signaling pathways (Montagne *et al.*, 2010; see Chapter 2). This nuclear receptor is thus a major link between ecdysone and insulin pathways. A role for HR3 in oogenesis and vitellogenesis has been described for *Aedes* (Kapitskaya *et al.*, 2000; Li *et al.*, 2000; Raikhel *et al.*, 2002), *Bombyx* (Eystathioy *et al.*, 2001), *Tr.* (Xu *et al.*, 2010), and *Caenorhabditis* (Kostrouchova *et al.*, 1998). Like E75 and FTZ-F1, HR3 seems to act as a central

link in the interplay between ecdysone and JH pathways (Siaussat *et al.*, 2004). HR3 is also expressed in embryos of *Drosophila* (Carney *et al.*, 1997) and *Caenorhabditis* (Kostrouchova *et al.*, 1998). Furthermore, *Drosophila* HR3 mutants have defects in their tracheal and nervous systems and die during embryogenesis (Carney *et al.*, 1997; Ruaud *et al.*, 2010). Just as for E75, the possible implication of HR3 in circadian rhythm is still unknown. Given the wide variety of tissues and stages of expression for HR3, it is clear that other functions will be identified by detailed analysis of mutants. This is nicely illustrated by two studies showing a specific role for HR3 during the formation of wings in *Drosophila* and in *Bombyx*. Clonal analysis reveals requirements for *Drosophila* HR3 in the development of adult wings, bristles, and cuticle, but no apparent function in eye or leg development (Lam *et al.*, 1999). In a *Bombyx* wing-deficient mutant called *flügellos*, HR3 expression is reduced only in wing discs, and not in the testis and fat body, while ECR, USP, E75, and HR38 are not affected (Matsuoka and Fujiwara, 2000). The wings are one of many organs where HR3 is expressed, and the role of this gene in their development in two different insects could only be revealed with appropriate genetic manipulations.

6.2.1.4. HR96 The *HR96* gene (NR1J1; CG11783) has been discovered in *Drosophila* (Fisk and Thummel, 1995). This gene is present in all insect genomes sequenced, except in the pea aphid (**Figure 4**). This absence of HR96 in *Acyrthosiphon* is difficult to interpret: Is it a real lineage-specific loss or is it due to a gap in this version of the sequence? Note that in *Daphnia* three genes DpHR97a, b, and g (group NR1L) are divergent duplicates of the single orthologue DpHR96 (Thomson *et al.*, 2009). In *Caenorhabditis*, three homologues of HR96 are known: DAF-12, NHR-8, and NHR-48 (Sluder and Maina, 2001). All of these ecdysozoan receptors are related to the vertebrate receptors NR1I: VDR, which is the receptor for vitamin D, PXR, and CAR. These receptors bind a wide variety of xenobiotics (Moore *et al.*, 2006). In *Drosophila*, the analysis of HR96 mutants has shown that, by regulating metabolic and stress-response genes, this protein plays a role in the response to xenobiotics such as phenobarbital and DDT (King-Jones *et al.*, 2006). A similar function is known for one of the *Caenorhabditis* homologues, NHR-8 (Lindblom *et al.*, 2001). *Drosophila* HR96 binds cholesterol and is a central regulator of triacylglycerol and cholesterol homeostasis (Horner *et al.*, 2009; Sieber and Thummel, 2009; Bujold *et al.*, 2010). Interestingly, the ligand of the *Caenorhabditis* homologue DAF-12 is dafachronic acid, which is a cholesterol derivative (Motola *et al.*, 2006; Wang *et al.*, 2009). DAF-12, which functions downstream of the insulin and TGF-β signaling pathways, regulates diapause, developmental age, and adult longevity (Antebi *et al.*, 1998, 2000; Gerisch *et al.*, 2001; Jia *et al.*, 2002).

Common regulatory mechanisms control developmental timing in *Caenorhabditis* and *Drosophila*, with a central role for the ecdysone pathway in insects (Thummel, 2001; see Chapters 4 and 5). It seems that the regulation of lipid metabolism and xenobiotics is an ancestral function of the NRIJ group of nuclear receptors. All of these receptors respond to and bind sterols, like members of the closely related NR1H group (ECR, LXR, FXR). Both HR96 and ECR link environmental conditions (nutriments, xenobiotics) to physiology (metabolism, obesity, life span) and development (dauer, molts, metamorphosis). As such, their characteristics fit very well with our current understanding of the universal and ancestral function of nuclear receptors.

6.2.2. Subfamily NR2: A Large Collection of Old and New Genes

Only subfamily NR2 has been found in all metazoans, suggesting that it is probably at the origin of the family (**Figure 3**). Most of the NR2 proteins are orphan repressors (except HNF4 and USP/RXR) that do not form heterodimers (except USP/RXR). In insects, subfamily NR2 contains nine genes, including the most conserved nuclear receptors: HNF4, *seven-up* (SVP/COUP), and *tailless* (TLL). The NR2E group contains five different genes: TLL, HR51, DSF, HR83, and NR2E6. They probably all share a primary function in the developing nervous system. Three of them are apparently missing in the genome of some species: DSF in pea aphid, HR83 in *Bombyx*, and NR2E6 in *Drosophila* (**Figure 4**).

6.2.2.1. HNF4

HNF4 (NR2A; CG9310) is one of the best conserved nuclear receptors among animals. This gene was identified in *Drosophila* (Zhong *et al.*, 1993), *Aedes* (Kapitskaya *et al.*, 1998), *Bombyx* (Swevers and Iatrou, 1998), and in all insect genomes sequenced (**Figure 4**). In nematodes, an explosive burst of duplications of HNF4 produced most of the (10 to 250) supplementary nuclear receptors found in these species (Robinson-Rechavi *et al.*, 2005). The reasons for this expansion are still unknown, as well as the function of most of the supplementary genes. Three homologous HNF4 genes exist in vertebrates (only two in mammals). HNF4 may have an important role in the reproduction of insects. In the mosquito, three isoforms have been identified, each having a specific temporal pattern during the vitellogenesis that takes place in the female fat body (Kapitskaya *et al.*, 1998). Furthermore, the reduction of HNF4 expression in *Tr.* leads to a significant negative effect on egg laying and maturation (Tan and Palli, 2008a). The embryonic expression pattern of this receptor is remarkably well conserved between insects and vertebrates. HNF4 is transcribed zygotically in several analogous organs: the midgut, Malpighian tubules, and fat body of insects (Zhong *et al.*, 1993;

Hoshizaki *et al.*, 1994; Kapitskaya *et al.*, 1998; Swevers and Iatrou, 1998; Palanker *et al.*, 2009) and the intestine, kidney, and liver of vertebrates (Laudet and Gronemeyer, 2002). Furthermore, analyses of mutants show that the normal development of these organs requires HNF4 activity in *Drosophila* (Zhong *et al.*, 1993) as well as in the mouse (Chen *et al.*, 1994). Knocking-down the expression of *Tr.* HNF4 using RNA interference causes a small decrease of viability (Tan and Palli, 2008a; Xu *et al.*, 2010). Similarly, *Drosophila* HNF4 null mutants are viable. However, *Drosophila* mutant larvae are sensitive to starvation, because they are unable to use stored fat, due to a reduced lipid catabolism (Palanker *et al.*, 2009). A positive feedback loop may imply the activation of HNF4 by the fatty acids produced by starvation. This activation could be direct, since *Drosophila* HNF4 LBD acts as a ligand-dependent sensor induced by metabolic signals and exogenous fatty acids (Palanker *et al.*, 2006, 2009). Furthermore, mammalian HNF4 receptors bind fatty acids constitutively (Dhe-Paganon *et al.*, 2002; Wisely *et al.*, 2002). Finally, analysis of *nhr-49* and *nhr-64*, two of the numerous HNF4 homologues in *C. elegans*, has revealed similar functions in fat storage and response to starvation (Van Gilst *et al.*, 2005a,b; Liang *et al.*, 2010). Thus it is a plausible hypothesis to consider HNF4 as a fatty acid sensor that adapts the transcriptional regulation of target genes to nutritional conditions in all metazoans. This function is very close to the putative function of the ancestral nuclear receptor, which fits well with the phylogenetic position of HNF4/NR2A (**Figure 3**).

6.2.2.2. HR78

The *HR78* gene (NR2D1; CG7199) has been identified in *Drosophila* (Fisk and Thummel, 1995; Zelhof *et al.*, 1995b), *Bombyx* (Hirai *et al.*, 2002), *Te.* (Mouillet *et al.*, 1999), and in all insect genomes sequenced (**Figure 4**). These genes are distantly related to the vertebrate's orphan receptors TR2 and TR4 (NR2C), and to the very divergent nematode genes *nhr-41* (*Caenorhabditis*; Sluder and Maina, 2001) and *nhr-2* (filarial nematode *Brugia malayi*; Moore and Devaney, 1999). HR78 is required for ecdysteroid signaling during the onset of metamorphosis of *Drosophila* (Fisk and Thummel, 1995, 1998; Zelhof *et al.*, 1995b). This receptor is inducible by 20-hydroxyecdysone (20HE) and binds to over 100 sites on polytene chromosomes, many of which correspond to ecdysteroid regulated puff loci. *Drosophila* HR78 is a repressor and an obligate partner for the cofactor Moses (Baker *et al.*, 2007). By contrast to the important sequence divergence of the proteins, the temporal profile of expression of HR78 is very well conserved between *Drosophila* and *Te.* with an early activation at the end of larval stages followed by a rather constant level during pre-pupal stages (Fisk and Thummel, 1995; Zelhof *et al.*, 1995b; Mouillet *et al.*, 1999). *Drosophila* HR78 null mutants die during the second and third larval instar with tracheal defects, showing that despite a uniform expression in the embryo this receptor is not essential for early development (Fisk and Thummel, 1998; Astle *et al.*, 2003). A similar conclusion can be drawn from RNA interference experiments on *Tr*. In this species, silencing the expression of HR78 results in a decrease in egg production and a 40% larval lethality, but has no effects on embryogenesis and metamorphosis (Tan and Palli, 2008a; Xu *et al.*, 2010). *Drosophila* HR78 mutant larvae are small, while hypomorphic alleles of its obligate partner *moses* display overgrowth (Baker *et al.*, 2007). Similar defects in TR4 mutant mice suggest that regulation of growth may be an evolutionary conserved function of the NR2CD group of nuclear receptors. A surprising result was obtained by studying the *Bombyx* homologue. Yeast two-hybrid and pull down assays showed that *Bombyx* HR78 could interact with USP-RXR (Hirai *et al.*, 2002). If we consider that HR78 and ECR bind to the same DNA sites *in vitro* (Fisk and Thummel, 1995; Zelhof *et al.*, 1995b), and have the same USP-RXR partner, it is possible that competition occurs between these two proteins during the onset of metamorphosis.

6.2.2.3. SVP The *seven-up* gene (SVP, NR2F3; CG11502) is a member of the COUP-TFs group, one of the most conserved nuclear receptors. It has been studied in various insects, notably *Drosophila* (Henrich *et al.*, 1990; Mlodzik *et al.*, 1990), *Aedes* (Miura *et al.*, 2002), *Bombyx* (Togawa *et al.*, 2001), Coleopterans (Mouillet *et al.*, 1999; Tan and Palli, 2008a; Xu *et al.*, 2010; Ahn *et al.*, 2010), in the honeybee (Velarde *et al.*, 2006), and in the grasshopper *Schistocerca gregaria* (Broadus and Doe, 1995). One homologue (UNC-55) was studied in *Caenorhabditis* (Zhou and Walthall, 1998), and three COUP-TF genes are present in vertebrates (Pereira *et al.*, 2000). These orphan receptors are repressors that have a very broad pattern of expression and are essential for development in all of the organisms studied. In the embryos of *Drosophila* and *Schistocerca*, SVP was found in the developing eyes, the central and peripheral nervous systems, the Malpighian tubules, and the fat body (Mlodzik *et al.*, 1990; Hoshikazi *et al.*, 1994; Broadus and Doe, 1995; Kerber *et al.*, 1998; Sudarsan *et al.*, 2002; Kanai *et al.*, 2005). Analysis of null mutants demonstrated the vital role of these expressions in *Drosophila*. In addition, SVP participates in the diversification of cardioblasts in the heart tube of *Drosophila* embryos (Lo and Frasch, 2001; Ponzielli *et al.*, 2002). The SVP protein is regulated by different pathways, such as Ras in eyes, EGF in Malpighian tubules, and hedgehog in heart, and thus plays multiple roles from cell fate determination to regulation of cell cycle and differentiation. Interestingly, SVP/COUP-TF inhibits USP-RXR signaling pathways, both in vertebrates and in insects. This has been tested *in vivo* with *Drosophila* in which the overexpression of

SVP that leads to lethality at the onset of metamorphosis can be rescued by a concomitant overexpression of USP-RXR (Zelhof *et al.*, 1995a). Similarly, in *Tr.*, SVP is required to complete the metamorphic process (Tan and Palli, 2008a). It is of note that SVP is expressed at the beginning of metamorphosis, when the ecdysteroid concentration is low, both in *Drosophila* and in *Te.* (Mouillet *et al.*, 1999). A similar inhibition of the 20HE response occurs during vitellogenesis in the mosquito *Aedes*. Furthermore, in this species, it was shown *in vitro* that the repression of the ecdysone pathway is probably due to a direct interaction between SVP and USP-RXR. The replacement of ECR by SVP in USP-RXR heterodimers would be essential for the termination of vitellogenesis when the ecdysteroid concentration declines (Miura *et al.*, 2002; Zhu *et al.*, 2003). However, no physical interaction between SVP and USP-RXR was detected in *Drosophila* (Zelhof *et al.*, 1995a). Another nice example of SVP's repressive role was discovered in the cowpea bruchid *Callosobruchus maculatus* (Ahn *et al.*, 2007). In this herbivorous coleopteran, an adaptive strategy to minimize the effects of plant protease inhibitors is to activate inhibitor-insensitive digestive enzymes. A major activator in this process turned out to be HNF4, whose transcriptional activity is repressed by protein-protein interactions with SVP (Ahn *et al.*, 2010). Thus, coordination between the activator HNF-4 and the repressor SVP may be important for the counter-defense gene regulation in insects. One idea that is emerging from all these studies is that SVP may be viewed as a major developmental timer that negatively regulates important temporal switches during the life of animals.

6.2.2.4. TLL The *tailless* gene (TLL, NR2E2; CG1378) is one of the most conserved nuclear receptors. It was first identified in *Drosophila* as a terminal gap gene determining embryo segmentation (Jürgens *et al.*, 1984). Homologues have been studied in *Drosophila virilis* (Liaw and Lengyel, 1993), the house fly *Musca domestica* (Sommer and Tautz, 1991), the coleopteran *T. castaneum* (Schröder *et al.*, 2000), the wasp *Nasonia vitripennis* (Lynch *et al.*, 2006), and the honeybee (Wilson and Dearden, 2009). There is also a single *tailless* gene in Nematodes (*nhr-67*) and in vertebrates (*tlx*). Genetic interaction analyses have suggested that this orphan nuclear receptor has a dual function as both a repressor and activator, according to the recruitment of different cofactors (Laudet and Gronemeyer, 2002; Haecker *et al.*, 2007). In *Drosophila* embryos, *tailless* may act exclusively as a transcriptional repressor (Moran and Jimenèz, 2006). Inactivating the *tailless* gene, either by mutation or by RNA interference, leads to embryonic lethality in all the species tested. Tailless function in posterior patterning is conserved in holometabolous insects. Despite this conservation, the early regulation of *tailless* involves a surprising diversity of mechanisms (Wilson and Dearden, 2009). The function of the early anterior expression is more variable than the posterior one. *Drosophila* and honeybee embryos show similar segmentation defects at the anterior pole, while the anterior patterning seems unaffected by *tailless* inactivation in *Tr.* and in *Nasonia* (Schröder *et al.*, 2000; Lynch *et al.*, 2006; Wilson and Dearden, 2009). In general, the role of *tailless* in the developing brain is very similar in insects (Daniel *et al.*, 1999; Hartmann *et al.*, 2001; Kurusu *et al.*, 2009) and in vertebrates (Monaghan *et al.*, 1997; Yu *et al.*, 2000). In *C. elegans*, the tailless homologue nhr-67 is required for the development of many neuronal lineages (Sarin *et al.*, 2009). In conclusion, the primary conserved function for *tailless* would be in the development of the anterior nervous system, whereas its role in patterning segmentation was probably acquired during the evolution of insects.

6.2.2.5. HR51

The *unfulfilled* gene (HR51, NR2E3; CG16801) was discovered thanks to genome annotation of insects (**Figure 4**). Although it was called Hormone Receptor 51 (band 51F7 on chromosome II), this repressor, which is not responsive to ecdysone, is probably not a hormone receptor (King-Jones and Thummel, 2005). Rather, HR51 was shown to contain a heme iron and to bind NO and CO similar to E75 (de Rosny *et al.*, 2008). This nuclear receptor is present in all bilaterians (**Figure 3**). The expression of the vertebrate homologue, photoreceptor-specific nuclear receptor (PNR), is restricted to the retina (Kobayashi *et al.*, 1999). PNR plays a critical role in the development of photoreceptors (Haider *et al.*, 2000; Schorderet and Escher, 2009). In *Drosophila*, *unfulfilled* is expressed in the central nervous system (CNS) at all developmental stages, but not in the visual system like PNR (Sung *et al.*, 2009). The hypomorphic mutant alleles did not reveal any function related to the embryonic expression in *Drosophila*, whereas RNA interference experiments have shown that HR51 is required for egg production and embryogenesis in *Tr.* (Xu *et al.*, 2010). In *Drosophila* pupae, there is a notable accumulation of transcripts in the mushroom bodies, which are brain structures required for complex behaviors in insects, such as olfaction, memory, and learning. Consistent with this pattern, *unfulfilled* mutants fail to develop normal neurons in the mushroom bodies, showing defects in axon morphogenesis and axon guidance (Bates *et al.*, 2010; Lin *et al.*, 2009). Although ECR and USP work together in the same neurons, it seems that HR51 acts independently of the ecdysone cascade. Homozygous mutants are also affected by problems of emergence from the pupae, poor coordination, failure of wing expansion, and sterility (Sung *et al.*, 2009). All of these phenotypes are possibly related to neural defects, which suggest that HR51 is important to regulate neuronal identity and remodeling during metamorphosis. RNAi experiments in

Tr. suggest that the role of HR51 during metamorphosis may be conserved in holometabolous insects (Tan and Palli, 2008a).

6.2.2.6. DSF

The *dissatisfaction* gene (DSF, NR2E4; CG9019) has been identified in all insect genomes sequenced (**Figure 4**). NR2E4 is one of the two bilaterian nuclear receptors (along with NR2E5 = HR83) that were lost in nematodes and chordates (**Figure 3**). In *Drosophila*, DSF is necessary for appropriate sexual behavior and sex-specific neural development. Mutant females resist male courtship and fail to lay eggs, while mutant males are bisexual and mate poorly (Finley *et al.*, 1997; O'Kane and Asztalos, 1999). DSF is also required for egg production in *Tr.* (Xu *et al.*, 2010). In *Drosophila*, the DSF protein acts as a repressor, downstream of the genes *Sex lethal* and *transformer*, in the sex determination cascade (Finley *et al.*, 1997; Pitman *et al.*, 2002). It will be very interesting to test whether DSF can also act as a ligand-dependent activator, since no sex hormones are known in insects (De Loof and Huybrechts, 1998). The *dsf* gene is expressed at all stages of development both in *Drosophila* and in *Tr.* In *Drosophila*, it is expressed in a very limited set of neurons in the brain of larvae, pupae, and adults with no sex specificity (Finley *et al.*, 1998). In *Tr.*, RNAi e experiments have shown that DSF is required for embryogenesis, but not for post-embryonic development (Tan and Palli, 2008a; Xu *et al.*, 2010).

6.2.2.7. HR83

The *HR83* gene (NR2E5; CG10296) was discovered thanks to genome annotation of insects (King-Jones and Thummel, 2005). It has been identified in all insect genomes sequenced, except in the silkworm *Bombyx mori* and in the human body louse *Pediculus humanus* where HR83 may have been lost during evolution. This gene is also absent from the genome of a crustacean, *Daphnia pulex* (**Figure 4**). NR2E5 is one of the two bilaterian nuclear receptors, along with NR2E4 (DSF), that were lost in nematodes and in chordates (**Figure 3**). The *Caenorhabditis* protein FAX-1, a regulator of neuron identity, was originally classified as a member of the NR2E5 group (Nuclear Receptor Nomenclature Committee, 1999). However, FAX-1 has no LBD and its DBD is actually related to the NR2E3 group. Thus, it should no longer be considered as a homologue of HR83 (DeMeo *et al.*, 2008; G. Markov, personal communication). In insects, an LBD can be identified in the HR83 protein sequence of the hymenopterans *Nasonia* and honeybee but not in *Drosophila* and *Tr.* These results show that the LBD of HR83 is a fast evolving domain in ecdysozoans. Therefore, if this nuclear receptor is liganded, we could expect a significant variation in the nature of its ligand. The only evidence on this topic comes from *in vitro* experiments showing that *Drosophila* HR83 LBD has a significant affinity for heme iron (de Rosny *et al.*, 2008). However, when compared to E75 and HR51, this affinity is very weak and we cannot conclude that heme is a natural ligand for HR83 (the same is true for *Drosophila* HNF-4 LBD). The expression of the *HR83* gene is very low at all stages in *Drosophila* (modENCODE RNAseq, FlyBase), in the mosquito *Ae. aegypti* (Cruz *et al.*, 2009) and in *Tr.*, where its inactivation, using RNAi i, has no effect on embryogenesis, molting, or metamorphosis (Tan and Palli, 2008a; Xu *et al.*, 2010). HR83 is probably an ecdysone-independent repressor of transcription, but its function remains unknown.

6.2.2.8. NR2E6

The *NR2E6* gene was first identified in the genome sequence of the honeybee *Apis mellifera* (Velarde *et al.*, 2006). It was later found that, among insects, only *Drosophila* species and the pea aphid lack this gene in their genome (**Figure 4**). As for HR96, it is difficult to interpret the absence of NR2E6 in *Acyrthosiphon*: Is it real lineage-specific loss or incomplete sequence? The NR2E6 is actually an ancient nuclear receptor that was probably present in the ancestor of all bilaterians and maybe in cnidarians as well (**Figure 3**). Several secondary losses occurred in vertebrates, nematodes, and *Drosophila*; in other words the losses occurred in the major model organisms, which delayed the identification and functional analysis of NR2E6. In the honeybee this gene is expressed in the brain and in the compound eye of pupa and adult (Velarde *et al.*, 2006; see also Chapter 11). Because this pattern is reminiscent of the retina-specific pattern of PNR, NR2E6 is sometimes named PNR-like or PNR. However, the true insect homologue of vertebrate PNR is HR51 (NR2E3) and the phylogenetic relationships between the highly divergent NR2E are still unresolved (Bonneton *et al.*, 2008). The expression of NR2E6 is constant and not regulated by ecdysone within fat body and ovary, which are the two main reproductive tissues of the adult female mosquito *Ae. aegypti* (Cruz *et al.*, 2009). Inactivation of *Tr.* NR2E6 using RNAi has no effect on embryogenesis, molting, or metamorphosis (Tan and Palli, 2008a; Xu *et al.*, 2010).

6.2.3. Subfamily NR3: ERR (NR3B4; CG7404) a Classical Hormone Receptor in Insects?

The subfamily NR3 appeared very early in animals (one NR3 in Placozoa) and later diversified into two main groups in Bilateria (**Figure 3**). The steroid receptors (NR3C: AR, PR, GR, and MR) are specific to chordates, where they experienced a fast rate of evolution. The estrogen receptors (ER, NR3A) are found in chordates and in some lophotrochozoans. The estrogen-related receptors (NR3B, ERR) are found in chordates, in arthropods, and in some lophotrochozoans. In insects, the only member of the NR3 subfamily is ERR, which is present in all insect

genomes sequenced (**Figure 4**). In vertebrates, the three ERR genes have broad expression patterns and influence a wide range of physiological processes. Furthermore, they share many structural and functional properties with the ERs (hence their name), including binding to synthetic estrogenic ligands, such as the inverse agonists 4-hydroxytamoxifen (OHT) and diethylstilbestrol (DES; Giguère, 2002; Bardet *et al.*, 2006). Vertebrate ERRs are not real orphans, but their natural ligands (possibly steroids) are unknown. Several lines of evidence suggest that ERR in insects may be a classical hormone receptor as well. The ligand sensor system, set up by Palanker *et al.* (2006) revealed that ERR LBD acts as an activator with a widespread and dynamic ligand-dependent activity. GAL4-ERR displays a switch in activity during mid-embryogenesis from strong activation in the myoblasts to specific and strong activation in the CNS (Sullivan and Thummel, 2003; Palanker *et al.*, 2006). Muscles and CNS also display GAL4-ERR activity in third instar larvae. Moreover, ERR appears to be responding to a widespread, temporally restricted activating signal that occurs in the mid-third instar transition one day before the ecdysteroid pulse that triggers entry into metamorphosis. This response might reflect a systemic mid-third instar pulse of an ERR hormone. In *Drosophila*, a triple mutant in the LBD confers a sensitivity to OHT and DES on ERR, similar to the mammalian type (Östberg *et al.*, 2003). Bisphenol A (BPA), an endocrine disruptor that interferes with the estrogenic response of vertebrates, binds weakly to ER and strongly to ERRγ. This suggests that the low-dose effects of BPA are due to a disruption of the unknown ERR signaling system. Interestingly, in the midge *Chironomus riparius*, ERR gene expression is increased by BPA and other xenoestrogens (Park *et al.*, 2010). All these data show that ERR might well be a conserved classical hormone receptor. It seems that ERR has an important function in the nervous system, as shown by its identification in a screen for *Drosophila* embryonic neural precursor genes (Brody *et al.*, 2002) and by its expression in the adult brain of honeybee (Velarde *et al.*, 2006). The GAL4-ERR pattern observed by Palanker *et al.* (2006) with activity in muscles and CNS, is reminiscent of what is known in vertebrates (Bardet *et al.*, 2006). Inactivation of *Tr.* ERR using RNA interference has no effect on embryogenesis, molting, or metamorphosis (Tan and Palli, 2008a; Xu *et al.*, 2010). More extensive genetic analyses are required to understand the biological role of ERR in insects.

6.2.4. Subfamily NR4: HR38 is an Alternative Partner for USP

The subfamily NR4 is a small group of orphan nuclear receptors highly conserved among bilaterians (**Figure 3**). The *HR38* gene (NR4A4; CG1864) has been identified simultaneously in *Drosophila* and *Bombyx* (Fisk and Thummel, 1995; Sutherland *et al.*, 1995) and later in all insect genomes sequenced (**Figure 4**). Like their vertebrate homologues, NGFIBα and NURR1, insect HR38 can bind DNA either as a monomer or through an interaction with USP/RXR (Sutherland *et al.*, 1995; Crispi *et al.*, 1998; Zhu *et al.*, 2000). In *Drosophila*, HR38 is expressed in ovaries and in various organs during all stages of development (Kozlova *et al.*, 1998; Komonyi *et al.*, 1998; Palanker *et al.*, 2006). A null mutant allele leads to pharate adult lethality, showing a role in metamorphosis and adult epidermis formation (Kozlova *et al.*, 1998, 2009). This result is consistent with the fact that HR38 is a direct regulator of genes that are important in the formation of adult epidermis and cuticle (Bruey-Sedano *et al.*, 2005; Davis *et al.*, 2007; Kozlova *et al.*, 2009). In *Tr.*, HR38 is important for successful embryogenesis and for both larval–pupal as well as pupal–adult metamorphosis (Tan and Palli, 2008a; Xu *et al.*, 2010). There is a competition between HR38 and ECR for the interaction with USP (Sutherland *et al.*, 1995). As a result, HR38 has an indirect repressive role on the ecdysone response. Surprisingly, the *Drosophila* heterodimer HR38-USP responds to several insect ecdysteroids (not only 20HE) independently of ECR and without direct binding of the ligand to either HR38 or USP (Baker *et al.*, 2003). Crystal structure analysis reveals that the *Drosophila* HR38 LBD lacks both a conventional ligand-binding pocket and a bona fide AF2 transactivation domain (Baker *et al.*, 2003). Interestingly, the vertebrate NR4 receptors are ligand-independent transcriptional activators considered to be real orphans (Wang *et al.*, 2003; Flaig *et al.*, 2005). These data show that NR4 receptors operate through an atypical mechanism of activation. How can *Drosophila* HR38 respond to ecdysteroids? One possibility is by interacting with a hormone-binding protein, maybe through the A/B domain of HR38, which has a partially unfolded conformation with regions of secondary structures (Dziedzic-Letka *et al.*, 2011). This domain is pliable and can adopt more ordered conformations in response to changes in environmental conditions, all characteristics that are important for multiple protein-protein interactions. The HR38-USP heterodimer could play the role of a physiological switch in the mode of ecdysteroid-signaling from the ECR to the HR38-mediated pathway. A nice example of the indirect repressive role of HR38 on the ecdysone response is found in the female mosquito *Ae. aegypti*. In this insect, vitellogenesis requires a blood meal that triggers a 20HE cascade to activate yolk synthesis. Before a blood meal, this regulation is inhibited despite the presence of the hormone. A key factor of this inhibition is the competitive binding of HR38 to USP, which prevents the formation of the ECR-USP heterodimer, the functional ecdysone receptor (Zhu *et al.*, 2000; Raikhel *et al.*, 2002). Although HR38 is not directly regulated by 20HE, it can participate in the 20HE pathway as an alternative partner to USP.

6.2.5. Subfamily NR5: Competition in the Ecdysone Pathway

The subfamily NR5 is composed of two closely related groups of nuclear receptors present in all bilaterians except for NR5B, which was lost in nematodes and vertebrates, but is present in the cephalochordate amphioxus (**Figure 3**). In insects, the two NR5 genes, *FTZ-F1* and *HR39*, have opposing roles in the 20HE pathway.

6.2.5.1. FTZ-F1   The *FTZ-F1* gene (NR5A3; CG4059) was first identified as a cofactor of fushi-tarazu in *Drosophila* (Ueda *et al.*, 1990). It was later found in all insect genomes sequenced (**Figure 4**). In vertebrates, the NR5A group contains the Steroidogenic Factor-1 (SF-1; NR5A1) and the Liver Receptor Homologue-1 (LRH-1; NR5A2), which are both essential regulators of development and steroidogenesis (Parker *et al.*, 2002; Fayard *et al.*, 2004). One orthologue (*nhr-25*) is present in the genome of *Caenorhabditis* (Asahina *et al.*, 2000; Gissendanner and Sluder, 2000). The *FTZ-F1* gene has two promoters that drive two protein isoforms, α and β, differing in their A/B domain and specific roles. The αFTZ-F1 isoform acts during segmentation of the early embryo, a function that is an innovation of holometabolous insects (see Chapter 5).

The βFTZ-F1 isoform is crucial for reproduction, embryogenesis, molting, and metamorphosis in all insects studied (Yamada *et al.*, 2000; Tan and Palli, 2008a; Xu *et al.*, 2010; Ruaud *et al.*, 2010). Similar functions are shared by the *C. elegans* homologue, *nhr-25*, suggesting that the basic roles of this NR5A3 isoform may be conserved in all ecdysozoans (Asahina *et al.*, 2000; Gissendanner and Sluder, 2000). In *Drosophila*, βFTZ-F1 is repressed by 20HE and its expression always occurs when the concentration of 20HE is low (Yamada *et al.*, 2000; Sullivan and Thummel, 2003; Palanker *et al.*, 2006). This pattern of induction is conserved in insects, as shown by studies on *B. mori* (Sun *et al.*, 1994), *Manduca sexta* (Hiruma and Riddiford, 2001; Weller *et al.*, 2001), *Ae. aegypti* (Li *et al.*, 2000; Cruz *et al.*, 2009), *Te. molitor* (Mouillet *et al.*, 1999), and *B. germanica* (Cruz *et al.*, 2008). Activation of βFTZ-F1 requires HR3 at all stages of development in *Drosophila* (Kageyama *et al.*, 1997; White *et al.*, 1997; Ruaud *et al.*, 2010), but also during vitellogenesis of *Ae. aegypti* (Li *et al.*, 2000) and molting of *B. germanica* (Cruz *et al.*, 2008). This functional interaction may be conserved in *C. elegans* where phenocopies of *nhr-25* (FTZ-F1) and *nhr-23* (HR3) have similar molting defects (Kostrouchova *et al.*, 1998; Asahina *et al.*, 2000; Gissendanner and Sluder, 2000). In *Drosophila*, βFTZ-F1 activates the late pre-pupal genes and controls the pre-pupal–pupal transition (Lavorgna *et al.*, 1993; Woodard *et al.*, 1994; Fortier *et al.*, 2003). Analysis of *Drosophila* mutants shows that βFTZ-F1 provides competence for stage-specific responses to 20HE throughout the organism (Broadus *et al.*, 1999; Yamada *et al.*, 2000). This role was also identified during the larva–pupa transition in *Bombyx* (Cheng *et al.*, 2008) and during vitellogenesis in the mosquito *Ae. aegypti*. In this latter species, we know interesting details about the molecular mechanisms by which βFTZ-F1 acts in the acquisition of competence to 20HE in the fat body. First, this competence is achieved through a post-transcriptional activation of βFTZ-F1 by JH (Zhu *et al.*, 2003). Second, the protein-protein interaction between βFTZ-F1 and the co-activator FISC (a bHLH-PAS protein) increases recruitment of FISC to ECR-USP in a 20HE-dependent manner and facilitates binding of the ecdysone receptor on the target promoters (Zhu *et al.*, 2006). These mechanisms have not yet been identified in other insects. Another important role of βFTZ-F1 is to regulate the timing of ecdysteroid synthesis, both in *Drosophila* (Parvy *et al.*, 2005; see Chapter 5) and in *B. germanica* (Cruz *et al.*, 2008; Martin, 2010). This may be an ancestral function of NR5, because both SF-1 and LRH-1 are involved in the control of steroidogenesis in mammals (Fayard *et al.*, 2004). Finally, βFTZ-F1 plays an important role in the adult brain of insects, such as neuron remodeling in *Drosophila* mushroom body (Boulanger *et al.*, 2011) or prothoracic gland degeneration in the cockroach *B. germanica* (Mané-Padrós *et al.*, 2010).

6.2.5.2. HR39

Identification of NR5B homologues in lophotrochozoans and in invertebrate chordates suggests that the NR5A/NR5B duplication occurred very early during bilaterian evolution (Bertrand *et al.*, 2004). This gene was subsequently lost in nematodes and in vertebrates, but is present in the cephalochordate amphioxus (**Figure 3**). In insects, the *HR39* gene (NR5B1; CG8676) was identified in *Drosophila* (Ohno and Petkovich, 1993; Ayer *et al.*, 1993), in *Bombyx* (Niimi *et al.*, 1997), and in all insect genomes sequenced (**Figure 4**). Since HR39 was isolated by cross-hybridization with αFTZ-F1, it was initially described with the misleading name of FTZ-F1β but should not be confused with the β-isoform of FTZ-F1 (NR5A3). The *HR39* gene of *Drosophila* is induced by 20HE and expressed at every stage of development, with a maximum at the end of third larval and pre-pupal stages (Horner *et al.*, 1995). This pattern of expression, which is inversely related to that of βFTZ-F1, is very similar in *Bombyx* (Niimi *et al.*, 1997; Cheng *et al.*, 2008). Therefore, contrary to β*FTZ-F1*, *HR39* behaves like an early gene during the onset of metamorphosis. Both HR39 and βFTZ-F1 proteins bind as monomers to the same response elements (Ayer and Benyajati, 1992; Ayer et al., 1993; Ohno and Petkovich, 1993; Crispi *et al.*, 1998). Furthermore, the overexpression of HR39

or βFTZ-F1 has opposite effects on promoter activity (Ayer *et al.*, 1993; Ohno *et al.*, 1994). Finally, the ligand sensor system has shown that, whereas βFTZ-F1 LBD behaves as an activator, HR39 LBD probably acts as a repressor (Palanker *et al.*, 2006). All of these data suggest competition between these two nuclear receptors. In *Drosophila*, *HR39* null mutants are viable (Horner and Thummel, 1997). By contrast, RNAi experiments have shown that HR39 is required for embryonic and post-embryonic development of *Tr.* (Tan and Palli, 2008a; Xu *et al.*, 2010). Finally, HR39 controls reproductive functions in insects. *Drosophila* mutant females are sterile because of defects in the sperm-storing spermathecae and glandular parovaria (Allen and Spradling, 2008). In *Tr.*, the eggs laid by females injected with HR39 dsRNA produced no offspring (Xu *et al.*, 2010). Thus, both βFTZ-F1 and HR39 control the sexual reproduction of insects similar to SF-1 and LRH-1 in vertebrates.

6.2.6. Subfamily NR6: HR4, an Early-late Ecdysone-Inducible Gene

The *HR4* gene (NR6A2; CG16902) is homologous to Germ Cell Nuclear Factor (GCNF; NR6A1), an orphan nuclear receptor of vertebrates. It was first identified as a GCNF-Related Factor (GRF) in *Te.* (Mouillet *et al.*, 1999) and in *Bombyx* (Charles *et al.*, 1999) and was later found in all insect genomes sequenced (**Figure 4**). One single NR6A homologue is found in all bilaterians and in cnidarians (**Figure 3**). As for many nuclear receptors, the most conserved functions of NR6A are probably in embryogenesis and gametogenesis. In adult vertebrates, GCNF is predominantly expressed in the germ cells of gonads, and loss of GCNF function causes embryonic lethality (Chung and Cooney, 2001). In *Drosophila*, the *HR4* gene is expressed in embryonic gut and ovary, and a P-insertion (PL78) located in this gene causes embryonic lethality (Bourbon *et al.*, 2002; Sullivan and Thummel, 2003). In *Tr.*, RNAi experiments showed that HR4 is required for vitellogenesis and oogenesis (Xu *et al.*, 2010). Later during development, *HR4* acts as an early-late gene involved in the onset of metamorphosis. HR4 is directly inducible by 20HE in *Manduca* (Hiruma and Riddiford, 2001), in a *Trichoplusia* cell line (Chen *et al.*, 2002), and in *Drosophila* (King-Jones *et al.*, 2005). Its activation during molting and metamorphosis starts after the early gene *HR3* and before the mid-pre-pupal gene *FTZ-F1* in all insect species analyzed (Charles *et al.*, 1999; Mouillet *et al.*, 1999; Hiruma and Riddiford, 2001; Weller *et al.*, 2001; Sullivan and Thummel, 2003; King-Jones *et al.*, 2005). Loss of function mutations in *Drosophila* HR4 result in growth defects and premature pupariation. In addition, HR4 is both a repressor of early ecdysone regulatory genes and an inducer of βFTZ-F1, which provides competence for stage-specific responses to 20HE throughout the organism (King-Jones *et al.*, 2005). This metamorphic function is well conserved in holometabolous insects, since HR4 inactivation blocks the larval–pupal transition in *Tr.* (Tan and Palli, 2008a). Finally, HR4 is required to complete ecdysis in the hemimetabolous insect *B. germanica* (Martin, 2010). In conclusion, it seems that HR4 might coordinate growth and maturation by mediating the 20HE response that controls molting and critical weight achievement during postembryonic development of insects.

6.2.7. Subfamily 0: Half Nuclear Receptors

The paraphyletic subfamily NR0 was artificially created to encompass all the proteins that contain only one of the two conserved domains of the nuclear receptors (Nuclear Receptors Nomenclature Committee, 1999). Without an LBD or a DBD, proteins of this subfamily are not complete nuclear receptors. The group NR0A contains proteins that lack an LBD but contains a DBD similar to the one found in the family. As such, they function like classical ligand-independent transcription factors. In insects, the group NR0A contains three genes: *knirps* (KNI, NR0A1; CG4717), *knirps-related* (KNRL, NR0A2; CG4761), and *eagle* (EG, NR0A3; CG7383). No homologue can be found in animals other than arthropods. The origin and the phylogeny of this group are not fully resolved because the C-terminal part of these proteins evolved particularly fast. However, their DBD is related to the DBD of NR1HJ (Laudet, 1997). Insects lack proteins of the NR0B group (DAX-1, SHP), which has an LBD but no DBD. The NR0 illustrates how protein domains are reshuffled during evolution, each module following its own evolutionary path, which is not necessarily identical to the one of the whole protein. In the NR family, this remains an exception.

The gap gene *knirps* has been identified in various cyclorrhaphan Diptera, such as *Drosophila* species (Nauber *et al.*, 1988; Gerwin *et al.*, 1994; Wittkopp *et al.*, 2003), the house fly *M. domestica* (Sommer and Tautz, 1991), and the hover fly *Episyrphus balteatus* (Lemke *et al.*, 2010). However, it is absent from the genome of other insects, including nematoceran Diptera such as the moth midge *Clogmia albipunctata* (García-Solache *et al.*, 2010) and mosquitoes (**Figure 4**). By contrast, the closely related gene *knrl* is found in almost all insect genomes sequenced. *Knrl* is expressed at all stages of development, whereas *knirps* is expressed only during embryonic stages (Oro *et al.*, 1988; Rothe *et al.*, 1989). The chromosomal location and the pattern of expression of both genes suggest that *knirps* is probably the result of a duplication of an ancestral *knrl* gene that occurred during the evolution of brachyceran Diptera. This hypothesis is reinforced by the strong conservation of function between *Drosophila knirps* and *knrl*. Both genes are functionally redundant

for their role in determining the anterior head structures at the blastoderm stage (Gonzalez-Gaitan *et al.*, 1994). Furthermore, they control together the development of several other organs, such as trachea (Chen *et al.*, 1998; Myat *et al.*, 2005), wing veins (Lunde *et al.*, 1998, 2003), foregut, and hindgut (Fuss *et al.*, 2001). In the short germ insect *Tr.*, the single orthologue *knrl* plays a minor role in abdominal segmentation. By contrast, it is crucial for head segmentation, a process during which it does not function as a canonical gap gene (Cerny *et al.*, 2008). Note that in *Tr.* RNAi experiments have shown that *knrl* is required not only for embryogenesis (Xu *et al.*, 2010), but also for larval, pupal, and adult viability (Tan and Palli, 2008a).

The *eagle* gene (initially named *egon* for embryonic gonads) is found in all insect genomes sequenced, except in mosquitoes, which have only *knrl* as an NR0A (**Figure 4**). A null mutant allele of *Drosophila eagle* results in late embryonic/early larval lethality (Dittrich *et al.*, 1997; Lundell and Hirsch, 1998). This gene is required for the specification of serotonergic neurons and other neuroblasts in the embryonic and larval CNS of *Drosophila* (Higashijima *et al.*, 1996; Dittrich *et al.*, 1997; Lundell and Hirsch, 1998; Couch *et al.*, 2004; Lee and Lundell, 2007). The pattern of expression of *Tr. eagle* is very different from that of the *Drosophila* homologue. In *Tr.*, *eagle* transcripts are maternally localized at the anterior pole of the early embryo and later in a segmented pattern, but not in the nervous system (Bucher *et al.*, 2005). Inactivation of *Tr. eagle* using RNAi has partial effects on embryogenesis (Bucher *et al.*, 2005; Xu *et al.*, 2010) and no effect at all on molting or metamorphosis (Tan and Palli, 2008a). These data support the idea that the *eagle* gene underwent a major functional shift during the divergence between *Drosophila* and *Tr.*

6.3. Evolution of Insect Nuclear Receptors

Despite their functional importance, the evolution of insect nuclear receptors has been largely overlooked (Henrich and Brown, 1995; Laudet and Bonneton, 2005). Three reasons may account for this disappointing situation. First, for a long time most of the functional analyses were restricted to Diptera (*Drosophila*, malaria vector *Anopheles gambiae*, yellow fever mosquito *Ae. aegypti*) and Lepidoptera (silk worm *B. mori*, tobacco hornworm *M. sexta*) (Raikhel *et al.*, 2002; Riddiford *et al.*, 2003; King-Jones and Thummel, 2005). Yet, these two highly derived holometabolous orders could not provide a satisfactory understanding of the evolution of insect nuclear receptors. Fortunately, the situation was improved recently, thanks to the achievement of genome projects and to the power of RNAi methodologies. Now we have the complete set of nuclear receptors in many species, including other holometabolous orders, such as Coleoptera and Hymenoptera, but also several heterometabolous insects (**Figure 4**). Most notably, important functional results have already been obtained for the coleopteran *T. castaneum* (Tan and Palli, 2008a; Xu *et al.*, 2010) and for the hemimetabolous model, the cockroach *B. germanica* (Martin, 2010). Second, these proteins have not been studied in insects as a homogeneous family of transcription factors, such as the products of homeotic genes. Rather, they were analyzed separately through their involvement in various developmental pathways, like embryonic segmentation (*knirps, tailless, αFtz-F1*), molting and metamorphosis (*EcR, Usp, βFtz-F1, E75*), or eye morphogenesis (*svp*). However, their common structure and mode of action requires an integrated view of their evolution. Third, natural ligands still remain to be identified for most of the insect nuclear

receptors (**Figure 5**). Are they all real orphans? This is still an open question and an essential issue to understanding their function and their evolution. Nevertheless, we can summarize the present knowledge to suggest the orientations of future work that should help to understand both the evolution of nuclear receptors and their role during the diversification of insects.

6.3.1. Evolutionary Trends

The set of nuclear receptors is strongly conserved in pterygote insects, ranging from 19 to 21 genes that are distributed into 22 groups (**Figure 4**). Unlike nematodes and chordates, insects did not experience any lineage-specific expansions within the nuclear receptors family.

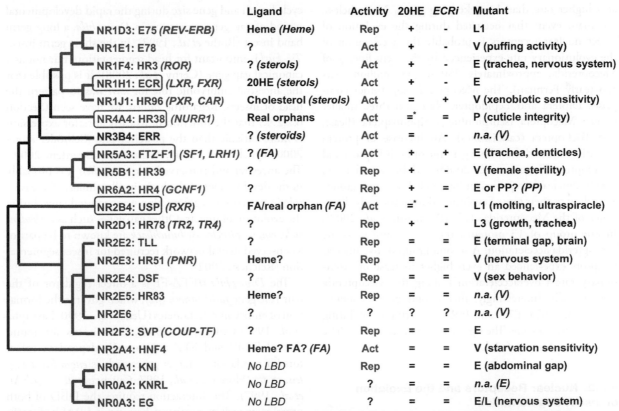

Figure 5 Functional characteristics of insect nuclear receptors. The phylogeny of the proteins is the same as in **Figure 4**. The four insect nuclear receptors with at least one 3D structure are boxed (FTZ-F1: PDB 2IZ2, Yoo and Cho, personal communication). The name of clear vertebrate orthologues is written in italics and in brackets. The column labeled Ligand indicates the known ligand for insect nuclear receptors. Ligands for vertebrate homologues are shown in italics and in brackets. The column labeled Activity indicates whether a given *Drosophila* nuclear receptor is an activator (Act, in green) or a repressor (Rep, in red), according to the ligand sensor system developed by Palanker *et al.* (2006). The column labeled 20HE indicates whether the expression of a given *Drosophila* nuclear receptor is induced (+) or unchanged (=) by 20-hydroxyecdysone (Beckstead *et al.*, 2005). * indicates that in *Drosophila*, two important players of the ecdysone response have special characteristics. HR38 is induced by ecdysone, not by 20HE (Baker *et al.*, 2003), and USP is the obligatory heterodimeric partner for the 20HE receptor, ECR (see Section 6.3.5.). The column labeled *ECRi* indicates whether the expression of a given *Drosophila* nuclear receptor is upregulated (+), downregulated (-), or not regulated (=) by disruption of *EcR* function using RNA interference at puparium formation (Beckstead *et al.*, 2005). The column labeled Mutant indicates the stage of lethality of the *Drosophila* null mutants: E, embryo; L1, first instar larva; L3, third instar larva; PP: pre-pupa; P, pupa. In addition, the viable phenotypes (V) are also indicated. When mutants are not available (*n.a.*) in *Drosophila*, the phenocopy obtained using RNA interference in *Tr.* is indicated in brackets (Tan and Palli, 2008a; Xu *et al.*, 2010).

Consequently, only small variations in the number of genes are found, notably NR2E5/6 (0–2 genes) and NR0 (1–3 genes). Further analysis of recent and upcoming genomes should clarify the origin and evolution of insect nuclear receptors through comparisons with related arthropods, notably Crustacea, the sister group of hexapods. Quantitative analysis of the evolutionary rate has revealed that the majority of nuclear receptors underwent high selective pressure during the radiation of holometabolous insects (Bonneton et al., 2008). In this clade, nuclear receptors can be distributed along a gradient from slow-evolving to fast-evolving proteins (NR0 were not considered). Genes with the most constrained evolution are encoding proteins whose structure and function are highly conserved throughout animals (SVP/COUP-TF, FTZ-F1, HR38/NURR1). In contrast, HR83, E78, ERR, and HR78 show a less constrained evolution and evolved at a higher rate than housekeeping genes. Nevertheless, the main event that occurred during the evolution of insect nuclear receptors is probably the acceleration of the evolutionary rate experienced by the stem lineage of Mecopterida, approximately 280 to 300 million years ago (early Permian). The Mecopterida superorder comprises Diptera and Lepidoptera, but also three smaller orders: Mecoptera (scorpionflies), Siphonaptera (fleas), and Trichoptera (caddishflies). An increase of protein evolution affected the whole genome of both Diptera and Lepidoptera (Savard et al., 2006b; Zdobnov and Bork, 2007). Nuclear receptors can be divided into two groups according to their rate of evolution during the early divergence of the Mecopterida clade (Bonneton et al., 2008). In one group of 13 proteins, the rate is similar to the Mecopterida acceleration. A second group of five nuclear receptors experienced an even higher increase of evolutionary rate ("overacceleration") along the Mecopterida stem branch. Interestingly, this second group contains ECR, USP, HR3, E75, and HR4, all acting early during the ecdysone cascade. The significance of this result will be discussed in Section 6.3.4.).

6.3.2. Nuclear Receptors and the Evolution of Insect Segmentation

Three of the most studied nuclear receptors in insects (kni, tll, and ftz-f1) are involved in the control of segmentation of the Drosophila embryo. Since the understanding of this developmental process played a seminal role in the emergence of evo-devo, the function of these genes has been analyzed in other species. The goal was to know whether the regulatory network identified in the long-germ band embryo of the fly was functionally conserved in short-germ band insects.

Both knirps and tailless were identified as gap genes in Drosophila (Nauber et al., 1988; Jürgens et al., 1984). As seen in Section 6.2.7. knirps is actually a new gene that

appeared after the duplication of an ancestral knrl gene during the evolution of brachyceran Diptera. Therefore, like the anterior morphogen bicoid, knirps is an essential gene only for true flies, but not for all the other species of insects where it is absent. However, its function is not totally dispensable, but is carried out by knrl. The two duplicates have partly redundant functions during the embryogenesis of Drosophila (Gonzalez-Gaitan et al., 1994). Thus it appears that Drosophila knirps and knrl, which are very close together on the third chromosome, have conserved most of their embryonic regulatory regions after gene duplication. An interesting difference, however, is the size of their intron sequences — one kb in knirps and 19 kb in knrl. The consequence of this difference in intron size is that knrl can complement knirps mutations only as an artificial intron-less transgene. This effect is explained by the required coordination of mitotic cycle length and gene size during the rapid developmental period of gap gene expression in Drosophila, a long-germ band insect (Rothe et al., 1992). In the short germ insect Tr., knrl is important for head segmentation, but not as a canonical gap gene (Cerny et al., 2008). It is possible that the ancestral embryonic role for knrl was to pattern the head of insects. Regarding tailless, we have seen (Section 6.2.2.4.) that the function of its early anterior expression is more variable than the posterior one (Schröder et al., 2000; Lynch et al., 2006; Wilson and Dearden, 2009). The ancestral and conserved role for tailless was probably in the development of the brain in all animals, whereas its role in patterning segmentation was acquired later during the evolution of insects. Other gap genes such as orthodenticle, empty spiracles, or hunchback are known to be part of a conserved neural network recruited for insect segmentation (Reichert, 2002).

The Drosophila αFTZ-F1 is a direct regulator of the pair-rule gene fushi-tarazu (ftz), which governs the formation of embryonic metameres (Ueda et al., 1990; Lavorgna et al., 1991; Ohno et al., 1994). Later, it was also found that αFTZ-F1 and FTZ are mutually dependent cofactors for regulation of target genes such as engrailed, wingless, or ftz (Florence et al., 1997; Guichet et al., 1997; Yu et al., 1997). This interaction requires the DBD of both proteins as well as a contact between a LRALL domain of the FTZ protein with the AF-2 activation domain of αFTZ-F1 (Schwartz et al., 2001; Suzuki et al., 2001; Yussa et al., 2001). The LRALL domain matches the consensus box LxxLL found in many cofactors of nuclear receptors (Laudet and Gronemeyer, 2002). Therefore, FTZ acts as a co-activator of αFTZ-F1, as confirmed by the crystal structure of the αFTZ-F1 LBD bound to the FTZ region (9 residues) that contains the LRALL motif (PDB: 2IZ2). The ftz gene is a derived Hox gene that has lost its homeotic function in insects and acquired new roles in neurogenesis and segmentation (Hughes and Kaufman, 2002). The LRALL domain was stably acquired in holometabolous insects long after the loss of homeotic function (Alonso et al., 2001; Lohr et al., 2001; Heffer et al., 2010). The recruitment of *ftz* in the holometabolous pair-rule network required regulatory changes both in gene expression and in cofactor interaction motifs, such as the one that allows the cooperation of αFTZ-F1.

These results, obtained with nuclear receptors, support the notion that the gap and pair-rule functions, as defined in *Drosophila*, are evolutionarily flexible and may be specific to long-germ band embryos. By contrast, the subsequent establishment of parasegment boundaries by segment polarity genes is well conserved among arthropods (Damen, 2007).

6.3.3. Nuclear Receptors and Phenotypic Plasticity in Insects

"Phenotypic plasticity is the ability of a given genotype to produce different phenotypes in different environments" (Gibert et al., 2007). Since nuclear receptors are involved in the integration of genomic and environmental processes, they are likely essential to understand the molecular basis of phenotypic plasticity. A few promising pieces of evidence in favor of this hypothesis have been provided by the study of four insect nuclear receptors.

The hormonal modulation of βFTZ-F1 activity during oogenesis might be one of the factors controlling honeybee caste development. In this social insect, JH and ecdysteroids induce a dramatic reduction in ovariole number during the fourth larval instar of workers but not in queens (see Chapter 11). In a search for genes controlling this phenotypic plasticity, the βFTZ-F1 of *A. mellifera* was identified by differential display of RT-PCR as a 20HE downregulated gene in the larval ovary (Hepperle and Hartfelder, 2001). This pioneering work highlights the potential role of nuclear receptors in regulating polyphenism, a major issue in the evo-devo field (Nijhout, 2003). Interestingly, βFTZ-F1 also plays important roles in the adult brain of insects, such as neuron remodeling in the *Drosophila* mushroom body (Boulanger et al., 2011) or prothoracic gland degeneration in the cockroach *B. germanica* (Mané-Padrós et al., 2010). Once again, the hormonal modulation of βFTZ-F1 activity (along with other nuclear receptors) could be involved in the regulation of caste polyphenism in the honeybee, as suggested by its expression and hormonal regulation in the adult mushroom bodies (Velarde et al., 2006, 2009).

We have seen that the HR38-USP heterodimer can switch the mode of ecdysteroid-signaling from the ECR to the HR38-mediated pathway (Section 6.2.4.). Such an HR38-mediated functional switch might be used for the regulation of age-polyethism in the honeybee. In this social insect, young workers are nurse bees and older ones are foragers. Expression of the HR38 gene is higher in the mushroom body of foragers as compared to nurse

bees and queens (Yamazaki et al., 2006). This brain structure is required for complex behaviors in insects, such as olfaction, memory, and learning. It is thus tempting to speculate that the enhanced expression of HR38 in the forager brain might contribute to the switch in the mode of ecdysteroid-signaling in the mushroom body from the ECR to the HR38-mediated pathway in association with age-polyethism of the workers.

The last example illustrates how changing the regulation of a key developmental gene can create new functions during the course of evolution. Alterations in the timing of ovarian ECR and USP expression characterize the evolution of larval reproduction in two species of gall midges, *Heteropeza pygmaea* and *Mycophila speyeri* (Hodin and Riddiford, 2000). In these Diptera, pedogenesis involves the precocious growth and differentiation of the ovary and subsequent parthenogenetic reproduction in an otherwise larval form. Genetic analysis has shown that ECR and USP regulate the timing of ovarian morphogenesis during *Drosophila* metamorphosis (Hodin and Riddiford, 1998). These data suggest that heterochronic shifts in the ovarian regulation of both genes may play an important role in the evolution of dipteran pedogenesis. From this example, we can speculate that an ancestral plasticity in the timing of ECR and USP ovarian expression was a determining step toward the establishment of heterochrony. In a larger perspective, a growing amount of evidence suggests that the genetic cascade triggered by ecdysteroids may play a central role in the control of developmental timing in Ecdysozoa (Thummel, 2001).

6.3.4. Evolution of the Ecdysone Pathway

Insects are ecdysozoans or animals that molt (Aguinaldo et al., 1997). Molting is required to replace the cuticular exoskeleton and to allow the organism to grow. During this process, the epidermis grows and separates from the cuticle. A new exoskeleton is secreted, which expands and hardens after the old cuticle is shed at ecdysis (see Chapter 7). If we consider the number of described species, the two main groups of ecdysozoans are arthropods and nematodes. In addition, this clade includes Onychophora (velvet worm) and Tardigrada (water bears) that share segmentation and appendages with Arthropoda, whereas Nematomorpha (horsehair worms), Priapulida (Penis worm), Kinorhyncha, and Loricifera are worms like Nematoda (Telford et al., 2008). In insects, the process of molting is initiated by neurohormones in response to signals that integrate physiological cues such as the photoperiod and the weight of the animal. Relatively little is known about the molecular mechanisms of molting in nematodes and other ecdysozoans (Frand et al., 2005; Ewer, 2005). On the other hand, it is well established that ecdysteroids are the steroid hormones that regulate molting in insects. This signal has undoubtedly acquired a central role in the developmental timing controlled by nuclear receptors (Riddiford *et al.*, 2003; King-Jones and Thummel, 2005; Nakagawa and Henrich, 2009). The importance of this role is reinforced by cross-talks with JH and insulin pathways involved in growth and metabolism (Spindler *et al.*, 2009). In *Drosophila*, the principal active compound is 20HE, which triggers a genetic and biochemical cascade (**Figure 6**). More than half of the nuclear receptors participate, directly or indirectly, in the upstream part of the 20HE pathway (**Figure 5**; see also Chapter 5). It is thus tempting to speculate that the ecdysone-signaling cascade would be an ecdysozoan synapomorphy. Indeed, several homologues of these nuclear receptors are known to control molting in *C. elegans* (Gissendanner *et al.*, 2004; **Figure 3**). This occurs for NHR-23 (HR3), NHR-25 (βFTZ-F1), and NHR-41 (HR78). Homologues of E75 (NHR-85), E78 (SEX-1), HR4 (NHR-91), and HR38 (NHR-6) are also found in this species, although their role in molting is not known (Magner and Antebi, 2008). However, strikingly, *EcR* and *usp* genes are not present in the genome of *C. elegans* (Sluder and Maina, 2001).

Therefore, if ecdysteroids control ecdysis in nematodes, then *C. elegans* has to use a novel receptor that is not like ECR-USP. Good evidence for the existence of such a receptor has been reported in *Drosophila* (Costantino *et al.*, 2008). Nevertheless, recent genomic data revealed the presence of genes from the NR1H and NR2B groups in several species of nematodes, such as *Dirofilaria immitis* (Shea *et al.*, 2004, 2010), *B. malayi* (Tzertzinis *et al.*, 2010), *Haemonchus contortus* (Graham *et al.*, 2010), and *Pristionchus pacificus* (Parihar *et al.*, 2010). Whether they act *in vivo* as ecdysteroid receptors, like their insect homologues ECR and USP, remains an open question. A simple hypothesis would consider ECR as a synapomorphy for Ecdysozoa. It may be wiser to consider the whole network of ecdysteroid response, and not the receptor alone, as a putative hallmark of ecdysozoans. This kind of approach proved to be fruitful in the case of the evolution of sex determination in animals where the upstream part of the genetic cascade is highly variable in contrast to strong similarity at the bottom (Zarkower, 2001). Furthermore, solving the problem of the origin of

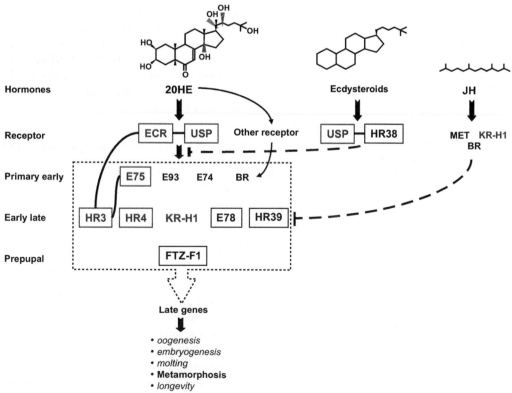

Figure 6 Simplified summary of the ecdysone regulatory cascade with 12 transcription factors known to act as classic early regulators during the onset of *Drosophila* metamorphosis (Thummel, 2001; King-Jones *et al.*, 2005; King-Jones and Thummel, 2005) and other insects (Martin, 2010). The heterodimer USP-HR38 is a sensor of ecdysteroids, including 20HE (Baker *et al.*, 2003). In *Drosophila*, there is a novel, uncharacterized 20HE receptor that is not ECR-USP (Costantino *et al.*, 2008). The receptor for JH is still unknown. However, *Methoprene tolerant* (*Met*; Konopova and Jindra, 2007), *Kruppel-homologue 1* (*Kr-H1*; Riddiford, 2008) and *Broad Complex* (*BR-C*: Wilson *et al.*, 2006; Konopova and Jindra, 2008) mediate the repressive action of JH on the ecdysone pathway. Nuclear receptors are boxed. The six proteins that overaccelerated in Mecopterida are in red. Large black bonds indicate the known protein-protein interactions. At the bottom of this figure, different developmental transitions that require the ecdysone pathway have been listed.

ecdysozoan innovations surely requires a better knowledge of the endocrine and developmental mechanisms of molting in nematodes. Currently, when the monophyly of Ecdysozoa is a robust fact, we have to admit that most of the important questions raised by this discovery are still unanswered. Is molting controlled by ecdysteroids in all ecdysozoans? Did molting evolve one or several times independently in Ecdysozoa? Are nuclear receptors linked to the emergence of ecdysozoans?

At a smaller evolutionary scale, we have learnt that the ecdysone pathway is well conserved in pterygote insects (Truman and Riddiford, 2002; Lafont et al., 2005; Martin, 2010; see Chapter 4). This is particularly true for the genetic core of the cascade, where 9 nuclear receptors transduce the 20HE response at the onset of *Drosophila* metamorphosis (**Figure 6**). Thanks to developmental genetics experiments carried out on dipteran, lepidopteran, and coleopteran species, it appears that these regulators have roughly the same role during metamorphosis in all holometabolous insects. Six of the same receptors also control ecdysis in the hemimetabolous insect *B. germanica*, suggesting that the transition from hemimetaboly to holometaboly did not involve major changes at the top of the ecdysone cascade (Martin, 2010). Actually, theories trying to explain the evolution of insect metamorphosis usually consider that JH is the key player, rather than 20HE (Sehnal et al., 1996; Truman and Riddiford, 1999, 2002). They proposed that a heterochronic shift in embryonic JH secretion, with an earlier appearance of this hormone, could have been the major event in the transition from hemimetabolous to holometabolous insects.

Is it possible that subtle modifications have occurred in the ecdysone pathway, although the regulatory logic remains unchanged? Yes, this kind of evolutionary pattern, where a function is maintained despite the divergence of its effectors, has actually been found for the core of the ecdysone response in Mecopterida (Bonneton et al., 2008). As explained in Section 6.3.1., an acceleration of evolutionary rate occurred during the emergence of the superorder Mecopterida (Diptera, Lepidoptera, Mecoptera, Siphonaptera, and Trichoptera). Five nuclear receptors, ECR, USP, HR3, E75, and HR4, plus Kr-h1, underwent an overacceleration that did not affect the other transcriptional regulators involved in this cascade, such as HR39, FTZ-F1, E78, E74, E93, and BR (**Figure 6**). The ecdysone receptor ECR-USP directly regulates the expression of E75, HR3, and HR4, which cross-regulatory interactions converge on FTZ-F1 to initiate metamorphosis or ecdysis (King-Jones et al., 2005; Martin, 2010). The *Kr-h1* (*Kruppel-homologue 1*) gene is regulated by both ecdysone and JH and encodes a zinc finger protein, which modulates the pre-pupal response in *Drosophila* (Pecasse et al., 2000; Beck et al., 2005; Riddiford, 2008). Therefore, it seems that closely interacting genes and proteins acting

in the upstream part of the ecdysone cascade constitute a small network of co-evolving regulators (Bonneton et al., 2008). Molecular evidence in favor of this co-evolution has been provided for by the heterodimerization interface between ECR and USP (Iwema et al., 2009; see Section 6.3.5. and Chapter 5). It would be particularly interesting to know whether co-evolution has also affected the protein-protein interactions that exist between ECR and HR3 on the one hand, and E75 and HR3 on the other hand. In conclusion, an important evolutionary transition established a separation within holometabolous insects between the Mecopterida and the non-Mecopterida species. To our knowledge, this molecular divergence is not correlated with any obvious phenotypic change. However, it is possible that, contrary to Mecopterida, non-Mecopterida species may never use ecdysteroids just JH to regulate vitellogenesis (Raikhel et al., 2005).

Two important issues should be addressed in the future to fully understand the evolution of the ecdysone pathway in insects. First, is the ecdysone cascade similar at every developmental transition? Second, is the ecdysone cascade similar in all pterygote insects? In other words, it is necessary to move away from the study of *Drosophila* metamorphosis. Fortunately, the trend toward a wider developmental and evolutionary analysis of ecdysone response has already started (Ruaud et al., 2010; Martin, 2010). A crucial point is to go past the gene candidate approach, where the *Drosophila* metamorphosis genetic network drives the research on other stages and other species. Only direct and reverse genetic methods applied at a large scale in appropriate model insect species are able to test the assumed universality of *Drosophila* mechanisms. The advent of the beetle *Tr.* and of the wasp *Nasonia* as model organisms with a complete genetic toolbox should facilitate this expected progress (Bonneton, 2010). In addition, a systems biology approach including exhaustive data on target genes (Gauhar et al., 2009), co-regulators (Francis et al., 2010), and expression (modENCODE consortium) in various species with a sequenced genome will help to understand the regulatory logic and the evolutionary ability of the ecdysone pathway.

6.3.5. Evolution of the Ecdysone Receptor

To understand the evolution of the ecdysone pathway, the first and simplest way was to start with the evolution of the ecdysone receptor (Bonneton et al., 2003). The functional ecdysone receptor is a heterodimer of the nuclear receptor ECR (ecdysone receptor, NR1H1) and USP (ultraspiracle, NR2B4). The requirement of heterodimerization between ECR and USP, which was originally found in *Drosophila* (Koelle et al., 1991; Oro et al., 1992; Yao et al., 1993), is actually conserved in all of the other arthropod species studied (Nakagawa and Henrich, 2009). At least two other ecdysteroid pathways independent of ECR have been identified in *Drosophila* (Baker *et al.*, 2003; Costantino *et al.*, 2008). However, the responsible receptors are not characterized. Before addressing this issue of ECR-USP evolution, it is necessary to recall that nuclear receptors are modular proteins with five to six regions, designated A to F, from the N- to the C-terminal end (**Figure 1**). Ideally, one would like to compare the complete and intact 3D structure of a nuclear receptor to understand how the different regions cooperate to fulfill receptor activity. Unfortunately, these kinds of data are currently available only for the human heterodimer PPAR-γ/RXR-α bound to DNA, ligands, and co-activator peptides (Chandra *et al.*, 2008). However, regarding ECR and USP, we have numerous insect sequences and several 3D structures of isolated DBD or LBD. As for other nuclear receptors, the regions A/B, D, and F of ECR and USP are poorly conserved. By contrast, their DBD and LBD are highly conserved, which allowed detailed comparative analysis (**Figure 7**).

In nuclear receptors, the A/B domain contains the ligand-independent activation function region (AF-1), which interacts with other transcription factors. This highly flexible region has a partially unfolded conformation, which is characteristic of intrinsically disordered proteins (Chandra *et al.*, 2008; Nocula-Lugowska *et al.*, 2009; Dziedzic-Letka *et al.*, 2011). Furthermore, phosphorylation of the A/B domain can modulate the activity of the receptor (Rochette-Egly, 2003). In insects, differential promoter usage and alternative splicing of the *EcR* gene produce two isoforms with specific regions in the A/B domain: ECR-A and ECR-B1 (*Drosophila* has an additional B2

isoform; Talbot *et al.*, 1993). Each ECR isoform has a distinct transactivation activity, expression, and function during the development of insects (Mouillet *et al.*, 2001; Wang *et al.*, 2002; Hu *et al.*, 2003; Davis *et al.*, 2005; Cruz *et al.*, 2006; Tan and Palli, 2008b). A structural comparison has shown that each isoform-specific region of ECR contains evolutionary conserved microdomain structures (Watanabe *et al.*, 2010). The ECR-A isoform has four conserved microdomains, whereas the ECR-B1 isoform has three. Interestingly, holometabolous insects are characterized by a specific (K/R)RRW motif at the N-terminal end of the EcR-B1 isoform, suggesting that this protein has acquired novel interactions with transcriptional co-regulators (Watanabe *et al.*, 2010). It should be noted that ECR isoforms were not affected by the Mecopterida acceleration, showing that the evolutionary paths followed by the A/B domain on one hand and those followed by the LBD on the other hand were independent of one another. Several isoforms have been found for USP as well, but they were not studied as intensively as the ECR isoforms, probably because the *Drosophila usp* gene is intron-less and encodes only one form (Nakagawa and Henrich, 2009).

One of the clear evolutionary acquisitions of Mecopterida insects is the presence of a divergent F domain at the C-terminal end of ECR (**Figure 7**). In Mecopterida, the length of this domain ranges from 18 residues in *Choristoneura* to 226 in *Drosophila*, whereas other insect ECRs have only 2 to 4 amino acids (Bonneton *et al.*, 2003). Most nuclear receptors do not contain this domain, including the mammalian homologues LXRs. It is known that, when present (ERα, HNF-4), the F domain

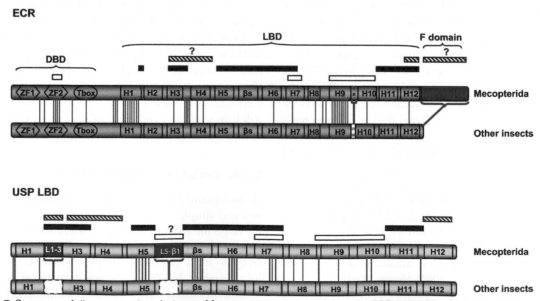

Figure 7 Summary of divergent regions between Mecopterida and other insects for ECR and USP. The Mecopterida divergences are indicated by vertical pink lines for single amino acid substitutions and by a dark pink color for large Mecopterida insertions (Bonneton *et al.*, 2003, 2006). LBD: ligand-binding domain; H: helix; L: loop; βs: β-strand; ZF: zinc finger; and DBD: DNA-binding domain. The bars above the proteins summarize the functional domains, with white for heterodimerization, black for ligand binding, and stripes for transcriptional activation. Question marks indicate putative function.

of nuclear receptors can modulate different functions of the LBD (Montano *et al.*, 1995; Nichols *et al.*, 1998; Peters and Khan, 1999). The F domain of *Drosophila* ECR contributes modestly to the activation function of the LBD, at least *in vitro* (Thornmeyer *et al.*, 1999; Hu *et al.*, 2003). Altogether, these data suggest that, in Mecopterida, the F domain of ECR may have evolved in cooperation with the adjacent LBD.

The most important changes for the ecdysone receptor have been identified in the Mecopterida sequences of ECR and USP LBDs (Bonneton *et al.*, 2003, 2006, 2008). The LBD of the nuclear receptor is a modular domain that fulfills three functions: ligand binding, dimerization, and recruitment of co-regulators. For ECR and USP, the analysis of primary structures suggests that every functional property of the LBDs might have changed (**Figure 7**).

Regarding ligand binding, although the LBD of ECR underwent a significant increase of substitution rate in Mecopterida, its structure remained unchanged (Billas *et al.*, 2003; Carmichael *et al.*, 2005; Iwema *et al.*, 2007). In all insects, and presumably in all arthropods, ECR LBD can bind ecdysteroids, especially 20HE (Riddiford *et al.*, 2000; **Figure 8**). This fundamental interaction is probably the primary selective constraint acting on this robust

domain. However, the structural and functional conservation of ECR LBD does not prevent a significant plasticity. Comparative tests of toxicity and receptor affinity of non-steroid insecticides revealed that the *Chironomus* ecdysone receptor behaves more like a lepidopteran than like *Drosophila* (Smagghe *et al.*, 2002). Furthermore, the *Aedes* ECR is more sensitive to ecdysone than the *Drosophila* receptor, whereas both proteins exhibit the same sensitivity to 20HE, the natural active hormone (Wang *et al.*, 2000). Finally, comparison of *Heliothis* ECR crystal structures bound to two different ligands revealed an extreme structural flexibility and adaptability of the ligand-binding pocket. This property allows the molding of the receptor around its ligand (Billas *et al.*, 2003).

By contrast to ECR, the rapid evolution of USP LBD in Mecopterida is associated with important changes of its ligand-binding affinity (**Figure 8**). The USP of hemipteran and coleopteran insects lack a ligand-binding pocket (Carmichael *et al.*, 2005; Iwema *et al.*, 2007). In these species, USP is unable to be activated by RXR ligands and, consequently, it acts as a constitutively silent partner of ECR. By contrast, in Mecopterida, a large ligand-binding pocket contains bacterial phospholipids with high affinity for USP (Billas *et al.*, 2001; Clayton *et al.*, 2001). The

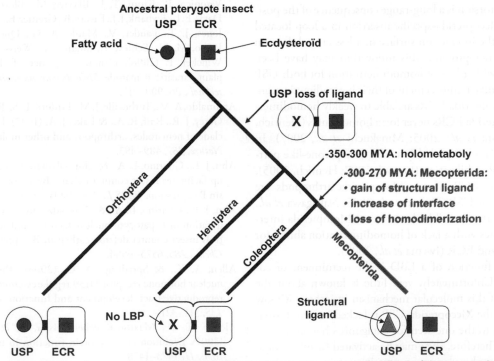

Figure 8 Evolution of the ecdysone receptor in insects. In this simplified scheme, the heterodimer of the ancestral pterygote insect is shown as an ecdysteroid (red square) liganded ECR (flat top) associated to a fatty acid (blue circle) binding USP (round). This type of receptor is conserved in *Locusta* (Nowickyj *et al.*, 2008). Later during evolution, USP lost its ligand-binding pocket (cross), as found in current Hemiptera (Carmichael *et al.*, 2005) and Coleoptera (Iwema *et al.*, 2007). During the emergence of Mecopterida (-300-270 mya), USP acquired a large ligand-binding pocket (white round), partly filled with phospholipids (yellow triangle), and a larger dimerization interface (black lines). In non-Mecopterida, both ECR and USP can form homodimers, whereas the new Mecopterida structure is associated with a loss of homodimerization and a reinforcement of heterodimerization (for details, see Iwema *et al.*, 2009).

presence of these ligands is probably not an artifact, since the use of cultured insect cells for expression leads to a similar result (Billas *et al.*, 2003). Phospholipids might thus function as structural cofactors for USP, just as fatty acids do for HNF-4 (Benoit *et al.*, 2004). However, the mode of activation of the Mecopterida USPs remains obscure, since the loop between helices H1 and H3 is located inside the hydrophobic furrow of the LBD, preventing interactions of helix H12 with co-activators. Consequently, these USPs are locked in a conformation similar to the one seen for antagonists-bound nuclear receptors (Billas *et al.*, 2001, 2003; Clayton *et al.*, 2001). In addition, USP of the basal insect *Locusta* (Orthoptera) binds 9-cis retinoic acid like the RXRs of vertebrates, whose natural ligands are fatty acids (Nowickyj *et al.*, 2008). In conclusion, the structure of the ligand-binding pocket of USP has shown an extreme plasticity during the evolution of insects, adopting three of the four states that are known for nuclear receptors: nutritional sensor (basal insects), real orphan (Hemiptera, Coleoptera), and receptor with a constitutive activity (Mecopterida; **Figure 8**).

Nuclear receptor LBDs are also involved in heterodimerization activity. The ECR-USP heterodimerization surface is symmetric and 30% larger in Mecopterida (Iwema *et al.*, 2009). This novel surface originates from a 15 degree torsion of a subdomain of USP LBD toward ECR. This torsion is a long-range consequence of the position of a Mecopterida-specific insertion in a loop located outside of the interaction surface in a less crucial domain of the partner protein. This innovation may have been associated with a loss of homodimerization for both USP and ECR and a reinforcement of their heterodimerization. If non-Mecopterida ECRs are able to weakly homodimerize, Mecopterida ECRs never form homodimers (Henrich, 2005; Ogura *et al.*, 2005; Minakuchi *et al.*, 2007). In Mecopterida, USP probably acts as a chaperone-like partner for ECR stabilization (Li *et al.*, 1997; Henrich, 2005). By contrast, ECR of non-Mecopterida arthropods are active without USP (Ogura *et al.*, 2005; Nakagawa *et al.*, 2007). It is thus possible that the new Mecopterida interface correlates with a lack of homodimerization ability for both USP and ECR (Iwema *et al.*, 2009).

The last function of a LBD is the recruitment of co-regulators. Unfortunately, very little is known about the evolution of this molecular mechanism in insects. We saw earlier that the Mecopterida USP is locked in a conformation similar to the one seen for antagonists-bound nuclear receptors. Therefore, they may be activated by other ways, such as phosphorylation or protein-protein interactions. In that respect, we know that small changes in the LBD of USP can lead to important functional consequences in whole organisms. For example, replacing the domains D and E (LBD) of the *Drosophila* USP protein with the equivalent domains of another dipteran, *Chironomus*, allows the rescue of *usp* mutants from first instar larvae lethality, but

does not restore a normal metamorphosis (Henrich *et al.*, 2000). Therefore, despite the 72% similarity between the exchanged domains, this chimeric USP fails to function normally in *Drosophila*. This result could be interpreted as evidence for the importance of protein-protein contacts in the activity of USP. We can speculate that, in a heterolog cellular context, the *Chironomus* USP LBD might not be able to establish efficient interactions with its partners. The resulting mismatches may be buffered by chaperones until the larval stages, but their accumulation would subsequently prevent the complex and tightly regulated process of metamorphosis. Similar studies involving chimeric regions and site-directed mutations should help in the future to understand the functional evolution of ECR and USP in insects.

References

Abad, P., Gouzy, J., Aury, J. M., Castagnone-Sereno, P., Danchin, E. G., Deleury, E., Perfus-Barbeoch, L., Anthouard, V., Artiguenave, F., Blok, V. C., Caillaud, M. C., Coutinho, P. M., Dasilva, C., De Luca, F., Deau, F., Esquibet, M., Flutre, T., Goldstone, J. V., Hamamouch, N., Hewezi, T., Jaillon, O., Jubin, C., Leonetti, P., Magliano, M., Maier, T. R., Markov, G. V., McVeigh, P., Pesole, G., Poulain, J., Robinson-Rechavi, M., Sallet, E., Segurens, B., Steinbach, D., Tytgat, T., Ugarte, E., van Ghelder, C., Veronico, P., Baum, T. J., Blaxter, M., Bleve-Zacheo, T., Davis, E. L., Ewbank, J. J., Favery, B., Grenier, E., Henrissat, B., Jones, J. T., Laudet, V., Maule, A. G., Quesneville, H., Rosso, M. N., Schiex, T., Smant, G., Weissenbach, J., & Wincker, P. (2008). Genome sequence of the metazoan plant-parasitic nematode *Meloidogyne incognita*. *Nat. Biotechnol.*, *26*, 909–915.

Aguinaldo, A. M., Turbeville, J. M., Linford, L. S., Rivera, M. C., Garey, J. R., Raff, R. A., & Lake, J. A. (1997). Evidence for a clade of nematodes, arthropods and other molting animals. *Nature*, *387*, 489–493.

Ahn, J. E., Guarino, L. A., & Zhu-Salzman, K. (2007). Seven-up facilitates insect counter-defense by suppressing cathepsin B expression. *FEBS J.*, *274*, 2800–2814.

Ahn, J. E., Guarino, L. A., & Zhu-Salzman, K. (2010). Coordination of hepatocyte nuclear factor 4 and seven-up controls insect counter-defense cathepsin B expression. *J. Biol. Chem.*, *285*, 6573–6584.

Allen, A. K., & Spradling, A. C. (2008). The Sf1-related nuclear hormone receptor Hr39 regulates *Drosophila* female reproductive tract development and function. *Development*, *135*, 311–321.

Alonso, C. R., Maxton-Kuechenmeister, J., & Akam, M. (2001). Evolution of Ftz protein function in insects. *Curr. Biol.*, *11*, 1473–1478.

Antebi, A., Culotti, J. G., & Hedgecock, E. M. (1998). daf-12 regulates developmental age and the dauer alternative in *Caenorhabditis elegans*. *Development*, *125*, 1191–1205.

Antebi, A., Yeh, W. H., Tait, D., Hedgecock, E. M., & Riddle, D. L. (2000). daf-12 encodes a nuclear receptor that regulates the dauer diapause and developmental age in *C. elegans*. *Genes Dev.*, *14*, 1512–1527.

Arnosti, D. N., Gray, S., Barolo, S., Zhou, J., & Levine, M. (1996). The gap protein knirps mediates both quenching and direct repression in the *Drosophila embryo. EMBO J.*, *15*, 3659–3666.

Asahina, M., Ishihara, T., Jindra, M., Kohara, Y., Katsura, I., & Hirose, S. (2000). The conserved nuclear receptor Ftz-F1 is required for embryogenesis, moulting and reproduction in *Caenorhabditis elegans. Genes Cells*, *5*, 711–723.

Astle, J., Kozlova, T., & Thummel, C. S. (2003). Essential roles for the Dhr78 orphan nuclear receptor during molting of the *Drosophila* tracheal system. *Insect Biochem. Mol. Biol.*, *33*, 1201–1209.

Ayer, S., & Benyajati, C. (1992). The binding site of a steroid hormone receptor-like protein within the *Drosophila* Adh adult enhancer is required for high levels of tissue-specific alcohol dehydrogenase expression. *Mol. Cell Biol.*, *12*, 661–673.

Ayer, S., Walker, N., Mosammaparast, M., Nelson, J. P., Shilo, B. Z., & Benyajati, C. (1993). Activation and repression of *Drosophila* alcohol dehydrogenase distal transcription by two steroid hormone receptor superfamily members binding to a common response element. *Nucleic Acids Res.*, *21*, 1619–1627.

Baker, K. D., Beckstead, R. B., Mangelsdorf, D. J., & Thummel, C. S. (2007). Functional interactions between the Moses corepressor and DHR78 nuclear receptor regulate growth in *Drosophila. Genes Dev.*, *21*, 450–464.

Baker, K. D., Shewchuk, L. M., Kozlova, T., Makishima, M., Hassell, A., Wisely, B., Caravella, J. A., Lambert, M. H., Reinking, J. L., Krause, H., Thummel, C. S., Willson, T. M., & Mangelsdorf, D. J. (2003). The *Drosophila* orphan nuclear receptor DHR38 mediates an atypical ecdysteroid signaling pathway. *Cell*, *113*, 731–742.

Bardet, P. L., Laudet, V., & Vanacker, J. M. (2006). Studying non-mammalian models? Not a fool's ERRand! *Trends Endocrinol. Metab.*, *17*, 166–171.

Bates, K. E., Sung, C. S., & Robinow, S. (2010). The unfulfilled gene is required for the development of mushroom body neuropil in *Drosophila. Neural. Dev.*, *5*, 4.

Beck, Y., Dauer, C., & Richards, G. (2005). Dynamic localisation of KR-H during an ecdysone response in *Drosophila. Gene. Expr. Patterns*, *5*, 403–409.

Beckstead, R. B., Lam, G., & Thummel, C. S. (2005). The genomic response to 20-hydroxyecdysone at the onset of *Drosophila* metamorphosis. *Genome Biol.*, *6*, R99.

Benoit, G., Malewicz, M., & Perlmann, T. (2004). Digging deep into the pockets of orphan nuclear receptors: insights from structural studies. *Trends Cell Biol.*, *14*, 369–376.

Bertrand, S., Belgacem, M.R., Escriva, H., (2011). Nuclear hormone receptors in chordates. *Mol. Cell Endocrinol.*, *334*, 67–75.

Bertrand, S., Brunet, F. G., Escriva, H., Parmentier, G., Laudet, V., & Robinson-Rechavi, M. (2004). Evolutionary genomics of nuclear receptors: from twenty-five ancestral genes to derived endocrine systems. *Mol. Biol. Evol.*, *21*, 1923–1937.

Bialecki, M., Shilton, A., Fichtenberg, C., Segraves, W. A., & Thummel, C. S. (2002). Loss of the ecdysteroid-inducible E75A orphan nuclear receptor uncouples molting from metamorphosis in *Drosophila. Dev. Cell*, *3*, 209–220.

Billas, I. M., Iwema, T., Garnier, J. M., Mitschler, A., Rochel, N., & Moras, D. (2003). Structural adaptability in the ligand-binding pocket of the ecdysone hormone receptor. *Nature*, *426*, 91–96.

Billas, I. M.L., Moulinier, L., Rochel, N., & Moras, D. (2001). Crystal structure of the ligand-binding domain of the ultraspiracle protein USP, the orthologue of retinoid X receptors in insects. *Journal of Biological Chemistry*, *276*, 7465–7474.

Bonneton, F. (2010). [When Tr. complements the genetics of *Drosophila*]. *Med. Sci. (Paris)*, *26*, 297–303.

Bonneton, F., Brunet, F. G., Kathirithamby, J., & Laudet, V. (2006). The rapid divergence of the ecdysone receptor is a synapomorphy for Mecopterida that clarifies the Strepsiptera problem. *Insect Mol. Biol.*, *15*, 351–362.

Bonneton, F., Chaumot, A., & Laudet, V. (2008). Annotation of *Tr.* nuclear receptors reveals an increase in evolutionary rate of a network controlling the ecdysone cascade. *Insect Biochem. Mol. Biol.*, *38*, 416–429.

Bonneton, F., Zelus, D., Iwema, T., Robinson-Rechavi, M., & Laudet, V. (2003). Rapid divergence of the ecdysone receptor in Diptera and Lepidoptera suggests coevolution between ECR and USP-RXR. *Mol. Biol. Evol.*, *20*, 541–553.

Boulanger, A., Clouet-Redt, C., Farge, M., Flandre, A., Guignard, T., Fernando, C., Juge, F., & Dura, J. M. (2011). ftz-f1 and Hr39 opposing roles on EcR expression during *Drosophila* mushroom body neuron remodeling. *Nat. Neurosci.*, *14*, 37–44.

Bourbon, H. M., Gonzy-Treboul, G., Peronnet, F., Alin, M. F., Ardourel, C., Benassayag, C., Cribbs, D., Deutsch, J., Ferrer, P., Haenlin, M., Lepesant, J. A., Noselli, S., & Vincent, A. (2002). A P-insertion screen identifying novel X-linked essential genes in *Drosophila. Mech. Dev.*, *110*, 71–83.

Bourguet, W., Vivat, V., Wurtz, J. M., Chambon, P., Gronemeyer, H., & Moras, D. (2000). Crystal structure of a heterodimeric complex of RAR and RXR ligand-binding domains. *Mol. Cell.*, *5*, 289–298.

Brelivet, Y., Kammerer, S., Rochel, N., Poch, O., & Moras, D. (2004). Signature of the oligomeric behaviour of nuclear receptors at the sequence and structural level. *EMBO Rep.*, *5*, 423–429.

Bridgham, J. T., Eick, G. N., Larroux, C., Deshpande, K., Harms, M. J., Gauthier, M. E., Ortlund, E. A., Degnan, B. M., & Thornton, J. W. (2010). Protein evolution by molecular tinkering: diversification of the nuclear receptor superfamily from a ligand-dependent ancestor. *PLoS Biol.*, *8*.

Broadus, J., & Doe, C. Q. (1995). Evolution of neuroblast identity: seven-up and prospero expression reveal homologous and divergent neuroblast fates in *Drosophila* and *Schistocerca. Development*, *121*, 3989–3996.

Broadus, J., McCabe, J. R., Endrizzi, B., Thummel, C. S., & Woodard, C. T. (1999). The *Drosophila* beta FTZ-F1 orphan nuclear receptor provides competence for stage-specific responses to the steroid hormone ecdysone. *Mol. Cell*, *3*, 143–149.

Brody, T., Stivers, C., Nagle, J., & Odenwald, W. F. (2002). Identification of novel *Drosophila* neural precursor genes using a differential embryonic head cDNA screen. *Mech. Dev.*, *113*, 41–59.

Bruey-Sedano, N., Alabouvette, J., Lestradet, M., Hong, L., Girard, A., Gervasio, E., Quennedey, B., & Charles, J. P. (2005). The *Drosophila* ACP65A cuticle gene: deletion scanning analysis of cis-regulatory sequences and regulation by DHR38. *Genesis*, *43*, 17–27.

Bryant, Z., Subrahmanyan, L., Tworoger, M., LaTray, L., Liu, C. R., Li, M. J., van den Engh, G., & Ruohola-Baker, H. (1999). Characterization of differentially expressed genes in purified *Drosophila* follicle cells: toward a general strategy for cell type-specific developmental analysis. *Proc. Natl. Acad. Sci. U S A, 96,* 5559–5564.

Bucher, G., Farzana, L., Brown, S. J., & Klingler, M. (2005). Anterior localization of maternal mRNAs in a short germ insect lacking bicoid. *Evol. Dev., 7,* 142–149.

Bujold, M., Gopalakrishnan, A., Nally, E., & King-Jones, K. (2010). Nuclear receptor DHR96 acts as a sentinel for low cholesterol concentrations in *Drosophila melanogaster. Mol. Cell Biol., 30,* 793–805.

Carmi, I., & Meyer, B. J. (1999). The primary sex determination signal of *Caenorhabditis elegans. Genetics, 152,* 999–1015.

Carmichael, J. A., Lawrence, M. C., Graham, L. D., Pilling, P. A., Epa, V. C., Noyce, L., Lovrecz, G., Winkler, D. A., Pawlak-Skrzecz, A., Eaton, R. E., Hannan, G. N., & Hill, R. J. (2005). The X-ray structure of a hemipteran ecdysone receptor ligand-binding domain - Comparison with a Lepidopteran ecdysone receptor ligand-binding domain and implications for insecticide design. *Journal of Biological Chemistry, 280,* 22258–22269.

Carney, G. E., Wade, A. A., Sapra, R., Goldstein, E. S., & Bender, M. (1997). DHR3, an ecdysone-inducible early-late gene encoding a *Drosophila* nuclear receptor, is required for embryogenesis. *Proc. Natl. Acad. Sci. U S A, 94,* 12024–12029.

Cerny, A. C., Grossmann, D., Bucher, G., & Klingler, M. (2008). The *Tr.* orthologue of knirps and knirps-related is crucial for head segmentation but plays a minor role during abdominal patterning. *Dev. Biol., 321,* 284–294.

Chan, S. M. (1998). Cloning of a shrimp (Metapenaeus ensis) cDNA encoding a nuclear receptor superfamily member: an insect homologue of E75 gene. *Febs. Letters, 436,* 395–400.

Chandra, V., Huang, P., Hamuro, Y., Raghuram, S., Wang, Y., Burris, T. P., & Rastinejad, F. (2008). Structure of the intact PPAR-gamma-RXR-alpha nuclear receptor complex on DNA. *Nature,* 350–356.

Charles, J. P., Shinoda, T., & Chinzei, Y. (1999). Characterization and DNA-binding properties of GRF, a novel monomeric binding orphan receptor related to GCNF and betaFTZ-F1. *Eur. J. Biochem., 266,* 181–190.

Chen, C. K., Kuhnlein, R. P., Eulenberg, K. G., Vincent, S., Affolter, M., & Schuh, R. (1998). The transcription factors KNIRPS and KNIRPS RELATED control cell migration and branch morphogenesis during *Drosophila* tracheal development. *Development, 125,* 4959–4968.

Chen, J. H., Turner, P. C., & Rees, H. H. (2002). Molecular cloning and induction of nuclear receptors from insect cell lines. *Insect Biochem. Mol. Biol., 32,* 657–667.

Chen, W. S., Manova, K., Weinstein, D. C., Duncan, S. A., Plump, A. S., Prezioso, V. R., Bachvarova, R. F., & Darnell, J. E., Jr. (1994). Disruption of the HNF-4 gene, expressed in visceral endoderm, leads to cell death in embryonic ectoderm and impaired gastrulation of mouse embryos. *Genes Dev., 8,* 2466–2477.

Cheng, D., Xia, Q., Duan, J., Wei, L., Huang, C., Li, Z., Wang, G., & Xiang, Z. (2008). Nuclear receptors in *Bombyx mori*: Insights into genomic structure and developmental expression. *Insect Biochemistry and Molecular Biology, 38,* 1130–1137.

Cheung, E., & Kraus, W. L. (2010). Genomic analyses of hormone signaling and gene regulation. *Annu. Rev. Physiol., 72,* 191–218.

Christiaens, O., Iga, M., Velarde, R. A., Rouge, P., & Smagghe, G. (2010). Halloween genes and nuclear receptors in ecdysteroid biosynthesis and signalling in the pea aphid. *Insect Mol. Biol., 2*(Suppl. 19), 187–200.

Chung, A. C., & Cooney, A. J. (2001). Germ cell nuclear factor. *Int. J. Biochem. Cell Biol., 33,* 1141–1146.

Clayton, G. M., Peak-Chew, S. Y., Evans, R. M., & Schwabe, J. W. (2001). The structure of the ultraspiracle ligand-binding domain reveals a nuclear receptor locked in an inactive conformation. *Proc. Natl. Acad. Sci. U S A, 98,* 1549–1554.

Costantino, B. F., Bricker, D. K., Alexandre, K., Shen, K., Merriam, J. R., Antoniewski, C., Callender, J. L., Henrich, V. C., Presente, A., & Andres, A. J. (2008). A novel ecdysone receptor mediates steroid-regulated developmental events during the mid-third instar of *Drosophila. PLoS Genet., 4,* e1000102.

Couch, J. A., Chen, J., Rieff, H. I., Uri, E. M., & Condron, B. G. (2004). robo2 and robo3 interact with eagle to regulate serotonergic neuron differentiation. *Development, 131,* 997–1006.

Crispi, S., Giordano, E., D'Avino, P. P., & Furia, M. (1998). Cross-talking among *Drosophila* nuclear receptors at the promiscuous response element of the ng-1 and ng-2 intermolt genes. *J. Mol. Biol., 275,* 561–574.

Crossgrove, K., Laudet, V., & Maina, C. V. (2002). Dirofilaria immitis encodes Di-hhr-7, a putative orthologue of the *Drosophila* ecdysone-regulated E78 gene. *Molecular and Biochemical Parasitology, 119,* 169–177.

Cruz, J., Mane-Padros, D., Belles, X., & Martin, D. (2006). Functions of the ecdysone receptor isoform-A in the hemimetabolous insect *Blattella germanica* revealed by systemic RNAi in vivo. *Developmental Biology, 297,* 158–171.

Cruz, J., Martin, D., & Belles, X. (2007). Redundant ecdysis regulatory functions of three nuclear receptor HR3 isoforms in the direct-developing insect *Blattella germanica. Mech. Dev., 124,* 180–189.

Cruz, J., Nieva, C., Mane-Padros, D., Martin, D., & Belles, X. (2008). Nuclear receptor BgFTZ-F1 regulates molting and the timing of ecdysteroid production during nymphal development in the hemimetabolous insect *Blattella germanica. Dev. Dyn., 237,* 3179–3191.

Cruz, J., Sieglaff, D. H., Arensburger, P., Atkinson, P. W., & Raikhel, A. S. (2009). Nuclear receptors in the mosquito *Aedes aegypti*: annotation, hormonal regulation and expression profiling. *FEBS J., 276,* 1233–1254.

Damen, W. G. (2007). Evolutionary conservation and divergence of the segmentation process in arthropods. *Dev. Dyn., 236,* 1379–1391.

Daniel, A., Dumstrei, K., Lengyel, J. A., & Hartenstein, V. (1999). The control of cell fate in the embryonic visual system by atonal, tailless and EGFR signaling. *Development, 126,* 2945–2954.

Davis, M. B., Carney, G. E., Robertson, A. E., & Bender, M. (2005). Phenotypic analysis of EcR-A mutants suggests that EcR isoforms have unique functions during *Drosophila* development. *Dev. Biol., 282,* 385–396.

Davis, M. M., Yang, P., Chen, L., O'Keefe, S. L., & Hodgetts, R. B. (2007). The orphan nuclear receptor DHR38 influences transcription of the DOPA decarboxylase gene in epidermal and neural tissues of *Drosophila melanogaster*. *Genome, 50,* 1049–1060.

De Loof, A., & Huybrechts, R. (1998). Insects do not have sex hormones: a myth? *Gen. Comp. Endocrinol., 111,* 245–260.

de Rosny, E., de Groot, A., Jullian-Binard, C., Borel, F., Suarez, C., Le Pape, L., Fontecilla-Camps, J. C., & Jouve, H. M. (2008). DHR51, the *Drosophila melanogaster* homologue of the human photoreceptor cell-specific nuclear receptor, is a thiolate heme-binding protein. *Biochemistry, 47,* 13252–13260.

de Rosny, E., de Groot, A., Jullian-Binard, C., Gaillard, J., Borel, F., Pebay-Peyroula, E., Fontecilla-Camps, J. C., & Jouve, H. M. (2006). *Drosophila* nuclear receptor E75 is a thiolate hemoprotein. *Biochemistry, 45,* 9727–9734.

Debernard, S., Bozzolan, F., Duportets, L., & Porcheron, P. (2001). Periodic expression of an ecdysteroid-induced nuclear receptor in a lepidopteran cell line (IAL-PID2). *Insect Biochem. Mol. Biol., 31,* 1057–1064.

DeMeo, S. D., Lombel, R. M., Cronin, M., Smith, E. L., Snowflack, D. R., Reinert, K., Clever, S., & Wightman, B. (2008). Specificity of DNA-binding by the FAX-1 and NHR-67 nuclear receptors of *Caenorhabditis elegans* is partially mediated via a subclass-specific P-box residue. *BMC Mol. Biol., 9,* 2.

Dhe-Paganon, S., Duda, K., Iwamoto, M., Chi, Y. I., & Shoelson, S. E. (2002). Crystal structure of the HNF4 alpha ligand binding domain in complex with endogenous fatty acid ligand. *J. Biol. Chem., 277,* 37973–37976.

Dittrich, R., Bossing, T., Gould, A. P., Technau, G. M., & Urban, J. (1997). The differentiation of the serotonergic neurons in the *Drosophila* ventral nerve cord depends on the combined function of the zinc finger proteins Eagle and Huckebein. *Development, 124,* 2515–2525.

Dubrovskaya, V. A., Berger, E. M., & Dubrovsky, E. B. (2004). Juvenile hormone regulation of the E75 nuclear receptor is conserved in Diptera and Lepidoptera. *Gene., 340,* 171–177.

Dubrovsky, E. B., Dubrovskaya, V. A., & Berger, E. M. (2004). Hormonal regulation and functional role of *Drosophila* E75A orphan nuclear receptor in the juvenile hormone signaling pathway. *Dev. Biol., 268,* 258–270.

Dziedzic-Letka, A., Rymarczyk, G., Kaplon, T. M., Gorecki, A., Szamborska-Gbur, A., Wojtas, M., Dobryszycki, P., & Ozyhar, A. (2011). Intrinsic disorder of *Drosophila melanogaster* hormone receptor 38 N-terminal domain. *Proteins, 79,* 376–392.

Escriva, H., Delaunay, F., & Laudet, V. (2000). Ligand binding and nuclear receptor evolution. *BioEssays, 22,* 717–727.

Ewer, J. (2005). How the ecdysozoan changed its coat. *PLoS Biol., 3,* e349.

Eystathioy, T., Swevers, L., & Iatrou, K. (2001). The orphan nuclear receptor BmHR3A of *Bombyx mori*: hormonal control, ovarian expression and functional properties. *Mech. Dev., 103,* 107–115.

Fayard, E., Auwerx, J., & Schoonjans, K. (2004). LRH-1: an orphan nuclear receptor involved in development, metabolism and steroidogenesis. *Trends Cell Biol., 14,* 250–260.

Feigl, G., Gram, M., & Pongs, O. (1989). A member of the steroid hormone receptor gene family is expressed in the 20-OH-ecdysone inducible puff 75B in *Drosophila melanogaster*. *Nucleic Acids Res., 17,* 7167–7178.

Finley, K. D., Edeen, P. T., Foss, M., Gross, E., Ghbeish, N., Palmer, R. H., Taylor, B. J., & McKeown, M. (1998). Dissatisfaction encodes a tailless-like nuclear receptor expressed in a subset of CNS neurons controlling *Drosophila* sexual behavior. *Neuron, 21,* 1363–1374.

Finley, K. D., Taylor, B. J., Milstein, M., & McKeown, M. (1997). dissatisfaction, a gene involved in sex-specific behavior and neural development of *Drosophila melanogaster*. *Proc. Natl. Acad. Sci. U S A, 94,* 913–918.

Fisk, G. J., & Thummel, C. S. (1995). Isolation, regulation, and DNA-binding properties of three *Drosophila* nuclear hormone receptor superfamily members. *Proc. Natl. Acad. Sci. U S A, 92,* 10604–10608.

Fisk, G. J., & Thummel, C. S. (1998). The DHR78 nuclear receptor is required for ecdysteroid signaling during the onset of *Drosophila* metamorphosis. *Cell, 93,* 543–555.

Flaig, R., Greschik, H., Peluso-Iltis, C., & Moras, D. (2005). Structural basis for the cell-specific activities of the NGFI-B and the Nurr1 ligand-binding domain. *J. Biol. Chem., 280,* 19250–19258.

Florence, B., Guichet, A., Ephrussi, A., & Laughon, A. (1997). Ftz-F1 is a cofactor in Ftz activation of the *Drosophila* engrailed gene. *Development, 124,* 839–847.

Folkertsma, S., van Noort, P. I., Brandt, R. F., Bettler, E., Vriend, G., & de Vlieg, J. (2005). The nuclear receptor ligand-binding domain: a family-based structure analysis. *Curr. Med. Chem., 12,* 1001–1016.

Fortier, T. M., Vasa, P. P., & Woodard, C. T. (2003). Orphan nuclear receptor betaFTZ-F1 is required for muscle-driven morphogenetic events at the prepupal-pupal transition in *Drosophila melanogaster*. *Dev. Biol., 257,* 153–165.

Francis, V. A., Zorzano, A., & Teleman, A. A. (2010). dDOR is an EcR coactivator that forms a feed-forward loop connecting insulin and ecdysone signaling. *Curr. Biol., 20,* 1799–1808.

Frand, A. R., Russel, S., & Ruvkun, G. (2005). Functional genomic analysis of *C. elegans* molting. *PLoS Biol., 3,* e312.

Furness, S. G., Lees, M. J., & Whitelaw, M. L. (2007). The dioxin (aryl hydrocarbon) receptor as a model for adaptive responses of bHLH/PAS transcription factors. *FEBS Lett., 581,* 3616–3625.

Fuss, B., Meissner, T., Bauer, R., Lehmann, C., Eckardt, F., & Hoch, M. (2001). Control of endoreduplication domains in the *Drosophila* gut by the knirps and knirps-related genes. *Mech. Dev., 100,* 15–23.

Gampe, R. T., Jr., Montana, V. G., Lambert, M. H., Miller, A. B., Bledsoe, R. K., Milburn, M. V., Kliewer, S. A., Willson, T. M., & Xu, H. E. (2000). Asymmetry in the PPARgamma/RXRalpha crystal structure reveals the molecular basis of heterodimerization among nuclear receptors. *Mol. Cell, 5,* 545–555.

García-Solache, M., Jaeger, J., & Akam, M. (2010). A systematic analysis of the gap gene system in the moth midge *Clogmia albipunctata*. *Dev. Biol., 344,* 306–318.

Gauhar, Z., Sun, L. V., Hua, S. J., Mason, C. E., Fuchs, F., Li, T. R., Boutros, M., & White, K. P. (2009). Genomic mapping of binding regions for the Ecdysone receptor protein complex. *Genome Research, 19,* 1006–1013.

Gerisch, B., Weitzel, C., Kober-Eisermann, C., Rottiers, V., & Antebi, A. (2001). A hormonal signaling pathway influencing *C. elegans* metabolism, reproductive development, and life span. *Dev. Cell, 1,* 841–851.

Gerwin, N., LaRosee, A., Sauer, F., Halbritter, H. P., Neumann, M., Jackle, H., & Nauber, U. (1994). Functional and conserved domains of the *Drosophila* transcription factor encoded by the segmentation gene knirps. *Mol. Cell Biol., 14,* 7899–7908.

Gibert, J. M., Peronnet, F., & Schlotterer, C. (2007). Phenotypic plasticity in *Drosophila* pigmentation caused by temperature sensitivity of a chromatin regulator network. *PLoS Genet., 3,* e30.

Giguère, V. (2002). To ERR in the estrogen pathway. *Trends Endocrinol. Metab., 13,* 220–225.

Gissendanner, C. R., Crossgrove, K., Kraus, K. A., Maina, C. V., & Sluder, A. E. (2004). Expression and function of conserved nuclear receptor genes in *Caenorhabditis elegans*. *Dev. Biol., 266,* 399–416.

Gissendanner, C. R., & Sluder, A. E. (2000). nhr-25, the *Caenorhabditis elegans* orthologue of ftz-f1, is required for epidermal and somatic gonad development. *Dev. Biol., 221,* 259–272.

Gonzalez-Gaitan, M., Rothe, M., Wimmer, E. A., Taubert, H., & Jackle, H. (1994). Redundant functions of the genes knirps and knirps-related for the establishment of anterior *Drosophila* head structures. *Proc. Natl. Acad. Sci. U S A, 91,* 8567–8571.

Graham, L. D., Kotze, A. C., Fernley, R. T., & Hill, R. J. (2010). An orthologue of the ecdysone receptor protein (EcR) from the parasitic nematode *Haemonchus contortus*. *Mol. Biochem. Parasitol., 171,* 104–107.

Grimaldi, D., & Engel, M. S. (2005). *Evolution of the insects*: Cambridge University Press.

Guichet, A., Copeland, J. W., Erdelyi, M., Hlousek, D., Zavorszky, P., Ho, J., Brown, S., Percival-Smith, A., Krause, H. M., & Ephrussi, A. (1997). The nuclear receptor homologue Ftz-F1 and the homeodomain protein Ftz are mutually dependent cofactors. *Nature, 385,* 548–552.

Haecker, A., Qi, D., Lilja, T., Moussain, B., Andrioli, L. P., Luschnig, S., & Mannervik, M. (2007). *Drosophila* brakeless interacts with atrophin and is required for tailless-mediated transcriptional repression in early embryos. *PLoS Biol., 5,* e145.

Haider, N. B., Jacobson, S. G., Cideciyan, A. V., Swiderski, R., Streb, L. M., Searby, C., Beck, G., Hockey, R., Hanna, D. B., Gorman, S., Duhl, D., Carmi, R., Bennett, J., Weleber, R. G., Fishman, G. A., Wright, A. F., Stone, E. M., & Sheffield, V. C. (2000). Mutation of a nuclear receptor gene, NR2E3, causes enhanced S cone syndrome, a disorder of retinal cell fate. *Nat. Genet., 24,* 127–131.

Hartmann, B., Reichert, H., & Walldorf, U. (2001). Interaction of gap genes in the *Drosophila* head: tailless regulates expression of empty spiracles in early embryonic patterning and brain development. *Mech. Dev., 109,* 161–172.

Heffer, A., Shultz, J. W., & Pick, L. (2010). Surprising flexibility in a conserved Hox transcription factor over 550 million years of evolution. *Proc. Natl. Acad. Sci. U S A, 107,* 18040–18045.

Henrich, V. C. (2005). The ecdysteroid receptor. In L. I. Gilbert, K. Iatrou, & S. S. Gill (Eds.), *Comprehensive Molecular Insect Science*, Vol. 3. (pp. 243–285). Elsevier.

Henrich, V. C., & Brown, N. E. (1995). Insect Nuclear Receptors - a Developmental and Comparative Perspective. *Insect Biochemistry and Molecular Biology, 25,* 881–897.

Henrich, V. C., Sliter, T. J., Lubahn, D. B., MacIntyre, A., & Gilbert, L. I. (1990). A steroid/thyroid hormone receptor superfamily member in *Drosophila melanogaster* that shares extensive sequence similarity with a mammalian homologue. *Nucleic Acids Res., 18,* 4143–4148.

Henrich, V. C., Vogtli, M. E., Antoniewski, C., Spindler-Barth, M., Przibilla, S., Noureddine, M., & Lezzi, M. (2000). Developmental effects of a chimeric ultraspiracle gene derived from *Drosophila* and *Chironomus*. *Genesis, 28,* 125–133.

Hepperle, C., & Hartfelder, K. (2001). Differentially expressed regulatory genes in honey bee caste development. *Naturwissenschaften, 88,* 113–116.

Higashijima, S., Shishido, E., Matsuzaki, M., & Saigo, K. (1996). eagle, a member of the steroid receptor gene superfamily, is expressed in a subset of neuroblasts and regulates the fate of their putative progeny in the *Drosophila* CNS. *Development, 122,* 527–536.

Hirai, M., Shinoda, T., Kamimura, M., Tomita, S., & Shiotsuki, T. (2002). *Bombyx mori* orphan receptor, BmHR78: cDNA cloning, testis abundant expression and putative dirnerization partner for *Bombyx* ultraspiracle. *Molecular and Cellular Endocrinology, 189,* 201–211.

Hiruma, K., & Riddiford, L. M. (2001). Regulation of transcription factors MHR4 and betaFTZ-F1 by 20-hydroxyecdysone during a larval molt in the tobacco hornworm, *Manduca sexta*. *Dev. Biol., 232,* 265–274.

Hodin, J., & Riddiford, L. M. (1998). The ecdysone receptor and ultraspiracle regulate the timing and progression of ovarian morphogenesis during *Drosophila* metamorphosis. *Dev. Genes. Evol., 208,* 304–317.

Hodin, J., & Riddiford, L. M. (2000). Parallel alterations in the timing of ovarian ecdysone receptor and ultraspiracle expression characterize the independent evolution of larval reproduction in two species of gall midges (Diptera: Cecidomyiidae). *Dev. Genes. Evol., 210,* 358–372.

Horner, M. A., Chen, T., & Thummel, C. S. (1995). Ecdysteroid regulation and DNA binding properties of *Drosophila* nuclear hormone receptor superfamily members. *Dev. Biol., 168,* 490–502.

Horner, M. A., Pardee, K., Liu, S., King-Jones, K., Lajoie, G., Edwards, A., Krause, H. M., & Thummel, C. S. (2009). The *Drosophila* DHR96 nuclear receptor binds cholesterol and regulates cholesterol homeostasis. *Genes. Dev., 23,* 2711–2716.

Horner, M. A., & Thummel, C. S. (1997). Mutations in the DHR39 orphan receptor gene have no effect on viability. *D.I.S, 80,* 35–37.

Hoshizaki, D. K., Blackburn, T., Price, C., Ghosh, M., Miles, K., Ragucci, M., & Sweis, R. (1994). Embryonic fat-cell lineage in *Drosophila melanogaster*. *Development, 120,* 2489–2499.

Hu, X., Cherbas, L., & Cherbas, P. (2003). Transcription activation by the ecdysone receptor (EcR/USP): identification of activation functions. *Mol. Endocrinol., 17,* 716–731.

Hughes, C. L., & Kaufman, T. C. (2002). Hox genes and the evolution of the arthropod body plan. *Evol. Dev., 4,* 459–499.

Iwema, T., Billas, I. M., Beck, Y., Bonneton, F., Nierengarten, H., Chaumot, A., Richards, G., Laudet, V., & Moras, D. (2007). Structural and functional characterization of a novel type of ligand-independent RXR-USP receptor. *EMBO J.*, 26, 3770–3782.

Iwema, T., Chaumot, A., Studer, R. A., Robinson-Rechavi, M., Billas, I. M.L., Moras, D., Laudet, V., & Bonneton, F. (2009). Structural and Evolutionary Innovation of the Heterodimerization Interface between USP and the Ecdysone Receptor ECR in Insects. *Molecular Biology and Evolution*, 26, 753–768.

Jakob, M., Kolodziejczyk, R., Orlowski, M., Krzywda, S., Kowalska, A., Dutko-Gwozdz, J., Gwozdz, T., Kochman, M., Jaskolski, M., & Ozyhar, A. (2007). Novel DNA-binding element within the C-terminal extension of the nuclear receptor DNA-binding domain. *Nucleic Acids Res.*, 35, 2705–2718.

Jetten, A. M. (2009). Retinoid-related orphan receptors (RORs): critical roles in development, immunity, circadian rhythm, and cellular metabolism. *Nucl. Recept. Signal.*, 7, e003.

Jia, K., Albert, P. S., & Riddle, D. L. (2002). DAF-9, a cytochrome P450 regulating *C. elegans* larval development and adult longevity. *Development*, 129, 221–231.

Jindra, M., Sehnal, F., & Riddiford, L. M. (1994a). Isolation, characterization and developmental expression of the ecdysteroid-induced E75 gene of the wax moth *Galleria mellonella*. *Eur. J. Biochem.*, 221, 665–675.

Jindra, M., Sehnal, F., & Riddiford, L. M. (1994b). Isolation and developmental expression of the ecdysteroid-induced GHR3 gene of the wax moth *Galleria mellonella*. *Insect Biochem. Mol. Biol.*, 24, 763–773.

Jürgens, G., Wieschaus, E., Nusslein-Volhard, C., & Kluding, H. (1984). Mutations affecting the pattern of the larval cuticle in *Drosophila melanogaster*. *Roux's Arch. Devel. Biol*, 193, 283–295.

Kageyama, Y., Masuda, S., Hirose, S., & Ueda, H. (1997). Temporal regulation of the mid-prepupal gene FTZ-F1: DHR3 early late gene product is one of the plural positive regulators. *Genes Cells*, 2, 559–569.

Kanai, M. I., Okabe, M., & Hiromi, Y. (2005). Seven-up Controls switching of transcription factors that specify temporal identities of *Drosophila* neuroblasts. *Dev. Cell*, 8, 203–213.

Kapitskaya, M. Z., Dittmer, N. T., Deitsch, K. W., Cho, W. L., Taylor, D. G., Leff, T., & Raikhel, A. S. (1998). Three isoforms of a hepatocyte nuclear factor-4 transcription factor with tissue- and stage-specific expression in the adult mosquito. *Journal of Biological Chemistry*, 273, 29801–29810.

Kapitskaya, M. Z., Li, C., Miura, K., Segraves, W., & Raikhel, A. S. (2000). Expression of the early-late gene encoding the nuclear receptor HR3 suggests its involvement in regulating the vitellogenic response to ecdysone in the adult mosquito. *Mol. Cell Endocrinol.*, 160, 25–37.

Kerber, B., Fellert, S., & Hoch, M. (1998). Seven-up, the *Drosophila* homologue of the COUP-TF orphan receptors, controls cell proliferation in the insect kidney. *Genes. Dev.*, 12, 1781–1786.

Keshan, B., Hiruma, K., & Riddiford, L. A. (2006). Developmental expression and hormonal regulation of different isoforms of the transcription factor E75 in the tobacco hornworm *Manduca sexta*. *Developmental Biology*, 295, 623–632.

Khorasanizadeh, S., & Rastinejad, F. (2001). Nuclear-receptor interactions on DNA-response elements. *Trends Biochem. Sci.*, 26, 384–390.

Kim, H. W., Lee, S. G., & Mykles, D. L. (2005). Ecdysteroid-responsive genes, RXR and E75, in the tropical land crab, *Gecarcinus lateralis*: Differential tissue expression of multiple RXR isoforms generated at three alternative splicing sites in the hinge and ligand-binding domains. *Molecular and Cellular Endocrinology*, 242, 80–95.

King, N., Westbrook, M. J., Young, S. L., Kuo, A., Abedin, M., Chapman, J., Fairclough, S., Hellsten, U., Isogai, Y., Letunic, I., Marr, M., Pincus, D., Putnam, N., Rokas, A., Wright, K. J., Zuzow, R., Dirks, W., Good, M., Goodstein, D., Lemons, D., Li, W., Lyons, J. B., Morris, A., Nichols, S., Richter, D. J., Salamov, A., Sequencing, J. G., Bork, P., Lim, W. A., Manning, G., Miller, W. T., McGinnis, W., Shapiro, H., Tjian, R., Grigoriev, I. V., & Rokhsar, D. (2008). The genome of the choanoflagellate *Monosiga brevicollis* and the origin of metazoans. *Nature*, 451, 783–788.

King-Jones, K., Charles, J. P., Lam, G., & Thummel, C. S. (2005). The ecdysone-induced DHR4 orphan nuclear receptor coordinates growth and maturation in *Drosophila*. *Cell*, 121, 773–784.

King-Jones, K., Horner, M. A., Lam, G., & Thummel, C. S. (2006). The DHR96 nuclear receptor regulates xenobiotic responses in *Drosophila*. *Cell Metab.*, 4, 37–48.

King-Jones, K., & Thummel, C. S. (2005). Nuclear receptors—a perspective from *Drosophila*. *Nat. Rev. Genet.*, 6, 311–323.

Kirkness, E. F., Haas, B. J., Sun, W., Braig, H. R., Perotti, M. A., Clark, J. M., Lee, S. H., Robertson, H. M., Kennedy, R. C., Elhaik, E., Gerlach, D., Kriventseva, E. V., Elsik, C. G., Graur, D., Hill, C. A., Veenstra, J. A., Walenz, B., Tubio, J. M., Ribeiro, J. M., Rozas, J., Johnston, J. S., Reese, J. T., Popadic, A., Tojo, M., Raoult, D., Reed, D. L., Tomoyasu, Y., Krause, E., Mittapalli, O., Margam, V. M., Li, H. M., Meyer, J. M., Johnson, R. M., Romero-Severson, J., Vanzee, J. P., Alvarez-Ponce, D., Vieira, F. G., Aguade, M., Guirao-Rico, S., Anzola, J. M., Yoon, K. S., Strycharz, J. P., Unger, M. F., Christley, S., Lobo, N. F., Seufferheld, M. J., Wang, N., Dasch, G. A., Struchiner, C. J., Madey, G., Hannick, L. I., Bidwell, S., Joardar, V., Caler, E., Shao, R., Barker, S. C., Cameron, S., Bruggner, R. V., Regier, A., Johnson, J., Viswanathan, L., Utterback, T. R., Sutton, G. G., Lawson, D., Waterhouse, R. M., Venter, J. C., Strausberg, R. L., Berenbaum, M. R., Collins, F. H., Zdobnov, E. M., & Pittendrigh, B. R. (2010). Genome sequences of the human body louse and its primary endosymbiont provide insights into the permanent parasitic lifestyle. *Proc. Natl. Acad. Sci. U S A*, 107, 12168–12173.

Kobayashi, M., Takezawa, S., Hara, K., Yu, R. T., Umesono, Y., Agata, K., Taniwaki, M., Yasuda, K., & Umesono, K. (1999). Identification of a photoreceptor cell-specific nuclear receptor. *Proc. Natl. Acad. Sci. U S A*, 96, 4814–4819.

Koelle, M. R., Segraves, W. A., & Hogness, D. S. (1992). Dhr3-a *Drosophila* Steroid-Receptor Homologue. *Proceedings of the National Academy of Sciences of the United States of America, 89*, 6167–6171.

Koelle, M. R., Talbot, W. S., Segraves, W. A., Bender, M. T., Cherbas, P., & Hogness, D. S. (1991). The *Drosophila* EcR gene encodes an ecdysone receptor, a new member of the steroid receptor superfamily. *Cell, 67*, 59–77.

Komonyi, O., Mink, M., Csiha, J., & Maroy, P. (1998). Genomic organization of DHR38 gene in *Drosophila*: presence of Alu-like repeat in a translated exon and expression during embryonic development. *Arch. Insect. Biochem. Physiol., 38*, 185–192.

Konopova, B., & Jindra, M. (2007). Juvenile hormone resistance gene Methoprene-tolerant controls entry into metamorphosis in the beetle *Tr. castaneum. Proc. Natl. Acad. Sci. USA, 104*, 10488–10493.

Konopova, B., & Jindra, M. (2008). Broad-Complex acts downstream of Met in juvenile hormone signaling to coordinate primitive holometabolan metamorphosis. *Development, 135*, 559–568.

Kostrouch, Z., Kostrouchova, M., & Rall, J. E. (1995). Steroid/thyroid harmone receptor genes in *Caenorhabditis elegans. Proc. Natl. Acad. Sci. USA, 92*, 156–159.

Kostrouchova, M., Krause, M., Kostrouch, Z., & Rall, J. E. (1998). CHR3: a *Caenorhabditis elegans* orphan nuclear hormone receptor required for proper epidermal development and molting. *Development, 125*, 1617–1626.

Kostrouchova, M., Krause, M., Kostrouch, Z., & Rall, J. E. (2001). Nuclear hormone receptor CHR3 is a critical regulator of all four larval molts of the nematode *Caenorhabditis elegans. Proc. Natl. Acad. Sci. USA, 98*, 7360–7365.

Kozlova, T., Lam, G., & Thummel, C. S. (2009). *Drosophila* DHR38 nuclear receptor is required for adult cuticle integrity at eclosion. *Dev. Dyn., 238*, 701–707.

Kozlova, T., Pokholkova, G. V., Tzertzinis, G., Sutherland, J. D., Zhimulev, I. F., & Kafatos, F. C. (1998). *Drosophila* hormone receptor 38 functions in metamorphosis: a role in adult cuticle formation. *Genetics, 149*, 1465–1475.

Kristensen, N. P. (1981). Phylogeny of insect orders. *Annu. Rev. Entomol., 26*, 135–157.

Kurusu, M., Maruyama, Y., Adachi, Y., Okabe, M., Susuki, E., & Furukubo-Tokunaga, K. (2009). A conserved nuclear receptor, Tailless, is required for efficient proliferation and prolonged maintenance of mushroom body progenitors in the *Drosophila* brain. *Dev. Biol., 326*, 224–236.

Lafont, R., Dauphin-Villemant, C., Warren, J. T., & Rees, H. (2005). Ecdysteroid chemistry and biochemistry. In L. I. Gilbert, K. Iatrou, & S. S. Gill (Eds.), *Comprehensive Molecular Insect Science Vol. 3*. (pp. 125–195): Elsevier.

Lam, G., Hall, B. L., Bender, M., & Thummel, C. S. (1999). DHR3 is required for the prepupal-pupal transition and differentiation of adult structures during *Drosophila* metamorphosis. *Dev. Biol., 212*, 204–216.

Lam, G. T., Jiang, C., & Thummel, C. S. (1997). Coordination of larval and prepupal gene expression by the DHR3 orphan receptor during *Drosophila* metamorphosis. *Development, 124*, 1757–1769.

Lan, Q., Hiruma, K., Hu, X., Jindra, M., & Riddiford, L. M. (1999). Activation of a delayed-early gene encoding MHR3 by the ecdysone receptor heterodimer EcR-B1-USP-1 but not by EcR-B1-USP-2. *Mol. Cell Biol., 19*, 4897–4906.

Lan, Q., Wu, Z., & Riddiford, L. M. (1997). Regulation of the ecdysone receptor, USP, E75 and MHR3 mRNAs by 20-hydroxyecdysone in the GV1 cell line of the tobacco hornworm, *Manduca sexta. Insect Mol. Biol., 6*, 3–10.

Langelan, R. E., Fisher, J. E., Hiruma, K., Palli, S. R., & Riddiford, L. M. (2000). Patterns of MHR3 expression in the epidermis during a larval molt of the tobacco hornworm *Manduca sexta. Dev. Biol., 227*, 481–494.

Laudet, V. (1997). Evolution of the nuclear receptor superfamily: early diversification from an ancestral orphan receptor. *J. Mol. Endocrinol., 19*, 207–226.

Laudet, V., & Adelmant, G. (1995). Nuclear receptors. Lonesome orphans. *Curr. Biol., 5*, 124–127.

Laudet, V., & Bonneton, F. (2005). Evolution of nuclear hormone receptors in insects. In L. I. Gilbert, K. Iatrou, & S. S. Gill (Eds.), *Comprehensive Molecular Insect Science Vol. 3*. (pp. 287–318): Elsevier.

Laudet, V., & Gronemeyer, H. (2002). *The Nuclear Receptor FactsBook*: Academic Press.

Laudet, V., Hanni, C., Coll, J., Catzeflis, F., & Stehelin, D. (1992). Evolution of the nuclear receptor gene superfamily. *EMBO J., 11*, 1003–1013.

Lavorgna, G., Karim, F. D., Thummel, C. S., & Wu, C. (1993). Potential Role for a Ftz-F1 Steroid-Receptor Superfamily Member in the Control of *Drosophila* Metamorphosis. *Proceedings of the National Academy of Sciences of the United States of America, 90*, 3004–3008.

Lavorgna, G., Ueda, H., Clos, J., & Wu, C. (1991). FTZ-F1, a steroid hormone receptor-like protein implicated in the activation of fushi tarazu. *Science, 252*, 848–851.

Lee, H. K., & Lundell, M. J. (2007). Differentiation of the *Drosophila* serotonergic lineage depends on the regulation of Zfh-1 by Notch and Eagle. *Mol. Cell Neurosci., 36*, 47–58.

Lemke, S., Busch, S. E., Antonopoulos, D. A., Meyer, F., Domanus, M. H., & Schmidt-Ott, U. (2010). Maternal activation of gap genes in the hover fly *Episyrphus. Development, 137*, 1709–1719.

Li, C., Kapitskaya, M. Z., Zhu, J., Miura, K., Segraves, W., & Raikhel, A. S. (2000). Conserved molecular mechanism for the stage specificity of the mosquito vitellogenic response to ecdysone. *Dev. Biol., 224*, 96–110.

Li, C., Schwabe, J. W.R., Banayo, E., & Evans, R. M. (1997). Coexpression of nuclear receptor partners increases their solubility and biological activities. *Proceedings of the National Academy of Sciences of the United States of America, 94*, 2278–2283.

Li, Y., Lambert, M. H., & Xu, H. E. (2003). Activation of nuclear receptors: a perspective from structural genomics. *Structure, 11*, 741–746.

Liang, B., Ferguson, K., Kadyk, L., & Watts, J. L. (2010). The role of nuclear receptor NHR-64 in fat storage regulation in *Caenorhabditis elegans. PLoS One, 5*, e9869.

Liaw, G. J., & Lengyel, J. A. (1993). Control of tailless expression by bicoid, dorsal and synergistically interacting terminal system regulatory elements. *Mech. Dev., 40*, 47–61.

Lin, S., Huang, Y., & Lee, T. (2009). Nuclear receptor unfulfilled regulates axonal guidance and cell identity of *Drosophila* mushroom body neurons. *PLoS One, 4,* e8392.

Lindblom, T. H., Pierce, G. J., & Sluder, A. E. (2001). A C. elegans orphan nuclear receptor contributes to xenobiotic resistance. *Curr. Biol., 11,* 864–868.

Lo, P. C., & Frasch, M. (2001). A role for the COUP-TF-related gene seven-up in the diversification of cardioblast identities in the dorsal vessel of *Drosophila. Mech. Dev., 104,* 49–60.

Lohr, U., Yussa, M., & Pick, L. (2001). *Drosophila fushi tarazu:* a gene on the border of homeotic function. *Current Biology, 11,* 1403–1412.

Lunde, K., Biehs, B., Nauber, U., & Bier, E. (1998). The knirps and knirps-related genes organize development of the second wing vein in *Drosophila. Development, 125,* 4145–4154.

Lunde, K., Trimble, J. L., Guichard, A., Guss, K. A., Nauber, U., & Bier, E. (2003). Activation of the knirps locus links patterning to morphogenesis of the second wing vein in *Drosophila. Development, 130,* 235–248.

Lundell, M. J., & Hirsh, J. (1998). eagle is required for the specification of serotonin neurons and other neuroblast 7-3 progeny in the *Drosophila* CNS. *Development, 125,* 463–472.

Lynch, J. A., Olesnicky, E. C., & Desplan, C. (2006). Regulation and function of tailless in the long germ wasp *Nasonia vitripennis. Dev. Genes. Evol., 216,* 493–498.

Magner, D. B., & Antebi, A. (2008). Caenorhabditis elegans nuclear receptors: insights into life traits. *Trends Endocrinol. Metab., 19,* 153–160.

Mane-Padros, D., Cruz, J., Vilaplana, L., Nieva, C., Urena, E., Belles, X., & Martin, D. (2010). The hormonal pathway controlling cell death during metamorphosis in a hemimetabolous insect. *Dev. Biol., 346,* 150–160.

Mane-Padros, D., Cruz, J., Vilaplana, L., Pascual, N., Belles, X., & Martin, D. (2008). The nuclear hormone receptor BgE75 links molting and developmental progression in the direct-developing insect *Blattella germanica. Dev. Biol., 315,* 147–160.

Markov, G. V., & Laudet, V. (2011). Origin and evolution of the ligand-binding ability of nuclear receptors. *Mol. Cell Endocrinol., 334,* 21–30.

Martin, D. (2010). Functions of nuclear receptors in insect development. In C. M. Bunce, & M. J. Campbell (Eds.), *Nuclear Receptors. Current concepts and future challenges* (Vol. 8): Springer.

Marvin, K. A., Reinking, J. L., Lee, A. J., Pardee, K., Krause, H. M., & Burstyn, J. N. (2009). Nuclear receptors *Homo sapiens* Rev-erbbeta and *Drosophila melanogaster* E75 are thiolate-ligated heme proteins which undergo redox-mediated ligand switching and bind CO and NO. *Biochemistry, 48,* 7056–7071.

Matsuoka, T., & Fujiwara, H. (2000). Expression of ecdysteroid-regulated genes is reduced specifically in the wing discs of the wing-deficient mutant (fl) of *Bombyx mori. Dev. Genes. Evol., 210,* 120–128.

Minakuchi, C., Ogura, T., Miyagawa, H., & Nakagawa, Y. (2007). Effects of the structures of ecdysone receptor (EcR) and ultraspiracle (USP) on the ligand-binding activity of the EcR/USP heterodimer. *J. Pestic. Sci., 32,* 379–384.

Miura, K., Zhu, J., Dittmer, N. T., Chen, L., & Raikhel, A. S. (2002). A COUP-TF/Svp homologue is highly expressed during vitellogenesis in the mosquito *Aedes aegypti. J. Mol. Endocrinol., 29,* 223–238.

Mlodzik, M., Hiromi, Y., Weber, U., Goodman, C. S., & Rubin, G. M. (1990). The *Drosophila* seven-up gene, a member of the steroid receptor gene superfamily, controls photoreceptor cell fates. *Cell, 60,* 211–224.

Monaghan, A. P., Bock, D., Goss, P., Schwager, A., Wolfer, D. P., Lipp, H. P., & Schutz, G. (1997). Defective limbic system in mice lacking the tailless gene. *Nature, 390,* 515–517.

Montagne, J., Lecerf, C., Parvy, J. P., Bennion, J. M., Radimerski, T., Ruhf, M. L., Zilbermann, F., Vouilloz, N., Stocker, H., Hafen, E., Kozma, S. C., & Thomas, G. (2010). The nuclear receptor DHR3 modulates dS6 kinase-dependent growth in *Drosophila. PLoS Genet., 6,* e1000937.

Montano, M. M., Muller, V., Trobaugh, A., & Katzenellenbogen, B. S. (1995). The carboxy-terminal F domain of the human estrogen receptor: role in the transcriptional activity of the receptor and the effectiveness of antiestrogens as estrogen antagonists. *Mol. Endocrinol., 9,* 814–825.

Moore, J., & Devaney, E. (1999). Cloning and characterization of two nuclear receptors from the filarial nematode *Brugia pahangi. Biochem. J., 344*(Pt 1), 245–252.

Moore, J. T., Collins, J. L., & Pearce, K. H. (2006). The nuclear receptor superfamily and drug discovery. *Chem. Med. Chem., 1,* 504–523.

Moore, T. W., Mayne, C. G., & Katzenellenbogen, J. A. (2010). Minireview: Not picking pockets: nuclear receptor alternate-site modulators (NRAMs). *Mol. Endocrinol., 24,* 683–695.

Moran, E., & Jimenez, G. (2006). The tailless order receptor acts as a dedicated repressor in the early *Drosophila* embryo. *Mol. Cell Biol., 26,* 3446–3454.

Motola, D. L., Cummins, C. L., Rottiers, V., Sharma, K. K., Li, T., Li, Y., Suino-Powell, K., Xu, H. E., Auchus, R. J., Antebi, A., & Mangelsdorf, D. J. (2006). Identification of ligands for DAF-12 that govern dauer formation and reproduction in *C. elegans. Cell, 124,* 1209–1223.

Mouillet, J. F., Bousquet, F., Sedano, N., Alabouvette, J., Nicolai, M., Zelus, D., Laudet, V., & Delachambre, J. (1999). Cloning and characterization of new orphan nuclear receptors and their developmental profiles during *Te.* metamorphosis. *Eur. J. Biochem., 265,* 972–981.

Mouillet, J. F., Henrich, V. C., Lezzi, M., & Vogtli, M. (2001). Differential control of gene activity by isoforms A, B1 and B2 of the *Drosophila* ecdysone receptor. *Eur. J. Biochem., 268,* 1811–1819.

Myat, M. M., Lightfoot, H., Wang, P., & Andrew, D. J. (2005). A molecular link between FGF and Dpp signaling in branch-specific migration of the *Drosophila* trachea. *Dev. Biol., 281,* 38–52.

Näär, A. M., & Thakur, J. K. (2009). Nuclear receptor-like transcription factors in fungi. *Genes Dev., 23,* 419–432.

Nakagawa, Y., & Henrich, V. C. (2009). Arthropod nuclear receptors and their role in molting. *FEBS. J., 276,* 6128–6157.

Nakagawa, Y., Sakai, A., Magata, F., Ogura, T., Miyashita, M., & Miyagawa, H. (2007). Molecular cloning of the ecdysone receptor and the retinoid X receptor from the scorpion *Liocheles australasiae. Febs. J., 274,* 6191–6203.

Nauber, U., Pankratz, M. J., Kienlin, A., Seifert, E., Klemm, U., & Jackle, H. (1988). Abdominal segmentation of the *Drosophila* embryo requires a hormone receptor-like protein encoded by the gap gene knirps. *Nature, 336,* 489–492.

Nichols, M., Rientjes, J. M., & Stewart, A. F. (1998). Different positioning of the ligand-binding domain helix 12 and the F domain of the estrogen receptor accounts for functional differences between agonists and antagonists. *Embo. J., 17,* 765–773.

Niimi, T., Morita, S., & Yamashita, O. (1997). The profiles of mRNA levels for *BHR39,* a *Bombyx* homologue of *Drosophila* hormone receptor 39, and *Bombyx* FTZ-F1 in the course of embryonic development and diapause. Devel. *Genes. Evol., 207,* 410–412.

Nijhout, H. F. (2003). Development and evolution of adaptive polyphenisms. *Evol. Dev., 5,* 9–18.

Nocula-Lugowska, M., Rymarczyk, G., Lisowski, M., & Ozyhar, A. (2009). Isoform-specific variation in the intrinsic disorder of the ecdysteroid receptor N-terminal domain. *Proteins, 76,* 291–308.

Nowickyj, S. M., Chithalen, J. V., Cameron, D., Tyshenko, M. G., Petkovich, M., Wyatt, G. R., Jones, G., & Walker, V. K. (2008). Locust retinoid X receptors: 9-Cis-retinoic acid in embryos from a primitive insect. *Proceedings of the National Academy of Sciences of the United States of America, 105,* 9540–9545.

Nuclear Receptors Nomenclature Committee. (1999). A unified nomenclature system for the nuclear receptor superfamily. *Cell, 97,* 161–163.

O'Kane, C. J., & Asztalos, Z. (1999). Sexual behaviour: Courting dissatisfaction. *Curr. Biol., 9,* R289–R292.

Ogura, T., Minakuchi, C., Nakagawa, Y., Smagghe, G., & Miyagawa, H. (2005). Molecular cloning, expression analysis and functional confirmation of ecdysone receptor and ultraspiracle from the Colorado potato beetle *Leptinotarsa decemlineata. Febs. Journal, 272,* 4114–4128.

Ohno, C. K., & Petkovich, M. (1993). FTZ-F1 beta, a novel member of the *Drosophila* nuclear receptor family. *Mech. Dev., 40,* 13–24.

Ohno, C. K., Ueda, H., & Petkovich, M. (1994). The *Drosophila* nuclear receptors FTZ-F1 alpha and FTZ-F1 beta compete as monomers for binding to a site in the fushi tarazu gene. *Mol. Cell Biol., 14,* 3166–3175.

Oro, A. E., McKeown, M., & Evans, R. M. (1992). The *Drosophila* retinoid X receptor homologue ultraspiracle functions in both female reproduction and eye morphogenesis. *Development, 115,* 449–462.

Oro, A. E., Ong, E. S., Margolis, J. S., Posakony, J. W., McKeown, M., & Evans, R. M. (1988). The *Drosophila* gene knirps-related is a member of the steroid-receptor gene superfamily. *Nature, 336,* 493–496.

Östberg, T., Jacobsson, M., Attersand, A., Mata de Urquiza, A., & Jendeberg, L. (2003). A triple mutant of the *Drosophila* ERR confers ligand-induced suppression of activity. *Biochemistry, 42,* 6427–6435.

Palanker, L., Necakov, A. S., Sampson, H. M., Ni, R., Hu, C., Thummel, C. S., & Krause, H. M. (2006). Dynamic regulation of *Drosophila* nuclear receptor activity in vivo. *Development, 133,* 3549–3562.

Palanker, L., Tennessen, J. M., Lam, G., & Thummel, C. S. (2009). *Drosophila* HNF4 regulates lipid mobilization and beta-oxidation. *Cell Metab., 9,* 228–239.

Palli, S. R., Hiruma, K., & Riddiford, L. M. (1992). An ecdysteroid-inducible *Manduca* gene similar to the *Drosophila* DHR3 gene, a member of the steroid hormone receptor superfamily. *Dev. Biol., 150,* 306–318.

Palli, S. R., Hormann, R. E., Schlattner, U., & Lezzi, M. (2005). Ecdysteroid receptors and their applications in agriculture and medicine. *Vitamins and Hormones - Advances in Research and Applications, 73,* 59–100.

Palli, S. R., Ladd, T. R., & Retnakaran, A. (1997a). Cloning and characterization of a new isoform of *Choristoneura* hormone receptor 3 from the spruce budworm. *Arch. Insect. Biochem. Physiol., 35,* 33–44.

Palli, S. R., Ladd, T. R., Ricci, A. R., Sohi, S. S., & Retnakaran, A. (1997b). Cloning and development expression of *Choristoneura* hormone receptor 75: a homologue of the *Drosophila* E75A gene. *Dev. Genet., 20,* 36–46.

Palli, S. R., Ladd, T. R., Sohi, S. S., Cook, B. J., & Retnakaran, A. (1996). Cloning and developmental expression of *Choristoneura* hormone receptor 3, an ecdysone-inducible gene and a member of the steroid hormone receptor superfamily. *Insect Biochem. Mol. Biol., 26,* 485–499.

Palli, S. R., Sohi, S. S., Cook, B. J., Lambert, D., Ladd, T. R., & Retnakaran, A. (1995). Analysis of ecdysteroid action in *Malacosoma disstria* cells: cloning selected regions of E75- and MHR3-like genes. *Insect Biochem. Mol. Biol., 25,* 697–707.

Parihar, M., Minton, R. L., Flowers, S., Holloway, A., Morehead, B. E., Paille, J., & Gissendanner, C. R. (2010). The genome of the nematode *Pristionchus pacificus* encodes putative homologues of RXR/Usp and EcR. *Gen. Comp. Endocrinol., 167,* 11–17.

Park, K., & Kwak, I. S. (2010). Molecular effects of endocrine-disrupting chemicals on the *Chironomus riparius* estrogen-related receptor gene. *Chemosphere, 79,* 934–941.

Parker, K. L., Rice, D. A., Lala, D. S., Ikeda, Y., Luo, X., Wong, M., Bakke, M., Zhao, L., Frigeri, C., Hanley, N. A., Stallings, N., & Schimmer, B. P. (2002). Steroidogenic factor 1: an essential mediator of endocrine development. *Recent Prog. Horm. Res., 57,* 19–36.

Parvy, J. P., Blais, C., Bernard, F., Warren, J. T., Petryk, A., Gilbert, L. I., O'Connor, M. B., & Dauphin-Villemant, C. (2005). A role for betaFTZ-F1 in regulating ecdysteroid titers during post-embryonic development in *Drosophila melanogaster. Dev. Biol., 282,* 84–94.

Pecasse, F., Beck, Y., Ruiz, C., & Richards, G. (2000). Kruppel-homologue, a stage-specific modulator of the prepupal ecdysone response, is essential for *Drosophila* metamorphosis. *Dev. Biol., 221,* 53–67.

Pereira, F. A., Tsai, M. J., & Tsai, S. Y. (2000). COUP-TF orphan nuclear receptors in development and differentiation. *Cell Mol. Life Sci., 57,* 1388–1398.

Peters, G. A., & Khan, S. A. (1999). Estrogen receptor domains E and F: role in dimerization and interaction with coactivator RIP-140. *Mol. Endocrinol., 13,* 286–296.

Phelps, C., Gburcik, V., Suslova, E., Dudek, P., Forafonov, F., Bot, N., MacLean, M., Fagan, R. J., & Picard, D. (2006). Fungi and animals may share a common ancestor to nuclear receptors. *Proc. Natl. Acad. Sci. U S A, 103,* 7077–7081.

Pierceall, W. E., Li, C., Biran, A., Miura, K., Raikhel, A. S., & Segraves, W. A. (1999). E75 expression in mosquito ovary and fat body suggests reiterative use of ecdysone-regulated hierarchies in development and reproduction. *Molecular and Cellular Endocrinology, 150,* 73–89.

Pitman, J. L., Tsai, C. C., Edeen, P. T., Finley, K. D., Evans, R. M., & McKeown, M. (2002). DSF nuclear receptor acts as a repressor in culture and in vivo. *Dev. Biol., 245,* 315–328.

Ponzielli, R., Astier, M., Chartier, A., Gallet, A., Therond, P., & Semeriva, M. (2002). Heart tube patterning in *Drosophila* requires integration of axial and segmental information provided by the Bithorax Complex genes and hedgehog signaling. *Development, 129,* 4509–4521.

Preitner, N., Damiola, F., Lopez-Molina, L., Zakany, J., Duboule, D., Albrecht, U., & Schibler, U. (2002). The orphan nuclear receptor REV-ERBalpha controls circadian transcription within the positive limb of the mammalian circadian oscillator. *Cell, 110,* 251–260.

Priya, T. A., Li, F., Zhang, J., Yang, C., & Xiang, J. (2010). Molecular characterization of an ecdysone inducible gene E75 of Chinese shrimp *Fenneropenaeus chinensis* and elucidation of its role in molting by RNA interference. *Comp. Biochem. Physiol. B Biochem. Mol. Biol., 156,* 149–157.

Raikel, A. S., Brown, M. R., & Belles, X. (2005). Hormonal control of reproductive processes. In: L. I. Gilbert, K. Iatrou & S. S. Gill (Eds.), *Comprehensive Molecular Insect Science* Vol. 3, (pp. 433–491). Elsevier.

Raikhel, A. S., Kokoza, V. A., Zhu, J., Martin, D., Wang, S. F., Li, C., Sun, G., Ahmed, A., Dittmer, N., & Attardo, G. (2002). Molecular biology of mosquito vitellogenesis: from basic studies to genetic engineering of antipathogen immunity. *Insect Biochem. Mol. Biol., 32,* 1275–1286.

Reichert, H. (2002). Conserved genetic mechanisms for embryonic brain patterning. *Int. J. Dev. Biol., 46,* 81–87.

Reinking, J., Lam, M. M., Pardee, K., Sampson, H. M., Liu, S., Yang, P., Williams, S., White, W., Lajoie, G., Edwards, A., & Krause, H. M. (2005). The *Drosophila* nuclear receptor e75 contains heme and is gas responsive. *Cell, 122,* 195–207.

Riddiford, L. M. (2008). Juvenile hormone action: A 2007 perspective. *Journal of Insect Physiology, 54,* 895–901.

Riddiford, L. M., Cherbas, P., & Truman, J. W. (2000). Ecdysone receptors and their biological actions. *Vitam. Horm., 60,* 1–73.

Riddiford, L. M., Hiruma, K., Zhou, X., & Nelson, C. A. (2003). Insights into the molecular basis of the hormonal control of molting and metamorphosis from *Manduca sexta* and *Drosophila melanogaster. Insect Biochem. Mol. Biol., 33,* 1327–1338.

Robinson-Rechavi, M., Maina, C. V., Gissendanner, C. R., Laudet, V., & Sluder, A. (2005). Explosive lineage-specific expansion of the orphan nuclear receptor HNF4 in nematodes. *J. Mol. Evol., 60,* 577–586.

Rochette-Egly, C. (2003). Nuclear receptors: integration of multiple signalling pathways through phosphorylation. *Cell Signal, 15,* 355–366.

Rothe, M., Nauber, U., & Jackle, H. (1989). Three hormone receptor-like *Drosophila* genes encode an identical DNA-binding finger. *EMBO J., 8,* 3087–3094.

Rothe, M., Pehl, M., Taubert, H., & Jackle, H. (1992). Loss of gene function through rapid mitotic cycles in the *Drosophila* embryo. *Nature, 359,* 156–159.

Ruaud, A. F., Lam, G., & Thummel, C. S. (2010). The *Drosophila* nuclear receptors DHR3 and betaFTZ-F1 control overlapping developmental responses in late embryos. *Development, 137,* 123–131.

Russell, S. R., Heimbeck, G., Goddard, C. M., Carpenter, A. T., & Ashburner, M. (1996). The *Drosophila* Eip78C gene is not vital but has a role in regulating chromosome puffs. *Genetics, 144,* 159–170.

Sarin, S., Antonio, C., Tursun, B., & Hobert, O. (2009). The C. elegans Tailless/TLX transcription factor nhr-67 controls neuronal identity and left/right asymmetric fate diversification. *Development, 136,* 2933–2944.

Savard, J., Tautz, D., Richards, S., Weinstock, G. M., Gibbs, R. A., Werren, J. H., Tettelin, H., & Lercher, M. J. (2006b). Phylogenomic analysis reveals bees and wasps (Hymenoptera) at the base of the radiation of Holometabolous insects. *Genome Res., 16,* 1334–1338.

Schorderet, D. F., & Escher, P. (2009). NR2E3 mutations in enhanced S-cone sensitivity syndrome (ESCS), Goldmann-Favre syndrome (GFS), clumped pigmentary retinal degeneration (CPRD), and retinitis pigmentosa (RP). *Hum. Mutat., 30,* 1475–1485.

Schröder, R., Eckert, C., Wolff, C., & Tautz, D. (2000). Conserved and divergent aspects of terminal patterning in the beetle Tr. castaneum. *Proc. Natl. Acad. Sci. U S A, 97,* 6591–6596.

Schwartz, C. J., Sampson, H. M., Hlousek, D., Percival-Smith, A., Copeland, J. W., Simmonds, A. J., & Krause, H. M. (2001). FTZ-Factor1 and Fushi tarazu interact via conserved nuclear receptor and coactivator motifs. *EMBO J., 20,* 510–519.

Segraves, W. A., & Hogness, D. S. (1990). The E75 ecdysone-inducible gene responsible for the 75B early puff in *Drosophila* encodes two new members of the steroid receptor superfamily. *Genes. Dev., 4,* 204–219.

Segraves, W. A., & Woldin, C. (1993). The E75 gene of *Manduca sexta* and comparison with its *Drosophila* homologue. *Insect Biochem Mol. Biol., 23,* 91–97.

Sehnal, F., Svàcha, P., & Zrzavy, J. (1996). Evolution of insect metamorphosis. In L. I. Gilbert, J. R. Tata, & B. G. Atkinson (Eds.), *Metamorphosis: postembryonic reprogramming of gene expression in amphibian and insect cells* (pp. 3–58). San Diego: Academic Press.

Shea, C., Hough, D., Xiao, J., Tzertzinis, G., & Maina, C. V. (2004). An rxr/usp homologue from the parasitic nematode, *Dirofilaria immitis. Gene., 324,* 171–182.

Shea, C., Richer, J., Tzertzinis, G., & Maina, C. V. (2010). An EcR homologue from the filarial parasite, *Dirofilaria immitis* requires a ligand-activated partner for transactivation. *Mol. Biochem. Parasitol., 171,* 55–63.

Shigenobu, S., Bickel, R. D., Brisson, J. A., Butts, T., Chang, C. C., Christiaens, O., Davis, G. K., Duncan, E. J., Ferrier, D. E., Iga, M., Janssen, R., Lin, G. W., Lu, H. L., McGregor, A. P., Miura, T., Smagghe, G., Smith, J. M., van der Zee, M., Velarde, R. A., Wilson, M. J., Dearden, P. K., & Stern, D. L. (2010). Comprehensive survey of developmental genes in the pea aphid, *Acyrthosiphon pisum*: frequent lineage-specific duplications and losses of developmental genes. *Insect Mol. Biol., 2*(Suppl. 19), 47–62.

Siaussat, D., Mottier, V., Bozzolan, F., Porcheron, P., & Debernard, S. (2004). Synchronization of *Plodia interpunctella* lepidopteran cells and effects of 20-hydroxyecdysone. *Insect Molecular Biology, 13*, 179–187.

Sieber, M. H., & Thummel, C. S. (2009). The DHR96 nuclear receptor controls triacylglycerol homeostasis in *Drosophila*. *Cell Metab., 10*, 481–490.

Sladek, F. M. (2011). What are nuclear receptor ligands? *Mol. Cell Endocrinol., 334(1–2)*, 3–13.

Sluder, A. E., & Maina, C. V. (2001). Nuclear receptors in nematodes: themes and variations. *Trends Genet., 17*, 206–213.

Smagghe, G., Dhadialla, T. S., & Lezzi, M. (2002). Comparative toxicity and ecdysone receptor affinity of non-steroidal ecdysone agonists and 20-hydroxyecdysone in *Chironomus tentans. Insect Biochem. Mol. Biol., 32*, 187–192.

Sommer, R., & Tautz, D. (1991). Segmentation gene expression in the housefly *Musca domestica. Development, 113*, 419–430.

Spindler, K. D., Honl, C., Tremmel, C., Braun, S., Ruff, H., & Spindler-Barth, M. (2009). Ecdysteroid hormone action. *Cell Mol. Life Sci., 66*, 3837–3850.

Stein, L. D., Bao, Z., Blasiar, D., Blumenthal, T., Brent, M. R., Chen, N., Chinwalla, A., Clarke, L., Clee, C., Coghlan, A., Coulson, A., D'Eustachio, P., Fitch, D. H., Fulton, L. A., Fulton, R. E., Griffiths-Jones, S., Harris, T. W., Hillier, L. W., Kamath, R., Kuwabara, P. E., Mardis, E. R., Marra, M. A., Miner, T. L., Minx, P., Mullikin, J. C., Plumb, R. W., Rogers, J., Schein, J. E., Sohrmann, M., Spieth, J., Stajich, J. E., Wei, C., Willey, D., Wilson, R. K., Durbin, R., & Waterston, R. H. (2003). The genome sequence of *Caenorhabditis briggsae*: a platform for comparative genomics. *PLoS Biol., 1*, E45.

Sudarsan, V., Pasalodos-Sanchez, S., Wan, S., Gampel, A., & Skaer, H. (2002). A genetic hierarchy establishes mitogenic signalling and mitotic competence in the renal tubules of *Drosophila. Development, 129*, 935–944.

Sullivan, A. A., & Thummel, C. S. (2003). Temporal profiles of nuclear receptor gene expression reveal coordinate transcriptional responses during *Drosophila* development. *Mol. Endocrinol., 17*, 2125–2137.

Sun, G. C., Hirose, S., & Ueda, H. (1994). Intermittent expression of BmFTZ-F1, a member of the nuclear hormone receptor superfamily during development of the silkworm *Bombyx mori. Dev. Biol., 162*, 426–437.

Sung, C., Wong, L. E., Chang Sen, L. Q., Nguyen, E., Lazaga, N., Ganzer, G., McNabb, S. L., & Robinow, S. (2009). The unfulfilled/DHR51 gene of *Drosophila melanogaster* modulates wing expansion and fertility. *Dev. Dyn., 238*, 171–182.

Sutherland, J. D., Kozlova, T., Tzertzinis, G., & Kafatos, F. C. (1995). *Drosophila* Hormone-Receptor-38-a 2nd Partner for *Drosophila* Usp Suggests an Unexpected Role for Nuclear Receptors of the Nerve Growth Factor-Induced Protein B-Type. *Proceedings of the National Academy of Sciences of the United States of America, 92*, 7966–7970.

Suzuki, T., Kawasaki, H., Yu, R. T., Ueda, H., & Umesono, K. (2001). Segmentation gene product Fushi tarazu is an LXXLL motif-dependent coactivator for orphan receptor FTZ-F1. *Proc. Natl. Acad. Sci. USA, 98*, 12403–12408.

Swevers, L., Eystathioy, T., & Iatrou, K. (2002a). The orphan nuclear receptors BmE75A and BmE75C of the silkmoth *Bombyx mori*: hornmonal control and ovarian expression. *Insect Biochem. Mol. Biol., 32*, 1643–1652.

Swevers, L., & Iatrou, K. (1998). The orphan receptor BmHNF-4 of the silkmoth *Bombyx mori*: ovarian and zygotic expression of two mRNA isoforms encoding polypeptides with different activating domains. *Mech. Dev., 72*, 3–13.

Swevers, L., Ito, K., & Iatrou, K. (2002b). The BmE75 nuclear receptors function as dominant repressors of the nuclear receptor BmHR3A. *J. Biol. Chem., 277*, 41637–41644.

Talbot, W. S., Swyryd, E. A., & Hogness, D. S. (1993). *Drosophila* tissues with different metamorphic responses to ecdysone express different ecdysone receptor isoforms. *Cell, 73*, 1323–1337.

Tan, A., & Palli, S. R. (2008a). Identification and characterization of nuclear receptors from the red flour beetle, *Tr. castaneum. Insect Biochem. Mol. Biol., 38*, 430–439.

Tan, A., & Palli, S. R. (2008b). Edysone receptor isoforms play distinct roles in controlling molting and metamorphosis in the red flour beetle, *Tr. castaneum. Molecular and Cellular Endocrinology, 291*, 42–49.

Telford, M. J., Bourlat, S. J., Economou, A., Papillon, D., & Rota-Stabelli, O. (2008). The evolution of the Ecdysozoa. *Philos. Trans. R Soc. Lond. B Biol. Sci., 363*, 1529–1537.

Terashima, J., & Bownes, M. (2006). E75A and E75B have opposite effects on the apoptosis/development choice of the *Drosophila* egg chamber. *Cell Death Differ., 13*, 454–464.

Thakur, J. K., Arthanari, H., Yang, F., Pan, S. J., Fan, X., Breger, J., Frueh, D. P., Gulshan, K., Li, D. K., Mylonakis, E., Struhl, K., Moye-Rowley, W. S., Cormack, B. P., Wagner, G., & Naar, A. M. (2008). A nuclear receptor-like pathway regulating multidrug resistance in fungi. *Nature, 452*, 604–609.

Thomson, S. A., Baldwin, W. S., Wang, Y. H., Kwon, G., & Leblanc, G. A. (2009). Annotation, phylogenetics, and expression of the nuclear receptors in *Daphnia pulex. BMC Genomics, 10*, 500.

Thormeyer, D., Tenbaum, S. P., Renkawitz, R., & Baniahmad, A. (1999). EcR interacts with corepressors and harbours an autonomous silencing domain functional in both *Drosophila* and vertebrate cells. *J. Steroid Biochem. Mol. Biol., 68*, 163–169.

Thummel, C. S. (2001). Molecular mechanisms of developmental timing in *C. elegans* and *Drosophila. Dev. Cell, 1*, 453–465.

Togawa, T., Shofuda, K., Yaginuma, T., Tomino, S., Nakato, H., & Izumi, S. (2001). Structural analysis of gene encoding cuticle protein BMCP18, and characterization of its putative transcription factor in the silkworm, *Bombyx mori. Insect Biochem. Mol. Biol., 31*, 611–620.

Truman, J. W., & Riddiford, L. M. (1999). The origins of insect metamorphosis. *Nature, 401*, 447–452.

Truman, J. W., & Riddiford, L. M. (2002). Endocrine insights into the evolution of metamorphosis in insects. *Annu. Rev. Entomol., 47*, 467–500.

Tzertzinis, G., Egana, A. L., Palli, S. R., Robinson-Rechavi, M., Gissendanner, C. R., Liu, C., Unnasch, T. R., & Maina, C. V. (2010). Molecular evidence for a functional ecdysone signaling system in *Brugia malayi. PLoS Negl. Trop. Dis., 4*, e625.

Ueda, H., & Hirose, S. (1990). Identification and purification of a *Bombyx mori* homologue of FTZ-F1. *Nucleic Acids Res.*, *18*, 7229–7234.

Unnasch, T. R., Bradley, J., Beauchamp, J., Tuan, R., & Kennedy, M. W. (1999). Characterization of a putative nuclear receptor from *Onchocerca volvulus*. *Mol. Biochem. Parasitol.*, *104*, 259–269.

Van Gilst, M. R., Hadjivassiliou, H., Jolly, A., & Yamamoto, K. R. (2005a). Nuclear hormone receptor NHR-49 controls fat consumption and fatty acid composition in *C. elegans*. *PLoS Biol.*, *3*, e53.

Van Gilst, M. R., Hadjivassiliou, H., & Yamamoto, K. R. (2005b). A *Caenorhabditis elegans* nutrient response system partially dependent on nuclear receptor NHR-49. *Proc. Natl. Acad. Sci. U S A*, *102*, 13496–13501.

Velarde, R. A., Robinson, G. E., & Fahrbach, S. E. (2006). Nuclear receptors of the honey bee: annotation and expression in the adult brain. *Insect Mol. Biol.*, *15*, 583–595.

Velarde, R. A., Robinson, G. E., & Fahrbach, S. E. (2009). Coordinated responses to developmental hormones in the Kenyon cells of the adult worker honey bee brain (*Apis mellifera* L.). *J. Insect. Physiol.*, *55*, 59–69.

Wang, S. F., Li, C., Sun, G., Zhu, J., & Raikhel, A. S. (2002). Differential expression and regulation by 20-hydroxyecdysone of mosquito ecdysteroid receptor isoforms A and B. *Mol. Cell Endocrinol.*, *196*, 29–42.

Wang, S. F., Li, C., Zhu, J. S., Miura, K., Miksicek, R. J., & Raikhel, A. S. (2000). Differential expression and regulation by 20-hydroxyecdysone of mosquito ultraspiracle isoforms. *Developmental Biology*, *218*, 99–113.

Wang, Z., Benoit, G., Liu, J., Prasad, S., Aarnisalo, P., Liu, X., Xu, H., Walker, N. P., & Perlmann, T. (2003). Structure and function of Nurr1 identifies a class of ligand-independent nuclear receptors. *Nature*, *423*, 555–560.

Wang, Z., Zhou, X. E., Motola, D. L., Gao, X., Suino-Powell, K., Conneely, A., Ogata, C., Sharma, K. K., Auchus, R. J., Lok, J. B., Hawdon, J. M., Kliewer, S. A., Xu, H. E., & Mangelsdorf, D. J. (2009). Identification of the nuclear receptor DAF-12 as a therapeutic target in parasitic nematodes. *Proc. Natl. Acad. Sci. U S A*, *106*, 9138–9143.

Watanabe, T., Takeuchi, H., & Kubo, T. (2010). Structural diversity and evolution of the N-terminal isoform-specific region of ecdysone receptor-A and-B1 isoforms in insects. *Bmc Evolutionary Biology*, *10*.

Weller, J., Sun, G. C., Zhou, B., Lan, Q., Hiruma, K., & Riddiford, L. M. (2001). Isolation and developmental expression of two nuclear receptors, MHR4 and betaFTZ-F1, in the tobacco hornworm, *Manduca sexta*. *Insect Biochem. Mol. Biol.*, *31*, 827–837.

White, K. P., Hurban, P., Watanabe, T., & Hogness, D. S. (1997). Coordination of *Drosophila* metamorphosis by two ecdysone-induced nuclear receptors. *Science*, *276*, 114–117.

Whiting, M. F., Carpenter, J. C., Wheeler, Q. D., & Wheeler, W. C. (1997). The streptisera problem: phylogeny of the holometabolous insect orders inferred from 18S and 28S ribosomal DNA sequences and morphology. *Syst. Biol.*, *46*, 1–68.

Wiegmann, B. M., Trautwein, M. D., Kim, J. W., Cassel, B. K., Bertone, M. A., Winterton, S. L., & Yeates, D. K. (2009). Single-copy nuclear genes resolve the phylogeny of the holometabolous insects. *BMC Biol.*, *7*, 34.

Wilson, M. J., & Dearden, P. K. (2009). Tailless patterning functions are conserved in the honeybee even in the absence of Torso signaling. *Dev. Biol.*, *335*, 276–287.

Wilson, T. G., Yerushalmi, Y., Donnell, D. M., & Restifo, L. (2006). Interaction between hormonal signaling pathways in *Drosophila melanogaster* as revealed by genetic interaction between Methoprene-tolerant and Broad-Complex. *Genetics*, *172*, 253–264.

Wisely, G. B., Miller, A. B., Davis, R. G., Thornquest, A. D., Jr., Johnson, R., Spitzer, T., Sefler, A., Shearer, B., Moore, J. T., Willson, T. M., & Williams, S. P. (2002). Hepatocyte nuclear factor 4 is a transcription factor that constitutively binds fatty acids. *Structure*, *10*, 1225–1234.

Wittkopp, P. J., Williams, B. L., Selegue, J. E., & Carroll, S. B. (2003). *Drosophila* pigmentation evolution: divergent genotypes underlying convergent phenotypes. *Proc. Natl. Acad. Sci. U S A*, *100*, 1808–1813.

Woodard, C. T., Baehrecke, E. H., & Thummel, C. S. (1994). A molecular mechanism for the stage specificity of the *Drosophila* prepupal genetic response to ecdysone. *Cell*, *79*, 607–615.

Wu, W., Niles, E. G., El-Sayed, N., Berriman, M., & LoVerde, P. T. (2006). *Schistosoma mansoni* (Platyhelminthes, Trematoda) nuclear receptors: sixteen new members and a novel subfamily. *Gene.*, *366*, 303–315.

Wu, W., Niles, E. G., Hirai, H., & LoVerde, P. T. (2007). Identification and characterization of a nuclear receptor subfamily I member in the Platyhelminth *Schistosoma mansoni* (SmNR1). *FEBS J.*, *274*, 390–405.

Wu, W., Tak, E. Y., & LoVerde, P. T. (2008). Schistosoma mansoni: SmE78, a nuclear receptor orthologue of *Drosophila* ecdysone-induced protein 78. *Exp. Parasitol.*, *119*, 313–318.

Xu, J., Tan, A., & Palli, S. R. (2010). The function of nuclear receptors in regulation of female reproduction and embryogenesis in the red flour beetle, *Tr. castaneum*. *J. Insect Physiol.*, *56*, 1471–1480.

Yamada, M., Murata, T., Hirose, S., Lavorgna, G., Suzuki, E., & Ueda, H. (2000). Temporally restricted expression of transcription factor betaFTZ-F1: significance for embryogenesis, molting and metamorphosis in *Drosophila melanogaster*. *Development*, *127*, 5083–5092.

Yamazaki, Y., Shirai, K., Paul, R. K., Fujiyuki, T., Wakamoto, A., Takeuchi, H., & Kubo, T. (2006). Differential expression of HR38 in the mushroom bodies of the honeybee brain depends on the caste and division of labor. *FEBS Lett.*, *580*, 2667–2670.

Yao, T. P., Forman, B. M., Jiang, Z., Cherbas, L., Chen, J. D., McKeown, M., Cherbas, P., & Evans, R. M. (1993). Functional ecdysone receptor is the product of EcR and Ultraspiracle genes. *Nature*, *366*, 476–479.

Yin, L., Wu, N., Curtin, J. C., Qatanani, M., Szwergold, N. R., Reid, R. A., Waitt, G. M., Parks, D. J., Pearce, K. H., Wisely, G. B., & Lazar, M. A. (2007). Rev-erbalpha, a heme sensor that coordinates metabolic and circadian pathways. *Science*, *318*, 1786–1789.

Yu, R. I., Chiang, M. Y., Tanabe, I., Kobayashi, M., Yasuda, K., Evans, R. M., & Umesono, K. (2000). The orphan nuclear receptor Tlx regulates Pax2 and is essential for vision. *Proc. Natl. Acad. Sci. USA*, *97*, 2621–2625.

Yu, Y., Li, W., Su, K., Yussa, M., Han, W., Perrimon, N., & Pick, L. (1997). The nuclear hormone receptor Ftz-F1 is a cofactor for the *Drosophila* homeodomain protein Ftz. *Nature, 385*, 552–555.

Yussa, M., Lohr, U., Su, K., & Pick, L. (2001). The nuclear receptor Ftz-F1 and homeodomain protein Ftz interact through evolutionarily conserved protein domains. *Mech. Dev., 107*, 39–53.

Zarkower, D. (2001). Establishing sexual dimorphism: conservation amidst diversity? *Nat. Rev. Genet., 2*, 175–185.

Zdobnov, E. M., & Bork, P. (2007). Quantification of insect genome divergence. *Trends Genet., 23*, 16–20.

Zelhof, A. C., Yao, T. P., Chen, J. D., Evans, R. M., & McKeown, M. (1995). Seven-up inhibits ultraspiracle-based signaling pathways in vitro and in vivo. *Mol. Cell Biol., 15*, 6736–6745.

Zelhof, A. C., Yao, T. P., Evans, R. M., & McKeown, M. (1995). Identification and characterization of a *Drosophila* nuclear receptor with the ability to inhibit the ecdysone response. *Proc. Natl. Acad. Sci. U S A, 92*, 10477–10481.

Zhao, X. F., Wang, J. X., Xu, X. L., Li, Z. M., & Kang, C. J. (2004). Molecular cloning and expression patterns of the molt-regulating transcription factor HHR3 from *Helicoverpa armigera. Insect Mol. Biol., 13*, 407–412.

Zhong, W., Sladek, F. M., & Darnell, J. E., Jr. (1993). The expression pattern of a *Drosophila* homologue to the mouse transcription factor HNF-4 suggests a determinative role in gut formation. *EMBO J., 12*, 537–544.

Zhou, B., Hiruma, K., Jindra, M., Shinoda, T., Segraves, W. A., Malone, F., & Riddiford, L. M. (1998). Regulation of the transcription factor E75 by 20-hydroxyecdysone and juvenile hormone in the epidermis of the tobacco hornworm, *Manduca sexta*, during larval molting and metamorphosis. *Dev. Biol., 193*, 127–138.

Zhou, H. M., & Walthall, W. W. (1998). UNC-55, an orphan nuclear hormone receptor, orchestrates synaptic specificity among two classes of motor neurons in *Caenorhabditis elegans. J. Neurosci., 18*, 10438–10444.

Zhu, J., Chen, L., & Raikhel, A. S. (2003). Posttranscriptional control of the competence factor betaFTZ-F1 by juvenile hormone in the mosquito *Aedes aegypti. Proc. Natl. Acad. Sci. U S A, 100*, 13338–13343.

Zhu, J., Chen, L., Sun, G., & Raikhel, A. S. (2006). The competence factor beta Ftz-F1 potentiates ecdysone receptor activity via recruiting a p160/SRC coactivator. *Mol. Cell Biol., 26*, 9402–9412.

Zhu, J., Miura, K., Chen, L., & Raikhel, A. S. (2000). AHR38, a homologue of NGFI-B, inhibits formation of the functional ecdysteroid receptor in the mosquito *Aedes aegypti. EMBO J., 19*, 253–262.

Zhu, J. S., Miura, K., Chen, L., & Raikhel, A. S. (2003). Cyclicity of mosquito vitellogenic ecdysteroid-mediated signaling is modulated by alternative dimerization of the RXR homologue Ultraspiracle. *Proceedings of the National Academy of Sciences of the United States of America, 100*, 544–549.

7 Neuroendocrine Regulation of Ecdysis

D. Zitnan
Slovak Academy of Sciences, Bratislava, Slovakia
M.E. Adams
University of California, Riverside, CA, USA

© 2012 Elsevier B.V. All Rights Reserved

7.1. Introduction

Insects are the dominant terrestrial life forms on earth. Their impressive evolutionary success is in no small measure a consequence of unique developmental and reproductive strategies that have facilitated efficient resource exploitation and radiation into a wide range of ecological niches (Truman and Riddiford, 2002). Developmental strategies are most remarkable in the holometabolous insects where, as vermiform larvae, they are essentially mobile digestive systems, optimized for a wide range of food resources. The complete morphological transformation from larva to winged reproductive adult, or metamorphosis, in many ways represents the birth of a new organism within the same life cycle, and enables the same species to minimize intraspecific competition by occupying separate ecological niches. The remarkable ability of insects to change form during the life cycle, to enter dormancy at various stages of development, and to adapt to environmental changes via polyphenism, is made possible by the processes of molting and ecdysis.

Each stage of insect development is characterized by an alternation between feeding and molting, the latter culminating in ecdysis. During this intricate process, insects shed the old cuticle surrounding the external surface, and the lining of the foregut, hindgut, and tracheal tubes of the respiratory system. While ecdysis is the gateway to the next developmental stage, it is also effectively a life-threatening process, since failure to execute the sequence on schedule can result in death of the organism. Its success depends on coordinated gene expression controlled by ecdysteroids and juvenile hormones (JH) responsible for the formation of new cuticle and other tissues appropriate for the next stage. Ecdysteroids also participate in activation of a peptide signaling cascade that regulates the ecdysis sequence at the end of the molt.

DOI: 10.1016/B978-0-12-384749-2.10007-X

7.1.1. The Molt Cycle

Upon hatching from the egg, growth through successive immature stages culminates in the winged, reproductive adult, exquisitely adapted for dispersal and proliferation of new young. At the beginning of each larval instar, the animal feeds and grows until it acquires the proper size and nutritional state. The feeding stage or "intermolt" proceeds until integrative processes in the brain produce a decision to move on to the next stage, whereupon production and release of ecdysteroids occurs. The appearance of elevated ecdysteroid levels brings an end to the intermolt and onset of the molt associated with production of new cuticle (**Figure 1**). At the end of last larval instar, holometabolous insects empty the gut and engage in wandering behavior aimed at finding an optimal location for pupation. Metamorphosis is initiated in the pharate pupa (pupal development) and continues after pupation (adult development). This complex process involves remodeling or programmed death of larval organs and proliferation of new cells and tissues specific for pupal and adult stages. In spite of these dramatic changes, it seems likely that hormonal mechanisms controlling larval, pupal, and adult ecdyses may be similar or only slightly modified.

The critical endocrine signal that switches the animal from feeding to molting is elevation of ecdysteroids, including ecdysone, 20-hydroxyecdysone (20E), and other hydroxylated analogs of ecdysone, depending on the insect (see Chapters 4–6). Epidermal cells respond by changing shape and expressing a set of genes that prepare the animal for passage to the next stage. An early step of this process is detachment of the epidermal cell layer from the existing cuticle, or apolysis, whereupon synthesis of a new cuticle layer appropriate for the next stage begins. Secretion of enzymatic molting fluid by the epidermal cells dissolves unsclerotized old cuticle so that its chitin and protein components are recovered and re-cycled into new cuticle or other tissues. Similarly, the old cuticular lining of tracheal tubes detaches from underlying epidermal cells and the gap between these layers is filled with molting fluid. This fluid is later reabsorbed from the old head capsule and space between the old and new cuticles, while the new tracheal tubes become inflated with air.

Apolysis begins the "pharate" stage, defined as the incipient new animal covered by its old "coat." Thus, a pharate fifth instar larva has already produced a new cuticle, but is still wearing its old coat from the fourth instar. A pharate pupa is a developing pupa still encased in its old larval cuticle. A pharate adult has developed all adult organs, but is still covered by its old pupal cuticle.

As described in this chapter, the rise and fall of ecdysteroid levels induce sequential bouts of gene expression, resulting in the production of peptides and their receptors, peptide processing enzymes, and signal transduction proteins. These events ensure the activation and proper timing of processes that underlie the upcoming ecdysis sequence, which involves air swallowing, tracheal dynamics, and a series of behaviors that culminates in cuticle shedding. The order of these events and their precise timing are determined by a peptide and neurotransmitter signaling cascade.

The molt cycle can be considered according to the following sequential steps:

1. Feeding and growth, which begins shortly after ecdysis of the previous stage.
2. Initiation of molt induced by increased ecdysteroid levels in concert with juvenile hormones to regulate gene expression important for development of the next stage phenotype, including the formation of new cuticle and tissues and synthesis of regulatory molecules.
3. Termination of the molt controlled by a decline in steroid levels and secretion of peptide hormones that control the ecdysis sequence.

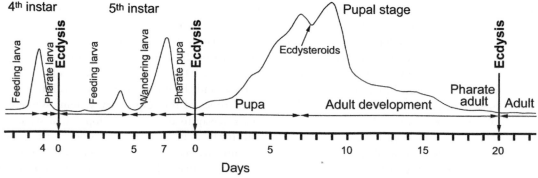

Figure 1 Ecdysteroid levels in hemolymph during development of *M. sexta*. Ecdysteroid levels are low during larval feeding stages, but surge to initiate the molt. This occurs ~48 h before ecdysis during the fourth instar, whereupon larvae stop feeding and develop a new cuticle to become pharate fifth instar larvae. After ecdysteroid decline, pharate larvae ecdyse to fifth instar (day 0) and resume feeding. A small ecdysteroid peak on day 4 or 5 induces wandering (larvae stop feeding and seek a suitable pupation site). A large ecdysteroid peak induces pupal development on days 7–9 (pharate pupa), and after ecdysteroids decline animals perform the pupal ecdysis sequence. Subsequent elevated ecdysteroid levels control adult development. (Normalized ecdysteroid levels are adapted from *Zitnan et al.*, 1999; *Zitnanova et al.*, 2001; *Bollenbacher et al.*, 1981 in the pupal stage.)

4. Post-ecdysial processes characterized by expansion, sclerotization, and melanization of the new cuticle important for a definitive shape of the new developmental stage.

7.1.2. Neuroendocrine Basis of Ecdysis

The role of endocrine signals in control of ecdysis was demonstrated 40 years ago through a series of classic brain transplantation experiments in giant silkmoths (Truman and Riddiford, 1970). These animals display "gated" circadian eclosion rhythms, meaning that adults eclose only during a certain time of day when the gate is "open." Once the gate is closed, eclosion (adult ecdysis) is delayed until the next day. Eclosion gating is abolished by surgical removal of the brain, but is re-established by implantation of a loose brain into the body cavity of the animal. Even more astounding, swapping brains between two species of moths that eclose at different times of day causes recipient moths to follow the eclosion schedule of the donor brain. This rather elegantly demonstrated two things: first, the existence of a timing device for eclosion control in the brain and second, a role for the brain as an endocrine organ by secreting a hormonal factor(s) that initiates eclosion. These experiments were reminiscent of the historic Kopec experiments, where transplantation of brains in moths was used to demonstrate an endocrine basis for the molt (Kopec, 1917, 1922; see Chapter 1). Eventually, a 62 amino acid peptide was identified as eclosion hormone (EH), which induces eclosion upon injection into pharate adults (Truman, 1992).

7.1.3. Brain Neurosecretory Cells, Corazonin, and Eclosion Hormone

Early accounts of EH neurons focused on brain neurosecretory cells identified as Ia neurons in *Manduca*, which also express the circadian clock protein PER (Copenhaver and Truman, 1986a; Homberg *et al.*, 1991; Wise *et al.*, 2002; Zitnan *et al.*, 1995). These cells are involved in regulation of circadian rhythms in the silkmoth *Antheraea pernyi* (Sauman and Reppert, 1996) and could be involved in the circadian gating of adult eclosion. "EH activity" was detected in bioassays of nine ipsilateral brain neurosecretory cells (Type Ia; **Figure 2A**), which project to the corpus cardiacum-corpus allatum (CC-CA). Extracts of these cells induced premature ecdysis when injected into *Manduca* pharate larvae. Moreover, electrical stimulation of CC nerves (NCC 1,2) led to release of EH activity from the CC (Copenhaver and Truman, 1986a,b). Since a subset of Ia cells (five pairs of Ia$_2$ cells; **Figure 2B**) reacted with an antiserum to EH, it was concluded that the EH activity originated from these neurons. However, subsequent studies revealed that EH gene expression is absent in Ia$_1$ neurons and is confined to brain ventromedial neurons (VM; type V; **Figure 2C**),

which project to the CC-CA complex only in pharate adults, not in larvae or pupae (Horodyski *et al.*, 1989; Kamito *et al.*, 1992; Riddiford *et al.*, 1994). It is now clear that Type Ia$_1$ produces corazonin and Ia$_2$ neurons produce ion transport peptide (ITP; **Figure 2B**). Interestingly, ITP shows some sequence homology with EH in one part of its primary structure (see **Figure 19**), so it is therefore possible that the EH antiserum used by Copenhaver and Truman may have cross-reacted with sequence regions common to EH and ITP. Because extracts of ipsilateral cells contained both cell types (Ia$_1$ and Ia$_2$), the observed EH activity was very likely induced by corazonin. Evidence for corazonin as an initiator of the ecdysis behavioral sequence is reviewed in Section 7.4.1.

Our understanding of the complexity of ecdysis signaling has grown considerably in recent years, but it remains clear that neuroendocrine signals are critical for the initiating and scheduling of the ecdysis sequence. Current evidence shows that corazonin and EH act sequentially to regulate release of ecdysis triggering hormones (ETHs) from endocrine Inka cells of the epitracheal endocrine system. These peptides, called PETH and ETH in moths and ETH1 and ETH2 in other insects, act directly on the central nervous system (CNS) to initiate and schedule pre-ecdysis, ecdysis, and post-ecdysis behaviors (Ewer *et al.*, 1997; Kingan *et al.*, 1997; Zitnan *et al.*, 1996, 1999). The scheduling of behavioral steps occurs through release of numerous neuropeptides within the CNS to coordinate the ecdysis sequence and associated processes (Gammie and Truman, 1999; Kim *et al.*, 2006a,b; Zitnan and Adams, 2000; Zitnan *et al.*, 2007).

Many peptides involved in ecdysis signaling were discovered years ago by using facile bioassays such as heartbeat or gut contraction. The names corazonin, crustacean cardioactive peptide, kinin, myoinhibitory peptides, and myosuppressins reflect the biological activities of peptides discovered in this manner. However, it is now clear that these names do not necessarily represent their authentic physiological functions. Many of these peptides are central neuromodulators of circuits underlying behavioral control (e.g., pre-ecdysis, ecdysis, post-ecdysis) and physiological functions associated with ecdysis, such as air swallowing, tracheal inflation, gut contractions, cuticle sclerotization, and pigmentation. In addition, many peptides are multifunctional, carrying out some signaling functions both as circulatory hormones and as neurotransmitters and/or neuromodulators through release at or near synapses within the CNS or at neuromuscular junctions.

Publication of the *Drosophila* genome (Adams *et al.*, 2000) and subsequently those of many other insects has opened a new era of opportunity for elucidation of the peptide cascade controlling ecdysis. These genome databases will enable much more rapid identification of both ligand and receptor orthologues and their mechanisms of action in a variety of species across wide evolutionary distances.

7.1.4. Chapter Overview

In this chapter, we review current knowledge of the neuroendocrine basis for ecdysis, focusing first on cellular and chemical signaling mechanisms that control gene expression prior to its onset, followed by events leading to initiation and proper scheduling of its steps. With regard to ecdysteroid-regulated gene expression, we concentrate on gene expression events that create the peptidergic signaling framework underlying activation and execution of the ecdysis sequence. Regarding the ecdysis sequence, emphasis is placed on peptide hormones and neuromodulators, which coordinate steps leading to shedding of the cuticle, including swallowing behaviors, tracheal dynamics, and pre-ecdysis, ecdysis, and post-ecdysis behaviors. Ultimately, a truly mechanistic analysis of ecdysis will depend on understanding its molecular basis. Because emphasis here is placed on the neuroendocrine basis of ecdysis, many details of ecdysis in different stages and orders of insects are not covered in depth. Most of the discussion is devoted to the moths *Manduca* and *Bombyx*, and the fruit fly *Drosophila*. Where possible, immunohistochemical demonstration of homologous cellular substrates for ecdysis in different insect phyla is presented, since such information may help to guide future research in these groups. For aspects of ecdysis not covered in this chapter, the reader is referred to recent reviews (Ewer and Reynolds, 2002; Roller *et al.*, 2010; Truman, 2005; Zitnan *et al.*, 2007). Section 7.2. characterizes peptides of the epitracheal endocrine system implicated in ecdysis control and their receptors in the CNS. Section 7.3. summarizes the roles of rising and falling ecdysteroids in programming the neural and endocrine signaling substrates that underlie ecdysis control and how early and late gene expression ensures proper timing of ecdysis initiation. Section 7.4. describes two phases of peptide release from Inka cells controlled by the brain peptides corazonin and eclosion hormone. Section 7.5. details what is known regarding the peptide signaling cascade within the CNS that initiates sequentially central pattern generators driving pre-ecdysis, ecdysis, and post-ecdysis steps in the behavioral sequence. Finally, Section 7.6. presents mechanistic models for endocrine control of the ecdysis sequence and summarizing statements.

7.2. Epitracheal Glands, Inka Cells, and Ecdysis Triggering Hormones

7.2.1. Epitracheal Glands and Inka Cells

The first observations of epitracheal glands were published in the early part of the last century using the silkworm, *Bombyx mori* (Ikeda, 1913). These glands, which appear to have both endocrine and exocrine functions, remained unnoticed for almost 80 years until they were rediscovered on histological sections of pharate larvae and pupae of the waxmoth *Galleria mellonella* during a search for peptidergic cells using an antiserum to FMRFamide (Zitnan, 1989). Nine pairs of segmentally distributed glands were found to be closely associated with the tracheal system and spiracles. One pair of glands is attached to tracheae near each prothoracic spiracle and eight pairs of glands are located on tracheae close to each abdominal spiracle (**Figure 3A**). Only the largest cell of each epitracheal gland shows strong FMRFamide-like immunoreactivity (IR), while two to three other cells are not stained (**Figure 3B,C**; Zitnan, 1989). Ultrastructural studies of epitracheal glands revealed numerous electron dense droplets in the largest gland cell of pharate pupal *Bombyx*, which degenerate in freshly ecdysed pupae, but their function was not determined (Akai, 1992). Further immunohistochemical studies with antisera to horseradish peroxidase (HRP), and small cardioactive peptide B (SCP$_B$) showed that each epitracheal gland of *Manduca sexta* is composed of one prominent peptidergic cell and three smaller cells. The smaller cells are present in all larval stages, but degenerate after pupation, leaving only nine pairs of large peptidergic cells in developing adults. These cells show strong SCP$_B$-IR in pharate larvae, pupae, and adults, but peptide staining disappears after each ecdysis, indicating that they release their contents into the hemolymph (Zitnan *et al.*, 1996). These remarkable peripheral endocrine cells were named Inka cells in honor of a beautiful fairy from the Tatra mountains. Subsequent investigations demonstrated that Inka cells produce pre-ecdysis triggering hormone (PETH) and ETH, which are released into the hemolymph to activate pre-ecdysis and ecdysis motor programs in the CNS (**Figure 3D–G**).

Ultrastructural studies of epitracheal glands in *Lymantria dispar* showed that each gland contains four cells: type I (Inka cell homologue) and type II endocrine cells, exocrine cells, and canal cells (Klein *et al.*, 1999). Inka cells release their hormonal content at ecdysis, but secretion of type II endocrine cells was not observed and their endocrine function requires confirmation by other techniques. As described in *Lymantria*, *Manduca* epitracheal glands contain an exocrine cell, which projects a duct through the canal cell into the lumen between new and old tracheae (Zitnanova *et al.*, 2001). The apparent size decrease of the exocrine cell in freshly ecdysed larvae and pupae indicates that its content is released into the tracheal lumen, possibly to aid shedding of the old tracheae during ecdysis.

The widespread occurrence of Inka cells in insects has been demonstrated by immunohistochemistry using an antiserum against PETH. Using this approach, Inka cells were observed to occur in ~40 representatives of diverse insect orders (Roller *et al.*, 2010; Zitnan *et al.*, 2003). Surprisingly, two general patterns of Inka cells have been described. In most insect orders, hundreds or thousands of Inka cells of various sizes and shapes are scattered throughout the tracheal surface (**Figure 4A–F, I**). This contrasts sharply with the pattern observed in certain holometabolous insects (*Manduca, Bombyx, Aedes, Drosophila*), where a well-defined system of 8–9 pairs of segmental epitracheal glands occurs near the spiracles, each with a single, prominent Inka cell associated with 3–4 other cells (**Figure 4 G, H, J, K**; (Zitnan *et al.*, 2002a, 2003). In the dragonfly, *Sympetrum sp.*, numerous small, round Inka cells were stained (**Figure 4A**), and similar cells were found in the apterygote silverfish *Lepisma saccharina* and primitive aquatic insects such as mayflies and damselflies. Two distinct cell types were found at different tracheal locations of the cockroach *Nauphoeta cinerea*; oval Inka cells occur on the major segmental tracheae (**Figure 4B**), while cells with cytoplasmic processes were observed on thin tracheae of the gonads and gut. Simple oval Inka cells also are found in the cockroach *Periplaneta americana* and the locust *Locusta migratoria*, whereas cells with prominent cytoplasmic processes are detected in the cockroaches *Blabera craniifera* and *Phylodromica sp.*, the cricket *Acheta domestica* (**Figure 4C**), the stonefly *Perla sp.*, and the bugs *Pyrrhocoris apterus* (**Figure 4D**) and *Triatoma infestans*.

Inka cells of holometabolous insects also can be quite variable. Numerous, single Inka cells dispersed throughout the tracheal system occur in *Sialis* and the antlion *Myrmeleon*. Variability of Inka cells in representatives of several beetle species reflects the enormous diversity of this large insect group. For example, the water beetle *Laccophilus sp.* contains coupled elongated Inka cells with short cytoplasmic processes (**Figure 4E**), but the mealworm beetle *Te. molitor* has two distinct types of Inka cells: groups of 2–6 larger cells are attached to tracheae near each spiracle, while small cells are scattered throughout the tracheal surface (**Figure 4F**). Numerous small Inka cells of various shapes are found in other beetles, but only 9 pairs of large Inka cells associated with epitracheal glands are present in the Colorado potato beetle, *Leptinotarsa decemlineata* (for details see Zitnan *et al.*, 2003). Similar epitracheal glands occur in most investigated hymenopterans and all lepidopterans and dipterans. The only exception is the honeybee *Apis mellifera*, which contains a large number of small Inka cells (**Figure 4I**). Variability of epitracheal glands containing prominent Inka cells is depicted in the moths *Manduca*, *Bombyx*, mosquito *Aedes*, and fruit fly *Drosophila* (**Figure 4 G, H, J, K**).

Strong PETH-IR detected in representatives of both hemi- and holometabolous insects before ecdysis disappears after each larval, pupal, and adult ecdysis. This suggests that Inka cells release their entire hormonal content into the hemolymph to initiate and regulate shedding of the old cuticle. Inka cells of hemimetabolous insects appear

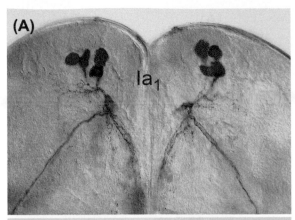

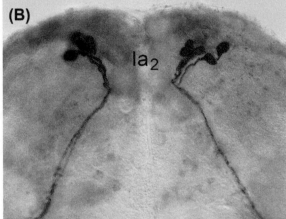

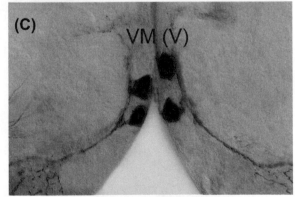

Figure 2 Brain neurons in moths implicated in ecdysis regulation. (Top) Neurosecretory cells la$_1$ produce corazonin, implicated in the early phase of ETH release. (Middle) la$_2$ cells produce ITP. (Bottom) Ventromedial cells (VM) type V release EH, implicated in the second phase of ETH release.

to degenerate after adult ecdysis as indicated by permanent disappearance of PETH-IR on tracheal surfaces, whereas Inka cells in some representative holometabolous insects (beetles, moths, and dipterans) persist in adults and continue to produce ETH. Physiological roles for these peptides in adults have not been assigned.

Similar PETH-IR in peripheral paired cells has been detected in pharate nymphs of the ticks *Rhipicephalus appendiculatus* and *Ixodes ricinus* (**Figure 4L**). These putative endocrine cells located on lateral sides of each pedal

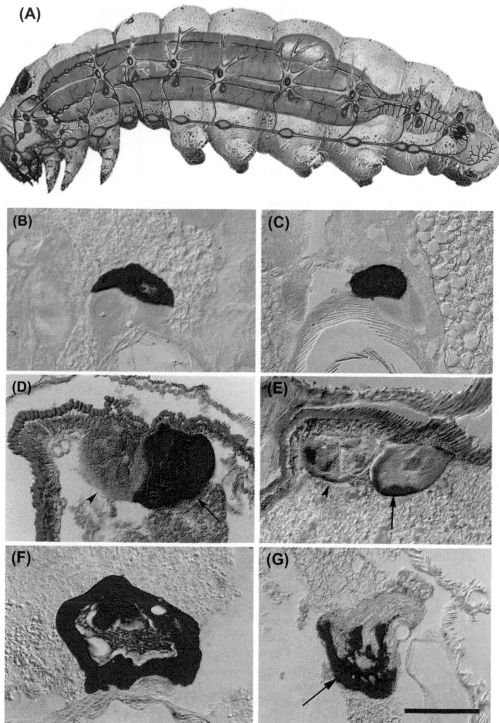

Figure 3 Distribution and peptide immunoreactivity (IR) of Inka cells in the moths *B. mori*, *G. mellonella*, and *M. sexta*. (A) Schematic drawings of Inka cells and other important nervous and endocrine organs of *Bombyx* pharate pupa (D. Zitnan and A. Macková). Bilaterally paired prothoracic and abdominal Inka cells (red) are attached to lateral tracheae near each spiracle. Different colors depict the following organs: CNS ganglia (light brown), CC and enteric nervous system (blue) composed of the frontal ganglion and enteric plexus innervating the fore- and midgut (yellow), epiproctodeal ganglia (dark blue) innervating the hindgut (ochre), peripheral link neurons (violet) innervating the prothoracic glands or heart, ecdysteroid-producing prothoracic glands (dark green) and gonads (light blue), and JH-producing CA (light green). (B, C) The peptidergic nature of Inka cells was first identified with an antiserum to FMRFamide in pharate pupae of the waxmoth *Galleria*. (D, E) Strong ETH-IR in Inka cells (D, arrow) of a *Manduca* pharate pupa disappears at onset of ecdysis and only traces of staining are present in the cytoplasm at its conclusion (E, arrow). Three associated cells with likely exocrine functions also decrease in size (arrowheads). (F, G) Accumulated ETH-IR in a *Bombyx* pharate adult (F) is greatly reduced after eclosion (G, arrow). Scale bar, 200 μm.

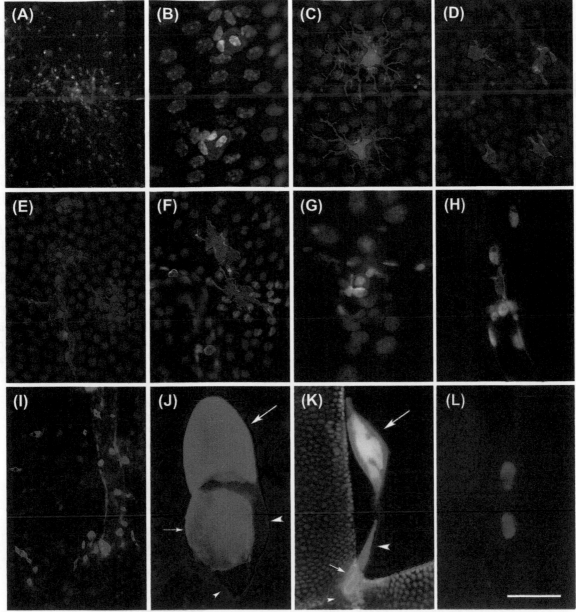

Figure 4 Variability of Inka cells in different insect species and similar cells in the tick. Immunohistochemical staining with PETH antiserum shows that all hemimetabolous and most holometabolous insects contain numerous small Inka cells (stained orange-red or yellow), which are scattered throughout the entire surface of the tracheal system (nuclei are stained blue with DAPI). (A) PETH-IR in thousands of small Inka cells in the dragonfly *Sympetrum sp.* (B) Inka cells of the cockroach *N. cinerea* are simple and oval, whereas the cricket *A. domestica* (C) contains cells with prominent cytoplasmic processes. (D) Numerous Inka cells with narrow processes are distributed on tracheae of the bug *P. apterus.* (E) The water beetle *Laccophilus sp.* contains groups of 2–4 Inka cells with thick processes. (F) Two types of cells are present in the mealworm beetle *T. molitor*; groups of larger single or coupled Inka cells are located on major tracheae close to spiracles, whereas single small cells are distributed throughout narrower branching tracheae. (G, H) Individual Inka cells of the mosquito *Aedes* (G) and *Drosophila* (H). (I) A large group of small Inka cells on a major tracheal branch in the honeybee, *Apis.* (J, K) Large epitracheal glands containing Inka cells in *Manduca* (J) and *Bombyx* (K). Double immunohistochemical staining with antisera to PETH and HRP reveals that each epitracheal gland is composed of a prominent Inka cell (yellow, large arrow) that produces peptide hormones, narrow cells (large arrowhead), exocrine cells (green/yellow, small arrow) and canal cells (small arrowhead). (L) PETH-IR in the pedal endocrine cells of the tick *Ixodes ricinus.* Two pairs of these cells are located in each pedal segment and may represent Inka cell homologues. Scale bar, 200 μm in A, J, and K; 50 μm in B–I; and 10 μm in L.

segment have been named "pedal endocrine cells" and may represent tick homologues of insect Inka cells (Roller *et al.*, 2010).

7.2.2. Ecdysis Triggering Hormones in Lepidoptera

A physiological function for epitracheal glands and identification of an active peptide hormone were first described in *Manduca* (Zitnan *et al.*, 1996). Injection of epitracheal gland extracts into *Manduca* pharate larvae induced the ecdysis behavioral sequence within 5–10 min. Exposure of the isolated CNS *in vitro* to the same extracts produced corresponding fictive motor bursts (see Section 7.5.1.). Peptides in saline extracts of 50 glands from four pharate pupae were separated by a single step of high-performance liquid chromatography (HPLC). Biological activity identical to that of the gland extract was found associated with a linear, 26 amino acid, C-terminally amidated peptide (mol. wt. 2940; **Figure 5**). This peptide was named *Manduca* ecdysis triggering hormone or Mas-ETH (Zitnan *et al.*, 1996). Injection of synthetic ETH mimicked the effects of the native peptide. A similar peptide

(mol. wt. 2656), with a three amino acid deletion near the N-terminus, subsequently was isolated from Inka cells of pharate pupal *B. mori* and called Bom-ETH (**Figure 5**; Adams and Zitnan, 1997). Either peptide is sufficient to induce the ecdysis sequence upon injection into *Manduca* or *Bombyx* larvae.

Identification of the cDNA precursor and gene encoding ETH in *Manduca* revealed two additional peptides produced by Inka cells (**Figure 5**). One of these, an 11 amino acid, C-terminally amidated peptide (mol. wt. 1269), was subsequently isolated from epitracheal gland extracts by HPLC and synthesized. Since this peptide induced only pre-ecdysis behavior in *Manduca*, it was named pre-ecdysis triggering hormone or PETH (Zitnan *et al.*, 1999). The third peptide, composed of 47 amino acids with a free carboxyl C-terminus was also synthesized, but showed no obvious biological function upon injection into pharate larvae. It was therefore named ETH-associated peptide, or ETH-AP (mol. wt. 5658). Significant levels of partially processed precursor peptides corresponding to PETH-ETH (PE; mol. wt. 4407) and ETH-ETH-AP (EA; mol. wt. 8953), along with the entire, unprocessed precursor PETH-ETH-ETH-AP (PEA; mol. wt. 10419), were also

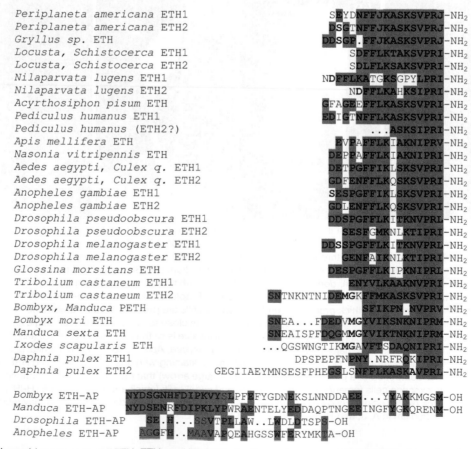

Figure 5 Amino acid sequences of PETH, ETH, and ETH-AP of arthropods — sequence identities (green) and sequence similarities (yellow). Since MS data cannot distinguish leucine (L) from isoleucine (I) in *Periplaneta* and *Gryllus* ETHs, J is used to denote the ambiguous L/I residue assignment. Modified from *Roller et al.* (2010).

isolated directly from Inka cell extracts by HPLC. The ETH precursor thus contains one copy each of PETH, ETH, and ETH-AP (Zitnan *et al.*, 1999). A similar precursor structure and corresponding peptides occur in *Bombyx* (Zitnan *et al.*, 2002a).

Immunohistochemical staining with antisera specific to ETH and ETH-AP show that these peptides are produced only in Inka cells of *Manduca* and *Bombyx* (Zitnan *et al.*, 2002a; Zitnan *et al.*, 1999). Release of PETH and ETH into the hemolymph at each ecdysis was demonstrated by enzyme immunoassays and immunohistochemistry using antisera raised against PETH and ETH (Zitnan *et al.*, 2002a; Zitnan *et al.*, 1999; Zitnanova *et al.*, 2001). These studies demonstrated that, during the initial 20 min of pre-ecdysis, low levels of these peptides (up to 5 nM) are present in the hemolymph. These levels then rise sharply, reaching ~30 nM about 20 min later. This is correlated with elevated cyclic GMP levels in Inka cells. The sharp rise in hemolymph ETHs corresponds to virtually complete depletion of these peptides in Inka cells. Hemolymph levels of PETH and ETH have decreased to ~15 nM by the onset of ecdysis (Zitnan *et al.*, 1999). Low levels of a partially processed precursor peptide (ETH-ETH-AP) are also co-released into the bloodstream along with fully processed peptides. The physiological significance of this precursor co-release, if any, is unclear.

Candidate regulatory elements contributing to ETH expression were revealed through identification of promoter regions in the ETH gene of moths and flies (**Figure 6**). Among the elements observed are direct repeat motifs thought to function as ecdysone response elements (EcRE). In *Manduca*, the direct repeat AGGTCATTAGGTCA and elements of Broad Complex are present in the ETH gene promoter (Zitnan *et al.*, 1999). A similar sequence occurs in the *Drosophila eth* promoter: AGGTCAggttAGTTCA (Park *et al.*, 1999). Such EcRE are known to bind the ecdysone receptor-ultraspiracle (EcR-USP) heterodimer in *Drosophila* and mosquitoes (Vogtli *et al.*, 1998; Wang *et al.*, 1998; see Chapter 5), and seem likely to function in the transcriptional regulation of *eth* gene expression.

7.2.3. Ecdysis Triggering Hormones in Diptera

The *eth* gene in *Drosophila* was identified through *"in silico"* screening using the nucleotide sequence encoding the *Manduca* ETH precursor (Park *et al.*, 1999). This approach subsequently was used to identify a similar gene in the mosquito *Anopheles gambiae* (Zitnan *et al.*, 2003). The *eth* gene of both dipterans encodes two related amidated peptides, ETH1 and ETH2, and a different non-amidated peptide ETH-AP (Park *et al.*, 1999, 2002a; Zitnan *et al.*, 2003). ETH-AP has been isolated from tracheal extracts of *Drosophila* (Schoofs, Clynen, and Zitnan, unpublished). Sequences of all peptide hormones and organization of the ETH gene of moths and flies are shown in **Figures 5** and **6**.

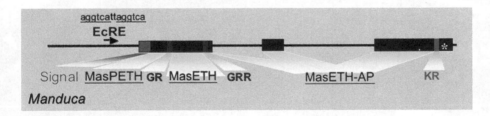

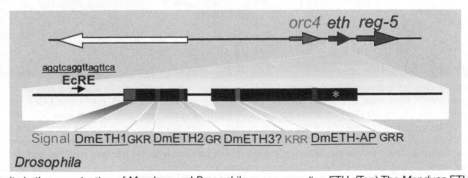

Figure 6 Similarity in the organization of *Manduca* and *Drosophila* genes encoding ETH. (Top) The *Manduca* ETH gene consists of a signal sequence, followed by one copy each of MasPETH, MasETH, and MasETH-AP. An upstream direct repeat ecdysone response element (EcRE) is a putative steroid receptor binding site for regulation of the gene. The Mas-ETH-AP sequence is interrupted by two introns. (Bottom) The signal sequence of *Drosophila eth* is followed by one copy each of DmETH1 and DmETH2. An upstream EcRE also is present in this gene. The sequence immediately downstream of DmETH2 is designated DmETH-AP1. Its sequence is followed by a second larger peptide designated DmETH-AP2. The gene *eth* occurs on the second chromosome right arm 60E1-2. See *Zitnan et al.* (1999) and *Park et al.* (1999) for details.

The N-terminus of ETH1 was deduced through algorithms designed for prediction of signal peptide cleavage sites in eukaryotic cells (Nielsen *et al.*, 1997). This prediction led to the assignment of ETH1 as an 18 amino acid peptide with N-terminal amino acids DDSSP, but an alternative sequence consisting of 23 amino acids containing the N-terminal sequence AISQADDSSP was also considered (Park *et al.*, 1999, 2002a; Zitnan *et al.*, 2003). These peptides were both prepared by chemical synthesis and compared in bioassays for potency in triggering adult eclosion in flies. These data showed ETH1 (N-terminus DDSSP) was at least 100 times more potent in triggering adult eclosion than the alternative peptide (N-terminus AISQADDSSP). Subsequent mass spectometric analysis provided confirmatory evidence for the original sequence assignments for both ETH1 and ETH2 (**Figure 5**; Hermesman and Adams, unpublished).

The *eth* gene in *Drosophila* is expressed only in Inka cells. This was demonstrated by producing a transgenic line of flies expressing the transgene *2eth3-egfp* (Park *et al.*, 2002a). The construct contains the upstream promoter and coding regions of the *eth* gene followed by the chimeric *egfp* gene encoding Enhanced Green Fluorescent Protein (EGFP). These flies show co-localization of ETH-IR and EGFP-IR in Inka cells (**Figure 7**); no other cells in the entire organism were stained in larvae and adults. Secretory activity of Inka cells was observed as a sharp decline of cellular EGFP fluorescence just before tracheal inflation and initiation of the ecdysis behavioral sequence in larvae (Park *et al.*, 2002a). These findings confirm that the ETH gene is expressed in Inka cell-specific fashion, and that the peptides are released at the appropriate time just preceding ecdysis in flies.

Peritracheal myomodulin cells (PM cells) also were described on the trachea of six insect species using an antiserum to molluscan myomodulin (O'Brien and Taghert, 1998). Double labeling with myomodulin and SCP_B antibodies indicated that PM cells are Inka cell homologues. O'Brien and Taghert speculated that myomodulin-like

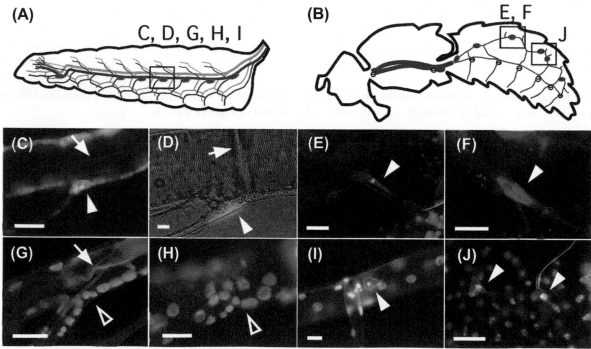

Figure 7 *Drosophila* Inka cell immunohistochemistry, EGFP expression in the *2eth3egfp* fly line, and loss of immunoreactivity in the deletion mutant *eth²⁵ᵇ*. Red, blue, and green colors in photomicrographs are from Cy3, DAPI, and EGFP, respectively. Scale bars in all panels indicate 10 μm. (A) Diagram showing the larval tracheal system and positions of Inka cells (red). (B) Diagram showing adult tracheal system and positions of the Inka cells (red). Letters associated with boxed areas in the diagrams are shown in panels C–J below. (C) Cy3 staining in the late first instar using the DrmETH1 antibody. Old intima (arrow) is already separated from new intima. The Inka cell (arrowhead) is located along the main dorsal tracheal trunk at each branchpoint of the transverse connectives. (D) Expression of the *eth3-egfp* transgene in late third instar. The Inka cell expressing EGFP (arrowhead) at a node (arrow) is shown with background illuminated by low intensity transmitted light. (E) Cy3 staining in the adult stage using the DrmETH1 antibody. (F) Expression of the *eth3-egfp* transgene in the adult stage. (G) The *eth* deletion mutant *eth²⁵ᵇ* Cy3 (late first instar) shows no immunohistochemical staining of the Inka cell. The collapsed old tracheal tube (arrow) is separated from new intima. The open arrowhead indicates the normal location of the Inka cell in wild-type flies. (H) Depletion of ETH immunoreactivity in the Inka cell of wild-type flies (open arrowhead) immediately after ecdysis to early second instar. (I) Immunoreactive Inka cell in the late third instar using an antibody against MasPETH. (J) Inka cell immunoreactivity in the adult stage using the MasPETH antiserum. Data from *Park et al.* (2002a).

peptide(s) were co-released with ETHs to control different functions at ecdysis. However, no function for myomodulin has been determined and no myomodulin-like peptide has been identified in insects. Inka cells of moths and flies also show strong immunohistochemical staining with antibodies to several other neuropeptides (e.g., FMRFamide, PBAN, vasopressin (Zitnan *et al.*, 2003). These neuropeptides share conserved amidated sequence motifs (RXamide or PRXamide; X = I, L, M, V) with PETH and ETH, indicating that antisera to these neuropeptides cross-react with amidated C-termini of PETH and ETH. Indeed, cross-reactivity of myomodulin and FMRFamide antisera with PETH and ETH was confirmed by specific enzyme immunoassays in HPLC-isolated extracts of *Manduca* epitracheal glands and in *Drosophila* and *Musca* tracheal extracts (Zitnan *et al.*, 2003). It was further demonstrated that Inka cells of mutant larvae lacking the *eth* gene failed to react with myomodulin antiserum, whereas strong staining was detected in Inka cells of control *Drosophila* larvae (Canton S) producing ETH (Zitnan *et al.*, 2003). These data demonstrate that Inka cells are highly specialized for production of peptide hormones expressed by the ETH gene, and do not produce myomodulin- or FMRFamide-related peptides in moths and flies.

7.2.4. Conservation of ETHs in Arthropods

The presence of ETH-related peptide hormones in Inka cells from various insects, including the cockroach *Nauphoeta*, cricket *Acheta*, bug *Pyrrhocoris*, beetle *Te.*, fly *Drosophila*, and mosquito *Aedes* was tested by injection of tissue extracts into *Bombyx* pharate larvae (Zitnan *et al.*, 2003). Injection of extracts induced within a few minutes pre-ecdysis followed in most instances by ecdysis behavior in *Bombyx* larvae. These data suggest that the ETH signaling system is conserved in diverse insect groups and perhaps all insects.

Further searches for Inka cell peptides in hemimetabolous insects resulted in detection of PETH-immunoreactive peaks in HPLC-purified tracheal extracts of *Periplaneta* and *Pyrrhocoris*. Four immunopositive peaks were detected in extracts of *Periplaneta*, whereas only two peaks were found in *Pyrrhocoris* (Zitnan *et al.*, 2003). These techniques were later used for isolation of ETHs from pharate adults of the cockroach (*P. americana*), locust (*Schistocerca americana*) and cricket (*Gryllus sp.*) and their amino acid sequences were determined by liquid secondary ion mass spectrometry and MALDI/MS-MS (Roller *et al.*, 2010). *S. americana* peptides are identical to ETH1 and ETH2 encoded by the cDNA precursor of *L. migratoria* (Clynen *et al.*, 2006). Additional ETH sequences have been predicted from cDNA clones or genomes of several hemi- and holometabolous insects, the water flea *Daphnia pulex* and the tick *Ixodes scapularis* (Roller *et al.*,

2010). Comparison of all identified ETHs revealed that FFLKASKSVPRI-NH$_2$ is the most conserved sequence motif in arthropods (**Figure 5**).

7.2.5. ETH Receptors

ETH receptors were first identified in *Drosophila*. The *Drosophila* gene *CG5911* encodes two ETH receptor subtypes by alternative splicing of the 3′ exon; they are referred to as CG5911a and CG5911b (**Figure 8A**). These gene products are GPCRs that are highly sensitive and selective for ETH ligands of *Drosophila* (Iversen *et al.*, 2002; Park *et al.*, 2003a). CG5911 is related to the vertebrate TRH receptor and more distantly to the neuromedin U (NMU) receptor group.

Drosophila receptors show strikingly different sensitivities to ETHs when expressed in CHO cells (Park *et al.*, 2003a); CG5911a is ~400-fold less sensitive to ETH1 than CG5911b. It is not yet clear whether this difference in sensitivity of the two receptors is real, or if it is a consequence of the heterologous expression system used. At higher concentrations, related peptides having the –PRXamide sequence motif also activate CG5911b, the most active of these being SCP$_B$, CAP2b-1, hugγ, and HezPBAN (**Figure 8B**). At still higher concentrations, the invertebrate peptide myomodulin and vertebrate peptides NMU and thyrotropin-releasing hormone (TRH) activate CG5911b. This pharmacological analysis of CG5911 supports their hypothesized evolutionary relationships with TRH receptors and the NMU receptor groups predicted by phylogenetic analysis.

Subsequent work on the cloning and expression of ETH receptors in *Manduca* identified functional orthologues of CG5911 in the CNS (Kim *et al.*, 2006a). The *Manduca* receptors also occur as two subtypes through alternative splicing and exhibit pharmacological profiles that are generally consistent with those of CG5911a and CG5911b. The *Manduca* receptor subtypes, which are equally sensitive to PETH and ETH, show higher ligand specificity compared to their *Drosophila* counterparts. Whereas CG5911b of *Drosophila* responds to both PBAN-like *hugin* peptides, its *Manduca* orthologue is completely insensitive to PBAN or *hugin* peptides at up to 10 μM concentration. The high specificity and sensitivity of the *Manduca* receptors provides further evidence that CG5911 and its orthologues in other insects function as the physiological ETH receptors.

Expression of ETH receptors is observed throughout the nervous system in both *Drosophila* and *Manduca*. ETHR-A and ETHR-B are expressed in separate populations of neurons, although small numbers of cells appear to express both receptor subtypes (Kim *et al.*, 2006a,b). Most ETHR-A neurons have been identified and are found to be peptidergic. Peptides expressed in ETHR-A neurons are members of the peptidergic signaling cascade

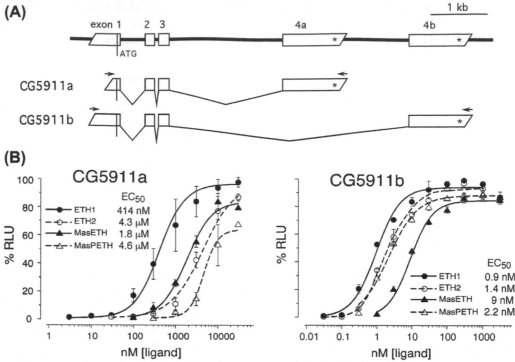

Figure 8 Organization of the ETH receptor gene and pharmacology of the ETH receptor. (A) Genomic structure of ETH receptors (CG5911a and CG5911b) arising from two alternative transcripts with different exons 4a and 4b at the 3' end of the gene. Transcripts were confirmed experimentally by RT-PCR with the primers (arrows) located on the boundary of start and stop codons in the predicted open reading frames. Each 5' and 3' prime end of the transcript is not determined and left as a slanted rectangle of the exon boxes. (B) Pharmacology of ETH receptor subtypes A and B. Transcripts were expressed in CHO cells with the G protein Gα16 (see *Park et al.*, 2003a for details). Using this heterologous expression system, CG5911a is less sensitive to ETH peptides from *Drosophila* and *Manduca* as compared to CG5911b, perhaps due to suboptimal receptor coupling with the G protein. Note that both *Drosophila* and *Manduca* ETHs are active on both receptors. Reproduced with permission from *Park et al.* (2003a).

that schedules successive steps in the ecdysis sequence. Notably, in *Manduca* larvae ETHR-A transcripts are found abundantly in the VM neurons (EH), $L_{3,4}$ neurons (kinin/DH), 27/704 neurons (CCAP/MIPs), neurons that produce FGLa, NPF, sNPF, Cal-DH31, and others (**Figure 9**). The $L_{3,4}$ neurons are implicated in initiation of pre-ecdysis I and IN-704 neurons are thought to be initiators of ecdysis behavior (see Sections 7.5.1.4. and 7.5.1.7. for details). While most ETHR-A neurons have been identified, the identities of ETHR-B neurons have

yet to be elucidated (**Figure 10**). Quantitative RT-PCR and *in situ* hybridizations provided the first demonstration of ETHRs in the CA of *Bombyx* and *Manduca* (Yamanaka *et al.*, 2008). This may help to explain the surge in JH levels at ecdysis, previously reported in *Manduca* (Kinjoh *et al.*, 2007).

The peptide signaling cascade identified in *Manduca* is largely conserved in *Drosophila* (Kim *et al.*, 2006b). In pharate pupae, *in situ* hybridization identified ETHR-A transcripts in neurons producing kinin, FMRFa, EH,

Figure 9 Expression of ETH receptor subtype A (ETHR-A) in central peptidergic neurons of *Manduca* larvae before and after ecdysis. (A) *In situ* hybridization detected ETHR-A expression in numerous neurons in the CNS of pharate third instar larva. (A') Subsequent immunohistochemical staining of the same CNS with neuropeptide and cGMP antibodies revealed the peptidergic nature of all receptor neurons. In the brain, two pairs of ventromedial (VM) neurons produce EH (green color, arrowheads), whereas small lateral interneurons express MIPs (red color, arrows). Dorsolateral cells in the SG and TG1-3 are L_1/704 neurons producing cGMP and bursicon (arrowheads; Dai *et al.*, 2008), whereas small anterior neurons produce cGMP only (arrows). Neuropeptide F (NPF) is produced by posterior dorsolateral interneurons in TG1-3 (green color, arrows). Segmental $L_{3,4}$ cells in AG3-7 express kinins and DHs (green color, arrowheads), whereas IN-704 in AG1-7 produce CCAP and MIPs (yellow color, arrows). Note that TAG contains fused AG7 and AG8. Small anteroventral interneurons shown in AG5 produce FGLa (allatostatin A; arrows). Visceral modulatory neurons (VMN) in the posterior TAG (AG8) co-express calcitonin-like DH31 and MIPs (red color, arrows) and short neuropeptide F (sNPF) and MIPs (orange color, arrows). These neurons arborize on the hindgut surface and posterior part of the last abdominal segment, respectively. (B) After ecdysis ETHR-A mRNA disappears in most neurons and only traces of expression are detected in VM and L_3 neurons (arrowheads). Brain: SG, subesophageal ganglion; TG3, thoracic ganglion 3; AG5, abdominal ganglion 5; and TAG, terminal abdominal ganglion. Scale bars, 100 μm. Adapted from *Kim et al.* (2006a).

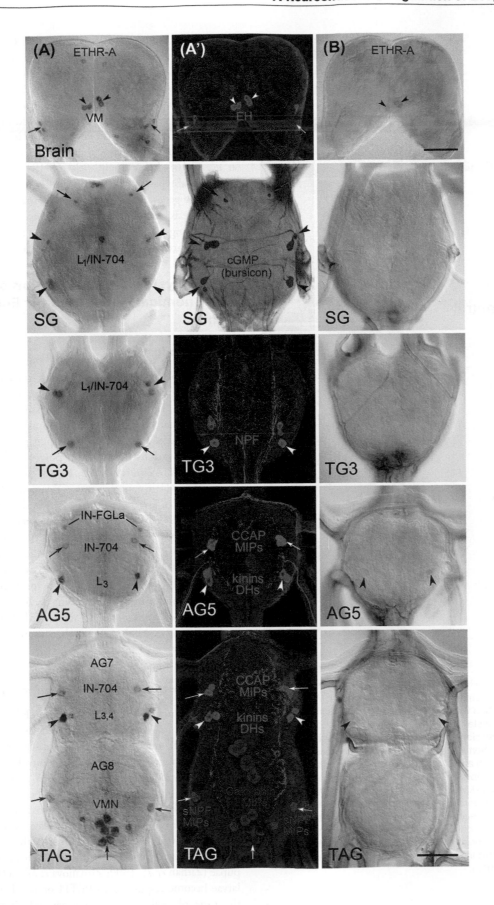

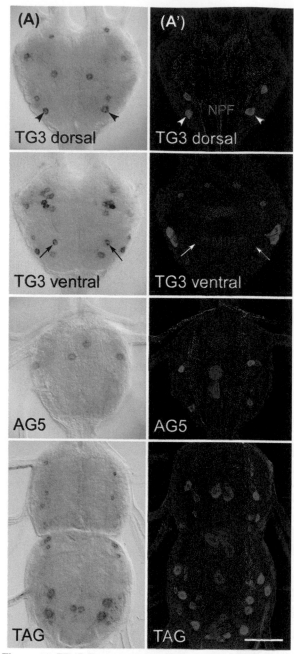

Figure 10 ETHR-B expression and neuropeptide immunohistochemistry in the CNS of *Manduca*. (A) *In situ* hybridization with probes for ETHR-B transcripts. In pharate third instar larvae, numerous neurons showed ETHR-B expression in the ventral ganglia, but subsequent immunohistochemical staining (A') with antibodies to RFamide (green) and MIP (red) revealed co-localization with ETHR-B transcripts in TG3 only. These neurons produce NPF (arrowheads) and MIPs (arrows) in each TG1-3. We speculate that abdominal dorsal unidentified neurons shown in AG5 may control pre-ecdysis II. In addition, we propose that thoracic clusters of neurons shown in TG3 are inhibitory and delay ecdysis onset. NPF in each TG1-3 and AG1-7 neurons may release this inhibition (see text for details). Adapted from *Kim et al.* (2006a).

bursicon, CCAP, and MIPs (**Figure 11**). Calcium imaging studies indicate that these "peptidergic ensembles" are activated sequentially, suggesting they schedule pre-ecdysis and ecdysis behaviors (see Section 7.5.3.2.1. for details).

Subsequent surveys of genomes show that sequences predictive of ETH receptors are widespread in the Insecta. Putative ETH receptor genes also are found in arachnids (the tick, *Ixodes*) and crustaceans (the water flea, *Daphnia*) (Roller *et al.*, 2010). Interestingly, tick ETHR-A and ETHR-B receptor subtypes are predicted to arise through expression of separate genes instead of through alternative splicing of a single gene as has been reported for all insect species thus far. These findings suggest that ETH signaling in ecdysis control is highly conserved evolutionarily (**Figure 12**).

7.3. Ecdysteroid Specification of Neural and Chemical Substrates for Ecdysis Control

7.3.1. Introduction

The transition from feeding to pharate stage during each stadium of development is initiated by a rise in ecdysteroid levels. This event triggers successive bouts of gene expression that have been referred to as early, early-late, and late phases, according to the appearance of sequential puffs at various positions on polytene chromosomes of *Drosophila* (Ashburner *et al.*, 1974). In this section, we note early and late gene expression events in *Manduca* that are involved in assembly of the peptide signaling cascade that regulates ecdysis. Early events that occur around the peak of ecdysteroid levels include acquisition of CNS sensitivity to ETH and increased production of ETH in Inka cells. Although Inka cells accumulate large amounts of peptides during this phase, they are incapable of releasing their contents until ecdysteroids decline to low levels. Acquisition of competence to release ETH is associated with late gene expression events that occur once ecdysteroids decline to <0.1 µg/ml.

7.3.2. Effects of Elevated Ecdysteroid Levels

7.3.2.1. Induction of CNS sensitivity to ETH and ETHR expression   In *Manduca* (and indeed in all insects thus far examined), only pharate stages are sensitive to PETH and ETH. Injection of these peptides into freshly ecdysed, feeding, or wandering larvae, or ecdysed pupae and adults never results in pre-ecdysis or ecdysis behaviors. Animals become sensitive and show specific behavioral responses to PETH and ETH approximately 1–2 days prior to ecdysis, depending on the particular stage. Onset of this sensitivity has been determined for pharate fifth instar larvae and pharate pupae (Zitnan *et al.*, 1999; Zitnanova *et al.*, 2001). Pharate larvae become responsive to PETH or ETH injections at the time of peak ecdysteroid levels, which corresponds

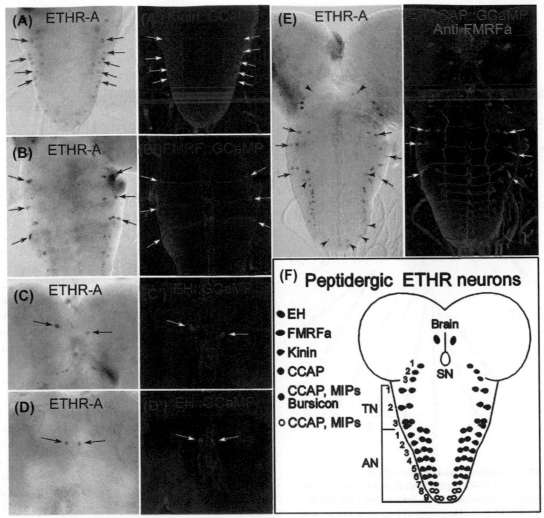

Figure 11 Identification of central peptidergic neurons expressing ETHR-A in *Drosophila* (A–E) Neurons were identified using a combination of *in situ* hybridization (A–E) with a specific ETHR-A DNA probe followed by immunohistochemical staining (A'–E') with antibodies to neuropeptides or GFP/GCaMP targeted by the binary neuropeptide promoter GAL4/UAS system. (A, A') ETHR-A is expressed in seven pairs of lateral abdominal neurons, not all of which are visible in this preparation (A, arrows) producing kinin (A'). (B, B') ETHR-A is detected in three pairs of thoracic ventrolateral neurosecretory cells (B, arrows) producing FMRFamides (B'). (C, D) ETHR-A expression in a pair of brain EH cells of pharate pupae (C, C') and adults (D, D'). (E, E') ETHR-A expression (E) in a network of 27/704 neurons co-expressing CCAP, MIPs, and/or bursicon (E', red) in the subesophageal neuromere (SN), thoracic neuromeres (TN) 1–3, and abdominal neuromeres (AN) 1–9. Arrows indicate position of FMRFamide neurons described earlier (green). Arrowheads in (E) point to a few ETHR-A cells that do not show CCAPGCaMP staining. (F) Schematic diagram showing peptidergic ensembles producing ETHR-A. Neurons 27/704 are subdivided into three subgroups on the basis of cotransmitter expression: CCAP in SNs, TNs, and AN5–7; CCAP/MIPs/bursicon in AN1–4; CCAP/MIPs in AN8,9. Adapted from *Kim et al.* (2006b).

to initiation of head capsule slip (~30 h prior to ecdysis). These animals respond to ETH injection by performing pre-ecdysis behavior only, but after the head capsule is completely slipped 2 h later, ETH injection induces the entire sequence in most animals. ETH treatment of pharate pupae elicits ecdysis behavior about 48 h prior to expected ecdysis (**Figure 13A**). These animals do not show pre-ecdysis contractions. The latency from ETH injection to the onset of ecdysis behavior is progressively shorter as animals approach the time of normal ecdysis (Zitnanova *et al.*, 2001). In *Bombyx*, sensitivity to ETH begins ~30 h prior to ecdysis of

fifth instar larvae, which again corresponds to head capsule slip, 24 h before ecdysis of non-spinning pharate pupae, and in developing adults already 6–8 days prior to eclosion (Zitnan *et al.*, 2002b). These studies showed that sensitivity to PETH and ETH is correlated with appearance of the highest ecdysteroid titers in the hemolymph (**Figure 13A**). The stage-specific differences in appearance of sensitive periods are remarkable and reflect occurrence of the ecdysteroid peak at different times in the hemolymph during development. These results also suggest that CNS circuits for pre-ecdysis and ecdysis behaviors are intact far

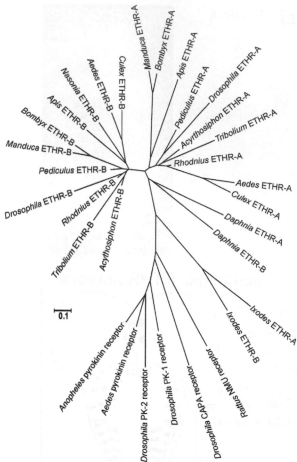

Figure 12 Phylogenetic relationships of alternatively spliced exons of ETHR-A and ETHR-B and related GPCRs (transmembrane domains 4 to 7). Sequences of ETHR-A and ETHR-B subtypes place them in two separate groups in all instances, except for *Daphnia* and *Ixodes*. Alternatively spliced exons for ETHR-A and ETHR-B apparently branched before the diversification of insect taxa. The tree was constructed by a minimum evolution method (Gaps/Missing Data-Pairwise deletion) in MEGA4. Reproduced with permission from *Roller et al.* (2010).

earlier than previously indicated by experiments with EH (Zitnanova *et al.*, 2001).

These experiments lead to the hypothesis that the behavioral response to ETH is enabled by exposure to high ecdysteroid levels. To test this, *Manduca* fifth instar larvae were examined for ability to respond to ETH injection at various stages following treatment with ecdysone, 20E, or the ecdysteroid agonist (tebufenozide, RH-5992). The results of these experiments showed that pre-treatment with steroid or steroid agonist induced ETH sensitivity in freshly ecdysed intact larvae or isolated abdomens of *Manduca* within 1–2 days (Zitnanova *et al.*, 2001). ETH sensitivity was detected by injection of the peptide, which induced pre-ecdysis and in some cases ecdysis behaviors in steroid-treated animals. Indeed, repeated injections of 20E followed by ETH into isolated abdomens induced

the behavioral response (pre-ecdysis) three times within six days (**Figure 13B**; Zitnanova *et al.*, 2001). Direct action of 20E on the CNS was demonstrated by *in vitro* experiments. The CNS dissected from freshly ecdysed or feeding larvae was incubated with physiological levels of 20E for 24–28 h. Addition of ETH into the incubation medium elicited typical fictive ecdysis bursts in dorsal nerves of these isolated nerve cords (Zitnan *et al.*, 1999; Zitnanova *et al.*, 2001). These experiments suggest that ecdysteroids act directly on the CNS to induce sensitivity to ETH.

Recent data from quantitative PCR experiments demonstrate that ETH receptor expression is induced by elevated ecdysteroids (**Figure 14**; Kim *et al.*, 2006a). Furthermore, exposure of freshly ecdysed fifth instar larvae or the isolated CNS from these animals to 20E results in expression of ETHR-A (Cho and Adams, unpublished). Expression of ETH receptors in the yellow fever mosquito *Aedes aegypti* also is under the control of ecdysteroids (see Section 7.5.4.; Dai and Adams, 2009). These data support the idea that steroid regulation of ETHR gene expression is an evolutionarily conserved event in ecdysis control.

7.3.2.2. Regulation of ETH gene expression in *Manduca* and *Bombyx*

During ecdysis, epitracheal glands release most, if not all, of their peptide stores. Depletion of peptide stores is accompanied by a corresponding decrease in cell volume. Following ecdysis, cell volume and PETH and ETH content increase gradually during the subsequent feeding stage. However, a sharp increase in PETH and ETH production occurs upon appearance of ecdysteroids to signal onset of the next molt. The concomitant rise of hemolymph ecdysteroid levels and Inka cell peptides have been quantified by enzyme immunoassays following the natural ecdysteroid pulse that initiates the molt in *Manduca* fourth instar larvae. Elevation of peptide content in Inka cells was blocked by pre-treatment with the transcription inhibitor actinomycin D (AcD), indicating the involvement of new gene expression (Zitnan *et al.*, 1999). Likewise, onset of the molt in fifth instar *Manduca* larvae involves sharp increases in the rate of PETH and ETH synthesis immediately after each of two successive ecdysteroid pulses (**Figure 15**) (Zitnanova *et al.*, 2001). Interestingly, levels of peptide precursors also increased sharply following each steroid pulse, but their levels quickly decreased as they were enzymatically processed to PETH and ETH (**Figure 15**). Consistent with this, injection of 20E or the steroid analog tebufenozide (RH5992) induced expression of the ETH gene and increased production of peptide hormones and their precursors in Inka cells (Zitnanova *et al.*, 2001). These results show that steroid pulses induce sharply increased rates of Inka cell peptide production. This increased expression sets the stage for ecdysis initiation at the end of the molt.

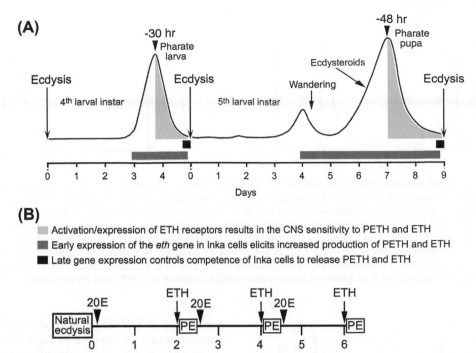

Figure 13 Ecdysteroid action on the CNS and Inka cells prior to larval and pupal ecdyses of *Manduca*. (A) Freshly ecdysed, feeding or wandering larvae, which contain low ecdysteroid levels in the hemolymph are insensitive to ETH. Peak ecdysteroid levels in the hemolymph coincide with onset of CNS sensitivity to ETH. Light gray areas illustrate time periods of CNS sensitivity to ETH, where its injection induces the ecdysis sequence. High ecdysteroid levels induce expression of the ETH gene, whereas declining ecdysteroids confer secretory competence of Inka cells. (B) Injection of 20-hydroxyecdysone (20E) into isolated abdomens induces CNS sensitivity to ETH within 1–2 days. Ecdysteroid treatments (20E, arrowheads) followed by ETH injections (arrows) 2 days later induced repeated onset of pre-ecdysis behavior (PE). Modified from *Zitnanova et al.* (2001).

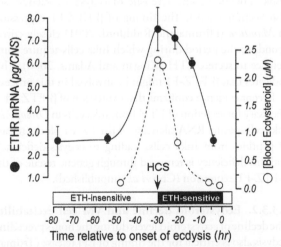

Figure 14 Elevated level of ETHR transcripts coincides with increased ecdysteroid levels and CNS sensitivity to ETH in *Manduca*. Temporal expression of ETHR-A,B transcripts (both subtypes), quantified by qPCR analysis, is shown by black filled circles, and changing ecdysteroid levels in the hemolymph are shown by white circles at different time points prior to ecdysis of pharate fifth instar larvae at time zero. Each point represents a mean value ±S.E. for ETHR mRNA in pg/CNS ($n = 3$). (Data on ETHR transcript levels taken from *Kim et al.*, 2006a; blood ecdysteroid levels and CNS sensitivity to ETH are taken from *Zitnan et al.*, 1999).

7.3.3. Effects of Declining Ecdysteroids

7.3.3.1. Declining ecdysteroids enable secretory competence in Inka cells High ecdysteroid levels that initiate the molt and coordinate gene expression in preparation for the next stage also serve to inhibit onset of the ecdysis sequence. This was first demonstrated by experiments showing that high steroid levels delay ecdysis (Curtis *et al.*, 1984; Slama, 1980; Truman *et al.*, 1983). However, the CNS is capable of responding to ETH by playing out the ecdysis sequence as early as 1–2 days prior to natural ecdysis, which marks the peak of ecdysteroid levels (Zitnan *et al.*, 1999). It is now clear that high ecdysteroid levels inhibit ecdysis by preventing ETH release from Inka cells (Kingan and Adams, 2000).

A decline in ecdysteroid levels is necessary for ecdysis to proceed (**Figure 16**). In *Manduca*, the onset of ecdysteroid decline coincides with the release of MIPs from the epiproctodeal glands (Davis *et al.*, 2003). When ecdysteroids drop to below 0.1 µg/ml, Inka cells become competent to release ETH. Inhibition of ecdysis by elevated ecdysteroid levels may be a fail-safe method that prevents life-threatening premature ecdysis. In pharate pupae of *Manduca*, Inka cells challenged with EH acquire competence to respond with ETH release about 8 h before pupation. Injection of 20E into pre-competent insects delays this acquisition of

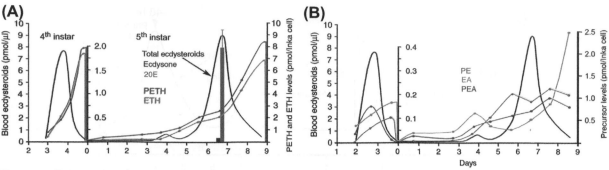

Figure 15 Ecdysteroids stimulate production of PETH, ETH, and their precursors in Inka cells. (A, B) Only small amounts of peptides and their precursors are present in Inka cells during low ecdysteroid levels in fourth and fifth instar larvae. Elevation in ecdysteroid levels is associated with increased production of PETH, ETH, and all precursor forms (PE, EA, PEA). Fully processed peptides and EA reached the highest levels on the last day of each instar, whereas levels of PE and PEA decreased prior to each ecdysis due to precursor processing. The ratio of ecdysone and 20E in the pre-pupal peak (fifth instar) was 1:20. Each peptide determination represents the mean ± SD of 3–4 sets of 20–40 epitracheal glands. Each ecdysteroid determination represents the mean ± SD of 5–11 hemolymph samples. Adapted from *Zitnan et al.* (1999). Steroid induction of a peptide hormone gene leads to orchestration of a defined behavioral sequence. *Zitnan et al.* (1999) and *Zitnanova et al.* (2001).

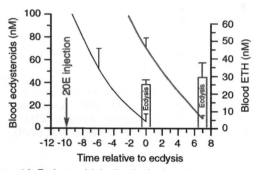

Figure 16 Ecdysteroid decline in the hemolymph is required for ETH release from *Manduca* Inka cells. In normal larvae, ecdysteroid decline to very low levels (black line), which leads to ETH secretion (black bar) and onset of the ecdysis sequence at 0 h. Injection of 10 μg 20-hydroxyecdysone (20E) 10 h prior to ecdysis (red arrow) delays ETH release and ecdysis for 6–8 h (red bar). Reprinted with permission from *Zitnan et al.* (1999).

competence, even though EH-evoked accumulation of the second messenger cyclic 3′,5′-guanosine monophosphate (cGMP) occurs. This demonstrates that the EH receptor is present in the Inka cell at an early stage, and that secretory competence involves a signaling step downstream of receptor activation. Pre-competent glands transferred from a high ecdysteroid environment into steroid-free culture medium acquire competence within about 12 h, but this can be prevented by inclusion of 20E above 0.1 μg/ml (Kingan and Adams, 2000). AcD completely inhibits acquisition of competence, showing its dependence on transcriptional events. These findings indicate that declining ecdysteroids trigger late transcriptional events in Inka cells, the product of which is an ETH secretion-enabling process independent of EH receptor activation and cGMP accumulation (Kingan and Adams, 2000).

The low levels of 20E (≤ 0.1 μg/ml) permissive for acquisition of competence correspond to those that allow

expression of β-FTZ-F1, a transcription factor whose expression occurs in response to high followed by low ecdysteroid levels (Hiruma and Riddiford, 2001; Lavorgna *et al.*, 1993). β-FTZ-F1 is implicated in acquisition of competence of various tissues to respond to hormonal cues, including competence of salivary glands to degenerate at the end of the pre-pupal period and in the formation of normal cuticle at the end of each molt (Yamada *et al.*, 2000). Interestingly, flies that are either deficient in β-FTZ-F1 or express it prematurely are unable to undergo successful ecdysis. These flies show a "double mouth hook" phenotype characteristic of ecdysis deficiency (see also Section 7.5.3.). The timing of β-FTZ-F1 expression in *Manduca* (Hiruma and Riddiford, 2001) closely corresponds to the period during which Inka cells acquire competence to secrete ETH (Kingan and Adams, 2000). This indicates that β-FTZ-F1 might be involved in this process.

Recent evidence confirms that expression of β-FTZ-F1 is necessary for secretion of ETH from Inka cells in *Drosophila*. Inka cell specific RNA silencing of β-FTZ-F1 blocks release of peptides from Inka cells, leading to ecdysis deficiency, and this deficiency is removed through genetic rescue of the β-FTZ-F1 expression (Cho *et al.*, unpublished).

7.3.3.2. Ecdysteroid decline and VM cell excitability

The decline of ecdysteroid levels during the hours preceding ecdysis also is critical for the timing of EH release (Truman *et al.*, 1983). Electrophysiological experiments showed that the ecdysteroid decline is followed by elevated excitability of the VM neurosecretory cells that produce EH (Hewes and Truman, 1994). Intracellular recordings from these cells showed that they are not spontaneously active either during the feeding stage, when steroid levels are low, or during the early pharate stage, when steroids rise to peak levels. In contrast, subsequent ecdysteroid decline at the end of pharate stage was followed by increased excitability of VM neurons. Excitability of these neurons was detected about 1 h prior to expected ecdysis. Prolongation of elevated ecdysteroid levels by injection of 20E delayed the appearance of excitability in VM neurons. Likewise, treatment with the RNA synthesis inhibitor, AcD at the appropriate time (about 6 h prior to ecdysis) blocked VM activity and the ability to release EH. These experiments suggest that late gene transcription induced by ecdysteroid decline is responsible for excitability of VM cells and consequent release of EH (Hewes and Truman, 1994).

A critical factor in the onset of VM excitability is the discovery that circulating ETH triggers the onset of the ecdysis sequence about 1 h prior to ecdysis through direct actions on the CNS (Zitnan *et al.*, 1996, 1999). CNS targets of ETH in the CNS include the VM neurons, since their exposure to ETH increases excitability and causes spike broadening (Gammie and Truman, 1999). Recent evidence confirms that VM neurons are primary targets of ETH, since they express ETH receptors in both *Manduca* and *Drosophila* and they respond to ETH by calcium mobilization (Kim *et al.*, 2006a,b).

7.4. Regulation of Inka Cell Secretion: Two Phases of Peptide Release

7.4.1. Early Release of PETH and ETH by Corazonin

Inka cells trigger pre-ecdysis and ecdysis through two modes of peptide release (Kim *et al.*, 2004b). The early phase of Inka cell release is initiated by corazonin release into the hemolymph by neurosecretory cells of the brain that project to the CA. This early, low-level phase of release leads to appearance of PETH and ETH in the hemolymph at levels between 1 and 5 nM (Zitnan *et al.*, 1999).

Corazonin originally was identified as a potent cardioaccelerator from the CC of the cockroach *P. americana* (**Figure 17**; Veenstra, 1989). Its recognized physiological actions now include a role in pigmentation in locusts (Tawfik *et al.*, 1999), and possible roles in nutritional stress (Veenstra, 2009) in addition to its role as an ecdysis initiator (Kim *et al.*, 2004b). The amino acid sequence of corazonin is remarkably conserved across distantly related phylogenetic groups. Slightly modified versions of corazonin have been predicted from genome databases of certain insects, as well as in crustaceans and ticks (Veenstra, 2009). Corazonin is produced by ipsilateral-projecting neurosecretory cells in the brain and paired segmental interneurons in the ventral nerve cord (Cantera *et al.*, 1994; Roller *et al.*, 2003; Veenstra and Davis, 1993). Corazonin neurosecretory cells have been identified as Ia₁ neurons in *Manduca* (**Figure 2**), which also express the circadian clock protein PER (Copenhaver and Truman, 1986a; Homberg *et al.*, 1991; Wise *et al.*, 2002; Zitnan *et al.*, 1995). These cells are involved in regulation of circadian rhythms in

Figure 17 Amino acid sequences of corazonin and FLRFamide-related peptides — sequence identity (green), sequence similarity (yellow).

the silkmoth *A. pernyi* (Sauman and Reppert, 1996) and could be involved in the circadian gating of adult eclosion described by Truman and Riddiford (1970).

Corazonin association with ecdysis is indicated by *in vivo* and *in vitro* experiments in *Manduca* and *Bombyx* pharate larvae and pupae. These experiments showed that low concentrations of corazonin (25–50 pM) induce the release of PETH and ETH to initiate the ecdysis sequence (Kim *et al.*, 2004; Daubnerová and Zitnan, unpublished). Corazonin release into the hemolymph just prior to pre-ecdysis initiation was demonstrated by a sensitive assay using a CHO cell line expressing corazonin receptor (Kim *et al.*, 2004). Using this assay, hemolymph concentrations of 30–80 pM were detected 20–30 min prior to the onset of pre-ecdysis in pharate fifth instar *Manduca* larvae. *In situ* hybridizations and Northern blots showed that strong corazonin receptor expression is restricted to the Inka cells and CNS (**Figure 18**). These data provide evidence that corazonin is the initial signal for PETH/ETH release that leads to activation of a complex neuroendocrine cascade essential for ecdysis.

The corazonin receptor in *Manduca* was cloned, expressed, and localized in Inka cells of the epitracheal endocrine system following its identification in *Drosophila* (Kim *et al.*, 2004b). Corazonin receptor cDNA in *Manduca* encodes a protein of 436 amino acids with seven putative transmembrane domains. The *Manduca* corazonin receptor (Mas-CRZR) appears to bind corazonin with a higher affinity than the *Drosophila* CRZR, with an EC_{50} value of ~200 pM when expressed in *Xenopus* oocytes and ~75 pM when expressed in CHO cells. Mas-CRZR exhibits high sensitivity and selectivity for corazonin.

7.4.2. EH Elicits a Second Phase of Massive Peptide Release from Inka Cells

7.4.2.1. Eclosion hormone (EH) The role of EH in insect ecdysis has been reviewed in detail (Ewer and Reynolds, 2002; Truman, 2005; Zitnan *et al.*, 2007); a brief summary of this peptide and its newly identified receptor is provided here. Release of an eclosion signal from transplanted brains in giant silkmoths demonstrated the role of a circulatory factor(s) released from the CC-CA

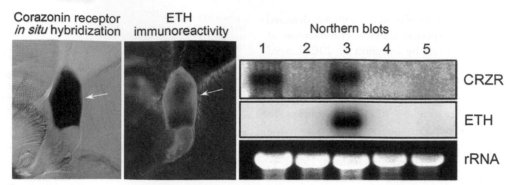

Figure 18 Expression of the corazonin receptor (*CRZR*) in the Inka cells of *Bombyx* and *Manduca*. *In situ* hybridization shows strong CRZR expression restricted to the Inka cell as indicated by ETH immunoreactivity (arrow) in *Bombyx* pharate pupa ~3 h prior to ecdysis. ^{32}P-labeled CRZR and ETH probes were used for Northern blot analysis of various tissues extracted from *Manduca* pharate pupae. Strong CRZR expression is detected in the CNS and Inka cells, while ETH mRNA is expressed only in Inka cells. Each lane contains 10 µg of total RNA extracted from the CNS (lane 1), epidermis (lane 2), Inka cells (lane 3), muscles (lane 4), and tracheae (lane 5). rRNA serves as loading control. (Northern blot data modified from *Kim et al.*, 2004).

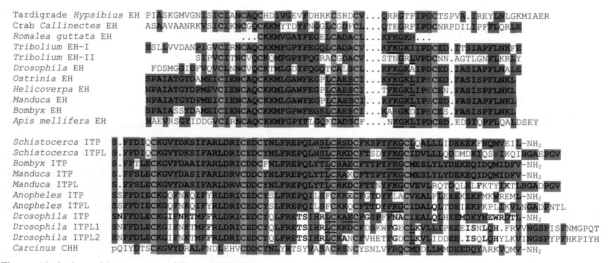

Figure 19 Amino acid sequences of EH and ITP/CHH of insects, crustaceans, and a tardigrade. Conserved sequence regions in EH and ITP/ITPL are underlined — conserved cysteines (turquoise), sequence identities (green), sequence similarities (yellow).

(Truman, 1971; Truman and Riddiford, 1970). Further studies showed that CC-CA extracts from pharate adults triggered premature ecdysis in pharate larvae, pupae, and adults upon injection into the hemocoel (Copenhaver and Truman, 1982; Reynolds *et al.*, 1979; Truman, 1980).

After initial isolation and partial sequencing of eclosion hormone in *Bombyx* (Kono *et al.*, 1987; Nagasawa *et al.*, 1987), the complete sequence of the hormone was simultaneously identified in two labs from CC/CA or heads of *Manduca* pharate adults (Kataoka *et al.*, 1987; Marti *et al.*, 1987). EH of both species is a closely related 62 amino acid peptide with three internal disulfide bonds (**Figure 19**; Kataoka *et al.*, 1992; Kono *et al.*, 1990b; Terzi *et al.*, 1988). The cDNA and gene encoding EH were cloned or predicted in several insects, crab, and tardigrade, indicating that this neuropeptide is widespread in arthropods (Horodyski *et al.*, 1989, 1993; Kamito *et al.*, 1992; Zitnan *et al.*, 2007).

In *Manduca*, EH is produced by two pairs of the brain ventromedial neurosecretory cells (VM cells, type V, **Figure 2C**). These neurons project axons along the entire length of the ventral nerve cord, which exit the CNS via the terminal and proctodeal nerves to make neurohemal release sites for EH on the hindgut surface (Truman and Copenhaver, 1989). During adult development they project additional axons to the CC, while axons running through the ventral nerve cord remain intact (Riddiford *et al.*, 1994). Similar pairs of VM cells have been identified in *Bombyx* with a monoclonal antibody to *Bombyx* EH (Kono *et al.*, 1990a). *In situ* hybridizations of the CNS in embryos, larvae, and developing adults, and Northern blots of different larval ganglia confirmed that the EH gene is only expressed in the brain VM cells (Horodyski *et al.*, 1989; Kamito *et al.*, 1992; Riddiford *et al.*, 1994).

An antiserum against *Manduca* EH was used to reveal VM neurons in diverse insect species (**Figure 20**).

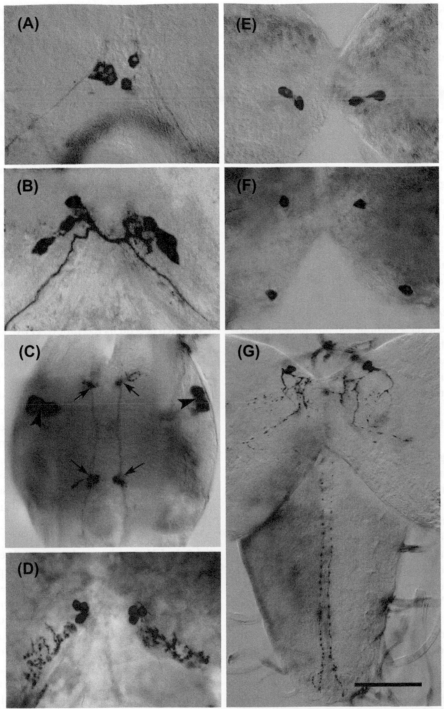

Figure 20 EH-IR in various insect species. Hemimetabolous insects usually show a variable number of EH-IR cells in the brain. (A, B) Two brain specimens of the cockroach *Nauphoeta* show EH-IR in a cluster of six (A) and ten (B) ventromedial (VM) cells. (C) VM cells of the cricket *Acheta* project axons throughout the entire nerve cord and make prominent arborizations in each ventral ganglion (AG4 is depicted; arrows). Two pairs of additional dorsolateral neurons are stained in these ventral ganglia (AG4, arrowheads). (D) Six EH-IR neurons were detected in the brain of the bug *Pyrrhocoris*, but other specimens show 4–9 immunopositive cells. (E) Two pairs of VM cells are stained in the brain of alderfly *Sialis* and (F) antlion *Myrmeleon*. Note that a second pair of VM neurons in *Myrmeleon* is located more posteriorly. (G) Only one pair of EH-producing neurons is stained in the dorsomedial brain region of the fruitfly *Drosophila*. These cells arborize in the brain and project two axons along entire ventral nerve cord (arrows).

EH-immunoreactivity (EH-IR) was detected in variable numbers of neurosecretory cells in the brain, dependig on the species. For example, two pairs of VM cells are present in the dragonfly *Sympetrum sp*, while a cluster of 4–10 EH-IR neurons is stained in the cockroach *N. cinerea* and bug *P. apterus*, and one pair of cells is found in locust *L. migratoria* and cricket *A. domestica* (**Figure 20A–D**). VM neurons of most species project non-arborizing axons posteriorly along the entire ventral nerve cord. Interestingly, axons of *Acheta* VM cells show obvious arborizations in the anterior and posterior regions of each ventral ganglion, and two additional pairs of lateral neurons are stained in these ganglia (**Figure 20C**). EH-producing neurons in most holometabolous insects examined resemble those described in *Manduca* (Truman and Copenhaver, 1989). Two pairs of VM neurons are present in the alderfly *Sialis sp.*, antlion *Myrmeleon sp.*, mealworm beetle *Te. molitor* and silkmoth *B. mori*, while only one pair of dorsomedial neurons shows EH-IR in the fruitfly *Drosophila* (**Figure 20E** and **G**).

Immunohistochemical staining and physiological experiments suggest that EH is released centrally from axons within ventral ganglia to participate in ecdysis activation (Hewes and Truman, 1991; Zitnan *et al.*, 2002a), as well as peripherally to regulate release of PETH and ETH from Inka cells and other functions (Ewer *et al.*, 1997; Kim *et al.*, 2004b; Kingan *et al.*, 1997).

7.4.2.2. Timing of EH release and action on Inka cells

Induction of pre-ecdysis and ecdysis behaviors originally was attributed to actions of circulating EH on the CNS, first examined in the context of circadian-gated adult eclosion (Truman and Sokolove, 1972). Thus, injection of EH initiates adult eclosion and also ecdysis in earlier instars, but only during a relatively narrow window of sensitivity that arises close to the time of natural ecdysis (Copenhaver and Truman, 1982; Reynolds *et al.*, 1979; Truman, 1980; Weeks and Truman, 1984a,b). Evidence for action of EH on the CNS came from experiments demonstrating that semi-isolated ventral nerve cords, when treated with CC extracts containing EH, generate patterned motor output consistent with pre-ecdysis and ecdysis behaviors (Weeks and Truman, 1984a). However, the success of these experiments depended on the presence of an intact tracheal system, ostensibly to oxygenate the highly aerobic nervous system.

It is now recognized that the requirement of the tracheal system for EH action on the CNS *in vitro* is due to the presence of segmentally distributed endocrine Inka cells attached to the tracheal tubes near each spiracle (Zitnan *et al.*, 1996). Furthermore, this effect of EH is observed only during the 7–8 h prior to pupation, when Inka cells are competent to release (Kingan *et al.*, 1997). Blood-borne EH acts on a newly described receptor guanylyl cyclase in Inka cells to induce the release of PETH

and ETH (Chang *et al.*, 2009). The beginning of sensitivity to EH and the latencies from injection to behavioral onset in *Manduca* pharate second and fifth instar larvae, pharate pupae, and pharate adults have been described previously (Copenhaver and Truman, 1982; Ewer *et al.*, 1997; Reynolds *et al.*, 1979; Truman, 1980). Further studies have established that EH actions on the Inka cell provide a mechanism for the complete depletion of Inka cells necessary for the transition from pre-ecdysis to ecdysis behavior (Kim *et al.*, 2004b; Kingan *et al.*, 1997).

The natural release of EH into the hemolymph during the ecdysis sequence is difficult to measure directly, given its very low, biologically active concentration (~2–10 pM). However, elevation of cGMP in Inka cells approximately 30 min after onset of pre-ecdysis in *Manduca* larvae is a good marker of EH appearance in the hemolymph (Kingan *et al.*, 1997). A similar elevation of cGMP in *Drosophila* Inka cells has been documented (Clark *et al.*, 2004). As described earlier, pre-ecdysis is initiated by low levels of PETH/ETH induced by corazonin. ETH then acts on its receptor (ETHR-A) in VM neurons to elicit the EH release (Ewer *et al.*, 1997; Gammie and Truman, 1999; Kim *et al.*, 2006a). Blood-borne EH acts back on Inka cells to cause cGMP production and PETH/ETH depletion in a positive feedback loop (Ewer *et al.*, 1997). The appearance of cGMP in Inka cells ~30 min into pre-ecdysis corresponds to the steep elevation of PETH/ETH in the hemolymph, known to occur upon exposure of Inka cells to EH *in vitro* (Kingan *et al.*, 1997). Furthermore loss of EH-IR in larval proctodeal nerves, which are neurohemal organs for EH, suggests that this peptide is released at this time (Novicki and Weeks, 1996; Zitnan and Adams, 2000).

7.4.2.3. EH signal transduction and receptors in Inka cells

EH causes release of ETH from epitracheal glands *in vitro* at threshold concentrations of ~3–10 pM (Kingan *et al.*, 1997). EH also elevates cGMP in epitracheal glands at concentrations similar to those that cause ETH release, and both cGMP and 8-Br-cGMP actually mimic the action of EH. The *in vitro* secretory response to EH occurs during a narrow window of development, beginning ~8 h prior to pupal ecdysis, which corresponds to the behavioral sensitivity observed previously *in vivo*. However, EH can cause cGMP elevation in epitracheal glands long before they acquire competence to release ETH, suggesting that the EH receptor and initial steps in the signal transduction cascade are in place early in development, well before the behavioral response appears (Kingan *et al.*, 1997). The absence of a downstream step in the cascade apparently prevents secretion. During natural ecdysis, cGMP levels in epitracheal glands show an abrupt elevation ~30 min after onset of pre-ecdysis, well after ETH secretion has been initiated. ETH secretion therefore can be viewed as a two-step process, beginning at pre-ecdysis prior to cGMP

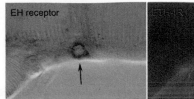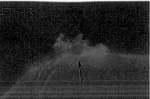

Figure 21 Expression of the EH receptor in an Inka cell of a *Drosophila* pharate pupa. (Left) *In situ* hybridization with an EH receptor probe showed a specific reaction in the Inka cell, which was subsequently stained with antibody to *Drosophila* ETH (arrows).

elevation, followed by a massive release resulting from a logarithmic elevation of cGMP.

Recently, mediators of EH action in Inka cells have been identified as receptor guanylate cyclases in epitracheal glands of the oriental fruit fly, *Bactrocera dorsalis* (Chang *et al.*, 2009). Two forms were described: a high-affinity receptor named BdmGC-1 and its isoform BdmGC-1B. The B form exhibits the same conserved domains and putative N-glycosylation sites found in BdmGC-1, but it possesses an additional 46-amino acid insertion in the extracellular domain and lacks the C-terminal tail of BdmGC-1. Immunolabeling revealed that BdmGC-1 is expressed in Inka cells and possibly also in adjacent cells of the epitracheal gland. Preliminary evidence from *in situ* hybridization in *Drosophila* confirms this finding (**Figure 21**). Thus it could be that EH regulates not only release of ETH from Inka cells, but also signaling among neighboring cells in the epitracheal gsland.

Heterologous expression of BdmGC-1 in HEK cells leads to robust increases in cGMP following exposure to low picomolar concentrations of EH. The B-isoform responds only to higher EH concentrations, suggesting different physiological roles of these cyclases (Chang *et al.*, 2009).

7.5. Orchestration of the Ecdysis Sequence by a Peptide Signaling Cascade

Insect development from larva to adult in all insects involves repeated shedding of old cuticle via a set of hormonally regulated motor movements called the ecdysis sequence. The endocrinology and physiology underlying ecdysis have been extensively studied in several species of moths, which pass through several larval stages and a pupal stage prior to reaching the winged, reproductive adult stage. These include the tobacco hornworm, *Manduca* (Copenhaver and Truman, 1982; Truman, 1980; Zitnan *et al.*, 1996, 1999), giant silkmoths *Hyalophora* and *Antherea* (Truman, 1972; Truman and Riddiford, 1974; Truman and Sokolove, 1972), and the silkworm *Bombyx* (Adams and Zitnan, 1997; Zitnan *et al.*, 2002b). Several studies also have described the ecdysis sequence in *Drosophila, Aedes,* and *Tr.*, where available

genetic tools have permitted studies of ETH-, EH-, and CCAP/MIP/bursicon-deficient flies (Arakane *et al.*, 2008; Baker *et al.*, 1999; Clark *et al.*, 2004; Dai and Adams, 2009; Kim *et al.*, 2006b; McNabb *et al.*, 1997; Park *et al.*, 2002a, 2003b).

Depending on the stage in question, ecdysis involves the initiation and scheduling of a behavioral sequence that includes sequential behaviors referred to as pre-ecdysis, ecdysis, and post-ecdysis. What follows is a summary of the roles played by endocrine signaling molecules in the initiation and scheduling of these events. Evidence is presented that ETH orchestrates a series of downstream peptides that activate sequentially neural circuits in the CNS to generate motor behaviors for the successive behavioral steps of the ecdysis sequence.

7.5.1. Ecdysis Sequence in *Manduca*

7.5.1.1. Embryonic ecdysis sequence During normal development, *Manduca* larvae undergo one embryonic and four post-embryonic ecdyses. PETH and ETH production in Inka cells is detectable in embryos at about 45% of embryonic development by enzyme immunoassay and immunohistochemical staining. Peptide levels in these cells reach a peak at about 65% of development. Embryonic ecdysis at 65–70% coincides with depletion of PETH and ETH immunoreactivity in Inka cells indicating that these peptides have been released into the hemolymph. Dissected embryos respond to ETH injection by showing rhythmic contractions that result in sloughing of the embryonic cuticle. These experiments suggest that ETH controls embryonic ecdysis in *Manduca*. By contrast, larval hatching behavior is not associated with ETH release (Hermesman and Adams, unpublished).

7.5.1.2. Natural and peptide-induced larval ecdysis sequence Studies of ecdysis regulation are assisted by the presence of external morphological markers that allow for recognition of time relative to onset of behaviors. Such morphological markers have been described for pharate second instar, pharate fifth instar, and pharate pupa *Manduca* (Copenhaver and Truman, 1982; Zitnanova *et al.*, 2001). Markers for pharate fifth and pharate pupa stages are summarized in **Table 1**. The natural ecdysis behavioral sequence in *Manduca* larvae is composed of three distinct patterns initiated in the following order: pre-ecdysis I, pre-ecdysis II and ecdysis (Miles and Weeks, 1991; Weeks and Truman, 1984a,b; Zitnan *et al.*, 1999). Pre-ecdysis I is characterized by synchronous dorsoventral contractions of abdominal segments (**Figure 22**), and lasts for 60–70 min (**Figure 23**). Observations of natural behavior indicate that larvae show occasional weak, intermittent pre-ecdysis I contractions for 2–5 min interrupted by ~5 min quiet periods for ~30 min before launching into robust, continuous behavior. Intermittent pre-ecdysis behavior coincides with the first appearance of corazonin in the hemolymph and low levels of circulating PETH and ETH (Kim *et al.*, 2004b).

About 20 min after onset of pre-ecdysis I, larvae initiate pre-ecdysis II (posteroventral and proleg contractions) and continue performing both behaviors simultaneously for about 40 min (**Figures 22** and **23**). During this time, posteroventral and proleg contractions alternate with dorsoventral contractions. About 1 h after initiation of pre-ecdysis I, a switch from pre-ecdysis to ecdysis behavior occurs, characterized by successive anteriorly directed peristaltic movements of abdominal segments and proleg retractions (**Figure 22**). Ecdysis movements last for about 10 to 12 min (**Figure 23**), serving first to rupture the old cuticle along the anterior midline of the thorax

Table 1 Timelines of morphological markers in *Manduca sexta*

Pharate 5th Instar Larva	
-30h	Molting fluid accumulation in the prothorax followed by slippage of the old head capsule
-28h	"Head capsule slip" (HCS), slippage of the old head capsule is completed
-10h	"Yellow mandibles", yellow pigmentation of new mandibles
-8h	"Brown mandibles", darkening of brown mandibles
-4–6h	Resorption of molting fluid and appearance of air in the old head capsule
-1h	Natural pre-ecdysis
0	Natural ecdysis
Pharate Pupa	
-48h	Yellowish dorsal thorax
-24h	Light yellow bars on metathorax "yellow bars"
-12h	"Brown bars"; brown bars on dorsal metathorax
-4h	Resorption of molting fluid and cuticle shrinkage on ventral side of first abdominal segment
-3.5h	"Anterior shrink"; distinct fold on lateral and ventral sides of first abdominal segment
-2.5h	"Posterior shrink"; shrinkage of the old cuticle covering posterior abdominal segments
-1.5h	Tanning of the anterior dorsal region of first abdominal segment
-1.0h	Natural pre-ecdysis
-0.5h	Tanning of the anterior dorsal region of second abdominal segment
0	Natural ecdysis

Table 2 Timelines of morphological markers in *Bombyx mori*

Pharate 5th Instar Larva	
-28h	Head capsule slip (HCS)
-15h	Edges of new light yellow spiracles show grey pigmentation
-12h	Edges of new spiracles turn black
-8h	"Yellow mandibles", yellow pigmentation of new mandibles
-6h	"Brown mandibles", darkening of brown mandibles
Pharate Pupa	
-24h	Pharate pupae stop spinning the cocoon
-4h	Shrinking of the old larval cuticle between all thoracic and abdominal segments
Pharate Adult	
6-7 days	Yellow-brown pigmentation of eyes
5-6 days	Black pigmentation of eyes
1-2 days	Dark pigmentation of antennae, legs and lines on the wings

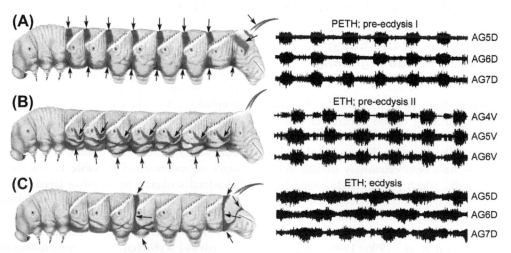

Figure 22 The ecdysis sequence of *Manduca* pharate larvae *in vivo* and *in vitro*. (A) PETH-induced pre-ecdysis I composed of synchronous dorsoventral contractions of abdominal segments (arrowheads) and corresponding synchronous bursts in dorsal nerves of abdominal ganglia 5–7 (AG5, 6, 7D) of the isolated nerve cord. (B) ETH-induced pre-ecdysis II characterized by posteroventral and proleg contractions (arrows) and corresponding bursts *in vitro* in ventral nerves (AG4, 5, 6V). (C) Ecdysis peristaltic abdominal contractions associated with proleg retractions (arrows), which result in shedding of the old cuticle and tracheae. Isolated nerve cord shows corresponding ecdysis bursts in dorsal ganglia (AG5, 6, 7D). Modified from *Zitnan et al.* (2000).

and subsequently to shed the cuticle with attached old tracheae.

Injection of PETH and ETH elicits pre-ecdysis and ecdysis behaviors identical to those observed under natural conditions. Likewise, exposure of the isolated CNS to these peptides elicits fictive motor patterns that correspond closely to *in vivo* behaviors (Zitnan and Adams, 2000; Zitnan *et al.*, 1999). These studies confirm that PETH and ETH activate pre-ecdysis and ecdysis behaviors organized within the CNS by central pattern generators.

In *Manduca*, there are clear differences in the precise actions of PETH and ETH. Injection of PETH into pharate larvae elicits pre-ecdysis I only (**Figures 22 and 23**). PETH-injected animals perform pre-ecdysis I behavior for about 35 to 60 min, but then fail to advance to pre-ecdysis II or ecdysis, regardless of the dose administered (Zitnan *et al.*, 1999). Corresponding pre-ecdysis I synchronous bursts in AGs are recorded *in vitro* from the isolated CNS exposed to PETH (**Figure 22**). Once PETH-induced

pre-ecdysis I behavior subsides, larvae injected again with PETH (100–500 pmol) are unresponsive. However injection of these animals with ETH elicits pre-ecdysis II followed by ecdysis behaviors (**Figures 22** and **23**). In contrast, naïve larvae injected with ETH alone initiate pre-ecdysis I and II simultaneously, followed by ecdysis behavior. Similar results are obtained upon exposure of the isolated CNS *in vitro* (Zitnan *et al.*, 1999; **Figure 22**). Thus in *Manduca*, ETH elicits the entire sequence consisting of pre-ecdysis I, II, and ecdysis, whereas PETH produces pre-ecdysis I only.

7.5.1.3. Natural and peptide-induced pupal and adult behaviors

7.5.1.3.1. Pupal ecdysis The ecdysis sequence during pupation was described *in vivo* and *in vitro* (Truman, 1980; Weeks and Truman, 1984a; *Zitnan et al.*, 1996). Pre-ecdysis of pharate pupae always occurs with the animal positioned ventral side down, and is visible as weak rhythmic contractions of the abdomen. Although these contractions are often difficult to discern visually, blood pressure measurements revealed clear rhythmic pulses during pre-ecdysis (Zitnan *et al.*, 1996). Robust ecdysis peristaltic contractions begin approximately 1 h later in the most posterior abdominal segment and move anteriorly. After rupture of the old cuticle along the dorsal thoracic midline, animals turn ventral side up and with the aid of strong peristaltic movements the cuticle is shed. During post-ecdysis, the proboscis, wing pads, and legs of the fresh pupa are then expanded and sclerotized to their typical shapes.

PETH injection has either no effect or causes weak rhythmic contractions likely representing pre-ecdysis in animals injected 3–4 h prior to ecdysis. ETH treatment 3–48 h prior to natural ecdysis induces within 35–60 min ecdysis behavior; weak pre-ecdysis behavior preceding ecdysis is only observed in pharate pupae injected 3–40 h prior to natural ecdysis. Removal of the head and thorax or CO_2 treatment during pre-ecdysis accelerates the onset of ecdysis behavior as observed in larvae (Fuse and Truman, 2002; Zitnan and Adams, 2000). ETH also induces strong ecdysis bursts in dorsal nerves of the isolated nerve cord *in vitro* (Zitnan *et al.*, 1996).

7.5.1.3.2. Adult eclosion The adult eclosion sequence is quite different from larval and pupal ecdysis sequences and is composed of pre-eclosion and eclosion behaviors, as well as digging from the underground pupation site (Mesce and Truman, 1988; Reynolds *et al.*, 1979). Pre-eclosion behavior is initiated by occasional rotations of the abdomen, which may be missing in some animals. The onset of eclosion behavior is characterized by strong movements of wing bases that break anterior parts of the pupal cuticle while abdominal retractions and extensions help the adult to emerge from the pupal case and to escape

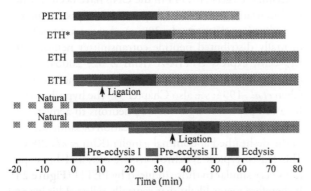

Figure 23 Time lines of natural and peptide-induced pre-ecdysis and ecdysis behaviors in *Manduca*. PETH injection into pharate fifth instar larvae induces pre-ecdysis I behavior only, which lasts for ~30 min, but in some animals it lasts for up to 1 h (stippled line). These larvae do not respond to further PETH treatment, but subsequent ETH injection (ETH*) induces pre-ecdysis II and ecdysis behaviors. However, naïve larvae injected only with ETH initiate pre-ecdysis I and II simultaneously, followed by ecdysis movements about 40 min later. These animals cannot shed their cuticle because hormone injection elicited behaviors prematurely, so ecdysis contractions last for up to 1 h. Ligation (arrow) between abdominal segments 1–2 and removal of all anterior segments at 10–15 min into pre-ecdysis greatly accelerates the initiation of ecdysis. The natural ecdysis sequence starts with occasional weak dorsoventral contractions in posterior abdominal segments (broken stippled squares). Continuous strong dorsoventral contractions are considered as the initiation of normal pre-ecdysis I is indistinguishable from PETH-induced behavior. About 20 min later, animals initiate posterolateral and proleg contractions (pre-ecdysis II) alternating with pre-ecdysis I compressions exactly like ETH-injected larvae. Ecdysis behavior is initiated about 1 h later and lasts 12–15 min. After the cuticle is completely shed, ecdysis movements immediately stop. Ligation of larvae 35–40 min into natural pre-ecdysis accelerates ecdysis onset as described earlier.

to the surface. Interestingly, adults retain the larval ecdysis motor program. Typical larval-like peristaltic abdominal movements can be induced in eclosing pharate adults by transecting connectives between the pterothoracic and abdominal ganglia (Mesce and Truman, 1988).

Pharate adults are sensitive to ETH beginning about 24 h prior to natural eclosion. ETH injections during this time period induce within 3–10 min occasional abdominal rotations, followed by a relatively quiescent phase. Eclosion behavior is initiated 2–3 h after ETH treatment and adults emerge in 10–20 min, but are usually wet and do not completely spread their wings (Zitnan *et al.*, 1996). These animals, nevertheless, climb to a suitable perch and assume the posture normally associated with wing expansion.

Since ETH-AP is processed from the ETH precursor and released at ecdysis (Zitnan *et al.*, 1999), its function was tested by injection parallel with PETH and ETH. However, no obvious effects were observed following injections of ETH-AP into pharate larvae, pupae, or adults and the isolated larval CNS shows no specific response to ETH-AP treatment *in vitro* (Zitnan *et al.*, 1999; Zitnan, unpublished observations).

7.5.1.4. Central regulation of pre-ecdysis (kinins, DHs)

Initiation of pre-ecdysis I and II involves direct actions of blood-borne PETH and ETH on abdominal ganglia. Pre-ecdysis I behavior is driven by a single pair of ascending interneurons whose cell bodies are located in the posterior region of the TAG, the IN-402 cells. They make monosynaptic connections with MN-2,3 in each abdominal ganglion to synchronize bursting of these motor neurons in all abdominal segments (Miles and Weeks, 1991; Novicki and Weeks, 1993, 1995). Transection of the ventral nerve cord during natural pre-ecdysis abolishes patterned bursting of MN-2,3 rostral to the cut, consistent with removal of ascending input from these interneurons (Novicki and Weeks, 1993). Likewise, PETH injection into pharate larvae already transected between AG4-5 or AG6-7 leads to normal rhythmic pre-ecdysis I contractions caudal to the cut, but abnormal, prolonged dorsoventral contractions rostral to the cut. Similar results are obtained from the isolated CNS following PETH exposure *in vitro*: transection of connectives leads to unpatterned, prolonged motor bursts in ganglia anterior to the cut, whereas posterior ganglia retain normal synchronous rhythmic bursts (Zitnan and Adams, 2000).

IN-402 cells probably produce acetylcholine to provide the necessary input for normal synchronized rhythmic contractions (Novicki and Weeks, 1995). IN-402 and MN2,3 retain their intrinsic properties and strength of synaptic connections in pharate pupae, indicating that weakening of pre-ecdysis behavior in this stage is caused by other unknown components of this behavior (Novicki and Weeks, 2000).

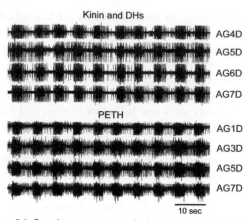

Figure 24 Synchronous pre-ecdysis I bursts induced by exposure of the desheathed CNS of a *Manduca* pharate larva to a mixture of kinin and DHs. PETH elicits very similar bursts indicating that Inka cell peptides cause the central release of kinins/DHs to control pre-ecdysis I behavior. Adapted from *Kim et al.* (2006a).

Cellular targets of ETH in the CNS have been revealed by localization of ETH receptor transcripts by *in situ* hybridization (Kim *et al.*, 2006a). An ensemble of segmentally distributed peptide cotransmitter neurons, the $L_{3,4}$ cells, express ETHR-A (Kim *et al.*, 2006a). These neurons co-express kinins and diuretic hormones (DHs; Chen *et al.*, 1994; see also Chapter 9). The hypothesis that Inka cell peptides target the $L_{3,4}$ neurons to activate pre-ecdysis was tested by bath co-application of kinin and DH to the desheathed abdominal ganglia (Kim *et al.*, 2006a). This led to typical synchronous pre-ecdysis I burst patterns very similar to those induced by PETH (**Figure 24**). It therefore seems likely that centrally released kinins and DHs are proximal signaling molecules for recruitment of a specific network of inter- and motoneurons that give rise to pre-ecdysis I behavior. $L_{3,4}$ neurons also release these neuropeptides into the hemolymph, presumably to control diuretic activity of Malphigian tubules. The amino acid sequences of *Bombyx* and *Drosophila* kinins and DHs are shown in **Figure 25**. Structure and peripheral roles of all insect neuropeptides in regulation of water balance has been described in detail (see Chapter 9).

In contrast to pre-ecdysis I, where motor neurons in each abdominal ganglion (AG) are slaves to ascending interneurons from the TAG, the pacemakers for pre-ecdysis II reside in each ganglion, such that motor bursts occur independent of inputs from other ganglia. Ligation or transection of connectives between various AGs does not affect proleg and posteroventral contractions during pre-ecdysis II, and isolated AG produces patterned bursting similar in duration and frequency to that recorded in intact nerve cords (Zitnan and Adams, 2000). Neurons and neurotransmitters controlling pre-ecdysis II have not been identified, but it seems likely that the principal and accessory planta retractor motoneurons (PPR, APR)

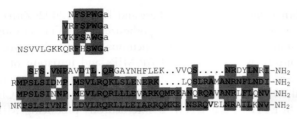

Bombyx kinin I   NFSPWGa
Bombyx kinin II   VRFSPWGa
Bombyx kinin III   KVKFSAWGa
Drosophila kinin NSVVLGKKQRFHSWGa

Bombyx DH34   SFS.VNEAVDTL.QRGAYNHFLEK..VVQS.....NRDYLNRI-NH₂
Bombyx DH41   RMPSLSIDMP.MSVLRQKLSLENERK....LQSLRAMANRNFLNDI-NH₂
Bombyx DH45   MPSLSINNP.MEVLRQRLLLEVARKQMREANQRQAVANRLFLQNV-NH₂
Drosophila DH44 NKPSLSIVNP.LDVLRQRLLLEIARRQMKE.NSRQVELNRAILKNV-NH₂

Figure 25 Alignment of kinins and DHs identified in *Bombyx* and *Drosophila*. *Bombyx* neuropeptides are identical or very similar to those identified in other moths including *Manduca* — sequence identity (green), sequence similarity (yellow).

involved in proleg retractions during ecdysis also participate in similar contractions during pre-ecdysis II (Weeks and Truman, 1984b). The ability of ETH to elicit pre-ecdysis II contractions in individually isolated abdominal ganglia indicates the presence of command ETHR neurons in each ganglion. Unidentified dorsolateral neurons expressing ETHR-B in AG1-7 are prime candidates for modulation of the pre-ecdysis II circuit (**Figure 10**; Kim *et al.*, 2006a).

7.5.1.5. Central regulation of ecdysis activation (EH)

In *Manduca* larvae, ETH "activates" neurosecretory cells 27 (also known as L₁) and interneurons 704 (27/704) in the AG1-7 through elevation of cGMP (Ewer *et al.*, 1994, 1997; Zitnan and Adams, 2000; Zitnan *et al.*, 2002a). We use the term "activation" here not to indicate behavioral activation, but an elevation of the second messenger cGMP, which is associated with increased excitability that precedes onset of the behavior. Cells 27 are classical neurosecretory cells that project to several peripheral neurohemal sites for release of peptides into the circulatory system. Intracellular recordings from *Manduca* cells 27 showed that production of cGMP is associated with excitability of these neurons (Gammie and Truman, 1997b). The production of cGMP in a homologous network of 27/704 neurons also was detected at ecdyses of a wide range of phylogenetically diverse insects, including silverfish, locust, cricket, mealworm beetle, and mosquito, but not in the fruitfly *Drosophila* (Ewer and Truman, 1996). These data indicate that cGMP elevation may be a conserved mechanism for activation of ecdysis neurons in most insects.

It appears that ETH elevates cGMP in 27/704 cells of ventral ganglia via multiple downstream pathways. Intracellular recordings showed that either epitracheal gland extracts or synthetic ETH elicits increased electrical activity in the brain VM neurons *in vitro* within 4–13 min. Individually isolated VM somata also show increased excitability within 10 min of ETH exposure (Ewer *et al.*, 1997; Gammie and Truman, 1999). Direct action of ETH on VM cells is supported by the presence ETHR-A transcripts in these EH-producing neurons (Kim *et al.*, 2006a). These data suggest that ETH action on VM neurons results in EH release. Centrally released EH may then participate in activation of the 27/704 network through cGMP elevation.

Although VM neurons are one primary target of ETH, this peptide may act through additional pathways to elevate cGMP in ventral ganglia. Evidence for this comes from ETH action on the debrained larval nerve cord, which does not release EH in response to ETH exposure (Zitnan and Adams, 2000). In spite of this, ETH exposure results in normal cGMP elevation and ecdysis bursts as observed in control intact CNS preparations. Moreover, direct EH application to the intact CNS does not result in elevation of cGMP. However, exposure of isolated ventral nerve cords with desheathed SG and TG1-3 to EH causes strong cGMP elevation in all intact (non-desheathed) abdominal ganglia. Conversely, EH treatment of desheathed abdominal ganglia results in increased cGMP production in intact (non-desheathed) SG and TG1-3 (Zitnan and Adams, 2000). Since EH does not penetrate the blood–brain barrier, these data indicate that additional downstream neurons may participate in the activation of the IN-704 network in *Manduca* through cGMP elevation. Which neurons may be responsible for this remains a question, but it is known that ETHR-A and ETHR-B transcripts appear in a large number of additional neurons in the brain and ventral ganglia, including the entire network of IN-704 in the SG, TG1-3, and AG1-7 (Kim *et al.*, 2006a). We propose that ETH action on its receptor (ETHR-A) in the descending IN-704 in the SG and TG1-3 elicits release of unknown neuropeptide(s) participating in cGMP elevation and activation of abdominal network 27/704.

It is also important to note that cGMP elevation may not be obligatory for ecdysis activation in all instances. For example, during pupal and adult ecdyses, only cells 27 show cGMP elevation, while IN704 neurons do not (Ewer and Truman, 1997). Also, debrained nerve cords exposed to ETH show normal ecdysis activity, but little or no cGMP production (Asuncion-Uchi *et al.*, 2009; Ewer *et al.*, 1997). This suggests that the IN-704 network does not require elevation of cGMP, at least in these stages (Asuncion-Uchi *et al.*, 2009; Ewer *et al.*, 1997). In these instances, the role of EH may be limited to post-ecdysis/eclosion events.

7.5.1.6. Descending inhibition delays ecdysis onset (MIPs, FGLamides)

The transition from pre-ecdysis to ecdysis behavior in *Manduca* requires ETH activation of centers in the brain and SG, from which descending inputs to the ventral nerve cord activate the ecdysis neuronal circuitry (Novicki and Weeks, 1996; Zitnan and Adams, 2000). During the natural ecdysis sequence in larval *Manduca*, activation of ecdysis circuitry, indicated by cGMP elevation in abdominal 27/704 neurons, is complete ~20–30 min after the start of pre-ecdysis, but ecdysis is initiated only after an additional 30 min delay, apparently due to descending inhibition. In ETH-injected animals, activation of ecdysis circuitry already is complete 10–15 min into pre-ecdysis, but the transition to ecdysis behavior is similarly delayed for ~30 min. Once the ecdysis circuitry is activated, the continued presence of ETH is not necessary for ecdysis initiation, indicating that ETH is a "triggering" peptide that sets in motion a sequence of events culminating in ecdysis.

The delay (~30 min) between ETH activation of the ecdysis network and initiation of behavior is controlled by a balance of excitation and inhibition descending from the SG cephalic and especially thoracic ganglia (TG1-3). Evidence for this comes from experiments showing that ligations posterior to the head or thorax or brief CO_2 exposure accelerates the onset of ecdysis behavior during larval, pupal, and adult ecdyses (Fuse and Truman, 2002; Zitnan and Adams, 2000; **Figure 23**). These experiments indicate that descending inhibitory input from the SG and TG1-3 delays the central release of CCAP and MIPs from the 27/704 network in AG1-7 and initiation of ecdysis. An independent clock in each abdominal ganglion controls the timing of disinhibition, allowing animals to switch from pre-ecdysis to ecdysis behavior.

Excitation and inhibition leading to initiation of ecdysis behavior may to be related to sequential phases of cGMP elevation in different groups of central neurons. About six hours prior to the onset of natural pre-ecdysis, cGMP immunoreactivity (cGMP-IR) is detected in 27/704 neurons of the SG and TG1-3, but not in AG1-7 (Zitnan and Adams, 2000). This is well before ETH is released into the hemolymph. Following the onset of pre-ecdysis, cGMP staining in the SG and TG1-3 gradually increases. At 30 min into pre-ecdysis, cGMP staining in 27/704 neurons of AG1-7 appears and increases in intensity over the next 15 min. At this time additional neurosecretory cells $L_{2,5}$ show weaker cGMP-IR in each AG1-7 (Fuse and Truman, 2002; Zitnan and Adams, 2000). It appears that cGMP elevation in AG1-7 neurons coincides with activation of ecdysis circuitry, since ligations or removal of the head or thorax at this time accelerates initiation of ecdysis behavior.

In ETH-injected larvae, cGMP is elevated in 27/704 neurons of AG1-7 within 10–15 min. Surgery or ligations applied after this time accelerates initiation of ecdysis behavior, whereas their application before this time prevents ecdysis. Similar, but less obvious results are obtained from measurements of fictive behavior in isolated nerve cords exposed to ETH (Zitnan and Adams, 2000).

Surgical and ligation experiments showed that the principal source of inhibitory input resides in the SG

and TG1-3 (Fuse and Truman, 2002; Zitnan and Adams, 2000). We hypothesize that ETH activates descending inhibitory interneurons during early pre-ecdysis to delay release of CCAP/MIPs and initiation of ecdysis. Most of these inhibitory neurons and compounds (probably neuropeptides) have not been identified. Clusters of ventral ETHR-B interneurons producing MIPs and other unknown factors in the SG, TG1-3, as well as small ventrolateral interneurons expressing ETHR-A,B and FGLa (AST-A) in each AG1-7 are probably involved in inhibition of the ecdysis network (**Figure 10**).

7.5.1.7. Central regulation of ecdysis initiation (CCAP, MIPs)

7.5.1.7.1. Crustacean cardioactive peptide
Crustacean cardioactive peptide (CCAP) was originally identified from the CNS of the crab *Carcinus maenas* based on its heart stimulatory activity (Stangier *et al.*, 1987). An identical neuropeptide and the gene encoding a single copy of CCAP later were identified or annotated in many other insects (**Figure 26**; Cheung *et al.*, 1992; Furuya *et al.*, 1993; Lehman *et al.*, 1993; Loi *et al.*, 2001; Stangier *et al.*, 1989).

Using immunohistochemistry and *in situ* hybridization techniques, developmental changes in expression of CCAP and its mRNA precursor were described in *Manduca* (Davis *et al.*, 1993; Loi *et al.*, 2001). In early larval stages, strong CCAP-IR is present only in two pairs of neurons in each AG1-7 and in the peripheral neurons of abdominal segments 2–7. The paired neurons were identified as neurosecretory cells 27 (or NS-L$_1$) and interneurons 704 (IN-704) (Davis *et al.*, 1993; Ewer *et al.*, 1994; Sandstrom and Weeks, 1991; Taghert and Truman, 1982a). Curiously, SG and TG1-3 homologues of 27/704 cells do not produce CCAP in *Manduca* larvae; instead these cells express IgG-like protein and bursicon (Klukas *et al.*, 1996). CCAP expression appears in the entire network of 27/704 neurons in the pupal stage, but vanishes after adult emergence. In late larval stage and in pupae,

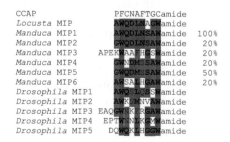

```
CCAP                      PFCNAFTGCamide
Locusta  MIP              AWQDLNAGWamide
Manduca  MIP1             AWQDLNSAWamide       100%
Manduca  MIP2             GWQDLNSAWamide        20%
Manduca  MIP3       APEKWAAFHGSWamide           20%
Manduca  MIP4             GWNDMSSAWamide         20%
Manduca  MIP5             GWQDMSSAWamide         50%
Manduca  MIP6             AWSALHGAWamide         20%
Drosophila MIP1           AWQSLQSSWamide
Drosophila MIP2           AWKSMNVAWamide
Drosophila MIP3      EAQGWNKFRGAWamide
Drosophila MIP4         EPTWNNLKGMWamide
Drosophila MIP5          DQWQKLHGGWamide
```

Bombyx cDNA encoding MIPs

```
MRWCLFALWVFGVATVVTAAEEPHHDAAPQTDNEVDLTEDDKRAWSSLHSGWAKR
AWQDMSSAWGKRAWQDLNSAWGKRGWQDLNSAWGKRAWQDLNSAWGKRGWQDLNSAW
GKRDDDEAMEKKSWQDLNSVWGKRAWQDLNSAWGKRAWQDLNSAWGKRGWNDISSVW
GKRAWQDLNSAWGKRAWQDMSSAWGKRAPEKWAAFHGSWGKRSSIEPDYEEIDAVEQ
LVPYQQAPNEEHIDAPEKKAWSALHGTWGKRPVKPMFNNEHSATTNEA
```

Figure 26 Amino acid sequences of CCAP and MIPs from different insect species — conserved tryptophans (turquoise), sequence identities (green), sequence similarities (yellow).

CCAP-IR is also detected in a pair of abdominal moto-neurons MN1, two medial neurosecretory cells NC₄, and two ventral unpaired neurons VUM (Davis *et al.*, 1993; Loi *et al.*, 2001).

Further studies with CCAP antisera revealed a conserved pattern of paired lateral neurons in the CNS in many apterygote, hemi- and holometabolous insects (Breidbach and Dircksen, 1991; Dircksen *et al.*, 1991; Ewer and Truman, 1996; Helle *et al.*, 1995; Jahn *et al.*, 1991; Kostron *et al.*, 1996). Examples of CCAP-IR neurons in various insects are shown in **Figures 27** and **28**.

Homologues of cells 27 in *Gryllus* and *Periplaneta* originally were described as lateral white (LW) cells (Adams and O'Shea, 1981; O'Shea and Adams, 1981) and type I and II neurons in *Locusta* (Davis *et al.*, 2003; Dircksen *et al.*, 1991). For the sake of clarity, we refer to them here as 27 cells and IN-704.

CCAP was also detected in the CNS of several crustaceans, spiders, and an opilionid (Breidbach *et al.*, 1995; Dircksen and Keller, 1988; Stangier *et al.*, 1988). High levels of CCAP also were detected in the hemolymph during ecdysis of decapod crustaceans, *C. maenas* and

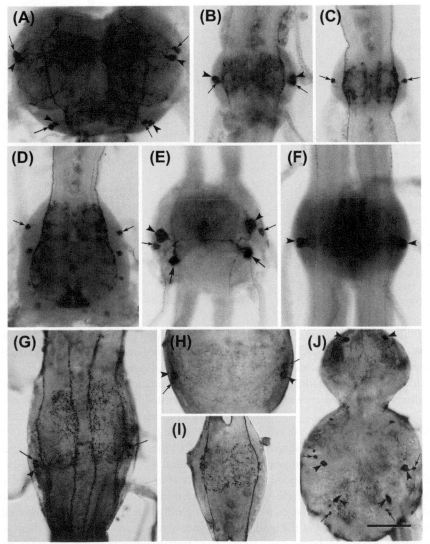

Figure 27 Variability of CCAP-IR in the CNS of hemimetabolous insects. All insects show CCAP-IR in putative homologues of neurosecretory cells 27 (arrowheads) and interneurons 704 (IN-704; small arrows). (A–D) In *P. americana* both cell types are stained in the fused TG3-AG1 (A) and unfused AG2-4 (B), whereas AG5, 6 (C) and TAG (D) lack cells 27 and show only staining in IN-704 and several other small cells. (E) In addition to cells 27/704 in AG1-7, *L. migratoria* shows a very strong CCAP-IR in a pair of large neurosecretory cells in AG1-6 (large arrows). These cells probably show cGMP elevation at ecdysis (*Truman et al.*, 1996). (F) CCAP-IR in *Acheta* is restricted to a pair of 27 cells in AG1-7; the dragonfly *Sympetrum sp.* shows a very similar pattern (not shown). (G) One pair of 27/704 homologues is stained in AG1-7 of *Lepisma*. (H, I) In *Carausius* cells 27/704 are detected in the TG1-3 (H), but no cell bodies are stained in AG1-7 (I). (J) Similarly, TG1-3 of *Pyrrhocoris* show CCAP-IR in putative 27/704 cells, but no immunopositive cells are found in the fused abdominal ganglion. Immunoreactive axons in AG of the two latter species probably originate from 27/704 cells in the SG and TG1-3.

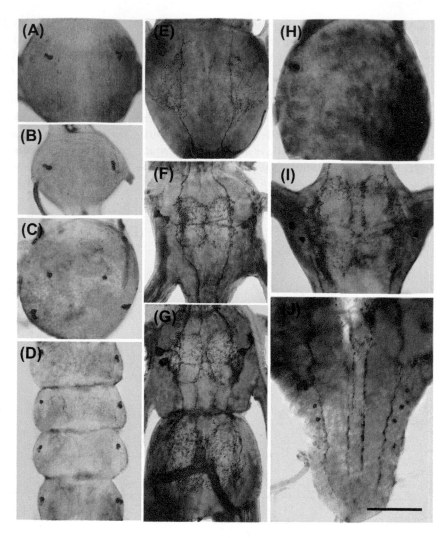

Figure 28 CCAP-IR in the ventral nerve cord of holometabolous larvae. A network of CCAP-IR neurons 27/704 is more conserved in holometabolous insects. (A, B) In *Sialis* one pair of putative cells 27 and IN-704 is detected in TG1-3 (A) and AG1-7 (B). (C, D) A similar distribution of CCAP-IR cells 27/704 is detected in the TG1-3 (C) and AG1-7 (D) of *Myrmeleon*, but some additional neurons are stained in the TG1-3. (E–G) In *Bombyx* larvae no cell bodies are stained in SG/TG1-3 (E), whereas strong immunoreactivity is found in cells 27/704 in AG1-7 (F, G). (H, I) A pair of cells 27/704 is detected in the TG1-3 (H) and AG1-7 (I) of *Te.* (J) *Drosophila* shows CCAP-IR in one pair of lateral cells probably homologous with cell 27 in each subesophageal, thoracic, and first six abdominal neuromeres.

Orconectes limosus (Phlippen *et al.*, 2000). The identical structure and widespread occurrence of CCAP in a conserved network of neurons provide increasing evidence that this neuropeptide plays an important role in regulation of cuticle shedding in insects and crustaceans.

7.5.1.7.2. Myoinhibitory peptides
This family of neuropeptides has been identified in widely divergent insects using various bioassays and referred to as myoinhibitory peptides (MIP), allatostatin B, or prothoracicostatic peptides. The first of these was identified in cephalic ganglia of *L. migratoria*, where it was found to inhibit contractions of the isolated hindgut and oviduct and of the hindgut of the cockroach *Leucophaea maderae* (Schoofs *et al.*, 1991). Another MIP inhibiting visceral muscle contractions was isolated from the CC-CA extracts of the cockroach *P. americana* (Predel, 2001). Several MIP-related peptides were isolated from brains of *Gryllus bimaculatus* and *Carausius morosus*. Because they suppress JH biosynthesis by the CA of adult crickets *in vitro*, they were named allatostatins type B. However, they failed to suppress JH production in *Carausius* (Lorenz *et al.*, 1995, 2000).

Six related peptides (MIP-I-VI) suppressing adult ileum contractions were identified from extracts of adult *Manduca* abdominal ganglia (Blackburn *et al.*, 1995, 2001). A "prothoracicostatic hormone" (PSTH) identical to Mas-MIP1 was isolated from larval brains of *B. mori* (see Chapter 1). This peptide inhibits production of ecdysteroids from prothoracic glands *in vitro* (Hua *et al.*, 1999). In *Bombyx* the cDNA precursor encodes thirteen copies of seven peptides identical or closely related to *Manduca* MIPs (**Figure 26**; Hua *et al.*, 1999). Numerous related peptides were predicted from available insect genomes. Since these peptides originally were named on the basis of inhibitory activity in visceral muscle, we refer to all of them as MIPs.

Immunohistochemistry has been used to detect neurons producing MIPs in the CNS of several insect (Predel, 2001; Schoofs *et al.*, 1996; Veelaert *et al.*, 1995). In the moths *Manduca* and *Bombyx*, MIPs are expressed in numerous small interneurons throughout the CNS, but prominent expression is detected in IN-704 in AG1-7 and clusters of large medial neurons in AG5, 6, and TAG (**Figures 9** and **10**; Davis *et al.*, 2003; Kim *et al.*, 2006a; Yamanaka *et al.*, 2008).

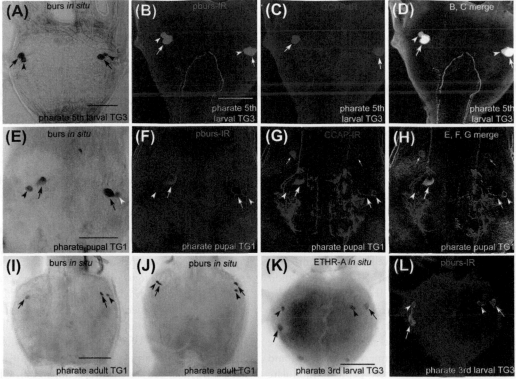

Figure 29 Bursicon expression in 27/704 cells in the thoracic ganglia of *Manduca*. Pharate larvae (A–D, K, L), pharate pupae (E–H), and pharate adults (I, J). (A–D) Cells 27/704 of TG3 contain burs transcripts and show pburs-IR. Cell 27 (arrows) co-expresses bursicon and CCAP in pharate fifth instar larvae (B–D). (E–H) In pharate pupa, bursicon, and CCAP are co-expressed in both cells 27/704. Two anterior bilateral neurons show CCAP-IR (small arrows in G and H). (I, J) In the pharate adult stage, only weak burs/pburs expression is detected in 27/704 neurons. (K, L) ETHR-A is expressed in 27/704 neurons (arrows and arrowheads) as revealed by *in situ* hybridization (K) and pburs antibodies (L). Scale bars: 50 μm. Reproduced with permission from *Dai et al.* (2008a).

7.5.1.7.3. Roles of CCAP and MIPs in ecdysis initiation

Recent studies provide increased evidence that central release of CCAP and MIPs from IN-704 provide the necessary signaling for immediate initiation of ecdysis behavior. Immunohistochemical and *in situ* hybridization studies show that IN-704 expresses both types of neuropeptides and ETHR-A (**Figure 9**; Davis *et al.*, 2003; Ewer and Truman, 1996; Loi *et al.*, 2001). Central neuropeptide release is indicated by decreased CCAP/MIP-IR in elaborate axonal arborizations in the ventral nerve cord after larval, pupal, and adult ecdysis/eclosion of *Manduca* (Davis *et al.*, 2003; Gammie and Truman, 1997a). This neuropeptide release is associated with the disappearance of ETHR-A expression in IN-704 after ecdysis (Kim *et al.*, 2006a). Furthermore, application of synthetic CCAP on the isolated, desheathed chain of abdominal ganglia AG1-8 induces the ecdysis motor program (Gammie and Truman, 1999; Zitnan and Adams, 2000). However, CCAP only causes occasional non-patterned or synchronous bursts when applied on the entire desheathed CNS. On the other hand, application of a mixture of CCAP and MIPs induces typical ecdysis bursts in these nerve cords in a few minutes (Kim *et al.*, 2006a). These data suggest that MIPs suppress activity of pre-ecdysis neurons, while CCAP controls activity of the ecdysis network.

It is not clear what factors regulate release of CCAP/MIPs. We speculate that direct ETH action on its receptors in thoracic ganglia relieve descending inhibition and induce the release of CCAP/MIPs. Paired dorsolateral interneurons that express ETHR-A and ETHR-B along with neuropeptide F (NPF) in TG-1-3 and AG1-7 are suspected candidates for this function (**Figures 9 and 10**).

7.5.1.7.4. Motoneurons controlling ecdysis contractions

Electrophysiological and cobalt backfilling techniques were used to identify motoneurons controlling muscle contractions during larval and pupal ecdyses of *Manduca*. These motoneurons innervate specific muscles of the body wall or prolegs to control specific contractions during ecdysis behavior (Levine and Truman, 1982, 1983; Taylor and Truman, 1974; Weeks and Truman, 1984a,b). Peristaltic contractions of intersegmental muscles are regulated by medial ISM motoneurons, which exit from the dorsal nerves of the following ganglion. Proleg retractions at larval ecdysis are controlled by principal and accessory planta retractor muscles innervated by the posterolateral PPR and APR motoneurons, respectively. Interestingly, these neurons retain their activity even if prolegs degenerate at pupation (Weeks and Truman, 1984b). Dorsoventral contractions of tergopleural muscles during peristaltic movements are controlled by contralateral motoneurons 2 and 3 (MN-2,3), which also control synchronous compressions during pre-ecdysis I behavior (Weeks and Truman, 1984a).

7.5.1.8. Regulation of muscle contractions and ecdysis termination (FRamides, sNPF, MIPs)

Quantitative immunoassays (ELISA) suggest participation of *Manduca* FLRFamides in ecdysis. ELISA of three FLRFamides (F7G, F7D, F10; **Figure 17**) originally isolated from the *Manduca* CNS (Kingan et al., 1990, 1996) showed that larval, pupal and adult ecdyses are associated with decline of these peptides in ventral ganglia (Miao et al., 1998). Consistently, elevated levels of F7D and F10 were detected in the hemolymph 10 min after adult emergence (Kingan et al., 1996). The release of F10 also is suggested by the considerable reduction of its mRNA precursor levels in the ventral nerve cord following each larval, pupal, and adult ecdyses (Lu et al., 2002). Presence of FLRFamide-IR in fine fibers innervating skeletal and visceral muscles indicates that these peptides modulate muscle contractions during pre-ecdysis and ecdysis behaviors (Miao et al., 1998).

Other RFamides along with MIPs are probably involved in regulation of visceral muscle contractions and ecdysis termination. After the old cuticle is completely shed, animals immediately stop ecdysis behavior. However, if the old cuticle cannot be shed (especially from the last abdominal segments), ecdysis contractions continue for a much longer time (up to 1 h). Similarly, isolated nerve cords lacking sensory connections to the cuticle show continuous ecdysis bursts for hours. These observations suggest that cessation of ecdysis behavior is controlled by sensory input from the last abdominal segment, which probably activates inhibitory neurons in the CNS to irreversibly stop the ecdysis motor program after shedding is completed. In search for possible neurons controlling ecdysis termination, we found several large visceral neuromodulatory ETHR-A neurons in AG8 producing short neuropeptides F (sNPF) and MIPs (Kim et al., 2006a). These neurons project axons through the terminal nerves to produce elaborate innervation on the hindgut surface and posterior regions of the last abdominal segment (**Figures 9** and **10**; Davis et al., 1997). We speculate that successful shedding of the old cuticle is associated with the central and peripheral release of these inhibitory peptides, which results in irreversible suppression of visceral muscle contractions and termination of the ecdysis behavior.

7.5.1.9. Regulation of water balance during ecdysis (ITP/CHH, kinins, DHs)

Insect ion transport peptides (ITP) belong to a large family of crustacean hyperglycemic hormones (CHH) and molt inhibiting hormones (MIH) that are prominent in crustacean physiology (for reviews see de Kleijn et al., 1998; Van Herp, 1998; Webster, 1998). Since some actions of CHH are associated with ecdysis, it is given some emphasis here. CHH originally was identified in sinus gland extracts of *C. maenas* as an amidated 72 amino acid peptide (Kegel et al., 1989). Subsequently, numerous related peptides and corresponding precursors and genes were identified in several crustaceans (de Kleijn et al., 1994, 1995; Gu and Chan, 1998; Gu et al., 2000; Ohira et al., 1997; Tensen et al., 1991; Weidemann et al., 1989).

Immunohistochemistry revealed that CHH is produced by the crustacean eyestalk neuroendocrine system composed of neurosecretory cells in the X-organ and neurohemal sinus gland (Dircksen et al., 1988). Subsequent studies showed that CHH is also produced by a novel endocrine system comprising thousands of endocrine cells in the fore- and hindgut of *Carcinus* (Chung et al., 1999; Webster et al., 2000). Massive release of CHH from gut endocrine cells was detected during ecdysis of *Carcinus*, causing water and ion uptake necessary for swelling of the animal and rupture of the epidermal lines, followed by ecdysis of the old cuticle and the subsequent increase in size during post-ecdysis (Chung et al., 1999). CHH/ITP-related peptides likely control water and ion balance in other arthropods as well, as first described in the locust.

Ion transport peptide originally was identified in the storage lobe of locust CC and shown to stimulate salt and water reabsorption and to inhibit acid secretion by the hindgut of *Schistocerca gregaria* (Audsley et al., 1992a,b). Precursors encoding peptides related to ITP and CHH have been identified in numerous insects. These precursors encode the amidated 72 amino acid ITP and additional longer free-carboxy terminal peptides called ITP-like peptides, or ITPL (**Figure 19**; Dai et al., 2007a; Dircksen, 2009).

In situ hybridization and immunohistochemical staining with ITP, ITPL, and CHH antisera were used to detect 5–6 pairs of Ia_2 neurosecretory cells in the *Bombyx*, *Manduca*, and *Locusta* brain (**Figure 2**). Smaller paired neurons also were stained in each ventral ganglion and in abdominal peripheral neurons (Dai et al., 2007a; Dircksen, 1998). Immunohistochemistry and especially ELISA suggest the release of ITP/ITPL-IR peptides into the hemolymph several hours prior to ecdysis of *Manduca* larvae (Drexler et al., 2007). However, specific ITP/ITPL roles in insect ecdysis are not understood. Based on analogy with gut CHH in crustaceans (Chung et al., 1999) and ITP role in gut ion transport in locusts (Phillips et al., 1998), we speculate that blood-borne ITP/ITPL may control functions associated with water reabsorption, which is a necessary preparatory step for shedding of the old cuticle. Other neuropeptides with diuretic functions (kinins, CRF-like DHs, and calcitonin-like DH31) are apparently produced by ETHR-A neurons and may participate in regulation of water balance during pre-ecdysis and ecdysis.

7.5.1.10. Regulation of air swallowing behaviors

Air swallowing behavior is associated with larval and adult ecdyses and has been studied in the locust *Schistocerca* (Bernays, 1972b), the cricket *Teleogryllus* (Carlson, 1977a), the butterfly *Pieris brassicae* (Cottrell, 1964), and the blowfly *Calliphora erythrocephala* (Cottrell, 1962a). Air swallowing prior or during ecdysis is required for expanding the new exoskeleton, which is important for splitting and shedding the old cuticle. Freshly ecdysed insects need air swallowing for inflation and expansion of the new soft cuticle (e.g., wings and body appendages) to achieve the final shape of the new developmental stage.

In *Manduca* the frontal ganglion (FG) controls behaviors associated with air swallowing before adult eclosion (Miles and Booker, 1998). FG innervates dilator and compressor muscles of a cibarial pump in the foregut that develops during metamorphosis. Motoneurons in the FG regulate contractions of the cibarial pump required for molting fluid swallowing many hours before eclosion and air swallowing necessary for wing and abdomen expansion after adult emergence. The pump is later used for feeding of emerged moths. Anterior and posterior dilator moto-neurons (AD1, PD1) have been identified during adult feeding in the FG. Since the activity of the cibarial pump is regulated by the FG neurons at feeding and later during swallowing of molting fluid and air, AD1 and PD1may participate in both behaviors.

Application of EH into isolated heads of pharate adults induced within 7 to 105 min of the irregular and later rhythmic activity of the cibarial pump muscles corresponds to the air swallowing behavior (Miles and Booker, 1998). This indicates that EH may control this behavior, but it is unclear how this occurs, since natural air swallowing is initiated several hours before EH is known to be released into the hemolymph. Also mechanisms of EH penetration into the FG are not clear. Further studies are needed to elucidate hormonal regulation of air swallowing.

7.5.1.11. Regulation of cuticle plasticization and sclerotization (bursicon)

Important post-ecdysis events include expansion, sclerotization (hardening), and melanization (darkening) of the cuticle. These processes are critical for the normal development of many structures. Some examples include expansion of the proboscis, antennae, legs, and wings in newly ecdysed moth pupae or adults. The term "tanning" often has been used to describe the combined processes of sclerotization and melanization. Somewhat paradoxically, bursicon also has been implicated in the induction of cuticle extensibility or "plasticization" necessary for wing expansion prior to sclerotization (see Chapter 3). Tanning in blowflies could be delayed for hours by forcing newly eclosed flies to dig continuously

through artificial substrate following eclosion (Fraenkel, 1935). Only when animals were allowed to break free of substrate would expansion and tanning proceed. It was also observed that mild anesthesia could abolish the delay, confirming sensory input is important in the control of tanning. After a 30-year hiatus, a series of papers demonstrated release of a hormonal factor that is critical in the tanning process (Cottrell, 1962a,b,c; Fraenkel, 1965; Fraenkel and Hsiao, 1962, 1963). This factor was coined "bursicon," a term derived from the Greek works *bursa* pertaining a hide or purse and *bursikos* meaning a tanning process. Bursicon-like biological activity has been detected in brain, CC, and ventral ganglia, and release sites are known to be both in the CC and in the transverse nerves of ventral ganglia, that is, the perivisceral (perisympathetic) organs (Grillot *et al.*, 1976; Honegger *et al.*, 2002; Taghert and Truman, 1982b).

Various bioassays have been devised to detect bursicon-like bioactivity. The most famous involves use of sarcophagid flies, but the assay works equally well with all cyclorraphan flies (Fraenkel, 1965). Flies in the process of emergence are ligated between head and thorax. Because descending input from the brain is essential for bursicon release, only the head tans, while the thorax and abdomen remain unsclerotized. Injection of tissue extracts containing bursicon into thorax or abdomen result in full tanning within 2 h. Other assays have utilized vermiform locust larvae (Bernays, 1971), pupal *Te.* (Grillot *et al.*, 1976), and wings from *Manduca* pharate adults (Truman, 1973).

Neurons containing bursicon activity first were identified in segmental ganglia of *Manduca* larvae using bioassays (Taghert and Truman, 1982a). Due to the unknown molecular nature of bursicon at the time, this work was done by dissecting single cell bodies of neurosecretory cells displaying the Tyndall effect and assaying for bursicon activity using the wing assay of Truman. This elegant work demonstrated bursicon activity in neurosecretory cells 23-27 (L_{1-5}) in unfused abdominal ganglia of *Manduca* larvae. Some of these cells exhibit elevated cGMP levels during the ecdysis behavioral sequence and cell 27 also expresses CCAP (Davis *et al.*, 1993; Ewer and Truman, 1996). These cells project to the perisympathetic organs in the transverse nerve, a known neurohemal release site for bursicon. However, *in situ* hybridization and immunohistochemical studies showed clearly that bursicon expression in larvae is restricted to cells 27/704 in the SG, TG1-3, and AG1 (**Figure 29**) (Dai *et al.*, 2008a; Honegger *et al.*, 2008). Therefore bursicon-like activity in other AGs may be mediated by a different, as yet unidentified hormone.

In *Manduca*, the number of bursicon neurons changes during development from larva to adult (Dai *et al.*, 2008a). Expression of bursicon in pharate larvae is confined to paired neurons (Cells 27 and 704) in each of the subesophageal and thoracic ganglia 1–3 and in the first unfused abdominal ganglion (**Figure 29**). The number of bursicon neurons increases in the pharate pupa to include all abdominal ganglia. In pharate adults, expression in the subesophageal and thoracic ganglia disappears and bursicon is confined to all abdominal ganglia.

One of the interesting aspects of bursicon biology relates to its reported induction of both cuticular extensibility and sclerotization. These seemingly opposite effects on the same tissue were reported using tissue extracts, and in the absence of pure synthetic material, it was impossible to rule out the involvement of multiple hormones. Following identification of bursicon, pure preparations of the hormone became available through its expression in HEK cells (Luo et al., 2005). Use of this biosynthetic material in the Manduca wing stretch assay demonstrated that bursicon indeed promotes both cuticle plasticization and sclerotization (Dai et al., 2008a). The precise molecular mechanisms underlying these diverse biological actions remains to be elucidated.

Cell 27 is a paired, serially homologous neurosecretory cell in thoracic and abdominal ganglia of Manduca, whose homologues also have been described in many other insects (**Figures 11** and **12**; Ewer and Truman, 1996; Honegger et al., 2002). A network of Cell 27-like neurons originally was described as "lateral A" cells, characterized their large size, pronounced white appearance due to the Tyndall effect, and strong staining with paraldehyde-fuchsin. They also are characterized by association with large vacuole-like structures adjacent to the cell body (Gaude, 1975). More detailed descriptions of these LW cells in crickets and cockroaches showed an association with proctolin-like biological activity, but this was never confirmed (O'Shea and Adams, 1981). Rather, work from various laboratories provided definitive evidence for co-localization of bursicon and CCAP in these neurons (Adams and Phelps, 1983; Davis et al., 1993; Taghert and Truman, 1982a). The recent availability of antisera specific for bursicon confirmed that Cell 27 contains bursicon, and also shows intense bursicon staining in vacuoles associated with these cell bodies (Honegger et al., 2002). The presence of bursicon staining in these vacuoles, together with early ultrastructural work showing that the lumen of the vacuole is continuous with cisternae of the endoplasmic reticulum, shows that the vacuole is a large storage depot for secretory product, specialized for large-scale release of hormonal substances into the hemolymph (Adams and O'Shea, 1981). In the cockroach, the cell 27 homologues project axons not only to the perisympathetic organs, but also the heart, most likely for general liberation of bursicon into the hemolymph. The co-release of CCAP likely enhances distribution of bursicon to distant locations through its cardioacceleratory activity. CCAP in cells 27 is apparently co-localized with bursicon in Periplaneta, Gryllus, Manduca, and probably other insects (Honegger et al., 2002; Kostron et al., 1996). These peptides are very likely co-released at ecdysis to control post-ecdysis

processes; CCAP increases heartbeat and blood pressure for cuticle expansion, while bursicon controls sclerotization and tanning of expanded new cuticle.

7.5.1.12. Control of cuticle melanization and pigmentation (PBAN/MRCH, JH)

Various colorations of lepidopteran larvae are achieved by a combination of melanin in the cuticle and ommochrome in the epidermis. Melanization and ommochrome synthesis is under hormonal control and requires the activity of both peptides and JHs (Hiruma et al., 1984, 1993). There are two known types of mechanisms required for pigmentation of lepidopteran larvae. In Mamestra, Leucania, and Spodoptera, melanization and reddish coloration hormone (MRCH) is the only factor controlling cuticle melanization (Matsumoto et al., 1990), whereas JH and MRCH are required for ommochrome synthesis (Hiruma et al., 1984). MRCH is identical with pheromonotropic and diapause hormones (PBAN/DH), which are produced along with other related PRLamide peptides by three clusters of neurosecretory cells in the SG (Davis et al., 1996; Hagino et al., 2010; Predel and Wegener, 2006). Surgical experiments showed that MRCH is probably released during the initiation of pre-ecdysis to control cuticle melanization several hours after completion of ecdysis, while high or low JH levels have no apparent influence on this process (Hiruma et al., 1984).

By contrast, in Manduca and Bombyx, the absence or decreased JH levels in early pharate stages induce increased deposition of premelanin granules in the cuticle and stronger melanization (Curtis et al., 1984; Kiguchi, 1972). MRCH has no effect in this process, but deposition of these granules containing phenoloxidase is under control of a neuroendocrine factor from the abdominal nerve cord (Hiruma and Riddiford, 1993). Therefore, low JH levels and a humoral factor (neuropeptide?) released at the end of the molt are required for cuticle melanization in these species.

Interestingly, strong ETHR-A expression was detected in the CA of pharate larvae in Manduca and Bombyx (Yamanaka et al., 2008); Zitnan, unpublished), while ETHR-B staining was observed in ventromedial neurons of the SG (Kim et al., 2006a). These neurons probably produce MRCH/PBAN and other PRLamides (Daubnerova, Roller, and Zitnan, unpublished). The presence of ETHR-A, B in appropriate targets and timing of JH and neuropeptide release at ecdysis indicate that ETH signaling may be directly involved in the initiation of cuticle pigmentation.

7.5.2. Natural and Peptide-induced Behaviors in Bombyx

7.5.2.1. Natural and ETH-induced pre-ecdysis and ecdysis in larvae

Developmental markers and behavioral sequences during larval and pupal ecdyses and adult eclosion have been recently described (Zitnan et al.,

2002b; see Table 2). Natural larval behavior is initiated by pre-ecdysis composed of asynchronous dorsoventral, ventral, and posterolateral body wall contractions, plus leg and proleg retractions. After about 1 h of pre-ecdysis contractions, larvae switch to anteriorly directed peristaltic movements (ecdysis behavior), which result in shedding of the old cuticle in 10–12 min (**Figure 30**).

Comparisons of *Manduca* pre-ecdysis and ecdysis behaviors with those in *Bombyx* revealed substantial differences.

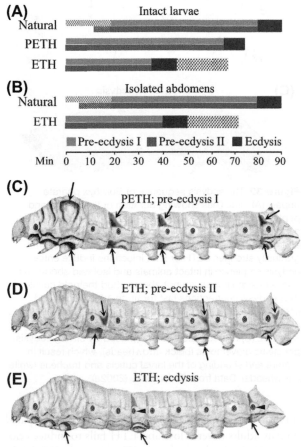

Figure 30 The ecdysis sequence of *Bombyx* pharate larvae. (A) Time lines showing duration of natural and peptide-induced pre-ecdysis and ecdysis in intact pharate larvae. Note that injection of PETH or ETH (50 pmol) induces the entire sequence and ecdysis onset is accelerated when compared with natural behavior. ETH is more effective than PETH and most larvae injected with ETH continue to show ecdysis contractions for up to 30 min (stippled line). (B) Time of the initiation and duration of natural and ETH-induced ecdysis sequences are very similar in isolated abdomens compared with intact pharate larvae of the same stage. (C) PETH-induced asynchronous dorsoventral and leg contractions in thorax and abdomen (pre-ecdysis I, arrows pointing at shaded areas indicate direction of muscle contractions). (D) Subsequent ETH injection induces asynchronous ventral, posterolateral and proleg contractions (pre-ecdysis II, shaded areas, arrows). (E) These animals then switch to anteriorly directed peristaltic contractions accompanied with proleg retractions (ecdysis, shaded areas, arrowheads). Data from *Zitnan et al.* (2002a).

For instance, *Bombyx* pre-ecdysis I and II are not synchronized and body wall or proleg contractions can occur randomly on the left or right side of any segment (Zitnan *et al.*, 2002a). Ligation and transection experiments of *Bombyx* larvae *in vivo* and *in vitro* confirmed that each ganglion contains the entire circuitry for pre-ecdysis I and II, and that connections to the TAG or other distal ganglia are not necessary for generation of these motor patterns. This indicates that IN-402 necessary for synchronous pre-ecdysis I contractions in *Manduca* are probably missing or are not activated in *Bombyx*. Thus pre-ecdysis contractions in *Bombyx* and *Manduca* are quite different, while ecdysis movements are very similar in both species.

Peptide-induced behaviors in *Bombyx* also can differ substantially from those induced in *Manduca*. For example, injection of either PETH or ETH into pharate *Bombyx* larvae (10–15 h prior to ecdysis) induces both pre-ecdysis and ecdysis behaviors *in vivo*. Likewise, peptide treatments of the isolated CNS *in vitro* elicit corresponding burst patterns (**Figure 31**). On the other hand, injections of pharate *Bombyx* larvae 1 day prior to expected ecdysis elicit responses similar to those induced in *Manduca*: PETH elicits only pre-ecdysis I, whereas ETH induces the entire sequence. These data indicate that CNS sensitivity to PETH and ETH becomes progressively stronger as animals approach the onset of natural ecdysis behavior (Zitnan *et al.*, 2002b).

Peptide-induced behaviors in *Bombyx* differ from those observed in *Manduca* in another notable way. Injection of isolated abdomens with ETH elicits the entire ecdysis sequence associated with normal elevation of cGMP in abdominal ganglia, but no detectable release of EH. This means that descending input from the *Bombyx* brain to abdominal ganglia is not required for activation of ecdysis circuitry as was observed in *Manduca*. Finally, although ETH-AP has no detectable behavioral action in *Manduca*, *Bombyx* larvae respond to ETH-AP injection with non-synchronized proleg and ventral contractions resembling pre-ecdysis II (Zitnan *et al.*, 2002b).

7.5.2.2. Natural and ETH-induced pupal ecdysis and adult eclosion In contrast to *Manduca* (see Section 7.5.1.), *Bombyx* pharate pupae show well-defined and pronounced pre-ecdysis behavior composed of dorsoventral, leg, and proleg contractions for about 1–1.5 h and then switch to robust ecdysis peristaltic movements (**Figure 32**). During anteriorly directed ecdysis contractions larval cuticle of the head capsule and thorax is ruptured along the dorsal line, and the entire old cuticle is shed with attached larval foregut, hindgut, and tracheae in 10–12 min. As described in pharate larvae, injections of either PETH or ETH elicit the entire behavioral sequence (**Figure 32**). ETH-AP treatment induced weaker proleg and ventral contractions for about 1 h followed by weak and occasional ecdysis movements for up to 1 h. Isolated abdomens show

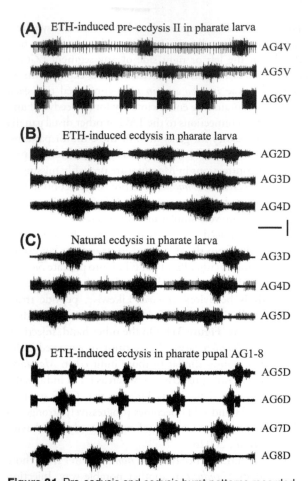

Figure 31 Pre-ecdysis and ecdysis burst patterns recorded in the isolated CNS of *Bombyx* pharate larvae and pharate pupae *in vitro*. (A, B) ETH-induced pre-ecdysis II and ecdysis in the entire CNS of pharate fifth instar larvae. (A) Note asynchronous pre-ecdysis II bursts in ventral nerves of abdominal ganglia 4–6 (AG4-6V). (B) ETH-induced ecdysis bursts in dorsal nerves of abdominal ganglia 2–5 (AG2-5D). These burst patterns are very similar to natural ecdysis bursts (C). (D) ETH-induced ecdysis bursts in dorsal nerves of the isolated chain of abdominal ganglia (AG1-8) of pharate pupa. Ecdysis burst patterns in intact CNS are very similar (not shown). Calibration bars: horizontal 5 s, vertical 10 μV. Modified from *Zitnan et al.* (2002a).

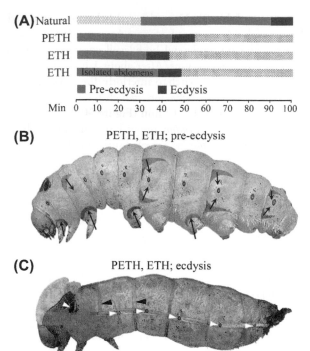

Figure 32 The ecdysis sequence of *Bombyx* pharate pupae. (A) Time lines showing duration of the natural and peptide-induced ecdysis sequences. Stippled green line during the first 30 min of pre-ecdysis indicates initial weak and occasional dorsoventral contractions, which become gradually stronger. PETH or ETH injections induce entire ecdysis sequence in intact animals and isolated abdomens. These pharate pupae are not able to shed their old cuticle, thus ecdysis contractions last for up to 1 h (stippled red lines). (B) Shaded areas indicate dorsoventral, leg, and proleg contractions during pre-ecdysis induced by PETH or ETH injection. (C) Pre-ecdysis is later replaced by strong ecdysis peristaltic movements (black arrowheads), which result in rupture and shedding of the larval cuticle and tracheae (white arrowheads). Data from *Zitnan et al.* (2002a).

natural pre-ecdysis and ecdysis at the expected time and both behaviors could be prematurely induced by ETH injection (**Figure 32**).

The natural eclosion sequence of pharate adults consists of three distinct phases: regular rotations of the abdomen, a quiescent phase, and eclosion peristaltic contractions of the abdomen (**Figure 33**). During peristaltic eclosion movements, adults partially emerge from the pupal cuticle and, after dissolving the cocoon around the head with a salivary secretion containing cocoonase, they escape and spread their wings.

Injection of PETH or ETH induces the entire behavioral sequence in most pharate adults. The importance of the brain in ecdysis activation was demonstrated by surgical experiments. Decapitated or brain-extirpated pharate adults never eclose and ETH fails to induce eclosion behavior in these animals (Zitnan *et al.*, 2002b). In summary, these data showed that both PETH and ETH induce the entire behavioral sequence, although latencies for the onset of ecdysis or eclosion movements by PETH were more variable and generally longer than for ETH.

The neuronal network expressing ETHR-A is almost identical in *Bombyx* and *Manduca*, the number of ETHR-B neurons in the SG and TG1-3 is lower and in AG1-8 higher in *Bombyx* compared to *Manduca*. This may reflect much higher autonomy of regulation of pre-ecdysis and ecdysis by individual ganglia in *Bombyx*.

7.5.2.3. Transgenesis approaches for analysis of peptide functions in *Bombyx* ecdysis In recent years, the baculovirus expression system and *piggyBac* germ-line transformation have been successfully employed to elucidate mechanisms of hormonal regulation of *Bombyx*

(A)

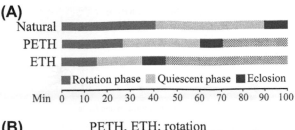

(B)

PETH, ETH; rotation

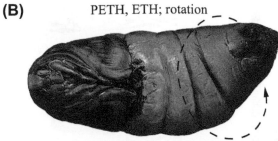

(C)

PETH, ETH; eclosion

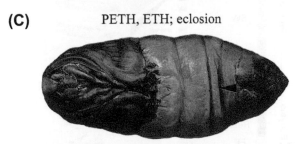

Figure 33 The eclosion sequence of *Bombyx* pharate adult. (A) Time lines showing rotation, quiescent, and eclosion behavioral phases during natural and peptide-induced eclosion sequence. Animals injected with PETH or ETH show accelerated response compared to that expressed during natural eclosion. Since peptide-injected animals fail to escape from the pupal cuticle, eclosion contractions last for up to 1 h. (B) Both PETH and ETH elicit vigorous rotations of abdomen (arrow), interrupted by short quiet intervals. (C) After a prolonged quiescent phase, the animal initiates eclosion peristaltic contractions of abdominal segments (arrowhead). Data from *Zitnan et al.* (2002a).

development and ecdysis (Daubnerova *et al.*, 2009; Tamura *et al.*, 2000; Tanaka, 2000). The baculovirus Bac-to-Bac system proved useful for identification of ETH promoter (Daubnerova *et al.*, 2009) and upstream regulatory regions of other neuropeptide genes (Shiomi *et al.*, 2003). Germline transgenesis with *piggyBac* and the Gal4-UAS binary system were used to evaluate the role of ETH at ecdysis. Double-stranded RNA targeting the ETH gene was expressed under control of the actin promoter to suppress its expression in Inka cells. The resulting decrease in ETH production led to severe defects in ecdysis behavior and lethality in pharate second instar larvae (Dai *et al.*, 2008b). This study confirms the essential role of ETH in activation of the ecdysis sequence in *Bombyx*.

The promoter of the *Drosophila* heat shock protein was used to suppress EH expression in the *Bombyx* brain with RNA interference (RNAi). Heat shock effectively reduced EH expression and consequently delayed adult eclosion

for 2–3 days and decreased egg fertility to 50% (Dai *et al.*, 2007b). The RNAi approach also was employed successfully for assessment of the role of bursicon after eclosion. As expected, RNAi suppression of bursicon transcripts disrupted expansion of wings after adult eclosion, but surprisingly had no effect on tanning of the cuticle (Huang *et al.*, 2007).

7.5.3. Ecdysis Sequence in the Fruitfly, *Drosophila*

7.5.3.1. Larval ecdysis The first detailed study of the larval ecdysis sequence in *Drosophila* focused on a series of stereotypic physiological and behavioral events occurring at the transition from first to second instar (Park *et al.*, 2002a). About 60 min prior to ecdysis, a new set of mouthparts for the second instar become visible adjacent to those of the first instar, leading to the appearance of "double mouth hooks." The appearance of "double vertical plates" (dVP) 30 min later was used as reference point, or "time zero," in the ecdysis sequence. Subsequent events in the ecdysis sequence are referred to this time point (**Figure 34**).

Using a transgenic fly line in which the nucleotide sequence of the fluorescent enhanced green jellyfish protein (*egfp*) was fused to the 3′ end of the *eth* gene, release of ETH from Inka cells was observed as a loss of fluorescence shortly after dVP. Almost immediately after ETH release, old tracheae collapse and the new, second instar tracheae are inflated. This occurs 10 min after dVP. About 5 min later, pre-ecdysis begins with a series of "anterior-posterior" (A-P) contractions. Five minutes later, another pattern of muscle contractions referred to as "squeezing waves" (SW) occurs. During A-P and SW movements, referred to as pre-ecdysis behaviors, vigorous muscle contractions are observed to pull the mouthparts alternately in anterior and posterior directions. These movements may be critical to the separation of old from new mouthparts during subsequent ecdysis behavior.

Ecdysis behavior is initiated ~20 min after dVP. It begins with one or two forward head thrusts, where old mouth hooks and vertical plates are detached from the new and planted onto the surface of the substrate. The forward thrust also detaches and extricates parts of the old tracheal system through lateral segmental spiracular pits, which are normally closed but become functionally open during ecdysis. Subsequent backward thrust detaches the old from the new posterior spiracles and pulls the remaining old tracheal linings out through the new spiracles (**Figure 34**). Ecdysis is completed after forward and turning movements help the larva to escape from the old cuticle. The sequence of events leading up to ecdysis from second to third instar follows a very similar pattern. The only difference is that the appearance of double mouth hooks occurs considerably sooner (104 min before dVP) as compared to the first ecdysis (30 min before dVP).

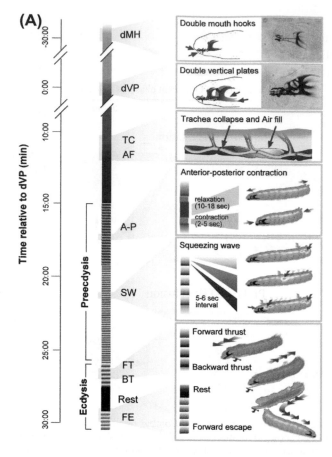

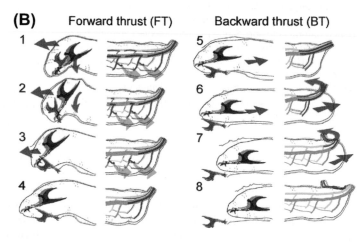

Figure 34 Schematic drawings depicting the time lines and behavioral phases during *Drosophila* larval ecdysis sequence. (A) About 60 min prior to ecdysis, a new set of mouth hooks for the second instar larva becomes visible adjacent to those of the first instar "double mouth hooks" (dMH). The appearance of "double vertical plates" (dVP) 30 min later was used as "time zero" in the ecdysis sequence. Upon ETH release from Inka cells 10 min after dVP, old tracheae collapse and the new, second instar tracheae are inflated. About 5 min later, pre-ecdysis begins with a series of "anterior-posterior" contractions (A-P) followed by a different pattern of muscle contractions called "squeezing waves" (SW). During A-P and SW pre-ecdysis movements, old and new mouthparts are separated. (B) About 25 min after dVP, ecdysis behavior is initiated with one or two forward head thrusts (FT). These movements help to plant old mouthparts onto the substrate and to extricate old tracheae through lateral segmental spiracular pits, which are normally closed but become functional during ecdysis. The following backward thrust (BT) detaches and extricates remaining old posterior spiracles and tracheal linings out through the new spiracles. Ecdysis is completed following forward escape (FE) and turning movements.

7.5.3.1.1. Roles of ETH and EH in larval ecdysis

As mentioned in the previous section, ETH is released from Inka cells just after the dVP stage and just before tracheal inflation. If ETH is responsible for tracheal inflation and initiation of the ecdysis sequence, injection of the peptide prematurely should elicit the steps of natural ecdysis already described. Premature injection of ETH1 accelerates all the events shown in **Figure 34**, including tracheal inflation, A-P contractions, squeezing waves, and ecdysis behavior (Park *et al.*, 2002a). Injection of the peptide (≥ 0.1 fmol) into *Drosophila* larvae at the double

mouth hook stage induces within a 3–4 min collapse of old tracheae and inflation of new tracheae, followed by a complete set of pre-ecdysis and ecdysis behaviors. ETH2 also induces tracheal and behavioral responses, but higher concentrations are needed (≥ 1 fmol for tracheal dynamics; ≥ 10 fmol or ecdysis behavior). Interestingly, pre-ecdysis behaviors are absent; tracheal inflation is followed by a quiescent period, after which animals initiate ecdysis behavior (Park *et al.*, 2002a). These studies illuminated a previously unrecognized role for ETH in eliciting tracheal dynamics, although it is likely that

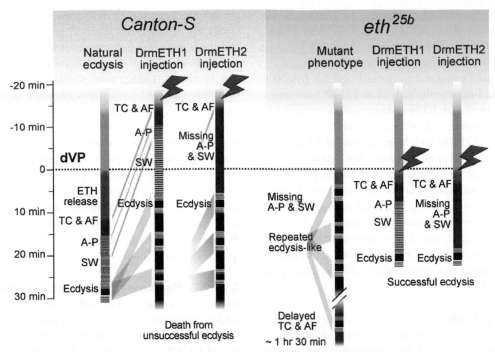

Figure 35 Time lines showing the ecdysis sequence in wild-type Canton-S and in the *eth* deletion mutant (line *eth*25b) During natural ecdysis, ETH release begins shortly after double vertical plates (dVP), followed by collapse of old tracheal tubes (TC) and air inflation (AF) of new, pre-ecdysis consisting of anterior-posterior (A-P) contractions and squeezing waves (SW), and finally ecdysis. Premature DrmETH1 injection accelerates all events that normally follow natural ETH release, but DrmETH2 elicits only accelerated tracheal dynamics and ecdysis. When ETH1 or ETH2 is injected prematurely, ecdysis behavior ensues but is unsuccessful due to failure to separate old and new mouth hooks and vertical plates. The mutant phenotype *eth*25b exhibits delayed tracheal dynamics, fails to perform pre-ecdysis and normal ecdysis behaviors, and shows high mortality (>98%) at the first larval ecdysis. Injection of ETH1 or ETH2 at the dVP stage reverses the mortality observed in this mutant phenotype. Data from *Park et al.* (2002a).

ETH acts through other downstream signals on the tracheal system.

ETH is sufficient to elicit the ecdysis sequence, but is it necessary? To address this question, the gene *eth* encoding ETH1 and ETH2 was deleted by imprecise P-element excision (Park *et al.*, 2002a). To excise *eth*, a fly line having a P-element insertion 1427 bp downstream of the gene was obtained. The P-element was mobilized by crossing this line with the Δ2-3 transposase line. Deletion of *eth* resulted in lethal behavioral and physiological deficits. Homozygous null mutants (*eth-*) develop through embryogenesis and the first instar, but 98% die at the first larval ecdysis. These animals show considerable delays in tracheal collapse and air inflation, lack pre-ecdysis behavior, exhibit abnormal ecdysis behavior, and are unable to shed the old cuticle completely. Precisely timed injection of ETH1 into *eth* null mutants rescues both respiratory and behavioral defects and animals progress to the second instar (**Figure 35**). The gene encoding ETH1 therefore plays an obligatory role in the regulation of ecdysis, and absence of ETHs leads to defects in air filling of the tracheal system, unsuccessful ecdysis, and lethality (Park *et al.*, 2002a).

While the experiments of Park *et al.* (2002a) focused exclusively on the roles of ETH in larval ecdysis, a more recent study examined the interacting roles of ETH, EH, and CCAP in the scheduling of larval ecdysis from second to third instar (Clark *et al.*, 2004). Effects of these peptides were evaluated in both wild-type and EH-KO flies. Results of the study indicate that, as in moths, EH and ETH engage in a positive feedback signaling loop that involves elevation of cGMP in Inka cells and massive release of ETH. Furthermore, both ETH and EH may be involved in the regulation of tracheal inflation, because EH-knockout (KO) flies also have tracheal inflation defects. These flies can be rescued by injection of either ETH or membrane-permeant cGMP analogs, which mimic the action of EH. An interesting point raised by Clark *et al.* (2004) is the inability of ETH injection to prematurely initiate the ecdysis in EH-KO flies, although ETH does trigger premature tracheal inflation, and all animals eventually display ecdysis behavior with some delay. This was interpreted as an inability of ETH to trigger the ecdysis sequence in the absence of EH. This is a difficult result to explain mechanistically because ETH KO flies show virtually 100% lethality at the first larval ecdysis (Park *et al.*, 2002a), whereas 30–34% of EH KO flies survive to become viable, fertile adults. The lethality observed in the remaining 66–70% animals occurs mostly during larval development because of defects in tracheal inflation, but not behavioral deficits. Also, pharate EH-KO adults show all components of eclosion behavior, although individual behavioral phases are prolonged and less coordinated compared to normal controls (McNabb et al., 1997). Importantly, ecdysis in EH-KO mutants is always associated with ETH release.

Calcium imaging and other experimental approaches showed that ETH acts on a large number of ETHR neurons that include a complex ecdysis/post-ecdysis network (Kim et al., 2006b). This clearly demonstrates the ability of ETH to activate the entire ecdysis/post-ecdysis circuitry, but in the absence of EH, resulting behaviors may be delayed and less synchronized. We conclude that ETH is essential for activation of all components of the ecdysis sequence, while ETH-EH interactions are important for proper timing and coordination of ecdysis and post-ecdysis behaviors.

7.5.3.2. Pupal ecdysis

A detailed account of morphological markers and behaviors that occur during metamorphosis was published by Bainbridge and Bownes (1981). These authors identified numerous stages from wandering larva to adult subdivided into 52 distinct events. Here we focus on those events associated with pupariation and the pupal behavioral sequence.

Pupal ecdysis in *Drosophila* is unique in that it is the only ecdysis during its life history that does not involve actual escape from the old cuticle. Instead, the wandering larva finds a suitable pupation site and uses the old third instar cuticle to form a puparium ~12 h prior to initiating the pupal ecdysis sequence. This process allows the animal to separate from the old cuticle, but it remains inside the puparium until metamorphosis is completed and the adult emerges during eclosion. Formation of a white puparium can be considered as the reference point or time zero for pupal ecdysis. A gas bubble becomes visible in the dorsal part of the abdomen ~3 h later and apolysis is initiated at 4–6 h. This stage is called "cryptocephalic pupa." Major lateral tracheal trunks are obscured at 6.5–8 h followed by initiation of dorsomedial abdominal contractions at 10–12 h. The abdominal bubble disappears at 12 h, while another bubble appears in the posterior part of puparium. The abrupt appearance of the posterior air bubble initiates the pupal ecdysis sequence, which consists of three centrally patterned behavioral steps: pre-ecdysis, (~10 min), ecdysis (~5 min), and post-ecdysis (~60–70 min) (Park et al., 2003b). Details of these behaviors were re-examined in both puparium-intact and puparium-free preparations by Kim et al. (2006b).

Pre-ecdysis involves anteriorly directed rolling contractions that occur along the lateral edges of the abdomen, alternating on the left and right sides of the animal. These contractions move the air bubble anteriorly to separate the pupal cuticle from the puparium. After ~10 min of pre-ecdysis behavior, ecdysis contractions commence, consisting of anteriorly directed peristaltic contractions and lateral swinging movements of the abdomen. Within 1 min, head eversion occurs (Ward et al., 2003; Zdarek and Friedman, 1986), after which ecdysis contractions continue, leading to extension of pupal legs and wing pads posteriorly along the abdomen and shedding of the larval tracheae. Ecdysis contractions gradually give way to post-ecdysis behaviors, consisting of peristaltic contractions proceeding in the posterior direction; these alternate with longitudinal "stretch-compression" movements of the abdomen. The frequency and intensity of post-ecdysis contractions wane over a period of 60–110 min. These contractions compress the pupa in the posterior end of the puparium (Park et al., 2003b). This stage is named "phanerocephalic pupa."

7.5.3.2.1. Peptidergic signaling in pupal ecdysis

Studies in *Manduca* implicated CCAP as a proximal signal for activation of larval ecdysis behavior (see Section 7.5.1.7.3.; Gammie and Truman, 1997a). The possible role of CCAP in *Drosophila* ecdyses was tested by ablation of CCAP neurons using the cell death gene *reaper* (Park et al., 2003b). Ablation of CCAP neurons produces a severe ecdysis phenotype at pupation, but only subtle effects on larval ecdysis and adult eclosion. The pupal ecdysis phenotype consists of a normal, but prolonged pre-ecdysis phase and a failure of ecdysis. Ecdysis defects include head eversion failure, absence of typical ecdysis associated contractions, incomplete shedding of larval tracheae, and failure to extend legs and wings. These observations suggest that CCAP neurons play important roles in pupal ecdysis.

With the identification of ETH receptors in *Drosophila*, (Iversen et al., 2002; Park et al., 2003a) it became possible to identify neural targets of this peptide in the CNS and to monitor cellular activity in response to ETH exposure. Using *in situ* hybridization, it was demonstrated that ETH receptor subtype A (ETHR-A) expression is widespread in groups or "ensembles" of peptidergic neurons, including those expressing EH, FMRFamide, kinin, CCAP, MIPs, and bursicon. ETH receptor subtype B neurons were also visualized and comprise a population largely non-overlapping with the ETHR-A neurons. The identities of ETHR-B neurons remain to be elucidated.

Using cell-specific promoters to drive the calcium reporter GCaMP in specific ETHR-A ensembles, it was shown that they respond to ETH application by mobilization of calcium at specific times (**Figure 36**). FMRFamide neurons exhibit calcium dynamics within a few minutes of ETH exposure and remain active during the entire ecdysis sequence. EH, CCAP, and CCAP/MIP neurons become active just before ecdysis, whereas CCAP/MIP/bursicon neurons are activated during ecdysis and post-ecdysis. These results indicate that specific peptidergic ensembles express ETHR-A, are direct targets of ETH, and likely schedule each step of the ecdysis sequence in this stage of fly development. Mechanisms underlying peptidergic ensemble-specific timing of calcium mobilization remain to be elucidated.

Ablation of EH neurons causes prolonged pre-ecdysis, but normal ecdysis and post-ecdysis occur. However, ablation of CCAP/MIP/bursicon neurons caused greatly prolonged pre-ecdysis and failure of ecdysis and post-ecdysis, as reported previously (Park *et al.*, 2003b).

7.5.3.3. Adult eclosion Time lines of the eclosion behavioral sequence and morphological markers useful in predicting ecdysis events in *Drosophila* pharate adults have been described (Kimura and Truman, 1990; McNabb *et al.*, 1997; Park *et al.*, 1999). Resorption of molting fluid is initiated about 6–7 h before expected eclosion. The pupal cuticle then passes through a series of changes in appearance referred to as smooth (S), smooth/grainy (S/G), grainy (G), and finally, white (W). Certain steps differ in duration between males and females; namely that the smooth stage lasts for ~5.5 h before females transition to the smooth/grainy stage, whereas males

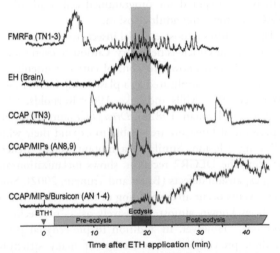

Figure 36 ETH-induced calcium mobilization in central peptidergic ensembles that express ETHR-A in pharate pupal *Drosophila*. The timing of calcium mobilization occurs with ensemble-specific latencies. ETH1 application at time 0 (green arrow) elicits the ecdysis sequence starting with pre-ecdysis after ~2.5 min. Within a few minutes, FMRFa neurons in thoracic neuromeres 1–3 (TN1-3) mobilize calcium and remain active during the entire behavioral sequence. After ~10 min, CCAP neurons in TN3 and EH neurons become active, followed at ~12.5 min by calcium dynamics in CCAP/MIPs neurons of abdominal neuromeres 8 and 9 (AN8, 9) shortly before onset of ecdysis behavior. CCAP/MIPs/bursicon neurons in AN1–4 become active at the beginning of ecdysis and reach peak levels of calcium mobilization at ~40 min. The ethogram at the bottom shows durations of the pre-ecdysis, ecdysis, and post-ecdysis behaviors induced by injection of 0.4 pmol ETH1 into puparium-free pharate pupa. Adapted from *Kim et al.* (2006b).

show the smooth stage for only a matter of minutes before advancing to the smooth/grainy stage. Commencement of the white stage (W) occurs upon inflation of the head tracheal sacs and indicates that pre-eclosion behavior has been initiated.

Pre-eclosion starts about 1 h prior to eclosion and is composed of dorsoventral contractions of the first abdominal tergum and ptilinium extension resulting in inflation/expansion and tracheal filling of the head (W). Pre-eclosion is followed by a quiescent period, but the entire behavior may be absent in up to 1 in 4 animals (Park *et al.*, 1999). Eclosion behavior consists of four consecutive events, including head expansion to open the operculum, forward head thrusts, lateral thoracic contractions, and forward peristaltic abdominal contractions helping the fly to emerge from puparium. Interestingly, neck ligation of pharate adults during pre-eclosion (extended ptilinum stage) causes immediate onset of ecdysis (Baker *et al.*, 1999). This indicates that inhibitory neurons and factor(s) in the cephalic ganglia delay the initiation of ecdysis, similar to that observed in *Manduca* (Ewer *et al.*, 1997; Fuse and Truman, 2002; Zitnan and Adams, 2000).

As described in *Drosophila* larvae, injections of ETHs into pharate adults accelerate onset of pre-eclosion and eclosion behaviors, and again, ETH1 is more effective than ETH2 (Park *et al.*, 1999). Injection of *Manduca* ETH also induces the entire behavioral sequence (Baker *et al.*, 1999; McNabb *et al.*, 1997; Park *et al.*, 1999), but animals injected with PETH show no response.

7.5.3.4. Roles for EH in the *Drosophila* eclosion sequence *Drosophila* EH is expressed in a single pair of dorsomedial neurons projecting from the brain into the CC as well as posteriorly within the fused ventral ganglion (Baker *et al.*, 1999; Horodyski *et al.*, 1993). There is a good deal of evidence generated from studies of EH in *Manduca* (Truman, 1992), *Bombyx* (Kono *et al.*, 1990b), and *Heliothis virescens* (Kataoka *et al.*, 1987) that EH is sufficient to trigger pre-ecdysis and ecdysis behaviors. In the first study to test for the necessity of EH in eclosion, the cell-specific promoter of the EH gene in *Drosophila* was used to drive the cell death genes *reaper* and *head involution defective* to kill EH synthesizing cells (McNabb *et al.*, 1997). In spite of EH cell ablation, ~50% of developing flies reached the puparium stage, and ~80% of these successfully performed eclosion.

Nevertheless, several defects in the coordination of eclosion-related events are evident in EH-KO flies. First, tracheal inflation is delayed. Second, while bouts of head expansion, head thrusts, thoracic contractions, and abdominal contractions that comprise eclosion behavior are synchronized and alternate with quiescent periods in wild-type flies, they are disorganized and performed continuously with no quiescent periods in EH-KO flies. Furthermore, eclosion behaviors are weaker in the EH-KO flies, and thus the entire eclosion sequence is prolonged. Finally, two-thirds of the mutant flies failed to undergo normal post-eclosion expansion of wings and hardening of the adult cuticle.

These observations indicate that, despite the ability of flies to undergo larval, pupal, and adult ecdysis behaviors, they fail to perform many of the steps on schedule and in a coordinated fashion. Nevertheless, it is remarkable that almost one-half of the mutant fly population survives to the adult stage in the absence of EH. This study indicates that EH contributes to the coordination and overall robustness of the ecdysis/eclosion sequence. In its absence, survival is possible, but occurs at a lower rate than in wild-type flies (McNabb et al., 1997).

Other interesting outcomes emerged from the study of EH-KO flies. First, based on Truman's original discoveries of EH activity in giant silkmoths, it has been assumed that EH is involved in the circadian gating of adult eclosion (Truman and Riddiford, 1974). Second, EH has been associated for many years with programmed cell death of larval abdominal body wall muscles during metamorphosis (Schwartz and Truman, 1984). However studies of EH-KO flies demonstrated both a normal free-running circadian rhythm of eclosion as well as programmed cell death of intersegmental muscles (McNabb et al., 1997). These findings indicate that other hormonal signals are likely involved in these processes.

7.5.3.5. Identification of bursicon and its receptor

Many years of struggle by countless investigators failed to identify bursicon, although it was recognized as a 30–40 kDa protein in many insects, including various species of flies, moths, and cockroaches (summarized in Seligman, 1980; Reynolds, 1983). Finally, during the past 8 years, a number of excellent papers have been published that define bursicon and its receptor (reviewed by Honegger et al., 2008).

The first breakthrough came with publication of information on the primary structure of bursicon by Honegger and colleagues, who reported partial amino acid sequences of a protein(s) with bursicon-like bioactivity from the cockroach CNS (Honegger et al., 2002). Several sequences obtained in this manner were then used to identify one of the likely genes encoding bursicon (CG13419) in Drosophila through a BLAST search of the fly genome (Dewey et al., 2004). This gene eventually was recognized to encode burs, one of two monomers that form the bioactive bursicon heterodimer. The next year, the other piece of the ligand puzzle, "partner of burs" or "pburs" (CG15284) was identified as well as the bursicon receptor (CG) (Luo et al., 2005). Independent discoveries of bursicon and its receptor were published the same year (Mendive et al., 2005); these authors refer to burs as "burs-α" and pburs as "bursβ". Subsequent reports indicate that bursicon genes are ubiquitous and highly

conserved in the Arthropoda, including classes Insecta, Crustacea, and Arachnida (Honegger et al., 2008).

In Drosophila larvae, bursicon expression is confined to abdominal neuromeres 1–4. These neurons express both burs (bursα) and pburs (bursβ). Interestingly, a number of additional neurons express only burs cells located in the SEG and all thoracic and remaining abdominal neuromeres (Luo et al., 2005; Peabody et al., 2008). The functional significance of burs expression alone is unclear.

As the Fraenkel bioassay demonstrates, decapitation or ligation behind the head immediately after eclosion inhibits the tanning process. Thus, while most bursicon-containing neurons reside in ventral ganglia, they depend on descending input from the cephalic ganglia for activation. In pharate adult Drosophila, bursicon expression is observed in 14 abdominal neurons referred to as B(AG) and 2 neurons in the SEG, the B(SEG) neurons (Peabody et al., 2008). Using targeted suppression of excitability, Peabody et al. (2008) showed that activity in a pair of bursicon neurons in the B(SEG) cells, which arborize widely in the CNS, controls release of bursicon from the B(AG) neurons and the physiological consequences of bursicon release, which include cuticle hardening and darkening and wing expansion. Interestingly, activity of the B(SEG) cells is also required for programmed cell death of the B(AG) neurons after adult eclosion.

For structures in post-ecdysial insects to expand, plasticization of the cuticle is necessary. Curiously, even though bursicon is well known to control cuticle hardening, it also seems to be implicated as a plasticizing factor as well (Fraenkel, 1966; Mills and Lake, 1966; Reynolds, 1977; Taghert and Truman, 1982b). Drosophila mutants lacking expression of bursicon are unable to expand their wings following eclosion. Similarly, a mutant called "rickets" targeting the DLGR2 receptor, shows melanization and wing expansion defects (Baker and Truman, 2002). Since these events occur after shedding of the cuticle, bursicon has been known almost exclusively as a "post-eclosion hormone." However, in addition to pre-ecdysial plasticization, pre-ecdysial sclerotization of many structures also occurs well before ecdysis (Cottrell, 1962a). These include tanning of tarsal claws, parts of head capsules, and mandibles in moths, and vertical plates, mouth hooks, and sensory structures in flies. It is likely that bursicon is involved in these pre-ecdysial tanning events, but the mechanism(s) of selective tanning of these structures is unknown.

Recent evidence from studies of Drosophila provides evidence that bursicon is released in two temporal waves, the first prior to ecdysis and the second after. In larvae, immunohistochemical evidence demonstrated loss of bursicon staining from nerve terminals at the neuromuscular junction between the double mouth hook stage and the double vertical plate stage. Although the authors speculate that the function of this release could be to tan mouthparts, RNA silencing of either bursicon or its receptor did not inhibit mouthpart tanning. Another explanation suggested is that release of the burs monomer alone may be involved in some as yet unknown function. It will be interesting to further investigate the function roles of bursicon or its subunits prior to ecdysis.

A putative bursicon receptor in *Drosophila* was reported to be the GPCR DLGR2, as flies carrying a mutation in gene encoding this protein ("rickets") failed to perform wing expansion, sclerotization, and melanization following ecdysis (Baker and Truman, 2002). These investigators showed that flies carrying the rickets mutation failed to respond to injections of bursicon. However, defects could be rescued by injection of cyclic AMP analogs, indicating that bursicon operates via the cAMP signal transduction pathway. The identity of this protein as the authentic bursicon receptor was confirmed subsequently (Luo *et al.*, 2005; see Chapter 3). Independent confirmation of the bursicon receptor was published that same year (Mendive *et al.*, 2005). The pburs/burs heterodimer from *D. melanogaster* binds with high affinity and specificity to activate the G protein-coupled receptor DLGR2, leading to the stimulation of cAMP signaling *in vitro* and tanning in neck-ligated blowflies. Native bursicon of the cockroach *P. americana* also is a heterodimer. In *D. melanogaster* levels of pburs, burs, and DLGR2 transcripts are increased before ecdysis, consistent with their role in post-ecdysial cuticle changes. Immunohistochemical analyses in diverse insect species revealed the co-localization of pburs- and burs-immunoreactivity in some of the neurosecretory neurons that also express crustacean cardioactive peptide. Forty-three years after its initial description, the elucidation of the molecular identities of bursicon and its receptor allow for further elucidation of bursicon actions in regulating cuticle tanning, wing expansion, and as yet unknown functions, including possible actions of its monomeric subunits. Because bursicon subunit genes are homologous to the vertebrate bone morphogenetic protein antagonists, these findings also may facilitate investigations of functions for these proteins during vertebrate development.

7.5.4. Ecdysis Sequence in the Yellow Fever Mosquito, *Ae. Aegypti*

A detailed description of ecdysis regulation in the mosquito, *Ae. aegypti* has been reported (Dai and Adams, 2009). Morphological markers and ecdysis behaviors were described for three stages: larval (pharate fourth instar) ecdysis, pupal ecdysis, and adult eclosion. Two peptides (AeaETH1, AeaETH2) extracted from tracheae of pharate fourth instar larvae were purified by HPLC and sequenced by mass spectrometry. These peptides show high sequence similarity to the previously described ETH ligands. Comparison of these sequences to the *Ae.*

aegypti genome database provided confirmation of processing sites inferred in the pre-propeptide precursor and in particular, pinpointed the N-terminal amino acid of AeaETH1. Injection of AeaETH1 and AeaETH2 into pharate pupae ~9–12 h prior to initiation of the normal ecdysis sequence elicited typical ecdysis contractions, but no pre-ecdysis behavior.

Aedes ETH receptors, identified as two subtypes (AeaETHR-A, AeaETHR-B), were cloned and functionally expressed in CHO cells for pharmacological profiling using aequorin flash assays. As in previous accounts of ETH receptors in moths and flies, the two mosquito ETH receptor subtypes arise by alternative splicing of 3' exons. Both receptor subtypes were most sensitive to mosquito ETH peptides, but also responded to *Drosophila* and *Manduca* ETHs, with the *Drosophila* ligands proving to be more potent than those of the moth. One exception was MasPETH, which was ~100-fold less active against AeaETHR-A and completely inactive against AeaETHR-B.

Further analysis of hormone dynamics in the pharate pupa showed that ecdysteroid levels rise ~25 h prior to initiation of the pupal ecdysis sequence to initiate the molt. The total AeaETHR transcript number rises in register with ecdysteroids, as does behavioral sensitivity to AeaETH injection. These data provide further confirmatory evidence that, as observed in *Manduca*, *Bombyx*, and *Drosophila*, steroids regulate expression of genes responsible for the ecdysis signaling axis.

7.5.5. Eclosion Sequence in Beetles and Honeybee

7.5.5.1. Natural eclosion sequence in the red flour beetle, *Tr. castaneum* In an effort to gain a better understanding of neuroendocrine regulation of ecdysis behaviors, the natural eclosion sequence was examined in the red flour beetle, *Tr. castaneum*, a model organism that provides an exceptional experimental platform for RNA silencing (Arakane *et al.*, 2008). Adult eclosion is initiated 5–6 six days after pupation and is composed of pre-eclosion, eclosion, and post-eclosion behaviors. Pre-eclosion lasts for up to 10 h and is characterized by rhythmic dorsoventral contractions of the abdomen. Subsequent eclosion behavior lasts for 22 min and is composed of several consecutive steps that include bouts of body bendings, tracheal air inflation, detachment of antennae, stretching of legs and wings, and A-P movements essential for complete shedding of the pupal cuticle. During post-eclosion, the exuvium is detached from the tip of the abdomen by robust shaking movements and the fully stretched hindwings and abdomen gradually retract under the non-sclerotized forewings (elytrae). The elytrae and exoskeleton are tanned over a period of 5 days (**Figure 37**).

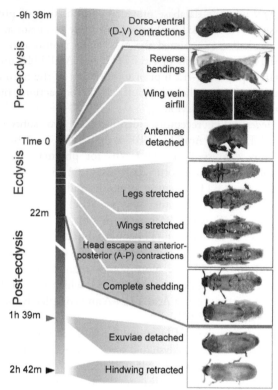

Figure 37 Schematic drawings depicting the time lines and behavioral phases during adult eclosion in the red flour beetle, *T. castaneum*. Arrows indicate morphological changes and behavioral patterns of movements Reproduced with permission from *Arakane et al.* (2008).

7.5.5.2. Roles of neuropeptides and their receptors in *Tr.* eclosion

A wide array of RNAi experiments was performed to determine specific roles of neuropeptides and their receptors during *Tr.* eclosion (Arakane *et al.*, 2008). This thorough analysis revealed essential functions for signaling pathways of ETH, EH, CCAP, and bursicon. Injections of dsRNA led to strong and specific suppression of transcripts for these neuropeptides and their receptors, and resulted in severe behavioral and developmental defects. Injections of double-stranded RNA targeting eclosion hormone (ds-*eh*), ETH (ds-*eth*), and ETH receptor subtype A (ds-*ethr-a*) disrupted the entire eclosion sequence including pre-ecdysis, whereas ds-*ethr-b* elicited no discernible phenotype. Silencing of CCAP (ds-*ccap*) and its receptor subtype 2 (ds-*ccapr-2*) completely disrupted eclosion behavior, while ds-*ccapr-1* had no effect. RNAi targeting bursicon subunits and the bursicon receptor "rickets" suppresses wing expansion, but surprisingly did not affect cuticle tanning or viability. In addition, RNAi of the latter genes also weakens pre-eclosion movements. These results show clearly that ETH and EH signaling are essential for activation and initiation of the entire eclosion sequence, whereas CCAP is necessary for eclosion behavior only. Bursicon is required for post-eclosion expansion of wings, but not for cuticle tanning.

Bursicon also may participate in regulation of pre-ecdysis behavior. These findings indicate that while the basic regulatory components for ecdysis signaling are similar to those previously described in fly and moths, specific functions of endocrine signals may be considerably modified, depending on the species.

7.5.5.3. Eclosion sequence in the mealworm beetle, *T. molitor*

In a related species, the mealworm beetle *T. molitor*, the eclosion sequence is similar to that described earlier (Roller *et al.*, 2010). Natural pre-eclosion starts 7–8 h before eclosion and is characterized by dorsoventral movements (up and down twitches) of the entire abdomen. Pre-eclosion movements develop gradually into strong and regular twitches of the head, thorax, and abdomen, and contractions of the mandibles and all tarsi. Stretched legs and a strong abdominal rotation indicate a switch to eclosion behavior, which lasts for 16–18 min. Eclosion behavior is characterized by anteriorly directed peristaltic movements combined with 3–7 robust abdominal rotations, resulting in rupture and shedding of the old cuticle. Injection of *Tr.* ETH1 into pharate adults of *Te.* induces pre-eclosion only, while injection of the much longer peptide *Tr.* ETH2 elicits the entire eclosion sequence (Roller *et al.*, 2010).

7.5.5.4. Eclosion sequence in the honeybee, *Apis mellifera*

Pre-eclosion behavior in the honeybee *Apis mellifera* is initiated 2–2.5 h prior to eclosion and is characterized by various movements of the head capsule and head/thoracic appendages (the antennae, proboscis, mandibles, and leg tarsi). These initially weak pre-eclosion movements gradually become more obvious and later include robust crossing and scratching of the entire legs. These strong and regular pre-eclosion movements of appendages on the head and thorax are displayed for 1 h and are interrupted by a quiet period for 20–25 min. Animals then switch to eclosion behavior characterized by anteriorly directed peristaltic contractions of the abdomen and grooming-like movements of legs. Peristaltic contractions are essential for rupturing the old cuticle along the dorsal midline, whereas strong leg movements are important for peeling of the pupal cuticle from the head and all appendages. Animals require sticky walls of honeycomb cells for successful eclosion. Injection of *Apis* ETH (a single copy of ETH is produced in bees) accelerates these behaviors (Roller *et al.*, 2010).

7.5.6. Ecdysis Sequence in Orthopterans

7.5.6.1. Hatching and ecdysis of locust nymphs

Behavior and neurophysiology during hatching and the first ecdysis have been described in nymphs of *L. migratoria* and *S. gregaria* (Bernays, 1971, 1972b,c; Truman *et al.*,

1996). Three behavioral phases are observed: (1) hatching behavior of undulatory vermiform larva with legs held close to the body to escape from the eggshell and dig through the soil to the surface, (2) ecdysis behavior initiated upon reaching the surface to shed the embryonic cuticle, and (3) expansional behavior to assume a posture and locomotory patterns typical of a hexapod nymph (Bernays, 1971, 1972b). Immunohistochemical techniques showed that the second phase of these behaviors (ecdysis) is associated with elevation of cGMP in a set of neurons producing CCAP. Hatching and digging animals do not produce cGMP in these neurons, but cGMP appears in the CCAP neuronal network immediately before the free larva initiates ecdysis behavior associated with air swallowing and tracheal air filling (Truman *et al.*, 1996). The cGMP immunoreactivity (cGMP-IR) reaches a peak about 5 min after initiation of ecdysis and then continually decreases. By 20 min, cGMP-IR is considerably reduced and disappears 10 min later. If the nymphs hatch under the moist cheesecloth, they do not ecdyse but continue digging behavior. These larvae do not show cGMP elevation or only traces of cGMP-IR after 10–60 min of digging. However, when ecdysing nymphs with elevated cGMP levels are covered with moist cheesecloth, they resume digging. These "digging ecdysing larvae" produce cGMP in CCAP neurons for prolonged time, ranging from 40–100 min. Once ecdysed nymphs assume the hopper form, covering with the cheesecloth does not resume digging behavior and cGMP elevation; hence, sensory input from the cuticle appears to be important in the initiation as well as cessation of the ecdysis behavior and associated cGMP production in CCAP neurons.

7.5.6.2. Adult ecdysis of crickets and locusts

Detailed behavioral analysis and associated muscle contractions recorded by electromyography were described during adult ecdysis of the cricket *Teleogryllus oceanicus* (Carlson, 1977a,b; Carlson and Bentley, 1977). The entire ecdysis sequence was divided into four phases: preparatory (2 h), ecdysial (20 h), expansional (1 h), and exuvial (30 min) and a total of 48 motor programs were enumerated. The preparatory phase corresponds to pre-ecdysis in other insects and consists of contractions of various muscles to loosen, anchor, and split the old cuticle. We will refer to this phase as pre-ecdysis to be consistent with observations in other insects. Cricket adult pre-ecdysis is initiated with grooming, locomotion, and pushups, later accompanied with synchronous longitudinal contractions of the abdomen to loosen and split the old cuticle along the dorsal ecdysial line. These synchronous abdominal contractions resemble pre-ecdysis I in *Manduca*. Other motor programs are later activated; for example, alternating contractions of coxal muscles help to anchor the claws to the substrate and anntenal beating and scissoring result in loosening the antennae. The onset of ecdysis behavior is preceded by a quiescent phase for 10 min. As

in other insects, cricket ecdysis behavior is characterized by anteriorly directed bouts of abdominal contractions and air swallowing. Additional motor programs are recruited later and involve either synchronous or alternating contractions of muscles in the head, thorax, or legs to extricate entire body from the exuvium.

The post-ecdysis expansional phase includes rhythmic contractions of abdominal and leg muscles, as well as air swallowing to expand cuticle of the body and appendages (legs and wings). After the new cuticle is expanded to its definitive shape, the expansional posture is abandoned and the animal enters the exuvial phase, involving ingestion of its shed cuticle (exuvium) accompanied by sporadic fluttering, spreading, and refolding of wings (Carlson, 1977a).

Removal of the old cuticle prior to onset of the ecdysis sequence results in delayed initiation of behaviors by 2 h. After this delay, normal pre-ecdysis and ecdysis behaviors are initiated, but the animal becomes arrested during the first half of the ecdysis motor programs. When the old cuticle is peeled at ecdysis, the expansional and exuvial phases follow after some delay. These observations indicate that ecdysis motor programs can be initiated without sensory input in *Teleogryllus*. However, sensory information from certain structures (e.g., cerci) is necessary for initiation, termination, or reactivation of some specific programs. For example, if cerci are peeled prior to ecdysis sequence, animals never initiate pre-ecdysis and ecdysis behaviors, but show abdominal tetanic contractions characteristic for the post-ecdysis expansional phase. By comparison, *Manduca* larvae never initiate the ecdysis sequence if the old cuticle is manually peeled, but adults usually inflate their wings and abdomen after they are removed from the pupal cuticle several hours prior to expected eclosion (Miles and Booker, 1998; Zitnan, unpublished).

The cephalic ganglia appear to be important for inhibiting the ecdysis peristaltic program in *Teleogryllus*. Removal of the head at any time during the behavioral sequence results in the initiation of ecdysis abdominal peristaltic movements in combination with the original behavior. For example, if animals are decapitated during pre-ecdysis, they continue to show pre-ecdysis synchronous longitudinal pumping alternating with peristaltic ecdysis contractions, whereas decapitation during the expansional phase results in a combination of tetanic and peristaltic contractions (Carlson, 1977a). Decapitation of *Manduca* and *Drosophila* suppresses pre-ecdysis within a few minutes and accelerates ecdysis onset (see Sections 7.5.1.4. and 7.5.3.3.).

The desert locust, *S. gregaria* displays similar behaviors during adult ecdysis as described in crickets. The locust ecdysis sequence has been divided to several stages consisting of pre-emergence (stages 1, 2), emergence (stages 3, 4), expansional (stage 5), and post-expansional (stage 6) behaviors (Hughes, 1980a). The ecdysis sequence starts after the fully grown fifth instar nymphs cease feeding and enter the pre-emergence behavior characterized by various synchronous or alternating contractions of skeletal muscles. This set of behaviors corresponds to pre-ecdysis in other animals. Emergence or ecdysis behaviors are composed of air swallowing to inflate the gut, as well as peristaltic, anteriorly directed movements of the abdomen to push the animal out of the cuticle and other motor programs to extricate legs and appendages on the head. Air swallowing is driven by neural circuits in the frontal ganglion. It has been reported recently that rhythmic bursting output of the frontal ganglion is initiated and modulated by exposure to ETH and CCAP (Zilberstein *et al.*, 2006). Interestingly, these authors found that the effects of CCAP were influenced by pre-treatment with ETH. After the ecdysis is completed, the new cuticle is expanded within 1 h. The last post-expansional phase is characterized by dorsoventral abdominal contractions during which the gut is deflated and tracheal sacs are filled with air (Hughes, 1980b,c,d).

7.5.6.3. Control of cuticle melanization in locusts — corazonin

His[7] corazonin was isolated from the CC of two locusts as a dark-pigment-inducing agent controlling phase polymorphism (melanization) in an albino mutant and in normal nymphs of *L. migratoria* (Tanaka and Pener, 1994; Tawfik *et al.*, 1999). But while corazonin also induces cuticle melanization in other locust species,

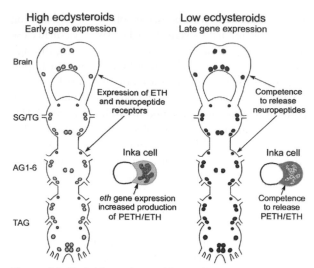

Figure 38 Model for ecdysteroid specification of neural and chemical ecdysis substrates in *Manduca*. High ecdysteroid levels induce CNS sensitivity to PETH/ETH associated with expression of ETH receptors (ETHR) in specific peptidergic neurons and in CA. They also stimulate *eth* expression in Inka cells to increase production of PETH/ETH. These processes are probably mediated by ecdysone receptor (EcR) and other ecdysteroid-inducible transcription factors. At this time, Inka cells are not competent to release PETH/ETH. Competence and timing to release peptides from the CNS and Inka cells are regulated by late gene expression after ecdysteroid levels decrease. These processes may be mediated by βFTZ-F1 and other late transcription factors.

Figure 39 Model for regulation of the moth ecdysis sequence by PETH/ETH and central neuropeptides. (A) The ecdysis sequence is initiated when sensory input to the CNS conveys readiness to ecdyse. Corazonin is released into hemolymph from Ia₁ cells projecting to the CC-CA, eliciting low-level release of PETH and ETH from Inka cells. Low levels of these peptides act on $L_{3,4}$ neurons (green), promoting release of kinins and DHs in AG2-7 to initiate pre-ecdysis I. Kinins and DHs act on IN-402 and MN2,3 to drive dorsoventral pre-ecdysis I contractions. Pre-ecdysis II may be initiated through PETH and ETH action on unidentified neurons in AG1-8 (stippled neurons). During pre-ecdysis, ETH activates the ecdysis network, which includes VM neurons producing EH and the entire network of IN-704 neurons that produce bursicon, CCAP, MIPs, and other neuropeptides. However, inhibitory descending interneurons in the SG and TG1-3 delay onset of ecdysis behavior through release of MIPs and other factors, including FLGa in AG1-8, to delay ecdysis onset. At the appropriate time, ecdysis behavior is initiated when segmental interneurons producing NPF suppress descending inhibition and evoke central release of CCAP and MIPs. These neuropeptides modulate a whole set of motoneurons (MN2,3, PPR, APR, and ISMM) to control peristaltic ecdysis movements. After the old cuticle is shed, ecdysis behavior is terminated by visceral modulatory neurons in AG8, which innervates the hindgut and the last abdominal segment. Associated events include JH production from the CA and tracheal inflation during pre-ecdysis, while the ETH/EH-activated network of neurosecretory cells $L_{1,2,5}$ and M_{Lb} producing bursicon, CCAP, PRLamides, and other factors control post-ecdysis processes, which include cuticle plasticization, expansion, sclerotization, and pigmentation (see text for details). Inka cells are labeled dark red, neurosecretory cells green, excitatory interneurons red and pink, inhibitory interneurons violet, visceral modulatory neurons orange and yellow, and motoneurons blue. (B) Time courses of cGMP elevation in subsets of central neurons, neuropeptide release, and behavior performance during the ecdysis sequence. cGMP is elevated in L_1/704 cells (L_1 is synonymous with cell 27) of the SG and TG1-3 up to 6 h prior to ecdysis of *Manduca* larvae. At this time, bursicon could control sclerotization of mouthparts and crochets. About 4–6 h later, corazonin release initiates the PETH/ETH signaling cascade. Low hemolymph concentrations of PETH/ETH act through ETHR-A in the CA to stimulate production of JHs, and through $L_{3,4}$ neurons to release kinins and DHs for initiation of pre-ecdysis I. During this time, ETH action on ETHR-B neurons leads to three events: (1) peripheral secretion of PBAN/PRLamides from M_{Lb} cells in the SG to control subsequent cuticle pigmentation and diapauses; (2) activation of inhibitory neurons producing MIPs, FGLa, and other factors in the SG, TG1-3, AG1-8 to delay ecdysis onset; and (3) initiation of pre-ecdysis II by unidentified neurons in the AG1-8. About 20 min later, ETH acts on ETHR-A in VM cells and L_1/IN-704 to activate the ecdysis/post-ecdysis network via ETHR-A and EH receptor. This activation is indicated by EH-stimulated cGMP production in abdominal cells L_1/IN-704. Then, 20–30 min later, a switch to ecdysis behavior is probably regulated by NPF from interneurons co-expressing ETHR-A,B, immediately followed by CCAP/MIPs release from abdominal cells L_1/IN-704. This is associated with cGMP appearance in cells $L_{2,5}$ of AG1-7. Within 10–15 min, peristaltic ecdysis contractions are terminated by a mixture of inhibitory peptides (MIPs, sNPF) from visceral neurons expressing ETHR-A,B. After ecdysis, a second surge of bursicon, CCAP, and other peptides from cells $L_{1,2,5}$ is mediated by ETH/EH signaling through both ETHR-A and EHR to control cuticle expansion, sclerotization, and pigmentation.

this effect does not appear to extend beyond this group (Tanaka, 2000). Implantation of CNS ganglia, or injection of their extracts from representatives of 13 insect orders induced cuticle melanization in the albino locust, but failed to evoke melanization in donor species (Roller *et al.*, 2003; Tanaka, 2000). These data indicate that most insects produce corazonin, but its role in cuticle melanization is specific for locusts (Tanaka, 2001).

7.6. Proposed Models for Ecdysis Behavior Control

7.6.1. Regulation of Larval Ecdysis in Manduca sexta

Based on our findings and those from various laboratories, we propose models for hormonal regulation of the ecdysis

sequence in *Manduca* and *Drosophila* (**Figures 38–40**). Models depicted in **Figures 38** and **39** are based heavily on results emanating from studies on moth larval ecdysis, but general principles may apply to other insects. Nevertheless, several important differences exist in regulation of *Drosophila* ecdysis, as discussed in Section 7.5.3.).

Successful ecdysis depends on a signaling cascade consisting of ecdysteroid and peptide hormones. In *Manduca* fourth instar larvae, a surge in ecdysteroid levels initiates the molt about 48 h prior to initiation of the ecdysis sequence. Elevated ecdysteroids exert both positive and negative effects on a variety of cellular targets. One target is the Inka cell, where steroids upregulate expression of their own receptors (EcR) and the ETH gene; the latter effect results in increased production of the peptides PETH and ETH. Elevated ecdysteroid levels also target specific neurons in the CNS, inducing sensitivity to Inka cell peptides through expression

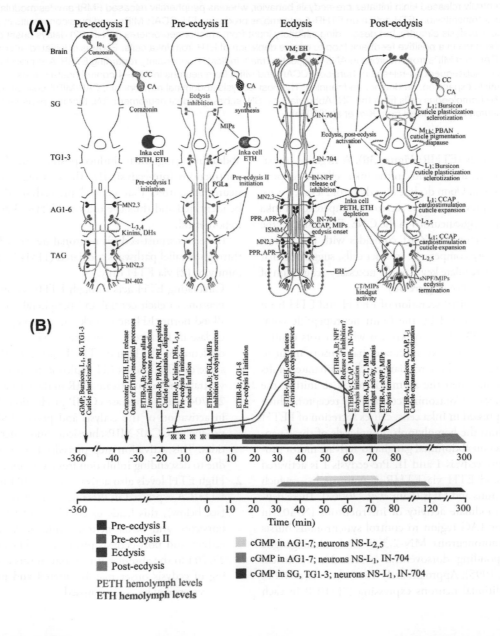

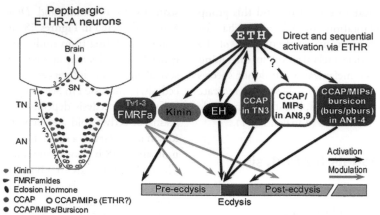

Figure 40 Model of *Drosophila* ecdysis. This model is based on pupal ecdysis, which is accomplished by activation of three major behaviors: pre-ecdysis (0–10 min), ecdysis (10–15 min), and post-ecdysis (15–100 min). Each behavior is programmed in the CNS and sequentially activated by direct actions of ETH from Inka cells. Several min before the onset of the ecdysis sequence ~50% of Inka cells start to release ETHs into the hemolymph, whereas the remaining cells initiate their secretion after onset of pre-ecdysis. Released ETH activates several ETHR-A-expressing peptidergic ensembles that produce FMRFamides and kinin. Centrally released kinin initiates pre-ecdysis behavior, whereas peripherally released FMRFamides modulate neuromuscular transmission. ETH action on ETHR-A ensembles producing EH, CCAP, MIPs, and bursicon results in activation of ecdysis/post-ecdysis circuitry, but descending inhibitory input from anterior neuromeres (ETHR-B?) delays onset of ecdysis. ETH and EH engage in a positive feedback loop to ensure depletion of ETH from Inka cells. Ecdysis is initiated after central release of CCAP and MIPs from neurons in AN8, 9 (question mark indicates uncertainty regarding ETHR-A expression in these neurons), while subsequent co-release of bursicon, CCAP, and MIPs from neurons in AN1–4 control post-ecdysis contractions as well as cuticle expansion, hardening, and tanning. The post-ecdysis abdominal network of CCAP/MIP/bursicon neurons is activated by homologous cells from the SN. Abbreviations: SN, subesophageal neuromere; TN, thoracic neuromere; AN, abdominal neuromere. Modified from *Kim et al.* (2006b).

of ETH receptors (ETHR; **Figure 38**). A negative regulatory effect of elevated steroids in the Inka cell is to suppress peptide secretion. Upon decline of ecdysteroid levels, a wave of late gene expression confers secretory competence in Inka cells and peptidergic neurons in the CNS. Expression of the transcription factor β-FTZ-F1 coincides with the appearance of secretory competence in Inka cells, suggesting that this factor may regulate key event(s) necessary acquisition of secretory competence (**Figure 38**).

In *Manduca*, initial secretion of PETH and ETH from Inka cells is regulated by the brain neuropeptide corazonin (**Figure 39**). It is not known what factors control corazonin release, but we speculate that sensory input to the larval brain or the circadian clock in pharate adults may determine when the animal is ready to initiate the ecdysis sequence. Corazonin action on its receptor, which is highly expressed in Inka cells, causes secretion of PETH and ETH into the hemolymph. Low levels of these peptides then act on abdominal ganglia to activate motor programs for pre-ecdysis I and II. Pre-ecdysis I is activated by PETH and ETH via ETHR-A neurons $L_{3,4}$, which co-release kinins and DHs (Kim *et al.*, 2006a). These neuropeptides modulate activity of interneurons IN-402 in the posterior TAG region to control synchronous bursts of paired motoneurons MN-2,3 in abdominal ganglia and corresponding dorsoventral contractions (Novicki and Weeks, 1995). Approximately 15–20 min later, ETH acts on additional neurons expressing ETHR-B in each

abdominal ganglion to induce posteroventral and proleg contractions (pre-ecdysis II). The neuronal network and neurotransmitters that control pre-ecdysis II remain to be identified and thus the roles of ETHR-B are currently hypothetical.

The ecdysis/post-ecdysis neuronal network is activated through parallel pathways mediated (1) via ETH and EH jointly or (2) via ETH only.

1. Circulating ETH acts through ETHR-A on brain VM neurons to elicit central and peripheral release of EH. Blood-borne EH acts back on its receptor guanylyl cyclase in Inka cells to induce cGMP production and depletion of PETH and ETH (**Figure 39**). The resulting surge of ETH causes EH depletion from VM axons in ventral ganglia to activate EHR in the entire network of 27/704 cells to increase cGMP production. This results in activation of the ecdysis and post-ecdysis circuitry producing CCAP, MIPs, bursicon, and other neuropeptides. This activation step precedes behavior initiation due to descending inhibition from anterior ganglia.

2. High ETH levels also activate the 27/704 network via ETHR-A expressed in these neurons. After some delay (see below), this leads to co-release of CCAP, MIPs, bursicon, and other neuropeptide(s), which control ecdysis and post-ecdysis processes. ETH can activate 27/704 in the absence of EH, but in this case the timing of ecdysis onset may be altered and post-ecdysis processes may be compromised.

Although cGMP elevation and activation of the ecdysis/post-ecdysis network in abdominal ganglia are apparent already 30 min into natural pre-ecdysis, the switch to ecdysis behavior occurs only after an additional 30 min delay. This delay from activation of the CCAP/MIP network to initiation of ecdysis behavior may be controlled by putative descending inhibitory ETHR-B neurons in the SG and TG1-3 and abdominal neurons expressing both ETHR-A and ETHR-B (ETHR-A/B) and FGLa. Segmental interneurons expressing ETHR-A,B and NPF may be responsible for disinhibition of IN-704 and consequent central co-release of CCAP and MIPs. CCAP and MIPs initiate the ecdysis motor circuit, while MIPs suppress activities of pre-ecdysis neurons. CCAP probably modulates activities of several identified motoneurons (MN2,3, ISMM, PPR, and APR) controlling ecdysis movements (**Figure 39**). After the old cuticle is shed, the ecdysis motor program and hindgut contractions are terminated irreversibly by visceral modulatory neurons expressing ETHR-A,B and inhibitory neuropeptides — MIPs and sNPF in the TAG.

Post-ecdysis is mediated by cells 27 expressing both ETHR-A and EHR. These cells release bursicon and CCAP after ecdysis to regulate plasticization, expansion, and hardening of the cuticle. Activation of ETHR-A in CA and ETHR-B in M_{Lb} cells in the SG results in the release of JH and PBAN/MRCH, respectively, which participate in regulation of cuticle pigmentation. Peripheral release of kinins and DHs from $L_{3,4}$ cells and visceral modulatory neurons may be important for regulation of tracheal inflation and water balance during and after ecdysis (**Figure 39**).

7.6.2. Regulation of *Drosophila* Pupal Ecdysis

Calcium imaging, RNA silencing, and cell killing experiments in *Drosophila* showed that a neuropeptide signaling cascade is involved in the proper timing and execution of the ecdysis sequence. The ecdysis sequence, composed of pre-ecdysis, ecdysis, and post-ecdysis behaviors, is initiated by direct ETH action on the CNS. Regulation of ETH release from Inka cells has not been demonstrated clearly, so we hypothesize that EH and possibly other humoral factor controls proper timing of this process. Released ETHs act on their receptors (ETHR-A, B) in the CNS to activate multiple peptidergic targets responsible for initiation of the ecdysis sequence. ETH activation of ETHR-A in kinin neurons leads to initiation of pre-ecdysis associated with tracheal inflation. ETH also elicits release of FMRFamides from thoracic VLT (Tv) cells expressing ETHR-A to modulate neuromuscular efficacy in the periphery. Other unidentified ETHR-B neurons may participate in modulation of pre-ecdysis. Meanwhile, ETH acts through ETHR-A on EH neurons in the brain and the 27/704 network in the ventral nerve cord to activate the entire ecdysis network. Putative inhibitory neurons

producing ETHR-B may delay ecdysis onset. The release of CCAP and MIPs and possibly bursicon initiates ecdysis and post-ecdysis behaviors (**Figure 40**).

Abbreviations:

AG – abdominal ganglion
APR – anterior planta retractor moneuron
CCAP – crustacean cardioactive peptide
CHH – crustacean hyperglycemic hormone
CT – calcitonin-like DH
DAPI – diamino phenyl indole
DH – diuretic hormone
dsRNA – double-stranded RNA
EH – eclosion hormone
EHR – eclosion hormone receptor
ETH – ecdysis triggering hormone
ETH-AP – ETH-associated peptide
ETHR-A – ETH receptor subtype A
ETHR-B – ETH receptor subtype B
FLRFamide – Phe-Leu-Arg-Phe-amide
FGLamide – cockroach allatostatin (type A)
HPLC – high-performance liquid chromatography
HRP – horseradish peroxidase
IN-402 – interneurons 402
IN-704 – interneurons 704
ISMM – intersegmental muscle motoneuron
ITP – ion transport peptide
MIP – myoinhibitory peptide
MN-2,3 – motoneurons 2, 3
MRCH – melanization and reddish coloration hormone
NPF – neuropeptide F
PBAN – pheromone biosynthesis activating hormone
PETH – pre-ecdysis triggering hormone
PPR – principal planta retractor motoneuron
RNAi – RNA interference
SCP_B – small cardioactive peptide B
SG – subesophageal ganglion
sNPF – short neuropeptide F
TG – thoracic ganglion
TAG – terminal abdominal ganglion
VM – ventromedial neurosecretory cells
20E – 20-hydroxyecdysone

Acknowledgments

We thank Ivana Daubnerova for comments on the manuscript and technical assistance and Larry Gilbert for infinite patience. This work was supported by grants from the National Institutes of Health (AI40555 and GM067310) and Vedecká grantová agentúra (95/5305/800 and 2/7168/20).

References and further reading may be available for this article. To view references and further reading you must purchase this article.

References

Adams, M. D., Celniker, S. E., Holt, R. A., Evans, C. A., Gocayne, J. D., et al. (2000). The genome sequence of *Drosophila melanogaster*. *Science, 287*, 2185–2195.

Adams, M. E., & O'Shea, M. (1981). Vacuolation of an identified peptidergic (proctolin-containing) neuron. *Brain Res., 230*, 439–444.

Adams, M. E., & Phelps, M. N. (1983). Co-localization of bursicon bioactivity and proctolin in identified neurons. *Society for Neuroscience Abstracts, 9*, 313.

Adams, M. E., & Zitnan, D. (1997). Identification of ecdysis-triggering hormone in the silkworm *Bombyx mori*. *Biochemical and Biophysical Research Communications, 230*, 188–191.

Akai, H. (1992). Ultrastructure of Epitracheal Gland During Larval-Pupal Molt of *Bombyx mori*. *Cytologia, 57*, 195–201.

Arakane, Y., Li, B., Muthukrishnan, S., Beeman, R. W., Kramer, K. J., et al. (2008). Functional analysis of four neuropeptides, EH, ETH, CCAP and bursicon, and their receptors in adult ecdysis behavior of the red flour beetle, *Tr. castaneum. Mech. Dev., 125*, 984–995, Epub 2008 Sep 19.

Ashburner, M., Chihara, C., Meltzer, P., & Richards, G. (1974). Temporal control of puffing activity in polytene chromosomes. *Cold Spring Harbor Symposia on Quantitative Biology, 38*, 655–662.

Asuncion-Uchi, M., El Shawa, H., Martin, T., & Fuse, M. (2009). Different actions of ecdysis-triggering hormone on the brain and ventral nerve cord of the hornworm, *Manduca sexta. Gen. Comp. Endocrinol., 166*, 54–65.

Audsley, N., McIntosh, C., & Phillips, J. E. (1992a). Isolation of a neuropeptide from locust corpus cardiacum which influences ileal transport. *Journal of Experimental Biology, 173*, 261–274.

Audsley, N., McIntosh, C., & Phillips, J. E. (1992b). Actions of ion-transport peptide from locust corpus cardiacum on several hindgut transport processes. *Journal of Experimental Biology, 173*, 275–288.

Bainbridge, S. P., & Bownes, M. (1981). Staging the metamorphosis of *Drosophila melanogaster*. *Journal of Embryology and Experimental Morphology, 66*, 57–80.

Baker, J. D., McNabb, S. L., & Truman, J. W. (1999). The hormonal coordination of behavior and physiology at adult ecdysis in *Drosophila melanogaster*. *J. Exp. Biol., 202*, 3037–3048.

Baker, J. D., & Truman, J. W. (2002). Mutations in the *Drosophila* glycoprotein hormone receptor, rickets, eliminate neuropeptide-induced tanning and selectively block a stereotyped behavioral program. *J. Exp. Biol., 205*, 2555–2565.

Bernays, E. A. (1971). The vermiform larva of *Schistocerca gregaria* form and activity Insecta Orthoptera. *Zeitschrift Fuer Morphologie der Tiere, 70*, 183–200.

Bernays, E. A. (1972). The Intermediate Molt 1st Ecdysis of *Schistocerca gregaria* Insecta Orthoptera. *Zeitschrift Fuer Morphologie der Tiere, 71*, 160–179.

Blackburn, M. B., Jaffe, H., Kochansky, J., & Raina, A. K. (2001). Identification of four additional myoinhibitory peptides (MIPs) from the ventral nerve cord of *Manduca sexta*. *Arch Insect Biochem. Physiol., 48*, 121–128.

Blackburn, M. B., Wagner, R. M., Kochansky, J. P., Harrison, D. J., Thomas-Laemont, P., et al. (1995). The identification of two myoinhibitory peptides, with sequence similarities to the galanins, isolated from the ventral nerve cord of *Manduca sexta. Regul. Pept., 57*, 213–219.

Bollenbacher, W. E., Smith, S. L., Goodman, W., & Gilbert, L. I. (1981). Ecdysteroid titer during larval–pupal–adult development of the tobacco hornworm, *Manduca sexta. Gen. Comp. Endocrinol., 44*(3), 302–306.

Breidbach, O., & Dircksen, H. (1991). Crustacean Cardioactive Peptide-Immunoreactive Neurons in the Ventral Nerve Cord and the Brain of the Meal Beetle T *enebrio molitor* During Postembryonic Development. *Cell & Tissue Research, 265*, 129–144.

Breidbach, O., Dircksen, H., & Wegerhoff, R. (1995). Common general morphological pattern of peptidergic neurons in the arachnid brain: Crustacean cardioactive peptide-immunoreactive neurons in the protocerebrum of seven arachnid species. *Cell & Tissue Research, 279*, 183–197.

Cantera, R., Veenstra, J. A., & Nassel, D. R. (1994). Postembryonic development of corazonin-containing neurons and neurosecretory cells in the blowfly, *Phormia terraenovae*. *J. Comp. Neurol., 350*, 559–572.

Carlson, J. R. (1977a). The Imaginal Ecdysis of the Cricket *Teleogryllus oceanicus* Part 1 Organization of Motor Programs and Roles of Central and Sensory Control. *Journal of Comparative Physiology A Sensory Neural & Behavioral Physiology, 115*, 299–317.

Carlson, J. R. (1977b). The Imaginal Ecdysis of the Cricket *Teleogryllus oceanicus* Part 2 the Roles of Identified Motor Units. *Journal of Comparative Physiology A Sensory Neural & Behavioral Physiology, 115*, 319–336.

Carlson, J. R., & Bentley, D. (1977). Ecdysis: neural orchestration of a complex behavioral performance. *Science, 195*, 1006–1008.

Chang, J. C., Yang, R. B., Adams, M. E., & Lu, K. H. (2009). Receptor guanylyl cyclases in Inka cells targeted by eclosion hormone. *Proc. Natl. Acad. Sci. USA., 106*, 13371–13376, Epub 2009 Jul 28.

Chen, Y., Veenstra, J. A., Hagedorn, H., & Davis, N. T. (1994). Leucokinin and diuretic hormone immunoreactivity of neurons in the tobacco hornworm, *Manduca sexta*, and colocalization of this immunoreactivity in lateral neurosecretory cells of abdominal ganglia. *Cell & Tissue Research, 278*, 493–507.

Cheung, C. C., Loi, P. K., Sylwester, A. W., Lee, T. D., & Tublitz, N. J. (1992). Primary structure of a cardioactive neuropeptide from the tobacco hawkmoth, *Manduca sexta. FEBS Lett., 313*, 165–168.

Chung, J. S., Dircksen, H., & Webster, S. G. (1999). A remarkable, precisely timed release of hyperglycemic hormone from endocrine cells in the gut is associated with ecdysis in the crab *Carcinus maenas. Proc. Natl. Acad. Sci. USA, 96*, 13103–13107.

Clark, A. C., del Campo, M. L., & Ewer, J. (2004). Neuroendocrine control of larval ecdysis behavior in *Drosophila*: complex regulation by partially redundant neuropeptides. *J. Neurosci., 24*, 4283–4292.

Clynen, E., Huybrechts, J., Verleyen, P., De Loof, A., & Schoofs, L. (2006). Annotation of novel neuropeptide precursors in the migratory locust based on transcript screening of a public EST database and mass spectrometry. *BMC Genomics.*, *7*, 201.

Copenhaver, P. F., & Truman, J. W. (1982). The Role of Eclosion Hormone in the Larval Ecdyses of *Manduca sexta*. *Journal of Insect Physiology*, *28*, 695–702.

Copenhaver, P. F., & Truman, J. W. (1986a). Identification of the cerebral neurosecretory cells that contain eclosion hormone in the moth *Manduca sexta*. *Journal of Neuroscience*, *6*, 1738–1747.

Copenhaver, P. F., & Truman, J. W. (1986b). Control of neurosecretion in the moth *Manduca sexta*: physiological regulation of the eclosion hormone cells. *Journal of Comparative Physiology. a, Sensory, Neural, and Behavioral Physiology*, *158*, 445–455.

Cottrell, C. B. (1962a). General observations on the imaginal ecdysis of blowflies. *Transactions of the Royal Entomology Society of London*, *114*, 317–333.

Cottrell, C. B. (1962b). The imaginal ecdysis of blowflies. The control of cuticular hardening and darkening. *J. Exp. Biol.*, *39*, 395–411.

Cottrell, C. B. (1962c). The imaginal ecdysis of blowflies. Detection of the blood-borne darkening factor and determination of some of its properties. *J. Exp. Biol.*, *39*.

Cottrell, C. B. (1964). Insect ecdysis with particular emphasis on cuticular hardening and darkening. *Advances in Insect Physiology*, *2*, 175–218.

Curtis, A. T., Hori, M., Gren, J. M., Wolfgang, W. J., Hiruma, K., et al. (1984). Ecdysteroid regulation of the onset of cuticular melanization in the allatectomized and *black* mutant *Manduca sexta* larvae. *J. Insect. Physiol.*, *30*, 597–606.

Dai, H., Jiang, R., Wang, J., Xu, G., Cao, M., et al. (2007b). Development of a heat shock inducible and inheritable RNAi system in silkworm. *Biomol Eng.*, *24*, 625–630, Epub 2007 Oct 23.

Dai, H., Ma, L., Wang, J., Jiang, R., Wang, Z., et al. (2008b). Knockdown of ecdysis-triggering hormone gene with a binary UAS/GAL4 RNA interference system leads to lethal ecdysis deficiency in silkworm. *Acta. Biochim. Biophys. Sin. (Shanghai)*, *40*, 790–795.

Dai, L., & Adams, M. E. (2009). Ecdysis triggering hormone signaling in the yellow fever mosquito *Aedes aegypti. Gen. Comp. Endocrinol.*, *162*, 43–51, Epub 2009 Mar 17.

Dai, L., Dewey, E. M., Zitnan, D., Luo, C. W., Honegger, H. W., et al. (2008a). Identification, developmental expression, and functions of bursicon in the tobacco hawkmoth, *Manduca sexta. J. Comp. Neurol.*, *506*, 759–774.

Dai, L., Zitnan, D., & Adams, M. E. (2007a). Strategic expression of ion transport peptide gene products in central and peripheral neurons of insects. *J. Comp. Neurol.*, *500*, 353–367.

Daubnerova, I., Roller, L., & Zitnan, D. (2009). Transgenesis approaches for functional analysis of peptidergic cells in the silkworm *Bombyx mori. Gen. Comp. Endocrinol.*, *162*, 36–42, Epub 2008 Dec 10.

Davis, N. T., Blackburn, M. B., Golubeva, E. G., & Hildebrand, J. G. (2003). Localization of myoinhibitory peptide immunoreactivity in *Manduca sexta* and *Bombyx mori*, with indications that the peptide has a role in molting and ecdysis. *Journal of Experimental Biology*, *206*, 1449–1460.

Davis, N. T., Homberg, U., Dircksen, H., Levine, R. B., & Hildebrand, J. G. (1993). Crustacean cardioactive peptide-immunoreactive neurons in the hawkmoth *Manduca sexta* and changes in their immunoreactivity during postembryonic development. *J. Comp. Neurol.*, *338*, 612–627.

Davis, N. T., Homberg, U., Teal, P. E. A., Altstein, M., Agricola, H., et al. (1996). Neuroanatomy and Immunocytochemistry of the median neuroendocrine cells of the subesophageal ganglion of the tobacco hawk moth, *Manduca sexta*; Immunoreactivities to PBAN and other neuropeptides. *Microscopy Research Techniques*, *35*, 201–229.

Davis, N. T., Veenstra, J. A., Feyereisen, R., & Hildebrand, J. G. (1997). Allatostatin-like-immunoreactive neurons of the tobacco hornworm, *Manduca sexta*, and isolation and identification of a new neuropeptide related to cockroach allatostatins. *J. Comp. Neurol.*, *385*, 265–284.

de Kleijn, D. P., de Leeuw, E. P., van den Berg, M. C., Martens, G. J., & van Herp, F. (1995). Cloning and expression of two mRNAs encoding structurally different crustacean hyperglycemic hormone precursors in the lobster *Homarus americanus. Biochim. Biophys. Acta*, *1260*, 62–66.

de Kleijn, D. P., Janssen, K. P., Martens, G. J., & Van Herp, F. (1994). Cloning and expression of two crustacean hyperglycemic-hormone mRNAs in the eyestalk of the crayfish *Orconectes limosus. Eur. J. Biochem.*, *224*, 623–629.

de Kleijn, D. P., Janssen, K. P., Waddy, S. L., Hegeman, R., Lai, W. Y., et al. (1998). Expression of the crustacean hyperglycaemic hormones and the gonad-inhibiting hormone during the reproductive cycle of the female American lobster *Homarus americanus. J. Endocrinol.*, *156*, 291–298.

Dewey, E. M., McNabb, S. L., Ewer, J., Kuo, G. R., Takanishi, C. L., et al. (2004). Identification of the gene encoding bursicon, an insect neuropeptide responsible for cuticle sclerotization and wing spreading. *Curr. Biol.*, *14*, 1208–1213.

Dircksen, H. (1998). Conserved crustacean cardioactive peptide (CCAP) neuronal networks and functions in arthropod evolution. In G. M. Coast, & S. G. Webster (Eds.), *Recent advances in arthropod endocrinology* (pp. 302–333). Cambridge University Press.

Dircksen, H. (2009). Insect ion transport peptides are derived from alternatively spliced genes and differentially expressed in the central and peripheral nervous system. *J. Exp. Biol.*, *212*, 401–412.

Dircksen, H., & Keller, R. (1988). Immunocytochemical localization of CCAP, a novel crustacean cardioactive peptide, in the nervous system of the shore crab, *Carcinus maenas. Cell & Tissue Research*, *254*, 347–360.

Dircksen, H., Mueller, A., & Keller, R. (1991). Crustacean Cardioactive Peptide in the Nervous System of the Locust Locusta-Migratoria an Immunocytochemical Study on the Ventral Nerve Cord and Peripheral Innervation. *Cell & Tissue Research*, *263*, 439–458.

Dircksen, H., Webster, S. G., & Keller, R. (1988). Immunocytochemical demonstration of the neurosecretory systems containing putative moult-inhibiting and hyperglycemic hormone in the eyestalk of brachyuran crustaceans. *Cell and Tissue Research*, *251*, 3–12.

Drexler, A. L., Harris, C. C., dela Pena, M. G., Asuncion-Uchi, M., Chung, S., et al. (2007). Molecular characterization and cell-specific expression of an ion transport peptide in the tobacco hornworm, *Manduca sexta*. *Cell Tissue Res., 329*, 391–408, Epub 2007 Apr 21.

Ewer, J., De Vente, J., & Truman, J. W. (1994). Neuropeptide induction of cyclic GMP increases in the insect CNS: resolution at the level of single identifiable neurons. *J. Neurosci., 14*, 7704–7712.

Ewer, J., Gammie, S. C., & Truman, J. W. (1997). Control of insect ecdysis by a positive-feedback endocrine system: roles of eclosion hormone and ecdysis triggering hormone. *J. Exp. Biol.*, 869–881.

Ewer, J., & Reynolds, S. (2002). Neuropeptide control of molting in insects. In D. W. Pfaff, et al. (Ed.), *Hormones, Brain and Behavior* (pp. 1–92). Boston: Elsevier Science.

Ewer, J., & Truman, J. W. (1996). Increases in cyclic 3′, 5′-guanosine monophosphate (cGMP) occur at ecdysis in an evolutionarily conserved crustacean cardioactive peptide-immunoreactive insect neuronal network. *J. Comp. Neurol., 370*, 330–341.

Ewer, J., & Truman, J. W. (1997). Invariant association of ecdysis with increases in cyclic 3′,5′-guanosine monophosphate immunoreactivity in a small network of peptidergic neurons in the hornworm, *Manduca sexta. J. Comp. Physiol. [A], 181*, 319–330.

Fraenkel, G. (1935). Observations and experiments on the blowfly (*Calliphora erythrocephala*) during the first day after emergence. *Proceedings of the Zoological Society of London, 87*, 893–904.

Fraenkel, G. (1965). Bursicon, a hormone which mediates tanning of the cuticle in the adult fly and other insects. *Journal of Insect Physiology, 11*, 513–556.

Fraenkel, G. (1966). Properties of bursicon: an insect protein hormone that controls cuticular tanning. *Science, 151*, 91–93.

Fraenkel, G., & Hsiao, C. (1962). Hormonal and nervous control of tanning in the fly. *Science, 138*, 27–29.

Fraenkel, G., & Hsiao, C. (1963). Tanning in the adult fly: A new function of neurosecretion in the brain. *Science, 141*, 1057–1058.

Furuya, K., Liao, S., Reynolds, S. E., Ota, R. B., Hackett, M., et al. (1993). Isolation and identification of a cardioactive peptide from *Te. molitor* and *Spodoptera eridania*. *Biol. Chem. Hoppe. Seyler, 374*, 1065–1074.

Fuse, M., & Truman, J. W. (2002). Modulation of ecdysis in the moth *Manduca sexta*: the roles of the suboesophageal and thoracic ganglia. *J. Exp. Biol., 205*, 1047–1058.

Gammie, S. C., & Truman, J. W. (1997a). Neuropeptide hierarchies and the activation of sequential motor behaviors in the hawkmoth, Manduca sexta. *J. Neurosci., 17*, 4389–4397.

Gammie, S. C., & Truman, J. W. (1997b). An endogenous elevation of cGMP increases the excitability of identified insect neurosecretory cells. *J. Comp. Physiol. [A], 180*, 329–337.

Gammie, S. C., & Truman, J. W. (1999). Eclosion hormone provides a link between ecdysis-triggering hormone and crustacean cardioactive peptide in the neuroendocrine cascade that controls ecdysis behavior. *Journal of Experimental Biology, 202*, 343–352.

Gaude, H. (1975). Histological Studies on the Structure and Function of the Neuro Secretory System of the House Cricket *Acheta domesticus*. *Zoologischer Anzeiger, 194*, 151–164.

Grillot, J.-P., Delachambre, J., & Provansal, A. (1976). Roles des organes perisympathetiques et dynamique de la secretion de la bursicon chez Te. molitor. *Journal of Insect Physiology, 22*, 763–780.

Gu, P. L., & Chan, S. M. (1998). The shrimp hyperglycemic hormone-like neuropeptide is encoded by multiple copies of genes arranged in a cluster. *FEBS Lett., 441*, 397–403.

Gu, P. L., Yu, K. L., & Chan, S. M. (2000). Molecular characterization of an additional shrimp hyperglycemic hormone: cDNA cloning, gene organization, expression and biological assay of recombinant proteins. *FEBS Lett., 472*, 122–128.

Hagino, A., Kitagawa, N., Imai, K., Yamashita, O., & Shiomi, K. (2010). Immunoreactive intensity of FXPRL amide neuropeptides in response to environmental conditions in the silkworm, *Bombyx mori*. *Cell & Tissue Research, 342*, 459–469.

Helle, J., Dircksen, H., Eckert, M., Naessel Dick, R., Spoerhase-Eichmann, U., et al. (1995). Putative neurohemal areas in the peripheral nervous system of an insect, *Gryllus bimaculatus*, revealed by immunocytochemistry. *Cell & Tissue Research, 281*, 43–61.

Hewes, R. S., & Truman, J. W. (1991). The Roles of Central and Peripheral Eclosion Hormone Release in the Control of Ecdysis Behavior in *Manduca sexta*. *Journal of Comparative Physiology A Sensory Neural & Behavioral Physiology, 168*, 697–708.

Hewes, R. S., & Truman, J. W. (1994). Steroid regulation of excitability in identified insect neurosecretory cells. *J. Neurosci., 14*, 1812–1819.

Hiruma, K., Matsumoto, S., Isogai, A., & Suzuki, A. (1984). Control of ommochrome synthesis by both juvenile hormone and melanization hormone in the cabbage armyworm, *Mamestra brassicae. J. Comp. Physiol. B, 154*, 13–21.

Hiruma, K., Norman, A., & Riddiford, L. M. (1993). A neuroendocrine factor essential for cuticular melanization in the tobacco hornworm, *Manduca sexta J. Insect Physiol., 39*, 353–360.

Hiruma, K., & Riddiford, L. M. (1993). Molecular mechanisms of cuticular melanization in the tobacco hornworm, *Manduca sexta*. (L.) (*Lepidoptera: Sphingidae*) International Journal of Insect Morphology & Embryology, 22, 103–117.

Hiruma, K., & Riddiford, L. M. (2001). Regulation of transcription factors MHR4 and betaFTZ-F1 by 20-hydroxyecdysone during a larval molt in the tobacco hornworm, *Manduca sexta. Developmental Biology, 232*, 265–274.

Homberg, U., Davis, N. T., & Hildebrand, J. G. (1991). Peptide-immunocytochemistry of neurosecretory cells in the brain and retrocerebral complex of the sphinx moth *Manduca sexta. J. Comp. Neurol., 303*, 35–52.

Honegger, H. W., Dewey, E. M., & Ewer, J. (2008). Bursicon, the tanning hormone of insects: recent advances following the discovery of its molecular identity. *J. Comp. Physiol. A Neuroethol. Sens. Neural. Behav. Physiol., 194*, 989–1005, Epub 2008 Nov 13.

Honegger, H. W., Market, D., Pierce, L. A., Dewey, E. M., Kostron, B., et al. (2002). Cellular localization of bursicon using antisera against partial peptide sequences of this insect cuticle-sclerotizing neurohormone. *J. Comp. Neurol.*, *452*, 163–177.

Horodyski, F. M., Ewer, J., Riddiford, L. M., & Truman, J. W. (1993). Isolation, characterization and expression of the eclosion hormone gene of *Drosophila melanogaster. European Journal of Biochemistry*, *215*, 221–228.

Horodyski, F. M., Riddiford, L. M., & Truman, J. W. (1989). Isolation and expression of the eclosion hormone gene from the tobacco hornworm, *Manduca sexta. Proceedings of the National Academy of Sciences of the United States of America*, *86*, 8123–8127.

Hua, Y. J., Tanaka, Y., Nakamura, K., Sakakibara, M., Nagata, S., et al. (1999). Identification of a prothoracicostatic peptide in the larval brain of the silkworm, *Bombyx mori. J. Biol. Chem.*, *274*, 31169–31173.

Huang, J., Zhang, Y., Li, M., Wang, S., Liu, W., et al. (2007). RNA interference-mediated silencing of the bursicon gene induces defects in wing expansion of silkworm. *FEBS Letters*, *581*, 697–701, Epub 2007 Jan 24.

Hughes, T. D. (1980a). The Imaginal Ecdysis of the Desert Locust *Schistocerca gregaria* 1. Description of the Behavior. *Physiological Entomology*, *5*, 47–54.

Hughes, T. D. (1980b). The Imaginal Ecdysis of the Desert Locust *Schistocerca gregaria* 2. Motor Activity Underlying the Preemergence and Emergence Behavior. *Physiological Entomology*, *5*, 55–72.

Hughes, T. D. (1980c). The Imaginal Ecdysis of the Desert Locust *Schistocerca gregaria* 3. Motor Activity Underlying the Expansional and Post Expansion Behavior. *Physiological Entomology*, *5*, 141–152.

Hughes, T. D. (1980d). The Imaginal Ecdysis of the Desert Locust *Schistocerca gregaria* 4. The Role of the Gut. *Physiological Entomology*, *5*, 153–164.

Ikeda, E. (1913). Kimon Rimensen. In E. Ikeda (Ed.), *Experimental Anatomy and Physiology of Bombyx mori* (pp. 242–243). Tokyo: Meibundo.

Iversen, A., Cazzamali, G., Williamson, M., Hauser, F., & Grimmelikhuijzen, C. J. (2002). Molecular identification of the first insect ecdysis triggering hormone receptors. *Biochem. Biophys. Res. Commun.*, *299*, 924–931.

Jahn, G., Käuser, G., & Koolman, J. (1991). Localization of the crustacean cardioactive peptide (CCAP) in the central nervous system of larvae of the blowfly, *Calliphora vicina. General and Comparative Endocrinology*, *82*, 287.

Kamito, T., Tanaka, H., Sato, B., Nagasawa, H., & Suzuki, A. (1992). Nucleotide sequence of cDNA for the eclosion hormone of the silkworm, *Bombyx mori*, and the expression in a brain. *Biochem. Biophys. Res. Commun.*, *182*, 514–519.

Kataoka, H., Li, J. P., Lui, A. S., Kramer, S. J., & Schooley, D. A. (1992). Complete structure of eclosion hormone of *Manduca sexta*. Assignment of disulfide bond location. *Int. J. Pept. Protein Res.*, *39*, 29–35.

Kataoka, H., Troetschler, R. G., Kramer, S. J., Cesarin, B. J., & Schooley, D. A. (1987). Isolation and primary structure of the eclosion hormone of the tobacco hornworm, *Manduca sexta. Biochem. Biophys. Res. Commun.*, *146*, 746–750.

Kegel, G., Reichwein, B., Weese, S., Gaus, G., Peter-Katalinic, J., et al. (1989). Amino acid sequence of the crustacean hyperglycemic hormone (CHH) from the shore crab, *Carcinus maenas. FEBS Lett.*, *255*, 10–14.

Kiguchi, K. (1972). Hormonal control of the coloration of larval body and the pigmentation of larval markings in *Bombyx mori*. (1) Endocrine organs affecting the coloration of larval body and the pigmentation of markings. *J. Sericult. Sci. Jpn.*, *41*, 407–412.

Kim, Y. J., Spalovska-Valachova, I., Cho, K. H., Zitnanova, I., Park, Y., et al. (2004). Corazonin receptor signaling in ecdysis initiation. *Proc. Natl. Acad. Sci. U S A.*, *101*, 6704–6709, Epub 2004 Apr 19.

Kim, Y. J., Zitnan, D., Cho, K. H., Schooley, D. A., Mizoguchi, A., et al. (2006a). Central peptidergic ensembles associated with organization of an innate behavior. *Proc. Natl. Acad. Sci. U S A.*, *103*, 14211–14216, Epub 2006 Sep 12.

Kim, Y. J., Zitnan, D., Galizia, C. G., Cho, K. H., & Adams, M. E. (2006b). A command chemical triggers an innate behavior by sequential activation of multiple peptidergic ensembles. *Curr. Biol.*, *16*, 1395–1407.

Kimura, K. I., & Truman, J. W. (1990). Postmetamorphic Cell Death in the Nervous and Muscular Systems of *Drosophila melanogaster. Journal of Neuroscience*, *10*, 403–411.

Kingan, T. G., & Adams, M. E. (2000). Ecdysteroids regulate secretory competence in Inka cells. *J. Exp. Biol.*, *203*, 3011–3018.

Kingan, T. G., Gray, W., Zitnan, D., & Adams, M. E. (1997). Regulation of ecdysis-triggering hormone release by eclosion hormone. *J. Exp. Biol.*, *200*, 3245–3256.

Kingan, T. G., Shabanowitz, J., Hunt, D. F., & Witten, J. L. (1996). Characterization of two myotropic neuropeptides in the FMRFamide family from segmental ganglia of the moth, *Manduca sexta*: candidate neurohormones and neuromodulators. *J. Exp. Biol.*, *199*.

Kingan, T. G., Teplow, D. B., Phillips, J. M., Riehm, J. P., Rao, K. R., et al. (1990). A new peptide in the FMRF amide family isolated from the CNS of the hawkmoth *Manduca sexta. Peptides*, *11*, 849–856.

Kinjoh, T., Kaneko, Y., Itoyama, K., Mita, K., Hiruma, K., et al. (2007). Control of juvenile hormone biosynthesis in *Bombyx mori*: cloning of the enzymes in the mevalonate pathway and assessment of their developmental expression in the corpora allata. *Insect. Biochem. Mol. Biol.*, *37*, 808–818, Epub 2007 Mar 19.

Klein, C., Kallenborn, H. G., & Radlicki, C. (1999). The 'Inka cell' and its associated cells: ultrastructure of the epitracheal glands in the gypsy moth, *Lymantria dispar. J. Insect. Physiol.*, *45*, 65–73.

Klukas, K. A., Brelje, T. C., & Mesce, K. A. (1996). Novel mouse IgG-like immunoreactivity expressed by neurons in the moth *Manduca sexta*: developmental regulation and colocalization with crustacean cardioactive peptide. *Microsc. Res. Tech.*, *35*, 242–264.

Kono, T., Mizoguchi, A., Nagasawa, H., Ishizaki, H., Fugo, H., et al. (1990a). A monoclonal antibody against a synthetic carboxyl-terminal fragment of the eclosion hormone of the silkworm, *Bombyx mori*: characterization and application to immunohistochemistry and affinity chromatography. *Zoological Science*, *7*, 47–54.

Kono, T., Nagasawa, H., Isogai, A., Fugo, H., & Suzuki, A. (1987). Amino Acid Sequence of Eclosion Hormone of the Silkworm *Bombyx mori. Agricultural & Biological Chemistry, 51*, 2307–2308.

Kono, T., Nagasawa, H., Kataoka, H., Isogai, A., Fugo, H., et al. (1990b). Eclosion hormone of the silkworm *Bombyx mori*. Expression in Escherichia coli and location of disulfide bonds. *FEBS Lett., 263*, 358–360.

Kopec, S. (1917). Experiments on metamorphosis of insects. *Bull. Int. Acad. Cracovie B*, 57–60.

Kopec, S. (1922). Studies on the necessity of the brain for the inception of insect metamorphosis. *Biol. Bull., 42*, 322–342.

Kostron, B., Kaltenhauser, U., Seibel, B., Bräunig, P., Honegger, H., et al. (1996). Localization of bursicon in CCAP-immunoreactive cells in the thoracic ganglia of the cricket *Gryllus bimaculatus. J. Exp. Biol., 199*, 367–377.

Lavorgna, G., Karim, F. D., Thummel, C. S., & Wu, C. (1993). Potential role for a FTZ-F1 steroid receptor superfamily member in the control of *Drosophila* metamorphosis. *Proceedings of the National Academy of Sciences of the United States of America, 90*, 3004–3008.

Lehman, H. K., Murguic, C. M., Miller, T. A., Lee, T. D., & Hildebrand, J. G. (1993). Crustacean cardioactive peptide in the sphinx moth, *Manduca sexta. Peptides, 14*, 735–741.

Levine, R. B., & Truman, J. W. (1982). Metamorphosis of the insect nervous system: changes in morphology and synaptic interactions of identified neurones. *Nature, 299*, 250–252.

Levine, R. B., & Truman, J. W. (1983). Peptide activation of a simple neural circuit. *Brain Res., 279*, 335–338.

Loi, P. K., Emmal, S. A., Park, Y., & Tublitz, N. J. (2001). Identification, sequence and expression of a crustacean cardioactive peptide (CCAP) gene in the moth *Manduca sexta. J. Exp. Biol., 204*, 2803–2816.

Lorenz, M. W., Kellner, R., & Hoffmann, K. H. (1995). A family of neuropeptides that inhibit juvenile hormone biosynthesis in the cricket, *Gryllus bimaculatus. J. Biol. Chem., 270*, 21103–21108.

Lorenz, M. W., Kellner, R., Hoffmann, K. H., & Gade, G. (2000). Identification of multiple peptides homologous to cockroach and cricket allatostatins in the stick insect *Carausius morosus. Insect. Biochem. Mol. Biol., 30*, 711–718.

Lu, D., Lee, K. Y., Horodyski, F. M., & Witten, J. L. (2002). Molecular characterization and cell-specific expression of a *Manduca sexta* FLRFamide gene. *J. Comp. Neurol., 446*, 377–396.

Luo, C. W., Dewey, E. M., Sudo, S., Ewer, J., Hsu, S. Y., et al. (2005). Bursicon, the insect cuticle-hardening hormone, is a heterodimeric cystine knot protein that activates G protein-coupled receptor LGR2. *Proc. Natl. Acad. Sci. U S A., 102*, 2820–2825, Epub 2005 Feb 9.

Marti, T., Takio, K., Walsh, K. A., Terzi, G., & Truman, J. W. (1987). Microanalysis of the amino acid sequence of the eclosion hormone from the tobacco hornworm *Manduca sexta. FEBS Lett., 219*, 415–418.

Matsumoto, S., Kitamura, A., Nagasawa, H., Kataoka, H., Orikasa, C., et al. (1990). Functional diversity of a neurohormone produced by the suboesophageal ganglion: molecular identity of melanization and reddish colouration hormone and pheromone biosynthesis activating neuropeptide. *J. Insect. Physiol., 36*, 427–432.

McNabb, S. L., Baker, J. D., Agapite, J., Steller, H., Riddiford, L. M., et al. (1997). Disruption of a behavioral sequence by targeted death of peptidergic neurons in *Drosophila. Neuron, 19*, 813–823.

Mendive, F. M., Van Loy, T., Claeysen, S., Poels, J., Williamson, M., et al. (2005). *Drosophila* molting neurohormone bursicon is a heterodimer and the natural agonist of the orphan receptor DLGR2. *FEBS Lett., 579*, 2171–2176.

Mesce, K. A., & Truman, J. W. (1988). Metamorphosis of the Ecdysis Motor Pattern in the Hawkmoth *Manduca sexta. Journal of Comparative Physiology A Sensory Neural & Behavioral Physiology, 163*, 287–300.

Miao, Y., Waters, E. M., & Witten, J. L. (1998). Developmental and regional-specific expression of FLRFamide peptides in the tobacco hornworm, *Manduca sexta*, suggests functions at ecdysis. *J. Neurobiol., 37*, 469–485.

Miles, C. I., & Booker, R. (1998). The role of the frontal ganglion in the feeding and eclosion behavior of the moth *Manduca sexta. J. Exp. Biol., 201*, 1785–1798.

Miles, C. I., & Weeks, J. C. (1991). Developmental attenuation of the pre-ecdysis motor pattern in the tobacco hornworm, *Manduca sexta. J. Comp. Physiol. [A], 168*, 179–190.

Mills, R. R., & Lake, C. R. (1966). Hormonal control of tanning in the American cockroach. IV. Preliminary purification of the hormone. *Journal of Insect Physiology, 12*, 1395–1401.

Nagasawa, H., Mikogami, T., Kono, T., Fugo, H., & Suzuki, A. (1987). Molecular Heterogeneity of Eclosion Hormone in Adult Heads of the Silkworm *Bombyx mori. Agricultural & Biological Chemistry, 51*, 1741–1743.

Nielsen, H., Engelbrecht, J., Brunak, S., & von Heijne, G. (1997). Identification of prokaryotic and eukaryotic signal peptides and prediction of their cleavage sites. *Protein Engineering, 10*, 1–6.

Novicki, A., & Weeks, J. C. (1993). Organization of the larval pre-ecdysis motor pattern in the tobacco hornworm, *Manduca sexta. J. Comp. Physiol. [A], 173*, 151–162.

Novicki, A., & Weeks, J. C. (1995). A single pair of interneurons controls motor neuron activity during pre-ecdysis compression behavior in larval *Manduca sexta. J. Comp. Physiol. [A], 176*, 45–54.

Novicki, A., & Weeks, J. C. (1996). The initiation of pre-ecdysis and ecdysis behaviors in larval *Manduca sexta*: The roles of the brain, terminal ganglion and eclosion hormone. *Journal of Experimental Biology, 199*, 1757–1769.

Novicki, A., & Weeks, J. C. (2000). Developmental attenuation of *Manduca* pre-ecdysis behavior involves neural changes upstream of motoneurons and relay interneurons. *J. Comp. Physiol. [A], 186*, 69–79.

O'Brien, M. A., & Taghert, P. H. (1998). A peritracheal neuropeptide system in insects: release of myomodulin-like peptides at ecdysis. *J. Exp. Biol., 201*, 193–209.

O'Shea, M., & Adams, M. E. (1981). Pentapeptide (proctolin) associated with an identified neuron. *Science, 213*, 567–569.

Ohira, T., Watanabe, T., Nagasawa, H., & Aida, K. (1997). Cloning and sequence analysis of a cDNA encoding a crustacean hyperglycemic hormone from the Kuruma prawn *Penaeus japonicus. Mol. Mar. Biol. Biotechnol., 6*, 59–63.

Park, J. H., Schroeder, A. J., Helfrich-Forster, C., Jackson, F. R., & Ewer, J. (2003b). Targeted ablation of CCAP neuropeptide-containing neurons of *Drosophila* causes specific defects in execution and circadian timing of ecdysis behavior. *Development, 130,* 2645–2656.

Park, Y., Filippov, V., Gill, S. S., & Adams, M. E. (2002a). Deletion of the ecdysis-triggering hormone gene leads to lethal ecdysis deficiency. *Development, 129,* 493–503.

Park, Y., Kim, Y. J., Dupriez, V., & Adams, M. E. (2003a). Two subtypes of ecdysis-triggering hormone receptor in *Drosophila melanogaster. J. Biol. Chem., 278,* 17710–17715.

Park, Y., Zitnan, D., Gill, S. S., & Adams, M. E. (1999). Molecular cloning and biological activity of ecdysis-triggering hormones in *Drosophila melanogaster. FEBS Letters, 463,* 133–138.

Peabody, N. C., Diao, F., Luan, H., Wang, H., Dewey, E. M., et al. (2008). Bursicon functions within the *Drosophila* CNS to modulate wing expansion behavior, hormone secretion, and cell death. *J. Neurosci., 28,* 14379–14391.

Phillips, J. E., Meredith, J., Audsley, N., Richardson, N., Macins, A., et al. (1998). Locust ion transport peptide (ITP): A putative hormone controlling water and ionic balance in terrestrial insects. *American Zoologist, 38,* 461–470.

Phlippen, M. K., Webster, S. G., Chung, J. S., & Dircksen, H. (2000). Ecdysis of decapod crustaceans is associated with a dramatic release of crustacean cardioactive peptide into the haemolymph. *J. Exp. Biol., 203,* 521–536.

Predel, R. (2001). Peptidergic neurohemal system of an insect: mass spectrometric morphology. *J. Comp. Neurol., 436,* 363–375.

Predel, R., & Wegener, C. (2006). Biology of the CAPA peptides in insects. *Cell. Mol. Life. Sci., 63,* 2477–2490.

Reynolds, S. E. (1977). Control of cuticle extensibility in the wings of adult *Manduca* at the time of eclosion: effects of eclosion hormone and bursicon. *Journal of Experimental Biology, 70,* 27–39.

Reynolds, S. E. (1983). Bursicon. In: R. G. H. Downer, H. Laufer (Eds.), *Endocrinology of insects.* Alan R. Liss, NY, (pp. 235–348).

Reynolds, S. E., Taghert, P. H., & Truman, J. W. (1979). Eclosion Hormone and Bursicon Titers and the Onset of Hormonal Responsiveness During the Last Day of Adult Development in *Manduca sexta. Journal of Experimental Biology, 78,* 77–86.

Riddiford, L. M., Hewes, R. S., & Truman, J. W. (1994). Dynamics and metamorphosis of an identifiable peptidergic neuron in an insect. *Journal of Neurobiology, 25,* 819–830.

Roller, L., Tanaka, Y., & Tanaka, S. (2003). Corazonin and corazonin-like substances in the central nervous system of the Pterygote and Apterygote insects. *Cell Tissue Res., 312,* 393–406.

Roller, L., Zitnanova, I., Dai, L., Simo, L., Park, Y., et al. (2010). Ecdysis triggering hormone signaling in arthropods. *Peptides, 31,* 429–441.

Sandstrom, D. J., & Weeks, J. C. (1991). Reidentification of larval interneurons in the pupal stage of the tobacco hornworm, *Manduca sexta. J. Comp. Neurol., 308,* 311–327.

Sauman, I., & Reppert, S. M. (1996). Circadian clock neurons in the silkmoth *Antheraea pernyi*: novel mechanisms of Period protein regulation. *Neuron, 17,* 889–900.

Schoofs, L., Holman, G. M., Hayes, T. K., Nachman, R. J., & De Loof, A. (1991). Isolation, identification and synthesis of locustamyoinhibiting peptide (LOM-MIP), a novel biologically active neuropeptide from *Locusta migratoria. Regul. Pept., 36,* 111–119.

Schoofs, L., Veelaert, D., Broeck, J. V., & De Loof, A. (1996). Immunocytochemical distribution of locustamyoinhibiting peptide (Lom-MIP) in the nervous system of *Locusta migratoria. Regul. Pept., 63,* 171–179.

Schwartz, L. M., & Truman, J. W. (1984). Hormonal Control of Muscle Atrophy and Degeneration in the Moth Antheraea polyphemus. *Journal of Experimental Biology, 111,* 13–30.

Seligman, I. M. (1983). Bursicon. In: I. A. Miller, (Ed.), *Neurohormonal techniques in insects.* Springer, Berlin, (pp. 137–153).

Shiomi, K., Kajiura, Z., Nakagaki, M., & Yamashita, O. (2003). Baculovirus-mediated efficient gene transfer into the central nervous system of the silkworm, *Bombyx mori. J. Insect Biotechnol. Sericol., 72,* 149–155.

Slama, K. (1980). Homeostatic function of ecdysteroids in ecdysis and oviposition. *Acta entmologica bohemoslovaca, 77,* 145–168.

Stangier, J., Hilbich, C., Beyruether, K., & Keller, R. (1987). Unusual Crustacean Cardioactive Peptide Ccap from Pericardial Organs of the Shore Crab *Carcinus maenas. Proceedings of the National Academy of Sciences of the United States of America, 84,* 575–579.

Stangier, J., Hilbich, C., Dircksen, H., & Keller, R. (1988). Distribution of a novel cardioactive neuropeptide (CCAP) in the nervous system of the shore crab *Carcinus maenas. Peptides, 9,* 795–800.

Stangier, J., Hilbich, C., & Keller, R. (1989). Occurrence of Crustacean Cardioactive Peptide Ccap in the Nervous System of an Insect Locusta migratoria. *Journal of Comparative Physiology B, Biochemical, Systemic, & Environmental Physiology, 159,* 5–12.

Taghert, P. H., & Truman, J. W. (1982a). Identification of the bursicon containing neurons in abdominal ganglia of the tobacco hornworm *Manduca sexta. Journal of Experimental Biology, 98,* 385–402.

Taghert, P. H., & Truman, J. W. (1982b). The Distribution and Molecular Characteristics of the Tanning Hormone Bursicon in the Tobacco Hornworm *Manduca sexta. Journal of Experimental Biology, 98,* 373–384.

Tamura, T., Thibert, C., Royer, C., Kanda, T., Abraham, E., et al. (2000). Germline transformation of the silkworm *Bombyx mori* L. using a piggyBac transposon-derived vector. *Nat. Biotechnol., 18,* 81–84.

Tanaka, S. (2000). Induction of darkening by corazonins in several species of Orthoptera and their possible presence in ten orders. *Appl. Ent. Zool., 35,* 509–517.

Tanaka, S. (2001). Endocrine mechanisms controlling body-color polymorphism in locusts. *Arch. Insect. Biochem. Physiol., 47,* 139–149.

Tanaka, S., & Pener, M. P. (1994). A neuropeptide controlling the dark pigmentation in color polymorphism of the migratory locust. *J. Insect. Physiol., 40,* 997–1005.

Tawfik, A. I., Tanaka, S., De Loof, A., Schoofs, L., Baggerman, G., et al. (1999). Identification of the gregarization-associated dark-pigmentotropin in locusts through an albino mutant. *Proc. Natl. Acad. Sci. U S A, 96,* 7083–7087.

Taylor, H. M., & Truman, J. W. (1974). Metamorphosis of the Abdominal Ganglia of the Tobacco Hornworm *Manduca sexta* Changes in Populations of Identified Motor Neurons. *Journal of Comparative Physiology, 90*, 367–388.

Tensen, C. P., Verhoeven, A. H., Gaus, G., Janssen, K. P., Keller, R., et al. (1991). Isolation and amino acid sequence of crustacean hyperglycemic hormone precursor-related peptides. *Peptides, 12*, 673–681.

Terzi, G., Truman, J. W., & Reynolds, S. E. (1988). Purification and Characterization of Eclosion Hormone from the Moth *Manduca sexta. Insect Biochemistry, 18*, 701–708.

Truman, J., Ewer, J., & Ball, E. (1996). Dynamics of cyclic GMP levels in identified neurones during ecdysis behaviour in the locust *Locusta migratoria. J. Exp. Biol., 199*, 749–758.

Truman, J. W. (1971). Physiology of Insect Ecdysis Part 1 the Eclosion Behavior of Saturniid Moths and Its Hormonal Release. *Journal of Experimental Biology, 54*, 805–814.

Truman, J. W. (1972). Physiology of Insect Rhythms Part II. The Silk Moth Brain as the Location of the Biological Clock Controlling Eclosion. *Journal of Comparative Physiology, 81*, 99–114.

Truman, J. W. (1973). Physiology of insect ecdysis. III. Relationship between the hormonal control of eclosion and of tanning in the tobacco hornworm, *Manduca sexta. J. Exp. Biol., 58*, 821–829.

Truman, J. W. (1980). In M. Locke & D. S. Smith (Ed.), *Eclosion Hormone: Its Role in Coordinating Ecdysial Events in Insects* (pp. 385–402).

Truman, J. W. (1992). The eclosion hormone system in insects. *Prog. Brain Res., 92*, 361–373.

Truman, J. W. (2005). Hormonal control of insect ecdysis: endocrine cascades for coordinating behavior with physiology. *Vitam. Horm., 73*, 1–30.

Truman, J. W., & Copenhaver, P. F. (1989). The larval eclosion hormone neurones in *Manduca sexta*: Identification of the brain-proctodeal neurosecretory system. *J. Exp. Biol., 147*, 457–470.

Truman, J. W., & Riddiford, L. M. (1970). Neuro Endocrine Control of Ecdysis in Silk Moths. *Science, 167*, 1624–1626.

Truman, J. W., & Riddiford, L. M. (1974). Hormonal Mechanisms Underlying Insect Behavior. In J. E. Treherne, & M. J. Berridge (Eds.), *Advances in Insect Physiology* (pp. 297–352). New York: Academic Press.

Truman, J. W., & Riddiford, L. M. (2002). Endocrine insights into the evolution of metamorphosis in insects. *Annu. Rev. Entomol, 47*, 467–500.

Truman, J. W., Rountree, D. B., Reiss, S. E., & Schwartz, L. M. (1983). Ecdysteroids regulate the release and action of eclosion hormone in the tobacco hornworm, *Manduca sexta* (L.). *J. Insect Physiol., 29*, 895–900.

Truman, J. W., & Sokolove, P. G. (1972). Silkmoth eclosion: hormonal triggering of a centrally programmed pattern of behavior. *Science, 175*, 1490–1493.

Van Herp, F. (1998). Molecular, cytological and physiological aspects of the crustacean hyperglycemic hormone family. In G. M. Coast, & S. G. Webster (Eds.), *Recent advances in arthropod endocrinology* (pp. 53–70). Cambridge: Cambridge University Press.

Veelaert, D., Schoofs, L., Tobe, S. S., Yu, C. G., Vullings, H. G., et al. (1995). Immunological evidence for an allatostatin-like neuropeptide in the central nervous system of *Schistocerca gregaria, Locusta migratoria* and *Neobellieria bullata. Cell Tissue Res., 279*, 601–611.

Veenstra, J. A. (1989). Isolation and structure of corazonin, a cardioactive peptide from the American cockroach. *FEBS Lett., 250*, 231–234.

Veenstra, J. A. (2009). Does corazonin signal nutritional stress in insects? *Insect. Biochem. Mol. Biol., 39*, 755–762, Epub 2009 Oct 6.

Veenstra, J. A., & Davis, N. T. (1993). Localization of corazonin in the nervous system of the cockroach *Periplaneta americana. Cell Tissue Res., 274*, 57–64.

Vogtli, M., Elke, C., Imhof, M. O., & Lezzi, M. (1998). High level transactivation by the ecdysone receptor complex at the core recognition motif. *Nucleic Acids Research, 26*, 2407–2414.

Wang, S. F., Miura, K., Miksicek, R. J., Segraves, W. A., & Raikhel, A. S. (1998). DNA binding and transactivation characteristics of the mosquito ecdysone receptor-Ultraspiracle complex. *Journal of Biological Chemistry, 273*, 27531–27540.

Ward, R. E., Reid, P., Bashirullah, A., D'Avino, P. P., & Thummel, C. S. (2003). GFP in living animals reveals dynamic developmental responses to ecdysone during *Drosophila* metamorphosis. *Dev. Biol., 256*, 389–402.

Webster, S. G. (1998). Molecular, cytological and physiological aspects of the crustacean hyperglycemic hormone family. In G. M. Coast, & S. G. Webster (Eds.), *Recent advances in arthropod endocrinology* (pp. 33–52). Cambridge: Cambridge University Press.

Webster, S. G., Dircksen, H., & Chung, J. S. (2000). Endocrine cells in the gut of the shore crab *Carcinus maenas* immunoreactive to crustacean hyperglycaemic hormone and its precursor-related peptide. *Cell Tissue Res., 300*, 193–205.

Weeks, J. C., & Truman, J. W. (1984a). Neural organization of peptide-activated ecdysis behaviors during the metamorphosis of *Manduca sexta*. I. Conservation of the peristalsis motor pattern at the larval-pupal transformation. *J. Comp. Physiol. A, 155*, 407–422.

Weeks, J. C., & Truman, J. W. (1984b). Neural organization of peptide-activated ecdysis behaviors during the metamorphosis of *Manduca sexta*. II. Retention of the proleg motor pattern despite the loss of prolegs at pupation. *J. Comp. Physiol., 155*, 423–433.

Weidemann, W., Gromoll, J., & Keller, R. (1989). Cloning and sequence analysis of cDNA for precursor of a crustacean hyperglycemic hormone. *FEBS Lett., 257*, 31–34.

Wise, S., Davis, N. T., Tyndale, E., Noveral, J., Folwell, M. G., et al. (2002). Neuroanatomical studies of period gene expression in the hawkmoth, *Manduca sexta. J. Comp. Neurol., 447*, 366–380.

Yamada, M., Murata, T., Hirose, S., Lavorgna, G., Suzuki, E., et al. (2000). Temporally restricted expression of transcription factor betaFTZ-F1: significance for embryogenesis, molting and metamorphosis in *Drosophila melanogaster. Development, 127*, 5083–5092.

Yamanaka, N., Yamamoto, S., Zitnan, D., Watanabe, K., Kawada, T., et al. (2008). Neuropeptide receptor transcriptome reveals unidentified neuroendocrine pathways. *PLoS One, 3*, e3048.

Zdarek, J., & Friedman, S. (1986). Pupal ecdysis in flies: mechanisms of evagination of the head and expansion of the thoracic appendages. *Journal of Insect Physiology, 32,* 917–923.

Zilberstein, Y., Ewer, J., & Ayali, A. (2006). Neuromodulation of the locust frontal ganglion during the moult: a novel role for insect ecdysis peptides. *J. Exp. Biol., 209,* 2911–2919.

Zitnan, D. (1989). *Regulatory peptides and peptidergic organs of insects. Ph.D Dissertation.* Bratislava: Slovak Academy of Sciences.

Zitnan, D., & Adams, M. E. (2000). Excitatory and inhibitory roles of central ganglia in initiation of the insect ecdysis behavioural sequence. *Journal of Experimental Biology, 203,* 1329–1340.

Zitnan, D., Hollar, L., Spalovska, I., Takac, P., Zitnanova, I., et al. (2002a). Molecular cloning and function of ecdysis-triggering hormones in the silkworm *Bombyx mori. J. Exp. Biol., 205,* 3459–3473.

Zitnan, D., Kim, Y. J., Zitnanova, I., Roller, L., & Adams, M. E. (2007). Complex steroid-peptide-receptor cascade controls insect ecdysis. *Gen. Comp. Endocrinol., 153,* 88–96.

Zitnan, D., Kingan, T. G., Hermesman, J., & Adams, M. E. (1996). Identification of ecdysis-triggering hormone from an epitracheal endocrine system. *Science, 271,* 88–91.

Zitnan, D., Kingan, T. G., Kramer, S. J., & Beckage, N. E. (1995). Accumulation of neuropeptides in the cerebral neurosecretory system of *Manduca sexta* larvae parasitized by the braconid wasp Cotesia congregata. *Journal of Comparative Neurology, 356,* 83–100.

Zitnan, D., Ross, L. S., Zitnanova, I., Hermesman, J. L., Gill, S. S., et al. (1999). Steroid induction of a peptide hormone gene leads to orchestration of a defined behavioral sequence. *Neuron, 23,* 523–535.

Zitnan, D., Zitnanova, I., Spalovska, I., Takac, P., Park, Y., et al. (2003). Conservation of ecdysis-triggering hormone signalling in insects. *J. Exp. Biol., 206,* 1275–1289.

Zitnanova, I., Adams, M. E., & Zitnan, D. (2001). Dual ecdysteroid action on the epitracheal glands and central nervous system preceding ecdysis of *Manduca sexta. J. Exp. Biol., 204,* 3483–3495.

8 The Juvenile Hormones

W.G. Goodman
University of Wisconsin-Madison, Madison, WI, USA
M. Cusson
Canadian Forest Service, Natural Resources Canada,
Quebec City, QC Canada

© 2012 Elsevier B.V. All Rights Reserved

Summary

The juvenile hormones, a family of acyclic sesquiterpenoids, are essential to insect development and reproduction. This family of hormones has been extensively studied because of its central role in regulating development and value as an insect pest control agent. Recent advances have been made in the areas of chemistry, biosynthesis, transport, catabolism, and biological roles the hormones play in the immature insect. This chapter examines the new compounds that have been added to the family. Recent structural elucidation of the hemolymph transport and catabolic proteins is examined. Molecular biology surrounding the long-sought JH receptor and potential candidates are discussed.

DOI: 10.1016/B978-0-12-384749-2.10008-1

8.1. Introduction

The absence of metamorphosis in normal nymphs before the fifth stage must therefore be due to an inhibitory factor or hormone in the blood

V.B. Wigglesworth (1934)

Nearly eight decades have passed since V.B. Wigglesworth made the insightful observations that led to the discovery of a metamorphosis-inhibiting factor that we now call the juvenile hormones (JHs). While Wigglesworth's initial observations focused on growth and development in immature insects, we now know the hormones are involved in reproduction, caste determination, behavior, stress response, diapause, and various polyphenisms (Nijhout, 1994). The naturally occurring JHs are a family of acyclic sesquiterpenoids primarily limited to insects. At least one or more JH homologues have been identified in approximately 100 insect species spanning at least 10 insect orders. This chapter will present the current status of our knowledge about JHs, focusing primarily on their role in the immature insect. This chapter is based on an earlier review (Goodman and Granger, 2005) and will focus primarily on advances made since that time.

8.2. Chemistry of the Juvenile Hormones

8.2.1. Homologues

Two common features characterize the naturally occurring JHs: their site of synthesis, the corpora allata (CA), and their farnesoid structure. Through insect evolution, the farnesol backbone has been chemically modified to generate a homologous series of hormones. Röller and colleagues (1967) identified the first homologue from lipid extracts of the wild silkmoth, *Hyalophora cecropia*. This JH, methyl (2E,6E,10-*cis*)-10,11-epoxy-7-ethyl-3, 11-dimethyl-2,6-tridecadienoate, was termed *cecropia* JH or C_{18}JH in the older literature, but it is now identified as JH I (**Figure 1**). This structure was confirmed as the 2E,6E,10-*cis* isomer (Dahm *et al.*, 1968), while the absolute configuration of JH I at its chiral centers (C10 and C11) was determined to be 10R,11S (Faulkner and Petersen, 1971; Nakanishi *et al.*, 1971; Meyer *et al.*, 1971).

Meyer *et al.* (1968) identified a minor component in the *H. cecropia* extracts that differed from JH I by a methyl group at C7 (**Figure 1**). Termed JH II, methyl (2E,6E, 10-*cis*)-10,11-epoxy-3,7,11-trimethyl-2,6-tridecadienoate,

like JH I, displays an *E, E* configuration at C2, C3, and C6, C7. The absolute configuration at the C10, C11 positions for the naturally occurring JH II has not been rigorously determined (Baker, 1990).

A third JH homologue, JH III, methyl 10,11-epoxyfarnesoate, was identified from medium in which the CA of the tobacco hornworm, *Manduca sexta*, had been maintained (Judy *et al.*, 1973; **Figure 1**). JH III differs from the higher homologues in that all three branches of the carbon skeleton, at C3, C7, and C11, are methyl groups; however, it displays the same 2E,6E geometry. The hormone has only one chiral carbon, C10, which in the naturally occurring hormone displays the 10R configuration. JH III appears to be the most common homologue among the species studied (Schooley *et al.*, 1984).

The trihomosesquiterpenoids JH 0 and its isomer 4-methyl JH I (iso-JH 0) were identified in *M. sexta* eggs (Bergot *et al.*, 1981a), but nothing is currently known of their functions (**Figure 1**). JH III with a second epoxide substitution at C6, C7 was isolated and identified from *in vitro* cultures of larval ring glands of *Drosophila melanogaster* (Richard *et al.*, 1989a,b; **Figure 1**). This homologue, termed JH III bisepoxy (JHB₃), is active in a *D. melanogaster* bioassay and *Drosophila* S2 cells (Wang *et al.*, 2009) and has been found in various dipteran species (Lefevre *et al.*, 1993; Moshitzky and Applebaum, 1995; Yin *et al.*, 1995; Jones and Jones, 2007).

It is a curious fact that the JH of the triatomid *Rhodnius prolixus*, the hemipteran used by Wigglesworth (1934) in his pioneering experiments on the nature of the hormones controlling growth and reproduction, does not correspond to any known JH. Wigglesworth (1961) suggested that farnesol could mimic the morphogenic as well as gonadotrophic roles of JH. Later attempts to identify the hemipteran JHs suggested that JH III was the primary JH in a pyrrhocorid, *Dysdercus fasciatus* (Bowers *et al.*, 1983); however, studies on the lygaeid *Oncopeltus fasciatus* did not meet with success (Baker *et al.*, 1988). Gelman *et al.* (2007) reported that nymphs of the greenhouse (silverleaf) whitefly, *Bemisia tabaci* (Biotype B), contained JH III only.

Another report suggested that the alydid *Riptortus clavatus* produces JH I (Numata *et al.*, 1992). Unfortunately, this identification seems doubtful in light of the results of Kotaki (1993, 1996). These studies revealed that while the CA products of the pentatomid *Plautia stali* have a sesquiterpenoid skeleton similar to JH III, the addition of the JH III precursors farnesoic acid or farnesol to CA *in vitro* does not stimulate the production of JH III or JHB₃, but

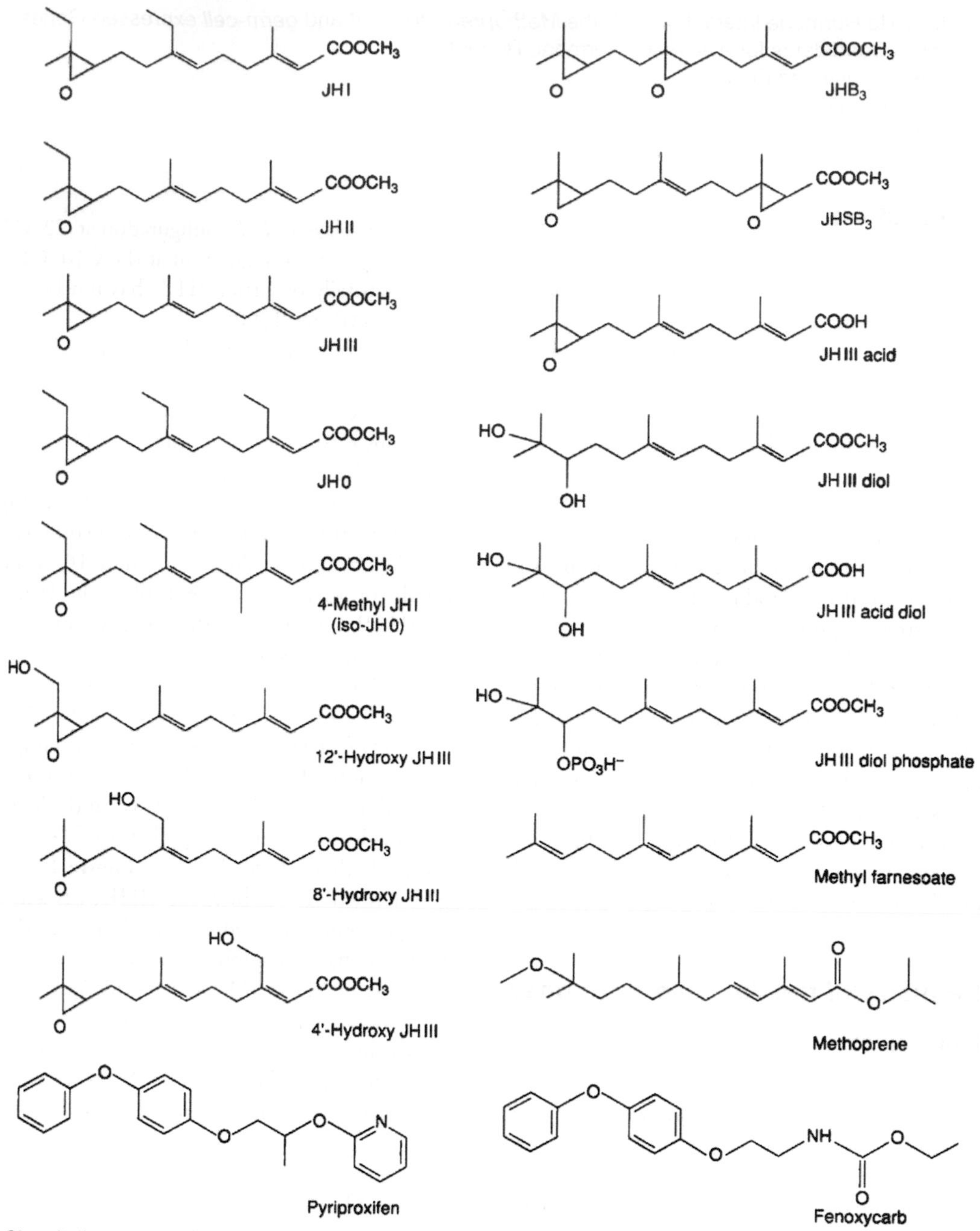

Figure 1 Chemical structures of the naturally occurring juvenile hormone (JH) homologues, hydroxylated JHs, JH metabolites, and commonly used JH agonists.

it does stimulate a molecule that differs chromatographically from any known JHs (Kotaki, 1997). Kotaki *et al.* (2009) identified the molecule as JH III skipped bisepoxide (JHSB$_3$; **Figure 1**). The term "skipped" refers to the fact that a second epoxide substitution occurs at C2, C3 rather than at C6, C7 as in JHB$_3$.

Elucidation of the JHSB$_3$ structure was performed in a novel fashion. Radiolabeled precursors were incubated *in vitro* with isolated CA to generate radiolabeled JH products. These products were then separated by thin-layer chromatography and compared to known JH

homologues. Because the radiolabeled product did not match the known homologues, Kotaki *et al.* (2009) subjected the CA product to gas chromatography-mass spectrometry (GC-MS). GC-MS indicated that the molecular formula was JHB$_3$, suggesting an isomer. A library of all possible stereoisomers for JHB$_3$ was synthesized and these compounds used as standards to identify the CA product; the results, in combination with bioassays, confirmed that (2R,3S,10R) JH III bisepoxide is a hemipteran JH. Curiously, biological activity depended most upon the (2R,3S) configuration whereas chirality around C10, C11 appeared less important. The opposite is true in JHs containing a single chiral center at this position.

Another structural group in the family of JH homologues is the hydroxylated JHs (HJHs; **Figure 1**). These HJHs — 4-OH, 8-OH, and 12-OH JH III — are synthesized and released by the CA of the African locust, *Locusta migratoria* (Darrouzet *et al.*, 1997, 1998; Mauchamp *et al.*, 1999). The biological activity of synthetic 12-OH JH III was analyzed by bioassay utilizing *Te. molitor* and was found to be 100-fold more active than JH III (Darrouzet *et al.*, 1997). Little work has been done on the HJHs during the past decade, so the pathways of synthesis and the physiological roles of the HJHs remain a fertile area for investigation.

The naturally occurring JHs discussed thus far contain one or more epoxide moieties. Increasing evidence, however, suggests that the farnesoid derivative methyl farnesoate (MF; **Figure 1**) may also serve as a hormone in some insects. MF is a precursor in the JH biosynthetic pathway (see Section 8.5.2.1.) but it is also secreted from the CA. MF, long thought to be the crustacean form of JH, is also present during embryogenesis in the cockroach *Nauphoeta cinerea* (Brüning *et al.*, 1985; Bürgin and Lanzrein, 1988) and is biosynthesized by larval *D. melanogaster* ring glands (Richard *et al.*, 1989a,b), the embryonic CA of *Diploptera punctata*, the larval CA of *Pseudaletia unipuncta* (Cusson *et al.*, 1991), and the adult CA of *Phormia regina* (Yin *et al.*, 1995) and *Aedes aegypti* (Jones *et al.*, 2010). Most remarkable is the level of MF in the hemolymph of *D. melanogaster*, where it is 50 times higher than the level of JH III (Jones and Jones, 2007).

Bioassays have been employed to better understand the biological role of MF in *D. melanogaster*. Two types of bioassays were performed: a feeding/continual contact assay and a topical application assay (Riddiford and Ashburner, 1991). Both of these have developmental delay or morphological aberration as the end point. While it is difficult to extrapolate from bioassay data to *in vivo* functions, patterns of biological activity as defined by the bioassay provide insight. MF activity in continually exposed *D. melanogaster* larvae was much greater than that of JH III or JHB$_3$ (Harshman *et al.*, 2010). When tested using the white puparial bioassay, however, MF was the least active of the three compounds. These findings led the authors to assert that MF, when continuously applied during the larval stage, is more active in blocking adult development than JH III or JHB$_3$. Conversely, MF is less active than JH III or JHB$_3$ once pupariation has been initiated. Similar results were reported by Jones *et al.* (2010).

8.2.2. Metabolites of Juvenile Hormones as Hormones?

A thought-provoking essay by Davey (2000) posed the intriguing question: How many JHs are there? The answer is elusive, because JHs appear to play a multifaceted role during larval development. Using the simple definition presented earlier (i.e., JHs are farnesoid molecules secreted by the CA) one can count but a handful of compounds. However, if a JH is defined by its biological activity, the number of molecules increases significantly, creating the difficult problem of sorting out which molecules actually evolved the function of modifying insect development.

An open mind is needed when considering whether metabolites of JHs serve as a substrate for conversion to the biologically active hormone or perhaps act as hormones themselves. The older literature speculates that JH conjugates may also play a role (Roe and Venkatesh, 1990). A major metabolite resulting from the injection or application of JH II into wandering larvae and pre-pupae of the lepidopteran *Trichoplusia ni* was found to be an unidentified, water-soluble, polar product of the hormone (Kallapur *et al.*, 1996). Identification by MS confirmed that the molecule is a novel polar metabolite. *In vitro* studies on the biosynthetic products of the CA from *M. sexta* and two other lepidopteran species indicate the glands are synthesizing a JH conjugate that is not recognized by antibodies highly specific for the JHs and their acids; nevertheless, when this compound is hydrolyzed by esterases, JH III acid is formed (Granger *et al.*, 1995). This product also appears to circulate in the hemolymph. Results of this study intimated the moiety is either a glucuronide of JH III or an acylglycerol with JH III as the side chain(s).

The increased sensitivity afforded by modern analytical tools should ease the task of deciphering the biological roles of these JH-like molecules. Are they biologically relevant or are they the result of promiscuous biochemical reactions that yield end products with no real biological activity? There is good evidence from vertebrate endocrinology that this question poses a real problem. A very early study on corticosteroid biosynthesis discovered more than 40 corticosteroids present in the human adrenal glands, even with the limited technology of that time (Heftmann and Mosettig, 1960). Of those, only a handful has ever been detected in the blood. A similar situation pertains to the insect ecdysteroids (Rees, 1985).

8.2.3. Juvenile Hormone Agonists

That the insect endocrine system could be exploited as a potential target for control of pest insects has not gone unnoticed by insect biologists. Williams (1956, 1957) proposed that the JHs might become the "third generation pesticides" that would offer a safer alternative to the conventional pesticides of the era. By 1975, the JH agonist methoprene was registered for use. JH agonists (analogs) have been employed as control agents for mosquitoes, flies, stored-product pests, fleas, and fire ants, and are used to increase silk production in *Bombyx mori* (Henrick, 2007). There are reports that JH agonists may even control human microbial pathogens (Esteva *et al.*, 2002, 2005).

The numbers of analogs that have been synthesized and bioassayed are in the thousands; most interesting are the strikingly diverse chemical structures that give rise to biological activity. As Henrick (2007) noted, structure–activity relationships are difficult to make, in part due to the diversity in biological response of insects tested. Often, a molecule with exceptional biological activity in one insect order will be inactive in another order (Sláma et al., 1974; Sehnal, 1976). Molecules with biological activity include derivatives of the farnesoid backbone (Sláma et al., 1974; Minakuchi and Riddiford, 2006; Henrick, 2007), derivatives containing amino acids (Sláma et al., 1974), peptides (Hlavacek et al., 1993), and the family of carbamate pesticides (Rejzek, 1992; Minakuchi and Riddiford, 2006). The most active — those active in the widest range of insect species and at very low doses — are molecules that contain a carbon backbone of 14 to 17 units. There must be limited branching and a balance of hydrophilic and hydrophobic sites. That balance is often observed in a lipophilic backbone with polar substituents at the ends of the molecule. Some of the most active agonists contain one or more asymmetric carbons (Henrick, 1982).

One group of JH agonists that displays exceptional biological activity in many orders of insects is the alkyl and alkynyl 3,7,11-trimethyl-2,4-dodecadienoates, more commonly referred to as methoprene, hydroprene, and kinoprene (**Figure 1**). Methoprene's high level of biological activity is undoubtedly due to its ability to bind to JH receptors; moreover, the agonist lacks the unstable epoxide and methyl ester (Henrick, 2007). These unstable groups, while vexing for researchers manipulating JHs, allow the insect to rapidly clear peripheral tissue of this potent biological signal. Substituting less susceptible groups significantly increases half-life.

Methoprene's commercial success is based on its low mammalian toxicity, wide range of insect pest targets, and short environmental half-life (Cusson and Palli, 2000; Minakuchi and Riddiford, 2006; Henrick, 2007). Yet the prediction by Williams (1957) that insects would not become resistant to their own hormones (agonists) has proven wrong. Under field conditions, resistance to methoprene has been observed in mosquitoes (Cornel et al., 2000, 2002) and the aleyrodid B. tabaci (Wilson et al., 2007; Crowder et al., 2007). In laboratory strains of methoprene-resistant D. melanogaster, where a target gene has been isolated, the elevated variety of amino acid changes that result in resistance suggests continued use of JH-type insecticides will ultimately lead to resistance in the field (Wilson et al., 2006). It is worth noting that in areas where methoprene has been judiciously used for better than 20 years as a mosquito control agent, resistance has not been observed (Henrick, 2007).

Despite the positive qualities of the JH agonists, there remains one environmental question — the inadvertent disruption of non-target organisms. Methoprene in micromolar concentrations reduced growth and viability of the thermophilic eubacterium Bacillus stearothermophilus through perturbation of the cellular respiratory system (Monteiro et al., 2005, 2008). It remains to be seen whether other microorganisms are affected in the same fashion. Since the JH agonists are extensively employed as mosquito larvicides, non-target arthropods, especially aquatic species, are seriously threatened. Henrick (2007) provided a list of non-target species that were studied prior to 2000. Non-target species examined more recently include crustaceans (Cripe et al., 2003; McKenney et al., 2004; McKenney, 2005; Raimondo and McKenney, 2005; Tuberty and McKenney, 2005; Wang et al., 2005; Oda et al., 2007). The hexapod Folsomia candida (Campiche et al., 2006, 2007) and the non-neopterous mayflies Rhithrogena semicolorata and Ephemerella ignita (Licht et al., 2004) have also been shown to be susceptible. It should be noted that in some of the studies, the amounts of agonist used far exceeded the rates recommended for control.

A novel approach to modifying delivery, uptake, and stability of JH agonists has been developed in the juvenogens — chimeric molecules containing an agonist covalently attached to a biologically inert compound (Sláma and Romaňuk, 1976). The juvenogens have low biological activity until they are activated through in vivo biochemical pathways or through environmental factors such as humidity, UV, or light. Once cleaved from its inert partner, the agonist is free to act. Several types of juvenogens have been developed by the Wimmer group (2007) that take advantage of the chemical characteristics of the inert partner. One group of juvenogens attaches a carbohydrate moiety, which makes the compound more hydrophilic and thus soluble in aqueous solvents. Another group of juvenogens complexes the agonist with a fatty acid ester (Jurcek et al., 2009). What makes these compounds so attractive is the ability to modify a given agonist using different inert partners to target different insects and/or their host plants.

8.3. Corpora Allata

8.3.1. Gland Anatomy and Fine Structure

During the larval stages, the primary endocrine organs responsible for JH synthesis are the CA. The CA, typically found in the head posterior to the brain, are innervated via multiple nerves from the brain and subesophageal ganglion (Cassier, 1979, 1990; Tobe and Stay, 1985; Nijhout, 1994; Goodman et al., 2005). The location of the retrocerebral complex, the CA and the corpora cardiaca (CC), varies significantly in relation to the brain/digestive tract, as does the CA in relation to the CC. Cassier (1979, 1990) suggested the brain and retrocerebral complex together can be classified into six morphological types (**Figure 2**).

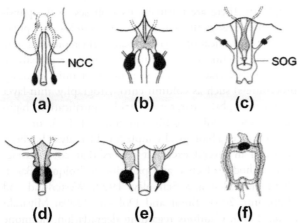

Figure 2 Morphological types of retrocerebral glands in insects. (a) Laterized type, (b) semicentralized type, (c) ventral type, (d) centralized type, (e) distal laterized type, and (f) annular or ring type. Solid black areas represent the corpora allata, shaded areas represent the corpora cardiaca, (NCC) indicates the nervi corporis cardiaci, and (SOG) indicates the subesophageal ganglion. Adapted from *Cassier* (1979).

In the lateralized type, the CA lie behind the brain and on either side of the esophagus, a condition seen in hexapods as well as some non-neopterous insects. The lateralized CA do not fuse with the CC. In the ventral arrangement, the CA are found beneath the aorta and the CC are not fused. A modified version of this is seen in the Ephemeroptera, where the CA are ventrolateral to the digestive tract and the major allatal nerves arise from the subesophageal ganglion rather than the brain. The semicentralized type is characterized by a connection via nerve fibers of the CA to the CC, but there is no fusion. The CC are directly attached to the aorta and the CA just posterior. According to Cassier (1979), this form is common in many species. The centralized arrangement is a fusion of CA and CC that rests beneath the aorta. The distal lateralized type, while similar to the lateralized, is distinct owing to its fusion of the CC with the CA. The annular retrocerebral complex results from the fusion of the CA, above the aorta, with the CC below to form the characteristic ring gland.

In any discussion of insects, variation is a predominant theme and this is true for the anatomy of the CA. Variation is underscored since the CA undergo significant cytological fluctuations during development. In general, the glands range from strongly ovoid to spherical and are approximately the diameter of the aorta or smaller (Tobe and Stay, 1985). The CA display closely interdigitating cells that are surrounded by a non-cellular sheath or basal lamina that encapsulates the glands (Sedlak, 1985; Tobe and Stay, 1985). The presence of smooth endoplasmic reticulum (SER), a predominant organelle in the CA, is associated with periods of high JH biosynthesis (Sedlak, 1985). SER is noticeable in *M. sexta* CA that are actively engaged in JH biosynthesis and its presence declines as JH titers fall prior to pupation. SER is completely absent in pupal CA, where JH titers are below detectable levels (Sedlak *et al.*, 1983).

Ultrastructural analysis of the CA indicates that JH is released immediately upon synthesis, since lipid droplets or cellular inclusions are not apparent during periods of active JH biosynthesis. Where biosynthetic studies have been conducted, precursors are rapidly incorporated into the hormone and the newly synthesized molecule is secreted (Tobe and Stay, 1985). Thus, in the few insects that have been thoroughly studied, there appears to be no accumulation of newly synthesized JH in the CA.

8.3.2. Innervation

Innervation of the CA is a vital component in the regulation of CA activity. In addition to neural fibers that transport prothoracicotrophic hormone (PTTH) from the brain to the CA for release (Agui *et al.*, 1979), the CA are served by inputs from the subesophageal ganglion. The evolutionary diversity of insects has led to a number of variations in innervation (Tobe and Stay, 1985; Nijhout, 1994; Shiga, 2003). Three nerves emanate from the brain and enter the CC: nervi corporis cardiaci (NCC) I, NCC II, and NCC III. In some species the neural tracts that appear to innervate the CC pass through the gland and terminate in or on the CA. These nerves that exit the CC and enter the CA are collectively termed the nervi corporis allata (NCA) I, whereas tracts from the subesophageal ganglion are termed NCA II. This appears to be the basic template, but there are significant variations among species.

In *M. sexta,* for example, NCC I and II innervate the CA and fuse outside the brain. They have been labeled NCC I + II (Nijhout, 1975). Upon entering the brain, the fused neural tract splits, with one branch running along the ventral side of the brain and bending sharply upward upon crossing into the contralateral lobe. This bundle of axons connects with the lateral neurosecretory cells L-NSC III (= IIa of Carrow *et al.*, 1984) of the brain, which are responsible for the synthesis of PTTH. The CA are also innervated by fibers from the contralateral medial cells, M-NSC Ia, that travel to the surface of the gland. This may explain the innervations seen in fine-structure examination of the basal lamina (Sedlak, 1985). Terminal varicosities in M-NSC I suggest that neurosecretory material may be released from this area. Fibers leading away from the group of cells labeled IIb (Carrow *et al.*, 1984) run ipsilaterally to the CA via NCC II.

Using monoclonal antibodies to PTTH, O'Brien *et al.* (1988) identified an extensive dendritic field in the protocerebrum in addition to the long fibers that extend into and arborize in the CA. Given this extensive field and probable overlap with dendritic fields from different neurosecretory axons, neuronal intercommunication is likely.

8.4. Analyses of Juvenile Hormone Titers

The lipophilic nature of JH, its occurrence in exceptionally low concentrations in biological samples (parts per billion or lower), its lability, and its tendency to bind non-specifically make it one of the most difficult hormones to accurately measure. These problems are compounded by relatively rapid changes in JH titers during development; precision staging of the insect under study is essential in generating a meaningful JH titer. Three methods have been employed to quantitatively assess JH titers: bioassays, physicochemical assays, and radioimmunoassays. Relative JH titers based on biosynthetic capacity of the CA *in vitro* can be determined using radiochemical assays.

8.4.1. Bioassays

The first techniques used to measure JH titers were an assortment of bioassays that quantified JH effect on a morphological or physiological process (see Sláma *et al.*, 1974). Initially, these bioassays were used to assess the activity of transplanted CA or tissue extracts (Gilbert and Schneiderman, 1958), but are now more often used to quantify JHs indirectly by comparing the biological response of an unknown preparation to a known amount of JH. These types of assays have been well described and critiqued in previous reviews (Feyereisen, 1985; Baker, 1990). Their use has declined for a number of reasons: (1) they are time-consuming and require maintenance of a test insect colony, (2) they are subjective in their scoring, (3) they lack specificity because they measure only total JH, and (4) they show differential sensitivity to the various homologues. Nevertheless, several bioassays are still employed, especially when the biologically active compound is thought to have a non-JH-like structure, or when monitoring activity of a JH is done during its purification (Harshman *et al.*, 2010; Jones *et al.*, 2010; Jurcek *et al.*, 2009; Kotaki *et al.*, 2009).

8.4.2. Physicochemical Assays

With technology for studying organic compounds becoming more sophisticated, a number of physicochemical methods have been developed for measuring JH in biological samples. Unfortunately, none of these methods are rapid, simple, or inexpensive. Moreover, the literature surrounding the physicochemical detection of JH is awash in methods that can be bewildering for a biologist. The current physicochemical methods reflect several decades of experimentation that emphasized specificity and ultrasensitivity at the expense of

simplicity. There are normally two phases to the physicochemical assay: the sample extraction/clean-up phase and the quantification phase. Because JH represents only a minute fraction of the extractable lipids, especially in whole-body extracts, protocols generally require clean-up procedure(s) such as column chromatography, thin-layer chromatography, separatory-cartridge purification, high-performance liquid chromatography (HPLC), or GC. Preparation of a biological sample for JH analysis has been discussed in several excellent papers that cover extraction methods, solvents, and partition techniques (Baker, 1990; Halarnkar and Schooley, 1990; Westerlund and Hoffmann, 2004; Brent and Dolezal, 2009a; Miyazaki *et al.*, 2009). Cautions regarding degradation, hormone volatility, and non-specific adsorption have also been extensively outlined by Goodman *et al.* (1995).

Once the clean-up procedure is completed, JHs can be quantified by a number of different techniques that employ a GC interfaced with an electron capture detection unit or mass spectrometer (Baker, 1990). To enhance sensitivity and specificity, methods were developed that generate JH derivatives with unique chemical tags. The first derivatives to be developed were the organohalides, which are detected by a GC interfaced with an electron capture detection (ECD) system. This method greatly increased sensitivity of the assay, but it had serious drawbacks in uniformity of derivatization and in identification of the homologues. To overcome these difficulties, Rembold *et al.* (1980) and Bergot *et al.* (1981b) developed GC- MS methods which, while still requiring derivatization of the hormones, were far more efficient and sufficiently selective to allow each JH homologue to be identified using selected ion monitoring. GC-MS-EI (electron impact) and GC-MS-CI (chemical ionization) analyses are now performed with slight variations on these procedures (Neese *et al.*, 2000; Smith *et al.*, 2000; Cole *et al.*, 2002; Park and Raina, 2004; Brent and Dozel, 2009a).

The assay of choice for future studies may be one of the recently developed physicochemical techniques that do not require derivatization to quantify JHs in biological samples. Teal *et al.* (2000), Westerlund *et al.* (2004), and Teal and Proveaux (2006) developed sensitive methods of this kind to titer JHs in hemolymph (Burns *et al.*, 2002) and culture medium (Bede *et al.*, 2001; Teal and Proveaux, 2006) in which lipid content is relatively low. Juvenile hormone titers from whole-body extracts of the Formosan subterranean termite, *Coptotermes formosanus*, have been analyzed employing liquid chromatography-electrospray ionization-mass spectrometry (LC-ESI-MS; Miyazaki *et al.*, 2009). Derivatization of the hormone is not required in this method. The authors report that the limit of detection is in the low picogram range and have provided very convincing evidence that this particular method deserves attention by those interested in titering JHs from whole-body extracts. This procedure is sufficiently robust so that a JH titer from a single worker termite can be derived.

The advantage of the physicochemical assays is their sensitivity and their unequivocal identification of the different JHs and metabolites. The disadvantages include accessibility to the costly equipment and the relatively low throughput; advances in automation may make this approach routine in the near future.

8.4.3. Radioimmunoassays

Radioimmunoassays (RIAs) are a rapid, sensitive, and inexpensive alternative to measuring JHs, both in biological samples and in medium from incubations of CA. The JH RIA is a competitive protein-binding assay in which JH from a biological sample competes with a fixed amount of radiolabeled JH for a limited number of binding sites on JH-directed antibodies. The radiolabeled JH bound to the antibody in the presence of the unknown is compared to a standard curve generated with known amounts of radioinert JH.

As in the physicochemical methods, JHs from whole-body or hemolymph extracts must be partially purified prior to the RIA. Partial purification is essential for a successful assay, since a number of lipids can interfere non-specifically. Misidentification of the JH homologues and overestimation of titers can also occur when chromatographic systems contaminated with high levels of JH standards are used for the purification of biological samples (Baker *et al.*, 1984). It is for these reasons that RIA technology has rightfully been criticized (Feyereisen, 1985; Tobe and Stay, 1985). A number of excellent clean-up procedures have been reported (Strambi *et al.*, 1981; Goodman *et al.*, 1995; Niimi and Sakurai, 1997; Noriega *et al.*, 2001). It is important to note that despite differences in extraction methods and JH-directed antibodies, reasonable agreement between RIAs can be achieved (Goodman *et al.*, 1993; Chen *et al.*, 2007).

8.4.4. Radiochemical Assays

The radiochemical assay (RCA), originally developed by Pratt and Tobe (1974), has been employed for many years to measure *in vitro* JH biosynthesis by the CA of a wide variety of insects (Feyereisen and Tobe, 1981; Tobe and Feyereisen, 1983; Tobe and Stay, 1985; Yagi and Tobe, 2001; Li *et al.*, 2006; Li, 2007; Burtenshaw *et al.*, 2008; Brent and Dolezal, 2009b). The RCA measures the rate of incorporation of the methyl group from either [^{14}C-methyl]-methionine or [^{3}H-methyl]-methionine. The availability of assay components, the relative simplicity of the protocol, its sensitivity (0.1 pmol), and the short incubation time (1–3 h) have made this assay popular among insect endocrinologists.

There has been a general assumption that isolated glands lacking neural connections function in the same fashion as glands *in vivo*. This assumption has been challenged by Horseman *et al.* (1994), who demonstrated that nerve transection in adult *L. migratoria* led to a rapid, 16-fold increase in JH production. Thus, the RCA may, at certain times, seriously overestimate the expected *in vivo* production of JH (Pratt *et al.*, 1990). Another potential problem is the use of [^{3}H-methyl]-methionine as the methyl donor (Yagi and Tobe, 2001). The lack of definitive data from the manufacturers on the purity and specific activity of this compound can compromise the RCA.

8.5. Biosynthesis

8.5.1. Biosynthetic Routes

The JH biosynthetic pathway may be viewed as two distinct biosynthetic units or "arms" (**Figure 3**). In the first arm, biosynthesis proceeds through the mevalonate pathway (MP), which insects share with other organisms and for which cholesterol is the best known end product. The MP uses acetyl-CoA as starting material to generate mevalonate, which is then converted to isopentenyl disphosphate (IPP), the C-5 isoprene unit used by prenyltransferases to build prenyl chains whose carbon atom numbers are typically in multiples of five. Biogenesis of the sesquiterpene (i.e., C-15) precursor of JHs, farnesyl diphosphate (FPP), is effected by an FPP synthase (FPPS, a prenyltransferase), which catalyzes the head-to-tail condensation of three isoprene units. In the latter reaction, the chain initiator is the allylic isomer of IPP, dimethylallyl diphosphate (DMAPP), the production of which is catalyzed by an IPP isomerase (IPPI).

Insects and other arthropods have lost the ability to synthesize cholesterol and depend on dietary sources for acquisition of this essential compound. Insects lack the genes encoding the enzymes required for the production of cholesterol from FPP, including squalene synthase, which catalyzes the reductive condensation of two FPP molecules (Clark and Bloch, 1959). Although insects utilize FPP in other pathways (e.g., ubiquinone synthesis, dolichol synthesis, protein prenylation, etc.), its use as a JH precursor is not only well documented, but is a hallmark feature of insect biochemistry. Thus, the second arm of JH biosynthesis, the JH-specific arm, comprises enzymatic steps unique to JH-producing organisms. FPP is first converted to farnesol by a farnesyl phosphatase (FP), while farnesol undergoes two sequential oxidation reactions that generate farnesal and farnesoic acid (FA). The order of the two last steps, epoxidation and methyl esterification, catalyzed by an epoxidase and a JH acid methyltransferase (JHAMT), may vary between species (Schooley and Baker, 1985).

Although all insects produce JHs using the same general biosynthetic scheme, variations are evident for taxa known to produce JHs other than JH III. For example, production

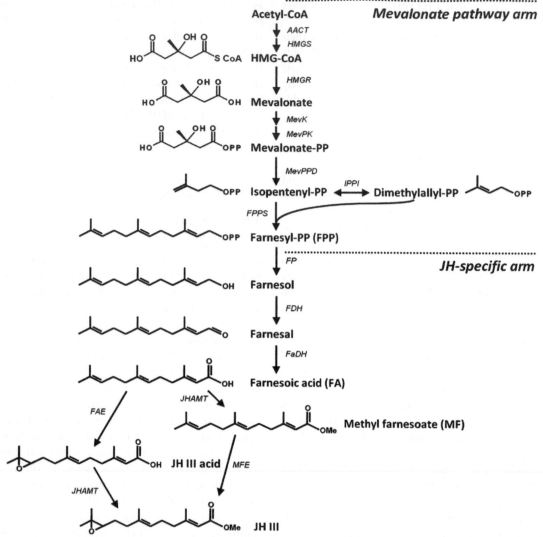

Figure 3 The JH biosynthetic pathway. This pathway is comprised of two biosynthetic units or "arms": the mevalonate pathway (MP) arm that insects share with most organisms, and the JH-specific arm, whose sequential enzymatic steps are unique to JH biosynthesis. The order of the last two enzymatic steps, methylation and epoxidation, may vary. Abbreviations: AACT, acetoacetyl-CoA thiolase; HMGS, HMG-CoA synthase; HMGR, HMG-CoA reductase; MevK, mevalonate kinase; MevPK, phosphomevalonate kinase; MevPPD, diphosphomevalonate decarboxylase; IPPI, isopentenyl diphosphate isomerase; FPPS, farnesyl diphosphate synthase; FP, farnesyl phosphatase; FDH, farnesol dehydrogenase; FaDH, farnesal dehydrogenase; JHAMT, JH acid methyltransferase; FAE, farnesoic acid epoxidase; and MFE, methyl farnesoate epoxidase.

of ethyl-branched JHs by Lepidoptera involves the substitution of propionyl-CoA for acetyl-CoA in an early step of JH synthesis. This substitution leads to the production of the C-6 homologue of IPP (homo-IPP), which may then be used as substrate, with its allylic isomer homo-DMAPP, by FPPS to generate the various FPP homologue precursors of ethyl-branched JHs (Schooley *et al.*, 1973). Propionyl-CoA is generated by the metabolism of leucine and valine and depends on a branched-chain amino acid transaminase found only in the CA of lepidopterans (Brindle *et al.*, 1987, 1988, 1992). Variations in the activity of this enzyme are correlated with changes in the ratio of JH homologues released by the CA *in vitro* (Cusson *et al.*, 1996). In addition, other lepidopteran JH biosynthetic

enzymes appear to display adaptations with respect to their selectivity toward ethyl-branched substrates. In the higher Diptera and the Hemiptera that produce two different bisepoxy forms of JH III, catalytic features of the epoxidases are expected to differ from those found in other insects (Helvig *et al.*, 2004). Characterization of these enzymes must await their isolation. Finally, as noted earlier, the order in which methylation and epoxidation take place, in the last two steps of JH biosynthesis, varies among different groups of insects. In Lepidoptera, in which JH III acid (i.e., epoxyfarnesoic acid) is secreted by larval CA during the second half of the last stadium (Bhaskaran *et al.*, 1986) and is the only secreted product of adult male CA (Peter *et al.*, 1981; for an exception to this rule see Cusson *et al.*, 1999), epoxidation precedes esterification (Reibstein *et al.*, 1976). The same appears to be true for the production of JHB_3 by the higher Diptera (Moshitzky and Applebaum, 1995). In cockroaches, however, esterification precedes epoxidation because the epoxidase only accepts MF as substrate (Helvig *et al.*, 2004).

Recent work on the isolation and characterization of JH biosynthetic enzymes has greatly benefited from genomics projects, including whole genome sequencing (Kinjoh *et al.*, 2007; Ueda *et al.*, 2009) and CA transcriptomics studies (Shinoda and Itoyama, 2003; Helvig *et al.*, 2004; Noriega *et al.*, 2006). The latter approach has proven particularly useful with respect to identifying enzymes involved in the JH-specific arm of the biosynthetic pathway. Unlike those of the MP, the genes encoding these highly specific enzymes cannot be isolated by homology cloning. The present overview focuses on these recent advances. For earlier work, the reader is invited to consult the previous edition of this chapter as well as the review by Schooley and Baker (1985) and more recent reviews by Bellés *et al.* (2005) and Palli and Cusson (2007).

8.5.2. Mevalonate Pathway Enzymes

Taking advantage of the sequenced *B. mori* genome (BAG, 2004; Mita *et al.*, 2004), Kinjoh *et al.* (2007) cloned the cDNAs of all silkworm MP enzymes and monitored their transcription in various tissues during development. All enzymes are encoded by a single-copy gene, except for FPPS, for which they identified three paralogues. With the exception of FPPS1, all enzymes are expressed predominantly in the CA, a finding that was later corroborated by *in situ* hybridization (Ueda *et al.*, 2009). CA-specific transcriptional profiles throughout development (from day 1 of the fourth stadium to day 4 of adult stage) display coordinated variations in transcript abundance for all genes, with the exception of the pupal stage, where transcript levels are low for most genes except for acetoacetyl-CoA thiolase (AACT), HMG-CoA reductase (HMGR), HMG-CoA synthase (HMGS), FPPS1, and FPPS3. Interestingly, levels of HMGR transcripts tended to be relatively low and almost undetectable in tissues other than the CA. Similarly, Express sequence tags for this enzyme were not identified in a study focusing on cockroach and mosquito CA transcriptomics, pointing to the low abundance of its transcripts in these insects as well (Noriega *et al.*, 2006). This observation may be relevant to the putative regulatory role of HMGR in JH biosynthesis, and may parallel that observed for mammalian steroid biosynthesis (Brown and Goldstein, 1980).

8.5.2.1. Farnesyl diphosphate synthase

Since the first cloning of an insect FPPS cDNA by Castillo-Garcia and Couillaud (1999), additional insect FPPSs have been isolated (Vandermoten *et al.*, 2009). Most insects have a single copy of this gene, with the notable exception of *Apis mellifera*, for which scanning of the genome revealed the presence of seven copies (HGSC, 2006). However, no information is available about which homologues are involved in JH biosynthesis. Aphids have two FPPS genes that encode very similar proteins (Vandermoten *et al.*, 2008; Lewis *et al.*, 2008). Lepidopterans produce two very distinct forms of this enzyme, designated type-1 and type-2 FPPSs (Cusson *et al.*, 2006), with *B. mori* displaying two slightly different isoforms of the type-2 enzyme (i.e., BmFPPS2 and BmFPPS3; Kinjoh *et al.*, 2007).

Early prenyltransferase studies conducted using *M. sexta* CA homogenates suggest selectivity of the endogenous enzyme toward homologous substrates (Sen *et al.*, 1996). This finding has since been corroborated by additional studies using substrate analogs. These studies demonstrated that, compared to pig liver FPPS, the lepidopteran enzyme displays greater steric latitude around the C-3 and C-7 alkyl positions of DMAPP and geranyl diphosphate (GPP), the C-10 intermediate generated by the first condensation step catalyzed by FPPS (Sen *et al.*, 2006). These discoveries generated interest in the isolation and further characterization of lepidopteran FPPSs, with the aim of identifying an FPPS homologue with structural adaptations favoring the biosynthesis of ethyl-branched FPPs. Unlike the type-2 enzyme (FPPS2), type-1 lepidopteran FPPS (FPPS1) features several active-site substitutions that were predicted to increase the size of the catalytic cavity and thus favor the condensation of the bulkier C-6 substrates used in the biosynthesis of ethyl-branched JHs (Cusson *et al.*, 2006). However, tissue-specific transcriptional analysis of the two FPPS types in *B. mori* revealed ubiquitous expression of FPPS1 and CA-specific expression of FPPS2, pointing to the latter as the more likely JH-specific FPPS (Cusson *et al.*, 2006; Kinjoh *et al.*, 2007). Whereas recombinant *Choristoneura fumiferana* FPPS1 (CfFPPS1) failed to display any FPPS activity *in vitro*, rCfFPPS2 was active and showed slightly higher activity toward HDMAPP than DMAPP (Sen *et al.*, 2007). Experimental association of rCfFPPS1 and rCfF-PPS2 enhanced prenyl coupling as compared to results obtained using rCfFPPS2 alone, but whether these two enzymes form heterodimers *in vivo* has not been verified experimentally (FPPSs are typically active as homodimers; Sen *et al.*, 2007). Work is presently underway to determine which features of the type-2 enzyme favor the condensation of homologous substrates.

8.5.2.2. Isopentenyl diphosphate isomerase

This is another JH biosynthetic enzyme suspected of displaying substrate selectivity toward a homologous substrate, that is, HIPP. Both *M. sexta* CA homogenates (Baker *et al.*, 1981) and a partially purified IPPI from *B. mori* (Koyama *et al.*, 1985) can convert HIPP to the correct isomer HDMAPP, whereas pig liver IPPI, which generates products unusable for synthesis of homologous JHs (Koyama et al., 1973). IPPI has now been cloned from many insects and is reported to be present as a single-copy gene in the genome of B. mori (Kinjoh et al., 2007). However, lepidopteran IPPIs are very similar to those of other eukaryotes, and sequence comparisons have so far failed to reveal obvious lepidopteran-specific substitutions that could account for the ability of this enzyme to isomerize HIPP to HDMAPP.

8.5.3. Juvenile Hormone-specific Enzymes

Until recently, no cDNA encoding enzymes from the JH-specific portion of the pathway had been cloned and their transcripts characterized using recombinant proteins. However, cDNAs have now been obtained from at least one insect species for all of the enzymes except farnesal dehydrogenase.

8.5.3.1. Farnesyl phosphatase

In an effort to isolate an insect FP, which hydrolyzes FPP to farnesol, Cao et al. (2009) screened the D. melanogaster genome for phosphatases and identified eight genes, three of which were found to be expressed in the ring gland. Two of these genes were cloned and their transcripts used to produce the recombinant proteins, which were then submitted to phosphatase assays. Although both proteins display phosphatase activity in the presence of p-nitrophenyl phosphate, only one displays high activity using FPP; the same protein is inactive in the presence of the diterpene (C-20) precursor geranylgeranyl diphosphate. Transcript levels in whole animals correlate well with JH titers throughout development, but it remains to be determined whether these transcripts are produced by the ring gland or by other tissues. FP has not been isolated from other insect species.

8.5.3.2. Farnesol dehydrogenase

Farnesol oxidation is typically catalyzed by nicotinamide-dependent dehydrogenases (Sperry and Sen, 2001). One such alcohol dehydrogenase gene, *JGW*, was cloned from two African *Drosophila* species (Zhang et al., 2004). Although the recombinant protein displays high substrate specificity toward geraniol and farnesol, as would be expected for an enzyme involved in JH biosynthesis, its expression appears confined to the testes (Zhang et al., 2004), thus ruling it out as a JH biosynthetic enzyme. Work on *M. sexta* CA homogenates suggests that the enzyme responsible for the oxidation of farnesol to farnesal may not be a dehydrogenase, but a specific metal-dependent alcohol oxidase, given that the conversion was oxygen-dependent (Sperry and Sen, 2001). However, it has not been possible to isolate this alcohol oxidase from other insect species. More recently, Mayoral et al. (2009a) reported cloning and characterization of a farnesol dehydrogenase from the CA of *Ae. aegypti*; Noriega et al.

(2006) had identified an EST candidate for this enzyme in an earlier study. The recombinant protein was active as a homodimer and oxidized farnesol to farnesal in the presence of NADP⁺, for which farnesol dehydrogenase displays an absolute requirement. This protein has a typical short chain dehydrogenase (SDR) fold, with orthologues found in other species, including *B. mori*. Surprisingly, mRNA levels were highest in the midgut and the brain, as opposed to the CA, where transcript abundance was relatively low. On the other hand, transcript levels in adult female CA were well-correlated with *in vitro* JH biosynthesis (Mayoral et al., 2009a). It remains to be determined whether mosquitoes and moths use different enzymes (i.e., an alcohol dehydrogenase *versus* an alcohol oxidase) to oxidize farnesol to farnesal.

8.5.3.3. Juvenile hormone acid methyltransferase

In Lepidoptera, methylation of the precursor is effected by an O-methyltransferase that uses JH acid as substrate; in most other groups of insects, FA is the putative substrate of this enzyme. The lepidopteran JHAMT has been cloned from *B. mori*, using a fluorescent differential display approach (Shinoda and Itoyama, 2003). PCR amplicons whose abundance displayed developmental changes matching those expected for this enzyme were submitted to sequence analysis. One of the cDNAs was found to code for a protein of 278 amino acids with no clear homologue; however, it contained a conserved motif found on several *S*-adenosyl methionine (SAM)-dependent methyltransferases. Northern blot analysis showed that this transcript could be detected only in the CA, and q-RT-PCR analysis indicated that its transcription declines at the spinning stage of the last stadium, concomitant with the documented loss of JHAMT activity. Its levels remain very low until the pharate adult stage, when it rises to high levels in female adult CA (Kinjoh et al., 2007). The recombinant His-tagged JHAMT could convert both JH acids and FA to the expected products, although it showed greater conversion rates with JH I and II acids as substrates when compared to JH III acid and FA (measured at 100 µM substrate concentration); no conversion was observed with several saturated and unsaturated fatty acids. This enzyme is clearly rate-limiting in *B. mori*. Orthologues of *JHAMT* have now been cloned and characterized in *Tr. castaneum* (Minakuchi et al., 2008a), *D. melanogaster* (Niwa et al., 2008) and *Ae. aegypti* (Mayoral et al., 2009b). In all three species, the enzyme is expressed predominantly in the CA and the recombinant protein can methylate JH III acid and FA at similar rates. In *T. castaneum*, RNAi-mediated *JHAMT* gene silencing in third instars induces a precocious metamorphosis, pointing to the key regulatory role of this enzyme in the red flour beetle (Minakuchi et al., 2008a). A similar approach used in *D. melanogaster* did not disrupt metamorphosis, but JHAMT overexpression caused a pharate adult lethal phenotype (Niwa et al., 2008). In

Ae. aegypti, JHAMT transcript levels did not always correlate well with *in vitro* JH biosynthesis, suggesting that it may not be a rate-limiting enzyme in this species (Mayoral *et al.*, 2009b). It appears that the enzymes used by Lepidoptera to methylate JH acid and by other groups of insects to methylate FA belong to the same family and display similar catalytic properties. As such, these enzymes are unrelated to the putative FA methyltransferase (FAMeT) isolated from crustacean mandibular organs (Holford *et al.*, 2004). Interestingly, while most recombinant FAMeT orthologues isolated from either crustaceans or insects have failed to display methyltransferase activity, a JHAMT orthologue has recently been identified in the water flea *Daphnia pulex* (Hui *et al.*, 2010), suggesting that the crustacean enzyme responsible for the conversion of FA to MF may be a JHAMT orthologue.

8.5.3.4. Methyl farnesoate epoxidase

With the possible exception of what is observed in Lepidoptera (Bhaskaran *et al.*, 1986), *Drosophila* (Moshitzky and Applebaum, 1995) and Hemiptera (Kotaki *et al.*, 2009), the last step of JH biosynthesis is the epoxidation of MF. It has been known for some time that this epoxidase is a microsomal cytochrome P450 enzyme in cockroach and locust (Feyereisen *et al.*, 1981; Hammock, 1975), but its cDNA was cloned only recently from a cockroach, following the construction of a *D. punctata* CA cDNA library and the 5' end sequencing of 1056 clones from it (Helvig *et al.*, 2004). The recombinant protein, designated CYP15A1, showed high affinity for MF, which it converted to JH III. The enzyme cannot metabolize other MF-related compounds such as farnesol, farnesoic acid, and farnesyl methyl ether as substrates, and shows selectivity for the natural geometric isomer as well as for the 10*R* enantiomer. CYP15A1 is expressed only in the CA and only during peak JH production. Interestingly, selected 1,5-disubstituted imidazoles, known to inhibit JH biosynthesis and cause accumulation of MF in cockroach CA, display parallel effectiveness in inhibiting the activity of recombinant CYP15A1 and *in vitro* JH biosynthesis by isolated CA (Helvig *et al.*, 2004). Surprisingly, no clear orthologues of CYP15A1 are found in *D. melanogaster*, perhaps because this insect produces a bisepoxide of JH III, the production of which may require a different epoxidase. However, an orthologue of this enzyme was identified in an *Ae. aegypti* CA EST collection (Noriega *et al.*, 2006). The high specificity of the cockroach epoxidase for MF suggests that the corresponding enzyme in other insects that use FA as substrate will show distinct structural and catalytic features.

8.5.4. Regulation of Juvenile Hormone Biosynthesis

Regulation of JH biosynthesis is a complex process involving many allatoregulating factors that differ among insect species and act at various points of the JH biosynthetic pathway. Unlike mammalian systems where sterols regulate MP enzymes through the action of a sterol regulatory element binding protein (SREBP), control of MP enzyme transcription in insects is insensitive to sterols due to the loss of the homologous SREBP, SREBP-2 (Bellés *et al.*, 2005). Juvenile hormones have been shown to have a regulatory effect on the expression of several MP enzymes in tissues other than the CA, most notably in the midgut of scolytid beetles where a monoterpenoid (C-10) pheromone is synthesized (Seybold and Tittiger, 2003; Tittiger *et al.*, 1999, 2003). It has also been suggested that JHs can act on MP enzymes by modulating the translatability and/or stability of their transcripts (Bellés *et al.*, 2005), but whether this occurs in the CA is not clear. Although the transcriptional analysis of MP enzymes in *B. mori* CA suggests a relatively well-coordinated regulation of this biosynthetic branch (Kinjoh *et al.*, 2007), some enzymes involved in precursor supply (i.e., upstream of the MP; Sutherland and Feyereisen, 1996) and in the conversion of FPP to JH (e.g., farnesol dehydrogenase, Mayoral *et al.*, 2009a; JHAMT, Kinjoh *et al.*, 2007) have been proposed as rate-limiting steps and could therefore be the target of allatoregulatory factors and signals.

8.5.5. Allatoregulatory Factors

A wide variety of factors have been identified as being involved in the regulation of JH biosynthesis by the CA, including inhibitory (allatostatins) and stimulatory (allatotropins) neuropeptides, biogenic amines, second messengers, 20-hydroxyecdysone (20E), and JH itself. These factors were reviewed in detail in the previous edition of this chapter, and allatoregulatory neuropeptides have been the subject of three recent reviews (Stay and Tobe, 2007; Audsley *et al.*, 2008; Weaver and Audsley, 2009) to which the reader is referred for background information.

The most important recent advances in this area of research have been in the identification of allatostatin (AST) and allatotropin (AT) receptors. G protein coupled receptors (GPCRs) for both A-type (FGLamide) and C-type (PISCF) ASTs have now been isolated and characterized (reviewed in Weaver and Audsley, 2009). Important breakthroughs in our understanding of neuropeptide-mediated regulation of JH biosynthesis in the CA came from a single study of the neuropeptide receptor transcriptome of *B. mori* (Yamanaka *et al.*, 2008). The authors used the 40 known *D. melanogaster* neuropeptide GPCRs to scan the *B. mori* genome for homologues. The cDNAs of the silkworm GPCRs were then cloned and used to assess their tissue-specific expression in larvae. Six were found to be expressed in the CC-CA complex, including BNGRA1, an orthologue of the AST-C receptor identified earlier in *D. melanogaster* (Kreienkamp *et al.*, 2002). Employing a cell-based assay, the cells responded to AST-C in a dose-dependent manner. Using the same type of assay, another receptor, BNGRA16, responded specifically and in a dose-dependent fashion to AT, clearly identifying this protein as the first AT receptor to be isolated. Among the other GPCRs expressed in the CC-CA complex, two were suspected of being receptors for the poorly characterized "short neuropeptide F" (sNPF) family of peptides, based on their homology to the corresponding receptors in *D. melanogaster*. sNPF peptides were thus purified from *B. mori* brains and the cDNA encoding the precursor protein was cloned. The latter was found to contain three distinct sNPF peptides for which the proteins BNGRA10 and BNGRA11 were shown to be specific receptors in a cell-based assay. When these peptides were assayed for their effect on *in vitro* JH biosynthesis, they were found to have a strong inhibitory effect, identifying sNPF peptides as a new class of allatostatins. Strikingly, the AT receptor was found to be expressed in the CC, as opposed to the CA, and in the same CC cells where the sNPF precursor protein is produced. This observation led the authors to suggest that AT may regulate CA activity by suppressing the production or release of sNPF (i.e., allatostatin) from the CC. This putative role of AT remains to be experimentally verified.

8.6. Hemolymph Transport Proteins for the Juvenile Hormones

Upon release from the CA, JH is dispersed into the hemolymph to act at distant peripheral sites. The physicochemical nature of JHs presents a quandary — dispersing a lipophilic hormone in an aqueous circulatory system. While the JH homologues are water-soluble at levels that far exceed physiological titers (Kramer *et al.*, 1976), their amphiphilic nature promotes surface binding (Law, 1980). When dispersed in an aqueous medium, JH displays a marked propensity for non-specific binding to nearly any surface, making its distribution problematic. Early in the evolution of insects and crustaceans, hemolymph proteins arose that interacted with JHs in a noncovalent fashion yet allowed them to be dispersed in the aqueous environment of the hemolymph (Li and Borst, 1991; King *et al.*, 1995). These transport molecules have evolved into highly specific hormone carriers that have been extensively studied since their discovery in the 1970s (Trautmann, 1972; Whitmore and Gilbert, 1972).

8.6.1. Categories of Hemolymph Juvenile Hormone Binding Proteins

The hemolymph of insect species contains one and possibly two classes of JH-binding macromolecules: (1) non-specific binding molecules characterized by a high equilibrium dissociation constant (K_d = >10^{-6} M; low affinity) and (2) specific molecules exhibiting a low equilibrium dissociation constant (K_d = <10^{-6} M; high affinity;

Goodman, 1990). Juvenile hormone binding proteins of both types have been reported in approximately 50 species (Trowell, 1992). Examples of low-affinity interactions between JHs and proteins can be observed when photoaffinity-labeled JH probes are incubated *in vitro* with hemolymph proteins (Prestwich *et al.*, 1994). In contrast to the low-affinity binders, the high-affinity JH binding proteins of the hemolymph have captured considerable attention and it is these that will be discussed.

8.6.2. High-Affinity, High Molecular Weight Hemolymph Juvenile Hormone Binding Proteins

The circulatory system of a number of insect species contains high-affinity, high molecular weight JH binding proteins, here termed hemolymph JH binding proteins (hJHBPs) to distinguish them from intracellular hormone binding proteins. Their high molecular weight appellation stems from the fact that the native molecular weight of these binding proteins routinely exceeds 300 kDa. The high molecular weight hJHBPs can be divided into two subgroups, the lipophorins and the storage proteins or hexamerins. With the appropriate homologues and enantiomers, the affinities are reasonably high (see Trowell, 1992; deKort and Granger, 1996).

8.6.3. Larval Lipophorins as Juvenile Hormone Transporters

Lipophorins are multimeric hemolymph proteins that transport dietary lipids, pheromones, and cuticular lipids to their sites of utilization (Soulages and Wells, 1994; Canavoso *et al.*, 2004; Fan *et al.*, 2004; Cheon *et al.*, 2006; Weers and Ryan, 2006). These multifaceted transport proteins also transport JHs in insect species belonging to an evolutionarily diverse array of insect orders, including Dictyoptera, Coleoptera, Diptera, and Hymenoptera.

Much of the work on the JH–lipophorin interaction has centered on adult insects, and as noted by King and Tobe (1993), the information derived from the adult stage may not be applicable to the larval stage. The situation in adults becomes more complicated, as vitellogenins as well as lipophorins interact with JHs during the adult stage (Engelmann and Mala, 2000). Nevertheless, if one assumes that immunological identity between lipophorins of the nymphal and adult stage implies these proteins have a close if not identical structure (King and Tobe, 1993), various characteristics of the larval JH-binding lipophorin (JHBL) can be extrapolated from those of the adult. The JHBL from the adult female *D. punctata* is a multimer of 680 kDa composed of subunits of apolipophorin I (230 kDa) and apolipophorin II (80 kDa; King and Tobe, 1988). Covalent labeling with a JH III photoaffinity probe indicates that the binding site resides on the larger subunit only. The protein appears to be specific for the naturally occurring enantiomer (10*R*) JH III and displays a dissociation constant of 3 nM (King and Tobe, 1988). These binding characteristics, high affinity and selectivity toward the correct enantiomer, are in keeping with the JHBLs from other species (Trowell, 1992; deKort and Granger, 1996). Since lipophorin levels in the hemolymph are typically high, ranging from 6 to 16 mg/ml during the last nymphal stadium (10–40% of the total hemolymph protein), titers of this transporter are significant and yield a large excess of unoccupied binding sites (King and Tobe, 1993). These investigators confirmed that JHBL offered protection from enzymatic degradation.

Trowell *et al.* (1994) characterized a JHBL in the hemolymph of larval *Lucilia cuprina*, the Australian sheep blowfly. As is the case with other JHBLs, this protein is composed of two types of subunits, apolipophorin I (228 kDa) and apolipophorin II (70 kDa). Binding studies demonstrate that JH III is the preferred homologue, followed sequentially by JH II, JH I, JH III acid, and JH diol. Curiously, no displacement is observed with JHB_3 even though it is the predominant *in vitro* product of the blowfly CA (Trowell *et al.*, 1994). The dissociation constant for the JH-JHBL complex is approximately 30 nM and, as with the other JHBLs studied to date, the fat body is the site of synthesis.

8.6.4. Larval Hexamerins as Juvenile Hormone Transporters

A second type of high-affinity, high molecular weight JH transport molecule, one having the characteristics of a hexamerin, has been discovered. The hexamerins, composed of six 70–80 kDa subunits, are widely distributed throughout the phylum Arthropoda and have been found in insects, crustaceans, and certain chelicerates (Burmester, 2002; Hagner-Holler *et al.*, 2007; Hathaway *et al.*, 2009); however, they are not typically employed as hemolymph JH transporters. Only species in the order Orthoptera, including *L. migratoria* (Koopmanschap and deKort, 1988; Braun and Wyatt, 1996) and *Melanoplus sanguinipes* (Ismail and Gillott, 1995), are known to exploit hexamerins as JH transport proteins.

The most thoroughly studied of the JH-binding hexamerins (JHBHs) is that of *L. migratoria*. This protein is 74.4 kDa as deduced by cDNA analysis (Braun and Wyatt, 1996) and contains 15% lipid (Koopmanschap and deKort, 1988). Binding analysis shows the dissociation constant for the (10*R*) JH III–JHBH complex is 1–4 nM. According to Koopmanschap and deKort (1988), this hexamerin is present at relatively low concentrations, never exceeding 2% of the total hemolymph protein. Yet its hexameric structure allows a single molecule to bind up to 6 molecules of hormone; as a result, this JHBH contains a very large number of unoccupied hormone-binding sites. Its 4.3 kb mRNA encodes a protein

of 668 amino acids and contains 2 kb of 3' untranslated region (GenBank Account Number: U74469; Braun and Wyatt, 1996). Functional assays to locate the JH binding site suggest that it resides in the N-terminus, since a truncated JHBH lacking this region does not bind the hormone. A comparison of its amino acid sequence with other members of the hexamerin superfamily indicates that the JHBH from *L. migratoria* represents a new form that is most closely aligned with the hemocyanins (Braun and Wyatt, 1996).

More recent studies have found that the hJHBPs of *Schistocerca gregaria*, the desert locust, and the field cricket *Gryllus bimaculatus* are hexamers composed of identical subunits (Tawfik *et al.*, 2006). The JHBH of *S. gregaria* has a M_r of 480 kDa with subunits of 77 kDa, while the *G. bimaculatus* hJHBH has a M_r of 510 kDa with subunits of 81 kDa. The locust hJHBH displays a K_d of 19 nM for JH III; that of the cricket displays a K_d of 28 nM for JH III. Tawfik *et al.* (2006) report that the N-terminal sequence of the cricket hJHBP shows approximately 56% homology with a hexamerin from the dipteran *Calliphora vicina*.

8.6.5. High-Affinity, Low Molecular Weight Hemolymph Juvenile Hormone Binding Proteins

While the high-affinity, low molecular weight hJHBPs are limited to only Lepidoptera and Diptera, they have, nevertheless, been extensively characterized since relatively large quantities of larval hemolymph can be easily obtained from just a few insects. The low molecular weight hJH-BPs are usually monomeric and range in molecular weight from 25 to 35 kDa. Although low molecular weight hJH-BPs have been identified in a number of lepidopterans, only those from *M. sexta*, *B. mori*, and *Galleria mellonella* have been extensively studied.

8.6.5.1. Chemical and physical characteristics The lepidopteran hJHBPs are monomers composed of 220 to 240 amino acid residues (**Figure 4**). In species where the primary sequence of hJHBP has been deduced, including *M. sexta* (GenBank Account Number: AF226857), *B. mori* (GenBank Account Number: NM_001043609), *G. mellonella* (GenBank Account Number: AY 579371), and *Heliothis virescens* (GenBank Account Number: U22515), 4 to 6 cysteine residues have been reported with each protein containing two disulfide bridges (Park and Goodman, 1993; Wojtasek and Prestwich, 1995; Vermunt *et al.*, 2001; Debski *et al.*, 2004). The results of CNBr cleavage (Park and Goodman, 1993), together with crystallography data on *G. mellonella* hJHBP and *B. mori* hJHBP, indicate that the disulfide bridges link Cys^{9-10} to Cys^{16-17} and $Cys^{151-152}$ to $Cys^{194-195}$ (**Figure 4**). The cysteines forming the disulfide bridges are highly conserved in the proteins studied. The role of the disulfide bridge between Cys^{9-10} and Cys^{16-17} in

```
M. sexta       MNGFKVFLFLLYAKCVLSDQGAL---FEPCSTQDIACLSRATQQFLEKACRGVPEYDIRP  57
B. mori        MASLKVFLVFVFARYVASDGDAL---LKPCKLGDMQCLSSATEQFLEKTSKGIPQYDIWP  57
H. virescens   MAAYTSFLLLAFASCVLSEGGVF---FNPCYKSDIKCLSNATETFLEKTCNGYPNTEIKA  57
G. mellonella  MITLNIFLVLVIYQCALSDGSKLNLSTEPCDVSDIECISKATQVFLDNTYQGIPEYNIKK  60
               *   .  **.:    .  *:  .  :    :**   *:  *:*  **:  **:::  .*  *:  :*
```

```
M. sexta       IDPLIIPSLDVAAYDDIGLIFHFKNLNVTGLKNQKISDFRMDTTRKSVLLKTQADLNVVA  117
B. mori        IDPLVVTSLDVIAPNDAGVVIRFKNLNITGLKNQQISDFQMDTKAKTVLLKTKADLHIVG  117
H. virescens   IDPLVIPELKVVVDESMGLVFDFKNININVGLKNQQISDFKMDTDKKSVVLKTKAILNIVA  117
G. mellonella  LDPITIPSLEKSIEK-INLNVRYNNLKVTGFKNQKISHFTLVRDTKAVNFKTKVNFTAEG  119
               :**:  :..*.       .    .:   .  ::*:::.*:***:**.*  :      *:*  :**:.  :    .
```

```
M. sexta       DVVIELSKQSKSFAGVMNIQASIIGGAKYSYDLQDDSKGVKHFEVGQETISCESIGEPAV  177
B. mori        DIVIELTEQSKSFTGLYTADTNVIGAVRYGYNLKNDDNGVQHFEVQPETFTCESIGEPKV  177
H. virescens   DLKIEFTKQNKVFNGPYIAKATALGSSQYGYSFTKKDD-KEYFVVGSEENACEIIGEPDV  176
G. mellonella  KLVIELPKSSKTYTGEVTIEASAEGGAAYSYSVKTDDKGVEHYEAGPETVSCEIFGEPTL  179
               .:  **:.:..*  :  *      .:.  *.  *.*..  ...  ::: .   *   :** :*** :
```

```
M. sexta       NLNPELADALLKDPDTTHYRKDYEAHRVSIRQRSLCKIVELCYVDVVHNIRAVAKILPST  237
B. mori        TLSSDLSSALEKDSGNNSLEPDMEPLKT-LRQAAICKIAEACYISVVHNIRASAKILPAS  236
H. virescens   EIGEELQKALLNDADAKAMKPDYEANKVALRKKTLCHIVEAAYVTVIHNIRAVAKLFPKE  236
G. mellonella  SVSSTLEDALKLDSDFKKIFTEYGKQLTEGRKQTACRIVETVYAVSVHNIRAAARILPKS  239
               :. * .** *.. .    :      *: : *:*.*  *  :***** *:::*
```

```
M. sexta       AFFTDVN  244
B. mori        SFFENLN  243
H. virescens   AFFLDI-  242
G. mellonella  AYFKNV-  245
               ::* ::
```

Figure 4 Alignment of the sequenced hemolymph JH binding proteins (hJHBPs) from Lepidoptera. Amino acid sequences of *M. sexta* (Madison wild-type) hJHBP (GenBank Accession Number: AAF45309), *B. mori* hJHBP 2 (AAF19268), *H. virescens* hJHBP (AAA68242), and *G. mellonella* hJHBP (AAN06604). Alignment was performed using the CLUSTALW algorithm. Letters in black represent amino acids in the signal peptide. Symbols beneath the columns of amino acids represent the degree of conservation in each column: (*) indicates residues completely conserved; (:) indicates residues with only very conservative substitutions; and (.) indicates residues with mostly conservative substitutions. The sequences are colored according to the physiochemical characteristics of the amino acids.

JH binding can be inferred from studies on the *H. virescens* hJHBP. Using the *H. virescens* hJHBP construct, Wojtasek and Prestwich (1995) generated mutant hJHBPs in which an alanine was substituted at each of the cysteine positions. They discovered that Cys^{9-10} and Cys^{16-17} are critical for JH binding, but cysteine residues at other sites are much less important. This is particularly interesting since the N-terminus region surrounding the disulfide bridge does not appear to be directly involved in the putative ligand-binding domain (Suzuki *et al.*, 2010a,b). It may well be that the bridge adds to the N-terminus the rigidity required for binding site integrity.

MALDI-TOF measurements have determined the molecular mass of *M. sexta* hJHBP and the mass attributable to amino acids has been deduced; carbohydrates account for approximately 10% of the native protein's mass (W. Goodman, unpublished observations). Debski *et al.* (2004), in their study on the *G. mellonella* hJHBP, report that of the potential glycosylation sites only Asn^{94} is glycosylated. The function of these carbohydrate moieties is unknown; however, glycosylation of hormone transport proteins in vertebrates is thought to limit proteolytic cleavage of the protein increasing its half-life in the circulatory system (Westphal, 1986).

Two recent studies have shown that despite only a moderate level of sequence similarity, hJHBP structure in *G. mellonella* (Kolodziejczyk *et al.*, 2008) and *B. mori* (Suzuki *et al.*, 2010a,b) is quite similar (**Figure 5**). The proteins display an unusual fold, consisting of a long α-helix wrapped in a curved antiparallel β-sheet, and belong to the α- and β-protein family exhibiting a "super roll" architecture. This folding pattern is observed in several other lipid binding proteins: Takeout (Sarov-Blat *et al.*, 2000), a potential ubiquinone binding protein (Hamiaux *et al.*, 2009); a bactericidal permeability-increasing protein (Kolodziejczyk *et al.*, 2008); and a cholesteryl ester transfer protein (Kolodziejczyk *et al.*, 2008). The cylindrical hJHBP appears to have two binding sites, one that binds JHs and one with no known function. In Takeout, the lack of a disulfide bridge corresponding to Cys^{151} and Cys^{195} in the *G. mellonella* hJHBP creates a larger contiguous hydrophobic tunnel instead of two smaller pockets.

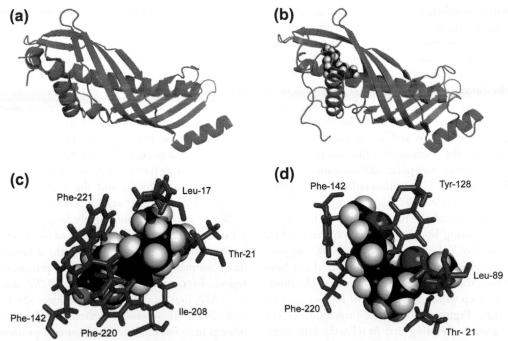

Figure 5 Structure of the *Bombyx mori* hemolymph JH binding protein (hJHBP). (a) apo-hJHBP deduced using crystallographic analyses; (b) holo-hJHBP with JH bound to the putative binding site; (c) binding site amino acid residues involved in interaction with the "hydrophobic face" of JH III; and (d) binding site amino acid residues involved in interaction with the methyl ester at C1 and the epoxide at C10, C11 of JH III. Adapted from *Suzuki et al.* (2010a,b). (Protein Data Base ID: apo-hJHBP = 2RQF; holo-hJHBP = 3A1Z.)

The hormone-binding domain lacks an obvious entrance/exit channel, which suggests that upon loading hJHBP it may change shape to enclose the hormone (**Figure 5**). Wieczorek and Kochman (1991) demonstrated a shift in the sedimentation coefficient between the apo-hJHBP (2.30 s) and holo-hJHBP (2.71 s) forms of the hJHBP from *G. mellonella*, as well as slight differences in electrophoretic mobility. The Kochman group used circular dichroism analyses to confirm their earlier discovery that hormone binding leads to changes in the secondary structure of *G. mellonella* hJHBP (Krzyzanowska *et al.*, 1998). Kolodziejczyk *et al.* (2008) speculated that the N-terminal arm of the protein, consisting of the first 9 residues, may be sufficiently flexible to form a lid that closes over the ligand binding site. Such hormone-induced conformational changes have been observed in certain vertebrate serum steroid hormone binding proteins (Grishkovskaya *et al.*, 2002; Avvakumov *et al.*, 2010).

The lepidopteran hJHBPs interact with the naturally occurring JHs in the low nanomolar range (Goodman and Granger, 2005); however, the homologues are not all bound with the same affinity. It was originally postulated (Goodman *et al.*, 1978) that the polarity rule prevails, that is, within a series of small non-polar homologous hormones, the least polar of the group will be bound with the highest affinity (Westphal, 1986). Park *et al.* (1993), in an extensive kinetics study of *M. sexta* hJHBP binding, demonstrated that the absolute polarity of the JH homologues

was not as important to the binding equilibrium as first proposed. This research revealed that the dissociation constants for the binding protein–hormone complex were approximately the same for JH I and II but, as previously observed, the interaction with JH III was considerably weaker. Although the polarity rule may not apply to the equilibrium constants, it is quite clear that the more polar the homologue, the shorter the half-life of the hormone–protein complex (Park *et al.*, 1993). Thus, complexes containing JH I, the least polar of the homologues tested, had the longest half-life (29 sec), while those containing JH III, the most polar homologue, had the shortest (13 sec). These half-times of dissociation are consistent with vertebrate hormone transport proteins (Mendel, 1989).

The hJHBP of *M. sexta* preferentially binds the naturally occurring enantiomer of (10*R*) JH III (Schooley *et al.*, 1978) and the (10*R*,11*S*) enantiomers of JH I and II (Park *et al.*, 1993). This specificity appears universal among the high-affinity JH binding proteins, including lipophorins and hexamerins (Trowell, 1992; deKort and Granger, 1996), indicating the binding site(s) for the hormone is selective. In addition to binding analyses using both enantiomers of each JH homologue, studies have been conducted using geometrical isomers of the hormones, hormone metabolites, and biologically active analogs (Goodman *et al.*, 1976; Peterson *et al.*, 1977, 1982). The results indicate that highly active JH agonists and JH metabolites do not interact with hJHBP.

Armed with this information, several investigators have speculated on the nature of the ligand-binding domain and its constituent residues. Goodman *et al.* (1978) presumed the primary interactions between the hormone and the binding site would occur along the alkyl side chains of the hormone and the methyl ester moiety at C1, which together form a distinct hydrophobic surface. Interruption of that surface, especially by the introduction of highly polar groups at C1, significantly reduces binding. However, the hormone-binding site is not sterically constrained, since non-polar additions such as ethyl and propyl groups at the C1 position reduce binding only moderately (Peterson *et al.*, 1977).

The recent monumental structural analyses by Suzuki *et al.* (2010a,b) using both crystallography and NMR technology and by Kolodziejczyk *et al.* (2008) support the earlier hypotheses about the interaction of the hormone with the protein. NMR studies indicate the binding site is very deep within the hJHBP of *B. mori* (Suzuki *et al.*, 2010a,b). **Figure 5** shows the interaction of JH III with the putative binding site. *In silico* docking indicates the site contains a number of hydrophobic amino acids that interact with the hormone and, as predicted, the side chains at C3, C7, and C11 play a key role in that interaction. The C1 methyl ester interacts with Leu[89]. Tyr[128] is involved in JH III binding in two ways: Van der Waals interactions between Tyr[128] and the side chains at C3 and C7 and hydrogen bonding between the hydroxyl group of Tyr[128] and both the oxygen at C1 and the epoxide at C10,11. The residues that interact with the hydrophobic "face" of the hormone are Leu[17], Ile[208], and Phe[142, 221, 222].

8.6.5.2. Genomic structures

The genomic structure of several *hJHBPs* have been examined (Orth *et al.*, 2003a; Sok *et al.*, 2005). Both *M. sexta* and *G. mellonella* are derived from five exons and are flanked by typical eukaryotic and insect-specific transcriptional control elements (**Figure 6**). In *M. sexta*, the first exon encodes the 5'-untranslated region, a signal peptide plus the first two amino acids of the open reading frame, while the other exons encode the mature secreted protein and the 3'-untranslated region. The proximal promoter region contains a TA-rich region similar in position to but slightly longer than that of the JH esterase gene from *Trichoplusia ni* (Jones *et al.*, 1998); however, both of these genes lack the canonical TATAA motif normally associated with the proximal promoter region. In contrast, the *Galleria hJHBP* does contain the TATA box extending from position -29 to -24 (Sok *et al.*, 2005). These workers believe that the core promoter belongs to a class of TATA-initiator core promoters.

Computational analysis indicates numerous potential transcription factor binding sites in the 5'-flanking region of the gene. Most intriguing are two closely spaced ecdysteroid receptor response elements approximately 5 kb upstream from the start site (Orth *et al.*, 2003a). The JHs are presumed to act on the ecdysteroid signaling pathway (Riddiford, 1994), regulating gene expression by modulating the transcriptional activity of the ecdysteroid receptor and its heterodimeric partner, Ultraspiracle (Jones and

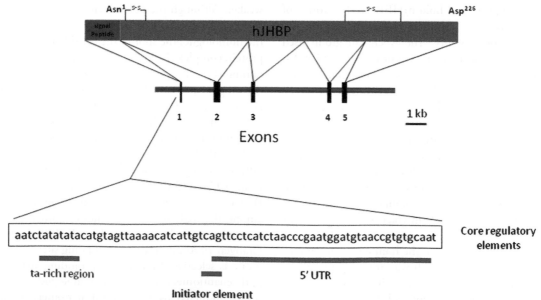

Figure 6 Structure of *M. sexta* hemolymph JH binding protein (hJHBP) and its corresponding genomic region. The top figure displays the protein and relative positions of the signal peptide and disulfide bridges. The center figure represents the genomic region encoding the hJHBP and the relative distances between each of the encoding exons. The span of residues encoded by each exon is represented by the connecting lines. The core regulatory region upstream from the 5' end of the gene is shown and each of the deduced functional regions is outlined below it.

Sharp, 1997). Considering that both the protein (Hidayat and Goodman, 1994) and *hJHBP* gene expression (Orth *et al.*, 1999) undergo a significant decline during a larval-to-larval molt, it is not unreasonable to speculate that ecdysteroids, acting via these ecdysteroid response elements, may be involved in the regulation of *hJHBP* expression and indirectly modulate JH titers. Sok *et al.* (2005) have noted a number of *Broad complex* regulatory elements upstream from the transcriptional start site and within the *G. mellonella hJHBP*.

8.6.5.3. Regulation

The discovery of the high-affinity JHBPs naturally led to the question of whether these proteins have a direct influence on JH titers. In most insects, the exceedingly high ratio of hJHBP to JH makes it unlikely that even major shifts in hJHBP levels would play a direct role in hormone regulation (Hidayat and Goodman, 1994). Nevertheless, the hJHBP titers do fluctuate significantly during development (Goodman, 1990; Hidayat and Goodman, 1994), and these changes do not mirror the changes in total hemolymph protein levels. Assessment of bound and unbound hormone levels concludes that virtually the entire hormone pool is bound at any given time and that the level of holo-hJHBP is a tiny fraction of apo-hJHBP.

Why should the fat body, the site of synthesis of hJHBP (Nowock *et al.*, 1975; Orth *et al.*, 2003a), be engaged in excessive synthesis of this protein, and why should its levels be so precisely controlled? The presence of excess hJHBP remains a mystery, but one intriguing hypothesis asserts that the protein acts as a high-affinity scavenger to sequester the hormone in the hemolymph, where it can be metabolized by hemolymph enzymes (Touhara *et al.*, 1996). This scavenger hypothesis has also been proposed for the role of hJHBP in the embryo, where maternal JHs may need to be compartmentalized to prevent interference with embryonic development (Orth *et al.*, 2003b). While this hypothesis is provocative, it assumes that cellular degradation of the hormone is minimal; as noted below, this may not be the case (Section 8.7.2.).

Nijhout and Reed (2008) attempted to mathematically model JH titers employing binding, rate, and enzymatic constants of the known factors controlling JH titers. Their model demonstrates, among other things, that the role of hJHBP in maintaining JH titers is highly underestimated. Furthermore, the model shows more JH present than can be accounted for by synthesis, sequestration, and degradation. They posit that the peripheral pool of hormone, typically limited to hormone associated with hemolymph proteins, must also include a large pool of tissue-adsorbed JH. Unfortunately, the binding and rate constants employed were generated using highly purified proteins and *in vitro* conditions for CA hormone production. Under *in vivo* conditions, with the appropriate neural connections, ions, pH, and potential interacting partners, the JH titers might be different. The work by Nijhout and Reed (2008) challenges those who work with hJHBP and JH esterases to re-examine their kinetic data in light of the model and bring resolution to this important issue.

Although it is unclear why the levels of hJHBP must be controlled so precisely during development, a picture of the elements controlling the expression of its gene is beginning to emerge. As previously noted, the presence of ecdysteroid response elements in the 5'-flanking region of *hJHBP* suggests ecdysteroids may be involved in the regulation of *hJHBP* expression. During the molting period, the elevated levels of ecdysteroids required for molting reduce expression of *hJHBP* (Hidayat and Goodman, 1994). In addition, there is reasonably good evidence that JH itself has a role in regulating the expression of this gene. It has been observed that an inverse relationship exists between JH and hJHBP titers in *M. sexta* (Hidayat and Goodman, 1994). Hormone titers at the beginning of the penultimate stadium are high, but hJHBP titers are low. When JH titers drop during the late intermolt period, levels of hJHBP rise (Hidayat and Goodman, 1994). The same inverse relationship can be seen when the insect is subjected to mild stress (Tauchman *et al.*, 2007) or parasitism (Goodman and Beckage, unpublished observations): JH titers rise and the levels of hJHBP fall. Elevated titers of the hormone downregulate hJHBP mRNA and protein levels, whereas reduced JH titers elevate both message and protein. If the sole function of hJHBP is hormone delivery, then the developmental pattern and excess titer remain a mystery, but, as with the vertebrate serum hormone-binding proteins, hJHBP may be more than a transport macromolecule.

8.6.5.4. Functions

Facilitating the transport and dispersal of JH to distant target sites is the most obvious function of hJHBP, yet other, equally important roles for the binding protein have been hypothesized or experimentally determined (Westphal, 1986; Goodman, 1990). These functions include: (1) reducing the levels of promiscuous binding to non-target tissues, (2) reducing enzymatic degradation of the hormone (Hammock *et al.*, 1975), (3) acting as a scavenger to enhance JH metabolism in larval hemolymph (Touhara *et al.*, 1996) or the developing embryo (Orth *et al.*, 2003b), (4) providing a peripheral reservoir of hormone that is readily available at the target tissue, (5) enhancing JH synthesis in the CA by acting as a sink to eliminate a short feedback loop, and (6) aiding hormone movement from the hemolymph into the target cell. The first four functions are either well established or have at least gained a level of scientific respect.

Preliminary evidence now suggests that hJHBP may promote the synthesis of JH by dispersing the hormone from the immediate vicinity of the CA. Using the RCA to assess the effect of several proteins that interact with JH (bovine serum albumin, JH-directed antiserum, and *M. sexta* hJHBP) on JH biosynthesis *in vitro*, it was observed that only hJHBP significantly increased the incorporation of radiolabeled tracer into JH (Granger and Goodman, unpublished observations). *M. sexta* CA incubated in medium containing physiological levels of hJHBP synthesized nearly 60% more JH than CA incubated with other JH-interacting proteins or controls. While the results are preliminary, they do suggest that hJHBP aids the diffusion of JH from the CA and may prevent hormone from accumulating non-specifically in the vicinity of the glands. The binding proteins, in essence, create a longer feedback loop and thus block short-circuiting of JH biosynthesis.

The most intriguing aspect of possible hJHBP functions has yet to be explored: its action at the surface of a target cell. A body of evidence has convinced some vertebrate endocrinologists that hormone binding proteins release their ligands into the circulatory system, thus allowing them to enter the cell in the unbound state (Westphal, 1986; Mendel, 1989; Grasberger *et al.*, 2002). For some hormones, this is a reasonable conclusion; however, several protein-bound hormones and vitamins have specific cell surface receptors for their respective transport proteins (Kahn *et al.*, 2002; Willnow and Nykjaer, 2010). Retinol binding protein (RBP), for example, binds to a plasma membrane receptor that participates in transfer of the ligand from this serum transport protein to the cytoplasmic RBP (Sundaram *et al.*, 1998, 2002). As levels of apo-RBP levels far exceed those of the holo-RBP, a conformational change in the protein must occur so that cell surface receptors are not interacting with ligand-free RBP (Chau *et al.*, 1999).

Even more striking are the roles of sex hormone binding globulin (SHBG), a vertebrate serum protein that binds and transports the sex steroids. Target tissues requiring the sex steroids display a plasma membrane receptor for SHBG parallel to RBP. Unlike the RBP receptor, the SHBG receptor binds only the unloaded form of the transport protein; the hormone-loaded SHBG complex does not bind (Kahn *et al.*, 2002; see Hammes *et al.*, 2005 for a different viewpoint). Sex steroids then bind to the receptor-bound SHBG on the cell surface which, in turn, activates adenylyl cyclase. The role of the activated second messenger cascade remains unknown. It seems that the extracellular transport protein is more than just a vehicle for hormone dispersal; anchoring at the cell surface may be required to facilitate uptake. Moreover, the receptor–protein complex may have its own intrinsic signaling ability.

Recently Zalewska *et al.* (2009) reported that *G. mellonella* hJHBP interacts with the high-abundance hemolymph proteins apolipophorin, arylphorin, and hexamerin. While each of these proteins is known to transport hydrophobic molecules, the concept that they may also form complexes

with proteins such as hJHBP is novel. Since there are cell surface receptors for the hexamerins (Hansen *et al.*, 2003), JH may enter the target cell bound to a complex of hJHBP and hexamerin (or apolipophorin or arylphorin). It should be noted that the interaction of hJHBP with these proteins is speculative, since size separation studies on hemolymph proteins routinely characterize the hJHBPs as molecules of 25 to 35 kDa. Association with high molecular weight proteins has not been detected immunologically (Goodman *et al.*, 1990). These investigators also noted that ATP synthase, a protein normally associated with mitochondria, was present on the plasma membrane of fat body cells and interacted with the *G. mellonella* hJHBP. Importantly, ATP synthase binds the apo-hJHBP form only; holo-hJHBP was not bound. The authors suggest this is part of the secretory pathway for newly synthesized hJHBP, but tethering apo-hJHBP to a membrane may be a method of concentrating the hormone in close proximity to the target cell to aid uptake.

8.6.5.5. Future studies   Given its high affinity and selectivity for JH, the concentration of hJHBP in insect hemolymph makes the transporter a major factor in determining the levels of unbound hormone at the target site. It is clear that the role of hJHBP in JH biosynthesis and mode of molecular action requires further attention. For example, what is the role of hJHBP in CA feedback loops? Can changes in hJHBP levels actually influence peripheral distribution or is the concentration of transporter so large in relation to JH that reductions are not that important? What role do the lipophorins play in distribution of JH, especially in species that have high-affinity, low molecular weight transport proteins? What is the molecular interaction between hJHBP and hemolymph JH esterases? What is the role of the apparent second binding cavity of hJHBP? Are there unknown ligands that activate the protein or play a part in uptake? The role of hJHBP in uptake at the target site remains enigmatic. Clearly, studies using primary *in vitro* tissue are not convincing as long as the tissues are saturated with endogenous hJHBP. Models that employ cell lines responsive to JH must be developed to clarify how a rapidly metabolized hormone can be delivered to its target and retain its biological activity.

8.7. Catabolism of the Juvenile Hormones

The maintenance of effective hormone titers at the target site is a delicate balancing act among various processes: synthesis, delivery, metabolism, and cellular uptake. Endocrine signals, by their very nature, must be transitory to elicit the exquisite control of the target response. Soon after the chemical structures of JHs were elucidated in the late 1960s, a search for hormone-inactivating enzymes began. That search has continued unabated with nearly 400 articles describing some aspect of JH metabolism (Kamita and Hammock, 2010). This prodigious number of publications reflects the agricultural interest in catabolism as a means of insect pest control. It is assumed that dramatically increasing or decreasing catabolism of JH may lead to developmental derailment (Bonning and Hammock, 1996; Hammock, 1998; Guerrero and Rosell, 2005; Tan et al., 2005). While this novel means of pest control has yet to be commercially exploited, the biological information gleaned from studies on JH metabolism is of considerable importance in understanding hormone titer regulation.

JH catabolism may be initiated either in the hemolymph or in target cells, resulting in different products; however, the end point is the same: biological inactivation of the hormone. Inactivation of JH in the hemolymph is the result of enzymatic activity of several different esterases capable of hydrolyzing the hormone at the C1 position to form the metabolite JH acid (Roe and Venkatesh, 1990; **Figure 1**). Although cellular inactivation of JH is primarily the result of epoxide hydrolases that hydrate the epoxide at the C10, C11 position to form the JH diol, esterase activity that yields JH acid is also associated with cells (Roe and Venkatesh, 1990; Lassiter et al., 1995). Our examination of JH catabolism begins in the hemolymph compartment, where esterases represent the primary mechanism of hormone inactivation.

8.7.1. Juvenile Hormone Esterases

The hemolymph of a number of species contains esterases that are capable of hydrolyzing the ester at the C1 position to form JH acid. While the early literature suggested two classes of esterases responsible for JH hydrolysis, JH-specific esterase (JHE) and non-specific or general esterases (Sanburg et al., 1975), more recent work has been less clear about the role of the non-specific enzyme(s) (Gilbert et al., 2000). General esterases were originally defined as those enzymes that could metabolize both a general substrate, α-naphthyl acetate, and JH (Sanburg et al., 1975) and were inhibited by O,O-diisopropyl phosphorofluoridate (DFP). More recent evidence suggests that the general esterases are less important in JH metabolism than first thought (Roe and Venkatesh, 1990; Gilbert et al., 2000).

Because conversion to JH acid can be carried out by a number of enzymes, Hammock (1985) proposed a working definition for the JHEs. From a biochemical standpoint, such an enzyme must have a low apparent Michaelis constant (K_m) for JH and it should therefore hydrolyze JH with a high k_{cat}/K_m ratio. Furthermore, the enzyme must be able to hydrolyze JH in the presence or absence of a JH binding protein. From a biological standpoint, JHE must have activity that correlates with

a decline in JH titer. The premise that these enzymes are vital to hormone titer regulation has received considerable attention.

8.7.1.1. Chemical and physical properties JH-specific esterases (EC 3.1.1.1) are members of the carboxylesterase family and have been studied in at least eight orders of insects, including Thysanura, Orthoptera, Blattodea, Hemiptera, Coleoptera, Lepidoptera, Diptera, and Hymenoptera. The best characterized of the JHEs are those from the hemolymph of Lepidoptera. The native JHEs of Lepidoptera appear to be composed of a single polypeptide chain with a relative molecular mass of approximately 63–67 kDa (Hinton and Hammock, 2003). A recent study indicates that the relative molecular mass of the JHE from the coleopteran *T. castaneum* is approximately 63 kDa (Tsubota et al., 2010). The hemolymph of another coleopteran, *Leptinotarsa decimlineata*, contains a JHE with a native relative molecular mass of 120 kDa, and appears to be composed of two polypeptide chains of approximately 57 kDa each (Vermunt et al., 1997). A similar situation exists in the orthopteran *G. assimilis*, where the native JHE displays a relative molecular mass of 98 kDa and is composed of two identical subunits of 52 kDa (Zera et al., 2002). The isoelectric points of all the JHEs studied are acidic, being in the range of 5.2 to 6.3. The JHE of one lepidopteran, *T. ni*, is glycosylated (Hanzlik and Hammock, 1987); however, the JHE of another, *M. sexta*, lacks glycosylation, even though computational analysis of its sequence suggests three potential glycosylation sites (Kamita et al., 2003).

8.7.1.2. Protein structure and catalytic site Since our last review in 2005, two important papers have emerged that clarify the structure and the catalytic site of *M. sexta* JHE. Wogulis et al. (2006) generated a 2.7 Å crystal structure of JHE complexed to the potent inhibitor 3-octylthio-1,1,1-trifluoropropan-2-one (OTFP). Structural analyses indicated the enzyme belongs to the α/β hydrolase family (**Figure 7**). *M. sexta* JHE displays 14 β-strands and 16 α-helices that closely match those of the well-characterized esterase acetylcholine esterase. The deeply buried catalytic site is formed by loops and helices that are widely dispersed within the primary sequence. The long narrow tunnel leading to the site is unusual for the α/β hydrolases and it makes the catalytic site nearly inaccessible to surface solvents when a substrate/inhibitor is bound. There is a ring of polarity at the entrance to the tunnel and about a third of the way into the catalytic site. Due to their hydroxyl groups, Thr[314], Tyr[416], and Tyr[424] are key to this domain. As might be expected, the catalytic site is formed largely from hydrophobic residues; however, at the base of the pocket are the hydrophilic residues of the catalytic His[471], Ser[226], and Glu[357], which are involved in hydrolysis of JH. Of the 24 non-catalytic residues in

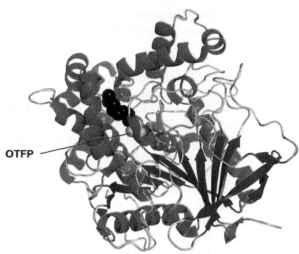

OTFP

Figure 7 Structure of the *M. sexta* JH esterase with the esterase inhibitor OTFP covalently attached to the catalytic site. Adapted from *Wogulis et al.* (2006); Protein Data Base ID 2FJ0.

the binding site, 14 are conserved in lepidopteran JHEs (Kamita *et al.*, 2010).

Two essential residues that are not part of the catalytic triad, Phe[259] and Thr[314], are vital in generating the low K_m for JH. The phenyl group of Phe[259] appears to form β-stacking interactions with the ester of JH and the Thr[314] may form a hydrogen bond with the epoxide group of JH. Their position in the pocket creates a tight fit for JH and thus explains the specificity exhibited for various homologues and inhibitors; however, the tight fit comes at the cost of reduced JH metabolism. Recent studies using site-directed mutagenesis indicate that removing either of the residues will increase the K_m approximately three- to six-fold. The exceptional conservation of Phe[259] and Thr[314] among lepidopteran species suggests that these residues play an important role in hydrolyzing JH at nanomolar concentrations.

One might reasonably expect the catalytic site of JHEs of other species to show some sequence similarity and, not surprisingly, they do. A comparison of sequences among the known lepidopteran JHEs, *B. mori* (Hirai *et al.*, 2002; GenBank Account Number: AF287267), *C. fumiferana* (Feng *et al.*, 1999; GenBank Account Number: AF153367), *H. virescens* (Hanzlik *et al.*, 1989; GenBank Account Number: 037197), *Sesamia nonagrioides* (GenBank Account Number: EU178813), *M. sexta* (Hinton and Hammock, 2001; GenBank Account Number: AF327882), and *Omphisa fuscidentalis* (GenBank Account Number: EU523701) reveals approximately 25% identity. The residues associated with the catalytic site are in complete agreement and display the appropriate alignment; however, with the exception of the putative catalytic site residues and a sequence of G-Q-S-A-G surrounding Ser[226] of *M. sexta* JHE, the JHEs as a family are not highly

conserved. This might be expected since only the cleft into which the substrate fits needs to be lined with hydrophobic residues (Kamita *et al.*, 2003).

8.7.1.3. Kinetic parameters and specificity Kinetics and substrate specificity have long been a focus of JHE studies because of the obvious agricultural potential of this information. Elucidation of JHE kinetic constants permits prediction of the rate limits of JH metabolism as well as provides valuable insight into how peripheral JH levels are regulated. A number of reports from the 1980s indicate that among lepidopterans the apparent K_m for the naturally occurring JHs ranges from 10^{-8} to 10^{-6} M (Roe and Venkatesh, 1990). Juvenile hormone titers in the species tested are at least 10 to 100 times lower than the estimated K_m concentration, indicating that the enzyme is very sensitive to changes in JH concentration, but its catalytic potential is wasted. Since the K_ms seemed unduly high compared to substrate concentration, investigators examined the individual rate components of the enzymatic reaction to better explain the observed data (Sparks and Rose, 1983; Abdel-Aal and Hammock, 1986).

Several important lessons can be drawn from the kinetic data. First, JHE has a very high affinity for JH. Second, it has a relatively low turnover number or k_{cat}, where k_{cat} is the maximum number of substrate JH molecules converted to JH acid per active site per unit time. These factors translate into a very effective enzyme scavenger that can "find" JH at physiological concentrations and convert it to JH acid (Abdel-Aal and Hammock, 1986). The structural elucidation of the protein now provides a rationale for the low turnover (Kamita *et al.*, 2010). The rate-determining event may be release of the lipophilic JH acid, which is buried deep within the enzyme.

Substrate specificity of JHE represents another characteristic that seems counterintuitive. The K_m and V_{max} that the JHEs display toward the JH homologues is surprisingly low when compared to other substrates, such as α-naphthyl acetate. For example, recombinant *M. sexta* JHE displays a K_m (410 μM) and V_{max} (21 μmol/min/mg protein) for α-naphthyl acetate that is considerably higher than for its natural substrate, JH III ($K_m = 0.052$ μM, $V_{max} = 1.4$ μmol/min/mg protein; Hinton and Hammock, 2003). As noted by Fersht (1985), when discriminating between two competing compounds, specificity should be determined by the k_{cat}/K_m ratio and not K_m alone. The k_{cat}/K_m ratios for JH III and α-naphthyl acetate are 27 and 0.04, respectively, underscoring the high degree of specificity of JHE for JH III (Hinton and Hammock, 2003). While these kinetic parameters suggest that the enzyme has a high degree of specificity, it appears that JHEs from several species also hydrolyze methyl and ethyl esters of JH I and III at similar rates (Grieneisen *et al.*, 1997). Moreover, even *n*-propyl and *n*-butyl esters of these homologues can serve effectively as substrates, albeit at a lower rate of hydrolysis. One of the most unexpected groups of JHE substrates is found in the naphthyl and p-nitrophenyl series (Rudnicka and Kochman, 1984; Hanzlik and Hammock, 1987; Kamita *et al.*, 2003). Once again, structural analysis provides clues as to how the catalytic site may accommodate certain bulky molecules such as α-naphthyl acetate. If α-naphthyl acetate enters the catalytic site in the acid-first orientation, there is sufficient space to carry out hydrolysis. Computational modeling posits that the catalytic site is flexible enough to accommodate larger molecules such as the ethyl esters of JH I and III or α-naphthyl acetate (Kamita *et al.*, 2010).

Another counterintuitive observation regarding JHE activity is that while its major role is the conversion of JH to JH acid, both native and recombinant JHEs from several species can, under the appropriate conditions, transesterify JH to form the higher ester homologues, such as JH ethyl, JH n-propyl, and JH n-butyl esters (Grieneisen *et al.*, 1997). Although JHE-mediated JH transesterification may be a curiosity limited to the test tube, Debernard *et al.* (1995) demonstrated that when JH III was dissolved in ethanol (10 μl) and injected into *L. migratoria*, it was converted to JH III acid and also to JH III ethyl ester. Thus, it should be noted that care must be taken to avoid artifacts when alcohols are used as carrier solvents for the hormone in JHE assays. Once again, flexibility of the catalytic site is key. Kamita *et al.* (2010) suggest that low concentrations of detergents and alcohols distort the binding site and allow JH and various substrates better access to the catalytic residues or improve the release of JH acid from the catalytic site. While JHE in a biological milieu clearly serves as an esterase, these new findings imply that the enzyme may have other physiological roles that are as yet undiscovered (Anspaugh *et al.*, 1995).

8.7.1.4. Inhibitors The development of effective JHE inhibitors has been ongoing for more than 30 years. While JHE inhibitors have not yet been commercialized for agricultural use, they have, nevertheless, been major tools in discovering the role of JH catabolism *in vivo*. Two important series of JHE inhibitors have emerged from these studies: the trifluoromethyl ketones and the phosphoramidothiolates (Roe and Venkatesh, 1990). The trifluoromethyl ketones (TFKs), such as 1,1,1-trifluorotetradecan-2-one (TFT), act as transitional state analogs, mimicking the α,β saturation of JH (Hammock *et al.*, 1982). Several important JHE inhibitors have been developed, including OTFP (Abdel-Aal and Hammock, 1985) and 1-octyl[1-(3,3,3-trifluoropropan-2,2-dihydroxy)] sulfone (OTFP-sulfone; Roe *et al.*, 1997; Wheelock *et al.*, 2001). The attractiveness of the TFKs lies in their reversible binding to the enzyme, which permits use in affinity-matrix purification of JHEs. Moreover, the TFKs are useful for *in vivo* studies, since they are highly specific and appear to have little effect on non-specific esterases (Roe *et al.*, 1997). In contrast

to the TFKs, the organophosphate inhibitors of JHE — most importantly O-ethyl S-phenyl phosphoramidothiolate (EPPAT) — bind irreversibly to JHE (Hammock, 1985). Like the TFKs, EPPAT appears to be highly selective for JHE and can be used *in vivo*. Although EPPAT has only a moderate inhibitory effect, it has a much longer life *in vitro* than the TFKs.

8.7.1.5. Genomic structures The genomic structure of several *JHE*s has been described, including that of *H. virescens* (Harshman *et al.*, 1994), *D. melanogaster* (Campbell *et al.*, 2001), and *T. castaneum* (Tsubota *et al.*, 2010). Genomic characterization of the *JHE* from *T. castaneum* has demonstrated that the gene has five splicing variants that differ only in the 5'-untranslated region (Tsubota *et al.*, 2010). All splice variants have eight exons, with the variations occurring in the first exon. In *D. melanogaster*, the gene is composed of six exons and contains an open reading frame of 1.7 kb. Immediately adjacent to the *JHE* gene in the 5'-position is another carboxylesterase gene (FlyBase Identification Number: CG 8424). It displays 42% identity with *JHE* and contains the putative catalytic site (Kamita *et al.*, 2003). This gene is nearly identical in length and is thought to be the result of a gene duplication event (Campbell *et al.*, 2001).

8.7.1.6. Tissue distribution and regulation The importance of JHE in regulating JH titers has prompted a number of investigators to examine its physiological and genetic regulation. The primary site of synthesis for the hemolymph JHEs, the fat body (Whitmore *et al.*, 1974; Hammock *et al.*, 1975; Wing *et al.*, 1981), has been the focus of many studies and has been reviewed in some detail (Roe and Venkatesh, 1990; Roe *et al.*, 1993; deKort and Granger, 1996; Gilbert *et al.*, 2000). Tissues other than the fat body can also express the enzyme (see Roe and Venkatesh, 1990, for a review; Feng *et al.*, 1999). Most studies conclude that the JHEs of different tissues present at different time points are similar, if not identical, to the hemolymph form of JHE; however, their regulation may be different (Jesudason *et al.*, 1992). Sparks *et al.* (1989) made the novel suggestion that the isoforms of hemolymph JHE may be the result of JHE production and release into the hemolymph from a variety of tissues. Likewise, a recently characterized *JHE* gene from *Tr.* displayed variants (Tsubota *et al.*, 2010), although it is unclear if the variants are products of different tissues.

To best understand JHE regulation, a brief description of hemolymph JHE levels is in order. In *M. sexta*, a peak in hemolymph JHE is observed prior to ecdysis in larval stadia two through five (Roe and Venkatesh, 1990), with activity in each stadium increasing in direct proportion to the growing insect's weight (Roe *et al.*, 1993). The precise timing in the appearance of JHE and its developmental profile in relation to total hemolymph protein suggests specific regulation. However, the role of JHE in early instars is unclear, since hemolymph JH titers remain relatively high even in the presence of enzyme (Hidayat and Goodman, 1994).

In contrast to the earlier stadia, during the last stadium of *M. sexta* the developmental profiles of JHE activity and of JH titers reveal two peaks (Vince and Gilbert, 1977; Baker *et al.*, 1987; Roe and Venkatesh, 1990). Titers of JH are highest in the hours immediately following the molt from the fourth to the fifth stadium (**Figure 8**). They then drop dramatically, becoming undetectable by 48 h after ecdysis. JH titers rise a second time, peaking at 156 h and then declining prior to metamorphosis. JHE activity begins to climb midway through the feeding period and peaks shortly before wandering. A smaller rise in JHE activity appears after the second peak of JH but drops at pupation (Baker *et al.*, 1987). Similar developmental profiles are seen in the last stadium of *Spodoptera littoralis* (Zimowska *et al.*, 1989) and *Leptinotarsa decemlineata* (deKort, 1990). Most studies have pinpointed the fat body as the primary source of JHE in the hemolymph (Wing *et al.*, 1981; Wroblewski *et al.*, 1990). However, Jesudason *et al.* (1992) suggested that the first peak of JHE is of fat body origin, while the second peak, during the pre-pupal phase, is not. It could be that in this study, the absence of protease inhibitors led to the destruction of JHE during preparation of pre-pupal fat body samples. While the JHE profile in hemolymph appears to be similar for most lepidopterans, it appears to differ outside of this order. Some species display a single burst of JHE activity that declines either midway through the last stadium or at the very end of larval life (Tsubota *et al.*, 2010; see Roe and Venkatesh, 1990, for a review).

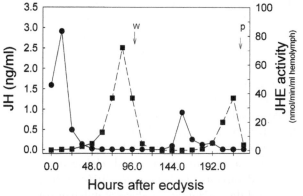

Figure 8 Hemolymph JH titer and JH esterase (JHE) activity during the fifth stadium of *M. sexta* larval development. Circles represent the total titer for all homologues detected (JH 0, I, II, III). JH acids are not shown. JH titers were combined for both females and males and expressed as the average. Squares represent the JHE activity. JHE activities were combined for both females and males and expressed as the average. (W) indicates the onset of wandering behavior and (P) indicates the time of pupal ecdysis. Adapted from *Baker et al.* (1987).

That JHE activity is relatively low until after the JH titer drops (**Figure 8**) does not support the concept that the enzyme is directly responsible for the decline in JH titers. Nevertheless, the enzyme displays exceptional scavenging properties, and it may be that the increase in enzyme activity occurs to ensure that no detectable JH is present during commitment to pupation. More problematic is the observation that JH is not totally eliminated when JHE is present. For example, a comparison of JH levels in the penultimate stadium of *M. sexta* (Hidayat and Goodman, 1994) with JHE levels during this same period (Roe and Venkatesh, 1990) reveals measurable amounts of JH in the presence of JHE. Browder *et al.* (2001) suggested the pre-wandering peak of JHE has no functional relationship to metamorphosis. Although not stated explicitly, the same conclusion can be drawn from studies of *T. ni* (Sparks *et al.*, 1979; Hanzlik and Hammock, 1988; Jones *et al.*, 1990b). JH levels appear to be unaffected by JHE in diapausing larvae of *Ostrinia nubilalis* (Bean *et al.*, 1982) and *S. nonagrioides* (Schafellner *et al.*, 2008). Yet, it is clear from the work of Tan *et al.* (2005) that JHE is capable of reducing JH titers *in vivo*. Taken together, hemolymph JHE levels follow a developmental pattern that suggests the enzyme may be directly involved in JH catabolism, yet there remains a degree of uncertainty about the assumed role of JHE in eliminating the hormone from the circulatory system.

The fact that JHE levels do not follow a developmental pattern typical of other hemolymph proteins prompted a search for possible regulatory mechanisms modulating its activity. It would be logical for JHE to be regulated by JH, but regulation of JHE by JH in lepidopteran larvae, especially during the last larval stadium, depends on whether the insect is in the pre- or post-wandering phase. In the pre-wandering phase of the last larval stadium, JH regulation of hemolymph JHE levels is ambiguous. The brain appears to play an inhibitory role in the control of JHE activity, and other factors, such as tissue competence, seem to be involved (Venkatesh and Roe, 1988; Jesudason *et al.*, 1992; Roe *et al.*, 1993). In contrast, manipulation of JH titers by surgical and chemical means during the post-wandering period of the last stadium has a direct impact on JHE activity (Roe *et al.*, 1993; see Gilbert *et al.*, 2000, for a review). Treatment of larvae with JH or JH agonists leads to an increase in JHE activity in the hemolymph, whereas allatectomy leads to a decline in JHE activity that can be reversed by supplying exogenous hormone. At the molecular level, nuclear run-on experiments using the fat body of *T. ni* have confirmed that exogenous JH or JH analogs stimulate *JHE* expression and that the response can be detected within three hours of treatment (Venkataraman *et al.*, 1994). Kethidi *et al.* (2004) examined the upstream region of the *C. fumiferana JHE* gene and identified a 30 bp region, located between -604 and -574, that can support both JH

I induction and 20E suppression. This 30 bp region contains two conserved hormone-response element half-sites separated by a 4-nucleotide spacer.

Using a cell line derived from the midgut of *C. fumiferana*, Feng *et al.* (1999) demonstrated that JH can increase the abundance of JHE mRNA within one hour, suggesting a potential upregulation of *JHE* expression by JH. However, these investigators rightfully caution that the apparent rise in message could instead be due to stabilization of JHE mRNA. Curiously, the potent protein synthesis inhibitor cyclohexamide mimicked the action of JH, prompting the suggestion that an inhibition of protein synthesis leads to an increase in transcription of JHE message or, alternatively, stabilization of the message. In contrast to the apparent stimulation of message levels by JH, the presence of 20E in the medium leads to a decrease in JHE mRNA in a dose-dependent manner (Feng *et al.*, 1999). A similar situation has been observed in *D. melanogaster* (Kethidi *et al.*, 2005). This discovery underscores the complexity surrounding regulation of *JHE* expression and it indicates that the role of ecdysteroids must be considered.

The endocrine system is an important bridge between the target cell and the environment, and it is axiomatic that environmental factors are major players in regulation. Factors such as stress (Gruntenko *et al.*, 2000), circadian cues that induce diapause (Bean *et al.*, 1982, 1983), nutrition (Cymborowski *et al.*, 1982; Sparks *et al.*, 1983; Venkatesh and Roe, 1988, 1990; Anspaugh and Roe, 2005), and parasitism (Hayakawa, 1990; Edwards *et al.*, 2006; Schafellner *et al.*, 2007) are all thought to play a role in regulating JHE levels.

8.7.1.7. Catabolism of juvenile hormone esterases

After a rapid increase midway through the feeding period of the last stadium in Lepidoptera, JHE levels drop dramatically just prior to the onset of wandering. Experimental evidence corroborates this observation; the half-life of recombinant JHE, when injected into *M. sexta* second instars, is approximately 20 min, and control proteins of similar molecular mass display half-lives in days (Ichinose *et al.*, 1992). Since no cleavage products of JHE can be detected in the hemolymph, these authors hypothesized that the enzyme is removed from the circulatory system by cellular uptake. They found that the pericardial cells were responsible for uptake and destruction of the hemolymph JHE (Ichinose *et al.*, 1992). Pericardial cells are a collection of cells that surround the insect heart in various configurations, depending upon the order (Wigglesworth, 1972; Crossley, 1985). In *M. sexta*, the pericardial mass is punctuated with numerous labyrinthine channels that increase exposure of the cells to the hemolymph (Brockhouse *et al.*, 1999), and it is thought that this tissue mass is responsible for the removal of hemolymph proteins and small colloidal particles from the hemolymph (Crossley, 1985).

The rapid removal of JHE from the hemolymph by the pericardial cells was initially speculated to occur via receptor-mediated endocytosis, followed by transport to lysosomes for destruction (Bonning *et al.*, 1997). During the passage from the endosome to the target, JHE appears to interact with pericardial cell-specific proteins that cross-react immunologically with a universal antiserum directed toward heat shock protein70 (Shanmugavelu *et al.*, 2000). In addition to these heat shock proteins, a novel 29 kDa protein, termed P29, also appeared to bind JHE. This protein has been detected in fat body and pericardial cells of insects from all five larval stadia in *M. sexta*. The *P29* gene has been identified using a phage display library and has been sequenced (GenBank Account Number: AF153450). Liu *et al.* (2007) examined a P29 cognate in *D. melanogaster* (GenBank Account Number: NM_138145) and discovered that it is targeted to the mitochondria. The significance of this localization is unknown and, as noted by the authors, may be an artifact.

8.7.2. Juvenile Hormone Epoxide Hydrolase

The preponderance of literature on JH catabolism has focused on the role of JHE in metamorphosis, yet the less well-studied pathway (i.e., epoxide hydration via JH epoxide hydrolase; JHEH), may actually be more biologically relevant in some species (Gilbert *et al.*, 2000). For example, when JH I is injected into early fifth instar *M. sexta*, the major metabolite is not JH acid or JH acid diol, but a JH diol phosphate conjugate (Halarnkar *et al.*, 1993). Further evidence that JHEH plays a significant role in the reduction of JH titers at critical times can be found in the mosquito *Culex quinquefasciatus*, in which JHEH activity is two to four times higher than JHE activity throughout most of the life cycle (Lassiter *et al.*, 1994). Moreover, two peaks of JHEH are seen during the last stadium, suggesting that the enzyme is involved in JH metabolism during critical periods of development (Lassiter *et al.*, 1995). Even in *T. ni*, a species in which so many of the JHE studies have been performed, JH diol is a major metabolite or intermediate. It should be noted that JHE activity is still a factor in JH metabolism in this species (Kallapur *et al.*, 1996). Thus, it appears that the contribution of JHEH to the overall catabolism of JH varies extensively, even within the same order (Hammock, 1985).

8.7.2.1. Physical properties

The JHEHs (EC.3.3.2.3) belong to a large family of proteins that display the α/β hydroxylase fold (Debernard *et al.*, 1998) and are responsible for hydration of the JH epoxide to a diol. Since hydrolases are involved in a wide range of enzymatic reactions and utilize multiple substrates, it has been justifiably cautioned that JHEHs may not be as specific as they are portrayed (Harris *et al.*, 1999). Nevertheless, the term JHEH will be employed as it is commonly used in the literature.

A number of species have been catalogued as having JHEH activity but in only a few species have the JHEHs been well characterized (Hammock, 1985). Because a single species may have more than one form of JHEH (Harshman *et al.*, 1991; Keiser *et al.*, 2002), it has been suggested that the different JHEHs represent tissue-specific enzymes (Harshman *et al.*, 1991). Epoxide hydrolases (EHs) in general are present in both soluble (cytoplasmic) and insoluble (microsomal) forms, but JHEH is found associated mainly with the microsomal fraction (Wisniewski *et al.*, 1986a,b; Touhara and Prestwich, 1993; Wojtasek and Prestwich, 1996; Harris *et al.*, 1999; Keiser *et al.*, 2002). The enzyme appears to be a monomer, with a M_r ranging from 46 to 53 kDa (Harshman *et al.*, 1991; Touhara and Prestwich, 1993; Harris *et al.*, 1999; Keiser *et al.*, 2002; Zhang *et al.*, 2005). In general, the pH range for JHEH activity is broad, extending from approximately pH 5 to 9, and probably reflects the involvement of the reactive histidine (pKa 6.5 for the imidazole group) in the catalytic site (Debernard *et al.*, 1998). The JHEHs of *D. melanogaster* are heat- and organic solvent-tolerant, withstanding incubation at 55°C and concentrations of ethanol exceeding 40% (Harshman *et al.*, 1991). Moreover, the recombinant enzyme from *M. sexta* retains full activity in the presence of low levels of the reducing agent dithiothreitol and the sulfhydryl modifying reagent iodoacetamide (Debernard *et al.*, 1998).

8.7.2.2. Kinetic parameters, specificity, and the catalytic site of the juvenile hormone epoxide hydrolases
Determinations of the kinetic parameters of the JHEH from the microsomal fraction of *M. sexta* eggs revealed K_ms for JH I, II, and III of 0.61, 0.55, and 0.28 μM, respectively (Touhara and Prestwich, 1993). Interestingly, the V_{max}/K_m ratio for JH III, the least abundant JH homologue in eggs (Bergot *et al.*, 1981a), was 25 times greater than that for JH I, suggesting that the enzyme displays considerably more specificity for JH III than the higher homologues (Touhara and Prestwich, 1993). This observation was confirmed by Debernard *et al.* (1998), using recombinant JHEH from *M. sexta*. These investigators observed that while the enzyme could hydrolyze JH I, II, and several synthetic substrates, JH III was the most favored substrate. JH III is also a better substrate for JHEH than JH acid (Touhara and Prestwich, 1993; Kallapur *et al.*, 1996). In *D. melanogaster*, JH III is a more suitable substrate for JHEH than JHB$_3$ (Casas *et al.*, 1991). By contrast, Zhang *et al.* (2005) found that recombinant JHEH from *B. mori* hydrolyzed JH I better than JH II and III. However, as noted by Harris *et al.* (1999), more direct evidence, such as a correlation between inhibition of the enzyme *in vivo* and a decrease in JH metabolism, is required before assuming the enzyme is

JH-directed. A cautionary example is seen in *A. mellifera*, where a putative JHEH has negligible participation in JH degradation (Mackert *et al.*, 2010).

There is remarkable sequence conservation of the active-site residues among both insect and mammalian epoxide hydrolases. All JHEHs, like their mammalian counterparts, contain a tryptophan residue at positions 150–155, forming part of the oxyanion hole that may stabilize the hydroxyl-alkyl intermediate (Lacourciere and Armstrong, 1993; Wojtasek and Prestwich, 1996; Keiser *et al.*, 2002; Zhang *et al.*, 2005). The high degree of conservation in the catalytic site suggests that the mechanism of epoxide hydration in insects is similar to that found in mammals (Wojtasek and Prestwich, 1996; Roe *et al.*, 1996; Debernard *et al.*, 1998; Zhang *et al.*, 2005). Using the JHEH of *M. sexta* as a model, Debernard *et al.* (1998) proposed a two-step catalytic process. The first step involves a nucleophilic attack on the epoxide at the least hindered position, C10, by Asp[227]. This, in turn, opens the ring, leading to the formation of a hydroxyl-alkyl intermediate covalently attached to the enzyme. The neighboring Trp[228] is thought to activate the epoxide for a nucleophilic attack (Harris *et al.*, 1999). There is exceptional conservation of this residue in all the insect species examined to date. The second step involves the hydrolysis of this covalent intermediate by a water molecule that is activated by His[428] and Asp[350]. The histidine residue activates water, with the aspartate or glutamate residue acting as the proton scavenger from the histidine to reactivate the enzyme. The resulting product is (10*S*,11*S*) JH diol. Linderman *et al.* (2000) reported a variation on this putative reaction, in which Glu[403] instead of Asp[350] acts as the charge relay partner.

8.7.2.3. Inhibitors
In contrast to the early discoveries of highly effective JHE inhibitors, the search for JHEH inhibitors has been less productive (Hammock, 1985; Casas *et al.*, 1991; Harshman *et al.*, 1991; Roe *et al.*, 1996, 2005). Investigations have taken one of two directions, examining either compounds that mimic the JH backbone (Roe *et al.*, 1996; Linderman *et al.*, 2000) or compounds based on urea and amide pharmacophores that are not subject to metabolism through epoxide degradation (Severson *et al.*, 2002).

Roe *et al.* (1996) demonstrated that methyl 10,11-epoxy-11-methyldodecanoate (MEMD), a long chain aliphatic epoxide, displayed an I_{50} (molar concentration needed to inhibit 50% of the enzyme's activity) in the low nanomolar range. A promising group of MEMD analogs was investigated by Linderman *et al.* (2000), who examined the structure–activity relationship of MEMD epoxide substitution and enantioselectivity. Two classes of MEMD analogs were synthesized: a glycidol-ester series and an epoxy-ester series. As a group, the glycidol-esters were more potent inhibitors than the corresponding epoxy-esters by an order of magnitude. The inhibitory activity in both classes was dependent upon the absolute configuration of the epoxide at C10, with the R configuration displaying the higher degree of inhibition. The inhibitory activity of the most potent compound of the series ($I_{50} = 1.2 \times 10^{-8}$ M) is thought to be due to the hydroxyl group in the active site, which forms an additional hydrogen bond. This bond may stabilize the enzyme-inhibitor complex by inducing a conformational change, or it could reduce the rate at which the diol product dissociates from the enzyme's active site.

The other class of JHEH inhibitors consists of analogs of the urea and amide pharmacophores that are potent inhibitors of mammalian soluble and microsomal epoxide hydrolases (Severson et al., 2002). To date, none of the nearly 60 compounds tested are as active as the glycidol-esters in the inhibition of recombinant M. sexta JHEH — a surprising result, given their action on mammalian enzymes. The most potent of the series, N-[(Z)-9-octadecenyl]-N'-propyl urea (NOPU), has an $I_{50} = 8.0 \times 10^{-8}$ M. Severson et al. (2002) suggested that when the inhibitor enters the catalytic site, the carbonyl group interacts with two tyrosine residues in the oxyanion hole. Since the inhibitor lacks an epoxide, it does not covalently bind to the reactive Asp^{227} residue, but it does block the catalytic site from further activity.

8.7.2.4. Genomic structures At the present time, the genomic structure of only one insect JHEH, D. melanogaster, has been well documented. The gene cluster, approximately 8.6 kb long, is located on chromosome 2 in region 55F7-8, near the JHE locus at 52F1. As suggested by Harshman et al. (1991) and confirmed through genomic studies (FlyBase), there are multiple forms of JHEH in D. melanogaster. Annotation of the JHEH locus identified three putative genes that display approximately 37% identity in their amino acid sequence. The genomic region containing JHEH1 (GenBank Account Number: NM_137541; CG15101) is approximately 1.8 kb long and contains four exons. Approximately 0.7 kb downstream from JHEH1 lies JHEH2 (GenBank Account Number: NM_137542; CG15102), which is 2.2 kb in length and has two potential transcripts. One transcript has four exons (GenBank Account Number: NM_137524; CG15102) while the other has three (GenBank Account Number: NM_176233; CG15102). These discrepancies in the annotation of JHEH2 will require future resolution. JHEH3 (GenBank Account Number: NM_137543; CG15106) is composed of three exons. JHEH1 and JHEH2 are more related to each other than to JHEH3, yet all three putative genes display the residues of the catalytic site in the correct sequence, and all are of about the same length. Computational analysis using CLUSTALW indicates that JHEH1 and JHEH3 are most divergent in their first 70 residues, which might be

expected if the N-terminal is needed for attachment to cell membranes. Since JHEH activity has been demonstrated in other cell fractions besides microsomes, it may be that the first exons of JHEH1 and JHEH2 code for sequences that target these enzymes to different locations within the cell (Casas et al., 1991; Harshman et al., 1991).

Recent evidence suggests that the N-terminal of EHs may play another role. The soluble human EH, EPXH2, also possesses phosphatase activity, so the enzyme cannot only transform epoxy fatty acids to their corresponding diols but also dephosphorylate dihydroxy lipid phosphates (Newman et al., 2003, 2005). The phosphatase activity localized to the N-terminal domain is unaffected by a number of classic phosphatase inhibitors. Alignment of EPXH2 with the JHEH of several insects shows less than 25% sequence similarity under non-stringent conditions, and regions of similarity are limited to the C-terminal domain. Nevertheless, an insect JHEH with phosphatase activity presents an interesting twist in the metabolic pathway for JH catabolism.

8.7.3. Secondary Metabolism of Juvenile Hormone: Juvenile Hormone Diol Kinase

Investigations of secondary metabolism of JH (Roe and Venkatesh, 1990; Halarnkar et al., 1993; Grieneisen et al., 1995) may have been misleading, since the enzymes used may have contained multiple hydrolytic activities (Halarnkar et al., 1993). In only one instance has an actual JH conjugate been unequivocally identified. The conjugate (10S,11S) JH diol phosphate (**Figure 1**) is the product of a two-step enzymatic process: conversion of JH to JH diol and then addition of a phosphate group to C10 (Halarnkar et al., 1993). The enzyme responsible for the phosphorylation of JH diol is JH diol kinase (JHDK), which was first characterized from the Malpighian tubules of early fifth instars of M. sexta (Grieneisen et al., 1995; Maxwell et al., 2002a,b). JHDK (EC 2.1.7.3) was discovered when an analysis of JH I metabolites in vivo yielded, in addition to the expected metabolites, a very polar JH I conjugate that was subsequently identified as JH I diol phosphate (Halarnkar and Schooley, 1990).

8.7.3.1. Physical properties JHDK from M. sexta Malpighian tubules is a cytosolic protein composed of two identical subunits of 20 kDa, as determined by MS (Maxwell et al., 2002a). Gel filtration studies indicate it has a molecular mass of approximately 43 kDa. JHDK displays a K_m in the nanomolar range for JH I diol, which is appropriate for an enzyme responsible for clearance of a hormone whose titers rarely exceeds 10 nM. Most significantly, the catalytic activity of JHDK parallels developmentally that of JHEH, a requisite if JH diol phosphate is a legitimate terminal metabolite. Analysis of the k_{cat}/K_m ratio for the diols of JH I, II, and III indicates that JH I diol is the preferred substrate, suggesting a preference for an ethyl group at the C7 position. JHDK requires both Mg^{2+} and ATP for activity (Grieneisen *et al.*, 1995; Maxwell *et al.*, 2002a), although excess Mg^{2+} and Ca^{2+} inhibit its activity (Maxwell *et al.*, 2002a).

The specificity of JHDK for JH I diol is relatively high, considering the multitude of potential phosphate acceptor groups present in a cell. The enzyme does not recognize methyl geranoate diol (one isoprenyl unit shorter than JH) nor methyl geranylgeranoate diol (one isoprenyl group longer than JH), yet it does recognize JH I ethyl ester diol. It also recognizes both JH diol enantiomers, indicating that the absolute stereospecificity of the hydroxyl groups is of minor importance. Most surprising is the enzyme's inability to recognize JH acid diols. Because JH acid diol cannot be phosphorylated by JHDK, the generally accepted pathway for JH catabolism (JH acid is converted to JH acid diol) must be reconsidered. However, JH acid diol could undergo further catabolism by cytochrome p450s, such as the one found in the CA and midgut of the cockroach *D. punctata* (Sutherland *et al.*, 1998). Still, the role of cellular JHE becomes problematic if the pathway catalyzed by JHEH and JHDK is the major pathway for JH catabolism in the cell. The fact that JH diol phosphate is a significant metabolite (Halarnkar *et al.*, 1993) certainly weakens the long-held dogma about JH catabolism.

8.7.3.2. Genomic structures

The sequence and hypothetical structures of *M. sexta*, *D. melanogaster* (Maxwell *et al.*, 2002b), and *B. mori* JHDK (Li *et al.*, 2005) have been analyzed. A partial characterization of JHDK from whole-body homogenates of *D. melanogaster* indicates that it is similar to the enzyme in *M. sexta*, with the exception of its subunit structure. The active *D. melanogaster* JHDK is a monomer of ~20 kDa, while the active *M. sexta* (GenBank Account Number: AJ430670) and *B. mori* JHDKs (GenBank Account Number: AY363308) are composed of two identical 20 kDa subunits. Similarities in chromatographic properties, isoelectric point, and enzyme activity led Maxwell *et al.* (2002b) to conclude that sarcoplasmic calcium-binding protein 2 (dSCP2) is the probable *D. melanogaster* homologue of *M. sexta* JHDK. The *M. sexta* gene codes for an enzyme that has 59% sequence identity and >80% similarity to dSCP2 of *D. melanogaster* (GenBank Account Number: AF093240; CG14904). Li *et al.* (2005) reported that the *B. mori JHDK* is composed of a single exon of 637 bp. The *B. mori* JHDK is expressed most prevalently in the gut, as determined by Northern blot analyses, and is not under the direct control of JH at the transcriptional level.

Maxwell *et al.* (2002b) generated a 3D model that they used for *in silico* docking simulations. They capitalized on the facts that the catalytic site of JHDK must contain a purine (GTP) binding site and hydrophobic pocket for JH diol, and that the scaffolding for dSCP2 is known. Surrounding the putative substrate-binding site, both the *M. sexta* and *D. melanogaster* JHDKs contain the three conserved nucleotide-binding elements common to nucleotide binding proteins. The model further demonstrates that the protein contains four domains that form two pairs of a helix-loop-helix motif (EF-hand; Branden and Tooze, 1999). Charge interactions in the hydrophobic binding pocket, as well as its depth (19 Å), are complementary to the extended conformation of the diol. Moreover, the hydrophobic nature of the binding pocket complements the C1 ester of the substrate and supports the observation that JH diol is the only substrate for this enzyme (Maxwell *et al.*, 2002a).

8.7.4. Catabolism and Caveats

The field of JH catabolism is changing, with the application of sophisticated analytical tools and the use of *in vivo* metabolism studies to uncover potential catabolic pathways. Moreover, details about the structure of the enzymes via NMR and crystallographic analyses have added considerable insight into the molecular action of these molecules. The commonly used JHE phase-separation assay can be misleading, because it yields polar metabolites resulting from enzymatic activity other than that generated by JHE. The JHE phase-separation assay should be employed only as the first step. It should then be followed by the use of advanced chromatographic tools and detection systems. In addition, it is critical that the correct controls be employed. This point is driven home by the fact that JHEs from several species can, under the appropriate conditions, transesterify JH (Debernard *et al.*, 1995; Grieneisen *et al.*, 1997). The surprising discovery that JH diol phosphate conjugates are major JH catabolites underscores the fact that there is still much to learn about JH metabolism (Halarnkar *et al.*, 1993; Gilbert *et al.*, 2000). The need to demonstrate biological relevance applies not only to the newly discovered JH-like molecules, but also to the enzymes involved in JH catabolism. It may be that some of the enzymes are really not JH-directed, but will, under experimental conditions, generate metabolites not seen under physiological conditions.

Analysis of JH catabolism is further complicated when *in vitro* studies do not reflect physiological conditions. Because of their high affinity and homologue selectivity, the JHBPs have a significant influence on JH catabolism (Hammock *et al.*, 1975; Halarnkar *et al.*, 1993). Studies have overlooked the fact that multiple homologues may be present and have focused on the metabolism of just a single homologue. Moreover, a lack of attention to enantiomeric specificity of catabolic enzymes may skew results (Peter, 1990). The use of chromatographic procedures to separate racemic preparations of JH should alleviate this problem (Cusson *et al.*, 1997).

Finally, in the drive to discover a unifying theory to explain JH catabolism, the diversity of the class Insecta is often overlooked. To assume that all insect species use (a) common pathway(s) for JH catabolism would be a gross oversimplification. One need only compare JH metabolism in the flea *Chlamydophila. felis,* which utilizes JHE (Keiser *et al.*, 2002), the mosquito *C. quinquefasciatus* and *D. melanogaster*, which utilize JHEH (Casas *et al.*, 1991; Lassiter *et al.*, 1994), and the moth *T. ni,* which utilizes both (Kallapur *et al.*, 1996), to understand that multiple pathways for JH catabolism have evolved in this diverse class.

8.8. Juvenile Hormones in Pre-metamorphic Development

8.8.1. Juvenile Hormone Titers and Potential Problems

The pre-metamorphic titers of JH appear to vary greatly among the insect orders, with some orders, such as Blattodea, displaying levels 100-fold greater than Lepidoptera (Gilbert *et al.*, 2000). Yet one pattern remains constant: JH titers are high, on average, while the larva is growing and feeding but drop at a well-defined point to permit metamorphosis. That pattern can be further refined to distinguish between hemimetabolous and holometabolous insects (Riddiford, 1994). In hemimetabolous insects, JH titers are low to undetectable during the final stadium (Treiblmayr *et al.*, 2006). In holometabolous insects, JH titers are relatively high at the beginning of the final larval stadium but decline to undetectable levels prior to the cessation of feeding. In contrast to the hemimetabolous insects, a second increase in the hemolymph JH titer is observed after the insect has found a suitable pupation site. This peak in the titer is thought to prevent precocious adult differentiation of imaginal discs and other imaginal precursors (Riddiford, 1994).

One of the major impediments to better understanding the role(s) of JH is the lack of precise titers during biologically relevant periods. Virtually all hormones studied in vertebrates display daily oscillations in their titer and release and are in many cases linked to critical homeostatic events (Nader *et al.*, 2010). Early studies demonstrated a daily bimodal fluctuation in the size of the nuclei in CA of adult *D. melanogaster* (Rensing, 1964). While size of the CA cells does not necessarily indicate whether a gland is active in JH biosynthesis (Tobe and Stay, 1985), there is ample evidence that enlarged glands are biosynthetically active to suggest that fluctuations in JH synthesis may be circadian (Chiang *et al.*, 1989, 1991).

The importance of examining titer fluctuations in detail can be seen in work from Steel's lab, which demonstrates a circadian rhythmicity of ecdysteroid titers in nymphal *R. prolixus* (Steel and Ampleford, 1984; Vafopoulou and Steel, 2001). While the work is beyond the scope of this review, in adult *Gryllus firmus* JH titers rise 10- to 20-fold during the photophase of the flight-capable long-wing morph (Zhao and Zera, 2004; Zera and Zhao, 2009). A single-point JH determination every 24 h may not be a serious problem for the student of metamorphosis where JH levels are constant over several days. However, infrequent sampling does make a difference when relatively rapid changes in JH titer regulate critical cellular events. Recent studies demonstrating that JH titers fluctuate rapidly in adult insects underscore the need for detailed and precisely timed studies of JH titers during the premetamorphic stages. Elekonich *et al.* (2001) found that honeybee foragers show a diurnal increase in hemolymph JH titers, which rise from approximately 100 ng/ml in the late morning to >350 ng/ml by late evening, a period spanning less than 12 h. A similar change is seen in the fourth stadium of *M. sexta*. The initial rise and then significant drop of JH titers over an 8 h period may have some influence on levels of its hJHBP (Fain and Riddiford, 1975; Orth *et al.*, 1999).

Obtaining meaningful JH titers can be problematic. Precise synchronization of development cannot be presumed even in laboratory-reared populations. Even if precise staging is possible, the measurement of JH titers in species or stages displaying low levels of hormone, or from smaller-sized species, requires tissue pooled from a number of individuals. Valuable information about population variability is inevitably lost. Ideally, a single individual should be sampled sequentially over time, but this process leads to wound-induced effects on JH titers (Caveney, 1970). In our hands, wounding has rapid and significant effects on hJHBP levels, which potentially modulate JH titers (Tauchman *et al.*, 2007). Infections with sublethal doses of pathogenic bacteria have been shown to regulate genes that are involved in JH biosynthesis and catabolism, such that JH titers are increased (Huang *et al.*, 2009). Even moderate handling prior to sample collection may have an effect on JH titers (Varjas *et al.*, 1992).

8.8.2. Pre-metamorphic Roles

The roles of JH in the stages preceding metamorphosis are ubiquitous, affecting behavior, organs and tissues, cellular organelles, and biochemical pathways. Several excellent reviews have covered these areas (Nijhout, 1994; Riddiford, 1994, 1996), and the focus here is on research since their publication.

8.8.2.1. Behavioral and neuronal responses There are a number of behavioral phenomena ascribed to JH during the adult stage, including pheromone production and calling (see Cusson *et al.*, 1994; Tillman *et al.*, 2004; Grozinger and Robinson, 2007, for reviews), aggression and display (Emlen *et al.*, 2006; Scott, 2006), migration (see Dingle and Winchell, 1997; Min *et al.*, 2004; Zhu *et al.*, 2009), phonotaxis (Stout *et al.*, 1992, 1998; Bronsert *et al.*, 2003), caste determination (Cristino *et al.*, 2006), and neuronal remodeling (see Robinson and Vargo, 1997; Elekonich and Robinson, 2000; Spiess and Rose, 2004; Jeffery *et al.*, 2005; Verma, 2007; Hewes, 2008, for reviews). However, far less is known about the effect of JH on behavior during the pre-metamorphic period.

In many of the studies that have been conducted on larval insects, exogenous JH is applied and then behavior monitored in the adult stage. Late in the third stadium, *A. mellifera* larvae destined to become queens have JH titers five times higher than larvae destined to become workers (Rachinsky and Hartfelder, 1991; Rembold *et al.*, 1992). When worker-destined larvae were treated with JH, there appeared to be no influence on adult behavior; however, when larvae destined to be queens were treated, time to adult emergence was shortened (Elekonich *et al.*, 2003). In another hymenopteran, the ant *Phiedole bicarinata*, large doses of methoprene, if applied during a critical period in the last stadium, induced worker-destined larvae to become soldiers (Wheeler and Nijhout, 1981). Higher JH titers during the last stadium of certain migratory species appear to elicit a stationary, rather than migratory, adult stage (Yagi and Kuramochi, 1976; Nijhout and Wheeler, 1982). Yin *et al.* (1987) demonstrated that methoprene can affect the circadian rhythm in the lepidopteran *Diatraea grandiosella*, inducing a dose-dependent phase shift in adult eclosion.

At the neuronal level, most endocrine studies have focused on the effect of ecdysteroids on the neural circuitry during metamorphosis, although JH appears to play a role as well. Truman and Reiss (1988) and Williams and Truman (2005) demonstrated that reorganization of neurons during metamorphosis of *M. sexta* is, in part, under the control of JH, but the regulatory elements involved are still undefined. JH has been demonstrated to affect larval neurons innervating the prothoracic gland of the cockroach *Periplaneta americana* (Richter and Gronert, 1999). Exposure of the insect to exogenous JH III or methoprene both *in vivo* and *in vitro* induces a short-term depression of spike activity in neurons innervating the prothoracic gland, but has no effect on the nervus connectivus of the stomatogastric nervous system. This response to JH III and methoprene is very rapid, occurring within three minutes of treatment and reaching maximum spike depression (75% reduction) within 15 minutes. Curiously, fenoxycarb, a potent JH analog, was only half as active as JH III and methoprene. The authors speculate that JH is acting directly on membrane receptors to reduce neurotropic activity. This observation relates to an earlier hypothesis that JH can influence inhibitory neurotransmitter receptors such as GABA (Stout *et al.*, 1992). Unfortunately, no further work on this novel research has been reported.

8.8.2.2. Epidermal responses It has long been known that JH has a marked influence on epidermal and cuticular structure. The early work of Williams (1952) demonstrated that extracts containing JH, when applied at a developmentally sensitive time during the pupal period, can induce formation of a second pupal cuticle. This discovery prompted Williams to suggest that JH is a "status quo" hormone, a label that is still in use (Riddiford *et al.*, 2010).

The integument of the insect is composed of an outer layer of secreted proteins, lipids, pigments, and complex carbohydrates termed the cuticle. The single layer of cells immediately beneath the cuticle, the epidermis, is responsible for synthesis and secretion of most, but not all, of the proteins found in the cuticle (Willis, 1996). The epidermis contains another subset of proteins, pigment-associating proteins, whose expression and position in the cells appear to be regulated by JH. The radical changes in morphology at metamorphosis led Wigglesworth (1959) to predict that each stage of development expressed its own set of unique genes. While there does appear to be a limited group of stage-specific cuticular proteins, especially in the higher flies, many cuticular proteins can be found in more than one stage of the life cycle (Willis, 1996; see Willis and Muthukrishnan, 2010, for a review).

Hormonal regulation of epidermal gene expression has been examined in a number of insect species, but the most extensive studies have been carried out on the epidermal genes of *M. sexta* (Riddiford, 1994, 1996; Riddiford *et al.*, 2003). Charles (2010) and Hiruma and Riddiford (2010) have recently reviewed research on the genes under the control of JH. The best-studied of these include the gene for larval cuticular protein 14 (*LCP14*; Rebers and Riddiford, 1988; GenBank Account Number: 813279), *LCP14.6* (Rebers *et al.*, 1997; GenBank Account Number: U65902), *LCP16/17* (Horodyski and Riddiford, 1989; GenBank Account Number: M25486), the gene for the biliverdin-associating proteins insecticyanin a and b (*INS*; Li and Riddiford, 1994; GenBank Account Numbers: 864714 and 864715), the gene for dopa decarboxylase (*DDC*; Hiruma and Riddiford, 1988; Hiruma *et al.*, 1995; GenBank Account Number: U03909), and *JP29* (Shinoda *et al.*, 1997; GenBank Account Number: U05270). Studies using the epidermis are particularly compelling because the results of *in vitro* manipulations mirror those observed *in vivo*, and because the genomic structure and flanking regions of the genes in question have been determined.

LCP14 is a larval-specific gene that encodes a 14 kDa protein expressed during the feeding period of each stadium (Riddiford, 1994). At the onset of a larval molt, mRNA for LCP14 rapidly becomes undetectable and remains so until the insect molts to the next stadium, at which time the mRNA levels rise again. In the last stadium, levels of *LCP14* message rise during the first several days, but fall sharply prior to wandering. *In vitro* manipulation indicates that 20E suppresses the expression of this gene. If JH is present when 20E titers rise, suppression by 20E is only transient; if JH is absent, the gene is permanently silenced. It is uncertain how JH acts at the molecular level to maintain the expression of *LCP14*, but its function may involve the transcription factor ßFTZ-F1. It has been shown that there are three potential binding sites for ßFTZ-F1 on the *LCP14* gene, one approximately 2 kb upstream from the translational start site and two in the first intron (Q. Lan, personal communication). The expression of *ßFTZ-F1* begins about 16 h before molting, peaks about 8 h later, and then ceases approximately 3 h before molting (Weller *et al.*, 2001). Functional analysis of the putative *LCP14* promoter indicates that the *ßFTZ-F1* response elements in the first intron may be involved in downregulating *LCP14* expression (Q. Lan, personal communication). The decline in *ßFTZ-F1* expression coincides with the rise in JH at the very beginning of the fifth stadium and, with it, the increase in *LCP14* expression. Whether JH acts directly on the gene or is involved in modulating the ecdysteroid effect remains unclear.

LCP14.6 is another hormonally controlled cuticular gene that, like *LCP14*, is downregulated by 20E. In contrast to *LCP14*, *LCP14.6* is suppressed *in vitro* by large doses of methoprene (Riddiford, 1986). Its expression is temporally and spatially complex and occurs in the larval, pupal, and adult stages. A comparison of its expression pattern with the profile of JH titers suggests that, *in vivo*, *LCP14.6* expression may be suppressed by JH. *LCP16/17* encodes a multigene family of three proteins that appear midway through the feeding period of the fifth stadium (Horodyski and Riddiford, 1989). The developmental appearance of these proteins correlates with a thinning of cuticular lamellae and a corresponding increase in stiffness. Expression of *LCP16/17*, like *LCP14.6*, is suppressed by large doses of methoprene.

One of the more striking effects elicited by JH is its action on pigmentation of the immature insect (Applebaum *et al.*, 1997; Suzuki and Nijhout, 2006, 2008; Pener and Simpson, 2009; Nijhout, 2010). Nowhere is this more evident than in larval *M. sexta*. Mutants have been discovered that are more highly melanized than the wild type (*black* or *bl* mutant; Safranek and Riddiford, 1975) or are nearly white (white mutant; Panchapakesan *et al.*, 1994). To maintain the normal wild-type phenotype following a molt, that is, a transparent cuticle and an epidermis with intracellular vesicles containing the biliverdin-associating protein insecticyanin (Goodman *et al.*, 1985), JH titers must be appropriate at the time of head capsule slippage (Hiruma and Riddiford, 1988; Riddiford, 1994). If JH titers are too high at the time of head capsule slippage, the resulting cuticle will be transparent and the underlying epidermal cells will lack insecticyanin vesicles, causing the insect to appear white (Panchapakesan *et al.*, 1994).

Conversely, if JH titers are too low at head capsule slippage, the resulting cuticle will be highly melanized and the epidermis devoid of insecticyanin vesicles, causing the insect to appear black (Goodman *et al.*, 1987). It should be noted that not all color variants may be linked to deviations in JH titer (Bear *et al.*, 2010).

Insecticyanin (INS) is a 21 kDa protein that associates with biliverdin IXγ to yield an intensely blue protein essential to larval camouflage (Goodman *et al.*, 1985). INS is found in epidermal cells sequestered in 1 μm vesicles and as a soluble protein in the hemolymph. At commitment to pupation, when JH is no longer present and ecdysteroid levels rise, the entire population of INS-containing vesicles is secreted from the epidermal cell (Sedlak *et al.*, 1983) and the epidermis ceases production of the protein (Goodman *et al.*, 1987; Riddiford *et al.*, 1990). Interestingly, the *bl* mutant of *M. sexta*, which has a low JH titer at head capsule slippage, lacks the INS-containing vesicles. Topical application of JH to *bl* mutants just prior to head capsule slippage will induce the appearance of INS vesicles in the next stadium, similar in number and position to those in the wild type (Goodman *et al.*, 1987; Goodman and Granger, 2005). Moreover, JH applied at head capsule slippage prevents cuticular melanization in the next stadium (Safranek and Riddiford, 1975). The role of JH in this process is still unclear, but ultrastructural data suggest that the cytoskeleton is involved in retaining INS vesicles in the apical portion of the epidermal cell (W. Goodman, unpublished observations).

JH regulation of *INS* expression is complex due to the fact that the gene is expressed in more than one tissue. In addition to its presence in the epidermis of wild-type larvae, *INS* mRNA has been found in the fat body (Li and Riddiford, 1994; Li, 1996). JH appears to upregulate epidermal *INS* mRNA abundance, while it downregulates *INS* mRNA in the fat body (Li, 1996). As in *LCP14* expression, 20E downregulates epidermal *INS* mRNA during the molting period (Riddiford *et al.*, 1990).

DDC is the enzyme responsible for catalyzing the conversion of dopa to dopamine (Hiruma and Riddiford, 1984, 2009; Arakane *et al.*, 2009) while granular phenoloxidase (PO) catalyzes the oxidative dehydrogenation of diphenols to quinones (Hopkins and Kramer, 1992). For cuticular melanization to occur, vesicles containing dopamine, the granular proenzyme form of PO, and other enzymes necessary for melanin synthesis must be secreted into the new endocuticle following head capsule slippage. If JH titers are low or absent at the time of head capsule slippage, the proenzyme form of PO will be synthesized by the epidermis and deposited in the endocuticle for a period of 12–14 h. Approximately 3 h before ecdysis, PO is activated and melanin begins to appear in the new cuticle (Jiang *et al.*, 2003). Topical application of JH to larvae that have been neck-ligated at head capsule slippage will block phenoloxidase activity and a new transparent cuticle will be formed (Riddiford *et al.*, 2003). While JH suppresses cuticular melanization in a striking fashion, it remains unclear whether the hormone is acting on transcription or translational events that regulate PO activity.

Hormonal regulation of epidermal DDC (EC 4.1.1.28) in *D. melanogaster* and *M. sexta* has been extensively studied by the Hodgett (see Chen *et al.*, 2002a,b, for reviews) and Riddiford laboratories (Riddiford, 1994; Riddiford *et al.*, 2003). In *M. sexta*, 20E works directly on epidermis to block DDC synthesis; however, as the ecdysteroid titers decline during the head capsule slippage period, levels of DDC begin to rise and peak near the time of the larval molt. While much is known about *DDC* and its regulation via ecdysteroids, the role of JH in its control remains unclear (Hiruma *et al.*, 1995; Chen *et al.*, 2002a, b; Riddiford *et al.*, 2003). It has been demonstrated that the epidermis of allatectomized *M. sexta* larvae displays approximately 50% more DDC activity than epidermis from sham-operated insects, and that activity of the enzyme can be suppressed by moderate levels of JH I. While this observation is certainly provocative, it may well be that JH has no direct regulatory action on *DDC* expression (K. Hiruma, personal communication).

JP29 is an abundant epidermal protein whose presence is developmentally regulated by JH (Palli *et al.*, 1990, 1991). Although it was originally thought to be a JH receptor, its affinity for JH is too low for it to be physiologically important as a receptor (Charles *et al.*, 1996). It has been suggested that due to its relatively high abundance in the epidermis, it may act as a hydrophobic sink for JH: The higher number of low-affinity binding sites afforded by JP29 can compete successfully with the lower number of high-affinity binding sites provided by hJHBP.

8.8.2.3. Fat body responses

The pre-metamorphic fat body is responsible for metabolism of nutrients, synthesis of most hemolymph proteins, and detoxification of xenobiotics (Locke, 1984; Haunerland and Shirk, 1995; Aguila *et al.*, 2007). Given the central roles this tissue plays, it is not surprising that an extensive body of literature has focused on the role of JH in regulating cellular and molecular events occurring in the fat body.

Numerous studies involving hormonal regulation of ultrastructural changes in adult fat body, especially those regarding vitellogenesis, have been conducted, but few have focused on the ultrastructural changes induced by JH in the larval fat body. Such experiments are problematic: JH titers are already relatively high during most of larval life, and challenging the fat body with additional JH may result in pharmacological responses that mask the actual underlying events. Thus, the most comprehensive studies are developmental, correlating JH titers with ultrastructural changes occurring during the pre-metamorphic and metamorphic periods (Locke, 1984; Dean *et al.*, 1985). During the last stadium of both hemimetabolous

and holometabolous insects, when JH titers are dropping, the fat body synthesizes and secretes several hexameric proteins collectively termed storage proteins (Levenbook, 1985). These proteins are retrieved from the hemolymph by the fat body and stored in relatively large vesicles for later use during the pupal–adult transformation (Locke, 1984; Dean *et al.*, 1985).

It has been shown that a JH agonist affects fat body nuclei. Treatment of nymphal *L. migratoria* with high levels of methoprene leads to an increase in the size of fat body nuclei, which could be linked to either increased transcription rates or to increased ploidy levels (Cotton and Anstee, 1991). The latter option appears more probable, since Jensen and Brasch (1985) demonstrated that allatectomy of adult *L. migratoria* prevents polyploidization while methoprene restores the process.

The role of JH in fat body gene expression has been studied by a number of investigators in several different insect species. Certain gene targets are discussed elsewhere in this series and a comprehensive discussion of the various fat body genes thought to be under the control of JH has been presented by Riddiford (1994; Hiruma and Riddiford, 2010). One of the better studied groups of pre-metamorphic fat body genes codes for the hexamerins (Burmester *et al.*, 1998; Burmester, 2002). The insect hexamerins, which belong to the superfamily of hexamerins, encompass at least five different subgroups based on amino acid sequence. They include the coleopteran and dictyopteran JH-suppressible arylphorins, the lepidopteran JH-suppressible hexamerins, the lepidopteran aromatic amino acid-rich arylphorins, the lepidopteran methionine-rich hexamerins, and the dipteran arylphorins (Beintema *et al.*, 1994). The products encoded by the hexamerin genes are expressed predominantly during late larval life, and thus appear to be negatively regulated by JH.

Another group of proteins in the hexamerin superfamily are the hemolymph cyanoproteins (Miura *et al.*, 1998) and the orthopteran hemolymph JH binding proteins (Koopmanschap and deKort, 1988; Braun and Wyatt, 1996). As first demonstrated by Jones *et al.* (1988, 1990a), the expression of certain fat body genes is downregulated by JH and JH analogs; however, exceedingly high doses of JH analogs were used to elicit the responses. More recently, Hwang *et al.* (2001) and Cheon *et al.* (2002), using the fall webworm, *Hyphantria cunea*, have isolated two genes encoding hexamerins that are JH-suppressible at low doses of JH analog. Expression of these genes is not dose-dependent, but can be downregulated within a 6h period, suggesting that JH is acting directly on transcription. The major hemolymph protein from the fat body of *B. mori*, the "30K protein," also appears to be negatively regulated by JH (Ogawa *et al.*, 2005). Surgical extirpation of the CA led to a rapid accumulation of 30K protein in the hemolymph of fourth instars and was repressed in primary cultured fat body cells treated with methoprene. The general pattern in Lepidoptera appears to be that JH represses the high-abundance hexamerins during the early larval stages, but as the insect enters the last larval stadium, when JH titers drop, the titers of hexamerins increase. By contrast, expression of the hexamerin genes *hex 70b* and *hex 70c* in *A. mellifera* larvae appears to be under positive JH control (Cunha *et al.*, 2005; Martins *et al.*, 2010). Exogenous treatment with JH agonists prevents the normal decline in expression of the hexamerin genes.

Technical difficulties in working with the fat body have limited the progress in exploring its response to JH. Its cellular heterogeneity, massive tracheal penetration, and apparent sensitivity to wounding and to culture *in vitro* make experimental manipulation of the pre-metamorphic fat body problematic. Moreover, many of the studies lack a dose-response curve, relying on data based on a single dose of hormone or analog. As demonstrated by Orth *et al.* (1999) in their study on JH regulation of *hJHBP* expression, the effect of increasing doses of JH on the fat body is stimulatory only within a limited range; after peak stimulation is reached, further increases in dose result in a negative response. Furthermore, microgram quantities of JH can form a monolayer at the surface of the medium, hindering gaseous exchange and compromising the physiology of the fat body in culture. A more physiological approach to hormone treatment, coupled with technological advances in molecular analysis, should ultimately reveal how JH functions with respect to the fat body.

8.8.2.4. Responses in other tissues

There are several other tissues that appear to be the target of JH action, including larval muscles and the prothoracic glands (see Goodman and Granger, 2005; Sakurai, 2005, for reviews). JH involvement in the regulation of muscle development and fate during metamorphosis is well documented by an impressive body of work from a number of laboratories (Schwartz, 1992; Riddiford, 1994; Hegstrom *et al.*, 1998; Roy and Vijayraghavan, 1999; Buszczak and Segraves, 2000; Cascone and Schwartz, 2001; Lee *et al.*, 2002; Rose, 2004; Mamatha *et al.*, 2008).

It has long been known that the prothoracic glands, the site of synthesis of ecdysteroids, are in turn regulated by ecdysteroids via short positive and negative feedback loops (Williams, 1952; Sakurai and Williams, 1989; Marchal *et al.*, 2010). Given the close relationship between JH and ecdysteroids in regulation of metamorphosis, the role of JH on ecdysteroid biosynthesis has been extensively studied. That interaction is now known to include the degeneration of the gland at the nymphal–adult molt in the roach *Blattella germanica* via 20E-regulated transcription factors (Mane-Padros *et al.*, 2010). The degeneration of the gland can be prevented by the topical application of a JH agonist. A review of the older literature can be found in Goodman and Granger (2005).

Recent studies from the Palli lab have focused on the metamorphic restructuring of the midgut in several insects, including *Ae. aegypti* (Wu *et al.*, 2006) and *H. virescens* (Parthasarathy and Palli, 2007). Using nuclear staining technology and analysis of mRNA expression levels in genes involved with in 20E action, programmed cell death of *Ae. aegypti* larval midgut cells was examined. Quantitative PCR analyses indicated that methoprene blocked midgut remodeling by modulating expression of *EcRB*, *ultraspiracle*, *Broad complex*, *E93*, *ftz-f1*, *dronc*, and *drice* genes, which are critical to 20E action and programmed cell death. In *H. virescens*, several genes involved in remodeling the midgut were examined, including *caspase-1*, *caspase ICE*, and *inhibitor of apoptosis* (IAP). Application of high doses of methoprene blocked programmed cell death by maintaining high levels of IAP expression while downregulating the caspases, markers for programmed cell death.

8.9. Molecular Mode of Action of the Juvenile Hormones

Even as the chemical structure of JH I was being reported by Röller *et al.* (1967), one of the first of many reviews on JH action at the molecular level was going to press (Kroeger, 1968). Since that time, there have been a number of major reviews focusing on the topic (Riddiford, 1996, 2008; Gade *et al.*, 1997; Gilbert *et al.*, 2000; Lafont, 2000; Wheeler and Nijhout, 2003; Wilson, 2004; Berger and Dubrovsky, 2005; De Loof, 2008). Despite the number of words written about the subject, the molecular actions of JH remain poorly understood. Our assessment of the literature suggests that some of the problems in deciphering the molecular action of JH lie not in unfathomable molecular mechanisms, but rather in the quality of the hormone, choice of experimental model, and development of experimental design. Nevertheless, a number of promising JH target genes have been identified (see Berger and Dubrovsky, 2005, for a review), renewing hope that these targets may offer insight into the molecular action(s) of the hormone.

A number of experimental approaches and targets have been explored, including high-affinity transcription factors, signal transduction mechanisms, and membrane receptors. At present there are several potential leads in this very dynamic field.

8.9.1. Juvenile Hormone Interaction with Ultraspiracle

One of the more intriguing ideas concerning the molecular action of JH was suggested by Jones and Sharp (1997), who demonstrated that the nuclear receptor Ultraspiracle (USP) binds JH III. USP, an orphan receptor for which no endogenous ligand has been unambiguously established (Billas *et al.*, 2001; Iwema *et al.*, 2007), is a heterodimeric partner that interacts with the ecdysone receptor (EcR) to form a USP:EcR complex (Thomas *et al.*, 1993; Yao *et al.*, 1993). USP is an orthologue of the vertebrate retinoid X receptor, RXR (Oro *et al.*, 1990; Henrich and Beatty, 2010), which is responsive to very high levels of methoprene (Harmon *et al.*, 1995). The DNA-binding domain of USP displays a high degree of identity with RXR, but the ligand-binding domains of these proteins show less than 50% similarity (Riddiford, 2008). Importantly, this form of USP is found only in Lepidoptera and Diptera; other insect orders and arthropods display an orthologue that is more similar to RXR (Iwema *et al.*, 2007).

Using a fluorescence assay that detects changes in protein conformation induced by ligand binding, Jones and Sharp (1997) demonstrated that JH III and JH III acid change the conformation of recombinant *D. melanogaster* USP, whereas farnesol and 20E do not. The USP-JH III acid interaction generates anomalous spectra, suggesting that it may not trigger a conformational change identical to that obtained with JH III. These investigators estimated the dissociation constant of the USP:JH III complex to be approximately 1 μM. This value is considerably higher than might be expected for a nuclear receptor that must sequester hormone directly from the circulatory system. Further refinement of the original studies indicated that methoprene competes with JH III for a binding site on USP. More recently, the Jones group has focused on the potential juvenilizing ability of MF in *D. melanogaster* and *Ae. aegypti* (Jones *et al.*, 2010). Their binding studies indicate that an isoform of USP from *Ae. aegypti* bound methyl farnesoate far better that JH III; however, displacement as measured by fluorescence quenching still indicates a relatively low affinity for MF.

Despite the low binding affinity constants, evidence from functional transcription assays supports USP as a JH binding transcription factor. Using modified *USP* response elements coupled to a *JHE* core promoter (-61 to +28 relative to the start site) and a reporter gene, Jones *et al.* (2001) demonstrated that expression could be induced by JH III in a dose-dependent fashion. Functional transcriptional assays also demonstrated that point mutations in the putative JH III binding site of USP, which abolish JH binding, caused the mutant receptor to act as a dominant negative and suppress JH III activation (Xu *et al.*, 2002; Fang *et al.*, 2005; Jones *et al.*, 2006).

Jones *et al.* (2001) suggested that high levels of JH III induce a conformational change that stabilizes dimeric/oligomeric forms of USP. This interaction may be important in dimerization of USP with its ecdysteroid receptor partner, but how it occurs and the role of USP-DNA binding in stabilization of the complex is not yet understood (Iwema *et al.*, 2009). A possible model for the role of USP in JH action can be constructed that takes into account larval growth, larval molting, and the metamorphic molt: (1) JH binds to homodimeric USP, which in turn interacts with specific JH response elements to drive expression related to larval growth; (2) JH binds to USP:EcR, and when 20E is present acts in concert with ecdysteroid-specific response elements to coordinate a larval molt; and (3) in the absence of JH, 20E binds to USP:EcR to coordinate a metamorphic molt.

Until USP was demonstrated to interact with JH III, it was not clear that this orphan receptor had a naturally occurring ligand. The initial crystallographic analyses of the USP ligand-binding domains from *H. virescens* (Billas *et al.*, 2001) and *D. melanogaster* (Clayton *et al.*, 2001) indicated that the putative ligand-binding site is locked in an antagonist conformation for JH III. Sasorith *et al.* (2002) reinvestigated the site with recombinant *H. virescens* USP. Their computer algorithms indicated that JH homologues and agonists may fit the putative site. However, the percentage of occupancy of the ligand-binding sites by these ligands lies in the bottom range of values for classical nuclear receptors, thus raising concern about the validity of USP as the JH receptor. It is worth noting that the binding site of the recombinant *H. virescens* USP contains a phospholipid inserted by the expression host. It may be that JH III competes with this endogenous ligand, thus skewing the binding analyses.

USP, with its large ligand-binding site and a low level of occupancy, behaves more like a sensor than a classical high-affinity receptor. Using the sensor model (Chawla *et al.*, 2001; Handschin and Meyer, 2005), one might envision an entirely different role for USP based on nutritional levels in the pre-metamorphic larva. Its role may be similar to an element of the target of rapamycin pathway, where JH plays an important role in nutrition (Shiao *et al.*, 2008). When sufficient levels of key dietary lipids have been attained, the USP binding site becomes saturated with that nutritionally important ligand, which, in turn, alters the conformation of the protein. Ligand-activated USP might then enhance 20E binding by its partner, EcR. Interestingly, developmental changes in larval *M. sexta* epidermis USP-1 and USP-2 mRNA are contrary to what might be expected if they were indeed directly related to JH levels (Hiruma and Riddiford, 2010); on the other hand, USP-1 and USP-2 may have roles as nutritional sensors. Ligand activation of the appropriate EcR isoforms is known to lead to changes in neuronal activity (Hewes and Truman, 1994; Truman, 1996) that may trigger activity in the prothoracicotropes or dendritic fields enervating them. While the hypothesis of USP as a sensor is simplistic, it could account for the very early signaling events that initiate ecdysis. It now seems clear that USP binds JH, albeit with low affinity, as well as other lipids. It may be that refinement of experiments presented by the Jones lab will shed new light on how the hormone acts at the molecular level.

8.9.2. Juvenile Hormone Interaction with the Methoprene-tolerant and Germ-cell Expressed Genes

Using mutagenesis studies to dissect the molecular action of JH, Wilson and his group examined a locus in *D. melanogaster*, the *Methoprene-tolerant* (*Met*) locus, which encodes a protein similar in sequence to proteins in the basic-helix-loop-helix-PAS (bHLH-PAS) family of transcriptional regulators (Wilson and Ashok, 1998). Treating wild-type flies with high levels of methoprene at certain times during larval development leads to a distinct array of developmental abnormalities, including aberrations in the central nervous system, salivary glands, and muscles (Restifo and Wilson, 1998). In contrast to wild-types, flies carrying the *Met* mutation are resistant to high levels of methoprene and appear to develop and reproduce normally (Wilson and Fabian, 1986).

The *Met* gene (GenBank Account Number: AF034859) codes a 716 residue protein, MET, which in wild-type flies displays a dissociation constant of 6 nM for JH III (Shemshedini *et al.*, 1990). Flies carrying the *Met* mutation exhibit a significantly reduced binding affinity for JH III (38 nM). The high affinity of a recombinant MET for JH III was supported by the studies of Miura *et al.* (2005), although these investigators did not rule out the contribution of other proteins in the binding analysis. Competitive displacement studies comparing the JH homologues and agonists indicated that JH I, JH II, and JH III acid as well as methoprene are weak competitors when compared with JH III; methoprene competition is 100-fold less than that of JH III. The reduced affinity of MET for methoprene translates into a 50- to 100-fold increase in resistance to both its toxic and morphogenetic effects. Flies carrying the *Met* mutation are also resistant to toxic levels of the naturally occurring hormones JH III and JHB_3, but not to various other classes of insecticides. This important distinction demonstrates that *Met* is not a general insecticide-resistance gene, but it is specific for JH and JH analogs (Wilson *et al.*, 2003).

Immunolocalization studies suggest that MET is limited to the nuclei of cells expressing *Met* (Pursley *et al.*, 2000), thus ruling out the possibility that MET acts as a cytosolic "sink" (Shemshedini *et al.*, 1990). It is present in all cells of the *Drosophila* embryo from the 256-cell stage until early gastrulation, when the signal begins to decline. In second and third stadium larvae MET is found in salivary glands, fat body, imaginal discs, and gut primordium. MET is also present in pupal histoblasts and in adult reproductive tissues, including the ovarian follicle cells and spermatheca and the male accessory glands (Pursley *et al.*, 2000).

Analysis of the *D. melanogaster Met* sequence indicates it contains three regions displaying similarity to the bHLH-PAS family of transcriptional regulators. These regions appear to be conserved in *Ae. aegypti*, *Anopheles gambiae* (Wang *et al.*, 2007), *T. castaneum* (Konopova and Jindra, 2007), and *B. mori* (Li *et al.*, 2010), which indicates a significant degree of evolutionary relatedness. It is instructive to note that the bHLH-PAS gene family was named for three important members of the group: the *Period* gene, the *Aryl hydrocarbon receptor* (*Ahr*) gene, and the *Single-minded* gene. Of particular importance is the similarity of *Met* to the *Ahr* gene, which encodes a xenobiotic binding protein (Ashok *et al.*, 1998). If MET resembles a member of the AHR family of receptors, it may explain why so many structurally distant compounds have JH-like activity (Stahl, 1975). Equally interesting is that the vertebrate AHR partners with another protein, the aryl hydrocarbon nuclear translocator, to form an active transcriptional regulator that activates genes responsible for mediating the response to xenobiotics.

Over a dozen alleles of *Met* have been recovered in genetic screens (Wilson *et al.*, 2003), but one in particular, Met^{27}, poses a perplexing question about the role of MET in JH action. On the basis of Northern and RT-PCR analyses, flies homozygous for Met^{27} lack *Met* transcripts, yet these insects survive and develop into adults. The lack of *Met* expression impacts only the adult female by an 80% reduction in oogenesis (Wilson and Ashok, 1998). If flies carrying a null mutation in *Met* can undergo seemingly normal development and limited oogenesis, the role of MET in JH action is open to question. Li *et al.* (2010) also noted that RNAi silencing of the *B. mori Met* gene in the fourth stadium did not affect the ability of larvae to molt and pupate successfully. Pursley *et al.* (2000) suggested that "genetic redundancy" (Krakauer and Plotkin, 2002), that is, back-up systems for biological phenomena, may explain this situation. Apparently MET is superfluous during the larval stages (however, see Konopova and Jindra, 2007) yet required for reproduction (Wilson and Ashok, 1998; Wilson *et al.*, 2003). This suggests that JH utilizes several distinct molecular pathways to control development and reproduction.

The discovery that MET is not essential to JH action during larval development led Wilson's group to continue the search for critical partners in JH action (Godlewski *et al.*, 2006; Baumann *et al.*, 2010). A critical partner may be germ cell expressed (GCE). The *gce* gene, a paralogous PAS domain, has approximately 60% sequence similarity to *Met* (Baumann *et al.*, 2010). The *gce* gene is expressed in the embryo and early larval stages of *D. melanogaster*, but expression declines during day 3 (Bauman *et al.*, 2010). Expression of *gce* increases during the wandering and puparium stages, declines in the pupa, and reappears in the adult stage; it does not follow the pattern of *Met* expression precisely.

Prior to recent studies, the role of *gce* was only poorly understood. Liu *et al.* (2009), employing genetic ablation procedures to eliminate ring gland CA cells, demonstrated that the ensuing JH deficiency led to significant upregulation of fat body caspases that are involved in pro-grammed cell death. In *Met-* and *gce*-deficient animals, these caspases were downregulated, leading to a reduction in fat body cell death. In animals where MET was overex-pressed, fat body caspases were significantly upregulated; when exposed to methoprene, programmed cell death in the fat body was suppressed. In over- and underexpression studies in *D. melanogaster, gce* was shown to play an impor-tant role in pupal development and it was demonstrated that GCE can substitute *in vivo* for MET (Baumann *et al.*, 2010). Moreover, knockdown of *gce* expression in larvae was demonstrated to result in pre-adult lethality in the absence of *Met* expression.

It is clear that GCE and MET are involved in the action of JH in *D. melanogaster*, and appear near the top of the hierarchy of critical events in several other insect orders (Konopova and Jindra, 2008), but there is no direct evi-dence that either of the molecules are bona fide JH recep-tors (Baumann *et al.*, 2010). Baumann *et al.* (2010) outlined some of the key components that make *gce* and *Met* important targets for study. It was previously sug-gested that MET and GCE may heterodimerize to initi-ate JH activity (Godlewski *et al.*, 2006); however, given recent data, mandatory formation of a heterodimer is less obvious. Wilson and his group speculate that ligand bind-ing may be the sole responsibility of MET in the MET-GCE complex; the loss of GCE does affect JH action but not due to a failure of JH binding. How MET and GCE interact in insect species other than the higher dipterans (Konopova and Jindra, 2007) is of considerable interest.

8.9.3. Other Candidates for a Juvenile Hormone Receptor

Studies indirectly point toward JHs acting through signal transduction mechanisms including cyclic nucleotides, calcium ions, or other small bioactive molecules. While JH may be sufficiently soluble to traverse a plasma mem-brane (see Davey, 2000, for an alternative view), its labile nature in an unbound state guarantees a very short half-life in the cytoplasm. A number of lipophilic compounds as diverse as lysophospholipids; arachidonic acid metabo-lites; and short-, medium-, and long-chain fatty acids as well as steroid-like molecules exert their effects as extracel-lular mediators through a family of transmembrane pro-teins, GCPRs. The signaling paradigm involving GPCRs includes: (1) the capacity of one GPCR to couple and ini-tiate signaling through multiple pathways, (2) the ability of one G protein to activate many effectors, and (3) the ability of a GPCR to transduce signals through G protein-independent pathways (Woehler and Ponimaskin, 2009). "Cross-talk" between JHs and other biologically active molecules such as ecdysteroids (Henrich *et al.*, 2003; Wheeler and Nijhout, 2003; Berger and Dubrovsky, 2005; Dubrovsky, 2005) and insulin-like peptides

(Mirth *et al.*, 2005; Emlen *et al.*, 2006; Shingleton *et al.*, 2007) may originate with the GPCRs.

JH-membrane interactions have been implicated with regards to membrane perturbation (Baumann, 1969; Barber *et al.*, 1981) and shifts in cellular potassium levels, potentially through the Na^+/K^+ ATPase (Kroeger, 1968). Signal transduction via second messenger regulation has been suggested (Everson and Feir, 1976; Kensler *et al.*, 1978; Yamamoto *et al.*, 1988; Pszczolkowski *et al.*, 2005; Wang *et al.*, 2009). Davey (2000) reported that JH binds to a membrane protein that initiates a cascade where a 100 kDa protein, the α-subunit of the Na^+/K^+ ATPase, is phosphorylated. Each of these studies provides a tempting glimpse into interaction of JH with membrane proteins. The potential scenarios become even more com-plicated considering that different life stages and tissues may not use the same receptors or signal transduction mechanisms.

8.9.4. The Downstream Picture

Having described the molecules that initially respond to JH, we turn our attention to the downstream molecules involved in JH action. Since our last review, several new downstream genes and their expression products have been placed into the JH signaling cascade. It is clear that we are only just beginning to populate the pathways with the interacting players.

8.9.4.1. FKBP39 and Chd64 Li *et al.* (2007), using microarray technology, examined JH-induced genes in *A. mellifera* and the *D. melanogaster* L57 cell line. After identifying a suite of genes that JH upregulated in both species, putative *cis*-acting regulatory elements were located in the promoter region of the target genes. These *cis*-acting elements, or JH response elements (JHREs), were covalently attached to a matrix to capture proteins that interacted with the tethered 29 bp oligonucleotide. It should be noted that this JHRE may not be a universal JHRE, as different JHRE sequences have been identified in other species (Li *et al.*, 2007). Two proteins were found associated with this JHRE: a 39 kDa member of the FK506 binding protein family, FKBP39 (GenBank Account Number: Z46894.1), and a 21 kDa calponin-like protein, Chd64 (GenBank Account Number: AF217286). The authors suggest that FKBP39 may aid kinase function by maintaining the proper protein-protein interactions in the kinase complex, and that complex may ultimately play a role in signal transduction. The developmental profile for expression of the two genes is interesting. *FKBP39* is expressed primarily at larval molts, at pupariation, and during the time corresponding to the ecdysteroid peak required for the adult molt. *Chd64* expression is present at periods corresponding to the larval molts but not present in the third instar or through metamorphosis.

Using double-stranded RNA technology, the Palli group (Li *et al.*, 2007) demonstrated that downregulation of either *FKBP39* or *Chd64* expression prevented the JH III upregulation of a JHRE reporter construct, suggesting that both proteins are necessary for JH action. Studies using a two-yeast hybrid assay indicated that both proteins interacted with EcR, USP, and MET and may generate a transcription complex capable of modifying target gene expression. Further elaboration of the model comes from incubating nuclear proteins isolated from L57 cells with a JHRE and ATP (Kethidi *et al.*, 2006). This resulted in a reduction in nuclear protein binding to DNA. Either JH III or calf intestinal alkaline phosphatase restored binding of nuclear proteins to the JHRE. In addition, inhibitors of protein kinase C (PKC) increased and activators of PKC reduced the binding of nuclear proteins to the JHRE. These data suggest that PKC-mediated phosphorylation prevents binding of nuclear proteins to JHREs and leads to suppression of JH action. The involvement of PKC indicates that JH may act via signal transduction as well as by activating specific transcription factors.

8.9.4.2. Epac Another gene that JH upregulates in *D. melanogaster* cells is *Epac* (*E*xchange *p*rotein directly *a*ctivated by *c*yclic AMP; GenBank Account Number: NM 001103732), a guanine nucleotide exchange factor for Rap1 (Wang *et al.*, 2009; Willis *et al.*, 2010). Addition of as little as 100 ng/ml 10*R* JH III led to a rapid (1 h) increase in *Epac* mRNA. While early studies ascribed the activation of protein kinase A (PKA) by cyclic AMP (cAMP) as the primary link between GPCRs and their targets, more recent evidence suggests that Epac also plays an important role (Bos, 2006). Epac, upon activation by cAMP, aids Rap1 as it cycles between an inactive GDP-bound and an active GTP-bound conformation (de Rooij *et al.*, 1998). In turn, Rap1 modifies cell adhesion processes through cadherin-mediated cell junction formation and integrin-mediated cell adhesion (Bos, 2005). In studies on *Drosophila* development, Asha *et al.* (1999) demonstrated that Rap1 plays a critical role in regulating normal morphogenesis in eye discs, ovaries, and embryos. In addition, Rap1 mutations disrupt cell migration and induce abnormalities in cell shape, further implicating Rap1 as a regulator of morphogenesis *in vivo*. Expression of *Epac* occurs during embryogenesis and episodically during the larval and pupal stages. Late third instars exposed to a diet containing methoprene (500 ng/g diet) expressed significantly higher levels of *Epac* 12 to 18 h after exposure. Where this protein fits into the cascade of early events regulated by JHs remains unclear, but its role in activation of Rap1 certainly provides insight into the potential mechanism of JH action.

8.9.4.3. Krüpel-homologues Another gene family involved in the JH signaling cascade is the zinc finger transcription factor group *Krüpel-homologues* (*Kr-h1*; FlyBase Identification Number: FBgn0028420). In *D. melanogaster* Kr-h1 has three isoforms displaying different N-terminal sequences; the β isoform is involved in remodeling of the central nervous system late in larval life (Pecasse *et al.*, 2000), whereas Kr-h1α is necessary for metamorphosis and may modify ecdysteroid-regulated processes (Pecasse *et al.*, 2000; Beckstead *et al.*, 2005; Minakuchi *et al.*, 2008b). Genetic studies indicate that mutants lacking Kr-h1α die in the early stages of pupation. Conversely, overexpression of *Kr-h1α* in the epidermis leads to missing or shortened bristles in the dorsal midline of adults, a phenotype similar to that observed in wild-type animals treated with JH. Topical application of a JH agonist to *methoprene-tolerant* mutants did not block the outgrowth of abdominal bristles, suggesting that Kr-h1 functions downstream from MET (Minakuchi *et al.*, 2008b). Misexpression of *Kr-h1* during early adult development leads to the abnormal re-expression of *Broad* (*br*, see below) in the abdomen and results in the formation of a second pupal cuticle (Zhou *et al.*, 2002; Minakuchi *et al.*, 2008b). When *br* was overexpressed, *Kr-h1* expression was not detected, suggesting that in abdominal epidermis *Kr-h1* is upstream of *br*. In the larger picture, JH-induced *Kr-h1* expression may be preventing the permanent cessation of *br* expression in imaginal abdominal epidermal cells during the onset of adult development; however, it is still unclear whether Kr-h1 acts directly on *br* transcription (Minakuchi *et al.*, 2008b).

Kr-h1α is regulated in part by 20E and in turn modifies the ecdysteroid-regulated processes (Pecasse *et al.*, 2000; Beckstead *et al.*, 2005). In *Kr-h1α* mutants, ecdysteroid regulation of transcription factors at pupariation appears disrupted (Pecasse *et al.*, 2000), which may be linked to USP (Shi *et al.*, 2007). It is curious that *Kr-h1* null mutants exhibit normal larval development, and it is only at metamorphosis that the mutation is noted (Riddiford, 2008). Clearly, other factors, both in the ecdysteroid and the JH signaling pathways, must be compensating for the loss of *Kr-h1*.

A similar story is emerging in *T. castaneum*, where treatment with a JH agonist promotes a rapid and significant induction in *Kr-h1* transcription and the formation of a second pupa (Minakuchi *et al.*, 2009). Knockdown of *Kr-h1* expression in larvae results in precocious metamorphosis. Moreover, suppressing JH biosynthesis by using double-stranded RNA-JHAMT leads to a significant reduction in the expression of *Kr-h1*. Taken together, these results show that Kr-h1 is an important element in the JH signaling cascade. A developmental profile of *Kr-h1* expression using quantitative RT-PCR analyses demonstrated that the transcript is continuously present during embryogenesis and larval development; it was undetectable in the pupal stage but reappeared in the adult.

8.9.4.4. Broad complex One of the major molecular targets in the JH signaling cascade is the group of transcription factors termed Broad complex (FlyBase Identification Number: FBgn0010011), a key component in the initiation of metamorphosis (see Riddiford *et al.*, 2003; Riddiford, 2008, for reviews). The Broad isoforms, in conjunction with other early-appearing transcription factors and ecdysteroids, direct a large suite of downstream genes involved in initiating and orchestrating metamorphosis (Hiruma and Riddiford, 2010).

The *Broad* gene is large (100 kb) and in *Drosophila* encodes at least six isoforms of the protein, all of which share a common N-terminal of approximately 425 amino acids (Bayer *et al.*, 1996). Through alternative splicing, the C-terminal can be represented by any one of four pairs of C_2H_2-type zinc finger domains, Z1, Z2, Z3, and Z4. The N- and C-terminals are connected by linkage domains of varying lengths. Based on sequence comparisons, it has been inferred that the different isoforms of *br* arose through a series of duplication events (Spokony and Restifo, 2007). The temporal and tissue abundance of the isoforms changes in a hormonally defined sequence during the course of metamorphosis. This choreographed response has led to the suggestion that the various members (or combination of members) of the isoform family function in different developmental pathways (Bayer *et al.*, 1996; Suzuki *et al.*, 2008).

Broad has been examined in several insect orders. It has been shown to be expressed in one of the most basal orders, Thysanura, as well as in the highly derived dipterans (Erezyilmaz *et al.*, 2009). In *D. melanogaster*, mutations in *br* allow normal larval development to occur but prevent pupation (Kiss *et al.*, 1988; Bayer *et al.*, 1996; Riddiford *et al.*, 2003). Ingestion of the JH agonist pyriproxifen by first instar *D. melanogaster* prolongs the third stadium by several days (Riddiford *et al.*, 2003) and induces a 12h delay in the appearance of Broad in abdominal epidermis. By the time of pupation, however, Broad levels are similar to those of controls, indicating that the JH agonist cannot prevent the appearance of this protein. The adult abdominal epidermis, formed from nests of histoblasts, begins to proliferate and spread over the abdomen after puparium formation, with the new cells displacing the existing larval cells. Broad is found in the nuclei of larval cells that are destined to die, but cannot be found in imaginal cells once they have spread across the abdomen. Increased levels of the Z1 isoform through misexpression at the onset of adult cuticle formation mimic the effects of JH, causing the reappearance of mRNAs encoding pupal cuticular proteins and the suppression of mRNAs encoding adult cuticular proteins (Zhou and Riddiford, 2002). These investigators suggest that JH prolongs the expression of the Z1 isoform, which suppresses the onset of adult cuticle formation.

In *M. sexta*, only three Broad isoforms have been reported, Z2, Z3, and Z4, with Z4 being most prominent in epidermal tissue (Zhou *et al.*, 1998; Zhou and Riddiford, 2001). Although *br* is not expressed until pupal commitment, its expression can be prevented by JH. *In vitro*, *br* expression in the epidermis was upregulated within 6h of 20E treatment. Once *br* appears, JH can no longer suppress it and levels remain elevated, even during the rise in JH titers later in the stadium. Then *br* expression disappears early in the pupal stage at the initiation of adult cuticular synthesis (Zhou *et al.*, 1998; Zhou and Riddiford, 2001). Juvenile hormone treatment of pupae prior to the onset of the adult molt leads to an upregulation of *br* expression and the formation of a second pupal cuticle. Silencing of *br* in *B. mori* using RNAi results in disruption of metamorphosis and developmental arrest (Uhlirova *et al.*, 2003). Thus, in these species, JH levels must decline prior to the initial onset of *br* expression. However, once *br* is expressed, that expression in response to 20E can be maintained in the presence of JH.

A slightly different pattern emerges in the coleopteran *T. castaneum*. Elimination of *br* in *T. castaneum* leads to precocious development with a mix of larval and adult traits (Parthasarathy *et al.*, 2008; Suzuki *et al.*, 2008). The expression pattern in *T. castaneum* is different than in *M. sexta* and *D. melanogaster*; all isoforms of *br* are expressed at reduced levels in the penultimate stadium and then increase in the pre-pupal period of the last stadium (Parthasarathy *et al.*, 2008; Suzuki *et al.*, 2008). Treatment with a JH agonist during the penultimate stadium led to a repeat of the penultimate *br* expression patterns in the next stadium, followed by a supernumerary larval stadium (Suzuki *et al.*, 2008).

Throughout embryonic and nymphal development in the hemimetabolous insect *O. fasciatus br* is expressed and is thought to aid the morphogenetic changes that occur between the different nymphal stadia (Erezyilmaz *et al.*, 2006, 2009). Then *br* disappears during the last nymphal stadium-when ecdysteroid levels rise in the absence of JH. Topical application of JH reduces *br* expression and induces supernumerary nymphal stages. Precocious adult development induced by precocene II, a compound that reduces CA activity, leads to a loss of *br* expression. From these observations it was concluded that JH is necessary to maintain expression of *br* during the nymphal stage.

While there are variations in the developmental pattern among insect species, it is clear that Broad is a major pupal specifier in holometabolous insects. It is responsible for the shift from larva to pupa in epidermal commitment and for the onset of metamorphic differentiation of imaginal discs. Juvenile hormone delays or prevents *br* expression from increasing in response to 20E. Hemimetabolous insects, by contrast, express *br* throughout pre-adult development until the last nymphal stage. This difference suggests that the expression of *br* prominent during embryonic and nymphal development of ametabolous and hemimetabolous insects was lost evolutionarily with the emergence of holometabolous insects (Erezyilmaz *et al.*, 2009).

8.9.5. New Directions

From the earliest studies, insect endocrinologists have used surgical ablation or ligation procedures that remove or compartmentalize specific hormone-secreting organs to better understand their endocrine functions. However, little is known about surgically induced stress or inadvertent removal of non-target cells and neural connectives that often accompany these procedures. With the development of genetic ablation methods that eliminate the CA cells of the *D. melanogaster* ring gland, surgically induced stress is eliminated and off-target damage is significantly reduced. This ability to manipulate the output of the CA in a non-invasive fashion will certainly provide new approaches to understanding the role of JH in *D. melanogaster* and hopefully provide working models for other species. Several recent studies using this extraordinary genetic technology examined the role of JH in programmed cell death of the fat body (Liu *et al.*, 2009) and in adult eye morphogenesis (Riddiford *et al.*, 2010).

To develop the tools necessary for genetic ablation, Siegmund and Korge (2001) generated fly lines with labeled neurons specific for either the prothoracic or CA cells of the ring gland. One line, *Aug21*, contains a GAL4 driver that specifically targets gene expression to the ring gland (Colombani *et al.*, 2005; Mirth *et al.*, 2005). An upstream activator sequence (UAS) was added to *grim*, which encodes a protein that induces programmed cell death (Liu *et al.*, 2009). *UAS-grim*, when expressed in the *Aug21* line, leads to ablation of the CA during the early wandering period in third instars and to death after normal pupariation. It should be noted that CA cells are still present during the earlier larval stadia, which may be why the larva survives. In addition to the depletion of CA cells, expression of *UAS-grim* in the *Aug21* line induces cell death in the salivary glands (Riddiford *et al.*, 2010).

The fat body in the *Aug21-UAS-grim* line displayed a precocious and enhanced programmed cell death that interfered with the remodeling process (Liu *et al.*, 2009). Central to the remodeling process are two caspases: Dronc, an initiator caspase necessary to activate pro-forms of effector caspases, and Drice, an effector caspase necessary to trigger the apoptotic process (Hay and Guo, 2006). These important markers of apoptotic activity are upregulated in the fat body when JH titers are low and 20E levels are elevated. *Dronc* and *Drice* expression is upregulated early in the JH-deficient *Aug21-UAS-grim* line (Liu *et al.*, 2009). These data suggest that JHs in the wild-type insect antagonize the 20E-induced appearance of the two caspases. The authors assert the JH-induced suppression of

fat body programmed cell death does not appear to act on the generally accepted 20E-triggered transcriptional cascade; rather, it acts in a yet unexplained fashion.

Contrary to what might be expected, mutants deficient in *Met/gce* did not display precocious fat body programmed cell death; rather, they displayed a lethal phenotype similar to that seen when JHAMT is globally overexpressed (i.e., JH levels are elevated). Moreover, when *Met* is overexpressed, *Dronc* and *Drice* are upregulated just as in JH-deficient animals. When mutants overexpressing MET are treated with methoprene in the early wandering stage, these caspases are significantly downregulated. In other words, JH somehow counteracts the overexpression of its putative receptor complex MET/GCE by downregulating several caspases important in programmed cell death. Thus, the search for a JH receptor outside of the MET/GCE complex must continue.

In *D. melanogaster*, JH treatment at the time of pupariation affects the normal reorganization of the central nervous and muscular systems and, as might be expected, the effects of JHs are blocked in the *Met*-null mutants unless given at very high doses (Restifo and Wilson, 1998). More recent work by Wilson (2006) noted severely malformed ommatidial defects in the compound eye in *Met* mutants. Using the *Aug21-UAS-grim* line of *D. melanogaster*, Riddiford *et al.* (2010) re-examined eye development and found premature expression of *EcR-B*1 in both photoreceptors and the optic lobe, indicating that the early appearance of ecdysteroid receptor is due to the lack of JH during the late third stadium.

Several morphological changes in the visual system were observed in the *Aug21-UAS-grim* line. First, cell proliferation in the outer proliferation zone of the optic lobe is significantly reduced and can be rescued by feeding the genetically allatectomized larvae a diet containing a JH agonist. Second, ommatidial connections with the optic lobe of the brain are temporally modified. An ommatidium contains eight photoreceptor cells, of which six send neural projections to terminate in the lamina region of the optic lobe. Photoreceptors 7 and 8 send axon projections past the lamina to the medulla. The projection of these neurons, from their external location to the brain, depends upon highly coordinated spatial and temporal signals in their growth cones and surrounding tissue (Birkholz *et al.*, 2009). The growth cones for photoreceptors 7 and 8 are initially bundled but then separate as they innervate the different layers of the medulla. In genetically allatectomized flies, the photoreceptors begin the separation process about 12 h earlier than in controls, but JH agonists delay the process.

8.9.6. Questions

Better than a decade ago, Davey (2000) posed a series of prescient questions surrounding the future of JH research. Among the questions to be answered were (1) How many

JHs are present in insects? (2) Is the product of the CA the effective hormone? (3) How does JH enter the target cell? (4) How many modes of action are there? and (5) How many receptors are there? Riddiford (2008) added an additional question: Does JH have a positive role in promoting larval development or does it act only to prevent metamorphosis? With emerging genetic technology and expanding genomic databases covering many species, the prospects for answering these questions in the next decade look bright.

Acknowledgements

The authors would like to thank Ms. Hedda Goodman for her editorial comments. Work from our laboratories was supported by the NIH (WG), Wisconsin Agricultural Experiment Station (WG), the Natural Sciences and Engineering Research Council of Canada (MC), the Canadian Forest Service (MC), and the Canadian Biotechnology Strategy (MC). In searching the literature it came to our attention that the senior editor of this series, Professor Lawrence Gilbert, has been publishing in the field of insect endocrinology for better than 50 years (1958 to 2010). Thus, in appreciation for Professor Gilbert's outstanding contributions for over better than a half-century, this chapter is dedicated to him.

References

Abdel-Aal, Y. A. I., & Hammock, B. D. (1985). 3-Octylthio-1, 1,1-trifluoro-2-propanone, a high affinity and slow binding inhibitor of juvenile hormone esterase from *Trichoplusia ni*. *Insect Biochem.*, *15*, 111–122.

Abdel-Aal, Y. A. I., & Hammock, B. D. (1986). Transition state analogs as ligands for affinity purification of juvenile hormone esterase. *Science*, *233*, 1073–1076.

Agui, N., Granger, N. A., Gilbert, L. I., & Bollenbacher, W. E. (1979). Cellular localization of the insect prothoracicotropic hormone: In vitro assay of a single neurosecretory cell. *Proc. Natl. Acad. Sci. USA*, *76*, 5694–5698.

Aguila, J. R., Suszko, J., Gibbs, A. G., & Hoshizaki, D. K. (2007). The role of larval fat cells in adult *Drosophila melanogaster*. *J. Exp. Biol.*, *210*, 956–963.

Anspaugh, D. D., Kennedy, G. G., & Roe, R. M. (1995). Purification and characterization of a resistance-associated esterase from the Colorado potato beetle, *Leptinotarsa decemlineata*. *Pest. Biochem. Physiol.*, *53*, 84–96.

Anspaugh, D. D., & Roe, R. M. (2005). Regulation of JH epoxide hydrolase versus JH esterase activity in the cabbage looper, *Trichoplusia ni*, by juvenile hormone and xenobiotics. *J. Insect Physiol.*, *51*, 523–535.

Applebaum, S. W., Avisar, E., & Heifetz, Y. (1997). Juvenile hormone and locust phase. *Arch. Insect Biochem. Physiol.*, *35*, 375–391.

Arakane, Y., Lomakin, J., Beeman, R. W., Muthukrishnan, S., Gehrke, S. H., Kanost, M. R., & Kramer, K. J. (2009). Molecular and functional analyses of amino acid decarboxylases involved in cuticle tanning in *Tr. castaneum*. *J. Biol. Chem.*, *284*, 16584–16594.

Asha, H., de Ruiter, N. D., Wang, M. G., & Hariharan, I. K. (1999). The Rap1 GTPase functions as a regulator of morphogenesis *in vivo*. *EMBO J.*, *18*, 605–615.

Ashok, M., Turner, C., & Wilson, T. G. (1998). Insect juvenile hormone resistance gene homology with the bHLH-PAS family of transcriptional regulators. *Proc. Natl. Acad. Sci. USA*, *95*, 2761–2766.

Audsley, N., Matthews, H. J., Price, N. R., & Weaver, R. J. (2008). Allatoregulatory peptides in Lepidoptera, structures, distribution and functions. *J. Insect Physiol.*, *54*, 969–980.

Avvakumov, G. V., Cherkasov, A., Muller, Y. A., & Hammond, G. L. (2010). Structural analyses of sex hormone-binding globulin reveal novel ligands and function. *Mol.Cell Endocrinol.*, *316*, 13–23.

[BAG] Biology Analysis Group. (2004). A draft sequence for the genome of the domesticated silkworm (*Bombyx mori*). *Science*, *306*, 1937–1940.

Baker, F. C., Lee, E., Bergot, B. J., & Schooley, D. A. (1981). Isomerization of isopentenyl pyrophosphate and homoisopentenyl pyrophosphate by *Manduca sexta* corpora cardiaca-corpora allata homogenates. In G. E. Pratt, & G. T. Brooks (Eds.), *Juvenile Hormone Biochemistry* (pp. 67–80). Amsterdam: Elsevier/North Holland Biomedical Press.

Baker, F. C., Lanzrein, B., Miller, C. A., Tsai, L. W., Jamieson, G. C., & Schooley, D. A. (1984). Detection of only JH III in several life-stages of *Nauphoeta cinerea* and *Thermobia domestica*. *Life Sci.*, *35*, 1553–1560.

Baker, F. C., Tsai, L. W., Reuter, C. C., & Schooley, D. A. (1987). *In vivo* fluctuation of JH, JH acid, and ecdysteroid titer, and JH esterase activity, during development of fifth stadium *Manduca sexta*. *Insect Biochem.*, *17*, 989–996.

Baker, F. C., Tsai, L. W., Reuter, C. C., & Schooley, D. A. (1988). The absence of significant levels of the known juvenile hormones and related compounds in the milkweed bug, *Oncopeltus fasciatus*. *Insect Biochem.*, *18*, 453–462.

Baker, F. C. (1990). Techniques for identification and quantification of juvenile hormones and related compounds in arthropods. In A. P. Gupta (Ed.), *Morphogenetic Hormones of Arthropods* (pp. 389–444). New Brunswick, NJ: Rutgers University Press.

Barber, R. F., Downer, R. G. H., & Thompson, R. G. H. (1981). Perturbation of phospholipid membranes by juvenile hormone. *Biochim. Biophys. Acta*, *643*, 593–600.

Baumann, A., Barry, J., Wang, S., Fujiwara, Y., & Wilson, T. G. (2010). Paralogous genes involved in juvenile hormone action in *Drosophila melanogaster*. *Genetics*, *185*, 1327–1336.

Baumann, G. (1969). Juvenile hormone: Effects on bimolecular lipid membranes. *Nature*, *223*, 316–317.

Bayer, C. A., Holley, B., & Fristrom, J. W. (1996). A switch in broad-complex zinc-finger isoform expression is regulated posttranscriptionally during the metamorphosis of *Drosophila* imaginal discs. *Dev. Biol.*, *177*, 1–14.

Bean, D. W., Beck, S. D., & Goodman, W. G. (1982). Juvenile hormone esterases in diapause and nondiapause larvae of the european corn borer, *Ostrinia nubilalis*. *J. Insect Physiol.*, *6*, 485–492.

Bean, D. W., Goodman, W. G., & Beck, S. D. (1983). Regulation of juvenile hormone esterase activity in the european corn borer, *Ostrinia nubilalis*. *J. Insect Physiol.*, *29*, 877–883.

Bear, A., Simons, A., Westerman, E., & Monteiro, A. (2010). The genetic, morphological, and physiological characterization of a dark larval cuticle mutation in the butterfly, *Bicyclus anynana*. *PLoS One*, *5*, e11563.

Beckstead, R. B., Lam, G., & Thummel, C. S. (2005). The genomic response to 20-hydroxyecdysone at the onset of *Drosophila* metamorphosis. *Genome Biol.*, *6*, R99.

Bede, J. C., Teal, P. E., Goodman, W. G., & Tobe, S. S. (2001). Biosynthetic pathway of insect juvenile hormone III in cell suspension cultures of the sedge *Cyperus iria*. *Plant Physiol.*, *127*, 584–593.

Bellés, X., Martín, D., & Piulachs, M.-D. (2005). The mevalonate pathway and the synthesis of juvenile hormone in insects. *Ann. Rev. Entomol.*, *50*, 181–199.

Beintema, J. J., Stam, W. T., Hazes, B., & Smidt, M. P. (1994). Evolution of arthropod hemocyanins and insect storage proteins (hexamerins). *Mol. Biol. Evol.*, *11*, 493–503.

Berger, E. M., & Dubrovsky, E. B. (2005). Juvenile hormone molecular actions and interactions during development of *Drosophila melanogaster*. *Vitam. Horm.*, *73*, 175–215.

Bergot, B. J., Baker, F. C., Cerf, D. C., Jamieson, G., & Schooley, D. A. (1981a). Qualitative and quantitative aspects of juvenile hormone titers in developing embryos of several insect species: discovery of a new JH-like substance extracted from eggs of *Manduca sexta*. In G. E. Pratt, & G. T. Brooks (Eds.), *Juvenile Hormone Biochemistry* (pp. 33–45). Amsterdam: Elsevier/North Holland Biomedical Press.

Bergot, B. J., Ratcliff, M., & Schooley, D. A. (1981b). A method for quantitative determination of juvenile hormones by mass spectroscopy. *J. Chromat.*, *204*, 231–244.

Bhaskaran, G., Sparagana, S. P., Barrera, P., & Dahm, K. H. (1986). Change in corpus allatum function during metamorphosis of the tobacco hornworm *Manduca sexta*: regulation at the terminal step in juvenile hormone biosynthesis. *Arch. Insect Biochem. Physiol.*, *3*, 321–338.

Billas, I. M. L., Moulinier, L., Rochel, N., & Moras, D. (2001). Crystal structure of the ligand-binding domain of the ultraspiracle protein USP, the orthologue of retinoid x receptors in insects. *J. Biol. Chem.*, *276*, 7465–7474.

Birkholz, D. A., Chou, W., Phistry, M. M., & Britt, S. G. (2009). *Rhomboid* mediates specification of blue- and green-sensitive R8 photoreceptor cells in *Drosophila*. *J. Neurosci.*, *29*, 2666–2675.

Bonning, B. C., & Hammock, B. D. (1996). Development of recombinant baculoviruses for insect control. *Ann. Rev. of Entomol.*, *41*, 191–210.

Bonning, B. C., Booth, T. F., & Hammock, B. D. (1997). Mechanistic studies of the degradation of juvenile hormone esterase in *Manduca sexta*. *Arch. Insect Biochem. Physio.*, *34*, 275–286.

Bos, J. L. (2005). Linking Rap to cell adhesion. *Curr. Op. Cell Biol.*, *17*, 123–128.

Bos, J. L. (2006). Epac proteins: Multi-purpose cAMP targets. *Trends Biochem. Sci.*, *31*, 680–686.

Bowers, W. S., Marsella, P. A., & Evans, P. H. (1983). Identification of an hemipteran juvenile hormone: In vitro biosynthesis of JH III by *Dysdercus fasciatus*. *J. Exp. Zool.*, *228*, 555–559.

Branden, C., & Tooze, J. (1999). *Introduction to Protein Structure* (2nd ed.). New York: Garland Publishing, Inc.

Braun, R. P., & Wyatt, G. R. (1996). Sequence of the hexameric juvenile hormone binding protein from the hemolymph of *Locusta migratoria*. *J. Biol. Chem.*, *271*, 31756–31762.

Brent, C., & Dolezal, A. (2009a). Juvenile hormone extraction, purification, and quantification in ants. *CSH Protoc*, *2009*, pdb prot5246.

Brent, C., & Dolezal, A. (2009b). Radiochemical assay of juvenile hormone biosynthesis rate in ants. *CSH Protoc*, *2009*, pdb prot5248.

Brindle, P. A., Baker, F. C., Tsai, L. W., Reuter, C. C., & Schooley, D. A. (1987). Sources of propionate for the biogenesis of ethyl-branched insect juvenile hormones: role of isoleucine and valine. *Proc. Natl. Acad. Sci. USA*, *84*, 7906–7910.

Brindle, P. A., Baker, F. C., Tsai, L. W., & Schooley, D. A. (1992). Comparative metabolism of isoleucine by corpora allata of nonlepidopteran insects versus lepidopteran insects, in relation to juvenile hormone biosynthesis. *Arch. Insect Biochem. Physiol.*, *19*, 1–15.

Brindle, P. A., Schooley, D. A., Tsai, L. W., & Baker, F. C. (1988). Comparative metabolism of branched-chain amino acids to precursors of juvenile hormone biogenesis in corpora allata of lepidopterous vs nonlepidopterous insects. *J. Biol. Chem.*, *263*, 10653–10657.

Brockhouse, A. C., Horner, H. T., Booth, T. F., & Bonning, B. C. (1999). Pericardial cell ultrastructure in the tobacco hornworm, *Manduca sexta*. *Internat. J. Insect Morph. Embryol.*, *28*, 261–271.

Bronsert, M., Bingol, H., Atkins, G., & Stout, J. (2003). Prolonged response to calling songs by the L3 auditory interneuron in female crickets (*Acheta domesticus*): Possible roles in regulating phonotactic threshold and selectiveness for call carrier frequency. *J. Exp. Zool. Part A*, *296A*, 72–85.

Browder, M. H., D'Amico, L. J., & Nijhout, H. F. (2001). The role of low levels of juvenile hormone esterase in the metamorphosis of *Manduca sexta*. *J. Insect Science*, *1*, 11.

Brown, M. S., & Goldstein, J. L. (1980). Multivalent feedback regulation of HMG CoA reductase, a control mechanism coordinating isoprenoid synthesis and cell growth. *J. Lipid Res.*, *21*, 505–517.

Brüning, E., Saxer, A., & Lanzrein, B. (1985). Methyl farnesoate and juvenile hormone III in normal and precocene treated embryos of the ovoviviparous cockroach *Nauphoeta cinerea*. *Int. J. Invert. Reprod. Dev.*, *8*, 269–278.

Bürgin, C., & Lanzrein, B. (1988). Stage dependent biosynthesis of methyl farnesoate and juvenile hormone III and metabolism of juvenile hormone III in embryos of the cockroach, *Nauphoeta cinera*. *Insect Biochem.*, *18*, 3–9.

Burmester, T., Massey, H. C., Zakharkin, S. O., & Benes, H. (1998). The evolution of hexamerins and the phylogeny of insects. *J. Mol. Evol.*, *47*, 93–108.

Burmester, T. (2002). Origin and evolution of arthropod hemocyanins and related proteins. *J. Comp. Physiol. Part B*, *172*, 95–107.

Burns, S. N., Teal, P. E. A., Meer, R. K. V., Nation, J. L., & Vogt, J. T. (2002). Identification and action of juvenile hormone III from sexually mature alate females of the red imported fire ant, *Solenopsis invicta*. *J. Insect Physiol.*, *48*, 357–365.

Burtenshaw, S. M., Su, P. P., Zhang, J. R., Tobe, S. S., Dayton, L., & Bendena, W. G. (2008). A putative farnesoic acid O-methyltransferase (FAMeT) orthologue in *Drosophila melanogaster* (CG10527): relationship to juvenile hormone biosynthesis? *Peptides, 29*, 242–251.

Buszczak, M., & Segraves, W. A. (2000). Insect metamorphosis: Out with the old, in with the new. *Curr. Biol., 10*, R830–R833.

Campbell, P. M., Harcourt, R. L., Crone, E. J., Claudianos, C., Hammock, B. D., Russell, R. J., & Oakeshott, J. G. (2001). Identification of a juvenile hormone esterase gene by matching its peptide mass fingerprint with a sequence from the *Drosophila* genome project. *Insect Biochem. Mol. Biol., 31*, 513–520.

Campiche, S., BeckerVanSlooten, K., Ridreau, C., & Tarradellas, J. (2006). Effects of insect growth regulators on the nontarget soil arthropod Folsomia candida (Collembola). *Ecotox. Environ. Saf., 63*, 216–225.

Campiche, S., L'Ambert, G., Tarradellas, J., & Becker-van Slooten, K. (2007). Multigeneration effects of insect growth regulators on the springtail *Folsomia candida*. *Ecotoxicol. Environ. Saf., 67*, 180–189.

Canavoso, L. E., Yun, H. K., Jouni, Z. E., & Wells, M. A. (2004). Lipid transfer particle mediates the delivery of diacylglycerol from lipophorin to fat body in larval *Manduca sexta*. *J. Lipid Res., 45*, 456–465.

Cao, L., Zhang, P., & Grant, D. F. (2009). An insect farnesyl phosphatase homologous to the N-terminal domain of soluble epoxide hydrolase. *Biochem. Biophys. Res. Comm., 380*, 188–192.

Carrow, G. M., Calabrese, R. L., & Williams, C. M. (1984). Architecture and physiology of insect cerebral neurosecretory cells. *J. Neurosci., 4*, 1034–1044.

Casas, J., Harshman, L. G., Messeguer, A., Kuwano, E., & Hammock, B. D. (1991). *Invitro* metabolism of juvenile hormone-III and juvenile hormone-III bisepoxide by *Drosophila melanogaster* and mammalian cytosolic epoxide hydrolase. *Arch. Biochem. Biophys., 286*, 153–158.

Cascone, P. J., & Schwartz, L. M. (2001). Post-transcriptional regulation of gene expression during the programmed death of insect skeletal muscle. *Dev. Genes Evol., 211*, 397–405.

Cassier, P. (1979). The corpora allata of insects. *Int. Rev. Cytol., 57*, 1–73.

Cassier, P. (1990). Morphology, histology, and ultrastructure of JH-producing glands in insects. In A. Gupta (Ed.), *Morphogenetic Hormones of Arthropods* (pp. 83–194). New Brunswick, NJ: Rutgers University Press.

Castillo-Gracia, M., & Couillaud, F. (1999). Molecular cloning and tissue expression of an insect farnesyl diphosphate synthase. *Eur. J. Biochem., 262*, 365–370.

Caveney, S. (1970). Juvenile hormone and wound modelling of Te. cuticle architecture. *J. Insect Physiol., 16*, 1087–1107.

Charles, J. P., Wojtasek, H., Lentz, A. J., Thomas, B. A., Bonning, B. C., Palli, S. R., Parker, A. G., Dorman, G., Hammock, B. D., Prestwich, G. D., & Riddiford, L. M. (1996). Purification and reassessment of ligand binding by the recombinant, putative juvenile hormone receptor of the tobacco hornworm, *Manduca sexta*. *Arch. Insect Biochem. Physiol., 31*, 371–393.

Charles, J. P. (2010). The regulation of expression of insect cuticle protein genes. *Insect Biochem. Mol. Biol., 40*, 205–213.

Chau, P. L., Vanaalten, D. M. F., Bywater, R. P., & Findlay, J. B. C. (1999). Functional concerted motions in the bovine serum retinol-binding protein. *J. Comp. Aided Mol. Des., 13*, 11–20.

Chawla, A., Repa, J. J., Evans, R. M., & Mangelsdorf, D. J. (2001). Nuclear receptors and lipid physiology: Opening the X-files. *Science, 294*, 1866–1870.

Chen, L., OKeefe, S. L., & Hodgetts, R. B. (2002a). Control of Dopa decarboxylase gene expression by the Broad-Complex during metamorphosis in *Drosophila*. *Mech. Dev., 119*, 145–156.

Chen, L., Reece, C., OKeefe, S. L., Hawryluk, G. W. L., Engstrom, M. M., & Hodgetts, R. B. (2002b). Induction of the early-late *Ddc* gene during *Drosophila* metamorphosis by the ecdysone receptor. *Mech. Dev., 114*, 95–107.

Chen, Z., Linse, K. D., Taub-Montemayor, T. E., & Rankin, M. A. (2007). Comparison of radioimmunoassay and liquid chromatography tandem mass spectrometry for determination of juvenile hormone titers. *Insect Biochem. Mol. Biol., 37*, 799–807.

Cheon, H. M., Hwang, S. J., Kim, H. J., Jin, B. R., Chae, K. S., Yun, C. Y., & Seo, S. J. (2002). Two juvenile hormone suppressible storage proteins may play different roles in *Hyphantria cunea*. *Arch. Insect Biochem. Physiol., 50*, 157–172.

Cheon, H. M., Shin, S. W., Bian, G. W., Park, J. H., & Raikhel, A. S. (2006). Regulation of lipid metabolism genes, lipid carrier protein lipophorin, and its receptor during immune challenge in the mosquito *Aedes aegypti*. *J. Biol. Chem., 281*, 8426–8435.

Chiang, A. S., Gadot, M., & Schal, C. (1989). Morphometric analysis of corpus allatum cells in adult females of three cockroach species. *Mol. Cell. Endocrinol., 67*, 179–184.

Chiang, A. S., Gadot, M., Burns, E. L., & Schal, C. (1991). Developmental regulation of juvenile hormone synthesis: ovarian synchronization of volumetric changes of corpus allatum cells in cockroaches. *Mol. Cell. Endocrinol., 75*, 141–147.

Clark, A. J., & Bloch, K. (1959). The absence of sterol synthesis in insects. *J. Biol. Chem., 254*, 2578–2582.

Clayton, G. M., Peak-Chew, S. Y., Evans, R. M., & Schwabe, J. W.R. (2001). The structure of the ultraspiracle ligand-binding domain reveals a nuclear receptor locked in an inactive conformation. *Proc. Natl. Acad. Sci. USA, 98*, 1549–1554.

Cole, T. J., Beckage, N. E., Tan, F. F., Srinivasan, A., & Ramaswamy, S. B. (2002). Parasitoid-host endocrine relations: self-reliance or co-optation? *Insect Biochem. Mol. Biol., 32*, 1673–1679.

Colombani, J., Bianchini, L., Layalle, S., Pondeville, E., DauphinVillemant, C., Antoniewski, C., Carre, C., Noselli, S., & Leopold, P. (2005). Antagonistic actions of ecdysone and insulins determine final size in *Drosophila*. *Science, 310*, 667–670.

Cornel, A. J., Stanich, M. A., Farley, D., Mulligan, F. S., & Byde, G. (2000). Methoprene tolerance in *Aedes nigromaculis* in Fresno County, California. *J. Amer. Mosq. Control Assoc., 16*, 223–238.

Cornel, A. J., Stanich, M. A., McAbee, R. D., & Mulligan, F. S., 3rd (2002). High level methoprene resistance in the mosquito *Ochlerotatus nigromaculis* (Ludlow) in central California. *Pest. Manag. Sci., 58*, 791–798.

Cotton, G., & Anstee, J. H. (1991). A biochemical and structural study on the effects of methoprene on fat body development in *Locusta migratoria* L. *J. Insect Physiol.*, *37*, 525–539.

Cripe, G. M., McKenney, C. L., Jr., Hoglund, M. D., & Harris, P. S. (2003). Effects of fenoxycarb exposure on complete larval development of the xanthid crab, *Rhithropanopeus harrisii*. *Environ. Pollut.*, *125*, 295–299.

Cristino, A. S., Nunes, F. M.F., Lobo, C. H., Bitondi, M. M. G., Simoes, Z. L. P., Costa, L. D., Lattorff, H. M. G., Moritz, R. F. A., Evans, J. D., & Hartfelder, K. (2006). Caste development and reproduction: a genome-wide analysis of hallmarks of insect eusociality. *Insect Mol. Biol.*, *15*, 703–714.

Crossley, A. C. (1985). Nephrocytes and pericardial cells. In G. A. Kerkut, & L. I. Gilbert (Eds.), *Comprehensive Insect Physiology, Biochemistry and Pharmacology* (Vol. 3), (pp. 487–515). New York: Pergamon Press.

Crowder, D. W., Dennehy, T. J., Ellers-Kirk, C., Yafuso, L. C., Ellsworth, P. C., Tabashnik, B. E., & Carriere, Y. (2007). Field evaluation of resistance to pyriproxyfen in *Bemisia tabaci* (B biotype). *J. Econ. Entomol.*, *100*, 1650–1656.

Cunha, A. D., Nascimento, A. M., Guidugli, K. R., Simoes, Z. L. P., & Bitondi, M. M. G. (2005). Molecular cloning and expression of a hexamerin cDNA from the honey bee, *Apis mellifera*. *J. Insect Physiol.*, *51*, 1135–1147.

Cusson, M., Yagi, K. J., Ding, Q., Duve, H., Thorpe, A., McNeil, J. N., & Tobe, S. S. (1991). Biosynthesis and release of juvenile hormone and its precursors in insects and crustaceans: The search for a unifying arthropod endocrinology. *Insect Biochem. Mol. Biol.*, *21*, 1–6.

Cusson, M., Tobe, S. S., & Mcneil, J. N. (1994). Juvenile hormones - Their role in the regulation of the pheromonal communication system of the armyworm moth, *Pseudaletia unipuncta*. *Arch. Insect Biochem. Physiol.*, *25*, 329–345.

Cusson, M., Le Page, A., McNeil, J. N., & Tobe, S. S. (1996). Rate of isoleucine metabolism in lepidopteran corpora allata: regulation of the proportion of juvenile hormone homologues released. *Insect Biochem. Mol. Biol.*, *26*, 195–201.

Cusson, M., Delisle, J., & Miller, D. (1999). Juvenile hormone titers in virgin and mated *Choristoneura fumiferana* and *C. rosaceana* females: assessment of the capacity of males to produce and transfer JH to the female during copulation. *J. Insect Physiol.*, *45*, 637–646.

Cusson, M., Miller, D., & Goodman, W. G. (1997). Characterization of antibody 444 using chromatographically purified enantiomers of juvenile hormones I, II, and III: Implications for radioimmunoassays. *Analyt. Biochem.*, *249*, 83–87.

Cusson, M., & Palli, S. (2000). Can juvenile hormone research help rejuvenate integrated pest management? *Can. Entomol.*, *132*, 263–280.

Cusson, M., Béliveau, C., Sen, S. E., Vandermoten, S., Rutledge, R. J., Stewart, D., Francis, F., Haubruge, É., Rehse, P., Huggins, D. J., Dowling, A. P.G., & Grant, G. H. (2006). Characterization and tissue-specific expression of two lepidopteran farnesyl diphosphate synthase homologues: implications for the biosynthesis of ethyl-substituted juvenile hormones. *Proteins*, *65*, 742–758.

Cymborowski, B., Bogus, M., Beckage, N. E., Williams, C. M., & Riddiford, L. M. (1982). Juvenile hormone titers and metabolism during starvation-induced supernumerary larval moulting of the tobacco hormworm, *Manduca sexta*. *J. Insect Physiol.*, *28*, 129–135.

Dahm, K. H., Röller, H., & Trost, B. M. (1968). The JH: IV. Stereochemistry of JH and biological activity of some of its isomers and related compounds. *Life Sci.*, *7*, 129–137.

Darrouzet, E., Mauchamp, B., Prestwich, G. D., Kerhoas, L., Ujvary, I., & Couillaud, F. (1997). Hydroxy juvenile hormones: New putative juvenile hormones biosynthesized by locust corpora allata *in vitro*. *Biochem. Biophys. Res. Commun.*, *240*, 752–758.

Darrouzet, E., Rossignol, F., & Couillaud, F. (1998). The release of isoprenoids by locust corpora allata *in vitro*. *J. Insect Physiol.*, *44*, 103–111.

Davey, K. G. (2000). The modes of action of juvenile hormones: some questions we ought to ask. *Insect Biochem. Mol. Biol.*, *30*, 663–669.

De Loof, A. (2008). Ecdysteroids, juvenile hormone and insect neuropeptides: Recent successes and remaining major challenges. *Gen. Comp. Endocrinol.*, *155*, 3–13.

de Rooij, J., Zwartkruis, F. J., Verheijen, M. H., Cool, R. H., Nijman, S. M., Wittinghofer, A., & Bos, J. L. (1998). Epac is a Rap1 guanine-nucleotide-exchange factor directly activated by cyclic AMP. *Nature*, *396*, 474–477.

Dean, R. L., Locke, M., & Collins, J. V. (1985). Structure of the fat body. In G. A. Kerkut, & L. I. Gilbert (Eds.), *Comprehensive Insect Physiology, Biochemistry and Pharmacology* (Vol. 3), (pp. 155–210). Oxford: Pergamon Press.

Debernard, S., Rossignol, F., Malosse, C., Mauchamp, B., & Couillaud, F. (1995). Transesterification of juvenile hormone occurs in vivo in locust when injected in alcoholic solvents. *Experientia.*, *51*, 1220–1224.

Debernard, S., Morisseau, C., Severson, T. F., Feng, L., Wojtasek, H., Prestwich, G. D., & Hammock, B. D. (1998). Expression and characterization of the recombinant juvenile hormone epoxide hydrolase (JHEH) from *Manduca sexta*. *Insect Biochem. Mol. Biol.*, *28*, 409–419.

Debski, J., Wyslouch-Cieszynska, A., Dadlez, M., Grzelak, K., Kludkiewicz, B., Kolodziejczyk, R., Lalik, A., Ozyhar, A., & Kochman, M. (2004). Positions of disulfide bonds and N-glycosylation site in juvenile hormone binding protein. *Arch. Biochem. Biophys.*, *421*, 260–266.

deKort, C. A. (1990). Thirty-five years of diapause research with the Colorado potato beetle *Entomol. Exp. Appl.*, *56*, 1–13.

deKort, C. A. D., & Granger, N. A. (1996). Regulation of JH titers: The relevance of degradative enzymes and binding proteins. *Arch. Insect Biochem. Physiol.*, *33*, 1–26.

Dingle, H., & Winchell, R. (1997). Juvenile hormone as a mediator of plasticity in insect life histories. *Arch. Insect Biochem. Physiol.*, *35*, 359–373.

Dubrovsky, E. B. (2005). Hormonal cross talk in insect development. *Trends Endocrinol. Metabol.*, *16*, 6–11.

Edwards, J. P., Bell, H. A., Audsley, N., Marris, G. C., Kirkbride-Smith, A., Bryning, G., Frisco, C., & Cusson, M. (2006). The ectoparasitic wasp *Eldophus pennicornis* (Hymenoptera: Eulophiclae) uses instar-specific endocrine disruption strategies to suppress the development of its host *Lacanobia oleracea* (Lepidoptera: Noctuidae). *J. Insect Physiol.*, *52*, 1153–1162.

Elekonich, M. M., & Robinson, G. E. (2000). Organizational and activational effects of hormones on insect behavior. *J. Insect Physiol.*, *46*, 1509–1515.

Elekonich, M. M., Schulz, D. J., Bloch, G., & Robinson, G. E. (2001). Juvenile hormone levels in honey bee (*Apis mellifera* L.) foragers: foraging experience and diurnal variation. *J. Insect Physiol.*, *47*, 1119–1125.

Elekonich, M. M., Jez, K., Ross, A. J., & Robinson, G. E. (2003). Larval juvenile hormone treatment affects pre-adult development, but not adult age at onset of foraging in worker honey bees (*Apis mellifera). J. Insect Physiol.*, *49*, 359–366.

Emlen, D. J., Szafran, Q., Corley, L. S., & Dworkin, I. (2006). Insulin signaling and limb-patterning: candidate pathways for the origin and evolutionary diversification of beetle 'horns'. *Heredity.*, *97*, 179–191.

Engelmann, F., & Mala, J. (2000). The interactions between juvenile hormone (JH), lipophorin, vitellogenin, and JH esterases in two cockroach species. *Insect Biochem. Mol. Biol.*, *30*, 793–803.

Erezyilmaz, D. F., Riddiford, L. M., & Truman, J. W. (2006). The pupal specifier broad directs progressive morphogenesis in a direct-developing insect. *Proc. Natl. Acad. Sci. USA*, *103*, 6925–6930.

Erezyilmaz, D. F., Rynerson, M. R., Truman, J. W., & Riddiford, L. M. (2009). The role of the pupal determinant broad during embryonic development of a direct-developing insect. *Dev. Genes Evol.*, *219*, 535–544.

Esteva, M., Ruiz, A. M., & Stoka, A. M. (2002). *Trypanosoma cruzi*: methoprene is a potent agent to sterilize blood infected with trypomastigotes. *Exp. Parasitol.*, *100*, 248–251.

Esteva, M., Maidana, C., Sinagra, A., Luna, C., Ruiz, A. M., & Stoka, A. M. (2005). Effect of a juvenile hormone analogue on *Leishmania amazonensis* and *Leishmania braziliensis*. *Exp. Parasitol.*, *110*, 162–164.

Everson, R. D., & Feir, D. (1976). Juvenile hormone regulation of cyclic AMP and cAMP phosphodiesterase activity in *Oncopeltus fasciatus*. *J. Insect Physiol.*, *22*, 781–784.

Fain, M., & Riddiford, L. M. (1975). Juvenile hormone titers in the hemolymph during late larval development of the Tobacco Hornworm, *Manduca sexta*. *Biol. Bull.*, *149*, 506–521.

Fan, Y. L., Schal, C., Vargo, E. L., & Bagneres, A. G. (2004). Characterization of termite lipophorin and its involvement in hydrocarbon transport. *J. Insect Physiol.*, *50*, 609–620.

Fang, F., Xu, Y., Jones, D., & Jones, G. (2005). Interactions of ultraspiracle with ecdysone receptor in the transduction of ecdysone- and juvenile hormone-signaling. *FEBS J.*, *272*, 1577–1589.

Faulkner, D. J., & Petersen, H. R. (1971). Synthesis of C18 cecropia juvenile hormone to obtain optically active forms of known absolute configuration. *J. Amer. Chem. Soc.*, *93*, 3766–3767.

Feng, Q. L., Ladd, T. R., Tomkins, B. L., Sundaram, M., Sohi, S. S., Retnakaran, A., Davey, K. G., & Palli, S. R. (1999). Spruce budworm (*Choristoneura fumiferana*) juvenile hormone esterase: hormonal regulation, developmental expression and cDNA cloning. *Mol. Cell. Endocrinol.*, *148*, 95–108.

Fersht, A. (1985). *Enzyme Structure and Mechanisms* (2nd ed.). San Francisco: W.H. Freeman and Co.

Feyereisen, R., & Tobe, S. S. (1981). A rapid partition assay for routine analysis of juvenile hormone release by insect corpora allata. *Analyt. Biochem.*, *111*, 372–375.

Feyereisen, R., Pratt, G. E., & Hamnett, A. F. (1981). Enzymic synthesis of juvenile hormone in locust corpora allata: evidence for a microsomal cytochrome P-450 linked methyl farnesoate epoxidase. *Eur. J. Biochem.*, *118*, 231–238.

Feyereisen, R. (1985). Regulation of juvenile hormone titer: synthesis. In G. A. Kerkut, & L. I. Gilbert (Eds.), *Comprehensive Insect Physiology Biochemistry and Pharmacology* (Vol. 7), (pp. 391–430). Oxford: Pergamon Press.

Gade, G., Hoffmann, K. H., & Spring, J. H. (1997). Hormonal regulation in insects: Facts, gaps, and future directions. *Physiol. Rev.*, *77*, 963–1032.

Gelman, D. B., Pszczolkowski, M. A., Blackburn, M. B., & Ramaswamy, S. B. (2007). Ecdysteroids and juvenile hormones of whiteflies, important insect vectors for plant viruses. *J. Insect Physiol.*, *53*, 274–284.

Gilbert, L. I., & Schneiderman, H. A. (1958). On the specificity of juvenile hormone biosynthesis in the male *cecropia*. *Science*, *128*, 844.

Gilbert, L. I., Granger, N. A., & Roe, R. M. (2000). The juvenile hormones: historical facts and speculations on future research directions. *Insect Biochem. Mol. Biol.*, *30*, 617–644.

Godlewski, J., Wang, S., & Wilson, T. G. (2006). Interaction of bHLH-PAS proteins involved in juvenile hormone reception in *Drosophila. Biochem. Biophys. Res. Commun.*, *342*, 1305–1311.

Goodman, C. L., Wagner, R. M., Nabli, H., Wright-Osment, M. K., Okuda, T., & Coudron, T. A. (2005). Partial morphological and functional characterization of the corpus allatum-corpus cardiacum complex from the two-spotted stinkbug, *Perillus bioculatus* (Hemiptera: Pentatomidae). *In vitro Cell Dev. Biol.-Anim.*, *41*, 71–76.

Goodman, W. G., Bollenbacher, W. E., Zvenko, H. L., & Gilbert, L. I. (1976). A competitive binding protein assay for juvenile hormone. In L. I. Gilbert (Ed.), *The Juvenile Hormones* (pp. 75–95). New York: Plenum Press.

Goodman, W. G., Schooley, D. A., & Gilbert, L. I. (1978). Specificity of the juvenile hormone binding protein: The geometrical isomers of juvenile hormone I. *Proc. Natl. Acad. Sci. USA*, *75*, 185–189.

Goodman, W. G., Adams, B., & Trost, J. T. (1985). Purification and characterization of a biliverdin-associated protein from the hemolymph of *Manduca sexta*. *Biochemistry*, *24*, 1168–1175.

Goodman, W. G., Tatham, G., Nesbit, D. J., Bultmann, H., & Sutton, R. D. (1987). The role of juvenile hormone in endocrine control of pigmentation in *Manduca sexta*. *Insect Biochem.*, *17*, 1065–1070.

Goodman, W. G. (1990). Biosynthesis, titer regulation and transport of juvenile hormones. In A. P. Gupta (Ed.), *Morphogenetic Hormones of Arthropods: Discoveries, Syntheses, Metabolism, Evolution, Mode of Action and Techniques* (pp. 83–124). New Brunswick, NJ: Rutgers University Press.

Goodman, W. G., Huang, Z. H., Robinson, G. E., Strambi, C., & Strambi, A. (1993). Comparison of 2 Juvenile Hormone Radioimmunoassays. *Arch. Insect Biochem. Physiol.*, *23*, 147–152.

Goodman, W. G., Orth, A. P., Toong, Y. C., Ebersohl, R., Hiruma, K., & Granger, N. A. (1995). Recent advances in radioimmunoassay technology for the juvenile hormones. *Arch. Insect Biochem. Physiol., 30*, 295–305.

Goodman, W. G., & Granger, N. A. (2005). Comprehensive Molecular Insect Science. In L. I. Gilbert, K. Iatrou, & S. S. Gill (Eds.), *The Juvenile Hormones* (Vol. 3), (pp. 319–409). Oxford: Elsevier/Pergamon.

Granger, N. A., Janzen, W. P., & Ebersohl, R. (1995). Biosynthetic products of the corpus allatum of the tobacco hornworm, *Manduca sexta. Insect Biochem. Mol. Biol., 25*, 427–439.

Grasberger, H., Golcher, H. M. B., Fingerhut, A., & Janssen, O. E. (2002). Loop variants of the serpin thyroxine binding globulin: implications for hormone release upon limited proteolysis. *Biochem. J., 365*, 311–316.

Grieneisen, M. L., Kieckbusch, T. D., Mok, A., Dorman, G., Latli, B., Prestwich, G. D., & Schooley, D. A. (1995). Characterization of the juvenile hormone epoxide hydrolase (JHEH) and juvenile hormone diol phosphotransferase (JHDPT) from *Manduca sexta* Malpighian tubules. *Arch. Insect Biochem. Physiol., 30*, 255–270.

Grieneisen, M. L., Mok, A., Kiechbusch, I. D. & Schooley, D. A. (1997). The specificity of juvenile hormone esterase revisited. *Insect Biochem. Mol. Biol., 27*, 365–376.

Grishkovskaya, I., Avvakumov, G. V., Hammond, G. L., Catalano, M. G., & Muller, Y. A. (2002). Steroid ligands bind human sex hormone binding globulin in specific orientations and produce distinct changes in protein conformation. *J. Biol. Chem., 277*, 32086–32093.

Grozinger, C. M., & Robinson, G. E. (2007). Endocrine modulation of a pheromone-responsive gene in the honey bee brain. *J. Comp. Physiol. Part A., 193*, 461–470.

Gruntenko, N. E., Wilson, T. G., Monastirioti, M., & Rauschenbach, I. Y. (2000). Stress-reactivity and juvenile hormone degradation in *Drosophila melanogaster* strains having stress-related mutations. *Insect Biochem. Mol. Biol., 30*, 775–783.

Guerrero, A., & Rosell, G. (2005). Biorational approaches for insect control by enzymatic inhibition. *Curr. Med. Chem., 12*, 461–469.

Hagner-Holler, S., Pick, C., Girgenrath, S., Marden, J. H., & Burmester, T. (2007). Diversity of stonefly hexamerins and implication for the evolution of insect storage proteins. *Insect Biochem. Mol. Biol., 37*, 1064–1074.

Halarnkar, P. P., & Schooley, D. A. (1990). Reversed-phase liquid chromatographic separation of juvenile hormone and its metabolites and its application for an *in vivo* juvenile hormone catabolism study in *Manduca sexta. Analyt. Biochem., 188*, 394–397.

Halarnkar, P. P., Jackson, G. P., Straub, K. M., & Schooley, D. A. (1993). Juvenile hormone catabolism in *Manduca sexta* - Homologue selectivity of catabolism and identification of a diol-phosphate conjugate as a major end product. *Experientia., 49*, 988–994.

Hamiaux, C., Stanley, D., Greenwood, D. R., Baker, E. N., & Newcomb, R. D. (2009). Crystal structure of *Epiphyas postvittana* takeout 1 with bound ubiquinone supports a role as ligand carriers for takeout proteins in insects. *J. Biol. Chem., 284*, 3496–3503.

Hammes, A., Andreassen, T. K., Spoelgen, A., Raila, J., Hubner, N., Schulz, H., Metzger, J., Schweigert, F. J., Luppa, P. B., Nykjaer, A., & Willnow, T. E. (2005). Cellular uptake of sex steroid hormones. *Cell, 122*, 751–762.

Hammock, B. D., Nowock, J., Goodman, W. G., Stamoudis, V., & Gilbert, L. I. (1975). The influence of hemolymph binding protein on juvenile hormone stability and distribution in *Manduca sexta* fat body and imaginal discs *in vitro. Mol. Cell. Endocrinol., 3*, 167–184.

Hammock, B. D. (1975). NADPH dependent epoxidation of methyl farnesoate to juvenile hormone in the cockroach *Blaberus giganteus* L. *Life Sci., 17*, 323–328.

Hammock, B. D., Wing, K. D., McLaughlin, J., Lovell, V. M., & Sparks, T. C. (1982). Trifluoromethylketones as possible transition state analog inhibitors of juvenile hormone esterase. *Pest. Biochem. Physiol., 17*, 76–88.

Hammock, B. D. (1985). Regulation of juvenile hormone titer: degradation. In G. A. Kerkut, & L. I. Gilbert (Eds.), *Comprehensive Insect Physiology, Biochemistry and Pharmacology* (Vol. 7), (pp. 431–472). New York: Pergamon Press.

Hammock, B. D. (1998). *Status of recombinant baculoviruses in insect pest control.* Amsterdam, Netherlands: I O S Press.

Handschin, C., & Meyer, U. A. (2005). Regulatory network of lipid-sensing nuclear receptors: roles for CAR, PXR, LXR, and FXR. *Arch. Biochem. Biophys., 433*, 387–396.

Hansen, I. A., Gutsmann, V., Meyer, S. R., & Scheller, K. (2003). Functional dissection of the hexamerin receptor and its ligand arylphorin in the blowfly *Calliphora vicina. Insect Mol. Biol., 12*, 427–432.

Hanzlik, T. N., & Hammock, B. D. (1987). Characterization of affinity purified JHE from *Trichoplusia ni. J. Biol. Chem., 262*, 13584–13589.

Hanzlik, T. N., & Hammock, B. D. (1988). Characterization of juvenile hormone hydrolysis in early larval development of *Trichoplusia ni. Arch. Insect Biochem. Physiol., 9*, 135–156.

Hanzlik, T. N., Abdel-aal, Y. A.I., Harshman, L. G., & Hammock, B. D. (1989). Isolation and sequencing of cDNA clones coding for juvenile hormone esterase from *Heliothis virescens* – Evidence for a catalytic mechanism for the serine-carboxylesterases cifferent from that of the serine proteases. *J. Biol. Chem., 264*, 12419–12425.

Harmon, M. A., Boehm, M. F., Heyman, R. A., & Mangelsdorf, D. J. (1995). Activation of mammalian retinoid X receptors by the insect growth regulator methoprene. *Proc. Natl. Acad. Sci. USA, 92*, 6157–6160.

Harris, S. V., Thompson, D. M., Linderman, R. J., Tomalski, M. D., & Roe, R. M. (1999). Cloning and expression of a novel juvenile hormone-metabolizing epoxide hydrolase during larval-pupal metamorphosis of the cabbage looper, *Trichoplusia ni. Insect Mol. Biol., 8*, 85–96.

Harshman, L. G., Casas, J., Dietze, E. C., & Hammock, B. D. (1991). Epoxide hydrolase activities in *Drosophila melanogaster. Insect Biochem., 21*, 887–894.

Harshman, L. G., Ward, V. K., Beetham, J. K., Grant, D. F., Grahan, L. J., Zraket, C. A., Heckel, D. G., & Hammock, B. D. (1994). Cloning, characterization, and genetics of the juvenile hormone esterase gene from *Heliothis virescens. Insect Biochem. Mol. Biol., 24*, 671–676.

Harshman, L. G., Song, K. D., Casas, J., Shuurmans, A., Kuwano, E., Kachman, S. D., Riddiford, L. M., & Hammock, B. D. (2010). Bioassays of compounds with potential juvenoid activity on *Drosophila melanogaster*: Juvenile hormone III, bisepoxide JH III and methyl farnesoates. *J. Insect Physiol.*, *56*, 1465–1470.

Hathaway, M., Hatle, J., Li, S., Ding, X., Barry, T., Hong, F., Wood, H., & Borst, D. (2009). Characterization of hexamerin proteins and their mRNAs in the adult lubber grasshopper: The effects of nutrition and juvenile hormone on their levels. *Comp. Biochem. Physiol. - Part A: Mol. Integ. Physiol.*, *154*, 323–332.

Haunerland, N. H., & Shirk, P. D. (1995). Regional and functional differentiation in the insect fat body. *Ann. Rev. Entomol.*, *40*, 121–145.

Hay, B. A., & Guo, M. (2006). Caspase-dependent cell death in *Drosophila*. *Ann. Rev. Cell Dev. Biol.*, *22*, 623–650.

Hayakawa, Y. (1990). Juvenile hormone esterase activity repressive factor in the plasma of parasitized insect larvae. *J. Biol. Chem.*, *265*, 10813–10816.

Heftmann, E., & Mosettig, E. (1960). *Biochemistry of Steroids*. New York: Reinhold Publishing.

Hegstrom, C. D., Riddiford, L. M., & Truman, J. W. (1998). Steroid and neuronal regulation of ecdysone receptor expression during metamorphosis of muscle in the moth, *Manduca sexta*. *J. Neurosci.*, *18*, 1786–1794.

Helvig, C., Koener, J. F., Unnithan, G. C., & Feyereisen, R. (2004). CYP15A1, the cytochrome P450 that catalyzes epoxidation of methyl farnesoate to juvenile hormone III in cockroach corpora allata. *Proc. Natl. Acad. Sci. USA*, *101*, 4024–4029.

Henrich, V. C., Burns, E., Yelverton, D. P., Christensen, E., & Weinberger, C. (2003). Juvenile hormone potentiates ecdysone receptor-dependent transcription in a mammalian cell culture system. *Insect Biochem. Mol. Biol.*, *33*, 1239–1247.

Henrich, V. C., & Beatty, J. M. (2010). Nuclear receptors in *Drosophila* melanogaster. In *Handbook of Cell Signaling* (2nd ed.). (pp. 2027–2037). San Diego: Academic Press.

Henrick, C. A. (1982). Juvenoid-structure activity. In J. R. Coats (Ed.), *Insecticide Mode of Action* (pp. 315–402). New York: Academic Press.

Henrick, C. A. (2007). Methoprene. *J. Amer. Mosq. Control Assoc.*, *23*, 225–239.

Hewes, R. S., & Truman, J. W. (1994). Steroid regulation of excitability in identified insect neurosecretory cells. *J. Neurosci.*, *14*, 1812–1819.

Hewes, R. S. (2008). The buzz on fly neuronal remodeling. *Trends Endocrinol. Metab*, *19*, 317–323.

[HGSC] Honeybee Genome Sequencing Consortium (2006). Insights into social insects from the genome of the honeybee *Apis mellifera*. *Nature*, *443*, 931–949.

Hidayat, P., & Goodman, W. G. (1994). Juvenile hormone and hemolymph juvenile hormone binding protein titers and their interaction in the hemolymph of fourth stadium *Manduca sexta*. *Insect Biochem. Mol. Biol.*, *24*, 709–715.

Hinton, A. C., & Hammock, B. D. (2001). Purification of juvenile hormone esterase and molecular cloning of the cDNA from *Manduca sexta*. *Insect Biochem. Mol. Biol.*, *32*, 57–66.

Hinton, A. C., & Hammock, B. D. (2003). In vitro expression and biochemical characterization of juvenile hormone esterase from *Manduca sexta*. *Insect Biochem. Mol. Biol.*, *33*, 317–329.

Hirai, M., Kamimura, M., Kikuchi, K., Yasukochi, Y., Kiuchi, M., Shinoda, T., & Shiotsuki, T. (2002). cDNA cloning and characterization of *Bombyx mori* juvenile hormone esterase: an inducible gene by the imidazole insect growth regulator KK-42. *Insect Biochem. Mol. Biol.*, *32*, 627–635.

Hiruma, K., & Riddiford, L. M. (1984). Regulation of melanization of tobacco hornworm larval cuticle *in vitro*. *J. Exp. Zool.*, *230*, 393–403.

Hiruma, K., & Riddiford, L. M. (1988). Granular phenoloxidase involved in cuticular melanization in the tobacco hornworm: regulation of its synthesis in the epidermis by juvenile hormone. *Dev. Biol.*, *130*, 87–97.

Hiruma, K., Carter, M. S., & Riddiford, L. M. (1995). Characterization of the dopa decarboxylase gene of *Manduca sexta* and its suppression by 20-hydroxyecdysone. *Dev. Biol.*, *169*, 195–209.

Hiruma, K., & Riddiford, L. M. (2009). The molecular mechanisms of cuticular melanization: the ecdysone cascade leading to dopa decarboxylase expression in *Manduca sexta*. *Insect Biochem. Mol. Biol.*, *39*, 245–253.

Hiruma, K., & Riddiford, L. M. (2010). Developmental expression of mRNAs for epidermal and fat body proteins and hormonally regulated transcription factors in the tobacco hornworm, *Manduca sexta*. *J. Insect Physiol.*, *56*, 1390–1395.

Hlavacek, J., Koudelka, J., & Jarv, J. (1993). Structure activity relationships in peptide juvenoids. *Bioorg. Chem.*, *21*, 7–13.

Holford, K. C., Edwards, K. A., Bendena, W. G., Tobe, S. S., Wang, Z., & Borst, D. W. (2004). Purification and characterization of a mandibular organ protein from the American lobster, *Homarus americanus*: a putative farnesoic acid O-methyltransferase. *Insect Biochem. Mol. Biol.*, *34*, 785–798.

Hopkins, T. E., & Kramer, K. J. (1992). Insect cuticle sclerotization. *Ann. Rev. Entomol.*, *37*, 273–302.

Horodyski, F. M., & Riddiford, L. M. (1989). Expression and hormonal control of a new larval cuticular multigene family at the onset of metamorphosis of the tobacco hornworm. *Dev. Biol.*, *132*, 292–303.

Horseman, G., Hartmann, R., Virantdoberlet, M., Loher, W., & Huber, F. (1994). Nervous control of juvenile hormone biosynthesis in *Locusta migratoria*. *Proc. Natl. Acad. Sci. USA*, *91*, 2960–2964.

Huang, L., Cheng, T., Xu, P., Cheng, D., Fang, T., & Xia, Q. (2009). A genome-wide survey for host response of silkworm, *Bombyx mori* during pathogen *Bacillus bombysepieus* infection. *PLoS One*, *4*, e8098.

Hui, J. H. L., Hayward, A., Bendena, W. G., Takahashi, T., & Tobe, S. S. (2010). Evolution and functional divergence of enzymes involved in sesquiterpenoid hormone biosynthesis in crustaceans and insects. *Peptides*, *31*, 451–455.

Hwang, S. J., Cheon, H. M., Kim, H. J., Chae, K. S., Chung, D. H., Kim, M. O., Park, J. S., & Seo, S. J. (2001). cDNA sequence and gene expression of storage protein-2. A juvenile hormone-suppressible hexamerin from the fall webworm, *Hyphantria cunea*. *Comp. Biochem. Physiol. Part B.*, *129*, 97–107.

Ichinose, R., Nakamura, A., Yamoto, T., Booth, T. F., Maeda, S., & Hammock, B. D. (1992). Uptake of juvenile hormone esterase by pericardial cells of *Manduca sexta*. *Insect Biochem. Mol. Biol.*, *22*, 893–904.

Ismail, S. M., & Gillott, C. (1995). Identification, characterization, and developmental profile of a high molecular weight, juvenile hormone binding protein in the hemolymph of the migratory grasshopper, *Melanoplus sanguinipes. Arch. Insect Biochem. Physiol., 29*, 415–430.

Iwema, T., Billas, I. M., Beck, Y., Bonneton, F., Nierengarten, H., Chaumot, A., Richards, G., Laudet, V., & Moras, D. (2007). Structural and functional characterization of a novel type of ligand-independent RXR-USP receptor. *EMBO J., 26*, 3770–3782.

Iwema, T., Chaumot, A., Studer, R. A., Robinson-Rechavi, M., Billas, I. M., Moras, D., Laudet, V., & Bonneton, F. (2009). Structural and evolutionary innovation of the heterodimerization interface between USP and the ecdysone receptor ECR in insects. *Mol. Biol. Evol., 26*, 753–768.

Jeffery, J., Navia, B., Atkins, G., & Stout, J. (2005). Selective processing of calling songs by auditory interneurons in the female cricket, *Gryllus pennsylvanicus*: Possible roles in behavior. *J. Exp. Zool. Part A Comp. Exp. Biol., 303A*, 377–392.

Jensen, A. L., & Brasch, K. (1985). Nuclear development in locust fat body: the influence of juvenile hormone on inclusion bodies and the nuclear matrix. *Tiss. Cell, 17*, 117–130.

Jesudason, P., Anspaugh, D. D., & Roe, R. M. (1992). Juvenile hormone metabolism in the plasma, integument, midgut, fat body, and brain during the last instar of the tobacco hornworm, *Manduca sexta. Arch. Insect Biochem. Physiol., 20*, 87–105.

Jiang, H., Wang, Y., Yu, X.-Q., & Zhu, Y. (2003). Prophenoloxidase-activating proteinase-3 from *Manduca sexta* hemolymph: a clip-domain serine proteinase regulated by serpin-1 and serine proteinase homologues. *Insect Biochem. Mol. Biol., 33*, 10491060.

Jones, D., & Jones, G. (2007). Farnesoid secretions of dipteran ring glands: What we do know and what we can know. *Insect Biochem. Mol. Biol., 37*, 771–798.

Jones, G., Hiremath, S. T., Hellmann, G. M., & Rhoads, R. E. (1988). Juvenile hormone regulation of mRNA levels for a highly abundant hemolymph protein in larval *Trichoplusia ni. J. Biol. Chem., 263*, 1089–1092.

Jones, G., Brown, N., Manczak, M., Hiremath, S., & Kafatos, F. C. (1990a). Molecular cloning, regulation, and complete sequence of a hemocyanin-related, juvenile hormone-suppressible protein from insect hemolymph. *J. Biol. Chem., 265*, 8596–8602.

Jones, G., Hanzlik, T., Hammock, B. D., Schooley, D. A., Miller, C. A., Tsai, L. W., & Baker, F. C. (1990b). The juvenile hormone titre during the penultimate and ultimate larval stadia of *Trichoplusia ni. J. Insect Physiol., 36*, 77–83.

Jones, G., & Sharp, P. A. (1997). Ultraspiracle: An invertebrate nuclear receptor for juvenile hormones. *Proc. Natl. Acad. Sci. USA, 94*, 13499–13503.

Jones, G., Manczak, M., Schelling, D., Turner, H., & Jones, D. (1998). Transcription of the juvenile hormone esterase gene under the control of both an initiator and AT-rich motif. *Biochem. J., 335*, 79–84.

Jones, G., Wozniak, M., Chu, Y. X., Dhar, S., & Jones, D. (2001). Juvenile hormone III-dependent conformational changes of the nuclear receptor ultraspiracle. *Insect Biochem. Mol. Biol., 32*, 33–49.

Jones, G., Jones, D., Teal, P., Sapa, A., & Wozniak, M. (2006). The retinoid-X receptor orthologue, ultraspiracle, binds with nanomolar affinity to an endogenous morphogenetic ligand. *FEBS J., 273*, 4983–4996.

Jones, G., Jones, D., Li, X., Tang, L., Ye, L., Teal, P., Riddiford, L. M., Sandifer, C., Borovsky, D., & Martin, J. R. (2010). Activities of natural methyl farnesoids on pupariation and metamorphosis of *Drosophila melanogaster. J. Insect Physiol., 56*, 1456–1464.

Judy, K. J., Schooley, D. A., Dunham, L. L., Hall, M. S., Bergot, J., & Siddall, J. B. (1973). Isolation, structure, and absolute configuration of a new natural insect juvenile hormone from *Manduca sexta. Proc. Natl. Acad. Sci. USA, 70*, 1509–1513.

Jurcek, O., Wimmer, Z., Bennettova, B., Moravcova, J., Drasar, P., & Saman, D. (2009). Novel juvenogens (insect hormonogenic agents): preparation and biological tests on *Neobellieria bullata. J. Agric. Food Chem., 57*, 10852–10858.

Kahn, S. M., Hryb, D. J., Nakhla, A. M., Romas, N. A., & Rosner, W. (2002). Sex hormone binding globulin is synthesized in target cells. *J. Endocrinol., 175*, 113–120.

Kallapur, V. L., Majumder, C., & Roe, R. M. (1996). *In vivo* and *in vitro* tissue specific metabolism of juvenile hormone during the last stadium of the cabbage looper, *Trichoplusia ni. J. Insect Physiol., 42*, 181–190.

Kamita, S. G., Hinton, A. C., Wheelock, C. E., Wogulis, M. D., Wilson, D. K., Wolf, N. M., Stok, J. E., Hock, B., & Hammock, B. D. (2003). Juvenile hormone (JH) esterase: why are you so JH specific? *Insect Biochem. Mol. Biol., 33*, 1261–1273.

Kamita, S. G., Wogulis, M. D., Law, C. S., Morisseau, C., Tanaka, H., Huang, H., Wilson, D. K., & Hammock, B. D. (2010). Function of phenylalanine 259 and threonine 314 within the substrate binding pocket of the juvenile hormone esterase of *Manduca sexta. Biochemistry, 49*, 3733–3742.

Kamita, S. G., & Hammock, B. D. (2010). Juvenile hormone esterase: biochemistry and structure. *J. Pest. Sci., 35*, 265–274.

Keiser, K. C. L., Brandt, K. S., Silver, G. M., & Wisnewski, N. (2002). Cloning, partial purification and in vivo developmental profile of expression of the juvenile hormone epoxide hydrolase of *Ctenocephalides felis. Arch. Insect Biochem. Physiol., 50*, 191–206.

Kensler, T. W., Verma, A. K., Boutwell, R. K., & Mueller, G. C. (1978). Effects of retinoic acid and juvenile hormone on the induction of ornithine decarboxylase activity by 12-O-tetradecanoylphorbol-13-acetate. *Cancer Res., 38*, 2896–2899.

Kethidi, D. R., Perera, S. C., Zheng, S., Feng, Q. L., Krell, P., Retnakaran, A., & Palli, S. R. (2004). Identification and characterization of a juvenile hormone (JH) response region in the JH esterase gene from the spruce budworm, *Choristoneura fumiferana. J. Biol. Chem., 279*, 19634–19642.

Kethidi, D. R., Xi, Z., & Palli, S. R. (2005). Developmental and hormonal regulation of juvenile hormone esterase gene in *Drosophila melanogaster. J. Insect. Physiol., 51*, 393–400.

Kethidi, D. R., Li, Y., & Palli, S. R. (2006). Protein kinase C mediated phosphorylation blocks juvenile hormone action. *Mol. Cell. Endocrinol., 247*, 127–134.

King, L. E., & Tobe, S. S. (1988). The Identification of an enantioselective JH III binding protein from the haemolymph of the cockroach, *Diploptera punctata*. *Insect Biochem.*, *18*, 793–805.

King, L. E., & Tobe, S. S. (1993). Changes in the titre of a juvenile hormone III binding lipophorin in the hemolymph of *Diploptera punctata* during development and reproduction. *J. Insect Physiol.*, *39*, 241–252.

King, L. E., Ding, Q., Prestwich, G. D., & Tobe, S. S. (1995). The characterization of a haemolymph methyl farnesoate binding protein and the assessment of methyl farnesoate metabolism by the haemolymph and other tissues from *Procambrus clarkii*. *Insect Biochem. Mol. Biol.*, *25*, 495–501.

Kinjoh, T., Kaneko, Y., Itoyama, K., Mita, K., Hiruma, K., & Shinoda, T. (2007). Control of juvenile hormone biosynthesis in *Bombyx mori*: Cloning of the enzymes in the mevalonate pathway and assessment of their developmental expression in the corpora allata. *Insect Biochem. Mol. Biol.*, *37*, 807–818.

Kiss, I., Beaton, A. H., Tardiff, J., Fristrom, D., & Fristrom, J. W. (1988). Interactions and developmental effects of mutations in the Broad-Complex of *Drosophila melanogaster*. *Genetics*, *118*, 247–259.

Kolodziejczyk, R., Bujacz, G., Jakob, M., Ozyhar, A., Jaskolski, M., & Kochman, M. (2008). Insect juvenile hormone binding protein shows ancestral fold present in human lipid-binding proteins. *J. Mol. Biol.*, *377*, 870–881.

Konopova, B., & Jindra, M. (2007). Juvenile hormone resistance gene *Methoprene-tolerant* controls entry into metamorphosis in the beetle *Tr. castaneum*. *Proc. Natl. Acad. Sci. USA*, *104*, 10488–10493.

Konopova, B., & Jindra, M. (2008). Broad-Complex acts downstream of Met in juvenile hormone signaling to coordinate primitive holometabolan metamorphosis. *Development*, *135*, 559–568.

Koopmanschap, A. B., & de Kort, C. A. D. (1988). Isolation and characterization of a high molecular weight JH-III transport protein in the hemolymph of *Locusta migratoria*. *Arch. Insect Biochem. Physiol.*, *7*, 105–118.

Kotaki, T. (1993). Biosynthetic products by heteropteran corpora allata *in vitro*. *Appl. Entomol. Zool.*, *28*, 242–245.

Kotaki, T. (1996). Evidence for a new juvenile hormone in a stink bug, *Plautia stali*. *J. Insect Physiol.*, *42*, 279–286.

Kotaki, T. (1997). A putative juvenile hormone in a stink bug, *Plautia stali*: The corpus allatus releases a JH-active product different from any known JHs in vitro. *Invert. Reprod. Dev.*, *31*, 225–230.

Kotaki, T., Shinada, T., Kaihara, K., Ohfune, Y., & Numata, H. (2009). Structure determination of a new juvenile hormone from a heteropteran insect. *Org. Lett.*, *11*, 5234–5237.

Koyama, T., Ogura, K., & Seto, S. (1973). Studies on isopentenyl pyrophosphate isomerase with artificial substrates. *J. Biol. Chem.*, *248*, 8043–8051.

Koyama, T., Matsubara, M., & Ogura, K. (1985). Isoprenoid enzyme systems of silkworm. II. Formation of the juvenile hormone skeletons by farnesyl pyrophosphate synthase II. *J. Biochem.*, *98*, 457–463.

Krakauer, D. C., & Plotkin, J. B. (2002). Redundancy, antiredundancy, and the robustness of genomes. *Proc. Natl. Acad. Sci. USA*, *99*, 1405–1409.

Kramer, K. J., Dunn, P. E., Peterson, R., & Law, J. H. (1976). Interaction of juvenile hormone with binding proteins in insect hemolymph. In L. I. Gilbert (Ed.), *The Juvenile Hormones* (pp. 327–341). New York: Plenum Press.

Kreienkamp, H.-J., Larusson, H. J., Witte, I., Roeder, T., Birgül, N., Hönck, H.-H., Harder, S., Ellinghausen, G., Buck, F., & Richter, D. (2002). Functional annotation of two orphan G protein-coupled receptors, Drostar1 and -2, from *Drosophila melanogaster* and their ligands by reverse pharmacology. *J. Biol. Chem.*, *277*, 39937–39943.

Kroeger, H. (1968). Gene activities during insect metamorphosis and their control by hormones. In W. Etkin, & L. I. Gilbert (Eds.), *Metamorphosis: A Problem in Developmental Biology* (pp. 185–220). New York: Appleton-Century-Crofts.

Krzyzanowska, D., Lisowski, M., & Kochman, M. (1998). UV-difference and CD spectroscopy studies on juvenile hormone binding to its carrier protein. *J. Peptide Res.*, *51*, 96–102.

Lacourciere, G. M., & Armstrong, R. N. (1993). The catalytic mechansim of microsomal epoxide hydrolase involves an ester intermediate. *J. Amer. Chem. Soc.*, *115*, 10466–10467.

Lafont, R. (2000). Understanding insect endocrine systems: molecular approaches. *Entomol. Exp. Appl.*, *97*, 123–136.

Lassiter, M. T., Apperson, C. S., Crawford, C. L., & Roe, R. M. (1994). Juvenile hormone metabolism during adult development of *Culex quinquefasciatus*. *J. Med. Entomol.*, *31*, 586–593.

Lassiter, M. T., Apperson, C. S., & Roe, R. M. (1995). Juvenile hormone metabolism during the fourth stadium and pupal stage of the southern house mosquito, *Culex quinquefasciatus*. *J. Insect Physiol.*, *41*, 869–876.

Law, J. H. (1980). Lipid-protein interactions in insects. In M. Locke, & D. S. Smith (Eds.), *Insect Biology in the Future: "VBW80"* (pp. 295–310). New York: Academic Press.

Lee, C. Y., Simon, C. R., Woodard, C. T., & Baehrecke, E. H. (2002). Genetic mechanism for the stage- and tissue-specific regulation of steroid triggered programmed cell death in *Drosophila*. *Dev. Biol.*, *252*, 138–148.

Lefevre, K. S., Lacey, M. J., Smith, P. H., & Roberts, B. (1993). Identification and quantification of juvenile hormone biosynthesized by larval and adult Australian sheep blowfly *Lucilia cuprina*. *Insect Biochem. Mol. Biol.*, *23*, 713–720.

Levenbook, L. (1985). Insect storage proteins. In G. A. Kerkut, & L. I. Gilbert (Eds.), (pp. 307–346). Oxford: Pergamon Press.

Lewis, M. J., Prosser, I. M., Mohib, A., & Field, L. M. (2008). Cloning and characterisation of a prenyltransferase from the aphid *Myzus persicae* with potential involvement in alarm pheromone biosynthesis. *Insect. Mol. Biol.*, *17*, 437–443.

Li, H., & Borst, D. W. (1991). Characterization of a methyl farnesoate binding protein in hemolymph from *Libinia emarginata*. *Gen. Comp. Endocrinol.*, *81*, 335–342.

Li, S., Zhang, Q. R., Xu, W. H., & Schooley, D. A. (2005). Juvenile hormone diol kinase, a calcium-binding protein with kinase activity, from the silkworm, *Bombyx mori*. *Insect Biochem. Mol. Biol.*, *35*, 1235–1248.

Li, W. C., & Riddiford, L. M. (1994). The two duplicated insecticyanin genes, ins-a and ins-b, are differentially expressed in the tobacco hornworm, *Manduca sexta*. *Nuc. Acids Res.*, *22*, 2945–2950.

Li, W. C. (1996). Differential expression of the two duplicated insecticyanin genes, ins-a and ins-b, in the black mutant of *Manduca sexta. Arch. Biochem. Biophys., 330*, 65–70.

Li, X. (2007). Juvenile hormone and methyl farnesoate production in cockroach embryos in relation to dorsal closure and the reproductive modes of different species of cockroaches. *Arch. Insect Biochem. Physiol., 66*, 159–168.

Li, Y., Hernandez-Martinez, S., Fernandez, F., Mayoral, J. G., Topalis, P., Priestap, H., Perez, M., Navare, A., & Noriega, F. G. (2006). Biochemical, molecular, and functional characterization of PISCF-allatostatin, a regulator of juvenile hormone biosynthesis in the mosquito *Aedes aegypti. J. Biol. Chem., 281*, 34048–34055.

Li, Y., Zhang, Z., Robinson, G. E., & Palli, S. R. (2007). Identification and characterization of a juvenile hormone response element and its binding proteins. *J. Biol. Chem., 282*, 37605–37617.

Li, Z. Q., Cheng, D. J., Wei, L., Zhao, P., Shu, X., Tang, L., Xiang, Z. H., & Xia, Q. Y. (2010). The silkworm homologue of *Methoprene-tolerant (Met)* gene reveals sequence conservation but function divergence. *Insect Sci., 17*, 313–324.

Licht, O., Jungmann, D., Ludwichowski, K. U., & Nagel, R. (2004). Long-term effects of fenoxycarb on two mayfly species in artificial indoor streams. *Ecotox. Environ. Saf., 58*, 246–255.

Linderman, R. J., Roe, R. M., Harris, S. V., & Thompson, D. M. (2000). Inhibition of insect juvenile hormone epoxide hydrolase: asymmetric synthesis and assay of glycidol-ester and epoxy-ester inhibitors of *Trichoplusia ni* epoxide hydrolase. *Insect Biochem. Mol. Biol., 30*, 767–774.

Liu, Y., Sheng, Z., Liu, H., Wen, D., He, Q., Wang, S., Shao, W., Jiang, R. J., An, S., Sun, Y., Bendena, W. G., Wang, J., Gilbert, L. I., Wilson, T. G., Song, Q., & Li, S. (2009). Juvenile hormone counteracts the bHLH-PAS transcription factors MET and GCE to prevent caspase-dependent programmed cell death in *Drosophila. Development, 136*, 2015–2025.

Liu, Z. Y., Ho, L. D., & Bonning, B. (2007). Localization of a *Drosophila melanogaster* homologue of the putative juvenile hormone esterase binding protein of *Manduca sexta. Insect Biochem. Mol. Biol., 37*, 155–163.

Locke, M. (1984). The structure and development of the vacuolar system in the fat body of insects. In R. C. King, & H. Akai (Eds.), *Insect Ultrastructure* (pp. 151–197). New York: Plenum Press.

Mackert, A., Hartfelder, K., Bitondi, M. M., & Simoes, Z. L. (2010). The juvenile hormone (JH) epoxide hydrolase gene in the honey bee *(Apis mellifera)* genome encodes a protein which has negligible participation in JH degradation. *J. Insect. Physiol., 56*, 1139–1146.

Mamatha, D. M., Kanji, V. K., Cohly, H. H., & Rao, M. R. (2008). Juvenile hormone analogues, methoprene and fenoxycarb dose-dependently enhance certain enzyme activities in the silkworm *Bombyx mori* (L). *Int. J. Environ. Res. Public Health, 5*, 120–124.

Mane-Padros, D., Cruz, J., Vilaplana, L., Nieva, C., Urena, E., Belles, X., & Martin, D. (2010). The hormonal pathway controlling cell death during metamorphosis in a hemimetabolous insect. *Dev. Biol., 346*, 150–160.

Marchal, E., Vandersmissen, H. P., Badisco, L., Van de Velde, S., Verlinden, H., Iga, M., Van Wielendaele, P., Huybrechts, R., Simonet, G., Smagghe, G., & Vanden Broeck, J. (2010). Control of ecdysteroidogenesis in prothoracic glands of insects: a review. *Peptides, 31*, 506–519.

Martins, J. R., Nunes, F. M., Cristino, A. S., Simoes, Z. L., & Bitondi, M. M. (2010). The four hexamerin genes in the honey bee: structure, molecular evolution and function deduced from expression patterns in queens, workers and drones. *BMC Mol. Biol., 11*, 23.

Mauchamp, B., Darrouzet, E., Malosse, C., & Couillaud, F. (1999). 4′-OH-JH-III: an additional hydroxylated juvenile hormone produced by locust corpora allata in vitro. *Insect Biochem. Mol. Biol., 29*, 475–480.

Maxwell, R. A., Welch, W. H., Horodyski, F. M., Schegg, K. M., & Schooley, D. A. (2002a). Juvenile hormone diol kinase. II. Sequencing, cloning, and molecular modeling of juvenile hormone-selective diol kinase from *Manduca sexta. J. Biol. Chem., 277*, 21882–21890.

Maxwell, R. A., Welch, W. H., & Schooley, D. A. (2002b). Juvenile hormone diol kinase. I. Purification, characterization, and substrate specificity of juvenile hormone-selective diol kinase from *Manduca sexta. J. Biol. Chem., 277*, 21874–21881.

Mayoral, J. G., Nouzova, M., Navare, A., & Noriega, F. G. (2009a). NADP+-dependent farnesol dehydrogenase, a corpora allata enzyme involved in juvenile hormone synthesis. *Proc. Natl. Acad. Sci. USA, 106*, 21091–21096.

Mayoral, J. M., Nouzova, M., Yoshiyama, M., Shinoda, T., Hernandez-Martinez, S., Dolghih, E., Turjanski, A. G., Roitberg, A. E., Priestap, H., Perez, M., Mackenzie, L., Li, Y., & Noriega, F. G. (2009b). Molecular and functional characterization of a juvenile hormone acid methyltransferase expressed in the corpora allata of mosquitoes. *Insect Biochem. Mol. Biol., 39*, 31–37.

McKenney, C. L., Cripe, G. M., Foss, S. S., Tuberty, S. R., & Hoglund, M. (2004). Comparative embryonic and larval developmental responses of estuarine shrimp *(Palaemonetes pugio)* to the juvenile hormone agonist fenoxycarb. *Arch. Environ. Contam. Toxicol., 47*, 463–470.

McKenney, C. L. (2005). The influence of insect juvenile hormone agonists on metamorphosis and reproduction in estuarine crustaceans. *Integ. Comp. Biol, 45*, 97–105.

Mendel, C. M. (1989). The free hormone hypothesis - A physiologically based mathematical model. *Endocrinol. Rev., 10*, 232–274.

Meyer, A. S., Schneiderman, H. A., Hanzmann, E., & Ko, J. (1968). The two juvenile hormones from the *cecropia* silk moth. *Proc. Natl. Acad. Sci. USA, 60*, 853–860.

Meyer, A. S., Hanzmann, E., & Murphy, R. C. (1971). Absolute configuration of *cecropia* juvenile hormone. *Proc. Natl. Acad. Sci. USA, 68*, 2312–2315.

Min, K. J., Jones, N., Borst, D. W., & Rankin, M. A. (2004). Increased juvenile hormone levels after long-duration flight in the grasshopper, *Melanoplus sanguinipes. J. Insect Physiol., 50*, 531–537.

Minakuchi, C., & Riddiford, L. M. (2006). Insect juvenile hormone action as a potential target of pest management. *J. Pest. Sci., 31*, 77–84.

Minakuchi, C., Namiki, T., Yoshiyama, M., & Shinoda, T. (2008a). RNAi-mediated knockdown of juvenile hormone acid O-methyltransferase gene causes precocious metamorphosis in the red flour beetle *Tr. castaneum. FEBS J., 275*, 2919–2931.

Minakuchi, C., Zhou, X., & Riddiford, L. M. (2008b). Kruppel homologue 1 (Kr-h1) mediates juvenile hormone action during metamorphosis of *Drosophila melanogaster. Mech. Dev., 125*, 91–105.

Minakuchi, C., Namiki, T., & Shinoda, T. (2009). Kruppel homologue 1, an early juvenile hormone-response gene downstream of Methoprene-tolerant, mediates its anti-metamorphic action in the red flour beetle *Tr. castaneum*. *Dev. Biol.*, *325*, 341–350.

Mirth, C., Truman, J. W., & Riddiford, L. M. (2005). The role of the prothoracic gland in determining critical weight to metamorphosis in *Drosophila melanogaster*. *Curr. Biol.*, *15*, 1796–1807.

Mita, K., Kasahara, M., Sasaki, S., Nagayasu, Y., Yamada, T., Kanamori, H., Namiki, N., Kitagawa, M., Yamashita, H., Yasukochi, Y., Kadono-Okuda, K., Yamamoto, K., Ajimura, M., Ravikumar, G., Shimomura, M., Nagamura, Y., Shin-I, T., Abe, H., Shimada, T., Morishita, S., & Sasaki, T. (2004). The genome sequence of silkworm, *Bombyx mori*. *DNA Research*, *11*, 27–35.

Miura, K., Shinoda, T., Yura, M., Nomura, S., Kamiya, K., Yuda, M., & Chinzei, Y. (1998). Two hexameric cyanoprotein subunits from an insect, *Riptortus clavatus*, sequence, phylogeny and developmental and juvenile hormone regulation. *Eur. J. Biochem.*, *258*, 929–940.

Miura, K., Oda, M., Makita, S., & Chinzei, Y. (2005). Characterization of the *Drosophila Methoprene-tolerant* gene product. Juvenile hormone binding and ligand-dependent gene regulation. *FEBS J.*, *272*, 1169–1178.

Miyazaki, M., Mao, L., Henderson, G., & Laine, R. A. (2009). Liquid chromatography-electrospray ionization-mass spectrometric quantitation of juvenile hormone III in whole body extracts of the Formosan subterranean termite. *J. Chromat. Part B*, *877*, 3175–3180.

Monteiro, J. P., Jurado, A. S., Moreno, A. J., & Madeira, V. M. (2005). Toxicity of methoprene as assessed by the use of a model microorganism. *Toxicol. In Vitro*, *19*, 951–956.

Monteiro, J. P., Videira, R. A., Matos, M. J., Dinis, A. M., & Jurado, A. S. (2008). Non-selective toxicological effects of the insect juvenile hormone analogue methoprene. A membrane biophysical approach. *Appl. Biochem. Biotech.*, *150*, 243–257.

Moshitzky, P., & Applebaum, S. W. (1995). Pathway and regulation of JH III-bisepoxide biosynthesis in adult *Drosophila melanogaster* corpus allatum. *Arch. Insect Biochem. Physiol.*, *30*, 225–238.

Nader, N., Chrousos, G. P., & Kino, T. (2010). Interactions of the circadian CLOCK system and the HPA axis. *Trends Endocrinol. Metab.*, *21*, 277–286.

Nakanishi, K., Schooley, D. A., Koreeda, M., & Dillon, J. (1971). Absolute configuration of the C18-juvenile hormone: application of a new circular dichroism method using tris(dipivaloylmethanato) praseodymium. *Chem. Commun.*, 1235–1236.

Neese, P. A., Sonenshine, D. E., Kallapur, V. L., Apperson, C. S., & Roe, R. M. (2000). Absence of insect juvenile hormones in the American dog tick, *Dermacentor variabilis* (Say) (Acari:Ixodidae), and in *Ornithodoros parkeri* Cooley (Acari:Argasidae). *J. Insect Physiol.*, *46*, 477–490.

Newman, J. W., Morisseau, C., Harris, T. R., & Hammock, B. D. (2003). The soluble epoxide hydrolase encoded by EPXH2 is a bifunctional enzyme with novel lipid phosphate phosphatase actvity. *Proc. Natl. Acad. Sci. USA*, *100*, 1558–1563.

Newman, J. W., Morisseau, C., & Hammock, B. D. (2005). Epoxide hydrolases: their roles and interactions with lipid metabolism. *Prog. Lipid Res.*, *44*, 1–51.

Niimi, S., & Sakurai, S. (1997). Development changes in juvenile hormone and juvenile hormone acid titers in the hemolymph and in vitro juvenile hormone synthesis by corpora allata of the silkworm, *Bombyx mori*. *J. Insect Physiol.*, *43*, 875–884.

Nijhout, H. F. (1975). Axonal pathways in the brain-retrocerbral neuroendocrine complex of *Manduca sexta*. *Int. J. Insect Morph. Embryol.*, *4*, 529–538.

Nijhout, H. F., & Wheeler, D. E. (1982). Juvenile hormone and the physiological basis of insect polymorphisms. *Quar. Rev. Biol.*, *57*, 109–133.

Nijhout, H. F. (1994). *Insect Hormones*. Princeton, N.J: Princeton University Press.

Nijhout, H. F., & Reed, M. C. (2008). A mathematical model for the regulation of juvenile hormone titers. *J. Insect Physiol.*, *54*, 255–264.

Nijhout, H. F. (2010). Molecular and physiological basis of colour pattern formation. *Adv. Insect Physiol.*, *38*, 219–265.

Niwa, R., Niimi, T., Honda, N., Yoshiyama, M., Itoyama, K., Kataoka, H., & Shinoda, T. (2008). Juvenile hormone acid *O*-methyltransferase in *Drosophila melanogaster*. *Insect Biochem. Mol. Biol.*, *38*, 714–720.

Noriega, F. G., Edgar, K. A., Goodman, W. G., Shah, D. K., & Wells, M. A. (2001). Neuroendocrine factors affecting the steady-state levels of early trypsin mRNA in *Aedes aegypti*. *J. Insect Physiol.*, *47*, 515–522.

Noriega, F. G., Ribeiro, J. M. C., Koener, J. F., Valenzuela, J. G., Hernandez-Martinez, S., Pham, V. M., & Feyereisen, R. (2006). Genomic endocrinology of insect juvenile hormone biosynthesis. *Insect Biochem. Mol. Biol.*, *36*, 366–374.

Nowock, J., Goodman, W. G., Bollenbacher, W. E., & Gilbert, L. I. (1975). Synthesis of juvenile hormone binding proteins by the fat body of *Manduca sexta*. *Gen. Comp. Endocrinol.*, *27*, 230–239.

Numata, H., Numata, A., Takahashi, C., Nakagawa, Y., Iwatani, K., Takahashi, S., Miura, K., & Chinzei, Y. (1992). Juvenile hormone I is the principal juvenile hormone in a hemipteran insect, *Riptortus clavatus*. *Experientia*, *48*, 606–610.

O'Brien, M. A., Katahira, E. J., Flanagan, T. R., Arnold, L. W., Haughton, G., & Bollenbacher, W. E. (1988). A monoclonal antibody to the insect prothoracicotropic hormone. *J. Neurosci.*, *8*, 3247–3257.

Oda, S., Tatarazako, N., Dorgerloh, M., Johnson, R. D., Kusk, K. O., Leverett, D., Marchini, S., Nakari, T., Williams, T., & Iguchi, T. (2007). Strain difference in sensitivity to 3,4-dichloroaniline and insect growth regulator, fenoxycarb, in *Daphnia magna*. *Ecotox. Environ. Saf.*, *67*, 399–405.

Ogawa, N., Kishimoto, A., Asano, T., & Izumi, S. (2005). The homeodomain protein PBX participates in JH-related suppressive regulation on the expression of major plasma protein genes in the silkworm, *Bombyx mori*. *Insect Biochem. Mol. Biol.*, *35*, 217–229.

Oro, A. E., McKeown, M., & Evans, R. M. (1990). Relationship between the product of the *Drosophila* ultraspiracle locus and the vertebrate retinoid X receptor. *Nature, 347,* 298–301.

Orth, A. P., Lan, Q., & Goodman, W. G. (1999). Ligand regulation of juvenile hormone binding protein mRNA in mutant *Manduca sexta. Mol. Cell.. Endocrinol., 149,* 61–69.

Orth, A. P., Doll, S. C., & Goodman, W. G. (2003a). Sequence, structure and expression of the hemolymph juvenile hormone binding protein gene in the tobacco hornworm, *Manduca sexta. Insect Biochem. Mol. Biol., 33,* 93–102.

Orth, A. P., Tauchman, S. J., Doll, S. C., & Goodman, W. G. (2003b). Embryonic expression of juvenile hormone binding protein and its relationship to the toxic effects of juvenile hormone in *Manduca sexta. Insect Biochem. Mol. Biol., 33,* 1275–1284.

Palli, S. R., Osir, E. O., Eng, W. S., Boehm, M. F., Edwards, M., Kulcsar, P., Ujvary, I., Hiruma, K., Prestwich, G. D., & Riddiford, L. M. (1990). Juvenile hormone receptors in insect larval epidermis - Identification by photoaffinity labeling. *Proc. Natl. Acad. Sci. USA, 87.*

Palli, S. R., Riddiford, L. M., & Hiruma, K. (1991). Juvenile hormone and retinoic acid receptors in *Manduca* epidermis. *Insect Biochemistry, 21,* 5–15.

Palli, S. R., & Cusson, M. (2007). Future insecticides targeting genes involved in the regulation of molting and metamorphosis. In I. Ishaaya, R. Nauen, & A. R. Horowitz (Eds.), *Insecticide Design Using Advanced Technologies* (pp. 105–134). Berlin: Springer-Verlag.

Panchapakesan, K., Lampert, E. P., Granger, N. A., Goodman, W. G., & Roe, R. M. (1994). Biology and physiology of the white mutant of the tobacco hornworm, *Manduca sexta. J. Insect Physiol., 40,* 423–429.

Park, Y. C., & Goodman, W. G. (1993). Analysis and modification of thiols in the hemolymph juvenile hormone binding protein of *Manduca sexta. Arch. Biochem. Biophys., 302,* 12–18.

Park, Y. C., Tesch, M. J., Toong, Y. C., & Goodman, W. G. (1993). Affinity purification and binding analysis of the hemolymph juvenile hormone binding protein from *Manduca sexta. Biochemistry, 32,* 7909–7915.

Park, Y. I., & Raina, A. K. (2004). Juvenile hormone III titers and regulation of soldier caste in *Coptotermes formosanus* (Isoptera: Rhinotermitidae). *J. Insect Physiol., 50,* 561–566.

Parthasarathy, R., & Palli, S. R. (2007). Developmental and hormonal regulation of midgut remodeling in a lepidopteran insect, *Heliothis virescens. Mech. Dev., 124,* 23–34.

Parthasarathy, R., Tan, A., Bai, H., & Palli, S. R. (2008). Transcription factor broad suppresses precocious development of adult structures during larval-pupal metamorphosis in the red flour beetle, *Tr. castaneum. Mech. Dev., 125,* 299–313.

Pecasse, F., Beck, Y., Ruiz, C., & Richards, G. (2000). Kruppel-homologue, a stage-specific modulator of the prepupal ecdysone response, is essential for *Drosophila* metamorphosis. *Dev. Biol., 221,* 53–67.

Pener, M. P., & Simpson, S. J. (2009). Locust phase polyphenism: An update. *Adv. Insect Physiol., 36,* 1–272.

Peter, M. G., Shirk, P. D., Dahm, K. H., & Röller, H. (1981). On the specificity of juvenile hormone biosynthesis in the male *cecropia. Zeitschrift. Naturforsch., 36c,* 579–585.

Peter, M. G. (1990). Chiral recognition in insect juvenile hormone metabolism. In B. Testa (Ed.), *Chirality and Biological Activity* (pp. 111–117). New York: Alan R. Liss.

Peterson, R. C., Reich, M. F., Dunn, P. E., Law, J. H., & Katzenellenbogen, J. A. (1977). Binding specificity of the juvenile hormone carrier protein from the hemolymph of the tobacco hornworm *Manduca sexta. Biochemistry, 16,* 2305–2311.

Peterson, R. C., Dunn, P. E., Seballos, H. L., Barbeau, B. K., Keim, P. S., Riley, C. T., Heinrickson, R. L., & Law, J. H. (1982). Juvenile hormone carrier protein of *Manduca sexta* haemolymph. Improved purification procedure: protein modification studies and sequence of the amino terminus of the protein. *Insect Biochem., 12,* 643–650.

Pratt, G. E., & Tobe, S. S. (1974). Juvenile hormone radiobiosynthesized by *corpora allata* of adult female locusts *in vitro. Life Sci., 14,* 576–586.

Pratt, G. E., Farnsworth, D. E., & Feyereisen, R. (1990). Changes in the sensitivity of adult cockroach corpora allata to a brain allatostatin. *Mol. Cell. Endocrinol., 70,* 185–195.

Prestwich, G. D., Touhara, K., Riddiford, L. M., & Hammock, B. D. (1994). Larva lights: a decade of photoaffinity labeling with juvenile hormone analogues. *Insect Biochem. Mol. Biol., 24,* 747–761.

Pszczolkowski, M. A., Peterson, A., Srinivasan, A., & Ramaswamy, S. B. (2005). Pharmacological analysis of ovarial patency in *Heliothis virescens. J. Insect Physiol., 51,* 445–453.

Pursley, S., Ashok, M., & Wilson, T. G. (2000). Intracellular localization and tissue specificity of the *Methoprene-tolerant (Met)* gene product in *Drosophila melanogaster. Insect Biochem. Mol. Biol., 30,* 839–845.

Rachinsky, A., & Hartfelder, K. (1991). Differential production of juvenile hormone and its deoxy precursor by corpora allata of honeybees during a critical period of caste development. *Naturwissenschaften, 78,* 270–272.

Raimondo, S., & McKenney, C. L. (2005). Projecting population-level responses of mysids exposed to an endocrine disrupting chemical. *Integ. Comp. Biol., 45,* 151–157.

Rebers, J. E., & Riddiford, L. M. (1988). Structure and expression of a *Manduca sexta* larval cuticle gene homologous to *Drosophila* cuticle genes. *J. Mol. Biol., 203,* 411–423.

Rebers, J. E., Niu, J., & Riddiford, L. M. (1997). Structure and spatial expression of the *Manduca sexta MSCP14.6* cuticle gene. *Insect Biochem. Mol. Biol., 27,* 229–240.

Rees, H. H. (1985). Biosynthesis of ecdysones. In G. A. Kerkut, & L. I. Gilbert (Eds.), *Comprehensive Insect Physiology Biochemistry and Pharmacology* (Vol. 7), (pp. 249–293). Oxford: Pergamon Press.

Reibstein, D., Law, J. H., Bowlus, S. B., & Katzenellenbogen, J. A. (1976). Enzymatic synthesis of juvenile hormone in *Manduca sexta.* In L. I. Gilbert (Ed.), *The Juvenile Hormones* (pp. 131–146). New York: Plenum Press.

Rejzek, M. (1992). Carbamate series of juvenoids. *Bioorg. Med. Chem. Lett., 29,* 963–966.

Rembold, H., Hagenguth, H., & Rascher, J. (1980). A sensitive method for detection and estimation of juvenile hormones from biological samples by glass capillary combined gas chromatography-selected ion monitoring mass spectrometry. *Analyt. Biochem., 101,* 356–363.

Rembold, H., Czoppelt, C., Grune, M., Lackner, B., Pfeffer, J., & Woker, E. (1992). Juvenile horomone titers during honeybee embryogenesis and metamorphosis. In B. Mauchamp, F. Couillaud, & J. C. Baehr (Eds.), *Insect Juvenile Hormone Research: Chemistry, Biochemistry and Mode of Action* (pp. 37–43). Paris: INRA.

Rensing, L. (1964). Daily rhythmicity of corpus allatum and neurosecretory cells in *Drosophila melanogaster. Science, 144,* 1586–1587.

Restifo, L. L., & Wilson, T. G. (1998). A juvenile hormone agonist reveals distinct developmental pathways mediated by ecdysone-inducible broad complex transcription factors. *Dev. Gen., 22,* 141–159.

Richard, D. S., Applebaum, S. W., & Gilbert, L. I. (1989a). Developmental regulation of juvenile hormone biosynthesis by the ring gland of *Drosophila melanogaster. J. Comp. Physiol. Part B, 159,* 383–387.

Richard, D. S., Applebaum, S. W., Sliter, T. J., Baker, F. C., Schooley, D. A., Reuter, C. C., Henrich, V. C., & Gilbert, L. I. (1989b). Juvenile hormone bisepoxide biosynthesis in vitro by the ring gland of *Drosophila melanogaster* – A putative juvenile hormone in the higher Diptera. *Proc. Natl. Acad. Sci. USA, 86,* 1421–1425.

Richter, K., & Gronert, M. (1999). Neurotropic effect of juvenile hormone III in larvae of the cockroach, *Periplaneta americana. J. Insect Physiol., 45,* 1065–1071.

Riddiford, L. M. (1986). Hormonal regulation of sequential larval cuticular gene expression. *Arch. Insect Biochem. Physiol., 3*(Suppl. 1), 75–86.

Riddiford, L. M., Palli, S. R., Hiruma, K., Li, W., Green, J., Hice, R. H., Wolfgang, W. J., & Webb, B. A. (1990). Developmental expression, synthesis, and secretion of insecticyanin by the epidermis of the tobacco hornworm, *Manduca sexta. Arch. Insect Biochem. Physiol., 14,* 171–190.

Riddiford, L. M., & Ashburner, M. (1991). Effects of juvenile hormone mimics on larval development and metamorphosis of *Drosophila melanogaster. Gen. Comp. Endocrinol., 82,* 172–183.

Riddiford, L. M. (1994). Cellular and molecular actions of the juvenile hormones 1. General considerations and premetamorphic actions. *Adv. Insect Physiol., 24,* 213–274.

Riddiford, L. M. (1996). Juvenile hormone: the status of its "status quo" action. *Arch. Insect Biochem. Physiol., 32,* 271–286.

Riddiford, L. M., Hiruma, K., & Zhou, X. (2003). Insights into the molecular basis of the hormonal control of molting and metamorphosis from *Manduca sexta* and *Drosophila melanogaster. Insect Biochem. Mol. Biol., 33,* 1327–1338.

Riddiford, L. M. (2008). Juvenile hormone action: A 2007 perspective. *J. Insect Physiol., 54,* 895–901.

Riddiford, L. M., Truman, J. W., Mirth, C. K., & Shen, Y. C. (2010). A role for juvenile hormone in the prepupal development of *Drosophila melanogaster. Development, 137,* 1117–1126.

Robinson, G. E., & Vargo, E. L. (1997). Juvenile hormone in adult eusocial Hymenoptera: Gonadotropin and behavioral pacemaker. *Arch. Insect Biochem. Physiol., 35,* 559–583.

Roe, R. M., & Venkatesh, K. (1990). Metabolism of juvenile hormones: degradation and titer regulation. In A. P. Gupta (Ed.), *Morphogenetic Hormones of Arthropods* (pp. 126–179). New Brunswick, NJ: Rutgers University Press.

Roe, R. M., Jesudason, P., Venkatesh, K., Kallapur, V. L., Anspaugh, D. D., Majumder, C., Linderman, R. J., & Graves, D. M. (1993). Developmental role of juvenile hormone metabolism in Lepidoptera. *Amer. Zool., 33,* 375–383.

Roe, R. M., Kallapur, V., Linderman, R. J., Viviani, F., Harris, S. V., Walker, E. A., & Thompson, D. M. (1996). Mechanism of action and cloning of epoxide hydrolase from the cabbage looper, *Trichoplusia ni. Arch. Insect Biochem. Physiol., 32,* 527–535.

Roe, R. M., Anspaugh, D. D., Venkatesh, K., Linderman, R. J., & Graves, D. M. (1997). A novel geminal diol as a highly specific and stable in vivo inhibitor of insect juvenile hormone esterase. *Arch. Insect Biochem. Physiol., 36,* 165–179.

Roe, R. M., Kallapur, V., Linderman, R. J., & Viviani, F. (2005). Organic synthesis and bioassay of novel inhibitors of JH III epoxide hydrolase activity from fifth stadium cabbage loopers, *Trichoplusia ni. Pest. Biochem. Physiol., 83,* 140–154.

Röller, H., Dahm, D. H., Sweeley, C. C., & Trost, B. M. (1967). The structure of the juvenile hormone. *Angew. Chem. Int. Ed., 6,* 179–180.

Rose, U. (2004). Morphological and functional maturation of a skeletal muscle regulated by juvenile hormone. *J. Exp. Biol., 207,* 483–495.

Roy, S., & Vijayraghavan, K. (1999). Muscle pattern diversification in *Drosophila:* the story of imaginal myogenesis. *Bioessays, 21,* 486–498.

Rudnicka, M., & Kochman, M. (1984). Purification of the juvenile hormone esterase from the haemolymph of the wax moth *Galleria mellonella,* (Lepidoptera). *Insect Biochem., 14,* 189–198.

Safranek, L., & Riddiford, L. M. (1975). The biology of the black larval mutant of the tobacco hornworm, *Manduca sexta. J. Insect Physiol., 21,* 1931–1938.

Sakurai, S., & Williams, C. M. (1989). Short-loop negative and positive feedback on ecdysone secretion by prothoracic gland in the tobacco hornworm, *Manduca sexta. Gen. Comp. Endocrinol., 75,* 204–216.

Sakurai, S. (2005). Feedback regulation of prothoracic gland activity. In L. I. Gilbert, I. Katrou, & S. S. Gill (Eds.), *Comprehensive Molecular Insect Science* (Vol. 3), (pp. 19–408). Oxford: Elsevier.

Sanburg, L. L., Kramer, K. J.F.J.K., & Law, J. H. (1975). Juvenile hormone-specific esterases in the haemolymph of the tobacco hornworm, *Manduca sexta. J. Insect Physiol., 21,* 873–887.

Sarov-Blat, L., So, W. V., Liu, L., & Rosbash, M. (2000). The *Drosophila takeout* gene is a novel molecular link between circadian rhythms and feeding behavior. *Cell, 101,* 647–656.

Sasorith, S., Billas, I. M.L., Iwema, T., Moras, D., & Wurtz, J. M. (2002). Structure-based analysis of the ultraspiracle protein and docking studies of putative ligands. *J. Insect Sci., 2,* 25.

Schafellner, C., Marktl, R. C., & Schopf, A. (2007). Inhibition of juvenile hormone esterase activity in *Lymantria dispar* (Lepidoptera, Lymantriidae) larvae parasitized by *Glyptapanteles liparidis* (Hymenoptera, Braconidae). *J. Insect Physiol., 53,* 858–868.

Schafellner, C., Eizaguirre, M., Lopez, C., & Sehnal, F. (2008). Juvenile hormone esterase activity in the pupating and diapausing larvae of *Sesamia nonagrioides. J. Insect Physiol., 54,* 916–921.

Schooley, D. A., Judy, K. J., Bergot, B. J., Hall, M. S., & Siddall, J. B. (1973). Biosynthesis of juvenile hormones of *Manduca sexta:* labeling pattern from mevalonate, propionate and acetate. *Proc. Natl. Acad. Sci. USA, 70,* 2921–2925.

Schooley, D. A., Bergot, B. J., Goodman, W. G., & Gilbert, L. I. (1978). Synthesis of both optical isomers of insect juvenile hormone JH III and their affinity for the JH-specific binding protein of *Manduca sexta. Biochem. Biophys. Res. Commun., 81,* 743–749.

Schooley, D. A., Baker, F. C., Tsai, L. W., Miller, C. A., & Jamieson, G. C. (1984). Juvenile hormones 0, I, II exist only in Lepidoptera. In J. Hoffmann, & M. Porchet (Eds.), *Biosynthesis, Metabolism and Mode of Action of Invertebrate Hormones* (pp. 373–383). Berlin: Springer Verlag.

Schooley, D. A., & Baker, F. C. (1985). Juvenile hormone biosynthesis. In G. A. Kerkut, & L. I. Gilbert (Eds.), *Comprehensive Insect Physiology Biochemistry and Pharmacology* (Vol. 7), (pp. 363–389). Oxford: Pergamon Press.

Schwartz, L. M. (1992). Insect muscle as a model for programmed cell death. *J. Neurobiol., 23,* 1312–1326.

Scott, M. P. (2006). Resource defense and juvenile hormone: the "challenge hypothesis" extended to insects. *Horm. Behav., 49,* 276–281.

Sedlak, B. J., Marchione, L., Devorkin, B., & Davino, R. (1983). Correlations between endocrine gland ultrastructure and hormone titers in the fifth larval instar of *Manduca sexta. Gen. Comp. Endocrinol., 52,* 291–310.

Sedlak, B. J. (1985). Structure of endocrine glands. In G. A. Kerkut, & L. I. Gilbert (Eds.), *Comprehensive Insect Physiology Biochemistry and Pharmocology* (Vol. 7), (pp. 25–60). Oxford: Pergamon Press.

Sehnal, F. (1976). Action of juvenoids on different groups of insects. In L. I. Gilbert (Ed.), *The Juvenile Hormones* (pp. 301–322). New York: Plenum.

Sen, S. E., Ewing, G. J., & Thurston, N. (1996). Characterization of lepidopteran prenyltransferase in *Manduca sexta* corpora allata. *Arch. Insect Biochem. Physiol., 32,* 315–332.

Sen, S. E., Hitchcock, J. R., Jordan, J. L., & Richard, T. (2006). Juvenile hormone biosynthesis in *Manduca sexta:* substrate specificity of insect prenyltransferase utilizing homologous diphosphate analogs. *Insect Biochem. Mol. Biol., 36,* 827–834.

Sen, S. E., Cusson, M., Trobaugh, C., Béliveau, C., Richard, T., Graham, W., Mimms, A., & Roberts, G. (2007). Purification, properties and heteromeric association of type-1 and type-2 lepidopteran farnesyl diphosphate synthases. *Insect Biochem. Mol. Biol., 37,* 819–828.

Severson, T. F., Goodrow, M. H., Morisseau, C., Dowdy, D. L., & Hammock, B. D. (2002). Urea and amide-based inhibitors of the juvenile hormone epoxide hydrolase of the tobacco hornworm *Manduca sexta. Insect Biochem. Mol. Biol., 32,* 1741–1756.

Seybold, S. J., & Tittiger, C. (2003). Biochemistry and molecular biology of de novo isoprenoid pheromone production in the Scolitidae. *Ann. Rev. Entomol., 48,* 425–453.

Shanmugavelu, M., Baytan, A. R., Chesnut, J. D., & Bonning, B. C. (2000). A novel protein that binds juvenile hormone esterase in fat body tissue and pericardial cells of the tobacco hornworm *Manduca sexta. J. Biol. Chem., 275,* 1802–1806.

Shemshedini, L., Lanoue, M., & Wilson, T. G. (1990). Evidence for a juvenile hormone receptor involved in protein synthesis in *Drosophila melanogaster. J. Biol. Chem., 265,* 1913–1918.

Shi, L., Lin, S., Grinberg, Y., Beck, Y., Grozinger, C. M., Robinson, G. E., & Lee, T. (2007). Roles of *Drosophila* Krüppel-homologue 1 in neuronal morphogenesis. *Dev. Neurobiol., 67,* 1614–1626.

Shiao, S. H., Hansen, I. A., Zhu, J., Sieglaff, D. H., & Raikhel, A. S. (2008). Juvenile hormone connects larval nutrition with target of rapamycin signaling in the mosquito *Aedes aegypti. J. Insect. Physiol., 54,* 231–239.

Shiga, S. (2003). Anatomy and functions of brain neurosecretory cells in Diptera. *Microsc. Res. Tech., 62,* 114–131.

Shingleton, A. W., Frankino, W. A., Flatt, T., Nijhout, H. F., & Emlen, D. J. (2007). Size and shape: the developmental regulation of static allometry in insects. *Bioessays, 29,* 536–548.

Shinoda, T., Hiruma, K., Charles, J. P., & Riddiford, L. M. (1997). Hormonal regulation of JP29 in the epidermis during larval development and metamorphosis in the tobacco hornworm, *Manduca sexta. Arch. Insect Biochem. Physiol., 34,* 409–428.

Shinoda, T., & Itoyama, K. (2003). Juvenile hormone acid methyltransferase: a key regulatory enzyme for insect metamorphosis. *Proc. Natl. Acad. Sci. USA, 100,* 11986–11991.

Siegmund, T., & Korge, G. (2001). Innervation of the ring gland of *Drosophila melanogaster. J. Comp. Neurol., 431,* 481–491.

Sláma, K., Romaňuk, M., & Šorm, F. (1974). *Insect Hormones and Bioanalogues.* New York: Springer-Verlag.

Sláma, K., & Romaňuk, M. (1976). Juvenogens, biochemically activated juvenoid complexes. *Insect Biochem., 6,* 579–586.

Smith, P. A., Clare, A. S., Rees, H. H., Prescott, M. C., Wainwright, G., & Thorndyke, M. C. (2000). Identification of methyl farnesoate in the cypris larva of the barnacle, *Balanus amphitrite,* and its role as a juvenile hormone. *Insect Biochem. Mol. Biol., 30,* 885–890.

Sok, A. J., Czajewska, K., Ozyhar, A., & Kochman, M. (2005). The structure of the juvenile hormone binding protein gene from *Galleria mellonella. Biol. Chem., 386,* 1–10.

Soulages, J. L., & Wells, M. A. (1994). Metabolic fate and turnover rate of hemolymph free fatty acids in adult *Manduca sexta. Insect Biochem. Mol. Biol., 24,* 79–86.

Sparks, T. C., Willis, W. S., Shorey, H. H., & Hammock, B. D. (1979). Haemolymph juvenile hormone esterase activity in synchronous last instar larvae of the cabbage looper, *Trichoplusia ni. J. Insect Physiol., 25,* 125–132.

Sparks, T. C., Hammock, B. D., & Riddiford, L. M. (1983). The haemolymph juvenile hormone esterase of *Manduca sexta* – inhibition and regulation. *Insect Biochem., 13,* 529–541.

Sparks, T. C., & Rose, R. L. (1983). Inhibition and substrate specificity of the haemolymph juvenile hormone esterase of the cabbage looper, *Trichoplusia ni. Insect Biochem., 13,* 633–640.

Sparks, T. C., Allen, L. G., Schneider, F., & Granger, N. A. (1989). Juvenile hormone esterase activity from *Manduca sexta* corpora allata in vitro. *Arch. Insect Biochem. Physiol., 11,* 93–108.

Sperry, A. E., & Sen, S. E. (2001). Farnesol oxidation in insects: evidence that the biosynthesis of insect juvenile hormone is mediated by a specific alcohol oxidase. *Insect Biochem, Mol. Biol., 31*, 171–178.

Spiess, R., & Rose, U. (2004). Juvenile hormone-dependent motor activation in the adult locust *Locusta migratoria. J. Comp. Physiol. Part A, 190*, 883–894.

Spokony, R. F., & Restifo, L. L. (2007). Anciently duplicated *Broad Complex* exons have distinct temporal functions during tissue morphogenesis. *Dev. Genes. Evol., 217*, 499–513.

Stahl, G. B. (1975). Insect growth regulators with juvenile hormone activity. *Ann. Rev Entomol., 20*, 417–460.

Stay, B., & Tobe, S. S. (2007). The role of allatostatins in juvenile hormone synthesis in insects and crustaceans. *Ann. Rev. Entomol., 52*, 277–299.

Steel, C. G.H., & Ampleford, E. J. (1984). Circadian control of haemolymph ecdysteroid titers and ecdysis rhythms in *Rhodnius prolixus*. In R. Porter, & G. M. Collins (Eds.), *Photoperiodic Regulation of Insect and Molluscan Hormones* (pp. 150–169). London: Pitman Publishing.

Stout, J., Hayes, V., Zacharias, D., Henley, J., Stumper, A., Hao, J., & Atkins, G. (1992). Juvenile hormone controls phonotactic responsivness of female crickets by genetic regulation of the response properties of identified auditory interneurons. In B. Mauchamp, F. Couillaud, & J. C. Baehr (Eds.), *Insect Juvenile Hormone Research: Fundamental and Applied Approaches* (pp. 265–283). Paris: INRA.

Stout, J., Hao, J., Kim, P., Mbungu, D., Bronsert, M., Slikkers, S., Maier, J., Kim, D., Bacchus, K., & Atkins, G. (1998). Regulation of the phonotactic threshold of the female cricket, *Acheta domesticus*: juvenile hormone III, allatectomy, l1 auditory neuron thresholds and environmental factors. *J. Comp. Physiol. Part A, 182*, 635–645.

Strambi, C., Strambi, A., De Reggi, M. L., Hirn, M. H., & DeLagge, M. A. (1981). Radioimmunoassay of insect juvenile hormones and of their diol derivatives. *Eur. J. Biochem., 118*, 401–406.

Sundaram, M., Sivaprasadarao, A., Desousa, M. M., & Findlay, J. B.C. (1998). The transfer of retinol from serum retinol binding protein to cellular retinol binding protein is mediated by a membrane receptor. *J. Biol. Chem., 273*, 3336–3342.

Sundaram, M., vanAalten, D. M.F., Findlay, J. B.C., & Sivaprasadarao, A. (2002). The transfer of transthyretin and receptor binding properties from the plasma retinol binding protein to the epididymal retinoic acid-binding protein. *Biochem. J., 362*, 265–271.

Sutherland, T. D., Unnithan, G. C., Andersen, J. F., Evans, P. H., Murataliev, M. B., Szabo, L. Z., Mash, E. A., Bowers, W. S., & Feyereisen, R. (1998). A cytochrome P450 terpenoid hydroxylase linked to the suppression of insect juvenile hormone synthesis. *Proc. Natl. Acad. Sci. USA, 95*, 12884–12889.

Sutherland, T. D., & Feyereisen, R. (1996). Target of cockroach allatostatin in the pathway of juvenile hormone biosynthesis. *Mol. Cell. Endocrinol., 120*, 115–123.

Suzuki, R., Fujimoto, Z., Shiotsuki, T., Momma, M., Tase, A., Miyazawa, M., & Yamazaki, T. (2010a). Solution structure of juvenile hormone binding protein from silkworm in complex with JH III. *PDB ID 2RQF.*

Suzuki, R., Fujimoto, Z., Shiotsuki, T., Momma, M., Tase, A., Miyazawa, M., & Yamazaki, T. (2010b). Crystal structure of juvenile hormone binding protein from silkworm. *PDB ID 3AIZ.*

Suzuki, Y., & Nijhout, H. F. (2006). Evolution of a polyphenism by genetic accommodation. *Science, 311*, 650–652.

Suzuki, Y., & Nijhout, H. F. (2008). Constraint and developmental dissociation of phenotypic integration in a genetically accommodated trait. *Evol. Dev., 10*, 690–699.

Suzuki, Y., Truman, J. W., & Riddiford, L. M. (2008). The role of Broad in the development of *Tr. castaneum*: implications for the evolution of the holometabolous insect pupa. *Development, 135*, 569–577.

Tan, A., Tanaka, H., Tamura, T., & Shiotsuki, T. (2005). Precocious metamorphosis in transgenic silkworms overexpressing juvenile hormone esterase. *Proc. Natl. Acad. Sci. USA, 102*, 11751–11756.

Tauchman, S. J., Lorch, J. M., Orth, A. P., & Goodman, W. G. (2007). Effects of stress on the hemolymph juvenile hormone binding protein titers of *Manduca sexta. Insect Biochem. Mol. Biol., 37*, 847–854.

Tawfik, A. I., Kellner, R., Hoffmann, K. H., & Lorenz, M. W. (2006). Purification, characterisation and titre of the haemolymph juvenile hormone binding proteins from *Schistocerca gregaria* and *Gryllus bimaculatus. J. Insect Physiol., 52*, 255–268.

Teal, P. E.A., Gomezsimuta, Y., & Proveaux, A. T. (2000). Mating experience and juvenile hormone enhance sexual signaling and mating in male Caribbean fruit flies. *Proc. Natl. Acad. Sci. USA, 97*, 3708–3712.

Teal, P. E.A., & Proveaux, A. T. (2006). Identification of methyl farnesoate from *in vitro* culture of the retrocerebral complex of adult females of the moth, *Heliothis virescens* (Lepidoptera: Noctuidae) and its conversion to juvenile hormone III. *Arch. Insect Biochem. Physiol., 61*, 98–105.

Thomas, H. E., Stunnenberg, H. G., & Stewart, A. F. (1993). Heterodimerization of the *Drosophila* ecdysone receptor with retinoid-x receptor and Ultraspiracle. *Nature, 362*, 471–475.

Tillman, J. A., Lu, F., Goddard, L. M., Donaldson, Z. R., Dwinell, S. C., Tittiger, C., Hall, G. M., Storer, A. J., Blomquist, G. J., & Seybold, S. J. (2004). Juvenile hormone regulates de novo isoprenoid aggregation pheromone biosynthesis in pine bark beetles, *Ips spp.*, through transcriptional control of HMG-CoA reductase. *J. Chem. Ecol., 30*, 2459–2494.

Tittiger, C., Blomquist, G. J., Ivarsson, P., Borgeson, C. E., & Seybold, S. J. (1999). Juvenile hormone regulation of HMG-R gene expression in the bark beetle *Ips paraconfusus* (Coleoptera: Scolytidae): Implications for male aggregation pheromone biosynthesis. *Cell. Mol. Life Sci., 55*, 121–127.

Tittiger, C., Barkawi, L. S., Bengoa, C. S., Blomquist, G. J., & Seybold, S. J. (2003). Structure and juvenile hormone-mediated regulation of the HMG-CoA reductase gene from the Jeffrey pine beetle, *Dendroctonus jeffreyi. Mol. Cell. Endocrinol., 199*, 11–21.

Tobe, S. S., & Feyereisen, R. (1983). Juvenile hormone biosynthesis:Regulation and assay. In R. G.H. Downer, & H. Laufer (Eds.), *Endocrinology of Insects* (pp. 161–178). New York: Alan R. Liss, Inc.

Tobe, S. S., & Stay, B. (1985). Structure and function of the corpus allatum. *Adv. Insect Physiol, 18*, 303–438.

Touhara, K., & Prestwich, G. D. (1993). Juvenile hormone epoxide hydrolase – photoaffinity labeling, purification, and characterization from tobacco hornworm eggs. *J. Biol. Chem., 268*, 19604–19609.

Touhara, K., Wojtasek, H., & Prestwich, G. D. (1996). In vitro modeling of the ternary interaction in juvenile hormone metabolism. *Arch. Insect Biochem. Physiol, 32*, 399–406.

Trautmann, K. (1972). In vitro Studium der Tragerproteine von 3H-markierten Juvenilhormonwirksamen verbindungen in der Hamolymphe von *Te. molitor* L. *Larvae. Zeitschreift Naturforsch, 27*, 263–273.

Treiblmayr, K., Pascual, N., Piulachs, M. D., Keller, T., & Belles, X. (2006). Juvenile hormone titer versus juvenile hormone synthesis in female nymphs and adults of the German cockroach, *Blattella germanica. J. Insect Sci, 6*, 1–7.

Trowell, S. C. (1992). High affinity juvenile hormone carrier proteins in the haemolymph of insects. *Comp. Biochem. Physiol. Part B, 103*, 795–808.

Trowell, S. C., Hines, E. R., Herlt, A. J., & Rickards, R. W. (1994). Characterization of a juvenile hormone binding lipophorin from the blowfly, *Lucilia cuprina. Comp. Biochem. Physiol. Part B, 109*, 339–359.

Truman, J. W., & Reiss, S. E. (1988). Hormonal regulation of the shape of identified motorneurons in the moth *Manduca sexta. J. Neurosci, 8*, 765–775.

Truman, J. W. (1996). Steroid receptors and nervous system metamorphosis in insects. *Dev. Neurobiol., 18*, 87–101.

Tsubota, T., Minakuchi, C., Nakakura, T., Shinoda, T., & Shiotsuki, T. (2010). Molecular characterization of a gene encoding juvenile hormone esterase in the red flour beetle, *Tr. castaneum. Insect Mol. Biol., 19*, 527–535.

Tuberty, S. R., & McKenney, C. L. (2005). Ecdysteroid responses of estuarine crustaceans exposed through complete larval development to juvenile hormone agonist insecticides. *Integr. Comp. Biol., 45*, 106–117.

Ueda, H., Shinoda, T., & Hiruma, K. (2009). Spatial expression of the mevalonate enzymes involved in juvenile hormone biosynthesis in the corpora allata in *Bombyx mori. J. Insect Physiol., 55*, 798–804.

Uhlirova, M., Foy, B. D., Beaty, B. J., Olson, K. E., Riddiford, L. M., & Jindra, M. (2003). Use of Sindbis virus-mediated RNA interference to demonstrate a conserved role of Broad-Complex in insect metamorphosis. *Proc. Natl. Acad. Sci. USA, 100*, 15607–15612.

Vafopoulou, X., & Steel, C. G.H. (2001). Induction of rhythmicity in prothoracicotropic hormone and ecdysteroids in *Rhodnius prolixus*: roles of photic and neuroendocrine Zeitgebers. *J. Insect Physiol., 47*, 935–941.

Vandermoten, S., Charloteaux, B., Santini, S., Sen, S. E., Béliveau, C., Vandenbol, M., Francis, F., Brasseur, R., Cusson, M., & Haubruge, É (2008). Characterization of a novel aphid prenyltransferase displaying dual geranyl/farnesyl diphosphate synthase activity. *FEBS Letters, 582*, 1928–1934.

Vandermoten, S., Haubruge, É., & Cusson, M. (2009). New insights into short-chain prenyltransferases: structural features, evolutionary history and potential for selective inhibition. *Cell. Mol. Life Sci., 66*, 3685–3695.

Varjas, L., Kulcsar, P., Fekete, J., Bihatsi-Karsai, E., & Lelik, L. (1992). JH titres measured by GC-MS in the hemolymph of *Mamestra oleracea* larvae reared under different photoperiodic conditions. In B. Mauchamp, F. Couillaud, & J. C. Baehr (Eds.), *Insect Juvenile Hormone Research: Fundamental and Applied Approaches* (pp. 45–50). Paris: INRA.

Venkataraman, V., Omahony, P. J., Manzcak, M., & Jones, G. (1994). Regulation of juvenile hormone esterase gene transcription by juvenile hormone. *Dev. Gen., 15*, 391–400.

Venkatesh, K., & Roe, R. M. (1988). The role of juvenile hormone and brain factor(s) in the regulation of plasma juvenile hormone esterase activity during the last larval stadium of the tobacco hornworm, *Manduca sexta. J. Insect Physiol., 34*, 415–426.

Verma, K. K. (2007). Polyphenism in insects and the juvenile hormone. *J. Biosci., 32*, 415–420.

Vermunt, A. M.W., Vermeesch, A. M.G., & dekort, C. A.D. (1997). Purification and characterization of juvenile hormone esterase from hemolymph of the Colorado potato beetle. *Arch. Insect Biochem. Physiol., 35*, 261–277.

Vermunt, A. M.W., Kamimura, M., Hirai, M., Kiuchi, M., & Shiotsuki, T. (2001). The juvenile hormone binding protein of silkworm haemolymph: gene and functional analysis. *Insect Mol. Biol., 10*, 147–154.

Vince, R., & Gilbert, L. I. (1977). Juvenile hormone esterase activity in precisely timed instar larvae and pharate pupae of *Manduca sexta. Insect Biochem., 7*, 115–120.

Wang, J., Lindholm, J. R., Willis, D. K., Orth, A., & Goodman, W. G. (2009). Juvenile hormone regulation of *Drosophila* Epac–a guanine nucleotide exchange factor. *Mol. Cell. Endocrinol., 305*, 30–37.

Wang, S., Phong, T. V., Tuno, N., Kawada, H., & Takagi, M. (2005). Sensitivity of the larvivorous copepod species, *Mesocyclops pehpeiensis* and *Megacyclops viridis*, to the insect growth regulator, pyriproxyfen. *J. Amer. Mosq. Control. Assoc., 21*, 483–488.

Wang, S., Baumann, A., & Wilson, T. G. (2007). *Drosophila melanogaster Methoprene-tolerant (Met)* gene homologues from three mosquito species: Members of PAS transcriptional factor family. *J. Insect Physiol., 53*, 246–253.

Weaver, R. J., & Audsley, N. (2009). Neuropeptide regulators of juvenile hormone synthesis. *Trends Comp. Endocrinol. Neuro., 1163*, 316–329.

Weers, P. M.M., & Ryan, R. O. (2006). Apolipophorin III: Role model apolipoprotein. *Insect Biochem. Mol. Biol., 36*, 231–240.

Weller, J., Sun, G. C., Zhou, B. H., Lan, Q., Hiruma, K., & Riddiford, L. M. (2001). Isolation and developmental expression of two nuclear receptors, MHR4 and beta FTZ-F1, in the tobacco hornworm, *Manduca sexta. Insect Biochem. Mol. Biol., 31*, 827–837.

Westerlund, S. A., & Hoffmann, K. H. (2004). Rapid quantification of juvenile hormones and their metabolites in insect haemolymph by liquid chromatography-mass spectrometry (LC-MS). *Analyt. Bioanalyt. Chem., 379*, 540–543.

Westphal, U. (1986). *Steroid-Protein Interactions II*. Berlin: Springer-Verlag.

Wheeler, D. E., & Nijhout, H. F. (1981). Soldier determination in ants: new role for juvenile hormone. *Science, 213*, 361–363.

Wheeler, D. E., & Nijhout, H. F. (2003). A perspective for understanding the modes of juvenile hormone action as a lipid signaling system. *Bioessays, 25*, 994–1001.

Wheelock, C. E., Severson, T. F., & Hammock, B. D. (2001). Synthesis of new carboxylesterase inhibitors and evaluation of potency and water solublity. *Chem. Res. Toxicol., 14*, 1536–1572.

Whitmore, E., & Gilbert, L. I. (1972). Haemolymph lipoprotein transport of juvenile hormone. *J. Insect Physiol., 18*, 1153–1167.

Whitmore, D., Gilbert, L. I., & Ittycheriah, P. I. (1974). The origin of hemolymph carboxylesterases 'induced' by the insect juvenile hormone. *Mol. Cell. Endocrinol., 1*, 37–54.

Wieczorek, E., & Kochman, M. (1991). Conformational change of the haemolymph juvenile hormone binding protein from *Galleria mellonella*. *Eur. J. Biochem., 201*, 347–353.

Wigglesworth, V. B. (1934). The physiology of ecdysis in *Rhodnius prolixus* (Hemiptera). II. Factors controlling molting and 'metamorphosis'. Quart. *J. Microsc. Sci., s2–77*, 191–222.

Wigglesworth, V. B. (1959). Metamorphosis, polymorphism, differentiation. *Sci. Amer., 200*, 100–106.

Wigglesworth, V. B. (1961). Some observations on the JH effect of farnesol in *Rhodnius prolixus*. *J. Insect Physiol., 7*, 73–78.

Wigglesworth, V. B. (1972). *Principles of Insect Physiology* (7th ed.). London: Chapman and Hall.

Williams, C. M. (1952). Physiology of insect diapause IV. The brain and prothoracic glands as an endocrine system in the cecropia silkworm. *Biol. Bull., 103*, 120–138.

Williams, C. M. (1956). The juvenile hormone of insects. *Nature, 178*, 212–213.

Williams, C. M. (1957). The juvenile hormone. *Sci. Amer., 108*, 67–74.

Williams, D. W., & Truman, J. W. (2005). Remodeling dendrites during insect metamorphosis. *J. Neurobiol., 64*, 24–33.

Willis, D. K., Wang, J., Lindholm, J. R., Orth, A., & Goodman, W. G. (2010). Microarray analysis of juvenile hormone response in *Drosophila melanogaster* S2 cells. *J. Insect Sci., 10*.

Willis, J. H. (1996). Metamorphosis of the cuticle, its proteins, and their genes. In L. I. Gilbert, J. R. Tata, & B. G. Atkinson (Eds.), *Metamorphosis* (pp. 253–282). San Diego: Academic Press Inc.

Willis, J. H., & Muthukrishnan, S. (2010). Special issue on the insect cuticle - Foreword. *Insect Biochem. Mol. Biol., 40*, 165.

Willnow, T. E., & Nykjaer, A. (2010). Cellular uptake of steroid carrier proteins-Mechanisms and implications. *Mol. Cell. Endocrinol., 316*, 93–102.

Wilson, M., Moshitzky, P., Laor, E., Ghanim, M., Horowitz, A. R., & Morin, S. (2007). Reversal of resistance to pyriproxyfen in the Q biotype of *Bemisia tabaci* (Hemiptera: Aleyrodidae). *Pest. Manag. Sci., 63*, 761–768.

Wilson, T. G., & Fabian, J. (1986). A *Drosophila melanogaster* mutant resistant to a chemical analog of juvenile hormone. *Dev. Biol., 118*, 190–201.

Wilson, T. G., & Ashok, M. (1998). Insecticide resistance resulting from an absence of target-site gene product. *Proc. Natl. Acad. Sci. USA, 95*, 14040–14044.

Wilson, T. G., DeMoor, S., & Lei, J. (2003). Juvenile hormone involvement in *Drosophila melanogaster* male reproduction. *Insect Biochem. Mol. Biol., 33*, 1167–1175.

Wilson, T. G. (2004). The molecular site of action of juvenile hormone and juvenile hormone insecticides during metamorphosis: how these compounds kill insects. *J. Insect Physiol., 50*, 111–121.

Wilson, T. G., Wang, S., Beno, M., & Farkas, R. (2006). Wide mutational spectrum of a gene involved in hormone action and insecticide resistance in *Drosophila melanogaster*. *Mol. Genet. Genomics, 276*, 294–303.

Wimmer, Z., Jurcek, O., Jedlicka, P., Hanus, R., Kuldova, J., Hrdy, I., Bennettova, B., & Saman, D. (2007). Insect pest management agents: hormonogen esters (juvenogens). *J. Agric. Food Chem., 55*, 7387–7393.

Wing, K. D., Sparks, T. C., Lovell, V. M., Levinson, S. O., & Hammock, B. D. (1981). The distribution of juvenile hormone esterase and its interrelationship with other proteins influencing juvenile hormone metabolism in the cabbage looper, *Trichoplusia ni*. *Insect Biochem., 11*, 473–485.

Wisniewski, J. R., Muszynska-Pytel, M., & Kochman, M. (1986a). Juvenile hormone degradation in brain and corpora cardiaca-corpora allata complex during the last larval instar of *Galleria mellonella*. *Experientia, 42*, 167–168.

Wisniewski, J. R., Rudnicka, M., & Kochman, M. (1986b). Tissue specific juvenile hormone degradation in *Galleria mellonella*. *Insect Biochem., 16*, 843–849.

Woehler, A., & Ponimaskin, E. G. (2009). G protein–mediated signaling: same receptor, multiple effectors. *Curr. Mol. Pharmacol., 2*, 237–248.

Wogulis, M., Wheelock, C. E., Kamita, S. G., Hinton, A. C., Whetstone, P. A., Hammock, B. D., & Wilson, D. K. (2006). Structural studies of a potent insect maturation inhibitor bound to the juvenile hormone esterase of *Manduca sexta*. *Biochemistry, 45*, 4045–4057.

Wojtasek, H., & Prestwich, G. D. (1995). Key disulfide bonds in an insect hormone binding protein: cDNA cloning of a juvenile hormone binding protein of *Heliothis virescens* and ligand binding by native and mutant forms. *Biochemistry, 34*, 5234–5241.

Wojtasek, H., & Prestwich, G. D. (1996). An insect juvenile hormone-specific epoxide hydrolase is related to vertebrate microsomal epoxide hydrolases. *Biochem. Biophys. Res. Commun., 220*, 323–329.

Wroblewski, V. J., Harshman, L. G., Hanzlik, T. N., & Hammock, B. D. (1990). Regulation of juvenile hormone esterase gene expression in the tobacco budworm *(Heliothis virescens)*. *Arch. Biochem. Biophys., 278*, 461–466.

Wu, Y., Parthasarathy, R., Bai, H., & Palli, S. R. (2006). Mechanisms of midgut remodeling: juvenile hormone analog methoprene blocks midgut metamorphosis by modulating ecdysone action. *Mech. Dev., 123*, 530–547.

Xu, Y., Fang, F., Chu, Y. X., Jones, D., & Jones, G. (2002). Activation of transcription through the ligand-binding pocket of the orphan nuclear receptor ultraspiracle. *Eur. J. Biochem., 269*, 6026–6036.

Yagi, K. J., & Tobe, S. S. (2001). The radiochemical assay for juvenile hormone biosynthesis in insects: problems and solutions. *J. Insect Physiol., 47*, 1227–1234.

Yagi, S., & Kuramochi, K. (1976). The role of juvenile hormone in larval duration and spermiogenesis in relation to phase variation in the tobacco cutworm, *Spodoptera litura*. *Appl. Entomol. Zool.*, *11*, 133–138.

Yamamoto, C. K., Chadarevian, A., & Pellegrini, M. (1988). Juvenile hormone action mediated in male accessory glands of *Drosophila* by calcium and kinase C. *Science*, *239*, 916–919.

Yamanaka, N., Yamamoto, S., Zitnan, D., Watanabe, K., Kawada, T., Satake, H., Kaneko, Y., Hiruma, K., Tanaka, Y., Shinoda, T., & Kataoka, H. (2008). Neuropeptide receptor transcriptome reveals unidentified neuroendocrine pathways. *PLoS One*, *3*, e3048.

Yao, T. P., Forman, B. M., Jiang, Z. Y., Cherbas, L., Chen, J. D., Mckeown, M., Cherbas, P., & Evans, R. M. (1993). Functional ecdysone receptor is the product of *EcR* and *ultraspiracle* genes. *Nature*, *366*, 476–479.

Yin, C.-M., Takeda, M., & Wang, Z. -S. (1987). Juvenile hormone analogue, methoprene as a circadian and development modulator in *Diatraea grandiosella*. *J. Insect Physiol.*, *33*, 95–102.

Yin, C. M., Zou, B. X., Jiang, M. G., Li, M. F., Qin, W. H., Potter, T. L., & Stoffolano, J. G. (1995). Identification of juvenile hormone III bisepoxide (JHB3), JH III and methyl farnesoate secreted by the corpus allatum of *Phormia regina*, in vitro and function of JHB3 either appplied alone or as a part of a juvenoid blend. *J. Insect Physiol.*, *41*, 473–488.

Zalewska, M., Kochman, A., Esteve, J. P., Lopez, F., Chaoui, K., Susini, C., Ozyhar, A., & Kochman, M. (2009). Juvenile hormone binding protein traffic – Interaction with ATP synthase and lipid transfer proteins. *Biochim. Biophys. Acta*, *1788*, 1695–1705.

Zera, A. J., Sanger, T., Hanes, J., & Harshman, L. (2002). Purification and characterization of hemolymph juvenile hormone esterase from the cricket, *Gryllus assimilis*. *Arch. Insect Biochem. Physiol.*, *49*, 41–55.

Zera, A. J., & Zhao, Z. (2009). Morph-associated JH titer diel rhythm in *Gryllus firmus*: Experimental verification of its circadian basis and cycle characterization in artificially selected lines raised in the field. *J. Insect Physiol.*, *55*, 450–458.

Zhang, J., Dean, A. M., Brunet, F., & Long, M. (2004). Evolving protein functional diversity in new genes of *Drosophila*. *Proc. Natl. Acad. Sci. USA*, *101*, 16246–16250.

Zhang, Q. R., Xu, W. H., Chen, F. S., & Li, S. (2005). Molecular and biochemical characterization of juvenile hormone epoxide hydrolase from the silkworm, *Bombyx mori*. *Insect Biochem. Mol. Biol.*, *35*, 153–164.

Zhao, Z., & Zera, A. J. (2004). A morph-specific daily cycle in the rate of JH biosynthesis underlies a morph-specific daily cycle in the hemolymph JH titer in a wing-polymorphic cricket. *J. Insect Physiol.*, *50*, 965–973.

Zhou, B., & Riddiford, L. M. (2001). Hormonal regulation and patterning of the broad-complex in the epidermis and wing discs of the tobacco hornworm, *Manduca sexta*. *Dev. Biol.*, *231*, 125–137.

Zhou, B. H., Hiruma, K., Shinoda, T., & Riddiford, L. M. (1998). Juvenile hormone prevents ecdysteroid-induced expression of broad complex RNAs in the epidermis of the tobacco hornworm, *Manduca sexta*. *Dev. Biol.*, *203*, 233–244.

Zhou, X. F., & Riddiford, L. M. (2002). Broad specifies pupal development and mediates the 'status quo' action of juvenile hormone on the pupal-adult transformation in *Drosophila* and *Manduca*. *Development*, *129*, 2259–2269.

Zhu, H., Gegear, R. J., Casselman, A., Kanginakudru, S., & Reppert, S. M. (2009). Defining behavioral and molecular differences between summer and migratory monarch butterflies. *BMC Biol.*, *7*, 14.

Zimowska, G., Rembold, H., & Bayer, G. (1989). Juvenile hormone identification, titer, and degradation during the last larval stadium of *Spodoptera littoralis*. *Arch. Insect Biochem. Physiol.*, *12*, 1–14.

9   Hormones Controlling Homeostasis in Insects

D.A. Schooley
University of Nevada, Reno, NV, USA
F.M. Horodyski
Ohio University, Athens, OH, USA
G.M. Coast
Birkbeck College, London, UK

© 2012 Elsevier B.V. All Rights Reserved

9.1. Introduction

Insects, like other animals, face a variety of challenges to their survival in the varying conditions of their environment and food availability. Many changes in their physiological status must be controlled to achieve homeostasis. In this chapter, the hormonal controllers of the supply of endogenous metabolic energy and the hormonal control of fluid and electrolyte balance are discussed.

Due to the wide variety of environmental niches inhabited by insects, the mechanisms for control of homeostasis may show considerable variation between genera, although basic underlying themes are conserved. Insects are an extremely ancient grouping of organisms, with short life spans and frequently multiple generations per year. On completion of sequencing of the *Anopheles gambiae* genome, the genome and proteome of this dipteran insect were compared with that of the related dipteran *Drosophila* melanogaster (Zdobnov *et al.*, 2002). One of the salient conclusions is that these two "closely related" dipterans show greater genetic divergence than the genomes of the human and the puffer fish, although the estimated time of divergence between these two dipterans (~250 million years ago) is considerably more recent than that between the higher organisms (Zdobnov *et al.*,

2002). This rapid rate of evolution is reflected in differences in the nature of main substrates used as metabolic fuels for flight; in considerable interspecific diversity in structure between hormones in certain families that regulate homeostasis, especially diuretic hormones; and also in differences in which family of diuretic hormones appears to dominate as the major controllers of fluid excretion between different genera. It has also become apparent that signal transduction pathways for a given family of homeostatic hormones may differ between genera.

It is interesting to compare and contrast the depth of our current knowledge in this area to that when the predecessor to this book (Mordue and Morgan, 1985) was published. At that time, the sequences of only two insect neuropeptides were known: proctolin and adipokinetic hormone (AKH) from *Locusta migratoria* (herein called Locmi-AKH-I). Technical advances in peptide isolation and analysis made in the early 1980s were responsible for an explosive growth in the isolation and identification of new insect neuropeptides. The availability of synthetic peptide hormones then allowed detailed studies of their mechanism of physiological action. It has also become clear, as will be seen in this chapter, that insect peptide hormones, like their vertebrate homologues, frequently have more than one function. Well over 200

DOI: 10.1016/B978-0-12-384749-2.10009-3

insect neuropeptides have been identified, belonging to a number of families. The majority of these were isolated on the basis of their myotropic activity, because of the ease and rapidity of bioassays for such effects. Owing to a huge literature, only those myotropic peptides are covered (certain forms of AKH and the myokinins) that are known to be important homeostatic regulators.

9.2. Hormonal Control of Energy Stores

It is difficult to discuss this section without considering comparative endocrinology. In vertebrates the storage of fats in adipose tissue and storage of glycogen in muscles and liver are increased by high levels of insulin. Mobilization of fats from adipocytes and glycogen from liver and muscles is triggered by high levels of the counter-regulatory hormone glucagon. Glucagon acts through the cyclic AMP (cAMP) signal transduction pathway to activate hormone-sensitive lipase in adipocytes, to activate glycogen phosphorylase and inactivate glycogen synthase in muscle and liver, and to activate gluconeogenesis and inactivate glycolysis in the liver and kidney. Insulin, by activating cAMP phosphodiesterase and phosphoprotein phosphatase I, has the opposite effects on these pathways.

9.2.1. Adipokinetic Hormone and Hypertrehalosemic Hormone

9.2.1.1. Biological role of AKH Peptides in the AKH family stimulate metabolism by mobilizing energy stores in the fat body. The actions mediated by the AKH peptides are critical to supply energy to tissues, such as the flight muscles, with substrates necessary to maintain long distance flight. Depending on the insect species, either lipids, carbohydrates, proline, or a combination thereof are released from the fat body during times of high physical activity. For example, the energy substrate released into the hemolymph of cockroaches is trehalose, a disaccharide of glucose and the major hemolymph carbohydrate of insects (Friedman, 1985). When energy demands are high, locusts release predominantly diacylglycerols from the fat body, which are transported by lipophorins in the hemolymph (Chino and Gilbert, 1964, 1965; Soulages and Wells, 1994), and also release carbohydrate under some conditions (Goldsworthy, 1969). Proline is released into the hemolymph to provide fuel for the contraction of flight muscles in the tsetse fly (*Glossina morsitans;* Bursell, 1963) and other dipteran insects (*Phormia regina;* Sacktor and Childress, 1967 and *Aedes aegypti:* Scaraffia and Wells, 2003), and in coleopteran insects (De Kort *et al.*, 1973; Gäde and Auerswald, 2002). Peptides that primarily mobilize lipids in the species in which they were isolated are referred to as AKH (Mayer and Candy, 1969a), whereas peptides whose predominant role is to mobilize carbohydrates are known as hypertrehalosemic

hormones (HrTH), since the increase in total carbohydrate is accounted for by an increase in trehalose. The adipokinetic and hypertrehalosemic functions of AKH peptides are similar to the metabolic responses induced by the vertebrate hormone glucagon.

AKH peptides are synthesized by, and released from, the cells of the glandular lobe of the corpus cardiacum (CC; Goldsworthy *et al.*, 1972a), an endocrine organ attached to the brain. The effect of the CC on metabolism was first described in *Periplaneta americana*, where injection of a CC extract into adult cockroaches caused an increased level of trehalose in the hemolymph and a decrease in fat body glycogen content (Steele, 1961). In the locusts *Schistocerca gregaria* and *L. migratoria*, injection of a CC extract into the hemolymph increases levels of diacylglycerol (Beenakkers, 1969; Mayer and Candy, 1969a), resembling the changes in lipid composition that occur during flight (Beenakkers, 1965; Mayer and Candy, 1967). This elevation of diacylglycerol reflects the oxidation of lipids as the primary fuel during prolonged flight in locusts (Weis-Fogh, 1952). The onset of flight is rapidly followed by the release of AKH into the hemolymph to mobilize energy substrates from the fat body (Cheeseman and Goldsworthy, 1979; Orchard and Lange, 1983a; Candy, 2002). The AKH concentration in the hemolymph during flight (Cheeseman and Goldsworthy, 1979; Candy, 2002) is similar to the concentration necessary to induce lipid release from the fat body *in vivo* (Goldsworthy *et al.*, 1986a). In addition to the adipokinetic effect in *L. migratoria*, injection of a CC extract also causes an increase in hemolymph carbohydrate levels in young male locusts whose fat body contains sufficient glycogen stores (Goldsworthy, 1969). In the moth *Manduca sexta*, AKH stimulates glycogen breakdown in larval insects and lipid mobilization in adults (Siegert and Ziegler, 1983; Ziegler and Schulz, 1986; Ziegler *et al.*, 1990).

The similarity of AKH and HrTH was demonstrated before their structures were known; a *P. americana* CC extract induced an adipokinetic response in *L. migratoria* and an *L. migratoria* CC extract induced a hyperglycemic response in *P. americana* (Goldsworthy *et al.*, 1972a). The differences in the metabolic responses reflect the strategy of the insect in the nature of the fuel stored in the fat body. An AKH was first identified in *L. migratoria* (Stone *et al.*, 1976). The structural similarity of the *L. migratoria* and *P. americana* peptides was confirmed when three groups isolated two octapeptides in the same year from *P. americana* based on myotropic assays (Baumann and Penzlin, 1984; Scarborough *et al.*, 1984; Witten *et al.*, 1984). However, Scarborough *et al.* (1984) also showed that the less abundant peptide from *P. americana*, Peram-CAH-II, has a very significant sequence identity to glucagon (**Figure 1C**). Later Siegert and Mordue (1986a) isolated these factors as specific hypertrehalosemic agents from cockroach. Because of the functional relatedness of

AKH peptides, heterologous bioassays, such as the lipid mobilization assay in *L. migratoria* and carbohydrate mobilization assay in *P. americana*, are often used for the isolation of metabolic neuropeptides in a large number of insect species (Gäde, 1980).

The biochemical effect of a CC extract in the *P. americana* fat body is the activation of glycogen phosphorylase (Steele, 1963). Similarly in *L. migratoria*, glycogen phosphorylase is activated by a CC extract (Van Marrewijk *et al.*, 1980) or by synthetic AKH (Gäde, 1981). The effect of AKH on lipid mobilization in *M. sexta* adults is mediated by the activation of triacylglycerol lipase (TAG lipase), which converts TAG into diacylglycerol that is subsequently released into the hemolymph (Arrese *et al.*, 1996). Injection of Locmi-AKH-I into *Sc. gregaria* resulted in a twofold increase of TAG lipase activity in the fat body (Ogoyi *et al.*, 1998).

AKH possesses numerous biological activities in addition to its well-known metabolic functions. Some of these actions include the acceleration of heart rate in *P. americana* (Baumann and Gersch, 1982; Scarborough *et al.*, 1984), *Blaberus discoidalis* (Keeley *et al.*, 1991), and *Drosophila melanogaster* (Noyes *et al.*, 1995), the stimulation of myotropic contractions in *P. americana* (O'Shea *et al.*, 1984; Witten *et al.*, 1984), and the stimulation of heme synthesis in *B. discoidalis* (Keeley *et al.*, 1991). Analogous to the action of glucagon, AKH inhibits anabolic pathways such as the synthesis of lipids (Gokuldas *et al.*, 1988; Lee and Goldsworthy, 1995; Lorenz, 2001, 2003; Anand and Lorenz, 2010), proteins (Carlisle and Loughton, 1979; Cusinato *et al.*, 1991), and RNA (Kodrik and Goldsworthy, 1995) and a neuromodulatory effect in *P. americana* (Wicher *et al.*, 1994), the stimulation of locomotory activity in *M. sexta* (Milde *et al.*, 1995) and *Pyrrhocoris apterus* (Socha *et al.*, 1999; Kodrík *et al.*, 2000), and a lipogenic effect in which lipids are translocated from the hemolymph to the fat body in *P. americana* (Oguri and Steele, 2003).

Although AKH is widely viewed as an energy mobilization hormone synthesized by and released from the CC, effects on the nervous system have been documented. Peram-AKH-I and, to a lesser extent, Peram-AKH-II accelerate the spike frequency in DUM neurons in *P. americana* through the increase of a Ca^{2+} current (Wicher *et al.*, 1994, 2006). The result is an increase in locomotor activity in response to injection by Peram-AKH-I. Injection of Manse-AKH into the mesothoracic neuropil increases motor activity in *M. sexta*, similar to the effects of octopamine (Milde *et al.*, 1995). A stimulation of locomotory activity in the firebug, *P. apterus*, was observed after injection of Locmi-AKH-I or the endogenous peptide, Pyrap-AKH (Socha *et al.*, 1999; Kodrík *et al.*, 2000). In *D. melanogaster*, AKH mediates hyperactivity in response to starvation, an effect that is not seen in flies devoid of AKH neurons (Lee and Park, 2004). Thus, one of the functions of AKH is to integrate food-searching behavior in response to starvation.

Although the endocrine cells of the CC are considered the sole source of AKH (Lewis *et al.*, 1997; Schooneveld *et al.*, 1983; Kim and Rulifson, 2004; Lee and Park, 2004;

Figure 1 (A) Alignment of sequences of the confirmed AKH/HrTH from the Insecta from the known 51 peptides with a unique structure. Sequences were aligned with ClustalW (http://npsa-pbil.ibcp.fr/cgi-bin/npsa_automat.pl?page=/NPSA/npsa_clu stalw.html), as were all peptide and protein sequences in this chapter, using the Blosum matrix unless otherwise specified. Alignments were plotted with ESPript at this site. White letters in red boxes denote sequence identity and red letters in yellow boxes indicate sequence similarity. Peptide sequences were compiled from Gäde (2009) with the inclusion of additional sequences found subsequently (Gäde *et al.*, 2008; Gäde and Simek, 2010). The sequence of one peptide from *C. morosus* (Carmo-HrTH-I*) is mannosylated on the Trp8 residue (Munte *et al.*, 2008), a peptide from *Trichostetha fascicularis* (Trifa-CC*) is phosphorylated on the Thr6 residue (Gäde *et al.*, 2006), and an additional peptide from *Platypleura capensis* (Placa-HrTH-I*) contains a modification that has not been characterized (Gäde and Janssens, 1994). pQ denotes pyroglutamate, **a** at the C-terminus denotes amidation of the residue. Species names are as follows: Aedae, *Aedes aegypti*; Anaim, *Anax imperator*; Anoga, *Anopheles gambiae*; Bladi, *Blaberus discoidalis*; Bommo, *Bombyx mori*; Carmo, *Carausius morosus*; Corpu, *Corixa punctata*; Declu, *Decapotoma lunata*; Emppe, *Empusa pennata*; Erysi, *Erythemis simplicicollis*; Galyu, *Galloisiana yuasai*; Grybi, *Gryllus bimaculatus*; Helze, *Heliothis zea*; Letin, *Lethocerus indicus*; Libau, *Libellula auripennis*; Locmi, *Locusta migratoria*; Manse, *Manduca sexta*; Manto, unidentified species from the order Mantophasmatodea; Melci, *Melittea cinxia*; Melme, *Melolontha melolontha*; Micvi, *Microhodotermes viator*; Nepci, *Nepa cinerea*; Oniay, *Onitis aygulus*; Panbo, *Pandalus borealis*; Peram, *Periplaneta americana*; Phote, *Phormia terranovae*; Phymo, *Phymateus morbillosus*; Phyle, *Phymateus leprosus*; Placa, *Platypleura capensis*; Polae, *Polyphaga aegyptiaca*; Psein, *Pseudagrion inconspicuum*; Pyrap, *Pyrrhocoris apterus*; Rommi, *Romalea microptera*; Scade, *Scarabaeus deludens*; Schgr, *Schistocerca gregaria*; Tabat, *Tabanus atratus*; Tenar, *Tenthredo arcuata*; Tenmo, *Te. molitor*; Trica, *Tr. castaneum*; Trifa, *Trichostetha fascicularis*; Vanca, *Vanessa cardui*. (B). Alignment of the predicted (not chemically identified) unique AKH/HrTH/RPCH from Insecta and Crustacea from the following species: *Tr. castaneum* (Trica-AKH-II; Amare and Sweedler, 2007; Weaver and Audsley, 2008), *Daphnia pulex* (Dappu-RPCH; Christie *et al.*, 2008), *Bombyx mori* (Bommo-AKH-II; Roller *et al.*, 2007), *Spodoptera frugiperda* (Spofr-AKH-IV; Abdel-Latief and Hoffmann, 2007), and *Glossina morsitans* (Glomo-AKH; Kaufmann *et al.*, 2009). (C) Alignment of sequences of selected AKH/HrTH peptides from the Insecta with an AKH/corazonin-related peptide (ACP; Hansen *et al.*, 2010), corazonin (Veenstra, 1989), an AKH-GnRH-like peptide (AKH-GnRH; Lindemans *et al.*, 2009), a gonadotropin-releasing hormone (GnRH), and human glucagon. Species names are as follows: Caeel, *Caenorhabditis elegans* and Aplca, *Aplysia californica*. The 15 N-terminal residues are shown (16–29 deleted).

Wicher et al., 2006), there is immunological and mass spectrometric evidence that some AKHs are also present in the brain (Siegert, 1999; Clynen et al., 2001; Kaufmann and Brown, 2006; Kaufmann et al., 2009). AKHs could also gain access to the brain through intracerebral projections of AKH-producing CC cells (Lee and Park, 2004) or by penetration of the peptide across the ganglionic sheath (Wicher et al., 2006).

The interaction of AKH with both the humoral and cellular immune system was demonstrated in *L. migratoria* (Goldsworthy et al., 2002, 2003a). One component of the insect defense response to wounding and infection is

the activation of phenoloxidase via a serine proteinase cascade triggered by the recognition of bacterial and fungal cell wall components resulting in the production of toxic quinones (Gillespie et al., 1997). Injection of Locmi-AKH-I or -II enhanced the activation of phenoloxidase in response to a challenge with laminarin (a β-1,3-glucan) and activated phenoloxidase in the presence of bacterial lipopolysaccharide (LPS; Goldsworthy et al., 2002). The different potencies of Locmi-AKH-I and -II in phenoloxidase activation are consistent with their relative potencies in the locust lipid mobilization assay (Goldsworthy et al., 1986b), suggesting a direct link between the immune and lipid mobilization responses (Goldsworthy et al., 2002). The enhancement of phenoloxidase activation by AKH is age dependent, and restricted to the mature adult stage (Mullen and Goldsworthy, 2003). This time dependence correlates with the lipid mobilization response to AKH (Mwangi and Goldsworthy, 1977a) and to the concentration of apolipophorin III (ApoLp-III) in the hemolymph (Mwangi and Goldsworthy, 1977b; Mullen and Goldsworthy, 2003). In addition to the role of ApoLp-III in lipid metabolism, the lipid associated form of ApoLp-III stimulates antimicrobial activity in the hemolymph of *Galleria mellonella* (Wiesner et al., 1997; Detloff et al., 2001). These data suggest that AKH may be exerting its effect on phenoloxidase activation in part through its effects on ApoLp-III metabolism (Mullen and Goldsworthy, 2003). An effect on phenoloxidase activation by LPS in the absence of AKH was observed in starved locusts, presumably because of the effects of starvation on lipid mobilization (Goldsworthy et al., 2003a).

An important component of the cellular immune response is the formation of nodules by hemocyte aggregation to engulf or entrap foreign bodies (Lavine and Strand, 2002). AKH increases nodule formation when coinjected with LPS (Goldsworthy et al., 2003a). Unlike the effects of AKH on phenoloxidase activation, which is specific to mature locusts (Goldsworthy et al., 2002), its effects on nodule formation are apparent from the fifth instar nymphal stage through the mature adult stage. AKH may be exerting its effect on the humoral immune system (activation of prophenoloxidase) and the cellular immune system (nodule formation) via separate mechanisms, since they exhibit differing sensitivities to pharmacological agents and show different correlations to the nutritional state and lipid composition (Goldsworthy et al., 2003b).

9.2.1.2. Structural features of AKH

Evidence for multiple AKHs within a species was first found in locusts (Carlsen et al., 1979), and has been extended to include many insect species (Gäde, 2009), with three biologically active peptides present in some insects (Oudejans et al., 1991; Siegert et al., 2000; Köllisch et al., 2003). Each AKH is derived from a unique mRNA (Schulz-Aellen

(A)

```
                    1         10
                    .         .
Placa-HrTH-I*    pQVNFSHSWGNa
Placa-HrTH-II    pQVNFSHSWGNa
Anaim-AKH        pQVNFSHSWa..
Letin-AKH        pQVNFGPYWa..
Galyu-AKH        pQVNFGPTWa..
Peram-CAH-I      pQVNFSPNWa..
Bladi-HrTH       pQVNFSPGWGTa
Manto-CC         pQVNFSPGWa..
Locmi-HrTH       pQVTFSRDWSPa
Anoga-AKH-II     pQVTFSRDWNAa
Phote-HrTH       pQLTFSPDWa..
Scade-CC-I       pQFNYSPDWa..
Scade-CC-II      pQFNYSPVWa..
Melme-CC         pQLNYSPDWa..
Oniay-CC         pQYNFSTGWa..
Grybi-AKH        pQVNFSTGWa..
Schgr-AKH-II     pQLNFSTGWa..
Tenar-HrTH       pQLNFSTGWGGa
Trica-AKH        pQLNFSTDWa..
Helze-HrTH       pQLTFSSGWGNa
Melci-AKH        pQLTFSSGWa..
Nepci-AKH        pQLNFSSGWa..
Locmi-AKH-II     pQLNFSAGWa..
Panbo-RPCH       pQLNFSPGWa..
Declu-CC         pQLNFSPNWGNa
Tenmo-HrTH       pQLNFSPNWa..
Corpu-AKH        pQLNFSPSWa..
Phymo-AKH        pQINFTPNWGSa
Pyrap-AKH        pQLNFTPNWa..
Locmi-AKH-I      pQLNFTPNWGTa
Carmo-HrTH-I*    pQLTFTPNWGTa
Carmo-HrTH-II    pQLTFTPNWGTa
Phyle-CC         pQLTFTPNWGSa
Peram-CAH-II     pQLTFTPNWa..
Polae-HrTH       pQITPTPNWa..
Tabat-HoTH       pQLTFTPGWGYa
Tabat-AKH        pQLTFTPGWa..
Bommo-AKH        pQLTFTPGWGQa
Anoga-HrTH       pQLTFTPAWa..
Vanca-AKH        pQLTFTSSWGGD
Manse-AKH        pQLTFTSSWGa.
Aedae-AKH-I      pQLTFTSSWa..
Locmi-AKH-III    pQLNFTPWWa..
Erysi-AKH        pQLNFTPSWa..
Micvi-CC         pQINFTPNWa..
Phymo-AKH-III    pQINFTPWWa..
Trifa-CC*        pQINMTTGWa..
Rommi-CC         pQVNFTPNWGTa
Emppe-AKH        pQVNFTPNWa..
Libau-AKH        pQVNFTPSWa..
Psein-AKH        pQVNFTPGWa..
```

(B)

```
                  1         10
                  .         .
Trica-AKH-II   pQVIFSRDWNPa..
Bommo-AKH-II   pQLIFSRDWSGa..
Dappu-RPCH     pQVNFSTSWa....
Spofr-AKH-IV   pQLIFSSGWGNCTS
Glomo-AKH      pQLIFSPGWa....
```

(C)

```
                  1         10
                  .         .
Locmi-HrTH     .pGVFSNSWSPa..
Anoga-ACP      .pVVFSRSWNAa..
Anoga-HrTH     .pLLFSPAWa...
Caeel-AKH-GnRH .pQMTFTDCYT....
Human-glucagon HSQGTFTSDYSKYLD
Peram-Corazonin pQTFQYSRGWTNa..
Aplca-GnRH     pQNYHFSNGWYAa..
```

et al., 1989; Noyes and Schaffer, 1990; Fischer-Lougheed et al., 1993; Bogerd et al., 1995). In L. migratoria, a fourth member of the AKH family (Locmi-HrTH) was characterized based on its hypertrehalosemic activity in P. americana, but no function has yet been identified in L. migratoria (Siegert, 1999). The three Locmi-AKHs differ in amount, with Locmi-AKH I the most abundant and Locmi-AKH-III the least abundant (Oudejans et al., 1993; Clynen et al., 2001). One rationale for multiplicity of peptides is their varying potencies in different biological assays, which may enable the fine-tuning of an energy-demanding response such as long-distance flight (Vroemen et al., 1998a). Locmi-AKH-I is more effective than Locmi-AKH-II in the lipid mobilization assay, the ability to alter circulating lipoproteins, and the ability to activate glycogen phosphorylase, while Locmi-AKH-II is more effective in increasing cAMP levels in the fat body (Orchard and Lange, 1983a; Goldsworthy et al., 1986b). These data suggest that Locmi-AKH-I is the predominant hyperlipemic hormone, and Locmi-AKH-II is largely responsible for carbohydrate mobilization. Locusts employ carbohydrate as an energy source for the initial stages of flight and lipid for more prolonged flight (Weis-Fogh, 1952; Mayer and Candy, 1969b; Jutsum and Goldsworthy, 1976), suggesting that Locmi-AKH-II is most important at the onset of flight and Locmi-AKH-I assumes a major role in prolonged flight activity (Vroemen et al., 1997). Locmi-AKH-III is present at low levels and has a high turnover, suggesting it may be a modulatory entity that provides the locust with energy when not flying.

The peptides that comprise the AKH family are the most thoroughly characterized insect peptides with respect to their evolution and structural diversity (Gäde et al., 1997). At least 51 distinct peptides have been structurally characterized from at least 214 species that include representatives from all major insect orders (**Figure 1A**). Five additional unique members of the AKH/RPCH family were predicted from known nucleotide sequence information (Abdel-Latief and Hoffmann, 2007; Amare and Sweedler, 2007; Roller et al., 2007; Weaver and Audsley, 2008; Kaufmann et al., 2009; **Figure 1B**), but some of these predicted peptides were not detected by mass spectrometry (MS) indicating that their genome presence does not guarantee expression of the peptide (Gäde et al., 2008; Weaver and Audsley, 2008). This remarkable structural diversity in insects is not seen in crustaceans, in which only one peptide (Panbo-RPCH) has been identified (Fernlund and Josefsson, 1972) and whose presence was subsequently confirmed in insects (Gäde et al., 2003). However, a second unique peptide was predicted in two Daphnia species from screening crustacean EST databases (Christie et al., 2008, 2010b; **Figure 1B**). Panbo-RPCH has been shown to induce hyperlipemia in locusts (Herman et al., 1977; Mordue and Stone, 1977).

In contrast to the insect peptides whose roles include the mobilization of metabolic stores, the role of RPCH in crustaceans is to induce aggregation of pigment granules in chromatophores (Rao, 2001). Thus, these orthologous peptides found in insects and crustacea do not retain a common function.

Several common structural features define the AKH/RPCH family (Gäde, 2009; **Figure 1**). The peptides are from 8 and 11 amino acids in length and are characterized by a pyroglutamate residue at their N-terminus, an aromatic residue (Phe or Tyr) at position 4 with one exception (Gäde et al., 2006), a Trp residue at position 8, a Gly residue at position 9 with two exceptions (Kaufmann and Brown, 2006; Siegert, 1999), and a C-terminal amidation with one exception (Köllisch et al., 2000).

Members of the AKH-HrTH family share a lesser degree of identity with various invertebrate and vertebrate peptides, which sheds light on the evolution of their structures and biological functions. It was first observed that Peram-CAH-II shares significant sequence identity to glucagon (**Figure 1C**) and that both peptides elevate hemolymph carbohydrate in an assay where trehalose is hydrolyzed to glucose (Scarborough et al., 1984). Corazonin was shown to have limited sequence identity to Helze-HrTH (Veenstra, 1989; (**Figure 1C**), and the structures of the predicted AKH and corazonin precursors are similar suggesting a common evolutionary origin for these peptides. During evolution, this peptide family may have split into two branches, one (AKH) expressed in the CC, and the other (corazonin) expressed in the nervous system (Veenstra, 1994a). Each peptide may have specialized in the control of different aspects of stress (Veenstra, 2009; Hansen et al., 2010).

A number of recent studies have shed significant light on the evolution of the AKH and corazonin peptide families. The An. gambiae genomic database was screened with various insect AKH and corazonin peptides, and a preprohormone was identified that predicted the presence of a peptide that was related to both AKH and corazonin designated ACP (AKH/corazonin-related peptide) (Hansen et al., 2010; **Figure 1C**). ACPs were predicted in eight additional insect species, but were absent in some, including Drosophila species, Apis mellifera, Acyrthosiphon pisum, Pediculus humanus, and the crustacean Daphnia pulex. It was argued that two members of the AKH peptide family, Locmi-HrTH and Anoga-AKH-II, are actually insect ACPs based on their identity with ACPs, their presence in the brain (unlike the majority of AKH peptides), and the lack of a described AKH-like function thus far (Hansen et al., 2010). The presence of a peptide related to AKH was recently predicted in Caenorhabditis elegans and seven additional species in the Nematoda (Lindemans et al., 2009; **Figure 1C**). This predicted non-amidated peptide was designated AKH-GnRH (AKH-GnRH-like) and the injection of this peptide into P. americana resulted in a hypertrehalosemic response, suggesting a common evolutionary origin for AKH and AKH-GnRH precursors. Gene silencing of Caeel-AKH-GnRH or its receptor in the nematode resulted in a delay of egg laying but did not affect fat content, which supports the hypothesis that the common ancestor for this family of peptides may have had a reproductive role (Lindemans et al., 2009). Also, the limited sequence similarity between the APRPs (located on the AKH precursor) with members of the growth hormone-releasing factor (GRF) family suggests that AKHs may be members of the GRF superfamily (Clynen et al., 2004).

All AKH precursor proteins share a common organization — an N-terminal signal peptide followed by AKH and the AKH precursor-related peptide (APRP; Martínez-Pérez et al., 2002). The AKH precursor also shares a similar organization with the red pigment-concentrating hormone (RPCH; Linck et al., 1993), corazonin (Veenstra, 1994a), the APGW peptide family of mollusks (Martínez-Pérez et al., 2002), AKH-GnRH (Lindemans et al., 2009), and ACP (Hansen et al., 2010) suggesting that they could have come from a common ancestral gene.

The most critical residues for bioactivity are probably the most conserved (pGlu1, Phe4, Trp8) and might be involved in binding to the receptor, stability of the peptide, or formation of an intrapeptide interaction stabilizing the conformation needed to achieve optimal activity (Hayes and Keeley, 1990; Gäde, 1992). For example, the conserved Trp8 was proposed to form a hydrogen bond with Ser5 or Phe4 (Hayes and Keeley, 1990), and the hydrophobic nature and relative spacing of residues 2, 4, and 8 may be important in determining receptor binding.

The biological activity of neuropeptides is mediated by their interaction with a membrane-bound receptor on the target cell that alters the concentration of second messengers and results in biochemical changes that ultimately lead to a physiological response. Small peptides, such as AKH, are flexible structures that adopt many conformations in solution, and their potency is related to their structural conformation upon interaction with the receptor (Nachman et al., 1993). The identification of AKH receptors and their expression allows the direct assay of peptide (or analog) binding to the receptor and subsequent response of the target cell to identify the critical structural features of the peptides that are required for receptor binding and activation. Considerable information on the importance of individual amino acids or parts of the peptide for binding to the receptor has been acquired using bioassays in structure–activity studies in several insect species, but these results are also influenced by such factors as the relative stability of the peptides in vivo. The overall conclusion that can be drawn from structure–activity studies is that the endogenous peptides are usually the most potent in the respective bioassays, and

co-evolution of the peptide and the receptor took place to achieve optimal binding (Hayes and Keeley, 1990; Fox and Reynolds, 1991a).

Stone et al. (1978) predicted that many members of the AKH/RPCH family have the potential to adopt a β-turn conformation. Even though AKH peptides do not form an ordered structure in aqueous solution, the presence of sodium dodecylsulfate (SDS) causes a change in the circular dichroism (CD) spectra characteristic of a β-turn (Goldsworthy and Wheeler, 1989; Cusinato et al., 1998). However, the CD spectra of Locmi-AKH-II and Manse-AKH, peptides that lack a proline residue in position 6, are not affected by the presence of SDS and lack the capacity to form a β-turn. The capacity of some AKH peptides to form a β-turn between residues 4 and 8 was also confirmed using nuclear magnetic resonance (NMR) spectroscopy (Zubrzycki and Gäde, 1994, 1999; Nair et al., 2001). It was suggested that the activity of AKH in the L. migratoria lipid mobilization assay is correlated with the ability to form a β-structure in SDS micelles, which in turn quantitatively affects its interactions with the receptor (Goldsworthy et al., 1997; Cusinato et al., 1998). Locmi-AKH-II and Manse-AKH, which lack the ability to form a β-structure, exhibit low potency in the L. migratoria lipid mobilization assay (Cusinato et al., 1998). Analogs with substitutions in residues 6–8 of AKHs exhibit reduced potency since the substitutions apparently hinder the formation of a β-turn (Lee et al., 1996). However, the capacity to form a β-turn does not correlate with potency when assayed for inhibition of acetate uptake into fat body in vitro, since Acheta domesticus AKH and Locmi-AKH-II are equally active using this assay (Cusinato et al., 1998). This pleiotropic response may reflect the heterogeneity of receptors, which exhibit differing structural preferences for their ligand. Structural analysis of -peptides from Melolontha melolontha (Melme-CC), Te. molitor (Tenmo-HrTH), and Decapotoma lunata (Declu-CC) using NMR supports the hypothesis that the β-turn is important for receptor binding in the L. migratoria lipid mobilization assay (Nair et al., 2001). The Asn residue at position 7 (Asn7) and the Phe residue at position 4 (Phe4) are essential for a potent response, and it was shown that the Asn7 projects outward from the β-turn, and the orientation of Asn7 and the tightness of the β-turn are influenced by the aromatic residue at position 4.

Structure–activity studies were also carried out using the activation of glycogen phosphorylase in M. sexta fat body to assay several naturally occurring AKH peptides and synthetic analogs (Ziegler et al., 1991, 1998). Manse-AKH lacks the ability to form a β-turn (Goldsworthy and Wheeler, 1989), and among the peptides tested, the most inactive were the ones with the highest probability of forming a β-turn (Ziegler et al., 1991, 1998). The substitution of Ser6 of Manse-AKH with Pro, which favors the formation of a β-turn, drastically reduced its abilities to activate glycogen phosphorylase and to bind to the receptor in a fat body membrane preparation. This indicates that the *M. sexta* AKH receptor does not bind peptides that contain a β-turn (Ziegler *et al.*, 1998).

Removal of pGlu1 from Locmi-AKH-I abolishes activity (Gäde, 1990), but replacement of pGlu1 by an unblocked amino acid partially restored lipid mobilization activity *in vivo* to varying degrees depending on the identity of the substituted amino acid, and activity was further enhanced by N-terminal blockage (Lee *et al.*, 1997). Replacement of pGlu1 in Manse-AKH, however, did not restore potency in a bioassay measuring activity of glycogen phosphorylase unless the N-terminus was blocked (Herman *et al.*, 1977; Ziegler *et al.*, 1998). These data imply that pGlu1 is not absolutely essential for activity, and that one of the roles of the blocked N-terminus might be to impart increased stability to the peptide by preventing aminopeptidase attack (Lee *et al.*, 1997; Ziegler *et al.*, 1998).

9.2.1.3. AKH synthesis and release

Like most neuropeptides, AKH is derived by processing a larger precursor protein (Hekimi and O'Shea, 1987). The steps involved in AKH synthesis have been most thoroughly characterized in *Sc. gregaria* (O'Shea and Rayne, 1992), and subsequent studies in other insects have shown that the basic features of AKH biosynthesis are conserved. AKH precursor proteins share a similar organization, an N-terminal signal peptide, followed by AKH and the AKH precursor related peptide (APRP; O'Shea and Rayne, 1992), as first shown in *Sc. gregaria* (Schulz-Aellen *et al.*, 1989) and *M. sexta* (Bradfield and Keeley, 1989). The locust AKH precursor exists as a dimer formed by oxidation of the Cys residues present in the APRP. This occurs prior to proteolytic processing in the trans-Golgi at basic amino acid residues and C-terminal amidation (Rayne and O'Shea, 1994). The dimeric structure of the precursor might confer a conformation that facilitates or allows the correct processing of the precursor. Since multiple AKHs (Diederen *et al.*, 1987; Hekimi *et al.*, 1991) and their mRNAs (Bogerd *et al.*, 1995) are present in the same glandular cells, random formation of dimeric precursors gives rise to both homodimers and heterodimers (Hekimi *et al.*, 1991; Huybrechts *et al.*, 2002). Further processing of *L. migratoria* APRPs from the AKH-I and -II precursors takes place to yield additional peptides designated the adipokinetic hormone joining peptides (AKH-JP-I and A-II), but their release was not demonstrated (Baggerman *et al.*, 2002). Biological roles for the APRPs or the AKH-JPs have not been found (Oudejans *et al.*, 1991; Hatle and Spring, 1999; Baggerman *et al.*, 2002). Relative peptide levels in the glandular cells are controlled by multiple mechanisms. The 4.5:1 ratio of AKH-I to -II present in the glandular cells (Hekimi *et al.*, 1991) is regulated by a 1.7:1 ratio of

AKH-I to -II mRNA and the more efficient translation of AKH-I mRNA (Fischer-Lougheed *et al.*, 1993).

Antisera specific to AKH-I or AKH-II have demonstrated their co-localization in the same secretory granules in the glandular cells of the CC (Diederen *et al.*, 1987). Since an AKH-III-specific antiserum is not available, the co-localization of each APRP implies that AKH-III is also co-localized with AKH-I and -II (Harthoorn *et al.*, 1999). The amount of AKHs increases dramatically in the CC during development (Siegert and Mordue, 1986b) due to their continued synthesis (Fischer-Lougheed *et al.*, 1993; Oudejans *et al.*, 1993) and to the increase in the number of cells in the glandular lobe of the CC (GCC; Kirschenbaum and O'Shea, 1993). The continued synthesis of AKH generates a large pool of secretory granules in the glandular cells (Diederen *et al.*, 1992), and newly synthesized AKH is preferentially released over AKHs stored in older secretory granules (Sharp-Baker *et al.*, 1995), which constitutes a pool of peptides that cannot be released (Sharp-Baker *et al.*, 1996; Harthoorn *et al.*, 2002). Therefore, only a small percentage of the total AKH present in the glandular cells is released during flight (Cheeseman *et al.*, 1976) or upon stimulation by CCAP (Harthoorn *et al.*, 2002). The AKHs are released in the same proportion as their levels in the CC (Harthoorn *et al.*, 1999), and the continuous synthesis of AKH is required for the secretion of peptides from the glandular cells in response to metabolic need (Harthoorn *et al.*, 2002).

The isolation of the cloned gene for AKH in several insect species has permitted the localization of AKH mRNA. *In situ* hybridization has demonstrated abundant levels of AKH mRNA in the CC (Noyes and Schaffer, 1990; Noyes *et al.*, 1995; Kim and Rulifson, 2004). Independent confirmation of AKH gene expression in the CC was obtained in *D. melanogaster* by showing expression of a green fluorescent protein (GFP) or LacZ reporter gene driven by the AKH promoter in the CC from late embryos to the adult stage (Kim and Rulifson, 2004; Lee and Park, 2004; Isabel *et al.*, 2005). However, expression of the AKH gene outside the CC was detected in other insects. The mRNAs for alternatively spliced mRNAs from the *Spodoptera frugiperda* AKH gene were present in ovaries, muscle, fat body, and midgut by RT-PCR, and detected in the ovaries by *in situ* hybridization (Abdel-Latief and Hoffmann, 2007). Also, AKH-like immunoreactivity was detected in the brain of *An. gambiae* (Kaufmann and Brown, 2006) and *Aedes aegypti* (Kaufmann *et al.*, 2009), and in the thoracic ganglia of *An. gambiae* (Kaufmann and Brown, 2006).

In *Sc. gregaria* and *L. migratoria*, the hemolymph trehalose levels during flight are about 50% that of preflight values (Mayer and Candy, 1969b; Jutsum and Goldsworthy, 1976), and injection of a high concentration of trehalose prevents AKH release assayed by quantifying lipid mobilizing activity in the hemolymph (Cheeseman et al., 1976). Trehalose and glucose exert a direct action on the glandular cells of the CC, since high trehalose concentrations decreased both spontaneous AKH release and AKH release induced by the neuropeptide Locmi-TK I, 3-isobutyl-l-methylxanthine (IBMX), or high potassium concentrations in vitro (Passier et al., 1997). It was suggested that trehalose exerts this effect, in part, after its conversion to glucose. The decrease in trehalose concentration in response to the energy demands of flight relieves this inhibition, and is one of the factors that contribute to AKH release observed during flight (Cheeseman and Goldsworthy, 1979; Orchard and Lange, 1983b). It was also suggested that high levels of diacylglycerol resulting from mobilization of lipid energy stores may exert negative feedback on AKH release (Cheeseman and Goldsworthy, 1979). The neural and hormonal factors that modulate AKH release were first characterized in L. migratoria (Vullings et al., 1999). Implanted CC do not show the ultrastructural signs of enhanced AKH release during flight, suggesting that AKH secretion is under neural control by cells that make direct contact with the glandular lobe cells (Rademakers, 1977a). Neuroanatomical studies defined the secretomotor neurons in the lateral part of the protocerebrum that project through the nervi corporis cardiaci II (NCC II) to innervate the glandular lobe of the GCC and make synaptic contact with the AKH cells, and the axon terminals of the GCC are derived solely from the secretomotor cells (Rademakers, 1977b; Konings et al., 1989). Electrical stimulation of the NCC II resulted in the release of AKH I and II from the GCC and was accompanied by an increase in cAMP levels in cells of the GCC (Orchard and Loughton, 1981; Orchard and Lange, 1983a). Whereas stimulation of the NCC I had no direct effect on AKH release, an enhancement of the NCC II-stimulated release of AKH from the CC was observed (Orchard and Loughton, 1981). Hormonally mediated lipid mobilization during flight is abolished in locusts in which the NCC I and II were severed (Goldsworthy et al., 1972b). Together, these data indicate that the secretomotor neurons are involved in the control of AKH release, and compounds present in the NCC I modulate AKH secretion.

An antiserum raised against locustatachykinin I (Locmi-TK I), a member of a family of structurally related neuropeptides (Schoofs et al., 1993), stained a subset of secretomotor neurons projecting to the GCC and made synaptoid contacts with AKH-immunoreactive glandular cells, suggesting a role for a Locmi-TK-I-like peptide in AKH release (Nässel et al., 1995). Locmi-TK I and II induced AKH release from the CC in vitro and increased cAMP levels in the GCC, but an effect on AKH release required concentrations in the 50–200 mM range (Nässel et al., 1995, 1999). Furthermore, four Locmi-TK isoforms were identified in L. migratoria CC

extracts (Nässel et al., 1999). The anatomical features of the Locmi-TK-containing secretomotor neurons suggest that these peptides might act on synaptic receptors, and the high concentration of peptide required for AKH release is consistent with this hypothesis, particularly since synaptic receptors might be less accessible for an in vitro effect and thus would require higher peptide concentrations for an effect to be observed. SchistoFLRFamide is a member of a large family of neuropeptides known as the FMRFamide-related peptides (FaRPs), which are widely found in insects (Orchard et al., 2001). Antisera to FMRFamide and SchistoFLRFamide label a subset of secretomotor neurons distinct from the Locmi-TK-containing cells that make synaptoid contacts with glandular cells of the CC (Vullings et al., 1998). FMRFamide and SchistoFLRFamide have no effect on the spontaneous release of AKH in vitro, but reduced IBMX-induced AKH release at a 10 mM peptide concentration. Like Locmi-TK neuropeptides, the high concentration of FaRPs required for an effect is consistent with a direct supply of peptides on the glandular cells. Thus, FaRPs are inhibitory neuromodulators that act to fine-tune the release of AKH to meet the energetic demands of the flying animal (Vullings et al., 1998).

A direct approach was used to isolate a compound, the neuropeptide CCAP, from an Sc. gregaria brain extract that potently stimulates release of AKH from L. migratoria and Sc. gregaria CC in vitro (Veelaert et al., 1997). CCAP is a multifunctional neuropeptide and is best known for its stimulatory activity on the heartbeat (Tublitz and Truman, 1985) and its effect as a trigger for ecdysis behavior (Gammie and Truman, 1997). Unlike the Locmi-TK and SchistoFLRFamide-like peptides, there are no CCAP containing fibers in the GCC (Dircksen and Homberg, 1995; Veelaert et al., 1997); CCAP is active in the nanomolar range indicating that it may act as a neurohormonal releasing factor for AKH (Veelaert et al., 1997). The CC of L. migratoria and Diploptera punctata contain extensive processes and varicosities that exhibit immunoreactivity to the neuropeptides proctolin (Clark et al., 2006) and Dippu-allatostatin (Clark et al., 2008), suggesting that these peptides may act as releasing factors for AKH. Proctolin stimulated the release of Locmi-AKH-I from excised L. migratoria CC, and the maximal stimulation occurred at 100 nM proctolin (Clark et al., 2006). Similarly, four Dippu-ASTs (Dippu-AST 2, 7, 12, and 13) stimulated Locmi-AKH-I release at 10 nM peptide concentration, with Dippu-AST 2 the most potent in which 10 fM peptide showed maximal effect (Clark et al., 2008). Exposure to both proctolin and Dippu-AST 2 increased the cAMP content of the glandular lobe of the CC (Clark et al., 2008), which is consistent with its known involvement in stimulating the release of AKH (Pannabecker and Orchard, 1986, 1987).

Flight activity is the only natural stimulus known for AKH release (Cheeseman and Goldsworthy, 1979), but neither AKH mRNA levels nor prohormone synthesis were altered by flight activity (Harthoorn et al., 2001). Therefore, the availability of releasable AKH is dependent on its continuous biosynthesis. Flight activity increases the octopamine levels in the hemolymph of Sc. gregaria (Goosey and Candy, 1980), and in L. migratoria, octopamine was shown to stimulate AKH release from the CC in the presence of IBMX (Pannabecker and Orchard, 1986). It was later found that in this experimental design, octopamine potentiates the stimulatory effect of IBMX-mediated cAMP elevation on AKH release, and has no effect on its own (Passier et al., 1995). This suggests that octopamine has a neurohormonal role in modulating AKH release (Veelaert et al., 1997).

9.2.1.4. AKH degradation After secretion of AKHs into the hemolymph, their degradation will have a significant effect on peptide levels, and the differential rates of degradation will affect the ratios of peptides and, ultimately, the physiological response. Since AKHs are blocked at both termini, they are not accessible to the actions of aminopeptidase and carboxypeptidase. Thus, the initial step in their degradation is by an endopeptidase that yields inactive peptide fragments (Siegert and Mordue, 1987; Fox and Reynolds, 1991b; Rayne and O'Shea, 1992; Isaac, 2010). The endopeptidase has been localized to the external surface of several tissues (Rayne and O'Shea, 1992), including the Malpighian tubules (MTs; Baumann and Penzlin, 1987; Siegert and Mordue, 1987), but in M. sexta, AKH is cleaved by an enzyme circulating in the hemolymph (Fox and Reynolds, 1991b).

The synthesis of high-specific activity tritiated AKHs allowed the quantification of the half-lives of AKH-I to -III in L. migratoria after injecting a physiological dose of 1 pmol (Oudejans et al., 1996). This study clearly demonstrated that each AKH has a different rate of breakdown, with AKH-III significantly less stable in the hemolymph. The half-lives of AKH-I to -III at rest are 51, 40, and 5 min, respectively, and AKH-I and -III are degraded more rapidly during flight. In contrast to the relatively long half-lives of Locmi-AKH-I and -II, injected Gryllus bimaculatus AKH has a half-life of about 3 min and is degraded by enzymes released from hemocytes (Woodring et al., 2002).

9.2.1.5. Signal transduction for AKH There is substantial evidence from a number of systems that inositol-1,4,5-trisphosphate (IP_3) serves as the second messenger to transduce the hormonal signal of AKH/HrTH neuropeptides in the fat body and, in some cases, cAMP is also involved (Van der Horst et al., 1997; Gäde and Auerswald, 2003). In L. migratoria, the involvement of cAMP in AKH-mediated lipid mobilization was shown

by demonstrating the accumulation of cAMP in the fat body following an injection of a CC extract, and the increased hemolymph lipid concentration after treatment with dibutyryl-cAMP (db-cAMP; Gäde and Holwerda, 1976). A similar effect of a CC extract on cAMP levels was also shown in vitro in Sc. gregaria (Spencer and Candy, 1976). The effect of the CC extract on cAMP levels was shown to be due to AKH (Gäde, 1979; Goldsworthy et al., 1986b).

Although the mobilization of lipids is a crucial role of AKH in L. migratoria, the mechanism of AKH action in the target cell has been studied most extensively using glycogen phosphorylase activation as the assay for AKH activity (Van Marrewijk et al., 1980). The potencies of each AKH in cAMP elevation and glycogen phosphorylase activation differed when assayed at physiological levels, decreasing in the order AKH-III > AKH-II > AKH-I (Vroemen et al., 1995a). These differences in potency support the hypothesis that the action of AKH-II is more directed to carbohydrate metabolism than that of AKH-I (Orchard and Lange, 1983a). The increase in cAMP levels is mediated by the G protein G_s, since cholera toxin, an irreversible activator of G_s, enhanced cAMP levels and glycogen phosphorylase activity (Vroemen et al., 1995a). Pertussis toxin, an irreversible inhibitor of G_i, had no effect on cAMP levels or glycogen phosphorylase activation.

In L. migratoria, the uptake of Ca^{2+} into the fat body was stimulated by the presence of AKH, and the effects of AKH on diacylglycerol levels and glycogen phosphorylase activation are dependent on extracellular Ca^{2+} and can be mimicked by the calcium ionophore A23187 (Van Marrewijk et al., 1991). A similar Ca^{2+} dependence for AKH-mediated lipid mobilization was demonstrated in vitro (Lum and Chino, 1990; Wang et al., 1990). In addition, the release of Ca^{2+} from intracellular stores leads to the subsequent influx of Ca^{2+} into the cell (Van Marrewijk et al., 1993) by a process known as capacitative Ca^{2+} entry (Berridge, 1995), confirming a requirement for both intracellular and extracellular Ca^{2+} for the full effect of AKH on the fat body. Although each of the three AKHs increase Ca^{2+} uptake with similar potency, the three AKHs also enhance the efflux of Ca^{2+} into the fat body cytosol, but the increase in efflux increased in the order AKH-I < AKH-II < AKH-III (Vroemen et al., 1995b). The mobilization of intracellular Ca^{2+} by each of the AKHs in locusts is mediated by the stimulation of inositol phosphate (InsPn) formation, most notably IP_3 (Stagg and Candy, 1996; Van Marrewijk et al., 1996), but differences in potency were documented for each AKH in L. migratoria, decreasing in the order AKHIII > AKHI > AKHII (Vroemen et al., 1997). IP_3, generated by activation of phospholipase C (PLC), plays a critical role in Ca^{2+} mobilization that leads to activation of protein kinase C (PKC) for a wide variety of neuropeptides (Berridge, 1993). The stronger effect of AKH-I on IP_3

combined with its weak effect on Ca^{2+} efflux relative to AKH-II suggests that AKH-I is the major lipid mobilizing hormone, particularly since Ca^{2+} is required for translocation of hormone-sensitive lipases from the cytosol to lipid droplets to elicit lipolysis (Clark et al., 1991; Egan et al., 1992). The activation of glycogen phosphorylase by each of the AKHs is mediated by PLC activation, since a PLC inhibitor, U73122, dampened the response to the peptides, and the residual activity might be due to a cAMP pathway not influenced by IP_3 (Vroemen et al., 1997). Elevation of IP_3 levels by AKH is mediated by the G protein G_q, since a G_q antagonist dampened AKH-I induced glycogen phosphorylase activity (Vroemen et al., 1998b). Stimulation of glycogen phosphorylase by cAMP is independent of extracellular Ca^{2+} (Van Marrewijk et al., 1993), and the elevation of cAMP by forskolin or db-cAMP did not affect the enhancement of InsPn levels by AKH-I, so a direct linkage between the AKH receptor and PLC activation must exist that is not dependent on the elevation of cAMP. These data have led to a model describing AKH signaling in L. migratoria (Van der Horst et al., 1997; **Figure 2**). The AKH receptor is coupled to at least two distinct G proteins, G_s and G_q, and the activation of both pathways results in a complete enhancement of glycogen phosphorylase activation by AKH. AKH-induced Ca^{2+} influx is mediated by voltage-independent channels, possibly calcium release activated channels, since La^{3+}, a universal Ca^{2+} channel blocker, prevents glycogen phosphorylase activation (Vroemen et al., 1998b).

In P. americana, the effects of HrTH on phosphorylase activation in the fat body and trehalose release are dependent on the presence of extracellular Ca^{2+} (McClure and Steele, 1981; Orr et al., 1985; Steele and Paul, 1985) and the entry of Ca^{2+} into intact fat body (Steele and Paul, 1985) or fat body trophocytes (Steele and Ireland, 1999), which is stimulated by HrTH. A key role for Ca^{2+}

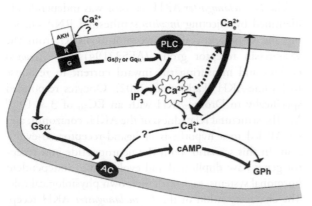

Figure 2 Proposed model for the coupling of AKH signaling pathways in L. migratoria fat body cells for carbohydrate mobilization. R, receptor; G, G protein; PLC, phospholipase C; IP_3, inositol-1,4,5-trisphosphate; AC, adenylate cyclase; GPh, glycogen phosphorylase. Reprinted with permission from Van der Horst et al. (1997).

in glycogen breakdown is its requirement, together with calmodulin, for the activity of phosphorylase kinase, the activator of glycogen phosphorylase (Pallen and Steele, 1988). The mode of HrTH action in P. americana was examined in disaggregated trophocytes, the fat body cells responsible for trehalose synthesis (Steele and Ireland, 1994). This system was used to provide the ability to test the effect of peptides or varying conditions on a large number of samples to better correlate changes in a discrete population of responsive cells. It was demonstrated that in P. americana HrTH I and II stimulated IP_3 formation with comparable potencies (Steele et al., 2001). The HrTH stimulated formation of IP_3 and the increase in intracellular Ca^{2+} concentration are probably involved in phosphorylase activation and trehalose efflux from trophocytes, since a PLC inhibitor blocked these effects. HrTH I and II increase Ca^{2+} levels in fat body trophocytes by the release of intracellular Ca^{2+} followed by the subsequent capacitative influx of extracellular Ca^{2+} (Sun and Steele, 2001). Capacitative Ca^{2+} entry is inhibited by activation of PKC or inhibition of calmodulin (Sun and Steele, 2001). HrTH stimulates the activity of phospholipase A_2 in trophocytes (Sun and Steele, 2002), which leads to the increased production of free fatty acids to provide an essential source of energy for trehalose synthesis (Ali and Steele, 1997). Phospholipase A_2 activation and the stimulation of trehalose efflux are dependent on G proteins, PKC, and calmodulin, and occur following the increase in cytosolic Ca^{2+} mediated by HrTH (Sun et al., 2002; Sun and Steele, 2002). In P. americana, a CC extract elevated cAMP levels in the fat body, but this effect was not due to the action of HrTH (Orr et al., 1985).

Transduction of the HrTH signal in B. discoidalis does not involve cyclic AMP, since HrTH does not elevate fat body cAMP levels (Park and Keeley, 1995), and elevating cAMP levels by IBMX or db-cAMP treatment does not lead to elevated trehalose synthesis (Lee and Keeley, 1994) or glycogen phosphorylase activity (Park and Keeley, 1995). Likewise, no effect of cAMP on trehalose production was observed in the cockroach P. americana (Orr et al., 1985) or Blaptica dubia (Becker et al., 1998).

The release of Ca^{2+} from intracellular stores by thimerosal or thapsigargin increased trehalose synthesis (Keeley and Hesson, 1995) and glycogen phosphorylase activation (Park and Keeley, 1996), even in the absence of extracellular Ca^{2+}. In contrast, stimulation of the influx of extracellular Ca^{2+} did not affect trehalose synthesis (Keeley and Hesson, 1995) and had only a modest effect on phosphorylase activation (Park and Keeley, 1996). However, a maximal response to hormone requires the presence of extracellular Ca^{2+} (Keeley and Hesson, 1995; Park and Keeley, 1996). Since HrTH was shown to increase IP_3 levels in the fat body, it was suggested that HrTH leads to the release of intracellular Ca^{2+} followed by an influx of extracellular Ca^{2+} to achieve a maximal effect on the fat body (Park and Keeley, 1996). A similar dependence on extracellular and intracellular Ca^{2+} was demonstrated for HrTH-mediated trehalose production in *B. dubia* (Becker *et al.*, 1998). However, the HrTH-mediated decrease in concentration of the metabolic signaling molecule, fructose 2,6-bisphosphate (Becker and Wegener, 1998), is largely dependent on Ca^{2+} entry, since this effect could be fully mimicked by the calcium ionophore A23187 (Becker *et al.*, 1998). This decrease in fructose 2,6-bisphosphate (Becker and Wegener, 1998) is a key regulatory step toward the inhibition of glycolysis and directing the products of glycogen breakdown toward the production of trehalose that results from HrTH action (Wiens and Gilbert, 1967).

Lipid mobilization by AKH in adult *M. sexta* involves both cAMP and Ca^{2+} and is accompanied by an increase in protein kinase A (PKA) activity (Arrese *et al.*, 1999). Although triglyceride lipase (TG-lipase) is phosphorylated by PKA *in vitro*, it does not enhance the activity of the enzyme (Patel *et al.*, 2004). It was found that the major PKA substrate that elicits the lipolytic effect is Lipid storage droplet protein-1 (Lsd1 or Lsdp1), a protein that is exclusively found in lipid droplets in the adult fat body, the site of stored lipids (Patel *et al.*, 2005). Lipolysis was reconstituted *in vitro* using triglyceride-labeled lipid droplets as a substrate for TG-lipase and PKA. Phosphorylation of Lsdp1 was the major change promoted by AKH (Patel *et al.*, 2006). Although Lsdp1 mRNA was present in both the larval and adult fat body, the protein was absent in the fat body of feeding larvae, which explains the lack of lipid mobilization by AKH in *M. sexta* larvae (Arrese *et al.*, 2008). Therefore, the majority of the lipolytic increase induced by AKH is due to changes in the substrate, rather than by increasing the activity of TG-lipase.

The fruit beetle, *Pachnoda sinuata*, mobilizes carbohydrate reserves and increases proline synthesis in response to AKH (Auerswald and Gäde, 1999). Injection of the endogenous AKH, Melme-CC, elevates cAMP levels in the fat body, and injection of a membrane-permeable cAMP analog or IBMX elevates proline levels in the hemolymph supporting the involvement of cAMP (Auerswald and Gäde, 2000). The induction of cAMP by AKH requires the presence of extracellular Ca^{2+} (Auerswald and Gäde, 2001a), and results in the rapid influx of Ca^{2+} into the fat body cell (Auerswald and Gäde, 2001b). The release of intracellular Ca^{2+} is required for the full stimulation of proline production by stimulating the influx of extracellular Ca^{2+} by capacitative Ca^{2+} entry (Auerswald and Gäde, 2001a). Injection of Melme-CC, or flight activity, stimulated TAG lipase activity. The oxidation of fatty acids yields acetyl-CoA, which combines with alanine (derived from oxidation of proline in the flight muscles) to synthesize proline (Auerswald *et al.*, 2005). In contrast to the effect of cAMP on proline production, it is not involved in carbohydrate mobilization or glycogen

phosphorylase activation (Auerswald and Gäde, 2000), which requires the presence of extracellular Ca^{2+} and the release of Ca^{2+} from intracellular stores (Auerswald and Gäde, 2001b). Injection of Melme-CC causes an increase in IP_3 levels in the fat body, but coinjection with a PLC inhibitor, U73122, affects only the carbohydrate mobilizing activity, and not the hyperprolinemic effect of the peptide (Auerswald and Gäde, 2002). These data suggest that two separate pathways exist for AKH action in *P. sinuata*, and that the release of the IP_3-dependent Ca^{2+} stores from the endoplasmic reticulum is not involved in proline production.

9.2.1.6. AKH receptor The *M. sexta* AKH receptor was initially characterized using tritium labeled Manse-AKH with a binding assay to larval fat body membrane preparations (Ziegler *et al.*, 1995). Specific and saturable binding that required the presence of Ca^{2+} was detected with a K_d of 7×10^{-10} M and was consistent with binding to one type of receptor. Specific binding to membrane preparations of *M. sexta* heart and muscle was not detected, but a low level of specific binding to the pterothoracic ganglion membrane preparations of adult *M. sexta* was observed (Ziegler *et al.*, 1995), consistent with the action of AKH observed on central nervous system neurons (Milde *et al.*, 1995).

Two independent methods were used to identify the AKH receptor cDNA from *D. melanogaster* (Park *et al.*, 2002; Staubli *et al.*, 2002). In one study, the *D. melanogaster* receptor cDNA related to the gonadotropin releasing hormone (GnRH) receptor (AF077299; Hauser *et al.*, 1998) was expressed in a CHO cell line expressing the promiscuous G protein $G\alpha16$ (CHO/G16) and aequorin, and Drome-AKH was identified as the ligand present in a third instar larval extract that specifically induced a bioluminescent response with an EC_{50} of 8×10^{-10} M (Staubli *et al.*, 2002).

The *D. melanogaster* AKH receptor was independently identified by injecting *in vitro* synthesized RNA encoding the G-protein-coupled receptor (GPCR) from the vasopressin receptor group (AF522194) into *Xenopus* oocytes and measuring the inward current in response to Drome-AKH (Park *et al.*, 2002). Oocytes responded specifically to Drome-AKH with an EC_{50} of 3×10^{-10} M. The structural relatedness of the AKH, corazonin, and CCAP led to a hypothesis of ligand-receptor co-evolution. In this scenario, both the neuropeptide and receptor genes have duplicated and evolved into independent hormonal systems, each with their own physiological role. The characterization of the *D. melanogaster* AKH receptor and the availability of genome sequences led to the subsequent identification and functional characterization of the AKH receptors in *B. mori* (Staubli *et al.*, 2002; Zhu *et al.*, 2009), *P. americana* (Hansen *et al.*, 2006), and *An. gambiae* (Belmont *et al.*, 2006). In addition, the receptor was identified by homology screening with known AKH receptors in *An. gambiae* (Hansen *et al.*, 2006; Kaufmann and Brown, 2006), *A. mellifera* (Hansen *et al.*, 2006), *Ae. aegypti* (Kaufmann *et al.*, 2009), *Tr. castaneum* (Hansen *et al.*, 2006; Hauser *et al.*, 2008), *Culex pipiens* (Hansen *et al.*, 2010), and *Nasonia vitripennis* (Hansen *et al.*, 2010). The *Ae. aegypti* AKH receptor consists of two splice variants, determined by RT-PCR, that differ in their C-terminal region (Kaufmann *et al.*, 2009). Phylogenetic analysis of the cloned receptors for AKH, ACP, corazonin, CCAP, and AVPL clearly demonstrates their evolutionary relatedness (Hansen *et al.*, 2010).

The presence of multiple AKH peptides in many insect species (Gäde, 2009), the differing rank order of potencies of these AKHs depending on the assay used (Goldsworthy *et al.*, 1986b), and the age-related changes in sensitivity to AKHs (Mwangi and Goldsworthy, 1977a; Ziegler, 1984; Woodring *et al.*, 2002) and biphasic responses to peptide analogs raise the question of whether AKH(s) interacts with multiple receptors (Gäde and Hayes, 1995). However, thus far, only a single receptor has been identified in any species, with the exception of two splice variants of the *Ae. aegypti* receptor (Kaufmann *et al.*, 2009). These receptors have not yet been functionally characterized, but it was suggested that the distinct C-termini might reflect differences in receptor internalization.

The G protein activated by peptide interaction with AKHR expressed in heterologous cells has been examined. The AKH-activated inward current observed in *Xenopus* oocytes transfected with *D. melanogaster* AKHR cRNA is characteristic of mobilization of intracellular Ca^{2+} stores and suggests coupling to G_q (Park *et al.*, 2002). In *P. americana*, the two AKHs (Peram-CAH-I and -II) couple with G_s in transfected HEK293 cells, but the two AKHs differ in their activation of G_q (Wicher *et al.*, 2006). This may reflect the differential effect of the two peptides to accelerate spiking of DUM cells and stimulate locomotion. Activation of the *B. mori* AKHR with the three Bommo-AKHs in HEK293 cells resulted in a differential activation of AKHR depending on the peptide ligand tested (Zhu *et al.*, 2009). Manse-AKH and Bommo-AKH resulted in cAMP accumulation characteristic of G_s activation and caused the transient phosphorylation of ERK1/2, components of the MAP kinase pathway, which are important for many GPCRs. Bommo-AKH-II was less effective. Also, Manse-AKH stimulated the release of Ca^{2+} from intracellular stores characteristic of G_q activation, and the other two AKHs were not tested.

The Akhr gene is expressed in all developmental stages in *D. melanogaster* (Hauser *et al.*, 1998), *An. gambiae* (Kaufmann and Brown, 2006), and *Ae. aegypti* (Kaufmann *et al.*, 2009). The Akhr mRNA was localized to the fat body by *in situ* hybridization (Grönke *et al.*, 2007) and by expression of a GFP marker driven by the

Akhr promoter (Bharucha *et al.*, 2008). This is consistent with the role of AKH as an energy-mobilizing hormone. In *D. melanogaster*, Akhr was also expressed in the subset of gustatory neurons in the subesophageal ganglion associated with attractive taste (Bharucha *et al.*, 2008). In *P. americana*, RT-PCR analysis demonstrated the presence of Akhr mRNA in additional tissues including the brain, ovaries, and digestive system (Wicher *et al.*, 2006). Significantly, the DUM cells from the sixth abdominal ganglion, which exhibit AKH-accelerated spiking, were shown to contain Akhr mRNA by single cell RT-PCR. An antibody against AKHR showed strong staining in the fat body, and AKHR-like immunoreactivity was also observed in neuronal somata and glial cells in all parts of the CNS, including the DUM neurons. In *Ae. aegypti*, one AKHR splice variant was expressed in the ovaries, suggesting that AKH may be involved in insect reproduction (Kaufmann *et al.*, 2009).

9.2.1.7. Effects on gene expression
The first example of the regulation of HrTH gene expression by nutritional status was obtained in *B. discoidalis* (Lewis *et al.*, 1998). It was demonstrated that starvation induced a twofold increase in HrTH mRNA levels in the CC, which returned to essentially normal levels after two days of re-feeding. This starvation-induced increase in HrTH mRNA level is accompanied by a similar increase in HrTH synthesis (Sowa *et al.*, 1996). The downregulation of HrTH mRNA levels by feeding is not simply a response to elevated hemolymph carbohydrate levels or a neural response to feeding, but is more likely due to the consumption of a complex of nutrients (Lewis *et al.*, 1998).

In the *B. discoidalis* fat body, the mRNA level of a cytochrome P450 (P4504C1) was increased directly in response to physiological levels of HrTH, and was also upregulated by starvation in a CC-dependent manner (Bradfield *et al.*, 1991; Lu *et al.*, 1995, 1996). It was suggested that P4504C1 is involved in mobilization of fat body resources in response to starvation either by the control of fatty acid oxidation to provide energy for the conversion of glycogen to trehalose in response to the release of HrTH (Bradfield *et al.*, 1991), or by providing an alternative route for carbohydrate synthesis during exposure to HrTH or starvation (Lu *et al.*, 1996).

9.2.2. A Counter-Regulatory Hormone for AKH?

The functional and structural similarities between AKH and glucagon are well established. Insects also contain a family of insulin-like peptides (ILPs), which was first described in *B. mori* and named bombyxin (Kawakami *et al.*, 1989). Contrary to expectations, injection of bombyxin into *B. mori* larvae lowered the trehalose concentration in the hemolymph, activated glycogen phosphorylase, and decreased glycogen content (Satake *et al.*, 1997), but bombyxin treatment of the adult moth did not cause the same effects (Satake *et al.*, 1999). Several groups have taken advantage of the genetic tools available in *D. melanogaster* to specifically manipulate gene expression or to ablate specific hormone-producing cells. In *D. melanogaster*, ablation of the cerebral insulin-producing neurosecretory cells (IPCs) by targeted expression of the apoptotic gene, *Reaper*, resulted in an increased level of hemolymph glucose and trehalose (Rulifson *et al.*, 2002). Heat-shock-inducible expression of the insulin-like peptide (dilp) reversed this phenotype, confirming that DILP is an essential regulator of metabolism in *D. melanogaster*. Targeted ablation of the AKH-expressing CC cells in *D. melanogaster* resulted in a decreased level of trehalose, and this effect was reversed by expression of an *akh* transgene (Kim and Rulifson, 2004; Lee and Park, 2004; Isabel *et al.*, 2005). In a separate study, the targeted ablation of AKH-producing cells resulted in an obese phenotype (Grönke *et al.*, 2007). Axonal processes of the IPCs terminate on the heart to facilitate DILP circulation, and on the CC suggesting that the AKH-producing cells of the CC contain DILP, suggesting that DILP is taken up by these cells (Rulifson *et al.*, 2002). Processes of the AKH-producing CC cells have extensive contact with IPC processes that terminate on the heart (Kim and Rulifson, 2004). The AKH cells express the homologues of the sulfonylurea receptor and the inward-rectifying K⁺ channel that makes up the ATP-sensitive K^+ channels that regulate hormone secretion in glucose-sensing cells. Exposure of larvae to sulfonylureas, which promotes depolarization and secretion, also leads to hyperglycemia, an effect that was not seen in CC-ablated animals (Kim and Rulifson, 2004). This suggests that hyperglycemic effects of AKH counter-regulate the activity of hormones such as the ILPs. To test whether CC cells exhibited glucose-sensing similar to that seen by mammalian pancreatic α-cells, cultured CC cells were marked with a Ca^{2+} indicator, and the Ca^{2+} level was used as a monitor for AKH secretion. It was demonstrated that low glucose and trehalose levels increased Ca^{2+} levels in CC cells, and high carbohydrate levels decreased Ca^{2+} levels (Kim and Rulifson, 2004). These results are consistent with the inhibitory effect of trehalose levels on AKH secretion in *L. migratoria* (Passier *et al.*, 1997). In *D. melanogaster*, mutation in the *Akhr* resulted in an obese phenotype, similar to that seen upon ablation of AKH-producing cells (Grönke *et al.*, 2007; Bharucha *et al.*, 2008) and overexpression of *Akhr* reduced fat storage (Grönke *et al.*, 2007). The obese phenotype of Akhr mutants was not altered by AKH, indicating that AKHR is the only receptor transmitting the signal induced by AKH. An independent confirmation of the role of AKH in carbohydrate mobilization was performed in *An. gambiae* in which the expression of *Akhr* was knocked down by RNA interference (Kaufmann and Brown, 2008). In these studies, the injection of Anoga-AKH-I into females

that were injected with *Akhr* dsRNA failed to mobilize glycogen reserves.

Together, these studies show that significant parallels exist between *D. melanogaster* and mammals with respect to the regulation of energy storage and utilization. The ability to perform genetic screens in *D. melanogaster* might uncover other components of the regulation of metabolism that could have relevance in treating human disorders, such as obesity and diabetes (Kim and Rulifson, 2004; Bharucha *et al.*, 2008).

9.3. Hormonal Control of Water and Electrolyte Homeostasis

Insects are small and hence have a high surface area to volume ratio, making them vulnerable to desiccation. Nevertheless, insects have evolved to be the most populous and diverse of all terrestrial animals. They have cuticles covered with water-impervious hydrocarbons to minimize water loss, and employ a variety of other behavioral, morphological, and physiological adaptations to keep water loss to a minimum, particularly in species that occupy very dry habitats. To maintain fluid homeostasis, water gained from the diet, metabolism and, in some species, from the atmosphere, must equal water lost by evaporation, respiration, and excretion. Evaporative and respiratory losses vary with environmental factors and the state of hydration, but the major site of regulation is the excretory system, and water loss via this route can change dramatically depending on the physiological status of the insect. Excretory water loss is determined by the rate at which fluid enters the hindgut from the midgut and MTs, and the rate of reabsorption therein. Fluid and ion homeostasis are under endocrine and possibly neural control, which allows the insect to regulate hemolymph volume and composition while permitting nitrogenous waste, toxic substances, and excess ions and/or water to be voided. The endocrine factors affecting the regulation are referred to as diuretic hormones (DHs) and antidiuretic hormones (ADHs), although this may not precisely describe their actions. Generally, DHs stimulate primary urine secretion by MTs, whereas ADHs increase fluid reabsorption from the hindgut, but there are exceptions to this.

9.3.1. Molecular Basis for Urine Excretion

With few exceptions (aglomerular fish), primary urine formation in vertebrates is driven by blood pressure (ultrafiltration), but there is a complex interplay between a number of hormones that control the volume and composition of the excreted urine, and hence fluid and ion homeostasis. Aldosterone and vasopressin (antidiuretic hormone) are largely responsible for maintaining electrolyte and water balance via reabsorption, whereas atrial natriuretic factor and angiotensin II primarily control blood volume (Beyenbach, 1993). In contrast, insects have low pressure circulatory systems and primary urine isosmotic to the hemolymph is secreted by the MT. Primary urine formation in insects must therefore be driven by active ion transport, not by blood pressure. Diuretic hormones increase MT secretion by stimulating ion transport into the lumen accompanied by osmotically obliged water (osmotic filtration), whereas ADHs either reduce MT secretion or stimulate fluid and vital solute reabsorption in the hindgut (Phillips, 1983; **Figure 3**).

A large literature exists on diuresis and its regulation in insects; early physiological studies have been reviewed comprehensively (Phillips, 1983; Spring, 1990). However, most early physiological studies on hormonal control of fluid homeostasis in insects utilized crude extracts of insect neuroendocrine tissues to affect target organs. It is now known that such studies dealt with a mixture of different active factors, so that unraveling the mode of action of any single factor was impossible. A fundamental

understanding of the control of fluid homeostasis requires that the controlling factors be identified, synthesized, and tested for their effects. Only the availability of synthetic factors can allow a detailed molecular understanding of the actions of factors controlling excretion and their physiological interplay, and whether prime control is due to diuresis or antidiuresis. There have been a number of reviews of the hormonal control of MTs and of hindgut function, focusing on identified hormones (Audsley *et al.*, 1994; Phillips and Audsley, 1995; Coast, 1996, 1998; Phillips *et al.*, 1998a; O'Donnell and Spring, 2000), and an extremely comprehensive review of both MT and hindgut function is available (Coast *et al.*, 2002b). The thrust of this chapter is to focus on advances in excretory physiology that concentrate on the effects of pure, defined factors.

An understanding of the molecular basis of primary urine production by the MTs of Insecta has emerged only during the last 20 years, and has been reviewed by Nicolson (1993) and Beyenbach (1995, 2003). Ion transport

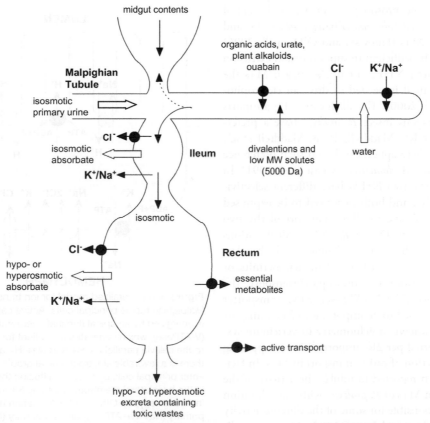

Figure 3 A generalized scheme of the classic view of the excretory process in insects. Malpighian tubule secretion is driven by the active transport of KCl and/or NaCl into the lumen, which draws water by osmosis. Other ions and metabolites enter the lumen by passive or active transporters. The primary urine enters the gut in most species at the midgut–hindgut junction and generally is directed posteriorly into the ileum and rectum where it is modified by isosmotic fluid reabsorption. The fluid entering the ileum and rectum is extensively modified by reabsorption of essential metabolites and fluid, which may be hypo- or hyperosmotic to the luminal contents. The driving force for ion transport is an apical electrogenic Cl⁻ pump, with cations entering passively, in contrast to the Malpighian tubules. Toxic wastes are retained in the hindgut lumen and are voided in the excreta, which can be strongly hypo- or hyperosmotic to hemolymph. Modified from *Coast et al.* (2002a).

leading to the secretion of primary urine is driven by an apical vacuolar-type H⁺-ATPase (V-ATPase) parallel to an apical monovalent cation/H⁺ antiporter (Grinstein and Wieczorek, 1994). The activity of the V-ATPase is stimulated by treatment of tubules with cAMP or peptides elevating cAMP. cGMP also stimulates its activity in *D. melanogaster* (O'Donnell *et al.*, 1996). The precise mechanism of stimulation is unknown, but activation of the V-ATPase in blowfly salivary glands involves the cAMP-dependent reassembly of the V_1 and V_0 subunits of the V_1V_0 holoenzyme (Dames *et al.*, 2006). The V_1 subunit C is a substrate for PKA and its phosphorylation most likely promotes assembly of the active holoenzyme (Voss *et al.*, 2007). There is also evidence for a Ca^{2+}-dependent activation of mitochondria in the apical region of principal cells in *D. melanogaster*, which will increase the availability of ATP to the V-ATPase and hence its activity (Terhzaz *et al.*, 2006).

The V-ATPase generates an electrochemical gradient favoring the entry of protons from the lumen in exchange for K^+ and/or Na^+ (Bertram *et al.*, 1991; Maddrell and O'Donnell, 1992; Weltens *et al.*, 1992; **Figure 4**). Measurements of the proton gradient across the apical membrane of *Formica polyctena* (Zhang *et al.*, 1994) and *Rhodnius prolixus* MTs (Ianowski and O'Donnell, 2006) indicate it would be sufficient to support cation extrusion via an electroneutral NHE, but in *Ae. aegypti* MTs the cell-negative potential is needed to drive an electrophoretic NHA (Petzel, 2000). The alkali cation-H⁺ antiports of the hematophagous insect *R. prolixus* have a preference for Na^+ over K^+ (Maddrell, 1978; Maddrell *et al.*, 1993b), whereas there appears to be no such preference in the house cricket, *A. domesticus* (Coast *et al.*, 1991). In *D. melanogaster*, the two NHAs have different selectivities for Na^+ and K^+, and both may need to be expressed on the apical membrane to allow transport of the two cations into the lumen (Day *et al.*, 2008). Alkali cations enter principal cells through ion channels in the basolateral membrane, if the electrochemical gradient permits, or by secondary (e.g., cation/Cl⁻ cotransport) transport. The basolateral primary (Na^+-K^+-ATPase) active transporter has long been proposed to be important in MT transport processes, yet its known stoichiometry in vertebrate systems (3 Na^+ exported per 2K^+ imported) is one opposing the observed direction of cation transport in MTs. In fact, serotonin has been reported to inhibit the activity of the Na^+-K^+-ATPase in MTs of *R. prolixus*, with this inhibition claimed to be responsible for some of the diuretic activity of this amine (Grieco and Lopes, 1997). However, indirect evidence suggests it may not play a significant role in *R. prolixus*, because ouabain, a potent inhibitor of this transporter, has no effect on fluid secretion (Ianowski *et al.*, 2002). The Na^+-K^+-ATPase does not appear to be of functional importance in MTs of *Ae. aegypti* (Weng *et al.*, 2003) or in *M. sexta* (A. Garrett, E. Chidembo, and D.A.

Schooley, unpublished data), as judged by a lack of vanadate or ouabain sensitivity of ATP hydrolase activity. In contrast, in *T. molitor* the Na^+-K^+-ATPase seems to play an important role in maintenance of cell K^+ concentration, and ouabain is an irreversible inhibitor of MT fluid secretion (Wiehart *et al.*, 2003). These species differences may be explained by the nature of cations secreted in primary urine; in the cryptonephric tubules of *T. molitor*, activities of K^+ exceeding 3mol $^{l-1}$ have been reported (O'Donnell and Machin, 1991), whereas *Ae. aegypti* secrete large amounts of Na^+ (Hegarty *et al.*, 1991). Species or order differences in the relative activity of the Na^+-K^+-ATPase may also underlie the high levels of hemolymph K^+ versus Na^+ in many Lepidoptera, Coleoptera, and Hymenoptera (see Sutcliffe, 1963 and Section 9.3.12.1.). A low activity of Na^+-K^+-ATPase in epithelial tissues would remove the driving force for keeping blood K^+ levels low versus Na^+ in hemolymph of most insect orders. The basal membrane Na^+-K^+-ATPase competes with apical membrane cation-H⁺ antiports for Na^+, and although ouabain has little effect on fluid secretion by cricket MTs, it markedly

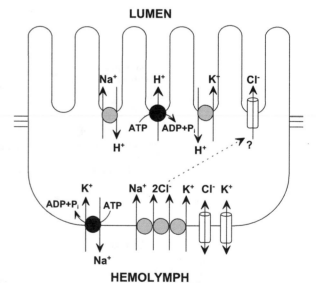

Figure 4 A generalized model for ion transport by Malpighian tubule principal cells. Active cation transport is energized by an apical (luminal) membrane H+ pump (V-ATPase), which generates a gradient for protons to return to the cell via parallel K+/H+ and Na+/H+ antiports. Where there is a favorable electrochemical gradient, cations may enter principal cells by passive diffusion through ion channels in the basolateral membrane (notably K+, but also Na+ in *Aedes aegypti*). Other routes for cation uptake include primary (Na+-K+-ATPase) and secondary (Na+-K+-2Cl⁻ and K+-Cl⁻ cotransport) active transport. Na+ entry may also be coupled to the transport of organic solutes (not shown). Cl⁻ that enters the cells via cation-Cl⁻ cotransporters probably exits through channels in the apical membrane. Arrows with circles indicate primary (black circles) or secondary (gray circles) active transport, while arrows through cylinders represent membrane ion channels. Modified from *Coast et al.* (2002a).

increases the Na⁺:K⁺ ratio of the secreted fluid from ~0.25 to unity (Coast, unpublished observations). The Na⁺-K⁺-ATPase is thus important in the excretion of K⁺, preserving the hemolymph Na⁺:K⁺ ratio (~22).

In general, the electrochemical gradient over the epithelium favors Cl⁻ diffusion into the lumen through a shunt pathway. The Cl⁻-selective shunt described by Pannabecker *et al.* (1993) in MTs of *Ae. aegypti* does not appear to be in the principal cells, and is either paracellular or through a second cell type, the stellate cells, which form thin (3–5 μm deep) windows between the hemolymph and tubule lumen. Not all species have stellate cells; they have been shown to have a different embryonic origin than principal cells (Denholm *et al.*, 2003).

9.3.2. A Multiplicity of Peptides Regulate Diuresis and Antidiuresis in Insects

A number of "model species" have been adopted by different investigators for studying diuresis and antidiuresis. The classic work of Ramsay (1954) was performed with a stick insect, *Carausius morosus*. A great deal of pioneering work on diuresis and its control has been done with the hematophagous hemipteran *R. prolixus* (Maddrell,

1963), and *D. melanogaster* has been proposed as a useful model for studying diuresis in insects (Dow *et al.*, 1998; MacPherson *et al.*, 2001). The Beyenbach group (2003) studied ion transport mechanisms in *Ae. aegypti* for decades. Locusts have been the preferred model for studying the role of the hindgut in fluid reabsorption, largely through the work of the Phillips group (Phillips and Audsley, 1995).

The first hint that more than two different classes of peptides acted on MTs was supplied by the Beyenbach group's studies of regulation of *Ae. aegypti* tubules by nervous system extracts separated on HPLC (Petzel *et al.*, 1985). The first undisputed identification of an insect DH was the identification from *M. sexta* of a 41 amino acid peptide with high sequence similarity to the sauvagine/urotensin/urocortin/corticotropin releasing factor (CRF) family (Kataoka *et al.*, 1989b). To date, 13 "CRF-related" DH structures have been identified by isolation and Edman degradation; others were identified by BLAST searches of the rapidly expanding list of arthropod genomes (**Figure 5**). Somewhat later, the leucokinins (myotropic peptides originally isolated based on their ability to cause contractions of the cockroach hindgut) were also shown to have potent diuretic effects (Hayes

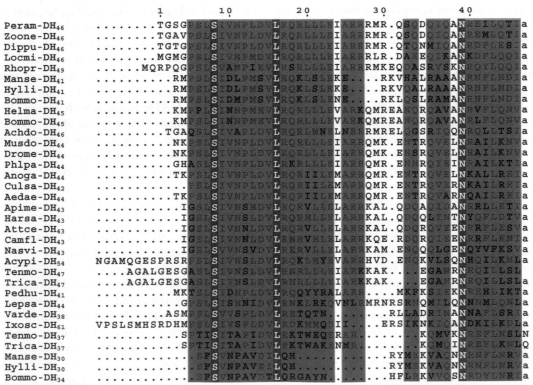

Figure 5 The CRF-related DH sequences were aligned with ClustalW at http://npsa-pbil.ibcp.fr/cgi-bin/npsa_automat.pl?page=/NPSA/npsa_clustalw.html using the Blosum matrix; the alignment was plotted with ESPript at this site. Identical residues are shown as white letters in red boxes, while similar residues are shown with red letters in yellow boxes. The CRF-related DHs are identified using the Swiss-Prot species abbreviations (defined in the text). All sequences are amidated (indicated by lower case a) at the carboxyl-terminus with the exception of three of the four DH from *T. molitor* and *T. castaneum*. For the tick DH Ixosc-DH61, 10 amino acids (GPLSLRSRGP) are deleted from the N-terminus to avoid resizing the alignment.

et al., 1989). Currently, 46 different myokinin sequences are known from 14 species of insects. In the last decade, it has become evident that a "cocktail" of neuropeptides is implicated in the regulation of MT secretion (Coast *et al.*, 2002a; Skaer *et al.*, 2002). The majority of these peptides have diuretic activity that stimulates fluid secretion. Others have antidiuretic activity and reduce the rate of secretion. From the mid-1990s onward, peptides belonging to four families in addition to the CRF-like DHs and kinins have been shown to influence tubule secretion: the calcitonin-like (CT-like) DHs, CAP_{2b}-like, tachykinin related peptides, and two antidiuretic factors, ADFa and ADFb. The greatest amount of research has been carried out on the first two families discovered, that is, the CRF-like and kinin families. They appear to be widespread, but not ubiquitous, among insects, although there is considerable interspecific variation in activity. For example, the CRF-like DH of *A. domesticus* stimulates maximal secretion, and is about threefold more active than the native kinins (Coast *et al.*, 1992). Alternatively, in *D. melanogaster* and in *Musca domestica* (Holman *et al.*, 1999), the CRF-like DHs have poor activity compared with the native kinin. The situation is even more extreme in *R. prolixus*, where CRF-like peptides can elicit maximal secretion and kinins appear to be inactive (Te Brugge *et al.*, 2002).

CT-like, CAP_{2b}-like, and tachykinin-related peptides may also be of widespread occurrence in insects, but have been investigated less extensively than the CRF-like DHs and kinins. CT-like peptides have only been identified in three species, *D. punctata*, *R. prolixus*, and *F. polyctena* (a partial sequence), but another 15 have been found in the genomes of Insecta of diverse orders (Dictyoptera, Diptera, Hemiptera, Lepidoptera and Hymenoptera), as well as in Acari and Crustacea (see Section 9.3.5.). Even less is known of the antidiuretic factors controlling MT secretion since these have only been described in *T. molitor* and *R. prolixus*. Intriguingly, peptides that have diuretic activity in one species have been shown to have antidiuretic activity in another. Manse-CAP_{2b} is a diuretic in Diptera (*D. melanogaster* and *M. domestica*), but an antidiuretic in *R. prolixus* and *T. molitor*. CAP_{2b} has no effect on secretion by MT of *M. sexta* at 1 mmol l^{-1} (Skaer *et al.*, 2002), the latter organism is the host from which it was identified. Tenmo-ADFb is an antidiuretic in *T. molitor*, but a diuretic in *A. domesticus* (Coast *et al.*, 2007). Even when comparing quite similar species, peptide activity can vary considerably, as for CAP_{2b} in *D. melanogaster* and *M. domestica*. In *D. melanogaster*, CAP_{2b} acts on principal cells to activate a calcium-calmodulin dependent NO synthase, and the rise in NO levels increases cGMP production by a soluble guanylate cyclase. Diuresis results from the cGMP-dependent stimulation of an apical membrane V-type ATPase that powers fluid secretion (Davies *et al.*, 1995). Conversely, CAP_{2b} has no effect on cGMP levels in *M. domestica* tubules or on the activity of the V-type

ATPase. Instead, diuresis results from the opening of a passive chloride shunt pathway, which has previously been attributed to kinin stimulation (C.S. Garside and G.M. Coast, unpublished data).

9.3.3. CRF-like DH

The first insect DH identified was Manse-DH_{41}, a 41 residue peptide isolated from 10,000 heads of pharate adult *M. sexta* using an *in vivo* assay (Kataoka *et al.*, 1989b). This assay utilized beheaded, newly eclosed adults of *Pieris rapae*, similar to a technique described for *P. brassicae* (Nicolson, 1976). Manse-DH_{41} has 41% sequence identity with sauvagine, a peptide isolated from skin of the frog *Phyllomedusa sauvagei* (Montecucchi and Henschen, 1981) originally believed to be the amphibian form of CRF. The direct sequence identity with human CRF is lower at 29%. Subsequently a smaller peptide, Manse-DH_{30} (also known as Manse-DPII), was isolated from dissected neuroendocrine cells of brains of *M. sexta*. Again, an *in vivo* bioassay was used to monitor the isolation using adults of *M. sexta*. Thanks to the "surgical purification" method, the peptide was obtained pure after a single step of reversed-phase liquid chromatography (RPLC; Blackburn *et al.*, 1991). The amount isolated was quite small, requiring both Edman degradation and mass spectral analysis. As aligned in **Figure 5**, Manse-DH_{30} has only nine residues identical with those in Manse-DH_{41}. Also in 1991, CRF-related DHs were isolated from *A. domesticus* and *L. migratoria*; the former based on its ability to increase levels of cAMP in MTs of *A. domesticus* maintained *in vitro* (Kay *et al.*, 1991a). This 46 amino acid peptide, Achdo-DH_{46}, is 41% identical with Manse-DH_{41}. Two publications reported the sequence of the same DH from *L. migratoria* within a few weeks of each other; both utilized purifications by multiple steps of RPLC. Lehmberg *et al.* (1991) isolated this peptide from 4600 locust brains + CC using a direct enzyme-linked immunosorbent assay (ELISA) to detect active fractions with antibodies raised against Manse-DH_{41}. Kay *et al.* (1991b) utilized 2000 whole heads of *L. migratoria* to isolate Locmi-DH_{46}, monitoring the isolation by *in vitro* assay for cAMP production with locust MTs. Locmi-DH_{46} is 49% identical with Manse-DH_{41}, but only 48% identical with Achdo-DH_{46}, another orthopteran peptide. This very low conservation of sequence contrasts with the CRF family in higher animals. Comparing sequences of CRF from a frog (*Xenopus laevis*), a fish (*Catostomus commersoni*), and seven mammals (rat, goat, sheep, pig, dog, horse, and human), 71% of residues are totally conserved (Lovejoy, 2009). Human and frog CRF are 95% identical. Thus, the surprisingly divergent sequences of the CRF-related DH are unusual.

In 1992 Kay *et al.* identified from *P. americana* a DH, Peram-DH_{46}, which differs from Locmi-DH_{46} at 13 residues, and from Achdo-DH_{46} at 18. Malpighian tubules of *L. migratoria* were used to monitor the isolation of this peptide. Audsley *et al.* (1995) advocated the use of *M. sexta* MTs for a general heterologous bioassay for isolation of DH from other species, monitoring the release of cAMP by the MTs, since *M. sexta* MTs cross-react more promiscuously with other CRF-related DH tested than do MTs of *L. migratoria* or *A. domesticus*, which respond poorly to CRF-related DHs of other species. Subsequently a number of new DHs were isolated using this assay, including Musdo-DH$_{44}$, isolated from both *M. domestica* and *Stomoxys calcitrans* (Clottens *et al.*, 1994). This peptide was accompanied by another, more hydrophobic active factor in *S. calcitrans*, whose existence was reported in a workshop (Schooley, 1994), but no details of its properties were ever published. A 42 amino acid DH, Culsa-DH$_{42}$, was isolated from the salt water mosquito, *C. salinarius*. The exact details of its isolation were not reported, although its sequence and biological properties are published (Clark *et al.*, 1998a). Two DHs were isolated from the white-lined Sphinx, *Hyles lineata* (Furuya *et al.*, 2000a); Hylli-DH$_{41}$ and Hylli-DH$_{30}$ differ from their *M. sexta* orthologues at only one residue. A 46 residue DH was isolated from a dampwood termite, *Zootermopsis nevadensis* (Baldwin *et al.*, 2001). Zoone-DH$_{46}$ was extracted with another DH, which is more basic on ion exchange LC, but less hydrophobic, than Zoone-DH$_{46}$. Two attempts to isolate this factor were unsuccessful, perhaps due to lability in the acidic solvents used for RPLC separation (Baldwin *et al.*, 2001).

An unusual CRF-related DH was identified in *T. molitor* (Furuya *et al.*, 1995); its isolation was monitored based on elevation of cAMP from MT of this species, as *T. molitor* extracts are without effect on *M. sexta* MTs. A 37 residue DH, Tenmo-DH$_{37}$, was isolated from 8400 pupal heads of *T. molitor*, accompanied by a less abundant, more hydrophobic factor isolated in insufficient quantity to sequence. After extracting an additional 20,000 heads, plus active fractions from the earlier isolation, a small amount of peptide (~50 pmol) was isolated and sequenced giving Tenmo-DH$_{47}$ (Furuya *et al.*, 1998). Structurally, both peptides differ from other CRF-related DHs as their C-terminal residues exist in the free carboxylate form rather than being amidated. In addition, Tenmo-DH$_{37}$ has a one-residue "extension" of its sequence. Most sequences from organisms other than *T. molitor* (or *Tr. castaneum*) have a C-terminal Ile-amide or Val-amide, whereas Tenmo-DH$_{37}$ has a Leu-Asn-OH C-terminus and Tenmo-DH$_{47}$ has a Leu-OH, lacking the Asn extension of its smaller congener. Tenmo-DH$_{37}$ is not active on the MTs of *M. sexta*; its activity is >10^4 lower than Manse-DH$_{41}$ in this assay. This is consistent with data on the free acid form of Manse-DH$_{41}$, which is ~1000-fold less active than its natural, amidated form in the *P. rapae* assay (Kataoka *et al.*, 1989b). Tenmo-DH$_{37}$ is ~600-fold more potent than its larger congener in an

in vitro assay measuring cAMP production with adult MTs (Furuya *et al.*, 1998), and ~200-fold more potent in a Ramsay fluid secretion assay with larval MTs (Wiehart *et al.*, 2002). Interestingly, Manse-DH$_{41}$ is only 17-fold less active than Tenmo-DH$_{37}$ in the cAMP assay with adult *T. molitor* MTs (Furuya *et al.*, 1995), although it is more similar to Tenmo-DH$_{47}$ than to -DH$_{37}$.

Subsequently Furuya *et al.* (2000b) isolated a 46 residue DH, Dippu-DH$_{46}$, from dissected brain-CC-CA complexes of the Pacific beetle roach, *D. punctata*, using *M. sexta* tubules to monitor its three-step isolation. Fractions from the first RPLC separation were also tested on MTs of *Schistocerca americana*. Two fractions stimulated *Sc. americana* tubules; one eluted much more quickly than Dippu-DH$_{46}$. The unknown factor active on *Sc. americana* tubules required two more RPLC steps to reach homogeneity. The 31 amino acid sequence was obviously unrelated to the CRF-related DH, but has six residues identical with chicken calcitonin (see Section 9.3.5.).

Only the 13 CRF-related DHs reviewed previously have been identified *de novo* from tissue extracts, but the availability of genome sequence data has allowed the identification of another 21 DHs from an increasingly wide range of species. The genomes currently available have few hemimetabolous species and are biased toward dipterans, including 12 *Drosophila* species, *An. gambiae*, *Ae. aegypti*, and *Cu. quinquefasciatus*, and hymenopterans (*A. mellifera*, *N. vitripennis*, *N. longicornis*, *N. giraulti*, and *Atta cephalotes*, *Camponotus floridanus*, and *Harpegnathos saltator*) with but a single lepidopteran (*B. mori*), a coleopteran (*Tr. castaneum*), a homopteran (*A. pisum*), a heteropteran (*R. prolixus*), an anopluran (*Pediculus humanus corporis*), and three non-insect arthropods: the Acari *Varroa destructor*, *Ixodes scapularis*, and the crustacean *Lepeophtheirus salmonis*. The quality of the genome data varies greatly from extensively annotated (*D. melanogaster*) to those containing incompletely assembled contigs from shotgun sequencing data. Alignment of these sequences with ClustalW (http://npsa-pbil.ibcp.fr/cgi-bin/npsa_automat.pl?page=/NPSA/npsa_server_html) gave the alignment shown in **Figure 5**. The Blosum matrix was used because it is better for this diverse family than the PAM or Gonnet matrices. There are at least two sets of paralogues in the CRF-related DH: the "long" DH are at the top of the figure and there is a "short" group at the bottom. The latter include five sequences from Lepidoptera and Coleoptera. Interestingly, the two sequences from Acari, Varde-DH$_{38}$ and Ixosc-DH$_{61}$, and the crustacean Lepsa-DH$_{44}$ are sorted by ClustalW as more similar to the short DHs, even though Ixosc-DH$_{61}$ is the largest CRF-related DH. The first 10 N-terminal amino acids (GPLSLRSRGP) of Ixosc-DH$_{61}$ are not shown in **Figure 5** to avoid resizing the alignment. Including members of the CRF superfamily (shown below the DH) in the alignment causes ClustalW to change the ordering of the paralogues.

In doing this, the three short lepidopteran DHs sort to the top of the group, adjacent to Peram-DH$_{46}$. Our alignment was adjusted to place these lepidopteran sequences adjacent to the coleopteran sequences, although the behavior with ClustalW suggests these groups are different paralogues. Three CRF-like DHs are known from *B. mori* (Roller *et al.*, 2008), and it seems likely that more paralogues remain to be discovered in other species.

Examination of the sequences of the 34 known CRF-related DHs (**Figure 5**) reveals only three amino acids that are completely conserved in this family. Two of these are a Ser and a Leu, separated by seven amino acids. The other is an Asn in the C-terminal domain. A Leu separated from the C-terminal residue by two amino acids (three in Tenmo-DH$_{37}$) is nearly completely conserved. The prospects for using degenerate probes to attempt to isolate additional new members of this family from cDNA or mRNA are poor, given that the most highly conserved domains are rich in amino acids with highly degenerate codons. In the CRF superfamily of peptides, again only four residues are completely conserved (Dautzenberg and Hauger, 2002) two are the Ser and Asn; also conserved in the CRF-related DH. Despite the exceptional sequence conservation of the various forms of CRF, the paralogues sauvagine, urotensin, and urocortin diverge considerably (Hauger *et al.*, 2003).

In 1991, two CRF sequences were identified by cDNA cloning in the white sucker *C. commersoni*. The sucker CRFs differ from the sequences of human, horse, rat, and dog CRFs (all identical) at only one and two residues (Morley *et al.*, 1991), yet are only 54% identical with urotensin I, the fish equivalent of sauvagine. The following year, a similar situation was discovered with *X. laevis*, which has two forms of CRF differing from the human sequence at only two and three residues (Stenzel-Poore *et al.*, 1992). Consequently a sauvagine-like (or urotensin-like) peptide was sought and found in rat brain; it was called urocortin 1, a vertebrate orthologue of urotensin and sauvagine (Vaughan *et al.*, 1995). Subsequently, two additional paralogues have been identified, urocortin 2/stresscopin related peptide and urocortin 3/stresscopin (Chang and Hsu, 2004). None of the urocortin sequences have been confirmed by peptide isolation. This is unfortunate, as several lack conventional dibasic amino acid cleavage sites at the N- and/or C-terminus. Although mouse urocortin III has only 26% identity with human CRF, there are no gaps in the alignment, which is a striking feature of aligning CRF-related DHs.

BLAST analysis of genome data at NCBI with a short CRF-related DH (Manse-DH$_{30}$) found the sequence of Bommo-DH$_{34}$, but no other DH. BLAST search with Tenmo-DH$_{37}$ as a query results in the discovery of only Trica-DH$_{37}$. BLAST search using Musdo-DH$_{44}$ as a query resulted in the discovery of Bommo-DH$_{45}$, Bommo-DH$_{41}$, fly DH, hymenopteran DH (Apime-DH$_{43}$,

Attce-DH$_{43}$, Harsa-DH$_{43}$, Camfl-DH$_{43}$, and Nasvi-DH$_{43}$; an identical DH is found in the genomes of the two other *Nasonia* species), Ixosc-DH$_{61}$, Acypi-DH$_{54}$, Pedhu-DH$_{41}$, Rhopr-DH$_{49}$, and Trica-DH$_{47}$. Use of Zoone-DH$_{46}$ as a query also located Varde-DH$_{38}$ and Lepsa-DH$_{44}$ from the crustacean *L. salmonis*. Thus, while BLAST searches locate orthologues of known peptides in genomic databases, they do not locate "novel" paralogous sequences. Only in Lepidoptera and Coleoptera do we know the sequence of multiple DH, although as discussed next, there is evidence for the existence of CRF-like DHs in insects from three hemimetabolous species and the stable fly.

A Pro residue lies between the totally conserved Ser and Leu residues in the N-terminal domain. The residue just upstream of the Pro is Asn in 13 of the DHs. The Asp-Pro bond is exceptionally labile to acid on standing in solution. A single base change to the Asn codon gives Asp. The inability to isolate a second CRF-related DH from *Z. nevadensis* may be due to the presence of an Asp-Pro bond (Baldwin *et al.*, 2001); this may hold for other species as well. Montuenga *et al.* (1996) showed the ampulla of *L. migratoria* MTs to contain endocrine cells that express not only Locmi-DH$_{46}$, but also a faster eluting CRF-like DH, which was not identified. *R. prolixus* also contains an unknown CRF-like DH in addition to Rhopr-DH$_{49}$ (D.A. Schooley, V.A. Te Brugge, and I. Orchard, unpublished). It is likely that identification of a DH paralogue from one of these species would allow BLAST location of its orthologues in other species.

B. mori is the first species for which three CRF-related DHs are found; the genome sequences have been experimentally verified and were shown to arise from alternative splicing of a single gene containing 11 exons (Roller *et al.*, 2008). In the precursor proteins of the three *B. mori* DHs, the first 70 amino acids are identical prior to the beginning of the unique sequences. This is in marked contrast to CRF and its paralogues in mouse and man, where each precursor protein is a unique sequence (SwissProt P06850, P55089, Q96RP3, and Q969E3 for human CRF and urocortins I–III, respectively). *Tr. castaneum* contains two CRF-related DHs (Li *et al.*, 2008) — Trica-DH$_{37}$ differs at ten residues from Tenmo-DH$_{37}$. However, Trica-DH$_{47}$ is 100% identical with Tenmo-DH$_{47}$, except that the C-terminus is amidated rather than being a carboxylate like the other three known beetle CRF-related DHs. As in *B. mori*, both peptides are encoded by the same five exon gene and result from alternative splicing (Cosme, 2009). Two additional DHs in **Figure 5** were located by searching ESTs of several organisms at NCBI: *Heliconius melpomene* DH (Helme-DH$_{45}$) was found using Bommo-DH$_{45}$ as a query, and the sand fly (*Phlebotomus papatasi*) ESTs were searched using Drome-DH$_{44}$ as a query, revealing Phlpa-DH$_{44}$.

The dipteran CRF-related DHs are unusual for their lack of diversity in structures. Seven *Drosophila* species contain Drome-DH$_{44}$ (*D. melanogaster, ananassae, erecta, sechellia, simulans, virilis,* and *yakuba*), and the five other species whose genomes are known (*D. pseudobscura, grimshawi, mojavensis, persimilis, willistoni*) contain Musdo-DH$_{44}$; this peptide differs from Drome-DH$_{44}$ only in conservative substitution of Ser31 in the latter for Thr in Musdo-DH$_{44}$. An idiosyncrasy of fly DH gene sequences is that they all have an intron in Val34. As a result, BLAST analysis with default parameters was unable to locate the C-terminal portion of the putative *Ae. aegypti* DH$_{44}$. Upon increasing the "expect" value to 100 the C-terminus was located on a different contig. There are 42.7 kb of sequence downstream of the splice site in *Ae. aegypti*; for example, the intron is larger than that. Recently, a gene for a SLC4-like anion exchanger was cloned in *Ae. aegypti* and found to contain an intron of 112 kb (Piermarini *et al.*, 2010). The precursor for Drome-DH$_{44}$ has been experimentally determined from one of three cDNAs in the GenBank (Cabrero *et al.*, 2002); two cDNAs were sequenced and both were found to be chimeric. However, one of these was sequenced and found to be apparently full length; it differed at six bases from the annotated genome sequence. There are two introns (3 and 1.6 kb) in the upstream portion of the gene. This gene sequence has an intron in Val34 of only 1842 bp, versus >42.7 kb in *Ae. aegypti*.

Of those DH predicted from genome sequences, only a few have been synthesized and functionally tested. This includes Drome-DH$_{44}$, which was shown to elevate cAMP and fluid transport from principal cells of *D. melanogaster* (Cabrero *et al.*, 2002). Anoga-DH$_{44}$ was shown to elevate cAMP and to stimulate Na$^+$ and K$^+$ transport by MTs of *An. gambiae* (Coast *et al.*, 2005: see also Section 9.3.12.4.). Very recently Te Brugge *et al.* (2010) isolated sufficient Rhopr-DH$_{49}$ to show it had identical mass spectrum to that of synthetic peptide, and found an EC$_{50}$ value for this peptide on MT of *R. prolixus* of ~3 nM, which is far more potent than the heterologous DHs tested: 670 nM EC$_{50}$ for Zoone-DH$_{46}$ and 550 nM for Dippu-DH$_{46}$.

Only the Manse-DH$_{41}$ cDNA sequence for a CRF-related DH has been determined by *de novo* experimental cloning (Digan *et al.*, 1992). The pre-propeptide has a 19 residue signal peptide for cellular secretion, followed by a 61 residue propeptide terminating in Lys-Arg (the cleavage site for a Kex-2 like enzyme), then the 41 residues of Manse-DH$_{41}$, followed by a Gly-Lys-Arg plus 15 additional amino acids of the propeptide. The Lys-Arg processing site results in a C-terminal Gly, which confers amidation of the C-terminal Ile upon action of peptidylglycine amidating monooxygenase (Prigge *et al.*, 2000) on the Gly.

There are few studies on structure–activity relationships with the CRF-related DHs because of their size and the expense of synthesis. Studies with Manse-DH$_{41}$ and Achdo-DH$_{46}$ show that a region near the N-terminus is essential for receptor activation while the remaining portion of the peptide is required for receptor binding. An analog Manse-DH(13–41), truncated by 12 residues at the N-terminus, binds Manse-DHR expressed in Sf9 cells with high affinity (IC$_{50}$ ~2.8 nmol l^{-1} vs. 0.16 nmol l^{-1} for the intact peptide) but does not stimulate cAMP production, whereas Manse-DH(3–41) has high receptor binding affinity and is a potent stimulant of adenylate cyclase (Reagan, 1995a). Manse-DH(21–41) and Manse-DH(26–41) have binding affinities reduced 100- and 1000-fold, respectively, while Manse-DH(31–41) has no binding activity at 1 mmol l^{-1} (Reagan *et al.*, 1993). Deletion of the first four residues from Manse-DH$_{41}$ decreases activity, while deletion of the next residue essentially abolishes it (P. Dey and D.A. Schooley, unpublished data). Taken together with the results of Reagan (1995a), this suggests that the receptor-activating domain of Manse-DH$_{41}$ lies very close to the N-terminus. Studies with N-terminal truncated analogs of Achdo-DH$_{46}$ in a cricket tubule fluid secretion assay confirm the importance of the N-terminus for receptor activation (Coast *et al.*, 1994). The activities of Achdo-DH(6–46) and Achdo-DH(7–46) are essentially indistinguishable from intact Achdo-DH, but the activity of Achdo-DH(11–46) is reduced by 60% and Achdo-DH(23–46) is inactive. These data suggest that the domain necessary for receptor activation is farther from the N-terminus in Achdo-DH$_{46}$ than in Manse-DH$_{41}$. However, results for peptides with low activity in a series of deletion peptides must be viewed with caution; unless the least active peptides were purified first, mere traces of active DH will bias the data. Intact Met residues are important for activity: one of the Met residues (at position 1, 3, or 13) in Locmi-DH$_{46}$ can become oxidized, with a loss of activity (I. Kay and G.M. Coast, unpublished data) and [Nle^{2,11}]-Manse-DH$_{41}$ (with Nle replacing Met^2 and Met^{11}) maximally stimulates cricket tubule secretion, whereas Manse-DH$_{41}$ gives only a 60% response (Coast *et al.*, 1992). In addition, Locmi-DH(1–23) and Locmi-DH(24–46) are inactive whether tested separately or together (Nittoli *et al.*, 1999; G.M. Coast, unpublished data); thus, the binding and activation domains must be bonded together to have a biologically active peptide. Manse-DH$_{41}$ with a free C-terminus (acid) stimulates cAMP production by adult *M. sexta* MTs (Audsley *et al.*, 1995) and post-eclosion diuresis in decapitated newly emerged *P. rapae* (Kataoka *et al.*, 1989b), but in both assays it was 1000-fold less potent than Manse-DH$_{41}$.

9.3.4. Diuretic and Myotropic Peptides: Kinins

Insect kinins (originally called myokinins) were isolated from whole head extracts of the Madeira cockroach, *Leucophaea maderae* (leucokinins; Leuma-Ks) (Holman *et al.*, 1986a,b, 1987a,b) based on their potent myotropic activity in a cockroach (*L. maderae*) hindgut assay (Holman *et al.*, 1991). The eight leucokinins differ in myotropic potency. They were later found to be potent stimulants of fluid secretion in *Ae. aegypti*, also causing a depolarization of the transepithelial potential (Hayes *et al.*, 1989). Soon thereafter, five kinins (achetakinins; Achdo-Ks) were isolated from *A. domesticus* using the *L. maderae* hindgut assay; they were shown to also possess high diuretic activity on MT of *A. domesticus* (Coast *et al.*, 1990). Similarly, locustakinin was isolated from 9000 brain-CA-CC-subesophageal complexes of *L. migratoria* using the *L. maderae* hindgut assay (Schoofs *et al.*, 1992). It is curious that only a single kinin, locustakinin (Locmi-K), was isolated from this species, whereas all other orthopterans and dictyopterans studied have 5–8 kinins. Using hindgut of *P. americana*, eight kinins were isolated from 800 CC-CA of this species (Predel *et al.*, 1997), five of which are unique sequences and three occur in other species: Locmi-K, Leuma-K-7, and Leuma-K-8. Availability of endogenous kinins is crucial to understanding their role in any species; for example, Musdo-K is about 10^6 more potent in *M. domestica* (Holman *et al.*, 1999) than Leuma-K-I.

The first kinins to be identified from a holometabolous species were those from *Ae. aegypti*, which were isolated using an ELISA rather than a functional assay (Veenstra, 1994b). Whereas these were termed "Aedes leucokinins I–III," this nomenclature suggests they may be the same sequence as those from *L. maderae*, but they are in fact unique. Here the Swiss-Prot-NCBI five-letter standards for species abbreviation, Aedae-K-1, Aedae-K-2, and Aedae-K-3, as previously recommended is used (Coast *et al.*, 2002a). Later, the sequence of the single cDNA encoding these kinins was determined, and their biological properties determined (Veenstra *et al.*, 1997). While all three Aedae-K peptides are capable of depolarizing the mosquito MT at doses between 0.1 and 1 nmol l^{-1}, only Aedae-K-3 gave good stimulation of fluid secretion, with an EC$_{50}$ near 10 nmol l^{-1}; Aedae-K-3 was appreciably less active and Aedae-K-2 was without effect on fluid secretion at 1 mmol l^{-1}. More recently, in a different laboratory, all three Aedae-Ks were found to stimulate MT secretion (Schepel *et al.*, 2010). The authors note that the MTs of some batches of mosquitoes do not respond to kinin stimulation and suggest this may be due to the nutritional, developmental, and/or physiological status of the donor insects. Accordingly, the failure of Aedae-K-II to stimulate secretion in the Veenstra *et al.*, (1997) study may be because it was tested on MT from developmentally or nutritionally compromised insects. All three Aedae-Ks stimulate *in vivo* urine excretion by adult female *Ae. aegypti* (Cady and Hagedorn, 1999a), and elevate inositol trisphosphate (IP$_3$) production by isolated MTs while having no effect on cAMP levels (Cady and Hagedorn, 1999b). Three kinins were isolated from wild-collected salt marsh mosquitoes, 94% of them *Cu. salinarius*, and the first kinin was christened "culekinin depolarizing peptide"

(Hayes *et al.*, 1994). They were isolated using the *L. maderae* hindgut assay, as well as an electrophysiological assay with tubules of *Ae. aegypti*. The sequence of only the heptamer Culsa-K-I was published (Hayes *et al.*, 1994). Some years later, the biological properties of Culsa-K-II and -III, along with the sequences, were published by other workers (Cady and Hagedorn, 1999a,b), but details of their isolation and identification remain unpublished. These peptides were also shown to stimulate *in vivo* urine production in adult female *Ae. aegypti* (Cady and Hagedorn, 1999a) and to increase (IP$_3$) concentrations in MTs of *Ae. aegypti*, but not cAMP levels in isolated MTs (Cady and Hagedorn, 1999b). Using the sequences of the Aedae-K as a BLAST query of the *An. gambiae* genome (Holt *et al.*, 2002), a single gene encoding three different kinins of 15, 10, and 21 residues was located: Anoga-K-1, -2, and -3. These peptides occur in the gene in the size order shown; the first and second are separated only by the amino acid residues Gly-Lys-Arg, as are the second and third (residues necessary for cleavage from the propeptide and formation of the amidated C-terminus). BLAST search of the insect genomes at NCBI resulted in discovery of kinin precursors from *A. pisum* (Acypi-K-1 to -3), *B. mori* (Bommo-K-1 to -3), *C. quinquefasciatus* (Culqu-K-1 to -3), *Mayetiola destructor* (Mayde-K), *Myzus persicae* (Myzpe-K-1 to -4), *P. humanus* (Pedhu-K), *R. prolixus* (Rhopr-K-1 through 4), and the tick *I. scapularis* (Ixosc-K-1 and -2), while the BLAST search of insect EST sequences at NCBI found a precursor for kinins from the sandfly *Lutzomyia longipalpis* (Lutzo-K-1 and -2). Certain sequences found in the BLAST search were identical with known peptides and are not shown in **Figure 6**.

Holman *et al.* (1999) isolated a 15 residue kinin from another fly species, *M. domestica*, using an ELISA with an antibody raised against Leuma-K-I. Musdo-K strongly stimulates fluid secretion in *M. domestica* MTs, whereas the hexapeptide Achdo-K-V, identical with the six C-terminal residues of Musdo-K, is 1000-fold less potent in this assay. Thus, the longer sequence is crucial for full activity in *M. domestica*. Musdo-K also stimulates the housefly hindgut, possibly aiding in excretion. Later that year, Terhzaz *et al.* (1999) reported the isolation of a single 15 residue kinin from *D. melanogaster*, using a purification strategy based on that for the Aedae-K peptides. The sequence differs from Musdo-K at the second residue, a Thr in Musdo-K versus a Ser in Drome-K. A BLAST search of the *D. melanogaster* genome confirms that this is the only kinin.

Studies of the effects of Drome-K on MTs showed that it elevates intracellular Ca^{2+}, but not cAMP. A BLAST search of the complete genome of this species verifies that the *pp* gene encodes but a single kinin. The isolation of kinins from a lepidopteran insect, the corn earworm *Helicoverpa zea*, is of interest because only the first step utilized a functional assay. Blackburn *et al.* (1995) used ~2000 abdominal ventral nerve cords (VNC) from adult moths as a tissue source. Homogenized tissues were pre-purified on

```
Leuma-K-V      ...............GSGFSSWGa
Pedhu-K        ...............GPGFRSWGa
Leuma-K-II     ..............DPGFSSWGa
Leuma-K-VII    ..............DPAFSSWGa
Acypi-K-2      ................PAFSSWGa
Locmi-K        .................AFSSWGa
Myzpe-K-1      ............RQKTVFSSWGa
Acypi-K-1      ............pQKTVFSSWGa
Myzpe-K-4      ......ASDKHGRPKQTFSSWGa
Rhopr-K-4      ..............KPIFSSWGa
Achdo-K-III    ..............ALPFSSWGa
Leuma-K-I      ..............DPAFNSWGa
Peram-K-5      ..............SPAFNSWGa
Peram-K-1      ..............RPSFNSWGa
Peram-K-3      ..............DPSFNSWGa
Leuma-K-III    ..............DQGFNSWGa
Musdo-K        .......NTVVLGKKQRFHSWGa
Drome-K        ......NSVVLGKKQRFHSWGa
Rhopr-K-3      ..............ARFNSWGa
Leuma-K-IV     ..............DASFHSWGa
Peram-K-2      ..............DASFSSWGa
Leuma-K-VIII   ..............GADFYSWGa
Achdo-K-1      .............SGADFYPWGa
Peram-K-4      ..............GAQFSSWGa
Rhopr-K-1      ...............AKFSSWGa
Achdo-K-V      ...............AFHSWGa
Lutlo-K-2      HQVNLDSGAYRIITRTPFHSWGa
Leuma-K-VI     ..............pQSSFHSWGa
Culsa-K-1      ...............NPFHSWGa
Culqu-K-3      .VSGRVHRQPKIVIRNPFHSWGa
Aedae-K-2      ...............NPFHAWGa
Anoga-K-3      .NMPRTHKQPKVVIRNPFHSWGa
Culsa-K-3      ........WKYVSKQKFFSWGa
Culqu-K-1      ........SKYVSKQKFFSWGa
Aedae-K-1      ........NSKYVSKQKFYSWGa
Anoga-K-1      .......DTPRYVSKQKFYSWGa
Helze-K-III    ..............KVKFSAWGa
Lutlo-K-1      ...............IKFHSWGa
Mayde-K        ..........NTIKKVKFHSWGa
Achdo-K-II     ................AYFSPWGa
Helze-K-I      ..................YFSPWGa
Helze-K-II     ................VRFSPWGa
Bommo-K-1      ................NFSPWGa
Achdo-K-IV     .............NFKFNPWGa
Ixosc-K-2      ..............ESGFNPWGa
Culsa-K-2      ...........NNANVFYPWGa
Aedae-K-3      ...........NNPNVFYPWGa
Culqu-K-2      .............NNVFYPWGa
Anoga-K-2      ...........NTAQVFYPWGa
Ixosc-K-1      ...............DTFGPWGa
```

Figure 6 A multiple sequence alignment (ClustalW) of 50 insect kinins from 19 species. Identical and similar residues are denoted as in the legend to **Figure 4**. At the C-terminus, a five amino acid motif is highly conserved: FX1X2WGamide. X1 is N, S, Y, or H in all known sequences. X2 is usually S, but may be P or A. The first kinins were isolated from *L. maderae* (Leuma). Two sequences have pyroglutamic acid as the N-terminal residue (denoted by pQ). For the definition of other species abbreviations, see text.

a Sep-Pak C_{18}, and then fractionated on RPLC. Fractions were assayed using a Ramsay assay with MTs of adult *M. sexta*. Three fractions with diuretic activity coincided with peaks having strong absorbance at 280 nm, which gave second derivative UV spectra characteristic of Trp, one of three amino acids totally conserved in the kinins. Subsequent purification steps were monitored based on the diagnostic UV spectral behavior of Trp, rather than by using a functional assay. The peak eluting first was purified to homogeneity in one more RPLC step, and sequenced by Edman degradation (Helze-K-III; see **Figure 6**). Two

peptides required additional purification. The Trp residue could not be detected by Edman sequencing in these two kinins, which were additionally sequenced by tandem MS methods. The deduced sequences of all Helze-Ks were synthesized and bioassayed on MTs of *M. sexta*. No EC_{50} values were determined, only "threshold values" — the lowest concentration at which a significant effect on fluid secretion can be measured. These were ~7 pmol l^{-1} for Helze-K-I, 0.6 pmol l^{-1} for Helze-K-II, and 6 pmol l^{-1} for Helze-K-III. None of these peptides had activity *in vivo* in *H. zea*, but they do not appear to have been tested *in vitro* in *H. zea*, or *in vivo* in *M. sexta*, so no valid comparison can be drawn. Partially purified extracts from neuroendocrine tissue of *R. prolixus*, which contain kinin-like immunoreactivity, do not stimulate tubule secretion, but are active on the hindgut (Te Brugge *et al.*, 2002). While four kinins are encoded by the *R. prolixus* genome, there are currently no data on their synthesis or assays. There is no evidence for a kinin peptide or receptor in the red flour beetle, *Tr. castaneum* (Hauser *et al.*, 2008), and selected Leuma-K, Aedae-K, and Musdo-K have no effect on secretion by MTs from the mealworm, *T. molitor* (Wiehart *et al.*, 2002). Beetles may therefore lack a kinin signaling pathway.

9.3.5. Calcitonin-like DH

The first member of this family was identified together with the CRF-related Dippu-DH$_{46}$ isolated from 1040 brains of *D. punctata* (Furuya *et al.*, 2000b). During the first step of RPLC purification of extracts, only one fraction (Dippu-DH$_{46}$) stimulated cAMP production by MTs of *M. sexta*, yet this fraction and a faster eluting fraction both elevated cAMP production by MTs of *Sc. americana*. The easier *M. sexta* assay was utilized to isolate Dippu-DH$_{46}$. Isolation to homogeneity of the second factor was accomplished using MTs of *Sc. americana*. Upon sequence analysis, it was apparent that this 31 residue peptide is not a member of the CRF-related DH. A BLAST search revealed similarities to certain proteins, but no significant resemblance to any bioactive peptide. Nevertheless, a manual search through the catalog of a commercial peptide supplier (Peninsula Laboratories) for the unusual GP-NH$_2$ C-terminal sequence revealed an interesting similarity to calcitonin; only the calcitonin family has a Pro-NH$_2$ at the C-terminus of all bioactive peptides listed in this catalog. In the alignment shown in **Figure 7**, only 6 of 31 residues of Dippu-DH$_{31}$ are well conserved in the calcitonin family. However, chicken calcitonin is only 30-fold less potent in a fluid secretion assay than Dippu-DH$_{31}$ using MTs of *D. punctata* (Furuya *et al.*, 2000b), suggesting a genuine homology: calcitonin lowers blood Ca^{2+} by increasing Ca^{2+} excretion by the kidneys in addition to increasing Ca^{2+} deposition in bone, which require active transport mechanisms. Interestingly, Dippu-DH$_{31}$ and Dippu-DH$_{46}$ have potent synergistic interaction in *D. punctata* (Furuya *et al.*, 2000b). The actions of Dippu-DH$_{31}$ in *L. migratoria* are

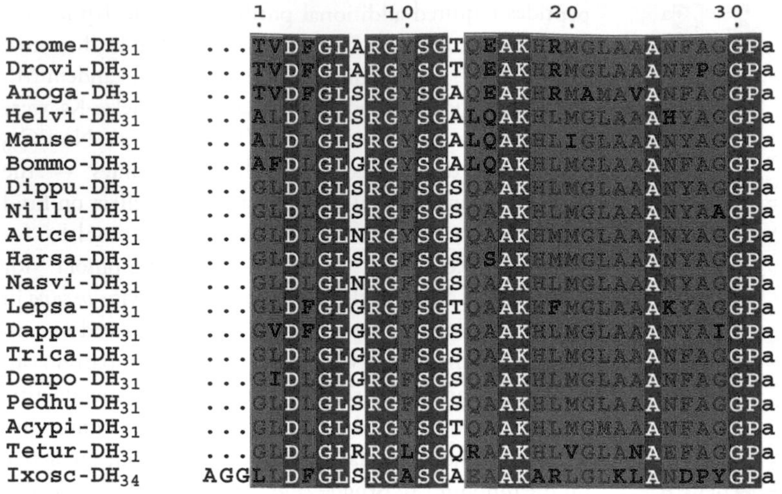

```
              1        10        20        30
              .         .         .         .
Drome-DH31  ...TVDFGLARGYSGTQEAKHRMGLAAANFAGGPa
Drovi-DH31  ...TVDFGLARGYSGTQEAKHRMGLAAANFPGGPa
Anoga-DH31  ...TVDFGLSRGYSGAQEAKHRMAMAVANFAGGPa
Helvi-DH31  ...AIDFGLSRGYSGALQAKHLMGLAAAHYAGGPa
Manse-DH31  ...AIDFGLSRGYSGALQAKHLIGLAAANYAGGPa
Bommo-DH31  ...AFDFGLGRGYSGALQAKHLMGLAAANFAGGPa
Dippu-DH31  ...GLDFGLSRGFSGSQAKHLMGLAAANYAGGPa
Nillu-DH31  ...GLDFGLSRGFSGSQAAKHLMGLAAANYAAGPa
Attce-DH31  ...GLDFGLNRGYSGSQAAKHMMGLAAANYAGGPa
Harsa-DH31  ...GLDFGLSRGFSGSLSAKHMMGLAAANYAGGPa
Nasvi-DH31  ...GLDFGLNRGFSGSQAAKHLMGLAAANYAGGPa
Lepsa-DH31  ...GLDFGLGRGFSGTQAAKHFMGLAAAKYAGGPa
Dappu-DH31  ...GVDFGLGRGFSGSQAAKHLMGLAAANYAIGPa
Trica-DH31  ...GIDFGLGRGFSGSQAAKHLMGLAAANFAGGPa
Denpo-DH31  ...GIDFGLGRGFSGSQAAKHLMGLAAANFAGGPa
Pedhu-DH31  ...GIDFGLSRGFSGSQAAKHLMGLAAANFAGGPa
Acypi-DH31  ...GIDFGLSRGVSGTQAAKHLMGMAAANFAGGPa
Tetur-DH31  ...GIDFGLRRGLSGQRAAKHLVGLANAEFAGGPa
Ixosc-DH34  AGGLIDFGLSRGASGAAAKARLGLKLANDPYGPa
```

Figure 7 The three chemically identified calcitonin-like DHs are shown using the Swiss-Prot species abbreviation convention, along with other sequences obtained from genome and EST data. All are amidated at the C-terminus (indicated with lower case **a**). Species abbreviations are given in the text.

more consistent with an elevation of intracellular Ca^{2+} in this species (Furuya *et al.*, 2000b). These apparent differences in signaling are not surprising in light of the fact that a cloned porcine calcitonin receptor has been shown to be capable of activating both the adenylate cyclase and PLC signal transduction systems (Chabre *et al.*, 1992; Force *et al.*, 1992).

Soon thereafter another DH_{31} was cloned "*in silico*" from the genome of *D. melanogaster* (gene *Dh31* CG13094); this peptide, Drome-DH_{31}, was synthesized and shown to be a potent diuretic in fruit fly tubules and to act via adenylate cyclase (Coast *et al.*, 2001). Similarly, Anoga-DH_{31} was identified from the *An. gambiae* genome, synthesized, and its activity compared with that of the CRF-related Anoga-DH_{44} (Coast *et al.*, 2005). The results of this study are consistent with the calcitonin-like DH being responsible for the large natriuresis required after a blood meal in the adult female mosquito (Section 9.3.12.4.).

Using an ELISA assay with antibodies to Dippu-DH_{31}, Te Brugge *et al.* (2008) isolated a peptide from 400 dissected CNS of *R. prolixus*, which was shown to have a sequence identical to Dippu-DH_{31} on Edman degradation and tandem MS analysis. The peptide was shown to have potent myotropic activity on the heart and hindgut, but had earlier been shown to have only a relatively small (14-fold) enhancement of fluid excretion (Te Brugge *et al.*, 2005) versus the 1000-fold stimulation of serotonin. Using a similar ELISA, a peptide has been isolated from *T. molitor*, which has a sequence identical to Trica-DH_{31} (**Figure 7**; D.A. Jensen and D.A. Schooley, unpublished). Moreover, Christie *et al.* (2010a) detected a cDNA corresponding to Trica-DH_{31} in an EST sequencing project in the lobster *Homarus americanus*; earlier they reported an EST encoding a different peptide, Dappu-DH_{31}, in the crustacean *D. pulex* (Gard *et al.*, 2009). The *H. americanus* peptide was shown by RT-PCR to be relatively widely expressed in the CNS and to have both inotropic and chronotropic effects on the heart, which were detectable down to 10–100 pM concentrations. We found

other CT-like DHs by searching genome data by NCBI BLAST: Acypi-DH_{31} from *A. pisum*, three hymenopteran species (Attce-DH_{31}, *A. cephalotes*; Harsa-DH_{31}, *H. saltator*; and Nasvi-DH_{31}, *N. vitripennis*), Pedhu-DH_{31} from *P. humanus corporis*, and non-insect sequences from the crustacean *L. salmonis* (Lepsa-DH_{31}), as well as novel extended form (Ixosc-DH_{34}) from the tick *I. scapularis*. The sequence for Tetur-DH_{31} from the Acari *Tetranychus urticae*, Denpo-DH_{31} from the beetle *Dendroctonus ponderosae*, Helvi-DH_{31} from *Heliothis virescens*, Manse-DH_{31} from *M. sexta*, and Nillu-DH_{31} from the planthopper *Nilaparvata lugens* was found by NCBI BLAST of EST data when choosing to BLAST by order. Drovi-DH_{31} from *D. virilis, erecta, persimilis, pseudoobscura, sechellia, simulans,* and *yakuba* has the unusual Pro28 substitution, differing from the other five *Drosophila* species.

Thus, it has become clear that the CT-like DHs constitute another family of DHs of near universal occurrence within the Insecta and other arthropods. The partial sequence of a similar peptide isolated from the Belgian forest ant, *F. polyctena*, was published in a Ph.D. thesis (Laenen, 1999). The isolation of the factor was monitored by observing its stimulation of MTs writhing in *L. migratoria*. Sequence analysis of this peptide was only possible out to 29 residues, but all are identical with the first 29 residues of Dippu-DH_{31} (see **Figure 7**). MS analysis revealed that the sequence was incomplete; the last two amino acids not detected must be different from those in Dippu-DH_{31} because the M_r is lower than that of the latter peptide (Laenen, 1999). Because the peptide lacked any diuretic activity on tubules of *F. polyctena*, from which it was isolated, it was not investigated further.

9.3.6. CAP$_{2b}$

Manse-CAP$_{2b}$ was isolated from *M. sexta* as a cardioactive factor (Huesmann *et al.*, 1995). Davies *et al.* (1995) found that it stimulates fluid secretion from MTs of

D. melanogaster by the unusual action (for a peptide hormone) of elevating production of nitric oxide (NO), which in turn activates a soluble guanylate cyclase. This peptide was reported, based exclusively on the retention time of biological activity from *D. melanogaster* tissue extracts on RPLC, to be an endogenous peptide in *D. melanogaster*. However, a BLAST search of the *D. melanogaster* genome (Kean *et al.*, 2002) revealed the existence of two homologues of CAP_{2b} (**Figure 8**). It was named the capability *(capa)* gene, because it was capable of encoding CAP_{2b} peptides that have a characteristic C-terminus of PRV-amide. The two CAP_{2b} homologues were named CAPA-1 and -2, whereas a third peptide encoded on capa was named CAPA-3, but it ends in PRLamide and therefore belongs to the pyrokinin (PK) family of peptides. This is an important distinction, because the C-terminal amidated valine residue of CAP_{2b} is critical for diuretic activity (Nachman and Coast, 2007), and there is a specific CAP_{2b} receptor that is not activated by pyrokinins (Iversen *et al.*, 2002). The homologue of capa in *M. sexta* also encodes two CAP_{2b} peptides, which have been named Manse-CAPA-1 (the peptide originally, and herein called Manse-CAP_{2b}) and Manse-CAPA-2 = Manse-CAP_{2b}-1 (**Figure 8**), along with a PK (Manse-PK-1; Loi and Tublitz, 2004). Homologues of capa are present in other insects and invariably encode one or two CAP_{2b} peptides and a PK. (The original nomenclature is preferred, because the name CAPA has been applied to peptides that lack the functionally important C-terminal motif, PRVamide). There has been a duplication of the *capa* gene in *R. prolixus*. The

genes encode pre-propeptides with 85% identity and are referred to as Rhopr-CAPA-α and -β (Paluzzi *et al.*, 2008, 2010). Only the Rhopr-CAP_{2b}-2 peptide(s) are identical and terminate in PRVamide, and is therefore a CAP_{2b}. Rhopr-CAP_{2b}-2 is biologically active on MTs (where it has antidiuretic activity; see the next section) and at the Rhopr-CAPA receptor, whereas the similar, but non-identical Rhopr-CAP_{2b}-1α and -1β (which end in LRAamide and LRAA, respectively) are inactive in both assays. Each gene also encodes a non-identical PRLamide peptide.

Predel *et al.* (1995) isolated a myotropin from perisympathetic organs of the cockroach *P. americana* christened periviscerokinin (PVK), which seemed to be a peptide unrelated to other families of peptides. However, this group (Predel *et al.*, 1998) subsequently isolated a second peptide from the same source, Peram-PVK-2, whose sequence is obviously related to both Peram-PVK-1 and CAP_{2b}. However, not all members of the so-called "PVK family" (Wegener *et al.*, 2002) are CAP_{2b}s, since they do not terminate in PRVamide, although they may be encoded on the *capa* gene. Partial sequences for two CAP_{2b}s were identified using highly sensitive matrix-assisted time of flight (MALDI-TOF/TOF) tandem MS (Clynen *et al.*, 2003) on single perisympathetic organs from two fly species, *Neobelleria bullata* and *M. domestica*. These sequences were incomplete because it was not possible to distinguish between Ile and Leu in peptides. Subsequent developments in MALDI-TOF/TOF MS with high-energy collision-induced dissociation revealed unique ions for Leu and Ile, and allowed unequivocal assignment of these residues (Nachman *et al.*, 2005). It is novel that this approach to isolation and identification does not require a bioassay, but the activities of Musdo-CAP_{2b}-1 and -2 were confirmed on housefly MTs. Interestingly, Garside and Coast (unpublished data) have found that, although CAP_{2b}s are diuretic on the MTs of houseflies, signal transduction is not via NO or cGMP, but appears to result from an increase in intracellular Ca^{2+}. MALDI-TOF MS was also used to identify two CAP_{2b}s in neurohemal organs of both the stable fly (*S. calcitrans*) and the horn fly (*Haematobia irritans*); the diuretic activity of the Stoca-CAP_{2b}-1 and -2 was confirmed on stable fly MTs (Nachman *et al.*, 2006). The same approach allowed CAP_{2b}s to be identified from the perisympathetic organs of several pentatomid plant sucking hemipteran bugs: *Nezara viridula*, *Acrosternum hilare*, *Banasa dimiata*, and *Euschistus servus* (Predel *et al.*, 2008). Confusingly, the authors refer to these as CAPA-PVK peptides. Identical CAP_{2b}-1 and -2 (**Figure 8**) are present in *N. viridula* and *A. hilare* (Nezvi-CAP_{2b}-1 and -2, **Figure 8**), whereas Bandi-CAP_{2b}-1 is slightly different, and the second peptide in *E. servus* is not a CAP_{2b} because it terminates in PRIamide. The biological activities of Acrhi/Nezvi-CAP_{2b}-1 and -2 were subsequently confirmed (Coast *et al.*, 2010) using *A. hilare* MTs, where both have antidiuretic activity.

```
Nezvi-CAP2b-2   ....EQLIPFPRVa
Bandi-CAP2b-1   ....DQLIPFPRVa
Nezvi-CAP2b-1   ....DQLIPFPRVa
Manse-CAP2b-1   .DGVLNLYPFPRVa
Pedhu-CAP2b-1   ..DVSGLFPFPRVa
Pedhu-CAP2b-2   ...pQGLIPFPRVa
Anoga-CAP2b-2   ...pQGLVPFPRVa
Acypi-CAP2b-1   ESAVAGLIPFPRVa
Peram-PVK-2     .GSSSGLISMPRVa
Leuma-PVK-2     ..GSSGLISMPRVa
Leuma-PVK-3     ..GSSGMIPFPRVa
Rhopr-CAP2b-2   ...EGGFISFPRVa
Drome-CAP2b-1   .GANMGLYAFPRVa
Manse-CAP2b     ....pQLYAFPRVa
Anoga-CAP2b-2   .GPTVGLEAFPRVa
Neobu-PVK-1     NGGTSGLEAFPRVa
Musdo-PVK-1     AGGTSGLYAFPRVa
Drome-CAP2b-2   ...ASGLVAFPRVa
Locmi-PVK-1     ...AAGLEQFPRVa
Musdo-PVK-2     ....ASGLNAFPRVa
```

Figure 8 The CAP_{2b} family of peptides. All are amidated at the C-terminus (indicated with lower case **a**), and three (Manse-CAP_{2b}, Anoga-CAP_{2b}, andPedhu-CAP_{2b}-2) have pyroglutamic acid as the N-terminal residue (denoted by pQ; Gln (Q) is the precursor of pGlu). Those sequences not described in the text are from *L. maderae* (Predel *et al.*, 2000), the locust (Predel and Gäde, 2002), the genome of *An. gambiae* (Riehle *et al.*, 2002), or from genome data (*P. humanus*).

9.3.7. Arginine Vasopressin-like Insect Diuretic Hormone

Proux *et al.* (1982) found that a factor in the subesophageal ganglia of *L. migratoria* stimulated *in vivo* clearance of injected amaranth from the hemolymph of locusts, an established assay for diuresis in this species (Mordue, 1969), and obtained evidence that this material was identical to a factor that cross-reacts with arginine vasopressin (AVP). This factor was isolated from 51,000 dissected subesophageal and thoracic ganglia using RPLC techniques and monitoring of fractions with a radioimmunoassay for AVP. In the first step of isolation two factors were separated, both of which were purified to homogeneity (Schooley *et al.*, 1987). Injection of the purified factors into *Sc. gregaria* revealed that only the less abundant, slower eluting factor (F2) increased amaranth excretion. The faster eluting F1 and slower eluting F2 were found to have identical amino acid compositions and identical primary sequences such as CLITNCPRGamide. This peptide was synthesized in both the amidated and free acid forms, and after reduction and carboxymethylation was compared with the similarly modified forms of F1 and F2 prepared for sequence analysis. This comparison established that F1 was identical to the nonapeptide whose sequence was shown earlier. Analysis of both materials by size exclusion chromatography suggested that F2 was a dimer of F1. Two separate, difficult specific syntheses were performed to prepare pure F2 as parallel and antiparallel dimers (Proux *et al.*, 1987). The synthetic antiparallel dimer proved to have retention properties identical on RPLC with F2 (Proux *et al.*, 1987). Natural and synthetic F2 were shown to promote fluid secretion in an unusual fluid secretion assay in which a group of *L. migratoria* MTs attached to the ampulla were removed, maintained in oxygenated saline, and the high combined flow of all tubules monitored (Proux *et al.*, 1988).

Later, this work was called into question by Coast *et al.* (1993), who reported that synthetic F2 had no stimulatory activity on a single *L. migratoria* MT in a conventional Ramsay assay, although Locmi-DH (a CRF-related peptide) had potent activity. Later, Montuenga *et al.* (1996) showed that there are endocrine cells in the ampulla of *L. migratoria* that contain Locmi-DH and an as yet uncharacterized peptide, which is recognized by a Locmi-DH antiserum but is faster eluting on RPLC than the latter peptide. It is conceivable that the AVP-like DH may stimulate the release of Locmi-DH from these endocrine cells. The localization of AVP-like immunoreactivity has been determined in the locust (Thompson *et al.*, 1991); the titer of AVP-like immunoreactivity changes in the hemolymph in response to unknown stimuli and neurons containing it innervate non-ocular photoreceptors (Thompson and Bacon, 1991). Analysis by RPLC

of AVP-like containing neurons showed them to contain not only the antiparallel dimer F2, but also the parallel dimer (D2) and the monomer F1 (Baines *et al.*, 1995). Recently Aikins *et al.* (2008) and Stafflinger *et al.* (2008) showed that the *Tr. castaneum* genome contains a gene homologous to the gene in vertebrates that encodes both vasopressin and its carrier protein neurophysin. This gene is also present in *N. vitripennis*, but absent in the genomes of the fruit fly, malaria mosquito, silkworm, and honeybee. Aikins *et al.* (2008) also showed that in *Tr. castaneum*, the monomer F1 has *in vivo* diuretic activity. Moreover, *in vitro* there is no direct activity on the MTs, only when CNS including the CC and CA are coincubated with the MT and F1. Thus, in both *L. migratoria* and *Tr. castaneum*, AVPL appears to act as a diuretic-releasing factor. Aikins *et al.* (2008) also cloned the receptor for this peptide, expressed in CHO cells with apoaequorin as a reporter. F1 gave a robust response (EC_{50} = 1.5 nM), while F2 gave ~7- to 12-fold lower response and D2, 60- to 80-fold lower.

9.3.8. Other Peptides with Diuretic Activity

Skaer *et al.* (2002) tested three different tachykinin-related peptides (TRP) on MTs of pharate adult *M. sexta*: locustatachykinin-1 from *L. migratoria* (Locmi-TK-1; GPSGFYGVRamide; Schoofs *et al.*, 1993) and two TRPs from *L. maderae* (Leuma TRP-1, APSGFLGVRamide and Leuma-TRP-4, APSGFMGMRamide; Muren and Nässel, 1996). Each TRP was tested at four concentrations from 1 nmol l^{-1} to 1 mmol l^{-1}. Each TRP exhibited a dose-dependent increase in the rate of fluid secretion, but a maximal response was reached only 30–40 min after application. While no EC_{50} values were measured, they appeared to be in the range of 10–100 nmol l^{-1}. Leuma-TRP-1 gave the highest maximal secretion of the three; at 1 mmol l^{-1} it increased the rate of fluid secretion 2.83-fold compared to that of the control. In contrast to the CAPs and leucokinin, TRP effects on tubule secretion activity were not long lasting. Two cardioactive peptides of as yet unknown sequence, CAP_{1a} and CAP_{1b} (Tublitz *et al.*, 1991), strongly stimulate *M. sexta* MTs; there is a distinct probability they could be kinins. The very slow response (20–30 min delay) observed as a result of treatment of tubules with all agonists may be attributable to their use of 8–12 cm long pieces of a single MT (Skaer *et al.*, 2002); *M. sexta* tubules tend to collapse on dissection and the delay in fluid secretion may result from the filling time of the tubule (G.M. Coast, unpublished data).

Recently, Locmi-TK-1 has been reported to have an EC_{50} of 1.2 nmol l^{-1} in *L. migratoria*, with an identical value reported for *Sc. gregaria* (Johard *et al.*, 2003). The response observed was about 75% of that observed in response to treatment with Locmi-DH; the latter was found to have synergistic activity with both Locmi-TK-1 and serotonin. In light of the response observed to this peptide in these two distantly related species (*M. sexta* vs. locusts), further studies of the effects of these peptides will be of great interest.

Three fractions were isolated from a head extract of *Ae. aegypti* by RPLC based on their effect on the transepithelial potential (TEP) of isolated perfused MTs (Petzel *et al.*, 1985). Each fraction contained a pronase-sensitive peptide with M_r values estimated by gel filtration chromatography to be 2400 (fraction I), 2700 (fraction II), and 1860 Da (fraction III; Petzel *et al.*, 1986). Fraction I depolarized the TEP. Although it has no effect on fluid secretion, it increases tritiated water (THO) loss and urine output from intact flies, possibly by inhibiting fluid uptake from the hindgut (Wheelock *et al.*, 1988). Fraction II also depolarized the TEP with a biphasic response — the TEP first depolarized, and then hyperpolarized. Fractions II and III both have diuretic activity and selectively stimulate secretion of NaCl-rich urine. Fraction III was the more potent of these two; its diuretic and natriuretic activity was indistinguishable from that of exogenous cyclic AMP, although the latter only hyperpolarized the TEP (Petzel *et al.*, 1985). Fraction III was named mosquito natriuretic peptide (MNP) and was shown subsequently to stimulate cyclic AMP production in isolated tubules (Petzel *et al.*, 1987). Hegarty *et al.* (1991) found that dibutyryl cAMP stimulated Na^+ secretion from *Ae. aegypti* MTs, with a concomitant decrease in K^+ secretion. MNP was shown to be the calcitonin-like DH of *An. gambiae* (Coast *et al.*, 2005).

Two groups have independently characterized, after partially isolating, insect homologues of the atrial natriuretic peptide (ANP) using antibodies against ANP. ANP is one of the important diuretics in vertebrates and causes vasorelaxation and Na^+ excretion (Beyenbach, 1993). Kim *et al.* (1994) reported such a peptide in *B. mori* eggs using an RIA with an antibody against atriopeptin III (SSCFGGRIDRIGAQSGLGCNSFRY). Egg extract was purified by gel filtration and RPLC, giving an RIA active fraction that elicited a dose-dependent relaxation of rat aortic strips, a highly selective assay for ANP. Chen *et al.* (1997) partially purified a stable fly (*S. calcitrans*) ANP by putting extracts through two RPLC steps guided by a commercial RIA kit for α-human ANP (SLRRSSCFGGRMDRIGAQSGLGCNSFRY) and supplemented with electrophysiological assays with mosquito tubules. Those fractions of semi-purified peptide that were immunoreactive were found to hyperpolarize *Ae. aegypti* MTs and to stimulate Na^+ secretion. These studies, in two distantly related species, show effects consistent with the effect of ANP on the vertebrate kidney, an effect mediated by cGMP, and different from those of the mosquito natriuretic peptide (Petzel *et al.*, 1985).

9.3.9. Serotonin and Other Biogenic Amines

The biogenic amine serotonin (5-hydroxytryptamine; 5HT) is known to stimulate diuresis in a variety of insects, including *R. prolixus* (Maddrell *et al.*, 1969), *L. migratoria* (Morgan and Mordue, 1984; cAMP independent), *P. brassicae* (Nicolson and Millar, 1983), and *M. sexta* (Skaer *et al.*, 2002). In most cases, levels required for stimulation are high. In *R. prolixus*, it stimulates distal tubules up to 1000-fold via a cAMP-dependent mechanism with EC_{50} ~30–40 nmol l^{-1} (Maddrell *et al.*, 1993a), whereas the response of *L. migratoria* tubules is only 25% of the maximum obtained with a CC extract and is mediated by a different second messenger, possibly Ca^{2+} (EC_{50} 10–100 nM; Morgan and Mordue, 1984). The response of *Ae. aegypti* tubules to serotonin varies with the stage of development. In larvae, it acts via cAMP to stimulate maximal secretion (EC_{50} ~100 nM; Clark and Bradley, 1996; Clark *et al.*, 1998a). In contrast, the response of adult tubules is ~25% of maximal (EC_{50} ~1 mM) and both cAMP and inositol trisphosphate (IP_3) production are increased (Veenstra, 1988; Cady and Hagedorn, 1999b). In *M. sexta* four concentrations were studied; while no EC_{50} value was reported, it appears to be in the range of 100–300 mmol l^{-1} (Skaer *et al.*, 2002). Octopamine also stimulates *M. sexta* tubules, but is much less potent. Fournier *et al.* (1994) reported that serotonin has antidiuretic effects in the rectum of *L. migratoria*.

Blumenthal (2003) showed tyramine to be a potent stimulant of *D. melanogaster* MTs, acting at levels as low as 1 nmol l^{-1}. Tyramine is several orders of magnitude more active than octopamine or dopamine, and the pharmacological profile of the response is consistent with it acting at a TA receptor (TAR). In support of this, the actions of TA on fluid secretion and V_{tep} are inhibited by yohimbine, a selective TAR antagonist. The TAR gene (CG7431) is abundantly expressed in MTs and mutation of this gene causes complete insensitivity to TA (Blumenthal, 2008). Interestingly, tyramine is produced in the MT from tyrosine taken up from the hemolymph, or culture medium. The immunocytochemical localization of TA indicates that it is synthesized from tyrosine in principal cells via tyrosine decarboxylase (TDC). MTs isolated from flies that are homozygous for a mutation of the non-neuronal form of TDC ($Tdc1^{f03311}$) do not respond to tyrosine, but tyrosine sensitivity can be rescued by expression of *Tdc1* in principal cells, not stellate cells (Blumenthal, 2009). Tyramine is believed to play an autocrine or paracrine role because it is produced in principal cells and is presumed to be released into the external medium to act at a TAR expressed on the stellate cell basolateral membrane (Blumenthal, 2003). In support of this, TA induces Cl^--dependent oscillations in the TEP, which have been linked to oscillations in the intracellular Ca^{2+}. This is the first demonstration of a physiological role for a TA receptor in

D. melanogaster (Blumenthal, 2003), and clearly the discovery that TA has a paracrine or autocrine role in fruit fly MTs merits studies in other species. For example, 10 nM TA has kinin-like activity in cricket MTs, but its synthesis from tyrosine has not been investigated (G.M. Coast, unpublished observation). Subsequently, Blumenthal (2005) showed that the sensitivity to both tyramine and a kinin are dependent on the osmolarity of the medium with the sensitivity decreasing as much as tenfold for TA with a relatively modest increase in osmolarity, from 200 to 280 mOsm. Decreases in osmolarity resulted in increased sensitivity to TA. These results may have broad implications for diuretic physiology, in that the response of MTs to TA (and perhaps other secretagogues) can be modulated to achieve hemolymph homeostasis under conditions of variable hydration/dehydration stress.

9.3.10. Antidiuretic Factors Inhibiting Malpighian Tubule Secretion

Spring *et al.* (1988) characterized an ADH from hemolymph of *A. domesticus* kept under very dry conditions; an extract of this hemolymph inhibited secretion by MT of control insects. Hemolymph extract from insects reared under normal humidity did not inhibit MT secretion. Lavigne *et al.* (2001) published a partial purification of a factor from *Leptinotarsa decemlineata* through five RPLC steps. While no peak was visible in the final separation step, they reported that the apparent molecular size of the factor (estimated by dialysis) was in the range of 25–50 amino acids. The first factor to be identified that inhibits MT secretion, Tenmo-ADFa, was isolated from pupal heads of *T. molitor*, based on its ability to elevate cyclic GMP (cGMP) levels in MTs of this species. Tenmo-ADFa was found to be a 14-mer with the sequence VVNTPGHAVSYHVY-OH, lacking the amidation usually found in regulatory peptides. It is exquisitely potent with an EC_{50} of 10 fmol l^{-1} (0.01 pmol l^{-1}; Eigenheer *et al.*, 2002), but receptor downregulation occurs at high concentrations (such as 1 nmol l^{-1}). A

second factor isolated from the same source, Tenmo-ADFb, is a 13-mer with the sequence YDDGSYKPHIYGF-OH, and is again non-amidated. It is 24,000 times less potent than Tenmo-ADFa, but still has EC_{50} = 0.24 nmol l^{-1} (Eigenheer *et al.*, 2003). These factors have poor solubility in the acidic solvents usually used for extracting neuropeptides from tissues.

Following typical isolation protocols resulted in low yields. However, upon discovering that the bulk of the activity remained in the pellet from acid extraction, the pellet was extracted with a neutral solvent, which by then had so many impurities removed that only three RPLC purification steps were required to isolate the factors. Curiously, these factors have extremely high sequence similarity to two cuticular proteins of this beetle. Tenmo-ADFa is identical to the C-terminus of the protein CAA03880 (Mathelin *et al.*, 1998) at all residues, except that Thr4 in the peptide is an Ala residue in the protein (**Figure 9**). Tenmo-ADFb is 100% identical with the 13 C-terminal residues of *T. molitor* putative cuticle protein 9.2 (TmPCP9.2; Baernholdt and Andersen, 1998). Consequently, the identification of ADFb was not published until immunohistochemical evidence showed specific staining for Tenmo-ADFb-like material in two pairs of bilaterally symmetrical neurosecretory cells of the protocerebrum of *T. molitor* (Eigenheer *et al.*, 2003). While the cuticle protein TmPCP9.2 lacks obvious enzymatic cleavage sites upstream of the portion identical with ADFb, this identity suggested the possibility that it could be a proteolysis fragment of the cuticle protein, and that this fragment might have coincidental biological activity. Tenmo-ADFa does have an interesting sequence identity to big endothelin 1 (**Figure 9**). The sequence identity to rabbit endothelin 1 (8 residues) and human endothelin 1 (7 residues) begins at exactly that bond where the endothelin-converting enzyme cleaves big endothelin into the potent vasoconstrictor endothelin 1. This cleavage site is marked with an arrow in **Figure 9**. Li *et al.* (2008) were unable to locate a gene encoding Tenmo-ADFa in the

Figure 9 Two antidiuretic factors from *T. molitor*: Tenmo-ADFa and Tenmo-ADFb. Tenmo-ADFa is identical at all but one residue with the C-terminus of the cuticle protein CAA03880 from this species (Mathelin *et al.*, 1998), and Tenmo-ADFb is 100% identical with the 13 C-terminal residues of another cuticle protein from this species, TmPCP9.2 (Baernholdt and Andersen, 1998). In addition, Tenmo-ADFa has an interesting sequence identity with big endothelin 1, a precursor of the potent vasoconstrictor endothelin 1. Endothelin converting enzyme cleaves big endothelin at the bond indicated by an arrow in the figure; this is exactly where the sequence identity with ADFa commences. Thus, the sequence identity is with the apparently inactive endothelin fragment removed in processing.

Tr. castaneum genome, but found a cluster of five genes encoding precursors of a peptide identical with Tenmo-ADFb, and four related peptides of the same length, but differing in sequence at 1, 3, or 4 residues.

Recently Coast (unpublished data) found that *M. sexta* allatotropin (Kataoka *et al.*, 1989a; Manse-AT) inhibits the Manse-DH$_{41}$ stimulated secretion of fluid by larval MT of this species at 50–200 nmol l^{-1} levels. This is the same concentration found earlier to inhibit fluid uptake from the anterior midgut (Lee *et al.*, 1998) of *M. sexta*. Incubation of 1 nmol l^{-1} Manse-DH$_{41}$ with and without 50 nmol l^{-1} Manse-AT (E. Chidembo and D.A. Schooley, unpublished data) shows that Manse-AT seems to act via lowering cAMP levels.

9.3.11. Antidiuretic Factors that Promote Fluid Reabsorption in the Hindgut

Hormonal control of fluid reabsorption in the hindgut has been heavily studied in locusts (*Sc. gregaria* and *L. migratoria*). Three antidiuretic neuropeptides have been either isolated or characterized from the CC of locusts (Phillips *et al.*, 1988, 1986): neuroparsins (Np), ion transport peptide (ITP), and chloride transport stimulating hormone (CTSH). Only the first two have been fully sequenced; CTSH is labile to the usual conditions used for RPLC isolation of peptides due to its instability in acidic conditions.

Herault *et al.* (1985) reported that the nervous and glandular lobes of the CC of *L. migratoria* each contain a factor acting as an antidiuretic on the rectum; these factors differ in size and extraction properties. The GCC factor was not purified, but Herault and Proux (1987) reported that GCC extracts cause a peak in rectal tissue cAMP levels coinciding with elevated short circuit current (I$_{sc}$). This stimulation is mimicked by forskolin, a stimulant of adenylate cyclase. The factor from the CC was reported to be an Np, because all antidiuretic activity in crude CC of *L. migratoria* was abolished by an antibody to this neuropeptide (Herault *et al.*, 1988). Neuroparsins are proteins (NpA, NpB) isolated and sequenced from the CC of *L. migratoria* (Girardie *et al.*, 1989, 1990). They have also been reported to have hypertrehalosemic (Moreau *et al.*, 1988) and antijuvenile hormone activity (Girardie *et al.*, 1987). NpB was reported to be a homodimer of a 78-residue polypeptide. NpA is identical to NpB except for having a heterogeneous N-terminus, the longest form of which has 83 residues. NpB is thought to be formed from NpA by enzymatic cleavage of the N-terminal sequence. Subsequently Hietter *et al.* (1991) confirmed the sequences of these small proteins, including the N-terminal heterogeneity, but revised the structural assignment for Nps involving three internal disulfide bridges in which the chains are monomeric. Fournier (1991) presented a body of data consistent with NpB acting on rectal fluid reabsorption by stimulating the IP$_3$ cascade with subsequent elevation

of cytosolic Ca^{2+}. They concluded that neuroparsins are the only antidiuretic in *L. migratoria* GCC and that these peptides act via the IP$_3$-Ca^{2+} second messenger system, whereas serotonin from the NCC acts on the rectum via a cAMP-mediated Ca^{2+} increase (Fournier *et al.*, 1994). They did not study actions of purified or synthetic Nps on rectal solute transport processes. Jeffs and Phillips (1996) found no effect of Locmi-Nps on either rectal ion or fluid transport in the locust *Sc. gregaria*. They attributed this either to fundamental differences in transport properties in these closely related insects or, more likely, to the pre-incubation of everted rectal sacs in Cl$^-$-free saline used by Fournier *et al.*, (1987). On return to normal saline, tissue swelling probably accounted for the gain in weight, rather than the Np-stimulated movement of water toward the hemolymph side.

Chloride transport stimulating hormone, a peptide occurring in the CC of *Sc. gregaria*, was characterized by its ability to increase short circuit current in *Sc. gregaria* rectal sheets maintained in Ussing chambers (Phillips *et al.*, 1980). In this assay, either cAMP or CC extracts give a tenfold increase in I$_{sc}$ for >8 h after stimulation. A single zone is seen on size exclusion chromatography with M$_r$ of about 8000 Da. The peptide nature is proved by its sensitivity to trypsin, and it is active at what appear to be nmol l^{-1} concentrations (Phillips *et al.*, 1986). The peptide can be extracted with water, saline, and aqueous ethanol, but biological activity is lost rapidly below pH 6. This may be consistent with the presence of an Asp-Pro bond in the primary sequence, which, as mentioned in Section 9.3.3., is suspected of causing an inability to isolate a CRF-related DH by RPLC techniques as well. Proux *et al.* (1985) provided some evidence that CTSH is produced in the pars intercerebralis and is transferred down nerves to the glandular lobe of the CC, after passing through the nervous lobe of the CC. CTSH is similar in size to Locmi-Nps, but is distinguished from the latter by the stability of Locmi-Nps to acidic conditions. Also, Locmi-Nps are reported to act via stimulation of PLC.

Audsley and Phillips (1990) used an assay similar to that for CTSH, substituting locust ileum for rectum, and found stimulants of ion transport in both the CC and the ventral ganglia. The effects elicited by CC and ganglia extracts differ in properties i.e., the factor in CC is stable to boiling and that in ganglia is not. The factor in the CC, unlike CTSH, is stable to acid. This stability permitted its successful isolation by RPLC, and a partial sequence was obtained to 31 amino acids (Audsley *et al.*, 1992b). This peptide was christened ion transport peptide (Schgr-ITP) to distinguish it from CTSH. It was found to be 44–59% identical to the crustacean hyperglycemic (CHH), molt-inhibiting (MIH), and vitellogenesis-inhibiting (VIH) hormones — a group of highly related 72 amino acid peptides (Audsley *et al.*, 1992b). This was the first direct evidence for the existence of a peptide related to this family outside of crustaceans. Audsley *et al.* (1992a) tested purified Schgr-ITP on the rectum and found that it had weak effects on short circuit current compared with CTSH, and no effect on the rate of fluid reabsorption. They concluded that separate neuropeptides act on the ileum and rectum of the locust.

Meredith *et al.* (1996) used degenerate primers designed from the partial protein sequence to clone by PCR a partial sequence of Schgr-ITP from a brain cDNA library. Use of 5′ and 3′ rapid amplification of cDNA ends (RACE) strategies led to isolation of a cDNA encoding a Schgr-ITP prepropeptide of 130 amino acid residues (Meredith *et al.*, 1996) with a 55-residue signal peptide and a dibasic cleavage site preceding the start of the partial ITP sequence (Audsley *et al.*, 1992b). The C-terminal Leu is followed by Gly-Lys-Lys-stop, consistent with C-terminal amidation and a second dibasic cleavage site to give the complete ITP sequence of 72 amino acid residues. Meredith *et al.* (1996) also used ITP cDNA sequence primers to probe a locust ileal mRNA library, and isolated an ITP-like clone (ITP-L), which was sequenced. It was identical to the brain cDNA for ITP except for an additional 121 bp insert at amino acid position 40 of ITP, suggesting alternative C-termini splicing of genomic DNA. The open reading frame for ITP-L (134 residues) is four residues longer than that of ITP (**Figure 10**), and the C-terminus differs because it is a free carboxylate. All six cysteines are conserved. The C-terminal 32 amino acid sequence of ITP differs from ITP-L at 17 positions with most of the difference over the last 20 residues. Using reverse transcriptase PCR (RT-PCR) techniques, ITP-L mRNA was detected in tissues (flight muscle, hindgut, and MTs) whose extracts have no stimulatory effect in the locust ileal I_{sc} bioassay, while ITP mRNA was found only in the brain and CC both of which do stimulate ileal I_{sc}. The ITP-L peptide does not stimulate I_{sc} in the locust ileal bioassay, but acts as a weak antagonist (Wang *et al.*, 2000). The Schgr-ITP synthesized by solid-phase peptide synthesis was oxidized to the appropriately folded form with three disulfide bonds (King *et al.*, 1999). This was

found to exhibit biological actions identical to that of the native peptide isolated from tissue.

ITP triggers reabsorption of NaCl and KCl by the ileum, and *reduces* H$^+$ secretion in the ileal bioassay. In contrast, cyclic AMP promotes KCl reabsorption by ileum, but *has no effect* on H$^+$ secretion, which Phillips *et al.* (1998) interpreted as requiring a different second messenger (perhaps cGMP) to stimulate H$^+$ secretion. In *H. americanus*, the CHH, an orthologue of ITP, was shown to act on target tissues via a membrane GCase (Goy, 1990). More recently, Lee *et al.* (2007) showed CHH and MIH to act via cGMP in the crab, and Nagai *et al.* (2009) showed that CHH acts via cGMP in the prawn *Marsupenaeus japonicus*.

In the 1990s there was concern over conflicts in results between the proponents of Nps versus Schgr-ITP. Three publications about crustacean neuroendocrinology strongly supported the role of ITP as the authentic osmoregulator. Chung *et al.* (1999) reported a large, precisely timed release of CHH from gut endocrine cells in the crab *Carcinus maenas* at ecdysis. This release appears to trigger the water and ion uptake required during molting, which allows the swelling necessary for successful ecdysis and the subsequent increase in size post-molt. The endogenous orthologue of CHH of the freshwater crayfish *Pachygrapsus marmoratus* (Spanings-Pierrot *et al.*, 2000) plays a crucial role in control of gill ion transport. In the crayfish *Astacus leptodactylus*, injection of CHH increases the hemolymph osmolality and Na$^+$ concentration 24 h after injection (Serrano *et al.*, 2003). Two other CHH-related peptides caused a smaller increase in Na$^+$ concentration.

Liao *et al.* (2000) characterized one of two factors that separate on RPLC of brain-CC-CA extracts of larval *M. sexta*. These factors trigger fluid reabsorption in an everted rectal sac bioassay with *M. sexta*. Owing to the cryptonephric anatomy of larvae of this species, the CRF-related peptide Manse-DH$_{41}$ also causes rectal fluid reabsorption. However, the effect of Manse-DH$_{41}$ is blocked by bumetanide (an inhibitor of the Na$^+$-K$^+$-2Cl$^-$ cotransporter), bafilomycin A$_1$, and amiloride, whereas that of

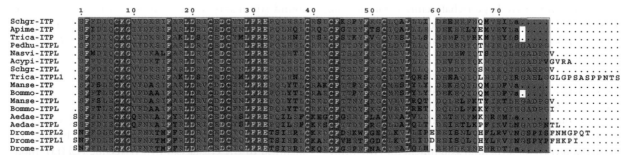

Figure 10 Clustal W alignment of Schgr-ITP and its splice variant Schgr-ITPL with their orthologues from *M. sexta*, *B. mori*, and *Ae. aegypti* (Dai *et al.*, 2007); *D. melanogaster* (Dircksen *et al.*, 2008); and *Tr. castaneum* (Begum *et al.*, 2008). All forms of ITP, with the exception of Trica-ITP, are amidated at the C-terminus (lower case **a**), while the ITPL peptides are not amidated. Because of gaps induced by the Drome-ITPL sequences, the default gap opening penalty in Clustal had to be decreased twofold.

the more potent of the two factors, Manse-ADFB, is not blocked by any of these inhibitors. The Cl⁻channel blockers 4,4'-diisothiocyanatostilbene-2,2'-disulfonic acid disodium salt (DIDS) and diphenylamine-2-carboxylic acid (DPC) both block the action of Manse-ADFB, as does H-89, a potent and specific inhibitor of PKA. These data suggest that this factor of unknown structure has a mechanism of action similar to Schgr-ITP, and may be the homologue of ITP identified in *M. sexta* by Dai *et al.* (2007).

Endo *et al.* (2000) isolated a cDNA encoding a "CHH family" peptide from *B. mori*. Dai *et al.* (2007) cloned gene products representing both ITP and ITP-L from *M. sexta*, *B. mori*, and *Ae. aegypti*, showing that the different isoforms result from alternative splicing of a precursor containing 3 exons. ITP is encoded by exons 1 and 3, and ITP-L by exons 1 and 2. Dircksen *et al.* (2008) found that in *D. melanogaster* a gene containing five exons encodes ITP and two ITP-L peptides. Begum *et al.* (2009) found that in *Tr. castaneum*, a gene containing four exons encodes ITP and two ITP-L peptides similar to *D. melanogaster*. Interestingly, the *Tr. castaneum* ITP is not amidated at the C-terminus; the sequence was checked multiple times to ensure this was not the result of an error. (In this context it is worth noting, from Section 9.3.3., that three of four known beetle CRF-related DHs are also not amidated.) Four sequences from a BLAST search of other insect genomes are shown in the alignment in **Figure 10**. In some species (*A. mellifera*) only an ITP was found, whereas in others (*N. vitripennis*, *P. humanus*, and *A. pisum*) only ITP-L peptide sequences were detected. This is likely attributable to the fact that the splice sites in ITP genes are poorly predicted using automated sequence analysis methods (Dircksen, 2009). A good portion of the C-terminus of the ITP-L peptides, beyond residue 72 in ITP and orthologues, is highly conserved in those sequences shown in **Figure 10**, suggesting an important, but unknown, function.

9.3.12. Cellular Mechanisms of Action

9.3.12.1. Introduction Does the diverse control of tubule secretion between different orders, and even between similar species belonging to the same order, have any rationale? Do differences in hemolymph composition, diet (and hence ion uptake), water availability, and the structure of the excretory system explain the diverse control strategies employed by different species? Primitive insect orders have a hemolymph characterized by a high concentration of Na⁺ relative to K⁺, and the sum of total anions and cations accounts for a considerable part of the total osmolarity (Sutcliffe, 1963). In contrast, more highly evolved insect orders have a greater proportion of hemolymph osmolarity attributable to high levels of amino acids and organic acids and, in some (Lepidoptera, Hymenoptera, and certain Coleoptera; Sutcliffe, 1963),

K^+ is the dominant cation, with elevated concentrations of Mg^{2+} and Ca^{2+}. Since tubule secretion is driven by secondary active cation transport, differences in hemolymph composition might place some constraint on how fluid secretion is regulated. The same could apply to the dietary intake of ions, most notably K^+ and Na^+. If the diet is K^+-rich, as in herbivorous insects, then it is appropriate for K^+ to be the dominant cation secreted by the MT and to control secretion by manipulating the rate of K^+ transport. Conversely, hematophagous insects take infrequent blood meals, which impose a considerable challenge on the excretory system for the removal of imbibed salt (NaCl) and water.

Not surprisingly, the MTs of mosquitoes (*Ae. aegypti*) respond to released diuretics with a dramatic increase in secretion driven by a switch from K^+ transport to Na^+ transport (Beyenbach, 2003). Hence, the CT-like DH of mosquitoes, which corresponds to the previously unidentified head factor, Mosquito Natriuretic Peptide (MNP; Beyenbach and Petzel, 1987) has both diuretic and natriuretic activity (Coast *et al.*, 2005). Interestingly, in the laboratory, the same mechanism has been shown to operate in the tubules of male mosquitoes (Plawner *et al.*, 1991), which do not feed on blood but on nectar (Clements, 1992), resulting in a diet that is not rich in NaCl. Periods of diuresis that do not require a concomitant natriuresis must therefore be initiated by the release of a different DH (Coast, 2009), possibly a kinin, which, because it stimulates anion movement, has a neutral effect on cation transport. Can a similar story be developed for other species? *D. melanogaster* feeds on fermentation products (the alcoholic of the Insecta), whereas the closely related *M. domestica* imbibes a mixture of saliva and the products of extracorporeal digestion. A major drawback in this analysis is that most researchers have studied effects on fluid secretion rather than on ion transport. Peptides with apparently similar diuretic activity could well differ in their effect on urine composition and hence on ion transport (the product of concentration and the rate of secretion). An early demonstration of the possibility of such an occurrence is in the MTs of *Ae. aegypti* treated with bumetanide, a drug that inhibits the Na^+-K^+-$2Cl^-$ cotransporter stimulated by cAMP. Bumetanide treatment has little effect on basal fluid secretion, but decreases K^+ secretion with a concomitant increase in Na^+ secretion (Hegarty *et al.*, 1991). Coast (1995) has shown that Locmi-DH₄₆ (CRF-related) and locustakinin have fairly similar diuretic potency, but have quite different effects on ion transport. The majority of physiological studies on the action of synthetic peptides affecting MT function have focused on cellular effects, such as elevation of second messengers, or on fluid transport. Relatively few have focused on the effects of synthetic factors on ion transport, which, given the complex interplay of factors controlling fluid and ion homeostasis in vertebrates, is a serious omission. Moreover, in the cricket

Teleogryllus oceanicus, the distal portion of the MTs have been reported to secrete a hyperosmotic urine containing 125 mM Mg^{2+}, together with a lesser concentration of Na^+ (Xu and Marshall, 1999). There are few studies on Mg^{2+} transporters in insect systems, yet in a number of insect species with high hemolymph K^+, the concentration of Mg^{2+} is actually higher (Sutcliffe, 1963; Cohen and Patana, 1982; Dow *et al.*, 1984). Thus, data on specific ion transport are under-investigated and are likely to shed some light on why so many peptides are involved in control of water and salt balance in insects.

Water availability may also play a role in determining the cocktail of diuretics and antidiuretics used to control tubule secretion. The simplistic view that diuretics stimulate tubule secretion whereas antidiuretics increase fluid reabsorption in the hindgut is no longer tenable (at least in some species). Antidiuretics that reduce MT secretion have been identified in few species. In *R. prolixus*, the antidiuretic activity of CAP_{2b} has been linked to the need to rapidly turn off tubule secretion (Quinlan *et al.*, 1997). Normally, the rapid degradation of a DH (probable half-life in the circulation measured in minutes; Li *et al.*, 1997) would suffice to end a period of diuresis. The situation in *R. prolixus* is very different, however, because the DH has such a dramatic effect on secretion, which at maximal rates would be sufficient to deplete the total store of water in the hemolymph within minutes. Under these conditions, it is essential that diuresis is switched off coincident with reduced fluid uptake from the midgut (Quinlan *et al.*, 1997), which is also inhibited by CAP_{2b} (Ianowski *et al.*, 2010). Now the insect conserves water until the next blood meal. Although this is an attractive hypothesis, CAP_{2b} has recently been shown to reduce secretion by MTs of two other hemipterans, the pentatomid stink bugs *A. hilare* (Coast *et al.*, 2010), which feeds on plants, and *Podisus maculiventris*, which feeds on caterpillars (G.M. Coast, unpublished data). The antidiuretic activity of CAP_{2b} on the MTs of hemipterans may therefore be a primitive feature of this group rather than a specialization associated with gorging on vertebrate blood.

In the period between its intermittent blood meals, *R. prolixus* resembles a xeric insect, because it conserves water and only rarely voids small quantities of urine. It is of considerable interest that another insect shown to use antidiuretic factors to control tubule secretion is *T. molitor*, an extreme example of a xeric species able to survive very dry conditions (O'Donnell and Machin, 1991). In both *T. molitor* and *R. prolixus*, antidiuresis appears to result from the activation of a cGMP-dependent cAMP phosphodiesterase, which lowers intracellular levels of cAMP, the second messenger used by both CRF-related and CT-like DH. Not surprisingly, the antidiuretic factors of *T. molitor* are extremely potent (Eigenheer *et al.*, 2002, 2003) and effectively counter the diuretic activity of native CRF-related DH (Wiehart *et al.*, 2002). To

grasp the importance of this, it is important to note that coleopteran larvae and adults have a cryptonephric system in which the distal ends of the MTs are closely associated with the rectum (Ramsay, 1964; Section 9.3.12.8.). Fluid is reabsorbed from the rectum into the cryptonephric region of the tubules and flows into the gut at the midgut–hindgut junction, where it is postulated to move anteriorly and to moisten the dry food for digestion prior to reabsorption by the midgut (Nicolson, 1991). To gain water (the cryptonephric system is used to take up water from subsaturated atmospheres), fluid absorption from the free portion of the tubules must exceed fluid secretion, although these two processes could be taking place in functionally different segments. Unfortunately, nothing is known of fluid reabsorption from *T. molitor* tubules or whether antidiuretic factors have any effect on urine composition. Unpublished work (G.M. Coast) with MT from *M. sexta* larvae, which also have a cryptonephric system, shows that one segment (the yellow segment) can display net secretion or net reabsorption depending on the composition of the bathing fluid and the complement of peptides present. The two processes appear to be proceeding in parallel with the overall balance tilted toward secretion or reabsorption by stimulants of each activity. The likelihood must be that similar mechanisms operate in *T. molitor* tubules.

9.3.12.2. CRF-related DH: Cellular actions
The CRF-related DHs studied all elevate cAMP in response to treatment of MT *in vitro*. As noted by Rafaeli *et al.* (1984), the cAMP is usually released from the tubule, which allows assay of the medium without having to homogenize and extract the tubules. This facilitates *in vitro* assays for isolating new members of this family. When using conspecific assays, CRF-related DH stimulate secretion by MTs of *A. domesticus* (Coast and Kay, 1994), *L. migratoria* (Patel *et al.*, 1995), *D. punctata* (Furuya *et al.*, 2000b), *M. domestica* (Iaboni *et al.*, 1998), *T. molitor* (both Tenmo-DH_{37} and Tenmo-DH_{47}; Wiehart *et al.*, 2002), and both CRF-related DHs of *M. sexta*, Manse-DH_{41} (Audsley *et al.*, 1993, 1995) and Manse-DH_{30} (Blackburn and Ma, 1994), at low nanomolar concentrations. Diuretic activity varies among species, from a maximal response in *A. domesticus* (Coast and Kay, 1994), equivalent to that obtained with a CC extract, to <25% of a maximal response in *M. domestica* (Iaboni *et al.*, 1998). This probably depends upon whether anion or cation transport is the rate-limiting step in tubule secretion.

The potency of CRF-related DH varies much more in heterologous assays. Tubules of *M. sexta* are extremely flexible in cross-reacting with all members of the CRF-related DH family tested (Audsley *et al.*, 1995) except the free acid form of Manse-DH_{41} and the two DHs of *T. molitor*, which have free acids at the carboxyl-terminus.

In contrast, tubules of *L. migratoria* seem to be very selective for their endogenous ligand; Peram-DH has an EC_{50} over 300-fold higher than Locmi-DH, and other ligands were less effective (Audsley *et al.*, 1995).

Elevated levels of cAMP stimulate ion transport, and hence fluid secretion, via a number of different actions that are not mutually exclusive. This is best illustrated with reference to *R. prolixus* MTs where elevated cAMP levels following stimulation with either a CRF-related DH (Zoone-DH) or serotonin cause a triphasic change in the transepithelial voltage (V_t; Donini *et al.*, 2008; O'Donnell and Maddrell, 1984). A detailed analysis of the voltage change (Ianowski and O'Donnell, 2001; Ianowski *et al.*, 2002) showed that the initial transient depolarization is caused by the opening of an apical Cl⁻ conductance, while the following hyperpolarization is due to activation of the V-ATPase. The voltage then depolarizes again due to stimulation of a bumetanide-sensitive Na⁺-K⁺-2Cl⁻ cotransporter, which brings more Cl⁻ into the cell and hence increases the electrochemical gradient favoring its diffusion into the lumen. The net result is a massive (up to 1000-fold) acceleration of fluid secretion driven by the transport of NaCl and KCl into the lumen of the distal MTs. Virtually all of the KCl is subsequently recovered by reabsorption across the water-impermeable lower tubule, so that a Na⁺-rich urine hypo-osmotic to the hemolymph is voided. It is noteworthy that the Na⁺-K⁺-2Cl⁻ cotransporter of *R. prolixus* MTs is unusual because Na⁺ competes for K⁺ binding sites (Ianowski *et al.*, 2004), which are believed to be part of an autonomous mechanism for regulating hemolymph K⁺ concentration.

In other insects, cAMP has a far less dramatic impact on fluid secretion, and targets just one or two transport mechanisms. Elevated cAMP levels are believed to activate a bumetanide-sensitive Na⁺-K⁺-2Cl⁻ cotransporter (**Figure 11**) in *M. sexta* (Audsley *et al.*, 1993), *A. domesticus* (Coast *et al.*, 2002a), and adults of *Ae. aegypti* (Beyenbach, 2003), but not in larval tubules (Clark and Bradley, 1996). The Na⁺-K⁺-2Cl⁻ cotransporter has been cloned from *M. sexta* (Reagan, 1995b). It is a large protein with 12 putative transmembrane helices resembling orthologous proteins from other species, and it contains a consensus site for phosphorylation by PKA (Reagan, 1995b). Similar transporters encoded in the *D. melanogaster* genome (CG2509 and CG31547) lack the PKA site (Ser1025) reported by Reagan (1995b), which might explain the relatively low sensitivity of this species to the CRF-related DH compared with kinins. Thr963 in the *An. gambiae* transporter (XM307782) is a likely PKA site (determined with NetPhos 2.0 server, http://www.cbs.dtu.dk/services/NetPhos/). In *D. melanogaster* MTs, cAMP appears to activate the apical membrane V-ATPase (Coast *et al.*, 2001; O'Donnell *et al.*, 1996), whereas in mosquito MTs it is believed to open a Na⁺ conductance in the basal membrane (Beyenbach, 2003), which

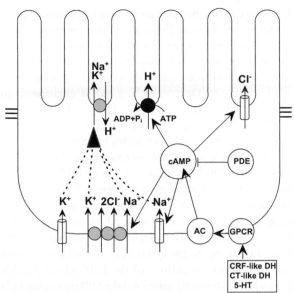

Figure 11 Proposed actions of cAMP in principal cells. CRF-related DH, CT-like DH, and serotonin stimulate adenylate cyclase (AC) activity and increased intracellular levels of cAMP. Diuresis results from the stimulation of ion transport by cAMP which, depending on species, activates the apical membrane V-ATPase (*D. melanogaster, T. molitor,* and *R. prolixus*), increases Na+-K+-2Cl⁻ cotransport (*R. prolixus, M. sexta, Ae. aegypti,* and *A. domesticus*) and opens a Na+ selective conductance in the basolateral membrane (*Ae. aegypti*). In addition, cAMP opens an apical membrane Cl⁻ conductance in *R. prolixus* tubules. Cyclic AMP levels are reduced by phosphodiesterase activity, which may account for the antidiuretic effects of Tenmo-ADFa,b and CAP2b (see text for details).

results in the secretion of Na⁺-rich urine (natriuresis). It is therefore surprising that, although both the CRF-related (Anoga-DH_{44}) and CT-like DH (Anoga-DH_{31}) stimulate cAMP production by the MTs of *An. gambiae*, only the latter has natriuretic activity (Coast *et al.*, 2005). Possibly the absence of a natriuretic response to Anoga-DH_{44} is because it also activates a Ca^{2+} signaling pathway (see next section).

In addition to elevating cAMP, there is evidence to suggest that in some species CRF-DH also activates a Ca^{2+} signaling pathway. The CRF-related peptide of *C. salinarius* (Culsa-DH) has weak diuretic and natriuretic activity in *A. aegypti* (Clark *et al.*, 1998b) and appeared to elevate intracellular Ca^{2+} at nmol l⁻¹ concentrations, but activated adenylate cyclase at ≥ 100 nmol l⁻¹ levels (Clark *et al.*, 1998b). These data are based on the biphasic effect of Culsa-DH on V_t, which transiently depolarizes to be followed by a short-lived hyperpolarization. Thapsigargin mimics the initial depolarization produced by nmol l⁻¹ concentrations of the peptide, while cAMP reproduces the subsequent hyperpolarization seen with ~100 nmol l⁻¹ levels of Culsa-DH. Anoga-DH_{44} has a similar effect on *An. gambiae* MTs (Coast *et al.*, 2005), with V_t transiently depolarizing then hyperpolarizing before finally stabilizing at a voltage that is more negative than that recorded prior to adding the peptide. The changes in V_t are mirrored by the voltage across the basolateral membrane (V_b), which first hyperpolarizes then depolarizes before repolarizing in the continued presence of the peptide. Hence the apical membrane voltage, V_a (= V_t - V_b), is unchanged. In contrast to the actions of Anoga-DH$_{44}$, cAMP and the CT-like peptide Anoga-DH$_{31}$ do not produce a transient depolarization in V_t, which remains hyperpolarized in the continued presence of the secretagogues. Interestingly, however, the triphasic effect of Anoga-DH$_{44}$ on V_t is mimicked with a combination of Anoga-DH$_{31}$ (or cAMP) and a kinin (Musdo-K), which uses Ca^{2+} as second messenger (Section 9.3.12.4.). Thus CRF-related DHs appear to stimulate both cAMP and Ca^{2+} second messenger pathways when acting on mosquito MTs. A similar conclusion was reached for the actions of the CRF-related peptide Achdo-DH$_{46}$ (formerly called Achdo-DP) on cricket MTs based upon a comparison of its diuretic activity with that of cAMP. Achdo-DH$_{46}$ stimulates maximal secretion (as defined by the maximum response to a crude extract of the CC), but the response to cAMP is only 80% of the maximum unless added in combination with thapsigargin or a kinin (Achdo-K-1). A dual activation of Ca^{2+} and cAMP signaling by Achdo-DH$_{46}$ could explain why the peptide does not act synergistically with Achdo-K-1 whereas cAMP does.

Zitnan *et al.* (2007) have shown that Manse-DH$_{41}$ and -DH$_{30}$, the calcitonin-like DH$_{31}$, and kinins are all expressed in specific cells in *M. sexta* abdominal ganglia, which also express the ecdysis-triggering hormone receptor subtype A. These cells are important in initiating pre-ecdysis I behavior; application of a mixture of these peptides to desheathed dorsal nerves of abdominal ganglia 2–8 mimics the firing pattern from application of pre-ecdysis triggering hormone. Application of the mixture of peptides is more effective than application of individual peptides (Zitnan *et al.*, 2007).

9.3.12.2.1. CRF-related DH receptors

Reagan (1994) utilized expression cloning to deduce the sequence of a receptor from *M. sexta* (Manse-DHR), and subsequently a similar one from *A. domesticus* (Reagan, 1996; Achdo-DHR). Two orthologues of these receptors are encoded in the *D. melanogaster* genome (CG8422 and CG12370); recently one of these (CG8422) was reported to be the receptor (Drome-DHR) for Drome-DH$_{44}$ (Johnson *et al.*, 2004). However, a later study showed that CG8422 is not found in the MT, but in the CNS (Johnson *et al.*, 2005). Recent work shows that CG12370 is expressed in the MT, but it is >100-fold less sensitive to Drome-DH$_{44}$ (EC$_{50}$ = 798 nM) than CG8422 (EC$_{50}$ = 5.1 nM; Hector *et al.*, 2009). The lower sensitivity of CG12370 compared with CG8422 is surprising, given that it appears to be a peripheral receptor and therefore exposed to DH levels in

the circulation. However, as discussed in Section 9.3.3., evidence for a factor different from Musdo-DH$_{44}$ was observed in *S. calcitrans*, and preliminary results suggest that *D. melanogaster* head extracts contain a material distinct from Drome-DH$_{44}$, which stimulates the CG12370 receptor and may represent a new paralogous ligand (D.A. Jensen and D.A. Schooley, unpublished data). Jagge and Pietrantonio (2008) reported orthologous receptors in MTs of *Ae. aegypti*; one of these was expressed in MTs (Aedae-DHR1 in **Figure 12**) while the other was isolated from head cDNA (Aedae-DHR2). Apparently due to lack of any synthetic Aedae-DH$_{44}$, no EC$_{50}$ values are available. Two DH receptors were isolated from the rice brown planthopper (*N. lugens*; Price *et al.*, 2004), but they are 96.1% identical, differing by a deletion of seven amino acids from the N-terminus of one and possessing but 16 other residues that differ; one may be a PCR artifact. The ligand is unknown in this species. The Park group has cloned two DH receptors from *Tr. castaneum* (Cosme, 2009), Trica-DHR1 and -DHR2. All are members of Family B of seven transmembrane domain GPCRs (a family that includes the CRF and calcitonin receptors of vertebrates; Dautzenberg and Hauger, 2002). As in most receptor families, the transmembrane domains (TMD) are highly conserved (**Figure 12**). Other domains differ in their degree of similarity; intracellular domain 1, domain 2, and domain 3 are quite highly conserved. In the CRF receptors, intracellular domain 3 is believed to be the site for G protein coupling (Hauger *et al.*, 2003). All of the CRF-related DH receptors (except Bommo-DHR) have six Cys residues in the N-terminal extracellular domain, a structural feature that is critical for ligand interaction in the CRF receptor (Qi *et al.*, 1997). The location of five Cys is completely conserved and the Cys closest to the N-terminus is variable in position and missing in Bommo-DHR (**Figure 12**). The disulfide pairings in the CRF receptor have been inferred (Qi *et al.*, 1997), and they provide conformational rigidity to the N-terminal domain, which is believed to be key in CRF binding (Dautzenberg *et al.*, 1998) together with the fourth extracellular loop (Sydow *et al.*, 1997, 1999). Curiously, extracellular loops EC2 and EC4 are poorly conserved in the DH receptors, although a portion of EC3 is well conserved (**Figure 12**). Certain amino acid residues in the N-terminal domain of *X. laevis* (Dautzenberg *et al.*, 1998) and human (Wille *et al.*, 1999) CRF1 receptors have been identified as crucial for ligand binding via site-directed mutagenesis. The intense research interest in these receptors has been driven by the development of small molecule receptor antagonists as drugs targeted at depression and anxiety (Kehne and De Lombaert, 2002); it is conceivable that a similar approach with agonists or antagonists of the insect receptors could lead to new insecticides.

The CRF1 receptor binds both CRF and urocortin 1 (Hauger *et al.*, 2003), but discriminates against urocortin

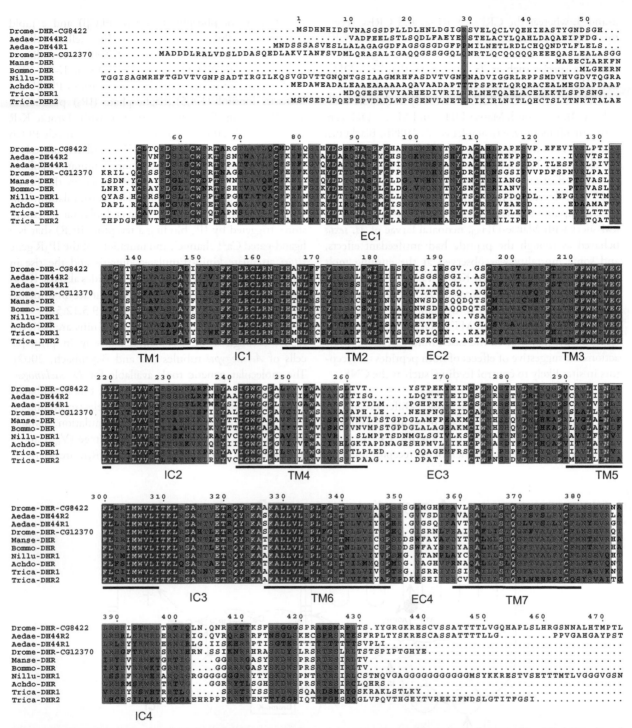

Figure 12 Experimentally cloned CRF-related DH receptors are shown; identical residues are indicated with white letters in red boxes; similar residues are in red letters in a yellow box. Sequences were aligned with Clustal W at http://npsa-pbil.ibcp.fr/cgi-bin/npsa_automat.pl?page=/NPSA/npsa_clustalw.html. The seven transmembrane domains (TMD) are indicated with a bold line and the domain number. Their approximate location was determined with the program TMHMM Server v. 2.0 at http://www.cbs.dtu.dk/index.shtml. This program locates transmembrane (TM) domains using hidden Markov models. The bold lines are the TMD determined for Manse-DHR, Bommo-DHR, and for Drome-DHR. There were disagreements of 2–3 amino acids in location of several of the TMD between these receptors; a consensus is shown. The Nillu-DHR shown is truncated by 90 residues from the N-terminus, and 31 residues from the C-terminus, for clarity, as are 28 and 27 residues from the C-terminus of one fruit fly and one mosquito receptor. Aedae-DHR1 and Drome-DHR-CG12370 are known to be localized in the MT, whereas Aedae-DHR2 and Drome-DHR-CG8422 are found in the CNS. The location of the four extracellular domains (EC1–EC4) and four intracellular domains (IC1–4) are indicated.

2 and 3; whereas the $CRF_{2\alpha}$, $CRF_{2\beta}$, and $CRF_{2\gamma}$ receptors show a preference for urocortin forms over CRF (Hauger *et al.*, 2003). Therefore, it seems highly likely that in a species such as *Tr. castaneum*, that Trica-DH_{37} and Trica-DH_{47} will show different preferences for Trica-DHR1 versus Trica-DHR2, but currently no data are available. In *M. sexta*, both Manse-DH_{41} and Manse-DH_{30} are active in adults *in vivo* (Kataoka *et al.*, 1989b; Blackburn *et al.*, 1991) as well as *in vitro* (Blackburn and Ma, 1994; Audsley *et al.*, 1995). However, Manse-DH_{30} is without apparent effect on the cryptonephric MT (Audsley *et al.*, 1995) and the ascending portion of the larval MT (G.M. Coast, unpublished data). Curiously, upon treatment of leaf slices with Manse-DH_{30}, neonatal larvae of *M. sexta* behaved as though the peptide had antifeedant effects, and some mortality was observed at the ensuing molt (Ma *et al.*, 2000). Similarly, Keeley *et al.* (1992) showed that injection of last instar larvae of *H. virescens* with synthetic Manse-DH_{41} had an antifeedant effect in addition to effects on fluid secretion; MT of larvae treated *in vitro* with this peptide showed increased fluid excretion. These actions are suggestive of effects of both peptides on receptors in sites likely to control feeding, such as the CNS.

9.3.12.3. Kinins: Cellular actions   Different models have been advanced for the mode of action of kinins in mosquitoes versus fruit flies (see **Figure 13**). In both models, kinins bind to a cell membrane receptor resulting in activation of phospholipase Cβ (PLCβ) and a rapid rise in intracellular Ca^{2+} ($[Ca^{2+}]_i$). The stellate cells of fruit fly tubules express a gene (CG10626) encoding a G-protein-coupled kinin receptor (Drome-K-R). The binding of Drome-K to this receptor activates PLCβ and increases inositol (1,4,5)-trisphosphate (IP_3) production and $[Ca^{2+}]_i$ in S2 cells transfected with Drome-K-R (Radford *et al.*, 2002). Two *Drosophila* genes encode PLCβ (*norpA* and *Plc21C*) and Drome-K-stimulated diuresis is greatly reduced in tubules from norpA mutants (Pollock *et al.*, 2003). An increase in $[Ca^{2+}]_i$ precedes the fastest physiological response of tubules to kinin stimulation. This results from the release of Ca^{2+} from endoplasmic reticulum stores triggered by IP_3 binding a receptor (IP_3R) that is a ligand-gated Ca^{2+} channel, and mutations of the IP_3R gene (*itpr*) attenuate kinin-stimulated diuresis and the rise in $[Ca^{2+}]_i$ (Pollock *et al.*, 2003). CAP_{2b} peptides also use IP_3 as a second messenger in fruit fly tubules, but act on principal cells rather than stellate cells (see Section 9.3.12.5.).

In marked contrast to the fruit fly, kinins are believed to act through a Ca^{2+} signaling pathway in principal cells of *Ae. aegypti* tubules (Yu and Beyenbach, 2002). The molecular genetic tools available for *D. melanogaster* are not yet available for mosquitoes, but Beyenbach's group has shown that the distinctive electrophysiological signature accompanying kinin stimulation, namely depolarization of the transepithelial voltage (V_{tep}) and a reduction in transepithelial resistance (R_t), are identical

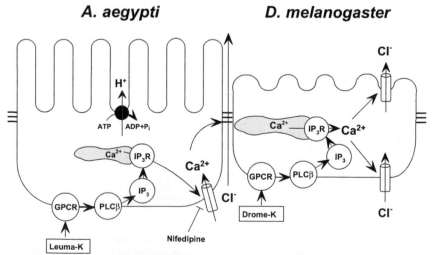

Figure 13 Kinin activity in Malpighian tubules of *Ae. aegypti* and *Drosophila melanogaster*. Kinins bind a PLCβ-coupled receptor and stimulate production of IP_3, which initiates Ca^{2+} release from endoplasmic reticulum IP_3-sensitive stores. The resultant increase in $[Ca^{2+}]_i$ opens a Cl^- selective shunt pathway that lies outside of the principal cells. In *D. melanogaster* tubules, Drome-K binds a receptor (Drome-K-R) on the stellate cells to stimulate production of IP_3. Release of Ca^{2+} from IP_3-sensitive stores elevates stellate cell $[Ca^{2+}]_i$, which opens a transcellular conductance pathway through Cl^- channels in the apical and basolateral membranes. In marked contrast, kinins (Aedae-K) were believed to act on principal cells in *Ae. aegypti* tubules. These cells have small IP_3-sensitive stores, but the opening of nifedipine-sensitive, store-operated Ca^{2+} Channels (SOCC) in the basolateral membrane allows Ca^{2+} entry from the external medium. The resultant increase in principal all $[Ca^{2+}]_i$ opens a poracellular conductance, permitting Cl^- to diffuse into the lumen through intercellular septate junctions. This model for the action of kinins on *A. aegypti* tubules is likely to need revising in light of recent work showing that, as in *D. Melanogaster*, the Aedae-K-R localizes to the basolateral membrane of stellate cells. Possibly it is the rise in stellate cells $[Ca^{2+}]_i$ that trigger the opening of the parcellular Cl^- conductance pathway.

in tubule segments with and without stellate cells (Yu and Beyenbach, 2004). Indeed, the fact that it is possible to measure levels of IP_3 in intact tubules of *Ae. aegypti* challenged with Aedae-Ks (Cady and Hagedorn, 1999b) suggests the response is not restricted to stellate cells, which are fewer in number and smaller in size than principal cells. The completion of the *An. gambiae* genome allowed identification of the kinin gene, which encodes Anoga-K-1 to -3, and the isolation of the cognate receptor from a related species of anopheline mosquito, *An. stephensi* (Radford *et al.*, 2004). All three kinins increase $[Ca^{2+}]_i$ in S2 cells transfected with the receptor. Importantly, the receptor was shown to be localized in stellate cells of *An. stephensi* MTs rather than principal cells (Radford *et al.*, 2004). These findings cast serious doubt over the assertion that kinins act on principal cells in *Ae. aegypti*, because *An. stephensi* is a closely related species (both belong to the family of Culicidae) and both feed on blood. It is possible, therefore, that the kinin response recorded by Yu and Beyenbach (2004) from short MT segments that appeared devoid of stellate cells was mediated by fine stellate cell processes that extend between neighboring principal cells. In a recent report, Lu *et al.* (2010) showed kinin receptors to be localized to stellate cells in *Ae. aegypti* as they are in the fruit fly.

Irrespective of the cell type to which the kinin activity is directed, V_{tep} depolarizes due to the opening of a Cl^- conductance pathway that lies outside of the principal cells. This pathway is believed to be through the stellate cells in fruit fly tubules because: (1) a high density of "maxi-Cl^-" channels are found in a small number of apical membrane patches, befitting the small number of stellate cells; and (2) vibrating probe analysis reveals Cl^- dependent current-density "hot spots" over stellate cells (O'Donnell *et al.*, 1998). Alternatively, Beyenbach's group argues that the Cl^- conductance pathway in *Ae. aegypti* tubules is paracellular through septate junctions (Wang *et al.*, 1996; Yu and Beyenbach, 2001). Although Cl^- channels are present in the apical membrane of stellate cells, they are not maxi channels and are not activated by Ca^{2+} (O'Connor and Beyenbach, 2001). The work of Yu and Beyenbach (2004; see previous section) would suggest that stellate cells neither signal nor mediate the kinin response, but as mentioned earlier these results should now be viewed with some caution. Malpighian tubules of the house cricket, *A. domesticus*, lack stellate cells, but respond to kinin stimulation with an increase in fluid secretion that is Cl^- dependent and probably mediated by a rise in $[Ca^{2+}]_i$ (Coast, 2001). As in dipteran tubules, kinins open a Cl^- conductance pathway causing V_{tep} to depolarize. Without stellate cells, this conductance can only be paracellular or through principal cells. Chloride uptake into principal cells is most likely via an electroneutral Na^+-K^+-$2Cl^-$ cotransporter in the basolateral membrane, and it moves into the lumen down a large electrochemical gradient through a Ca^{2+}-activated conductance pathway in the

apical membrane. In support of this model, the transepithelial Cl^- channel blocker diphenylamine-2-carboxylate significantly reduced kinin diuretic activity and decreased the accompanying depolarization of V_{tep} (Coast *et al.*, 2007). It is also worth noting that kinins have no effect on fluid secretion in *R. prolixus* tubules (Te Brugge *et al.*, 2002), which might be explained by the unusual lumen negative V_{tep} (Ianowski and O'Donnell, 2001) opposing passive diffusion of Cl^- through a paracellular conductance pathway.

9.3.12.3.1. Kinin receptors

One of a group of putative neuropeptide receptors from *D. melanogaster* (Hewes and Taghert, 2001) was identified as the probable receptor for Drome-K (Radford *et al.*, 2002). When heterologously expressed, the receptor shows an EC_{50} of 40 pmol l^{-1} and a $t(1/2) < 1$ s for Ca^{2+} elevation. Earlier, Cox *et al.* (1997) cloned a receptor from a *Lymnaea stagnalis* (pond snail) cDNA library (see **Figure 14**) and also identified the endogenous kinin, lymnokinin (PSFHSWS-amide), in this same paper. When heterologously expressed in CHO cells, this receptor was stimulated by lymnokinin with an EC_{50} for a Ca^{2+} response of 1.1 nmol l^{-1}. A "leucokinin-like" receptor has also been cloned from the cattle tick *Boophilus microplus* (Holmes *et al.*, 2003), which is clearly an orthologue to the previous two kinin receptors. However, no ligand is known for this receptor in the tick, so no affinity data are available. A kinin receptor was cloned from *Ae. aegypti* MT cDNAs using degenerate primers designed from conserved TMD sequences from the three kinin receptors already known (Pietrantonio *et al.*, 2005). When functionally expressed in CHO-K1 cells, it exhibited EC_{50} values for Aedae-K-I to -III of 49, 27, and 16 nM, respectively. The *An. stephensi* kinin receptor is also activated by "Anoga-LK-I to -III" with EC_{50} values of 2.0, 7.4, and 8.4 nM, respectively (Radford *et al.*, 2004). Unfortunately, these workers have likely misinterpreted the coding sequence of the *An. gambiae* kinin precursor: "Anoga-LK-II" (Anoga-K-3 in **Figure 6**) has a dibasic cleavage site upstream of a 21 mer, NMPRTHKQPKV-VIRNPFHSWGamide. Radford *et al.* (2004) synthesized a heptamer (from cleavage at an RN bond). Anoga-K-1 corresponds to their Anoga-LK-1, and Anoga-K-2 is their Anoga-LK-III (**Figure 6**). Orthologues of these receptors occur in other sequenced genomes, with the noted exceptions of *Tr. castaneum* and *N. vitripennis*.

Beyenbach's laboratory recently tested synthetic kinin analogs for effects on fluid secretion and tubule electrophysiology in *Ae. aegypti* (Schepel *et al.*, 2010). Surprisingly, some analogs increase secretion without an accompanying electrophysiological response, while others had the reverse effect, giving an electrophysiological response while having no effect on fluid secretion. These findings raise two important questions. First, since *Ae. aegypti* MTs apparently express only one kinin receptor (Pietrantonio *et al.*, 2005), how do the different responses

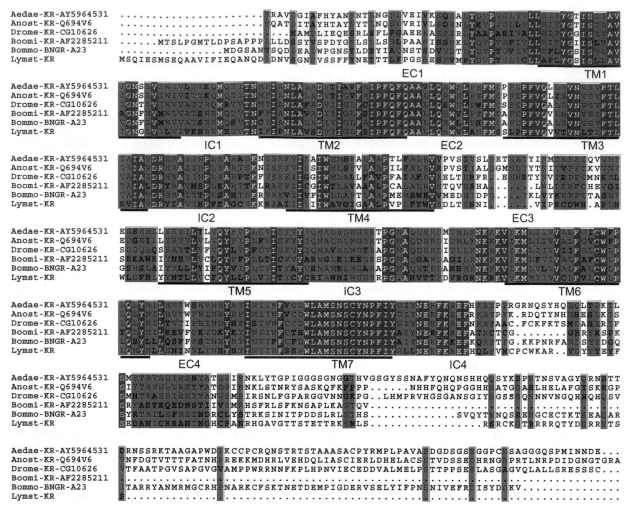

Figure 14 Experimentally cloned kinin receptors (KR) are shown, aligned with Clustal W. The seven transmembrane domains (TMD) are indicated with a bold line and the domain number. Their location was determined with the program TMHMM Server v. 2.0 at the Center for Biological Sequence Analysis, Technical University of Denmark (TMHMM), by analyzing the Aedae-K-R, Boomi-K-R, and Drome-K-R, and plotting the consensus of these. The final 44–76 C-terminal residues of the three fly (Aedae-, Anost-, and Drome-K-R) receptors are not shown for clarity and because of low sequence conservation. Accession numbers are shown after the abbreviations for four receptors.

arise? Secondly, how can the effects of kinin stimulation on tubule electrophysiology and fluid secretion occur independently of one another? The authors suggest two possible answers to the first question: agonist-directed signaling and glycosylation-dependent signaling. Both assume that different analogs activate different signaling pathways through the same receptor, although all the evidence shows that kinins act only through an increase in $[Ca^{2+}]_i$. Regarding the second question, the unstimulated MT is an example of a "tight" epithelium and maintains a large V_{tep}. In such epithelia, a small change in ion flow can have a pronounced effect on V_{tep}, but the net movement of ions is too small to have a detectable effect on the movement of osmotically obliged water. However, it is more difficult to explain why some analogs increase fluid secretion without producing an electrophysiological response. There is universal agreement that kinins open a

Cl⁻ conductance pathway and change the MT from a tight to a "leaky" epithelium, causing V_{tep} to depolarize. It may be significant that for these studies Schepel *et al.* (2010) did not record V_{tep}, instead they used a two-electrode voltage clamp to measure V_b in principal cells and their input resistance.

9.3.12.4. CT-like DH family: Cellular actions   The CT-like DH of *D. punctata* (Dippu-DH$_{31}$) is a potent stimulant of secretion by MTs from the cockroach (EC$_{50}$ = 9.8 nM) and the locust, *L. migratoria* (EC$_{50}$ = 0.56 nM), but in both assays its activity is <50% of that obtained with the CRF-related DH, Dippu-DH$_{46}$ (Furuya *et al.*, 2000b). The different activities of the CT-like and CRF-related DHs suggest they have different modes of action, which was unexpected, because both were isolated on the basis of their ability to stimulate cAMP production by isolated MTs, albeit from *S. americana* and *M. sexta*, respectively. There is unequivocal evidence to show Dippu-DH$_{46}$ acts via a cAMP-dependent signaling pathway in *D. punctata* MTs (Tobe *et al.*, 2005), but the same study failed to resolve the mode of action of Dippu-DH$_{31}$, which had no effect on cAMP or cGMP production by isolated MTs. Consistent with Dippu-DH$_{31}$ and Dippu-DH$_{46}$ having different modes of action, they act synergistically in stimulating secretion by *D. punctata* MTs, but this could not be demonstrated with *L. migratoria* MTs. On the other hand, the peptides clearly have different effects on locust MTs, because Dippu-DH$_{46}$ has a greater effect on the transport of Na$^+$ than K$^+$ (the secreted fluid [Na$^+$]:[K$^+$] ratio is increased from 0.16 to 0.43), whereas Dippu-DH$_{31}$ has a non-selective effect on cation transport (Furuya *et al.*, 2000b). The *L. migratoria* kinin (Locmi-K) also has a non-selective effect on cation transport (Coast, 1995), which suggests Dippu-DH$_{31}$ might act via a Ca^{2+} signaling pathway in locust MTs. In support of this, thapsigargin was shown to act additively with Dippu-DH$_{31}$, but synergistically with Dippu-DH$_{46}$ (Furuya *et al.*, 2000b). Surprisingly, however, Dippu-DH$_{31}$ (and Dippu-DH$_{46}$) acts synergistically with Locmi-K, although this is consistent with the CT-like DH stimulating cAMP production by the MT of another locust, *S. americana* (Furuya *et al.*, 2000b).

The CT-like peptide Drome-DH$_{31}$ is a potent stimulant of secretion by fruit fly MTs (EC$_{50}$ = 4.3 nM; Coast *et al.*, 2001), but its activity is <35% that of the kinin Musdo-K, which is virtually identical to Drome-K. The diuretic activity of Drome-DH$_{31}$ appears to be mediated solely by cAMP, and is associated with hyperpolarization of the apical membrane voltage and acidification of the secreted fluid. These actions are mimicked by exogenous cAMP (O'Donnell *et al.*, 1996), and are consistent with stimulation of the apical V-ATPase. Drome-DH$_{31}$ (and cAMP) appears to have no effect on cation uptake into principal cells and stimulated rates of secretion are not accompanied by any change in tubule fluid K$^+$ (and probably Na$^+$) concentration. In this respect, the activity of Drome-DH$_{31}$ on fruit fly MTs resembles that of Dippu-DH$_{31}$ on *L. migratoria* MTs, although the non-selective increase in cation transport in the latter is attributed to stimulation of the apical V-ATPase and the opening of a Cl$^-$ conductance pathway, respectively.

The CT-like DH (Anoga-DH$_{31}$) of the mosquitoes *An. gambiae* and *Ae. aegypti* also uses cAMP as a second messenger, but the stimulation of fluid secretion is accompanied by a dramatic increase in the [Na$^+$]:[K$^+$] ratio of the secreted fluid (Coast *et al.*, 2005). Previously, an unidentified peptide from the head was shown to act via cAMP to increase Na$^+$ transport at the expense of K$^+$ transport (Beyenbach and Petzel, 1987), and this so-called Mosquito Natriuretic Peptide is probably the CT-like DH. There is indirect evidence to show that MNP is released

into the circulation in response to blood feeding to stimulate the ensuing natriuresis, which is associated with an increase in tubule cAMP content. In support of this, passive immunization of female *An. gambiae* with an antiserum raised against Dippu-DH$_{31}$ reduced the amount of an injected Na$^+$ load excreted over 4 h from 33% to just 9% (Coast, 2009). Anoga-DH$_{31}$ (and cAMP) depolarizes the voltage across the basolateral membrane (Coast *et al.*, 2005), which is attributed to the opening of a Na$^+$ conductance pathway across this surface (Beyenbach, 2003), accounting for the accelerated transport of Na$^+$ into the tubule lumen. Surprisingly, the loop diuretic bumetanide blocks cAMP-stimulated Na$^+$ transport (Hegarty *et al.*, 1991), which may be due to some interaction between a bumetanide-sensitive Na$^+$-K$^+$-2Cl$^-$ cotransporter and Na$^+$ channels in the basolateral membrane (Scott *et al.*, 2004).

The CT-like DH of *R. prolixus* (Rhopr-DH$_{31}$) is identical to Dippu-DH$_{31}$. Tested at concentrations of 0.1 to 100 nM, it has very little effect on MT secretion (<1.5% maximal; Te Brugge *et al.*, 2005) and no effect on cAMP production (Te Brugge *et al.*, 2008). Likewise, the CT-like DH has no effect on fluid absorption from the expanded anterior midgut, although it does produce a small increase in cAMP content (Te Brugge *et al.*, 2009). Rhopr-DH$_{31}$ is therefore unlikely to have a direct role in the rapid diuresis that follows a blood meal. It may have an indirect role, however, by stimulating contractions of the anterior midgut, hindgut, and dorsal vessel, which would increase hemolymph circulation and the mixing of gut contents, reducing unstirred layers (Te Brugge *et al.*, 2008, 2009). In this context, it is noteworthy that the purification of a CT-like DH from the Belgium forest ant *F. polyctena* was guided by its effect on the spontaneous contractions of *L. migratoria* MTs, which were stimulated by doses of 0.001 head equivalents/μl (Laenen, 1999).

Holtzhausen *et al.* (2007) studied the activity of Bommo-DH$_{31}$, Anoga-DH$_{31}$, and Dippu-DH$_{31}$ on MTs of *T. molitor* and the dung beetle. They found EC$_{50}$ values of 0.6 and 14 nM for the silkworm and mosquito peptides, but Dippu-DH$_{31}$ was inactive even at 1 μM, just as in *M. sexta* (Furuya *et al.*, 2000b). The effect of these peptides was only about 50% of the maximal effect of 0.1 mM dibutyryl-cAMP. No synergistic effect was seen on co-treatment with cAMP, thapsigargin, or a CRF-like DH. Attempts to stain *T. molitor* tissues with affinity purified anti-Dippu-DH$_{31}$ were unsuccessful; this antibody has considerably lower cross-reactivity to Trica-DH$_{31}$ (D.A. Jensen and D.A. Schooley, unpublished).

9.3.12.4.1. CT-like DH receptors Johnson *et al.* (2005) reported the identification of a receptor for Drome-DH$_{31}$, which is encoded by the gene CG17415, homologous to the vertebrate calcitonin receptor. Like the calcitonin receptor, high-sensitivity signaling required the co-expression of a receptor component protein, or receptor activity modifying protein. The sensitivity of signaling did not seem to depend much on the subtype of protein co-expressed, with EC_{50} values of ~100 nM in each case (Johnson *et al.*, 2005). The receptor was expressed in principal cells of the MT and in ~7 neurons on each side of the brain, which co-express the Drome-DHR-CG8422 and the peptide corazonin. In addition, another neuron on this side expressed only the CG17415 receptor. A CT-like DH receptor, BNGR-B1, has been cloned from *B. mori*, but not expressed (Yamanaka, 2008). A CT-like DH receptor has been cloned from *Tr. castaneum*, but not expressed or characterized (Cosme, 2009).

9.3.12.5. CAP$_{2b}$ family: Cellular actions

The diuretic activity of peptides of the CAP$_{2b}$ family has been most intensively investigated in fruit fly tubules and how they act through a Ca^{2+}-NO-cGMP signaling pathway to stimulate cation transport by increasing V-ATPase activity in principal cells, as evidenced by hyperpolarization of V_{tep} and acidification of the urine (O'Donnell *et al.*, 1996; Kean *et al.*, 2002). The binding of CAP$_{2b}$ to a PLCβ-coupled receptor on principal cells in the main secretory segment stimulates the production of IP$_3$, which initiates the release of Ca^{2+} from endoplasmic reticulum IP$_3$-sensitive stores (see **Figure 15** for model). In support of this, the tubules express genes encoding PLCβ (*norpA* and *plc21*; Pollock *et al.*, 2003) and the IP$_3$ receptor (*itpr*; Blumenthal, 2001), and the CAP$_{2b}$-stimulated diuresis is significantly reduced in tubules from norpA and itpr mutant flies (Pollock *et al.*, 2003). The Ca^{2+} released from principal cell IP$_3$-sensitive stores is responsible for an initial short-lived spike in $[Ca^{2+}]_i$. Store emptying probably activates store-operated Ca^{2+} channels (SOCCs), and the influx of extracellular Ca^{2+} accounts for a second longer lasting increase in $[Ca^{2+}]_i$, which is essential for the CAP$_{2b}$ stimulated diuresis. Fruit fly tubules express genes that encode a variety of Ca^{2+} channels involved in Ca^{2+} transport and signaling. These include the genes *trp* and *trpl* that encode Trp and Trp-like (Trpl) channel proteins (MacPherson *et al.*, 2005). Analysis of *trp* and *trpl* mutants implicates TRPL in the diuretic response to CAP$_{2b}$, and they are expressed in principal cells of the main tubule segment. The CAP$_{2b}$ stimulated rise in $[Ca^{2+}]_i$ is sufficient to activate NO synthase, which is a Ca^{2+}/calmodulin-sensitive *D. melanogaster* nitric oxide synthase (dNOS) expressed only in principal cells (Davies *et al.*, 1997; Rosay *et al.*, 1997; Davies, 2000). The NO generated by dNOS activates a soluble guanylate cyclase (sGC) resulting in elevated levels of cGMP, which are confined to the principal cells (Kean *et al.*, 2002). Interestingly, fruit fly tubules express a gene encoding a cyclic nucleotide gated channel (CNG), and activation of this channel by cGMP contributes to the influx of extracellular Ca^{2+} and helps sustain the rise in $[Ca2+]_i$ (MacPherson *et al.*, 2001).

Other possible targets for the cGMP produced in response to CAP$_{2b}$ stimulation include cGMP-dependent protein kinases (cGKs), and two genes encoding cGK (*dg1* and *dg2*) are expressed in tubules. Tubules from *dg2* mutant flies are hypersensitive to cGMP, but this appears to be due to indirect modulation of the activity of a cGMP-specific phosphodiesterase (cG-PDE) (MacPherson *et al.*, 2004). Interestingly, CAP$_{2b}$ inhibits cGMP-PDE activity, which will contribute to the elevation of cGMP levels.

CAP$_{2b}$ increases V-ATPase activity by recruiting V_1 head-groups to the apical membrane from the basolateral membrane, and by activating apical mitochondria to increase the supply of ATP (Terhzaz *et al.*, 2006). Since there is only a modest redistribution of V_1 head-groups, the latter is believed to be more important for increasing V-ATPase activity. CAP$_{2b}$ stimulation triggers a fast (within a second) transient rise in $[Ca^{2+}]_i$, which is believed to be localized to the basolateral region of the cell, followed by a slower rise in the whole cell that peaks in 100–200 s.

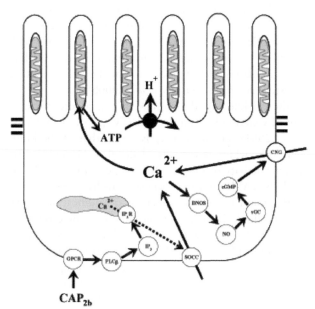

Figure 15 The nitridergic signaling pathway activated by CAP$_{2b}$ and orthologues in *Drosophila melanogaster* principal cells. The binding of CAP$_{2b}$ to a PLCβ-coupled receptor stimulates production of IP$_3$, which initiates Ca^{2+} release from endoplasmic reticulum stores by opening ligand gated Ca^{2+} channels (IP$_3$R). Emptying of these IP$_3$-sensitive stores activates store-operated Ca^{2+} channels (SOCC) and the resultant rise in $[Ca^{2+}]_i$ stimulates a Ca^{2+}-calmodulin nitric oxide synthase (DNOS) and elevates nitric oxide (NO) production. NO activates a soluble guanylate cyclase (sGC), increasing production of cGMP, which opens cyclic nucleotide-gated Ca^{2+} channels (CNG) in the basolateral membrane, augmenting the Ca^{2+} signal. Apical mitochondria are activated by the rise in $[Ca^{2+}]_i$, which increases the supply of ATP to the apical membrane V-ATPase. As a result, V-ATPase activity, the driving force for fluid secretion, is stimulated. The hydrolysis of cGMP by phosphodiesterase (PDE) appears to be important in modulating the stimulation of fluid secretion by CAP$_{2b}$.

Measurements of whole cell mitochondrial Ca^{2+} levels show these track the slow rise in $[Ca^{2+}]_i$, with the response largely restricted to mitochondria located within the apical microvilli. The time course of the slow rise in Ca^{2+} levels within apical mitochondria correlates with their activation (Terhzaz et al., 2006) as evidenced by hyperpolarization of the inner mitochondrial membrane recorded using a potential-sensitive dye. Although basal mitochondria also respond to CAP_{2b} stimulation, they are constitutively active, possibly because of the need to supply ATP to the Na^+-K^+-ATPase and organic solute transporters on the basolateral membrane. Results obtained using the proteomic technique of time series difference gel electrophoresis suggests that activation of the apical mitochondria involves extensive remodeling of the mitochondrial matrix. This correlates with an increase in tubule levels of ATP, which powers the apical V-ATPase. The causal link between the rise in mitochondrial Ca^{2+} levels and their activation has been demonstrated using Ru360 (the active component of ruthenium red), which is a selective blocker of the mitochondrial Ca^{2+} uniporter (Terhzaz et al., 2006). Ru360 blocks the mitochondrial Ca^{2+} response to CAP_{2b} stimulation and also abolishes the rise in ATP production. It is likely that the rise in mitochondrial Ca^{2+} produces long-term changes, because CAP_{2b}-stimulated secretion remains elevated even after Ca^{2+} levels return to their baseline values (Terhzaz et al., 2006).

The detailed dissection of the elaborate signal transduction pathway for CAP_{2b} in fruit fly tubules is testament to the advantages afforded by the powerful combination of molecular biology and genetic analysis in unraveling complex physiological processes. Nevertheless, it is important to remember that however elegant a model, it provides only a description of what occurs in the fruit fly and may be quite different in other insects. When CAP_{2b} peptides have been tested in other species their actions differ from those described in the fruit fly. For example, CAP_{2b} stimulates fluid secretion by housefly (M. domestica) tubules by an NO-cGMP-independent mechanism that results in V_{tep} depolarizing rather than hyperpolarizing and is not accompanied by urine acidification (C.S. Garside and G.M. Coast, unpublished data). The cellular actions of CAP_{2b} in the housefly are very similar to those of the kinins in that the depolarization of V_{tep} is due to the opening of a Cl^- conductance pathway that lies outside of the principal cells and may be paracellular or transcellular through the stellate cells (see earlier in this section). Moreover, CAP_{2b} activity is not dependent on Ca^{2+} influx from the bathing medium, but most likely involves Ca^{2+} release from intracellular stores, because prior treatment of tubules with thapsigargin blocks CAP_{2b} stimulated diuresis. If housefly principal cells have only small stores of releasable Ca^{2+}, as in fruit fly tubules, then the assumption is that CAP_{2b} acts on stellate cells. In this context, it is noteworthy that high concentrations (100 nmol l^{-1}) of

CAP_{2b} peptides increase $[Ca^{2+}]_i$ in the stellate cells of fruit fly tubules (Kean et al., 2002). The response to Drome-CAP_{2b}-1 (which the authors refer to as capa-1 because it is encoded on the capa gene) is large enough to open the stellate cell Cl^- conductance and, in combination with the stimulation of active cation transport by the principal cells, this might explain the more potent diuretic activity of this peptide compared with Drome-CAP_{2b}-2 and Manse-CAP_{2b}.

In contrast to the diuretic activity of CAP_{2b} peptides in the housefly and fruit fly, CAP_{2b} acts via cGMP to reduce secretion by MTs from R. prolixus and T. molitor larvae and Ae. aegypti adult females. CAP_{2b} (and cGMP) antagonizes the actions of cAMP and the diuretic hormones serotonin (R. prolixus; Quinlan et al., 1997) and Tenmo-DH_{37} (T. molitor; Wiehart et al., 2002), both of which act via cAMP-dependent mechanisms to stimulate tubule fluid secretion. CAP_{2b} also antagonizes the diuretic activity of the CRF-related Zoone-DH on stink bug MTs (Coast et al., 2010), although it is not known whether this is mediated by cGMP. The antagonist effects of CAP_{2b} and cGMP can be explained by the activation of a cGMP-dependent cAMP phosphodiesterase (Quinlan et al., 1997), which would lower intracellular levels of the diuretic second messenger cAMP and thus reduce tubule secretion, although there is as yet no direct evidence in support of this. Subsequently, they further studied the antagonistic effects of cGMP and cAMP on R. prolixus MTs; an inhibitory level of cGMP is without effect on the concentrations of K^+ and Na^+ of secreted fluid (Quinlan and O'Donnell, 1998). Transport of organic anions into upper tubules involves at least two different transporters: one for acylamides (e.g., p-aminohippuric acid) and another for sulfonates (e.g., amaranth and phenol red). Amaranth and phenol red blocked the actions of both cGMP and cAMP, whereas p-aminohippuric acid was without effect, suggesting that exogenous cyclic nucleotides are carried into the MT cell by the sulfonate transporter (Quinlan and O'Donnell, 1998).

CAP_{2b} also inhibits ion and fluid transport across the expanded anterior midgut (a functional crop) of R. prolixus, which is stimulated by both serotonin and Zoone-DH (Te Brugge et al., 2009). This is of critical importance, because during the rapid diuresis that follows a blood meal large amounts of NaCl-rich fluid are taken up into the hemolymph from the blood meal in the crop and transported into the lumen of the upper MT for elimination. The two processes must be very precisely coordinated to avoid any acute change in the volume and composition of the hemolymph. CAP_{2b} peptides are released into the circulation toward the end of the rapid diuresis (Paluzzi and Orchard, 2006), which lasts about 3 h, ensuring that fluid uptake from the crop and secretion by the upper MT are simultaneously shut down. Interestingly, CAP_{2b} does not stimulate cGMP production by the anterior midgut

(Ianowski *et al.*, 2010), and exogenous cGMP has no effect on fluid transport (Ianowski *et al.*, 2010). Moreover, the actions of CAP_{2b} on the crop are not dependent on extracellular Ca^{2+} or inhibited by preloading the gut with the intracellular Ca^{2+} chelator BAPTA-AM. Thus the actions of CAP_{2b} on the crop involve a different signaling pathway from that reported in MTs. Furthermore, CAP_{2b} has no effect on cAMP levels in the crop and inhibits fluid transport even when this is stimulated with a high concentration of the cell-permeant analog 8-bromo-cAMP, which is resistant to PDE hydrolysis (Ianowski *et al.*, 2010). Ion transport, mainly NaCl, by the crop has been little explored, but it is believed to be driven by a ouabain-sensitive Na^+-K^+-ATPase on the hemolymph side of the epithelium (Farmer *et al.*, 1981) rather than a V-ATPase. CAP_{2b} peptides might therefore act to inhibit this transporter or to modulate associated conductance pathways.

9.3.12.5.1. CAP_{2b} receptors Receptors for CAP_{2b} (CAP_{2b}-R) are evolutionarily related to vertebrate neuromedin receptors and are placed in a monophyletic clade along with receptors for pyrokinins and eclosion triggering hormones (Park *et al.*, 2002). CAP_{2b}-Rs have been cloned from *D. melanogaster* (Iversen *et al.*, 2002; Park *et al.*, 2002), *An. gambiae* (Olsen *et al.*, 2007), and *R. prolixus* (Paluzzi *et al.*, 2010), and are specific for CAP_{2b} peptides that have the C-terminal sequence PRVamide. A structure–activity study using *M. domestica* MTs (Nachman and Coast, 2007) revealed that the C-terminal valine residue, which differentiates them from pyrokinins ending in PRLa, is critical for diuretic activity. Rhopr-CAP_{2b}-R is expressed at high levels in the anterior midgut and the MT, consistent with the actions of CAP_{2b} peptides on these structures. The expression of CAP_{2b}-R is 60-fold higher in the upper secretory segment of the MT than in the lower reabsorptive segment, again consistent with the known action of CAP_{2b} peptides in inhibiting serotonin-stimulated secretion. Other CAP_{2b}-Rs have been identified from BLAST searches of sequenced genomes.

9.3.12.6. Cellular actions of antidiuretic factors inhibiting Malpighian tubule secretion

The first insect neuropeptide shown to inhibit MT secretion was Manse-CAP_{2b} (Quinlan *et al.*, 1997) as described in Section 9.3.12.5.

In *T. molitor*, both CAP_{2b} and the two *T. molitor* antidiuretics, Tenmo-ADFa and -ADFb, act by stimulating a guanylate cyclase, which is probably membrane bound. Wiehart *et al.* (2002) showed that Tenmo-ADFa and Tenmo-DH_{37} have antagonistic effects, the former elevating cGMP levels and the latter elevating cAMP levels. It also seems likely in this species that cGMP activates a cAMP-specific phosphodiesterase, as in vertebrate PDE2 (Bender and Beavo, 2006). Attempts to prove this by using *T. molitor* MT incubated with 1 mmol l^{-1}

erythro-9-(2-hydroxy-3-nonyl)adenine HCl (EHNA), a selective inhibitor of the cGMP-activated cAMP phosphodiesterase II (Podzuweit *et al.*, 1995), and 100 fmol l^{-1} Tenmo-ADFa, were unfortunately inconclusive (S.W. Nicolson, R.J. Eigenheer, and D.A. Schooley, unpublished data). Tenmo-ADFa also has antidiuretic action in *Ae. aegypti*; at 1 nmol l^{-1} concentrations in a Ramsay assay it significantly inhibits the rate of fluid secretion without significant effects on the concentrations of Na^+, K^+, and Cl^- in secreted fluid and also significantly increases intracellular cGMP concentration (Massaro *et al.*, 2004). Low doses of cGMP (20 µmol l^{-1}) inhibit isosmotic fluid secretion without inducing electrophysiological effects, whereas cGMP at 500 µmol l^{-1} significantly depolarized the basolateral membrane but hyperpolarized the transepithelial potential. Low concentrations of Tenmo-ADFa and cGMP inhibit isosmotic fluid secretion by reducing electroneutral transport, whereas high levels of cGMP increase an unknown conductance of the basolateral membrane and depolarize the basolateral membrane voltage supplemental to and independent of the electroneutral mechanism of action of low concentrations of cGMP (Massaro *et al.*, 2004).

In *T. molitor*, only two zones of material that elevated cGMP levels in MT were isolated and sequenced, even though ADFb (Eigenheer *et al.*, 2003) is 24,000 times less potent than ADFa (Eigenheer *et al.*, 2002). Manse-CAP_{2b} (EC_{50} ~ 85 nmol l^{-1}) is less potent than ADFb; the presence of CAP_{2b} or paralogue in *T. molitor* would probably have been detected. Li *et al.* (2008) found a *capa* gene in *Tr. castaneum*, but none of the three peptides it encodes has the vital PRVa required for CAP_{2b}-like activity. It is quite interesting that cGMP has thus far been found to be antidiuretic in two phylogenetically disparate blood feeders, *R. prolixus* and *Ae. aegypti*, and a starkly xeric species, *T. molitor*. Quinlan and O'Donnell (1997) argued that an antidiuretic factor could be of great importance in a blood feeder to cause a rapid decrease in post blood meal diuresis to avoid depleting the hemolymph volume. *T. molitor* is so highly evolved for water absorption that activities of K^+ in its cryptonephric MT exceeding 3 M have been measured (O'Donnell and Machin, 1991), allowing absorption of water vapor from the rectum above 88% relative humidity. (Despite this species having high hemolymph Na^+, the active transport mechanism in MT is quite selective for K^+.) Thus, antidiuretic factors could be very important in conserving water in this species.

It seems surprising that cGMP has the opposite effect of diuresis in *D. melanogaster* (Davies *et al.*, 1995) and *M. sexta* (Skaer *et al.*, 2002). However, cGMP is only weakly diuretic in the dipteran *M. domestica*, and Drome-CAP_{2b}-2 (one of three endogenous Manse-CAP_{2b}-like peptides encoded by the *D. melanogaster* genome) is diuretic with actions that resemble those of Musdo-K (G.M. Coast, unpublished data).

9.3.12.7. Cellular actions of ITP Schgr-ITP was reported to act via cAMP to stimulate salt (NaCl/KCl) and water uptake from the locust ileum (Phillips *et al.*, 1998b). Its actions have been studied extensively in flat sheet preparations of the locust ileum mounted in an Ussing chamber (this method is suitable only for large insects, even the hindgut of adult *M. sexta* is too small; S. Liao and D.A. Schooley, unpublished data). Addition of Schgr-ITP or cAMP to the hemolymph side results in a tenfold stimulation of the current (I_{sc}) needed to clamp the tissue at a transepithelial potential (V_{tep}) of 0 mV. The ITP stimulated I_{sc} is Cl⁻ dependent and reflects a tenfold increase in net active Cl⁻ transport from the luminal to the basal surface of the epithelium against an overall electrochemical potential gradient (see **Figure 16** for model). Extensive electrophysiological studies reveal that the active transport step takes place at the apical (luminal) membrane and involves an unusual (for vertebrates) electrogenic Cl⁻ pump stimulated by cAMP (Phillips *et al.*, 1998b). Chloride exits the cells to the hemolymph side by passive diffusion through a large basolateral membrane Cl⁻ conductance. The major cations (Na⁺ and K⁺) follow Cl⁻ passively across the apical membrane through separate conductance pathways, both of which are increased by cAMP. At the hemolymph side, K⁺ exits the cells passively via a large K⁺ conductance in the basolateral membrane, whereas Na⁺ is actively removed from the cells by a Na⁺-K⁺-ATPase pump. Fluid uptake from the locust ileum is driven by active salt (NaCl/KCl) transport and isosmotic fluid reabsorption is stimulated fourfold by Schgr-ITP and cAMP (Phillips *et al.*, 1998b). Interestingly, while both Schgr-ITP and crude extracts of CC inhibit active acid (H⁺) secretion into the lumen of the ileum, cAMP is without effect suggesting the involvement of another second messenger in the actions of this peptide (Phillips *et al.*, 1998a).

Schgr-ITP has been shown to have no effect on secretion of MT from *Sc. gregaria* (Coast *et al.*, 1999); similarly, Locmi-K and Locmi-DH have no effect on the ileum or rectum of *Sc. gregaria in vitro*. Locmi-K and Locmi-DH have a synergistic effect on the MT of *Sc. gregaria*, and this effect is much faster than the effect of Schgr-ITP on the ileum. These data would suggest that in early phases of feeding, urine secretion would commence before reabsorption, allowing voiding of excess water (Coast *et al.*, 1999).

Drexler *et al.* (2007) showed that ITP titer is elevated in blood of *M. sexta* at the time that molting fluid is reabsorbed. More recently, Begum *et al.* (2009) showed in *Tr. castaneum* that knockdown of *itp* and *itpl* by RNAi causes about 40% mortality as larvae, but high mortality and eclosion deficiency at the pupal–adult molt. Taken together, these observations may be consistent with ITP being required for molting fluid reabsorption, perhaps from the tracheae. ITP has also recently been detected in two pairs of clock neurons in the *D. melanogaster* brain (Johard *et al.*, 2009), but the role of these neurons is currently unclear.

9.3.12.8. The cryptonephric complex and cellular actions of factors acting on it The cryptonephric complex of Coleoptera and larval Lepidoptera is a composite tissue consisting of both excretory and reabsorptive tissue. In this anatomy, which differs between the two orders, the rectum is closely associated with the distal part of the MTs, which lie within a perinephric space separated from the hemolymph by a water-impermeable perinephric membrane. The cryptonephric tubules of *T. molitor* contact the perinephric membrane at regions called "boursouflures" (or "bulges"). Here the perinephric membrane is greatly reduced and forms a "blister" over an underlying leptophragma cell (Ramsay, 1964). Ramsay postulated that KCl is transported from the hemolymph into the cryptonephric tubules across the leptophragma cells without an accompanying movement of water, establishing a high osmotic concentration (as much as 6.8 Osm in *T. molitor* larvae, close to water-saturation for KCl) in the tubule lumen (O'Donnell and Machin, 1991). The perinephric space also has a high osmolarity consisting largely of non-ionic solutes (melting point depression up to -10°C) with a Na⁺-K⁺ ratio similar to that of hemolymph (Ramsay, 1964), unlike the distal parts of the cryptonephric tubules, which absorb K⁺ to

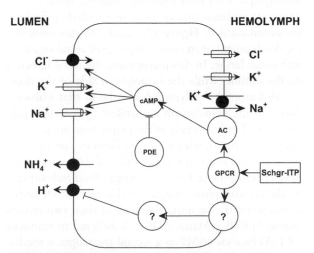

Figure 16 Proposed actions of Schgr-ITP on ion transport by locust ileum. Schgr-ITP stimulates ion transport and isosmotic fluid reabsorption from the lumen of the ileum. Acting via cAMP, it stimulates the apical electrogenic Cl⁻ pump, which is the prime driving force for ion and fluid transport. Cyclic AMP also opens conductance pathways for Na⁺ and K⁺ in the apical membrane. K⁺ and Cl⁻ exit to the hemolymph side of the epithelium through conductance pathways in the basolateral membrane, while Na⁺ is removed via Na⁺-K⁺-ATPase activity. Schgr-ITP also inhibits ileal acid (H⁺) secretion via a cAMP independent pathway. Abbreviations: AC, adenylate cyclase; PDE, phosphodiesterase; GPCR, Gsα-coupled receptor.

the near exclusion of Na^+ (O'Donnell and Machin, 1991). Using double-barreled microelectrodes and dye injections, O'Donnell and Machin (1991) also found that the perinephric space is an important functional part of the cryptonephric complex. They reported large pH (as much as 2 units, averaging 1.1 units) and electrochemical gradients (70 mV) across the apical membrane of the tubule, with the average K^+ activity across the membrane exceeding the Nernst equilibrium by 75-fold. They favor a model where cations enter the perinephric space through the anterior, more permeable part of the perinephric sheath. Anions may enter the perinephric space separately, possibly through the leptophragma cells, which are present in the *T. molitor* cryptonephric complex but absent from *M. sexta*. This could be consistent with the leptophragma cells being equivalent to stellate cells in other species. The highest activity of K^+ measured in their studies was apparently near the distal end of the cryptonephric MT. The high osmotic concentration in the tubule accounts for the osmotic withdrawal of water from first the perinephric space and ultimately the rectal lumen. This exceptional osmotic gradient may be necessary to drive water transport across at least two membranes and can account for the uptake of water vapor from air of >89% relative humidity (RH; Couchié and Machin, 1984). A microcalorimetric study on larvae of *T. molitor* exposed to a dry atmosphere showed that there was a doubling of metabolic activity prior to commencement of water vapor absorption from the air (Hansen *et al.*, 2004), which could indicate hormonal involvement in this metabolic activity. The consequence of this anatomy is that DH (at least CRF-related DH) promotes salt uptake from the hemolymph, but water uptake from the rectum. Ramsay (1964) hypothesized that the cryptonephric complex in beetles evolved to conserve water. Nicolson (1991) speculated that a DH in similar Coleoptera might serve as a clearance hormone instead of a diuretic hormone, with the fluid flow directed into the midgut to moisten the dry food for digestion, and the fluid reabsorbed in the midgut. This model raises the question of where reabsorption of electrolytes, at least other than KCl, from the hindgut occurs in these species. It is clear that the model developed for fluid reabsorption in locusts is not applicable for Coleoptera given the cryptonephric anatomy.

The cryptonephric complex of lepidopteran larvae differs significantly from that of *T. molitor* in that it is restricted to three longitudinal bands running the length of the rectum, corresponding to the three pairs of MTs (Ramsay, 1976). Between each band, a patch of "normal" rectal epithelium separates the luminal contents from the hemocoel. Additionally, boursouflures appear to be absent in the lepidopteran cryptonephric complex, which nevertheless is assumed to function in a manner similar to that of *T. molitor*, although high osmotic concentrations are not established. Ramsay (1976) argued that the

cryptonephric complex in Lepidoptera evolved to handle the osmotic challenge of ingesting massive quantities of salt, because he found that *M. sexta* larvae were able to feed on a diet containing sufficient NaCl to represent 1 kg of salt per day for an adult human; however, most lepidopterans feed on plants with high K^+ and low Na^+. Ramsay did not test KCl; perhaps this ability to excrete high Na^+ is more likely linked to the high hemolymph K^+ content of this species (Sutcliffe, 1963). About 90% of the fluid entering the rectum of day 1 fifth instar larvae is reabsorbed (Reynolds and Bellward, 1989), a quarter of which enters the cryptonephric tubules and is recycled to the midgut via the straight ascending and descending tubule segments (Moffett, 1994). At the same time, K^+ in the tubule lumen is returned to the midgut in exchange for Na^+ (Moffett, 1994). The remaining fluid is reabsorbed across the "normal" epithelium, which is composed of cells with extensive infoldings of the apical and basolateral membranes and associated mitochondria (Reynolds and Bellward, 1989). The latter group argued that the ileum of *M. sexta* has no ultrastructural characteristics of a reabsorptive tissue, suggesting that fluid reabsorption in this species is dependent on the "normal" rectal epithelium.

Audsley *et al.* (1993) presented evidence that Manse-DH$_{41}$ has a net antidiuretic effect on the cryptonephric complex of *M. sexta*. They showed that Manse-DH$_{41}$ stimulates a bumetanide-sensitive Na^+-K^+-$2Cl^-$ cotransporter, causing an influx of salt from the hemolymph, with fluid following from the rectal lumen. Transport processes that are consistent with their findings are summarized in **Figure 17**. Crucial to this work was the development of an assay using everted rectal sacs from fifth instar larvae. In this preparation, the rectal lumen is on the outside, while the hemolymph side of the tissue is sealed around a polypropylene tube so that reabsorption into the hemolymph is manifest by a weight gain over time. Further details of the preparation are given by Liao *et al.* (2000), who used it to demonstrate the existence of two factors in CC-CA of larval *M. sexta* that are distinct from Manse-DH$_{41}$ and trigger fluid reabsorption by the cryptonephric complex of this species (see earlier in this section). The more abundant of these two factors, Manse-ADF-B = Manse-ADF-2, is believed to stimulate a Cl^--ATPase via cAMP as a second messenger, a mechanism of action apparently identical to that of Schgr-ITP. Elucidating the mechanism of action of Manse-ADF-2 could only be accomplished via use of selective inhibitors, because Manse-DH$_{41}$ also exerts its antidiuretic effect on the tubules via cAMP. However, incubation of everted rectal sacs with bafilomycin A1 (an inhibitor of the V-ATPase) on the hemolymph side of the tissue was found to block the basal reabsorption by this tissue, masking the effect of Manse-DH$_{41}$. This is similar to incubation with amiloride on the lumen side, which is an inhibitor of the H^+-alkali metal antiporter. Addition of Manse-ADF-2

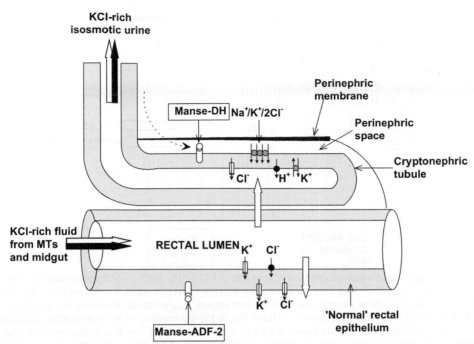

Figure 17 A model for the hindgut of *M. sexta* and similar lepidopteran larvae. At the top is one of three longitudinal segments of cryptonephric complex, and on the opposing side of the hindgut lies a longitudinal segment of 'normal' rectal epithelium. There are three radially symmetrical bands of cryptonephric tubules and 'normal' rectal epithelium. Fluid entering the rectal lumen from the midgut and Malpighian tubules is reabsorbed by the cryptonephric tubules (CNT) and by the 'normal' epithelium. The CNT lie in a perinephric space (PNS) isolated from the hemolymph by a water-impermeable perinephric membrane (PNM). Ion transport by CNT is similar to that described in principal cells, with cation uptake from the PNS occurring via Na^+-K^+-$2Cl^-$ cotransport. Manse-DH41 acts via cAMP to stimulate the cotransporter and hence increase secretion of KCl-rich fluid into the CNT lumen. The withdrawal of water from the PNS creates an osmotic gradient favoring the passive withdrawal of water from the rectal lumen. Ions may enter the PNS from the rectal lumen (not shown) or across a thinner and more permeable (to ions) anterior region of the PNM (dotted arrow). In contrast to the actions of Manse-DH41, Manse-ADF-2 acts via cAMP to stimulate ion transport and hence fluid reabsorption across the 'normal' rectal epithelium. Ion transport by the rectal epithelium appears to be driven by an apical electrogenic Cl^- pump (c.f., the locust ileum) stimulated by cAMP. Cl^- exits the cells by passive diffusion through Cl^- channels in the basolateral membrane. The active transport of Cl^- from lumen to hemocoel establishes a favorable gradient for passive reabsorption of K^+ through conductance pathways in the apical and basolateral membranes.

to these inhibited preparations caused a highly significant increase in fluid reabsorption, which was easier to measure in the presence of the inhibitors due to a strong decrease in basal reabsorption. Incubation of rectal sacs with H-89, a potent and selective inhibitor of PKA, blocks the effect of Manse-ADF-2. In addition, incubation of rectal sacs with the Cl^- channel blockers DIDS or DPC caused a significant decrease in reabsorption, and the reabsorption could not be relieved by treatment with Manse-ADF-2. The ionic transport believed to occur in the "normal" rectal epithelium as a result of this antidiuretic factor is summarized in **Figure 17**.

9.3.13. Synergism

In general, the transport of cations (K^+ and Na^+) and anions (Cl^-) by MTs appears to be regulated separately (O'Donnell *et al.*, 1996). Thus, depending on species, CRF-related, CT-like, and CAP_{2b} peptides (*D. melanogaster*) stimulate active cation transport, whereas kinins and

CAP_{2b} peptides (*M. domestica*) open a Cl^- selective shunt pathway to increase anion movement into the tubule lumen. By stimulating different transport processes, either through different second messengers or by acting on different cell types, combinations of these diuretics can have an effect on tubule fluid secretion that is greater than the sum of their separate effects. An electrophysiological basis for this synergism can be derived from an electrical equivalent circuit used to model transepithelial ion transport (Clark *et al.*, 1998a; **Figure 18**). The circuit comprises elements representing the active transport pathway for K^+ and Na^+ through the principal cells parallel with a passive transport pathway for Cl^- that lies outside of the principal cells. Within the active pathway, E_{cell} and R_{cell} represent the electromotive force and the resistance to cation transport, respectively, whereas R_{shunt} defines the passive transport pathway. The two pathways are coupled on the apical and basal sides of the epithelium by the conductances of the urine and hemolymph, respectively. It follows that any increase in ion transport through one pathway must

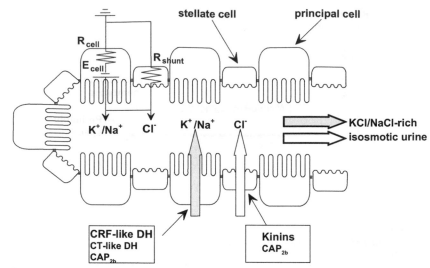

Figure 18 An electrical equivalent circuit model for the secretion of KCl and/or NaCl by Malpighian tubules. The circuit comprises parallel active cation (K+ and Na+) and passive anion (Cl-) transport pathways, the former through principal cells and the latter through either stellate cells or intercellular junctions (not shown). E_{cell} is the electromotive force of the active pathway and is primarily generated by the apical membrane V-ATPase. R_{cell} is the transcellular resistance to active transport, while R_{shunt} is the resistance of the anion pathway to the diffusion of Cl- into the lumen. Current flow (the equivalent of ion transport) through the separate pathways must be equal to preserve electrical neutrality. Synergism between factors that separately stimulate cation and anion transport can in part be explained by the combined effect of reducing R_{cell} and R_{shunt} on intraepithelial current flow, which is greater than the sum of their single effects. Adapted with permission from Ussing H.H., and Windhogen, E.E., (1964). Nature of short path and active sodium transport path through frog skin epithelium. *Acta Physiol. Scand. 61*, 484-504.

be matched by equivalent movement through the other pathway to maintain electrical neutrality. The current flow through the circuit (I_{loop}) is described by Ohm's law, where $I_{loop} = E_{cell}/R_t$, with R_t being the transepithelial resistance ($R_t = R_{cell}R_{shunt}/(R_{cell} + R_{shunt})$). Reducing either R_{cell} or R_{shunt} will increase transepithelial ion secretion as evidenced by the increase in I_{loop}. However, a reduction in both R_{cell} and R_{shunt} will increase ion secretion (I_{loop}) by far more than the sum of their separate effects (Clark *et al.*, 1998a).

Such synergistic control of tubule secretion has been demonstrated in tubules of *L. migratoria* (Coast, 1995) and *M. domestica* (Iaboni *et al.*, 1998) challenged with a combination of their CRF-related peptides and kinins to stimulate cation and anion transport, respectively. Interestingly, the CRF-related peptide (Dippu-DH$_{46}$) and CT-like peptide (Dippu-DH$_{31}$) act synergistically to stimulate secretion by *D. punctata* tubules, although in this instance the peptides appear to use the same second messenger, namely cAMP (Furuya *et al.*, 2000b). A possible explanation is that the peptides act on different cell types as seen in fruit fly tubules with the segregation of Ca^{2+} signaling pathways stimulated by CAP$_{2b}$ (principal cells) and Drome-K (stellate cells) (Rosay *et al.*, 1997). Synergism has also been demonstrated between serotonin and a peptide diuretic hormone from the mesothoracic ganglion mass (MTGM) of *R. prolixus* (Maddrell *et al.*, 1993a). The peptide has now been identified as Rhopr-DH$_{49}$ (Te Brugge *et al.*, 2010), which raises a number of intriguing

questions. First, why does Rhopr-DH$_{49}$ synergize with serotonin whereas CRF-DH from *L. migratoria* and *Z. nevadensis* do not? Secondly, how do two hormones using the same second messenger act synergistically? Possibly the answer to both questions lies in the fact that Rhopr-DH$_{49}$ acts via cAMP and another, as yet unknown, second messenger pathway, with the latter accounting for the synergism with serotonin. In this context, it is noteworthy that Rhopr-CAP$_{2b}$-2 antagonizes the diuretic activity of serotonin, but not that of the CRF-DH (Paluzzi and Orchard, personal communication). This would still curtail the postprandial diuresis if, as seems likely, high rates of urine flow are dependent on the synergistic actions of the two hormones.

Synergism between two (or more?) diuretics offers a number of advantages arising from the fact that less hormone needs to be released into the circulation to stimulate diuresis. Thus, diuresis can be turned on and off more rapidly, because less hormone needs to be released and later inactivated, and there will be energy savings in terms of the cost of hormone synthesis. For example, using data from Coast (1995) and Audsley *et al.* (1997a) it has been calculated that about 90% of the total store of Locmi-DH$_{46}$ must be released to increase MT secretion from 10 to 90% of the maximum rate, but this figure falls to just 5% in the presence of 0.05 nM Locmi-K (Coast *et al.*, 2002b). Unfortunately, although the hemolymph titer of Locmi-DH$_{46}$ is known to increase after feeding (Audsley *et al.*, 1997b), circulating levels of Locmi-K have not been measured in locusts. An added level of complexity is introduced by the reported synergism between Locmi-DH$_{46}$ and a tachykinin-related peptide, Locmi-TK-I (Johard et al., 2003), because such peptides are released into the circulation from gut endocrine cells during periods of starvation (Winther and Nässel, 2001). This could provide a potent mechanism for increasing MT secretion without a significant change in the titer of Locmi-DH$_{46}$ to accelerate the clearance of toxic metabolites. The steeper dose-response curve that characterizes synergism (Maddrell et al., 1993a; Coast, 1995) also means that a small change in hormone concentration can have a pronounced impact on fluid secretion, resulting in more precise control of diuresis. It is noteworthy that Locmi-DH and Locmi-K, which are co-localized in abdominal ganglion neurosecretory cells (Patel et al., 1994), are not co-localized in the brain and CC. To be effective hormone release from separate sites would need to be coordinated, and it is significant that the receptor for Drome-K (Drome-K-R) is present on neurosecretory cells in the pars intercerebralis that express Drome-DH$_{44}$ and in their axons in the retrocerebral complex (Radford et al., 2002).

9.3.14. Possible Utility of Research on Hormonal Control of Fluid Homeostasis

There is a long-term practical significance to this research. Over 30 years ago a paper from the journal *Nature* reported that insects poisoned with conventional insecticides were found to have undergone a profound diuresis, which probably contributed to their death (Maddrell and Casida, 1971). Control of fluid homeostasis has long been recognized as a point of great vulnerability of the Insecta, because for most species all water ingested comes from the diet. Discovery of low molecular weight agonists or antagonists of diuretic or antidiuretic hormones could lead to a new class of safe, insect-specific pesticides. Screening for such molecules will be possible with the implementation of rapid, simple receptor binding assays for important classes of neuropeptides crucial in controlling water balance. Such an assay has been described for the tachykinin receptor by Torfs et al. (2002), who engineered *D. melanogaster* S2 cells to express recombinant aequorin together with the tachykinin receptor, and also for the Drome-K receptor using the identical approach of Radford et al. (2002). Agents binding to this receptor, in particular agonists, trigger an easily measured bioluminescent response suitable for high-throughput screening. Such research is one of the major thrusts of pharmaceutical companies seeking to develop new drugs. Two classes of insecticides with hormonal activity are known: (1) juvenoids, agonists of the sesquiterpenoid juvenile hormones, are useful for controlling many insects where the adult is the pest and (2) ecdysone agonists, such as tefubenozide or MIMIC®, are useful for control of even larval stages of insects. One of the chief liabilities of these chemicals is that they work by altering gene transcription, so they are slow acting. Peptide hormone mimics should work within seconds to minutes, surmounting one of the principal shortcomings of prior hormonal pesticides.

Acknowledgments

D. Schooley thanks the Nevada Ag Experiment Station for financial support. F. Horodyski thanks NSF (IOS-0821930) for financial support. We also would like to thank numerous colleagues (Drs. C. Garside, R. Eigenheer, S. Nicolson, S. Liao, J.-P. Paluzzi, V. Te Brugge, and I. Orchard, as well as Mr. D.A. Jensen, Ms. E. Chidembo and P. Dey) for allowing us to cite unpublished data.

Relevant Web Sites

NetPhos — http://www.cbs.dtu.dk.
TMHMM — http://www.cbs.dtu.dk

References

Abdel-Latief, M., & Hoffmann, K. H. (2007). The adipokinetic hormones in the fall armyworm, Spodoptera frugiperda: cDNA cloning, quantitative real time RT-PCR analysis, and gene specific localization. *Insect Biochem. Molec. Biol*, 37, 999–1014.

Aikins, M. J., Schooley, D. A., Begum, K., Detheux, M., Beeman, R. W., et al. (2008). Vasopressin-like peptide and its receptor function in an indirect diuretic signaling pathway in the red flour beetle. *Insect Biochem. Mol. Biol.*, 38, 740–748.

Ali, I., & Steele, J. E. (1997). Evidence that free fatty acids in trophocytes of Periplaneta americana fat body may be regulated by the activity of phospholipase A$_2$ and cyclooxygenase. *Insect Biochem. Mol. Biol.*, 27, 681–692.

Amare, A., & Sweedler, J. V. (2007). Neuropeptide precursors in Tr. castaneum. *Peptides*, 28, 1282–1291.

Anand, A. N., & Lorenz, M. W. (2010). Age-dependent changes of fat body stores and the regulation of fat body lipid synthesis and mobilisation by adipokinetic hormone in the last larval instar of the cricket, Gryllus bimaculatus. *J. Insect Physiol.*, 54, 1404–1412.

Arrese, E. L., Flowers, M. T., Gazard, J. L., & Wells, M. A. (1999). Calcium and cAMP are second messengers in the adipokinetic hormone-induced lipolysis of triacylglycerols in Manduca sexta fat body. *J. Lipid Res.*, 40, 556–564.

Arrese, E. L., Mirza, S., Rivera, L., Howard, A. D., Chetty, P. S., & Soulages, J. L. (2008). Expression of lipid storage droplet protein-1 may define the role of AKH as a lipid mobilizing hormone in Manduca sexta. *Insect Biochem. Molec. Biol*, 38, 993–1000.

Arrese, E. L., Rojas-Rivas, B. I., & Wells, M. A. (1996). The use of decapitated insects to study lipid mobilization in adult Manduca sexta: effects of adipokinetic hormone and trehalose on fat body lipase activity. *Insect Biochem. Mol. Biol.*, 26, 775–782.

Audsley, N., Coast, G. M., & Schooley, D. A. (1993). The effects of Manduca sexta diuretic hormone on fluid transport by the Malpighian tubules and cryptonephric complex of Manduca sexta. *J. Exp. Biol.*, *178*, 231–243.

Audsley, N., Goldsworthy, G. J., & Coast, G. M. (1997a). Quantification of Locusta diuretic hormone in the central nervous system and corpora cardiaca: influence of age and feeding status, and mechanism of release. *Regulatory Peptides*, *69*, 25–32.

Audsley, N., Goldsworthy, G. J., & Coast, G. M. (1997b). Circulating levels of Locusta diuretic hormone: the effect of feeding. *Peptides*, *18*, 59–65.

Audsley, N., Kay, I., Hayes, T. K., & Coast, G. M. (1995). Cross reactivity studies of CRF-related peptides on insect Malpighian tubules. *Comp. Biochem. Physiol. A*, *110A*, 87–93.

Audsley, N., McIntosh, C., & Phillips, J. E. (1992a). Actions of ion-transport peptide from locust corpus cardiacum on several hindgut transport processes. *J. Exp. Biol.*, *173*, 275–288.

Audsley, N., McIntosh, C., & Phillips, J. E. (1992b). Isolation of a neuropeptide from locust corpus cardiacum which influences ileal transport. *J. Exp. Biol.*, *173*, 261–274.

Audsley, N., McIntosh, C., Phillips, J. E., Schooley, D. A., & Coast, G. M. (1994). Neuropeptide regulation of ion and fluid reabsorption in the insect excretory system. In K. G. Davey, R. E. Peter, & S. S. Tobe (Eds.), *Perspectives in Comparative Endocrinology* (pp. 74–80). Ottawa: National Research Council of Canada.

Audsley, N., & Phillips, J. E. (1990). Stimulants of ileal salt transport in neuroendocrine system of the desert locust. *Gen. Comp. Endocrinol.*, *80*, 127–137.

Auerswald, L., & Gäde, G. (1999). Effects of metabolic neuropeptides from insect corpora cardiaca on proline metabolism on the African fruit beetle, Pachnoda sinuata. *J. Insect Physiol.*, *45*, 535–543.

Auerswald, L., & Gäde, G. (2000). Cyclic AMP mediates the elevation of proline by AKH peptides in the cetoniid beetle, Pachnoda sinuata. *Biochim. Biophys. Acta*, *1495*, 78–89.

Auerswald, L., & Gäde, G. (2001b). The role of calcium in the activation of glycogen phosphorylase in the fat body of the fruit beetle, Pachnoda sinuata, by hypertrehalosemic hormone. *Biochim. Biophys. Acta*, *1499*, 199–208.

Auerswald, L., & Gäde, G. (2002). The role of Ins(1,4,5)P3 in signal transduction of the metabolic neuropeptide Mem-CC in the cetoniid beetle, Pachnoda sinuata. *Insect Biochem. Mol. Biol.*, *32*, 1793–1803.

Auerswald, L., Siegert, K. J., & Gäde, G. (2005). Activation of triacylglycerol lipase in the fat body of a beetle by adipokinetic hormone. *Insect Biochem. Molec. Biol.*, *35*, 461–470.

Auerswald, L., & Gäde, G. (2001a). Hormonal stimulation of proline synthesis in the fat body of the fruit beetle, Pachnoda sinuata, is calcium dependent. *Insect Biochem. Mol. Biol.*, *32*, 23–32.

Baernholdt, D., & Andersen, S. O. (1998). Sequence studies on post-ecdysial cuticular proteins from pupae of the yellow mealworm, Te. molitor. *Insect Biochem. Mol. Biol.*, *28*, 517–526.

Baggerman, G., Huybrechts, J., Clynen, E., Hens, K., Harthoorn, L., et al. (2002). New insights in adipokinetic hormone (AKH) precursor processing in Locusta migratoria obtained by capillary liquid chromatography-tandem mass spectrometry. *Peptides*, *23*, 635–644.

Baines, R. A., Thompson, K. S. J., Rayne, R. C., & Bacon, J. P. (1995). Analysis of the peptide content of the locust vasopressin-like immunoreactive (VPLI) neurons. *Peptides*, *16*, 799–807.

Baldwin, D., Schegg, K. M., Furuya, K., Lehmberg, E., & Schooley, D. A. (2001). Isolation and identification of a diuretic hormone from Zootermopsis nevadensis. *Peptides*, *22*, 147–152.

Baumann, E., & Gersch, M. (1982). Purification and identification of neurohormone D, a heart accelerating peptide from the corpora cardiaca of the cockroach Periplaneta americana. *Insect Biochem.*, *12*, 7–14.

Baumann, E., & Penzlin, H. (1984). Sequence analysis of neurohormone D, a neuropeptide of an insect Periplaneta americana. *Biomed. Biochim. Acta*, *43*, K13–K16.

Baumann, E., & Penzlin, H. (1987). Inactivation of neurohormone D by Malpighian tubules in an insect, Periplaneta americana. *J. Comp. Physiol. B*, *157*, 511–517.

Becker, A., Liewald, J. F., & Wegener, G. (1998). Signal transduction in isolated fat body from the cockroach Blaptica dubia exposed to hypertrehalosemic neuropeptide. *J. Comp. Physiol. B.*, *168*, 159–167.

Becker, A., & Wegener, G. (1998). Hypertrehalosemic neuropeptides decrease levels of the glycolytic signal fructose 2,6-bisphosphate in cockroach fat body. *J. Exp. Biol.*, *201*, 1939–1946.

Beenakkers, A. M. T. (1965). Transport of fatty acids in Locusta migratoria during sustained flight. *J. Insect Physiol.*, *11*, 879–888.

Beenakkers, A. M. T. (1969). The influence of corpus cardiacum on lipid metabolism in Locusta migratoria. *Gen. Comp. Endocrinol.*, *13*, 492.

Begum, K., Li, B., Beeman, R. W., & Park, Y. (2009). Functions of ion transport peptide and ion transport peptide-like in the red flour beetle Tr. castaneum. *Insect Biochem. Mol. Biol.*, *39*, 717–725.

Belmont, M., Cazzamali, G., Williamson, M., Hauser, F., & Grimmelikhuijzen, C. J. P. (2006). Identification of four evolutionarily related G protein-coupled receptors from the malaria mosquito *Anopheles gambiae*. *Biochem. Biophys. Res. Commun.*, *344*, 160–165.

Bender, A. T., & Beavo, J. A. (2006). Cyclic nucleotide phosphodiesterases: molecular regulation to clinical use. *Pharmacol. Rev.*, *58*, 488–520.

Berridge, M. J. (1993). Inositol trisphosphate and calcium signalling. *Nature*, *361*, 315–325.

Berridge, M. J. (1995). Capacitative calcium entry. *Biochem. J.*, *312*, 1–11.

Bertram, G., Schleithoff, L., Zimmermann, P., & Wessing, A. (1991). Bafilomycin A1 is a potent inhibitor of urine formation by Malpighian tubules of *Drosophila* hydei: is a vacuolar ATPase involved in ion and fluid secretion? *J. Insect Physiol.*, *37*, 201–209.

Beyenbach, K. W. (2003). Transport mechanisms of diuresis in Malpighian tubules of insects. *J. Exp. Biol.*, *206*, 3845–3856.

Beyenbach, K. W., & Petzel, D. H. (1987). Diuresis in mosquitoes: role of a natriuretic factor. *News Physiol. Sci., 2*, 171–175.

Beyenbach, K. W. (1993). Extracellular fluid homeostasis in insects? In K. W. Beyenbach (Ed.), *Structure and Function of Primary Messengers in Invertebrates: Insect Diuretic and Antidiuretic Peptides* (vol. 12, pp. 146–173). Basel: Karger.

Beyenbach, K. W. (1995). Mechanism and regulation of electrolyte transport in Malpighian tubules. *J. Insect Physiol., 41*, 197–207.

Beyenbach, K. W. (2003). Transport mechanisms of diuresis in Malpighian tubules of insects. *J. Exp. Biol., 206*, 3845–3856.

Bharucha, K. N., Tarr, P., & Zipursky, S. L. (2008). A glucagon-like endocrine pathway in *Drosophila* modulates both lipid and carbohydrate homeostasis. *J. Exp. Biol., 211*, 3103–3110.

Blackburn, M. B., Kingan, T. G., Bodnar, W., Shabanowitz, J., Hunt, D. F., et al. (1991). Isolation and identification of a new diuretic peptide from the tobacco hornworm, Manduca sexta. *Biochem. Biophys. Res. Commun., 181*, 927–932.

Blackburn, M. B., & Ma, M. C. (1994). Diuretic activity of Mas-DP II, an identified neuropeptide from Manduca sexta: an in vivo and in vitro examination in the adult moth. *Arch. Insect Biochem. Physiol., 27*, 3–10.

Blackburn, M. B., Wagner, R. M., Shabanowitz, J., Kochansky, J. P., Hunt, D. F., et al. (1995). The isolation and identification of three diuretic kinins from the abdominal ventral nerve cord of adult Helicoverpa zea. *J. Insect Physiol., 41*, 723–730.

Blumenthal, E. (2008). Molecular dissection of tyraminergic communication in the *Drosophila* Malpighian tubule. *Comp. Biochem. Physiol. A-Mol. Integr. Physiol., 150*, S138–S138.

Blumenthal, E. M. (2005). Modulation of tyramine signaling by osmolality in an insect secretory epithelium. *Am. J. Physiol. Cell. Physiol., 289*, C1261–C1267.

Blumenthal, E. M. (2009). Isoform- and cell-specific function of tyrosine decarboxylase in the *Drosophila* Malpighian tubule. *J. Exp. Biol., 212*, 3802–3809.

Blumenthal, E. M. (2001). Characterization of transepithelial potential oscillations in the *Drosophila* Malpighian tubule. *J. Exp. Biol., 204*, 3075–3084.

Blumenthal, E. M. (2003). Regulation of chloride permeability by endogenously produced tyramine in the *Drosophila* Malpighian tubule. *Am. J. Physiol. Cell Physiol., 284*, C718–C728.

Bogerd, J., Kooiman, F. P., Pijenburg, M. A. P., Hekking, L. H. P., Oudejans, R. C. H. M., et al. (1995). Molecular cloning of three distinct cDNAs, each encoding a different adipokinetic hormone precursor, of the migratory locust, Locusta migratoria. *J. Biol. Chem., 270*, 23038–23043.

Bradfield, J. Y., & Keeley, L. L. (1989). Adipokinetic hormone gene sequence from Manduca sexta. *J. Biol. Chem., 264*, 12791–12793.

Bradfield, J. Y., Lee, Y.-H., & Keeley, L. L. (1991). Cytochrome P450 family 4 in a cockroach: molecular cloning and regulation by hypertrehalosemic hormone. *Proc. Natl. Acad. Sci. USA, 88*, 4558–4562.

Bursell, A. (1963). Aspects of the metabolism of amino acids in the tsetse fly, Glossina (Diptera). *J. Insect Physiol., 9*, 439–452.

Cabrero, P., Radford, J. C., Broderick, K. E., Costes, L., Veenstra, J. A., et al. (2002). The DH gene of *Drosophila melanogaster* encodes a diuretic peptide that acts through cyclic AMP. *J. Exp. Biol., 205*, 3799–3807.

Cady, C., & Hagedorn, H. H. (1999a). The effect of putative diuretic factors on in vivo urine production in the mosquito, Aedes aegypti. *J. Insect Physiol., 45*, 317–325.

Cady, C., & Hagedorn, H. H. (1999b). Effects of putative diuretic factors on intracellular second messenger levels in the Malpighian tubules of Aedes aegypti. *J. Insect Physiol., 45*, 327–337.

Candy, D. J. (2002). Adipokinetic hormone concentrations in the haemolymph of Schistocerca gregaria measured by radioimmunoassay. *Insect Biochem. Mol. Biol., 32*, 1361–1367.

Carlisle, J. A., & Loughton, B. G. (1979). Adipokinetic hormone inhibits protein synthesis in Locusta. *Nature, 282*, 420–421.

Carlsen, J., Herman, W. S., Christensen, M., & Josefsson, L. (1979). Characterisation of a second peptide with adipokinetic and red pigment-concentrating activity from the locust corpora cardiaca. *Insect Biochem., 9*, 497–501.

Chabre, O., Conklin, B. R., Lin, H. Y., Lodish, H. F., Wilson, E., et al. (1992). A recombinant calcitonin receptor independently stimulates 3',5'-cyclic adenosine monophosphate and Ca^{2+}/inositol phosphate signaling pathways. *Mol. Endocrinol., 6*, 551–556.

Chang, C. L., & Hsu, S. Y. (2004). Ancient evolution of stress-regulating peptides in vertebrates. *Peptides, 25*, 1681–1688.

Cheeseman, P., & Goldsworthy, G. J. (1979). The release of adipokinetic hormone during flight and starvation in Locusta. *Gen. Comp. Endocrinol., 37*, 35–43.

Cheeseman, P., Jutsum, A. R., & Goldsworthy, G. J. (1976). Quantitative studies on the release of locust adipokinetic hormone. *Physiol. Ent., 1*, 115–121.

Chen, A. C., Pannabecker, T. L., & Taylor, D. (1997). Natriuretic and depolarizing effects of a stable fly (Stomoxys calcitrans) factor on Malpighian tubules. *J. Insect Physiol., 43*, 991–998.

Chino, H., & Gilbert, L. I. (1964). Diglyceride release from insect fat body. *Science, 143*, 359–361.

Chino, H., & Gilbert, L. I. (1965). Lipid release and transport in insects. *Biochim. Biophys. Acta, 98*, 94–110.

Christie, A. E., Stevens, J. S., Bowers, M. R., Chapline, M. C., Jensen, D. A., et al. (2010a). Identification of a calcitonin-like diuretic hormone that functions as an intrinsic modulator of the American lobster, Homarus americanus, cardiac neuromuscular system. *J. Exp. Biol., 213*, 118–127.

Christie, A. E., Cashman, C. R., Brennan, H. R., Ma, M., Sousa, G. L., Li, L., Stemmler, E. A., & Dickinson, P. S. (2008). Identification of putative crustacean neuropeptides using in silico analyses of publicly accessible expressed sequence tags. *Gen. Comp. Endocrinol., 156*, 246–264.

Christie, A. E., Durkin, C. S., Hartline, N., Ohno, P., & Lenz, P. H. (2010b). Bioinformatic analyses of the publicly accessible crustacean expressed sequence tags (ESTs) reveal numerous neuropeptide-encoding precursor proteins, including ones from members of several little studied taxa. *Gen. Comp. Endocrinol., 167*, 164–178.

Chung, J. S., Dircksen, H., & Webster, S. G. (1999). A remarkable, precisely timed release of hyperglycemic hormone from endocrine cells in the gut is associated with ecdysis in the crab Carcinus maenas. *Proc. Natl. Acad. Sci. USA, 96*, 13103–13107.

Clark, J. D., Lin, L.-L., Kriz, R. W., Ramesha, C. S., Sultzman, L. A., et al. (1991). A novel arachidonic acid-selective PLA2 contains a Ca^{2+}-dependent translocation domain with homology to PKC and GAP. *Cell, 65*, 1043–1051.

Clark, L., Lange, A. B., Zhang, J. R., & Tobe, S. S. (2008). The roles of Dippu-allatostatin in the modulation of hormone release in Locusta migratoria. *J. Insect Physiol., 54*, 949–958.

Clark, L., Zhang, J. R., Tobe, S., & Lange, A. B. (2006). Proctolin: A possible releasing factor in the corpus cardiacum/corpus allatum of the locust. *Peptides, 27*, 559–566.

Clark, T. M., & Bradley, T. J. (1996). Stimulation of Malpighian tubules from larval Aedes aegypti by secretagogues. *J. Insect Physiol., 42*, 593–602.

Clark, T. M., Hayes, T. K., & Beyenbach, K. W. (1998a). Dose-dependent effects of CRF-like diuretic peptide on transcellular and paracellular transport pathways. *Am. J. Physiol. Renal Physiol., 274*, F834–F840.

Clark, T. M., Hayes, T. K., Holman, G. M., & Beyenbach, K. W. (1998b). The concentration-dependence of CRF-like diuretic peptide: mechanisms of action. *J. Exp. Biol., 201*, 1753–1762.

Clements, A. N. (1992). *The Biology of Mosquitoes.* London: Chapman & Hall.

Clottens, F. L., Holman, G. M., Coast, G. M., Totty, N. F., Hayes, T. K., et al. (1994). Isolation and characterization of a diuretic peptide common to the house fly and stable fly. *Peptides, 15*, 971–979.

Clynen, E., Baggerman, G., Veelaert, D., Cerstiaens, A., Van der Horst, D., et al. (2001). Peptidomics of the pars intercerebralis-corpus cardiacum complex of the migratory locust, Locusta migratoria. *Eur. J. Biochem., 268*, 1929–1939.

Clynen, E., De Loof, A., & Schoofs, L. (2004). New insights into the evolution of the GRF superfamily based on sequence similarity between the locust APRPs and human GRF. *Gen. Comp. Endocrinol., 139*, 173–178.

Clynen, E., Huybrechts, J., De Loof, A., & Schoofs, L. (2003). Mass spectrometric analysis of the perisympathetic organs in locusts: identification of novel periviscerokinins. *Biochem. Biophys. Res. Commun., 300*, 422–428.

Coast, G. M. (2009). Neuroendocrine control of ionic homeostasis in blood-sucking insects. *J Exp. Biol., 212*, 378–386.

Coast, G. M., Garside, C. S., Webster, S. G., Schegg, K. M., & Schooley, D. A. (2005). Mosquito natriuretic peptide identified as a calcitonin-like diuretic hormone in *Anopheles gambiae* (Giles). *J. Exp. Biol., 208*, 3281–3291.

Coast, G. M., Nachman, R. J., & Schooley, D. A. (2007). An antidiuretic peptide (Tenmo-ADFb) with kinin-like diuretic activity on Malpighian tubules of the house cricket, Acheta domesticus (L.). *J. Exp. Biol., 210*, 3979–3989.

Coast, G. M., TeBrugge, V. A., Nachman, R. J., Lopez, J., Aldrich, J. R., et al. (2010). Neurohormones implicated in the control of Malpighian tubule secretion in plant sucking heteropterans: The stink bugs Acrosternum hilare and Nezara viridula. *Peptides, 31*, 468–473.

Coast, G. M. (1995). Synergism between diuretic peptides controlling ion and fluid transport in insect Malpighian tubules. *Regul. Pept., 57*, 283–296.

Coast, G. M. (1996). Neuropeptides implicated in the control of diuresis in insects. *Peptides, 17*, 327–336.

Coast, G. M. (1998). The regulation of primary urine production in insects. In S. G. Webster (Ed.), *Recent Advances in Arthropod Endocrinology* (pp. 189–209). Cambridge: Cambridge University Press.

Coast, G. M. (2001). The neuroendocrine regulation of salt and water balance in insects. *Zoology (Jena), 103*, 179–188.

Coast, G. M., Chung, J.-S., Goldsworthy, G. J., Patel, M., Hayes, T. K., et al. (1994). Corticotropin releasing factor related diuretic peptides in insects. In K. G. Davey, R. E. Peter, & S. S. Tobe (Eds.), *Perspectives in Comparative Endocrinology* (pp. 67–73). Ottawa: National Research Council of Canada.

Coast, G. M., Cusinato, O., Kay, I., & Goldsworthy, G. J. (1991). An evaluation of the role of cyclic AMP as an intracellular 2nd messenger in Malpighian tubules of the house cricket, Acheta domesticus. *J. Insect Physiol., 37*, 563–573.

Coast, G. M., Hayes, T. K., Kay, I., & Chung, J. S. (1992). Effect of Manduca sexta diuretic hormone and related peptides on isolated Malpighian tubules of the house cricket Acheta domesticus (L). *J. Exp. Biol., 162*, 331–338.

Coast, G. M., Holman, G. M., & Nachman, R. J. (1990). The diuretic activity of a series of cephalomyotropic neuropeptides, the achetakinins, on isolated Malpighian tubules of the house cricket, Acheta domesticus. *J. Insect Physiol., 36*, 481–488.

Coast, G. M., & Kay, I. (1994). The effects of Acheta diuretic peptide on isolated Malpighian tubules from the house cricket Acheta domesticus. *J. Exp. Biol., 187*, 225–243.

Coast, G. M., Meredith, J., & Phillips, J. E. (1999). Target organ specificity of major neuropeptide stimulants in locust excretory systems. *J. Exp. Biol., 202*, 3195–3203.

Coast, G. M., Orchard, I., Phillips, J. E., & Schooley, D. A. (2002a). Insect diuretic and antidiuretic hormones. *Adv. Insect Physiol., 29*, 279–409.

Coast, G. M., Rayne, R. C., Hayes, T. K., Mallet, A. I., Thompson, K. S. J., et al. (1993). A comparison of the effects of 2 putative diuretic hormones from Locusta migratoria on isolated locust Malpighian tubules. *J. Exp. Biol., 175*, 1–14.

Coast, G. M., Webster, S. G., Schegg, K. M., Tobe, S. S., & Schooley, D. A. (2001). The *Drosophila* melanogaster homologue of an insect calcitonin-like diuretic peptide stimulates V-ATPase activity in fruit fly Malpighian tubules. *J. Exp. Biol., 204*, 1795–1804.

Coast, G. M., Zabrocki, J., & Nachman, R. J. (2002b). Diuretic and myotropic activities of N-terminal truncated analogs of Musca domestica kinin neuropeptide. *Peptides, 23*, 701–708.

Cohen, A. C., & Patana, R. (1982). Ontogenetic and stress-related changes in hemolymph chemistry of Beet Armyworms. *Comp. Biochem. Physiol., 71A*, 193–198.

Cosme, L. V. (2009). *Diuretic hormones of Tr. castaneum (Herbst) (Coleoptera: Tenebrionidae). MSc. Department of Entomology.* Manhattan, KS: Kansas State University.

Couchié, P. A., & Machin, J. (1984). Allometry of water-vapor absorption in 2 species of tenebrionid beetle larvae. *Am. J. Physiol., 247*, R230–R236.

Cox, K. J. A., Tensen, C. P., Van der Schors, R. C., Li, K. W., Van Heerikhuizen, H., et al. (1997). Cloning, characterization, and expression of a G-protein-coupled receptor from Lymnaea stagnalis and identification of a leucokinin-like peptide, PSFHSWSamide, as its endogenous ligand. *J. Neurosci.*, *17*, 1197–1205.

Cusinato, O., Drake, A. F., Gäde, G., & Goldsworthy, G. J. (1998). The molecular conformations of representative arthropod adipokinetic peptides determined by circular dichroism spectroscopy. *Insect Biochem. Mol. Biol.*, *28*, 43–50.

Cusinato, O., Wheeler, C. H., & Goldsworthy, G. J. (1991). The identity and physiological actions of an adipokinetic hormone in Acheta domesticus. *J. Insect Physiol.*, *37*, 461–469.

Dai, L., Zitnan, D., & Adams, M. E. (2007). Strategic expression of ion transport peptide gene products in central and peripheral neurons of insects. *J. Comp. Neurol.*, *500*, 353–367.

Dames, P., Zimmermann, B., Schmidt, R., Rein, J., Voss, M., et al. (2006). cAMP regulates plasma membrane vacuolar-type H$^+$-ATPase assembly and activity in blowfly salivary glands. *Proc. Natl. Acad. Sci. USA*, *103*, 3926–3931.

Dautzenberg, F. M., & Hauger, R. L. (2002). The CRF peptide family and their receptors: yet more partners discovered. *Trends Pharmacol. Sci.*, *23*, 71–77.

Dautzenberg, F. M., Wille, S., Lohmann, R., & Spiess, J. (1998). Mapping of the ligand-selective domain of the Xenopus laevis corticotropin-releasing factor receptor 1: Implications for the ligand-binding site. *Proc. Natl. Acad. Sci. USA*, *95*, 4941–4946.

Davies, S. A. (2000). Nitric oxide signalling in insects. *Insect Biochem. Mol. Biol.*, *30*, 1123–1138.

Davies, S. A., Huesmann, G. R., Maddrell, S. H. P., O'Donnell, M. J., Skaer, N. J. V., et al. (1995). CAP$_{2b}$, a cardioacceleratory peptide, is present in *Drosophila* and stimulates tubule fluid secretion via cGMP. *Am. J. Physiol. Regul. Integr. Comp. Physiol.*, *269*, R1321–R1326.

Davies, S. A., Stewart, E. J., Huesmann, G. R., Skaer, N. J. V., Maddrell, S. H. P., et al. (1997). Neuropeptide stimulation of the nitric oxide signaling pathway in *Drosophila* melanogaster Malpighian tubules. *Am. J. Physiol. Regul. Integr. Comp. Physiol.*, *273*, R823–R827.

Day, J. P., Wan, S., Allan, A. K., Kean, L., Davies, S. A., et al. (2008). Identification of two partners from the bacterial Kef exchanger family for the apical plasma membrane V-ATPase of Metazoa. *J. Cell Sci.*, *121*, 2612–2619.

De Kort, C. A.D., Bartlink, A. K. M., & Schuurmans, R. R. (1973). The significance of L-proline for oxidative metabolism in the flight muscles of the Colorado beetle, Leptinotarsa decemlineata. *Insect Biochem.*, *3*, 11–17.

Denholm, B., Sudarsan, V., Pasalodos-Sanchez, S., Artero, R., Lawrence, P., et al. (2003). Dual origin of the renal tubules in *Drosophila*: mesodermal cells integrate and polarize to establish secretory function. *Curr. Biol.*, *13*, 1052–1057.

Detloff, M., Wittwer, D., Weise, C., & Wiesner, A. (2001). Lipophorin of lower density is formed during immune responses in the lepidopteran insect Galleria mellonella. *Cell Tissue Res.*, *306*, 449–458.

Diederen, J. H. B., Maas, H. A., Pel, H. J., Schooneveld, H., Jansen, W. F., et al. (1987). Co-localization of the adipokinetic hormones I and II in the same glandular cells and in the same secretory granules of corpus cardiacum of Locusta migratoria and Schistocerca gregaria. *Cell Tissue Res.*, *249*, 379–389.

Diederen, J. H. B., Peppelenbosch, M. P., & Vullings, H. C. B. (1992). Aging adipokinetic cells in Locusta migratoria: an ultrastructural morphometric study. *Cell Tissue Res.*, *268*, 117–121.

Digan, M. E., Roberts, D. N., Enderlin, F. E., Woodworth, A. R., & Kramer, S. J. (1992). Characterization of the precursor for Manduca sexta diuretic hormone Mas-DH. *Proc. Natl. Acad. Sci. USA*, *89*, 11074–11078.

Dircksen, H. (2009). Insect ion transport peptides are derived from alternatively spliced genes and differentially expressed in the central and peripheral nervous system. *J. Exp. Biol.*, *212*, 401–412.

Dircksen, H., & Homberg, U. (1995). Crustacean cardioactive peptide-immunoreactive neurons innervating brain neuropils, retrocerebral complex and stomatogastric nervous system of the locust, Locusta migratoria. *Cell Tissue Res.*, *279*, 495–515.

Dircksen, H., Tesfai, L. K., Albus, C., & Nässel, D. R. (2008). Ion transport peptide splice forms in central and peripheral neurons throughout postembryogenesis of *Drosophila* melanogaster. *J. Comp. Neurol.*, *509*, 23–41.

Donini, A., O'Donnell, M. J., & Orchard, I. (2008). Differential actions of diuretic factors on the Malpighian tubules of Rhodnius prolixus. *Journal of Experimental Biology*, *211*, 42–48.

Dow, J. A., Gupta, B. L., Hall, T. A., & Harvey, W. R. (1984). X-ray microanalysis of elements in frozen-hydrated sections of an electrogenic K$^+$ transport system: the posterior midgut of tobacco hornworm (Manduca sexta) in vivo and in vitro. *J. Membr. Biol.*, *77*, 223–241.

Dow, J. A. T., Davies, S. A., & Sözen, M. A. (1998). Fluid secretion by the *Drosophila* Malpighian tubule. *Am. Zool.*, *38*, 450–460.

Drexler, A. L., Harris, C. C., dela Pena, M. G., Asuncion-Uchi, M., Chung, S., et al. (2007). Molecular characterization and cell-specific expression of an ion transport peptide in the tobacco hornworm, Manduca sexta. *Cell Tissue Res.*, *329*, 391–408.

Egan, J. J., Greenberg, A. S., Chang, M.-K., Wek, S. A., Moos, M. C., Jr., et al. (1992). Mechanisms of hormone-stimulated lipolysis in adipocytes: translocation of hormone-sensitive lipase to the lipid storage droplet. *Proc. Natl. Acad. Sci. USA*, *89*, 8537–8541.

Eigenheer, R. A., Nicolson, S. W., Schegg, K. M., Hull, J. J., & Schooley, D. A. (2002). Identification of a potent antidiuretic factor acting on beetle Malpighian tubules. *Proc. Natl. Acad. Sci. USA*, *99*, 84–89.

Eigenheer, R. A., Wiehart, U. M., Nicolson, S. W., Schoofs, L., Schegg, K. M., et al. (2003). Isolation, identification and localization of a second beetle antidiuretic peptide. *Peptides*, *24*, 27–34.

Endo, H., Nagasawa, H., & Watanabe, T. (2000). Isolation of a cDNA encoding a CHH-family peptide from the silkworm Bombyx mori. *Insect Biochem. Mol. Biol.*, *30*, 355–361.

Farmer, J., Maddrell, S. H. P., & Spring, J. H. (1981). Absorption of fluid by the midgut of Rhodnius. *J. Exp. Biol.*, *94*, 301–316.

Fernlund, P., & Josefsson, L. (1972). Crustacean color-change hormone: amino acid sequence and chemical synthesis. *Science*, *177*, 173–175.

Fischer-Lougheed, J., O'Shea, M., Cornish, I., Losberger, C., Roulet, E., et al. (1993). AKH biosynthesis: transcriptional and translational control of two co-localized prohormones. *J. Exp. Biol.*, *177*, 223–241.

Force, T., Bonventre, J. V., Flannery, M. R., Gorn, A. H., Yamin, M., et al. (1992). A cloned porcine renal calcitonin receptor couples to adenylyl cyclase and phospholipase C. *Am. J. Physiol.*, *262*, F1110–F1115.

Fournier, B. (1991). Neuroparsins stimulate inositol phosphate formation in locust rectal cells. *Comp. Biochem. Physiol. B*, *99*, 57–64.

Fournier, B., Guerineau, N., Mollard, P., & Girardie, J. (1994). Effects of two neuronal antidiuretic molecules, neuroparsin and 5-hydroxytryptamine, on cytosolic free calcium monitored with indo-1 in epithelial and muscular cells of the African locust rectum. *BBA-Mol. Cell. Res.*, *1220*, 181–187.

Fournier, B., Herault, J. P., & Proux, J. (1987). Study of an antidiuretic factor from the nervous lobes of the migratory locust corpora cardiaca. Improvement of an existing bioassay. *Gen. Comp. Endocrinol.*, *68*, 49–56.

Fox, A. M., & Reynolds, S. E. (1991a). The pharmacology of the lipid-mobilising response to adipokinetic hormone family peptides in the moth, Manduca sexta. *J. Insect Physiol.*, *37*, 373–381.

Fox, A. M., & Reynolds, S. E. (1991b). Degradation of adipokinetic hormone family peptides by a circulating endopeptidase in the insect Manduca sexta. *Peptides*, *12*, 937–944.

Friedman, S. (1985). Carbohydrate metabolism. In G. A. Kerkut, & L. I. Gilbert (Eds.), *Comprehensive Insect Physiology, Biochemistry and Pharmacology* (vol. 10, pp. 43–70). Oxford: Pergamon Press.

Furuya, K., Harper, M., Schegg, K. M., & Schooley, D. A. (2000a). Isolation and characterization of CRF-related diuretic hormones from the whitelined sphinx moth Hyles lineata. *Insect Biochem. Mol. Biol.*, *30*, 127–133.

Furuya, K., Milchak, R. J., Schegg, K. M., Zhang, J., Tobe, S. S., et al. (2000b). Cockroach diuretic hormones: characterization of a calcitonin-like peptide in insects. *Proc. Natl. Acad. Sci. USA*, *97*, 6469–6474.

Furuya, K., Schegg, K. M., & Schooley, D. A. (1998). Isolation and identification of a second diuretic hormone from Te. molitor. *Peptides*, *19*, 619–626.

Furuya, K., Schegg, K. M., Wang, H., King, D. S., & Schooley, D. A. (1995). Isolation and identification of a diuretic hormone from the mealworm Te. molitor. *Proc. Natl. Acad. Sci. USA*, *92*, 12323–12327.

Gäde, G., & Simek, P. (2010). A novel member of the adipokinetic peptide family in a "living fossil", the ice crawler Galloisiana yuasai, is the first identified neuropeptide from the order Grylloblattodea. *Peptides*, *31*, 372–376.

Gäde, G. (1979). Studies on the influence of synthetic adipokinetic hormone and some analogs on cyclic AMP levels in different arthropod systems. *Gen. Comp. Endocrinol.*, *37*, 122–130.

Gäde, G. (1980). Further characteristics of adipokinetic and hyperglycaemic factor(s) of stick insects. *J. Insect Physiol.*, *26*, 351–360.

Gäde, G. (1981). Activation of fat body glycogen phosphorylase in Locusta migratoria by corpus cardiacum extract and synthetic adipokinetic hormone. *J. Insect Physiol.*, *27*, 155–162.

Gäde, G. (1990). Structure-function studies on hypertrehalosemic and adipokinetic hormones: activity of naturally occurring analogues and some N-and C-terminal modified analogues. *Physiol. Ent.*, *15*, 299–316.

Gäde, G. (1992). Structure-activity relationships for the carbohydrate-mobilizing action of further bioanalogues of the adipokinetic hormone/red pigment-concentrating hormone family of peptides. *J. Insect Physiol.*, *38*, 259–266.

Gäde, G. (2009). Peptides of the adipokinetic hormone/red pigment-concentrating hormone family. A new take on biodiversity. *Ann. NY Acad. Sci.*, *1163*, 125–136.

Gäde, G., & Auerswald, L. (2002). Beetles' choice – proline for energy output: control by AKHs. *Comp. Biochem. Physiol.*, *132B*, 117–129.

Gäde, G., & Auerswald, L. (2003). Mode of action of neuropeptides from the adipokinetic hormone family. *Gen. Comp. Endocrinol.*, *132*, 10–20.

Gäde, G., Auerswald, L., Simek, P., Marco, H. G., & Kodrík, D. (2003). Red pigment-concentrating hormone is not limited to crustaceans. *Biochem. Biophys. Res. Commun.*, *309*, 967–973.

Gäde, G., & Hayes, T. K. (1995). Structure-activity relationships for Periplaneta americana hypertrehalosemic hormone I: the importance of side chains and termini. *Peptides*, *16*, 1173–1180.

Gäde, G., Hoffmann, K.-H., & Spring, J. H. (1997). Hormonal regulation in insects: facts, gaps, and future directions. *Physiol. Rev.*, *77*, 963–1032.

Gäde, G., & Holwerda, D. A. (1976). Involvement of adenosine 3':5'-cyclic monophosphate in lipid mobilization in Locusta migratoria. *Insect Biochem.*, *6*, 535–540.

Gäde, G., & Janssens, M. P. (1994). Cicadas contain novel members of the AKH/RPCH family peptides with hypertrehalosemic activity. *Biol. Chem. Hoppe-Seyler*, *375*, 803–809.

Gäde, G., Marco, H. G., Simek, P., Audsley, N., Clark, K. D., & Weaver, R. J. (2008). Predicted versus expressed adipokinetic hormone, and other small peptides from the corpus cardiacum-corpus allatum: A case study with beetles and moths. *Peptides*, *29*, 1124–1139.

Gäde, G., Simek, P., Clark, K. D., & Auerswald, L. (2006). Unique translational modification of an invertebrate neuropeptide: a phosphorylated member of the adipokinetic hormone peptide family. *Biochem. J.*, *393*, 705–713.

Gammie, S. C., & Truman, J. W. (1997). Neuropeptide hierarchies and the activation of sequential motor behaviors in the hawkmoth, Manduca sexta. *J. Neurosci.*, *17*, 4389–4397.

Gard, A. L., Lenz, P. H., Shaw, J. R., & Christie, A. E. (2009). Identification of putative peptide paracrines/hormones in the water flea Daphnia pulex (Crustacea; Branchiopoda; Cladocera) using transcriptomics and immunohistochemistry. *Gen. Comp. Endocrinol.*, *160*, 271–287.

Gillespie, J. P., Kanost, M. R., & Trenczek, T. (1997). Biological mediators of insect immunity. *Annu. Rev. Entomol., 42,* 611–643.

Girardie, J., Boureme, D., Couilland, F., Tamarelle, M., & Girardie, A. (1987). Anti-juvenile effect of neuroparsin A, a new protein isolated from the locust corpora cardiaca. *Insect Biochem., 17,* 977–983.

Girardie, J., Girardie, A., Huet, J.-C., & Pernollet, J.-C. (1989). Amino acid sequence of locust neuroparsins. *FEBS Lett., 248,* 4–8.

Girardie, J., Huet, J. C., & Pernollet, J. C. (1990). The locust neuroparsin-A – sequence and similarities with vertebrate and insect polypeptide hormones. *Insect Biochem., 20,* 659–666.

Gokuldas, M., Hunt, P. A., & Candy, D. J. (1988). The inhibition of lipid synthesis in vitro in the locust Schistocerca gregaria by factors from the corpora cardiaca. *Physiol. Ent., 13,* 43–48.

Goldsworthy, G., Chandrakant, S., & Opoku-Ware, K. (2003a). Adipokinetic hormone enhances nodule formation and phenoloxidase activation in adult locusts injected with bacterial lipopolysaccharide. *J. Insect Physiol., 49,* 795–803.

Goldsworthy, G., Mullen, L., Opoku-Ware, K., & Chandrakant, S. (2003b). Interactions between the endocrine and immune systems in locusts. *Physiol. Ent., 28,* 54–61.

Goldsworthy, G., Opoku-Ware, K., & Mullen, L. (2002). Adipokinetic hormone enhances laminarin and bacterial lipopolysaccharide-induced activation of the prophenoloxidase cascade in the African migratory locust, Locusta migratoria. *J. Insect Physiol., 48,* 601–608.

Goldsworthy, G. J. (1969). Hyperglycaemic factors from the corpus cardiacum of Locusta migratoria. *J. Insect Physiol., 15,* 2131–2140.

Goldsworthy, G. J., Johnson, R. A., & Mordue, W. (1972b). vivo studies on the release of hormones from the corpora cardiaca of locusts. *J. Comp. Physiol., 79,* 85–96.

Goldsworthy, G. J., Lee, M. J., Luswata, R., Drake, A. F., & Hyde, D. (1997). Structures, assays and receptors for locust adipokinetic hormones. *Comp. Biochem. Physiol., 117B,* 483–496.

Goldsworthy, G. J., Mallison, K., & Wheeler, C. H. (1986b). The relative potencies of two known locust adipokinetic hormones. *J. Insect Physiol., 32,* 95–101.

Goldsworthy, G. J., Mallison, K., Wheeler, C. H., & Gäde, G. (1986a). Relative adipokinetic activities of members of the adipokinetic hormone/red pigment concentrating hormone family. *J. Insect Physiol., 32,* 433–438.

Goldsworthy, G. J., Mordue, W., & Guthkelch, J. (1972a). Studies on insect adipokinetic hormones. *Gen. Comp. Endocrinol., 18,* 545–551.

Goldsworthy, G. J., & Wheeler, C. H. (1989). Physiological and structural aspects of adipokinetic hormone function in locusts. *Pest. Sci., 25,* 85–95.

Goosey, M. W., & Candy, D. J. (1980). The D-octopamine content of the haemolymph of the locust Schistocerca americana gregaria and its elevation during flight. *Insect Biochem., 10,* 393–397.

Goy, M. F. (1990). Activation of membrane guanylate cyclase by an invertebrate peptide hormone. *J. Biol. Chem., 265,* 20220–20227.

Grieco, M. A. B., & Lopes, A. G. (1997). 5-Hydroxytryptamine regulates the (Na^+-K^+)ATPase activity in Malpighian tubules of Rhodnius prolixus: evidence for involvement of G-protein and cAMP-dependent protein kinase. *Arch. Insect Biochem. Physiol., 36,* 203–214.

Grinstein, S., & Wieczorek, H. (1994). Cation antiports of animal plasma membranes. *J. Exp. Biol., 196,* 307–318.

Grönke, S., Müller, G., Hirsch, J., Fellert, S., Andreou, A., Haase, T., Jäckle, H., & Kühnlein, R. P. (2007). Dual lipolytic control of body fat storage and mobilization in *Drosophila. PLoS Biol., 5,* 1248–1256.

Hansen, K. K., Hauser, F., Cazzamali, G., Williamson, M., & Grimmelikhuijzen, C. J. P. (2006). Cloning and characterization of the adipokinetic hormone receptor from the cockroach Periplaneta americana. *Biochem. Biophys. Res. Commun., 343,* 638–643.

Hansen, K. K., Stafflinger, E., Schneider, M., Hauser, F., Cazzamali, G., Williamson, M., Kollmann, M., Schächtner, J., & Grimmelikhuijzen, C. J. P. (2010). Discovery of a novel insect neuropeptide signaling system closely related to the insect adipokinetic hormone and corazonin hormonal systems. *J. Biol. Chem., 285,* 10736–10747.

Hansen, L. L., Ramlov, H., & Westh, P. (2004). Metabolic activity and water vapour absorption in the mealworm Te. molitor L. (Coleoptera, Tenebrionidae): real-time measurements by two channel microcalorimetry. *J. Exp. Biol., 207,* 545–552.

Harthoorn, L. F., Diederen, J. H. B., Oudejans, R. C. H. M., & Van der Horst, D. J. (1999). Differential location of peptide hormones in the secretory pathway of insect adipokinetic cells. *Cell Tissue Res., 298,* 361–369.

Harthoorn, L. F., Oudejans, R. C. H. M., Diederen, J. H. B., & Van der Horst, D. J. (2002). Coherence between biosynthesis and secretion of insect adipokinetic hormones. *Peptides, 23,* 629–634.

Harthoorn, L. F., Oudejans, R. C. H. M., Diederen, J. H. B., Van de Wijngaart, D. J., & Van der Horst, D. J. (2001). Absence of coupling between release and synthesis of peptide hormones in insect neurosecretory cells. *Eur. J. Cell Biol., 80,* 451–457.

Hatle, J. D., & Spring, J. H. (1999). Tests of potential adipokinetic hormone precursor related peptide (APRP) functions: Lack of responses. *Arch. Insect Biochem. Physiol., 42,* 163–166.

Hauger, R. L., Grigoriadis, D. E., Dallman, M. F., Plotsky, P. M., Vale, W. W., et al. (2003). International Union of Pharmacology. XXXVI. Current status of the nomenclature for receptors for corticotropin-releasing factor and their ligands. *Pharmacol. Rev., 55,* 21–26.

Hauser, F., Cazzamali, G., Williamson, M., Park, Y., Li, B., et al. (2008). A genome-wide inventory of neurohormone GPCRs in the red flour beetle Tr. castaneum. Front. *Neuroendocrinol., 29,* 142–165.

Hauser, F., Søndergaard, L., & Grimmelikhuijzen, C. J. P. (1998). Molecular cloning, genomic organization and developmental regulation of a novel receptor from *Drosophila melanogaster* structurally related to gonadotropin-releasing hormone receptors from vertebrates. *Biochem. Biophys. Res. Commun., 249,* 822–828.

Hayes, T. K., Holman, G. M., Pannabecker, T. L., Wright, M. S., Strey, A. A., et al. (1994). Culekinin depolarizing peptide: A mosquito leucokinin-like peptide that influences insect Malpighian tubule ion transport. Regul. *Pept., 52,* 235–248.

Hayes, T. K., & Keeley, L. L. (1990). Structure-activity relationships on hyperglycemia by representatives of the adipokinetic/hyperglycemic hormone family in Blaberus cockroaches. *J. Comp. Physiol.*, *160B*, 187–194.

Hayes, T. K., Pannabecker, T. L., Hinckley, D. J., Holman, G. M., Nachman, R. J., et al. (1989). Leucokinins, a new family of ion transport stimulators and inhibitors in insect Malpighian tubules. *Life Sci.*, *44*, 1259–1266.

Hector, C. E., Bretz, C. A., Zhao, Y., & Johnson, E. C. (2009). Functional differences between two CRF-related diuretic hormone receptors in *Drosophila*. *J. Exp. Biol.*, *212*, 3142–3147.

Hegarty, J. L., Zhang, B., Pannabecker, T. L., Petzel, D. H., Baustian, M. D., et al. (1991). Dibutyryl cAMP activates bumetanide-sensitive electrolyte transport in Malpighian tubules. *Am. J. Physiol.*, *261*, C521–C529.

Hekimi, S., Fischer-Lougheed, J., & O'Shea, M. (1991). Regulation of neuropeptide stoichiometry in neurosecretory cells. *J. Neurosci.*, *11*, 3246–3256.

Hekimi, S., & O'Shea, M. (1987). Identification and purification of two precursors of the insect neuropeptide adipokinetic hormone. *J. Neurosci.*, *7*, 2773–2784.

Herault, J. P. E., Girardie, J., & Proux, J. P. (1985). Separation and characteristics of antidiuretic factors from the glandular lobes of the migratory locust corpora cardiaca. *Int. J. Invert. Reprod. Devel.*, *8*, 325–335.

Herault, J. P. E., Girardie, J., & Proux, J. P. (1988). Further characterization of the antidiuretic factor from the glandular part of the corpora cardiaca of the migratory locust. *Int. J. Invert. Reprod. Devel.*, *13*, 183–192.

Herault, J. P. E., & Proux, J. P. (1987). Cyclic AMP, the second messenger of an antidiuretic hormone from glandular lobes of migratory locust CC. *J. Insect Physiol.*, *33*, 487–492.

Herman, W. S., Carlsen, J. B., Christensen, M., & Josefsson, L. (1977). Evidence for an adipokinetic function of the RPCH activity present in the desert locust neuroendocrine system. *Biol. Bull.*, *153*, 527–539.

Hewes, R. S., & Taghert, P. H. (2001). Neuropeptides and neuropeptide receptors in the *Drosophila* melanogaster genome. *Genome Res.*, *11*, 1126–1142.

Hietter, H., Van Dorsselaer, A., & Luu, B. (1991). Characterization of 3 structurally-related 8–9 kDa monomeric peptides present in the corpora cardiaca of Locusta – a revised structure for the neuroparsins. *Insect Biochem.*, *21*, 259–264.

Holman, G. M., Cook, B. J., & Nachman, R. J. (1986a). Isolation, primary structure and synthesis of two neuropeptides from Leucophaea maderae: members of a new family of cephalomyotropins. *Comp. Biochem. Physiol.*, *84C*, 205–211.

Holman, G. M., Cook, B. J., & Nachman, R. J. (1986b). Primary structure and synthesis of two additional neuropeptides from Leucophaea maderae: members of a new family of cephalomyotropins. *Comp. Biochem. Physiol.*, *84C*, 271–276.

Holman, G. M., Cook, B. J., & Nachman, R. J. (1987a). Isolation, primary structure and synthesis of leucokinins V and VI: myotropic peptides of Leucophaea maderae. *Comp. Biochem. Physiol.*, *88C*, 27–30.

Holman, G. M., Cook, B. J., & Nachman, R. J. (1987b). Isolation, primary structure and synthesis of Leucokinins VII and VIII: the final members of this new family of cephalomyotropic peptides isolated from head extracts of Leucophaea maderae. *Comp. Biochem. Physiol.*, *88C*, 31–34.

Holman, G. M., Nachman, R. J., & Coast, G. M. (1999). Isolation, characterization and biological activity of a diuretic myokinin neuropeptide from the housefly, Musca domestica. *Peptides*, *20*, 1–10.

Holman, G. M., Nachman, R. J., Schoofs, L., Hayes, T. K., Wright, M. S., et al. (1991). The Leucophaea maderae hindgut preparation – a rapid and sensitive bioassay tool for the isolation of insect myotropins of other insect species. *Insect Biochem.*, *21*, 107–112.

Holmes, S. P., Barhoumi, R., Nachman, R. J., & Pietrantonio, P. V. (2003). Functional analysis of a G protein-coupled receptor from the southern cattle tick Boophilus microplus (Acari: Ixodidae) identifies it as the first arthropod myokinin receptor. *Insect Mol. Biol.*, *12*, 27–38.

Holt, R. A., Subramanian, G. M., Hacpern, A., Sutton, G. G., Charlab, R., et al. (2002). The genome sequence of the malaria mosquito *Anopheles gambiae*. *Science*, *298*, 129–149.

Holtzhausen, W. D., & Nicolson, S. W. (2007). Beetle diuretic peptides: The response of mealworm (Te. molitor) Malpighian tubules to synthetic peptides, and cross-reactivity studies with a dung beetle (Onthophagus gazella). *J. Insect Physiol.*, *53*, 361–369.

Huesmann, G. R., Cheung, C. C., Loi, P. K., Lee, T. D., Swiderek, K. M., et al. (1995). Amino acid sequence of CAP_{2b}, an insect cardioacceleratory peptide from the tobacco hawkmoth Manduca sexta. *FEBS Lett.*, *371*, 311–314.

Huybrechts, J., Clynen, E., Baggerman, G., De Loof, A., & Schoofs, L. (2002). Isolation and identification of the AKH III precursor-related peptide from Locusta migratoria. *Biochem. Biophys. Res. Commun.*, *296*, 1112–1117.

Iaboni, A., Holman, G. M., Nachman, R. J., Orchard, I., & Coast, G. M. (1998). Immunocytochemical localisation and biological activity of diuretic peptides in the housefly, Musca domestica. *Cell Tissue Res.*, *294*, 549–560.

Ianowski, J. P., & O'Donnell, M. J. (2006). Electrochemical gradients for Na$^+$, K$^+$, $^-$ and H$^+$ across the apical membrane in Malpighian (renal) tubule cells of Rhodnius prolixus. *J. Exp. Biol.*, *209*, 1964–1975.

Ianowski, J. P., Christensen, R. J., & O'Donnell, M. J. (2004). Na$^+$ competes with K$^+$ in bumetanide-sensitive transport by Malpighian tubules of Rhodnius prolixus. *J. Exp. Biol.*, *207*, 3707–3716.

Ianowski, J. P., Paluzzi, J.-P., Te Brugge, V. A., & Orchard, I. (2010). The antidiuretic neurohormone RhoprCAPA-2 downregulates fluid transport across the anterior midgut in the blood-feeding insect Rhodnius prolixus. *Am. J. Physiol. Regul. Integr. Comp. Physiol.*, *298*, R548–R557.

Ianowski, J. P., Christensen, R. J., & O'Donnell, M. J. (2002). Intracellular ion activities in Malpighian tubule cells of Rhodnius prolixus: evaluation of Na$^+$–K$^+$–2Cl$^-$cotransport across the basolateral membrane. *J. Exp. Biol.*, *205*, 1645–1655.

Ianowski, J. P., & O'Donnell, M. J. (2001). Transepithelial potential in Malpighian tubules of Rhodnius prolixus: lumen-negative voltages and the triphasic response to serotonin. *J. Insect Physiol.*, *47*, 411–421.

Isaac, R. E. (2010). Neuropeptide-degrading endopeptidase activity of locust (Schistocerca gregaria) synaptic membranes. *Biochem. J.*, *255*, 843–847.

Isabel, G., Martin, J.-R., Chidami, S., Veenstra, J. A., & Rosay, P. (2005). AKH-producing neuroendocrine cell ablation decreases trehalose and induces behavioral changes in *Drosophila*. *Am. J. Physiol. Regul. Integr. Comp. Physiol., 288*, R531–R538.

Iversen, A., Cazzamali, G., Williamson, M., Hauser, F., & Grimmelikhuijzen, C. J. P. (2002). Molecular cloning and functional expression of a *Drosophila* receptor for the neuropeptides capa-1 and -2. *Biochem. Biophys. Res. Commun., 299*, 628–633.

Jagge, C. L., & Pietrantonio, P. V. (2008). Diuretic hormone 44 receptor in Malpighian tubules of the mosquito Aedes aegypti: evidence for transcriptional regulation paralleling urination. *Insect Mol. Biol., 17*, 413–426.

Jeffs, L. B., & Phillips, J. E. (1996). Pharmacological study of the second messengers that control rectal ion and fluid transport in the desert locust (Schistocerca gregaria). *Arch. Insect Biochem. Physiol., 31*, 169–184.

Johard, H. A., Yoishii, T., Dircksen, H., Cusumano, P., Rouyer, F., et al. (2009). Peptidergic clock neurons in *Drosophila*: ion transport peptide and short neuropeptide F in subsets of dorsal and ventral lateral neurons. *J. Comp. Neurol., 516*, 59–73.

Johard, H. A., Coast, G. M., Mordue, W., & Nassel, D. R. (2003). Diuretic action of the peptide locustatachykinin I: cellular localisation and effects on fluid secretion in Malpighian tubules of locusts. *Peptides, 24*, 1571–1579.

Johnson, E. C., Shafer, O. T., Trigg, J. S., Schooley, D. A., Dow, J. A., et al. (2005). A novel diuretic hormone receptor in *Drosophila*: Evidence for conservation of CGRP signaling. *J. Exp. Biol., 208*, 1239–1246.

Johnson, E. C., Bohn, L. M., & Taghert, P. H. (2004). *Drosophila* CG8422 encodes a functional diuretic hormone receptor. *J. Exp. Biol., 207*, 743–748.

Jutsum, A. R., & Goldsworthy, G. J. (1976). Fuels for flight in Locusta. *J. Insect Physiol., 22*, 243–249.

Kataoka, H., Toschi, A., Li, J. P., Carney, R. L., Schooley, D. A., et al. (1989a). Identification of an allatotropin from adult Manduca sexta. *Science, 243*, 1481–1483.

Kataoka, H., Troetschler, R. G., Li, J. P., Kramer, S. J., Carney, R. L., et al. (1989b). Isolation and identification of a diuretic hormone from the tobacco hornworm, Manduca sexta. *Proc. Natl. Acad. Sci. USA, 86*, 2976–2980.

Kaufmann, C., & Brown, M. R. (2006). Adipokinetic hormones in the African malaria mosquito, *Anopheles gambiae*: Identification and expression of genes for two peptides and a putative receptor. *Insect Biochem. Molec. Biol., 36*, 466–481.

Kaufmann, C., & Brown, M. R. (2008). Regulation of carbohydrate metabolism and flight performance by a hypertrehalosaemic hormone in the mosquito *Anopheles gambiae*. *J. Insect Physiol., 54*, 367–377.

Kaufmann, C., Merzendorfer, H., & Gäde, G. (2009). The adipokinetic hormone system in Culicinae (Diptera: Culicidae): Molecular identification and characterization of two adipokinetic hormone (AKH) precursors from Aedes aegypti and Culex pipiens and two putative AKH receptor variants from A. aegypti. *Insect Biochem. Mol. Biol., 39*, 770–781.

Kawakami, A., Iwami, M., Nagasawa, H., Suzuki, A., & Ishizaki, H. (1989). Structure and organization of 4 clustered genes that encode bombyxin, an insulin–related brain secretory peptide of the silkmoth Bombyx mori. *Proc. Natl. Acad. Sci. USA, 86*, 6843–6847.

Kay, I., Coast, G. M., Cusinato, O., Wheeler, C. H., Totty, N. F., et al. (1991a). Isolation and characterization of a diuretic peptide from Acheta domesticus – evidence for a family of insect diuretic peptides. *Biol. Chem. Hoppe-Seyler, 372*, 505–512.

Kay, I., Wheeler, C. H., Coast, G. M., Totty, N. F., Cusinato, O., et al. (1991b). Characterization of a diuretic peptide from Locusta migratoria. *Biol. Chem. Hoppe-Seyler, 372*, 929–934.

Kay, I., Patel, M., Coast, G. M., Totty, N. F., Mallet, A. T., & Gdds worthy G.J., (1992). Isolation, characterization and biological activity of a CRF-related diuretic peptide from *Peaplaneta americana. L. Regul. Pept., 42*, 111–122.

Kean, L., Cazenave, W., Costes, L., Broderick, K. E., Graham, S., et al. (2002). Two nitridergic peptides are encoded by the gene capability in *Drosophila* melanogaster. *Am. J. Physiol. Regul. Integr. Comp. Physiol., 282*, R1297–R1307.

Keeley, L. L., Chung, J. S., & Hayes, T. K. (1992). Diuretic and antifeedant actions by Manduca sexta diuretic hormone in lepidopteran larvae. *Experientia., 48*, 1145–1148.

Keeley, L. L., Hayes, T. K., Bradfield, J. Y., & Sowa, S. M. (1991). Physiological actions by hypertrehalosemic hormone and adipokinetic peptides in adult Blaberus discoidalis cockroaches. *Insect Biochem., 21*, 121–129.

Keeley, L. L., & Hesson, A. S. (1995). Calcium-dependent signal transduction by the hypertrehalosemic hormone in the cockroach fat body. *Gen. Comp. Endocrinol., 99*, 373–381.

Kehne, J., & De Lombaert, S. (2002). Non-peptidic CRF1 receptor antagonists for the treatment of anxiety, depression and stress disorders. *Curr. Drug Target CNS Neurol. Disord, 1*, 467–493.

Kim, S. H., Ryu, H., Kang, C. W., Kim, S. Z., Seul, K. H., et al. (1994). Atrial natriuretic peptide immunoreactivity in the eggs of the silkworm Bombyx mori. *Gen. Comp. Endocrinol., 94*, 151–156.

Kim, S. K., & Rulifson, E. J. (2004). Conserved mechanisms of glucose sensing and regulation by *Drosophila* corpora cardiaca cells. *Nature, 431*, 316–320.

King, D. S., Meredith, J., Wang, Y. J., & Phillips, J. E. (1999). Biological actions of synthetic locust ion transport peptide (ITP). *Insect Biochem. Mol. Biol., 29*, 11–18.

Kirschenbaum, S. R., & O'Shea, M. (1993). Postembryonic proliferation of neuroendocrine cells expressing adipokinetic hormone peptides in the corpora cardiaca of the locust. *Development, 118*, 1181–1190.

Kodrík, D., & Goldsworthy, G. J. (1995). Inhibition of RNA synthesis by adipokinetic hormones and brain factor(s) in adult fat body of Locusta migratoria. *J. Insect Physiol., 41*, 127–133.

Kodrík, D., Socha, R., Simek, P., Zemek, R., & Goldsworthy, G. J. (2000). A new member of the AKH/RPCH family that stimulates locomotory activity in the firebug, Pyrrhocoris apterus (Heteroptera). *Insect Biochem. Mol. Biol., 30*, 489–498.

Köllisch, G. V., Lorenz, M. W., Kellner, R., Verhaert, P. D., & Hoffmann, K. H. (2000). Structure elucidation and biological activity of an unusual adipokinetic hormone from corpora cardiaca of the butterfly, Vanessa cardui. *Eur. J. Biochem.*, *267*, 5502–5508.

Köllisch, G. V., Verhaert, P., & Hoffmann, K. H. (2003). Vanessa cardui adipokinetic hormone (Vanca-AKH) in butterflies and a moth. *Comp. Biochem. Physiol.*, *135A*, 303–308.

Konings, P. N. M., Vullings, H. G. B., Kok, O. J. M., Diederen, J. H. B., & Jansen, W. F. (1989). The innervation of the corpus cardiacum of Locusta migratoria: A neuroanatomical study with the use of Lucifer yellow. *Cell Tissue Res.*, *258*, 301–308.

Laenen, B. (1999). *Purification, Characterization and Mode of Action of Endogenous Neuroendocrine Factors in the Forest Ant, Formica polyctena*. Ph.D. thesis, Limburgs Universitair Centrum.

Lavigne, C., Embleton, J., Audy, P., King, R. R., & Pelletier, Y. (2001). Partial purification of a novel insect antidiuretic factor from the Colorado potato beetle, Leptinotarsa decemlineata (Say) (Coleoptera: Chrysomelidae), which acts on Malpighian tubules. *Insect Biochem. Mol. Biol.*, *31*, 339–347.

Lavine, M. D., & Strand, M. R. (2002). Insect hemocytes and their role in immunity. *Insect Biochem. Mol. Biol.*, *32*, 1295–1309.

Lee, G., & Park, P. (2004). Hemolymph sugar homeostasis and starvation-induced hyperactivity affected by genetic manipulations of the adipokinetic hormone-encoding gene in *Drosophila* melanogaster. *Genetics*, *167*, 311–332.

Lee, K. Y., Horodyski, F. M., & Chamberlin, M. E. (1998). Inhibition of midgut ion transport by allatotropin (Mas-AT) and Manduca FLRFamides in the tobacco hornworm Manduca sexta. *J. Exp. Biol.*, *201*, 3067–3074.

Lee, M. J., Cusinato, O., Luswata, R., Wheeler, C. H., & Goldsworthy, G. J. (1997). N-terminal modifications to AKH-I from Locusta migratoria: assessment of biological potencies in vivo and in vitro. *Reg. Peptides*, *69*, 69–76.

Lee, M. J., & Goldsworthy, G. J. (1995). Acetate uptake test: the basis of a rapid method for determining potencies of adipokinetics peptides for structure-activity studies. *J. Insect Physiol.*, *41*, 113–170.

Lee, M. J., Drake, A. F., & Goldsworthy, G. J. (1996). Locusta-AKH-III and related peptides containing two tryptophan residues have unusual CD spectra. *Biochem. Biophys. Res. Commun.*, *226*, 407–412.

Lee, S. G., Bader, B. D., Chang, E. S., & Mykles, D. L. (2007). Effects of elevated ecdysteroid on tissue expression of three guanylyl cyclases in the tropical land crab Gecarcinus lateralis: possible roles of neuropeptide signaling in the molting gland. *J. Exp. Biol.*, *210*, 3245–3254.

Lee, Y.-H., & Keeley, L. L. (1994). Intracellular transduction of trehalose synthesis by hypertrehalosemic hormone in the fat body of the tropical cockroach, Blaberus discoidalis. *Insect Biochem. Mol. Biol.*, *24*, 473–480.

Lehmberg, E., Ota, R. B., Furuya, K., King, D. S., Applebaum, S. W., et al. (1991). Identification of a diuretic hormone of Locusta migratoria. *Biochem. Biophys. Res. Commun.*, *179*, 1036–1041.

Lewis, D. K., Bradfield, J. Y., & Keeley, L. L. (1998). Feeding effects on gene expression of the hypertrehalosemic hormone in the cockroach, Blaberus discoidalis. *J. Insect Physiol.*, *44*, 967–972.

Lewis, D. K., Jezierski, M. K., Keeley, L. L., & Bradfield, J. Y. (1997). Hypertrehalosemic hormone in a cockroach: molecular cloning and expression. *Mol. Cell. Endocrinol.*, *130*, 101–108.

Li, B., Predel, R., Neupert, S., Hauser, F., Tanaka, Y., et al. (2008). Genomics, transcriptomics, and peptidomics of neuropeptides and protein hormones in the red flour beetle Tr. castaneum. *Genome Res.*, *18*, 113–122.

Li, H., Wang, H., Schegg, K. M., & Schooley, D. A. (1997). Metabolism of an insect diuretic hormone by Malpighian tubules studied by liquid chromatography coupled with electrospray ionization mass spectrometry. *Proc. Natl. Acad. Sci. USA*, *94*, 13463–13468.

Liao, S., Audsley, N., & Schooley, D. A. (2000). Antidiuretic effects of a factor in brain/corpora cardiaca/corpora allata extract on fluid reabsorption across the cryptonephric complex of Manduca sexta. *J. Exp. Biol.*, *203*, 605–615.

Linck, B., Klein, J. M., Mangerich, S., Keller, R., & Weidemann, W. M. (1993). Molecular cloning of crustacean red pigment concentrating hormone precursor. *Biochem. Biophys. Res. Commun.*, *195*, 807–813.

Lindemans, M., Liu, F., Janssen, T., Husson, S. J., Mertens, I., Gäde, G., & Schoofs, L. (2009). Adipokinetic hormone signaling through the gonadotropin-releasing hormone receptor modulates egg-laying in Caenorhabditis elegans. *Proc. Nat. Acad. Sci. USA*, *106*, 1642–1647.

Lorenz, M. W. (2003). Adipokinetic hormone inhibits the formation of energy stores and egg production in the cricket Gryllus bimaculatus. *Comp. Biochem. Physiol.*, *136B*, 197–206.

Lorenz, M. W. (2001). Synthesis of lipids in the fat body of Gryllus bimaculatus: age dependency and regulation by adipokinetic hormone. *Arch. Insect Biochem. Physiol.*, *47*, 148–214.

Lorenz, M. W., Kellner, R., Völkl, W., Hoffmann, K. H., & Woodring, J. (2001). A comparative study on hypertrehalosemic hormones in the Hymenoptera: sequence determination, physiological actions and biological significance. *J. Insect Physiol.*, *47*, 563–571.

Lovejoy, D. A. (2009). Structural evolution of urotensin-I: reflections of life before corticotropin releasing factor. *Gen. Comp. Endocrinol.*, *164*, 15–19.

Lu, H. -L., Kersch, C., & Pietrantonio, P. V. (2010). The kinin receptor is expressed in the Malpighian tubule stellate cells in the mosquito Aedes aegypti (L.): A new model needed to explain ion transport? *Insect Biochem. Mol. Biol*, doi: 10.1016/j.ibmb.2010.10.003.

Lu, K.-H., Bradfield, J. Y., & Keeley, L. L. (1995). Hypertrehalosemic hormone-regulated gene expression of cytochrome P$_{450}$4C1 in the fat body of the cockroach, Blaberus discoidalis. *Arch. Insect Biochem. Physiol.*, *28*, 79–90.

Lu, K.-H., Bradfield, J. Y., & Keeley, L. L. (1996). Age and starvation effects on hypertrehalosemic hormone-dependent gene expression of cytochrome P4504C1 in the cockroach, Blaberus discoidalis. *J. Insect Physiol.*, *42*, 925–930.

Lum, P. Y., & Chino, H. (1990). Primary role of adipokinetic hormone in the formation of low density lipophorin in insects. *J. Lipid. Res.*, *31*, 2039–2044.

Ma, M., Emery, S. B., Wong, W. K. R., & De Loof, A. (2000). Effects of Manduca diuresin on neonates of the tobacco hornworm, Manduca sexta. *Gen. Comp. Endocrinol.*, *118*, 1–7.

MacPherson, M. R., Broderick, K. E., Graham, S., Day, J. P., Houslay, M. D., et al. (2004). The dg2 (for) gene confers a renal phenotype in *Drosophila* by modulation of cGMP-specific phosphodiesterase. *J. Exp. Biol.*, *207*, 2769–2776.

MacPherson, M. R., Pollock, V. P., Broderick, K. E., Kean, L., O'Connell, F. C., et al. (2001). Model organisms: new insights into ion channel and transporter function. L-type calcium channels regulate epithelial fluid transport in *Drosophila* melanogaster. *Am. J. Physiol. Cell. Physiol.*, *280*, C394–C407.

Maddrell, S. H., & Casida, J. E. (1971). Mechanism of insecticide-induced diuresis in Rhodnius. *Nature*, *231*, 55–56.

Maddrell, S. H., Pilcher, D. E., & Gardiner, B. O. (1969). Stimulatory effect of 5-hydroxytryptamine (serotonin) on secretion by Malpighian tubules of insects. *Nature*, *222*, 784–785.

Maddrell, S. H. P. (1963). Excretion in the blood-sucking bug, Rhodnius prolixus Stål. I. The control of diuresis. *J. Exp. Biol.*, *40*, 247–256.

Maddrell, S. H. P. (1978). Transport in insect excretory epithelia. In H. H. Ussing (Ed.), *Membrane Transport in Biology* (pp. 239–271). Heidelberg: Springer-Verlag.

Maddrell, S. H. P., Herman, W. S., Farndale, R. W., & Riegel, J. A. (1993a). Synergism of hormones controlling epithelial fluid transport in an insect. *J. Exp. Biol.*, *174*, 65–80.

Maddrell, S. H. P., & O'Donnell, M. J. (1992). Insect Malpighian tubules: V-ATPase action in ion and fluid transport. *J. Exp. Biol.*, *172*, 417–429.

Maddrell, S. H. P., O'Donnell, M. J., & Caffrey, R. (1993b). The regulation of haemolymph potassium activity during initiation and maintenance of diuresis in fed Rhodnius prolixus. *J. Exp. Biol.*, *177*, 273–285.

Martínez-Pérez, F., Becerra, A., Valdés, J., Zinker, S., & Aréchiga, H. (2002). A possible molecular ancestor for mollusk APGWamide, insect adipokinetic hormone, and crustacean red pigment concentrating hormone. *J. Mol. Evol.*, *54*, 703–714.

Massaro, R. C., Lee, L. W., Patel, A. B., Wu, D. S., Yu, M.-J., et al. (2004). The mechanism of action of the antidiuretic pepide Tenmo ADFa in Malpighian tubules of Aedes aegypti. *J. Exp. Biol.*, *207*, 2877–2888.

Mathelin, J., Quennedey, B., Bouhin, H., & Delachambre, J. (1998). Characterization of two new cuticular genes specifically expressed during the post-ecdysial molting period in Te. molitor. *Gene*, *211*, 351–359.

Mayer, R. J., & Candy, D. J. (1967). Changes in haemolymph lipoproteins during locust flight. *Nature*, *215*, 987.

Mayer, R. J., & Candy, D. J. (1969a). Control of haemolymph lipid concentration during locust flight: an adipokinetic hormone from the corpora cardiaca. *J. Insect Physiol.*, *15*, 611–620.

Mayer, R. J., & Candy, D. J. (1969b). Changes in energy reserve during flight of the desert locust, Schistocerca gregaria. *Comp. Biochem. Physiol.*, *31*, 409–418.

McClure, J. B., & Steele, J. E. (1981). The role of extracellular calcium in hormonal activation of glycogen phosphorylase in cockroach fat body. *Insect Biochem.*, *11*, 605–613.

Meredith, J., Ring, M., Macins, A., Marschall, J., Cheng, N. N., et al. (1996). Locust ion transport peptide (ITP): primary structure, cDNA and expression in a baculovirus system. *J. Exp. Biol.*, *199*, 1053–1061.

Milde, J. J., Ziegler, R., & Wallstein, M. (1995). Adipokinetic hormone stimulates neurones in the insect central nervous system. *J. Exp. Biol.*, *198*, 1307–1311.

Moffett, D. F. (1994). Recycling of K^+, acid-base equivalents, and fluid between gut and hemolymph in lepidopteran larvae. *Physiol. Zool.*, *67*, 68–81.

Montecucchi, P. C., & Henschen, A. (1981). Amino acid composition and sequence analysis of sauvagine, a new active peptide from the skin of Phyllomedusa sauvagei. *Int. J. Pept. Protein Res.*, *18*, 113–120.

Montuenga, L. M., Zudaire, E., Prado, M. A., Audsley, N., Burrell, M. A., et al. (1996). Presence of Locusta diuretic hormone in endocrine cells of the ampullae of locust Malpighian tubules. *Cell Tissue Res.*, *285*, 331–339.

Mordue, W. (1969). Hormonal control of Malpighian tubule and rectal function in the desert locust, Schistocerca gregaria. *J. Insect Physiol.*, *15*, 273–285.

Mordue, W., & Morgan, P. J. (1985). Chemistry of Peptide Hormones. In G. A. Kerkut, & L. I. Gilbert (Eds.), *Comprehensive Insect Physiology, Biochemistry and Pharmacology* (pp. 153–183). Oxford: Pergamon Press.

Mordue, W., & Stone, J. V. (1977). Relative potencies of locust adipokinetic hormone and prawn red-pigment concentrating hormone in insect and crustacean systems. *Gen. Comp. Endocrinol.*, *33*, 103–108.

Moreau, R., Gourdeux, L., & Girardie, J. (1988). Neuroparsin: a new energetic neurohormone in the African locust. *Arch. Insect Biochem. Physiol.*, *8*, 135–145.

Morgan, P. J., & Mordue, W. (1984). 5-Hydroxytryptamine stimulates fluid secretion in locust Malpighian tubules independently of cAMP. *Comp. Biochem. Physiol.*, *79C*, 305–310.

Morley, S. D., Schonrock, C., Richter, D., Okawara, Y., & Lederis, K. (1991). Corticotropin-releasing factor (CRF) gene family in the brain of the teleost fish Catostomus commersoni (white sucker): molecular analysis predicts distinct precursors for two CRFs and one urotensin I peptide. *Mol. Mar. Biol. Biotechnol.*, *1*, 48–57.

Mullen, L., & Goldsworthy, G. (2003). Changes in lipophorins are related to the activation of phenoloxidase in the hemolymph of Locusta migratoria in response to injection of immunogens. *Insect Biochem. Mol. Biol.*, *33*, 661–670.

Munte, C. E., Gäde, G., Domogalla, B., Kremer, W., Kellner, R., & Kalbitzer, H. R. (2008). C-mannosylation in the hypertrehalosaemic hormone from the stick insect Carausius morosus. *FEBS J.*, *275*, 1163–1173.

Muren, J. E., & Nässel, D. R. (1996). Isolation of five tachykinin-related peptides from the midgut of the cockroach Leucophaea maderae: existence of N-terminally extended isoforms. *Regul. Pept.*, *65*, 185–196.

Mwangi, R. W., & Goldsworthy, G. J. (1977a). Age-related changes in the response to adipokinetic hormone in Locusta migratoria. *Physiol. Ent.*, *2*, 37–42.

Mwangi, R. W., & Goldsworthy, G. J. (1977b). Diglyceride-transporting lipoproteins in Locusta. *J. Comp. Physiol. B*, *114*, 177–190.

Nachman, R. J., & Coast, G. M. (2007). Structure-activity relationships for in vitro diuretic activity of CAP$_{2b}$ in the housefly. *Peptides, 28,* 57–61.

Nachman, R. J., Russell, W. K., Coast, G. M., Russell, D. H., & Predel, R. (2005). Mass spectrometric assignment of Leu/Ile in neuropeptides from single neurohemal organ preparations of insects. *Peptides, 26,* 2151–2156.

Nachman, R. J., Russell, W. K., Coast, G. M., Russell, D. H., Miller, J. A., et al. (2006). Identification of PVK/CAP2b neuropeptides from single neurohemal organs of the stable fly and horn fly via MALDI-TOF/TOF tandem mass spectrometry. *Peptides, 27,* 521–526.

Nachman, R. J., Holman, G. M., & Hadden, W. F. (1993). Leads for insect neuropeptide mimetic development. *Arch. Ins. Biochem. Physiol., 22,* 181–197.

Nagai, C., Asazuma, H., Nagata, S., & Nagasawa, H. (2009). Identification of a second messenger of crustacean hyperglycemic hormone signaling pathway in the kuruma prawn Marsupenaeus japonicus. *Ann. N.Y. Acad. Sci., 1163,* 478–480.

Nair, M. M., Jackson, G. E., & Gäde, G. (2001). Conformation study of insect adipokinetic hormones using NMR constrained molecular dynamics. *J. Computer-Aided Mol. Design, 15,* 259–270.

Nässel, D. R., Passier, P. C. C. M., Elekes, K., Dircksen, H., Vullings, H. G. B., et al. (1995). Evidence that locustatachykinin I is involved in release of adipokinetic hormone from locust corpora cardiaca. *Reg. Peptides, 57,* 297–310.

Nässel, D. R., Vullings, H. G. B., Passier, P. C. C. M., Lundquist, C. T., Schoofs, L., et al. (1999). Several isoforms of locustatachykinins may be involved in cyclic AMP-mediated release of adipokinetic hormones from the locust corpora cardiaca. *Gen. Comp. Endocrinol., 113,* 401–412.

Nicolson, S. W. (1976). Diuresis in the cabbage white butterfly, Pieris brassicae: fluid secretion by the Malpighian tubules. *J. Insect Physiol., 22,* 1347–1356.

Nicolson, S. W. (1991). Diuresis or clearance – is there a physiological role for the diuretic hormone of the desert beetle Onymacris? *J. Insect Physiol., 37,* 447–452.

Nicolson, S. W. (1993). The ionic basis of fluid secretion in insect Malpighian tubules – Advances in the last 10 years – Review. *J. Insect Physiol., 39,* 451–458.

Nicolson, S. W., & Millar, R. P. (1983). Effects of biogenic amines and hormones on butterfly Malpighian tubules: dopamine stimulates fluid secretion. *J. Insect Physiol., 29,* 611–615.

Nittoli, T., Coast, G. M., & Sieburth, S. M. (1999). Evidence for helicity in insect diuretic peptide hormones: computational analysis, spectroscopic studies, and biological assays. *J. Pept. Res., 53,* 99–108.

Noyes, B. E., Katz, F. N., & Schaffe, M. H. (1995). Identification and expression of the *Drosophila* adipokinetic hormone gene. *Mol. Cell.Endocrinal, 109,* 133–141.

Noyes, B. E., & Schaffer, M. H. (1990). The structurally similar neuropeptides adipokinetic hormone I and II are derived from similar, very small mRNAs. *J. Biol. Chem., 265,* 483–489.

O'Connor, K. R., & Beyenbach, K. W. (2001). Chloride channels in apical membrane patches of stellate cells of Malpighian tubules of Aedes aegypti. *J. Exp. Biol., 204,* 367–378.

O'Donnell, M. J., Dow, J. A. T., Huesmann, G. R., Tublitz, N. J., & Maddrell, S. H. P. (1996). Separate control of anion and cation transport in Malpighian tubules of *Drosophila* melanogaster. *J. Exp. Biol., 199,* 1163–1175.

O'Donnell, M. J., & Machin, J. (1991). Ion activities and electrochemical gradients in the mealworm rectal complex. *J. Exp. Biol., 155,* 375–402.

O'Donnell, M. J., & Maddrell, S. H. (1984). Secretion by the Malpighian tubules of Rhodnius prolixus stal: electrical events. *J. Exp. Biol., 110,* 275–290.

O'Donnell, M. J., Rheault, M. R., Davies, S. A., Rosay, P., Harvey, B. J., et al. (1998). Hormonally controlled chloride movement across *Drosophila* tubules is via ion channels in stellate cells. *Am. J. Physiol., 274,* R1039–R1049.

O'Donnell, M. J., & Spring, J. H. (2000). Modes of control of insect Malpighian tubules: synergism, antagonism, cooperation and autonomous regulation. *J. Insect Physiol., 46,* 107–117.

Ogoyi, D. O., Osir, E. O., & Olembo, N. K. (1998). Fat body triacylglycerol lipase in solitary and gregarious phases of Schistocerca gregaria (Forskal) (Orthoptera: Acrididae). *Comp. Biochem. Physiol., 119B,* 163–169.

Oguri, E., & Steele, J. E. (2003). A novel function of cockroach (Periplaneta americana) hypertrehalosemic hormone: translocation of lipid from hemolymph to fat body. *Gen. Comp. Endocrinol., 132,* 46–54.

Olsen, S. S., Cazzamali, G., Williamson, M., Grimmelikhuijzen, C. J. P., & Hauser, F. (2007). Identification of one capa and two pyrokinin receptors from the malaria mosquito *Anopheles gambiae. Biochem. Biophys. Res. Commun., 362,* 245–251.

Orchard, I., & Lange, A. B. (1983a). Release of identified adipokinetic hormones during flight and following neural stimulation in Locusta migratoria. *J. Insect Physiol., 29,* 425–429.

Orchard, I., & Lange, A. B. (1983b). The hormonal control of haemolymph lipid during flight in Locusta migratoria. *J. Insect Physiol., 29,* 639–642.

Orchard, I., Lange, A. B., Bendena, W. G., & Evans, P. D. (2001). FMRFamide-related peptides: a multifunctional family of structurally-related neuropeptides in insects. *Adv. Insect Physiol., 28,* 267–329.

Orchard, I., & Loughton, B. G. (1981). The neural control of release of hyperlipaemic hormone from the corpus cardiacum of Locusta migratoria. *Comp. Biochem. Physiol., 68A,* 25–30.

Orr, G. L., Gole, J. W. D., Jahagirdar, A. P., Downer, R. G. H., & Steele, J. E. (1985). Cyclic AMP does not mediate the action of synthetic hypertrehalosemic peptides from the corpus cardiacum of Periplaneta americana. *Insect Biochem., 15,* 703–709.

O'Shea, M., & Rayne, R. C. (1992). Adipokinetic hormones: Cell and molecular biology. *Experientia, 48,* 430–438.

O'Shea, M., Witten, J., & Schaffer, M. (1984). Isolation and characterization of two myoactive neuropeptides: Further evidence for an invertebrate peptide family. *J. Neurosci., 4,* 521–529.

Oudejans, R. C. H. M., Kooiman, F. P., Heerma, W., Versluis, C., Slotboom, A. J., et al. (1991). Isolation and structure elucidation of a novel adipokinetic hormone (Lom-AKH-III) from the glandular lobes of the corpus cardiacum of the migratory locust, Locusta migratoria. *Eur. J. Biochem., 195,* 351–359.

Oudejans, R. C. H. M., Mes, T. H. M., Kooiman, F. P., & Van der Horst, D. (1993). Adipokinetic peptide hormone content and biosynthesis during locust development. *Peptides, 14*, 877–881.

Oudejans, R. C. H. M., Vroemen, S. F., Jansen, R. F. R., & Van der Horst, D. J. (1996). Locust adipokinetic hormones: carrier-independent transport and differential inactivation at physiological concentrations during flight and rest. *Proc. Natl. Acad. Sci. USA, 93*, 8654–8659.

Pallen, C. J., & Steele, J. E. (1988). A putative role for calmodulin in corpus cardiacum stimulated trehalose synthesis in fat body of the American cockroach (Periplaneta americana). *Insect Biochem., 18*, 577–584.

Paluzzi, J. P., & Orchard, I. (2010). A second gene encodes the anti-diuretic hormone in the insect, Rhodnius prolixus. *Mol. Cell. Endocrinol., 317*, 53–63.

Paluzzi, J. P., Park, Y., Nachman, R. J., & Orchard, I. (2010). Isolation, expression analysis, and functional characterization of the first antidiuretic hormone receptor in insects. *Proc. Natl. Acad. Sci. USA, 107*, 10290–10295.

Paluzzi, J. P., Russell, W. K., Nachman, R. J., & Orchard, I. (2008). Isolation, cloning, and expression mapping of a gene encoding an antidiuretic hormone and other capa-related peptides in the disease vector, Rhodnius prolixus. *Endocrinology, 149*, 4638–4646.

Paluzzi, J.-P., & Orchard, I. (2006). Distribution, activity and evidence for the release of an anti-diuretic peptide in the kissing bug Rhodnius prolixus. *J. Exp. Biol., 209*, 907–915.

Ponnabecker, T., & Orchord, I. (1987). Regulater of adipokinetic hormone release from locustneuroendocrine tissue: Participation of calcium and cyclic Amp. *Brain Res, 423*, 13–22.

Pannabecker, T., & Orchard, I. (1986). Octopamine and cyclic AMP mediate release of adipokinetic hormone I and II from isolated neuroendocrine tissue. *Mol. Cell. Endocrinol., 48*, 153–159.

Pannabecker, T. L., Hayes, T. K., & Beyenbach, K. W. (1993). Regulation of epithelial shunt conductance by the peptide leucokinin. *J. Membr. Biol., 132*, 63–76.

Park, J. H., & Keeley, L. L. (1995). vitro hormonal regulation of glycogen phosphorylase activity in fat body of the tropical cockroach, Blaberus discoidalis. *Gen. Comp. Endocrinol., 98*, 234–243.

Park, J. H., & Keeley, L. L. (1996). Calcium-dependent action of hypertrehalosemic hormone on activation of glycogen phosphorylase in cockroach fat body. *Mol. Cell. Endocrinol., 116*, 199–205.

Park, Y., Kim, Y.-J., & Adams, M. E. (2002). Identification of G protein-coupled receptors for *Drosophila* PRXamide peptides, CCAP, corazonin, and AKH supports a theory of ligand-receptor coevolution. *Proc. Natl. Acad. Sci. USA, 99*, 11423–11428.

Passier, P. C. C. M., Vullings, H. G. B., Diederen, J. H. B., & Van der Horst, D. J. (1995). Modulatory effects of biogenic amines on adipokinetic hormone secretion from locust corpora cardiaca. *Gen. Comp. Endocrinol., 97*, 231–238.

Passier, P. C. C. M., Vullings, H. G. B., Diederen, J. H. B., & Van der Horst, D. J. (1997). Trehalose inhibits the release of adipokinetic hormones from the corpus cardiacum in the African migratory locust, Locusta migratoria, at the level of the adipokinetic cells. *J. Endocrinol., 153*, 299–305.

Patel, M., Chung, J. S., Kay, I., Mallet, A. I., Gibbon, C. R., et al. (1994). Localization of Locusta-DP in locust CNS and hemolymph satisfies initial hormonal criteria. *Peptides, 15*, 591–602.

Patel, M., Hayes, T. K., & Coast, G. M. (1995). Evidence for the hormonal function of a CRF-related diuretic peptide (Locusta-DP) in Locusta migratoria. *J. Exp. Biol., 198*, 793–804.

Patel, R., Soulages, J. L., Wells, M. A., & Arrese, E. L. (2004). cAMP-dependent protein kinase of Manduca sexta phosphorylates but does not activate the fat body triglyceride lipase. *Insect Biochem. Molec. Biol., 34*, 1269–1279.

Patel, R. T., Soulages, J. L., & Arrese, E. L. (2006). Adipokinetic hormone-induced mobilization of fat body triglyceride stores in Manduca sexta: Role of TG-lipase and lipid droplets. *Arch. Ins. Biochem. Physiol., 63*, 73–81.

Patel, R. T., Soulages, J. L., Hariharasundaram, B., & Arrese, E. L. (2005). Activation of the lipid droplet controls the rate of lipolysis of triglycerides in the insect fat body. *J. Biol. Chem., 280*, 22624–22631.

Petzel, D. H. (2000). Na^+/H^+ exchange in mosquito Malpighian tubules. *Am. J. Physiol.-Reg. Integr. Comp. Physiol., 279*, R1996–R2003.

Petzel, D. H., Berg, M. M., & Beyenbach, K. W. (1987). Hormone-controlled cAMP-mediated fluid secretion in yellow-fever mosquito. *Am. J. Physiol., 253*, R701–R711.

Petzel, D. H., Hagedorn, H. H., & Beyenbach, K. W. (1985). Preliminary isolation of mosquito natriuretic factor. *Am. J. Physiol., 249*, R379–R386.

Petzel, D. H., Hagedorn, H. H., & Beyenbach, K. W. (1986). Peptide nature of two mosquito natriuretic factors. *Am. J. Physiol., 250*, R328–R332.

Phillips, J. E. (1983). Endocrine control of salt and water balance: excretion. In R. G. H. Downer, & H. Laufer (Eds.), *Endocrinology of Insects* (pp. 411–425). New York: A.R. Liss.

Phillips, J. E., & Audsley, N. (1995). Neuropeptide control of ion and fluid transport across locust hindgut. *Am. Zool., 35*, 503–514.

Phillips, J. E., Audsley, N., Lechleitner, R., Thomson, B., Meredith, J., et al. (1988). Some major transport mechanisms of insect absorptive epithelia. *Comp. Biochem. Physiol. A, 90*, 643–650.

Phillips, J. E., Hanrahan, J., Chamberlin, M., & Thomson, B. (1986). Mechanisms and control of reabsorption in insect hindgut. *Adv. Insect Physiol., 19*, 329–422.

Phillips, J. E., Meredith, J., Audsley, N., Richardson, N., Macins, A., et al. (1998a). Locust ion transport peptide (ITP): a putative hormone controlling water and ionic balance in terrestrial insects. *Am. Zool., 38*, 461–470.

Phillips, J. E., Meredith, J., Audsley, N., Ring, M., Macins, A., et al. (1998b). Locust ion transport peptide (ITP): function, structure, cDNA and expression. In G. M. Coast, & S. G. Webster (Eds.) *Recent Advances in Arthropod Endocrinology* (pp. 210–226). Cambridge: Cambridge University Press.

Phillips, J. E., Mordue, W., Meredith, J., & Spring, J. (1980). Purification and characteristics of the chloride transport stimulating factor from locust corpora cardiaca: a new peptide. *Can. J. Zool., 58*, 1851–1860.

Piermarini, P. M., Grogan, L. F., Lau, K., Wang, L., & Beyenbach, K. W. (2010). A SLC4-like anion exchanger from renal tubules of the mosquito (Aedes aegypti): evidence for a novel role of stellate cells in diuretic fluid secretion. *Am. J. Physiol. Regul. Integr. Comp. Physiol., 298*, R642–R660.

Pietrantonio, P. V., Jagge, C., Taneja-Bageshwar, S., Nachman, R. J., & Barhoumi, R. (2005). The mosquito Aedes aegypti (L.) leucokinin receptor is a multiligand receptor for the three Aedes kinins. *Insect Mol. Biol., 14*, 55–67.

Plawner, L., Pannabecker, T. L., Laufer, S., Baustian, M. D., & Beyenbach, K. W. (1991). Control of diuresis in the yellow fever mosquito Aedes aegypti: evidence for similar mechanisms in the male and female. *J. Insect Physiol., 37*, 119–128.

Podzuweit, T., Nennstiel, P., & Mueller, A. (1995). Isozyme selective inhibition of cGMP-stimulated cyclic nucleotide phosphodiesterases by erythro-9-(2hydroxy-3-nonyl) adenine. *Cell. Signal., 7*, 733–738.

Pollock, V. P., Radford, J. C., Pyne, S., Hasan, G., Dow, J. A., et al. (2003). NorpA and itpr mutants reveal roles for phospholipase C and inositol (1,4,5)-trisphosphate receptor in *Drosophila* melanogaster renal function. *J. Exp. Biol., 206*, 901–911.

Predel, R., & Gäde, G. (2002). Identification of the abundant neuropeptide from abdominal perisympathetic organs of locusts. *Peptides, 23*, 621–627.

Predel, R., Kellner, R., Baggerman, G., Steinmetzer, T., & Schoofs, L. (2000). Identification of novel periviscerokinins from single neurohaemal release sites in insects-MS/MS fragmentation complemented by Edman degradation. *Eur. J. Biochem., 267*, 3869–3873.

Predel, R., Kellner, R., Rapus, J., Penzlin, H., & Gäde, G. (1997). Isolation and structural elucidation of eight kinins from the retrocerebral complex of the American cockroach, Periplaneta americana. *Regul. Pept., 71*, 199–205.

Predel, R., Linde, D., Rapus, J., Vettermann, S., & Penzlin, H. (1995). Periviscerokinin (Pea-PVK): a novel myotropic neuropeptide from the perisympathetic organs of the American cockroach. *Peptides., 16*, 61–66.

Predel, R., Rapus, J., Eckert, M., Holman, G. M., Nachman, R. J., et al. (1998). Isolation of periviscerokinin-2 from the abdominal perisympathetic organs of the American cockroach, Periplaneta americana. *Peptides., 19*, 801–809.

Predel, R., Russell, W. K., Russell, D. H., Lopez, J., Esquivel, J., et al. (2008). Comparative peptidomics of four related hemipteran species: pyrokinins, myosuppressin, corazonin, adipokinetic hormone, sNPF, and periviscerokinins. *Peptides., 29*, 162–167.

Price, D. R., Du, J., Dinsmore, A., & Gatehouse, J. A. (2004). Molecular cloning and immunolocalization of a diuretic hormone receptor in rice brown planthopper (Nilaparvata lugens). *Insect. Mol. Biol., 13*, 469–480.

Prigge, S. T., Mains, R. E., Eipper, B. A., & Amzel, L. M. (2000). New insights into copper monooxygenases and peptide amidation: structure, mechanism and function. *Cell. Mol. Life. Sci., 57*, 1236–1259.

Proux, J., Proux, B., & Phillips, J. (1985). Source and distribution of factors in locust nervous system which stimulate rectal Cl⁻ transport. *Can. J. Zool., 63*, 37–41.

Proux, J., Rougon, G., & Cupo, A. (1982). Enhancement of excretion across locust Malpighian tubules by a diuretic vasopressin-like hormone. *Gen. Comp. Endocrinol., 47*, 449–457.

Proux, J. P., Miller, C. A., Li, J. P., Carney, R. L., Girardie, A., et al. (1987). Identification of an arginine vasopressin-like diuretic hormone from Locusta migratoria. *Biochem. Biophys. Res. Commun., 149*, 180–186.

Proux, J. P., Picquot, M., Herault, J. P., & Fournier, B. (1988). Diuretic activity of a newly identified neuropeptide – the arginine-vasopressin-like insect diuretic hormone: use of an improved bioassay. *J. Insect. Physiol., 34*, 919–927.

Qi, L. J., Leung, A. T., Xiong, Y. T., Marx, K. A., & Abou-Samra, A. B. (1997). Extracellular cysteines of the corticotropin-releasing factor receptor are critical for ligand interaction. *Biochemistry., 36*, 12442–12448.

Quinlan, M. C., & O'Donnell, M. J. (1998). Anti-diuresis in the blood-feeding insect Rhodnius prolixus Stal: antagonistic actions of cAMP and cGMP and the role of organic acid transport. *J. Insect. Physiol., 44*, 561–568.

Quinlan, M. C., Tublitz, N. J., & O'Donnell, M. J. (1997). Anti-diuresis in the blood-feeding insect Rhodnius prolixus Stal: The peptide CAP$_{2b}$ and cyclic GMP inhibit Malpighian tubule fluid secretion. *J. Exp. Biol., 200*, 2363–2367.

Rademakers, L. H. P. M. (1977a). Effects of isolation and transplantation of the corpus cardiacum on hormone release from its glandular cells after flight in Locusta migratoria. *Cell. Tissue. Res., 184*, 213–224.

Rademakers, L. H.P.M. (1977b). Identification of a secreto-motor centre in the brain of Locusta migratoria controlling the secretory activity of the adipokinetic hormone producing cells of the corpus cardiacum. *Cell. Tissue. Res., 184*, 381–395.

Radford, J. C., Terhzaz, S., Cabrero, P., Davies, S.-A., & Dow, J. A. T. (2004). Functional characterisation of the Anopheles leucokinins and their cognate G-protein coupled receptor. *J. Exp. Biol., 207*, 4573–4586.

Radford, J. C., Davies, S. A., & Dow, J. A. (2002). Systematic G-protein-coupled receptor analysis in *Drosophila* melanogaster identifies a leucokinin receptor with novel roles. *J. Biol. Chem., 277*, 38810–38817.

Rafaeli, A., Pines, M., Stern, P. S., & Applebaum, S. W. (1984). Locust diuretic hormone-stimulated synthesis and excretion of cyclic-AMP: a novel Malpighian tubule bioassay. *Gen. Comp. Endocrinol., 54*, 35–42.

Ramsay, J. A. (1954). Active transport of water by the Malpighian tubules of the stick insect, Dixippus morosus (Orthoptera, Phasmidae). *J. Exp. Biol., 31*, 104–113.

Ramsay, J. A. (1964). The rectal complex of the meal-worm Te. molitor L. (Coleoptera, Tenebrionidae). *Phil. Trans. R. Soc. Lond. B. Biol. Sci., 248*, 279–314.

Ramsay, J. A. (1976). The rectal complex in the larvae of Lepidoptera. *Phil. Trans. R. Soc. Lond. B. Biol., 274*, 203–226.

Rao, K. R. (2001). Crustacean pigmentary-effector hormones: Chemistry and functions of RPCH, PDH, and related peptides. *Am. Zool., 41*, 364–379.

Rayne, R. C., & O'Shea, M. (1992). Inactivation of neuropeptide hormones (AKH I and AKH II) studied in vivo and in vitro. *Insect. Biochem. Mol. Biol., 22*, 25–34.

Rayne, R. C., & O'Shea, M. (1994). Reconstitution of adipokinetic hormone biosynthesis in vitro indicates steps in prohormone processing. *Eur. J. Biochem., 219,* 781–789.

Reagan, J. D., (1994). Expression cloning of an insect diuretic hormone receptor – a member of the calcitonin/secretin receptor family. *J. Biol. Chem. 269,* 9–12.

Reagan, J. D. (1995a). Functional expression of a diuretic hormone receptor in baculovirus-infected insect cells: evidence suggesting that the N-terminal region of diuretic hormone is associated with receptor activation. *Insect. Biochem. Mol. Biol., 25,* 535–539.

Reagan, J. D. (1995b). Molecular cloning of a putative Na+–K+–2Cl-cotransporter from the Malpighian tubules of the tobacco hornworm, Manduca sexta. *Insect. Biochem. Mol. Biol., 25,* 875–880.

Reagan, J. D. (1996). Molecular cloning and function expression of a diuretic hormone receptor from the house cricket, Acheta domesticus. *Insect. Biochem. Mol. Biol., 26,* 1–6.

Reagan, J. D., Li, J. P., Carney, R. L., & Kramer, S. J. (1993). Characterization of a diuretic hormone receptor from the tobacco hornworm, Manduca sexta. *Arch. Insect. Biochem. Physiol., 23,* 135–145.

Reynolds, S. E., & Bellward, K. (1989). Water balance in Manduca sexta caterpillars: water recycling from the rectum. *J. Exp. Biol., 141,* 33–45.

Riehle, M. A., Garczynski, S. F., Crim, J. W., Hill, C. A., & Brown, M. R. (2002). Neuropeptides and peptide hormones in *Anopheles gambiae. Science., 298,* 172–175.

Roller, L., Yamanaka, N., Watanabe, K., Daubnerova, I., Zitnan, D., et al. (2008). The unique evolution of neuropeptide genes in the silkworm Bombyx mori. *Insect. Biochem. Mol. Biol., 38,* 1147–1157.

Rosay, P., Davies, S. A., Yu, Y., Sozen, A., Kaiser, K., et al. (1997). Cell-type specific calcium signalling in a *Drosophila* epithelium. *J. Cell. Sci., 110,* 1683–1692.

Rulifson, E. J., Kim, S. K., & Nusse, R. (2002). Ablation of insulin-producing neurons in flies: growth and diabetic phenotypes. *Science., 296,* 1118–1120.

Sacktor, B., & Childress, C. (1967). Metabolism of proline in insect flight muscle and its significance in stimulating the oxidation of pyruvate. *Arch. Biochem. Biophys., 120,* 583–588.

Satake, S., Masumura, M., Ishizaki, H., Nagata, K., Kataoka, H., et al. (1997). Bombyxin, an insulin-related peptide of insects, reduces the major storage carbohydrates in the silkworm Bombyx mori. *Comp. Biochem. Physiol. [B]., 118B,* 349–357.

Satake, S., Nagata, K., Kataoka, H., & Mizoguchi, A. (1999). Bombyxin secretion in the adult silkmoth Bombyx mori: sex-specificity and its correlation with metabolism. *J. Insect. Physiol., 45,* 939–945.

Scaraffia, P. Y., & Wells, M. A. (2003). Proline can be utilized as an energy substrate during flight of Aedes aegypti females. *J. Insect. Physiol., 49,* 591–601.

Scarborough, R. M., Jamieson, G. C., Kalish, F., Kramer, S. J., McEnroe, G. A., et al. (1984). Isolation and primary structure of two peptides with cardioaccelatory and hyperglycemic activity from the corpora cardiaca of Periplaneta americana. *Proc. Natl. Acad. Sci. USA., 81,* 5575–5579.

Schepel, S. A., Fox, A. J., Miyauchi, J. T., Sou, T., Yang, J. D., et al. (2010). The single kinin receptor signals to separate and independent physiological pathways in Malpighian tubules of the yellow fever mosquito. *Am. J. Physiol. Regul. Integr. Comp. Physiol., 299,* R612–R622.

Schoofs, L., Holman, G. M., Proost, P., Van Damme, J., Hayes, T. K., et al. (1992). Locustakinin, a novel myotropic peptide from Locusta migratoria, isolation, primary structure and synthesis. *Regul. Pept., 37,* 49–57.

Schoofs, L., Vanden Broeck, J., & De Loof, A. (1993). The myotropic peptides of Locusta migratoria: structures, distribution, functions and receptors. *Insect. Biochem. Mol. Biol., 23,* 859–881.

Schooley, D. A. (1994). Peptide hormones regulating water balance: workshop report. *Arch. Insect. Biochem. Physiol., 27,* 323–324.

Schooley, D. A., Miller, C. A., & Proux, J. P. (1987). Isolation of two arginine vasopressin-like factors from ganglia of Locusta migratoria. *Arch. Insect. Biochem. Physiol., 5,* 157–166.

Schooneveld, H., Tesser, G. I., Veenstra, J. A., & Romberg-Privee, H. M. (1983). Adipokinetic hormone and AKH-like peptide demonstrated in the corpora cardiaca and nervous system of Locusta migratoria by immunocytochemistry. *Cell. Tiss. Res., 230,* 67–76.

Schulz-Aellen, M.-F., Roulet, E., Fischer-Lougheed, J., & O'Shea, M. (1989). Synthesis of homodimer neurohormone precursor of locust adipokinetic hormone studied by in vitro translation and cDNA cloning. *Neuron., 2,* 1369–1373.

Scott, B. N., Yu, M. -J., Lee, L. W., & Beyenbach, K. W. (2004). Mechanisms of K+ transport across basolateral membranes of principal cells in Malpighian tubules of the yellow fever mosquito, Aedes aegypti. *J. Exp. Biol., 207,* 1655–1663.

Serrano, L., Blanvillain, G., Soyez, D., Charmantier, G., Grousset, E., et al. (2003). Putative involvement of crustacean hyperglycemic hormone isoforms in the neuroendocrine mediation of osmoregulation in the crayfish Astacus leptodactylus. *J. Exp. Biol., 206,* 979–988.

Sharp-Baker, H. E., Diederen, J. H.B., Mäkel, K. M., Peute, J., & Van der Horst, D. J. (1995). The adipokinetic cell in the corpus cardiacum of Locusta migratoria preferentially release young secretory granules. *Eur J. Cell. Biol., 68,* 268–274.

Sharp-Baker, H. E., Oudejans, R. C. H. M., Kooiman, F. P., Diederen, J. H. B., Peute, J., et al. (1996). Preferential release of newly synthesized, exportable neuropeptides by insect neuroendocrine cells and the effect of ageing of secretory granules. *Eur. J. Cell. Biol., 71,* 72–78.

Siegert, K. J. (1999). Locust corpora cardiaca contain an inactive adipokinetic hormone. *FEBS. Letters., 447,* 237–240.

Siegert, K. J., Kellner, R., & Gäde, G. (2000). A third active AKH is present in the phygomorphid grasshoppers Phymateus morbillosus and Dictyophorus spumans. *Insect. Biochem. Mol. Biol., 30,* 1061–1067.

Siegert, K. J., & Mordue, W. (1986a). Elucidation of the primary structures of the cockroach hyperglycaemic hormones I and II using enzymatic techniques and gas phase sequencing. *Physiol. Ent., 11,* 205–211.

Siegert, K. J., & Mordue, W. (1986b). Quantification of adipokinetic hormones I and II in corpora cardiaca of Schistocerca gregaria and Locusta migratoria. *Comp. Biochem. Physiol.*, *84A*, 279–284.

Siegert, K. J., & Mordue, W. (1987). Breakdown of locust adipokinetic hormone I by Malpighian tubules of Schistocerca gregaria. *Insect. Biochem.*, *17*, 705–710.

Siegert, K. J., & Ziegler, R. (1983). A hormone from the corpora cardiaca controls fat body glycogen phosphorylase during starvation in tobacco hornworm larvae. *Nature.*, *301*, 526–527.

Skaer, N. J., Nassel, D. R., Maddrell, S. H., & Tublitz, N. J. (2002). Neurochemical fine tuning of a peripheral tissue: peptidergic and aminergic regulation of fluid secretion by Malpighian tubules in the tobacco hawkmoth M. sexta. *J. Exp. Biol.*, *205*, 1869–1880.

Socha, R., Kodrik, D., & Zemek, R. (1999). Adipokinetic hormone stimulates insect locomotory activity. *Naturwissenschaften.*, *86*, 85–86.

Soulages, J. L., & Wells, M. A. (1994). Lipophorin: the structure of an insect lipoprotein and its role in lipid transport in insects. *Adv. Prot. Chem.*, *45*, 371–415.

Sowa, S. M., Lu, K.-H., Park, J. H., & Keeley, L. L. (1996). Physiological effectors of hyperglycemic neurohormone biosynthesis in an insect. *Mol. Cell Endocrinol.*, *123*, 97–105.

Spanings-Pierrot, C., Soyez, D., Van Herp, F., Gompel, M., Skaret, G., et al. (2000). Involvement of crustacean hyperglycemic hormone in the control of gill ion transport in the crab Pachygrapsus marmoratus. *Gen. Comp. Endocrinol.*, *119*, 340–350.

Spencer, I. M., & Candy, D. J. (1976). Hormonal control of diacyl glycerol mobilization from fat body of the desert locust, Schistocerca gregaria. *Insect. Biochem.*, *6*, 289–296.

Spring, J. H. (1990). Endocrine regulation of diuresis in insects. *J. Insect. Physiol.*, *36*, 13–22.

Spring, J. H., Morgan, A. M., & Hazelton, S. R. (1988). A novel target for antidiuretic hormone in insects. *Science.*, *241*, 1096–1098.

Stafflinger, E., Hansen, K. K., Hauser, F., Schneider, M., Cazzamali, G., et al. (2008). Cloning and identification of an oxytocin/vasopressin-like receptor and its ligand from insects. *Proc. Natl. Acad. Sci. USA.*, *105*, 3262–3267.

Stagg, L. E., & Candy, D. J. (1996). The effect of adipokinetic hormones on the levels of inositol phosphates and cyclic AMP in the fat body of the desert locust Schistocerca gregaria. *Insect. Biochem. Mol. Biol.*, *26*, 537–544.

Staubli, F., Jørgensen, T. J. D., Cazzamali, G., Williamson, M., Lenz, C., et al. (2002). Molecular identification of the insect adipokinetic hormone receptors. *Proc. Natl. Acad. Sci. USA.*, *99*, 3446–3451.

Steele, J. E. (1961). Occurrence of a hyperglycaemic factor in the corpus cardiacum of an insect. *Nature.*, *192*, 680–681.

Steele, J. E. (1963). The site of action of insect hyperglycemic hormone. *Gen. Comp. Endocrinol.*, *3*, 46–52.

Steele, J. E., Garcha, K., & Sun, D. (2001). Inositol trisphosphate mediates the action of hypertrehalosemic hormone on fat body of the American cockroach, Periplaneta americana. *Comp. Biochem. Physiol.*, *130B*, 537–545.

Steele, J. E., & Ireland, R. (1994). The preparation of trophocytes from the disaggregated fat body of the cockroach (Periplaneta americana). *Comp. Biochem. Physiol.*, *107A*, 517–522.

Steele, J. E., & Ireland, R. (1999). Hormonal activation of phosphorylase in cockroach fat body trophocytes: a correlation with trans-membrane calcium flux. *Arch. Insect. Biochem. Physiol.*, *42*, 233–244.

Steele, J. E., & Paul, T. (1985). Corpus cardiacum stimulated trehalose efflux from cockroach (Periplaneta americana) fat body: Control by calcium. *Can. J. Zool.*, *63*, 63–66.

Stenzel-Poore, M. P., Heldwein, K. A., Stenzel, P., Lee, S., & Vale, W. W. (1992). Characterization of the genomic corticotropin-releasing factor (CRF) gene from Xenopus laevis: two members of the CRF family exist in amphibians. *Mol. Endocrinol.*, *6*, 1716–1724.

Stone, J. V., Mordue, W., Batley, K. E., & Morris, H. R. (1976). Structure of locust adipokinetic hormone, a neurohormone that regulates lipid utilisation during flight. *Nature.*, *263*, 207–211.

Stone, J. V., Mordue, W., Broomfield, C. E., & Hardy, P. M. (1978). Structure-activity relationships for the lipid-mobilizing action of locust adipokinetic hormone. Synthesis and activity of a series of hormone analogs. *Eur. J. Biochem.*, *89*, 195–202.

Sun, D., Garcha, K., & Steele, J. E. (2002). Stimulation of trehalose efflux from cockroach (Periplaneta americana) fat body by hypertrehalosemic hormone is dependent on protein kinase C and calmodulin. *Arch. Insect Biochem. Physiol.*, *50*, 41–51.

Sun, D., & Steele, J. E. (2001). Regulation of intracellular calcium in dispersed fat body trophocytes of the cockroach, Periplaneta americana, by hypertrehalosemic hormone. *J. Insect. Physiol.*, *47*, 1399–1408.

Sun, D., & Steele, J. E. (2002). Regulation of phospholipase A2 activity in cockroach (Periplaneta americana) fat body by hypertrehalosemic hormone: evidence for the participation of protein kinase C. *J. Insect. Physiol.*, *48*, 537–546.

Sutcliffe, D. W. (1963). Chemical composition of haemolymph in insects and some other arthropods, in relation to their phylogeny. *Comp. Biochem. Physiol.*, *9*, 121–135.

Sydow, S., Flaccus, A., Fischer, A., & Spiess, J. (1999). The role of the fourth extracellular domain of the rat corticotropin-releasing factor receptor type 1 in ligand binding. *Eur. J. Biochem.*, *259*, 55–62.

Sydow, S., Radulovic, J., Dautzenberg, F. M., & Spiess, J. (1997). Structure-function relationship of different domains of the rat corticotropin-releasing factor receptor. *Mol. Brain. Res.*, *52*, 182–193.

Te Brugge, V. A., Lombardi, V. C., Schooley, D. A., & Orchard, I. (2005). Presence and activity of a Dippu-DH$_{31}$-like peptide in the blood-feeding bug, Rhodnius prolixus. *Peptides.*, *26*, 1283, 29-42; Corrigendum.

Te Brugge, V. A., Schooley, D. A., & Orchard, I. (2008). Amino acid sequence and biological activity of a calcitonin-like diuretic hormone (DH$_{31}$) from Rhodnius prolixus. *J. Exp. Biol.*, *211*, 382–390.

Te Brugge, V., Ianowski, J., & Orchard, I. (2009). Biological activity of diuretic factors on the anterior midgut of the blood-feeding bug, Rhodnius prolixus. *Gen. Comp. Endocrinol.*, *162*, 105–112.

Te Brugge, V., Paluzzi, J.-P., Schooley, D. A., & Orchard, I. (2011). Identification of the elusive peptidergic diuretic hormone in the blood-feeding bug, Rhodnius prolixus: a CRF-related peptide. *J. Exp. Biol., 214,* 371–381.

Te Brugge, V. A., Schooley, D. A., & Orchard, I. (2002). The biological activity of diuretic factors in Rhodnius prolixus. *Peptides., 23,* 671–681.

Terhzaz, S., O'Connell, F. C., Pollock, V. P., Kean, L., Davies, S. A., et al. (1999). Isolation and characterization of a leucokinin-like peptide of *Drosophila* melanogaster. *J. Exp. Biol., 202,* 3667–3676.

Terhzaz, S., Southall, T. D., Lilley, K. S., Kean, L., Allan, A. K., et al. (2006). Differential gel electrophoresis and transgenic mitochondrial calcium reporters demonstrate spatiotemporal filtering in calcium control of mitochondria. *J. Biol. Chem., 281,* 18849–18858.

Thompson, K. S. J., & Bacon, J. P. (1991). The vasopressin-like immunoreactive (VPLI) neurons of the locust, Locusta migratoria. 2. Physiology. *J. Comp. Physiol. A., 168,* 619–630.

Thompson, K. S. J., Tyrer, N. M., May, S. T., & Bacon, J. P. (1991). The vasopressin-like immunoreactive (VPLI) neurons of the Locust, Locusta migratoria. 1. Anatomy. *J. Comp. Physiol. A., 168,* 605–617.

Tobe, S. S., Zhang, J. R., Schooley, D. A., & Coast, G. M. (2005). A study of signal transduction for the two divretic peptides of Diploptera punctata. *Peptides, 26,* 89–98.

Torfs, H., Poels, J., Detheux, M., Dupriez, V., Van Loy, T., et al. (2002). Recombinant aequorin as a reporter for receptor-mediated changes of intracellular Ca^{2+}-levels in *Drosophila* S2 cells. *Invert. Neurosci., 4,* 119–124.

Tublitz, N. J., Brink, D., Broadie, K. S., Loi, P. K., & Sylwester, A. W. (1991). From behaviour to molecules: an integrated approach to the study of neuropeptides. *Trends. Neurosci., 14,* 254–259.

Tublitz, N. J., & Truman, J. W. (1985). Insect cardioactive peptides II. Neurohormonal control of heart activity by two cardioacceleratory peptides in the tobacco hawkmoth, Manduca sexta. *J. Exp. Biol., 114,* 381–395.

Van der Horst, D. J., Vroemen, S. F., & Van Marrewijk, W. J.A. (1997). Metabolism of stored reserves in insect fat body: hormonal signal transduction implicated in glycogen mobilization and biosynthesis of the lipophorin system. *Comp. Biochem. Physiol., 117B,* 463–474.

Van Marrewijk, W. J. A., Van den Broek, A. T. M., & Beenakkers, A. M. T. (1980). Regulation of glycogenolysis in the locust fat body during flight. *Insect. Biochem., 10,* 675–679.

Van Marrewijk, W. J. A., Van den Broek, A. T. M., & Beenakkers, A. M. T. (1991). Adipokinetic hormone is dependent on extracellular Ca^{2+} for its stimulatory action on the glycogenolytic pathway in locust fat body in vitro. *Insect. Biochem., 21,* 375–380.

Van Marrewijk, W. J. A., Van den Broek, A. T. M., Gielbert, M.-L., & Van der Horst, D. J. (1996). Insect adipokinetic hormone stimulates inositol phosphate metabolism: roles for both Ins(1,4,5)P$_3$ and Ins(1,3,4,5)P$_4$ in signal transduction? *Mol. Cell Endocrinol., 122,* 141–150.

Van Marrewijk, W. J.A., Van den Broek, A. T.M., & Van der Horst, D. J. (1993). Adipokinetic hormone-induced influx of extracellular calcium into insect fat body cells is mediated through depletion of intracellular calcium stores. *Cell Signal., 5,* 753–761.

Vaughan, J., Donaldson, C., Bittencourt, J., Perrin, M. H., Lewis, K., et al. (1995). Urocortin, a mammalian neuropeptide related to fish urotensin I and to corticotropin-releasing factor. *Nature., 378,* 287–292.

Veelaert, D., Passier, P., Devreese, B., Vanden Broeck, J., Van Beeumen, J., et al. (1997). Isolation and characterization of an adipokinetic hormone release-inducing factor in locusts: the crustacean cardioactive peptide. *Endocrinology., 138,* 138–142.

Veenstra, J. A. (1988). Effects of 5-hydroxytryptamine on the Malpighian tubules of Aedes aegypti. *J. Insect. Physiol., 34,* 299–304.

Veensta, J. A. (1989). Isolation and structure of corozonin, a cardioactive peptide from the American cockroach. *FEBS Lett, 250,* 231–234.

Veensta, J. A. (1994a). Isolation and structure of the *Drosophila* corozonin gene. *Biophys.Res.Common.* 204, 292–296.

Veenstra, J. A. (1994b). Isolation and identification of three leucokinins from the mosquito Aedes aegypti. *Biochem. Biophys. Res. Commun., 202,* 715–719.

Veenstra, J. A. (2009). Does corazonin signal nutritional stress in insects?. *Insect. Biochem. Molec. Biol., 39,* 755–762.

Veenstra, J. A., Pattillo, J. M., & Petzel, D. H. (1997). A single cDNA encodes all three Aedes leucokinins, which stimulate both fluid secretion by the Malpighian tubules and hindgut contractions. *J. Biol. Chem., 272,* 10402–10407.

Voss, M., Vitavska, O., Walz, B., Wieczorek, H., & Baumann, O. (2007). Stimulus-induced phosphorylation of V-ATPase by protein kinase A. *J. Biol. Chem., 282,* 33735–33742.

Vroemen, S. F., De Jonge, H., Van Marrewijk, W. J.A., & Van der Horst, D. J. (1998b). The phospholipase C signaling pathway in locust fat body is activated via G$_q$ and not affected by cAMP. *Insect. Biochem. Mol. Biol., 28,* 483–490.

Vroemen, S. F., Van der Horst, D. J., & Van Marrewijk, W. J.A. (1998a). New insights into adipokinetic hormone signaling *Mol. Cell Endocrinol., 141,* 7–12.

Vroemen, S. F., Van Marrewijk, W. J.A., De Meijer, J., Van den Broek, A. T.M., & Van der Horst, D. J. (1997). Differential induction of inositol phosphate metabolism by three adipokinetic hormones. *Mol. Cell Endocrinol., 130,* 131–139.

Vroemen, S. F., Van Marrewijk, W. J.A., Schepers, C. C.J., & Van der Horst, D. J. (1995b). Signal transduction of adipokinetic hormones involves Ca^{2+} fluxes and depends on extracellular Ca^{2+} to potentiate cAMP-induced activation of glycogen phosphorylase. *Cell. Calcium., 17,* 459–467.

Vroemen, S. F., Van Marrewijk, W. J.A., & Van der Horst, D. J. (1995a). Stimulation of glycogenolysis by three locust adipokinetic hormones involves G$_s$ and cAMP. *Mol. Cell Endocrinol., 107,* 165–171.

Vullings, H. G.B., Diederen, J. H.B., Veelaert, D., & Van der Horst, D. J. (1999). The multifactorial control of the release of hormones from the locust retrocerebral glandular complex. *Micro. Res. Tech., 45,* 142–153.

Vullings, H. G.B., Ten Voorde, S. E.C.G., Passier, P. C.C.M., Diederen, J. H.B., Van der Horst, D. J., et al. (1998). A possible role of SchistoFLRFamide in inhibition of adipokinetic hormone release from locust corpora cardiaca. *J. Neurocytol.*, *27*, 901–913.

Wang, S., Rubenfeld, A. B., Hayes, T. K., & Beyenbach, K. W. (1996). Leucokinin increases paracellular permeability in insect Malpighian tubules. *J. Exp. Biol.*, *199*, 2537–2542.

Wang, Y. J., Zhao, Y., Meredith, J., Phillips, J. E., Theilmann, D. A., et al. (2000). Mutational analysis of the C-terminus in ion transport peptide (ITP) expressed in *Drosophila* Kc1 cells. *Arch. Insect Biochem. Physiol.*, *45*, 129–138.

Wang, Z., Hayakawa, Y., & Downer, R. G.H. (1990). Factors influencing cyclic AMP and diacylglycerol levels in fat body of Locusta migratoria. *Insect. Biochem.*, *20*, 325–330.

Weaver, R. J., & Audsley, N. (2008). Neuopeptides of the beetle, *Te. Moliton* identified using MALDI-TOF mass spectrometry and deduced sequence from the *Tr. Castaneum* genome. *Peptides*, *29*, 168–178.

Wegener, C., Herbert, Z., Eckert, M., & Predel, R. (2002). The periviscerokinin (PVK) peptide family in insects: evidence for the inclusion of CAP_{2b} as a PVK family member. *Peptides.*, *23*, 605–611.

Weis-Fogh, T. (1952). Fat combustion and metabolic rate of flying locusts (Schistocerca gregaria Forsk.). *Phil. Trans. R. Soc. Ser. B.*, *237*, 1–36.

Weltens, R., Leyssens, A., Zhang, S. L., Lohrmann, E., Steels, P., et al. (1992). Unmasking an apical electrogenic H pump in isolated Malpighian tubules (Formica polyctena) by the use of barium. *Cell. Physiol. Biochem.*, *2*, 101–116.

Weng, X. H., Huss, M., Wieczorek, H., & Beyenbach, K. W. (2003). The V-type H(+)-ATPase in Malpighian tubules of Aedes aegypti: localization and activity. *J. Exp. Biol.*, *206*, 2211–2219.

Wheelock, G. D., Petzel, D. H., Gillett, J. D., Beyenbach, K. W., & Hagedorn, H. H. (1988). Evidence for hormonal control of diuresis after a blood meal in the mosquito Aedes aegypti. *Arch. Insect Biochem. Physiol.*, *7*, 75–89.

Wicher, D., Agricoln, H. J., Söhler, S., Gundel, M., Heinemann, S. H., Wellweber, L., Stengl, M., & Derst, C. (2006). Differential receptor activation by cockroach adipkinetic hormones produces differential effects on ion currents, neuronal activity, and locomotion. *J. Neurophysiol.*, *95*, 2314–2325.

Wicher, D., Walther, C., & Penzlin, H. (1994). Neurohormone D induces ionic current changes in cockroach central neurones. *J. Comp. Physiol. A.*, *174*, 507–515.

Wiehart, U. I., Nicolson, S. W., & Van Kerkhove, E. (2003). K+ transport in Malpighian tubules of Te. molitor L.: a study of electrochemical gradients and basal K+ uptake mechanisms. *J. Exp. Biol.*, *206*, 949–957.

Wiehart, U. I.M., Nicolson, S. W., Eigenheer, R. A., & Schooley, D. A. (2002). Antagonistic control of fluid secretion by the Malpighian tubules of Te. molitor: effects of diuretic and antidiuretic peptides and their second messengers. *J. Exp. Biol.*, *205*, 493–501.

Wiens, A. W., & Gilbert, L. I. (1967). Regulation of carbohydrate mobilization and utilization in Leucophaea maderae. *J. Insect. Physiol.*, *13*, 779–794.

Wiesner, A., Losen, S., Kopacek, P., Weise, C., & Gotz, P. (1997). Isolated apolipophorin III from Galleria mellonella stimulates the immune reactions of this insect. *J. Insect. Physiol.*, *43*, 383–391.

Wille, S., Sydow, S., Palchaudhuri, M. R., Spiess, J., & Dautzenberg, F. M. (1999). Identification of amino acids in the N-terminal domain of corticotropinreleasing factor receptor 1 that are important determinants of high-affinity ligand binding. *J. Neurochem.*, *72*, 388–395.

Winther, Å., M. E., & Nässel, D. R. (2001). Intestinal peptides as circulating hormones: release of tachykinin-related peptide from the locust and cockroach midgut. *J. Exp. Biol.*, *204*, 1269–1280.

Witten, J. L., Schaffer, M. H., O'Shea, M., Cook, J. C., Hemling, M. E., et al. (1984). Structures of two cockroach neuropeptides assigned by fast atom bombardment mass spectrometry. *Biochem. Biophys. Res. Commun.*, *124*, 350–358.

Woodring, J., Lorenz, M. W., & Hoffmann, K. H. (2002). Sensitivity of larval and adult crickets (Gryllus bimaculatus) to adipokinetic hormone. *Comp. Biochem. Physiol.*, *133A*, 637–644.

Xu, W., & Marshall, A. T. (1999). Magnesium secretion by the distal segment of the Malpighian tubules of the black field cricket Teleogryllus oceanicus. *J. Insect. Physiol.*, *45*, 777–784.

Yamanaka, N., Yamamoto, S., Zitnan, D., Watanabe, K., Kawada, T., et al. (2008). Neuropeptide receptor transcriptome reveals unidentified neuroendocrine pathways. *PLoS. ONE.*, *3*, e3048.

Yu, M. J., & Beyenbach, K. W. (2001). Leucokinin and the modulation of the shunt pathway in Malpighian tubules. *J. Insect. Physiol.*, *47*, 263–276.

Yu, M. J., & Beyenbach, K. W. (2002). Leucokinin activates $Ca^{(2+)}$-dependent signal pathway in principal cells of Aedes aegypti Malpighian tubules. *Am. J. Physiol. Renal. Physiol.*, *283*, F499–F508.

Yu, M. J., & Beyenbach, K. W. (2004). Effects of leucokinin-VIII on Aedes Malpighian tubule segments lacking stellate cells. *J. Exp. Biol.*, *207*, 519–526.

Zdobnov, E. M., von Mering, C., Letunic, I., Torrents, D., Suyama, M., et al. (2002). Comparative genome and proteome analysis of *Anopheles gambiae* and *Drosophila* melanogaster. *Science.*, *298*, 149–159.

Zhang, S. L., Leyssens, A., Van Kerkhove, E., Weltens, R., Vandriessche, W., et al. (1994). Electrophysiological evidence for the presence of an apical H+-ATPase in Malpighian tubules of Formica polyctena - intracellular and luminal pH measurements. *Pflugers. Arch.-Eur. J. Physiol.*, *426*, 288–295.

Zhu, C., Huang, H., Hua, R., Li, G., Yang, D., Luo, J., Zhang, C., Shi, L., Benovic, J. L., & Zhou, N. (2009). Molecular and functional characterization of adipokinetichormone receptor and its peptide ligands in *Bombyx Mori*. *FEBS Lett*, *583*, 1463–1468.

Ziegler, R. (1984). Developmental changes in the response of the fat body of Manduca sexta to injections of corpora cardiaca extracts. *Gen. Comp. Endocrinol.*, *54*, 51–58.

Ziegler, R., Cushing, A. S., Walpole, P., Jasensky, R. D., & Morimoto, H. (1998). Analogs of Manduca adipokinetic hormone tested in a bioassay and in a receptor-binding assay. *Peptides*, *19*, 481–486.

Ziegler, R., Eckart, K., Jasensky, R. D., & Law, J. H. (1991). Structure-activity studies on adipokinetic hormones in Manduca sexta. *Arch. Insect. Biochem. Physiol., 18,* 229–237.

Ziegler, R., Eckart, K., & Law, J. H. (1990). Adipokinetic hormone controls lipid metabolism in adults and carbohydrate metabolism in larvae of Manduca sexta. *Peptides, 11,* 1037–1040.

Ziegler, R., Jasensky, R. D., & Morimoto, H. (1995). Characterization of the adipokinetic hormone receptor from the fat body of Manduca sexta. *Reg. Pept., 57,* 329–338.

Ziegler, R., & Schulz, M. (1986). Regulation of lipid metabolism during flight in Manduca sexta. *J. Insect. Physiol., 32,* 903–908.

Zitnan, D., Kim, Y. J., Zitnanova, I., Roller, L., & Adams, M. E. (2007). Complex steroid-peptide-receptor cascade controls insect ecdysis. *Gen. Comp. Endocrinol., 153,* 88–96.

Zubrzycki, I. Z., & Gäde, G. (1994). Conformational study on an insect neuropeptide of the AKH/RPCH-family by combined ^{1}H-NMR spectroscopy and molecular mechanics. *Biochem. Biophys. Res. Commun., 198,* 228–235.

Zubrzycki, I. Z., & Gäde, G. (1999). Conformational study on a representative member of the AKH/RPCH neuropeptide family, Emp-AKH, in the presence of SDS micelle. *Eur. J. Entomol., 96,* 337–340.

10 Hormonal Control of Diapause

D.L. Denlinger
Ohio State University, Columbus, OH, USA
G.D. Yocum and J.P. Rinehart
USDA-ARS Red River Valley Agricultural Research
Center, Fargo, ND, USA

© 2012 Elsevier B.V. All Rights Reserved

Summary

A critical feature of the insect life cycle is the ability to shut down development and enter a period of diapause during inimical seasons. This chapter briefly discusses features of the seasonal environment that program this developmental arrest and examines endocrine processes that preside over this decision. Although many links in the pathway leading from reception of environmental signals to expression of the diapause phenotype remain poorly known, recent advances have defined new elements. Ecdysteroids, juvenile hormones, and the neuropeptides that regulate their synthesis and release are well-known contributors to diapause regulation, but new roles for insulin signaling and newly discovered roles for additional neuropeptides have expanded our understanding of the regulatory schemes controlling diapause. Numerous physiological attributes are common to diapauses in different species and developmental stages, yet the diapause phenotype appears to have evolved numerous times using divergent molecular mechanisms.

10.1. Introduction

The evolution of diapause is arguably one of the most critical events bolstering the success of insects. Having the capacity to periodically shut down development has enabled insects and their arthropod relatives to invade environments that are seasonally hostile. Indeed, very few environments permit continuous insect development. Even in the tropics, where temperatures throughout the year may be compatible with ectotherm development, seasonal patterns of rainfall drive cycles of plant growth that favor insects with the ability to periodically become dormant.

Diapause is an arrest in development accompanied by a major shutdown in metabolic activity. Unlike a simple quiescence that is an immediate response to an unfavorable environmental condition, diapause is a genetically programmed response that occurs at a specific stage for each species. Sometimes the insect will enter diapause at this particular stage in each generation regardless of the environmental conditions it receives — a developmental program referred to as obligatory diapause. Much more frequently, the decision to enter diapause is determined by environmental factors, usually daylength, received by that individual or its mother at an earlier developmental stage. This is referred to as facultative diapause.

Development can effectively be interrupted at many points, as evidenced by the rich diversity of developmental stages used by different species for diapause. Embryonic diapauses have been documented for nearly all possible stages of embryonic development, ranging from early blastoderm formation to pharate larvae. Larval diapauses are most common in late instars, but early instar diapauses are not uncommon. Pupal diapause is well documented, and a few species diapause as fully developed pharate adults, but, not surprisingly, there appear to be no instances where development is halted midway through pharate adult development. The most prevalent cases of adult

DOI: 10.1016/B978-0-12-384749-2.10010-X

diapause involve newly emerged adults that have not yet reproduced, but there are also examples of diapause interrupting periods of reproduction. Most commonly, insects enter diapause only once during their life cycle, but some species, especially those living at higher latitudes, may extend their development over several years, spending successive winters in diapauses of different stages. Some long-lived adults may also go through multiple diapauses to coordinate egg laying with favorable conditions.

Diapause is programmed well in advance of its onset, which implies that there is a pre-diapause period that can be used to prepare the insect for the special conditions that lie ahead. Hence it is common for insects destined for diapause to have a somewhat prolonged period of pre-diapause development, attain a larger size, accumulate additional lipid reserves, and add additional waterproofing to their cuticle. Distinct coloration, often to match the insect's overwintering environment, may accompany the entry into diapause. Insects entering a winter diapause frequently undergo metabolic adjustments to enhance cold hardiness. In some cases the cold hardiness is a component of the diapause program, but in other cases cold hardiness is invoked by a different set of environmental cues.

The term diapause falsely suggests a period of arrest in which development comes to a complete halt. Diapause should not be regarded as a simple stop and restart of development. Andrewartha (1952) originally coined the term "diapause development" to refer to the ongoing progression of events that occurs during diapause and eventually results in its termination. It is evident that diapause is a dynamic process, as evidenced by characteristic changes in the utilization of energy reserves, systematic changes in patterns of oxygen consumption, changes in responsiveness to environmental stress and hormones, and distinct patterns and changes in gene expression. A progression of events must transpire before diapause can be terminated. In some cases, development is not completely halted. There are several examples in which developmental processes continue during diapause, albeit at much slower rates; for example, ovarian maturation during adult diapause in the mosquito *Culex pipiens* (Readio *et al.*, 1999) and embryonic development during diapause in the pea aphid *Acyrthosiphon pisum* (Shingleton *et al.*, 2003). In some cases there can be profound differences even within the body of a single individual. For example, during pupal diapause in the tobacco hornworm, *Manduca sexta*, most tissues of the body are arrested, but the testicular stem cells continue to undergo mitotic division (Friedlander and Reynolds, 1992). These unique events of diapause, coupled with pre-diapause distinctions in diapause-programmed individuals, suggest that it is most appropriate to view diapause as an alternate developmental pathway.

Several comprehensive reviews cover select features of diapause including environmental (Tauber *et al.*, 1986; Danks, 1987) and molecular (Denlinger, 2002; MacRae, 2010; Williams *et al.*, 2010) regulation, circadian mechanisms involved in timekeeping (Saunders, 2002, 2010a,b), photoreceptors and clock genes (Goto *et al.*, 2010; Kostal, 2011), the dynamics of diapause (Kostal, 2006), cold tolerance associated with diapause (Denlinger and Lee, 2010), impact of climate change (Bale and Hayward, 2010), and energy management during diapause (Hahn and Denlinger, 2007; 2011). The goal of this chapter is to complement the other reviews by highlighting major discoveries relevant to the hormonal control of diapause that have occurred subsequent to our previous reviews (Denlinger, 1985; Denlinger *et al.*, 2005). Earlier findings are briefly summarized in this chapter, with emphasis on new developments.

10.2. The Preparative Phase

During the preparative phase, the insect must make the decision to enter diapause (and sometimes how long to remain in diapause) and make the accompanying alterations that will result in the sequestration of additional energy reserves and waterproofing agents as well as initiate specific behavioral programs needed to find protected sites appropriate for overwintering. Some species, such as the Monarch butterfly, *Danaus plexippus*, engage in long-distance flights to reach their overwintering site, but more commonly, insects programmed for diapause migrate to closer, buffered sites. When diapause is viewed as an alternate developmental program, these phases unique to diapause, even if they occur prior to diapause, must also be viewed as part of the diapause syndrome.

These early distinctions underscore the fact that we cannot simply focus on the endocrine events that dictate the onset and termination of diapause. A comprehensive view of diapause also calls for an understanding of the control mechanisms regulating these early events, but regrettably we know very little about the events regulating early phases of this developmental pathway. For example, what endocrine signals direct a diapause-programmed insect to sequester more lipid reserves or coat its cuticle with additional waterproofing hydrocarbons? These observations suggest that the set points for lipid accumulation and synthesis of cuticular hydrocarbons differ in diapause- and non-diapause-destined individuals, distinctions most likely regulated by the endocrine system. The amounts of energy stores and the composition of those stores differ between diapause- and non-diapause-programmed individuals. A number of potential regulators of pre-diapause energy storage have been proposed (Hahn and Denlinger, 2011), including insulin signaling and its sister pathway, target of rapamycin (TOR) signaling, but few experiments have probed those pathways in relation to diapause.

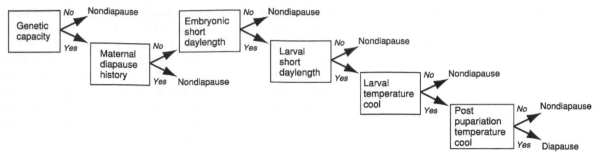

Figure 1 Sequence of signals required for the programming of pupal diapause in flesh flies (Sarcophaga). Reproduced with permission from *Denlinger* (1985).

10.2.1. Making the Decision to Enter Diapause

For an obligate diapause, the decision to enter diapause is hard-wired. Diapause will be entered at a specific stage regardless of the environmental conditions. This is an arrangement common to species that have a single generation each year, such as the gypsy moth and the Cecropia silkmoth, although in both of these examples a few individuals fail to enter diapause. By contrast, a facultative diapause enables the insect to complete several generations per year, and the individual "decides" its own developmental fate based on the environmental cues it perceives. The plasticity of this arrangement offers flexibility and enables the insect to more closely coordinate development with the appropriate seasons. An additional variation of this theme is when the individual does not directly decide and the decision is made by the mother.

The decision to diapause should not be viewed as a simple "yes/no" decision made at a single moment in time; it is the culmination of a series of events, as exemplified for the flesh fly example shown in **Figure 1**. This example shows that the programming of diapause can be averted at many points along the way, right up to the time that diapause is normally manifested. For the program to be enacted, all of the previous conditions must be fulfilled. Thus, the diapause program can be aborted at the last minute, but attempts to induce diapause by offering the correct signals only during the later phases of development are not sufficient to induce diapause.

10.2.1.1. Reading the environmental cues Short daylength, a reliable harbinger of winter, is the most common cue used for the programming of an overwintering diapause (**Figure 2**). The precision of this environmental token makes it a seasonal indicator of unparalleled reliability. The critical photoperiod, the transition point between photoperiods that elicit diapause and those that elicit non-diapause development, is usually quite sharp, implying that insects can readily distinguish daylength differences as small as 15 min (Denlinger, 1986). Insects that enter a summer diapause may use long daylengths instead of short daylengths for the programming of diapause (Masaki, 1980), and there

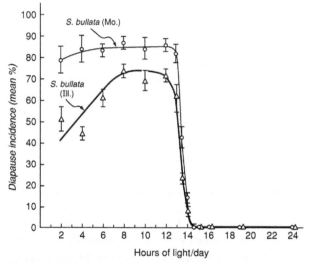

Figure 2 Critical photoperiod for diapause induction in populations of *Sarcophaga bullata* from Illinois (40° 15'N) and Missouri (38° 30'N) at 25°C. Reproduced with permission from *Denlinger* (1972).

are numerous examples of photoresponse curves showing restricted portions of the curve to be diapause inductive. Daylength appears to be measured as a threshold character, and only a few examples suggest that insects are capable of actually measuring changes in daylength (Tauber *et al.*, 1986). The widespread importance of photoperiodism in regulating diapause induction is reflected by the fact that it is used almost universally by insects in all geographic areas excluding areas within approximately 5 degrees of the equator. Diapause is still prevalent in the equatorial region, but other cues such as temperature, rainfall, or plant volatiles replace photoperiod as the cue for diapause induction in that region (Denlinger, 1986).

With a few exceptions, temperature also influences the decision to enter diapause. While it may act alone as the environmental cue used for diapause induction in some tropical species, temperature usually acts in concert with the photoperiod. As shown in **Figure 3a**, low temperature commonly increases the diapause incidence observed at short daylength without influencing the critical photoperiod, but in some cases the critical photoperiod

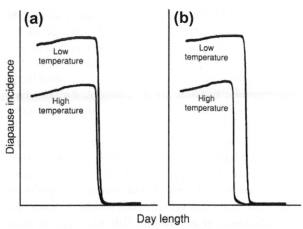

Figure 3 Low temperature can impact the photoperiodic response curve in two ways. (a) In some cases, the critical photoperiod remains the same but the incidence of diapause at short daylengths is higher. (b) In other instances, the critical photoperiod is also shifted toward longer daylengths at low temperature. Reproduced with permission from *Denlinger* (2001).

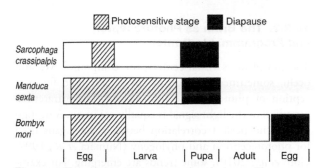

Figure 4 The relationship between the photosensitive stage and the stage of diapause in three insect examples. Reproduced with permission from *Denlinger* (1985).

will shift toward longer daylengths at low temperature (**Figure 3b**). Under natural conditions, insects are simultaneously subjected to thermoperiods as well as photoperiods. Thermoperiod can sometimes substitute for a photoperiodic signal, and it presumably does so by linking to a common mechanism (van Houten and Veerman, 1990). Other factors contributing to the diapause decision include food presence (Fielding, 1990), food quality (Hahn and Denlinger, 2011; Liu *et al.*, 2011), plant volatiles (Ctvrtecka and Zdarek, 1992), pheromones (Kipyatkov, 2001), and crowding (Desroches and Huignard, 1991; Saunders *et al.*, 1999).

The photosensitive stage, the stage that receives the photoperiodic cues used to program diapause, usually occurs fairly far in advance of the actual diapause stage, as noted in **Figure 4**. The photosensitive stage may be as brief as four days as observed in *Sarcophaga crassipalpis*, or it may encompass much of the pre-diapause period as noted in *M. sexta*. In the extreme example noted in the silkmoth *Bombyx mori,* the photosensitive period actually occurs in the previous generation. If diapause is to be programmed by daylength, it is imperative that the correct photoperiodic signals are received during this photosensitive period. At other times during development the length of the day is of no consequence for the programming of diapause.

Thus, insects that rely on short days to program their diapause need to answer two questions: (1) Was the day short? and (2) How many short days were received? Answering the first question requires a sophisticated clock that can measure daylength with precision, and answering the second question requires a counter capable of summing the number of short days that comprise the photosensitive period (Saunders, 2002, 2010a).

10.2.1.2. Letting mother decide In a number of parasitic Hymenoptera, higher Diptera, and some Lepidoptera it is the mother that decides the diapause fate of her progeny. This arrangement may be particularly useful when the mother has access to environmental cues not readily accessible to her progeny. Most of these cases involve the mother determining the diapause status of her eggs. The embryos that enter diapause are often in an early stage of development and are not yet equipped with the sophisticated neural systems required to receive and respond to photoperiodic cues. *Bombyx mori* is in this category, and since quite a bit is known about diapause in this species it will be discussed separately (see Section 10.3.1.1.). In other cases, however, the mother influences the diapause fate of her progeny in their larval (e.g., *Nasonia vitripennis*) or pupal (e.g., *S. bullata*) stages. This clearly implies that the mother is not simply exerting some direct effect on her progeny as might be expected in an embryonic diapause, but that she is initiating an effect expressed at a much later stage of development.

In spite of the intriguing nature of these maternal effects, very little is known about the mechanisms that regulate these transgenerational effects. In the flesh fly *S. bullata* a critical message that influences the maternal effect is transferred from the brain of the female to her ovary sometime between pupariation and shortly after adult emergence (Rockey *et al.*, 1989). Gamma aminobutyric acid (GABA) and octopamine are possibly involved in the transfer of information from mother to progeny (Webb and Denlinger,1998). Differential display of mRNA revealed a transcript unique to the ovaries of females that expresses the maternal effect, but the function of the protein it encodes is still unknown (Denlinger, 1998). How the mother can influence the diapause fate of her offspring offers an intriguing, and still unanswered, problem in development. This sort of transgenerational transfer of information (Ho and Burggren, 2010) appears to be ripe for investigating a potential epigenetic mechanism of regulation.

10.2.2. The Brain as Photoreceptor and Programmable Center

The conspicuous visual centers, the compound eyes and ocelli, sometimes function as the organs used for perception of photoperiodic information, but extraretinal reception of the light signal is equally common. Although there is no perfect correlation between the organs used for photoreception and phylogeny (Numata *et al.*, 1997; Goto *et al.*, 2010), some trends are emerging. For example, extraretinal reception seems to be particularly common among the Lepidoptera, whereas reliance upon both compound eyes and direct brain photoreception is evident among the Hemiptera, Coleoptera, and Diptera. In some cases both types of photoreception operate within the same species (Goto *et al.*, 2010).

In cases of extraretinal reception, blackening or ablating the compound eyes has no impact on the insect's ability to respond to photoperiod. A translucent cuticular window or lightened regions of cuticle sometimes overlay the brain, facilitating transfer of light to extraretinal receptors within the brain. In classic experiments with the giant silkmoth, *Antheraea pernyi*, Williams and Adkisson (1964) elegantly demonstrated that the brain is the direct recipient of the light signal. This species terminates pupal diapause in response to long daylength, and by using a divided photoperiod chamber that offered diapause-terminating conditions to one half of the body and diapause-maintaining conditions to the other half, they were able to demonstrate that diapause can be broken only when the portion of the body containing the brain is exposed to diapause-terminating conditions. This is true even when the brain is transplanted to the abdomen of the pupa. In yet another elegant demonstration of the brain's role in photoreception, Bowen *et al.* (1984) demonstrated that the brain of *M. sexta*, when exposed *in vitro* to short days, is capable of eliciting diapause when implanted into a debrained pupa. Likewise, larval brain–subesophageal ganglion complexes of *B. mori* can be photoperiodically programmed for diapause *in vitro* and elicit a diapause response when re-implanted into larvae (Hasegawa and Shimizu, 1987).

A number of more recent cases also demonstrate a role for the compound eyes (Goto *et al.*, 2010). Females of the fly *Protophormia terraenovae* (Shiga and Numata, 1997), the cricket *Modicogryllus siamensis* (Sakamoto and Tomioka, 2007), and the bugs *Plautia stali* (Morita and Numata, 1999), *Graphosoma lineatum* (Nakamura and Hodkova, 1998), and *Poecilocoris lewisi* (Miyawaki *et al.*, 2003) cannot distinguish short days from long days after surgical removal of the compound eyes. Painting the compound eyes of the bug *Riptortus clavatus* with phosphorescent paint effectively extends the photoperiod, resulting in termination of diapause

(Numata and Hidaka, 1983). By contrast, painting the head region above the pars intercerebralis fails to terminate diapause, suggesting that the compound eyes are the dominant photoreceptors for this species. Not all ommatidia of the compound eye are involved in the photoperiodic response of *R. clavatus*. By selectively removing ommatidia from various regions of the compound eyes, it is evident that only ommatidia in the central region of the compound eye are involved in this response (Morita and Numata, 1997b). Evidence from circadian rhythms suggests that in some cases (e.g., *P. terraenovae*) both retinal and extraretinal pathways contribute to rhythm entrainment within the same species (Hamasaka *et al.*, 2001). This also appears to be true for photoperiodism in some species such as *P. stali* (Morita and Numata, 1999).

The diapause program resides within the brain and can be transferred from one individual to another by transplantation of the programmed brain, as demonstrated in the embryonic diapause of *B. mori* (Hasegawa and Shimizu, 1987), pupal diapause of *M. sexta* (Safranek and Williams, 1980; Bowen *et al.*, 1984), and *S. crassipalpis* (Giebultowicz and Denlinger, 1986), and adult diapause of the linden bug *Pyrrhocoris apterus* (Hodkova, 1992). In *M. sexta* the duration of diapause is also programmed within the brain prior to the onset of diapause. In this species, the number of short days the larva receives determines the length of the pupal diapause. Short days throughout produce a high incidence of diapause, but such a diapause is of short duration. By contrast, exposure to only a few short days yields a low incidence of diapause, but a diapause of long duration. Brain transplantation experiments show that information about the duration of the diapause can be transplanted along with the brain (Denlinger and Bradfield, 1981).

Precisely which cells within the brain are involved in photoreception and storage of diapause information remains unknown in most cases, but future studies pinpointing photoreceptor pigments and specific clock genes involved in photoperiodism should reveal the correct target sites for this critical information. In the giant silkmoth *A. pernyi*, the response is localized within the dorsolateral region of the brain, the same region containing the lateral neurosecretory cells that produce prothoracicotropic hormone (PTTH), the neuropeptide that regulates diapause. Thus the cells involved in photoreception and the effector region that produces PTTH are together in the same region of the brain (Sauman and Reppert, 1996a). In the vetch aphid, *Megoura viciae*, antibodies directed against invertebrate and vertebrate opsins and phototransduction proteins consistently label an anterior ventral neuropil region of the protocerebrum (Gao *et al.*, 1999). Identification of this region is consistent with previous experiments using localized illumination and microlesions to identify the photoreceptor region.

10.2.3. Transduction of the Environmental Signals

When photoperiod is the primary cue leading to the induction of diapause, several key molecular components are essential. The first key component is a photoreceptor capable of distinguishing between the photophase and scotophase. Secondly, the insect must have the capacity to measure the length of each photophase (or scotophase), and thirdly, the insect must be able to sum the number of short (or long) days it has seen. Although our understanding of the molecular events involved in the transduction of the environmental signals remains in its infancy, several recent studies have proven insightful, and the early literature offers valuable clues about the processes involved. As discussed previously, the photoreceptive organ, and in some cases candidate cells, have been identified for several diapause models. Action spectra that elicit diapause have been described for several species, and in most cases the blue-green range of the visual spectrum is most sensitive to diapause induction (Truman, 1976), indicating that any putative photoreceptor involved in the diapause response must have features consistent with these early observations.

How temperature exerts its effect on the diapause response is poorly understood. One attractive model proposed by Saunders (1971) for flesh flies suggests that low temperature simply increases the number of short days the larvae receive. In this system, 14 short days are required for diapause induction, and lowering the temperature retards development so a larger portion of the population experiences the required number of short days. Although this model is attractive for species with a long photosensitive period, such a system seems less likely to operate in other insects (including some closely related flesh fly species) that have a very short photosensitive period. In some cases, thermoperiod can completely substitute for photoperiod as an environmental cue. Adult diapause is induced in the predatory mite *Amblyseius potentillae* by a short-day photoperiod or by rearing under constant darkness with a thermoperiod of 16 h at 15°C and 8 h at 27°C (van Houten et al., 1987). Exposing *A. potentillae* to short-day photoperiods or to short-day thermoperiods either concurrently or in succession is successful for inducing diapause, suggesting that a single physiological mechanism is integrating photoperiodic and thermoperiodic information to regulate diapause (van Houten and Veerman, 1990). The argument that a single mechanism is involved is further strengthened by the observation that mites reared on a diet deficient in carotenoids/vitamin A (see Section 10.2.3.1.) lose their ability to respond to either photoperiod or thermoperiod (van Houten et al., 1987). By contrast, carotenoids/vitamin A are essential for the photoperiod response but not for the thermoperiodic response in the cabbage butterfly, *Pieris brassicae* (Claret, 1989). Clearly, species differences must again be evident in the integration of photoperiodic and thermoperiodic information.

10.2.3.1. Cryptochromes and opsins

One of the most important quests in diapause research is to identify the photoreceptor pigment involved in this response. A number of proteins have been identified, but most appear to be involved in visual perception. Only a few are candidates for involvement in photoperiodism, especially given the blue light sensitivity of the diapause response and the common involvement of extraretinal light reception. One such candidate is the flavin-based blue light photoreceptor cryptochrome (CRY). This protein has been directly linked to the circadian pacemaker of the fruit fly *Drosophila*, acting to daily reset the clock upon initiation of the photophase as detailed in the next section. Null mutants of CRY exhibit poor synchronization during light cycling, whereas rhythmicity is present in constant darkness (Stanewsky et al., 1998). These features support a role for CRY in entrainment of the circadian clock by photoperiod. The possible role of CRY in diapause induction has yet to be established, although *cry* expression is upregulated in adult *S. crassipalpis* when the flies are reared at short as opposed to long daylengths (Goto and Denlinger, 2002).

Another group of proteins linked to photoperiodism are the carotenoid/vitamin A associated opsin photoreceptors. A causal link with diapause has been established by rearing insects on carotenoid-free diets. Although several species, including *Drosophila melanogaster* (Zimmerman and Goldsmith, 1971) and *B. mori* (Shimizu et al., 1981), retain circadian behavioral patterns when reared on a carotenoid-free diet, the induction of diapause can be affected in a number of species, including adults of the predatory mite *A. potentillae* (van Zon et al., 1981; Veerman et al., 1985) and spider mite *Tetranychus urticae* (Bosse and Veerman, 1996), embryonic diapause of *B. mori* (Shimizu and Kato, 1984), and pupal diapause of *P. brassicae* (Claret, 1989; Claret and Volkoff, 1992). The opsin proteins have been linked to photoperiodism as well. Melanopsin has been implicated in circadian mechanisms in vertebrates. In mice, null mutants of melanopsin retain circadian rhythmicity in constant darkness, but lack photoperiodic entrainment (Panda et al., 2002; Ruby et al., 2002). Opsin 5 has also recently been identified as a deep brain photoreceptor functioning as an extraretinal receptor that regulates seasonal reproduction in birds (Nakane et al., 2010). Similar work has yet to be conducted in insects, although boceropsin, a novel opsin expressed specifically in the central nervous system (CNS), has been proposed as a photoreceptor protein in *B. mori* (Shimizu et al., 2001). A comparison of the phenotypes of null mutants of opsin- and carotenoid-deficient silkworms will be insightful.

10.2.3.2. Clock genes Regardless of the type of photoreceptor involved in diapause, the photoreceptor must be functionally connected to a timekeeping mechanism to exert its effect. Whether the clock mechanism that controls circadian rhythms utilizes the same timekeeping mechanism involved in photoperiodic time measurement (i.e., diapause) remains controversial (Emerson *et al.*, 2008, 2009a; Goto *et al.*, 2010; Saunders, 2010a,b; Kostal, 2011). The molecular basis for circadian timekeeping is particularly well described in *D. melanogaster* (Stanewsky, 2002; Price, 2004; Helfrich-Forster, 2009). In *Drosophila*, the clock is composed of a multiple component autoregulatory feedback loop. The best described components involve the interactive oscillations of the proteins period (PER), timeless (TIM), clock (CLK), and cycle (CYC). In this system, the proteins PER and TIM heterodimerize to enter the nucleus and transcriptionally repress expression of CLK and CYC. Conversely, CLK and CYC act to transcriptionally upregulate both PER and TIM. Thus, when levels of the proteins PER and TIM are high, they repress transcription of CLK and CYC. As levels of CLK and CYC diminish, transcription of PER and TIM are decreased until they are low enough to no longer repress CLK and CYC. Once the levels of CLK and CYC are restored, PER and TIM will again be transcribed. The genes in this autoregulatory loop, in addition to several others including vrille, doubletime, and CREB2, create two interacting phases of protein oscillations (PER with TIM and CLK with CYC), creating a circadian rhythm of gene expression. The connection of this circadian pacemaker to photoperiod is attained through the interaction of the proteins TIM and CRY. TIM is degraded in the photophase through the action of CRY, "resetting the clock" according to the daily light cycle. Although most of these genes appear to be conserved, it is evident that details of the clock mechanism may differ among species. For example, in *A. pernyi*, PER apparently does not enter the nucleus (Sauman and Reppert, 1996b).

These major clock genes have been studied intensely in association with circadian rhythms, yet it is not absolutely clear whether diapause uses the same timekeeping system. Although much evidence points to a role of circadian rhythms in the programming of diapause (Saunders, 2002, 2010b), several arguments defend the position that diapause does not have a circadian basis (Veerman, 2001; Veerman and Veenendaal, 2003). Even if diapause does have a circadian basis it is not clear whether the same clock genes and response mechanisms are involved in both circadian rhythmicity and photoperiodism (diapause). Several experiments suggest two distinct mechanisms. In both spider mites (Veerman and Veenendaal, 2003) and the blow fly *Calliphora vicina* (Saunders and Cymborowski, 2003), it is possible to segregate the photoperiodic response mechanism from the mechanism governing daily behavioral rhythms. Experiments utilizing mutants suggest that distinct mechanisms may operate in these two systems. *per* null mutants of *D. melanogaster* are arrhythmic, yet they retain the ability to enter adult diapause (Saunders *et al.*, 1989; Saunders, 1990). Similarly, in the drosophilid *Chymomyza costata*, mutants with reduced rhythmicity continue to enter larval diapause (Lankinen and Riihimaa, 1992), and *per* null mutants retain the capacity for diapause as well (Shimada, 1999; Kostal and Shimada, 2001). It is possible that there is enough redundancy in the system so that knocking out a single gene does not render the timekeeping mechanism involved in photoperiodism inoperative, and in *per* mutants, TIM could be degraded directly through CRY as nicely pointed out by Bradshaw and Holzapfel (2007).

The involvement of clock genes in diapause is likely as suggested by several types of experiments. In *S. bullata*, the circadian system that regulates adult eclosion and diapause appear to be linked: a mutant that lacks rhythmicity also fails to enter diapause (Goto *et al.*, 2006). In *S. crassipalpis*, *period*, *cycle*, *clock*, and *cryptochrome* show little differences in expression patterns when the flies are reared under short day or long day. However, the expression of *timeless* is greatly suppressed in long days as compared to short days (Goto and Denlinger, 2002b), suggesting a possible role for *tim* in the mechanism distinguishing short and long days. Although the developmental arrest in *D. melanogaster* is modest by comparison to other species and is probably more akin to quiescence (Emerson *et al.*, 2009b), several lines of evidence point to the involvement of *tim* in this dormancy (Tauber *et al.*, 2007; Sandrelli *et al.*, 2007), as well as the diapause of the pitcher plant mosquito, *Wyeomyia smithii* (Mathias *et al.*, 2005). In addition, RNAi directed against *tim* elicited a modest effect in reducing diapause in *C. costata* (Pavelka *et al.*, 2003), which suggests a possible role for *tim* as a component of the photoperiodic timekeeping mechanism. A role for *tim* in the diapause of *C. costata* is further bolstered by the interesting observation that, in the non-diapausing mutant, the protein TIM is not present in the two neurons in each brain hemisphere that contain this protein in wild-type flies (Stehlik *et al.*, 2008). Collectively, these results underscore the importance of *tim* in the diapause of this species. Recent RNAi results also yielded a particularly interesting effect on diapause in the bug, *R. pedestris*: RNAi directed against *per* caused the bug to avert diapause when reared under a diapause-inducing photoperiod, while RNAi directed against *cycle* induced diapause under a diapause-averting photoperiod (Ikeno *et al.*, 2010). A minimal interpretation of these profound effects is that both of these well-known clock genes participate in the diapause decision. In this experiment, altering the expression patterns of these two clock genes affected diapause as well as the circadian pattern of cuticle deposition, a nice demonstration that the same genes are involved in both of these timekeeping mechanisms. It is difficult, as pointed out by Bradshaw and Holzapfel (2010), to clearly distinguish between functions for the clock genes in photoperiodism versus pleiotropic effects that may be acting on sister pathways.

Results are emerging that implicate a number of the clock genes in the programming of diapause, yet there are also sufficient reports in the literature suggesting that the gene interplay described so elegantly for the control of circadian rhythms of behavior in *D. melanogaster* may not be fully relevant to diapause induction. Resolution of these disparities is pivotal for understanding diapause induction at the molecular level.

10.3. Endocrine Regulators

10.3.1. Hormonal Basis for Embryonic Diapause

Embryonic diapause is common among the Orthoptera, Hemiptera, Lepidoptera, and Diptera (especially *Aedes* mosquitoes) with occasional reports from other orders as well. The diverse stages in which embryonic diapause occurs suggest a rich diversity of regulatory schemes. Hormonal regulation has been examined only in a few species of Lepidoptera, and it is evident that no single regulatory scheme prevails. The most intensely studied embryonic diapause is that of *B. mori* (Yamashita and Hasegawa, 1985; Yamashita, 1996). Classic experiments, begun in the 1950s by Fukuda (1951, 1952) and Hasegawa (1952), set the stage for a wealth of papers documenting the regulatory mechanism controlling the early embryonic diapause in this commercially important species. Recent work has provided major contributions to the hormonal, biochemical, and molecular characterization of diapause in *B. mori* (Yamashita *et al.*, 2001).

Diapause in *B. mori* differs from that of most other species. While both temperature and daylength are key components contributing to the diapause decision, the specifics are reversed as compared to other diapause models. High temperature and long daylength prompt diapause induction, while low temperature and short daylength lead to non-diapause development. Diapause induction in this species is under strict maternal control. The photosensitive stage occurs far in advance of the actual diapause stage: exposure of developing female embryos and larvae to diapause-inducing conditions cause the adult female to produce eggs that are destined to enter diapause. This effect is mediated by diapause hormone (DH), a neuropeptide released by the female adult during her period of egg maturation, which acts upon the ovarioles and leads to production of diapause-destined eggs. After being oviposited, the silkworm embryos enter a diapause characterized by G2 cell cycle arrest (Nakagaki *et al.*, 1991). Once diapause has been established, this arrest can be broken after the embryos have been chilled for 1–3 months at 5°C,

after which development can be reinitiated by exposure to higher temperatures. As early as 1926, ovary transplantation experiments indicated that the factor controlling diapause induction was blood-borne (Umeya, 1926). Later studies by Fukuda and Hasegawa reported that the brain and subesophageal ganglion controlled the release of DH, and bioassays performed by injecting non-diapause-type individuals with subesophageal ganglion and brain extracts showed DH activity in these tissues.

10.3.1.1. Diapause hormone and its action in *Bombyx mori*
Although the activity of DH was established quite early, purification and sequencing of the peptide progressed slowly. An initial characterization was made from extracts of 15,000 isolated brain–subesophageal ganglion complexes (Hasegawa, 1957). Further analysis, including an assessment of amino acid composition, was made from extracts of 1 million isolated male adult heads (Isobe *et al.*, 1973). A final push for determining the amino acid sequence began in 1987, when extracts from 100,000 subesophageal ganglia generated enough material for sequencing and finally yielded a 24 amino acid peptide (Imai *et al.*, 1991). Subsequent cloning of the cDNA for DH confirmed the amino acid sequence (**Figure 5**) and led to further characterization of the hormone (Sato *et al.*, 1992, 1993).

DH is a member of the FXPRL-amide peptide family, sharing the same C-terminus five amino acid sequence with a variety of hormones including pheromone biosynthesis activating neuropeptide (PBAN) and pyrokinin. Although a sequence of PRL-NH2 is the minimum structure required for DH activity (Imai *et al.*, 1998), substantially greater activity is noted from the C-terminal pentapeptide. By synthesizing a series of DH analogs of different lengths, it was determined that two other regions of the peptide in the center and near the N-terminus, as well as the C-terminal amide group, also contributed to biological activity (Saito *et al.*, 1994; Suwan *et al.*, 1994). Cloning of the cDNA for DH resulted in an 800 bp nucleic acid sequence with a 192 amino acid open reading frame, much larger than the 24 amino acid DH peptide (Sato *et al.*, 1992, 1993). Analysis of the deduced amino acid sequence revealed that not only DH, but several additional neuropeptides, including PBAN and three other members of the FXPRL neuropeptide family are encoded

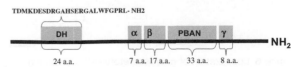

Figure 5 Amino acid sequence of diapause hormone (DH) from *Bombyx mori* and organization of the precursor polyprotein including locations of DH, PBAN, and three additional subesophageal ganglion neuropeptides. Adapted from *Sato et al.* (1993).

in the single open reading frame (**Figure 5**). The resulting polyprotein is a prohormone that is post-translationally processed to excise the active peptides. The presence of PBAN and other neuropeptides in the DH prohormone stresses the importance of post-translational control of hormone release, and helps to explain the formerly perplexing data indicating the presence of DH activity from a wide variety of developmental stages, from both sexes, and from both diapausing and non-diapausing individuals. For instance, it is now assumed that the large peak of activity seen in developing pharate adults (Sonobe *et al.*, 1977) can be attributed to the synthesis of PBAN as the moth completes its maturation and prepares for pheromone biosynthesis. Genomic characterization of DH has also been achieved: Southern blotting reveals a single copy of DH-PBAN in the *Bombyx* genome, and linkage analysis suggests that it is located on chromosome 11 (Pinyarat *et al.*, 1995). The gene contains six exons separating five introns with DH encoded by the first two exons. The region upstream of the open reading frame includes an ecdysone response element and five homeodomain TAAT binding core domains (Xu *et al.*, 1995a).

Even in the absence of purified hormone or the DH clone, early researchers were able to characterize DH control mechanisms and the sites of DH synthesis and release. Early transplantation experiments demonstrated DH activity in the subesophageal ganglion, and histological studies from Fukuda's laboratory identified a single pair of cells in the subesophageal ganglion as the likely site of DH release. More sensitive molecular methods have largely confirmed these results. Real-time PCR results indicate that DH is exclusively expressed in the subesophageal ganglion with no DH expression in the brain, thoracic or abdominal ganglia, or non-neural tissues (Sato *et al.*, 1994). *In situ* hybridization studies illustrate that expression is concentrated in 12 cells grouped into three clusters along the midline of the subesophageal ganglion (Sato *et al.*, 1994). Surgical manipulation indicates that the different clusters are responsible for different activities of the prohormone. Females retaining only the two cells in the posterior cluster continue to be capable of inducing diapause in their progeny, while females with only

the six cells in the medial cluster retain PBAN function. Removal of the anterior cluster of four cells does not substantially affect either function, suggesting that this cluster is involved in the function of other portions of the prohormone (Ichikawa *et al.*, 1996). Immunohistochemical studies using antibodies recognizing the FXPRL-amide further support the identification of cells with DH activity. Of the 12 cells in the subesophageal ganglion that exhibit antigenicity, only two show decreased staining in diapause-type as compared to non-diapause-type females (**Figure 6**), suggesting that the peptide is being released (Sato *et al.*, 1998; Kitagawa *et al.*, 2005). Immunocytochemistry combined with intracellular Lucifer Yellow injections reveals that axons from all three clusters pass through the brain and into the corpus cardiacum (Ichikawa *et al.*, 1995), the known site of DH release.

While early studies showed DH activity in larval, pupal, and adult stages (Hasegawa, 1964), more sensitive techniques led to further resolution of DH expression patterns. Use of RT-PCR determined that DH-PBAN transcription is controlled by both development and temperature (Xu *et al.*, 1995b; Morita *et al.*, 2004). When *B. mori* is raised under diapause-inducing conditions of 25°C, multiple expression peaks are noted throughout development: the first during embryogenesis, one during both the fourth and fifth instars, and two during pupal and adult development. In contrast, when the insects are reared under a non-diapausing temperature of 15°C, only the final peak of gene expression, during pupal and adult development, is observed (**Figure 7**). This final transcription peak is likely associated with PBAN, rather than DH, activity.

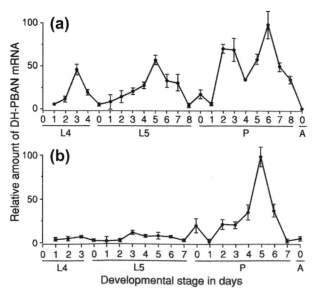

Figure 7 Levels of DH-PBAN mRNA in (a) diapause and (b) non-diapause type *Bombyx mori* during the course of development. L4, L5, fourth and fifth instars; P, pupa; A, adult. Adapted from *Xu et al.* (1995a).

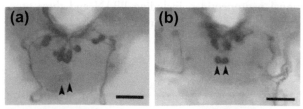

Figure 6 Localization of neurosecretory cells containing diapause hormone (DH) within the subesophageal ganglion of (a) diapause and (b) non-diapause type adults of *Bombyx mori*. Release of DH leads to a decrease of staining in diapause-type individuals. Arrows indicate the somata containing DH. Adapted from *Sato et al.* (1998).

The neural control of DH expression and release has also been characterized. Both *in vivo* and *in vitro* studies show that DH transcription is under control of the neurotransmitter dopamine. Dopamine levels in both the hemolymph and brain–subesophageal ganglion complexes of diapause-type individuals are significantly higher than in their non-diapause-type counterparts during the late larval and early pupal stages (Noguchi and Hayakawa, 2001), correlating with an increase in dopamine decarboxylase expression. Furthermore, experiments in which DOPA is fed to non-diapause-type larvae during the final larval instar or injected into non-diapause-type pupae resulted in the production of diapausing eggs by non-diapause-type females. Additionally, pupal injections of DOPA elevate DH mRNA in the brain–subesophageal ganglion complexes. *In vitro* studies further confirm the role of dopamine, as DH mRNA levels in isolated brain–subesophageal ganglion complexes from non-diapause-type pupae are elevated within 2 h after addition of either DOPA or dopamine to the medium. Expression of the dopamine receptors is not affected by diapause status and is concentrated in the mushroom bodies, suggesting that they contribute to the pathways leading to DH production (Mitsumasu *et al.*, 2008). Transcription factors involved in regulation of DH include a member of the pituitary homeobox gene family (Shiomi *et al.*, 2007), as well as at least two members of the POU transcription factor family (Xu *et al.*, 1995b; Zhang *et al.*, 2004a).

Neuronal control of DH is not restricted to the transcriptional level. The release of DH from the subesophageal ganglion is under control of GABAergic neurons. Culturing isolated brain–subesophageal complexes from diapause-egg-producing females in the presence of the inhibitory neurotransmitter GABA causes a substantial reduction in release of DH, while incubation with picrotoxin, the GABA inhibitor, promotes DH release from the cultured brain–subesophageal complexes (Hasegawa and Shimizu, 1990; Shimizu *et al.*, 1997). Further insight into the control of DH release may be gleaned from data for PBAN, the release of which is under circadian control, with PBAN-releasing neurons firing during the scotophase of a photoperiodic cycle. This pattern persists with a periodicity of 24 h (free runs) even in constant darkness, confirming the circadian nature of this pattern (Tawata and Ichikawa, 2001). At this point, however, there is no direct evidence indicating a circadian pattern of DH release.

The ovaries appear to be the sole target for DH. Many key components of the second messenger systems involved in DH signal transduction have been elucidated. The DH receptor has been identified as a 436 amino acid G-protein-coupled peptide, with affinity to several FXPRL-amides, but exhibiting highest affinity to DH (Homma *et al.*, 2006). Downstream components have been identified as well. *In vitro* experiments suggest

that early events, such as the upregulation of trehalase described later, are regulated by Ca^{2+} ion influx, since the addition of the chelating agent egtazic acid (EGTA) to the incubation medium suppresses DH activity, while the addition of Ca^{2+} restores this activity (Ikeda *et al.*, 1993). Later events may be mediated by cyclic nucleotide cascades. DH acts to reduce the concentration of cyclic GMP in diapause-destined eggs through the regulation of guanylate cyclase activity (Chen *et al.*, 1988; Chen and Yamashita, 1989).

Downstream of DH, regulation of diapause appears to involve accumulation of sorbitol (Horie *et al.*, 2000). A key indicator of DH activity has long been a metabolic switch, resulting in the accumulation of glycogen in diapause-destined eggs. This accumulation is due to the enzyme trehalase, which is transcriptionally upregulated after DH injection (Su *et al.*, 1994). Accumulated glycogen is then converted to sorbitol, which acts as a classic cryoprotectant, lowering the supercooling point of the diapausing eggs (Chino, 1958). Upon diapause termination, an ERK/MAPK-mediated pathway signals the reconversion of sorbitol to glycogen (Fujiwara *et al.*, 2006), which is subsequently used as an energy reserve. The accumulation of sorbitol not only offers cryoprotection but appears to exert a regulatory role as well (Horie *et al.*, 2000). When diapause-destined eggs are dechorionated and cultured in Grace's medium, diapause is averted. When sorbitol is added to the medium the ability to diapause is restored in a dose-dependent fashion. Substantial inhibition of development is noted at biologically relevant concentrations of sorbitol. Trehalose elicits similar inhibitory effects, but glycerol is less effective. Additionally, removal of sorbitol induces resumption of development, and addition of sorbitol to non-diapause-destined eggs leads to arrested development. Interestingly, the ERK/MAPK pathway appears to also be involved in regulation of embryonic diapause in the false melon beetle, *Atrachya menetriesi* (Kidokoro *et al.*, 2006b), even though there is no evidence for DH activity in this or any other non-lepidopteran species.

10.3.1.2. Other mechanisms for regulating embryonic diapause There is no evidence to suggest that the regulatory mechanisms carefully documented in *B. mori* are relevant to other forms of embryonic diapause, although DH has been implicated in the control of pupal diapause (see Section 10.3.2.). Embryonic diapauses are enormously diverse, as indicated by the wide range of stages at which development is arrested, and data suggest that the regulation is diverse as well. Consequently, a variety of regulatory mechanisms have been suggested for embryonic diapause as alternatives to what is seen in *B. mori*.

The gypsy moth, *Lymantria dispar*, diapauses as a pharate first instar larva. Several lines of evidence indicate that this diapause is induced and maintained by an elevated ecdysteroid titer. Using synthesis of a 55 kDa gut protein — now known to be the immune protein hemolin (Lee *et al.*, 2002) — as a distinct biomarker for diapause, Lee and Denlinger (1996, 1997) demonstrated that placing a ligature behind the prothoracic gland blocks hemolin synthesis, and co-culturing the gut with an active prothoracic gland promotes hemolin synthesis. Synthesis of hemolin can also be induced if an isolated abdomen is injected with 20E or the ecdysteroid agonist RH-5992. KK-42, an imidazole derivative thought to block ecdysteroid synthesis, averts diapause in this species (Suzuki *et al.*, 1993; Bell, 1996; Lee and Denlinger, 1996), but this effect can be reversed with the application of ecdysteroids (Lee and Denlinger, 1997). Additionally, at the end of the mandatory chilling period, diapause can be readily terminated if pharate larvae are transferred to high temperatures, but this response can be prevented with exogenous ecdysteroids. Experiments with a mutant, non-diapausing strain of *L. dispar* also support a role for ecdysteroids, as suggested by the restoration of diapause when ecdysteroid concentrations are artificially boosted in the mutant (Lee *et al.*, 1997). Although ecdysteroids of maternal origin are known to be incorporated into eggs in other species (Lagueux *et al.*, 1981), at this late stage of embryonic development the ecdysteroids that control diapause are likely synthesized in the pharate larva's prothoracic gland, which continuously synthesizes ecdysteroids throughout diapause but then shuts down synthesis as a first step in terminating diapause. A diapause-promoting effect for ecdysteroids may seem paradoxical, especially considering that a lack of ecdysteroids is the dominant feature of most larval and pupal diapauses (see Section 10.3.2.). Yet, it is well known that a drop in the ecdysteroid titer is an essential signal permitting the ongoing progression of development in other systems (Slama, 1980), and the gypsy moth appears to have exploited this function to bring about developmental arrest. The link between diapause induction and increased ecdysteroids is not exclusive to *L. dispar*. Metabolic depression has been linked to ecdysteroid peaks during embryogenesis in *Schistocerca gregaria* (Slama *et al.*, 2000). Additionally, gene regulation during pre-diapause in the ground cricket *Allonemobius socius* suggests that this embryonic diapause is regulated in a similar manner (Reynolds and Hand, 2009), and increased ecdysone levels induce cell cycle arrest in cultured mosquito cells (Gerenday and Fallon, 2004), further suggesting a link between increased ecdysone and the development of diapause traits.

Ironically, decreased ecdysteroid levels can also bring about embryonic diapause, as is the case for two locust species: the Australian plague locust *Chortoicetes terminifera* (Gregg *et al.*, 1987) and the migratory locust, *Locusta migratoria* (Tawfik *et al.*, 2002b). In both species, non-diapausing eggs contain approximately three times the amount of ecdysteroids as found in diapausing

eggs. In these species the diapauses occur early in embryonic development, and the ecdysteroids present are likely to be exclusively of maternal origin. Ecdysteroid titers increase in non-diapause eggs after the stage at which diapause normally occurs and in embryos that avert diapause in response to high temperature (Gregg *et al.*, 1987). Diapause termination also leads to an elevation in ecdysteroid titers (Tawfik *et al.*, 2002b), and exogenous application of ecdysteroids prompts diapause termination (Kidokoro *et al.*, 2006a), thus the responses of these two locusts consistently point to diapause being caused by the absence of ecdysteroids, much akin to the endocrine scenario evident in most larval and pupal diapauses. Parallels between embryonic and adult diapause are evident as well, as exogenous application of juvenile hormone analogs has been shown to terminate diapause in the orthopteran *Aulocara ellioti* (Neumann-Visscher, 1976), as well in the aforementioned *L. migratoria* (Kidokoro *et al.*, 2006a).

The pharate first instar larval diapause of the silkmoth *A. yamamai* represents yet another intriguing regulatory scenario (Suzuki *et al.*, 1990, 1993). The imidazole known to block ecdysteriod synthesis, KK-42, is highly efficacious for terminating diapause in intact pharate first instar larvae as well as in headless larvae. Still, diapause persists in an isolated head–thorax preparation, suggesting the presence of a repressive factor in the thorax. Isolated abdomens break diapause whether or not KK-42 is added. From these results, Suzuki *et al.* (1990) suggested the presence of a diapause-promoting factor produced in the mesothorax, and a development-promoting factor produced in the region of the second to fifth abdominal segments. During diapause the diapause-promoting factor prevails, halting development. Although the identities of these factors remain unknown, the effects cannot be mimicked by several of the obvious candidates: prothoracicotropic hormone, ecdysteroids, or JH.

10.3.1.3. Associated changes in gene expression

Although considerable effort has been devoted to the identification of diapause-specific genes, the majority of these genes are likely involved in maintaining the diapause state and in enhancing stress resistance rather than being specifically involved in diapause regulation. One exception may be the genes regulating the G2 cell cycle arrest characteristic of diapause in *B. mori* (Nakagaki *et al.*, 1991). Considerable interest has focused on the molecular basis of this cell cycle arrest. While there is little evidence for a connection between *cyclin B* transcription and diapause (Takahashi *et al.*, 1996), both *cyclin dependent kinase 2* and *cdc-2 related kinase* are substantially downregulated in diapausing individuals and are thought to be involved in mediating the diapause cell cycle arrest (Takahashi *et al.*, 1998). Another *B. mori* gene with possible regulatory implications is *BmEts*, a gene highly upregulated in early diapause but suppressed in a non-diapausing mutant strain (Suzuki et al., 1999). This gene is especially intriguing because, as the name suggests, *BmEts* belongs to the ETS gene family of transcription factors, which includes several genes such as Drosophila E74, a gene known to be important in the cellular ecdysone response. Additionally, the cold-induced gene *Samui* may be involved in the signal cascade of diapause termination. Downregulation of *Samui* occurs concurrently with activation of sorbitol dehydrogenase, the enzyme involved in converting sorbitol to glycogen (Moribe et al., 2001). As such it could be involved in removing the aforementioned sorbitol blockade to embryonic development.

Recent suppressive subtractive hybridization studies on the embryonic diapauses of the cricket *A. socius* (Reynolds and Hand, 2009) and the Asian tiger mosquito *Aedes albopictus* (Urbanski et al., 2010a,b) have also identified genes with unique diapause expression profiles. Specific links between these genes and the neuroendocrine control of diapause remain to be identified, but several of the cricket genes upregulated in the pre-diapause phase are associated with ecdysteroid synthesis and signaling, suggesting a possible role for ecdysteroids in promoting the onset of this diapause. Concurrently, genes involved in yolk mobilization and/or metabolism are consistently downregulated in the cricket embryos, as one might anticipate for the dormant state. The diapause-associated genes identified in *A. albopictus* appear to be involved mainly in downstream physiological components of the diapause syndrome such as desiccation resistance. The *A. albopictus* study focused on transcripts derived from oocytes dissected from the mother, thus in this case, the genes affecting the progeny's diapause are maternally derived.

10.3.2. Hormonal Basis for Larval and Pupal Diapauses

Larval diapauses are best known from the Lepidoptera, but there are good examples of larval (or nymphal) diapauses in other orders as well. Diapause in the last larval instar is reported most frequently, but some larvae diapause occurs in earlier instars or as pre-pupae. Pupal diapause has been examined extensively in Lepidoptera and the higher Diptera, but appears to be rare in other taxa. It is the true pupal stage that is usually involved in diapause, but occasional reports indicate a diapause late in pharate adult development. Larval and pupal diapauses share much in common. Central to both is a failure to progress to the next metamorphic stage. This suggests that one of the key endocrine events regulating such a diapause is likely to be a failure of the prothoracic gland to produce the ecdysteroids needed to initiate the next molt. Invariably, this has proven to be a key element of larval and pupal diapauses. One might also predict a shutdown until the end of diapause in the production of PTTH (see Chapter 1). This may often be the case but not always. The other major

metamorphic hormone, juvenile hormone (JH), may or may not play a role, depending on the species.

10.3.2.1. Role of the brain–prothoracic gland axis The older literature is replete with experiments showing that activation of the brain is essential for terminating diapause. Of special significance are the classic experiments by Carroll Williams (1946, 1947, 1952) demonstrating that the brain of the giant silkworm *Hyalophora cecropia*, when activated by chilling, is capable of terminating diapause when implanted into a brainless pupa. These experiments were critical not only for understanding the mechanism of diapause but also for deciphering the endocrine basis for insect metamorphosis. One can now be more specific and attribute these brain effects not to the whole brain, but to the production of PTTH by a pair of neurosecretory cells in the pars intercerebralis of each brain hemisphere (Agui et al., 1979; Sauman and Reppert, 1996).

In the pupal diapause of *M. sexta* intracellular electrical recordings of the cells that produce PTTH show all the characteristics of silent cells: high voltage thresholds, low input resistance, and few spontaneous action potentials (Tomioka et al., 1995). Electrical differences between the PTTH cells of diapausing and non-diapausing pupae are already evident within one day after pupation. As diapause progresses in chilled pupae, the cells gradually display more spontaneous action potentials, and more excitatory postsynaptic potentials are observed. In spite of the early curtailment of neurophysiological activity in the PTTH cells, there are no corresponding ultrastructural changes that would appear to reflect the electrical silencing of the cells in early diapause (Hartfelder et al., 1994).

One common method for monitoring PTTH activity is to culture prothoracic glands *in vitro* and monitor ecdysone production when brains or brain extracts are added to the culture medium. When this has been done, low PTTH activity associated with larval and pupal diapause has been observed in some cases. In *Diatraea grandiosella*, a species that diapauses late in the final larval instar, PTTH activity is far lower in brains and brain extracts from diapause-destined larvae and diapausing larvae than in non-diapausing larvae (Yin et al., 1985). Likewise, in the European corn borer, *Ostrinia nubilalis*, PTTH activity is lower in brains of diapausing larvae than in non-diapausing brains (Gelman et al., 1992). Similar low levels of PTTH activity are noted in pupal diapause of the cabbage armyworm, *Mamestra brassicae* (Endo et al., 1997). Yet, brains from all of these species clearly show some PTTH activity, and pupal brain extracts from *M. sexta* (Bowen et al., 1984) and *Sarcophagia argyrostoma* (Richard and Saunders, 1987) actually have as much or more PTTH activity than brain extracts from non-diapausing pupae. Likewise, PTTH activity is higher in larvae of *S. peregrina* that are destined for pupal diapause than in larvae not destined for diapause (Moribayashi *et al.*, 1992). These observations suggest that the hormone may be present during diapause but is simply not released.

In *Heliothis virescens*, expression of transcript encoding PTTH persists at relatively high levels throughout larval development, but at the onset of wandering behavior, expression in larvae programmed for pupal diapause drops and remains low during diapause, while the transcript continues to be highly expressed in non-diapause-destined larvae and non-diapausing pupae (Xu and Denlinger, 2003). These observations suggest that, in this species, PTTH is regulated at the level of transcription. Thus, there may be species variation in the mechanisms of PTTH regulation. Either a halt in PTTH synthesis or a failure of PTTH release could result in a developmental shutdown.

Several experiments suggest that high dopamine levels in the brain may be critical for diapause induction, possibly by preventing release of PTTH. High levels of dopamine in brains of diapause-destined larvae of *C. costata* (Kostal *et al.*, 1998), diapause-destined pupae of *M. brassicae* (Noguchi and Hayakawa, 1997), and pupae of *P. brassicae* (Puiroux *et al.*, 1990) suggest a causal relationship. In diapausing pupae of *A. pernyi*, a drop in brain dopamine is noted shortly before the surge of ecdysteroids that elicits diapause termination (Matsumoto and Takeda, 2002). But the most compelling evidence suggesting a role for dopamine is the fascinating observation that feeding L-DOPA to last instar larvae of *M. brassicae* reared under long daylengths (not diapause-inducing) elicits a diapause-like state in the pupae (Noguchi and Hayakawa, 1997). Other than DH in *Bombyx*, there are few examples of chemical agents capable of inducing diapause or a diapause-like state, thus this result with DOPA feeding is quite striking. A potential role is also discussed for dopamine in the induction of embryonic diapause of *B. mori* (Noguchi and Hayakawa, 2001; see Section 10.3.1.1.). Among genes that are upregulated by both short-day conditions and the feeding of DOPA in *M. brassicae*, the most interesting is an upregulated gene with high identity to a receptor for activated protein kinase C (RACK) from *H. virescens* (Uryu *et al.*, 2003). *In situ* hybridization with a RACK probe reveals expression of this gene within a few cells in the medial protocerebral neuropil. As a member of the protein kinase C (PKC) signaling cascade, RACK has the potential to be an important gene contributing to transduction of the signals involved in diapause induction. Noguchi and Hayakawa (1997) proposed that elevation of dopamine may be the key to preventing PTTH release.

Elevation of brain serotonin (5-hydroxytryptamine) has also been linked to diapause in several cases (Kostal *et al.*, 1999; L'Helias *et al.*, 2001), but not in others (Puiroux *et al.*, 1990). In *C. costata*, depletion of serotonin does not alter the fly's capacity to enter larval diapause (Kostal

et al., 1999), suggesting that the observed elevation of this monoamine is not essential for diapause induction.

The shutdown in the brain–prothoracic gland axis is not exclusively the result of a failure of the PTTH cells to produce or release PTTH. The prothoracic gland also becomes refractory to PTTH at this time, providing a double assurance that the neuroendocrine system will not provide the signal needed to promote development. Assays for PTTH activity indicate that prothoracic glands dissected from diapausing individuals are refractory to stimulation, as shown in larvae of *C. vicina* (Richard and Saunders, 1987) and *O. nubilalis* (Gelman *et al.*, 1992), and in pupae of *M. sexta* (Bowen *et al.*, 1984) and *S. argyrostoma* (Richard and Saunders, 1987). This loss of competency can be rather rapid (**Figure 8a**). In *S. argyrostoma* the prothoracic glands lose their competency to produce ecdysone within a 1- to 2-day period at the onset of diapause (Richard and Saunders, 1987). In the larval diapause of *C. vicina* loss of competency is more gradual, occurring over a 6-day period, and even during diapause the gland maintains limited competency to produce ecdysone (**Figure 8b**). At diapause termination, prothoracic gland competency is restored quickly. When diapausing larvae of *C. vicina* are transferred from 11 to 25°C, the prothoracic gland regains competency to produce ecdysone within 24 h (Richard and Saunders, 1987).

PTTH stimulation of the prothoracic gland in non-diapausing insects is mediated by cyclic AMP as a second messenger, but the diapausing glands of *M. sexta* (Smith *et al.*, 1986) and *S. argyrostoma* (Richard and Saunders, 1987) fail to respond to cyclic AMP or agents known to elevate cyclic AMP, implying that the block in gland function occurs beyond the stage of cyclic nucleotide action.

Reports from several species of Lepidoptera indicate that cells of the prothoracic gland are greatly reduced in size during diapause, both the cytoplasm and nuclei are shrunken, and the mitochondria take on a swollen configuration (Denlinger, 1985). In *S. crassipalpis*, there is a slight reduction in the size of the cells in the prothoracic gland component of the ring gland, but this difference is not as conspicuous as noted in the Lepidoptera (Joplin *et al.*, 1993). Results of an ultrastructural examination of the fly prothoracic gland during diapause are rather surprising. Instead of being inactive, the presence of extensive arrays of rough endoplasmic reticulum suggests that the glands remain active in protein synthesis. Pulse labeling of the ring gland further supports the contention that the gland continues to synthesize proteins during diapause. The function of this curious diapause activity remains unknown, but it would appear to be unrelated to ecdysone production since all evidence indicates that this function is shut down at this time.

Although there appears to be species variation as to whether the synthesis or release of PTTH is regulated, the net effect, a shutdown in the synthesis of ecdysteroids, is

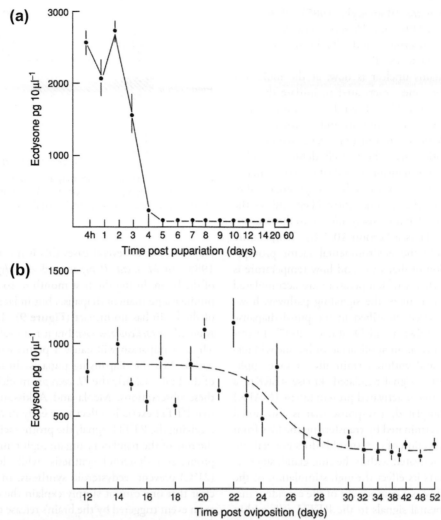

Figure 8 Decline in the competency of ring glands from diapause-destined (a) *Sarcophaga argyrostoma* and (b) *Calliphora vicina* to synthesize ecdysone in response to incubation with prothoracicotropic hormone (PTTH) producing brains. As *S. argyrostoma* enters pupal diapause and as *C. vicina* enters larval diapause their ring glands become refractory to PTTH stimulation. Adapted from *Richard et al.* (1987).

consistent. Earlier literature provides numerous examples documenting low or undetectable ecdysteroid titers in larvae or pupae during diapause and an elevation in the ecdysteroid titer at diapause termination. More recent studies continue to provide evidence that this is the case; for example, in larval diapauses of *O. nubilalis* (Gelman and Woods, 1983; Gelman and Brents, 1984; Peypelut *et al.*, 1990) and *Omphisa fuscidentalis* (Singtripop *et al.*, 2002) and pupal diapauses of *M. configurata* (Bodnaryk, 1985), *M. sexta* (Friedlander and Reynolds, 1992), *S. argyrostoma* (Richard *et al.*, 1987), and *S. peregrina* (Moribayashi *et al.*, 1988). As discussed in Section 10.3.2.4., some larvae periodically undergo a stationary molt during diapause, and in these cases a burst of ecdysteroid synthesis may occur during diapause and prompt the larva to molt. The surge in ecdysteroids that normally occurs at the end of diapause may come as two distinct waves. In *M. configurata* the release of ecdysone into the hemolymph is followed

several days later by the appearance of 20E (Bodnaryk, 1985). Since these two peaks are nearly identical in size, they may simply represent the conversion of ecdysone to 20E. In *O. nubilalis* a small transient peak of 20E is followed several days later by a major peak of 20E (Peypelut *et al.*, 1990). In this case, imaginal wing disc development is thought to be initiated by the small initial peak, but the later phases of post-diapause development require the sustained presence of 20E.

Consistent with the concept that the absence of ecdysteroids is central to maintaining the diapause state is the observation that an application of exogenous ecdysteroids or ecdysteroid mimics can terminate diapause. Again, the older literature includes numerous examples supporting this response, and more recent work continues to document this effect; for example, larval diapauses of *O. nubilalis* (Gadenne *et al.*, 1990) and *Diprion pini* (Hamel *et al.*, 1998), and pupal diapauses of *A. mylitta* (Mishra *et al.*,

2008), *M. configurata* (Bodnaryk, 1985), *P. brassicae* (Pullin and Bale, 1989), and *M. sexta* (Friedlander and Reynolds, 1992; Sielezniew and Cymborowski, 1997; Champlin and Truman, 1998).

What still remains unclear is how, at the molecular level, the brain becomes reactivated to initiate the process of promoting the PTTH synthesis and/or release that leads to ecdysteroid synthesis and diapause termination. Although the environmental events required for diapause termination have been well defined (Tauber *et al.*, 1986), the mechanism of activation still remains largely unknown. In the few cases that use photoperiod to terminate diapause, the response most likely utilizes the clock-gene neurons that are closely aligned to the neurons that produce PTTH (see Section 10.2.2.). Temperature is more commonly the environmental factor prompting the termination of diapause, and how temperature is perceived or the effects of temperature are accumulated is unclear. One element of the signaling pathway, however, has recently been identified in the pupal diapause of *S. crassipalpis* (Fujiwara and Denlinger, 2007). In this fly, high temperature or an application of hexane will terminate diapause, and within 10 min after hexane application, extracellular signal-regulated kinase (ERK), a member of the mitogen-activated protein kinase (MAPK) family, is phosphorylated, a response that is also noted when diapause is terminated by transferring the flies from 20 to 25°C. An injection of ecdysteroids does not activate ERK nor is ERK activated within the ring gland, suggesting that ERK exerts its effect through stimulation of the brain, quite possibly as a component of the cascade transmitting environmental signals to the PTTH cells within the brain. Two other members of the MAPK family, p38 MAPK and JNK, showed no response at diapause termination. A signaling cascade involving ERK emerges as a potential mediator of the environmental signal, resulting ultimately in ecdysteroid synthesis and diapause termination. A similar role for ERK in the termination of embryonic diapause was noted earlier (see Section 10.3.1.1.). From the *Drosophila* literature, it is evident that Torso, a receptor tyrosine kinase, functions as the PTTH receptor and its activation by PTTH stimulates the ERK pathway (Rewitz *et al.*, 2009; see also Chapter 1).

10.3.2.2. The missing pieces The role of the brain–prothoracic gland axis in diapause, as presented earlier, is reasonably cohesive. In a simplified version, the prothoracic gland fails to produce ecdysone because it has not been stimulated by PTTH from the brain. Yet, some pieces of information are not consistent with this simple story and remain unexplained. For *H. cecropia* the previous story is just fine. The brain, when chilled, becomes activated to release PTTH and this in turn prompts the prothoracic glands to secrete the ecdysteroids needed to initiate development. Without a brain, the pupa is locked into a permanent

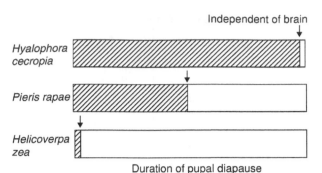

Figure 9 Time during pupal diapause at which several species are no longer dependent on the brain for initiation of adult development. Adapted from *Denlinger* (1985).

diapause, but in several cases this is not true (Denlinger, 1985). In *M. sexta*, *P. rapae*, and *A. polyphemus* removal of the brain during the first month or so of diapause will produce a permanent diapause, but in later stages removal of the brain has no impact (**Figure 9**). The extreme case is noted in *Helicoverpa zea*: brain removal within the first 4 h after pupation will cause a permanent diapause, but within 24 h the diapausing pupa is already independent of the brain. Clearly the *H. cecropia* model cannot explain these observations. Meola and Adkisson (1977) argued that PTTH exerts its effect very early in *H. zea*, and after receiving the PTTH signal, the prothoracic glands take on the role of the regulatory organ: high temperature (27°C) promotes ecdysteroid synthesis, while low temperature (21°C) prevents ecdysteroid synthesis. In these cases it is clear that one cannot simply explain the end of diapause as an event triggered by the brain's release of PTTH. What is lacking is a good explanation of this peculiar interaction between the brain and prothoracic gland. What does it mean for the brain, presumably via PTTH, to stimulate the prothoracic gland so early and for the gland to delay the onset of ecdysteroid synthesis by many months? An alternative explanation may be that our traditional view of PTTH playing a role in promoting the synthesis of ecdysteroids is not correct. This scenario gains credence from recent results in *Drosophila* showing that ablation of the PTTH-producing cells had no effect on molting or metamorphosis (McBrayer *et al.*, 2007), a result not too different from the diapause experiments showing that brain extirpation fails, in some cases, to block the termination of diapause. If, as shown in *Drosophila*, PTTH is not absolutely essential for stimulating ecdysteroid synthesis, the door is opened to considering alternative pathways for stimulating ecdysteroid synthesis.

Several lines of evidence suggest that there may be differences in the way the prothoracic gland is activated in non-diapausing individuals and at the termination of diapause. In *S. crassipalpis*, cyclic AMP levels in the brain and ring gland are high in non-diapausing pupae and low in diapausing pupae at the onset of diapause, and diapause can easily be averted in the diapause-programmed pupae if cyclic AMP levels are artificially boosted around the time of expected diapause entry (Denlinger, 1985). This observation is compatible with other evidence suggesting that cyclic AMP is used as a second messenger for PTTH (see Chapter 1). One might then assume it would also be possible to employ this same tactic of cyclic AMP elevation to initiate development in a pupa that has been in diapause for some time. Surprisingly, this does not work because cyclic AMP injected along with ecdysteroids actually retards the termination of diapause. By contrast, cyclic GMP, which is ineffective in averting diapause, can readily break diapause by itself and will enhance the diapause-breaking effect of ecdysteroids (Denlinger and Wingard, 1978). Richard and Saunders (1987) were unable to stimulate ecdysteroid production with PTTH or cyclic nucleotides in ring glands cultured from diapausing pupae of *S. argyrostoma* or diapausing larvae of *C. vicina*, which also suggests that the glands are somehow activated in a different way at the termination of diapause.

Although prothoracic glands dissected from *C. vicina* are competent to synthesize ecdysone within 24 h after a transfer of the diapausing larvae to a high temperature, brain–ring gland complexes cultured *in vitro* fail to gain competency when subjected to the same temperature switch (Richard and Saunders, 1987). This suggests that some additional component from another part of the body may contribute to reactivation of the prothoracic gland. The pieces of the fly puzzle include: cyclic GMP readily breaks diapause when injected *in vivo*, but fails to activate prothoracic glands cultured *in vitro*, and the brain–prothoracic gland can regain competency in response to high temperature *in vivo*, but not *in vitro*. A minimal interpretation of these observations is that some additional prothoracicotropic factor, acting through cyclic GMP, contributes to activation of the prothoracic gland at the termination of diapause.

Although our best evidence points to the ecdysteroids as the major coordinators of diapause in larvae and pupae, other hormones can be expected to play roles as well. Roles for JH (see Section 10.3.2.4.) and DH (see Section 10.3.2.5.) are also well established, but others beyond these may also be implicated. For example, neuropeptide-like precursor 4 is uniquely expressed in association with diapause in flesh fly pupae (Li *et al.*, 2009), but its function remains unknown. This potential regulatory agent, and others like it, suggest that we may not yet have identified all the players involved in regulating larval and pupal diapause.

10.3.2.3. Ecdysone receptor and Ultraspiracle

Two proteins, Ecdysone receptor (EcR) and Ultraspiracle (USP), dimerize to form the functional receptor for ecdysone (see Chapter 5). 20-Hydroxyecdysone binds to this complex, which in turn binds directly to ecdysone response elements to elicit specific gene action. The central role for ecdysteroids in the regulation of larval and pupal diapauses suggests that these proteins could contribute to the diapause control mechanism, but thus far few experiments have addressed this question. During pupal diapause of *S. crassipalpis* expression of EcR mRNA persists throughout diapause, but the expression of USP gradually declines at the onset of diapause and is undetectable after 20 days in diapause (Rinehart *et al.*, 2001). USP transcripts again appear in late diapause (beginning on day 50) and are further upregulated 9 h after pupae are artificially stimulated to break diapause. This 9 h time frame coincides precisely with the rise in the ecdysteroid titer following hexane application. It is interesting to note that upregulation of the USP transcripts at day 50 coincides with the pupa's increased sensitivity to injected ecdysteroids. This suggests the possibility that reappearance of the USP transcript is a preparatory step, perhaps a critical one, leading to the eventual termination of diapause. In the second instar larval diapause of *Choristoneura fumiferana*, genes encoding both isoforms of EcR are expressed throughout diapause, as is the gene encoding USP (Palli *et al.*, 2001). However, the ecdysone-inducible transcription factors (CHR75A and CHR75B) and hormone receptor 3 (CHR3) are not present during diapause, but reappear at diapause termination.

In *M. sexta* the transcript encoding EcR disappears at the onset of diapause and reappears when diapause is terminated (Fujiwara *et al.*, 1995). In the larval diapause of the midge *Chironomus tentans* (Imhof *et al.*, 1993) and the bamboo borer *O. fuscidentalis* (Tatun *et al.*, 2008), EcR transcripts are detectable throughout diapause and an increase is noted at diapause termination. The patterns of USP mRNA expression are unknown for these two species. Although the database remains small at this point, the results do not suggest the emergence of a universal pattern in relation to diapause, but the presence or absence of these receptor proteins could very well be an important component of the regulatory scheme for diapause. Turning off the expression of one or both of these genes could promote diapause, while the presence of these gene products is likely to be essential for the resumption of development.

10.3.2.4. Role of the corpora allata

The early views of larval and pupal diapause focused strictly on the brain–prothoracic gland axis, leaving little reason to suspect involvement of JH. This view was challenged by a set of experiments on *D. grandiosella* by Chippendale, Yin and their colleagues beginning in the early 1970s (Chippendale, 1977). What launched their work on JH was the surprising discovery that injection of ecdysteroids into a diapausing larva of *D. grandiosella* prompted a stationary larval molt rather than pupation. The stationary molt suggested the presence of JH because the larva would have been expected to pupate in the absence of JH. This suspicion of JH involvement was followed by verification that the hemolymph titer of JH remains high and the corpora allata (CA) are active during diapause. At the end of diapause, the JH titer drops, and the larva pupates. A similar regulatory scenario is evident in some of the *Chilo* stemborers. These two taxa both undergo stationary molts during diapause, and the only way to undergo a stationary molt rather than a progressive molt is to maintain a high JH titer. The high JH titer reported for larval diapause in the Mediterranean corn borer, *Sesamia nonagrioides* (Eizaguirre *et al.*, 1998, 2005) and the yellow-spotted longicorn beetle, *Psacothea hilaris* (Munyiri and Ishikawa, 2004) suggests that they too have a diapause maintained by JH.

Species that do not undergo stationary molts presumably do not need to maintain a high JH titer during diapause, and in a number of Lepidoptera larvae JH appears to play no role; for example, in *O. nubilalis* and *Laspeyresia pomonella* (Denlinger, 1985). In these species JH is actually high at the onset of diapause, but the titer drops to low levels shortly after diapause is initiated, and an injection of ecdysteroids elicits pupation, not a stationary molt. Why the JH titer is high at the onset of diapause is unclear, but it does not appear to be involved in the induction process. Application of JH to non-diapausing larvae at this stage does not cause the induction of larval diapause.

An application of JH can terminate larval diapause in *O. fuscidentalis* (Singtripop *et al.*, 2000), as reported previously for a number of other larval and pupal diapauses (Denlinger, 1985), and JH presumably exerts this effect by stimulating the prothoracic gland to produce ecdysone, as suggested by prothoracic gland transplantation studies in *O. fuscidentalis* (Singtripop *et al.*, 2008). There is no evidence to suggest that JH by itself is the natural trigger for diapause termination. A combined role for JH with ecdysteroids is, however, likely in some cases. In flesh flies, JH by itself has no effect in terminating pupal diapause, but dramatic synergism is noted when JH is applied in concert with a small dose of ecdysteroid (Denlinger, 1979). When diapause is terminated naturally in flesh flies, a small peak of JH activity precedes the sustained elevation of the ecdysteroid titer, thus the combined application of JH and ecdysteroid likely is an effective mimic of the natural reactivation process.

JH contributes to the regulation of metabolic events occurring during pupal diapause, but it does not appear to play a regulatory role in the induction or termination of diapause. In flesh flies, JH is involved in regulating metabolic cycles that persist during diapause (Denlinger *et al.*, 1984). In these flies the metabolic rate is not constant; it occurs in 4- to 6-day cycles. These cycles are driven by JH. The hemolymph JH titer progressively increases during the trough of the cycle, reaches a critical point,

triggers a bout of high oxygen consumption, and then drops precipitously. The rapid decline in JH titer is aided by a rise in JH esterase (Denlinger and Tanaka, 1989). When the ring gland containing the CA is removed, the diapausing pupae become acyclic. Application of a high dose of JH analog to the pupae just before diapause will not alter the diapause fate of the pupa, but such pupae fail to exhibit oxygen consumption cycles and instead consume oxygen at a high rate throughout diapause. Pupae treated this way have a much shorter diapause, presumably because their high metabolic rate has prematurely exhausted their energy reserves. This conclusion assumes a mechanism used by the pupae to monitor and respond to depleted energy reserves (Denlinger *et al.*, 1988; Hahn and Denlinger, 2011).

10.3.2.5. A role for diapause hormone? Diapause hormone is well known for its role in diapause induction in *B. mori* (see Section 10.3.1.), but attempts to induce diapause in other species with DH have failed. Thus, DH appeared to be a neuropeptide with some other primary function that was simply recaptured by *B. mori* as a diapause regulator (Denlinger, 1985). We now know, however, that a gene encoding a DH-like sequence is present and expressed in all Lepidoptera that have been examined, even those that lack an embryonic diapause. Diapause hormone and PBAN, along with several additional subesophageal ganglion neuropeptides (SGNPs), are encoded by a common precursor (see Section 10.3.1.1.). However, the function of this DH-like peptide in lepidopteran species other than *B. mori* has remained elusive.

Does DH contribute to diapause in other species? DH-like peptides from *H. virescens* (Xu and Denlinger, 2003), *H. armigera* (Zhang *et al.*, 2004a,b) and *H. zea* (Zhang *et al.*, 2008) appear to have no stimulatory effect on diapause induction, but quite surprisingly, diapausing pupae of these heliothine species readily respond to the DH-like peptides by terminating diapause. This could represent a non-specific effect, but non-amidated forms of DH and other injected proteins fail to terminate diapause, suggesting that the effect is specific for DH. The possibility that a drop in DH contributes to diapause induction is further supported by the observation that expression of the mRNA encoding DH-PBAN declines at the onset of larval wandering behavior and remains low throughout pupal diapause, whereas expression remains high in larvae and pupae not destined for diapause. When diapausing pupae of *H. armigera* are transferred to high temperatures to terminate diapause, the mRNA encoding DH-PBAN is quickly elevated (Wei *et al.*, 2005). In *H. virescens*, Northern blots indicate that the DH-PBAN message is most highly expressed in the subesophageal–ganglion complex (Xu and Denlinger, 2003), and this is supported by immunocytochemical evidence in *H. armigera*

(Wei *et al.*, 2005). The hemolymph titer of DH-like peptide is also much higher in non-diapausing pupae than in diapausing ones (Wei *et al.*, 2005). Injection of DH into diapausing pupae prompts a major increase in hemolymph ecdysteroids 3 to 7 days after DH injection. Current evidence from these noctuid moths is consistent with a drop in DH being essential for pupal diapause induction and DH elevation serving as a possible trigger for the ecdysteroid surge needed for diapause termination.

As shown in **Figure 5**, a single gene encodes DH, PBAN, and three SGNPs of unknown function. There is well-documented cross reactivity between DH and PBAN, and even the SGNPs are capable of terminating pupal diapause in *Helicoverpa assulta* (Zhao *et al.*, 2004) and *H. armigera* (Zhang *et al.*, 2004b), suggesting similarities in the receptors for members of this family. Yet, differences in efficacy of these different neuropeptides in bioassays and more recent evidence on the structures of the receptors for DH and PBAN (Homma *et al.*, 2006; Watanabe *et al.*, 2007) indicate that these neuropeptides use distinctly different receptors.

The 24-amino acid sequences of DH used by *H. zea* and *H. virescens* differ by only a single amino acid, but a greatly truncated version of the neuropeptide, C-terminal heptapeptide LWFGPRLa (the core sequence), is sufficient to terminate diapause (Zhang *et al.*, 2008). Identification of the active components of DH led to the development of several hyperpotent agonists (Zhang *et al.*, 2009) that could potentially disrupt diapause in this important group of agricultural pests.

How and why the heliothine moths use both DH and ecdysteroids for the termination of pupal diapause is not yet clear. Both hormones, by themselves, are capable of terminating diapause when injected into diapausing pupae, yet the responses are distinct. While an injection of ecdysteroids will break diapause at any temperature, DH is ineffective at 18°C but will readily break diapause at temperatures of 21°C or higher (Zhang *et al.*, 2008). The DH agonists that have been developed also show the same restricted range of temperature effectiveness as DH (Zhang *et al.*, 2009). Diapause hormone can stimulate the prothoracic glands of *H. armigera* to produce ecdysteroids (Zhang *et al.*, 2004c; Liu *et al.*, 2005), indicating that both PTTH and DH can promote steroidogenesis in heliothine prothoracic glands, presumably by activating the diazepam-binding inhibitor/acyl-CoA-binding pathway (Liu *et al.*, 2005). The temperature sensitivity of the DH response differs from that of the ecdysteroids, suggesting that the DH pathway for terminating diapause relies on high temperature and may trigger pupae to break diapause in response to high temperature. This scenario is consistent with the observation that the timing of diapause termination in heliothine moths is independent of the brain shortly after pupation (see Section 10.3.2.2.).

10.3.2.6. Associated changes in gene expression The endocrine mechanism that dictates the diapause fate can be viewed as a developmental switch that brings into play the upregulation and downregulation of huge suites of genes and their products as evidenced by recent large-scale studies on pupal diapause using microarrays (Emerson *et al.*, 2010; Ragland *et al.*, 2010), proteomics (Li *et al.*, 2007; Cheng *et al.*, 2009; Chen *et al.*, 2010; Lu and Xu, 2010), and metabolomics (Michaud and Denlinger, 2007). A large-scale proteomics approach has also been used to probe the larval diapause of *N. vitripennis* (Wolschin and Gadau, 2009). Some genes, such as the clock genes discussed in Section 10.2.3.2., are upstream of the endocrine regulatory switch, while others are involved with downstream events set in motion by the endocrine program. Downstream physiological responses commonly linked to the diapause program include metabolic depression, cell cycle arrest, and enhanced stress tolerance.

Transcriptomic (Ragland *et al.*, 2010) and metabolomic (Michaud and Denlinger, 2007) results with flesh flies suggest that pathways involving glycolysis and gluconeogensis are favored during pupal diapause. The expression patterns of several genes in the TCA cycle are altered by the diapause program, and the enrichment of genes involved in anaerobic metabolism during diapause is noted in flesh fly pupae as well as in diapausing larvae of the pitcher plant mosquito (Emerson *et al.*, 2010). This suggests that such a shift to anaerobic metabolism may be a common downstream component of diapause in larvae and pupae as well as in adult dormancy in *Drosophila melanogaster* and in the dauer state of the nematode *Caenorhabditis elegans* (Ragland *et al.*, 2010).

Another related consequence of favoring glycolysis and gluconeogenesis is the generation of carbohydrates, polyols, and amino acids that are known to have cryoprotective functions. Diapausing insects are well known to have enhanced tolerance to low temperatures, desiccation, hypoxia, and oxidative stress, and gene expression that supports the acquisition of these forms of stress protection are commonly noted during diapause. The upregulation of genes encoding heat shock proteins is also an attribute of diapause in quite a few species, and suppression of these genes in flesh flies results in impaired cold tolerance (Rinehart *et al.*, 2007).

Cell cycle arrest is another feature central to diapause. In diapausing flesh fly pupae this arrest in the brain occurs at the G_0/G_1 stage and downregulation of the cell cycle regulator, *proliferating cell nuclear antigen* (*pcna*), appears to be critical for bringing about the arrest (Tammariello and Denlinger, 1998). This response is quite similar to that observed in the CNS of diapausing larvae of the fly *C. costata* (Kostal *et al.*, 2009). A pentapeptide Yamamarin, extracted from diapausing pharate first instar larvae of the Japanese oak silkmoth *A. yamamai*, is a potent agent capable of arresting cell proliferation in insects and insect cell lines as well as rat leukemic cell lines (Sato *et al.*, 2010). This peptide can also induce embryonic diapause in *B. mori* when injected into pupae of silkmoths not programmed for diapause. How widespread such an agent may be remains unknown, but clearly some mechanism for halting the cell cycle can be expected to be a component common to many forms of diapause. It is not currently clear how the endocrine system is linked to the cell cycle arrests noted earlier, but a role for ecdysteroids in regulating the cell cycle is suggested from cell culture studies (Gerenday and Fallon, 2004).

Although most diapauses can be characterized by the same end points of metabolic depression, enhanced stress resistance, and cell cycle arrest, these end points can be attained by diverse tactics, and we can expect that different insect species will have targeted different components of the gene network to elicit a similar physiological response. We remain at a very early stage in sorting out what specifics genes are being regulated.

10.3.3. Hormonal Basis for Adult Diapause

The central feature of adult diapause is a cessation in reproduction. For females this implies an arrest in oocyte development and in males the most conspicuous feature is their failure to mate with receptive females. The status of the testes is not a consistently reliable indicator of diapause in males (Pener, 1992): in some cases the testes are underdeveloped, while in other cases they may be full of mature sperm cysts. In both males and females the accessory glands are usually reduced in size during diapause. Although a period of long-distance flight may precede the entry into diapause, once adults arrive at their diapause site, flight muscles may degenerate (especially common in Coleoptera) for the duration of diapause and then regenerate when diapause is terminated. Adults remain fairly inactive in diapause, but some local movements may be noted. Mating may occur either before the entry into diapause or after diapause has been completed. When mating occurs prior to diapause, the males usually die without entering diapause and only the females bridge the diapause period. When both sexes diapause, mating usually occurs only after diapause has been terminated.

Most early work on the hormonal control of adult diapause focused on JH. Pioneering work on adult diapause, initiated in the late 1950s by Jan de Wilde and his colleagues in the Netherlands, utilized the Colorado potato beetle as their model system (de Kort, 1990). They observed that the JH titer in diapause-programmed *Leptinotarsa decemlineata* decreases after adult emergence, remains low throughout diapause, and then rises after termination of diapause. Further evidence supporting a role for JH in adult diapause is provided by the responsiveness of diapausing adults to JH, histological studies of the CA, and experimental manipulations of the CA.

Application of synthetic JH and JH analogs to diapausing females prompts egg laying (although such periods are not sustained). The CA from diapausing *L. decemlineata* are small and inactive. Surgical removal of the CA induces a diapause-like state in long-day beetles (those not programmed for diapause): they stop feeding, leave the food, and burrow into the soil. Chemical destruction of the CA in long-day beetles with precocene II elicits the same diapause-like response. In non-diapausing adults, cauterizing the pars intercerebralis (the region of the brain that controls CA activity) also induces a diapause-like state. This same surgery performed with diapausing adults blocks the beetle's ability to become reproductively active when placed under long daylengths (the environmental signal used to promote reproduction). Severing the axons between the pars intercerebralis and the CA does not appear to alter the diapause decision, suggesting that the CA in this species is under hormonal, rather than neural, control.

In the early experimental years there would have been no reason to suspect a role for ecdysteroids as a regulator of adult diapause because the well-known larval source of ecdysteroid synthesis, the prothoracic gland, degenerates in adults. But the discovery in the late 1970s that adults can produce ecdysteroids from other tissues such as ovaries raised the possibility that ecdysteroids could also be involved in regulating reproductive events. In the Colorado potato beetle, the hemolymph ecdysteroid titer is nearly twice as high in diapause-programmed beetles as it is in non-diapause-programmed beetles, but the titer drops sharply a few days after adult eclosion. The functional significance of this elevated ecdysteroid titer associated with diapause has not been determined.

The scheme for the hormonal control of adult diapause that has been developed from experiments in *L. decemlineata* is applicable to many species that diapause as adults, but there are also clear species differences. These differences include whether both sexes enter diapause, sexual distinctions in the environmental cues used for regulating diapause, and the formation of stable intermediate phenotypes that lie somewhere between fully sexually active and completely inactive. The regulatory mechanisms controlling the CA during adult diapause vary among species. The brain's regulation of the CA can be either hormonal as in *L. decemlineata* or neural as in *P. apterus*, *Tetrix undulata*, and *L. migratoria*. Clear unifying themes are evident within adult diapause, yet species-specific distinctions are also apparent.

10.3.3.1. Role of the corpora allatal Research on the endocrine basis of adult diapause continues to focus primarily on the role of JH. In substantially more species we now have new or additional evidence that JH plays a major role; for example, the mosquito *C. pipiens* (Sim and Denlinger, 2008), the butterfly *Speyeria idalia* (Kopper

et al., 2001), the black rice bug *Scotinophara lurida* (Cho *et al.*, 2007), the stink bug *P. stali* (Kotaki *et al.*, 2011), the moth *Caloptilia fraxinella* (Evenden *et al.*, 2007), and the plum curculio *Conotrachelus nenuphar* (Hoffmann *et al.*, 2007).

Earlier results arguing a role for JHs in regulating the adult diapause of *C. pipiens* were based on the ability of JH applications to terminate diapause; more recently this evidence was supplemented by demonstrating that cultured CA dissected from diapausing females secrete very little JH (**Figure 10**), whereas those dissected from non-diapausing females are quite active, especially during the first week after adult eclosion (Readio *et al.*, 1999). Females of *C. pipiens* that are allatectomized and then transferred to diapause-terminating conditions remain in a diapause-like state and fail to initiate the blood feeding noted in control groups. Finally, the argument for JH regulation of adult diapause is further strengthened by gene silencing experiments (Sim and Denlinger, 2008), as discussed in Section 10.3.3.2. as well as with experiments demonstrating that a knockdown of two ribosomal genes, S2 and S3a, causes a diapause-like arrest in *C. pipiens* that can be reversed with application of JH III (Kim and Denlinger, 2010; Kim *et al.*, 2010).

The CA, of course, does not function autonomously. Activity of the CA is coordinated by the brain, and the brain exerts its control through both hormonal and neuronal conduits. Active suppression of the CA by the brain during diapause has been demonstrated in the following insect orders: Coleoptera: *L. decemlineata* (Khan *et al.*, 1983, 1988); Diptera: *P. terraenovae* (Matsuo *et al.*, 1997); Heteroptera, *P. stali* (Kotaki and Yagi, 1989), *R. clavatus* (Morita and Numata, 1997a), *and P. apterus* (Hodkova *et al.*, 2001); and Orthoptera: *L. migratoria* (Okuda and Tanaka, 1997). In the diapause of *P. terraenovae* inhibition is maintained by neurons that originate in the pars lateralis

and terminate within the CA (Shiga and Numata, 2000), but in several other species, including *P. apterus* (Hodkova, 1994), the inhibition appears to originate in the pars intercerebralis. The diapause program may actually influence the organization of the synapses. Rearing *L. decemlineata* throughout its life under short-day conditions boosts both the number and activity of neurosecretory synapses within the CA as compared to beetles reared under long-day conditions (Khan and Buma, 1985). Hormonal control of the CA is also evident in a number of species. CA activity in diapausing *L. migratoria* is suppressed by neuropeptide(s) stored in the corpora cardiaca (CC; Okuda and Tanaka, 1997). Neural regulation of the CA during diapause may not be as simple as the presence or absence of an allatostatin. Brain extracts from both diapausing and non-diapausing *P. apterus* stimulate increased JH production in CA from non-diapausing donors but have no effect on JH synthesis in CA from diapausing donors. This suggests that diapausing adults may lack receptors for allatotropins or the presence of an allato-inhibiting factor in the CA or CC that blocks the action of allatotropins (Hodkova *et al.*, 1996). Extracts of the brain–subesophageal ganglion-CC-CA complex from diapausing *P. apterus* inhibit the CA activity of *P. stali*, indicating that this neurosecretory complex from diapausing bugs may contain both allato-inhibiting and -stimulating factors (Hodkova *et al.*, 1996).

The JH hemolymph titer is a function of both JH synthesis and its degradation (de Kort and Granger, 1996). JH esterases (JHE) play a central role in regulating the JH titer in diapause-destined adults of *L. decemlineata*. There is an inverse relationship between the JH hemolymph titer and JHE activity in pre-diapausing adults, with peak esterase activity occurring just prior to diapause initiation. This peak in JHE activity contributes to the low JH titer noted during the first two days of adult life (Kramer, 1978). The effect of JH on JHE activity is indirect and appears to require a factor from the brain. JHE isolated from the hemolymph of fourth instar larvae of *L. decemlineata* is a dimer consisting of two 57 kDa subunits (Vermunt *et al.*, 1997a). cDNA clones of two different JHEs (A and B) have been isolated from fourth instar larvae (Vermunt *et al.*, 1997b, 1998). JHE-A appears to encode the protein first isolated from the hemolymph, whereas JHE-B encodes a protein that does not appear to be secreted into the hemolymph. Significant expression of the gene encoding JHE-A is present in brains of fourth instar larvae and fat body of both fourth instar larvae and short-day adults; low expression is noted in long-day adults (Vermunt *et al.*, 1999). Treating diapause-destined day 1 adults with the JH analog pyriproxyfen strongly suppresses JHE-A expression. JHE-A sensitivity to pyriproxyfen decreases as beetles near diapause initiation. Vermunt *et al.* (1999) concluded that JHE activity in the hemolymph is critical for the initiation of diapause in the Colorado potato beetle. However, a role for JHEs in regulation of the JH titer appears not to

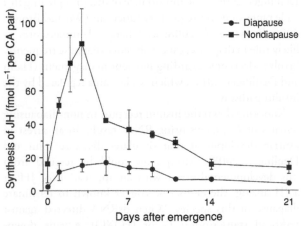

Figure 10 *In vitro* activity of the corpora allata (CA) from diapausing and nondiapausing females of the mosquito *Culex pipiens* on different days following adult emergence. Adapted from *Readio et al.* (1999).

be essential for some other cases of adult diapause. In the locusts *Nomadacris japonica* and *L. migratoria* there are no differences in JHE hemolymph activity between long- and short-day adults (Okuda *et al.*, 1996).

Although evidence supporting involvement of JH in reproductive diapause is compelling, results indicate that the absence of JH is not the exclusive cause of diapause induction and maintenance. Results, even in the Colorado potato beetle, suggest that lack of JH cannot account for all observations. Juvenile hormone application, at best, can stimulate only a short burst of oviposition in diapausing females, suggesting that the actual termination of diapause requires more than a single boost in the JH titer. This observation has usually been interpreted as the need for sustained stimulation of the CA by the brain (Denlinger, 1985). A minimal interpretation is that at the end of diapause the brain must become reactivated to promote JH synthesis by the CA. Brain activation clearly needs to precede activation of the CA and it is possible that the brain independently directs specific events associated with diapause termination. It is also possible that these events occur prior to activation of the CA. This is evident from the observation that adults of *L. decemlineata* actually dig their way out of the soil prior to activation of the CA (Lefevere *et al.*, 1989). This suggests that activation of the CA may be the consequence of diapause termination, rather than the cause.

The regulatory scheme for males is not always the same as that in females. In *P. apterus* (Hodkova, 1994) and *P. terraenovae* (Tanigawa *et al.*, 1999), regulation of diapause in females clearly involves the CA, but the CA does not appear to be involved in males. Allatectomy has little effect on the manifestation of diapause in males; they respond to photoperiod and display mating behavior even in the absence of a CA. The inhibitory center regulating mating behavior resides in the brain's pars intercerebralis, but this important inhibitory pathway does not involve the CA.

When the CA are involved in regulating diapause, they presumably preside over the whole diapause syndrome, not just the regulation of reproduction. This appears to be the case in *P. apterus*, a species in which photoperiod dictates not only the reproductive status of the female but also the profile of phospholipid molecules in cell membranes (Hodkova *et al.*, 2002). The unique winter pattern of phosphatidylethanolamines, associated with cold hardening, is programmed by short daylength, not low temperature. Non-diapausing females deprived of their CA assume the lipid profile normally associated with diapausing females. Similarly, earlier studies with the Colorado potato beetle demonstrated involvement of the CA in determining the status of flight muscles during diapause.

The occurrence of reproductive arrest in insects is not confined to diapause. Adults of *P. apterus* have two distinct forms of reproductive arrest, a photoperiod-controlled diapause and a non-diapause-starvation reproductive arrest. These two arrests are both controlled by the brain but by different mechanisms, resulting in distinct morphological changes in the corpus allatum (Hodkova *et al.*, 2001). Experiments using transplanted brain–CA complexes and application of JH analogs have clearly demonstrated that JH terminates both forms of reproductive arrest (Hodkova and Socha, 2006; Socha, 2007; Socha and Sula, 2008). Delineating commonalities and differences in the various forms of reproductive arrest will be helpful in defining convergent mechanisms for shutting down development.

Another intriguing connection between JH and adult diapause has been proposed by Hunt *et al.* (2007) for *Polistes* wasps. In this scenario, accumulation of a storage protein, hexamerin 1, may dictate the JH titer, determining the diapause fate of the emerging females. Low levels of hexamerin are suggested to elevate JH, which in turn promotes the non-diapause phenotype that is primed to reproduce, while high levels of hexamerin generate a JH-deficient diapause phenotype that fails to reproduce until the following year. Although pieces of this regulatory scheme remain to be defined, if this scheme is verified, this would be an unusual case where the JH titer is dictated by the presence of hexamerins, rather than the more usual assumption that JH and ecdysteroids regulate production and accumulation of the hexamerins (Denlinger *et al.*, 2005). Energy stores are enormously important for the success of diapause (Hahn and Denlinger, 2011), and the mechanism proposed for the paper wasps provides a nice link between nutrition and the endocrine mechanisms of diapause.

10.3.3.2. Insulin signaling Insulin signaling plays a critical role in dauer formation in the nematode *C. elegans* (Lee *et al.*, 2003), and the same appears to be true for reproductive arrest in *D. melanogaster* (Tatar *et al.*, 2001; Allen, 2007) and the mosquito *C. pipiens* (Sim and Denlinger, 2008). The importance of insulin signaling in a wide range of insect regulatory functions (Wu and Brown, 2006; Brown *et al.*, 2008; see Chapter 2) underscores its likely role in diapause, especially since diapause frequently involves decisions regarding nutrient management (Hahn and Denlinger, 2011), which is a key area regulated by the insulin pathway.

Knocking down the insulin receptor in non-diapausing females of *C. pipiens* using RNAi results in an arrest in ovarian development that simulates diapause (Sim and Denlinger, 2008). Mosquitoes could be rescued from this developmental arrest with an application of JH or a JH analog, the endocrine trigger known to terminate diapause in this species. When dsRNA directed against forkhead transcription factor (FOXO), a gene downstream of insulin, was injected into diapausing females, the females failed to accumulate the huge fat depositions that characterize diapause. This suggests that a shut down in insulin signaling halts JH production and simultaneously prompts activation of the downstream gene FOXO, leading to the diapause phenotype, as summarized in **Figure 11**. Although most invertebrates are thought to have just one insulin-like receptor, numerous insulin-like peptides (ILPs) are known from insects. Among these ILPs, ILP-1 appears to be the one of greatest importance for diapause regulation in *C. pipiens* (Sim and Denlinger, 2009a). The activation of FOXO is thus proposed to lead to the dramatic shift in lipid metabolism associated with diapause (Sim and Denlinger, 2009b). The scenario presented in **Figure 11** is consistent with the results currently on hand, but not all steps in the pathway between insulin signaling and JH are known at this time.

Key evidence pointing to a role for insulin in the reproductive arrest of *Drosophila* is presented by Williams *et al.* (2006). Genetic crosses of geographic strains of *D. melanogaster* exhibiting different potentials for ovarian arrest led to the isolation of *Dp110,* a gene that encodes phosphoinositol 3 OH kinase (PI3 K), a component of the insulin signaling pathway. Mutants with a *Dp110* deletion have an elevated incidence of reproductive arrest, whereas increased expression of *Dp110* in the nervous system elicits the opposite effect. Results from both *C. pipiens* and *D. melanogaster* are consistent with a model showing that reduced insulin signaling is linked to adult diapause.

10.3.3.3. Role of ecdysteroids The role of ecdysteroids in the regulation of adult diapause has been investigated most thoroughly in *D. melanogaster*, although this arrest appears to be a low-temperature induced halt, more akin to a quiescence than a photoperiodically induced diapause (Emerson *et al.*, 2009b). The arrest occurs at the pre-vitellogenic stage of ovarian follicular development

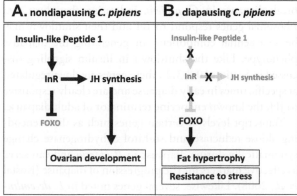

Figure 11 Model for the role of insulin signaling and FOXO in regulation of adult diapause in the mosquito *Culex pipiens.* (A) Insulin signaling in non-diapausing females results in ovarian development and suppression of FOXO. (B) Insulin signaling in diapausing females suppresses the insulin signaling pathway, resulting in arrested development. FOXO is activated, which leads to fat hypertrophy and enhanced stress resistance. Adapted from *Sim and Denlinger* (2008, 2009a).

(Saunders *et al.*, 1989). Yolk proteins (YPs) are present in the hemolymph of these females, but the oocytes fail to incorporate the YPs (Saunders *et al.*, 1990). Ecdysteroids are proposed to activate endocytosis at termination of the arrest allowing initiation of vitellogenesis (Richard *et al.*, 1998). The evidence for this intriguing scenario is compelling.

During this reproductive arrest ovaries produce low level of ecdysteroids; the rate of synthesis begins to increase 4 h after transferring the flies to 25 °C to terminate the arrest, and reaches maximum levels by 12 h (Richard *et al.*, 1998). Injection of 20E terminates the arrest in a dose-dependent manner as measured by ovarian maturation (Richard *et al.*, 2001b). Yolk protein uptake by the oocytes is by a receptor-mediated endocytotic mechanism (Bownes *et al.*, 1993) that involves at least three key proteins. Clathrin coats the pit, and adaptin binds to both clathrin and the transmembrane YP receptor (Gao *et al.*, 1991; Smythe and Warren, 1991). Once the vesicle is internalized the vesicle breaks down, and the clathrin and adaptin are recycled (Gao *et al.*, 1991). During the arrest, clathrin, adaptin, and YP receptor are undetectable by immunostaining. Terminating the arrest by transferring flies to 25°C or injecting them with 20E induces expression of clathrin, adaptin, and the YP receptor in nurse cells adjacent to the oocytes (Richard *et al.*, 2001a). Incubating ovaries from females in this reproductive arrest with 20E fails to elicit expression of clathrin, adaptin, or the YP receptor, indicating that dormancy termination is not regulated by the ovaries and that some external factor is needed (Richard *et al.*, 2001a).

How can these ecdysteroid results be reconciled with the huge body of data suggesting that vitellogenesis in higher Diptera is mediated by JH? Indeed application of JH III or JHB3 breaks this arrest in *D. melanogaster* (Saunders *et al.*, 1990), but this response is an indirect effect caused by JH triggering the ecdysteroid-mediated initiation of vitellogenesis (Richard *et al.*, 1998, 2001b). When flies are transferred to reproduction permissive conditions, it is the ecdysteroid titer that goes up quickly (within hours), while the JH titer remains low.

Roles for ecdysteroids are suggested in a number of other species as well, but the results are sometimes conflicting. Significantly lower ecdysteroid titers are reported for diapausing females of *L. migratoria* than for non-diapausing females (Tawfik *et al.*, 2002a), a result that is consistent with *D. melanogaster*. The ecdysteroid titer in *L. migratoria* fluctuates between 22 and 40 ng ml^{-1} in diapausing females, whereas the titer reaches a peak of 354 ng ml^{-1} in 18- to 22-day old non-diapausing females. Treating diapausing females with three consecutive applications of JH III induces ovarian development and increases ovarian and hemolymph titers of ecdysteroids to levels similar to those noted in sexually active females (Tawfik *et al.*, 2002a). This relationship between the ecdysteroid titer and diapause induction is, however, opposite of that seen in *L. decemlineata* (Briers and de Loof, 1981). In these beetles, high titers of ecdysteroids are noted in diapause-destined beetles shortly after adult eclosion, but the titers drop within the first week, whereas the titers are initially low in non-diapausing adults and then increase in females at the onset of reproduction. During diapause, the ecdysteroid titer gradually increases (Briers *et al.*, 1982). This may be essential for diapause termination. Interestingly, an injection of a combination of 20E and JH is much more effective in breaking diapause than either hormone by itself (Lefevere, 1989), a response also noted for pupal diapause termination in flesh flies (Denlinger, 1979). Lefevere *et al.* (1989) suggested that the ecdysteroid injection enhances diapause termination by blocking activation of JHE, facilitating elevation of the JH titer.

A role for ecdysteroids has also been proposed for diapause in *P. apterus* (Sauman and Sehnal, 1997). Under long-day conditions, maturation of the male accessory glands and testes of *P. apterus* occur during the fourth nymphal instar. Injecting an ecdysteroid or implanting brain–subesophageal ganglion complexes from long-day nymphs into early fourth instar nymphs reared at short days induces sexual development in the normally immature reproductive tract. Sauman and Sehnal (1997) concluded that the pre-diapause physiology in the fourth instar nymph of *P. apterus* is regulated by ecdysteroids and neurohormones, an observation that is consistent with other reports showing a low ecdysteroid titer contributing to the diapause fate of the adult.

Prothoracicotropic hormone, the neuropeptide that drives ecdysteroid production in the prothoracic gland of larvae and pupae, is generally considered not to play a role in adults because the prothoracic gland has degenerated by that time. Yet, ovarian-derived ecdysteroids are now well-known regulators of adult reproductive events, and a possible role for PTTH in adults cannot be discounted as evidenced by recent work on the mosquito *C. pipiens* (Zhang and Denlinger, 2011; see Chapter 1). Transcript abundance of the gene encoding PTTH drops quickly following adult emergence in non-diapausing female mosquitoes but persists at a high level for the first month in diapausing females. These results indicate that, at least at the transcript level, there are conspicuous PTTH distinctions between females programmed for diapause and non-diapause, suggesting possible differences in ecdysteroid titers as well.

It seems unlikely that adult diapause will prove to be based simply on either a JH or ecdysteroid deficiency. In all of the earlier examples showing evidence for ecdysteroid involvement, JH also has been implicated, suggesting roles for both of these hormones in orchestrating the regulation of adult diapause.

10.3.3.4. Associated changes in gene expression

As seen for larval and pupal diapauses the endocrine mechanisms controlling adult diapause set into motion complex changes in gene expression, as evidenced from a suppressive subtractive hybridization study in *C. pipiens* (Robich *et al.*, 2007), a targeted gene study in *P. apterus* (Kostal *et al.*, 2008), and differential display studies with *L. decemlineata* (Yocum *et al.*, 2009a,b). Thus far, the only large-scale genomic approach to adult diapause is based on the reproductive arrest noted in *D. melanogaster*, where a comparison was made between flies in arrested reproduction and those that terminated their arrest and initiated egg development (Baker and Russell, 2009). Such studies give a good perspective on the magnitude of gene expression changes associated with diapause. Numerous transcripts are altered in abundance. In the *Drosophila* study, 2786 transcripts were two- or threefold more abundant during diapause. As might be expected, genes involved in processes associated with follicle and egg maturation are downregulated, whereas many genes involved in response to environmental stress, lipid conservation, and protective structural modifications are upregulated.

Among diapause-associated genes that may be particularly important for regulating reproduction are ribosomal proteins S2 (Kim and Denlinger, 2010) and S3a (Kim *et al.*, 2010). Expression of both of these ribosomal proteins is shut down during early diapause in the mosquito *C. pipiens*, and RNAi directed against these genes results in a "diapause-like" ovarian state in mosquitoes programmed for continuous development. Both genes are known to play critical roles in regulating oogenesis in other insects, suggesting that their shutdown could be critical for the diapause response. Arrest in ovarian development observed in RNAi-treated females can be rescued with an application of JH III, the hormone known to break diapause in *C. pipiens*, indicating that these two ribosomal genes respond to JH and their shutdown may be an essential component for generating the diapause phenotype. Like the shutdown in insulin signaling discussed in Section 10.3.3.2., these genes are downregulated at specific times in early diapause and are clearly responsive to JH, the known endocrine terminator of adult diapause.

Transcript levels of certain genes such as those encoding aldose reductase and sorbitol dehydrogenase change systematically during diapause in *P. apterus* and can serve as effective markers for the progression of diapause (Kostal *et al.*, 2008). Likewise, several genes noted in *L. decemlineata* can be used as biomarkers to track the progression of diapause (Yocum *et al.*, 2009a,b).

As discussed in Section 10.2.3.2., clock genes have been examined in association with adult diapause in several species, but critical links between clock gene expression and downstream pathways, including the endocrine components that dictate the diapause fate, remain unknown.

10.4. Parallels to Other Dormancies

Diverse endocrine mechanisms are capable of halting development at different developmental stages and in different species that enter diapause at the same stage. Sometimes the same net effect, diapause, is elicited by contrasting responses to the very same hormone. For example, a shutdown in ecdysteroid production is key to most pupal diapauses, but an elevation in ecdysteroids elicits diapause in pharate first instar larvae of the gypsy moth. Similarly, the absence of JH characterizes many adult diapauses, while some larval diapauses are maintained by a high JH titer. Progression through the different stages of embryonic and post-embryonic development and the initiation of reproduction by the adult is orchestrated by a well-defined series of endocrine events requiring periods of hormone release and other equally important phases of hormone absence. This requirement for precisely timed periods of hormone presence or absence to elicit uninterrupted development suggests that nearly any interruption in this normal sequence could lock the insect into a developmental arrest at the point of interruption. This is presumably why we observe diverse endocrine mechanisms causing the same developmental shutdown we recognize as diapause. This means that the endocrine control mechanism for halting development has evolved independently numerous times. Certainly there are taxa in which the diapause stage and the control mechanism are highly conserved. For example, all of the flesh fly species (Sarcophagidae) from around the temperate and tropical world that have a diapause rely on a pupal diapause that represents a shutdown in the brain–prothoracic gland axis, suggesting a highly conserved trait within this family. By contrast, different members of the Drosophilidae are known to enter diapause as larvae, pupae, or adults, suggesting that the endocrine basis for diapause has evolved several times within this family.

The challenge to identify a unifying endocrine basis for diapause becomes even more difficult when one seeks parallels in the endocrine control mechanisms for dormancies in non-arthropods. A connection between the insulin-like receptor *daf-2* in the production of dauer larvae of the nematode *C. elegans* (Kimura *et al.*, 1997; Lee *et al.*, 2003) and the involvement of a homologous receptor that influences JH synthesis in *D. melanogaster* (Tatar *et al.*, 2001; Tatar and Yin, 2001) and *C. pipiens* (Sim and Denlinger, 2008) offers a tantalizing connection with some adult diapauses, but because JH is not an active player in many forms of insect diapause suggests that this connection will not be applicable to all insect diapauses. It is quite likely that many different non-arthropod species will have, like insects, tapped into diverse endocrine mechanisms to regulate development as well as developmental arrest.

This argument, however, does not preclude the possibility that at the molecular level diapause and other forms of dormancy may target the same or similar cellular processes to bring about an arrest in development. Several results already hint of some commonality to the molecular events associated with diapause. For example, the diapause-associated upregulation of genes encoding heat shock proteins was first noted in flesh fly pupae, but heat shock protein upregulation during insect diapause has now been documented in several pupal diapauses as well as embryonic, larval, and adult diapauses of other insects and in the dormancies of brine shrimp, nematodes, and some plants (Denlinger *et al.*, 2001; Rinehart *et al.*, 2007). This common feature represents a widespread contribution of heat shock proteins to cellular defense during diapause and/or a contribution to the mechanism for shutting down development. Although not all insects elevate heat shock proteins during diapause (Goto *et al.*, 1998), elevated stress responses of some sort are a component of most diapauses. Another feature that would appear to be an essential cellular event common to most diapauses is a shutdown of the cell cycle (Tammariello, 2001; Kostal *et al.*, 2009). Again, there may be diversity in the phase of the cell cycle arrested (e.g., G_0/G_1 in pupae of *S. crassipalpis* and larvae of *C. costata*, G_2 in pupae of *M. sexta*, and in embryos of *B. mori*) and in the cell cycle regulator that is targeted, but the overall targeting of the cell cycle is a likely ultimate target of whatever endocrine mechanism is utilized. Metabolic depression is yet another feature common to diapause, but again diverse mechanisms can be used to achieve this response. Diversity in the control mechanisms for diapause is most evident at the transcriptional level. In spite of common physiological end points such as metabolic depression, cell cycle arrest and enhanced stress tolerance in dauer larvae of *C. elegans*, adult dormancy in *D. melanogaster*, and pupal diapause in *S. crassipalpis*, there is remarkably little overlap in the genes generating the diapause response (Ragland *et al.*, 2010). This suggests that the key elements of diapause can be attained by different transcriptional strategies. This, in turn, suggests that we can expect equally diverse endocrine oversight directing diapause mechanisms.

In summary, we argue that the diverse patterns of diapause, although sharing many common physiological attributes, are attained through divergent molecular mechanisms that reflect an equal diversity of endocrine regulators. The ultimate physiological responses that characterize diapause can be turned on or off by influencing numerous targets in the gene network that generate the diapause phenotype.

References

Agui, N., Granger, N. A., Gilbert, L. I., & Bollenbacher, W. E. (1979). Cellular localization of insect prothoracicotropic hormone. *Proc. Natl. Acad. Sci. USA, 76,* 5684–5690.

Allen, M. J. (2007). What makes a fly enter diapause? *Fly, 1,* 307–310.

Andrewartha, H. G. (1952). Diapause in relation to the ecology of insects. *Biol. Rev., 27,* 50–107.

Baker, D. A., & Russell, S. (2009). Gene expression during *Drosophila melanogaster* egg development before and after reproductive diapause. *BMC Genomics, 10,* 242.

Bale, J. S., & Hayward, S. A.L. (2010). Insect overwintering in a changing climate. *J. Exp. Biol., 213,* 980–994.

Bell, R. A. (1996). Manipulation of diapause in the gypsy moth, *Lymantria dispar* L., by application of KK-42 and precocious chilling of eggs. *J. Insect Physiol., 42,* 557–563.

Bodnaryk, R. P. (1985). Ecdysteroid levels during postdiapause development and 20-hydroxyecdysoneinduced development in male pupae of *Mamestra configurata* Wlk. *J. Insect Physiol., 31,* 53–58.

Bosse, T. C., & Veerman, A. (1996). InvolvementofvitaminAin the photoperiodic induction of diapause in the spider mite *Tetranychus urticae* is demonstrated by rearing an albino mutant on a semi-synthetic diet and without b-carotene or vitamin A. *Physiol. Entomol., 21,* 188–192.

Bowen, M. F., Bollenbacher, W. E., & Gilbert, L. I. (1984). in vitro studies on the role of the brain and prothoracic glands in the pupal diapause of *Manduca sexta. J. Exp. Biol., 108,* 9–24.

Bownes, M., Ronaldson, E., Mauchline, D., & Martinez, A. (1993). Regulation of vitellogenesis in *Drosophila. J. Insect Morphol. Embryol., 22,* 349–367.

Bradshaw, W. E., & Holzapfel, C. M. (2007). Tantalizing *timeless. Science, 316,* 1851–1852.

Bradshaw, W. E., & Holzapfel, C. M. (2010). Circadian clock genes, ovarian development and diapause. *BMC Biol., 8,* 115.

Briers, T., & de Loof, A. (1981). Moulting hormone activity in the adult Colorado potato beetle, *Leptinotarsa decemlineata* Say in relation to reproduction and diapause. *Int. J. Invert. Reprod., 3,* 145–155.

Briers, T., Peferoen, M., & de Loof, A. (1982). Ecdysteroids and adult diapause in the Colorado potato beetle, *Leptinotarsa decemlineata. Physiol. Entomol., 7,* 379–386.

Brown, M. R., Clark, K. D., Gulia, M., Zhoa, Z., Garczynski, S. F., Crim, J. W., et al. (2008). An insulin-like peptide regulates egg maturation and metabolism in the mosquito *Aedes aegypti. Proc. Natl. Acad. Sci. USA, 105,* 5716–5721.

Champlin, D. T., & Truman, J. W. (1998). Ecdysteroid control of cell proliferation during optic lobe neurogenesis in the moth *Manduca sexta. Development, 125,* 269–277.

Chen, L., Ma, W., Xang, X., Niu, C., & Lei, C. (2010). Analysis of pupal head proteome and its alteration in diapausing pupae of *Helicoverpa armigera. J. Insect Physiol., 56,* 247–252.

Chen, J. H., Yaginuma, T., & Yamashita, O. (1988). Effect of diapause hormone on cyclic nucleotide metabolism in developing ovaries of the silkmoth, Bombyx mori. *Comp. Biochem. Physiol. B, 91,* 631–637.

Chen, J. H., & Yamashita, O. (1989). Activity changes of guanylate cyclase and cyclic GMP phosphodiesterase related to the accumulation of cyclic GMP in developing ovaries of the silkmoth, *Bombyx mori. Comp. Biochem. Physiol. B, 93,* 385–390.

Cheng, W. N., Li, X. L., Yu, F., Li, Y. P., Li, J. J., & Wu, J. X. (2009). Proteomic analysis or pre-diapause, diapause and post-diapause larvae of the wheat blossom midge, *Sitodiplosis mosellana* (Diptera: Cecidomyiidae). *Eur. J. Entomol., 106,* 29–35.

Chino, H. (1958). Carbohydrate metabolism in the diapause egg of the silkworm, *Bombyx mori.* II. Conversion of glycogen into sorbitol and glycerol during diapause. *J. Insect Physiol. 2,* 1–12.

Chippendale, G. M. (1977). Hormonal regulation of larval diapause. *Annu. Rev. Entomol., 22,* 121–138.

Cho, J. R., Lee, M., Kim, H. S., & Boo, K. S. (2007). Effect of the juvenile hormone analog, fenoxycarb on termination of reproductive diapause in *Scotinophara lurida* (Burmeister) (Heteroptera: Pentatomidae). *J. Asia_Pacific Entomol., 10,* 145–150.

Claret, J. (1989). Vitamine A et induction photopériodique ou thermopériodique de la diapause chez *Pieris brassicae* (Lepidoptera). *C.R. Acad. Sci. Paris, Se'rie III, 308,* 347–352.

Claret, J., & Volkoff, N. (1992). Vitamin A is essential for two processes involved in the photoperiodic reaction in *Pieris brassicae. J. Insect Physiol., 38,* 569–574.

Ctvrtecka, R., & Zdarek, J. (1992). Reproductive diapause and its termination in the apple blossom weevil *(Anthonomus pomorum)* (Coleoptera, Curculionidae). *Acta Entomol. Bohem., 89,* 281–286.

Danks, H. V. (1987). *Insect Dormancy: An Ecological Perspective.* Ottawa: Biological Survey of Canada.

de Kort, C. A.D. (1990). Thirty-five years of diapause research with the Colorado potato beetle. *Entomol.Exp. Applic., 56,* 1–13.

de Kort, C. A.D., & Granger, N. A. (1996). Regulation of JH titers: the relevance of degradative enzymes and binding proteins. *Arch. Insect Biochem. Physiol., 33,* 1–26.

Denlinger, D. L. (1979). Pupal diapause in tropical flesh flies: environmental and endocrine regulation, metabolic rate and genetic selection. *Biol. Bull. Woods Hole, 156,* 31–46.

Denlinger, D. L. (1985). Hormonal control of diapause. In G. A. Kerkut, & L. I. Gilbert (Eds.), *Comprehensive Insect Physiology, Biochemistry and Pharmacology* vol. 8. (pp. 353–412) Oxford: Pergamon Press.

Denlinger, D. L. (1986). Dormancy in tropical insects. *Annu. Rev. Entomol., 31,* 239–264.

Denlinger, D. L. (1998). Maternal control of fly diapause. In T. A. Mousseau, & C. A. Fox (Eds.), *Maternal Effects as Adaptations* (pp. 275–287). Oxford: Oxford University Press.

Denlinger, D. L. (2001). Interrupted development: the impact of temperature on insect diapause. In D. Atkinson, & M. Thorndyke (Eds.), *Environment and Animal Development: Genes, Life Histories and Plasticity* (pp. 235–250). Oxford: BIOS Scientific Publishers.

Denlinger, D. L. (2002). Regulation of diapause. *Annu. Rev. Entomol., 47,* 93–122.

Denlinger, D. L., & Bradfield, J. Y. (1981). Duration of pupal diapause in the tobacco hornworm is determined by number of short days received by the larva. *J. Exp. Biol., 91,* 331–337.

Denlinger, D. L., Giebultowicz, J. M., & Adedokun, T. A. (1988). Insect diapause: dynamics of hormone sensitivity and vulnerablility to environmental stress. In F. Sehnal,

A. Zabza, & D. L. Denlinger (Eds.), *Endocrinological Frontiers in Physiological Insect Ecology* (pp. 885–898). Wrocław: Wrocław Technical University Press.

Denlinger, D. L., & Lee, R. E., Jr. (2010). *Low Temperature Biology of Insects*. Cambridge: Cambridge University Press.

Denlinger, D. L., Shukla, M., & Faustini, D. L. (1984). Juvenile hormone involvement in pupal diapause of the flesh fly *Sarcophaga crassipalpis*: regulation of infradian cycles of O2 consumption. *J. Exp. Biol., 109*, 191–199.

Denlinger, D. L., & Tanaka, S. (1989). Cycles of juvenile hormone esterase activity during the juvenile hormone-driven cycles of oxygen consumption in pupal diapause of flesh flies. *Experientia, 45*, 474–476.

Denlinger, D. L., Yocum, G. D., & Rinehart, J. P. (2005). Hormonal control of diapause. In L. I. Gilbert, K. Iatrou, & S. S. Gill (Eds.), *Comprehensive Molecular Insect Science* vol. 3. (pp. 615–650). Oxford: Elsevier.

Denlinger, D. L., & Wingard, P. (1978). Cyclic GMP breaks pupal diapause in the flesh fly *Sarcophaga crassipalpis*. *J. Insect Physiol., 24*, 715–719.

Desroches, P., & Huignard, J. (1991). Effect of larval density on development and induction of reproductive diapause in *Bruchidius atrolineatus*. *Entomol. Exp. Applic., 61*, 255–263.

Eizaguirre, M., Prats, J., Abellana, M., Lopez, C., Llovera, M., et al. (1998). Juvenile hormone and diapause in the Mediterranean corn borer, *Sesamia nonagrioides*. *J. Insect Physiol., 44*, 419–425.

Eizaguirre, M., Schafellner, C., López, C., & Sehnal, F. (2005). Relationship between an increase of juvenile hormone titer in early instars and the induction of diapause in fully grown larvae of *Sesamia nonagrioides*. *J. Insect Physiol., 51*, 1127–1134.

Emerson, K. J., Letaw, A. D., Bradshaw, W. E., & Holzapfel, C. M. (2008). Extrinsic light:dark cycles, rather than endogenous circadian cycles, affect the photoperiodic counter in the pitcher-plant mosquito, *Wyeomyia smithii*. *J. Comp. Physiol. A, 194*, 611–615.

Emerson, K. J., Dake, S. J., Bradshaw, W. E., & Holzapfel, C. M. (2009a). Evolution of photoperiodic time measurement is independent of the circadian clock in the pitcher-plant mosquito, *Wyeomyia smithii*. *J. Comp. Physiol. A, 195*, 385–391.

Emerson, K. J., Uyemura, A. M., McDaniel, K. L., Schmidt, P. S., Bradshaw, W. E., & Holzapfel, C. M. (2009b). Environmental control of ovarian dormancy in natural populations of *Drosophila melanogaster*. *J. Comp. Physiol. A, 195*, 825–829.

Emerson, K. J., Bradshaw, W. E., & Holzapel, C. M. (2010). Microarrays reveal early transcriptional events during the termination of larval diapause in natural populations of the mosquito, *Wyeomyia smithii*. *PLoS ONE, 5*, e9574.

Endo, K., Fujimoto, Y., Kondo, M., Yamanaka, A., Watanabe, M., et al. (1997). Stage-dependent changes of the prothoracicotropic hormone (PTTH) activity of brain extracts and of the PTTH sensitivity of the prothoracic glands in the cabbage armyworm, *Mamestra brassicae,* before and during winter and festival pupal diapause. *Zool. Sci., 14*, 127–133.

Evenden, M. L., Armitage, G., & Lau, R. (2007). Effects of nutrition and methoprene treatment upon reproductive diapause in *Caloptilia fraxinella* (Lepidoptera: Gracillariidae). *Physiol. Entomol., 32*, 275–282.

Fielding, D. J. (1990). Photoperiod and food regulate termination of diapause in the squash bug, *Anasa tristis. Entomol. Exp. Applic., 55*, 119–124.

Friedlander, M., & Reynolds, S. (1992). Intratesticular ecdysteroid titres and the arrest of sperm production during pupal diapause in the tobacco hornworm, *Manduca sexta. J. Insect Physiol., 38*, 693–703.

Fujiwara, Y., & Denlinger, D. L. (2007). High temperature and hexane break pupal diapause in the flesh fly, *Sarcophaga crassipalpis*, by activating ERK/MAPK. *J. Insect Physiol., 53*, 1276–1282.

Fujiwara, H., Jindra, M., Newirtt, R., Palli, S. R., Hiruma, K., et al. (1995). Cloning of an ecdysone receptor homologue from *Manduca sexta* and the developmental profile of its mRNA in wings. *Insect Biochem. Mol. Biol., 25*, 845–856.

Fujiwara, Y., Tanaka, Y., Iwata, K. I., Rubio, R. O., Yaginuma, T., Yamashita, O., & Shiomi, K. (2006). ERK/MAPK regulates ecdysteroid and sorbitol metabolism for embryonic diapause termination in the silkworm, *Bombyx mori. J. Insect Physiol., 52*, 569–575.

Fukuda, S. (1951). Factors determining the production of nondiapause eggs in the silkworm. *Proc. Jap. Acad., 27*, 582–586.

Fukuda, S. (1952). Function of the pupal brain and subesophageal ganglion in the production of non-diapause and diapause eggs in the silkworm. *Annot. Zool. Japan, 25*, 149–155.

Gadenne, C., Varjas, L., & Mauchamp, B. (1990). Effects of the non-steroidal ecdysone mimic, RH-5849, on diapause and non-diapause larvae of the European corn borer, *Ostrinia nubilalis. J. Insect Physiol., 36*, 555–559.

Gao, B., Biosca, J., Craig, E. A., Greene, L. E., & Eisenberg, E. (1991). Uncoating of coated vesicles by yeast hsp70 proteins. *J. Biol. Chem., 266*, 19565–19571.

Gao, N., von Schantz, M., Foster, R. G., & Hardie, J. (1999). The putative brain photoperiodic photoreceptors in the vetch aphid, *Megoura viciae. J. Insect Physiol., 45*, 1011–1019.

Gelman, D. B., & Brents, L. A. (1984). Haemolymph ecdysteroid levels in diapause- and nondiapause-bound fourth and fifth instars and in pupae of the European corn borer, *Ostrinia nubilalis* (Hubner). *Comp. Biochem. Physiol. A, 78*, 319–325.

Gelman, D. B., Thyagaraja, B. S., Kelly, T. J., Masler, E. P., Bell, R. A., et al. (1992). Prothoracicotropic hormone levels in brains of the European corn borer, *Ostrinia nubilalis*: diapause vs. the non-diapause state. *J. Insect Physiol., 38*, 383–395.

Gelman, D. B., & Woods, C. W. (1983). Haemolymph ecdysteroid titers of diapause- and nondiapause-bound fifth instars and pupae of the European corn borer, *Ostrinia nubilalis* (Hubner). *Comp. Biochem. Physiol. A, 76*, 367–375.

Gerenday, A., & Fallon, A. M. (2004). Ecdysone-induced accumulation of mosquito cells in the G1 phase of the cell cycle. *J. Insect Physiol., 50*, 831–838.

Giebultowicz, J. M., & Denlinger, D. L. (1986). Role of the brain and ring gland in relation to pupal diapause in the flesh fly, *Sarcophaga crassipalpis. J. Insect Physiol., 32*, 161–166.

Goto, S. G., & Denlinger, D. L. (2002). Short-day and long-day expression patterns of genes involved in the flesh fly clock mechanism: period, timeless, cycle and cryptochrome. *J. Insect Physiol, 48*, 803–816.

Goto, S. G., Yoshida, K. M., & Kimura, M. T. (1998). Accumulation of Hsp70 mRNA under environmental stresses in diapausing and nondiapausing adults of *Drosophila triauraria. J. Insect Physiol., 44*, 1009–1015.

Goto, S. G., Han, B., & Denlinger, D. L. (2006). A nondiapausing variant of the flesh fly, *Sarcophaga bullata,* that shows arrhythmic adult eclosion and elevated expression of two circadian clock genes, *period and timeless. J. Insect Physiol., 52*, 1213–1218.

Goto, S. G., Shiga, S., & Numata, H. (2010). Photoperiodism in insects: perception of light and the role of clock genes. In R. J. Nelson, D. L. Denlinger, & D. E. Somers (Eds.), *Photoperiodism* (pp. 258–286). Oxford: Oxford University Press.

Gregg, P. C., Roberts, B., & Wentworth, S. L. (1987). Levels of ecdysteroids in diapause and non-diapause eggs of the Australian plague locust, Chortoicetes terminifera (Walker). *J. Insect Physiol., 33*, 237–242.

Hahn, D. A., & Denlinger, D. L. (2007). Meeting the energetic demands of insect diapause: nutrient storage and utilization. *J. Insect Physiol., 53*, 760–773.

Hahn, D. A., & Denlinger, D. L. (2011). Energetics of diapause. *Annu. Rev. Entomol., 56*, 103–121.

Hamasaka, Y., Watari, Y., Arai, T., Numata, H., & Shiga, S. (2001). Retinal and extraretinal pathways for entrainment of the circadian activity rhythm in the blow fly, *Protophormia terraenovae. J. Insect Physiol., 47*, 967–975.

Hamel, M., Geri, C., & Auger-Rozenberg, A. (1998). The effects of 20-hydroxyecdysone on breaking diapause of *Diprion pini* L. (Hym., Diprionidae). *Physiol. Entomol., 23*, 337–346.

Hartfelder, K., Hanton, W. K., & Bollenbacher, W. E. (1994). Diapause-dependent changes in prothoracicotropic hormone-producing neurons of the tobacco hornworm, *Manduca sexta. Cell Tissue Res., 277*, 69–78.

Hasegawa, K. (1952). Studies on the voltinism of the silkworm, *Bombyx mori* L., with special reference to the organs concerning determination of voltinism. *J. Fac. Agric. Tottori Univ, 1*, 83–124.

Hasegawa, K. (1957). The diapause hormone of the silkworm, *Bombyx mori. Nature, 179*, 1300–1301.

Hasegawa, K. (1964). Studies on the mode of action of the diapause hormone in the silkworm, *Bombyx mori* L. II. Content of diapause hormone in the subesophageal ganglion. *J. Exp. Biol., 41*, 855–863.

Hasegawa, K., & Shimizu, I. (1987). In vivo and in vitro photoperiodic induction of diapause using isolated brain–suboesophageal ganglion complexes of the silkworm, *Bombyx mori. J. Insect Physiol., 33*, 959–966.

Hasegawa, K., & Shimizu, I. (1990). Gabaergic control of the release of diapause hormone from the esophageal ganglion of the silkworm, *Bombyx mori. J. Insect Physiol., 36*, 909–915.

Helfrich-Forster, C. (2009). Does the morning and evening oscillator model fit better for flies or mice? *J. Biol. Rhythms, 24*, 259–270.

Ho, D. H., & Burggren, W. W. (2010). Epigenetics and transgenerationsl transfer: a physiological perspective. *J. Exp. Biol., 213*, 3–16.

Hodkova, M. (1992). Storage of the photoperiodic information within the implanted neuroendocrine complexes in females of the linden bug *Pyrrhocoris apterus* (L.) (Heteroptera). *J. Insect Physiol., 38*, 357–363.

Hodkova, M. (1994). Photoperiodic regulation of mating behaviour in the linden bug, *Pyrrhocoris apterus,* is mediated by a brain inhibitory factor. *Experientia, 50*, 742–744.

Hodkova, M., Berkova, P., & Zahradnickova, H. (2002). Photoperiodic regulation of the phospholipid molecular species composition in thoracic muscles and fat body of *Pyrrhocoris apterus* (Heteroptera) via an endocrine gland, corpus allatum. *J. Insect Physiol., 48*, 1009–1019.

Hodkova, M., Okuda, T., & Wagner, R. (1996). Stimulation of corpora allata by extract from neuroendocrine complex: comparison of reproducing and diapausing *Pyrrhocoris apterus* (Heteroptera: Pyrrhocoridae). *Eur. J. Entomol., 93*, 535–543.

Hodkova, M., Okuda, T., & Wagner, R. (2001). Regulation of corpora allata in females of *Pyrrhocoris apterus* (Heteroptera) (a mini-review). *In vitro Cell. Devel. Biol., 37*, 560–563.

Hodkova, M., & Socha, R. (2006). Endocrine regulation of the reproductive arrest in the long-winged females of a flightless bug, *Pyrrhocoris apterus* (Heteroptera: Pyrrhocoridae). *Eur. J. Entomol., 103*, 523–529.

Hoffmann, E. J., VanderJagt, J., & Whalon, M. E. (2007). Pyriproxyfen activates reproduction in pre-diapause northern strain plum curculio (*Conotrachelus nenuphar* Herbst). *Pest Manag. Sci., 63*, 835–840.

Homma, T., Watanabe, K., Tsurumaru, S., Kataoka, H., Imai, K., Kamba, M., Niimi, T., Yamashita, O., & Yaginuma, T. (2006). G protein-coupled receptor for diapause hormone, an inducer of *Bombyx* embryonic diapause. *Biochem. Biophy. Res. Comm., 344*, 386–393.

Horie, Y., Kanda, T., & Mochida, Y. (2000). Sorbitol as an arrester of embryonic development in diapausing eggs of the silkworm, *Bombyx mori. J. Insect Physiol., 46*, 1009–1016.

Hunt, J. H., Kensinger, B. J., Kossuth, J. A., Henshaw, M. T., Norberg, K., Wolschin, F., & Amdam, G. V. (2007). A diapause pathway underlies the gyne phenotype in *Polistes* wasps, revealing an evolutionary route to caste-containing insect societies. *Proc. Nat'l. Acad. Sci., USA, 104*, 14020–14025.

Ichikawa, T., Hasegawa, K., Shimizu, I., Katsuno, K., Kataoka, H., et al. (1995). Structure or neurosecretory cells with immunoreactive diapause hormone and pheromone biosynthesis activating neuropeptide in the silkworm, *Bombyx mori. Zool. Sci., 12*, 703–712.

Ichikawa, T., Shiota, T., Shimizu, I., & Kataoka, H. (1996). Functional differentiation of neurosecretory cells with immunoreactive diapause hormone and pheromone biosynthesis activating neuropeptide of the moth, *Bombyx mori. Zool. Sci., 13*, 21–25.

Ikeda, M., Su, Z. H., Saito, H., Imai, K., Sato, Y., et al. (1993). Induction of embryonic diapause and stimulation of ovary trehalase activity in the silkworm, *Bombyx mori*, by synthetic diapause hormone. *J. Insect Physiol.*, *39*, 889–895.

Ikeno, T., Tanaka, S. I., Numata, H., & Goto, S. G. (2010). Insect photoperiodism under control of circadian clock genes. *BMC Biol.*, *8*, 116.

Imai, K., Konno, T., Nakazawa, Y., Komiya, T., Isobe, M., et al. (1991). Isolation and structure of diapause hormone of the silkworm, *Bombyx mori*. *Proc. Jap. Acad. B*, *67*, 98–101.

Imai, K., Nomura, T., Katsuzaki, H., Komiya, T., & Yamashita, O. (1998). Minimum structure of diapause hormone required for biological activity. *Biosci. Biotechnol. Biochem.*, *62*, 1875–1879.

Imhof, M. O., Rusconi, S., & Lezzi, M. (1993). Cloning of the *Chironomus tentans* cDNA encoding a protein (cEcRH) homologous to the *Drosophila melanogaster* ecdysteroid receptor (dEcR). *Insect Biochem. Mol. Biol.*, *23*, 115–124.

Isobe, M., Hasegawa, K., & Goto, T. (1973). Isolation of the diapause hormone from the silkworm, *Bombyx mori*. *J. Insect Physiol.*, *19*, 1221–1240.

Joplin, K. H., Stetson, D. L., Diaz, J. G., & Denlinger, D. L. (1993). Cellular differences in ring glands of flesh fly pupae as a consequence of diapause programming. *Tissue and Cell*, *25*, 245–257.

Khan, M. A. (1988). Brain-controlled synthesis of juvenile hormone in adult insects. *Entomol. Exp. Applic.*, *46*, 3–17.

Khan, M. A., & Buma, P. (1985). Neural control of the corpus allatum in the Colorado potato beetle, *Leptinotarsa decemlineata*: an electron microscope study utilizing the in vitro tannic acid Ringer incubation method. *J. Insect Physiol.*, *31*, 639–645.

Khan, M. A., Koopmanschap, A. B., & de Kort, C. A.D. (1983). The relative importance of nervous and hormonal control of corpus allatum activity in the adult Colorado potato beetle, *Leptinotarsa decemlineata* (Say). *Gen. Comp. Endocrinol.*, *52*, 214–221.

Kidokoro, K., Iwata, K. i, Fujiwara, Y., & Takeda, M. (2006a). Effects of juvenile hormone analogs and 20-hydroxyecdysone on diapause termination in eggs of *Locusta migratoria* and *Oxya yezoensis*. *J. Insect Physiol.*, *52*, 473–479.

Kidokoro, K., Iwata, K. i, Takeda, M., & Fujiwara, Y. (2006b). Involvement of ERK/MAPK in regulation of diapause intensity in the false melon beetle, *Atrachya menetriesi*. *J. Insect Physiol.*, *52*, 1189–1193.

Kim, M., & Denlinger, D. L. (2010). A potential role for ribosomal protein S2 in the gene network regulating reproductive diapause in the mosquito *Culex pipiens*. *J. Comp. Physiol. B*, *180*, 171–178.

Kim, M., Sim, C., & Denlinger, D. L. (2010). RNA interference directed against ribosomal protein S3a suggests a link between this gene and arrested ovarian development during adult diapause in *Culex pipiens*. *Insect Mol. Biol.*, *19*, 27–33.

Kimura, K. D., Tissenbaum, H. A., Liu, Y., & Ruvkun, G. (1997). daf-2, an insulin receptor-like gene that regulates longevity and diapause in *Caenorhabditis elegans*. *Science*, *277*, 942–946.

Kipyatkov, V. E. (2001). A distantly perceived primer pheromone controls diapause termination in the ant *Myrmica rubra* L. (Hymenoptera, Formicidae). *J. Evol. Biochem. Physiol.*, *37*, 405–416.

Kitagawa, N., Shiomi, K., Imai, K., Niimi, T., Yamashita, O., & Yaginuma, T. (2005). Diapause hormone levels in subesophageal ganglia of uni-, bi- and poly-voltine races during pupal-adult development of *Bombyx mori*, and the effects of ouabain, an inhibitor of Na+-K+ ATPase, on the hormone levels. *J. Insect Biotech. Sericol.*, *74*, 57–62.

Kopper, B. J., Shu, S., Charlton, R. E., & Ramaswamy, S. B. (2001). Evidence for reproductive diapause in the fritillary *Speyeria idalia* (Lepidoptera: Nymphalidae). *Ann. Entomol. Soc. America*, *94*, 427–432.

Kostal, V. (2006). Eco-physiological phases of insect diapause. *J. Insect Physiol.*, *52*, 113–127.

Kostal, V. (2011). Insect photoperiodic calendar and circadian clock: independence, cooperation or unity? *J. Insect Physiol.*, in press.

Kostal, V., Noguchi, H., Shimada, K., & Hayakawa, Y. (1998). Developmental changes in dopamine levels in larvae of the fly *Chymomyza costata*: comparison between wild-type and mutant-nondiapause strains. *J. Insect Physiol.*, *44*, 605–614.

Kostal, V., Noguchi, H., Shimada, K., & Hayakawa, Y. (1999). Dopamine and serotonin in the larval CNS of a drosophilid fly, *Chymomyza costata*: are they involved in the regulation of diapause?. *Arch. Insect Biochem. Physiol.*, *42*, 147–162.

Kostal, V., & Shimada, K. (2001). Malfunction of circadian clock in the non-photoperiodic-diapause mutants of the drosophilid fly, *Chymomyza costata*. *J. Insect Physiol.*, *47*, 1269–1274.

Kostal, V., Simunkova, P., Kobelkova, A., & Shimada, K. (2009). Cell cycle arrest as a hallmark of insect diapause: changes in gene transcription during diapause induction in the drosophilid fly, *Chymomyza costata*. *Insect Biochem. Mol. Biol.*, *39*, 875–883.

Kostal, V., Tollarova, M., & Dolezel, D. (2008). Dynamism in physiology and gene transcription during reproductive diapause in a heteropteran bug, *Pyrrhocoris apterus*. *J. Insect Physiol.*, *54*, 77–88.

Kotaki, T., Shinada, T., Kaihara, K., Ohfune, Y., & Numata, H. (2011). Biological activities of juvenile hormone III skipped bisepoxide in last instar nymphs and adults of a stink bug, *Plautia stali*. *J. Insect Physiol.*, in press.

Kotaki, T., & Yagi, S. (1989). Hormonal control of adult diapause in the brown-winged green bug, *Plautia stali* Scott (Heteroptera: Pentatomidae). *Appl. Entomol. Zool.*, *24*, 42–51.

Kramer, S. J. (1978). Regulation of the activity of JH specific esterases in the Colorado potato beetle, *Leptinotarsa decemlineata*. *J. Insect Physiol.*, *24*, 743–747.

Lagueux, M., Harry, P., & Hoffman, J. A. (1981). Ecdysteroids are bound to vitellin in newly laid eggs of *Locusta*. *Mol. Cell. Endocrin.*, *24*, 325–338.

Lankinen, P., & Riihimaa, A. (1992). Weak circadian eclosion rhythmicity in *Chymomyza costata* (Diptera: Drosophilidae), and its independence of diapause type. *J. Insect Physiol.*, *38*, 801–811.

Lee, K.-Y., & Denlinger, D. L. (1996). Diapause-regulated proteins in the gut of pharate first instar larvae of the gypsy moth, *Lymantria dispar,* and the effects of KK-42 and neck ligation on expression. *J. Insect Physiol., 42,* 423–431.

Lee, K.-Y., & Denlinger, D. L. (1997). A role for ecdysteroids in the induction and maintenance of the pharate first instar diapause of the gypsy moth, *Lymantria dispar. J. Insect Physiol., 43,* 289–296.

Lee, K.-Y., Horodyski, F. M., Valaitis, A. P., & Denlinger, D. L. (2002). Molecular characterization of the insect immune protein hemolin and its high induction during embryonic diapause in the gypsy moth, *Lymantria dispar. Insect Biochem. Mol. Biol., 32,* 1457–1467.

Lee, K.-Y., Valaitis, A. P., & Denlinger, D. L. (1997). Further evidence that diapause in the gypsy moth, *Lymantria dispar,* is regulated by ecdysteroids: a comparison of diapause and nondiapause strains. *J. Insect Physiol., 43,* 897–903.

Lee, S. S., Kennedy, S., Tolonen, A. C., & Ruvkun, G. (2003). DAF-16 target genes that control *C. elegans* life-span and metabolism. *Science, 300,* 644–647.

Lefevere, K. S. (1989). Endocrine control of diapause termination in the adult female Colorado potato beetle, *Leptinotarsa decemlineata. J. Insect Physiol., 35,* 197–203.

Lefevere, K. S., Koopmanschap, A. B., & de Kort, C. A.D. (1989). Juvenile hormone metabolism during and after diapause in the female Colorado potato beetle, *Leptinotarsa decemlineata. J. Insect Physiol, 35,* 129–135.

L'Helias, C., Beaudry, P., Callebert, J., & Launay, J. M. (2001). Head tryptophan hydroxylase activity and serotonin content during premetamorphosis of *Pieris brassicae. Biogen. Amines, 16,* 497–522.

Li, A. Q., Popova-Butler, A., Dean, D. H., & Denlinger, D. L. (2007). Proteomics of the flesh fly brain reveals an abundance of upregulated heat shock proteins during pupal diapause. *J. Insect Physiol., 53,* 385–391.

Li, A., Rinehart, J. P., & Denlinger, D. L. (2009). Neuropeptide-like precursor 4 is uniquely expressed during pupal diapause in the flesh fly. *Peptides, 30,* 518–521.

Liu, Z., Gong, P., Li, D., & Wei, W. (2011). Pupal diapause of *Helicoverpa armigera* (Hubner)(Lepidoptera: Noctuidae) mediated by larval host plants: pupal weight is important. *J. Insect Physiol.,* in press.

Liu, M., Zhang, T. Y., & Xu, W. H. (2005). A cDNA encoding diazepam-binding inhibitor/acyl-CoA-binding protein in *Helicoverpa armigera*: Molecular characterization and expression analysis associated with pupal diapause. *Comparative Biochemistry and Physiology - C Toxicology and Pharmacology, 141,* 168–176.

Lu, Y.-X., & Xu, W.-H. (2010). Proteomic and phosphoproteomic analysis at diapause initiation in the cotton bollworm, *Helicoverpa armigera. J. Proteome Res.,* in press.

MacRae, T. H. (2010). Gene expression, metabolic regulation and stress tolerance during diapause. *Cell Mol. Life Sci., 67,* 2405–2424.

Masaki, S. (1980). Summer diapause. *Annu. Rev. Entomol., 25,* 1–25.

Mathias, D., Jacky, L., Bradshaw, W. E., & Holzapfel, C. M. (2005). Geographic and developmental variation in expression of the circadian rhythm gene, *timeless,* in the pitcher-plant mosquito, *Wyeomyia smithii. J. Insect Physiol., 51,* 661–667.

Matsumoto, M., & Takeda, M. (2002). Changes in brain monoamine contents in diapausing pupae of *Antheraea pernyi* when activated under long-day conditions. *J. Insect Physiol., 48,* 765–771.

Matsuo, J., Nakayama, S., & Numata, H. (1997). Role of the corpus allatum in the control of adult diapause in the blow fly, *Protophormia terraenovae. J. Insect Physiol., 43,* 211–216.

McBrayer, Z., Ono, H., Shimell, M. J., Parvy, J.-P., Beckstead, R. B., Warren, J. T., Thummel, C. S., Dauphin-Villemant, C., Gilbert, L. I., & O'Connor, M. B. (2007). Prothoracicotropic hormone regulates developmental timing and body size. *Dev. Cell, 13,* 857–871.

Meola, R. W., & Adkisson, P. L. (1977). Release of prothoracicotropic hormone and potentiation of developmental ability during diapause in the bollworm, *Heliothis zea. J. Insect Physiol., 23,* 683–688.

Michaud, M. R., & Denlinger, D. L. (2007). Shifts in carbohydrate, polyol, and amino acid pools during rapid cold-hardening and diapause-associated cold hardening in flesh flies (*Sarcophaga crassipalpis*): a metabolomics comparison. *J. Comp. Physiol. B, 177,* 753–763.

Mishra, P. K., Sharan, S. K., Kumar, D., Singh, B. M.K., Subrahmanyam, B., & Suryanarayana, N. (2008). Effect of ecdysone on termination of pupal diapause and egg production in *Antheraea mylita* Drury. *J. Adv. Zool., 29,* 128–136.

Mitsumasu, K., Ohta, H., Tsuchihara, K., Asaoka, K., Ozoe, Y., Niimi, T., Yamashita, O., & Yaginuma, T. (2008). Molecular cloning and characterization of cDNAs encoding dopamine receptor-1 and -2 from brain-suboesophageal ganglion of the silkworm, *Bombyx mori. Insect Mol. Biol., 17,* 185–195.

Miyawaki, R., Tanaka, S. I., & Numata, H. (2003). Photoperiodic receptor in the nymph of *Poecilocoris lewisi* (Heteroptera: Scutelleridae). *Eur. J. Entomol., 100,* 301–303.

Moribayashi, A., Kurahashi, H., & Ohtaki, T. (1988). Different profiles of ecdysone secretion and its metabolism between diapause- and nondiapause-destined cultures of the fleshfly, *Boettcherisca peregrina. Comp. Biochem. Physiol. A, 91,* 157–164.

Moribayashi, A., Kurahashi, H., & Ohtaki, T. (1992). Physiological differentiation of the ring glands in mature larvae of the flesh fly, *Boettcherisca peregrina,* programmed for diapause or non-diapause. *J. Insect Physiol., 38,* 177–183.

Moribe, Y., Niimi, T., Yamashita, O., & Yaginuma, T. (2001). Samui, a novel cold-inducible gene, encoding a protein with a BAG-domain similar to silencer of death domains (SODD/BAG-4), isolated from *Bombyx* diapause eggs. *Eur. J. Biochem., 268,* 3432–3442.

Morita, A., Niimi, T., & Yamashita, O. (2004). Physiological differentiation of DH-PBAN-producing neurosecretory cells in the silkworm embryo. *J. Insect Physiol., 49,* 1093–1102.

Morita, A., & Numata, H. (1997a). Role of the neuroendocrine complex in the control of adult diapause in the bean bug, *Riptotus clavatus. Arch. Insect Biochem. Physiol., 35,* 347–355.

Morita, A., & Numata, H. (1997b). Distribution of the photoperiodic receptor in the compound eyes of the bean bug, *Riptortus clavatus. J. Comp. Physiol. A, 180,* 181–185.

Morita, A., & Numata, H. (1999). Localization of the photoreceptor for photoperiodism in the stink bug, *Platutia crossota Stali. Physiol. Entomol., 24,* 189–195.

Munyiri, F. N., & Ishikawa, Y. (2004). Endocrine changes associated with metamorphosis and diapause induction in the yellow-spotted longicorn beetle, *Psacothea hilaris*. *J. Insect Physiol.*, *50*, 1075–1081.

Nakagaki, M., Takei, R., Nagashima, E., & Yaginuma, T. (1991). Cell-cycles in embryos of the silkworm, *Bombyx mori*: G2 arrest at diapause stage. *Roux's Arch. Devel. Biol.*, *200*, 223–229.

Nakamura, K., & Hodkova, M. (1998). Photoreception in entrainment of rhythms and photoperiodic regulation of diapause in a hemipteran, *Graphosoma lineatum*. *J. Biol. Rhythms*, *13*, 159–166.

Nakane, Y., Ikegami, K., Ono, H., Yamamoto, N., Yoshida, S., Hirunagi, K., Ebihara, S., Kubo, Y., & Yoshimura, T. (2010). A mammalian neural tissue opsin (Opsin 5) is a deep brain photoreceptor in birds. *Proc. Natl. Acad. Sci., USA*, *107*, 15264–15268.

Neumann-Visscher, S. (1976). The embryonic diapause of *Aulocara elliotti* (Orthoptera, Acrididae). Histological and morphometric changes during diapause development and following experimental termination with juvenile hormone analogue. *Cell Tissue Res.*, *174*, 433–452.

Noguchi, H., & Hayakawa, Y. (1997). Role of dopamine at the onset of pupal diapause in the cabbage armyworm, *Mamestra brassicae*. *FEBS Lett.*, *413*, 157–161.

Noguchi, H., & Hayakawa, Y. (2001). Dopamine is a key factor for the induction of egg diapause of the silkworm, *Bombyx mori*. *Eur. J. Biochem.*, *268*, 774–780.

Numata, H., & Hidaka, T. (1983). Compound eyes as the photoperiodic receptors in the bean bug. *Experientia*, *39*, 868–869.

Numata, H., Shiga, S., & Morita, A. (1997). Photoperiodic receptors in arthropods. *Zool. Sci.*, *14*, 187–197.

Okuda, T., & Tanaka, S. (1997). An allatostatic factor and juvenile hormone synthesis by corpora allata in *Locusta migratoria*. *J. Insect Physiol.*, *43*, 635–641.

Okuda, T., Tanaka, S., Kotaki, T., & Ferenz, H.-J. (1996). Role of the corpora allata and juvenile hormone in the control of imaginal diapause and reproduction in three species of locusts. *J. Insect Physiol.*, *42*, 943–951.

Palli, S. R., Kothapalli, R., Feng, Q., Ladd, T., Perera, S. C., et al. (2001). Molecular analysis of overwintering diapause. In D. L. Denlinger, J. M. Giebultowicz, & D. S. Saunders (Eds.), *Insect Timing: Circadian Rhythmicity to Seasonality* (pp. 133–144). Amsterdam: Elsevier.

Panda, S., Sato, T. K., Castrucci, A. M., Rollag, M. D., De Grip, W. J., et al. (2002). Melanopsin (Opn4) requirement for normal light-induced circadian phase shifting. *Science*, *298*, 2213–2216.

Pavelka, J., Shimada, K., & Kostal, V. (2003). Timeless: a link between fly's circadian and photoperiodic clocks? *Eur. J. Ent.*, *100*, 255–265.

Pener, M. P. (1992). Environmental cues, endocrine factors, and reproductive diapause in male insects. *Chronobiol. Int.*, *9*, 102–113.

Peypelut, L., Beydon, P., & Lavenseau, L. (1990). 20-Hydroxyecdysone triggers the resumption of imaginal wing disc development after diapause in the European corn borer, *Ostrinia nubilalis*. *Arch. Insect Biochem. Physiol.*, *15*, 1–19.

Pinyarat, W., Shimada, T., Xu, W. -H., Sato, Y., Yamashita, O., et al. (1995). Linkage analysis of the gene encoding precursor protein of diapause hormone and pheromone biosynthesis-activating neuropeptide in the silkworm, *Bombyx mori*. *Genet. Res.*, *65*, 105–111.

Price, J. L. (2004). *Drosophila melanogaster*: a model system for molecular chronobiology. In A. Sehgal (Ed.), *Molecular Biology of Circadian Rhythms* (pp. 33–74). Hoboken, NJ: Wiley-Liss.

Pullin, A. S., & Bale, J. S. (1989). Effects of ecdysone, juvenile hormone and haemolymph transfer on cryoprotectant metabolism in diapausing and non-diapausing pupae of *Pieris brassicae*. *J. Insect Physiol.*, *35*, 911–918.

Ragland, G. J., Denlinger, D. L., & Hahn, D. A. (2010). Mechanisms of suspended animation are revealed by transcript profiling of diapause in the flesh fly. *Proc. Natl. Acad. Sci., USA*, *107*, 14909–14914.

Readio, J., Chen, M.-H., & Meola, R. (1999). Juvenile hormone biosynthesis in diapausing and nondiapausing *Culex pipiens* (Diptera: Culicidae). *J. Med. Entomol.*, *36*, 355–360.

Reynolds, J. A., & Hand, S. C. (2009). Embryonic diapause highlighted by differential expression of mRNAs for ecdy-steroidogenesis, transcription and lipid sparing in the cricket *Allonemobius socius*. *J. Exp. Biol.*, *212*, 2075–2084.

Rewitz, K. F., Yamanaka, N., Gilbert, L. I., & O'Connor, M. B. (2009). The insect neuropeptide PTTH activates receptor tyrosine kinase torso to initiate metamorphosis. *Science*, *326*, 1403–1405.

Richard, D. S., Gilbert, M., Crum, B., Hollinshead, D. M., Schelble, S., et al. (2001a). Yolk protein endocytosis by oocytes in *Drosophila melanogaster*: immunofluorescent localization of clathrin, adaptin and the yolk protein receptor. *J. Insect Physiol.*, *47*, 715–723.

Richard, D. S., Jones, J. M., Barbarito, M. R., Cerula, S., Detweiler, J. P., et al. (2001b). Vitellogenesis in diapausing and mutant *Drosophila melanogaster*: further evidence for the relative roles of ecdysteroids and juvenile hormones. *J. Insect Physiol.*, *47*, 905–913.

Richard, D. S., & Saunders, D. S. (1987). Prothoracic gland function in diapause and non-diapause *Sarcophaga argyrostoma* and *Calliphora vicina*. *J. Insect Physiol.*, *33*, 385–392.

Richard, D. S., Watkins, N. L., Serafin, R. B., & Gilbert, L. I. (1998). Ecdysteroids regulate yolk protein uptake by *Drosophila melanogaster* oocytes. *J. Insect Physiol.*, *44*, 637–644.

Rinehart, J. P., Cikra-Ireland, R. A., Flannagan, R. D., & Denlinger, D. L. (2001). Expression of ecdysone receptor is unaffected by pupal diapause in the flesh fly, *Sarcophaga crassipalpis*, while its dimerization partner, USP, is down-regulated. *J. Insect Physiol.*, *47*, 915–921.

Rinehart, J. P., Li, A., Yocum, G. D., Robich, R. M., Hayward, S. A.L., & Denlinger, D. L. (2007). Up-regulation of heat shock proteins is essential for cold survival during insect diapause. *Proc. Natl. Acad. Sci., USA*, *104*, 11130–11137.

Robich, R. M., Rinehart, J. P., Kitchen, L. J., & Denlinger, D. L. (2007). Diapause-specific gene expression in the northern house mosquito, *Culex pipiens* L., identified by suppressive subtractive hybridization. *J. Insect Physiol.*, *53*, 235–245.

Rockey, S. J., Miller, B. B., & Denlinger, D. L. (1989). A diapause maternal effect in the flesh fly, *Sarcophaga bullata*: transfer of information from mother to progeny. *J. Insect Physiol.*, *35*, 553–558.

Ruby, N. F., Brennan, T. J., Xie, X. M., Cao, V., Franken, P., et al. (2002). Role of melanopsin in circadian responses to light. *Science*, *298*, 2211–2213.

Safranek, L., & Williams, C. M. (1980). Studies on the prothoraciotropic hormone in the tobacco hornworm, *Manduca sexta*. *Biol. Bull.*, *158*, 141–153.

Saito, H., Takeuchi, Y., Takeda, R., Hayashi, Y., Watanabe, K., et al. (1994). The core and complementary sequence responsible for biological activity of the diapause hormone of the silkworm, *Bombyx mori*. *Peptides*, *15*, 1173–1178.

Sakamoto, T., & Tomioka, K. (2007). Effects of unilateral compound-eye removal on the photoperiodic responses of nymphal development in the cricket *Modycorgryllus siamensis*. *Zool. Sci.*, *24*, 604–610.

Sandrelli, F., Tauber, E., Pegoraro, M., Mozzotta, G., Cisotto, P., Landskron, J., Stanewsky, R., JPiccin, A., Rosato, E., Zordan, M., Costa, R., & Kyriacou, C. P. (2007). A molecular basis for natural selection at the *timeless locus in Drosophila melanogaster*. *Science*, *316*, 1898–1900.

Sato, Y., Ikeda, M., & Yamashita, O. (1994). Neurosecretory cells expressing the gene for common precursor for diapause hormone and phermone biosynthesis-activating neuropeptide in the subesophageal ganglion of the silkworm, *Bombyx mori*. *Gen. Comp. Endocrinol.*, *96*, 27–36.

Sato, Y., Nakazawa, Y., Menjo, N., Imai, K., Komiya, T., et al. (1992). A new diapause hormone molecule of the silkworm, *Bombyx mori*. *Proc. Jap. Acad. B*, *68*, 75–79.

Sato, Y., Oguchi, M., Menjo, N., Imai, K., Saito, H., et al. (1993). Precursor polyprotein for multiple neuropeptides secreted from the subesophageal ganglion of the silkworm *Bombyx mori*: characterization of the DNA-encoding diapause hormone precursor and identification of additional peptides. *Proc. Natl Acad. Sci. USA*, *90*, 3251–3255.

Sato, Y., Shiomi, K., Saito, H., Imai, K., & Yamashita, O. (1998). Phe-X-Pro-Arg-Leu-NH2 peptide producing cells in the central nervous system of the silkworm, *Bombyx mori*. *J. Insect Physiol.*, *44*, 333–342.

Sauman, I., & Reppert, S. M. (1996a). Molecular characterization of prothoracicotropic hormone (PTTH) from the giant silkmoth *Antheraea pernyi*: Developmental appearance of PTTH-expressing cells and relationship to circadian clock cells in central brain. *Dev. Biol.*, *178*, 418–429.

Sato, Y., Yang, P., An, Y., Matsukawa, K., Ito, K., Imanishi, S., Matsuda, H., Uchiyama, Y., Imai, K., Ito, S., Ishida, Y., & Suzuki, K. (2010). A palmitoyl conjugate of insect pentapeptide Yamamarin arrests cell proliferation and respiration. *Peptides*, *31*, 827–833.

Sauman, I., & Reppert, S. M. (1996b). Circadian clock neurons in the silkmoth *Antheraea pernyi*: novel mechanisms of period protein regulation. *Neuron*, *17*, 889–900.

Sauman, I., & Sehnal, F. (1997). Immunohistochemistry of the products of male accessory glands in several hemimetabolous insects and the control of their secretion in *Pyrrhocoris apterus* (Heteroptera: Pyrrhocoridae). *Eur. J. Entomol.*, *94*, 349–360.

Saunders, D. S. (1971). The temperature-compensated photoperiodic clock "programming" development and pupal diapause in the flesh fly, *Sarcophaga argyrostoma*. *J. Insect Physiol.*, *17*, 801–812.

Saunders, D. S. (1990). The circadian basis of ovarian diapause regulation in *Drosophila melanogaster*: is the period gene causally involved in photoperiodic time measurement? *J. Biol. Rhythms*, *5*, 315–331.

Saunders, D. S. (2002). *Insect Clocks* (3rd ed.). Amsterdam: Elsevier.

Saunders, D. S. (2010a). Photoperiodism in insects: migration and diapause responses. In R. J. Nelson, D. L. Denlinger, & D. E. Somers (Eds.), *Photoperiodism* (pp. 218–257). Oxford: Oxford University Press.

Saunders, D. S. (2010b). Controversial aspects of photoperiodism in insects and mites. *J. Insect Physiol.*, *56*, 1491–1502.

Saunders, D. S., & Cymborowski, B. (2003). Selection for high diapause incidence in blow flies *(Calliphora vicina)* maintained under long days increases the maternal critical daylength: some consequences for the photoperiodic clock. *J. Insect Physiol.*, *49*, 777–784.

Saunders, D. S., Henrich, V. C., & Gilbert, L. I. (1989). Induction of diapause in *Drosophila melanogaster*: photoperiodic regulation and the impact of arrhythmic clock mutants on time measurement. *Proc. Natl Acad. Sci. USA*, *86*, 3748–3752.

Saunders, D. S., Richard, D. S., Applebaum, S. W., Ma, M., & Gilbert, L. I. (1990). Photoperiodic diapause in *Drosophila melanogaster* involves a block to juvenile hormone regulation of ovarian maturation. *Gen. Comp. Endocrinol.*, *79*, 174–184.

Saunders, D. S., Wheeler, I., & Kerr, A. (1999). Survival and reproduction of small blow flies *(Calliphora vicina;* Diptera: Calliphoridae) produced in severely overcrowded short-day larval cultures. *Eur. J. Entomol.*, *96*, 19–22.

Shiga, S., & Numata, H. (1997). Induction of reproductive diapause via perception of photoperiod through the compound eyes in the adult blow fly, *Protophormia terraenovae*. *J. Comp. Physiol. A*, *181*, 35–40.

Shiga, S., & Numata, H. (2000). The role of neurosecretory neurons in the pars intercerebralis and pars lateralis in reproductive diapause of the blowfly, *Protophormia terraenovae*. *Naturwissenschaften*, *87*, 125–128.

Shimada, K. (1999). Genetic linkage analysis of photoperiodic clock genes in *Chymomyza costata* (Diptera: Drosophilidae). *Entomol. Sci.*, *2*, 575–578.

Shimizu, I., Aoki, S., & Ichikawa, T. (1997). Neuroendocrine control of diapause hormone secretion in the silkworm, *Bombyx mori*. *J. Insect Physiol.*, *43*, 1101–1109.

Shimizu, I., & Kato, M. (1984). Carotenoid functions in photoperiodic induction in the silkworm, *Bombyx mori*. *Photochem. Photobiophys.*, *7*, 47–52.

Shimizu, I., Kitabatake, S., & Kato, M. (1981). Effect of carotenoid deficiency on photosensitvities in the silkworm, *Bombyx mori*. *J. Insect Physiol*, *27*, 593–599.

Shimizu, I., Yamakawa, Y., Shimazaki, Y., & Iwasa, T. (2001). Molecular cloning of *Bombyx* cerebral opsin (Boceropsin) and cellular localization of its expression in the silkworm brain. *Biochem. Biophys. Res. Commun.*, *287*, 27–34.

Shingleton, A. W., Sisk, G. C., & Stern, D. L. (2003). Diapause in the pea aphid *(Acyrthosiphon pisum)* is a slowing but not a cessation of development. *Devel. Biol., 3*, 7.

Shiomi, K., Fujiwara, Y., Yasukochi, Y., Kajiura, Z., Nakagaki, M., & Yaginuma, T. (2007). The Pitx homeobox gene in *Bombyx mori*: Regulation of DH-PBAN neuropeptide hormone gene expression. *Mol. Cell Neurosci., 34*, 209–218.

Sielezniew, M., & Cymborowski, B. (1997). Effects of ecdysteroid agonist RH-5849 on pupal diapause of the tobacco hornworm *(Manduca sexta). Arch. Insect Biochem. Physiol., 35*, 191–197.

Sim, C., & Denlinger, D. L. (2008). Insulin signaling and FOXO regulate eh overwintering diapause of the mosquito *Culex pipiens. Proc. Natl. Acad. Sci., USA, 105*, 6777–6781.

Sim, C., & Denlinger, D. L. (2009a). A shut-down in expression of an insulin-like peptide, ILP-1, halts ovarian maturation during overwintering diapause in the mosquito *Culex pipiens. Insect Mol. Biol., 18*, 325–332.

Sim, C., & Denlinger, D. L. (2009b). Transcription profiling and regulation of fat metabolism genes in diapausing adults of the mosquito *Culex pipiens. Physiol. Genomics, 39*, 202–209.

Singtripop, T., Manaboon, M., Tatun, N., Kaneko, Y., & Sakurai, S. (2008). Hormonal mechanisms underlying termination of larval diapause by juvenile hormone in the bamboo borer, *Omphisa fuscidentalis. J. Insect Physiol., 54*, 137–145.

Singtripop, T., Oda, Y., Wanichacheewa, S., & Sakurai, S. (2002). Sensitivities to juvenile hormone and ecdysteroid in diapause larvae of *Omphisa fuscidentalis* based on the hemolymph trehalose dynamics index. *J. Insect Physiol., 48*, 817–824.

Singtripop, T., Wanichacheewa, S., & Sakurai, S. (2000). Juvenile hormone-mediated termination of larval diapause in the bamboo borer, *Omphisa fuscidentalis. Insect Biochem. Mol. Biol., 30*, 847–854.

Slama, K. (1980). Homeostatic functions of ecdysteroids in ecdysis and oviposition. *Acta Entomol. Bohem, 77*, 145–168.

Slama, K. (2000). Correlation between metabolic depression and ecdysteroid peak during embryogenesis of the desert locust, *Schistocerca gregaria* (Orthoptera: Acrididae). *Eur. J. Entomol., 97*, 141–148.

Smith, W. A., Bowen, M. F., Bollenbacher, W. E., & Gilbert, L. I. (1986). Cellular changes in the prothoracic glands of diapausing pupae of *Manduca sexta. J. Exp. Biol., 120*, 131–142.

Smythe, E., & Warren, G. (1991). The mechanism of receptor mediated endocytosis. *Eur. J. Biochem, 202*, 689–699.

Socha, R. (2007). Factors terminating ovarian arrest in the long-winged females of a flightless bug, *Pyrrhocoris apterus* (Heteroptera: Pyrrhocoridae). *Eur. J. Entomol., 104*, 15–22.

Socha, R., & Sula, J. (2008). Regulation of development of flight muscles in long-winged females of a flightless bug, *Pyrrhocoris apterus* (Heteroptera: Pyrrhocoridae). *Eur. J. Entomol., 105*, 575–583.

Sonobe, H., Hiyama, Y., & Keino, H. (1977). Changes in the amount of the diapause factor in the subesophageal ganglion during development of the silkworm, *Bombyx mori. J. Insect Physiol., 23*, 633–637.

Stanewsky, R. (2002). Clock mechanisms in *Drosophila. Cell Tissue Res., 309*, 11–26.

Stanewsky, R., Kaneko, M., Emery, P., Beretta, B., Wager-Smith, K., et al. (1998). The cry(b) mutation identifies cryptochrome as a circadian photoreceptor in *Drosophila. Cell, 95*, 681–692.

Stehlík, J., Závodská, R., Shimada, K., Šauman, I., & Koštál, V. (2008). Photoperiodic induction of diapause requires regulated transcription of timeless in the larval brain of *Chymomyza costata. J. Biol. Rhythms, 23*, 129–139.

Su, Z. H., Ikeda, M., Sato, Y., Saito, H., Imai, K., et al. (1994). Molecular characterization of ovary trehalase of the silkworm, *Bombyx mori* and its transcriptional activation by diapause hormone. *Biochim. Biophys. Acta – Gene Struct. Expr, 1218*, 366–374.

Suwan, S., Isobe, M., Yamashita, O., Minakata, H., & Imai, K. (1994). Silkworm diapause hormone, structure activity-relationships, indispensable role of C-terminus amide. *Insect Biochem. Mol. Biol., 24*, 1001–1007.

Suzuki, K., Minagawa, T., Kumagai, T., Naya, S.-I., Endo, Y., et al. (1990). Control mechanism of diapause of the pharate first-instar larvae of the silkmoth *Antheraea yamamai. J. Insect Physiol, 36*, 855–860.

Suzuki, K., Nakamura, T., Yanbe, T., Kurihara, M., & Kuwano, E. (1993). Termination of diapause in pharate first-instar larvae of the gypsy moth *Lymantria dispar japonica* by an imidazole derivative KK-42. *J. Insect Physiol., 39*, 107–110.

Suzuki, M. G., Terada, T., Kobayashi, M., & Shimada, T. (1999). Diapause-associated transcription of BmEts, a gene encoding an ETS transcription factor homologue in *Bombyx mori. Insect Biochem. Mol. Biol., 29*, 339–347.

Takahashi, M., Iwasaki, H., Niimi, T., Yamashita, O., & Yaginuma, T. (1998). Changing profiles of mRNA levels of cdc2 and a novel cdc2-related kinase (Bcdrk) in relation to ovarian development and embryonic diapause of *Bombyx mori. Appl. Entomol. Zool., 33*, 551–559.

Takahashi, M., Niimi, T., Ichimura, H., Sasaki, T., Yamashita, O., et al. (1996). Cloning of a B-type cyclin homologue from *Bombyx mori* and the profiles of its mRNA level in non-diapause and diapause eggs. *Devel. Genes Evol., 206*, 288–291.

Tammariello, S. P. (2001). Regulation of the cell cycle during diapause. In D. L. Denlinger, J. M. Giebultowicz, & D. S. Saunders (Eds.), *Insect Timing: Circadian Rhythmicity to Seasonality* (pp. 173–183). Amsterdam: Elsevier.

Tammariello, S. P., & Denlinger, D. L. (1998). G0/G1 cell cycle arrest in the brains of *Sarcophaga crassipalpis* during pupal diapause and the expression pattern of the cell cycle regulator, proliferating cell nuclear antigen. *Insect Biochem. Mol. Biol., 28*, 83–89.

Tanigawa, N. A., Shiga, S., & Numata, H. (1999). Role of the corpus allatum in the control of reproductive diapause in the male blow fly, *Protophormia terraenovae. Zool. Sci., 16*, 639–644.

Tatar, M., Kopelman, A., Epstein, D., Tu, M.-P., Yin, C.-M., et al. (2001). A mutant *Drosophila* insulin receptor homologue that extends life-span and impairs neuroendocrine function. *Science, 292*, 107–110.

Tatar, M., & Yin, C.-M. (2001). Slow aging during insect reproductive diapause: why butterflies, grasshoppers and flies are like worms. *Exp. Gerontol, 36*, 723–738.

Tatun, N., Singtripop, T., & Sakurai, S. (2008). Dual control of midgut trehalase activity by 20-hydroxyecdysone and an inhibitory factor in the bamboo borer *Omphisa fuscidentalis* Hampson. *J. Insect Physiol., 54,* 351–357.

Tauber, M. J., Tauber, C. A., & Masaki, S. (1986). *Seasonal Adaptations of Insects.* Oxford: Oxford University Press.

Tauber, E., Zordan, M., Sandrelli, F., Pegoraro, M., Osterwalder, N., Breda, C., Daga, A., Selmin, A., Monger, K., Benna, C., Rosato, E., Kyriacou, C. P., & Costa, R. (2007). Natural selection favors a newly derived timeless allele in *Drosophila melanogaster. Science, 31,* 1895–1898.

Tawata, M., & Ichikawa, T. (2001). Circadian firing activities of neurosecretory cells releasing pheromonotropic neuropeptides in the silkmoth, *Bombyx mori. Zool. Sci., 18,* 645–649.

Tawfik, A. I., Tanaka, Y., & Tanaka, S. (2002a). Possible involvement of ecdysteroids in photoperiodically induced suppression of ovarian development in a Japanese strain of the migratory locust, *Locusta migratoria. J. Insect Physiol., 48,* 411–418.

Tawfik, A. I., Tanaka, Y., & Tanaka, S. (2002b). Possible involvement of ecdysteroids in embryonic diapause of *Locusta migratoria. J. Insect Physiol., 48,* 743–749.

Tomioka, K., Agui, N., & Bollenbacher, W. E. (1995). Electrical properties of the cerebral prothoracicotropic hormone cells in diapausing and nondiapausing pupae of the tobacco hornworm, *Manduca sexta. Zool. Sci., 12,* 165–173.

Truman, J. W. (1976). Extraretinal photoreception in insects. *Photochem. Photobiol., 23,* 215–225.

Urbanski, J. M., Aruda, A., & Armbruster, P. (2010a). A transcriptional element of the diapause program in the Asian tiger mosquito, *Aedes albopictus,* identified by suppressive subtractive hybridization. *J. Insect Physiol., 56,* 1147–1154.

Urbanski, J. M., Benoit, J. B., Michaud, R. M., Denlinger, D. L., & Armbruster, P. A. (2010b). The molecular physiology of increased egg desiccation resistance during diapause in the invasive mosquito, *Aedes albopictus. Pro. Roy. Soc. B, 277,* 2683–2692.

Umeya, Y. (1926). Experiments of ovarian transplantation and blood transfusions in silkworms, with special reference to the alternation of voltinism. (*Bombyx mori* L.) Bull. Sericult. *Exp. Station, Chosen, 1,* 1–26.

Uryu, M., Ninomiya, Y., Yokoi, T., Tsuzuki, S., & Hayakawa, Y. (2003). Enhanced expression of genes in the brains of larvae of *Mamestea brassicae* (Lepidoptera; Noctuidae) exposed to short daylengths or fed Dopa. *Eur. J. Entomol., 100,* 245–250.

van Houten, Y. M., Overmeer, W. P.J., & Veerman, A. (1987). Thermoperiodically induced diapause in a mite in constant darkness is vitamin A dependent. *Experientia, 43,* 933–935.

van Houten, Y. M., & Veerman, A. (1990). Photoperiodism and thermoperiodism in the predatory mite, *Amblyseius potentillae* are probably based on the same mechanism. *J. Comp. Physiol. A, 167,* 201–209.

van Zon, A. Q., Overmeer, W. P.J., & Veerman, A. (1981). Carotenoids function in photoperiodic induction of diapause in a predacious mite. *Science, 213,* 1131–1133.

Veerman, A. (2001). Photoperiodic time measurement in insects and mites: a critical evaluation of the oscillator clock hypothesis. *J. Insect Physiol., 47,* 1097–1109.

Veerman, A., Slagt, M. E., Alderlieste, M. F.J., & Veenendaal, R. L. (1985). Photoperiodic induction of diapause in an insect is vitamin A dependent. *Experientia, 41,* 1194–1195.

Veerman, A., & Veenendaal, R. L. (2003). Experimental evidence for a non-clock role of the circadian system in spider mite photoperiodism. *J. Insect Physiol., 49,* 727–732.

Vermunt, A. M.W., Koopmanschap, A. B., Vlak, J. M., & de Kort, C. A.D. (1997b). Cloning and sequence analysis of cDNA encoding a putative juvenile hormone esterase from the Colorado potato beetle. *Insect Biochem. Mol. Biol., 27,* 919–928.

Vermunt, A. M.W., Koopmanschap, A. B., Vlak, J. M., & de Kort, C. A.D. (1998). Evidence for two juvenile hormone esterase-related genes in the Colorado potato beetle. *Insect Mol. Biol., 7,* 327–336.

Vermunt, A. M.W., Koopmanschap, A. B., Vlak, J. M., & de Kort, C. A.D. (1999). Expression of the juvenile hormone esterase gene in the Colorado potato beetle, *Leptinotarsa decemlineata:* photoperiodic and juvenile hormone analog response. *J. Insect Physiol., 45,* 135–142.

Vermunt, A. M.W., Vermeesch, A. M.G., & de Kort, C. A.D. (1997a). Purification and characterization of juvenile hormone esterase from hemolymph of the Colorado potato beetle. *Arch. Insect Biochem. Physiol., 35,* 261–277.

Watanabe, K., Hull, J. J., Niimi, T., Imai, K., Matsumoto, S., Yaginuma, T., & Kataoka, H. (2007). FXPRL-amide peptides induce ecdysteroidogenesis through a G-protein coupled receptor expressed in the prothoracic gland of *Bombyx mori. Mol. Cell Endocrin., 273,* 51–58.

Webb, M.-L.Z., & Denlinger, D. L. (1998). GABA and picrotoxin alter expression of a maternal effect that influences pupal diapause in the flesh fly, *Sarcophaga bullata. Physiol. Entomol., 23,* 184–191.

Wei, Z.-J., Zhang, Q.-R., Kang, L., Xu, W.-H., & Denlinger, D. L. (2005). Molecular characterization and expression of prothoracicotropic hormone during development and pupal diapause in the cotton bollworm, *Helicoverpa armigera. J. Insect Physiol., 51,* 691–700.

Williams, C. M. (1946). Physiology of insect diapause: the role of the brain in the production and termination of pupal dormancy in the giant silkworm *Platysamia cecropia. Biol. Bull., 90,* 234–243.

Williams, C. M. (1947). Physiology of insect diapause. II. Interaction between the pupal brain and prothoracic glands in the metamorphosis of the giant silkworm, *Platysamia cecropia. Biol. Bull., 93,* 89–98.

Williams, C. M. (1952). Physiology of insect diapause. VI. The brain and prothoracic glands as an endocrine system in the cecropia silkworm. *Biol. Bull., 103,* 120–138.

Williams, C. M., & Adkisson, P. L. (1964). Physiology of insect diapause. XIV. An endocrine mechanism for the photoperiodic control of pupal diapause in the oak silkworm *Antheraea pernyi. Biol. Bull., 127,* 511–525.

Williams, K. D., Busto, M., Suster, M. L., So, A. K.-C., Ben-Shahar, Y., Leevers, S. J., & Sokolowski, M. B. (2006). Natural variation in *Drosophila melanogaster* diapause due to the insulin-regulated PI3-kinase. *Proc. Natl. Acad. Sci. USA, 103,* 15911–15915.

Williams, K. D., Schmidt, P. S., & Sokolowski, M. B. (2010). Photoperiodism in insects: molecular basis and consequences of diapause. In R. J. Nelson, D. L. Denlinger, & D. E. Somers (Eds.), *Photoperiodism* (pp. 287–317). Oxford: Oxford University Press.

Wolschin, F., & Gadau, J. (2009). Deciphering proteomic signatures of early diapause in *Nasonia*. *PLoS ONE, 4*(7), e6394.

Wu, Q., & Brown, M. (2006). Signaling and function of insulin-like peptides in insects. *Ann. Rev. Entomol, 51*, 1–24.

Xu, W.-H., & Denlinger, D. L. (2003). Molecular characterization of prothoracicotropic hormone and diapause hormone in *Heliothis virescens* during diapause, and a new role for diapause hormone. *Insect Mol. Biol., 12*, 509–516.

Xu, W.-H., Sato, Y., Ikeda, M., & Yamashita, O. (1995a). Stage-dependent and temperature controlled expression of the gene encoding the precursor protein of diapause hormone and pheromone biosynthesis activating neuropeptide in the silkworm, *Bombyx mori. J. Biol. Chem., 270*, 3804–3808.

Xu, W.-H., Sato, Y., Ikeda, M., & Yamashita, O. (1995b). Molecular characterization of the gene encoding the precursor protein of diapause hormone and pheromone biosynthesis activating neuropeptide (DH-PBAN) of the silkworm, *Bombyx mori* and its distribution in some insects. *Biochim. Biophys. Acta, 1261*, 83–89.

Yamashita, O. (1996). Diapause hormone of the silkworm, *Bombyx mori*: structure, gene expression and function. *J. Insect Physiol, 42*, 669–679.

Yamashita, O., & Hasegawa, K. (1985). Embryonic Diapause. In G. A. Kerkut, & L. I. Gilbert (Eds.), *Comprehensive Insect Physiology Biochemistry and Pharmacology* vol. 1. (pp. 407–343). Oxford: Pergamon Press.

Yamashita, O., Shiomi, K., Ishida, Y., Katagiri, N., & Niimi, T. (2001). Insights for future studies on embryonic diapause promoted by molecular analyses of diapause hormone and its action in *Bombyx mori*. In D. L. Denlinger, J. Giebultowicz, & D. S. Saunders (Eds.), *Insect Timing: Circadian Rhythmicity to Seasonality* (pp. 145–153). Amsterdam: Elsevier.

Yin, C.-M., Wang, Z. S., & Chaw, W. D. (1985). Brain neurosecretory cell and ecdysiotropin activity of the nondiapausing, pre-diapausing and diapausing southwestern corn borer, *Diatraea grandiosella* Dyar. *J. Insect Physiol, 31*, 659–667.

Yocum, G. D., Rinehart, J. P., Chirumamilla-Chapara, A., & Larson, M. L. (2009a). Characterization of gene expression patterns during the initiation and maintenance phases of diapause in the Colorado potato beetle, *Leptinotarsa decemlineata. J. Insect Physiol., 55*, 32–39.

Yocum, G. D., Rinehart, J. P., & Larson, M. (2009b). Down-regulation of gene expression between the diapause initiation and maintenance phases of the Colorado potato beetle, *Leptinotarsa decemlineata* (Coleoptera: Chrysomelidae). *Eur. J. Entomol., 106*, 471–476.

Zhang, T. Y., Kang, L., Zhang, Z. F., & Xu, W. H. (2004a). Identification of a POU factor involved in regulating the neuron-specific expression of the gene encoding diapause hormone and pheromone biosynthesis-activating neuropeptide in *Bombyx mori*. *Biochem. J., 380*, 255–263.

Zhang, T. Y., Sun, J. S., Zhang, Q. R., Xu, J., Jiang, R. J., & Xu, W. H. (2004b). The diapause hormone-pheromone biosynthesis activating neuropeptide gene of *Helicoverpa armigera* encodes multiple peptides that break, rather than induce, diapause. *J. Insect Physiol., 50*, 547–554.

Zhang, Q., Zdarek, J., Nachman, R. J., & Denlinger, D. L. (2008). Diapause hormone in the corn earworm, *Helicoverpa zea*: optimum temperature for activity, structure-activity relationships, and efficacy in accelerating flesh fly pupariation. *Peptides, 29*, 196–205.

Zhang, Q., Nachman, R. J., Zubrzak, P., & Denlinger, D. L. (2009). Conformational aspects and hyperpotent agonists of diapause hormone for termination of pupal diapause in the corn earworm. *Peptides, 30*, 596–602.

Zhang, Q., & Denlinger, D. L. (2011). Molecular structure and expression analysis of the prothoracicotropic hormone gene in the northern house mosquito, *Culex pipiens*, in association with diapause and blood feeding. *Insect Mol. Biol., 20*, 201–213.

Zhao, J. Y., Xu, W. H., & Kang, L. (2004). Functional analysis of the SGNP I in the pupal diapause of the oriental tobacco budworm, *Helicoverpa assulta* (Lepidoptera: Noctuidae). *Reg. Peptides, 118*, 25–31.

Zimmerman, W. F., & Goldsmith, T. H. (1971). Photosensitivity of the circadian rhythm and of visual receptors in carotenoid-depleted *Drosophila*. *Science, 171*, 1167–1169.

11 Endocrine Control of Insect Polyphenism

K. Hartfelder
Universidade de São Paulo, Ribeirão Preto,
São Paulo, Brazil
D.J. Emlen
University of Montana, Missoula, MT, USA

© 2012 Elsevier B.V. All Rights Reserved

Summary

Polyphenism in insects is associated with some of the most striking and successful life histories. Insect polyphenisms center around four major themes: wing length variation (including winglessness), differences in fertility or reproductive strategies, body and/or wing coloration and exaggerated morphologies. Hormones, especially juvenile hormone and ecdysteroids are important factors underlying the generation of distinct morphologies and reproductive strategies, and the insulin-signaling pathway is now also emerging as a major player. We present an overview of current knowledge regarding wing length dimorphism in crickets, the gradual phase shift from the solitary to gregarious syndrome in migrating locusts, the complex switching between wingless/winged forms and asexual/sexual reproduction in aphids, wing and body color variants in butterflies, male horn dimorphism in beetles, and caste polyphenism occurring in the hemimetabolous termites and holometabolous hymenopterans.

11.1. Polyphenism and Polymorphism: Discontinuity in the Variation of Insects

Historically, the terms polymorphism and polyphenism have the same meaning, namely intraspecific variation in sets of characters. This phenomenon is seen as an adaptive response that allows genotypes to track short-term cycles of environmental variation and, at the same time, maintain cohesiveness of adapted gene complexes. With recent advances in population genetics, enzymology, genomics, and proteomics the term polymorphism has now gained a much wider meaning. It has come to denote variation in nucleotide sequences in general. This genetic variation may or may not have consequences for phenotypic character states, depending on whether it occurs in coding regions, promoter and regulatory regions, or in selectively neutral DNA, such as microsatellites.

In this chapter we use the term polyphenism to describe, in a more narrow sense, the occurrence of intraspecific variation in phenotypes. Specifically, we focus on relatively discrete or discontinuous variation in phenotype expression. Polyphenic mechanisms can be described at two levels: (1) cues that serve as proximate triggers of developmental alternatives and (2) endogenous response cascades that drive the developmental responses. Triggering stimuli may consist of external environmental factors (e.g., photoperiod, crowding), internal genetic factors (e.g., alleles at polymorphic loci), or a mixture of the two. Endogenous response cascades involve a series of systems that relay, synchronize, and coordinate differentiated processes in target systems.

The most common form of polyphenism in animals is sexual dimorphism. This, in most cases, is triggered by genetic factors that generate distinct phenotypes that then act at the tissue level through a series of endocrine response cascades. Even though sexual polyphenism occurs to some degree in all insects, we consider it only peripherally in this chapter. Rather, and following the tradition already set forward in the review on this subject

DOI: 10.1016/B978-0-12-384749-2.10011-1

by Hardie and Lees (1985), and building on our prior version of this review (Hartfelder and Emlen, 2005), we focus on distinct phenotypes (in either or both sexes) that are triggered by environmental or other exogenous factors. When reviewing recent studies on this subject and comparing these to insights already presented in the 2005 version of this chapter, it became clear that considerable progress has been made in the elucidation of endogenous response cascades underlying polyphenic trait expression in some insect polyphenisms, and these now truly represent model systems. Major progress in the field can be attributed to genomic resources and high throughput platforms for both transcriptome and proteome analyses. These were championed for the honey bee, which became the first polyphenic (social) insect to have its genome sequenced (The Honey Bee Genome Sequencing Consortium, 2006). Next came the genome of the pea aphid *Acyrthosiphon pisum,* which has recently been concluded and published (International Aphid Genomics, 2010) and the road has been paved for several other genomes.

11.2. Polyphenism in the Hemimetabola

Wing length is the most predominant form of polyphenism in hemimetabolous insects. It involves differential investment in wings and wing muscles, and appears to reflect an ecophysiological trade-off in resource allocation between dispersal and reproduction. It is observed in species that colonize and/or exploit ephemeral resources, and may be integrated into life histories of considerable complexity, such as seen in locust and aphid phase polyphenism and in termite caste polyphenism. In the following sections, we will review the current status of endocrine regulation of wing, phase, and caste polyphenism encountered in the Hemimetabola.

11.2.1. Hemiptera/Homoptera Wing Polyphenism

11.2.1.1. Soapberry bugs A number of hemipteran species have polyphenic wing expression (Dingle and Winchell, 1997; Tanaka and Wolda, 1987). The hemipteran *Jadera haemotoloma* (the soapberry bug) exhibits a wing polymorphism as four morphs: three winged forms, one that always has viable flight muscle, one that histolyzes flight muscles part way through the adult stage, and a winged form that never produces viable flight musculature, as well as a completely wingless form that also has inviable flight musculature (Dingle and Winchell, 1997). *Jadera haemotoloma* has undergone rapid recent evolution following a host-shift that occurred within the past 50 years (Carroll and Boyd, 1992; Carroll *et al.*, 1997, 2001), and one characteristic that has

changed dramatically is the proportion of the population expressing wings: derived populations (on the new host) have significantly fewer winged animals than ancestral populations.

Dingle and Winchell (1997) demonstrated that this polyphenism is regulated by a threshold mechanism, and that this threshold behaves like a polygenic character with high levels of additive genetic variance. Manipulations of the amount of nymphal food predictably affected the percent of winged offspring, as did topical applications of methoprene (Dingle and Winchell, 1997). In laboratory animals sampled from an ancestral population, increases in the amount of food decreased the percent developing with wings. Similarly, experimental augmentation of juvenile hormone (JH) levels at the beginning of the final nymphal instar also decreased the proportion of winged animals, suggesting that high levels of JH result in a reduction of the relative amount of wing growth (Dingle and Winchell, 1997).

There is no information regarding the sensitivity of animals to JH during the penultimate or earlier instars, nor of the relative levels of ecdysteroids. Nevertheless, results available at this time show striking similarities with the polyphenic mechanism described for crickets (Section 11.2.2.1.). The timing of the sensitive period, the implicated role for JH, and the direction of the effect of JH all agree with the basic model for cricket wing polyphenism.

11.2.1.2. Planthoppers Planthoppers often occur in both winged and wingless forms (Denno and Perfect, 1994; Kisimoto, 1965; Morooka *et al.*, 1988), and dimorphism in these species results from a threshold mechanism with heritable variation for the threshold (Denno *et al.*, 1986, 1996; Matsumara, 1996; Morooka and Tojo, 1992; Peterson and Denno, 1997). Where studied, the environmental factor most relevant to wing expression appears to be population density, or crowding, and as with the wing polyphenism already described, the physiological response to nymphal crowding appears to be mediated by levels of JH.

To date, the endocrine regulation of wing polyphenism has been best studied in the brown planthopper, *Nilaparvata lugens* (Homoptera: Delphacidae). The default developmental pathway appears to involve wing production, but Iwanaga and Tojo (1986) showed that both exposure to solitary (low density) conditions and topical applications of JH could suppress wing development and result in a greater proportion of wingless individuals. They identified sensitive periods in the pre-penultimate (third) and penultimate (fourth) instars (Iwanaga and Tojo, 1986). The timing of these sensitive periods and the effect of JH were corroborated by Ayoade *et al.* (1999), who induced wingless morphs in a strain of planthoppers that had been selected to be entirely winged.

Bertuso *et al.* (2002) then induced expression of wings in a line selected to be entirely wingless by applying precocene to animals at these same sensitive periods. Dai *et al.* (2001) measured JH titers in presumptive winged and wingless nymphs and found that putative short-winged individuals had higher JH levels and lower juvenile hormone esterase (JHE) levels than putative long-winged individuals. Liu *et al.* (2008) showed that these patterns re-emerge in the fifth instar, with nymphs developing into short-winged adults having higher JH titers and lower JHE levels from 48 h onward during the fifth instar.

The two morphs also showed divergent titer levels in the adult stage, where a higher JH titer in the short-winged morph was associated with an earlier development of oocytes (Bertuso *et al.*, 2002), thus providing evidence for the dual (developmental and reproductive) role of JH in the divergent life history strategies of planthoppers. Tufail *et al.* (2010) measured vitellogenin (Vg) profiles and showed that levels increase earlier in short-winged adults than in long-winged adults (on days 3 and 4, respectively), and that topical application of JH could induce upregulation of vitellogenin transcription. The delayed rise in Vg exhibited by long-winged females is consistent with the hypothesis that this morph is capable of long-distance flights prior to reproduction.

Thus, wing polyphenism in planthoppers, as with wing polyphenisms in soapberry bugs and crickets, appears to be regulated by JH during brief sensitive periods late in nymphal life. Genomic resources are now available for the brown planthopper (Noda *et al.*, 2008), and this has permitted researchers to characterize the JHE gene (Liu *et al.*, 2008) and the farnesoic acid O-methyltransferase gene (an enzyme critical for JH biosynthesis; Liu *et al.*, 2008), as well as the vitellogenin gene (Tufail *et al.*, 2010), and studies on differential gene transcription between winged and wingless morphs are likely to be forthcoming.

11.2.1.3. Aphid phase and caste polyphenism   Aphids are a monophyletic group represented by around 4400 species worldwide which, in terms of life cycle complexity, are outstanding champions. Aphid life cycles can include at least two different forms of polyphenism, (1) cyclical switching between asexual reproduction (viviparous parthenogenesis) and sexual reproduction (associated with the production of haploid eggs capable of overwintering after fertilization), and (2) switching between a wingless sedentary morphology, and a winged morph capable of dispersal. This switching capacity enables aphids to successfully colonize and thrive on ephemeral habitats. In some cases, these polyphenic switches occur simultaneously (e.g., the switch from asexual to sexual reproduction may coincide with a switch from wingless to winged morphologies), or they may occur separately (the switch from wingless to winged morphologies can occur without coincident changes in mode of reproduction

or host plant preference). The large variation in species-specific aphid life cycles stems from the fact that these elements of polyphenism can be shuffled and integrated as seemingly independent modules in the evolutionary history of each species.

In this chapter we focus on the ontogenetic mechanisms that generate aphid polyphenism (for reviews of aphid life cycles and their ecological significance see Braendle *et al.*, 2006; Dixon, 1977, 1998; Moran, 1992). In an effort to facilitate the discussion of aphid polyphenism, and to place these mechanisms within the context of insect polyphenism in general, we present a simplified description of aphid life cycles and adapt a simplified terminology (Blackman, 1994).

11.2.1.3.1. Asexual versus sexual reproduction   All present day aphids reproduce parthenogenetically, a capacity which appears to have been acquired from a common ancestor approximately 250 million years ago. Yet, only a small number of derived species (i.e., those at the tips of the different branches in the phylogenetic tree) are strictly parthenogenetic (Simon *et al.*, 2002). Instead, the majority of aphid species facultatively switch between parthenogenetic and sexual modes of reproduction.

During the asexual phase, females give birth to parthenogenetic progeny that develop completely within the ovaries (for detailed reviews on the embryology of sexually versus asexually developing offspring see Le Trionnaire *et al.*, 2008; Miura *et al.*, 2003). While these progeny develop, their own ovaries become active and these embryos begin ovulation before being born. In this extreme telescoping of generations, both daughters and granddaughters develop simultaneously within a single adult female. This parthenogenetic/paedogenetic reproductive strategy of wingless female aphids is highly efficient for colonizing ephemeral habitats, because it can very rapidly generate large numbers of progeny. However, rapid production of parthenogenetic daughters can also lead to overcrowding, and with this, the need to disperse to new host plants (see Section 11.2.1.3.2.).

At the end of a parthenogenetic phase of reproduction, and generally at the end of the favorable season, female aphids begin producing sexually reproductive offspring. Often, this transition from asexual to sexual reproduction is coupled with a shift from the primary to a secondary host plant species. Aphids that live on a primary host plant with a short seasonal duration may switch to a secondary host plant, and this can occur in two ways, depending on the species. Asexual females may produce a generation of winged daughters that first disperse to the new host plant and then produce sexually reproductive (male and female) offspring. Alternatively, asexual females may produce winged males and females directly and these then disperse to the secondary host and reproduce sexually. In either case, the cycle is complete when daughters disperse back to the primary host the following season and begin parthenogenetic reproduction anew.

As may be expected, given the strict seasonality in both host plant alternation and cyclical parthenogenesis, photoperiod appears to be an important environmental cue for this polyphenism. The effects of photoperiod on these key elements in aphid life cycles have long been recognized (reviewed in Hardie and Lees, 1985), and have been studied extensively, especially in the black bean aphid, *Aphis fabae*, and the vetch aphid, *Megoura vicinae* (Hardie, 1981b,c; Hardie and Lees, 1983). The photoperiodic response in the switch from parthenogenetic to sexual forms depends primarily on the length of the scotophase (Hardie *et al.*, 1990), and immunocytochemical analyses of rhodopsins and phototransduction proteins suggest that photoperiodic receptors are located in the anterior dorsal region of the aphid protocerebrum (Hardie and Nunes, 2001).

Once the external trigger (changes in photoperiod, especially scotophase length, and to a lesser extent temperature) has been detected, the polyphenic switch between modes of reproduction is coordinated by circulating levels of hormones, in particular by JH. Specifically, in *A. fabae*, topical applications of JH inhibited the production of haploid eggs (and hence the switch from asexual to sexual reproduction), and promoted parthenogenetic development (Hardie, 1981b,c). JH also affected host plant preference behavior at this same stage (Hardie, 1980, 1981a; Hardie *et al.*, 1990), which makes sense, since these events occur simultaneously in *A. fabae*, and since both appear to involve a response to photoperiod. Based on these results, Hardie and Lees (1985) proposed that the endogenous JH titer should be high during long-day conditions, when animals feed on the primary host plant and reproduce parthenogenetically, and low later in the season when aphids switch from asexual to sexual reproduction and from primary to secondary host plants.

11.2.1.3.2. Wingless versus winged morphology The switch from wingless to winged morphologies can occur in two different situations. First, parthenogenetically reproducing aphids overcrowd their primary host plant and must disperse to other (also primary) host individuals. As conditions become crowded, females begin producing asexual daughters with wings. These winged females disperse to new plants and begin parthenogenetic production of wingless daughters all over again.

At the end of the favorable growing season, when conditions begin to deteriorate everywhere (e.g., as the primary host plants begin to die), aphids again produce a winged generation of offspring, except that this time the switch between wingless and winged morphologies may coincide with both the switch from primary to secondary host plants, and also with the switch from asexual to sexual reproduction. Thus, the first manifestation of the

wingless versus winged switch is not associated with a simultaneous switch in the mode of reproduction, but the second one is. This suggests that at least partially independent mechanisms may be involved. In some species the winged forms produced at these two times are morphologically distinct (Hille Ris Lambers, 1966), which is consistent with the idea that separate environmental cues and/or endocrine mechanisms may regulate wing expression at these two stages.

Early research on the mechanism of aphid wing polyphenism produced equivocal results, probably due to difficulties distinguishing between hormonally induced "juvenilization" of animals and polyphenism-specific effects on wing expression per se (reviewed in Hardie and Lees, 1985). The complex life cycles of aphids and the simultaneous involvement of several different forms of polyphenism made this problem still more difficult. However, several clever experiments with *A. fabae* finally resolved this dilemma and clearly established a role for JH in one form of aphid wing polyphenism (Hardie, 1980, 1981b,c; Lees, 1977, 1980). Juvenile hormone appears to regulate the switch from wingless to winged morphologies that occurs at the end of the growing season (and coincident with the switch from asexual to sexual reproduction). Animals harvested at a stage that would typically begin production of winged offspring (i.e., at the end of the season) could, if exposed to either long days or high temperatures (i.e., summer conditions), be induced to forego wing production and produce wingless daughters instead. This effect of environmental stimuli could be mimicked by topical applications of JH I at this same time, suggesting that high levels of JH are associated with continued production of wingless offspring, and that a seasonal decline in levels of JH may underlie the polyphenic switch from wingless to winged morphologies (as it did with the end-of-season switch between asexual and sexual reproduction).

In contrast, the most important environmental trigger of wing production in conditions of crowding is an increase in tactile stimulation between individuals (Johnson, 1965) or a response to the release of aphid alarm pheromone due to predator pressure (Hatano *et al.*, 2010; Kunert and Weisser, 2005). Thus, animals reared under crowded conditions produce winged progeny irrespective of photoperiod or temperature and even when exposed to topical applications of JH (Hardie, 1980). These results suggest that the two forms of wing polyphenism are regulated by independent endocrine mechanisms. In particular, they suggest that polyphenic production of winged morphologies during the period of parthenogenetic reproduction occurs by a mechanism other than circulating levels of JH, and one that can operate independent of levels of JH (i.e., by a mechanism capable of inducing wing expression even when environmental conditions stimulate high levels of JH).

Major obstacles to direct assessment of JH titers and corpora allata (CA) activity in aphids are their small size and telescoping of generations. Chemical allatectomy by precocenes turned out to be a valuable research tool in this context, even though not all aphid species were equally susceptible (Hardie, 1986), and not all precocene compounds acted in the same direction (Hardie *et al.*, 1995, 1996). Juvenile hormone rescue experiments on precocene-treated larvae of a pink *A. pisum* clone (Gao and Hardie, 1996) showed that the destruction of the CA by precocene II or precocene III resulted in precocious adult development. The alate-promoting property of these compounds appeared to be unrelated to the decreased JH titer, and instead depended heavily on population density (Hardie *et al.*, 1995). In contrast, flight muscle breakdown in alate adult *A. pisum* that had undertaken a migratory flight to a new host plant has been shown to be under JH control (Kobayashi and Ishikawa, 1993, 1994). Thus, there is substantial evidence for the role of JH in specific contexts of developmental regulation in aphids, but it is clearly not the general role that was attributed to it in the early days of aphid endocrinology. A role for JH in wing polyphenism has also been called into question by results of direct JH III titer determination using an LC-MS approach on pools of *A. pisum* aphids preferentially producing winged or unwinged offspring. As shown by Schwartzberg *et al.* (2008), there was no evidence for differences in JH titer between the two groups.

To our knowledge, little information exists on the role of ecdysteroids in wing phenotype expression of aphids, yet genomic resources have now enabled the annotation of genes involved in ecdysteroid biosynthesis as well as nuclear receptors, and modeled the possible binding properties of 20-hydroxyecdysone (20E) to a putative ecdysone receptor (Christiaens *et al.*, 2010). While there is little doubt that ecdysteroids play a role in the regulation of molting and metamorphosis in general, in aphids their role in wing polyphenism and/or cyclical parthenogenesis still warrants functional validation.

The proportion of winged females produced in response to a given environmental cue may vary between clonal genotypes, indicating genotype–environment interaction on this important trait of adaptive plasticity (Braendle *et al.*, 2005a,b). An even stronger influence of genotype has been detected in males of *A. pisum*, where wing production is under the control of a single locus on the X chromosome (Cauillaud *et al.*, 2002). Allelic variation in this locus, *aphicarus (api)*, is associated with clonal genotype differences in the propensity to produce winged males (Braendle *et al.*, 2005b). This finding is in line with earlier experiments that showed that precocene-mediated allatectomy was much more effective in apterizing presumptive alate females than males in this species (Christiansen-Weniger and Hardie, 2000).

Interestingly, apterization of presumptive alates can also occur as a consequence of parasitization during the early larval instars (Johnson, 1959). Parasitization effects on wing development appear to be regulated independently of metamorphosis, and in particular, seem to occur independent of JH (Hardie and Lees, 1985). In pea aphids, parasitization by *Aphidius pisum* caused apterization and other correlated changes in body shape (Christiansen-Weniger and Hardie, 2000), yet parasitization of the same species by *Aphidius ervi* had the opposite effect, resulting in a higher proportion of winged offspring (Sloggett and Weisser, 2002). The developmental mechanisms underlying the divergent responses to parasitization are not yet understood.

In conclusion, the mechanisms underlying aphid phase polyphenism are responses contingent on different environmental cues and may involve completely different endocrine mechanisms. Thus, wing polyphenism represents not one, but two developmental mechanisms. Again, superficially taken, reproductive polyphenism would appear to be very different from wing polyphenism, yet it occurs at the same time as one of the wing polyphenisms in response to the same environmental stimulus, and is regulated in the same direction by the same hormone (JH) (**Figure 1**). Thus, a useful description of aphid polyphenism may involve a classification scheme that distinguishes an "autumn" polyphenism that includes both wing morphology and mode of reproduction, and which is regulated by a seasonal change in photoperiod and a corresponding decline in levels of JH, from a "summer" polyphenism. This seasonal change only involves wing expression, which occurs in response to crowding and incorporates an as yet unidentified endocrine mechanism, independent from circulating levels of JH.

11.2.1.3.3. Soldier aphids It is worth noting that aphids, the masters of phase polyphenism, exhibit yet another form of polyphenism. Some aphid species facultatively produce a behaviorally and morphologically distinct soldier caste. Since its original description (Aoki, 1977), this phenomenon has been detected in over 50 aphid species in the families Pemphigidae and Hormaphididae (Stern *et al.*, 1997). Most studies of aphid soldiers have focused on clonal relatedness and kin selection (Abbot *et al.*, 2001; Carlin *et al.*, 1994; Pike and Foster, 2008), as well as on colony division of labor and defense against predators (Foster and Rhoden, 1998; Fukatsu *et al.*, 2005; Rhoden and Foster, 2002; Schutze and Maschwitz, 1991). Little is known about proximate mechanisms that induce soldier formation. Positive correlations have been found between soldier proportion and colony size (Ito *et al.*, 1995), which may be mediated through direct contact stimuli with non-soldier aphids in a colony and negative feedback from contact with soldier aphids (Shibao *et al.*, 2010). Ontogenetically,

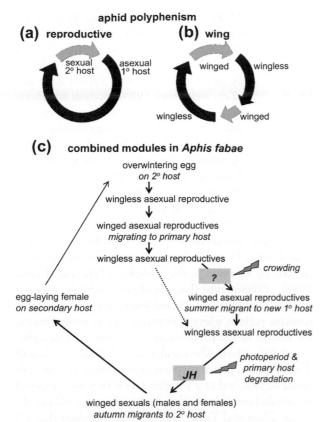

aphid polyphenism

(a) reproductive

sexual 2° host — asexual 1° host

(b) wing

winged — wingless

wingless — winged

(c) combined modules in *Aphis fabae*

overwintering egg on 2° host
↓
wingless asexual reproductive
↓
winged asexual reproductives *migrating to primary host*
↓
wingless asexual reproductives
? ⚡ *crowding*

egg-laying female *on secondary host*

winged asexual reproductives *summer migrant to new 1° host*
↓
wingless asexual reproductives
JH ⚡ *photoperiod & primary host degradation*

winged sexuals (males and females) *autumn migrants to 2° host*

Figure 1 Polyphenism in aphid life cycles can tentatively be assigned to two separate modules. (a) Module 1 (reproductive polyphenism) represents the switch from asexual (parthenogenetic) to sexual reproduction. This switch may be associated with a switch from a primary to a secondary host plant. (b) Module 2 (wing polyphenism) represents the switch from a wingless to a winged form. (c) These modules appear combined in complex life cycles such as that of the bean aphid, *Aphis fabae*. In the latter, three switches from wingless to winged can occur in the annual cycle. The first one occurs when wingless (parthenogenetic) females that arose from overwintered eggs produce winged (parthenogenetic) daughters that migrate to the primary host. The mechanisms underlying this switch are little understood. In a rapid sequence of generations, a large population of wingless (parthenogenetic) females then builds up on the primary host plant. When crowding reaches critical levels, some winged (parthenogenetic) females are produced. These colonize new primary host plants and initiate a new cycle of wingless generations. This switch from a wingless to a winged morph does not involve a switch in the mode of reproduction and appears to be independent of circulating (high) levels of JH. The third switch from a wingless to a winged morph occurs at the end of the favorable season when primary host plants degrade. This switch in wing expression now also involves a switch in reproductive mode, as the wingless asexual females start to produce winged males and females that mate and move to a secondary host plant. These either produce overwintering eggs themselves or produce a generation of egg-laying females. This third switch in wing expression is controlled by photoperiod (scotophase length) accompanied by a decrease in JH titers. Graph compiled from data by *Hardie and Lees* (1985) and *Hardie et al.* (1990, 1996).

the development of a soldier morphology appears to be linked to the replacement of the reproductive system by fat body cells and the lack of endosymbionts (Fukatsu and Ishikawa, 1992). Developmental trajectories have been mapped for *Pseuroregma bambucicola* (Ijichi *et al.*, 2004) and *Tuberaphis styraci* (Shibao *et al.*, 2010), inferring proximate cues that already act during embryonic stages and gradually separating developmental pathways in first instar nymphs.

11.2.1.3.4. Gene expression analyses and genomic resources A large collection of ESTs published in recent years (Ramsey *et al.*, 2007; Sabater-Muñoz *et al.*, 2006) and the establishment and use of microarray platforms (Brisson *et al.*, 2007; Le Trionnaire *et al.*, 2007; Wilson *et al.*, 2006) contributed to the establishment of a database of aphid genomic resources (Gauthier *et al.*, 2007). These efforts furthered the formation of a consortium for the sequencing and annotation of the first aphid genome, that of the pea aphid *A. pisum* (International Aphid Genomics Consortium, 2010). Furthermore, RNAi protocols have been devised for functional analyses of candidate genes or gene sets revealed by non-biased high throughput analyses (Jaubert-Possamai *et al.*, 2007; Mutti *et al.*, 2006).

Three noteworthy results from these novel approaches to aphid biology are summarized here, each hinting at significant insights to come. The first one concerns the perception of the photoperiod signal, which has long been attributed to an extraocular portion of the brain located within the region lateral to Group I neurosecretory cells (Steel and Lees, 1977). A transcriptome analysis of aphid heads revealed expression differences in cuticle protein encoding genes, suggesting a role in softening of the cuticle that overlies the photoreceptive area in short-day reared aphids (Le Trionnaire *et al.*, 2007). This approach also revealed alterations in the dopamine pathway (Gallot *et al.*, 2010; Le Trionnaire *et al.*, 2009), which may link both cuticle sclerotization/melanization in the head with neurotransmission, thus establishing a possible route for transgenerational signal transmission to the ovary.

A second important point is the annotation of a complete complement of DNA methyltransferases (Walsh *et al.*, 2010), bringing into focus the possible role of epigenetic alteration in the development of alternative phenotypes similar to findings in the honey bee (Kucharski *et al.*, 2008). The third point refers to the wing developmental gene regulatory network. Based on the *Drosophila* regulatory network, eleven genes were annotated in the aphid genome, and when assayed, six showed stage-specific variation and one gene, *apterous 1*, exhibited a significant difference in transcript levels between winged and unwinged morphs of *A. pisum* (for review see Brisson *et al.*, 2010) indicating that this gene may play a major role in polyphenic development.

11.2.2. Orthoptera

11.2.2.1. Wing polyphenism in crickets Facultative dispersal strategies in crickets involve the relative amounts of growth of the wings and associated wing musculature. Animals can have full-sized, functional wings (macropters), miniature wings (micropters), or be entirely wingless (brachypters). Wing expression and associated flight capability may differ among species (Harrison, 1980), among populations of a single species (Harrison, 1979), or even among time periods within a single individual's lifetime, with some crickets shedding their wings and histolyzing flight muscles after mating (Srihari *et al.*, 1975).

In a number of cricket species, wing expression is facultative and depends on environmental conditions encountered during nymphal development, such as temperature

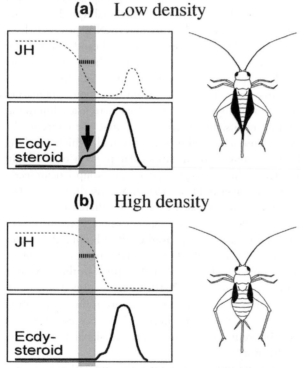

(a) **Low density**

JH

Ecdy-steroid

(b) **High density**

JH

Ecdy-steroid

Figure 2 Facultative wing-length polyphenism in the cricket *Gryllus rubens* is a response to environmental conditions, especially nymphal crowding, and is controlled by a hormonal threshold mechanism acting during critical periods (shaded bars) in the two final instars of post-embryonic development. At low population densities, the JH titer drops below a critical threshold level due to enhanced JHE activity, whereas the pre-molting ecdysteroid titer exceeds a threshold level. This endocrine situation permits wing development to a macropterous phenotype. In contrast, high population densities lead to an above-threshold JH titer and a low ecdysteroid titer during this critical period. Consequently, wing development is inhibited and brachypterous adults are being formed. Graph compiled from data and models by *Zera and Tiebel* (1988), *Zera and Holtmeier* (1992), and *Zera and Denno* (1997).

(Ghouri and McFarlane, 1958; McFarlane, 1962), photoperiod (Alexander, 1968; Masaki and Oyama, 1963; Mathad and McFarlane, 1968; Saeki, 1966a; Tanaka *et al.*, 1976), diet (McFarlane, 1962), and population density (Fuzeau-Braesch, 1961; Saeki, 1966b; Zera and Denno, 1997; Zera and Tiebel, 1988). In these taxa, wing expression appears to be regulated by a threshold mechanism and population comparisons, controlled-breeding and artificial selection studies all inferred considerable levels of genetic variation for the threshold of this polyphenism (Fairbairn and Yadlowski, 1997; Harrison, 1979; Roff, 1986, 1990; Zera *et al.*, 2007).

The physiology of wing polyphenism has been most thoroughly studied in *Gryllus rubens* and *G. firmus*. Wing expression in these species is sensitive to population density, in particular, the level of crowding experienced by nymphs as they develop (Zera and Tiebel, 1988). Experiments that transferred animals between experimentally staged high and low densities revealed two sensitive periods relevant to expression of wings, during the middle of the penultimate and in the final nymphal instar, respectively (Zera and Tiebel, 1988). The default developmental pattern appears to be winged, but animals could be switched to a wingless fate if they were exposed to crowded conditions or received an exogenous JH application (Zera and Tiebel, 1988). This suggested that JH titer differences might underlie wing polyphenism, and this hypothesis was corroborated by direct measurement of JH titers in presumptive winged and wingless animals (Zera *et al.*, 1989). In addition, a second titer difference was observed. Animals destined to produce wings had higher levels of ecdysteroids than animals destined not to produce wings (Zera *et al.*, 1989).

Surprisingly, when comparing rates of JH biosynthesis in presumptive winged and wingless animals, Zera and Tobe (1990) found no differences, suggesting that morph-specific variation in levels of JH result from differential clearance of JH from the hemolymph, rather than from differential rates of hormone biosynthesis. This was ascribed to the enzyme JHE, which was shown to attain higher levels in winged than in wingless juveniles (Zera and Tiebel, 1989). Furthermore, the timing of these morph-specific differences in levels of JHE coincided with the two already described sensitive periods and with the observed morph-specific differences in the JH titer. A JHE has been cloned and sequenced in *G. assimilis* (Crone *et al.*, 2007) revealing a 19 bp indel in an intron that was strongly associated with differences in enzyme activity among lines selected for increased or decreased wing expression. This indicates that genetic differences resulting in different JHE function do not reside in the coding sequence but in possibly regulatory intronic motifs, a finding in accordance with previous findings that showed JHEs of short- and long-winged morphs do not differ biochemically (Zera *et al.*, 2002; Zera and Zeisset, 1996).

Based on these results, the following model was proposed (Zera and Denno, 1997; Zera and Holtmeier, 1992). Crickets have two sensitive periods, one during the middle of the penultimate and a second in the final nymphal instars. During these periods, wing development is sensitive to external environmental factors (crowding), and these external stimuli appear to result in morph-specific differences in levels of two hormones, JH and ecdysone. Animals destined to produce wings have lower levels of JH, and higher levels of ecdysteroids, than animals destined to mature without wings (**Figure 2**). Morph-specific differences in levels of JH appear to result from subtle differences in the timing of the decline in JH titers during these sensitive periods, and these differences in JH titers result from prospective winged animals having higher levels of the degratory enzyme JHE than prospective wingless ones. Thus, the default pattern of development entails production of wings, but a subset of individuals (animals with levels of JH above a genetically mediated threshold level) experience an early rise in levels of ecdysteroids, and this pulse of ecdysteroids probably "re-programs" these individuals toward production of a wingless body form.

While developmentally established during the nymphal phase, the density-dependent wing polyphenism has its major implications in the reproductive physiology of the adults. In a fusion of resource allocation and metabolic biochemistry studies, Zera and Zhao (2003) and Zhao and Zera (2002) demonstrated a genetic bias for flight-fuel synthesis (mainly triglycerides) in the flight-capable genotype versus a bias for ovarian lipids (phospholipids) in the flightless one. When JH was topically applied, the adult flight-capable morph shifted its lipid metabolism toward that of the flightless morph. In addition to this expected metabolic biphenism a surprising difference between the short- and long-winged adults emerged when JH synthesis and JH titers were measured. The long-winged adults exhibited a strong diurnal modulation in these parameters in the photophase, but this was not so in the short-winged morph (Zera and Cisper, 2001; Zhao and Zera, 2004). Such diurnal variation was also found in natural populations (Zera *et al.*, 2007), and thus cannot be ascribed to laboratory conditions and selection. Furthermore, it persisted under constant darkness and was temperature compensated, thus qualifying as a genuine circadian rhythm within the endocrine system (Zera and Zhao, 2009). Allatostatin-like material was shown to also exhibit a diurnal modulation in long-winged crickets, but not in short-winged ones (Stay and Zera, 2010), thus qualifying as a potential upstream regulatory factor.

Genetic variation that could explain such threshold mechanisms has been found in laboratory strains and natural populations of *G. firmus*, making this a prime model system to investigate functional causes of adult life history evolution in the context of evolutionary endocrinology (Zera, 2006, 2007).

11.2.2.2. Phase polyphenism in locusts Locust phase polyphenism (a switch between solitary and gregarious forms) has dramatically impacted human history. The gregarization phenomenon can lead to staggering densities of animals, and these migratory swarms of locusts are one of the world's most devastating plagues (for historical and recent data on locust pest status see Pener and Simpson, 2009; Sword *et al.*, 2010). Phase polyphenism is common in several species of locusts, but is most clearly expressed and best studied in the migratory locust (*Locusta migratoria*) and the desert locust (*Schistocerca gregaria*).

Phase polyphenism is a complex phenomenon that involves changes in body coloration, wing morphology, reproductive physiology, energy metabolism, and behavior (Uvarov, 1921). Solitary phase animals (instead of solitary, the term "solitarious" is also frequently used in the scientific literature on locusts to resolve possible ambiguities in the term solitary, but for reasons of simplicity we prefer here to speak of solitary phase locusts, also when referring to laboratory animals reared in isolation) generally avoid conspecifics and have cryptic or green color patterns and reduced wing morphologies and musculature. Gregarious phase animals aggregate, actively seeking conspecifics, and have a dark background pigmentation with a frequently yellow or orange color pattern, as well as more fully developed wings and wing musculature. The complexity of this syndrome sets locust phase polyphenism apart from the apparently simpler wing polyphenisms already discussed for other hemimetabolans (including other orthopterans). Perhaps the most important difference between locust phase polyphenism and the other polyphenisms discussed so far is that the full transition from the solitary to the gregarious form generally requires multiple generations, and thus a transgenerational route of information transfer.

Gradual phase transition from solitary to gregarious locust forms was one of the earliest of the studied types of polyphenisms (Uvarov, 1921) and for decades the endocrine system, and in particular, JH, was advocated as the main control center (Couillaud *et al.*, 1987; Joly, 1954; Staal and de Wilde, 1962). However, results from many of these early hormone studies were equivocal, and the role of JH continues to be a controversial issue (Applebaum *et al.*, 1997; Dorn *et al.*, 2000; Pener, 1991; Pener and Yerushalmi, 1998). Fortunately, a series of hormone titer studies and their interpretation in an organismic context has now resolved several of the discrepancies, not only for JH but also as for other endocrine system functions.

Before going into further detail we call attention to two recent reviews, the one by Pener and Simpson (2009), which addresses the full complexity of locust phase polyphenism in an impressive 286 pages, and a more concise review by Verlinden *et al.* (2009). These reviews were helpful in revising some of the considerations on the hormonal control of locust phase polyphenism expressed in our previous version of this chapter (Hartfelder and Emlen, 2005).

11.2.2.2.1. Changes in color and morphology The most important cue leading to phase change is the level of crowding experienced by nymphs as they develop. Sensory excitation of hindleg mechanoreceptors has been identified as a powerful stimulus to elicit behavioral and gradually also other phase characteristic changes (Simpson *et al.*, 2001). Models built from particle physics properties showed that tactile stimuli to the hindlegs elicits self-organization of crowded locusts into marching hopper (nymph) bands and gregarious flight movement of adults (Buhl *et al.*, 2006). The driving force behind this self-organization is a simple rule, avoidance of cannibalism from behind (Bazazi *et al.*, 2008). A strong correlation was demonstrated between hindleg bristle mechanostimulation, gregarious behavior, and serotonin levels in thoracic

ganglia (Anstey *et al.*, 2009), which is evidence for the transformation of the population density signal into a fast (early) endogenous response, but first leading to a switch in behavioral state.

Color differences between gregarious and solitary locusts are the best known and most visible phase characteristic (**Figure 3**, lower panel). There are separate kinds of color polyphenism in acridids; namely a phase color polyphenism expressed as dark body pigmentation or not, a green-brown background color polyphenism, homochromy (i.e., adaptation of the body color to the respective background), and a yellow/orange background color which, in combination with dark body patterns, is frequently interpreted as having an aposematic function (Sword, 2002). Different locust species may vary in the degree of expression of these color variations. Furthermore, within a species they may vary with respect to environmental conditions, especially humidity. Also

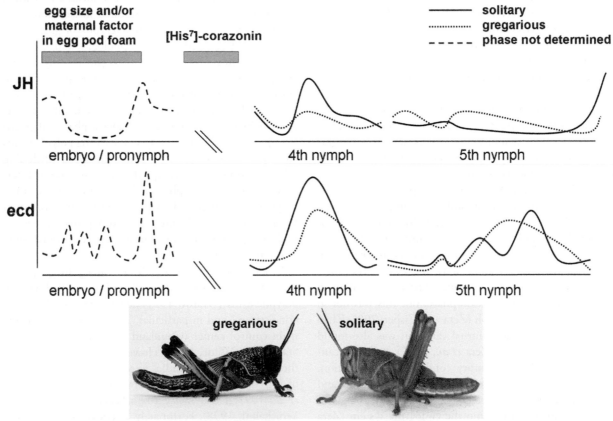

Figure 3 Juvenile hormone and ecdysteroid titers during embryonic and post-embryonic development of solitary and gregarious morphs of migratory locusts (*Locusta migratoria* and *Schistocerca gregaria*). Hormone titers during embryonic development were determined by *Lagueux et al.* (1977) and were not specified with respect to phase. The ecdysteroid titer peaks during embryonic development are associated with embryonic molts and the JH-free period in the middle of embryonic development is required for correct blastokinesis (*Truman and Riddiford*, 1999). Larval hormone titers compiled in this graph were determined by *Botens et al.* (1997), *Tawfik et al.* (1997b), and *Tawfik and Sehnal* (2003). JH application to gregarious/crowded fourth instar larvae was shown to shift several of the solitarious traits (especially green color), but did not shift the entire set that characterizes the gregarious morph (see text). Corazonin application to isolated/solitary second and third instar larvae induced the gregarious-phase dark foreground coloration (Tanaka, 2000a,b,c). Small egg size and/or exposure of eggs of isolated/solitarious females to a maternal factor present in egg pod foam from gregarious/crowded females shifts offspring to the gregarious phase (*Häegele et al.*, 2000; *Tanaka and Maeno*, 2006). Typical color variants for gregarious and solitarious morphs of *S. gregaria* are shown in the lower panel. Photo copyright Tom Fayle.

nymphs and adults may be different in these color combinations, in part due to the transgenerational gradual transition between the phases.

Exposure to crowded conditions induces a shift from the solitary (frequently green) to the gregarious (brown/black with yellow, *S. gregaria*) or orange (*L. migratoria*) form, and much of this shift is mediated by hormones. Functionally, the best characterized hormone is the neurohormone Lom-DCIN, originally identified in studies on an albino mutant of an *L. migratoria* strain (Tanaka, 1993). This neurohormone has since been chemically characterized as [His[7]]-corazonin (Tawfik *et al.*, 1999a) and shown to induce the expression of the black pigmentation and foreground color patterns typical of the gregarious form (Pener *et al.*, 1992; Tanaka, 2000a–c, 2001). Schoofs *et al.* (2000) found immunoreactive staining to a DCIN in lateral neurosecretory cells and in corpora cardiaca (CC), as well as in a few other distinct neurons. Subsequently, the detection of DCIN in both gregarious and solitary phase nymphs (Baggerman *et al.*, 2001) indicated that its release may be blocked in solitary ones.

Whereas [His7]-corazonin is established as the prime factor involved in the expression of the dark pigmentation of the gregarious form, JH is a major effector leading to the green (or better to say, not brown) background coloration. As stated by Pener and Simpson (2009):

In locust species that exhibit green-brown colour polyphenism, the green colour inducing effect of implantation of extra CA, or administration of JH or JH analogues (JHAs), has been repeatedly confirmed without any exception. These treatments induce green colour even in crowded locust hoppers that show a reduction or disappearance of the gregarious colouration with the increasing green colour.

It is now recognized that the solitary–gregarious phase transition is also associated with changes in a suite of morphological traits, including aspects of wing morphology and relative wing size (Dorn *et al.*, 2000), differences in morphometric ratios related to the hindleg femur (for review see Pener and Simpson, 2009), and especially the number of ovarioles. The number of ovarioles, which is already determined in the embryonic stage, is higher in solitary than in gregarious females (Pener, 1991) causing fertility differences between the morphs, contributing to a larger number of eggs per egg pod in the solitary morph. As this difference in ovariole number is also associated with the size of the eggs, which are smaller in solitary locusts, these traits are indicative of major phase differences related to female reproductive strategies.

Whereas JH and [His[7]]-corazonin certainly play major roles in pigmentation changes, their role in shaping the entire suite of phase characters is far from clear, despite a plethora of experimental studies (for a comprehensive review of discrepancies in results and interpretations see Pener and Simpson, 2009). It is noteworthy at this point that the solitary phase characteristics cannot be interpreted as an effect of juvenilization (or neotinization) simply due to higher JH titers. Nevertheless, measurements of endogenous hormone titers are crucial to any interpretation of results obtained by experimental manipulation of hormone levels, either by gland extirpations or transplantations, or topical application of hormones or pharmacological analogs (Zera, 2007).

11.2.2.2.2. Juvenile hormone and ecdysteroid titers of solitary and gregarious nymphs

In a pioneering study, Injeyan and Tobe (1981) showed that during the fourth nymphal instar, the CA were more active in animals reared under isolated conditions than they were in animals reared under crowded conditions. Subsequently, Botens *et al.* (1997) compared JH III titers measured for laboratory animals reared under isolated and crowded conditions. These authors noted that the JH hemolymph titer was higher in fourth (penultimate) instar nymphs reared in isolation (solitary) when compared to crowded ones, but such differences were no longer encountered in last instar nymphs. Furthermore, Botens *et al.* (1997) also measured JH III titers for wild-caught solitary and gregarious animals and found a similar result; JH titers during the middle of the fourth nymphal instar were higher in solitary animals than in gregarious ones. These findings are consistent with a critical window for JH action during nymphal development (**Figure 3**). In addition to JH III, which is the principal form of JH in *L. migratoria* (Bergot *et al.*, 1981), Darrouzet *et al.* (1997) found that the CA of *L. migratoria* also synthesize two hydroxy-juvenile hormones (12′-OH JH III and 8′-OH JH III). These products have not been detected by the standard radiochemical assays (Tobe and Pratt, 1974), and whether these are secreted into the hemolymph or whether they affect phase polyphenism remains to be explored.

Solitary and gregarious locusts also differ in their 20E levels (**Figure 3**). Animals reared under crowded conditions have a slightly lower but more prolonged premolting ecdysteroid peak than animals reared under solitary conditions in the penultimate and last nymphal instars (Tawfik *et al.*, 1996; Tawfik and Sehnal, 2003). As these differences coincide with the phase specifically modulated JH titer in the fourth instar, a synergistic interaction of these major morphogenetic hormones is feasible, even though a clear morphogenetic role for ecdysteroids in locust phase polyphenism has not yet been shown.

11.2.2.2.3. Changes in adult behavior and physiology

Adult solitary and gregarious locusts differ in behavior and physiology, as well as in morphology. For example, gregarious animals show an affinity for other locusts (aggregative behavior) and a strong propensity for long-duration migratory flights, in contrast to solitary animals. These behavioral differences are accompanied by corresponding differences in adult physiology related to the two different life-history strategies. Phase differences are noticeable in the strength of the adipokinetic reaction that mobilizes fat body lipids for long-distance flight (see Pener and Simpson, 2009 for a comprehensive discussion on the role of adipokinetic hormone in locusts) and at the onset of egg production, with crowded females generally beginning egg production slightly earlier than isolated ones (Tawfik *et al.*, 2000).

As with the morphological differences previously discussed, behavioral differences between solitary and gregarious forms also appear to be driven, at least in part, by hormones. In adult females, both solitary and gregarious animals show a major increase in JH levels within the first two weeks after eclosion (Dale and Tobe, 1986; Tawfik *et al.*, 2000). However, this rise in JH occurs earlier in solitary females than it does in gregarious ones (Dale and Tobe, 1986; Tawfik *et al.*, 2000). Similar differences in the timing of hormone secretion were observed for ecdysteroids, with ecdysone levels rising earlier in solitary females than in gregarious ones (Tawfik and Sehnal, 2003; Tawfik *et al.*, 1997b).

Even though the role of JH in the induction of vitellogenin synthesis in locusts is a well-established model in insect physiology (Nijhout, 1994), these phase polyphenism-related differences in the timing of adult female JH and ecdysteroid synthesis and titers are not easily reconciled with phase-specific differences in the timing of the first oogenic cycle (**Figure 4**), as females raised under crowded conditions begin producing eggs slightly sooner than females reared under isolated conditions (Tawfik *et al.*, 1997a, 2000). The hormone titer differences, however, may be related to migratory flight fuel metabolism (Wiesel *et al.*, 1996). Low levels of JH (as in gregarious females) stimulate utilization of stored lipids for flight, whereas high levels of JH (as in solitary females) reduce their use, and instead, result in vitellogenin synthesis and lipid allocation to oocytes (Wiesel *et al.*, 1996).

In adult males, the principal behavioral difference between solitary and gregarious phase individuals involves production and secretion of an aggregation pheromone that has phenylacetonitrile (PAN) as its main component (Mahamat *et al.*, 1993). Tawfik *et al.* (2000) showed that pheromone emission by crowded adult males starts between days 10 and 15 after fledging, coinciding with a conspicuous increase in the amount of JH III in the hemolymph. As pointed out by Pener and Simpson (2009), the term aggregation pheromone may not entirely be correct, and due to its repellent effects on conspecific males they

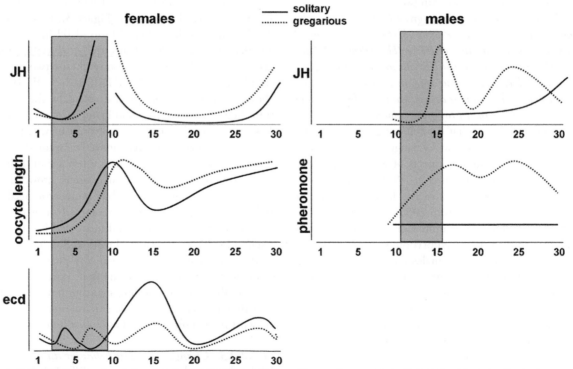

Figure 4 Hormone titers in relation to terminal oocyte length in solitary and gregarious-phase locust females (*Locusta migratoria* and *Schistocerca gregaria*), and the phase relationship between JH titer and pheromone production (phenylacetonitrile) in adult males. The phase differences in the JH and ecdysteroid titer of females are reflected in an earlier onset of growth of the terminal oocytes in solitarious females, nevertheless, laying of the first batch of eggs occurs somewhat earlier in gregarious females, possibly due to smaller egg size. In gregarious-phase males, the sharp increase in the JH titer seen between days 10 and 15 coincides but may not necessarily be causally related to the strongly enhanced production of a gregarization pheromone. Graph compiled from data by *Tawfik et al.* (1997a,b, 1999a, 2000), *Tawfik and Sehnal* (2003), and *Dorn et al.* (2000).

propose the term "rival male repelling pheromone." Since a chemically not yet defined maturation-accelerating effect of mature *S. gregaria* males on immature ones has long been noted and appears to be mediated via the CA (Loher, 1961), the transient steep rise in the hemolymph JH titer seen in young males (**Figure 4**) may be more related to the coordination of maturation among gregarious males, and less so with PAN production, especially since the latter has been shown to quickly respond to changes in population density (Deng *et al.*, 1996).

11.2.2.2.4. Maternal effects on offspring development

Schistocerca gregaria females reared in isolation do not only lay more eggs than those reared under crowded conditions, due to the larger number of ovarioles, but these also lay smaller ones. This was also observed for the first egg pod laid by a female reared under crowded conditions (Maeno and Tanaka, 2008). Egg size is therefore considered to be an important transgenerational cue (Tanaka and Maeno, 2010).

Egg development is influenced greatly by ovarian ecdysteroids, and by measuring these Tawfik and Sehnal (2003) and Tawfik *et al.* (1999b) found that ecdysteroid contents of ovaries of females (*S. gregaria*) reared under crowded conditions were up to four times higher than those in the ovaries of females reared in isolation (8.9 ng/mg vs. 2.3 ng/mg tissue before egg laying). These phase-specific differences in egg ecdysteroid levels persisted after the eggs were laid (89 ng vs. 14 ng per egg), and even were reflected in newly hatched larvae (Tawfik *et al.*, 1999b). High levels of ecdysteroids have long been detected in vitellogenic ovaries (Lagueux *et al.*, 1977), where they are synthesized by the follicle epithelial cells. These ecdysteroids are transferred to the developing eggs and, together with JH, they affect molting events during embryonic development (Lagueux *et al.*, 1979; Truman and Riddiford, 1999). These studies, thus, raise the possibility that endocrine differences in the mothers, resulting from their exposure to crowding as they developed, are carried over to their offspring. This may explain some aspects of the transgenerational nature of this phase transition.

Pheromones and semiochemicals transferred to eggs may also direct embryonic development. These compounds, when deposited on the eggs or egg pod material, not only attract other gravid females to the area resulting in clustered oviposition (Saini *et al.*, 1995), but they may also lead to an increase in the propensity of the hatchlings to express gregarious characteristics. Washing freshly laid eggs from gregarious *S. gregaria* females shifted phase characteristics from gregarious toward solitary, and application of female accessory gland products to these washed eggs restored expression of the gregarious characters (Hägele *et al.*, 2000; McCaffery *et al.*, 1998). These findings have been challenged by Tanaka and Maeno (2006), but a recent study on *L. migratoria* (Ben Hamouda *et al.*,

2009) observed an effect similar to that seen by Hägele *et al.* (2000). A possible reason given by Pener and Simpson (2009) for the discrepancy in the findings on *S. gregaria* egg foam activity could be a genetic difference in the strains used. Nevertheless, when testing egg foam effects in different strains, this hypothesis was not confirmed (Maeno and Tanaka, 2009), and instead, these authors proposed other factors for predetermination of the hatchling phase fate, such as the egg-size dependent amount of egg yolk. While the chemical nature of the egg-foam maternal agent has been tentatively identified as an alkylated L-DOPA analog (Miller *et al.*, 2008), its role and mode of action continues to be a matter of debate (Tanaka and Maeno, 2010). Nonetheless, the existence of a transgenerational transmission of phase characteristics is unquestionable.

A phase-related 6 kDa molecule has been identified from a proteomic screen on hemolymph of *S. gregaria* reared under crowded conditions (Rahman *et al.*, 2003). Interestingly, the strongest immunoreactivity for this peptide was found in follicle cells of the ovary and in seminal vesicles of the male accessory gland complex (Rahman *et al.*, 2008), thus opening the possibility that it may be transmitted to females during copulation and be part of the transgenerational phase determination process.

11.2.2.2.5. Genomic resources

Studies comparing patterns of protein and gene expression patterns between solitary and gregarious animals were initiated at the turn of the century (Clynen *et al.*, 2002; Rahman *et al.*, 2003; Wedekind-Hirschberger *et al.*, 1999), but only with the generation of a cDNA library and the sequencing of 76,012 ESTs clustered into 12,161 unique sequences (Kang *et al.*, 2004) did high throughput, large-scale analysis become feasible, The transcriptome comparisons of solitary and crowd-reared *L. migratoria* nymphs revealed over 500 differentially expressed genes, 70% of these represented novel transcripts without similarity in non-redundant databases. Two subsequent studies by this group (Guo *et al.*, 2010; Wei *et al.*, 2009) characterized differentially expressed small RNAs and transcripts of transposable elements, respectively. The most abundant one of the transposable element transcripts was cloned and turned out to be differentially expressed in the nervous system, making it a possible mediator in phase-specific neural responses.

An alternative to non-hypothesis-driven, high throughput studies are candidate gene approaches. Special attention has been given to neuroparsins, initially isolated from CC of *L. migratoria* (Girardie *et al.*, 1987) and genes encoding components of the insulin signaling pathway. Monitoring transcript levels of two neuroparsins, Scg-NPP3 and Scg-NPP4, in brain and abdominal tissue revealed phase-dependent modulation in adult locusts (Claeys *et al.*, 2006). Badisco *et al.* (2008) quantified mRNA levels of an insulin-related peptide (Scg-IRP) in adult *S. gregaria* and detected phase-dependent differences in the fat body. Furthermore, they showed that a recombinant Scg-NPP4 peptide was capable of binding Scg-IRP, inferring a cross-talk between these signaling pathways during sexual maturation.

11.2.3. "Isoptera" (Termites) — Caste Polyphenism in the Hemimetabola

The characterization of termites as "social cockroaches" (Korb, 2008) already hints at the very special position of termites within insects expressing caste polyphenism, even more so as the order Isoptera has recently literally been relinquished to the status of a clade nested within the cockroaches (Blattodea; Inward *et al.*, 2007). Caste polyphenism in termites differs from that of the holometabolous Hymenoptera in three very important ways. First, all isopteran species are social, whereas sociality has evolved in only a few branches of the Hymenoptera. Second, caste polyphenism in termites is a larval polyphenism, that is, it primarily affects the morphologies and behavior of larvae, rather than adults. Finally, in termite societies both males and females form castes, and thus equally contribute to the social organization. In contrast, hymenopteran societies are all female based.

Hemimetabolous development permits post-embryonic stages to actively participate in termite colony life, and these post-embryonic larval and nymphal stages constitute the major work force of termite colonies (in termites, the early post-embryonic instars that are frequently dependent on being cared for are called larvae, whereas later instars are called nymphs). The only true imagoes encountered in a termite colony are the sexuals, — primary reproductives (king and queen), and, at certain times of the year, the pre-dispersal sexual alates. In the adults, there is little visible dimorphism between the sexes, and the differences that exist primarily result from the high degree of ovarian activity leading to physogastry in the egg-laying queen (Bordereau, 1971). In contrast, many termites exhibit marked sexual dimorphism in the larval stages, especially in taxa that show a sex bias with respect to caste phenotypes. For termite sociality, the spotlight is thus on the polyphenism exhibited by the larval/nymphal stages.

Larval/nymphal polyphenism means that in subsequent developmental stages an individual can progressively specialize for different colony tasks ("temporal polyphenism" according to Noirot and Bordereau, 1988). As such, caste polyphenism is tightly linked to the molting process and to the endocrine factors regulating molting. Interestingly, molting in termites is no longer necessarily connected with growth (Noirot, 1989), and has instead been co-opted as a mechanism for polyphenic changes in animal shape.

Reviews on termite societies and trajectories of caste development within these (Korb and Hartfelder, 2008; Nalepa, 2009; Noirot, 1985a,b, 1989, 1990; Noirot and Pasteels, 1987; Roisin, 2000) emphasize the distinction between caste systems (and polyphenic mechanisms) in "lower" (Mastotermitidae, Kalotermitidae, Hodotermitidae, and Rhinotermitidae) and "higher" termites (Termitidae) and their relationship to nesting modality and nesting substrate types. Lower and higher termites differ primarily in the development of the worker caste. In the higher termites, workers represent a clear developmental trajectory, culminating in a terminal molt. In contrast, in the lower termites, a true worker caste is rare. Instead, most of the nymphal stages perform worker functions, such as colony maintenance. Workers in the lower termites do not undergo a terminal molt, and these individuals retain the capacity to subsequently develop into either soldiers or reproductives, or simply remain workers. In the lower termites, these late instar nymphs that comprise the major work force of a colony are generally described as false workers or pseudergates (for explanations on termite-specific terminology for developmental stages and their roles in termite societies see Korb and Hartfelder, 2008).

Caste polyphenism in termites involves several possible developmental switches, such as larva to pre-reproductive (with wingpads) to reproductive, larva to presoldier to soldier, and, in the higher termites, larva to worker. These developmental transformations may require several subsequent molts and several different critical periods (**Figure 5**). Several of these developmental transformations can be reversed midway through the process, permitting extraordinarily flexible adjustment of the production of castes in the regulation of colony structure. This is particularly common in the lower termites, which are famous for undergoing stationary, and even regressive molts.

11.2.3.1. The pathways leading to reproductive development Termite nymphs occasionally molt into reproductives. The most commonly encountered reproductives are the alates — the winged males and females that disperse from termite colonies to breed and found new colonies. After mating, each new royal pair sheds its wings and founds a colony. A second type of reproductive consists of a replacement king or queen within an existing colony. When colonies bud, or when one of the original members of the royal pair dies, replacement reproductives can be produced. In either case (dispersing alates or replacement reproductives), commitment toward a sexual fate is first evidenced by the appearance of rudimentary wingpads, which may appear several molts before the terminal, adult molt. However, in the developmental trajectory of "replacement" reproductives, these wingpads fail to fully develop, so that replacement kings or queens do not have functional wings. Termites

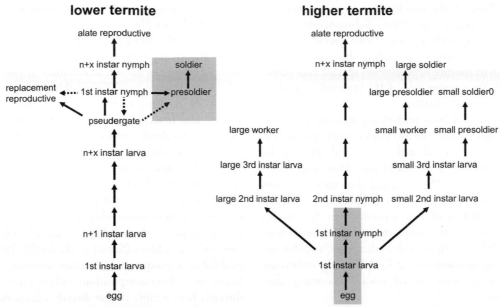

Figure 5 Generalized developmental pathways in lower and higher termites. In termite studies it is customary to denote all immatures stages without wingpads as larvae. Nymphs are stages exhibiting wingpads and thus potentially develop into sexual alates. Lower termites pass through a variable number of larval stages before the last nymphal stage (frequently referred to as the pseudergate stage), which is the main branching point for nymphal to alate sexual (imaginal) development, to neotenic replacement reproductives, or to soldiers. In most primitive termites there is no true worker caste, since tasks of colony maintenance are performed by immatures that still retain several developmental options. In higher termites, the branchpoint for developmental pathways is set early in larval development, leading alternatively to the nymphal/adult line, the soldier line, and to a true, definitive worker caste. Shading emphasizes nodes where hormone titers (JH and/or ecdysteroids) differ between castes, or where JH applications bias development, principally into presoldier/soldier differentiation. Modified from *Noirot* (1990), *Miura (2004), and Korb and Hartfelder (2008).*

that do not develop wingpads during any of these decisive molts are committed to remain in the work force or to become soldiers.

Interestingly, the number of larval/nymphal instars that an individual completes prior to becoming a reproductive or a neuter (worker or soldier) differs considerably among species. In the higher termites, this decision usually occurs during the first or second instar, while lower termites may pass through six or more pre-commitment molts (**Figure 5**).

Environmental triggers for termite reproductive caste development primarily involve demographic aspects of the colony, as communicated through chemical signals transmitted from workers to larvae (Noirot, 1990). Most early work on this polyphenism focused on the differentiation of *Kalotermes flavicollis* nymphs after removal of the primary reproductive pair (Lüscher, 1964; Wilson, 1971). Once the king or queen had been removed, nymphs began developing into replacement reproductives, suggesting that levels of chemical signals from the primary pair triggered the switch between non-reproductive and reproductive development. These studies led to the now classical model of negative feedback, where the king and queen each secrete compounds that repress the development of replacement reproductives of their respective sex

(Lüscher, 1964). These signals are thought to act via the neuroendocrine system (Lüscher, 1976).

While this classical model had its ups and downs during the last decades due to a lack of empirical evidence on the chemical nature of such pheromones, a major breakthrough has come from the termite *Reticulitermes speratus* for which a volatile inhibitory pheromone produced by female neotenics has been identified (Matsuura *et al.*, 2010). Interestingly, its active compounds are also released from eggs, inferring that reproductive status and inhibitory power are tightly linked.

Unfortunately, the hormonal control of termite reproductive development is not well understood. This is due in part to the complexity of the polyphenism (many different "switches" are involved, and several of these have multiple critical periods spread over several subsequent molts), and in part due to the lack of sensitive bioassays for the activity of termite endocrine glands. When monitoring *in vitro* CA activity for *Zootermopsis angusticollis*, the pheromonal secretions of the royal pair were denoted as inhibiting JH biosynthesis rates, inferring a link between pheromonal cues and polyphenic regulation of larval/nymphal development (Greenberg and Tobe, 1985). Similar findings come from the damp-wood termite *Hodotermes sjostedti*, where a JH peak normally observed during molting events was absent in the one leading up to imaginal differentiation (Cornette *et al.*, 2008). Subsequently, as non-physogastric nymphoids developed into queens, JH titers were observed to increase, preceding the progression of vitellogenesis in another *Reticulitermes* species, thus building evidence for a major role of JH in female reproductives (Maekawa *et al.*, 2010).

The physiological basis underlying seasonal alate production is more difficult to ascertain because production of alates depends on seasonal factors and not just the presence or absence of the social pair. However, studies of Lanzrein *et al.* (1985) suggested that termite queens may transfer hormones to their eggs, and in this fashion affect the production of alate reproductives. Specifically, they found that in two *Macrotermes* species, queens transferred both JH and ecdysteroids to their eggs, and elevated levels of these hormones were found during embryonic development in larvae biased toward becoming alate reproductives.

11.2.3.2. Mechanisms underlying soldier development

Much more effort has been put into studies of the endocrine regulation of soldier development. Soldier castes are the hallmark of termite societies and may be considered an evolutionary novelty characterizing the clade of "social cockroaches." Soldiers have heavily sclerotized exoskeletons and specialized structures for colony defense, including enlarged mandibles or enlarged heads that squirt sticky glandular secretions capable of entangling their main predators, ants.

Whereas most of the earlier studies investigated the role of exogenously applied JH as a means to pest control (promoted by offsetting the worker/soldier ratio in colonies), more recent efforts were directed to understanding the endocrinology of the presoldier and soldier molts. Most of these studies were done on diverse species of lower termites, and many of these studies connected field data to controlled laboratory experiment designs. JH III titers in presoldier stages of the Formosan subterranean termite *Coptotermes formosanus*, a major pest species, were significantly higher than those of workers or soldiers, denoting the important role of JH in this transition stage of soldier development both in field samples (Liu *et al.*, 2005a; Park and Raina, 2004) and under laboratory conditions varying temperature and nutrition (Liu *et al.*, 2005b). A subsequent study comparing CA activity in *R. flavicollis* showed that workers developing into presoldiers had 2.5-fold higher JH synthesis rates than those developing into neotenic reproductives (Elliott and Stay, 2008).

Most important, colony size and its worker/soldier ratio turned out to be crucial in stimulating or inhibiting competent larval stages to enter the presoldier–soldier route through modulating JH titers (Mao and Henderson, 2010). Furthermore, live soldiers or soldier head extracts were shown to inhibit soldier development

and the transcriptional profile associated with this pathway in *R. flavipes* (Tarver *et al.*, 2010).

The most thought-provoking results on a key mechanism underlying termite caste development, however, came from studies on the modulation of two major larval storage proteins hexamerin 1 (Hex1) and hexamerin 2 (Hex2) in *R. flavipes* (Scharf *et al.*, 2005a). Sequence analysis and functional assays indicated that (1) Hex1 is capable of covalently binding circulating JH, thus sequestering it from the biologically active hormone pool, and (2) Hex2 expression is contingent on JH levels (Zhou *et al.*, 2006a,b). This hexamerin-intrinsic circuitry favors soldier development when the Hex1/Hex2 ratio is low. In contrast, Hex1 accumulation in well-fed colonies and an appropriate worker/soldier ratio inhibited the development of new soldiers (Scharf *et al.*, 2007). This not only establishes a unique link between extrinsic conditions (nutrition, colony composition) and the intrinsic JH titer through hexamerins, it also directly affects downstream JH-responsive genomic networks, as shown by hexamerin RNAi (Zhou *et al.*, 2007).

11.2.3.3. Termite genomics and new frontiers

Taken together, these recent insights on endocrine system-mediated regulatory mechanisms underlying sexual (alates and secondary reproductives), worker (false/pseudergate or true workers), and soldier development now have little in common with hormone-application-derived models on caste development, which postulated distinct critical periods for polyphenic switching between termite castes, an early one for the nymphoid/alate pathway, and a later one for the worker/soldier decision.

Apart from a recent revival in endocrine studies, the quest for understanding caste differentiation in termites has gained new impetus from studies on differential gene expression. The first studies were directed at understanding the soldier developmental pathway (Miura, 2001; Miura *et al.*, 1999), leading to the identification of a gene with soldier-specific expression in the mandibular gland of the damp-wood termite *H. japonica*. This gene encodes a novel protein (SOL1) with a putative signal peptide, indicating that it may be a soldier-specific secretory product of this gland. The gland develops from a disc-like structure once a presoldier-differentiating molt has been induced by a high JH titer (Miura and Matsumoto, 2000; Ogino *et al.*, 1993). Subsequent DNA macroarray studies on *R. flavipes* set up to reveal the molecular underpinnings of reproductive caste development denoted 34 nymph-biased genes (Scharf *et al.*, 2005b), including those functionally related to vitellogenesis and JH sequestration. More recent differential gene expression screens identified gene sets related to the development of neotenics in the dry-wood termite *Cryptotermes secundus* (Weil *et al.*, 2007). With sequencing efforts on termite genomes underway, the results of these pioneering studies should address gene network questions built around environmental and endocrine factors in termite caste development.

Yet these certainly important genomic insights probably will do little to resolve a major enigma in termite development, namely the switches in molting types in lower termites, from progressive to stationary and even regressive molts. The latter are a major puzzle to insect physiologists. Endocrine signatures underlying these molting types are now emerging from mass hormone titer assaying in two species, *H. sjostedti* (Cornette *et al.*, 2008) and *C. secundus* (Korb *et al.*, 2009) indicating that relative JH and ecdysteroid titer dynamics during the late nymphal stages are crucial to predicting the outcome of the subsequent molt.

11.3. Polyphenism in the Holometabola

In contrast with the Hemimetabola, wing length polyphenism does not play a prominent role in the Holometabola. In this group, the prevalent forms of polyphenism are related either to camouflage, such as in color and wing pattern polyphenism in butterflies, to reproductive strategies, such as the development of weaponry in male stag and rhinoceros beetles, or to reproductive division of labor as seen in the female castes of many Hymenoptera.

11.3.1. Lepidoptera

11.3.1.1. Pupal color polyphenism Lepidopteran pupae display a remarkable crypsis; this immobile life stage is especially vulnerable to predators, and pupae rely on hiding for survival (Baker, 1970; Hazel, 1977; Hazel *et al.*, 1998; Sims, 1983; West and Hazel, 1982, 1985; Wicklund, 1975). Most butterflies pupate in the soil or leaf litter, and pupae are generally constitutively brown (West and Hazel, 1979, 1996). However, a number of species crawl out of the leaf litter to pupate on the undersides of leaves or branches. Wandering larvae in these species encounter a variety of substrate "backgrounds," and effective pupal crypsis requires a facultative mechanism of pigment production (Hazel *et al.*, 1998; Hazel and West, 1996; Jones *et al.*, 2007; West and Hazel, 1979, 1982, 1996).

Over a century ago it was observed that some butterfly species switch between light and dark pupal forms (e.g., green vs. brown), depending on the background substrate (Merrifield and Poulton, 1899; Poulton, 1887; Wood, 1867). Today, numerous representatives of four lepidopteran families (Danaidae, Nymphalidae, Papilionidae, and Pieridae) are known to exhibit facultative pupal-color polyphenism. Depending on the specific habitats and pupation-substrates of each species, larvae couple pupal color production with exposure to a variety of external stimuli, such as photoperiod (Ishizaki and Kato, 1956; Sheppard, 1958; West *et al.*, 1972; Yamanaka *et al.*, 2004,

2007), relative humidity (Ishizaki and Kato, 1956; Smith, 1978), background color (Gardiner, 1974; Smith, 1978; Wicklund, 1972), temperature (Yamanaka *et al.*, 2009), and background texture and substrate shape/size/geometry (Hazel, 1977; Hazel and West, 1979; Sevastopulo, 1975).

Pupal color polyphenism appears to be regulated by a threshold (Hazel, 1977; Sims, 1983). Although a continuous range of pupal phenotypes are possible (West *et al.*, 1972), natural populations tend to be dimorphic for pupal color, and genetic studies indicate heritable differences among populations for the sensitivity of animals to background substrate characteristics (Hazel, 1977; Hazel and West, 1979; Sims, 1983).

Pre-pupal larvae pass through a "sensitive period" (West and Hazel, 1985) when environmental cues associated with pupation substrate influence the release of a neuroendocrine factor (Awiti and Hidaka, 1982; Bückmann and Maisch, 1987; Hidaka, 1961a,b; Smith, 1978, 1980; Starnecker and Bückmann, 1997). Interestingly, the effect of this factor differs among the lepidopteran families studied.

In Papilionidae, release of this factor (called "browning hormone") results in the production of a dark (brown) pupa; inhibition of release of this factor results in a light (green) pupa, as seen in *Papilio xuthus*, *P. polytes*, *P. demoleus*, *P. polyxenes*, *P. glaucus*, *P. troilus*, *Eurytides marcellus*, and *Battus philenor* (Awiti and Hidaka, 1982; Hidaka, 1961a, 1961b; Smith, 1978; Starnecker and Hazel, 1999). In contrast, in Nymphalidae (*Inachis io*), Pieridae (*Pieris brassicae*), and Danaidae (*Danaus chrysippus*), the default pupal color appears to be relatively dark (green), and release of the neuroendocrine factor (pupal melanization-reducing factor; PMRF) stimulates production of a light (yellow) pupa (Maisch and Bückmann, 1987; Ohtaki, 1960, 1963; Smith *et al.*, 1988; Starnecker, 1997). Thus secretion of the neuroendocrine factor has opposite effects on the relative darkness of pupae.

Starnecker and Hazel (1999) extracted the respective neuroendocrine factor from ventral nerve cords of both a nymphalid (*I. io*) and a papilionid (*P. polyxenes*), and injected these neurohormones into sensitive-stage larvae of both the appropriate and the opposite species. In all cases, animals responded to neuroendocrine injections in the species-appropriate way. For example, nymphalid larvae responded to injection of either PMRF (appropriate) or browning factor (inappropriate) by production of light colored pupae, whereas papilionid larvae responded to injection of these same substances by production of dark pupae. Cross-reactivity of these neuroendocrine factors suggests that they are the same factor, and indicates that nymphalid and papilionid butterflies may have independently evolved the capacity to facultatively regulate pupal color (Starnecker and Hazel, 1999). Interestingly, although these butterfly lineages appear to have co-opted the same neuroendocrine factor, they coupled it with downstream processes of pigment synthesis that are very different (Jones *et al.*, 2007; Starnecker and Hazel, 1999; **Figure 6**).

11.3.1.2. Seasonal wing-pattern polyphenism in butterflies A multitude of butterfly species display seasonal polyphenisms for wing pattern and color. Here we review two of the more common and better characterized of these forms of wing pattern polyphenism: light versus dark hindwings, and presence or absence of ventral forewing eyespots.

11.3.1.2.1. Light versus dark hindwings The most common form of butterfly wing polyphenism involves the overall lightness or darkness of wings. Numerous species in at least four Lepidopteran families are characterized by distinct seasonal light and dark forms, and where it has been studied, this seasonal variation in wing pigmentation results from larval sensitivity to both photoperiod and temperature (Aé, 1957; Endo, 1984; Endo and Funatsu, 1985; Endo and Kamata, 1985; Endo *et al.*, 1988; Hoffman, 1973, 1978; Jacobs and Watt, 1994; Kingsolver, 1987; Kingsolver and Wiesnarz, 1991; Koch and Bückmann, 1987; Müller, 1955, 1956; Nylin, 1992; Reinhardt, 1969; Shapiro, 1976; Smith,

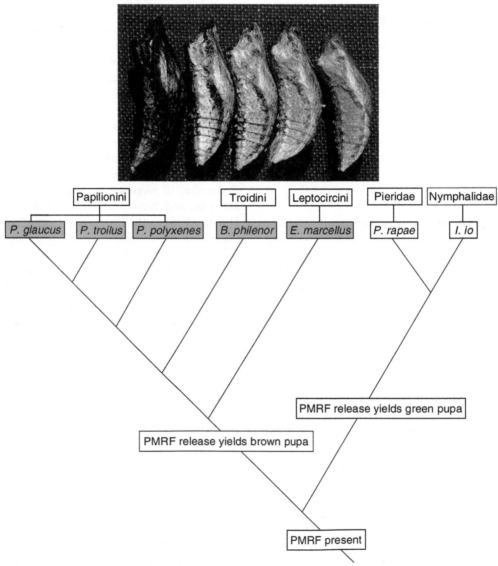

Figure 6 Color polyphenism in lepidopteran pupae. Pupal color polyphenism has arisen multiple times within the butterflies, such as green, orange, and orange-brown pupae of the swallowtail *Papilio xuthus*. (Reproduced with permission from *Yamanaka et al.*, 2004). Where studied, these mechanisms involve an endocrine signal (PMRF) that either darkens or lightens pupal color. A recent comparative study by *Jones et al.* (2007) showed that the Pieridae-Nymphalidae and the Papilionini-Troidini-Leptocircini lineage each evolved pupal color polyphenism independently through co-option of the same endocrine signal. Reproduced with permission from *Jones et al.* (2007). Photos courtesy of Wade Hazel.

1991; Süffert, 1924; Watt, 1968, 1969; Weismann, 1875).

Wing darkness is known to affect the solar absorption and thermal characteristics of butterflies (Jacobs and Watt, 1994; Kingsolver and Watt, 1983; Kingsolver and Wiesnarz, 1991; Watt, 1968, 1969), and natural selection for physiological performance under seasonally variable temperature regimes likely underlies many of these wing-darkness polyphenisms (Karl *et al.*, 2009; Kingsolver, 1987, 1995a,b; Kingsolver and Wiesnarz, 1991). Animals with dark wings are effective at absorbing solar radiation (Kingsolver, 1987; Kingsolver and Watt, 1983; Watt, 1968, 1969). These animals warm quickly, and perform well under cool conditions (e.g., spring), but these same animals are prone to lethal overheating under warmer (e.g., summer) conditions. Most butterfly species with light-dark polyphenisms produce dark forms when animals are reared under short days and cool temperatures (i.e., animals that will emerge in the spring), and lighter forms when they develop under longer days and warmer conditions.

A number of species have now been examined for endocrine regulation of wing melanism polyphenism: the lycaenid *Lycaena phlaeas* (Endo and Kamata, 1985), the papilionids *P. xuthus* (Endo and Funatsu, 1985; Endo *et al.*, 1985) and *P. glaucus* (Koch *et al.*, 2000), and the nymphalids *Polygonia c-aureum* (Endo, 1984; Endo *et al.*, 1988; Fukuda and Endo, 1966), *Araschnia levana* (Koch and Bückmann, 1985, 1987), and *Junonia (Precis) coenia* (Rountree and Nijhout, 1995).

Although these species represent three families, and they differ in the direction of the polyphenism (i.e., which morph is induced under each set of conditions) and in whether or not the polyphenism is tied with facultative induction of diapause, the underlying physiological mechanisms show many similarities. All of these polyphenisms respond to both photoperiod and temperature, and although each factor can influence wing pattern expression independently, the effects of these environmental variables are additive in all cases. All of these polyphenisms involve critical periods of hormone sensitivity that occur early in the pupal period, and all but one involve the relative timing of the rise in ecdysteroids: when ecdysteroid levels are low during the critical period (i.e., when the pupal peak in ecdysteroid levels starts *after* the polyphenism sensitive period), then the default wing morph develops (e.g., *L. phlaeas*: Endo and Kamata, 1985; *A. levana*: Koch and Bückmann, 1987; *P. coenia*: Rountree and Nijhout, 1995; and probably *P. c-aureum*: Endo *et al.*, 1988; see also Rountree and Nijhout, 1995). When the ecdysteroid pulse is induced to start earlier, then high levels of ecdysone are present during the critical period, and animals switch to production of the alternate morph (**Figure 7**).

The link between external stimulation (e.g., exposure to temperature) and the timing of ecdysteroid secretion involves the brain, and secretion of neurohormones — either PTTH (presumed for *A. levana*, Koch and Bückmann, 1987 and *P. coenia*, Rountree and Nijhout, 1995) or another neurohormone called summer morph producing hormone (SMPH; Endo and Funatsu, 1985; Endo and Kamata, 1985; Fukuda and Endo, 1966; Tanaka *et al.*, 2009), which has structural similarities to both bombyxin and small PTTH (Masaki *et al.*, 1988; Rountree and Nijhout, 1995). Extirpation of brains in animals prevented release of these neurohormones resulting in complete production of the default morph (Endo and Funatsu, 1985; Endo and Kamata, 1985; Endo *et al.*, 1988; Fukuda and Endo, 1966; Rountree and Nijhout, 1995). However, injection of ecdysteroids as these brainless animals pupate (i.e., prior to the critical period) can restore the polyphenism and switch animals to the alternate form (Endo and Kamata, 1985; Koch and Bückmann, 1987; Rountree and Nijhout, 1995).

(a)   Short days, Cool temperatures

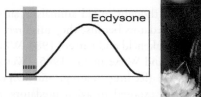

(b)   Long days, Warm temperatures

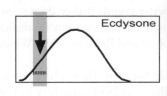

Figure 7 Seasonal wing color polyphenism in butterflies, as exemplified in *Junonia (Precis) coenia* represents the outcome of an interaction between allelic differences (genetic polymorphism for wing color) and environmental conditions. (a) Larvae exposed to a short-day photoperiod and low temperature regime secrete low levels of a small-PTTH-like neurohormone at the onset of the pupal phase. Consequently, the pupal ecdysteroid peak builds up late and reaches threshold levels only after the critical period (gray bar). This permits the expression of dark pigments in the developing wings which are, thus, adapted for higher solar absorption. (b) In contrast, exposure of larvae to a long-day photoperiod and high temperatures leads to an early and enhanced PTTH release, and consequently, above-threshold ecdysteroid levels during the critical period. These conditions favor the expression of the light colored, alternative wing phenotype. Based on data by *Masaki et al.* (1988), *Rountree and Nijhout* (1995), and other references cited in the text.

These results suggest the following model for light/dark polyphenism in butterflies (summarized for *P. coenia* in **Figure 7**): environmental stimuli affect the timing of ecdysteroid secretion indirectly via neurohormone secretion by the brain. Larvae exposed to warm temperatures and long days produce high levels of neurohormone. High neurohormone levels stimulate early secretion of ecdyseroids, and early secretion of ecdysteroids switches animals to production of the alternate (generally the lighter) morph.

11.3.1.2.2. Presence/absence of eyespots

Some butterfly species exhibit seasonal polyphenism for pattern elements, in addition to, or instead of, seasonal variation in overall levels of melanization (Brakefield and Larsen, 1984; Brakefield and Reitsma, 1991; Condamin, 1973; Roskam and Brakefield, 1999; Shapiro, 1976). A striking example occurs in the satyrid *Bicyclus anynana,* which exhibits a wet versus dry season polyphenism for presence (and size) of eyespots (Brakefield *et al.*, 1998; Brakefield and Reitsma, 1991; Kooi *et al.*, 1996; Windig *et al.*, 1994). Wet season animals express pronounced ventral forewing eyespots that are reduced or absent in dry season animals.

Ventral eyespots are thought to aid animals in surviving attacks by avian predators because they misdirect the aim of the predator (Brakefield and Larsen, 1984; Windig *et al.*, 1994; Wourms and Wassermann, 1985). Wet season animals are active (feeding, mating, and ovipositing), and are regularly exposed to avian predators, and it is presumed that these animals survive better with pronounced eyespots. Dry season animals are almost completely dormant, and escape predators through crypsis. These animals achieve better crypsis if they do not produce conspicuous markings, such as eyespots (Brakefield and Larsen, 1984; Brakefield and Reitsma, 1991; Windig *et al.*, 1994).

Eyespot polyphenism in Southern Hemisphere populations of *B. anynana* results from larval sensitivity to temperature (Brakefield and Mazotta, 1995; Brakefield and Reitsma, 1991; Kooi and Brakefield, 1999; Kooi *et al.*, 1994; Windig *et al.*, 1994). Late-stage larvae exposed to temperatures above 23°C produce the "wet season" pattern with large eyespots. Larvae exposed to temperatures below 19°C produce the "dry season" form with tiny or no eyespots (Brakefield *et al.*, 1998; Brakefield and Mazotta, 1995). Although natural populations of these butterflies are dimorphic, this developmental mechanism (like the light/dark polyphenism described in the previous section, incidentally) does *not* seem to involve a threshold: intermediate temperatures lead to intermediate eyespot sizes, and animals reared in the laboratory can be induced to produce a continuous range of wing patterns (Windig *et al.*, 1994). Instead, dimorphism appears to arise from distinct seasonal temperature regimes encountered by sequential generations of larvae in the wild (Oostra *et al.*, 2010; Windig *et al.*, 1994).

Temperature-induced variation in eyespot size appears to be mediated by the timing of the pupal pulse of ecdysteroids (Brakefield *et al.*, 1998; Koch *et al.*, 1996; Oostra *et al.*, 2010; Zijlstra *et al.*, 2004). Just as with the light/dark polyphenism in *A. levana* and *P. coenia*, animals have a hormone-sensitive period during the first few days of the pupal period. The default developmental pathway appears to be a pupal pulse of ecdysteroids that starts *after* this critical period. When ecdysteroid levels are low during the sensitive period, no eyespots form (the dry season form). Wet season (warm-temperature) animals have an earlier pulse of ecdysteroids that precedes the sensitive period, and presence of ecdysone during this period results in the production of large eyespots (Brakefield *et al.*, 1998; Koch *et al.*, 1996; Oostra *et al.*, 2010).

Genetic strains of *B. anynana* that have been selected for continuous expression of the no-eyespot morph have late pulses of ecdysteroids identical to those of the cool-temperature-induced dry season animals (Brakefield *et al.*, 1998; Koch *et al.*, 1996). However, these genetically eyespot-less animals can be induced to produce eyespots by injections of ecdysone prior to the sensitive period (Koch *et al.*, 1996). These results all point to a mechanism where polyphenic expression of butterfly eyespots results from a coupling of larval exposure to temperature with variation in the timing of secretion of ecdysone during the first 2 days of the pupal period (**Figure 8**), a model confirmed recently by ecdysteroid titer measures across a range of developmental temperatures (Oostra *et al.*, 2010).

Butterfly eyespots have served as a model system for characterizing the developmental control of pattern formation (Beldade and Brakefield, 2002; Brunetti *et al.*, 2001; French, 1997; French and Brakefield, 1995; Kühn and von Engelhardt, 1933; McMillan *et al.*, 2002; Nijhout, 1985, 1986, 1991; Wittkopp and Beldade, 2008), and recent genetic experiments provide exciting glimpses into the patterns of gene expression that underlie eyespot formation in general, and temperature-sensitive modulation of eyespot size in particular.

Butterfly eyespots result from a patterning mechanism that takes place in the wings of late larval and early stage pupae (Beldade and Brakefield, 2002; Brunetti *et al.*, 2001; French, 1997; French and Brakefield, 1995; Nijhout, 1985, 1986, 1991). Briefly, the eyespot consists of a series of concentric pigmented rings around an organizing center, or focus. Cells in the focus of an eyespot are critical for normal formation of the eyespot; if these cells are ablated at the end of the larval period, eyespots fail to form (Nijhout, 1980, 1991; Nijhout and Grunert, 1988). Likewise, if foci cells are transplanted, they can induce ectopic eyespots in other parts of the developing wing (Brakefield *et al.*, 1996; French and Brakefield,

(a) Dry Season, Cool temperatures

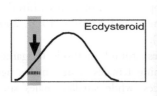

(b) Wet Season, Warm temperatures

Figure 8 Eyespot size polyphenism as exemplified in the tropical butterfly *Bicyclus anynana*. (a) The less active dry-season (low temperature) morph exhibits small eyespots on the ventral side of the forewings as a result of a low ecdysteroid titer during the critical period (shaded bar) at the onset of pupal development. (b) In contrast, the forewings of the wet-season (high temperature) morph develop large eyespots in response to an above-threshold ecdysteroid titer during this critical period. In the wing discs of last instar larvae, eyespot focal cells express the Distal-less protein. The earlier pupal ecdysteroid peak causes enhanced synthesis of dark pigment from this focal center and, consequently, generates large eyespots in the wet-season morph. Based on data and models by *Brakefield and Reitsma* (1991), *Kooi et al.* (1994), *Brakefield et al.* (1996, 1998), *Koch et al.* (1996, 2003), *Brunetti et al.* (2001), and *Beldade and Brakefield* (2002). For further references, see the text. Photos courtesy William Piel and Antonia Monteiro.

1995; Nijhout, 1991). It appears that foci establish the eyespot by secreting a diffusible chemical morphogen into the surrounding epidermis, and this signal is interpreted by the surrounding cells in a way that affects the color of the pigment synthesized; gradient contours in the diffusible substance interact with the relative sensitivities of the surrounding cells to generate the concentric rings of the eyespot (Dilao and Sainhas, 2004; Evans and Marcus, 2006).

Although the identity of the morphogen released from eyespot foci is not yet known, recent genetic studies using antibodies for *Drosophila* patterning genes have identified several signaling molecules and transcription factors expressed in the appropriate areas and at the appropriate times (Beldade and Brakefield, 2002; Brakefield *et al.*, 1996; Brunetti *et al.*, 2001; Carlin *et al.*, 1994; Keys *et al.*, 1999; Weatherbee *et al.*, 1999). For example, *distal-less*, *engrailed* and *spalt* all have circular regions of expression

that correspond with the position of eyespots (Beldade *et al.*, 2002; Brakefield *et al.*, 1996; Brunetti *et al.*, 2001; Keys *et al.*, 1999; Monteiro *et al.*, 2006) and their expression occurs at the time of eyespot pattern formation (late larval and early pupal period). All three genes are expressed together in the organizing focus of the eyespot, but they are expressed separately in the surrounding color rings, lending credence to the idea that they may in some way contribute to the patterning of the rings themselves (Beldade and Brakefield, 2002; Brunetti *et al.*, 2001).

Brakefield and colleagues have taken the next step by linking expression patterns of these genes — *distal-less* in particular — with quantitative variation in eyespot size. Genetic lines selected for large and small eyespots differ in the relative size of the eyespot focus, as measured by the region of *distal-less* expression (Beldade and Brakefield, 2002; Beldade *et al.*, 2002; Monteiro *et al.*, 1994, 1997), and these authors then showed that genetic variants of *distal-less* expression co-segregate with inter-individual variation in eyespot size.

The *wingless* gene is also expressed at appropriate times and locations in eyespot centers (Monteiro *et al.*, 2006). However, it is not yet clear how these gene products interact with each other during eyespot formation (Evans and Marcus, 2006; Saenko *et al.*, 2007) or how they regulate pigment synthesis (Koch *et al.*, 2000), and it is already apparent that different eyespots can utilize different combinations of these patterning elements (e.g., eyespots in Pieris spp. are patterned by different genes than eyespots in Nymphalid spp.; Monteiro *et al.*, 2006). Extensive genomic tools have now been developed for *Bicyclus*, including high-density genetic arrays, dense genetic maps, and genetic transformation techniques (Beldade *et al.*, 2006, 2008; Marcus *et al.*, 2004), which will facilitate further investigation of these patterning mechanisms.

Combined, these studies suggest an explicit model for the developmental basis of variation in eyespot size. According to Beldade and Brakefield (2002), differences in the relative amount of growth of the organizing focus lead to differences in the size of the focus at the time of pattern formation. This translates into the strength of the signal emitted from this organizing center and results in relatively larger or smaller eyespot diameters. Plasticity in the expression of eyespot size could then result from a coupling of the amount of growth of the organizing focus with the amount of ecdysteroid present during the patterning period (Koch *et al.*, 2003). Wet season (higher temperature) animals would have early rises in pupal ecdysteroid levels, stimulating growth of large eyespot foci, resulting in correspondingly large adult eyespots. Dry season (lower temperature) animals would have later rises in pupal ecdysteroid levels, less hormone present during the sensitive period, less growth of the eyespot foci, and relatively smaller final eyespots (**Figure 8**).

Interestingly, not all eyespots are plastic — even within the same individual. In *B. anynana* ventral forewing eyespots are exquisitely sensitive to the larval thermal environment, but dorsal eyespots are not. Dorsal eyespots are expressed in all individuals and display no detectable plasticity. Brakefield *et al.* (1998) proposed that this difference in the environmental sensitivity of expression of dorsal and ventral eyespots results from the presence of ecdysteroid receptors in the foci of ventral eyespots, and the absence of these same receptors in the focal cells of the dorsal eyespots. Thus, patterns of ecdysone receptor expression may underlie the coupling of an adult wing pattern element expression (eyespots) with seasonally variable components of the larval environment (temperature).

11.3.2. Coleoptera — Male Dimorphism for Weaponry

Males of many species face intense competition from rival males over access to reproduction (Andersson, 1994; Darwin, 1871; Thornhill and Alcock, 1983). In these species, it is not uncommon for dominant and subordinate individuals to adopt different behavioral tactics: dominant (generally large) males fight to guard territories frequented by females or display or sing to attract females, while subordinate (generally smaller) males of these same species adopt less aggressive alternative tactics, such as sneaking up to or mimicking females (Austad, 1984; Darwin, 1871; Dominey, 1984; Gross, 1996; Iguchi, 1998; Oliveira *et al.*, 2008; Shuster, 2002).

In the most extreme cases, large and small males differ in morphology as well as reproductive behavior. Rarely, these alternative male "morphs" result from allelic differences among the male types (Lank *et al.*, 1995; Ryan *et al.*, 1990; Shuster, 1989; Zimmerer and Kallmann, 1989). Instead, the majority of male dimorphisms appear to result from polyphenic developmental processes that switch among alternative phenotypic possibilities depending on larval growth, and/or the resulting body size attained by each individual (Clark, 1997; Eberhard, 1982; Goldsmith, 1985, 1987; Kukuk, 1966; Rasmussen, 1994; Tomkins, 1999). Males encountering favorable conditions grow large and produce one morphology, while genetically similar (e.g., sibling) individuals encountering poor conditions remain small and produce an alternative morphology. These male-dimorphic polyphenisms are characterized by a relatively abrupt switch between morphs that corresponds with a critical, or threshold, body size (Cook, 1987; Danforth, 1991; Diakonov, 1925; Eberhard and Gutierrez, 1991; Emlen *et al.*, 2005a; Iguchi, 1998; Kawano, 1995; Kukuk, 1966; Rasmussen, 1994; Tomkins and Simmons, 1996).

Only one male-dimorphism that we are aware of has been characterized physiologically — the alternative horn morphologies of the dung beetle *Onthophagus*

taurus (Coleoptera: Scarabaeidae). *Onthophagus taurus* is a European species that has been introduced into both Australia and the United States, where it is now an abundant inhabitant of horse and cow manure (Fincher and Woodruff, 1975; Tyndale-Biscoe, 1996).

Beetles fly into fresh manure pads and excavate tunnels into the soil below. Females dig the primary tunnels and spend a period of days pulling fragments of dung to the ends of these tunnels, where they pack them into oval masses called "brood balls" (Emlen, 1997; Fabre, 1899; Hallffter and Edmonds, 1982; Moczek and Emlen, 2000). A single egg is laid inside each brood ball, and larvae complete their development in isolation within these buried balls of dung (Emlen and Nijhout, 1999, 2001; Fabre, 1899; Main, 1922).

Male behavior revolves around methods of gaining entry to tunnels containing females. Large males fight to guard tunnel entrances, while smaller males sneak into these tunnels on the sly (Emlen, 1997; Moczek and Emlen, 2000). Large males produce a pair of long, curved horns that aid them in contests over tunnel occupancy (Moczek and Emlen, 1999). Smaller males dispense with horn production altogether.

Both overall body size and male horn length are sensitive to the larval nutritional environment (Emlen, 1994; Hunt and Simmons, 1997; Moczek and Emlen, 1999). Specifically, the amount and quality of larval food predictably influence the final body size and horn lengths of males (Emlen, 1994; Hunt and Simmons, 1997; Moczek and Emlen, 1999). Controlled breeding experiments, artificial selection experiments, population comparisons, and "common garden" experiments all suggest that horn expression is regulated by a threshold mechanism, and that natural populations contain measurable levels of additive genetic variation for the body size threshold, that is, the size associated with the switch between horned and hornless morphologies (Emlen, 1996; Moczek *et al.*, 2002). Somehow, then, the relative amount of growth of the horns must be influenced by the overall growth, or body size, attained by each animal. Specifically, this polyphenism appears to involve a reprogramming of animals that fall beneath a genetically mediated threshold body size, so that in these animals, growth of the horns is reduced (Emlen *et al.*, 2005a, 2006; Moczek, 2006, 2007).

Beetles pass through three larval instars before molting into a pupa (Emlen and Nijhout, 1999; Main, 1922). Larvae feed on dung supplies that are provided in the brood ball, and gain weight steadily. When these food supplies are depleted, a stereotyped series of events is commenced and this ultimately results in the metamorphic molt from larva to pupa (Emlen and Nijhout, 1999). The cessation of feeding appears to trigger a rapid drop in JH titers analogous to the drop that occurs with attainment of a critical size for metamorphosis in *Manduca sexta* (Nijhout, 1994). As animals stop feeding, they begin to purge their gut, they begin forming a mud/fecal shell that will protect them as pupae, and the imaginal structures (legs, wings, genitalia, horns) begin to grow.

A combination of hormone-application perturbation experiments (JH) and radioimmunoassays of hormone titer profiles (ecdysteroids) revealed two critical periods relevant to this developmental polyphenism. Perturbation of hormone levels during either of these sensitive periods influences the expression of male horns, although the effects of hormone application at these two times are distinct (Emlen et al., 2005b; Emlen and Nijhout, 1999, 2001). The first of these critical periods occurs at the end of the feeding period, before initiation of the metamorphic molt. Animals at this time have attained their largest sizes, and larval weight during this period exactly predicts patterns of male horn expression. Male larvae with sustained weights equal to or heavier than 0.12 g end up producing horns, whereas male larvae not sustaining this critical weight do not produce horns (Emlen and Nijhout, 2001). Furthermore, male larvae with weights beneath this critical size have a small pulse of ecdysteroids that is not present in larger males. These results suggest that body size is assessed at this time, and that animals beneath a critical larval weight are "reprogrammed" by a morph-specific pulse of ecdysteroids (also see Moczek, 2006). Levels of JH appear to be involved in this size-assessment process. Topical application of the JH analog methoprene during this critical period raised the threshold body size associated with horn expression, so that methoprene-treated animals needed to attain a larger body size (heavier larval weight) for horn production than acetone-treated control animals (Emlen and Nijhout, 2001).

A second critical period occurs after the metamorphic molt has been initiated, as animals purge their guts and enter the pre-pupa period. This is the period of horn growth. All of the imaginal structures, including the horns, are growing rapidly at this time, and all animals have very high levels of ecdysteroids irrespective of sex or morph (Emlen and Nijhout, 2001). Here again levels of JH appear correlated with body size, and perturbation of levels of JH influence the relative amount of growth of imaginal structures. In this case, augmentation of levels of JH (by topical application of methoprene) caused animals to produce disproportionately large horns. Specifically, it caused small, typically hornless males to produce horns (Emlen and Nijhout, 1999; Moczek et al., 2002).

In summary, experiments suggest that there are at least two periods during larval development when horn expression is sensitive to endocrine events. A first critical period occurs as animals attain their largest body sizes, and animals not attaining (or sustaining) a threshold size get reprogrammed by a small, morph-specific pulse of ecdysteroids at this time. Ecdysteroids affect patterns of gene transcription and are known to reprogram the developmental fates of specific structures (e.g., epidermis,

imaginal discs; reviewed in Nijhout, 1994). In this case, a pulse of ecdysteroids may reprogram the relative sensitivity of cells destined to become the horns, so that their growth is inhibited during the pre-pupal period when all of the imaginal structures grow to their full sizes.

Advances have now been made in understanding the genetic mechanisms underlying horn growth, such as how these structures arose and how they were subsequently modified in form as species in this genus diversified. The origin of beetle horns appears to have entailed the co-option of portions of traditional appendage patterning, in particular components of the network of genes responsible for patterning the proximo-distal axis (Emlen et al., 2006, 2007; Moczek and Nagy, 2005; Moczek and Rose, 2009; Moczek et al., 2006). Yet beetle horns are fantastically diverse, exhibiting stunning variety in shapes, sizes and types (e.g., head versus thorax). A phylogenetic analysis of 48 species of the genus Onthophagus revealed multiple origins of horns as well as losses (Emlen et al., 2005a,b). These structures clearly have been gained and lost repeatedly in the history of this genus, a pattern reflected in emerging studies of horn development: it is already evident that horns in different Onthophagus species — and even different horn types in the same species — utilize different subsets of the appendage patterning network in the regulation of their growth (Moczek and Nagy, 2005; Moczek and Rose, 2009; Moczek et al., 2006).

Polyphenic regulation of horn expression (horn dimorphism) also has been gained and lost repeatedly in these beetles, so that closely related species may differ in whether their horns are dimorphic and in the nature of their dimorphism (Emlen et al., 2005b). It has even been suggested that multiple threshold mechanisms may operate simultaneously on the same horn to yield populations with three distinct male types (facultative male trimorphism; Rowland and Emlen, 2009). This pattern of rapid and prolific evolution of polyphenic mechanisms is reflected in preliminary studies of their underlying developmental regulation. There clearly are several different ways that horn growth can be truncated in small males (and females). The most obvious entails the prevention of proliferation of horns in the first place — a process that occurs in the head horns of O. taurus, and to some extent, in the thoracic horns of O. nigriventris (Emlen et al., 2006). Preliminary studies of transcription of the O. nigriventris insulin receptor gene (InR) suggest that localized interruption of insulin signaling in the horn tissues may contribute to the truncation of horn cell proliferation in small males and females (Emlen et al., 2006; Lavine et al., unpublished results).

It is also possible for dimorphism to arise after this period of growth during the pupal period when epidermal tissues are remodeled prior to the formation of the adult cuticle (**Figure 9**). Moczek showed that significant remodeling of beetle horns can occur during this period,

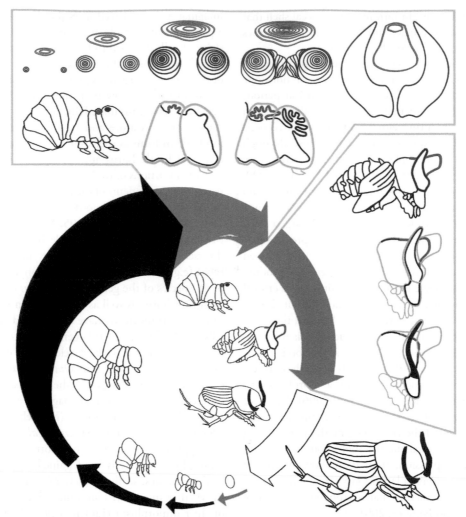

Figure 9 Polyphenic regulation of beetle horns can arise during either of two stages of horn development. Horns grow during a brief, non-feeding, period at the end of the third (final) larval instar and the pre-pupal period, and this process can be truncated to produce hornless individuals. After animals molt into pupae, the epidermal cells begin to produce the adult cuticle. At this time, significant remodeling of horn tissues can occur, probably through localized patterns of cell death. Dimorphism in horn expression can be generated or erased, depending on the amount of horn tissue loss that occurs in large males, small males, and females, respectively. Life cycle for *Onthophagus taurus* shown with arrow size roughly corresponding to animal size, and black arrows indicating feeding periods. In this species, thoracic horns (green) are grown during the pre-pupal period and removed completely in all individuals during the pupal period. Head horns (blue) proliferate in large males, but not in small males or females, and undergo only minor remodeling during the pupal period. Profile drawings modified from *Moczek* (2006). Reproduced with permission from *Emlen et al.* (2007).

and that this could generate dimorphism in horns that initially grow in all individuals of both sexes, which are subsequently reabsorbed in small males and females (Moczek, 2007; Wasik *et al.*, 2010). From this it is abundantly clear that the regulatory mechanisms are diverse, with either or both of these two processes occurring in each horn type; across different horn types (even within the same species), the mechanisms responsible for facultative horn loss vary. It appears that endocrine events near the end of the larval feeding period specify either a horned or a hornless trajectory for subsequent development, and that these regulatory events are implemented either during the pre-pupal period, through differential

amounts of horn tissue proliferation, or during the pupal period, through differential amounts of horn tissue reabsorption, to generate horned and hornless adult beetle morphologies.

Genomic tools are being developed for the dung beetle species *O. taurus* (Snell-Rood *et al.*, 2010) and *O. nigriventris* (Snell-Rood *et al.*, 2010; Warren, Lavine and Emlen unpublished results) as well as for the rhinoceros beetle *Trypoxylus dichotomus* (Warren, Lavine, and Emlen unpublished results), and these should permit the identification of suites of genes whose expression differs between horned and hornless morphs. Just this year, Snell-Rood *et al.* (2010) found that expression profiles for horn tissues in hornless males of both *O. taurus* and *O. nigriventris* were more similar to the profiles of females than they were to those of larger, horned males (**Figure 10**). These exciting results suggest that the magnitude of developmental reprogramming involved in male dimorphism rivals that which occurs between the sexes. They also are consistent with the observation that small, hornless males appear to have converged on a body morphology similar to that of the females.

11.3.3. Hymenoptera — Caste Polyphenism and Division of Labor

Except for termite caste polyphenism, most of the examples discussed so far can easily be accommodated within the framework of classic Darwinian fitness concepts, wherein alternative phenotypes prove to be intimately related to alternative reproductive strategies. In principle, each of the selective environments encountered by insects

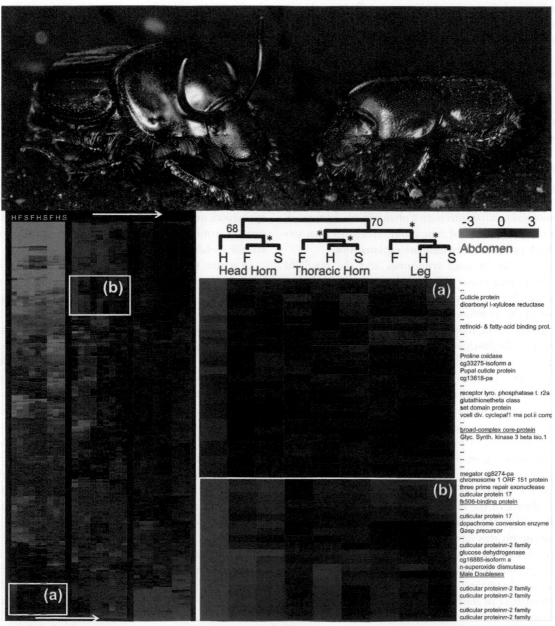

Figure 10 Polyphenic expression of horns in the dung beetle *Onthophagus taurus*. (a) Depending on the larval nutritional environment, males either develop into large adults with a pair of long, curved head horns (right), or into smaller adults with only rudimentary horns (left). (Photo courtesy of D. Emlen.) (b) Differences in patterns of horn growth are reflected by differential transcription of genes. Expression profiles for head tissues of small, hornless males (S) more closely resemble comparable tissues of females (F), than they do those of large, horned males (H). Photos D. Emlen; bottom panel reproduced with permission from *Snell-Rood et al.* (2010).

(e.g., spring vs. summer) favors a different optimal phenotype, and individuals maximize their reproductive success by expressing the appropriate morphology in the appropriate circumstance.

Caste polyphenisms, in contrast, are associated with a severe reduction in fertility of the subdominant (worker) caste. Workers in most social insect colonies never reproduce. Social insect societies with their complex caste systems, thus, do not obey the rules of classical Darwinian theory, a fact which has clearly and succinctly been stated by Darwin (1859). The dilemma of inserting social insect castes into a fully acceptable Darwinian framework was only resolved by the introduction of the theory of inclusive fitness and kin selection (Hamilton, 1964). Since then, a great deal of work has centered on ultimate (evolutionary) explanations of why individuals should refrain from reproducing. Numerous studies also address proximate (ontogenetic) mechanisms underlying developmental trajectories into reproductive individuals and non- or less-reproductive helpers (workers). A major issue that remains to be elucidated is the evolution of these proximate mechanisms, that is, the evolution of developmental pathways that generate the distinct queen/worker and also the soldier phenotypes. Only if it becomes possible to map these physiological mechanisms onto a phylogenetic framework may a unifying picture of caste evolution emerge that satisfies both proximate and ultimate explanations.

The diversity in caste syndromes and the manifold stimuli that trigger caste differentiation remain a challenge to any unifying view of developmental regulation. In part, the root of this problem lies in the conceptual framework of the caste. In the most general sense, the term caste is used to describe *functional* roles in reproduction and division of labor, such as the performance of different tasks by different members of an insect society, independent of whether these roles are a reflection of behavioral differences only, or whether there are also differences in morphological phenotype. In a more restricted sense, a caste is seen as a manifestation of pre-imaginal developmental diversification resulting in *morphologically distinct phenotypes*, which are then ever more prone to perform distinct functions in the division of labor. Distinct caste phenotypes are a hallmark of the highly eusocial insect societies, and in a discussion of insect polyphenism one might tend to concentrate on this latter aspect. Yet, in terms of evolutionary explanations, that is, how such morphologically distinct phenotypes may have emerged from modifications in developmental physiologies and gene expression, it is informative to start with incipient social systems that arise from individual differences in reproductive potential. In the subsequent sections, we will try to discern these two aspects of caste and the underlying endocrine mechanisms that govern behavioral and morphological diversification.

A first and major distinction in social insect organization and caste systems sets apart the hemimetabolous termites from the holometabolous Hymenoptera (wasps, bees, and ants). As previously outlined, termites are diplo-diploid hemimetabolans that descended from presocial cockroaches, and caste development in termites is essentially a problem of how endocrine regulation of post-embryonic development maintains immatures as a worker caste while permitting terminal differentiation of a soldier and a reproductive caste. Sociality in the Hymenoptera, in contrast, is built on asymmetries of genetic relationships generated by the haplo-diploid system of sex determination (Hamilton, 1964), and arose several times independently. Multiple evolutionary origins of sociality make hymenopterans, and in particular wasps and bees, useful objects to study the environmental, genetic, and endocrine background that set the stage for the development of the exclusively female caste phenotypes, including the primary ones — the queens and workers. Ants, in contrast, are all highly social, and thus are less ideally suited for comparative studies with such a focus. They are most valuable, however, when it comes to testing hypotheses on the evolution of multiple sterile female caste phenotypes, in particular, worker versus soldier development. It is important to emphasize at this point, that the soldier caste in ants is a highly derived phenotype that makes its appearance only in a few genera, in contrast to the soldier caste in termites, which is one of the essential features of termite sociality.

In general terms, holometabolous caste phenotypes can be seen as a progressive fixation of roles, first in reproduction, and secondarily in the specialization to specific tasks related to colony maintenance (Wilson, 1971). Evolutionarily, the monopolization of colony reproduction by a queen caste is a conflict-ridden situation that starts out with asymmetries in reproductive potential, either of females that are cofoundresses of a nest, or results from the repression of the reproductive potential of offspring daughters by the mother who persists in the nest. In the first case reproductive dominance should become established through a series of dominance interactions between adult and potentially reproductive females, whereas in the latter case, which requires an overlap of generations, the dominant egg-laying mother may repress reproductive activity in her daughters by manipulating a pre-imaginal caste bias.

11.3.3.1. Wasps As stated by West-Eberhard (1996): "wasps are a microcosm for the study of development and evolution" of insect sociality, and they serve as a logical baseline for endocrine studies of caste polyphenism. All social Hymenoptera evolved from wasps (Wilson, 1971), and most levels of sociality can be found within extant wasp taxa. Three of the subfamilies in the Vespidae, the Stenogastrinae, Polistinae, and Vespinae contain social species (Carpenter, 1991; Pickett and Carpenter, 2010) with the Polistinae being of special interest since they represent practically the entire range of social evolution. Many of the Old World Polistinae are open-nesting wasps, and colonies can be founded by one or more mated females. In such incipient colonies, a dominance hierarchy is gradually established by aggressive interactions among adult females (Sledge *et al.*, 2001). Thereafter, the dominant female signals her status to the subordinate females (Sledge *et al.*, 2001), and whenever she encounters an egg laid by another female she removes it. In such a system, foregoing reproduction with the perspective to eventually substitute the dominant female and taking over the nest has a relatively high value in the pay-off matrix, since survival chances of individual nest-founding females generally are quite low.

The endocrine basis of this reproductive dominance has initially been investigated in *Polistes dominulus* (formerly *P. gallicus*). These studies revealed a synergistic interaction between JH synthesis, ecdysteroid titer, and ovarian activity. Dominant females have higher CA activity and consequently a higher JH titer than subordinate females. In addition, the higher ovarian activity in dominant females also correlates with an elevated ecdysteroid titer. JH and/or ecdysteroid treatment of subordinate females resulted in a rise in their social rank (Barth *et al.*, 1975; Röseler *et al.*, 1984, 1985; Strambi, 1990). Such an interaction in hormonal control of physiology and behavior is a general trait in insect reproduction, and should be expected also in solitary or incipiently social Hymenoptera. These studies, however, also pointed out an important gap in our comprehension of reproductive dominance, that is, the distinction between the presence of an activated ovary and high social rank (Sledge *et al.*, 2001). Even when ovariectomized, some *Polistes* foundress females continued to maintain their dominant status (Röseler *et al.*, 1985). Generally, these were females with large CA. A large CA volume (and thus presumed high CA activity), however, was not consistently linked to dominance, since in other cases formerly subordinate females that became dominant over ovariectomized foundresses did not have larger CA. Recent results shed light on the multifactorial interplay in conflict and its resolution among *P. dominulus* females. Information on fighting ability is conveyed by facial signals that allow individual identification (Tibbetts and Huang, 2010), whereas cuticular hydrocarbon profiles indicate reproductive status (Izzo *et al.*, 2010). This multisignal situation is also reflected in JH titers. After queen removal, the JH titers are upregulated in workers as a response to this social stress and aggression-rich situation, whereas there is no relation between JH titer and aggression in queen-right colonies (Tibbetts and Huang, 2010). Furthermore, JH appears to play a role in division of labor among workers of stable colonies, advancing the onset of foraging activity in the lifetime of individual wasps (Tibbetts and Izzo, 2009).

With increasing social complexity the choice between alternative reproductive strategies evidently becomes restricted, and in parallel, morphological caste differences become implemented and increase in degree. This is nicely illustrated in the Ropalidiini, where a gradual shift toward a preimaginal caste bias has been convincingly demonstrated (Gadagkar *et al.*, 1988), together with an age-dependent pattern of task performance (temporal or age polyethism) in adult workers (Naug and Gadagkar, 1998).

In the Neotropical, swarm-founding Epiponini, morphological caste phenotypes are clearly expressed ranging from subtle differences between the extremes in a unimodal body size range, to clearly dimorphic castes resulting from non-isometric growth in the preimaginal phase. In the latter, queens are not necessarily always the largest individuals. Allometric analyses suggest size-independent allometries in many of these cases as seen by pre-imaginal caste determination marked by a resetting of growth parameters in critical phases of development (Jeanne *et al.*, 1995; Keeping, 2002; Noll *et al.*, 1997; O'Donnell, 1998). Furthermore, these analyses showed that the caste system in most genera of Epiponini appears to have originated from an ancestral lineage that did not have a queen caste (Noll and Wenzel, 2008).

Neither the Ropalidiini nor the Epiponini have been investigated regarding endocrine system functions in caste development, and only in two species has hormonal control in division of labor among the adult wasps been addressed. The results are of interest as they indicate a split in reaction patterns. In the primitively social *Ropalidia marginata*, an elevated JH titer stimulated egg development (Agrahari and Gadagkar, 2003), whereas in the socially more advanced *Polybia occidentalis*, treatment with the JH analog methoprene accelerated behavioral development in workers (O'Donnell and Jeanne, 1993), similar to what is observed in honey bees.

As in other insects, genomic studies are also gaining momentum in wasps, where a large EST library for *P. metricus* furthered the establishment of a platform for high throughput gene expression analyses (Toth *et al.*, 2007). Brain gene expression of workers that were rearing brood is similar to that of foundresses (that also rear brood) and markedly differs from queens. Among the differentially expressed genes were several related to the insulin signaling pathway, inferring a strong connection between nutritional and reproductive regulation in *Polistes* castes (Toth *et al.*, 2007). A second link of interest in developmental regulation of caste comes from a candidate gene analysis and proteomic screen showing that prospective queen and worker larvae differ in the expression pattern of diapause-related genes (Hunt *et al.*, 2010). The potential to become a reproductive (gyne) among *Polistes* female offspring, which becomes established during larval development, thus appears to be contingent on a diapause-related gene network. The latter represents a facultatively expressed trait in the life history of ancestral non-social wasps (Hunt et al., 2007b).

11.3.3.2. Ants   Like their hemimetabolan counterparts, the termites, ants are an apex in caste complexity, and thus, also a challenge to any unifying hypothesis on caste development and function. In this section we provide only a glimpse into the ant empire of behavioral and developmental diversity. Luckily, we can direct any reader interested in a more detailed presentation on ant biology to the masterpiece of scientific literature written by Hölldobler and Wilson (1990).

Prior to any further discussion it is important to emphasize that there are no solitary or primitively social ants. Possibly related to a major switch in foraging biology from flying to walking in search of prey, a mesozoic sphecoid wasp-like ancestor to ants appears to have been exposed to relaxed selective constraints on the aerodynamics of body shape, making possible a wide variation in worker ant morphologies. This resulted in an eminent wing polyphenism between queens and workers, which is evidenced already in the earliest ant fossils (Wilson, 1987). Many early studies on ant sociobiology capitalized on such variation in form, analyzing size allometries in relation to functions performed by colony members and in relation to phylogenetic patterns (reviewed in Hölldobler and Wilson, 1990; Wheeler, 1991; Wilson, 1971). Even though there is a wealth of data on complex trait allometries in ants, there still appears to be a lack of information linking these data to rules of *how* changes in shape are actually generated in development, and how these specific developmental pathways have evolved (Tschinkel, 1991; Tschinkel et al., 2003).

11.3.3.2.1. The Queen/worker decision   As in all highly social insects, ant queens differ markedly from workers. They are generally bigger, have a fully developed reproductive system, and have a well-developed flight apparatus for dispersal. Even though this queen/worker distinction is an ancestral character present in all ant species, we still have very limited knowledge as to how these caste differences arise ontogenetically. Many of the early studies on mechanisms governing queen production in ants were carried out on temperate climate species, which showed strong seasonal triggers for queen production (reviewed in Wheeler, 1986). In many species, only queens are produced from overwintered eggs or larvae, illustrating a requirement for a state of diapause in the development of queens.

For some ants, the critical period for the queen/worker decision has been studied by JH application experiments. For example, topical application of JH to larvae developing from queen-biased overwintered brood in *Myrmica rubra* increased the number of queens (Brian, 1974). In *Aphaenogaster senilis*, application of JH caused the appearance of queen-like workers (Ledoux, 1976), and this same pattern occurred in the Argentinian fire ant, *Solenopsis invicta* (Vinson and Robeau, 1974). In the latter species, metamorphosis was delayed significantly by the application of JH, and the "queen-like" appearance of workers was initially attributed to a simple increase in worker size (associated with the delay in metamorphosis, and extended period of larval feeding), rather than to a direct effect of JH on queen production (Wheeler, 1990).

Ecdysteroids also have been implicated in divergent queen/worker development. In *Plagiolepis pygmea*, worker-biased larvae have higher ecdysteroid titers during the last larval instar than queen-biased larvae (Suzzoni et al., 1983).

In all examples cited so far caste development is dependent on environmental factors. Yet, there is now increasing evidence for a genetic basis to the determination of queen/worker polyphenism in at least some species of ants. Such genetic factors (allelic differences between queens and workers) were originally proposed for the slave-making ant *Harpagoxenus sublaevis* (Winter and Buschinger, 1986), but have recently also been demonstrated for a *Camponotus* (Fraser et al., 2000), two *Pogonomyrmex* species (Julian et al., 2002; Volny and Gordon, 2002), and the very peculiar sex/caste determination system in the little fire ant *Wasmannia auropunctata* (Foucaud et al., 2010). A yet unresolved question is how the allelic combinations setting up the genetic basis for caste interact with the endocrine system.

Environmental triggers of ant caste development primarily involve pheromonal signals within the colony. Pheromonal regulation of ant reproduction has primarily been studied in *S. invicta*, which has completely sterile workers due to the lack of functional ovaries (Hölldobler and Wilson, 1990). Virgin *Solenopsis* queens normally leave to undertake a mating flight and thereafter shed their wings, activate their ovaries, and soon initiate egg laying. If prevented from flying, these gynes keep their wings and maintain their ovaries in an inactive state. Furthermore, gynes are prevented from maturing their ovaries as long as an egg-laying queen is present in the nest (Fletcher and Blum, 1981). This integrated behavioral and physiological response in *Solenopsis* virgin queens has led to the identification of a pheromone produced by the poison gland of the dominant queen, which inhibits precocious dealation and ovary activation in virgin queens as long as they are in the nest (Fletcher and Blum, 1983). The pheromone supposedly maintains high brain dopamine levels (Boulay et al., 2001) that are thought to suppress CA activity (Burns et al., 2002; Vargo and Laurel, 1994). Low CA activity in turn prevents vitellogenin uptake by the ovary. Vitellogenin synthesis appears to occur independent of JH (Vargo and Laurel, 1994), and the decision to initiate oogenesis or not seems to be taken by the activation of receptor-mediated vitellogenin uptake. In *Camponotus festinatus*, such a regulation has also been shown to prevent ovary activation in queenless workers (Martinez and Wheeler, 1991). Another interesting endocrine regulatory mechanism was found for the harvester ant *Pogonomyrmex californicus*, where queens can either found a nest alone or join into a group of cofounding queens exhibiting division of labor, especially with respect to foraging decisions. In both cases, JH titers were found to be elevated during the foraging stage (Dolezal *et al.*, 2009), revealing a foraging bias elicited by this hormone similar to what is known for honey bee workers.

The species considered so far all belong to a large group of ants that do not exhibit a marked polyphenism within the worker caste. Instead, they all have a worker caste with a unimodal or only slightly bimodal size frequency distribution. In these species, the principle polyphenism is between queens (reproductives) and workers (non-reproductives). In contrast, the genus *Pheidole* is characterized by multiple polyphenisms, such as between queens and workers and also between different types of workers (including true soldiers). The genus *Pheidole* exhibits the most spectacular polyphenism known for social insect worker castes (see the next section). In this genus, the queen/worker polyphenism occurs much earlier in development than the worker/soldier switch to the extent that queen determination in *Pheidole* appears to occur in embryonic development by maternal factors, primarily hormones deposited in the egg during oogenesis. Queens that laid worker-biased eggs had higher ecdysteroid levels than those laying queen-biased eggs, and, correspondingly, worker-biased eggs also exhibited a higher ecdysteroid content (Suzzoni *et al.*, 1980). Furthermore, JH application to eggs increased the proportion of females developing into queens (Passera and Suzzoni, 1979). These results suggest that during this early critical period high levels of ecdysteroids are associated with worker development, and high levels of JH stimulate queen development.

11.3.3.2.2. The worker/soldier decision In contrast with the other highly eusocial Hymenoptera, division of labor within ant worker castes does not seem to be governed by age-related shifts in task performance. Rather, it is the relatively large size range of workers that underlies task preference and morphological specialization. Developmental mechanisms underlying worker caste polyphenism have been extensively reviewed (Wheeler, 1991) and mathematical models have been proposed that explicitly address the problem of growth rule reprogramming (Nijhout and Wheeler, 1994). Worker/soldier reprogramming is thought to take place during late larval instars in response to different feeding conditions or other forms of social influence on larval growth. These, in turn, appear to impact the larval endocrine system, which regulates growth and the onset of metamorphosis. Studies

exploring this aspect were carried out on the red imported fire ant, *S. invicta*, which has a unimodal distribution of worker size (Wheeler, 1990), and on the strongly polyphenic *Pheidole bicarinata* (Wheeler, 1983, 1984).

Methoprene application to late larval instars led to an increase in size in *S. invicta* workers, probably due to a retarded onset of metamorphosis. In *P. bicarinata*, similar treatment resulted in the expression of soldier characters in the emerging brood, indicating a discrete JH-dependent developmental switch in the last larval instar. In species with a bi- or multimodal size distribution, larvae that reach a critical size, and consequently experience a different endocrine milieu, can thus be shunted into alternative developmental pathways leading to overt polyphenism, generally between a soldier and a worker caste (Wheeler, 1994).

While the role of JH seems to be fairly well established in *Pheidole* soldier/worker polyphenism, the primary triggering factors are still largely hypothetical. Two aspects could be relevant in this context. First is the larva/worker ratio, which determines the amount of nursing activity devoted to each larva and thus may affect body size (Porter and Tschinkel, 1985). The second factor is social pheromones, which inhibit the development of further soldiers once a soldier level appropriate to colony size has been reached (Wheeler, 1991). In terms of colony fitness, such regulatory mechanisms represent a generalized developmental basis for the concept of adaptive colony demographies postulated by Oster and Wilson (1978).

Similar to the queen/worker decision, a genetic basis has also been proposed for the observed polymorphism within the worker caste; for example, for the harvester ant *Pogonomyrmex badius* (Rheindt *et al.*, 2005; Smith *et al.*, 2008), the leafcutter ant *Acromyrmex echinator* (Hughes and Boomsma, 2007), and the army ant *Eciton burchelli* (Jaffe *et al.*, 2007). But there is a caveat in the proposals for genetic worker caste determination because these are all polyandric species, making it possible that the observed worker caste polyphenism may actually be the result of temporal variation in sperm genotypes that fertilize the eggs produced by the queen (Wiernasz and Cole, 2010).

Although our knowledge of the regulatory cascade from genetic, feeding, and/or pheromonal factors to endocrine activity and consequent programming of growth parameters is still fragmentary, a landmark study on gene networks underlying wing development in ants (Abouheif and Wray, 2002) has provided remarkable insight into downstream consequences of developmental reprogramming. In the wing buds of three ant species exhibiting different degrees of polyphenism, these authors investigated the expression patterns of six regulatory genes (*ultrabithorax, extradenticle, engrailed, wingless, scalloped,* and *spalt*), all of which are functionally conserved in wing development of holometabolous insects. Cessation of wing bud development in worker-determined larvae is a characteristic element in all ants, and in species of the genus *Pheidole* this cessation can occur in two steps. *Pheidole morrisi* soldier larvae develop large vestigial forewing discs but no visible hindwing discs, while worker-destined larvae develop neither of these wing discs. Of these six genes, all except for the most downstream one (*spalt*), showed correlated expression patterns for the large forewing disc in soldier-biased larvae. None of these genes, however, were expressed in the forewing disc of worker larvae or in the hindwing discs of both soldier and worker larvae. Tracing the gene expression patterns through earlier development suggested (1) a gradual shutdown of the gene regulatory network occurring between mid-embryogenesis and the last larval instar and (2) that this patterning network was not interrupted at a single step, as might be expected under the concept of a classical switch mechanism. Instead, this study suggested that many different points in the gene expression cascade were involved. Extending the study to other ant species with lesser degrees of polyphenism (*Neoformica nitidiventris* and *Crematogaster lineolata*) corroborated the findings from *P. morrisi*. This led (Abouheif and Wray, 2002) to the conclusion that natural selection may be playing an active role in determining the most efficient route to halting wing development, and that, within the network, it may operate directly (independently) on different genes in different species.

This hypothesis has been followed up in the red imported fire ant, *S. invicta*, where vestigial wings were also discovered in worker larvae. These discs express the patterning genes *extradenticle, ultrabithorax*, and *engrailed* in accordance with expectations for normal wing development, but the wing discs do not grow (Bowsher *et al.*, 2007), inferring that wing disc patterning and growth are evolutionarily dissociated. This is a view that opens up exciting perspectives for hormonal effects on the expression of individual genes in the regulatory networks, and should stimulate the search for putative hormone response elements in the upstream control region of these genes in cross-species comparisons.

11.3.3.2.3. Queen polyphenism, queen loss, and queenless ants and wing dimorphism in males

Notwithstanding its fascination, phenotypic diversity in the worker caste, which includes both continuous size variation and polyphenism, is restricted to only 15% of the ant genera and is mainly found in the highly derived groups. Yet, ants in the subfamilies Myrmeciinae and Ponerinae, which are characterized by a number of ancestral traits and are commonly believed to be socially less complex than the "higher" ants, are by no means lacking phenotypic variation. These subfamilies are of interest because they exhibit graded transitional series between queens and workers (Heinze, 1998), and workers in these ants are often not sterile, but may have large ovaries and possess developed spermathecae (Crossland *et al.*, 1988; Dietemann *et al.*, 2002; Ohkawara *et al.*, 1993). In some species, workers

attain a queen-like morphology but contribute relatively little to colony reproduction, whereas in others, workers contribute substantially (e.g., the colony founding queen may be substituted by mated intercastes; Heinze and Buschinger, 1987; Peeters and Hölldobler, 1995).

In the most extreme cases, the queen caste has been completely lost, and all reproduction is carried out by workers, which are denominated gamergates (Peeters, 1991). Because in such queenless ants all females can potentially mate and lay eggs, a reproductive conflict of interest among colony females (as is typical in primitively social insects) is a secondarily evolved consequence. These species resolve the reproductive conflict by employing a mixed strategy of aggression by a dominant female and chemical signaling (Cuvillier-Hot *et al.*, 2001; Liebig *et al.*, 2000; Monnin and Ratnieks, 2001; Peeters *et al.*, 1999).

JH plays an interesting role in social dominance and fertility in these queenless ants. In contrast to primitively social wasps, where the JH titer positively correlates with dominance, and especially so with fertility (see Section 11.3.3.1.), application of the JH analog pyriproxyfen in queenless ants resulted in a *decrease* in fertility in the alpha female and in a loss in dominance rank (Sommer *et al.*, 1993). This decline in social rank was accompanied by corresponding changes in the cuticular hydrocarbon profile (Cuvillier-Hot *et al.*, 2004). Consequently, the idea that queenless ants represent a reversion to a primitive social system appears to be a superficial oversimplification that does not withstand a critical analysis once underlying developmental mechanisms and hormonal regulation of reproduction are taken into account. An interesting aspect in this context are the wing-disc derived gemmae of queenless ants, which deserve to be studied in the context of gene regulatory networks underlying wing development in ants.

Shutdown in wing development may not be exclusive to the female sex in ants. Recently this has also been shown to occur in males of *Cardiocondyla obscurior*, where a winged and a wingless male morph was found (Cremer *et al.*, 2002). Just as in workers, the loss of wings is not an isolated character state, but is part of a correlated modification in other characters, including reduced eye size. These "worker-like" males have a much prolonged period of spermatogenesis compared with winged conspecifics and mate exclusively inside the nest. This finding is a complete novelty and should stimulate comparative analyses on physiological and genomic events underlying concurrent polyphenism in the two sexes. Interestingly, both sexes responded with wing development when treated with a JH analog in a critical larval stage (Schrempf and Heinze, 2006).

11.3.3.3. Social bees

Social systems with incipient caste differences between dominant egg-laying and subordinate females have originated multiply and apparently independently within the Apidea (Michener, 2000). This happened in the Halictinae, the Xylocopinae, and of course most notably, the Apinae, which contain the closely related social tribes Euglossini, Bombini, Apini, and Meliponini.

11.3.3.3.1. The primitively eusocial bumblebees
Sociality in the bumble bees is obligatory and colonies can become quite large. The Bombini are large bees adapted to nesting in colder climates. Virgin queens emerge in autumn, mate, and hibernate before founding a new nest in spring. These queens are long-lived and survive for the entire seasonal nest cycle. The first brood emerging in early spring consists of a few rather small females (workers) that aid their mother in rearing the subsequent batches of exclusively female brood. Workers can vary considerably in size, and there is even slight overlap with queens in size-frequency distributions. During midsummer, the queen starts to lay haploid (male-producing) eggs, in addition to further female eggs. Soon after this "switch point," a few workers of high social rank also begin to lay (haploid) eggs, contributing to male production. The onset of worker reproduction has been termed the "competition point" (Duchateau and Velthuis, 1988), and marks an important incision in the colony life cycle, since from this point onward social integration deteriorates. By the end of the season, the colony produces large queens, which disperse and subsequently mate and hibernate. These components of the annual nest cycle have best been studied in *Bombus terrestris* (Röseler, 1985), and efforts in breeding and colony-rearing programs have been made to successfully introduce this species as a pollinator in greenhouses.

Although females produced before the competition point can attain a relatively large size, they never become queens (Cnaani et al., 1997; Cnaani and Hefetz, 2001). This inhibition of queen production during the early stages of the colony cycle can theoretically be attributed either to "nutritional castration," that is, feeding of less or lower quality food to the respective larvae, or to a direct inhibitory effect from the egg-laying queen (Röseler, 1970). Long-term video recording of feeding acts, larval food analysis, and manipulation of feeding frequency (Pereboom, 2000; Pereboom et al., 2003; Ribeiro, 1999; Ribeiro et al., 1999) have now clearly established that nurse bees do not manipulate the larval feeding program; rather, nurses readily respond to hunger signals emitted by the larvae. Thus, it is the intrinsic feeding program of the larvae that is defined during the early larval instars.

The decision as to which feeding program the larvae will adopt has been attributed to a queen signal that inhibits larvae from developing into queens before the competition point in the colony cycle (Röseler, 1974). Although it is supposed to be a primer pheromone (Röseler et al., 1981), its source and nature have not yet been unambiguously determined (Bloch and Hefetz, 1999b). Meticulous

studies on the endocrine response elicited in the larvae pointed to the first and early second larval instar as the critical window for this queen signal (**Figure 11**). From the fifth day of the second instar, pre-competition-point larvae showed markedly reduced levels of JH synthesis (Cnaani et al., 1997) and lower JH titers (Cnaani et al., 2000b) when compared to queens. This response in the CA correlates with caste-specific differences in the ecdysteroid titer (Hartfelder et al., 2000), indicating a synergistic interaction of the CA and prothoracic gland during the second and third larval instars. Interestingly, these hormone titer differences vanish during the feeding phase of the last larval instar when both castes exhibit low JH and ecdysteroid titers (Cnaani et al., 2000b; Hartfelder et al., 2000). Hormonal caste differences make their reappearance during the spinning and pre-pupal stages (Strambi et al., 1984). In these early metamorphosis stages there is no quantitative modulation, only a temporal one in the position of the pre-pupal JH and ecdysteroid titer peaks (**Figure 11**).

Social conditions, especially the phase in the colony cycle, have a strong impact on the endocrine system, and thus on the switch from worker to queen development in the individual larvae. This was demonstrated when comparing JH release rates in larvae that were reared shortly before and shortly after the competition point (Cnaani et al., 2000a).

This cascade from pheromonal effects on the feeding program to associated endocrine events in queen/worker differentiation does not necessarily represent a generalized chain of events in bumblebee caste polyphenism. Of the few comparative studies carried out on other bumblebee species, the most noteworthy ones were done on *B. hypnorum*, which was chosen for its different strategy in provisioning the larvae (Röseler, 1970). In this species, the queen/worker polyphenism involves a much later critical period. Larval fate is not determined during the early instars (as it is in *B. terrestris*); it is the feeding conditions in the last larval instar that appear to be the decisive factor. Yet again, the feeding differences converge in a caste-specific differential response in the endocrine system, and, as in *B. terrestris*, the pre-pupal JH and ecdysteroid titer peaks of *B. hypnorum* workers precede the corresponding peaks in queens (Strambi et al., 1984).

As already mentioned, adult bumblebee workers and queens differ in size — despite some overlap in the bimodal size frequency distribution — and to some extent in their physiology, mainly in their energy metabolism (Röseler and Röseler, 1986), yet not in their capacity to lay eggs. Even in the presence of the queen, a considerable percentage of *B. terrestris* workers can exhibit signs of ovary activation. However, less than 40% of the workers complete oogenesis and these individuals do not lay eggs before the switch point in the colony cycle (Duchateau and Velthuis, 1989). Until this time point, egg laying in

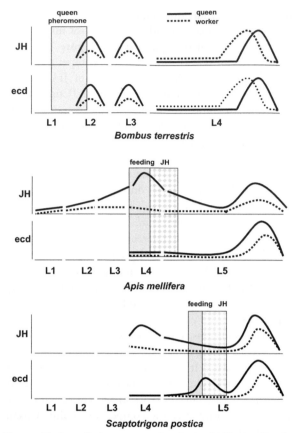

Figure 11 Juvenile hormone and ecdysteroid titer profiles in critical stages of caste development of the primitively eusocial bumblebee *Bombus terrestris* and the highly eusocial bees *Apis mellifera* and *Scaptotrigona postica*. In the primitively eusocial bumblebee, prospective queen and worker larvae differ in their JH and ecdysteroid titers during the early larval stages. Exposure during a critical period (shaded bar) to a repressor pheromone emitted by the egg-laying queen affects the hormone titers in the subsequent larval instars (L2 and L3). In the last larval instar (L4), the differences in hormone titers between queens and workers vanish and only a temporal shift in titer peaks is observed in the pre-pupal stage. In contrast, in the highly eusocial bees, *A. mellifera* and *S. postica*, the JH and ecdysteroid titers exhibit marked differences during the late larval stages. In *A. mellifera*, the major JH titer differences overlap with the nutritional switch from royal jelly to worker jelly in the fourth and early fifth instar (shaded bar). In *S. postica*, queen development is dependent on a prolonged feeding period (shaded bar) in the fifth instar. In *Apis* and in *Scaptotrigona* these critical stages overlap with a JH-sensitive period (dotted bar) established by JH application experiments. Morphological differences between queens and workers make their appearance in the highly eusocial bees but not in bumblebees. In the latter, queens differ from workers mainly in size and some physiological parameters. The expression of morphological caste differences thus appears to depend on sustained hormone titer differences in the last larval instar. Hormone titer profiles for *B. terrestris* were compiled from data by *Strambi et al.* (1984), *Cnaani et al.* (2000b), and *Hartfelder et al.* (2000). Data on *A. mellifera* were published by *Rembold* (1987) and *Rembold et al.* (1992), as well as by *Rachinsky et al.* (1990). Hormone titer curves for *S. postica* were adapted from *Hartfelder and Rembold* (1991). For data on critical periods and JH application experiments, see text.

workers is presumably inhibited by a queen-produced pheromone. The first studies pointing out a correlation between a queen inhibitory signal, JH production, and egg laying by workers were carried out by Röseler (1977) and Röseler and Röseler (1978). Reproductive workers had an elevated CA activity and JH titer relative to "non-activated" workers, and the queen inhibition of worker egg laying could be overcome by topical application of JH. Egg-laying inhibition by means of volatile pheromone was tested in a double-mesh screen assay and there was no evidence for such a signal (Alaux *et al.*, 2004), making it more likely that the workers autoregulate the onset of egg laying or eavesdrop on a queen signal that indicates that new queens will soon be produced in the colony (Alaux *et al.*, 2006).

A closer look at egg laying by workers revealed the existence of a dominance hierarchy within the worker caste in both queenright and queenless colonies. Dominant workers exhibit antagonistic behavior toward lower ranking workers, resulting in decreased CA activity and inhibition of egg laying in the latter ones (van Doorn, 1987). In small groups of queenless workers, a dominance hierarchy is quickly established, which is reflected in elevated JH titers in the high-ranking, egg-laying workers. The sequence in the response to queen removal was documented by monitoring JH synthesis, JH titer, ecdysteroid titer, and egg development during the first six days after queen removal and compared to queenright workers (Bloch *et al.*, 1996; Bloch *et al.*, 2000a,b). Levels of JH synthesis and JH titer were significantly elevated in workers three days after removal of the queen, followed by a similar increase in ecdysteroid titer, and larger terminal oocyte length at day six. This enhancement in JH release rates occurred independently of the colony cycle, that is, JH release was equally elevated in workers made queenless before and after the competition point (Bloch *et al.*, 1996). Yet, there was a strong correlation with worker age, since it was mainly the older workers that had elevated levels of JH release, and thus became dominant egg layers, both in queenless groups and in post-commitment point colonies. The absence of the queen is not the *sine qua non* for worker reproduction, since individually kept workers did not start egg laying and, correspondingly, maintained a low JH titer (Larrere and Couillaud, 1993). In accordance with previous studies on worker dominance hierarchies summarized by Röseler and van Honk (1990), it appears to be the high-ranking (older) workers that inhibit JH synthesis and ovary activation in lower ranking (younger) workers (Bloch and Hefetz, 1999a). Worker reproduction should be inhibited differentially during the two main phases of colony development, that is, prior to the competition point the queen inhibitory signal suppresses egg development in high-ranking workers, which acts synergistically with an inhibitory effect of dominant workers on ovary activation in the lower ranking workers. After the competition point, it is primarily the high-ranking workers that control egg production in their nestmates. Both inhibitory signals appear to be mediated through the neuroendocrine axis, particularly via JH release. The elevated ecdysteroid titer observed concomitantly with progressive oogenesis (Bloch et al., 2000b) should plausibly be interpreted as a consequence of the JH-stimulated ovary activation, yet one cannot exclude an ecdysteroid-mediated synergistic feedback effect on dominance as well (just as seen in *Polistes* wasps, see Section 11.3.3.1.).

Surprisingly, the strong correlation observed between JH titer and ovary activation in workers (Bloch et al., 2000a) was not as striking in queens, which have lower JH titers than egg-laying, queenless workers (Bloch et al., 2000a; Larrere et al., 1993). A high JH titer is therefore not an absolute requirement for egg production, at least in queens, and may be more related to a combination of dominance status and egg production in the workers than with queen reproduction. Ecdysteroids may play a more important role in queen reproduction than in worker reproduction, as seen by the elevated ecdysteroid titers in overwintered queens, especially in the ovaries (Bloch et al., 2000b; Geva et al., 2005). Thus it seems that in bumblebees we are confronted with a split in the role of JH and ecdysteroids in female reproduction, with JH playing the prominent role in workers and ecdysteroids in queens. An interesting question to ask is whether or not and how diapause physiology (see Chapter 10) may play a role in setting up this split in the endocrine system of reproductive physiology in the bumblebee castes, both ontogenetically and in evolutionary terms. Clear diapause effects on development and physiology are seen in butterfly phase (Section 11.3.1.2.1.) and other caste polyphenisms (Sections 11.3.3.1. and 11.3.3.2.).

11.3.3.3.2. The highly eusocial honey bees and stingless bees

Honey bees (Apini) and the closely related Meliponini, colloquially referred to as stingless bees, have been companions of mankind for almost as long as locusts. Their origin and social organization has been narrated in picturesque myths, and their biology is a well-represented element in ethnobiology. The common themes to these myths are the nature of social integration and the origin of the differences in form and function between queens and workers.

One of the main difficulties in explanations for queen/worker dimorphism, both for myths and the modern scientific approach, resides in the identification of the mode of action for the initial triggers in caste development. As in most social insects, this is a problem of larval nutrition. In the honey bees, larvae are continuously fed during larval development, and during the early larval instars, both queen and worker larvae receive mainly a glandular proteinaceous secretion (royal jelly). In the fourth and fifth larval instar this type of food is provided in large quantities

only to queen larvae, whereas worker-destined larvae are fed a mixture of glandular secretions, honey, and pollen. Apart from the switch in diet type, worker larvae are also visited and fed much less frequently than queen larvae (Beetsma, 1985). Attempts to identify specific queen-determining factors in royal jelly resulted in a partial purification of chemically labile factors (Rembold et al., 1974b), but unequivocal evidence for a "queen determinator" was not obtained, leading these authors to consider the requirement for a balanced diet. (But see Note added in proof).

This view is also consistent with results of larval food analyses in stingless bees (Hartfelder and Engels, 1989), which do not progressively feed their brood; instead, they mass provision the brood cells shortly after they are built. Due to mass provisioning of the brood cells and lack of further interaction of worker bees with the developing brood, other than regulation of the colony microclimate, there is no evidence for a nutritional switch in the stingless bees that could serve as a signal for queen/worker development. In the majority of the stingless bee species (i.e., those pertaining to the highly diverse trigonine genera), queen development depends on the quantity of larval food provided to the larvae, and this is reflected in the size of the cells. Large queen cells containing two- to threefold more larval food than the worker cells are often built at the margins of the horizontal brood combs (Engels and Imperatriz-Fonseca, 1990) or result from the fusion of two brood cells sitting on top of one another, thus providing a larva with a second portion of larval food. This strategy of queen production is also observed in the context of emergency queen rearing in some species when a colony has lost its queen (Faustino et al., 2002).

An exception to the queen/worker determining mechanism by modulation of larval food quantity is the genus *Melipona*. Their brood cells are of equal size, irrespective of whether queens, workers, or males are reared within them. Up to 25% of the female brood can be queens, which led Kerr (1950) to propose the hypothesis of a genetic predisposition with a two-locus, two-allele system where only double heterozygotes can become queens. Even in this system, larval food quality has a modulating effect on caste differentiation (Kerr et al., 1966), since maximal levels in queen production are only observed in strong colonies that have sufficiently large stocks of honey and pollen. Under suboptimal conditions, the proportion of queens in the female brood is generally well below 25%. An alternative hypothesis based on modeling optimal queen production frequencies in the queen/worker conflict attributes "self-determining" capacity to the larvae (Ratnieks, 2001). Although the two hypotheses are not mutually exclusive, because one addresses proximate mechanisms and the other ultimate causes, both are awaiting empirical validation. This may be obtained through genetic marker analyses, for example, by means of linkage in AFLP analyses (Hartfelder et al., 2006; Makert et al.,

2006). Irrespective of the distinct modalities in initial triggers (i.e., nutritional switch in honey bees, large food quantities in trigonine bees, or a genetic predisposition in *Melipona* species), all of these inputs eventually converge in caste-specific activity patterns in the endocrine system, which govern subsequent differentiation events in the target tissues.

11.3.3.3.2.1. The endocrine regulation of caste development in honey bees. This endocrine cascade has been best explored in the honey bee. JH application experiments carried out in the 1970s indicated an eminent role for JH in caste development, and established a critical window for its action in the fourth and early fifth larval instar (Dietz *et al.*, 1979; Rembold *et al.*, 1974a; and further references in Hartfelder, 1990). Subsequent analyses of endogenous JH titers by specific radioimmunoassays and highly sensitive GC-MS corroborated these experimental JH effects, revealing a marked difference in JH titers between queen- and worker-biased larvae (**Figure 11**) during larval, pre-pupal and the late pupal stages (Rachinsky *et al.*, 1990; Rembold, 1987; Rembold *et al.*, 1992). From these meticulous JH titer analyses it was possible to ask how the JH titer differences may be generated and especially, how they may affect caste-specific differentiation processes in target tissues. Analyses of JH biosynthesis and release rates by means of a radiochemical assay revealed a markedly elevated CA activity in prospective queen larvae between the fourth and the early fifth larval instar (Rachinsky and Hartfelder, 1990), inferring that the JH titer differences are primarily a result of differential CA activity. Rates of JH degradation were generally low in honey bee development (Mane and Rembold, 1977) and are primarily controlled by a recently identified honey bee JH esterase (Mackert *et al.*, 2008). A JH epoxy hydrolase-like gene was also identified in the honey bee genome, but functional assays showed that it does not degrade JH (Mackert *et al.*, 2010).

An interesting facet was the finding that the terminal steps in JH biosynthesis are the critical ones for the differential JH release rates in early fifth instar queen and worker larvae (Rachinsky and Hartfelder, 1991; Rachinsky *et al.*, 2000). Blocking the terminal steps in JH synthesis, from farnesoate to JH III, is an efficient means to generate and guarantee low JH titers in worker larvae during the fourth and early fifth instar. Furthermore, since this block is reversible, as required to generate the elevated JH titers in pre-pupae (Rachinsky and Hartfelder, 1990, 1991), the ortho-methylfarnesoate transferase and the JH-epoxidase are candidate targets for allatoregulatory factors. Radiochemical assays on CA of worker larvae showed that *M. sexta* allatotropin (Manse-AT) can stimulate JH synthesis in honey bees in a dose-dependent manner, yet does not overcome the block on the terminal steps (Rachinsky and Feldlaufer, 2000; Rachinsky *et al.*, 2000). Immunolocalization for Manse-AT-like material in honey

bee brains detected a discrete, small number of cells in pre-pupae, but not in the earlier stages of honey bee development, making it difficult to assert that a Manse-AT-like peptide is involved in the regulation of CA activity during the critical stages of caste determination. CA regulation becomes even more complex considering that a set of peptides from the brain and subesophageal ganglion (Rachinsky, 1996) as well as biogenic amines, especially serotonin (5HT) and octopamine (Rachinsky, 1994), may modulate JH biosynthesis to generate the observed caste-specific profile.

This search for allatoregulatory factors represents an upstream walk to close the gap between the initial nutritional signal that induces caste development and its translation into an endocrine response. The only pathways mapped in this context are the serotonergic (Boleli *et al.*, 1995; Seidel and Bicker, 1996) and the stomatogastric nervous system (Boleli *et al.*, 1998). Immunocytochemical mapping of serotonergic neurons in honey bee larvae and pupae revealed an apparent heterochrony in their development. Whereas serotonergic neurons in the ventral ganglia exhibited a seemingly mature architecture already in the larval stages, 5HT-neurons in the brain matured only during late pupal development (Boleli *et al.*, 1995). These results seemingly preclude an allatoregulatory function for protocerebral 5HT-neurons during pre-imaginal development. Instead, the detection of immunoreactive cell bodies in the vicinity of the CA, close to their connection to the *nervi corporis cardiaci III*, implicates the stomatogastric nervous system in the modulation of CA activity. Since the stomatogastric nervous system provides a direct link between sensory cells for food quality in the labral region of honey bee larvae (Goewie, 1978), and the food-intake-mediating neurons of the stomatogastric nervous system integrate with the retrocerebral endocrine complex (Boleli *et al.*, 1998), the nutritional switch information could certainly take this (alternative, or even more direct) pathway. Nevertheless, information on perception and processing of the nutritional switch signal in bees is still rudimentary and remains a major black box in our understanding of events upstream of the caste-specific endocrine responses.

One of the downstream effects resulting from differences in JH titers between honey bee queen and worker larvae involves the endocrine system functioning as an internal circuit between the CA and the prothoracic glands. The latter were identified in honey bee larvae as loose agglomerates of cells projecting from the foregut to the retrocerebral complex (Hartfelder, 1993), and their activity pattern closely reflects the ecdysteroid titer in the last larval instar (Rachinsky *et al.*, 1990). The ecdysteroid titer, with makisterone A as the main compound, is low at the beginning of the last instar (**Figure 11**), and exhibits a marked peak during the pre-pupal stage in both castes. Caste-related differences in the larval ecdysteroid titer are evident in the cocoon-spinning phase, with an earlier increase in queens than in workers. Mimicking a queen-like JH titer by an exogenous JH application to worker larvae in the fourth to early fifth instar resulted in a precocious increase in prothoracic gland activity (Rachinsky and Engels, 1995). This positive interaction between JH titer and prothoracic gland activity may involve a prothoracicotropic hormone (PTTH)-like factor, which was detected immunohistochemically in honey bee brain sections (Simões *et al.*, 1997).

Alternatively, JH could directly stimulate ecdysteroid synthesis and release in the prothoracic glands, as demonstrated by *in vitro* assays exposing these glands to different doses of methoprene (Hartfelder and Engels, 1998). The positive action of JH on the ecdysteroid titer in worker larvae seems to be restricted to the larval–pupal transition, since JH application to pupae inhibited and retarded the formation of the large pupal ecdysteroid peak (Zufelato *et al.*, 2000), which also mainly consists of makisterone A (Feldlaufer *et al.*, 1985).

In terms of target tissues, it is the ovary that has received the most attention, because it best reflects the functional caste differences. A typical queen ovary consists of 150–200 ovarioles, whereas a worker ovary most often contains between 2 and 12 of these serial units. These differences arise primarily during the final larval instars, even though some of the differentiation steps are already visible as early as the second or third instar (Dedej *et al.*, 1998; Reginato and Cruz-Landim, 2001). Caste-specific differentiation of the ovary consists primarily of a reduction in ovariole number in the worker caste, from an initially equal number of ovariole anlagen in the two castes — generally over 150 anlagen per ovary in the fourth instar — to the final set of ovarioles at pupation. Histologically, this process is marked by a reduced number of rosette-like cystocyte clusters and a large number of autophagic vacuoles in early fifth instar worker ovarioles (Hartfelder and Steinbrück, 1997).

The reduced number of cystocyte clusters is not due to a diminished mitotic activity in the ovaries of worker larvae (Schmidt Capella and Hartfelder, 1998); rather it results from a disintegrating actin cytoskeleton in the germ cells (Schmidt Capella and Hartfelder, 2002). These degradative processes initiated by actin-spectrin dissociation could be reverted by a single topical JH application to late fourth instar workers (Schmidt Capella and Hartfelder, 1998; Schmidt Capella and Hartfelder, 2002). Even though these results permit us to pinpoint a cell biological target for JH in an important caste differentiation process, we cannot yet establish whether JH acts directly on the affinity of actin to spectrin, or whether this is due to a transcriptional effect on factors that mediate the interaction between these cytoskeletal components.

A completely different approach toward understanding the development and evolution of the tremendous difference in ovary size between queen and worker honey bees comes from quantitative trait loci (QTL) mapping (Linksvayer *et al.*, 2009). This study identified a series of interesting candidate genes, such as *quail*, a gene involved in actin filament-dependent apoptosis in nurse cells in the *Drosophila* ovary, *cabut* an ecdysteroid-responsive transcriptional activator also associated with cell death, *delta*, the main player in Notch signaling, and *miro*, which is involved in mitochondrial homeostasis, apoptotic signal transduction, and cytoskeleton organization. Taken together, these results nicely illustrate the convergence of results coming from hypothesis-driven and non-biased research approaches, both pointing to the importance of cytoskeletal organization and cell signaling in the caste-specific differentiation of the honey bee ovary.

11.3.3.3.2.2. Honey bee caste development — Insights from genomic studies. As in functional genomics in general, complex transcriptional regulation can be expected to be a key factor in our comprehension of caste development, and in this respect, the honey bee stands out as a model organism among the social insects. The early studies on mRNA differences between queen and worker larvae (Severson *et al.*, 1989) have gained considerable depth and impetus from powerful molecular methods generating differentially expressed EST (Corona *et al.*, 1999; Hepperle and Hartfelder, 2001) and suppression-subtractive hybridization libraries (Evans and Wheeler, 1999, 2000). By clustering algorithms three major transcriptomic groupings became evident, a dichotomy for early versus late larval stages and a queen-worker dichotomy in gene expression that begins in the late larval stages. A major breakthrough in honey bee genomic analysis finally came with the publication of the complete and annotated honey bee genome (The Honey Bee Genome Sequencing Consortium, 2006). Against this database, the caste specifically expressed ESTs of the earlier studies could now be interpreted in context. A Gene Ontology-based annotation (Cristino *et al.*, 2006) inferred that metabolic regulation must be a major factor in caste development, and one gene that appears to be an important player herein encodes an ecdysone-responsive short chain dehydrogenase/reductase (Guidugli *et al.*, 2004). Motif searches for putative transcription factors in upstream control regions of differentially expressed genes then set the road map to the reconstruction of genomic networks acting in honey bee caste development (Cristino *et al.*, 2006). Such networks became comprehensive once larger data sets of caste- and stage-specific gene expression became available by means of microarray analyses (Barchuk *et al.*, 2007). As shown in **Figure 12**, the genomic regulatory networks derived for queen and worker development show strongly divergent topographies, denoting to the complexity in gene expression shifts during post-embryonic caste differentiation. Furthermore, network nodes are of heuristic value as they pinpoint genes at crucial connections within the network, which should be the focus of further in-depth studies.

(A) Bipartite networks with motifs and genes

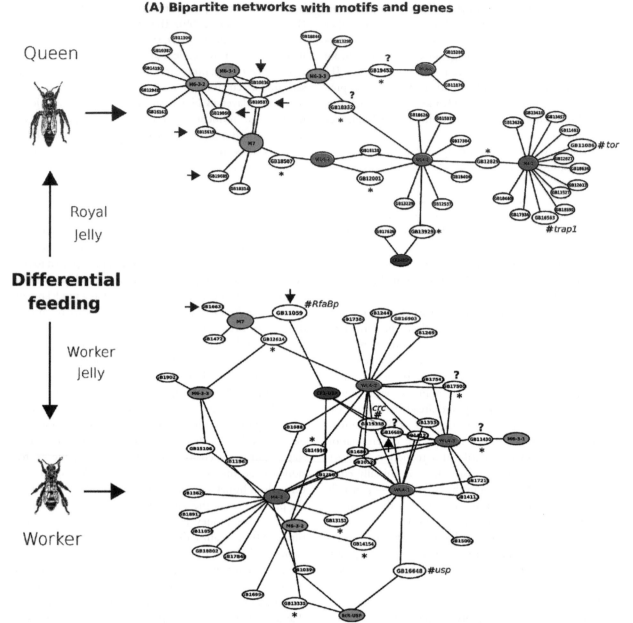

Figure 12 Network analysis depicting putative gene interaction in caste development of the honey bee, *Apis mellifera* (*Barchuk et al.*, 2007). Differential gene expression data were obtained by microarray analysis of third, fourth, and early fifth instar queen and worker larvae. After collecting Gene Ontology attributes for differentially expressed genes, their respective upstream control regions (UCRs) were retrieved from the genome sequence and were screened for putative regulatory motifs by means of a Gibbs sampling approach integrated into a motif discovery and analysis script (*Cristino et al.*, 2006). Functional ecdysone responsive elements and Ultraspiracle (USP) binding site information was obtained from studies on *Drosophila*. Motifs represented in blue were associated with JH responsive genes in caste development, those in green with cell death events. Yellow circles represent a set of ten genes that are the most overexpressed in workers compared to queens, and in magenta are motifs related to ecdysone response and USP. The black arrows point to genes consistently upregulated in caste development and in JH application assay. Genes with unknown function are marked by a question mark. Genes marked by an asterisk were not in the training data set for motif discovery. The worker differentially expressed genes marked by a hash (#) are *usp*, *crc*, and *RfaBp*, which are repressed by hormones. The queen differentially expressed genes marked by a hash (#) are *tor* and *trap1*, the latter being a negative regulator of apoptosis in response of nutrition. The more interconnected network obtained for worker larvae suggests that the differentially expressed genes share a larger number of *cis*-elements than the queen differentially expressed genes. Reproduced with permission from *Barchuk et al.* (2007).

Alternative approaches to the non-hypothesis-driven expression library or microarray analyses are candidate gene approaches. Because honey bee queens and workers greatly differ in size, factors controlling growth should be important. Consequently, several studies have now investigated the functionality of the insulin-insulin-like signaling (IIS) pathway together with its parallel branch, the target of rapamycin (TOR) pathway. The clearest results were obtained for *Apis mellifera* TOR gene function. *Amtor* expression is higher in young larvae destined to become queens, and a knockdown experiment on AmTOR function by feeding *Amtor* dsRNA to larvae reared *in vitro* inhibited the development of queen characters (Patel *et al.*, 2007). These results established a positive function for *Amtor* in queen development, in accordance with predictions form the *Drosophila* model. The role of the IIS pathway, on the other hand, turns out to be considerably more complex. This is because there are two insulin-like peptides (AmILP1 and AmILP2) and two insulin receptors (AmInR1 and AmInR2) predicted in the honey bee genome (Azevedo and Hartfelder, 2008; Watt, 1968; Wheeler *et al.*, 2006), and especially in the late larval stages, when the growth rates of queen larvae tremendously surpass those of workers, the expression of the AmILP2 encoding gene is, surprisingly, more expressed in worker than in queen larvae, and this ILP gene is the most expressed gene among the two predicted genes. Furthermore, the expression of both insulin receptor genes was downregulated in queen larvae while they exhibited enhanced growth (Azevedo and Hartfelder, 2008). Honey bees, thus, differ from the *Drosophila* model (Colombani *et al.*, 2005; Mirth *et al.*, 2005; Oldham and Hafen, 2003) when it comes to IIS-mediated growth control, thus representing an apparent paradox.

It is possible that the downregulation in IIS function may be contingent on the JH titer, which also shows a decrease during this phase in queen larvae, but searching for further interacting factors indicated that the hypoxia signaling pathway could be involved. A possible role for the involvement of oxidative metabolism in honey bee caste development had already been indicated from differential gene expression screens, but evidence dated back to very early studies on respiratory rates in honey bee larvae (Melampy and Willis, 1939) and investigations on mitochondrial functions, especially cytochrome c content (Eder *et al.*, 1983). Since the three hypoxia signaling core genes turned out to be highly conserved in the honey bee genome, it was possible to study their transcript levels during queen and worker development. It became clearly apparent that worker larvae strongly overexpress these genes (Azevedo *et al.*, 2011), even though ambient oxygen availability in the hive should be the same for queen and worker larvae. These results from non-biased microarray studies, as well as from candidate gene approaches, now pinpoint the importance of metabolic regulation to caste development.

A further intriguing insight into mechanisms underlying caste development comes from the observation that the honey bee genome, distinct from *Drosophila melanogaster*, has a full complement of DNA methylating enzymes (The Honey Bee Genome Sequencing Consortium, 2006). Using this genomic information, an RNAi-mediated knockdown of the DNA methyltransferase *Dnmt3* gene caused worker-destined larvae to express queen-like characters, such as large ovaries (Kucharski *et al.*, 2008). This was a surprising insight into the unexpected tremendous role that epigenetic regulation exerts on the honey bee transcriptome, and current studies are directed toward analyzing the genome-wide role of methylation in honey bee queens and workers (Foret *et al.*, 2009) and imprinting. Preliminary results indicate that large portions of the honey bee genome may be imprinted (G. Hunt, personal communication).

11.3.3.3.2.3. Honey bee reproduction and division of labor. The high rates of egg production in the honey bee queen require an enormous production of vitellogenin by the fat body. In contrast to most insects, hormonal regulation of vitellogenin expression in honey bees is rather puzzling. Juvenile hormone application experiments, allatectomy, and also JH titer analyses all test against a role for this hormone in the regulation of the vitellogenin titer in adult queens (Engels, 1974; other references cited in Hartfelder and Engels, 1998). Rather, the gonadotropic function of JH, especially the stimulation of vitellogenin expression, seems to have been shifted to the late pupal stage (Barchuk *et al.*, 2002), and strikingly, vitellogenin transcripts were already detected during larval development (Guidugli *et al.*, 2005b). As in other insects, much of the mysteries of JH action can be ascribed to a lack of insights into its receptor and downstream mode of action. In the honey bee, a homologue to Ultraspiracle, which is considered an insect JH receptor candidate, has been identified as a single copy gene that is upregulated in response to JH (Barchuk *et al.*, 2004). Furthermore, knockdown of this gene by RNAi delayed pupal development, but since it did not affect vitellogenin gene expression during this phase (Barchuk *et al.*, 2008), its function in the JH response cascade is far form clear. As indicated from functional studies on a JH response element (DmJRE1) that was found in the promoter region of some of the JH-induced honey bee genes, proteins interacting with this element were shown to also interact with the ecdysone receptor and Ultraspiracle gene products (Li *et al.*, 2007), thus inferring a cross-talk between JH and ecdysteroid signaling. This interaction could be especially relevant for vitellogenin induction in the late pupal phase.

In contrast to JH, no apparent role in reproduction and division of labor in female honey bees can be attributed to ecdysteroids. Except for a transient small peak early in the adult life cycle, ecdysteroid titers fluctuate at basal levels, independent of caste, reproductive status, and colony social conditions (Hartfelder *et al.*, 2002). This is all the more striking when considering that considerable levels of ecdysteroids were detected in the ovaries of adult queens (Feldlaufer *et al.*, 1986). Furthermore, a homologue of the ecdysteroid-regulated gene *E74* was shown to be selectively expressed in the ovaries of queens and in mushroom body interneurons of adult workers (Paul *et al.*, 2005).

Caste and reproduction in the highly eusocial insects are intimately related to life span, and contrary to standard animal models, high fertility is associated with extended life span (Heinze and Schrempf, 2008) or more accurately, with delayed aging, also making the honey bee a model system of particular interest to the medical field (Münch *et al.*, 2008). Notwithstanding the complexity of the aging and senescence syndrome, an apparently simple regulatory circuitry based on a mutual negative feedback of JH on the vitellogenin titer and vice versa emerged from initial modeling studies (Amdam and Omholt, 2002, 2003). These built on early evidence showing that the application of JH to young worker bees promotes a precocious switch from activities within the nest, especially brood care, to foraging activities typical of older workers (Jaycox, 1976). Subsequent analyses of JH titers and CA activity confirmed these data, showing that rates of JH synthesis and hemolymph JH titers in worker bees strongly increase as these individuals age and become foragers (Huang *et al.*, 1991; Robinson *et al.*, 1991, 1992). This switch from within-hive to foraging (out-of-hive) activity has long been known to be accompanied by a decrease in hemolymph vitellogenin levels (Engels, 1974), but the functional relevance of this mutual negative feedback only became apparent once novel properties of vitellogenin emerged, in addition to its primary role as a yolk protein precursor. Honey bee vitellogenin turned out to also be a major zinc transporting protein and to directly affect immunosenescence of hemocytes (Amdam *et al.*, 2005). Such age-related immunosenescence was further corroborated through an analysis of various stressor effects, showing that older workers were less resistant to stress than younger bees (Remolina *et al.*, 2007).

For proof of principle of the postulated mutual negative feedback between JH and vitellogenin (double repressor hypothesis), it was essential to show that not only does JH represses vitellogenin synthesis (Engels *et al.*, 1990), but also that downregulating vitellogenin expression may, in turn lead to an increase in the hemolymph JH titer. This has now been shown in several experiments using RNAi-mediated knockdown of vitellogenin gene function (Amdam *et al.*, 2005; Guidugli *et al.*, 2005a; Marco Antonio *et al.*, 2008; Nelson *et al.*, 2007). Since the interplay between JH and vitellogenin affected a whole suite of behavioral characters, including gustatory responses and preferences between pollen and nectar (Amdam *et al.*, 2006b; Tsuruda *et al.*, 2008), this circuitry is considered to represent a key module in the division of labor in a honey bee colony. Viewed against the background of social

insects having evolved from non-social ancestors, this circuitry has been formulated as the "reproductive ground plan hypothesis" (Amdam *et al.*, 2004).

A major open question herein is the molecular structure of the vitellogenin signaling pathway. A vitellogenin receptor candidate gene was shown to be expressed in honey bee fat body, ovaries, and also in the brain (Guidugli-Lazzarini *et al.*, 2008), consistent with findings for vitellogenin expression (Corona *et al.*, 2007). In addition, a signal originating from the ovaries was recently postulated based on ovary transplantation experiments (Wang *et al.*, 2010). Furthermore, a cross-talk of vitellogenin signaling with the IIS pathway has been observed (Corona *et al.*, 2007). In addition, high-throughput and quantitative RT-PCR gene expression analyses of worker honey bee brains further singled out the importance of nutrition-mediated signals acting through the IIS pathway (Ament *et al.*, 2008, 2010), and emphasized the role of the brood pheromone in orchestrating large-scale transcriptional responses in worker bees (Alaux *et al.*, 2009).

With all this evidence adding to the heuristic power of the apparently simple reproductive ground plan hypothesis, the complexity of its genetic architecture emerged from the genomic annotation of the pln QTL (Hunt *et al.*, 2007a) that contribute to pollen hoarding behavior and could be analyzed through a long-term selection program for this trait. In these QTL studies, genes related to ovarian development and the IIS pathway were detected as over-represented. Furthermore, evolutionary transitions in social insects, as related to the reproductive ground plan, do not seem to be restricted to the female sex, because vitellogenin and its receptor are also expressed in different tissues of honey bee drones (Colonello-Frattini and Hartfelder, 2009; Trenczek *et al.*, 1989).

To conclude, the reproductive ground plan hypothesis (Amdam *et al.*, 2004), which builds upon the prior ovarian ground plan hypothesis (West-Eberhard, 1996), hinges on the remodeling of (1) an ancestral JH-vitellogenin circuitry regulating oogenesis and (2) the segregation of behaviors — that ancestrally are associated with a reproductive phase (foraging and brood rearing with a protein/lipid-rich diet) and a non-reproductive phase (foraging for carbohydrate-rich food) in the life history of solitary wasps or bees — into two functionally and morphologically distinct castes, such as queens and workers. The hypothesis is well supported for temperate-climate paper wasps (Hunt *et al.*, 2007b) and bees, especially the honey-bee (Amdam *et al.*, 2006a; Amdam and Page, 2007) and has come to be a powerful framework in which to explain caste polyphenism in social Hymenoptera because these polyphenisms are all centered around reproduction. A possible extension to termites has not yet been explored.

11.3.3.3.2.4. Hormonal control of caste development and reproduction in stingless bees.
The stingless bees are not only closely related to the monogeneric tribe Apini, but

are also represented by an enormous number of species (Camargo and Pedro, 1992). Thus, they supply us with ample material for evolutionary insights into caste polyphenism and reproduction in highly social bees. As previously mentioned, it is the nutritional conditions that serve as the initial trigger in the divergence of the queen/worker developmental pathways, even though there is strong evidence for a genetic predisposition to caste fate in the genus *Melipona* (see Section 11.3.3.3.2.).

As in the honey bee, the initial investigations on the role of hormones in stingless bee caste differentiation all relied on JH application experiments. These comparative investigations on different species of stingless bees (Bonetti *et al.*, 1995; Buschini and Campos, 1994; Campos, 1978, 1979; Campos *et al.*, 1975) established the spinning stage of the last larval instar as the critical phase for JH-dependent induction of queen development. These findings subsequently received support from investigations on CA activity and JH titer measurements in *Scaptotrigona postica* (Hartfelder, 1987; Hartfelder and Rembold, 1991). These results indicated that JH-dependent differentiation steps in queen development occur much later in stingless bees when compared to the honey bee (Hartfelder, 1990; **Figure 11**). Also, queen development in the trigonine species takes longer than worker development, with the opposite rue for honey bees and the genus *Melipona*, indicating that the differences in nutritional programs among the three groups (*Apis*, *Melipona*, *Trigonini*) are reflected in larval and especially pupal ecdysteroid titers (Hartfelder and Rembold, 1991; Pinto *et al.*, 2002), arguing for a correlated regulation of caste development by JH and ecdysteroid in these groups of highly social bees.

The high species diversity in the stingless bees, their variation in colony size, nesting sites, and a large number of further aspects in social lifestyles, obviously provide ample material for variation on the basic themes of caste development, reproduction and division of labor, and possible hormonal regulation therein. An important difference between honey bees and stingless bees lies in the degree of morphological differences between the sexes and castes. The males of stingless bees are morphologically much more similar to workers than they are to queens (Kerr, 1987, 1990), even though growth rules for the different morphogenetic fields along the body axes appear to be adjusted rather independently (Hartfelder and Engels, 1992).

A striking phenomenon calling for functional explanations is the strong variation in queen size in some species. The occurrence and reproductive performance of miniature queens was closely studied and compared to normal-sized queens (Ribeiro *et al.*, 2006), indicating functional differences in reproductive performance related to colony conditions.

Since monopolization of reproduction by the queen is a key element in the social evolution of bees, the large variation in reproductive activities by the workers among

stingless bee species can provide insight into different evolutionary solutions to the queen–worker and also the worker–worker conflict over reproduction. Worker oviposition can take two forms: (1) trophic eggs that are laid shortly before the queen oviposits and (2) reproductive eggs that are laid after the queen's oviposition. Trophic eggs are unviable eggs that are specially produced as nutrition for the queen serving to maintain her high reproductive rates. Worker vitellogenin is, thus, directly shunted into egg production by the queen. Since the production of (trophic) worker eggs is clearly in the interest of the queen, she does not discourage workers to produce these unviable eggs and thus should keep the workers' ovaries in an active state. It is therefore not surprising that workers can also produce viable eggs that can make a significant contribution to male production in a colony (Cepeda, 2006; Engels and Imperatriz-Fonseca, 1990; Velthuis *et al.*, 2005). Some species, such as *Frieseomelitta varia* have, however, opted for a completely different solution to this conflict over reproduction, as their workers are completely sterile due to the complete degeneration of their ovariole anlagen during pupal development (Boleli *et al.*, 1999, 2000). All the more surprising, on first sight, but consistent with the evolutionary transitions implicit in the reproductive ground plan hypothesis, vitellogenin expression in this ovary-less stingless bee was practically constitutive both at the transcript and at the protein level (Dallacqua *et al.*, 2007; Hartfelder *et al.*, 2006).

The involvement of hormones in the regulation of reproduction and division of labor in adult stingless bee queens and workers has only marginally been investigated, and so far there is only negative evidence to this end in *Melipona quadrifasciata*. As in the honey bee, ecdysteroids do not seem to play any role in queen or worker reproduction in *M. quadrifasciata* (Hartfelder *et al.*, 2002). Interestingly, however, analyses of *ultraspiracle* gene homologues in *M. scutellaris* and *S. depilis* (Teles *et al.*, 2007) showed differences in pupal transcript levels that might be related to differences in reproductive strategies seen in adult workers of trigonine and *Melipona* species (Velthuis *et al.*, 2005).

Genomic resources are still scarce for stingless bees, with the exception of *M. quadrifasciata*, for which a suppression subtractive library analysis has revealed 337 unique sequences as differentially expressed between newly emerged queens and workers (Judice *et al.*, 2006). This situation is, however, bound to change as efforts are under way to generate 454 sequence data for at least two stingless bee species.

11.4. Synthesis and Perspectives

In the previous sections, we presented insect polyphenisms and detailed the underlying ontogenetic mechanism in a phylogenetic context to facilitate drawing parallels to general modes of development and reproduction in hemi- and holometabolous insects. The most common forms of polyphenism are built around the framework of dispersal and reproduction. Considering that the development of wings has played a major role in the extraordinary evolutionary and ecological radiation of insects, it appears that aspects of wing formation have proven especially amenable to the evolution of facultative or polyphenic patterns of wing expression. The switch between winged and wingless developmental pathways most often occurs during a late nymphal instar and frequently involves modulation of the JH titer during a critical physiological "sensitive period." Research on wing polyphenism in cricket species has not only firmly established a key role for JH in wing size variation, but also called attention to the fact that it is JH degradation by a JHE and not CA activity that is the controlling factor of the JH titer modulation in nymphs. In contrast, in adult crickets, wing length is associated with an interesting diurnal variation in JH synthesis in the long-winged morph, and genetic variation underlying this trait is a prime example of the interaction of genotype and environment (population density) in the expression of an adaptive life history polyphenism (Roff and Fairbairn, 2007; Zera et al., 2007).

When compared to crickets, phase polyphenism in locusts is a much more complex syndrome involving transgenerational transfer of information in the transition from the solitary to the gregarious morph. Whereas a role for the endocrine system is best established for changes in body pigmentation, with high levels of JH promoting expression of the green background color of the solitary morph and the neuropeptide [His7]-corazonin the dark foreground pattern, the physiological factors underlying the shift in other aspects of the syndrome (leg and wing morphometry, ovariole number, and behavior) are less well understood (Pener and Simpson, 2009). A much debated issue for understanding the phase shift in migratory locusts is the transfer of population density information to the next generation via priming during embryonic development. Both egg size and factors released by accessory glands and deposited on eggs during passage through the oviducts or added to the foam covering egg pods are considered crucial for this transgenerational effect (Pener and Simpson, 2009; Tanaka and Maeno, 2010). Cumulative effects on phase characteristics over the entire life cycle are the hallmark of locust polyphenism and have also been found in other groups that exhibit complex life cycle syndromes (Moran, 1994). Thus, elucidating the architecture of the multiple and synergistic developmental switches remains a challenging task for future work.

Aphids, with their highly complex life cycle shifts between sexual and asexual reproduction on the one hand, and winged and wingless morphs on the other, are still an enigma when it comes to mechanisms underlying these shifts, which can occur in different combinations. While crowding is a trigger for the shift from wingless asexual to winged asexual forms, it probably does not involve JH as an endocrine mediator. Juvenile hormone, however, seems to be involved in the photoperiod-induced shift from the winged asexually reproducing form to winged sexuals, a shift that frequently is accompanied by a host plant shift as well (Baker, 1970; Braendle et al., 2006; Le Trionnaire et al., 2008). Genomic analyses have now revealed a surprising alteration in cuticle proteins and the dopamine biosynthetic pathway that may contribute to photoperiod perception and signal transduction involved in this shift (Simon et al., 2010).

In termites, wing development is restricted to primary reproductives, with soldiers and workers (false or true ones) representing wingless phenotypes. Whereas the role of morphogenetic hormones in wing development is still unclear, hormonal regulation of caste development in this clade has been much exploited for pest control, because application of JH analogs can shift the caste-ratio balance toward soldiers (and away from workers and reproductives). Using this well-defined shift as a model, Scharf and colleagues identified hexamerins as hemolymph proteins capable of sequestering circulating JH from exerting its biological effects and showed that the quantitative balance between two hexamerins can be crucial to adjust soldier to worker ratios in a colony according to environmental and social conditions (Scharf et al., 2007; Zhou et al., 2007). The role of hexamerins emerged from differential gene expression screens (Scharf et al., 2003), similar to those performed earlier on genes underlying the formation and function of soldier-specific structures (Miura, 2004; Miura et al., 1999). In this context it will be important to address two sets of questions: (1) the relative immaturity of termite larvae, which, especially in the higher termites, need to be cared for by workers, and (2) the stationary and regressive molts in lower termites, an astonishing phenomenon in insect metamorphosis that emphasizes the importance of molting events in termite societies.

Wing polyphenism in holometabolans takes two very different forms: seasonal wing pattern differences in lepidopterans, and wing reduction in ant workers. Wing pattern polyphenism in butterflies is an allelic polymorphism in camouflage in response to predator pressure, which, in its expression, depends on the timing of the pupal ecdysteroid peak (Brakefield et al., 1998, 2007; Koch et al., 1996). Analysis of candidate genes in wing patterning identified Distal-less as a key factor, and the availability of genomic resources and mutants now sets the road map for deciphering the genomic regulatory network underlying wing eyespot formation (Saenko et al., 2007).

Winglessness in ant workers, in contrast, is an integral element of caste polyphenism in the Formicidae. It is a result of the shutdown of the gene expression cascade regulating wing disc development (Abouheif and Wray, 2002) in response to embryonic/larval hormone titers. While little progress has been made on the endocrine regulation of wing development suppression and worker development in general, a genetic bias to caste fate has now been identified in several ant species (Schwander et al., 2010).

While wing polyphenism is an ancestral character in the caste syndrome of ants, caste polyphenisms in the other social Hymenoptera (wasps and bees) are built on asymmetries in fecundity and, correspondingly, the development of the reproductive system. Such asymmetries are also crucial, and probably were the primary drivers in the evolution of sociality in ants. The focus on *A. mellifera* as a model organism in research on caste development in social Hymenoptera has certainly been promoted by the honey bee's economic importance, a factor that more recently has also stimulated much of the research on the bumblebee *B. terrestris*. Synthesis and titers of morphogenetic hormones have been monitored throughout the entire life cycle of these insects, and these studies have shown that a correlated modulation of JH and ecdysteroid titers drives caste development during the larval stages (Hartfelder and Engels, 1998). Furthermore, the full cascade from initial environmental triggers, through the endocrine response to these, and finally to downstream differentiative processes in target organs, are now starting to be mapped out due to the concentrated efforts in honey bee genomics (Barchuk *et al.*, 2007; Cristino *et al.*, 2006; Evans and Wheeler, 2001). Two emergent factors that may interact with JH and ecdysteroids in different ways are the insulin signaling pathway with TOR as a convergent entry (Azevedo and Hartfelder, 2008; Patel *et al.*, 2007; Wheeler *et al.*, 2006), and epigenetic effects impressively demonstrated through RNAi-mediated knockdown of a DNA methyltransferase (Kucharski *et al.*, 2008).

Another polyphenism related to reproduction, yet going in a completely different direction, is that of male weaponry in dung beetles. The development of these exaggerated morphologies appears to be contingent on at least two critical periods during larval development when horn expression is sensitive to endocrine events. Details of endocrine system functions still need to be worked out, but a great deal of progress has been made in understanding the genetic mechanisms underlying horn growth, which appears to have entailed the co-option of portions of traditional appendage patterning, especially along the proximo-distal axis (Emlen *et al.*, 2006, 2007; Moczek *et al.*, 2006).

The past decade has seen an exponential increase in genomic information, not only for classical model organisms in biology, but also for a wide range of species, including those expressing polyphenic traits. Two complete genomes have been sequenced and annotated for polyphenic insect species: the honey bee *A. mellifera* and the pea aphid *A. pisum*. These efforts were crucial for the interpretation of differential gene expression results and candidate gene analyses, and crystallized the following emergent topics: (1) the integration of signaling pathways through expression analyses of nuclear receptors related to the ecdysone and (tentative) JH response and insulin signaling, (2) the role of epigenetic modification through DNA methylation, and (3) the role of small and long noncoding RNAs in developmental regulation. The latter is

an emergent topic, especially in human genomics and no doubt will soon influence research underway in developmental biology in most multicellular organisms (Mattick, 2009, 2010).

Species exhibiting phenotypic plasticity, especially the existence of discrete, alternative phenotypes that go well beyond the more common forms of phenotypic plasticity seen in reaction norm variation, are natural experiments on how far genotype-environment interactions can be taken both in ontogeny and in evolution. Many of the polyphenic insects have been established as laboratory populations that can be exposed to different conditions, so that the molecular underpinnings of alternative phenotypes are accessible both at the organismal as well as at the tissue level. This is highly favorable to genome and transcriptome studies, which, without doubt, will come to dominate the field through the advance of new generation sequencing methodologies that will increase sequencing efforts at diminishing costs, not only for DNA and RNA analyses, but also in proteomics. The limits in all these analyses will be set by the advances in bioinformatics tools and computational capacity.

A critical issue to the interpretation of high throughput transcriptome or proteome data is, and will continue to be, the fact that most of the observed molecular level changes are downstream consequences of a "switch mechanism." It may be much more difficult to elucidate genes associated with the switch because there may be only a few genes, and these are harder to detect, and the critical time periods in question occur earlier, before there are obvious morphological differences between the forms. But here polyphenisms excel, because prior endocrine work in most cases has already identified the critical periods when these developmental "decisions" occur and the precise environmental conditions needed to generate animals with each form. Grasping the molecular underpinnings of switch mechanisms will, thus, require carefully devised experiments in combination with high throughput screens. We hope and predict that such approaches will shed light on the ways that hormones (including the elusive JH) alter gene expression to generate distinct alternative plastic phenotypes, and will illuminate the genetic bases of growth, morphology, behavior, and physiology in general (depending on the phenotypic differences relevant to each polyphenic species).

Perhaps the greatest challenge will be to conceptually link the results of such genomic/proteomic studies with organismic-level manifestations in physiology, behavior, and ecology. Here we expect two questions to become predominant. The first one is in the realm of developmental biology and addressing molecular and cellular mechanisms underlying reaction norms and phenotypic plasticity. Many aspects of phenotypic plasticity in insects can be ascribed to general size-related allometries. These have been mathematically formulated in growth models that assume competition for available nutrients between growing morphogenetic fields (Nijhout and Wheeler, 1994). Such intuitive trade-offs have, however, not yet been taken beyond genetically tractable model organisms, such as *Drosophila*, where gene regulatory networks for life histories and reaction norms are emergent themes (Colombani *et al.*, 2005; Debat *et al.*, 2009; Mirth *et al.*, 2005; Mirth and Riddiford, 2007). The second question is much broader and involves linking evolutionary aspects in the life histories of polyphenic insects to the evolution of regulatory mechanisms underlying the development of alternative phenotypes (West-Eberhard, 2003).

Note added in Proof: A recent landmark publication [KamaKura, M. (2011). Royalactin induces queen differentiation in honey bees. *Nature, Epub,* doi:10.1038/nature10093] convincingly demonstrated that royalactin, an MRJ P1 derived protein that easily degrades in stored royal jelly, is important for queen determination in the honey bee, *Apis mellifera.* It acts through an EGF receptor signaling pathay.

References

Abbot, P., Withgott, J. H., & Moran, N. A. (2001). Genetic conflict and conditional altruism in social aphid colonies. *Proc. Natl. Acad. Sci. USA, 98,* 12068–12071.

Abouheif, E., & Wray, G. A. (2002). Evolution of the genetic network underlying wing polyphenism in ants. *Science, 297,* 249–252.

Aé, S. A. (1957). Effects of photoperiod on Colias eurytheme. *Lepidopt. News, 11,* 207–214.

Agrahari, M., & Gadagkar, R. (2003). Juvenile hormone accelerates ovarian development and does not affect age polyethism in the primitively eusocial wasp, Ropalidia marginata. *J. Insect Physiol., 49,* 217–222.

Alaux, C., Jaisson, P., & Hefetz, A. (2004). Queen influence on worker reproduction in bumblebees (Bombus terrestris) colonies. *Insectes Soc., 51,* 287–293.

Alaux, C., Jaisson, P., & Hefetz, A. (2006). Regulation of worker reproduction in bumblebees (Bombus terrestris): workers eavesdrop on a queen signal. *Behav. Ecol. Sociobiol., 60,* 439–446.

Alaux, C., Le Conte, Y., Adams, H. A., Rodriguez-Zas, S., Grozinger, C. M., et al. (2009). Regulation of brain gene expression in honey bees by brood pheromone. *Genes Brain Behav., 8,* 309–319.

Alexander, R. D. (1968). Life cycle origins, speciation, and related phenomena in crickets. *Quart. Rev. Biol., 43,* 1–41.

Amdam, G. V., & Omholt, S. W. (2002). The regulatory anatomy of honey bee lifespan. *J. Theor. Biol., 21,* 209–228.

Amdam, G. V., & Omholt, S. W. (2003). The hive bee to forager transition in honey bee colonies: the double repressor hypothesis. *J. Theor. Biol., 223,* 451–464.

Amdam, G. V., Norberg, K., Fondrk, M. K., & Page, R. E. (2004). Reproductive ground plan may mediate colony-level selection effects on individual foraging behavior in honey bees. *Proc. Natl. Acad. Sci. USA, 101,* 11350–11355.

Amdam, G. V., Aase, A., Seehuus, S. C., Fondrk, M. K., Norberg, K., et al. (2005). Social reversal of immunosenescence in honey bee workers. *Exp. Gerontol., 40,* 939–947.

Amdam, G. V., Csondes, A., Fondrk, M. K., & Page, R. E. (2006a). Complex social behaviour derived from maternal reproductive traits. *Nature, 439,* 76–78.

Amdam, G. V., Norberg, K., Page, R. E., Erber, J., & Scheiner, R. (2006b). Downregulation of vitellogenin gene activity increases the gustatory responsiveness of honey bee workers (Apis mellifera). *Behav. Brain Res., 169,* 201–205.

Amdam, G. V., & Page, R. E. (2007). The making of a social insect: developmental architectures of social design. *Bioessays, 29,* 334–343.

Ament, S. A., Corona, M., Pollock, H. S., & Robinson, G. E. (2008). Insulin signaling is involved in the regulation of worker division of labor in honey bee colonies. *Proc. Natl. Acad. Sci. USA, 105,* 4226–4231.

Ament, S. A., Wang, Y., & Robinson, G. E. (2010). Nutritional regulation of division of labor in honey bees: toward a systems biology perspective. Wiley Interdisc. *Rev. Syst. Biol. Med., 2,* 566–576.

Andersson, M. (1994). *Sexual Selection.* Princeton University Press, Princetown.

Anstey, M. L., Rogers, S. M., Ott, S. R., Burrows, M., & Simpson, S. J. (2009). Serotonin mediates behavioral gregarization underlying swarm formation in desert locusts. *Science, 323,* 627–630.

Aoki, S. (1977). Colophina clematis, an aphid species with "soldiers". *Kontyu, 45,* 276–282.

Applebaum, S. W., Avisar, E., & Heifetz, Y. (1997). Juvenile hormone and locust phase. *Arch. Insect Biochem. Physiol., 35,* 375–391.

Austad, S. N. (1984). A classification of alternative reproductive behaviors and methods for field-testing ESS models. *Am. Zool., 24,* 309–319.

Awiti, L. R., & Hidaka, T. (1982). Neuroendocrine mechanisms involved in pupal color dimoprhism in swallowtail Papilio xuthus. *Insect Sci. Appl., 3,* 181–192.

Ayoade, O., Morooka, S., & Tojo, S. (1999). Enhancement of short wing formation and ovarian growth in the genetically defined macropterous strain of the brown planthopper, Nilaparvata lugens. *J. Insect Physiol., 45,* 93–100.

Azevedo, S. V., & Hartfelder, K. (2008). The insulin signaling pathway in honey bee (Apis mellifera) caste development - differential expression of insulin-like peptides and insulin receptors in queen and worker larvae. *J. Insect Physiol., 54,* 1064–1071.

Azevedo, S. V., Carantona, O. A.M., Oliveira, T. L., & Hartfelder, K. (2011). Differential expression of hypoxia pathway genes in honey bee (Apis mellifera L.) caste development. *J. Insect Physiol., 57,* 38–45.

Badisco, L., Claeys, I., Van Hiel, M., Clynen, E., Huybrechts, J., et al. (2008). Purification and characterization of an insulin-related peptide in the desert locust, Schistocerca gregaria: immunolocalization, cDNA cloning, transcript profiling and interaction with neuroparsin. *J. Mol. Endocrinol., 40,* 137–150.

Baggerman, G., Clynen, E., Mazibur, R., Veelaert, D., Breuer, M., et al. (2001). Mass spectrometric evidence for the deficiency in the dark color-inducing hormone, [His7]-corazonin in an albino strain of Locusta migratoria as well as for its presence in solitary Schistocerca gregaria. *Arch. Insect Biochem. Physiol., 47,* 150–160.

Baker, R. (1970). Bird predation as a selective pressure of the cabbage butterflies, Pieris rapae and P. brassicae. *J. Zool.*, *152*, 43–59.

Barchuk, A. R., Bitondi, M. M.G., & Simões, Z. L.P. (2002). Effects of juvenile hormone and ecdysone on the timing of vitellogenin appearance in hemolymph of queen and worker pupae of Apis mellifera. *J. Insect Sci.*, *2*, e8.

Barchuk, A. R., Maleszka, R., & Simoes, Z. L.P. (2004). Apis mellifera ultraspiracle: cDNA sequence and rapid up-regulation by juvenile hormone. *Insect Mol. Biol.*, *13*, 459–467.

Barchuk, A. R., Cristino, A. S., Kucharski, R., Costa, L. F., Simoes, Z. L.P., et al. (2007). Molecular determinants of caste differentiation in the highly eusocial honey bee Apis mellifera. *BMC Dev. Biol.*, *7*.

Barchuk, A. R., Figueiredo, V. L.C., & Simões, Z. L.P. (2008). Downregulation of ultraspiracle gene expression delays pupal development in honey bees. *J. Insect Physiol.*, *54*, 1035–1040.

Barth, R. H., Lester, L. J., Sroka, P., Kessler, T., & Hearn, R. (1975). Juvenile hormone promotes dominance behavior and ovarian development in social wasps (Polistes annularis). *Experientia*, *31*, 691–692.

Bazazi, S., Buhl, J., Hale, J. J., Anstey, M. L., Sword, G. A., et al. (2008). Collective motion and cannibalism in locust migratory bands. *Curr. Biol.*, *18*, 735–739.

Beetsma, J. (1985). Feeding behaviour of nurse bees, larval food composition and caste differentiation in the honey bee (Apis mellifera L.). In B. Hölldobler, & M. Lindauer (Eds.), *Experimental Behavioral Ecology and Sociobiology: in memoriam Karl von Frisch, 1886-1982* (pp. 407–410). Sunderland, Mass: Sinauer.

Beldade, P., & Brakefield, P. M. (2002). The genetics and evo-devo of butterfly wing patterns. *Nat. Rev. Genet.*, *3*, 442–452.

Beldade, P., Brakefield, P. M., & Long, A. D. (2002). Contribution of Distal-less to quantitative variation in butterfly eyespots. *Nature*, *415*, 315–318.

Beldade, P., Rudd, S., Gruber, J. D., & Long, A. D. (2006). A wing expressed sequence tag resource for Bicyclus anynana butterflies, an evo-devo model. *BMC Genomics*, *7*, e130.

Beldade, P., McMillan, W. O., & Papanicolaou, A. (2008). Butterfly genomics eclosing. *Heredity*, *100*, 150–157.

Ben Hamouda, A., Ammar, M., Ben Hamouda, M. H., & Bouain, A. (2009). The role of egg pod foam and rearing conditions of the phase state of the Asian migratory locust Locusta migratoria migratoria (Orthoptera, Acrididae). *J. Insect Physiol.*, *55*, 617–623.

Bergot, B. J., Schooley, D. A., & de Kort, C. A.D. (1981). Identification of JH III as the principal juvenile hormone in Locusta migratoria. *Experientia*, *37*, 909–910.

Bertuso, A. G., Morooka, S., & Tojo, S. (2002). Sensitive periods for wing development and precocious metamorphosis after Precocene treatment of the brown planthopper, Nilaparvata lugens. *J. Insect Physiol.*, *48*, 221–229.

Blackman, R. L. (1994). The simplification of aphid terminology. *Eur. J. Entomol.*, *91*, 139–141.

Bloch, G., Borst, D. W., Huang, Z. Y., Robinson, G. E., & Hefetz, A. (1996). Effects of social conditions on juvenile hormone mediated reproductive development in Bombus terrestris workers. *Physiol. Entomol.*, *21*, 257–267.

Bloch, G., & Hefetz, A. (1999a). Regulation of reproduction by dominant workers in bumblebee (Bombus terrestris) queenright colonies. *Behav. Ecol. Sociobiol.*, *45*, 125–135.

Bloch, G., & Hefetz, A. (1999b). Reevaluation of the role of mandibular glands in regulation of reproduction in bumblebee colonies. *J. Chem. Ecol.*, *25*, 881–896.

Bloch, G., Borst, D. W., Huang, Z. Y., Robinson, G. E., Cnaani, J., et al. (2000a). Juvenile hormone titers, juvenile hormone biosynthesis, ovarian development and social environment in Bombus terrestris. *J. Insect Physiol.*, *46*, 47–57.

Bloch, G., Hefetz, A., & Hartfelder, K. (2000b). Ecdysteroid titer, ovary status, and dominance in adult worker and queen bumble bees (Bombus terrestris). *J. Insect Physiol.*, *46*, 1033–1040.

Boleli, I. C., Hartfelder, K., & Simões, Z. L.P. (1995). Serotonin-like immunoreactivity in the central nervous and neuroendocrine system of honey bee (Apis mellifera) larvae. *Zool. Anal. Complex Syst.*, *99*, 58–67.

Boleli, I. C., Simões, Z. L.P., & Hartfelder, K. (1998). The stomatogastric nervous system of the honey bee (Apis mellifera) in a critical phase of caste development. *J. Morphol.*, *236*, 139–148.

Boleli, I. C., Simões, Z. L.P., & Bitondi, M. M.G. (1999). Cell death in ovarioles causes permanent sterility in Frieseomelitta varia workers bees. *J. Morphol.*, *242*, 271–282.

Boleli, I. C., Simões, Z. L.P., & Bitondi, M. M.G. (2000). Regression of the lateral oviducts during the larval-adult transformation of the reproductive system of Melipona quadrifasciata and Frieseomelitta varia. *J. Morphol.*, *243*, 141–151.

Bonetti, A. M., Kerr, W. E., & Matusita, S. H. (1995). Effects of juvenile hormones I, II and III, in single and fractionated doses in Melipona bees. *Rev. Bras. Biol.*, *55*(Suppl. 1), 113–120.

Bordereau, C. (1971). Dimorphisme sexuel du systéme trachéen chez les imago ailés de Bellicositermes natalensis Haviland (Isoptera, Termitidae) ; rapports avec la physogastrie de la reine. *Arch. Zool. Exp. Gen.*, *112*, 33–54.

Botens, F. F.W., Rembold, H., & Dorn, A. (1997). Phase-related juvenile hormone determinations in field catches and laboratory strains of different Locusta migratoria subspecies. In S. Kawashima, & S. Kikuyama (Eds.), *Advances in Comparative Endocrinology* (pp. 197–203). Bologna: Monduzzi.

Boulay, R., Hooper-Bui, L. M., & Woodring, J. (2001). Oviposition and oogenesis in virgin fire ant females Solenopsis invicta are associated with a high level of dopamine in the brain. *Physiol. Entomol.*, *26*, 294–299.

Bowsher, J. H., Wray, G. A., & Abouheif, E. (2007). Growth and patterning are evolutionarily dissociated in the vestigial wing discs of workers of the red imported fire ant, Solenopsis Invicta. *J. Exp. Zool. B*, *308*, 769–776.

Braendle, C., Friebe, I., Caillaud, M. C., & Stern, D. L. (2005a). Genetic variation for an aphid wing polyphenism is genetically linked to a naturally occurring wing polymorphism. *Proc. Soc. Lond. B Biol. Sci.*, *272*, 657–664.

Braendle, C., Caillaud, M. C., & Stern, D. L. (2005b). Genetic mapping of aphicarus - a sex-linked locus controlling a wing polymorphism in the pea aphid (Acyrthosiphon pisum). *Heredity*, *94*, 435–442.

Braendle, C., Davis, G. K., Brisson, J. A., & Stern, D. L. (2006). Wing dimorphism in aphids. *Heredity*, *97*, 192–199.

Brakefield, P. M., & Larsen, T. B. (1984). The evolutionary significance of dry and wet season forms in tropical butterflies. *Biol. J. Linn. Soc., 22*, 1–22.

Brakefield, P. M., & Reitsma, N. (1991). Phenotypic plasticity, seasonal climate and the population biology of Bicyclus butterflies (Satyridae) in Malawi. *Ecol. Entomol., 16*, 291–304.

Brakefield, P. M., & Mazotta, V. (1995). Matching field and laboratory experiments: effects of neglecting daily temperature variation in insect reaction norms. *J. Evol. Biol., 8*, 559–573.

Brakefield, P. M., Gates, J., Keys, D., Kesbeke, F., Wijngaarden, P. J., et al. (1996). Development, plasticity and evolution of butterfly eyespot patterns. *Nature, 384*, 236–242.

Brakefield, P. M., Kesbeke, F., & Koch, P. B. (1998). The regulation of phenotypic plasticity of eyespots in the butterfly Bicyclus anynana. *Am. Nat., 152*, 853–860.

Brakefield, P. M., Pijpe, J., & Zwaan, B. J. (2007). Developmental plasticity and acclimation both contribute to adaptive responses to alternating seasons of plenty and of stress in Bicyclus butterflies. *J. Biosci., 32*, 465–475.

Brian, M. V. (1974). Caste determination in Myrmica rubra: the role of hormones. *J. Insect Physiol., 20*, 1351–1365.

Brisson, J. A., Davis, G. K., & Stern, D. L. (2007). Common genome-wide patterns of transcript accumulation underlying the wing polyphenism and polymorphism in the pea aphid (Acyrthosiphon pisum). *Evol. Dev., 9*, 338–346.

Brisson, J. A., Ishikawa, A., & Miura, T. (2010). Wing development genes of the pea aphid and differential gene expression between winged and unwinged morphs. *Insect Mol. Biol., 19*, 63–73.

Brunetti, C., Selegue, J. E., Monterio, A., French, V., Brakefield, P. M., et al. (2001). The generation and diversification of butterfly eyespot patterns. *Curr. Biol., 11*, 1578–1585.

Bückmann, D., & Maisch, A. (1987). Extraction and partial purification of the pupal melanization reducing factor (PMRF) from Inachis io (Lepidoptera). *Insect Biochem., 17*, 841–844.

Buhl, J., Sumpter, D. J.T., Couzin, I. D., Hale, J. J., Despland, E., et al. (2006). From disorder to order in marching locusts. *Science, 312*, 1402–1406.

Burns, S. N., Teal, P. E.A., Vander Meer, R. K., & Vogt, J. T. (2002). Identification and action of juvenile hormone III from sexually mature alate females of the red imported fire ant, Solenopsis invicta. *J. Insect Physiol., 48*, 357–365.

Buschini, M. L.T., & Campos, L. A.O. (1994). Caste determination in Trigona spinipes (Hymenoptera: Apidae): influence of the available food and the juvenile hormone. *Rev. Bras. Biol., 55*(Suppl. 1), 121–129.

Camargo, J. M.F., & Pedro, S. R.M. (1992). Systematics, phylogeny and biogeography of the Meliponinae (Hymenoptera, Apidae): a mini-review. *Apidologie, 23*, 509–522.

Campos, L. A.O., Velthuis, H. H.W., & Velthuis-Kluppell, F. M. (1975). Juvenile hormone and caste determination in a stingless bee. *Naturwissenschaften, 62*, 98–99.

Campos, L. A.O. (1978). Sex determination in bees. VI: effect of a juvenile hormone analog in males and females of Melipona quadrifasciata (Apidae). *J. Kansas Entom. Soc., 51*, 228–234.

Campos, L. A.O. (1979). Determinação do sexo em abelhas. XIV. Papel do hormônio juvenil na diferenciação das castas na subfamilia Meliponinae (Hymenoptera: Apidae). *Rev. Bras. Biol., 39*, 965–971.

Carlin, N. F., Gladstein, D. S., Berry, A. J., & Pierce, N. E. (1994). Absence of kin discrimination behavior in a soldier-producing aphid, Ceratovacuna japonica (Hemiptera, Pemphigidae, Cerataphidini). *J.N.Y. Entomol. Soc., 102*, 287–298.

Carpenter, J. M. (1991). Phylogenetic relationships and the origin of social behaviour in the Vespidae. In K. G. Ross, & R. W. Matthews (Eds.), *The Social Biology of Wasps* (pp. 7–32). Ithaca, NY: Cornell University Press.

Carroll, S. P., & Boyd, C. (1992). Host radiation in the soapberry bug: Natural history, with the history. *Evolution, 51*, 1052–1069.

Carroll, S. P., Dingle, H., & Klassen, S. P. (1997). Genetic differentiation of fitnes-associated traits among rapidly evolving populations of the soapberry bug. *Evolution, 51*, 1182–1188.

Carroll, S. P., Dingle, H., Famula, T. R., & Fox, C. W. (2001). Genetic architecture of adaptive differentiation in evolving host races of the soapberry bug, Jadera hematoloma. *Genetica, 112*, 257–272.

Cauillaud, M. C., Boutin, M., Braendle, C., & Simon, J. C. (2002). A sex-linked locus controls wing polymorphism in males of the pea aphid, Acyrthosiphon pisum (Harris). *Heredity, 89*, 346–352.

Cepeda, O. I. (2006). Division oof labor during brood production in stingless bees with special reference to individual participation. *Apidologie, 37*, 175–190.

Christiaens, O., Iga, M., Velarde, R. A., Rouge, P., & Smagghe, G. (2010). Halloween genes and nuclear receptors in ecdysteroid biosynthesis and signalling in the pea aphid. *Insect Mol. Biol., 19*, 187–200.

Christiansen-Weniger, P., & Hardie, J. (2000). The influence of parasitism on wing development in male and female pea aphids. *J. Insect Physiol., 46*, 861–867.

Claeys, I., Breugelmans, B., Simonet, G., Van Soest, S., Sas, F., et al. (2006). Neuroparsin transcripts as molecular markers in the process of desert locust (Schistocerca gregaria) phase transition. *Biochem. Biophys. Res. Com., 341*, 599–606.

Clark, R. (1997). Dimorphic males display alternative reproductive strategies in the marine amphipod Jassa marmorata. *Ethology, 7*, 531–553.

Clynen, E., Stubbe, D., de Loof, A., & Schoofs, L. (2002). Peptide differential display: a novel approach for phase transition in locusts. *Comp. Biochem. Physiol. B, 132*, 107–115.

Cnaani, J., Borst, D. W., Huang, Z. Y., Robinson, G. E., & Hefetz, A. (1997). Caste determination in Bombus terrestris: differences in development and rates of JH biosynthesis between queen and worker larvae. *J. Insect Physiol., 43*, 373–381.

Cnaani, J., Robinson, G. E., Bloch, G., Borst, D., & Hefetz, A. (2000a). The effect of queen-worker conflict on caste determination in the bumblebee Bombus terrestris. *Behav. Ecol. Sociobiol., 47*, 346–352.

Cnaani, J., Robinson, G. E., & Hefetz, A. (2000b). The critical period for caste determination in Bombus terrestris and its juvenile hormone correlates. *J. Comp. Physiol. A-Sens. Neural Behav. Physiol., 186*, 1089–1094.

Cnaani, J., & Hefetz, A. (2001). Are queen Bombus terrestris giant workers or are workers dwarf queens? Solving the 'chicken and egg' problem in a bumblebee species. *Naturwissenschaften, 88*, 85–87.

Colombani, J., Bianchini, L., Layalle, S., Pondeville, E., Dauphin-Villemant, C., et al. (2005). Antagonistic actions of ecdysone and insulins determine final size in *Drosophila*. *Science*, *310*, 667–670.

Colonello-Frattini, N. A., & Hartfelder, K. (2009). Differential gene expression profiling of mucus glands of honey bee (Apis mellifera) drones during sexual maturation. *Apidologie*, *40*, 481–495.

Condamin, M. (1973). Monographie du genre Bicyclus (Lepidoptera, Satyridae). *Mém Inst. Fond. Afr. Noire*, *88*, 1–324.

Cook, D. (1987). Sexual selection in dung beetles I. A multivariate study of the morphological variation in two species of Ontophagus (Scarabeidae: Ontophagini). *Aust. J. Zool.*, *35*, 123–132.

Cornette, R., Gotoh, H., Koshikawa, S., & Miura, T. (2008). Juvenile hormone titers and caste differentiation in the damp-wood termite Hodotermopsis sjostedti (Isoptera, Termopsidae). *J. Insect Physiol.*, *54*, 922–930.

Corona, M., Estrada, E., & Zurita, M. (1999). Differential expression of mitochondrial genes between queens and workers during caste determination in the honey bee Apis mellifera. *J. Exp. Biol.*, *202*, 929–938.

Corona, M., Velarde, R. A., Remolina, S., Moran-Lauter, A., Wang, Y., et al. (2007). Vitellogenin, juvenile hormone, insulin signaling, and queen honey bee longevity. *Proc. Natl. Acad. Sci. USA*, *104*, 7128–7133.

Couillaud, F., Mauchamp, B., & Girardie, A. (1987). Biological, radiochemical and physicochemical evidence for the low activity of disconnected corpora allata in locust. *J. Insect Physiol.*, *33*, 223–228.

Cremer, S., Lautenschläger, B., & Heinze, J. (2002). A transitional stage between the ergatoid and winged male morph in the ant Cardiocondyla obscurior. *Insectes Soc.*, *49*, 221–228.

Cristino, A. S., Nunes, F. M.F., Lobo, C. H., Bitondi, M. M.G., Simoes, Z. L.P., et al. (2006). Caste development and reproduction: a genome-wide analysis of hallmarks of insect eusociality. *Insect Mol. Biol.*, *15*, 703–714.

Crone, E. J., Zera, A. J., Anand, A., Oakeshott, J. G., Sutherland, T. D., et al. (2007). JHE in Gryllus assimilis: cloning, sequence-activity associations and phylogeny. *Insect Biochem. Mol. Biol.*, *37*, 1359–1365.

Crossland, M. W.J., Crozier, R. H., & Jefferson, E. (1988). Aspects of the biology of the primitive ant genus Myrmecia F. (Hymenoptera: Formicidae). *J. Aust. Ent. Soc.*, *27*, 305–309.

Cuvillier-Hot, V., Cobb, M., Malosse, C., & Peeters, C. (2001). Sex, age and ovarian activity affect cuticular hydrocarbons in Diacamma ceylonense, a queenless ant. *J. Insect Physiol.*, *47*, 485–493.

Cuvillier-Hot, V., Lenoir, A., & Peeters, C. (2004). Reproductive monopoly enforced by sterile police workers in a queenless ant. *Behav. Ecol.*, *15*, 970–975.

Dai, H., Wu, X., & Wu, S. (2001). The change of juvenile hormone titer and its relation with wing dimorphism of brown planthopper, Nilaparvata lugens. *Acta Entom. Sin.*, *44*, 27–32.

Dale, J. F., & Tobe, S. S. (1986). Biosynthesis and titre of juvenile hormone during the first gonotrophic cycle in isolated and crowded Locusta migratoria females. *J. Insect Physiol.*, *32*, 763–769.

Dallacqua, R. P., Simões, Z. L.P., & Bitondi, M. M.G. (2007). Vitellogenin gene expression in stingless bee workers differing in egg-laying behavior. *Insectes Soc.*, *54*, 70–76.

Danforth, B. N. (1991). The morphology and behavior of dimorphic males in Perdita portalis (Hymenoptera: Andrenidae). *Behav. Ecol. Sociobiol.*, *29*, 235–247.

Darrouzet, E., Mauchamp, B., Prestwich, G. D., Kerhoas, L., Ujvary, I., et al. (1997). Hydroxy juvenile hormones: New putative juvenile hormones biosynthesized by locust corpora allata in vitro. *Biochem. Biophys. Res. Commun.*, *240*, 752–758.

Darwin, C. R. (1859). *On the Origin of Species by Means of Natural Selection*. London: John Murray.

Darwin, C. R. (1871). *The Descent of Man and Selection in Relation to Sex*. London: John Murray.

Debat, V., Debelle, A., & Dworkin, I. (2009). Plasticity, canalization, and developmental stability of the *Drosophila* wing: joint effects of mutations and developmental temperature. *Evolution*, *63*, 2864–2876.

Dedej, S., Hartfelder, K., Rosenkranz, P., & Engels, W. (1998). Caste determination is a sequential process: effect of larval age on ovariole number, hind leg size and cephalic volatiles in the honey bee (Apis mellifera carnica). *J. Apicult. Res.*, *37*, 183–190.

Deng, A. L., Torto, B., Hassanali, A., & Ali, E. E. (1996). Effects of shifting to crowded or solitary conditions on pheromone release and morphometrics of the desert locust, Schistocerca gregaria (Forskål) (Orthoptera: Acrididae). *J. Insect Physiol.*, *42*, 771–776.

Denno, F., & Perfect, T. J. (1994). *Planthoppers: Their Ecology and Management*. Chapman and Hall New York.

Denno, R. F., Douglass, L. W., & Jacobs, D. (1986). Effects of crowding and host plant nutrition on a wing dimorphic planthopper. *Ecology*, *67*, 116–123.

Denno, R. F., Roderick, G. K., Peterson, M. A., Huberty, A. F., Döbel, H., et al. (1996). Habitat persistence underlies intraspecific variation and dispersal strategies of planthoppers. *Ecol. Monogr*, *66*, 389–408.

Diakonov, D. M. (1925). Experimental and biometrical investigations on dimoprhic variability of Forficula. *J. Genet.*, *15*, 201–232.

Dietemann, V., Hölldobler, B., & Peeters, C. (2002). Caste specialization and differentiation in reproductive potential in the phylogenetically primitiva ant Myrmecia gulosa. *Insectes Soc.*, *49*, 289–298.

Dietz, A., Hermann, H. R., & Blum, M. S. (1979). The role of exogenous JH I, JH III and anti-JH (Precocene II) on queen induction in 4.5-day-old worker honey bee larvae. *J. Insect Physiol.*, *25*, 503–512.

Dilao, R., & Sainhas, J. (2004). Modelling butterfly wing eyespot patterns. *Proc. R. Soc. Lond. B Biol. Sci.*, *271*, 1565–1569.

Dingle, H., & Winchell, R. (1997). Juvenile hormone as a mediator of plasticity in insect life histories. *Arch. Insect Biochem. Physiol.*, *35*, 359–373.

Dixon, A. F.G. (1977). Aphid ecology - life cycles, polymorphism, and population regulation. *Annu. Rev. Ecol. Syst.*, *8*, 329–353.

Dixon, A. F.G. (1998). *Aphid Ecology*. London: Chapman & Hall.

Dolezal, A. G., Brent, C. S., Gadau, J., Holldobler, B., & Amdam, G. V. (2009). Endocrine physiology of the division of labour in Pogonomyrmex californicus founding queens. *Anim. Behav.*, *77*, 1005–1010.

Dominey, W. J. (1984). Alternative tactics and evolutionarily stable strategies. *Am. Zool.*, *24*, 385–396.

Dorn, A., Ress, C., Sickold, S., & Wedekind-Hirschberger, S. (2000). Arthropoda - Insecta: Endocrine control of phase polymorphism. In A. Dorn (Ed.) *Progress in Developmental Endocrinology, Vol. 10*. (pp. 205–253). Chichester: Wiley, part B.

Duchateau, M. J., & Velthuis, H. H.W. (1988). Development and reproductive strategies in Bombus terrestris colonies. *Behaviour*, *107*, 186–207.

Duchateau, M. J., & Velthuis, H. H.W. (1989). Ovarian development and egg laying in workers of Bombus terrestris. *Entomol. Exp. Appl.*, *51*, 199–213.

Eberhard, W. G. (1982). Beetle horn dimoprhism: making the best of a bad lot. *Am. Nat.*, *119*, 420–426.

Eberhard, W. G., & Gutierrez, E. E. (1991). Male dimorphism in beetles and earwigs and the question of developmental constraints. *Evolution*, *45*, 18–28.

Eder, J., Kremer, J. P., & Rembold, H. (1983). Correlation of cytochrome c titer and respiration in Apis mellifera: adaptive response to caste determination defines workers, intercastes and queens. *Comp. Biochem. Physiol. B*, *76*, 703–716.

Elliott, K. L., & Stay, B. (2008). Changes in juvenile hormone synthesis in the termite Reticulitermes flavipes during development of soldiers and neotenic reproductives from groups of isolated workers. *J. Insect Physiol.*, *54*, 492–500.

Emlen, D. J. (1994). Environmental control of horn length dimorphism in the beetle Onthophagus acuminathus (Coleoptera: Scarabaeidae). *Proc. Roy. Soc. Lond. Ser. B*, *256*, 131–136.

Emlen, D. J. (1996). Artificial selection on horn length-body size allometry in the horned beetle Onthophagus acuminatus (Coleoptera: Scarabaeidae). *Evolution*, *50*, 1219–1230.

Emlen, D. J. (1997). Alternative reproductive tactics and male dimorphism in the horned beetle Onthophagus acuminatus (Coleoptera: Scarabaeidae). *Behav. Ecol. Sociobiol.*, *41*, 335–341.

Emlen, D. J., & Nijhout, H. F. (1999). Hormonal control of male horn length dimorphism in the dung beetle Onthophagus taurus (Coleoptera: Scarabaeidae). *J. Insect Physiol.*, *45*, 45–53.

Emlen, D. J., & Nijhout, H. F. (2001). Hormonal control of male horn length dimorphism in Onthophagus taurus (Coleoptera: Scarabaeidae): a second critical period of sensitivity to juvenile hormone. *J. Insect Physiol.*, *47*, 1045–1054.

Emlen, D. J., Marangelo, J., Ball, B., & Cunningham, C. W. (2005a). Diversity in the weapons of sexual selection: horn evolution in the beetle genus Onthophagus (Coleoptera: Scarabaeidae). *Evolution*, *59*, 1060–1084.

Emlen, D. J., Hunt, J. H., & Simmons, L. W. (2005b). Evolution of sexual dimorphism and male dcimorphism in the expression of beetle horns: phylogenetic evidence for modularity, evolutionary lability, and constraint. *Am. Nat.*, *166*, S42–S68.

Emlen, D. J., Szafran, Q., Corley, L. S., & Dworkin, I. (2006). Insulin signaling and limb-patterning: candidate pathways for the origin and evolutionary diversification of beetle 'horns'. *Heredity*, *97*, 179–191.

Emlen, D. J., Lavine, L. C., & Ewen-Campen, B. (2007). On the origin and evolutionary diversification of beetle horns. *Proc. Natl. Acad. Sci. USA*, *104*(Suppl. 1), 8661–8668.

Endo, K. (1984). Neuroendocrine regulation of the development of seasonal forms of the Asian comma butterfly Polygonia c-aureum. *Dev. Growth Diff.*, *26*, 217–222.

Endo, K., & Funatsu, S. (1985). Hormonal control of seasonal morph determination in the swallowtail butterfly, Papilio xuthus L. (Lepidoptera: Papilionidae). *J. Insect Physiol.*, *31*, 669–674.

Endo, K., & Kamata, Y. (1985). Hormonal control of seasonal morph determination in the small copper butterfly, Lycaena phlaeas daimio Seitz. *J. Insect Physiol.*, *31*, 701–706.

Endo, K., Yamashita, I., & Chiba, Y. (1985). Effect of photoperiodic transfer and brain surgery on the photoperiodic control of pupal diapause and seasonal morphs in the swallowtail Papilio xuthus. *Appl. Entom. Zool.*, *20*, 470–478.

Endo, K., Masaki, T., & Kumagai, K. (1988). Neuroendocrine regulation of the development of seasonal morphs in the Asian comma butterfly, Polygonia c-aureum L. difference in activity of summer-morph-producing hormone from brain extracts of the long-day and short-day pupae. *Zool. Sci.*, *5*, 145–152.

Engels, W. (1974). Occurrence and significance of vitellogenins in female castes of social Hymenoptera. *Am. Zool.*, *14*, 1229–1237.

Engels, W., & Imperatriz-Fonseca, V. L. (1990). Caste development, reproductive strategies, and control of fertility in honey bees and stingless bees. In W. Engels (Ed.), *Social Insects - an Evolutionary Approach to Castes and Reproduction* (pp. 168–230). Heidelberg: Springer.

Engels, W., Kaatz, H., Zillikens, A., Simões, Z. L.P., Trube, A., et al. (1990). Honey bee reproduction: vitellogenin and caste-specific regulation of fertility. In M. Hoshi, & O. Yamashita (Eds.), *Advances in Invertebrate Reproduction, Vol. 5*. (pp. 495–502). Amsterdam: Elsevier.

Evans, J. D., & Wheeler, D. E. (1999). Differential gene expression between developing queens and workers in the honey bee, Apis mellifera. *Proc. Natl. Acad. Sci. USA*, *96*, 5575–5580.

Evans, J. D., & Wheeler, D. E. (2000). Expression profiles during honey bee caste determination. *Genome Biol.*, *2*, e6.

Evans, J. D., & Wheeler, D. E. (2001). Gene expression and the evolution of insect polyphenisms. *Bioessays*, *23*, 62–68.

Evans, T. M., & Marcus, J. M. (2006). A simulation study of the genetic regulatory hierarchy for butterfly eyespot focus determination. *Evol. Dev.*, *8*, 273–283.

Fabre, J. H. (1899). Souvenirs Entomologique, Paris, excerpts translated by A.T. de Mattos, 1922. In *More beetles*. London: Hodder and Stoughton.

Fairbairn, D. J., & Yadlowski, D. E. (1997). Coevolution of traits determining migratory tendency: correlated response of a critical enzyme, juvenile hormone esterase, to selection on wing morphology. *J. Evol. Biol.*, *10*, 495–513.

Faustino, C. D., Silva-Matos, E. V., Mateus, S., & Zucchi, R. (2002). First record of emergency queen rearing in stingless bees (Hymenoptera, Apinae, Meliponini). *Insectes Soc.*, *49*, 111–113.

Feldlaufer, M. F., Herbert, E. W.J., & Svoboda, J. A. (1985). Makisterone A: the major ecdysteroid from the pupae of the honey bee, Apis mellifera. *Insect Biochem.*, *15*, 597–600

Feldlaufer, M. F., Svoboda, J. A., & Herbert, E. W.J. (1986). Makisterone A and 24-methylenecholesterol from the ovaries of the honey bee, Apis mellifera L. *Experientia, 42,* 200–201.

Fincher, G. T., & Woodruff, R. E. (1975). A European dung beetle, Onthophagus taurus Schreber, new to the US (Coleoptera: Scarabeidae). *Coleopt. Bull., 29,* 349–350.

Fletcher, D. J.C., & Blum, M. S. (1981). Pheromonal control of dealation and oogenesis in virgin queen fire ants. *Science, 212,* 73–75.

Fletcher, D. J.C., & Blum, M. S. (1983). The inhibitory pheromone of queen fire ants: effects of disinhibition on dealation and oviposition by virgin queens. *J. Comp. Physiol. A, 153,* 467–475.

Foret, S., Kucharski, R., Pittelkow, Y., Lockett, G. A., & Maleszka, R. (2009). Epigenetic regulation of the honey bee transcriptome: unravelling the nature of methylated genes. *BMC Genomics, 14,* e472.

Foster, W. A., & Rhoden, P. K. (1998). Soldiers effectively defend aphid colonies against predators in the field. *Anim. Behav., 55,* 761–765.

Foucard, J., Estoup, A., Loiseau, A., Rey, O., & Orivel, J. (2010). Thelytokous parthenogenesis, male clonality, and genetic caste determination in the little fire ant: new evidence and insights from the lab. *Heredity, 105,* 205–212.

Fraser, V. S., Kaufmann, B., Oldroyd, B. P., & Crozier, R. H. (2000). Genetic influence on caste in the ant Camponotus consobrinus. *Behav. Ecol. Sociobiol., 47,* 188–194.

French, V., & Brakefield, P. M. (1995). Eyespot development in butterfly wings - the focal signal. *Dev. Biol., 168,* 112–123.

French, V. (1997). Pattern formation in colour on butterfly wings. *Curr. Opin. Genet. Dev., 7,* 524–529.

Fukatsu, T., & Ishikawa, H. (1992). Soldier and male of an eusocial aphid Colophina arma lack endosymbiont - implications for physiological and evolutionary interaction between host and symbiont. *J. Insect Physiol., 38,* 1033–1042.

Fukatsu, T., Sarjiya, A., & Shibao, H. (2005). Soldier caste with morphological and reproductive division in the aphid tribe Nipponaphidini. *Insectes Soc., 52,* 132–138.

Fukuda, S., & Endo, K. (1966). Hormonal control of the development of seasonal forms in the butterfly Polygonia c-aureum L. *Proc. Jpn. Acad., 42,* 1082–1987.

Fuzeau-Braesch, S. (1961). Variations dans la longeur des ailes en fonction de l'effect de groupes chez quelques especes de Gryllides. *Bull. Soc. Zool. Fr., 86,* 785–788.

Gadagkar, R., Vinutha, C., Shanubhogue, A., & Gore, A. P. (1988). Pre-imaginal biasing of caste in a primitively eusocial insect. *Proc. R. Soc., London B, 233,* 175–189.

Gallot, A., Rispe, C., Leterme, N., Gauthier, J. P., Jaubert-Possamai, S., et al. (2010). Cuticular proteins and seasonal photoperiodism in aphids. *Insect Biochem. Mol. Biol., 40,* 35–240.

Gao, N., & Hardie, J. (1996). Pre- and post-natal effects of precocenes on aphid morphogenesis and differential rescue. *Arch. Insect Biochem. Physiol., 32,* 503–510.

Gardiner, B. O.C. (1974). Observations on green pupae in Papilio machaon L. and Pieris brassicae L.W. Roux. *Arch. Entwicklungsmech., 176,* 13–22.

Gauthier, J. P., Legeai, F., Zasadzinski, A., Rispe, C., & Tagu, D. (2007). AphidBase: a database for aphid genomic resources. *Bioinformatics, 27,* 783–784.

Geva, S., Hartfelder, K., & Bloch, G. (2005). Reproductive division of labor, dominance, and ecdysteroid levels in hemolymph and ovary of the bumble bee Bombus terrestris. *J. Insect Physiol., 51,* 811–823.

Ghouri, A. S.K., & McFarlane, J. E. (1958). Occurrence of a macropterous form of Gryllodes sigillatus (Walker) (Othoptera:Gryllidae) in laboratory culture. *Can. J. Zool., 36,* 837–838.

Girardie, J., Boureme, D., Couillaud, F., Tamarelle, M., & Girardie, A. (1987). Anti–juvenile effect of neuroparsin A, a neuroprotein isolated from locust corpora allata. *Insect Biochem., 17,* 977–983.

Goewie, E. A. (1978). Regulation of caste differentiation in the honey bee (Apis mellifera). *Med. Landb. Wageningen, 1–75,* 78-15.

Goldsmith, S. K. (1985). Male dimorphism in Dendrobias mandibularis Audinet-Serville (Coleoptera: Cerambycidae). *J. Kans. Ent. Soc., 58,* 534–538.

Goldsmith, S. K. (1987). The mating system and alternative reproductive behaviors of Dendrobias mandibularis (Coleoptera: Cerambycidae). *Behav. Ecol. Sociobiol., 20,* 111–115.

Greenberg, S., & Tobe, S. S. (1985). Adaptation of a radiochemical assay for juvenile hormone biosynthesis to study caste differentiation in a primitive termite. *J. Insect Physiol., 31,* 347–352.

Gross, M. R. (1996). Alternative reproductie strategies and tactics: diversity within sexes. *Trends Ecol. Evol., 11,* 92–98.

Guidugli-Lazzarini, K. R., Nascimento, A. M., Tanaka, E. D., Piulachs, M. D., Hartfelder, K., et al. (2008). Expression analysis of putative vitellogenin and lipophorin receptors in honey bee (Apis mellifera L.) queens and workers. *J. Insect Physiol., 54,* 1138–1147.

Guidugli, K. R., Hepperle, C., & Hartfelder, K. (2004). A member of the short-chain dehydrogenase/reductase (SDR) superfamily is a target of the ecdysone response in honey bee (Apis mellifera) caste development. *Apidologie, 35,* 37–47.

Guidugli,K.R.,Nascimento,A.M.,Amdam,G.V.,Barchuk,A.R., Omholt, S., et al. (2005a). Vitellogenin regulates hormonal dynamics in the worker caste of a eusocial insect. *FEBS Lett., 579,* 4961–4965.

Guidugli, K. R., Piulachs, M. D., Belles, X., Lourenco, A. P., & Simoes, Z. L.P. (2005b). Vitellogenin expression in queen ovaries and in larvae of both sexes of Apis mellifera. *Arch. Insect Biochem. Physiol., 59,* 211–218.

Guo, W., Wang, X. H., Zhao, D. J., Yang, P. C., & Kang, L. (2010). Molecular cloning and temporal-spatial expression of I element in gregarious and solitary locusts. *J. Insect Physiol., 56,* 943–948.

Hägele, B., Oag, V., Bouaichi, A., McCaffery, A. R., & Simpson, S. J. (2000). The role of female accessory glands in maternal inheritance of phase in the desert locust Schistocerca gregaria. *J. Insect Physiol., 46,* 275–280.

Hamilton, W. D. (1964). The genetical theory of social behaviour I. & II. *J. Theor. Biol., 7,* 1–16, 17–52.

Hardie, J. (1980). Juvenile hormone mimics the photoperiodic apterization of the alate gynopara of aphid, Aphis fabae. *Nature, 286,* 602–604.

Hardie, J. (1981a). The effect of juvenile hormone on host plant preference in the black bean aphid, Aphis fabae. *Physiol. Entomol., 6,* 369–374.

Hardie, J. (1981b). Juvenile hormone and photoperiodically controlled polymorphism in Aphis fabae - prenatal effects on presumptive oviparae. *J. Insect Physiol.*, *27*, 257–265.

Hardie, J. (1981c). Juvenile hormone and photoperiodically controlled polymorphism in Aphis fabae - postnatal effects on presumptive gynoparae. *J. Insect Physiol.*, *27*, 347.

Hardie, J., & Lees, A. D. (1983). Photoperiodic regulation of the development of winged gynoparae in the aphid, Aphis fabae. *Physiol. Entomol.*, *8*, 385–391.

Hardie, J., & Lees, A. D. (1985). Endocrine control of polymorphism and polyphenism. In G. A. Kerkut, & L. I. Gilbert (Eds.), *Comprehensive Insect Physiology, Biochemistry and Pharmacology, Vol. 8.* (pp. 441–489). Oxford: Pergamon Press.

Hardie, J. (1986). Morphogenetic effects of precocenes on 3 aphid species. *J. Insect Physiol.*, *32*, 813–818.

Hardie, J., Mallory, A. C.L., & Quashiewilliams, C. A. (1990). Juvenile hormone and host plant colonization by the black bean aphid, Aphis fabae. *Physiol. Entomol.*, *15*, 331–336.

Hardie, J., Honda, K., Timar, T., & Varjas, L. (1995). Effects of 2,2-dimethylchromene derivatives on wing determination and metamorphosis in the pea aphid, Acyrthosiphon pisum. *Arch. Insect Biochem. Physiol.*, *30*, 25–40.

Hardie, J., Gao, N., Timar, T., Sebok, P., & Honda, K. (1996). Precocene derivatives and aphid morphogenesis. *Arch. Insect Biochem. Physiol.*, *32*, 493–501.

Hardie, J., & Nunes, M. V. (2001). Aphid photoperiodic clocks. *J. Insect Physiol.*, *47*, 821–832.

Harrison, R. G. (1979). Flight polymorphism in the field cricket Gryllus pennsylvannicus. *Oecologia*, *40*, 125–132.

Harrison, R. G. (1980). Dispersal polymorphism in insects. *Annu. Rev. Ecol. Syst.*, *11*, 95–118.

Hartfelder, K. (1987). Rates of juvenile hormone synthesis control caste differentiation in the stingless bee Scaptotrigona postica depilis. Rouxs. *Arch. Dev. Biol.*, *196*, 522–526.

Hartfelder, K., & Engels, W. (1989). The composition of larval food in stingless bees: evaluating nutritional balance by chemosystematic methods. *Ins. Soc.*, *36*, 1–14.

Hartfelder, K. (1990). Regulatory steps in caste development of eusocial bees. In W. Engels (Ed.), *Social Insects - an Evolutionary Approach to Castes and Reproduction* (pp. 245–264). Heidelberg: Springer.

Hartfelder, K., & Rembold, H. (1991). Caste-specific modulation of juvenile hormone-III content and ecdysteroid titer in postembryonic development of the stingless bee, Scaptotrigona postica depilis. *J. Comp. Physiol. B*, *160*, 617–620.

Hartfelder, K., & Engels, W. (1992). Allometric and multivariate analysis of sex and caste polymorphism in the neotropical stingless bee, Scaptotrigona postica. *Insectes Soc.*, *39*, 251–266.

Hartfelder, K. (1993). Structure and function of the prothoracic gland in honey bee (Apis mellifera L) development. *Invertebr. Reprod. Dev.*, *23*, 59–74.

Hartfelder, K., & Steinbrück, G. (1997). Germ cell cluster formation and cell death are alternatives in caste-specific differentiation of the larval honey bee ovary. *Invertebr. Reprod. Dev.*, *31*, 237–250.

Hartfelder, K., & Engels, W. (1998). Social insect polymorphism: hormonal regulation of plasticity in development and reproduction in the honey bee. *Curr. Topics Dev. Biol.*, *40*, 45–77.

Hartfelder, K., Cnaani, J., & Hefetz, A. (2000). Caste-specific differences in ecdysteroid titers in early larval stages of the bumblebee Bombus terrestris. *J. Insect Physiol.*, *46*, 1433–1439.

Hartfelder, K., Bitondi, M. M.G., Santana, W. C., & Simões, Z. L.P. (2002). Ecdysteroid titers and reproduction in queens and workers of the honey bee and of a stingless bee: loss of ecdysteroid function at increasing levels of sociality?. *J. Insect Physiol.*, *32*, 211–216.

Hartfelder, K., & Emlen, D. J. (2005). Endocrine control of insect polyphenisms. In L. I. Gilbert, K. Iatrou, & S. J. Gill (Eds.), *Comprehensive Insect Molecular Science, Vol. 3.* (pp. 651–703). Oxford: Elsevier.

Hartfelder, K., Makert, G. R., Judice, C. C., Pereira, G. A.G., Santana, W. C., et al. (2006). Physiological and genetic mechanisms underlying caste development, reproduction and division of labor in stingless bees. *Apidologie*, *37*, 144–163.

Hatano, E., Kunert, G., & Weisser, W. W. (2010). Aphid wing induction and ecological costs of alarm pheromone emission under field conditions. *PLoS One*, *5*, e6.

Hazel, W. N. (1977). The genetic basis of pupal colour dimorphism and its maintenance by natural selection in Papilio polyxenes (Papilionidae: Lepidoptera). *Heredity*, *38*.

Hazel, W. N., & West, D. A. (1979). Environmental control of pupal colour in swallowtail butterflies (Lepidoptera: Papilionidae: Battus philenor (L.) and Papilio polyxenes. *Fabr. Ecol. Entomol.*, *4*, 393–408.

Hazel, W. N., & West, D. A. (1996). Pupation site preference and environmentally-cued pupal colour dimorphism in the swallowtail butterflies Papilio polyxenes Fabr. (Lepidoptera: Papilionidae). *Biol. J. Linn. Soc.*, *57*, 81–87.

Hazel, W. N., Ante, S., & Strongfellow, B. (1998). The evolution of environmentally-cued pupal colour in swallowtail butterflies: natural selection for pupation site and pupation colour. *Ecol. Entomol.*, *23*, 41–44.

Heinze, J., & Buschinger, A. (1987). Queen polymorphism in a non-parasitic Leptothorax species (Hymenoptera, Formicidae). *Insectes Soc.*, *34*, 28–43.

Heinze, J. (1998). Intercastes, intermorphs and ergatoids: who is who in ant reproduction? *Ins. Soc.*, *45*, 113–124.

Heinze, J., & Schrempf, A. (2008). Aging and reproduction in social insects - a mini-review. *Gerontology*, *54*, 160–167.

Hepperle, C., & Hartfelder, K. (2001). Differentially expressed regulatory genes in honey bee caste development. *Naturwissenschaften*, *88*, 113–116.

Hidaka, T. (1961a). Mise en evidénce de l'activité sécrétoire du ganglion prothoracique dans l'adaptation chromatique de la nymphe du Papilio xuthus L. C. R. Soc. Biol. Paris, *154*, 1682–1685.

Hidaka, T. (1961b). Reserches sur le mechanisme endocrine de l'adaptation chromatique morphologique chez les nymphes du Papilio xuthus L. *J. Fac. Sci. Univ. Tokyo*, *9*(Sec. IV), 223–261.

Hille Ris Lambers, D. (1966). Polymorphism in Aphididae. *Annu Rev. Entom.*, *11*, 47–78.

Hoffman, R. J. (1973). Environmental control of seasonal variation in the butterfly Colias eurytheme. I. Adaptive aspects of a photoperiodic response. *Evolution*, *27*, 387–397.

Hoffman, R. J. (1978). Environmental uncertainty and evolution of physioplogical adaptation in Colias butterflies. *Am. Nat.*, *112*, 999–1015.

Hölldobler, B., & Wilson, E. O. (1990). *The Ants*. Cambridge, Mass: Belknap Press of Harvard University.

Huang, Z. Y., Robinson, G. E., Tobe, S. S., Yagi, K. J., Strambi, C., et al. (1991). Hormonal regulation of behavioral development in the honey bee is based on changes in the rate of juvenile hormone biosynthesis. *J. Insect Physiol.*, 37, 733–741.

Hughes, W. O. H., & Boomsma, J. J. (2007). Genetic polymorphism in leaf–cutting ants is phenotypically plastic. *Pcoc. R. Soc. B Biol.Sci.*, 274, 1625–1630.

Hunt, G. J., Amdam, G. V., Schlipalius, D., Emore, C., Sardesai, N., et al. (2007a). Behavioral genomics of honey bee foraging and nest defense. *Naturwissenschaften*, 94, 247–267.

Hunt, J., & Simmons, L. W. (1997). Patterns of fluctuating assymetry in beetle horns: an experimental examination of the honest signalling hypothesis. *Behav. Ecol. Sociobiol.*, 41, 109–114.

Hunt, J. H., Kensinger, B. J., Kossuth, J. A., Henshaw, M. T., Norberg, K., et al. (2007b). A diapause pathway underlies the gyne phenotype in Polistes wasps, revealing an evolutionary route to caste-containing insect societies. *Proc. Natl. Acad. Sci. USA*, 104, 14020–14025.

Hunt, J. H., Wolschin, F., Henshaw, M. T., Newman, T. C., Toth, A. L., et al. (2010). Differential gene expression and protein abundance evidence ontogenetic bias toward castes in a pimitively eusocial wasp. *PLoS Biology*, 5, e6.

Iguchi, Y. (1998). Horn dimorphism in Allomyrina dichotomia septentrionalis (Coleoptera: Scarabaeidae) affected by larval nutrition. *Ann. Ent. Soc. Amer.*, 91, 845–847.

Ijichi, N., Shibao, H., Miura, T., Matsumoto, T., & Fukatsu, T. (2004). Soldier differentiation during embryogenesis of a social aphid, Pseudoregma bambucicola. *Entomol. Sci.*, 7, 143–155.

Injeyan, H. S., & Tobe, S. S. (1981). Phase polymorphism in Schistocerca gregaria - assessment of juvenile hormone synthesis in relation to vitellogenesis. *J. Insect Physiol.*, 27, 203–210.

International Aphid Genomics Consortium, I. A. G. (2010). Genome sequence of the pea aphid Acyrthosiphon pisum. *PLoS Biology*, 8, e2.

Inward, D., Beccaloni, G., & Eggleton, (2007). Death of an order: a comprehensive molecular phylogenetic study confirms that termites are eusocial cockroaches. *Biol. Lett.*, 3, 331–335.

Ishizaki, H., & Kato, M. (1956). Environmental factors affecting the formation of orange pupae in Papilio xuthus. *Mem. Coll. Sci. Kyoto Univ. B*, 3, 11–18.

Ito, Y., Tanaka, S., Yukawa, J., & Tsuji, K. (1995). Factors affecting the proportion of soldiers in eusocial bamboo aphid, Pseudoregma bambucicola, colonies. *Ethol. Ecol. Evol.*, 7, 335–345.

Iwanaga, K., & Tojo, S. (1986). Effects of juvenile hormone and rearing density on wing dimorphism and oocyte development in the brown planthopper Niloparvata lugens. *J. Insect Physiol.*, 32, 585–590.

Izzo, A., Wells, M., Huang, Z., & Tibbetts, E. (2010). Cuticular hydrocarbons correlate with fertility, not dominance, in a paper wasp, Polistes dominulus. *Behav. Ecol. Sociobiol.*, 64, 857–864.

Jacobs, M. D., & Watt, W. B. (1994). Seasonal adaptation vs. physiological constraint: photoperiod, thermoregulation and flight in Colias butterflies. *Func. Ecol.*, 8, 366–376.

Jaffe, R., Kronaver, D. J. C., Kraus, F. B., Boomsma, J. J., & Moritz, R. F. A. (2007). Worker caste determination in the army ant Eciton burchellii. *Biol. Lett.*, 3, 513–516.

Jaubert-Possamai, S., Le Trionnaire, G., Bonhomme, J., Christophides, G. K., Rispe, C., et al. (2007). Gene knockdown by RNAi in the pea aphid Acyrthosiphon pisum. *BMC Biotech*, 7, e63.

Jaycox, E. R. (1976). Behavioral changes in worker honey bees (Apis mellifera) after injection with synthetic juvenile hormone (Hymenoptera: Apidae). *J. Kansas Entom. Soc.*, 49, 165–170.

Jeanne, R. L., Graf, C. A., & Yandell, B. S. (1995). Non-size-based morphological castes in a social insect. *Naturwissenschaften*, 82, 296–298.

Johnson, B. (1959). Effect of parasitization by Aphidius platensis Brèthes on the developmental physiology of its host, Aphis craccivora Koch. *Ent. Exp. Appl.*, 2, 82–99.

Johnson, B. (1965). Wing polymorphism in aphids II. Interaction between aphids. *Entomol. Exp. Appl.*, 8, 49–64.

Joly, L. (1954). Résultats d'implantations systématiques de coprora allata à de jeunes larves de Locusta migratoria. *C.R. Soc. Biol.*, 148, 579–583.

Jones, M., Rakes, L., Yochum, M., Dunn, G., Wurster, S., et al. (2007). The proximate control of pupal color in swallowtail butterflies: implications for the evolution of environmentally cued pupal color in butterflies (Lepidoptera: Papilionidae). *J. Insect Physiol.*, 53, 40–46.

Judice, C. C., Carazzole, M. F., Festa, F., Sogayar, M. C., Hartfelder, K., et al. (2006). Gene expression profiles underlying alternative caste phenotypes in a highly eusocial bee, Melipona quadrifasciata. *Insect Mol. Biol.*, 15, 33–44.

Julian, G. E., Fewell, J. H., Gadau, J., Johnson, R. A., & Lorrabee, D. (2002). Genetic determination of the queen caste in an ant hybrid zone. *Proc. Natl. Acad. Sci., USA*, 99, 8157–8160.

Kang, L., Chen, X. Y., Zhou, Y., Liu, B. W., Zheng, W., et al. (2004). The analysis of large-scale gene expression correlated to the phase changes of the migratory locust. *Proc. Natl. Acad. Sci. USA*, 101, 17611–17615.

Karl, I., Geister, T. L., & Fischer, K. (2009). Intraspecific variation in wing and pupal melanization in copper butterflies (Lepidoptera: Lycaenidae). *Biol. J. Linn. Soc.*, 98, 301–312.

Kawano, K. (1995). Horn and wing allometry and male dimorphism in giant rhinoceros beetles (Coleoptera: Scarabaeidae) of tropical Asia and America. *Ann. Ent. Soc. Amer.*, 88, 92–98.

Keeping, M. G. (2002). Reproductive and worker caste in the primitively eusocial wasp Belonogaster petiolata (DeGeer) (Hymenoptera: Vespidae): evidence for pre-imaginal differentiation. *J. Insect Physiol.*, 48, 867–879.

Kerr, W. E. (1950). Evolution of the mechanism of caste determination in the genus Melipona. *Evolution*, 4, 7–13.

Kerr, W. E., Stort, A. C., & Montenegro, M. S. (1966). Importância de alguns fatores ambientais na determinação das castas do gênero Melipona. *Anais Acad. Bras. Ciências*, 38, 149–168.

Kerr, W. E. (1987). Sex determination in bees. XVII. systems of caste determination in the Apinae, Meliponinae and Bombinae and their phylogenetic implications. *Rev. Bras. Genet.*, *10*, 685–694.

Kerr, W. E. (1990). Sex determination in bees. XXVI. masculinism of workers in the Apidae. *Rev. Bras. Genet.*, *13*, 479–489.

Keys, D. N., Lewis, D. L., Selegue, J. E., Pearson, B. J., Goodrich, L. V., et al. (1999). Recruitment of a hedgehog regulatory circuit in butterfly eyespot evolution. *Science*, *283*, 532–534.

Kingsolver, J. G., & Watt, W. B. (1983). Thermoregulatory strategies in Coliasbutterflies: thermal stress and the limits of adaptation in temporally varying environments. *Am. Nat.*, *121*, 32–55.

Kingsolver, J. G. (1987). Evolution and coadaptation of thermoregulatory behavior and wing pigmentation in pierid butterflies. *Evolution*, *41*, 472–490.

Kingsolver, J. G., & Wiesnarz, D. C. (1991). Seasonal polyphenism in wing-melanin pattern and thermoregulatory behavior pigmentation in Pieris butterflies. *Amer. Nat.*, *137*, 816–829.

Kingsolver, J. G. (1995a). Fitness consequences of seasonal polyphenism in western white butterflies. *Evolution*, *49*, 942–954.

Kingsolver, J. G. (1995b). Viability selection on seasonally polyphenic traits: wing melanin pattern in western white butterflies. *Evolution*, *49*, 932–941.

Kisimoto, R. (1965). Studies on the polymoprhism and its role playing in the population growth of the brown planthopper, Nilaparvata lugens Stal. *Bull. Shikoku Agric. Exp. Stn.*, *13*, 1–106.

Kobayashi, M., & Ishikawa, H. (1993). Breakdown of indirect flight muscles of alate aphids (Acyrthosiphon pisum) in relation to their flight, feeding and reproductive behavior. *J. Insect Physiol.*, *39*, 549–554.

Kobayashi, M., & Ishikawa, H. (1994). Involvement of juvenile hormone and ubiquitin-dependent proteolysis in flight muscle breakdown of alate aphid (Acyrthosiphon pisum). *J. Insect Physiol.*, *40*, 107–111.

Koch, P. B., & Bückmann, D. (1985). The seasonal dimorphism of Araschnia levana L. (Nymphalidae) in relation to hormonal controlled development. *Verh. Dt. Zool. Ges. 78*, 260.

Koch, P. B., & Bückmann, D. (1987). Hormonal control of seasonal morphs by the timing of ecdysteroid release in Araschnia levana L. (Nymphalidae: Lepidoptera). *J. Insect Physiol.*, *33*, 823–829.

Koch, P. B., Brakefield, P. M., & Kesbeke, F. (1996). Ecdysteroids control eyespot size and wing color pattern in the polyphenic butterfly Bicyclus anynana (Lepidoptera: Satyridae). *J. Insect Physiol.*, *42*, 223–230.

Koch, P. B., Bechnecke, B., & ffrench-Constant, R. H. (2000). The molecular basis of melanismand mimicry in a swallowtail butterfly. *Curr. Biol.*, *10*, 591–594.

Koch, P. B., Merk, R., & Reinhardt (2003). Localization of ecdysone receptor protein during colour pattern formation in wings of the butterfly Precis coenia (Lepidoptera: Nymphalidae) and co-expression with Distal-less protein. *Dev. Genes Evol.*, *212*, 571–584.

Kooi, R. E., Brakefield, P. M., & Schlatmann, E. G. M. (1994). Description of the larval sensitive period for polyphenic wing pattern induction in the tropical butterfly Bicyclus anynana (Satyrinae). *Proc. Exp. & Appl. Entomol. 5*, 47–52.

Kooi, R. E., Brakefield, P. M., & Rossie, W. E. M. (1996). Effects of food plant on phenotypic plasticity in the tropical butterfly Bicyclus anynana. *Entomol. Exp. Appl.*, *80*, 149–151.

Kooi, R. E., & Brakefield, P. M. (1999). The critical period for wing pattern induction in the polyphenic tropical butterfly Bicyclus anynana (Satyrinae). *J. Insect Physiol.*, *45*, 201–212.

Korb, J. (2008). Termites, hemimetabolous diploid white ants? *Frontiers Zool.*, *5*, e15.

Korb, J., & Hartfelder, K. (2008). Life history and development - a framework for understanding developmental plasticity in lower termites. *Biol. Rev.*, *83*, 295–313.

Korb, J., Hoffmann, K., & Hartfelder, K. (2009). Endocrine signatures underlying plasticity in postembryonic development of a lower termite, Cryptotermes secundus (Kalotermitidae). *Evol. Dev.*, *11*, 269–277.

Kucharski, R., Maleszka, J., Foret, S., & Maleszka, R. (2008). Nutritional control of reproductive status in honey bees via DNA methylation. *Science*, *319*, 1827–1830.

Kühn, A., & von Engelhardt, M. (1933). Über die Determination des Symmetriesystems auf dem Vorderfügel von Ephestia kühniella. W. Roux. *Arch. Entwicklungsmech. Org.*, *130*, 660.

Kukuk, P. F. (1966). Male dimorphism in Lasioglossum (Chilalictus) hemiochalceum: the role of larval nutrition. *J. Kansas Entom. Soc.*, *69*(suppl), 147–157.

Kunert, G., & Weisser, W. W. (2005). The importance of antennae for pea aphid wing induction in the presence of natural enemies. *Bull. Entomol. Res.*, *95*, 125–131.

Lagueux, M., Hirn, M., & Hoffmann, J. A. (1977). Ecdysone during ovarian development in Locusta migratoria. J. *Insect Physiol.*, *23*, 109–119.

Lagueux, M., Hetru, H., Goltzene, F., Kappler, C., & Hoffmann, J. A. (1979). Ecdysone titer and metabolism in relation to cuticulogenesis in embryos of Locusta migratoria. *J. Insect Physiol.*, *25*, 709–725.

Lank, D. B., Smith, C. M., Hanotte, O., Burke, T., & Cooke, F. (1995). Genetic polymoprhism for alternative mating behavior in lekking male ruff Philomachus pugnax. *Nature*, *378*, 59–62.

Lanzrein, B., Gentinetta, V., & Fehr, R. (1985). Titres of juvenile hormone and ecdysteroids in reproducion and eggs of Macrotermes michaelseni: relation to caste determination. In J. A.L. Watson, B. M. Okot-Kotber, & C. Noirot (Eds.),*Caste Differentiation in Social Insects* (pp. 307–327). Oxford: Pergamon Press.

Larrere, M., & Couillaud, F. (1993). Role of juvenile hormone biosynthesis in dominance status and reproduction of the bumblebee, Bombus terrestris. *Behav. Ecol. Sociobiol.*, *33*, 335–338.

Larrere, M., Lavenseau, L., Tasei, J. N., & Couillaud, F. (1993). Juvenile hormone biosynthesis and diapause termination in Bombus terrestris. *Invertebr. Reprod. Dev.*, *23*, 7–14.

Le Trionnaire, G., Jaubert, S., Sabater-Munoz, B., Benedetto, A., Bonhomme, J., et al. (2007). Seasonal photoperiodism regulates the expression of cuticular and signalling protein genes in the pea aphid. *Insect Biochem. Mol. Biol.*, *37*, 1094–1102.

Le Trionnaire, G., Hardie, J., Jaubert-Possamai, S., Simon, J. C., & Tagu, D. (2008). Shifting from clonal to sexual reproduction in aphids: physiological and developmental aspects. *Biol. Cell, 100,* 441–451.

Le Trionnaire, G., Francis, F., Jaubert-Possamai, S., Bonhomme, J., De Pauw, E., et al. (2009). Transcriptomic and proteomic analyses of seasonal photoperiodism in the pea aphid. *BMC Genomics, 10,* 1–14.

Ledoux, A. (1976). Action d'un dérivé du farnesol sur l'apparition des femelles ailées chez Aphaenogaster senilis (Hym. Formicoidea). *C.R. Acad. Sci., Paris, Ser. D, 228,* 569–570.

Lees, A. D. (1977). Action of juvenile hormone mimics on the regulation of larval-adult alary polymorphism in aphids. *Nature, 267,* 46–48.

Lees, A. D. (1980). Development of juvenile hormone sensitivity in alatae of the aphid Megoura viciae. *J. Insect Physiol., 26,* 143–151.

Li, Y., Zhang, Z., Robinson, G. E., & Palli, S. R. (2007). Identification and characterization of a juvenile hormone response element and its binding proteins. *J. Biol. Chem., 282,* 37605–37617.

Liebig, J., Peeters, C., Oldham, N. J., Markstädter, C., & Hölldobler, B. (2000). Are variations in cuticular hydrocarbons of queens and workers a reliable signal of fertility in the ant Harpegnathos saltator? *Proc. Natl. Acad. Sci., USA, 97,* 4124–4131.

Linksvayer, T. A., Rueppell, O., Siegel, A., Kaftanoglu, O., Page, R. E., et al. (2009). The genetic basis of transgressive ovary size in honey bee workers. *Genetics, 183,* 693–707.

Liu, S., Yang, B., Gu, J., Yao, X., Zhang, Y., et al. (2008). Molecular cloning and characterization of a juvenile hormone esterase gene from brown planthopper, Nilaparvata lugens. *J. Insect Physiol., 54,* 1495–1502.

Liu, Y. X., Henderson, G., Mao, L. X., & Laine, R. A. (2005a). Seasonal variation of juvenile hormone titers of the formosan subterranean termite, Coptotermes formosanus (Rhinotermitidae). *Environ. Entom., 34,* 557–562.

Liu, Y. X., Henderson, G., Mao, L. X., & Laine, R. A. (2005b). Effects of temperature and nutrition on juvenile hormone titers of Coptotermes formosanus (Isoptera: Rhinotermitidae). *Ann. Entom. Soc. Amer., 98,* 732–737.

Loher, W. (1961). The chemical acceleration of the maturation process and its hormonal control in the male of the desert locust. *Proc. R. Soc. Lond. Ser B Biol. Sci., 153,* 380–397.

Lüscher, M. (1964). Die spezifische Wirkung männlicher und weiblicher Ersatzgeschlechtstiere auf die Entstehung von Geschlechtstieren bei der Termite Kalotermes flavicollis (Fab.). *Insectes Soc., 11,* 79–90.

Lüscher, M. (1976). Evidence for endocrine control of caste determination in higher termites. In M. Lüscher (Ed.), *Phase and Caste Determination in Insects* (pp. 91–103). Oxford: Pergamon Press.

Mackert, A., do Nascimento, A. M., Bitondi, M. M. G., Hartfelder, K., & Simoes, Z. L.P. (2008). Identification of a juvenile hormone esterase-like gene in the honey bee, Apis mellifera L. - expression analysis and functional assays. *Comp. Biochem. Physiol. B, 150,* 33–44.

Mackert, A., Hartfelder, K., Bitondi, M. M. G., & Simões, Z. L. P. (2010). The juvenile hormone (JH) epoxide hydrolase gene in the honey bee (Apis mellifera) genome encodes a protein which has negligible participation in JH degradation. *J. Insect Physiol., 56,* 1139–1146.

Maekawa, K., Ishitani, K., Gotoh, H., Cornette, R., & Miura, T. (2010). Juvenile Hormone titre and vitellogenin gene expression related to ovarian development in primary reproductives compared with nymphs and nymphoid reproductives of the termite Reticulitermes speratus. *Physiol. Entomol., 35,* 52–58.

Maeno, K., & Tanaka, S. (2008). Maternal effects on progeny size, number and body color in the desert locust, Schistocerca gregaria: Density- and reproductive cycle-dependent variation. *J. Insect Physiol., 54,* 1072–1080.

Maeno, K., & Tanaka, S. (2009). Artificial miniaturization causes eggs laid by crowd-reared (gregarious) desert locusts to produce green (solitarious) offspring in the desert locust, Schistocerca gregaria. *J. Insect Physiol., 55,* 849–854.

Mahamat, H., Hassanali, A., Odongo, H., Torto, B., & El-Bashir, E.-S. (1993). Studies on the maturation-accelerating pheromone of the desert locust Schistocerca gregaria (Orthoptera: Acrididae). *Chemoecology, 4,* 159–164.

Main, H. (1922). Notes on the metamorphosis of Onthophagus taurus L. *Proc. Entom. Soc. London, 1922,* 14–16.

Maisch, A., & Bückmann, D. (1987). The control of cuticular melanin and lutein incorporation in the morphological colour adaptation of a nymphalid pupa, Inachis io L. *J. Insect Physiol., 33,* 393–402.

Makert, G. R., Paxton, R. J., & Hartfelder, K. (2006). An optimized method for the generation of AFLP markers in a stingless bee (Melipona quadrifasciata) reveals a high degree of genetic polymorphism. *Apidologie, 37,* 687–698.

Mane, S. D., & Rembold, H. (1977). Developmental kinetics of juvenile hormone inactivation in queen and worker castes of the honey bee, Apis mellifera. *Insect Biochem., 7,* 463–467.

Mao, L. X., & Henderson, G. (2010). Group size effect on worker juvenile hormone titers and soldier differentiation in Formosan subterranean termite. *J. Insect Physiol., 56,* 725–730.

Marco Antonio, D. S., Guidugli-Lazzarini, K. R., Nascimento, A. M., Simões, Z. L.P., & Hartfelder, K. (2008). RNAi-mediated silencing of vitellogenin gene function turns honey bee (Apis mellifera) workers into extremely precocious foragers. *Naturwissenschaften, 95,* 953–961.

Marcus, J. M., Ramos, D. M., & Monteiro, A. (2004). Germline transformation of the butterfly Bicyclus anynana. *Proc. R. Soc. Ser. B Biol. Sci., 271,* S263–S265.

Martinez, T., & Wheeler, D. E. (1991). Effect of the queen, brood annd worker caste on haemolymph vitellogenin titre in Camponotus festinatus workers. *J. Insect Physiol., 37,* 347–352.

Masaki, S., & Oyama, N. (1963). Photoperiodic control of growth and wing form in Nemobius yezoensis Shiraki. *Kontyu, 31,* 16–26.

Masaki, T., Endo, K., & Kumagai, K. (1988). Neuroendocrine regulation of development of seasonal morphs in the Asian comma butterfly, Polygonia c-aureum L.: is the factor producing summer morphs (SMPH) identical to the small prothoracicotropic hormone (4K-PTTH)? *Zool. Sci., 5,* 1051–1057.

Mathad, S. B., & McFarlane, J. E. (1968). Two effects of photoperiod on wing development in Grylloides sigillatus (Walk.). *Can. J. Zool.*, *46*, 57–60.

Matsumara, M. (1996). Genetic analysis of a threshold trait: density-dependent wing dimorphism in Sogatella furcifera(Horvath) (Hemiptera: Delphacidae), the white-backed planthopper. *Heredity*, *76*, 229–237.

Matsuura, K., Himuro, C., Yokoi, T., Yamamoto, Y., Vargo, E. L., et al. (2010). Identification of a pheromone regulating caste differentiation in termites. *Proc. Natl. Acad. Sci. USA*, *107*, 12963–12968.

Mattick, J. S. (2009). The genetic signatures of noncoding RNAs. *PLoS Genetics*, *5*, e1000459.

Mattick, J. S. (2010). RNA as the substrate for epigenome-environment interactions. *Bioessays*, *32*, 548–552.

McCaffery, A. R., Simpson, S. J., Islam, M. S., & Roessingh, P. (1998). A gregarizing factor in the egg pod foam of the desert locust Schistocerca gregaria. *J. Exp. Biol.*, *201*, 347–363.

McFarlane, J. E. (1962). Effect of diet and temperature on wing development of Gryllodes sigillatus (Walk.) (Orthoptera: Gryllidae). *Ann. Soc. Entom. Quebec.*, *7*, 28–33.

McMillan, W. O., Monterio, A., & Kapan, D. D. (2002). Development and evolution on the wing. *Trends Ecol. Evol.*, *17*, 125–133.

Melampy, R. M., & Willis, E. R. (1939). Respiratory metabolism during larval and pupal development of the female honey bee (Apis mellifica L.). *Physiol. Zool.*, *12*, 302–311.

Merrifield, F., & Poulton, E. B. (1899). The color relation between the pupae of Papilio machaon, Pieris napai and many other species, and the surroundings of the larvae preparing to pupate, etc. *Trans. Entom. Soc. London*, *1899*, 369–433.

Michener, C. D. (2000). *The Bees of the World*. Baltimore: John Hopkins University Press.

Miller, G. A., Islam, M. S., Claridge, T. D.W., Dodgson, T., & Simpson, S. J. (2008). Swarm formation in the desert locust Schistocerca gregaria: isolation and NMR analysis of the primary maternal gregarizing agent. *J. Exp. Biol.*, *211*, 370–376.

Mirth, C., Truman, J. W., & Riddiford, L. M. (2005). The role of the prothoracic gland in determining critical weight to metamorphosis in *Drosophila* melanogaster. *Curr. Biol.*, *15*, 1796–1807.

Mirth, C. K., & Riddiford, L. M. (2007). Size assessment and growth control: how adult size is determined in insects. *Bioessays*, *29*, 344–355.

Miura, T., Kamikouchi, A., Sawata, M., Takeuchi, H., Natori, S., et al. (1999). Soldier caste-specific gene expression in the mandibular glands of Hodotermopsis japonica (Isoptera: Termopsidae). *Proc. Natl. Acad. Sci. USA*, *96*, 13874–13879.

Miura, T., & Matsumoto, T. (2000). Soldier morphogenesis in a nasute termite: discovery of a disk-like structure forming a soldier nasus. *Proc. R. Soc. Lond. B Biol. Sci.*, *267*, 1185–1189.

Miura, T. (2001). Morphogenesis and gene expression in the soldier-caste differentiation of termites. *Insectes Soc.*, *48*, 216–223.

Miura, T., Braendle, C., Shingleton, A., Sisk, G., Kambhampati, S., et al. (2003). A comparison of parthenogenetic and sexual embryogenesis of the pea aphid Acyrthosiphon pisum (Hemiptera: Aphidoidea). *J. Exp. Zool. B*, *295*, 59–81.

Miura, T. (2004). Proximate mechanisms and evolution of caste polyphenism in social insects: From sociality to genes. *Ecol. Res.*, *19*, 141–148.

Moczek, A. (2006). Integrating micro- and macroevolution of development through the study of horned beetles. *Heredity*, *97*, 168–178.

Moczek, A. (2007). Pupal remodeling and the evolution and development of alternative male morphologies in horned beetles. *BMC Evol. Biol.*, *7*, e151.

Moczek, A. P., & Emlen, D. J. (1999). Proximate determination of male horn dimorphism in the beetle Onthophagus taurus (Coleoptera: Scarabaeidae). *J. Evol. Biol.*, *12*, 27–37.

Moczek, A. P., & Emlen, D. J. (2000). Male horn dimorphism in the scarab beetle Ontophagus taurus: do alternative reproductive tactics favor alternative phenotypes? *Anim. Behav.*, *59*, 459–466.

Moczek, A. P., Hunt, J., Emlen, D. J., & Simmons, L. W. (2002). Threshold evolution in exotic populations of a polyphenic beetle. *Evol. Ecol. Res.*, *4*, 587–601.

Moczek, A. P., & Nagy, L. M. (2005). Diverse developmental mechanisms contribute to different levels of diversity in horned beetles. *Evol. Dev.*, *7*, 175–185.

Moczek, A. P., Rose, D., Sewell, W., & Kesselring, B. R. (2006). Conservation, innovation, and the evolution of horned beetle diversity. *Dev. Genes Evol.*, *216*, 655–665.

Moczek, A. P., & Rose, D. (2009). Differential recruitment of limb patterning genes during development and diversification of beetle horns. *Proc. Natl. Acad. Sci. USA*, *106*, 8992–8997.

Monnin, T., & Ratnieks, F. L. (2001). Policing in queenless ants. *Behav. Ecol. Sociobiol.*, *50*, 97–108.

Monteiro, A., Glaser, G., Stockslager, S., Glansdorp, N., & Ramos, D. (2006). Comparative insights into questions of lepidopteran wing pattern homology. *BMC Dev. Biol.*, *6*, e52.

Monteiro, A. F., Brakefield, P. M., & French, V. (1994). The evolutionary genetics and developmental basis of wing pattern variation in the butterfly Bicyclus anynana. *Evolution*, *48*, 1147–1157.

Monteiro, A. F., Brakefield, P. M., & French, V. (1997). Butterfly eyespots: the genetics and development of the color rings. *Evolution*, *51*, 1207–1216.

Moran, N. A. (1992). The evolutionary maintenance of alternative phenotypes. *Am. Nat.*, *139*, 971–989.

Moran, N. A. (1994). Adaptation and constraint in complex life cycles of animals. *Annu. Rev. Ecol. Syst.*, *25*, 573–600.

Morooka, S., Ishibashi, N., & Tojo, S. (1988). Relationships between wing form response to nymphal density and black colouration in the brown planthopper Nilaparvata lugens (Homoptera: Delphacidae). *Appl. Ent. Zool.*, *23*, 449–458.

Morooka, S., & Tojo, S. (1992). Maintenance and selection of strains exhibiting specific wing form and body colour under high density conditions in the brown planthopper Nilaparvata lugens (Homoptera: Delphacidae). *Appl. Ent. Zool.*, *27*, 445–454.

Müller, H. J. (1955). Die Saisonformenbildung von Araschnia levana - ein photoperiodisch gesteuerter Diapauseeffekt. *Naturwissenschaften*, *42*, 134–135.

Müller, H. J. (1956). Die Wirkung verschiedener diurnaler Licht-Dunkel-Relationen auf die Saisonformenbildung von Araschnia levana. *Naturwissenschaften, 43,* 503–504.

Münch, D., Amdam, G. V., & Wolschin, F. (2008). Aging in a social insect: molecular and physiological characteristics of life span plasticity in the honey bee. *Funct. Ecol., 22,* 407–421.

Mutti, N. S., Park, Y., Reese, J. C., & Reeck, G. R. (2006). RNAi knockdown of a salivary transcript leading to lethality in the pea aphid (Acyrthosiphon pisum). *J. Insect Sci., 6,* e38.

Nalepa, C. (2009). Altricial development in subsocial cockroach ancestors: foundation for the evolution of phenotypic plasticity in termites. *Evol. Dev., 12,* 95–105.

Naug, D., & Gadagkar, R. (1998). The role of age in temporal polyethism in a primitively eusocial wasp. *Behav. Ecol. Sociobiol., 42,* 37–47.

Nelson, C. M., Ihle, K. E., Fondrk, M. K., Page, R. E., & Amdam, G. V. (2007). The gene vitellogenin has multiple coordinating effects on social organization. *PLoS Biology, 5,* e62.

Nijhout, H. F. (1980). Pattern formation of lepidopteran wings: determination of an eyespot. *Dev. Biol., 80,* 267–274.

Nijhout, H. F. (1985). The developmental physiology of color patterns in Lepidoptera. *Adv. Insect Physiol., 18,* 181–247.

Nijhout, H. F. (1986). Pattern and diversity of lepidopteran wings. *Bioscience, 36,* 527–533.

Nijhout, H. F., & Grunert, L. W. (1988). Color pattern regulation after surgery on the wing discs of Precis coenia (lepidoptera: Nymphalaidae). *Development, 102,* 377–385.

Nijhout, H. F. (1991). *The development and evolution of butterfly wing patterns.* Washington DC: Smithsonian Institution Press.

Nijhout, H. F. (1994). *Insect Hormones.* Princeton NJ: Princeton University Press.

Nijhout, H. F., & Wheeler, D. E. (1994). Growth models of complex allometries in holometabolous insects. *Am. Nat., 148,* 40–56.

Noda, H., Kawai, S., Koizumi, Y., Matsui, K., Zhang, Q., et al. (2008). Annotated ESTs from various tissues of the brown planthopper Nilaparvata lugens: a genomic resource for studying agricultural pests. *BMC Genomics, 9,* e1117.

Noirot, C. (1985a). Pathways of caste development in higher termites. In J. A.L. Watson, B. M. Okot-Kotber, & C. Noirot (Eds.), *Caste Differentiation in Social Insects* (pp. 75–86). Oxford: Pergamon Press.

Noirot, C. (1985b). Differentiation of reproductives in higher termites. In J. A.L. Watson, B. M. Okot-Kotber, & C. Noirot (Eds.), *Caste Differentiation in Social Insects* (pp. 177–186). Oxford: Pergamon Press.

Noirot, C., & Pasteels, J. M. (1987). Ontogenetic development and evolution of the worker caste in termites. *Experientia, 43,* 851–860.

Noirot, C., & Bordereau, C. (1988). Termite polymorphism and morphogenetic hormones. In A. P. Gupta (Ed.), *Morphogenetic hormones of Arthropods.* New Brunswick: Rutgers University Press.

Noirot, C. (1989). Social structure in termite societies. *Ethol., Ecol. Evol., 1,* 1–17.

Noirot, C. (1990). Sexual castes and reproductive strategies in termites. In W. Engels (Ed.), *Social Insects - an Evolutionary Approach to Castes and Reproduction* (pp. 5–35). Heidelberg: Springer.

Noll, F. B., Mateus, S., & Zucchi, R. (1997). Morphological caste differences in the neotropical swarm-founding and polygynous polistine wasps, Polybia scutellaris. *Stud. Neotrop. Fauna Environ., 21,* 76–80.

Noll, F. B., & Wenzel, J. W. (2008). Caste in the swarming wasps: 'queenless' societies in highly social insects. *Biol. J. Linn. Soc., 93,* 509–522.

Nylin, S. (1992). Seasonal plasticity in life history traits: growth and development in Polygonia c-album (Lepidoptera: Nymphalaidae). *Biol. J. Linn. Soc., 47,* 301–323.

O'Donnell, S., & Jeanne, R. L. (1993). Methoprene accelerates age polyethism in workers of a social wasp (Polybia occidentalis). *Physiol. Entomol., 18,* 189–194.

O'Donnell, S. (1998). Reproductive caste determination in eusocial wasps (Hymenoptera: Vespidae). *Annu. Rev. Entomol., 43,* 323–346.

Ogino, K. Y., Hirono, Y., Matsumoto, T., & Ishikawa, H. (1993). Juvenile hormone analogue, S-31183, causes a high level induction of presoldier differentiation in the Japanese damp-wood termite. *Zool. Sci., 10,* 361–366.

Ohkawara, K., Ito, F., & Higashi, S. (1993). Production and reproductive function of intercastes in Myrmecina graminicola nipponica colonies (Hymenoptera: Formicidae). *Insectes Soc., 40,* 1–10.

Ohtaki, T. (1960). Humoral control of pupal coloration in the cabbage white butterfly Pieris rapae crucivora. *Annot. Zool. Jap., 33,* 97–103.

Ohtaki, T. (1963). Further studies on the development of pupal colouration in the cabbage white butterfly Pieris rapae crucivora Bois. *Serio-Seiti, 9,* 84–89.

Oldham, S., & Hafen, E. (2003). Insulin/IGF and target of rapamycin signaling: a TOR de force in growth control. *Trends Cell Biol., 13,* 79–85.

Oliveira, R., Taborsky, M., & Brockmann, H. J. (2008). *Alternative Reproductive Tactics: An Integrative Approach.* Cambridge, MA: Cambridge University Press.

Oostra, V., de Jong, M., Invergo, B., Kesbeke, F., Wende, F., et al. (2010). Translating environmental gradients into discontinuous reaction norms via hormone signaling in a polyphenic butterfly. *Epub.*

Oster, G. F., & Wilson, E. O. (1978). *Caste and Ecology in the Social Insects.* Princeton: Princeton University Press.

Park, Y. I., & Raina, A. K. (2004). Juvenile hormone III titers and regulation of soldier caste in Coptotermes formosanus (Isoptera: Rhinotermitidae). *J. Insect Physiol., 50,* 561–566.

Passera, L., & Suzzoni, J. P. (1979). Le rôle de la reine de Pheidole pallidula (Nyl.) (Hymenoptera, Formicidae) dans la sexualisation du couvain après traitement per l'hormone juvénile. *Insectes Soc., 26,* 343–353.

Patel, A., Fondrk, M. K., Kaftanoglu, O., Emore, C., Hunt, G., et al. (2007). The making of a queen: TOR pathway is a key player in diphenic caste development. *PLoS One, 2,* e509.

Paul, R. K., Takeuchi, H., Matsuo, Y., & Kubo, T. (2005). Gene expression of ecdysteroid-regulated gene E74 of the honey bee in ovary and brain. *Insect Mol. Biol., 14,* 9–15.

Peeters, C. (1991). Ergatoid queens and intercastes in ants: two distinct adult forms which look morphologically intermediate between workers and winged queens. *Insectes Soc., 38,* 1–15.

Peeters, C., & Hölldobler, B. (1995). Reproductive cooperation between queens and their mated workers - the complex life history of an ant with a valuable nest. *Proc. Natl. Acad. Sci., U.S.A*, *92*, 10977–10979.

Peeters, C., Monnin, T., & Malosse, C. (1999). Cuticular hydrocarbons correlated with reproductive status in a queenless ant. *Proc. R. Soc. London B*, *266*, 1323–1327.

Pener, M. P. (1991). Locust phase polymorphism and Its endocrine relations. *Adv. Insect Physiol.*, *23*, 1–79.

Pener, M. P., Ayali, A., & Benm-Ami, E. (1992). Juvenile hormone is not a major factor in locust phase changes. In B. Mauchamp, F. Couillaud, & J. C. Baehr (Eds.), *Insect Juvenile Hormone Research, Fundamental and Applied Approaches* (pp. 125–134). Paris: INRA.

Pener, M. P., & Yerushalmi, Y. (1998). The physiology of locust phase polymorphism: an update. *J. Insect Physiol.*, *44*, 365–377.

Pener, M. P., & Simpson, S. J. (2009). Locust phase polyphenism: an update. *Adv. Insect Physiol.*, *36*, 1–272.

Pereboom, J. J.M. (2000). The composition of larval food and the significance of exocrine secretions in the bumblebee Bombus terrestris. *Insectes Soc.*, *47*, 11–20.

Pereboom, J. J.M., Velthuis, H. H.W., & Duchateau, M. J. (2003). The organisation of larval feeding in bumblebees (Hymenoptera, Apidae) and its significance to castre differentiation. *Insectes Soc.*, *50*, 127–133.

Peterson, M. A., & Denno, R. F. (1997). The influence of intraspecific variation in dispersal strategies on the genetic structure of planthopper populations. *Evolution*, *51*, 1189–1206.

Pickett, K. M., & Carpenter, J. (2010). Simultaneous analysis and the origin of eusociality in the Vespidae (Insecta: Hymenoptera). *Arthrop. Syst. Phyl.*, *68*, 3–33.

Pike, N., & Foster, W. A. (2008). The ecology of altruism in a clonal insect. In J. Korb, & J. Heinze (Eds.), *Ecology of Social Evolution* (pp. 37–56). Berlin: Springer.

Pinto, L. Z., Hartfelder, K., Bitondi, M. M.G., & Simões, Z. L.P. (2002). Ecdysteroid titers in pupae of highly social bees relate to distinct modes of caste development. *J. Insect Physiol.*, *48*, 783–790.

Porter, S. D., & Tschinkel, W. R. (1985). Fire ant polymorphisms: factors affecting worker size. *Ann. Ent. Soc. Amer.*, *78*, 381–386.

Poulton, E. B. (1887). An enquiry into the cause and extent of a special colour-relation between certain exposed lepidopterous pupae and the surfaces which immediately surround them. *Phil. Trans. R. Soc. London*, *178*, 311–441.

Rachinsky, A., & Hartfelder, K. (1990). Corpora allata activity, a prime regulating element for caste-specific juvenile hormone titre in honey bee larvae (Apis mellifera carnica). *J. Insect Physiol.*, *36*, 189–194.

Rachinsky, A., Strambi, C., Strambi, A., & Hartfelder, K. (1990). Caste and metamorphosis - hemolymph titers of juvenile hormone and ecdysteroids in last instar honey bee larvae. *Gen. Comp. Endocr.*, *79*, 31–38.

Rachinsky, A., & Hartfelder, K. (1991). Differential production of juvenile hormone and its deoxy precursor by corpora allata of honey bees during a critical period of caste development. *Naturwissenschaften*, *78*, 270–272.

Rachinsky, A. (1994). Octopamine and serotonin influence on corpora allata activity in honey bee (Apis mellifera) larvae. *J. Insect Physiol.*, *40*, 549–554.

Rachinsky, A., & Engels, W. (1995). Caste development in honey bees (Apis mellifera) - juvenile hormone turns on ecdysteroids. *Naturwissenschaften*, *82*, 378–379.

Rachinsky, A. (1996). Brain and suboesophageal ganglion extracts affect juvenile hormone biosynthesis in honey bee larvae (Apis mellifera carnica). *Zool.-Anal. Complex Syst.*, *99*, 277–284.

Rachinsky, A., & Feldlaufer, M. F. (2000). Responsiveness of honey bee (Apis mellifera L.) corpora allata to allatoregulatory peptides from four insect species. *J. Insect Physiol.*, *46*, 41–46.

Rachinsky, A., Tobe, S. S., & Feldlaufer, M. F. (2000). Terminal steps in JH biosynthesis in the honey bee (Apis mellifera L.): developmental changes in sensitivity to JH precursor and allatotropin. *Insect Biochem. Mol. Biol.*, *30*, 729–737.

Rahman, M. M., Vandigenen, A., Begum, M., Breuer, M., de Loof, A., et al. (2003). Search for phase specific genes in the brain of desert locust, Schistocerca gregaria (Orthoptera: Acrididae) by differential display polymerase chain reaction. *Comp. Biochem. Physiol. A*, *135*, 221–228.

Rahman, M. M., Breuer, M., Begum, M., Baggerman, G., Huybrechts, J., et al. (2008). Localization of the phase-related 6-kDa peptide (PRP) in different tissues of the desert locust Schistocerca gregaria - Immunocytochemical and mass spectrometric approach. *J. Insect Physiol.*, *54*, 543–554.

Ramsey, J. S., Wilson, A. C.C., de Vos, M., Sun, Q., Tamborindeguy, C., et al. (2007). Genomic resources for Myzus persicae: EST sequencing, SNP identification, and microarray design. *BMC Genomics*, *8*, e423.

Rasmussen, J. (1994). The influence of horn and body size on the reproductive behavior of the horned rainbow scarab beetle Phanaeus difformis (Coleoptera: Scarabaeidae). *J. Insect Behav.*, *7*, 67–82.

Ratnieks, F. L.W. (2001). Heirs and spares: caste conflict and excess queen production in Melipona bees. *Behav. Ecol. Sociobiol.*, *50*, 467–473.

Reginato, R. D., & Cruz-Landim, C. (2001). Differentiation of the worker's ovary in Apis mellifera L. (Hymenoptera, Apidae) during life of the larvae. *Invertebr. Reprod. Dev.*, *39*, 127–134.

Reinhardt, R. (1969). Über den Einfluss der Temperatur auf den Saisondimorphismus von Araschnia levana L. (Lepidopt. Nymphalidae) nach photoperiodischer Diapauseinduktion. *Zool. Jb. Physiol.*, *75*, 41–75.

Rembold, H., Czoppelt, C., & Rao, P. J. (1974a). Effect of juvenile hormone on caste differentiation in the honey bee, Apis mellifera. *J. Insect Physiol.*, *20*, 1193–1202.

Rembold, H., Lackner, B., & Geistbeck, J. (1974b). The chemical basis of queen bee determinator from royal jelly. *J. Insect Physiol.*, *20*, 307–314.

Rembold, H. (1987). Caste-specific modulation of juvenile hormone titers in Apis mellifera. *Insect Biochem.*, *17*, 1003–1006.

Rembold, H., Czoppelt, C., Grüne, M., Lackner, B., Pfeffer, J., et al. (1992). Juvenile hormone titers during honey bee embryogenesis and metamorphosis. In B. Mauchamp, F. Couillaud, & J. C. Baehr (Eds.), *Insect Juvenile Hormone Research* (pp. 37–43). Paris: INRA.

Remolina, S. C., Hafez, D. M., Robinson, G. E., & Hughes, K. A. (2007). Senescence in the worker honey bee Apis mellifera. *IJ. Insect Physiol.*, *53*, 1027–1033.

Rheindt, F. E., Strehl, C. P., & Gadau, J. (2005). A genetic component in the determination of worker polymorphism in the Florida harvester ant pogonomyrmex bodius. *Insectes Soc.*, *52*, 163–168.

Rhoden, P. K., & Foster, W. A. (2002). Soldier behaviour and division of labor in the aphid genus Pemphigius (Hemiptera: Aphididae). *Insectes Soc.*, *49*, 257–263.

Ribeiro, M. F. (1999). Long-duration feedings and caste differentiation in Bombus terrestris larvae. *Insectes Soc.*, *46*, 315–322.

Ribeiro, M. F., Velthuis, H. H.W., Duchateau, M. J., & van der Tweet, I. (1999). Feeding frequency and caste differentiation in Bombus terrestris larvae. *Insectes Soc.*, *46*, 306–314.

Ribeiro, M. F., Wenseleers, T., Santos Filho, P. S., & Alves, D. A. (2006). Miniature queens in stingless bees: basic facts and evolutionary hypotheses. *Apidologie*, *37*, 191–206.

Robinson, G. E., Strambi, C., Strambi, A., & Feldlaufer, M. F. (1991). Comparison of juvenile hormone and ecdysteroid hemolymph titers in adult worker and queen honey bees (Apis mellifera). *J. Insect Physiol.*, *37*, 929–935.

Robinson, G. E., Strambi, C., Strambi, A., & Huang, Z.-Y. (1992). Reproduction in worker honey bees is associated with low juvenile hormone titers and rates of biosynthesis. *Gen. Comp. Endocr.*, *87*, 471–480.

Roff, D. A. (1986). The genetic basis of wing dimorphism in the sand cricket, Gryllus firmus and its relevance to the evolution of wing dimorphism in insects. *Heredity*, *57*, 221–231.

Roff, D. A. (1990). Selection for changes in the incidence of wing dimorphism in Gryllus firmus. *Heredity*, *65*, 163–168.

Roff, D. A., & Fairbairn, D. J. (2007). The evolution and genetics of migration in insects. *Bioscience*, *57*, 155–164.

Roisin, Y. (2000). Diversity and evolution of caste patterns. In T. Abe, D. E. Bignell, & M. Higashi (Eds.), *Termites: Evolution, Sociality, Symbiosis, Ecology*. Dordrecht, Netherlands: Kluwer Academic Publishers.

Röseler, P.-F. (1970). Unterschiede in der Kastendetermination zwischen den Hummelarten Bombus hypnorum und Bombus terrestris. *Z. Naturforsch*, *25*, 543–548.

Röseler, P.-F. (1974). Grössenpolymorphismus, Geschlechtsregulation und Stabilisierung der Kasten im Hummelvolk. In G. H. Schmidt (Ed.), *Sozialpolymorphismus bei Insekten* (pp. 298–335). Stuttgart: Wissenschaftliche Verlagsgesellschaft.

Röseler, P.-F. (1977). Juvenile hormone control of oogenesis in bumblebee workers, Bombus terrestris. *J. Insect Physiol.*, *23*, 985–992.

Röseler, P.-F., & Röseler, I. (1978). Studies on the regulation of the juvenile hormone titre in bumblebee workers, Bombus terrestris. *J. Insect Physiol.*, *24*, 707–713.

Röseler, P.-F., Röseler, I., & van Honk, C. G.J. (1981). Evidence for inhibition of corpora allata activity in workers of Bombus terrestris by a pheromone from the queen's mandibular glands. *Experientia*, *37*, 348–351.

Röseler, P.-F. (1985). A technique for year-round rearing of Bombus terrestris (Apidae, Bombini) colonies in captivity. *Apidologie*, *16*, 165–170.

Röseler, P.-F., & Röseler, I. (1986). Caste-specific differences in fat body glycogen metabolism of the bumblebee, Bombus terrestris. *Insect Biochem.*, *16*, 501–508.

Röseler, P.-F., & van Honk, C. G.J. (1990). Castes and reproduction in bumblebees. In W. Engels (Ed.), *Social Insects - an Evolutionary Approach to Castes and Reproduction* (pp. 147–166). Heidelberg: Springer.

Röseler, P. F., Röseler, I., Strambi, A., & Augier, R. (1984). Influence of insect hormones on the establishment of dominance hierarchies among foundresses of the paper wasp, Polistes gallicus. *Behav. Ecol. Sociobiol.*, *15*, 133–184.

Röseler, P. F., Röseler, I., & Strambi, A. (1985). Role of ovaries and ecdysteroids in dominance hierarchy establishment among foundresses of the primitively social wasp, Polistes gallicus. *Behav. Ecol. Sociobiol.*, *18*, 9–13.

Roskam, J. C., & Brakefield, P. M. (1999). Seasonal polyphenism in Bicyclus (Lepidoptera: Satyridae) butterflies: different climates need different cues. *Biol. J. Linn. Soc.*, *66*, 345–356.

Rountree, D. B., & Nijhout, H. F. (1995). Genetic control of a seasonal morph in Precis coenia (Lepidoptera, Nymphalidae). *J. Insect Physiol.*, *41*, 1141–1145.

Rowland, J. M., & Emlen, D. J. (2009). Two thresholds, three male forms result in facultative male trimorphism in beetles. *Science*, *323*, 773–776.

Ryan, M. J., Hews, D. W., & Wagner, W. E.J. (1990). Sexual selection on alleles that determine body size in the swordtail Xiphophorus nigrensis. *Behav. Ecol. Sociobiol.*, *26*, 231–237.

Sabater-Muñoz, B., Legeai, F., Rispe, C., Bonhomme, J., Dearden, P. K., et al. (2006). Large scale gene discovery in the pea aphid Acyrthosiphon pisum (Hemiptera). *Genome Biol.*, *7*, e21.

Saeki, H. (1966a). The effect of day length on the occurrence of the macropterous form in a cricket, Scapsipedus aspersus Walker (Orthoptera: Gryllidae). *Jpn. J. Ecol.*, *16*, 49–52.

Saeki, H. (1966b). The effect of population density on the occurrence of the macropterous form in a cricket, Scapsipedus aspersus Walker (Orthoptera: Gryllidae). *Jpn. J. Ecol.*, *16*, 1–4.

Saenko, S. V., French, V., Brakefield, P. M., & Beldade, P. (2007). Conserved developmental processes and the formation of evolutionary novelties: examples from butterfly wings. *Phil. Trans. R. Soc. B*, *363*, 1549–1555.

Saini, R. K., Rai, M. M., Hassanali, A., Wawiye, J., & Odongo, H. (1995). Semiochemicals from froth of egg pods attract ovipositing female Schistocerca gregaria. *J. Insect Physiol.*, *41*, 711–716.

Scharf, M. E., Wu-Scharf, D., Pittendrigh, B. R., & Bennett, G. W. (2003). Caste- and development-associated gene expression in a lower termite. *Genome Biol.*, *4*, e10.

Scharf, M. E., Ratliff, C. R., Wu-Scharf, D., Zhou, X. G., Pittendrigh, B. R., et al. (2005a). Effects of juvenile hormone III on Reticulitermes flavipes: changes in hemolymph protein composition and gene expression. *Insect Biochem. Mol. Biol.*, *35*, 207–215.

Scharf, M. E., Wu-Scharf, D., Zhou, X., Pittendrigh, B. R., & Bennett, G. W. (2005b). Gene expression profiles among immature and adult reproductive castes of the termite Reticulitermes flavipes. *Insect Mol. Biol.*, *14*, 31–44.

Scharf, M. E., Buckspan, C. E., Grzymala, T. L., & Zhou, X. (2007). Regulation of polyphenic caste differentiation in the termite Reticulitermes flavipes by interaction of intrinsic and extrinsic factors. *J. Exp. Biol.*, *210*, 4390–4398.

Schmidt Capella, I. C., & Hartfelder, K. (1998). Juvenile hormone effect on DNA synthesis and apoptosis in caste-specific differentiation of the larval honey bee (Apis mellifera L.) ovary. *J. Insect Physiol.*, *44*, 385–391.

Schmidt Capella, I. C., & Hartfelder, K. (2002). Juvenile-hormone-dependent interaction of actin and spectrin is crucial for polymorphic differentiation of the larval honey bee ovary. *Cell Tissue Res.*, *307*, 265–272.

Schoofs, L., Baggerman, G., Veelaert, D., Breuer, M., Tanaka, S., et al. (2000). The pigmentotropic hormone [His7]-corazonin, absent in a Locusta migratoria albino strain, occurs in an albino strain of Schistocerca gregaria. *Mol. Cell Endocrinol.*, *168*, 101–109.

Schrempf, A., & Heinze, J. (2006). Proximate mechanisms of male, morph determination in the ant Cardiocondyla obscurior. *Evol. Dev.*, *8*, 266–272.

Schutze, M., & Maschwitz, U. (1991). Enemy recognition and defense within trophobiotic associations with ants by the soldier caste of Pseudoregma sundanica (Homoptera, Aphidoidea). *Entom. Gener.*, *16*, 1–12.

Schwander, T., Lo, N., Beekman, M., Oldroyd, B. P., & Keller, L. (2010). Nature versus nurture in social insect caste differentiation. *Trends Ecol. Evol.*, *25*, 275–282.

Schwartzberg, E. G., Kunert, G., Westerlund, S. A., Hoffmann, K. H., & Weisser, W. W. (2008). Juvenile hormone titres and winged offspring production do not correlate in the pea aphid, Acyrthosiphon pisum. *J. Insect Physiol.*, *54*, 1332–1336.

Seidel, C., & Bicker, G. (1996). The developmental expression of serotonin-immunoreactivity in the brain of the pupal honey bee. *Tissue Cell*, *28*, 663–672.

Sevastopulo, D. G. (1975). Dimorphism in Papilio pupae. *Entom. Rec. J. Var.*, *87*, 109–111.

Severson, D. W., Williamson, J. L., & Aiken, J. M. (1989). Caste-specific transcription in the female honey bee. *Insect Biochem.*, *19*, 215–220.

Shapiro, A. M. (1976). Seasonal polymorphism. *Evol. Biol.*, *9*, 259–333.

Sheppard, P. M. (1958). *Natural Selection and Heredity.* London: Hutchinson.

Shibao, S., Kutsukake, M., Matsuyama, S., Fukatsu, T., & Shimada, M. (2010). Mechanisms regulating caste differentiation in an aphid social system. *Commun. Integr. Biol.*, *4*, 1–5.

Shuster, S. M. (1989). Male alternative reproductive strategies in a marine isopod crustacean (Paracerceis sculpta): the use of genetic markers to measure diferences in fertilization success among alpha, beta, and gamma males. *Evolution*, *43*, 1683–1698.

Shuster, S. M. (2002). Mating strategies, alternative. In M. Pagel (Ed.), *Encyclopedia of Evolution Vol. 2* (pp. 688–693). Oxford: Oxford University Press.

Simões, Z. L.P., Boleli, I. C., & Hartfelder, K. (1997). Occurrence of a prothoracicotropic hormone-like peptide in the developing nervous system of the honey bee (Apis mellifera). *Apidologie*, *28*, 399–409.

Simon, J.-C., Rispe, C., & Sunnucks, P. (2002). Ecology and evolution of sex in aphids. *Trends Ecol. Evol.*, *17*, 34–39.

Simon, J. C., Stoeckel, S., & Tagu, D. (2010). Evolutionary and functional insights into reproductive strategies of aphids. *C.R. Biol.*, *333*, 488–496.

Simpson, S. J., Despland, E., Hägele, B. F., & Dodgson, T. (2001). Gregarious behavior in desert locusts is evoked by touching their back legs. *Proc. Natl. Acad. Sci. USA*, *98*, 3895–3897.

Sims, S. R. (1983). The genetic and environmental basis of pupal colour dimorphism in Papilio zelicaon (Lepidoptera: Papilionidae). *Heredity*, *50*, 159–168.

Sledge, M. F., Boscaro, F., & Turillazzi, S. (2001). Cuticular hydrocarbons and reproductive status in the social wasp Polistes dominulus. *Behav. Ecol. Sociobiol.*, *49*, 401–409.

Sloggett, J. J., & Weisser, W. W. (2002). Parasitoids induce production of the dispersal morph of the pea aphid, Acyrthosiphon pisum. *Oikos*, *98*, 323–333.

Smith, A. G. (1978). Environmental factors influencing pupal colour determination in Lepidoptera. I. Experiments with Papilio polytes, Papilio demoleus and Papilio polyxenes. *Proc. R. Soc. London*, *200*, 295–329.

Smith, A. G. (1980). Environmental factors influencing pupal colour determination in Lepidoptera. I. Experiments with Pieris rapae, Pieris napi and Pieris brassicae. *Proc. R. Soc. London*, *207*, 163–186.

Smith, C. R., Anderson, K. E., Tillberg, C. V., Gadau, J., & Suarez, A. V. (2008). Caste determination in a polymorphic social insect: nutritional, social and genetic factors. *Am. Nat.*, *172*, 497–507.

Smith, D. A., Shoesmith, E. A., & Smith, A. G. (1988). Pupal polyphenism in the butterfly Danaus chrysippus (L): environmental, seasonal and genetic influences. *Biol. J. Linn. Soc.*, *33*, 17–50.

Smith, K. C. (1991). The effects of temperature and daylength on the rosa polyphenism in the buckeye butterfly, Precis coenia (Lepidoptera: Nymphalidae). *J. Res. Lep.*, *30*, 225–236.

Snell-Rood, E. C., Cash, A., Han, M. V., Kijimoto, T., Andrews, J., et al. (2010). Developmental decoupling of alternative phenotypes: insights from the transcriptomes of horn-polyphenic beetles. *Evolution 65*, 231–245.

Sommer, K., Hölldobler, B., & Rembold, H. (1993). Behavioral and physiological aspects of reproductive control in a Diacamma species from Malaysia (Formicidae, Ponerinae). *Ethology*, *94*, 162–170.

Srihari, T., Gutmann, E., & Novak, V. J.A. (1975). Effect of ecdysterone and juvenoid on the developmental involution of flight muscles in Acheta domestica. *J. Insect Physiol.*, *21*, 1–8.

Staal, G. B., & de Wilde, J. (1962). Endocrine influences on the development of phase characters in Locusta. *Coll. Int. Centre Nat. Rech. Sci*, *114*, 89–105.

Starnecker, G. (1997). Hormonal control of lutein incorporation into pupal cuticle of the butterfly Inachis io and the pupal melanization reducing factor. *Physiol. Entomol.*, *22*, 73–78.

Starnecker, G., & Bückmann, D. (1997). Temporal occurrence of pupal melanization reducing factor during development of the butterfly Inachis io. *Physiol. Entomol.*, *22*, 79–85.

Starnecker, G., & Hazel, W. N. (1999). Convergent evolution of neuroendocrine control of phenotypic plasticity in pupal colour in butterflies. *Proc. R. Soc. London*, *266*, 2409–2412.

Stay, B., & Zera, A. J. (2010). Morph-specific diurnal variation in allatostatin immunostaining in the corpora allata of Gryllus firmus: implications for the regulation of a morph-specific circadian rhythm for JH biosynthetic rate. *J. Insect Physiol.*, *56*, 266–270.

Steel, C. G.H., & Lees, A. D. (1977). The role of neurosecretion in the photoperiodic control of polymorphism in the aphid Megoura viciae. *J. Exp. Biol.*, *67*, 117–135.

Stern, D. L., Whitfield, J. A., & Foster, W. A. (1997). Behavior and morphology of monomorphic soldiers from the aphid genus Pseudoregma (Cerataphidini, Hormaphididae): implications for the evolution of morphological castes in social aphids. *Insectes Soc.*, *44*, 379–392.

Strambi, A., Strambi, C., Röseler, P.-F., & Röseler, I. (1984). Simultaneous determination of juvenile hormone and ecdysteroid titers in the hemolymph of bumblebee prepupae (Bombus hypnorum and B. terrestris). *Gen. Comp. Endocr.*, *55*, 83–88.

Strambi, A. (1990). Physiology and reproduction in social wasps. In W. Engels (Ed.), *Social Insects - an Evolutionary Approach to Castes and Reproduction* (pp. 59–75). Heidelberg: Springer.

Süffert, F. (1924). Bestimmungsfaktoren des Zeichnungsmusters beim Saisondimorphismus von Araschnia levana-prorsa. *Biol. Zbl.*, *44*, 173–188.

Suzzoni, J. P., Passera, L., & Strambi, A. (1980). Ecdysteroid titer and caste determination in the ant, Pheidole pallidula (Nyl) (Hymenoptera, Formicidae). *Experientia*, *36*, 1228–1229.

Suzzoni, J. P., Passera, L., & Strambi, A. (1983). Ecdysteroid production during caste differentiation in larvae of the ant, Plagiolepis pygmaea. *Physiol. Entomol.*, *8*, 93–96.

Sword, G. A. (2002). A role for phenotypic plasticity in the evolution of aposematism. *Proc. R. Soc. Lond. Ser. B Biol. Sci.*, *269*, 1639–1644.

Sword, G. A., Lecoq, M., & Simpson, S. J. (2010). Phase polyphenism and preventative locust management. *J. Insect Physiol.*, *56*, 949–957.

Tanaka, A., Inoue, M., Endo, K., Kitazawa, C., & Yamanaka, A. (2009). Presence of a cerebral factor showing summer-morphproducing hormone activity in the brain of the seasonal nonpolyphenic butterflies Vanessa cardui, V. indica and Nymphalis xanthomelas japonica (Lepidoptera: Nymphalidae). *Insect Sci.*, *16*, 125–130.

Tanaka, S., Matsuka, M., & Sakai, T. (1976). Effect of change in photoperiod on wing form in Pteronemobius taprobanensis (Orthoptera: Gryllidae). *Appl. Ent. Zool.*, *11*, 27–32.

Tanaka, S., & Wolda, H. (1987). Seasonal wing dimoprhism in a tropical seed bug: ecological significance of the short-winged form. *Oecologia*, *73*, 559–565.

Tanaka, S. (1993). Hormonal deficiency causing albinism in Locusta migratoria. *Zool. Sci.*, *10*, 467–471.

Tanaka, S. (2000a). Induction of darkening by corazonins in several species of Orthoptera and their possible presence in ten insect orders. *Appl. Entomol. Zoolog.*, *35*, 509–517.

Tanaka, S. (2000b). The role of [His(7)]-corazonin in the control of body-color polymorphism in the migratory locust, Locusta migratoria (Orthoptera: Acrididae). *J. Insect Physiol.*, *46*, 1169–1176.

Tanaka, S. (2000c). Hormonal control of body-color polymorphism in Locusta migratoria: interaction between [His(7)]-corazonin and juvenile hormone. *J. Insect Physiol.*, *46*, 1535–1544.

Tanaka, S. (2001). Endocrine mechanisms controlling body-color polymorphism in locusts. *Arch. Insect Biochem. Physiol.*, *47*, 139–149.

Tanaka, S., & Maeno, K. (2006). Phase-related body-color polyphenism in hatchlings of the desert locust, Schistocerca gregaria: re-examination of the maternal and crowding effects. *J. Insect Physiol.*, *52*, 1054–1061.

Tanaka, S., & Maeno, K. (2010). A review of maternal and embryonic control of phase-dependent progeny characteristics in the desert locust. *J. Insect Physiol.*, *56*, 911–918.

Tarver, M. R., Zhou, X. G., & Scharf, M. E. (2010). Socio-environmental and endocrine influences on developmental and caste-regulatory gene expression in the eusocial termite Reticulitermes flavipes. *BMC Mol. Biol.*, *11*. e28.

Tawfik, A. I., Mathova, A., Sehnal, F., & Ismail, S. H. (1996). Haemolymph ecdysteroids in the solitary and gregarious larvae of Schistocerca gregaria. *Arch. Insect Biochem. Physiol.*, *31*, 427–438.

Tawfik, A. I., Osir, E. O., Hassanali, A., & Ismail, S. H. (1997a). Effects of juvenile hormone treatment on phase changes and pheromone production in the desert locust, Schistocerca gregaria (Forskal) (Orthoptera: Acrididae). *J. Insect Physiol.*, *43*, 1177–1182.

Tawfik, A. I., Vedrova, A., Li, W. W., Sehnal, F., & ObengOfori, D. (1997b). Haemolymph ecdysteroids and the prothoracic glands in the solitary and gregarious adults of Schistocerca gregaria. *J. Insect Physiol.*, *43*, 485–493.

Tawfik, A. I., Tanaka, S., De Loof, A., Schoofs, L., Baggerman, G., et al. (1999a). Identification of the gregarization-associated dark-pigmentotropin in locusts through an albino mutant. *Proc. Natl. Acad. Sci. USA*, *96*, 7083–7087.

Tawfik, A. I., Vedrova, A., & Sehnal, F. (1999b). Ecdysteroids during ovarian development and embryogenesis in solitary and gregarious Schistocerca gregaria. *Arch. Insect Biochem. Physiol.*, *41*, 134–143.

Tawfik, A. I., Treiblmayr, K., Hassanali, A., & Osir, E. O. (2000). Time-course haemolymph juvenile hormone titres in solitarious and gregarious adults of Schistocerca gregaria, and their relation to pheromone emission, CA volumetric changes and oocyte growth. *J. Insect Physiol.*, *46*, 1143–1150.

Tawfik, A. I., & Sehnal, F. (2003). A role for ecdysteroids in the phase polymoprhism of the desert locust. *Physiol. Entomol.*, *28*, 19–24.

Teles, A. C.A.S., Mello, T. R.P., Barchuk, A. R., & Simoes, Z. L.P. (2007). Ultraspiracle of the stingless bees Melipona scutellaris and Scaptotrigona depilis: cDNA sequence and expression profiles during pupal development. *Apidologie*, *38*, 462–471.

The Honey Bee Genome Sequencing Consortium (2006). Insights into social insects from the genome of the honey bee Apis mellifera. *Nature*, *443*, 931–949.

Thornhill, R., & Alcock, J. (1983). *The Evolution of Insect Mating Systems*. Cambridge Mass: Belknapp Press of Harvard University Press.

Tibbetts, E. A., & Izzo, A. S. (2009). Endocrine mediated phenotypic plasticity: Condition-dependent effects of juvenile hormone on dominance and fertility of wasp queens. *Horm. Behav.*, *56*, 527–531.

Tibbetts, E. A., & Huang, Z. Y. (2010). The challenge hypothesis in an insect: juvenile hormone increases during reproductive conflict following queen loss in Polistes wasps. *Am. Nat.*, *176*, 123–130.

Tobe, S. S., & Pratt, G. E. (1974). The influence of substrate concentrations on the rate of insect juvenile hormone biosynthesis by corpora allata of the desert locust *in vitro*. *Biochem. J.*, *144*, 107–113.

Tomkins, J. L., & Simmons, L. W. (1996). Dimorphism and fluctuating asymmetry in the forceps of male earwigs. *J. Evol. Biol.*, *9*, 753–770.

Tomkins, J. L. (1999). Environmental and genetic determinants of the male forceps length dimorphism in the European earwig Forficula auricularia L. *Behav. Ecol. Sociobiol.*, *47*, 1–8.

Toth, A. L., Varala, K., Newman, T. C., Miguez, F. E., Hutchison, S. K., et al. (2007). Wasp gene expression supports an evolutionary link between maternal behavior and eusociality. *Science*, *318*, 441–444.

Trenczek, T., Zillikens, A., & Engels, W. (1989). Developmental patterns of vitellogenin hemolymph titer and rate of synthesis in adult drone honey bees (Apis mellifera). *J. Insect Physiol.*, *35*, 475–481.

Truman, J. W., & Riddiford, L. M. (1999). The origin of insect metamorphosis. *Nature*, *401*, 447–452.

Tschinkel, W. R. (1991). Sociometry, a field in search of data. *Insectes Soc.*, *34*, 143–164.

Tschinkel, W. R., Mikheyev, A. S., & Storz, S. R. (2003). Allometry of workers of the fire ant, Solenopsis invicta. *J. Insect Sci.*, *3.2*, 11.

Tsuruda, J. M., Amdam, G. V., & Page, R. E. J. (2008). Sensory response system of social behavior tied to female reproductive traits. *PLoS One*, *3*, e3397.

Tufail, M., Naeemullah, M., Elmogy, M., Sharma, P. N., Takeda, M., et al. (2010). Molecular cloning, transcriptional regulation, and differential expression profiling of vitellogenin in two wing-morphs of the brown planthopper, Nilaparvata lugens Stål (Hemiptera: Delphacidae). *Insect Mol. Biol. Epub.*

Tyndale-Biscoe, M. (1996). *Australia's introduced dung beetles: original releases and redistribution*. Technical report No. 62 Canberra, ACT: CSIRO, Division of Entomology.

Uvarov, B. P. (1921). A revision of the genus Locusta L. (=Pachytylus Fieb.), with a new theory as to periodicity and migrations of locusts. *Bull. Ent. Res.*, *12*, 135–163.

van Doorn, A. (1987). Investigations into the regulation of dominace behaviour and the division of labour in bumblebee colonies (Bombus terrestris). *Neth. J. Zool.*, *37*, 255–276.

Vargo, E. L., & Laurel, M. (1994). Studies on the mode of action of a queen primer pheromone of the fire ant Solenopsis invicta. *J. Insect Physiol.*, *40*, 601–610.

Velthuis, H. H. W., Koedam, D., & Imperatriz-Fonseca, V. (2005). The males of Melipona and other stingless bees, and their mothers. *Apidologie*, *36*, 169–185.

Verlinden, H., Badisco, L., Marchal, E., Van Wielendaele, P., & Vanden Broeck, J. (2009). Endocrinology of reproduction and phase transition in locusts. *Gen. Comp. Endocr.*, *162*, 79–92.

Vinson, S. B., & Robeau, R. M., (1974). Insect growth regulators: effects on colonies of the imported fire ant. *J. Econ. Entom.*, *67*, 584–587.

Volny, V., & Gordon, D. M. (2002). Genetic basis for queen-worker dimorphism in a social insect. *Proc. Natl. Acad. Sci., USA*, *99*, 6108–6111.

Walsh, T. K., Brisson, J. A., Robertson, H. M., Gordon, K., Jaubert-Possamai, S., et al. (2010). A functional DNA methylation system in the pea aphid, Acyrthosiphon pisum. *Insect Mol. Biol.*, *19*, 215–228.

Wang, Y., Kaftanoglu, O., Siegel, A., Page, R. E.J., & Amdam, G. V. (2010). Surgically increased ovarian mass in the honey bee confirms link between reproductive physiology and worker behavior. *J. Insect Physiol.*, *56*, 1816–1824.

Wasik, B. R., Rose, D. J., & Moczek, A. P. (2010). Beetle horns are regulated by the Hox gene, Sex combs reduced, in a species- and sex-specific manner. *Evol. Dev.*, *12*, 353–362.

Watt, W. B. (1968). Adaptive significance of pigment polymorphisms in Colias butterflies. I. Variation of emlanin pigment in relation to thermoregulation. *Evolution*, *22*, 437–458.

Watt, W. B. (1969). Adaptive significance of pigment polymorphisms in Colias butterflies. II. Thermoregulation and photoperiodically controlled melanin variation in Colias eurytheme. *Proc. Natl. Acad. Sci., USA*, *63*, 767–774.

Weatherbee, S. D., Nijhout, H. F., Grunert, L. W., Haldeer, G., Galant, R., et al. (1999). Ultrabithorax function in butterfly wings and the evolution of insect wing patterns. *Curr. Biol.*, *9*, 109–115.

Wedekind-Hirschberger, S., Sickold, S., & Dorn, A. (1999). Expression of phase-specific haemolymph polypeptides in a laboratory strain and field catches of Schistocerca gregaria. *J. Insect Physiol.*, *45*, 1097–1103.

Wei, Y., Chen, S., Yang, P., Ma, Z., & Kang, L. (2009). Characterization and comparative profiling of the small RNA transcriptomes in two phases of locust. *Genome Biol.*, *10*, e6.

Weil, T., Rehli, M., & Korb, J. (2007). Molecular basis for the reproductive division of labour in a lower termite. *BMC Genomics*, *8*, e198.

Weismann, A. (1875). *Studien zur Deszendenztheorie 1. Über den Saisondimorphismus der Schmetterlinge*. Leipzig: Engelmann.

West-Eberhard, M. J. (1996). Wasp societies as microcosms for the study of development and evolution. In S. Turillazzi, & M. J. West-Eberhard (Eds.), *Natural History and Evolution of Paper Wasps* (pp. 290–317). Oxford: Oxford University Press.

West-Eberhard, M. J. (2003). *Developmental Plasticity and Evolution*. Oxford: Oxford University Press.

West, D. A., Snelling, W. N., & Herbeck, T. A. (1972). Pupal colour dimorphism and its environmental control in Papilio polyxenes asterias Stoll (Lep. Papilionidae). *J.N.Y. Entom. Soc.*, *80*, 205–211.

West, D. A., & Hazel, W. N. (1979). Natural pupation sites of swallowtail butterflies (Lepidoptera: Papilionidae): Papilio polyxenes Fabr., P. glaucus L., and Battus philenor L. *Ecol. Entomol.*, *4*, 387–392.

West, D. A., & Hazel, W. N. (1982). An experimental test of natural selection for pupation site in swallowtail butterflies. *Evolution*, *36*, 152–159.

West, D. A., & Hazel, W. N. (1985). Pupal colour dimorphism in swallowtail butterflies: timing of the sensitive period and environmental control. *Physiol. Entomol.*, *10*, 113–119.

West, D. A., & Hazel, W. N. (1996). Natural pupation sites of three North American swallowtail butterflies: Eurytides marcellus (Cramer), Papilio cresphontes Cramer, and P. troilus L. (Papilionindae). *J. Lepidopt. Soc., 50,* 297–302.

Wheeler, D. E. (1983). Soldier determination in Pheidole bicarinata: effect of methoprene on caste and size within castes. *J. Insect Physiol., 29,* 847–854.

Wheeler, D. E. (1984). Soldier determination in Pheidole bicarinata: inhibition by adult soldiers. *J. Insect Physiol., 30,* 127–135.

Wheeler, D. E. (1986). Developmental and physiological determinants of caste in social Hymenoptera: evolutionary implications. *Am. Nat., 128,* 13–34.

Wheeler, D. E. (1990). The developmental basis of worker polymorphism in fire ants. *J. Insect Physiol., 36,* 315–322.

Wheeler, D. E. (1991). The developmental basis of worker caste polymorphism in ants. *Am. Nat., 138,* 1218–1238.

Wheeler, D. E. (1994). Nourishment in ants: patterns in individuals and societies. In J. Hunt, & C. Nalepa (Eds.), *Nourishment & Evolution in Insect Societies* (pp. 245–278). Boulder, Colorado: Westview Press.

Wheeler, D. E., Buck, N., & Evans, J. D. (2006). Expression of insulin pathway genes during the period of caste determination in the honey bee, Apis mellifera. *Insect Mol. Biol., 15,* 597–602.

Wicklund, C. (1972). Pupal colour polymorphism in Papilio machaon L. in response to wavelenght of light. *Naturwissenschaften, 59,* 219.

Wicklund, C. (1975). Pupal colour polymorphism in Papilio machaon L., and the survival of cryptic versus non-cryptic pupae. *Trans. R. Ent. Soc. London, 127,* 73–84.

Wiesnasz, D. C., & Cole, B. J. (2010). Patriline shifting leads to apparent genetic caste determination in harvester ants. *Pcoc. Natl. Acad. Sci. USA, 107,* 12958–12962.

Wiesel, G., Tappermann, S., & Dorn, A. (1996). Effects of juvenile hormone and juvenile hormone analogues on the phase behaviour of Schistocerca gregaria and Locusta migratoria. *J. Insect Physiol., 42,* 385–395.

Wilson, A. C. C., Dunbar, H. E., Davis, G. K., Hunter, W. B., Stern, D. L., et al. (2006). A dual-genome microarray for the pea aphid, Acyrthosiphon pisum, and its obligate bacterial symbiont, Buchnera aphidicola. *BMC Genomics, 7,* e50.

Wilson, E. O. (1971). *The Insect Societies.* Cambridge, Mass: Belknapp Press of Harvard University Press.

Wilson, E. O. (1987). The earliest known ants: an analysis of the Cretaceous species and an inference on their social organization. *Paleobiology, 13,* 44–53.

Windig, J. J., Brakefield, P. M., Reitsma, N., & Wilson, J. G. M. (1994). Seasonal polyphenism in the wild: survey of wing patterns in five species of Bicyclus butterflies in Malawi. *Ecol. Entomol., 19,* 285–298.

Winter, U., & Buschinger, A. (1986). Genetically mediated queen polymorphism and caste determination in the slave-making ant, Harpagoxenus sublaevis (Hymenoptera: Formicidae). *Entom. Gener., 11,* 125–137.

Wittkopp, P., & Beldade, P. (2008). Development and evolution of insect pigmentation: genetic mechanisms and the potential consequences of pleiotropy. *Semin. Cell Dev. Biol., 20,* 65–71.

Wood, T. W. (1867). Remarks on the coloration of Chrysalides. *Proc. R. Soc. Lond., 1867,* 98–101.

Wourms, M. K., & Wassermann, F. E. (1985). Butterfly wing markings are more advantageous during handling than during the initial strike of an avian predator. *Evolution, 39,* 845–851.

Yamanaka, A., Imai, H., Adachi, M., Komatsu, M., Islam, A. T.M.F., et al. (2004). Hormonal control of orange coloration of diapause pupae in the swallowtail butterfly, Papilio xuthus L. (Lepidoptera: Papilionidae). *Zool. Sci., 21,* 1049–1055.

Yamanaka, A., Uchiyama, T., Naotori, M., Adachi, M., Inoue, M., et al. (2007). Effect of Bombyx mori central nervous system extracts on diapause pupal coloration in the swallowtail butterfly, Papilio xuthus L. (Lepidoptera, Papilionidae). *Chugoku Kontyu, 21,* 55–59.

Yamanaka, A., Kometani, M., Yamamoto, K., Tsujimura, Y., Motomura, M., et al. (2009). Hormonal control of pupal coloration in the painted lady butterfly Vanessa cardui. *J. Insect Physiol., 55,* 512–517.

Zera, A. J., & Tiebel, K. C. (1988). Brachypterizing effect of group rearing, juvenile hormone III and methoprene in the wing-dimorphic cricket, Gryllus rubens. *J. Insect Physiol., 34,* 489–498.

Zera, A. J., Strambi, C., Tiebel, K. C., Strambi, A., & Rankin, M. A. (1989). Juvenile hormone and ecdysteroid titers during critical periods of wing determination in Gryllus rubens. *J. Insect Physiol., 35,* 501–511.

Zera, A. J., & Tiebel, K. C. (1989). Differences in juvenile hormone esterase activity between presumptive macropterous and brachypterousGryllus rubens: implications for the hormonal control of wing polymorphism. *J. Insect Physiol., 35,* 7–17.

Zera, A. J., & Tobe, S. S. (1990). Juvenile hormone-III biosynthesis in presumtive long-winged and short-winged Gryllus rubens: implications for the endocrine regulation of wing polymorphism. *J. Insect Physiol., 36,* 271–280.

Zera, A. J., & Holtmeier, C. L. (1992). *In vivo* and *in vitro* degradation of juvenile hormone-III in presumtive long-winged and short-winged Gryllus rubens. *J. Insect Physiol., 38,* 61–74.

Zera, A. J., & Zeisset, M. (1996). Biochemical characterization of juvenile hormone esterases from lines selected for high and low enzyme activity in Gryllus assimilis. *Biochem. Genet., 34,* 421–435.

Zera, A. J., & Denno, R. F. (1997). Physiology and ecology of dispersal polymorphism in insects. *Annu. Rev. Entomol., 42,* 207–230.

Zera, A. J., & Cisper, G. (2001). Genetic and diurnal variation in the juvenile hormone titer in a wing-polymorphic cricket: implications for the evolution of life histories and dispersal. *Physiol. Biochem. Zool., 74,* 293–306.

Zera, A. J., Sanger, T., Hanes, J., & Harshman, L. G. (2002). Purification and characterization of hemolymph juvenile hormone esterase from the cricket, Gryllus assimilis. *Arch. Insect Biochem. Physiol., 49,* 41–55.

Zera, A. J., & Zhao, Z. W. (2003). Life-history evolution and the microevolution of intermediary metabolism: activities of lipid-metabolizing enzymes in life-history morphs of a wing-dimorphic cricket. *Evolution, 57,* 586–596.

Zera, A. J. (2006). Evolutionary genetics of juvenile hormone and ecdysteroid regulation in Gryllus: a case study in the microevolution of endocrine regulation. *Comp. Biochem. Physiol. A, 144,* 365–379.

Zera, A. J. (2007). Endocrine analysis in evolutionary-developmental studies of insect polymorphism: hormone manipulation versus direct measurement of hormonal regulators. *Evol. Dev.*, *9*, 499–513.

Zera, A. J., Harshman, L. G., & Williams, T. D. (2007). Evolutionary endocrinology: The developing synthesis between endocrinology and evolutionary genetics. *Annu. Rev. Ecol. Evol. Syst.*, *38*, 793–817.

Zera, A. J., & Zhao, Z. W. (2009). Morph-associated JH titer diel rhythm in Gryllus firmus: experimental verification of its circadian basis and cycle characterization in artificially selected lines raised in the field. *J. Insect Physiol.*, *55*, 450–458.

Zhao, Z. W., & Zera, A. J. (2002). Differential lipid biosynthesis underlies a tradeoff between reproduction and flight capability in a wing-dimorphic cricket. *Proc. Natl. Acad. Sci. USA*, *99*, 16829–16834.

Zhao, Z. W., & Zera, A. J. (2004). A morph-specific daily cycle in the rate of JH biosynthesis underlies a morph-specific daily cycle in the hemolymph JH titer in a wing-polymorphic cricket. *J. Insect Physiol.*, *50*, 965–973.

Zhou, X., Tarver, M. R., Bennett, G. W., Oi, F. M., & Scharf, M. E. (2006a). Two hexamerin genes from the termite Reticulitermes flavipes: sequence, expression, and proposed functions in caste regulation. *Gene.*, *376*, 47–58.

Zhou, X. G., Oi, F. M., & Scharf, M. E. (2006b). Social exploitation of hexamerin: RNAi reveals a major caste-regulatory factor in termites. *Proc. Natl. Acad. Sci. USA*, *103*, 4499–4504.

Zhou, X. G., Tarver, M. R., & Scharf, M. E. (2007). Hexamerin-based regulation of juvenile hormone-dependent gene expression underlies phenotypic plasticity in a social insect. *Development*, *134*, 601–610.

Zijlstra, W. G., Steigenga, M. J., Koch, P. B., Zwaan, B. J., & Brakefield, P. M. (2004). Butterfly selected lines explore the hormonal basis of interactions between life histories and morphology. *Am. Nat.*, *163*, E76–E87.

Zimmerer, E. J., & Kallmann, K. D. (1989). The genetic basis for alternative reproductive tactics in the pygmy swordtail, Xiphophorus nigrensis. *Evolution*, *43*, 1298–1307.

Zufelato, M. S., Bitondi, M. M.G., Simões, Z. L.P., & Hartfelder, K. (2000). The juvenile hormone analog pyriproxyfen affects ecdysteroid-dependent cuticle melanization and shifts the pupal ecdysteroid peak in the honey bee (Apis mellifera). *Arthrop. Struct. Dev.*, *29*, 111–119.

12   Pheromone Production: Biochemistry and Molecular Biology

G.J. Blomquist
University of Nevada, Reno, Reno, NV, USA
R. Jurenka
Iowa State University, Ames, IA, USA
C. Schal
North Carolina State University, Raleigh, NC, USA
C. Tittiger
University of Nevada, Reno, Reno, NV, USA

© 2012 Elsevier B.V. All Rights Reserved

12.1. Introduction and Overview

The elucidation of the structure of the first insect sex pheromone bombykol (*E,Z*)-10,12-hexadecadien-1-ol (Butenandt *et al.*, 1959; **Figure 1**), from the silkworm moth, *Bombyx mori* (L.) spanned more than 20 years and required a half million female abdomens. A few years later, (*Z*)-7-dodecenyl acetate (**Figure 1**) was identified as the sex pheromone of the cabbage looper, *Trichoplusia ni* (Berger, 1966). At about the same time Silverstein *et al.* (1966) identified three terpenoid alcohols, ipsenol, ipsdienol, and verbenol (**Figure 1**), as the pheromone of the bark beetle, *Ips paraconfusus*. This latter finding led to the recognition that most insect pheromones consisted of multicomponent blends. This has since been shown to be true for most insects, while single-component pheromones are rare. Rapid improvements in analytical instrumentation and techniques reduced the number of insects needed for

pheromone extracts from a half million or more to where now individual insects can sometimes provide sufficient material for chemical analysis. Over the last four decades, extensive research on insect pheromones has resulted in the chemical and/or behavioral elucidation of pheromone components from many thousands of insect species, with much of the work concentrating on sex pheromones from economically important pests.

An early issue addressed in pheromone production was the origin of pheromone components. Ultimately, all precursors for pheromone biosynthesis can be traced through dietary intake. A question asked in several systems was whether pheromone components were derived from dietary components only minimally altered, or whether they were synthesized *de novo*. This simple question proved surprisingly difficult to answer, and different answers were obtained for different groups

DOI: 10.1016/B978-0-12-384749-2.10012-3

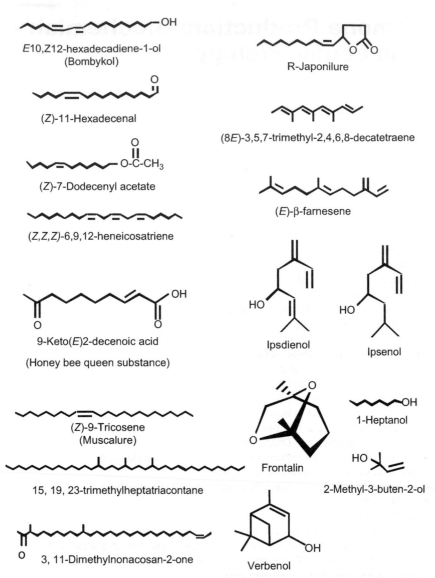

Figure 1 Selected pheromone components representing fatty acid-, hydrocarbon-, and isoprenoid-derived components. Components were selected based on historical interest and work performed on their biosynthesis.

of insects. It is now clear that most insect pheromone components are synthesized *de novo* by insect tissue, with a number of notable exceptions (Tillman *et al.*, 1999, Eisner and Meinwald, 2003; Blomquist *et al.*, 2005, 2010).

By the mid-1980s, it was apparent that the products of normal metabolism, particularly those of the fatty acid and isoprenoid pathways, were modified by a few pheromone-tissue specific enzymes to produce many of the myriad of pheromone molecules. The elegant work of the Roelofs Laboratory (Bjostad *et al.*, 1987; Jurenka, 2003) demonstrated that many of the lepidopteran pheromones could be formed by the appropriate interplay of highly selective chain-shortening and unique Δ11 and other desaturases, followed by modification of the carboxyl carbon. This work has been extended, and there now exists a clear understanding of the biosynthetic pathways for

many of the lepidopteran pheromones (Jurenka, 2003). The honeybee also uses highly specific chain shortening of fatty acids to produce the major component of the queen pheromone (Plettner *et al.*, 1996, 1998). In some insects, fatty acid elongation followed by oxidative decarbonylation produces the hydrocarbon and hydrocarbon-derived pheromones (Tillman *et al.*, 1999; Blomquist, 2010; Millar, 2010). Recent work in bark beetles has shown that *Ips* and *Dendroctonus* spp. produce their monoterpenoid-derived pheromones ipsenol, ipsdienol, and frontalin by modification of isoprenoid pathway products (Blomquist *et al.*, 2010).

The work on the biosynthesis and endocrine regulation of pheromone production has emphasized sex and aggregation pheromones in lepidopteran, coleopteran, dipteran, and blattodean models. Research on representative species from these orders was motivated by their economic importance, the relatively large amount of pheromone produced by some members, and extension of ongoing studies on hydrocarbon and fatty acid biosynthesis in some species.

The production and/or release of sex pheromones are influenced by a variety of environmental factors (Shorey, 1974). In general, insects do not release pheromones until they are reproductively competent, although exceptions occur. Pheromone production is usually age-related and coincides with the maturation of ovaries or testes, and in some cases with feeding. The observation that females of certain species have repeated reproductive cycles and that mating occurs only during defined periods of each cycle led to the proposal that pheromone production might be under hormonal control (Barth, 1965). Early work in cockroaches established that females require the presence of functional corpora allata (CA) to produce sex pheromone. It is now recognized that juvenile hormone (JH) regulates pheromone production in a number of species, especially among beetles (Seybold and Vanderwel, 2003) and cockroaches (Schal et al., 2003). A unifying theme of this work on cockroaches and beetles was that the same hormone that regulated ovarian maturation (JH) also regulated pheromone production, coordinating sexual maturity with mating. Thus, in retrospect, it was not surprising that ovarian-produced 20-hydroxyecdysone (20E), which plays an important role in reproduction in female Diptera, is also the key hormone inducing sex pheromone production in the female housefly, Musca domestica (Blomquist, 2003) and Drosophila (Wicker-Thomas, 1995a,b; Wicker-Thomas and Chertemps, 2010).

It was recognized by the mid-1980s that female moths regulated pheromone production through a different mechanism than flies, cockroaches, and beetles (Raina and Klun, 1984), but it was not until 1989 that the structure of the pheromone biosynthesis activating neuropeptide (PBAN) was elucidated (Raina et al., 1989), and work is ongoing deciphering its mode of action (Rafaeli and Jurenka, 2003; Rafaeli, 2009). In a few species, there does not appear to be endocrine regulation of pheromone production; instead, it is under developmental regulation.

12.2. Pheromone Chemistry

The pheromones of over three thousand insect species are now known. The Web site "pherobase" (http://www.pherobase.net/) is an up-to-date compilation of pheromones and other behavior-modifying chemicals found in insects. These numbers are a huge increase from the 80 pheromone components from about 120 lepidopteran species known in 1985 (Tamaki, 1985) and illustrate both the rapid growth in pheromone chemistry and the increased ease with which pheromones are identified.

12.3. Site of Pheromone Biosynthesis

There is a great deal of variability among insects in the anatomical location of pheromone production, just as there are many differences in the gross morphology and function of pheromone-producing tissue. Complexity varies from simple unicellular glands distributed throughout the integument to elaborate internal cellular aggregates connected to a reservoir. Of the orders emphasized in this chapter (Lepidoptera, Coleoptera, Blattodea, and Diptera), the most common location for pheromone production is the abdomen. There are a number of excellent reviews of the ultrastructure of exocrine cells in general (Ma and Ramaswamy, 2003; Quennedey, 1998; Percy-Cunningham and MacDonald, 1987) and social insects in particular (Billen and Morgan, 1998). Definitive proof that pheromone production and release occur in certain tissues comes from studies where the isolated tissue has been shown to incorporate labeled precursors into pheromone components.

12.3.1. Location of Pheromone Production in Lepidoptera

The oxygenated lepidopteran pheromone components are usually produced and released from extrudable glands located between the 8th and 9th abdominal segments (Percy-Cunningham and MacDonald, 1987; Ma and Ramaswamy, 2003). The secretory cells in these glands typically contain a well-developed endoplasmic reticulum that is involved in fatty acid metabolism. Extensive studies examining the pheromone products and precursors from these glands in a number of species show that unusual specific fatty acids that had the same carbon number, double bond positions, and stereochemistry as the acetate, alcohol, or acetate ester pheromone components were present. The role of these glands in pheromone production has been clearly demonstrated with radiochemical and stable isotope studies (Bjostad et al., 1987; Jurenka, 2003). Although a large number of moths utilize a gland located between the 8th and 9th segments, exceptions occur. It was shown in Theresimima ampelophaga (Zygaenidae) that the gland is located on the dorsal part of abdominal segments 3 to 5 (Hallberg and Subchev, 1997).

In contrast, the site of synthesis of the hydrocarbon and hydrocarbon-derived pheromone components is more complicated. 2-Methylheptadecane is not synthesized in the pheromone gland of Holomelina aurantiaca, but it is synthesized by epidermal tissue, presumably oenocytes, transported by lipophorin, and then sequestered into and released from the pheromone gland (Schal et al., 1998a). A similar situation exists for the Gypsy moth, Lymantria dispar, in which the hydrocarbon precursor is synthesized in epidermal tissue, then transported by lipophorin to the pheromone gland where it is epoxidized and released

(Jurenka *et al.*, 2003) (described in detail in Section 12.4.3.6.). The polyene hydrocarbons are also apparently biosynthesized in oenocytes by the elongation of linoleic and linolenic acid followed by loss of the carboxyl group (Millar, 2010).

12.3.2. Site of Pheromone Biosynthesis in Coleoptera

In many coleopteran species, pheromone production localizes to the abdomen, for example, among the Scarabaeidae (Tada and Leal, 1997), Anobiidae (Levinson *et al.*, 1983), Dermestidae (Barak and Burkholder, 1977), and others (Tillman *et al.*, 1999; Plarre and Vanderwel, 1999). In some cases there are defined glands, whereas in others, groups of cells in defined anatomical locations may not form a gland, but are, nevertheless, a localized site of pheromone biosynthesis. Other anatomical locations for pheromone production have also been identified for some species. These include putative pheromone glands on the antennae of *Batrisodes oculatus* (de Mazo and Vit, 1983) and on the forelimbs of *Tr. castaneum* (Faustini *et al.*, 1982).

Pheromone biosynthetic cells of many beetles are generally dermally derived, although pheromone biosynthesis in the fat body is also possible. There is evidence that fatty-acid-derived *exo*-brevicomin is synthesized in the male mountain pine beetle, *Dendroctonus ponderosae* fat body (M. Song, unpublished data), and monoterpenoid pheromone components have been reported to be synthesized in the fat body of male boll weevils, *Anthonomus grandis* (Wiygul *et al.*, 1990). In bark beetles, the isoprenoid aggregation pheromones are synthesized in midgut tissue (**Figure 2**; Hall *et al.*, 2002a,b; Nardi *et al.*, 2002; Barkawi *et al.*, 2003).

Why midgut tissue in bark beetles? In considering the evolution of pheromone components, it is generally thought that pheromone molecules, as other signaling chemicals, originally had another purpose and were co-opted for signal function. Components were then shaped by selective forces acting on pre-existing structures, and pheromones evolved so that the signal would be stronger and more easily discriminated. For bark beetles, it is thought that monoterpene components arose from the detoxification of oleo-resin tree defensive components, and that many of the detoxification reactions involved hydroxylation reactions (Wood, 1982; Vanderwel and Oehlschlager, 1987). To lessen dependence upon host chemicals and increase host range, bark beetles then apparently evolved the ability to synthesize many of their monoterpenoid pheromones *de novo*, and with this, evolved the JH III regulation of pheromone production. As the ancient beetles chewed through the bark and phloem, they encountered tree terpenoids. Gut tissue was the first line of defense, and detoxification enzymes (cytochrome P450 hydoxylases) could have evolved to detoxify monoterpenes in this tissue. As *de novo* synthesis evolved, it is possible that production of pheromone arose in midgut tissue so that the hydroxylation reactions that were already present could be used. Alternatively, midgut production may be an ancestral state because monoterpenoid components of the cotton boll weevil (*A. grandis*) are also synthesized in the midgut (Taban *et al.*, 2006).

12.3.3. Site of Pheromone Production in Diptera

The hydrocarbon pheromone components of many Diptera evolved from and still function as cuticular lipids. Thus, it is not unexpected that these pheromone

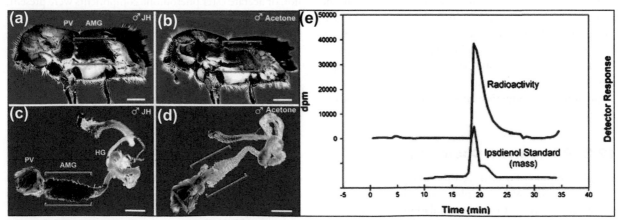

Figure 2 Tissue localization of pheromone production in *Ips pini*. "Exposed whole mounts" show that HMG-R mRNA is highly expressed in the midgut of JH-III-treated male *I. pini* (a) but not in untreated insects (b). Panels c and d show whole mount hybridizations of isolated *I. pini* alimentary canals. HMG-R expression in the anterior midgut (AMG, marked by yellow brackets) correlates with pheromone production in starved, JH-III-treated males (c), while starved and untreated males (d) that do not produce monoterpenoid pheromone components do not strongly express HMG-R. Isolated male midgut tissue from JH-III-treated males incorporated labeled acetate into a component that co-eluted on HPLC with ipsdienol. Modified from *Hall et al.* (2002a).

components are synthesized in the same specialized epidermal cells (oenocytes) that produce cuticular hydrocarbons. After synthesis, pheromone components are secreted into the hemolymph, where they are transported by lipophorin, as has been demonstrated in *Drosophila* (Ferveur *et al.*, 1997; Billeter *et al.*, 2009; Wicker-Thomas *et al.*, 2009) and the housefly (Schal *et al.*, 2001), before being deposited on the cuticular surface. During the process of grooming, relatively large amounts of (*Z*)-9-tricosene accumulate on the legs of female houseflies (Dillwith and Blomquist, 1982). The mechanism of how hydrocarbon pheromones, and hydrocarbons in general, are unloaded from lipophorin and transported across the cuticle is not known.

12.3.4. Site of Pheromone Production in Blattodea

As in Coleoptera, early reports considered various organs in the head, thorax, and abdomen as possible sites of pheromone production in cockroaches, but most of the more recent research implicates specialized abdominal glands. Periplanones-A and -B, the sex pheromone components of the American cockroach, *Periplaneta americana*, appear to be concentrated in the midgut and released in feces. However, female calling involves opening the genital vestibulum (= atrial gland), without excretion of feces, and this glandular tissue contains most of the periplanone-B, up to 60 ng (Abed *et al.*, 1993a). The epithelium of the atrial gland consists of class 1 glandular cells, in which the secretion passes directly to the cuticle and not through a duct. However, Yang *et al.* (1998) concluded that both periplanone-A and -B were most abundant in the colon and that this tissue produced the strongest electroantennogram (EAG) responses. Still, no definitive proof of the sites of pheromone biosynthesis is available through isotopic tracing of pheromone precursors in isolated tissues.

The volatile sex pheromones of other cockroaches have been localized to abdominal tergites and sternites. Females of the German cockroach, *Blattella germanica*, produce the pheromone in the 10th abdominal tergite (pygidium; Liang and Schal, 1993; Tokro *et al.*, 1993; Nojima *et al.*, 2005), *Supella longipalpa* in the 4th and 5th tergites (Schal *et al.*, 1992), and *Parcoblatta lata* (wood cockroach) females produce the pheromone in tergites 1–7 (Gemeno *et al.*, 2003). In all three species, the tergal glands are composed of multiple class 3 secretory units with each leading through a long unbranched duct to a single cuticular pore. The secretory cells are characterized by abundant mitochondria, SER, RER, a large nucleus, and numerous secretory vesicles that discharge through an end apparatus with numerous long microvilli into the duct and to the cuticular surface. In all three species pheromone is released during a calling behavior.

Male cockroaches also release volatile sex pheromones from tergal or sternal glands. In *Nauphoeta cinerea* (lobster cockroach), glands on sternites 3–7 are exposed during calling releasing a volatile male pheromone consisting of 3-hydroxy-2-butanone, 2-methylthiazolidine, 4-ethyl-2-methoxyphenol, and 2-methyl-2-thiazoline (Sreng, 1990; Sirugue *et al.*, 1992). In some species the male tergal glands form specialized regions on the cuticular surface and produce a blend of close-range attractants, phagostimulants, and nutrients that the male deploys to place the female in a pre-copulatory position. For example, the tergal glands of *B. germanica* consist of traverse cuticular depressions on the 7th and 8th tergites connected to numerous class 3 secretory cells (Sreng and Quennedey, 1976; Brossut and Roth, 1977).

Cuticular contact pheromones mediate species- and sex-recognition and, in most cases, they function as courtship-inducing pheromones. They are thought to be distributed throughout the epicuticular surface. The blend of hydrocarbon-derived ketones of *B. germanica* females is produced by oenocytes localized within the abdominal integument, separated from the hemocoel by a basal lamina (Liang and Schal, 1993). Fan *et al.* (2003) enzymatically dissociated the integument underlying the sternites into a cell suspension, and after further Percoll™ gradient centrifugation, assayed each fraction for incorporation of [1-^{14}C]propionate into methyl-branched hydrocarbons. Only oenocytes produced hydrocarbons and methyl ketone pheromones, whereas the much larger population of epidermal cells did not (Fan *et al.*, 2003). As in the dipterans *M. domestica* and *Drosophila* and the moth *Holomelina*, the hydrocarbon-derived pheromones of *B. germanica* are produced by oenocytes and transported by lipophorin (Section 12.5.5.3.).

12.4. Biochemistry of Pheromone Production

12.4.1. Modification of "Normal Metabolism"

By the mid-1980s (Prestwich and Blomquist, 1987), the biosynthetic pathways of pheromones for a limited number of species had been determined, and work was progressing toward the characterization of some of the unique enzymes involved. It became apparent that the products and intermediates of normal metabolism, particularly those of the fatty acid and isoprenoid pathways, were modified by a few specific enzymes in pheromone gland tissue to produce the myriad of pheromone molecules. Many of the lepidopteran pheromones could be formed by the appropriate interplay of highly selective chain shortening and a unique Δ11 and other desaturases followed by modification of the carboxyl carbon (Bjostad *et al.*, 1987). This work has been extended, and a clear understanding of the biosynthetic pathways for many of the lepidopteran pheromones is now known (Jurenka, 2003). The Δ11 and other pheromone-specific desaturases in Lepidoptera have been characterized at the molecular level (Knipple and Roelofs, 2003). Chain shortening of fatty acids is also involved in producing the queen pheromone in honeybees (Plettner *et al.*, 1996, 1998). In some insects, fatty acid elongation and reduction to the aldehyde followed by oxidative decarbonylation produce the hydrocarbon pheromones; for example, lepidopterans (Jurenka, 2003), dipterans (Blomquist, 2003; Jallon and Wicker-Thomas, 2003), the German cockroach (Schal *et al.*, 2003), and the social insects (Blomquist and Howard, 2003). More recent work in bark beetles showed that *Ips* and *Dendroctonus* spp. produce their monoterpenoid-derived pheromones ipsenol, ipsdienol, and frontalin by modification of isoprenoid pathway products (Seybold and Vanderwel, 2003; Blomquist *et al.*, 2010). Until that work, it was considered rare for animals to produce monoterpenoids (C10 isoprenoids).

It is worth noting that the production of pheromones is usually affected, either directly or indirectly, by feeding or starvation, even in insects that do not sequester pheromone precursors from the diet. Indirect effects may be through feeding-induced stimulation of endocrine factors (e.g., beetles, Vanderwel and Oehlschlager, 1987; German cockroach, Schal *et al.*, 2003). Foster (2009) showed in *Heliothis virescens* moths that *de novo* production of the fatty-acid-derived pheromone components is strongly influenced by sugar feeding. The pheromone titer rapidly declined in mated females, as did hemolymph trehalose, but ingestion of sugar restored the pheromone to nearvirgin levels. He suggested that redirection of carbon to the developing oocytes lowers blood sugar, and females can elevate both trehalose concentration and their pheromone titer with sugar feeding.

12.4.2. Pheromone Biosynthesis in Moths

Bjostad and Roelofs (1983) were the first to correctly determine how the major pheromone component for a particular moth was biosynthesized. They utilized the moth *T. ni* because the female produces a relatively large amount of pheromone (about 1 μg). They began by demonstrating that pheromone glands utilize acetate to produce the common fatty acids octadecanoate and hexadecanoate, which undergo Δ11 desaturation to produce *Z*11-18:acid and *Z*11-16:acid. However, the main pheromone component is *Z*7-12:OAc, which presumably is made from *Z*7-12:acid. To demonstrate how the fatty acid precursor *Z*7-12:acid was produced, [³H-16]-*Z*11-16:acid was applied to pheromone glands and found to be incorporated into both *Z*7-12:acid and *Z*7-12:OAc. They concluded that limited chain shortening of *Z*11-16:acid could account for this incorporation. The other minor components are produced in a similar way (**Figure 3**).

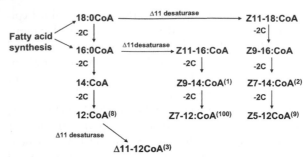

Figure 3 Biosynthetic pathways for producing the intermediate CoA derivatives of the pheromone blend of the cabbage looper, *Trichoplusia ni*. The CoA derivatives followed by the superscript number in parenthesis are reduced to an alcohol and acetylated to form the acetate esters that make up the pheromone blend. The superscript numbers indicate the approximate ratio of components found in the pheromone gland (*Bjostad et al.*, 1984).

Thus the pheromone components are produced through a fatty acid biosynthesis pathway involving a Δ11 desaturase and limited chain-shortening enzymes. The appropriate chain length fatty acid is then reduced and acetylated to form the acetate ester. Subsequent research demonstrated that similar pathways occur in a wide variety of female moth species (Roelofs and Wolf, 1988; Jurenka, 2003).

Most female moths produce their pheromone components through modifications of fatty acid synthesis pathways, thus techniques in fatty acid research were utilized to determine biosynthetic pathways (Bjostad *et al.*, 1987; Morse and Meighen, 1987b). The main techniques include thin-layer chromatography, gas chromatography (GC), and GC/mass spectrometry (MS) with the latter being the primary method utilized. The advantage of using GC-MS is that the label (stable isotopes deuterium or carbon-13) can be explicitly shown to be present in the compound of interest. By monitoring for diagnostic ions that correspond to unlabeled and labeled products, it can be determined with considerable certainty that the label is associated with a particular compound. By utilizing different proposed intermediate labeled fatty acids a biosynthetic pathway can be deduced. For example, it was determined that *Z*11-16:Ald is produced by Δ11 desaturation of 16:CoA followed by reduction in the moths *Helicoverpa zea* and *H. assulta* (Choi *et al.*, 2002). Whereas *Z*9-16:Ald, the major pheromone component in *H. assulta*, is produced by a Δ9 desaturase using 16:CoA as a substrate, in *H. zea Z*9-16:Ald is produced by Δ11 desaturation of 18:CoA to produce *Z*11-18:CoA that is then chain-shortened to *Z*9-16:CoA (Choi *et al.*, 2002). These types of studies have shown that the key enzymes of sex pheromone biosynthetic pathways are fatty acid biosynthetic enzymes, desaturases, chain-shortening enzymes, and specific enzymes to produce a functional group.

12.4.3. Enzymes Involved in Lepidopteran Pheromone Production

12.4.3.1. Fatty acid synthesis A combination of acetyl-CoA carboxylase and fatty acid synthase produce saturated fatty acids. Although no direct enzymatic studies have been conducted using pheromone gland cells, these enzymes are presumably similar to enzymes found in other cell types. Labeling studies conducted with acetate indicated that pheromone glands produce 16:acid and 18:acid saturated products (Bjostad and Roelofs, 1984; Jurenka *et al.*, 1991a, 1994; Tang *et al.*, 1989). Indirect evidence showed that when the activity of acetyl-CoA carboxylase was inhibited by herbicides, sex pheromone biosynthesis was also inhibited in *H. armigera* and *Plodia interpunctella* females (Eliyahu *et al.*, 2003; Tsfadia *et al.*, 2008).

12.4.3.2. Chain-shortening enzymes Insects in general have the ability to shorten long chain fatty acids to specific shorter chain lengths (Stanley-Samuelson *et al.*, 1988). This chain-shortening pathway has not been characterized at the enzymatic level in insects. It presumably is similar to the characterized pathway as it occurs in vertebrates and is essentially a partial β-oxidation pathway located in peroxisomes (Hashimoto, 1996). The evidence for limited chain-shortening enzymes in pheromone glands was originally demonstrated by Bjostad and Roelofs (1983) using the cabbage looper moth in which it was shown that $Z$11-16:acid labeled the intermediate fatty acid $Z$7-12:OAc. A similar study using *Argyrotaenia velutinana* demonstrated that deuterium labeled 16:acid was chain-shortened to 14:acid, which was used to make Z and E11-14:acid (Bjostad and Roelofs, 1984). Since then considerable evidence in a number of moths has accumulated to indicate that limited chain shortening occurs in a variety of pheromone biosynthetic pathways.

Radiolabeled or stable isotopes were topically applied directly to an intact pheromone gland to provide evidence for chain shortening. The position of the label was on the terminal methyl carbon making it difficult for any type of rearrangement to occur (Wolf and Roelofs, 1983; Rosell *et al.*, 1992). A more direct *in vitro* enzyme assay was utilized to demonstrate substrate preferences in a study using cabbage looper moths, *T. ni* (Jurenka *et al.*, 1994). This study was prompted by the finding of a mutant line of cabbage loopers that produced a greatly increased amount of $Z$9-14:OAc (Haynes and Hunt, 1990), which is a minor component of normal cabbage loopers. Increased amounts of $Z$9-14:OAc indicate that chain shortening was affected in the mutant cabbage loopers. To determine if chain shortening was affected, substrate specificities for both normal and mutant cabbage loopers were determined using an *in vitro* enzyme assay (Jurenka *et al.*,

1994). Pheromone glands from normal cabbage loopers preferred to chain-shorten $Z$11-16:CoA to $Z$7-12:CoA with two rounds of chain shortening, whereas pheromone glands from the mutant cabbage looper apparently have the ability to chain-shorten by only one round. Therefore, $Z$11-16:CoA was chain-shortened to $Z$9-14:CoA.

Changes in chain-shortening reactions have also been implicated in the alteration of pheromone ratios in several other species. In a laboratory selection pressure experiment using *A. velutinana*, the Z/E ratio of 11-14:OAc could not be changed much from a 92/8 ratio (Roelofs *et al.*, 1986). However, it was found that the ratio of $E$9-12:OAc/$E$11-14:OAc could be selected (Sreng *et al.*, 1989), and by comparing ratios of the 14-/12-carbon pheromone components and Z/E isomers of each chain length it was determined that chain-shortening enzymes were selective for the E isomer (Roelofs and Jurenka, 1996). Another example where a change in chain shortening can account for a new pheromone blend is the larch budmoth, *Zeiraphera diniana*. The pheromone was first identified as $E$11-14:OAc (Roelofs *et al.*, 1971), but it was later determined that some populations utilize $E$9-12:OAc, which is made by chain shortening $E$11-14:acid (Baltensweiler and Priesner, 1988; Guerin *et al.*, 1984). Another case where changes in chain shortening may have produced two populations of insects was found in the turnip moth, *Agrotis segetum*. A Swedish population has a ratio of $Z$9-14:OAc/$Z$7-12:OAc/$Z$5-10:OAc of 29/59/12, whereas a Zimbabwean population has a ratio of 2/20/78. After conducting labeling studies, it was determined that chain-shortening enzymes could be affected to produce the alteration in pheromone ratios (Wu *et al.*, 1998). These studies indicate that alteration in chain-shortening enzymes can have a major effect on pheromone blends.

12.4.3.3. Desaturases A double bond is introduced into the fatty acid chain by desaturases. A variety of desaturases have been described that are involved in the biosynthesis of female moth sex pheromones. The desaturases identified so far include enzymes that act on saturated and monounsaturated substrates. These include Δ5 (Foster and Roelofs, 1996), Δ9 (Löfstedt and Bengtsson, 1988; Martinez *et al.*, 1990), Δ10 (Foster and Roelofs, 1988), Δ11 (Bjostad and Roelofs, 1981, 1983), and Δ14 (Zhao *et al.*, 1990) desaturases that utilize saturated substrates. The combination of these desaturases along with chain shortening account for the majority of double bond positions in the various chain-length monounsaturated pheromones identified thus far (Roelofs and Wolf, 1988). **Figure 4** illustrates the large number of monounsaturated compounds that can be generated through a combination of desaturation and chain shortening. The addition of various functional groups — acetate esters, alcohols, and aldehydes — increases the potential number of pheromone components. Notice

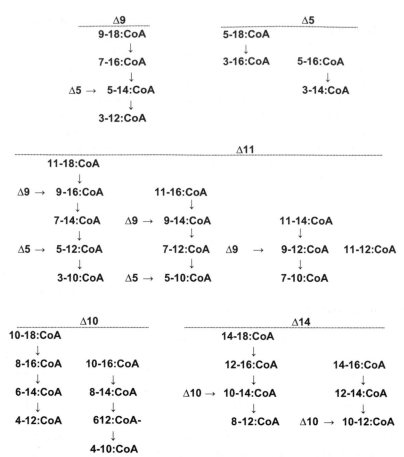

Figure 4 Combination of desaturation and chain shortening can produce a variety of monounsaturated acyl-CoA precursors that can be modified to form acetate esters, aldehydes, and alcohols. The number followed by the Δ sign indicates a desaturase that introduces a double bond into the first indicated chain length acyl-CoA. The arrow pointing down indicates limited chain shortening by two carbons. The arrow pointing to the right indicates that desaturation could produce the compound found within a chain-shortening pathway. This indicates that certain compounds could be produced in two different ways. Modification of all 16-, 14-, 12-, and 10-carbon acyl-CoA derivatives on the carbonyl carbon can account for the majority of monounsaturated acetate esters, aldehydes, and alcohols identified as sex pheromones.

that some of the intermediate compounds could be produced in two different ways. Therefore, although the desaturation and chain-shortening steps occur in a wide variety of moths, the order in which they occur and the type of desaturase must still be determined experimentally.

Some pheromone components are dienes and these can be produced by either the action of two desaturases or one desaturase and isomerization around the double bond. Some dienes with a 6,9-double bond configuration are produced using linoleic acid. Desaturases that utilize monounsaturated acyl-CoA substrates include Δ5 (Ono *et al.*, 2002), Δ9 (Martinez *et al.*, 1990), Δ11 (Foster and Roelofs, 1990), Δ12 (Jurenka, 1997), and Δ13 (Arsequell *et al.*, 1990). These can act sequentially to produce the diene (Foster and Roelofs, 1990; Jurenka, 1997) or conjugated dienes could be produced by the action of one desaturase followed by isomerization (Ando *et al.*, 1988; Löfstedt and Bengtsson, 1988; Fang *et al.*, 1995a). An apparently unique Δ6 desaturase has been found in *Antheraea pernyi* that couples with a Δ11 desaturase to produce the

(*E*,*Z*)-6,11-hexadecadienoic acid intermediate to the aldehyde and acetate ester pheromone (Wang *et al.*, 2010a).

The biosynthesis of triene pheromone components has not been extensively investigated. Pheromones with a triene double bond system that is n-3 (3,6,9-) are produced from linolenic acid (Millar, 2000, 2010; Choi *et al.*, 2007b). This was demonstrated in the saltmarsh caterpillar, *Estigmene acrea*, and the ruby tiger moth, *Phragmatobia fuliginosa* (Rule and Roelofs, 1989). Moths in the families Geometridae, Arctiidae, and Noctuidae apparently utilize linoleic and linolenic acid as precursors for their pheromones. Most of these pheromones are produced by chain elongation and loss of the carboxyl group to form hydrocarbons. Oxygen is added across one of the double bonds in the polyunsaturated hydrocarbon to produce an epoxide (Millar, 2000, 2010).

12.4.3.3.1. Molecular biology of the desaturases The first gene encoding a Δ11 desaturase was identified in the cabbage looper, *T. ni* (Knipple *et al.*, 1998). Since then a

number of desaturase genes have been cloned and functionally expressed (Knipple and Roelofs, 2003). Expression in a strain of yeast lacking an endogenous desaturase was utilized to demonstrate functionality. This strain of yeast will not grow in media lacking unsaturated fatty acids but will grow if a functional desaturase is inserted into the yeast genome. After growth, the fatty acids are analyzed to determine double bond positions and thus the desaturase can be characterized regarding double bond insertion and chain-length specificity.

Desaturase encoding cDNAs from pheromone glands have been identified that produce *Z*9, *Z*10, *Z/E*11, and *Z/E*14 double bonds in 14- and 16-carbon acids. The *Z*9 desaturase is comparable to the metabolic desaturase found in the fat body. A *Z*10 desaturase was characterized from *Planotortrix octo* that produces *Z*10-16:acid (Hao *et al.*, 2002), which would be chain-shortened to produce the precursor to the pheromone Z8-14:OAc (Foster and Roelofs, 1988). Several Δ11 desaturases have been characterized, including those from *T. ni* (Knipple *et al.*, 1998) and *H. zea* (Rosenfield *et al.*, 2001), which primarily produce *Z*11-16:acid. A single Δ11 desaturase was characterized from *A. velutinana* that produces both *Z*11- and *E*11-14:acid (Liu *et al.*, 2002a). This desaturase is unique in that it produces both isomers and uses 14:acid as a substrate. Another unique Δ11 desaturase was identified from *Epiphyas postvittana* that produced *E*11-14:acid and *E*11-16:acid from saturated precursors, but also *E*9,*E*11-14:acid from an *E*9-14:acid precursor (produced by chain-shortening *E*11-16:acid; Liu *et al.*, 2002b). Identification of desaturases in *Ostrinia nubilalis* and *O. furnicalis* produced the surprise finding that Δ11 and Δ14 desaturases are found in both moths (Roelofs *et al.*, 2002). Although three Δ14 desaturases and ten Δ11 desaturases have been found in *O. nubilalis*, only one Δ11 desaturase transcript is functional in this species, which produces both *Z*11- and *E*11-14:OAc (Xue *et al.*, 2007). *Ostrinia furnicalis* has two Δ14 desaturases and five Δ11 desaturases, but only protein products of a Δ14 desaturase gene were found in the pheromone gland, which makes *Z*12- and *E*12-14:OAc pheromone components (Roelofs and Rooney, 2003). These findings have implications regarding the evolution of pheromone blends in moths (Baker, 2002).

Based on the desaturases identified, phylogenetic relationships can be inferred and grouped according to structural activity (**Figure 5**; Liénard *et al.*, 2010). The Δ9 desaturases are apparently the most basal group and can be divided into those that prefer palmitic (C_{16}) or stearic (C_{18}) acids as substrates. A novel Δ9 desaturase was identified from the spotted fireworm moth (*Choristoneura parallela*), which produces primarily Z9-16 but also will utilize C_{14} to C_{26} fatty acids (Liu *et al.*, 2004). This is a more recently evolved desaturase related to the Δ14 lineage, although only a few of these desaturases have been

functionally characterized. The Δ11/Δ10 desaturases form a more recently evolved group that seems unique to the Lepidoptera.

12.4.3.4. Specific enzymes to produce functional group on carbonyl carbon

Once a specific chain-length pheromone intermediate with the appropriate double bonds is produced, the carbonyl carbon is modified to form a functional group. The majority of oxygenated pheromone components are acetate esters (or other esters), alcohols, and aldehydes. Production of these components requires the reduction of a fatty acyl precursor to an alcohol that is catalyzed by a fatty acyl reductase. The fatty acyl reductase has been identified in *B. mori* for production of bombykol (Moto *et al.*, 2003). It has been functionally characterized by expression in yeast cells and shown to preferentially reduce *E,Z*10,12—16:acid, which thus forms bombykol (Matsumoto, 2010).

Fatty acyl reductases have subsequently been identified in several other moths in which the alcohol serves as an intermediate for production of acetate esters. The reductase has been identified in the adzuki bean borer *O. scapulalis*, European corn borer *O. nubilalis*, and small ermine moths Yponomeuta spp. (Antony *et al.*, 2009; Lassance *et al.*, 2010; Liénard *et al.*, 2010). In the European corn borer two reductases were identified: one from each of the two strains that produce primarily *Z*11-14:OAc or *E*11-14:OAc. The reductase from each strain preferentially reduces *Z*11-14:acid or *E*11-14:acid resulting in a strain-specific pheromone blend (Lassance *et al.*, 2010). Sequences of the two desaturases exhibited a 3.8% nucleotide divergence and corresponding 7.5% amino acid divergence. These small differences apparently change the substrate preference for the enzyme. The reductase from the adzuki bean borer (Antony *et al.*, 2009) has a 99% identical amino acid composition to the *E*-strain reductase from the European corn borer. However, no substrate preference was conducted with the adzuki bean borer reductase. The reductases from three species of *Yponomeuta* have about 36% amino acid identity with the *Ostrinia* reductases, but they have a broad substrate specificity (Liénard *et al.*, 2010).

Formation of aldehydes requires the oxidation of primary alcohols and a cuticular oxidase has been characterized from pheromone glands of *H. zea* and *Manduca sexta* that produce aldehydes as pheromones (Fang *et al.*, 1995b; Teal and Tumlinson, 1988). In those insects that utilize both an alcohol and an aldehyde as part of their pheromone, it is unclear how the production of both components occurs. Luxova and Svatos (2006) isolated a membrane-bound alcohol oxidase from *M. sexta* pheromone glands with high specificity for primary alcohols, as occurs in yeast alcohol dehydrogenase.

Production of acetate ester pheromone components utilizes an enzyme called acetyl-CoA:fatty alcohol

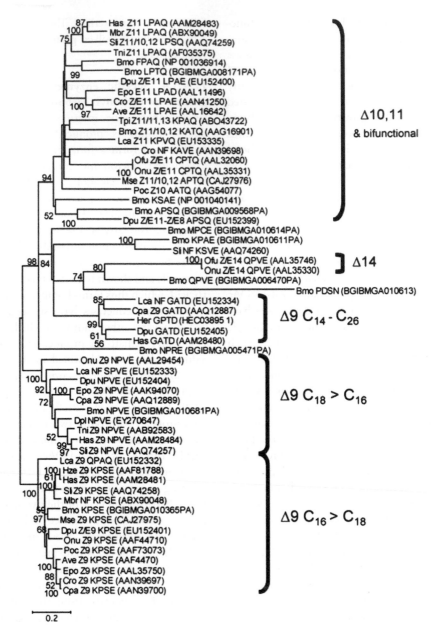

Figure 5 Phylogeny of desaturase genes of lepidopteran insects. Sequences are named according to the abbreviated species name, a desaturase catalytic activity (when assayed), and a four-amino-acid signature motif (SM; *Knipple et al., 2002*). The accession numbers are indicated in parentheses. The abbreviated species names correspond to Ave, *Argyrotaenia velutinana*; Bmo, *Bombyx mori*; Cpa, *Choristoneura parallela*; Cro, *Choristoneura rosaceana*; Dpl, *Danaus plexippus*; Dpu, *Dendrolimus punctatus*; Epo, *Epiphyas postvittana*; Has, *Helicoverpa assulta*; Her, *Heliconius erato*; Hze, *Helicoverpa zea*; Lca, *Lampronia capitella*; Mbr, *Mamestra brassicae*; Mse, *Manduca sexta*; Ofu, *Ostrinia furnicalis*; Onu, *Ostrinia nubilalis*; Poc, *Planotortrix octo*; Sli, *Spodoptera littoralis*; Tni, *Trichoplusia ni*, and Tpi, *Thaumetopoea pityocampa*. Reproduced with permission from *Liénard et al.* (2010).

acetyltransferase that converts a fatty alcohol to an acetate ester (Morse and Meighen, 1987a). Therefore, alcohols could be utilized as substrates for both aldehyde and acetate ester formation. Morse and Meighen (1987a) first demonstrated its presence in the spruce budworm, *Choristoneura fumiferana*, where it is involved in producing the acetate ester that serves as a precursor to the aldehyde pheromone (Morse and Meighen, 1987b). In some other tortricids, *A. velutinana*, *C. rosaceana*, and *Platynota*

idaeusalis, an *in vitro* enzyme assay was utilized to demonstrate specificity of the acetyltransferase for the *Z* isomer of 11-14:OH (Jurenka and Roelofs, 1989). This specificity contributes to the final ratio of pheromone components. These results indicate that the family Tortricidae has members that have an acetyltransferase specific for the *Z* isomer of monounsaturated fatty alcohols. In contrast, several studies have shown no substrate preference for the acetyltransferase in other moths (Bestmann *et al.*, 1987;

Teal and Tumlinson, 1987; Jurenka and Roelofs, 1989). Therefore this unique acetyltransferase apparently evolved within the Tortricidae.

The use of RNAi has illustrated the function of many of these enzymes in *B. mori*. Injecting dsRNA corresponding to the Z11/Δ10,12 desaturase, fatty acyl reductase, and acyl-CoA-binding protein into pupae resulted in a reduction of gene transcripts and sex pheromone production in the adult female (Ohnishi *et al.*, 2006; Matsumoto *et al.*, 2007).

Recent studies have examined the transcriptomes or sequenced genomes of insects. For example, Vogel *et al.* (2010) identified 8310 putative genes in the pheromone gland of *H. virescens*, 6435 of which were unique to the pheromone gland (by comparison with larval tissue). Comparison with EST databases from other moth species revealed 86 candidate genes encoding enzymes that could be involved in moth sex pheromone biosynthesis, including two Δ11 and six Δ9 desaturases. Matsumoto (2010) reviewed the molecular mechanisms of pheromone production in moths.

12.4.3.5. Production of specific pheromone blends

Most female moths utilize a blend of components produced in a specific ratio for pheromone attraction of conspecific males. A major question is how these species-specific ratios of components are produced. Research from several sources indicates that these ratios are produced by the inherent specificity of certain enzymes present in the biosynthetic pathways. The combination of these enzymes acting in concert produces the species-specific pheromone blend. Several examples will be utilized to illustrate this point.

The cabbage looper, *T. ni*, uses Z7-12:OAc as the major sex pheromone component, which is produced by Δ11 desaturation of 16:CoA followed by two rounds of chain shortening, reduction, and acetylation. The Δ11 desaturase has been characterized (Wolf and Roelofs, 1986) and the gene isolated (Knipple *et al.*, 1998). These studies indicate that 16:CoA and 18:CoA are substrates with 16:CoA, the preferred substrate, and Z11-16:acid is the most abundant monounsaturated fatty acid in pheromone glands (Bjostad *et al.*, 1984). The next step in the pathway is limited chain shortening, and it was shown that Z11-16:CoA is the preferred substrate for these enzymes (Jurenka *et al.*, 1994). After chain shortening the 14- and 12-carbon intermediates are reduced to an alcohol and acetylated. The acetyltransferase enzyme is not specific and will accept a variety of substrates (Jurenka and Roelofs, 1989). From these observations, it can be inferred that the final ratio of pheromone components is produced by the specificity found within the Δ11 desaturase and chain-shortening enzymes.

A blend of seven acetate esters used by *A. velutinana* is produced in a biosynthetic pathway similar to the one just described, except that the Δ11 desaturase starts with 14:CoA as the substrate and produces both Z and E isomers of 11-14:CoA in about a 60/40 ratio (Wolf and Roelofs, 1987). Cloning of the Δ11 desaturase from *A. velutinana* females indicates that the expressed enzyme produces a ratio of Z/E of about 6/1 (Liu *et al.*, 2002a). However, the final ratio of Z11- to E11-14:OAc is 92/8 (Miller and Roelofs, 1980). A selective increase in the Z isomer occurs within the biosynthetic pathway and it was determined that acetyl:CoA fatty alcohol acetyltransferase shows specificity for the Z isomer (Jurenka and Roelofs, 1989). Therefore, selective acetylation of Z11-14:OH and production of >60% Z11-14:CoA indicates that these enzymes have the inherent specificity to produce the 92:8 ratio of the major pheromone components Z11- and E11-14:OAc. Two minor pheromone components are produced by chain-shortening Z11- and E11-14:OAc. The ratio of Z9- to E9-12:OAc is about 1 to 2. This indicates that the chain-shortening enzymes may prefer E11-14:CoA or that very little Z11-14:CoA is available to chain-shorten. This combined information indicates that in *A. velutinana* pheromone glands the final ratio of pheromone components can be produced through the concerted action of a Δ11 desaturase that produces at least a 60:40 ratio of Z/E intermediate isomers. The final ratio of acetate esters (92/8) is produced through the specificity for the Z isomer by the acetyltransferase. The minor components are produced by specificity in chain shortening.

Another insect that utilizes specific ratios of Z11- and E11-14:OAc is the European corn borer, *O. nubilalis*. Two strains are known in which one produces a ratio of Z/E of about 97/3 (Z strain) and the other produces an opposite ratio of Z/E of about 1/99 (E strain). Hybridization studies between the two strains indicated that offspring have an acetate ester ratio of Z/E of about 30/70 (Klun and Maini, 1982). The Δ11 desaturase from both strains produced a product with about 30/70 Z/E in an *in vitro* enzyme assay (Wolf and Roelofs, 1987). These results indicate that the final ratio of acetate ester isomers is produced after the desaturation step. The enzymes that follow the desaturase are a reductase to make an alcohol and an acetyltransferase to produce the acetate esters. Two studies have shown that the acetyltransferase is similar between the two strains (Jurenka and Roelofs, 1989; Zhu *et al.*, 1996). However, labeled acids applied to glands *in vivo* were selectively incorporated into the correct pheromone ratio indicating that the reductase shows specificity (Zhu *et al.*, 1996). Therefore, the final pheromone ratios produced by females of the European corn borer are made through the action of a Δ11 desaturase that can produce both Z and E isomers. The final acetate ester ratio is strain dependent and is produced through the specificity found in the reductase enzyme, as was recently demonstrated by identification of the genes encoding the fatty acyl reductase as discussed earlier (Lassance *et al.*, 2010).

The previous three examples illustrate how a species-specific pheromone blend is produced by the concerted action of desaturases, chain-shortening enzymes, and a reductase and an acetyltransferase. The specificity inherent in certain enzymes in the pathway produces the final blend of pheromone components.

12.4.3.6. Hydrocarbon pheromones Moths in the families Geometridae, Arctiidae, Amatidae, Lymantriidae, Lyonetiidae, and some Noctuidae utilize hydrocarbons or epoxides of hydrocarbons as their sex pheromones (Millar, 2010). Biosynthesis of hydrocarbons occurs in oenocyte cells associated with either epidermal cells or fat body cells (Romer, 1991; Billeter *et al.*, 2009). Once the hydrocarbons are biosynthesized they are transported to the sex pheromone gland by lipophorin (Schal *et al.*, 1998a). When the transport of hydrocarbon sex pheromones in arctiid moths was investigated in detail by Schal *et al.* (1998a), it was found that a very specific uptake was occurring at pheromone glands. Lipophorin was shown to contain both the sex pheromone and cuticular hydrocarbons; however, only the pheromone gland had the sex pheromone. Other studies have shown similar pathways in other moths (Jurenka and Subchev, 2000; Subchev and Jurenka, 2001; Wei *et al.*, 2003; Matsuoka *et al.*, 2006).

Most moth sex pheromones that are straight chain hydrocarbons usually have an odd number of carbons. Most of these are polyunsaturated with double bonds in the 3,6,9- or 6,9-positions, indicating that they are derived from linolenic or linoleic acid, respectively (Rule and Roelofs, 1989; Millar, 2000; Matsuoka *et al.*, 2008). Linolenic and linoleic acid cannot be biosynthesized by moths so they must be obtained from the diet (Stanley-Samuelson *et al.*, 1988).

A few even-chain-length hydrocarbon sex pheromones have been identified that also have 3,6,9- or 6,9-double bond configurations (Millar, 2000, 2010), indicating they too are derived from linolenic or linoleic acids. A study using a winter moth, *Erannis bajaria*, which produces Z3,Z6,Z9-18:Hc, has demonstrated how these even-chained hydrocarbons are produced (Goller *et al.*, 2007). The pathway involves chain-elongating α-linolenic acid to Z11,Z14,Z17-20:acid followed by the key step of α-oxidation to produce Z10,Z13,Z16-19:acid. The 19C acid is then converted to the hydrocarbon Z3,Z6,Z9-18:Hc. The odd-chain-length pheromone component Z3,Z6,Z9-19:Hc is formed from Z11,Z14,Z17-20:acid as usual for odd-chain-length hydrocarbons.

A major class of sex pheromones derived from hydrocarbons are the polyene monoepoxides (Millar, 2010). These usually have double bonds in the 3,6,9-positions or 6,9-positions, again indicating they are biosynthesized from linolenic or linoleic acids, respectively. Although the production of hydrocarbon occurs in oenocytes, the

epoxidation step takes place in the pheromone gland. This has been demonstrated in several studies utilizing deuterium-labeled precursors. In a study on the Japanese giant looper, *Ascotis selenaria cretacea*, which uses 6,9-19:3,4Epox as a sex pheromone component, deuterium-labeled hydrocarbon precursor D3-3,6,9-19:Hc was topically applied to pheromone glands and found to be converted to the epoxide. This indicated that epoxidation takes place in pheromone glands (Miyamoto *et al.*, 1999). By using a variety of polyene precursors it was also determined that the monooxygenase regiospecifically attacked the n-3 double bond regardless of chain length or degree of unsaturation, indicating that the epoxidation enzyme is regiospecific in this insect (Miyamoto *et al.*, 1999).

A study using the gypsy moth, *L. dispar*, illustrates the overall pathways involved in production of epoxide pheromone components (**Figure 6**; Jurenka *et al.*, 2003). This insect uses disparlure, 2me-18:7,8Epox, as a pheromone component. Incubation of isolated abdominal epidermal tissue with deuterium-labeled valine resulted in incorporation into 2me-Z7-18:Hc. This indicates that the oenocyte cells associated with the epidermal tissues biosynthesize 2me-Z7-18:Hc using the carbons of valine to initiate the chain. The double bond is probably introduced by a Δ12 desaturase as determined by using specific deuterium-labeled intermediates. Hemolymph transport of 2me-Z7-18:Hc is indicated by the finding of this alkene in the hemolymph (Jurenka and Subchev, 2000). Demonstration that 2me-Z7-18:Hc is converted to the epoxide in the pheromone gland was shown by using deuterium-labeled 2me-Z7-18:Hc and incubation with isolated pheromone glands. Disparlure is a stereoisomer

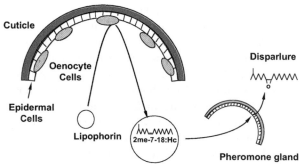

Figure 6 Production of the sex pheromone in the gypsy moth, *Lymantria dispar*. The oenocyte cells located in the abdomen biosynthesize the alkene hydrocarbon precursor to the pheromone, 2me-Z7-18:Hc. It is transported through the hemolymph by lipophorin. The alkene is taken up by pheromone gland cells where it is acted upon by an epoxidase to produce the pheromone disparlure, 2me-18:7,8Epox.

that has the 7*R*,8*S* or (+) configuration and chiral chromatography indicated that only the (+)-isomer was produced by pheromone glands (Jurenka *et al.*, 2003). These results indicate that hydrocarbon pheromones and their epoxides are produced through a pathway outlined in **Figure 6**.

A few moths utilize a combination of hydrocarbon- and fatty-acid-derived pheromone components (Ando *et al.*, 2004). The navel orangeworm moth is one of these and labeling studies supported different pathways producing the hydrocarbons and the aldehydes. The aldehydes were produced through a fatty acid biosynthetic route in pheromone glands while the hydrocarbons were produced by oenocytes and transported to the pheromone gland for release (Wang *et al.*, 2010b).

12.4.4. Pheromone Biosynthesis in Beetles

Pheromone biosynthesis in the Coleoptera is as diverse as the taxa and the pheromone structures, and the utilization of several types of pheromone biosynthetic pathways has been demonstrated (Vanderwel and Oehlschlager, 1987; Vanderwel, 1994; Seybold and Vanderwel, 2003; Tittiger, 2003; Blomquist *et al.*, 2010). Extensive work has been done on the biosynthesis of coleopteran pheromones, and the major systems that have been investigated are described in the following sections. Beetles can generate pheromone components by modification of dietary host compounds or by *de novo* biosynthesis, with the latter accounting for the majority of beetle pheromone components.

12.4.4.1. Isoprenoid pheromones from bark beetles
Most of our knowledge about beetle pheromone biosynthesis and endocrine regulation (see 12.5.3.) comes from studies of various bark beetles, especially *Ips* and *Dendroctonus* species (Scolytidae). Some bark beetles may modify fatty acyl or amino acid precursors (Vanderwel and Oehlschlager, 1987; Birgersson *et al.*, 1990); however, the majority of pheromone components are isoprenoid (Schlyter and Birgarson, 1999: Seybold *et al.*, 2000; Blomquist *et al.*, 2010).

While it was originally accepted that bark beetle pheromone components were produced by simple modification of host tree dietary precursors (Borden, 1985; Vanderwel and Oehlschlager, 1987; Vanderwel, 1994), it is now clear that most, but not all, monoterpenoid components are produced *de novo* (Seybold *et al.*, 1995; Tillman *et al.*, 1998; Lanne *et al.*, 1989; Barkawi *et al.*, 2003; Mortin *et al.*, 2003; Gilg *et al.*, 2005; Mitlin and Hedin, 1974; Thompson and Mitlin, 1979; Blomquist *et al.*, 2010).

12.4.4.2. Ipsdienol and ipsenol production
The final biosynthetic steps to produce ipsdienol have been extensively characterized in *I. pini*. Identification of terminal step enzymes was greatly facilitated by a functional genomics approach including a small-scale

expressed sequence tag (EST; Eigenheer *et al.*, 2003) and custom cDNA microarray analyses (Keeling *et al.*, 2004, 2006). The combination of sequence, microarray, biochemical, and other molecular data made candidate genes easier to identify. Subsequent "classical" biochemical and molecular analyses of gene regulation and enzyme activities confirmed candidate gene identities (Blomquist *et al.*, 2010). More is now known about the terminal steps to pheromonal ipsdienol biosynthesis in *I. pini* and *I. confusus* than any other coleopteran pheromone biosynthetic pathway.

Carbon flows first through the mevalonate pathway up to the isopentenyl diphosphate isomerase step, and is then diverted to ipsdienol by the dual function enzyme, geranyl diphosphate synthase/myrcene synthase (GPPS/MS; **Figure 7**). This enzyme is evolutionarily related to GGPPS, but has amino acid substitutions that make the active site smaller and less specific for isoprenoid-diphosphate substrates. This allows the enzyme to efficiently condense IPP and DMAPP to form GPP, which in turn acts as a substrate for myrcene production (Gilg *et al.*, 2005, 2009). Myrcene is then hydroxylated by the cytochrome P450, CYP9T2 (in *I. pini*), or CYP9T1 (in *I. confusus*; Sandstrom *et al.*, 2006, 2008). Surprisingly, these P450s produce the same ~85% 4*R*-(-)-ipsdienol enantiomeric blend, despite the fact that pheromonal ipsdienol for western *I. pini* populations is 95% (4*R*)-(-), while pheromonal ipsdienol for *I. confusus* is ~10% (4*R*)-(-) (**Figure 8**). Thus, the myrcene hydroxylases do not contribute to the final pheromonal ipsdienol blend (Sandstrom *et al.*, 2008). Interconversion of (-)- and (+)-ipsdienol is done via an ipsdienone intermediate, which is abundant in pheromone biosynthetic midguts (Ivarsson *et al.*, 1997). One enzyme known to be involved in this process is the short chain oxidoreductase IDOL DH, which readily oxidizes (4*R*)-(-)-ipsdienol to ipsdienone and stereospecifically catalyzes the reverse reaction (**Figure 9**; Figueroa-Teran, unpublished data). Genetic analyses of western and eastern *I. pini* populations suggest that one or a few loci regulate the enantiomeric composition of ipsdienol (Domingue *et al.*, 2006; Domingue and Teal, 2008). Other as yet uncharacterized enzymes may refine the final enantiomeric blend of ipsdienol and convert ipsdienol to ipsenol.

The capacity for *de novo* biosynthesis does not preclude the conversion of host precursors to pheromone components. Host myrcene ingested during feeding would enter the *de novo* pathway downstream of geranyl diphosphate. Similarly, cotton plant monoterpenes (myrcene and limonene) could enter a *de novo* biosynthetic pathway to grandlure in *A. grandis*. The question then arises as to which conversion — *de novo* biosynthesis or host precursor — is the preferred route to pheromone production. Male *I. pini* exposed to myrcene vapors produce a racemic mixture of ipsdienol (Lu, 1999), whereas

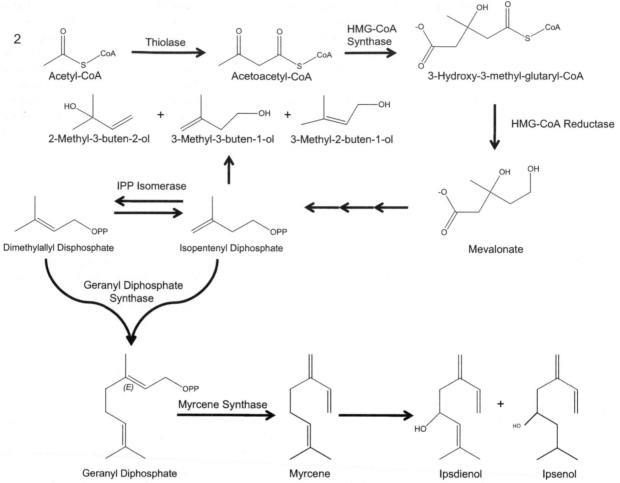

Figure 7 Proposed biosynthetic pathways for several of hemi- and monoterpenoid pheromone components in *Ips* spp.

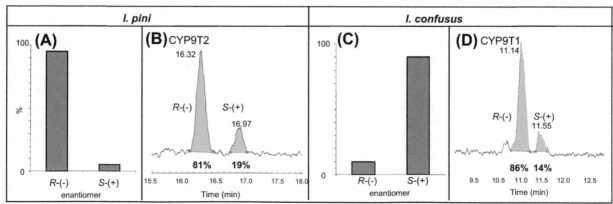

Figure 8 Ratios of the R-(-)/S-(+) ipsdienol used as pheromone in western *I. pini* (A) and *I. confusus* (C) and the products of their expressed CYP9T2/1 (B, D).

the naturally occurring pheromone of western *I. pini* is about 95:5 (-)/(+) ratio (Lu, 1999). The enzyme hydroxylating myrcene is CYP9T2, and its gene is coordinately regulated with mevalonate pathway genes in male midguts (Sandstrom *et al.*, 2006). A different P450 likely detoxifies exogenous myrcene because *CYP9T2* is not highly expressed in females (Sandstrom *et al.*, 2006), nor is it induced by its substrate (Griffith and Tittiger, unpublished data). Thus, *de novo* biosynthesis is clearly the most important route to ipsdienol and ipsenol in *I.*

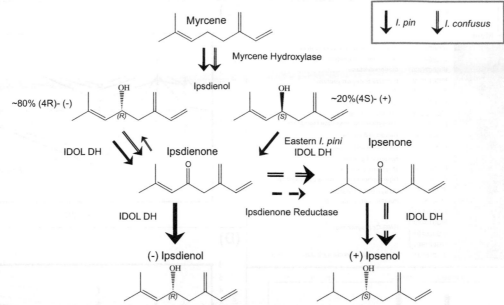

Figure 9 Current understanding of how the stereochemistry of ipsdienol and ipsenol are achieved in *I. pini* and *I. confusus*. Ipsdienol dehydrogenase (IDOL DH) converts ipsdienone stereospecifically to R-(-)-ipsdienol in *I. pini* (Figueroa-Teran, unpublished data).

pini and *I. paraconfusus*. Further examples should arise as other species are studied.

It is worth emphasizing the contribution of modern genomics techniques to our knowledge of pheromone biosynthesis in *Ips* spp. Without this approach, enzymes encoded by gene families, such as CYP9T2 and IDOL DH, would have been revealed at a much slower pace. The pheromone biosynthetic cluster contains many of the mevalonate pathway genes (**Figure 10A**), and the pheromone biosynthetic genes have a much higher basal level in males than in females (**Figure 10B**). Surprisingly, most of the genes involved in pheromone biosynthesis are upregulated by feeding and JH III in females as well as males (which produce the aggregation pheromone; **Figure 10C,D**) (Keeling *et al.*, 2004, 2006). Of the genes involved in pheromone production that were examined, only GPPS is downregulated in females. This approach has also facilitated gene identification in lepidopteran pheromone biosynthesis and is currently being applied to *Dendroctonus* spp. (Aw *et al.*, 2010; Keeling *et al.*, personal communication). Early genomic work in *Dendroctonus* spp. has revealed a surprising lack of coordinated regulation for pheromone biosynthetic genes; however, some candidate enzymes were still identified and are currently being pursued (Aw *et al.*, 2010).

12.4.4.3. Other pheromone components
Hemiterpene pheromone components of the bark beetles are similarly synthesized *de novo*. Lanne *et al.* (1989) demonstrated the incorporation of labeled acetate, glucose, and mevalonate into 2-methyl-3-buten-2-ol in *I. typographus*. This also argues for the *de novo* synthesis

of 3-methyl-3-buten-1-ol and 3-methyl-2-buten-1-ol. The mevalonate pathway intermediate, dimethylallyl diphosphate, likely provides the carbon skeleton for 3-methyl-2-buten-1-ol by dephosphorylation. The other 5-carbon intermediate, isopentenyl disphosphate, could be directly converted to 3-methyl-3-buten-1-ol and, perhaps through several steps, to 2-methyl-3-buten-2-ol.

While ipsdienol and ipsenol biosynthesis is relatively well characterized, the production of frontalin, and the male-produced pheromone of the Colorado potato beetle, (*S*)-3,7-dimethyl-2-oxo-oct-6-ene-1,3-diol (CPB1; Dickens *et al.*, 2002), is less clear. Barkawi *et al.* (2003) demonstrated that acetate and mevalonolactone are precursors to frontalin, demonstrating the isoprenoid origin of this common *Dendroctonus* pheromone component. The carbon skeleton of frontalin probably arises from geranyl disphosphate (GPP) or farnesyl- or geranylgeranyl-diphosphate via a putative dioxygenase, which converts GPP, FPP, or GGPP into the 8-carbon intermediate sulcatone. Sulcatone is an obvious precursor to sulcatol, which is a pheromone component of some bark beetles (*Gnathotrichus sulcatus*, Byrne *et al.*, 1974). In *Dendroctonus* spp., sulcatone may be converted to 6-methyl-6-hepten-2-one (6-MHO), a known intermediate to frontalin (Perez *et al.*, 1996). Alternative cyclizations of 6-MHO in other beetles could lead to the pheromone components pityol and vittatol. Significant amounts of sulcatone are also found in Colorado potato beetle males during their synthesis of the Colorado potato beetle pheromone compound 1 (Dickens *et al.*, 2002).

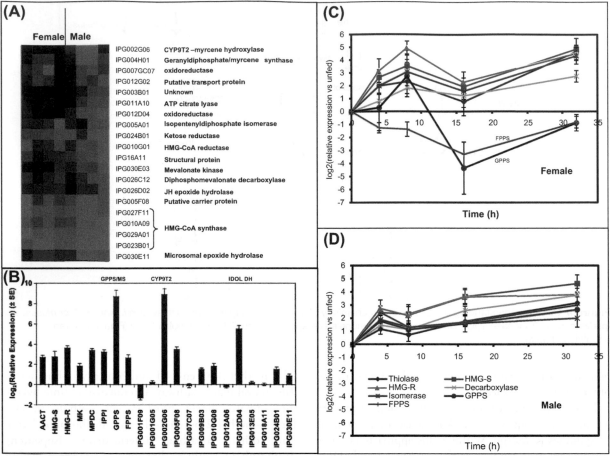

Figure 10 The "pheromone" biosynthetic gene cluster based on the analysis of microarray data showing the effect of JH III on mevalonate pheromone biosynthesis genes in *I. pini* (A) (*Eigenheer et al.*, 2003). qRT-PCR analyses show the effect of feeding on selected mevalonate genes from female (C) and male (D) midguts (*Keeling et al.*, 2004) and the basal levels of selected genes in males compared to females (B) (*Keeling et al.*, 2006).

The picture emerging from these studies is that isoprenoid pheromone component production in beetles is mostly *de novo*, with carbon being diverted from the mevalonate pathway at GPP (for linear monoterpenoids) or possibly later (e.g., for frontalin). Geranyl diphosphate may be directly modified through dephosphorylation and cyclizations (cotton boll weevil) or hydroxylations (*Ips* spp.). Alternatively, some bark beetles, such as *Dendroctonus* spp., may have a dioxygenase (although not yet characterized) that oxidizes GPP or geraniol (or C15 and/or C20 isoprenoids) to produce sulcatone, which is likely a precursor to the pheromone components sulcatol, pityol, and frontalin.

12.4.4.4. Fatty-acid-derived pheromones Numerous beetle genera use modified fatty acyl compounds as pheromone components. Less is known about their biosynthesis compared to isoprenoid pheromones, but the same general strategy of modifying or combining existing biosynthetic pathways is conserved.

For some beetles, the modifications are relatively minor. For example, *Attagenus* spp. (Dermestidae) myristic acid

may be desaturated at the Δ5 and Δ7 positions to produce tetradecadienoic acid pheromone components. The stereochemistries of the double bonds apparently provide specificity between species (Fukui *et al.*, 1977). It is unclear whether the short chain fatty acid precursors to these pheromone components are synthesized through normal fatty acid elongations, or are the β-oxidation products of longer fatty acids. For other beetles, modifications can become more complex. Female *Te. molitor* produce 4-methyl-1-nonanol from propionyl-, malonyl-, and methylmalonyl-precursors (Islam *et al.*, 1999). This is an example of carbon being shunted away from fatty acyl elongation before long fatty acids are completed. The use of methylmalonate to produce methyl-branched hydrocarbons is well established in other insect systems (Blomquist, 2003; Schal *et al.*, 2003), although it is unknown if beetles have a secondary fatty acyl synthase that, similar to houseflies, incorporates methyl-malonyl-CoA precursors efficiently.

The flexibility of the fatty acid biosynthetic pathway is extended in some nitidulid beetles (*Carpophilus* spp.), where males use propionate and butyrate (presumably as methylmalonyl-CoA and ethylmalonyl-CoA) to make methyl and ethyl-branched triene and tetraene pheromone components, apparently also via the fatty acid biosynthetic pathway (Bartelt *et al.*, 1992; Bartelt, 2010). The branched hydrocarbons generally have 10–12 carbon backbones with conjugated double bonds. In contrast to other systems, where pheromone component biosynthesis is highly specific, *Carpophilus* spp. males produce a mixture of related structures, some of which act as pheromones and some of which do not. Since di-substituted tetraenes are less abundant than mono- or unsubstituted tetraenes, it appears that non-acyl units placed in the growing hydrocarbon chains represent "mistakes" made by synthesis machinery with a low stringency for substrate selection (Bartelt, 1999). Such non-specific hydrocarbon biosynthesis may serve speciation, since changes in antennal receptivity may accommodate pre-existing compounds (Bartelt, 1999). Interestingly, the desaturated nature of these hydrocarbons is not due to fatty acyl desaturases, but to the inactivity of enoyl-ACP reductase during biosynthesis so that the enoyl-ACP intermediate formed during elongation is not reduced (Petroski *et al.*, 1994). This suggests that carbon is shunted out of the fatty acid biosynthetic pathway when the chains are the correct length, similar to the situation in *T. molitor*.

Rather than modifying the normal biosynthetic pathway to produce pheromone components, some beetles modify normal products of the pathway. For example, lactone pheromone components of some scarab beetles are produced by the stereospecific alterations of long chain fatty acids. Female *Anomala japonica* (Scarabaeidae) are perhaps best studied among scarab beetles for the biosynthesis of japonilure and buibuilactone, which involves the successive $\Delta 9$ desaturation, hydroxylation, and two rounds of β-oxidation to shorten the chain length, and cyclization of stearic and palmitic acids (Leal *et al.*, 1999). Of all these, only the hydroxylation step appears to be stereo-specific. This step is important because different enantiomers have different functions in different *Anomala* species (Leal *et al.*, 1999).

12.4.4.5. Host precursor modifications Whereas most bark beetle pheromones are clearly synthesized *de novo*, there is strong evidence that some pheromone components are the result of modifying host precursor molecules. α-Pinene is produced by pine trees, and can be hydroxylated to *cis-* and *trans*-verbenol by *Ips* beetles (Renwick *et al.*, 1976). A further oxidation of verbenol yields verbenone in *D. ponderosae* (Hunt *et al.*, 1989). Similarly, Jeffrey pine trees contain relatively low levels of monoterpenoids, but high levels of heptane. Female *D. jeffreyi* that attack these trees produce 1- and 2-heptanol, and 1-heptanol acts as an aggregation pheromone (Paine *et al.*, 1999).

12.4.5. Pheromone Biosynthesis in Diptera

12.4.5.1. Housefly pheromone biosynthesis: C23 sex pheromone components A combination of *in vivo* and *in vitro* studies using both radio- and stable-isotope techniques established the biosynthetic pathways for the major sex pheromone components in the housefly (**Figure 11**; Dillwith and Blomquist, 1982; Dillwith *et al.*, 1981, 1982; Blomquist *et al.*, 1984a; Vaz *et al.*, 1988). (*Z*)-9-Tricosene is formed by the microsomal elongation of 18:1-CoA to 24:1-CoA using malonyl-CoA and NADPH, and the elongated fatty acyl-CoA is then reduced to the aldehyde and converted to the hydrocarbon one carbon shorter (**Figure 11**; Reed *et al.*, 1994, 1995; Mpuru *et al.*, 1996). A cytochrome P450 enzyme is involved in the metabolism of the alkene to the corresponding epoxide and ketone (Ahmad *et al.*, 1987). It appears that the same enzyme that catalyzes formation of an epoxide from the double bond between carbons 9 and 10 of the alkene of (*Z*)-9-tricosene also hydroxylates it at position n-10. The secondary alcohol thus formed is then converted to the unsaturated ketone (Guo *et al.*, 1991).

12.4.5.2. Mechanism of hydrocarbon formation: Oxidative or reductive decarbonylation of an aldehyde? The mechanism of hydrocarbon formation has proven elusive. In an elegant set of experiments in the 1960s and early 1970s, Kolattukudy and co-workers demonstrated that fatty acyl-CoAs were elongated and then converted to hydrocarbon by the loss of the carboxyl group (reviewed in Kolattukudy *et al.*, 1976; Kolattukudy, 1980). In the 1980s and early 1990s, the hypothesis was put forward that very long chain fatty acyl-CoAs were reduced to the aldehyde and then decarbonylated to the hydrocarbon and carbon monoxide, and that this reaction did not require any cofactors or O_2. Evidence for this reductive decarbonylation mechanism was obtained from a plant (Cheesbrough and Kolattukudy, 1984), an alga (Dennis and Kolattukudy, 1991, 1992), a vertebrate (Cheesbrough and Kolattukudy, 1988), an insect (Yoder *et al.*, 1992), and has recently been confirmed in a cyanobacterium (Schirmer *et al.*, 2010).

In work on hydrocarbon formation in the housefly, it was found that the acyl-CoA is reduced to the aldehyde, with the conversion of the aldehyde to hydrocarbon requiring NADPH and molecular oxygen, and that the products were hydrocarbon and carbon dioxide (demonstrated by radio-GLC; Reed *et al.*, 1994, 1995). Antibodies to housefly cytochrome P450 and to P450 reductase inhibited hydrocarbon formation in microsomes, as did exposure to CO, and the latter could be partially reversed by white light. GC-MS analyses of specifically deuterated substrates showed that the protons on positions 2,2 and 3,3 of the acyl-CoA were retained during conversion to hydrocarbon, and that the proton on position 1 of the aldehyde

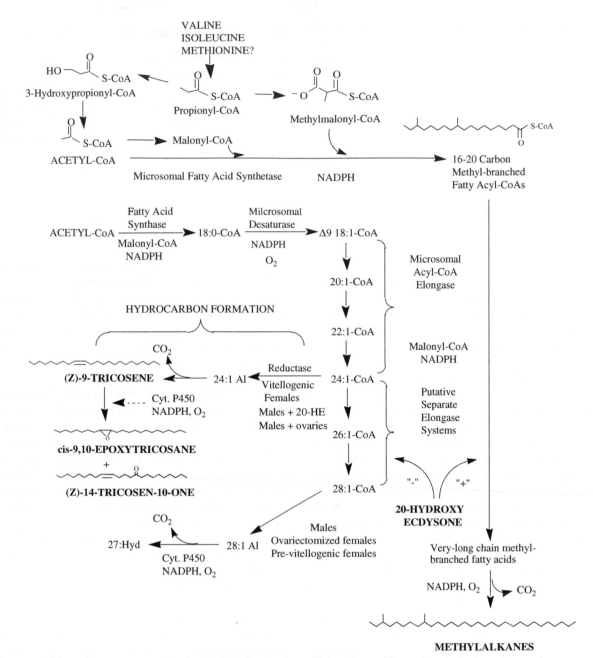

Figure 11 Biosynthetic pathways showing the putative steps at which ecdysteroids regulate the hydrocarbon and hydrocarbon-derived pheromone components of the female housefly, *Musca domestica*.

was transferred to the adjacent carbon and retained during hydrocarbon formation. Furthermore, several peroxides could substitute for O_2 and NADPH and support hydrocarbon production. All of this evidence strongly supports a cytochrome P450 involvement in hydrocarbon synthesis (Reed *et al.*, 1994, 1995).

It appears that both mechanisms of hydrocarbon formation exist, with strong evidence for the reductive decarbonylation occurring in microorganisms (Schirmer *et al.*, 2010) and presumably plants, and an oxidative decarbonylation to hydrocarbon and carbon dioxide occurring in insects.

12.4.5.3. Biosynthesis of the methylalkane pheromone components

Methylalkanes are formed by the substitution of methylmalonyl-CoA in place of malonyl-CoA at specific points during chain elongation. Carbon-13 NMR, mass spectrometry, and radiochemical studies (Dwyer *et al.*, 1981; Dillwith *et al.*, 1982; Chase *et al.*, 1990) demonstrated that the methylmalonyl-CoA was added during the initial steps of chain elongation in insects using what appears to be a novel microsomal fatty acid synthase (FAS; **Figure 8**). A microsomal FAS was first suggested from studies in the moth *T. ni* for the formation of methyl-branched very long chain alcohols (de Renobales *et al.*, 1989). In this insect, high rates of methyl-branched very long chain alcohol synthesis were observed in the mid-pupal stages at times when soluble FAS activity was very low or undetectable. The FAS of most organisms is soluble (cytoplasmic).

A microsomal FAS was also implicated in the biosynthesis of methyl-branched fatty acids and methyl-branched hydrocarbon precursors of the German cockroach contact sex pheromone (Juarez *et al.*, 1992; Gu *et al.*, 1993). A microsomal FAS present in the epidermal tissues of the housefly is a likely candidate responsible for methyl-branched fatty acid production (Blomquist *et al.*, 1994). The housefly microsomal and soluble FAS were purified to homogeneity (Gu *et al.*, 1997) and the microsomal FAS was shown to preferentially use methylmalonyl-CoA in comparison to the soluble FAS. GC-MS analyses showed that the methyl-branching positions of the methyl-branched fatty acids of the housefly (Blomquist *et al.*, 1994) were in positions consistent with them being the precursors of the methyl-branched hydrocarbons.

The methylmalonyl-CoA unit that is the precursor to methyl-branched fatty acids and hydrocarbons arises from the carbon skeletons of valine and isoleucine, but not succinate (Dillwith *et al.*, 1982). Propionate is also a precursor to methylmalonyl-CoA, and in the course of these studies, a novel pathway for propionate metabolism in insects was discovered. Many insect species, including the housefly, do not contain vitamin B12 (Wakayama *et al.*, 1984), and therefore cannot catabolize propionate via methylmalonyl-CoA to succinate. Instead, as first demonstrated in the housefly (Dillwith *et al.*, 1982), insects metabolize propionate to 3-hydroxypropionate and then to acetyl-CoA, with carbons 3 and 2 of propionate becoming carbons 1 and 2 of acetyl-CoA (Halarnkar *et al.*, 1986; **Figure 7**).

12.4.6. Pheromone Biosynthesis in *Drosophila*

The other major dipteran system in which pheromone production has been extensively studied is that of *Drosophila*. Wicker-Thomas and Chertemps (2010) provided a detailed review of the chemistry, biochemistry, molecular biology, and genetics of *Drosophila* pheromone production. *Drosophila*, as do *Musca*, use long chain, sex-specific hydrocarbons as close range and contact sex pheromones. *Drosophila melanogaster* Canton-S mature females have abundant (*Z,Z*)-7,11-heptacosadiene, which is absent in males and was shown to effectively stimulate wing vibration in conspecific males when applied to a dummy (Antony and Jallon, 1982; Ferveur and Sureau, 1996). In some *Drosophila*, 7-tricosene is involved in chemical communication. A series of studies established that the hydrocarbon pheromones are produced in the oenocytes (Ferveur *et al.*, 1997), transported through the hemolymph via lipophorin (Pho *et al.*, 1996), and then deposited on the cuticle.

A series of experiments assaying the incorporation of various precursors into hydrocarbons established that *Drosophila* use an elongation-decarboxylation pathway for hydrocarbon production (Pennanec'h *et al.*, 1997). An interesting question in *Drosophila* pheromone production is the origin of the 7,11 double bonds. Biochemical evidence suggested that either a Δ7 desaturase working on myristate or a Δ9 desaturase working on palmitate give rise to the double bond in the 7-position, which could then be elongated to produce vaccenate (n-7, 18:1). Vaccenate could then be elongated and decarboxylated to produce 7-tricosene and, with an additional desaturation step, to produce 7,11-27:2Hyd.

Wicker-Thomas *et al.* (1997) isolated a cDNA encoding a desaturase (*desat1*) in *D. melanogaster*. The expressed *desat1* protein is a Δ9 desaturase that preferentially used palmitate, resulting in n-7 fatty acids. A *desat2*, located close to *desat1*, appears to be responsible for the 5,9-dienes present in Tai strain females (Jallon and Wicker-Thomas, 2003). The complete sequencing of the *D. melanogaster* genome, as well as the genomes of other *Drosophila* species, has greatly contributed to our understanding of fruitfly desaturases. Shirangi *et al.* (2009) showed that the expression of *desatF*, which is responsible for the second desaturation step, correlates with the amount of long chain dienes. This desaturase has undergone six independent gene inactivations, three losses of expression without gene loss, two transitions in sex-specificity, and other changes, suggesting that frequent and rapid changes in the expression of *desatF* were responsible for evolutionary transitions in chemical communication and speciation in 24 *Drosophila* species.

12.4.7. Biosynthesis of Contact Pheromones in the German Cockroach

The female German cockroach attracts males with a volatile pheromone, gentisyl quinone isovalerate, or blattellaquinone (Nojima *et al.*, 2005). Upon antennal contact with a female, the male German cockroach rotates his body 180 degrees and raises his wings, thus exposing specialized tergal glands that attract the female and place her into a pre-copulatory position (Nojima *et al.*, 1999). The non-volatile contact pheromone responsible for this behavior was identified as (3*S*,11*S*)-dimethylnonacosan-2-one (**Figure 12**) (Nishida *et al.*, 1974) along with an alcohol (29-hydroxy-3*S*,11*S*-dimethylnonacosan-2-one) and aldehyde (29-oxo-3,11-dimethylnonacosan-2-one) derivative with the same 3,11-dimethylketone skeleton (Nishida and Fukami, 1983). A fourth pheromone component, 3,11-dimethylheptacosan-2-one, is less active than its C29 homologue (Schal *et al.*, 1990b). Recent studies have extended this pheromone blend to 6 biosynthetically related components, consisting of two homologous series of C27 and C29 methyl ketones, aldehydes, and alcohols (Eliyahu *et al.*, 2004, 2008; Mori, 2008).

The route of biosynthesis and its physiological regulation have been previously reviewed (Blomquist *et al.*, 1993; Tillman *et al.*, 1999; Schal *et al.*, 2003). Central to investigations of the biosynthetic pathway was the observation that the major cuticular hydrocarbon in all life stages of the German cockroach is an isomeric mixture of 3,7-, 3,9-, and 3,11-dimethylnonacosane (Jurenka *et al.*, 1989). The presence of only the 3,11-isomer in the cuticular dimethyl ketone fraction and only in adult females prompted Jurenka *et al.* (1989) to propose that production of the pheromone might result from the sex-specific oxidation of its hydrocarbon analog only in adult females. This scheme follows the well-established conversion of hydrocarbons to methyl ketone and epoxide pheromones in the housefly (see section 12.4.5.; Blomquist *et al.*, 1984a; Ahmad *et al.*, 1987).

This model has since been validated with several independent approaches. Biochemical studies on the biosynthesis of methyl-branched alkanes showed that the methyl branches are added during the early stages of chain elongation (Chase *et al.*, 1990). Using carbon-13 labeling and NMR analyses, Chase *et al.* (1990) showed that carbons 1 and 2 of acetate are incorporated as the chain initiator, and that the carbon skeleton of propionate serves as the methyl branch donor (**Figure 12**). Further, propionate and succinate labeled methyl-branched hydrocarbons and the methyl ketone pheromone, as did the amino acids valine, isoleucine, and methionine, all of which can be metabolized to propionate. NMR studies confirmed that these substrates were metabolized to methylmalonyl-CoA for incorporation into the methyl branch unit of hydrocarbons (Chase *et al.*, 1990), as in the housefly (Dillwith *et al.*, 1982; Halarnkar *et al.*, 1986), American cockroach (Halarnkar *et al.*, 1985), cabbage looper moth (de Renobales and Blomquist, 1983), and the termite *Zootermopsis* (Chu and Blomquist, 1980).

Methyl-branched fatty acids are intermediates in branched alkane biosynthesis (Juarez *et al.*, 1992). Thus, [1-^{14}C]propionate labeled methyl-branched fatty acids of 16–20 carbons, but did not label straight-chain saturated and monounsaturated fatty acids (Chase *et al.*, 1990).

Chase *et al.* (1992) investigated the hypothesis that the 3,11-dimethyl ketone sex pheromone arises from the insertion of an oxygen into the preformed 3,11-dimethyl alkane. When high-specific activity, tritiated 3,11-dimethylnonacosane (mixture of stereoisomers) was topically applied on the cuticle of *B. germanica* females, it readily penetrated the cockroach and radioactivity from the alkane was detected in both 3,11-dimethylnonacosan-2-ol and 3,11-dimethylnonacosan-2-one. Likewise, when tritiated 3,11-dimethylnonacosan-2-ol was applied to the cuticle it was readily and highly efficiently converted to the corresponding methyl ketone pheromone. Surprisingly, the dimethyl ketone pheromone was derived from the corresponding alcohol not only in females, as expected, but also in males. These results suggest that the sex pheromone of *B. germanica* arises via a female-specific hydroxylation of 3,11-dimethylnonacosane and a subsequent non-sex-specific oxidation, probably involving a polysubstrate monooxygenase system, to the (3*S*,11*S*)-dimethylnonacosan-2-one pheromone (**Figure 12**). Chase *et al.* (1992) also suggested that a similar hydroxylation and subsequent oxidation at the 29-position of 3,11-dimethylnonacosan-2-one might give rise to 29-hydroxy- and 29-oxo-(3,11)-dimethylnonacosan-2-one, the other components of the contact pheromone blend, but this hypothesis has yet to be tested. It is quite likely, as well, that the same mechanism converts 3,11-dimethylheptacosane to the corresponding methyl ketone pheromone, and its 27-hydroxy- and 27-oxo- analogs.

The contact sex pheromone of female *B. germanica* remains the only cockroach pheromone whose biosynthetic pathway has been investigated with radio- and stable-isotope tracers.

12.4.8. Biosynthesis of the Honeybee Queen Pheromone

The queen substance used for "queen control" inside the nest is also the substance used by virgin queens to attract drones for mating. It is the best understood of the sexual pheromones of the social insects. Callow and Johnston (1960) and Barbier and Lederer (1960) identified (*E*)-9-oxodec-2-enoic acid (9-ODA) in queen mandibular glands. 9-Hydroxy-2(*E*)-decenoic acid (9-HDA) is also present (Callow *et al.*, 1964) and together both attract drones. Keeling *et al.* (2003) identified a number of additional compounds that function synergistically with the 9-ODA and 9-HDA, making this the most complex pheromone blend known for any organism.

In an elegant set of experiments, Plettner *et al.* (1996, 1998) elucidated the biosynthetic pathways for the honeybee queen mandibular pheromone (QMP) components 9-ODA and 9-HDA and compared their biosynthesis to that of worker-produced 10-hydroxy-2(*E*)-decenoic acid and the corresponding diacid. Using carbon-13 and deuterated precursors, Plettner *et al.* (1996, 1998) demonstrated (1) the *de novo* synthesis of stearic acid in worker mandibular glands, (2) the hydroxylation of stearic acid at the n- (workers) and n-1 (queens) positions, (3) chain shortening through β-oxidation to the 10 and 8 carbon hydroxy acids, and (4) oxidation of n- and n-1 hydroxy groups to give diacids and 9-keto-2(*E*)-decenoic acid, respectively. Stearic acid was shown to be the main precursor of the pheromone molecules as it was converted to C10 hydroxy acids and diacids more efficiently than either 16 or 14 carbon fatty acids.

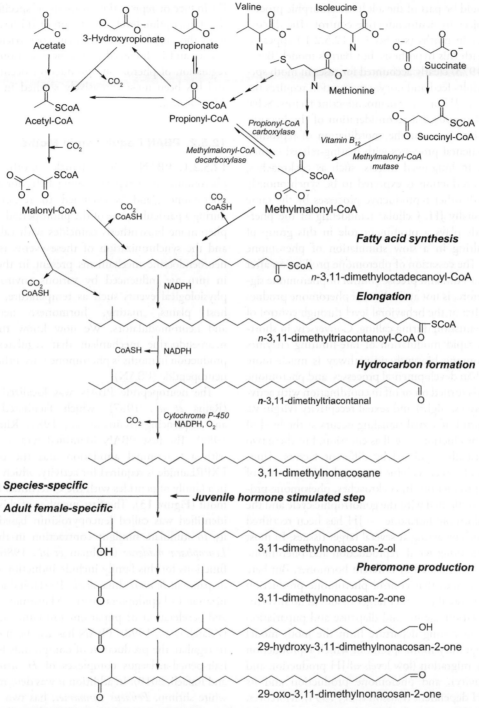

Figure 12 Proposed biosynthetic pathways for the major contact pheromone components of the German cockroach, *Blattella germanica*. The JH III regulated step appears to be the hydroxylation of the dimethylalkane.

12.5. Endocrine Regulation of Pheromone Production — Introduction

12.5.1. Barth's Hypothesis

The central role of JH in mate-finding was recognized in 1965 when Barth proposed that neuroendocrine control of pheromone production would be common in insects with a long-lived adult stage and with multiple reproductive cycles interrupted by periods during which sexual receptivity and mating are not appropriate or not even possible anatomically. Cockroaches and beetles are quintessential examples of this life history syndrome. Conversely, in insects that eclose with mature oocytes and live for only a few days as adults, Barth (1965) predicted that pheromone signaling would be part of the adult metamorphic process and not subject to neuroendocrine control. The discovery of PBAN in moths (see Section 12.5.2.1.) appeared in conflict with this hypothesis, but Barth's model (Barth and Lester, 1973) clearly accounted for cases in moth species where adults feed and oocyte maturation requires the participation of JH or other neuroendocrine factors. Schal *et al.* (2003) proposed a reconsideration of the hypothesis, taking into account the coordination of reproductive developmental processes with mating-related events. Accordingly, in long-lived insects, such as cockroaches, pheromone production is expected to be synchronously regulated with other reproductive processes by the same hormone, usually JH. Cellular remodeling of the pheromone glands plays a prominent role in this group of insects, resulting in a slow stimulation of pheromone production. The cessation of pheromone production after mating is also slow, and precise control of pheromone signaling, therefore, is not at the level of pheromone production, but rather at the behavioral level through control of pheromone emission during calling. Conversely, in short-lived moths rapid modulation of rate-limiting enzymes in the pheromone biosynthetic pathway is much more prominent than developmental processes, and pheromone biosynthesis is turned on or off in coordination with activity cycles (day vs. night) and sexual receptivity (virgin vs. mated). Control of sexual signaling occurs at the level of pheromone production as well as emission, but these two events are usually regulated by different factors. Thus, both groups of insects exhibit neuroendocrine control of pheromone production. In cockroaches, pheromone production is coordinated with the gonotrophic cycle and the major gonadotropic hormone — JH has been recruited to control both by acting at several target tissues. In most moths, on the other hand, reproduction and pheromone production are regulated by different hormones. But here also, the hormones that control pheromone production (e.g., PBAN) also affect other target tissues as do myotropins, melanization agents, and diapause and pupariation factors. An interesting departure from the moth model occurs in migratory moth species in which reproduction is delayed by migration (low levels of JH production and sexual inactivity), and pheromone production and its release are JH dependent (Cusson *et al.*, 1994). Moreover, even in PBAN-regulated pheromone production, JH appears to play an important role in inducing competence of the pheromone gland to respond to PBAN (Rafaeli *et al.*, 2003). All of these observations are consistent with our interpretation of Barth's model.

The three hormones that regulate pheromone production in insects are JH, 20E, and PBAN. PBAN has been studied in female moths and alters enzyme activity through second messengers at one or more steps during or subsequent to fatty acid synthesis during pheromone production (Rafaeli and Jurenka, 2003). In contrast, 20E and JH induce or repress the synthesis of specific enzymes at the transcription level. The action of JH has been studied most thoroughly in the German cockroach and in bark beetles and is discussed in the next section. Ecdysteroid regulation of pheromone production occurs in Diptera, and has been most extensively studied in the housefly, *M. domestica*.

12.5.2. PBAN Regulation in Moths

12.5.2.1. PBAN Most female moths release sex pheromones in a typical calling behavior in which the pheromone gland is extruded to release pheromone during a particular time of the photoperiod. In most cases pheromone biosynthesis coincides with calling behavior and the synchronization of these events is achieved by neuroendocrine mechanisms present in the female that in turn are influenced by various environmental and physiological events such as temperature, photoperiod, host plants, mating, hormones, neurohormones, and neuromodulators. We now know that the main neuroendocrine mechanism that regulates pheromone production in moths is pheromone biosynthesis activating neuropeptide (PBAN).

The neuropeptide PBAN was localized to the SEG (Raina *et al.*, 1987), which facilitated purification and sequencing (Raina *et al.*, 1989; Kitamura *et al.*, 1989). The first PBAN identified had 33 amino acids with a C-terminal amidation and the core sequence FXPRLamide is required for activity, which places PBAN in a family of peptides with the C-terminal FXPRLamide motif (**Figure 13**). The first member of this family to be identified was called leucopyrokinin based on its ability to stimulate hindgut contraction in the cockroach, *Leucophaea maderae* (Holman *et al.*, 1986). Additional functions for this family include induction of embryonic diapause in *B. mori* (Imai *et al.*, 1991), induction of melanization in Lepidoptera larvae (Matsumoto *et al.*, 1990), and acceleration of puparium formation in several flies (Zdarek *et al.*, 1998). PBAN has also been demonstrated to regulate the production of compounds found in male hair-pencil-aedeagus complexes of *H. armigera* (Bober and Rafaeli, 2010). In addition it was determined that the white shrimp, *Penaeus vannamei*, has two peptides that can induce myotropic activity (Torfs *et al.*, 2001). These results demonstrate the ubiquity and multifunctional nature of this family of peptides.

Localization of PBAN-like immunoreactivity in the central nervous system of adult moths indicated that several neurons in the SEG contain PBAN-like activity. These were found as clusters along the ventral midline, one each in the presumptive mandibular, maxillary, and labial neuromeres (Kingan *et al.*, 1992). All three groups of neurons have axons that project into the corpus cardiacum (CC; Kingan *et al.*, 1992; Davis *et al.*, 1996). Two

Function and Species		Peptide sequence	Reference
PBAN			
Helicoverpa zea		LSDDMPATPADQEMYRQDPEQIDSRTKY**FSPRL**amide[a]	(Raina *et al.*, 1989)
Helicoverpa assulta		LSDDMPATPADQEMYRQDPEQIDSRTKY**FSPRL**amide[b]	(Choi *et al.*, 1998)
Bombby mori		LSEDMPATPADQEMYQPDPEEMESRTRY**FSPRL**amide[a]	(Kitamura *et al.*, 1989)
Lymantria dispar		LADDMPATMADQEVYRPEPEQIDSRNKY**FSPRL**amide[a]	(Masler *et al.*, 1994)
Agrotis ipsilon		LADDTPATPADQEMYRPDPEQIDSRTKY**FSPRL**amide[b]	(Duportets *et al.*, 1999)
Mamestra brassicae		LADDMPATPADQEMYRPDPEQIDSRTKY**FSPRL**amide[b]	(Jacquin-Joly *et al.*, 1998)
Spodoptera littoralis		LADDMPATPADQELYRPDPDQIDSRTKY**FSPRL**amide[b]	(Iglesias *et al.*, 2002)
Pheromontropic peptides			
Bomby xmori	α	II**FTPKL**amide[b]	(Kawano *et al.*, 1992)
Helicoverpa zea	α	VI**FTPKL**amide[b]	(Ma *et al.*, 1994)
Helicoverpa assulta	α	VI**FTPKL**amide[b]	(Choi *et al.*, 1998)
Agrotis ipsilon	α	VI**FTPKL**amide[b]	(Duportets *et al.*, 1999)
Mamestra brassicae	α	VI**FTPKL**amide[b]	(Jacquin-Joly *et al.*, 1998)
Spodoptera littoralis	α	VI**FTPKL**amide[b]	(Iglesias *et al.*, 2002)
Bombyx mori	β	SVAKPQTHESLE**FIPRL**amide[b]	(Kawano *et al.*, 1992)
Helicoverpa zea	β	SLAYDDKSFENVE**FTPRL**amide[b]	(Ma *et al.*, 1994)
Helicoverpa assulta	β	SLAYDDKSFENVE**FTPRL**amide[b]	(Choi *et al.*, 1998)
Agrotis ipsilon	β	SLSYEDKMFDNVE**FTPRL**amide[b]	(Duportets *et al.*, 1999)
Mamestra brassicae	β	SLAYDDKVFENVE**FTPRL**amide[b]	(Jacquin-Joly *et al.*, 1998)
Pseudaletia separata	β	KLSYDDKVFENVE**FTPRL**amide[b]	(Matsumoto *et al.*, 1992a)
Spodoptera littoralis	β	SLAYDDKVFENVE**FTPRL**amide[b]	(Iglesias *et al.*, 2002)
Bombyx mori	γ	TMS**FSPRL**amide[b]	(Kawano *et al.*, 1992)
Helicoverpa zea	γ	TMN**FSPRL**amide[b]	(Ma *et al.*, 1994)
Helicoverpa assulta	γ	TMN**FSPRL**amide[b]	(Choi *et al.*, 1998)
Agrotis ipsilon	γ	TMN**FSPRL**amide[b]	(Duportets *et al.*, 1999)
Mamestra brassicae	γ	TMN**FSPRL**amide[b]	(Jacquin-Joly *et al.*, 1998)
Spodoptera littoralis	γ	TMN**FSPRL**amide[b]	(Iglesias *et al.*, 2002)
Diapause Hormone			
Bombyx mori		TDMKDESDRGAHSERGALC**FGPRL**amide[a]	(Imai *et al.*, 1991)
Helicoverpa zea		NDVKDGAASGAHSDRLGLW**FGPRL**amide[b]	(Ma *et al.*, 1994)
Helicoverpa assulta		NDVKDGAASGAHSDRLGLW**FGPRL**amide[b]	(Choi *et al.*, 1998)
Agrotis ipsilon		NDVKDGGADRGAHSDRGGMW**FGPR**lamide[b]	(Duportets *et al.*, 1999)
Spodoptera littoralis		NEIKDGGSDRGAHSDRAGLW**FGPRL**amide[b]	(Iglesias *et al.*, 2002)
Pyrokinins			
Leucophaea madera		ETS**FTPRL**amide[a]	(Hlolman *et al.*,1986)
Locusta migratoria	(I)	pEDSGDGWPQQP**FVPRL**amide[a]	(Schoofs *et al.*, 1991)
	(II)	pESVPT**FTPRL**amide[a]	(Schoofs *et al.*, 1993)
Myotropins			
Locusta migratoria	(I)	GAVPAAQ**FSPRL**amide[a]	(Schoofs *et al.*, 1990a)
	(II)	EGD**FTPRL**amide[a]	(Schoofs *et al.*, 1990b)
	(III)	RQQP**FVPRL**amide[a]	(Schoofs *et al.*, 1992)
	(IV)	RLHQNGMP**FSPRL**amide[a]	(Schoofs *et al.*, 1992)
Schistocerca gregaria	MT1	GAAPAAQ**FSPRL**amide[a]	(Veelaert *et al.*, 1997)
	MT2	TSSLFPH**PRL**amide[a]	(Veelaert *et al.*, 1997)
Periplaneta americana	PK1	HTAG**FIPRL**amide[a]	(Predel *et al.*, 1997)
	PK2	SPPFA**PRL**amide[a]	(Predel *et al.*, 1997)
	PK3	LVPFR**PRL**amide[a]	(Predel *et al.*, 1999)
	PK4	DHLPHDVY**SPRL**amide[a]	(Predel *et al.*,1999)
	PK5	GGGGSGETSGMW**FGPRL**amide[a]	(Predel *et al.*, 1999)
	PK6	SESEVPGMW**FGPRL**amide[a]	(Predel and Eckert, 2000)
Periplaneta fuliginosa	PK4	DHLSHDVY**SPRL**amide[a]	(Predel and Eckert, 2000)
Drosophila melanogaster	CAP2b-3	TGPSASSGLW**FGPRL**amide[b]	(Choi *et al.*, 2001)
	PK-22	SVP**FKPRL**amide[b]	(Choi *et al.*, 2001)
	ETH-1	DDSSPGFFLKITKNV**PRL**amide[b]	(Park *et al.*, 1999)
	hug γ	pELQSNGIPAYRVRT**PRL**amide[b]	(Meng *et al.*, 2002)
Penaeus vannamei (Crustacea)		DFA**FSPRL**amide[a]	(Torfs *et al.*, 2001)
		ADFA**FNPRL**amide[a]	(Torfs *et al.*, 2001)

Figure 13 Amino acid sequences of the pyrokinin/PBAN family of peptides and the species where identified. The FXPRLamide motif is shown in bold. X = S, T, G, or V. Five peptides are shown that have a PRLamide ending. Peptides are grouped together based on the primary function for which they were first identified. [a] - Identified from the amino acid sequence of a purified peptide. [b] - Deduced from the cloned gene sequence.

pairs of maxillary neurons send processes within the SEG that include posterior projections into the paired ventral nerve cord (VNC) and travel its entire length to terminate in the terminal abdominal ganglion (TAG). Arborizations arising from these paired projections were found in each segmental ganglion (Davis *et al.*, 1996). In addition the segmental ganglia have neurons that contain PBAN-like activity and these neurons can release peptides into the hemolymph (Davis *et al.*, 1996; Ma and Roelofs, 1995b; Ma *et al.*, 1996).

12.5.2.2. Molecular genetics of PBAN
The gene encoding PBAN was first characterized from *H. zea* and *B. mori* (Davis *et al.*, 1992; Imai *et al.*, 1991; Kawano *et al.*, 1992; Ma *et al.*, 1994; Sato *et al.*, 1993). The full-length cDNA was found to encode PBAN plus four

additional peptide domains with a common C-terminal FXPRL sequence motif including that of the diapause hormone of *B. mori*. Three additional peptides with the common C-termini and sequence homology to those of *H. zea* and *B. mori* have been deduced from cDNA isolated from pheromone glands of *Mamestra brassicae* (Jacquin-Joly and Descoins, 1996), *H. assulta* (Choi *et al.*, 1998), *Agrotis ipsilon* (Duportets *et al.*, 1999), *Spodoptera littoralis* (Iglesias *et al.*, 2002), *H. armigera* (Zhang *et al.*, 2001), and *Adoxophyes* sp. (Lee *et al.*, 2001). The post-translational processed peptides can be found in the SEG (Ma *et al.*, 1996; Sato *et al.*, 1993). The MALDI MS data indicated that PBAN was found to a greater extent in the mandibular and maxillary clusters than in the labial cluster (Ma *et al.*, 2000). The other neuropeptides were found in all clusters. In addition some larger peptide fragments were found indicating alternative processing of the precursor protein (Ma *et al.*, 2000). A study using *M. sexta* indicates that PBAN processing occurs in all three cell clusters of the SEG similar to *H. zea;* but in addition the *capa* gene, which produces a WFGPRLamide peptide, is also active in the labial cell cluster (Neupert *et al.*, 2009). The *capa* gene produces two periviscerokinin peptides in addition to the pyrokinin and is primarily expressed in the abdominal ganglia (Neupert *et al.*, 2009).

12.5.2.3. PBAN mode of action Through the development of a sensitive *in vitro* bioassay, studies on *H. armigera* and *H. zea* demonstrated that brain extracts and synthetic *H. zea* PBAN could stimulate the production of the main pheromone component (Soroker and Rafaeli, 1989; Rafaeli *et al.*, 1990, 1991, 1993; Rafaeli, 1994; Rafaeli and Gileadi, 1996). The response obtained was specific to the pheromone gland and independent of other abdominal tissues (Rafaeli, 1994; Rafaeli *et al.*, 1997b). Pharmacological evidence indicated that PBAN activates a G-protein-coupled receptor (GPCR) located in the cell membrane of pheromone glands (Rafaeli and Gileadi, 1996). Specific binding to a membrane receptor was demonstrated using a photoaffinity-labeled PBAN (Rafaeli and Gileadi, 1999; Rafaeli *et al.*, 2003). The first PBAN-receptor was cloned and sequenced from the moth, *H. zea* (Choi *et al.*, 2003). This was possible due to the sequencing of the *D. melanogaster* genome and annotation of the peptide GPCRs (Hewes and Taghert, 2001). Subsequently Park *et al.* (2002) functionally expressed several receptors that were similar to neuromedin U receptors of vertebrates and determined that two were activated by pyrokinin peptides. With the sequencing of additional insect genomes it is apparent that the PBAN/pyrokinin receptors are found in all insects based on sequence homology (Jurenka and Nusawardani, 2011). Within the Lepidoptera several receptors have been sequenced and functionally expressed in addition to the PBAN-receptor from *H. zea*. These

include the PBAN- and diapause hormone-receptors from *B. mori* (Hull *et al.*, 2004; Homma *et al.*, 2006) and the PBAN-receptor from *H. virescens* (Kim *et al.*, 2008). These receptors have high sequence homology in the transmembrane domains but little homology occurs in the N- and C-terminal domains (Stern *et al.*, 2007). The extracellular domains are important for ligand recognition as indicated by a study in which the extracellular loops were swapped between the *H. zea* PBAN-receptor and the pyrokinin receptor from *D. melanogaster* (Choi *et al.*, 2007a). Additional studies have shown that the PBAN-receptor is located in various tissues indicating the nature of the ligand (Rafaeli *et al.*, 2003, 2007). Based on sequence homology there appear to be two groups of receptors in insects: the PBAN or pyrokinin 2 receptors and the diapause hormone or pyrokinin 1 receptors (Jurenka and Nusawardani, 2011).

The signal transduction events that occur after PBAN binds to a receptor have been studied in several model moth species (**Figure 14**). The main difference found between these species so far is whether or not $3',5'$,cyclic-AMP (cAMP) is used as a second messenger. In the heliothines and several other moth species, cAMP is a second messenger. On the other hand, cAMP is thought not to act in pheromone gland cells of *B. mori* and *O. nubilalis* (Fónagy *et al.*, 1992; Ma and Roelofs, 1995a). Instead, in these insects, it is thought that an increase in cytosolic calcium directly activates downstream events leading to stimulation of the biosynthetic pathway.

Extracellular calcium is essential for pheromonotropic activity in all moth species studied to date (Fónagy *et al.*, 1992, 1999; Ma and Roelofs, 1995a; Matsumoto *et al.*, 1995). The influx of extracellular calcium is most likely mediated by store-operated calcium channels as demonstrated in *B. mori* (Hull *et al.*, 2007). Molecular and biochemical characterization of two proteins, STIM1 and Orai1, implicate them as essential components of the calcium channel (Hull *et al.*, 2009). The activation of calcium channels is most likely mediated by receptor-activated phospholipase C and the G protein α subunit Gq in pheromone glands of *B. mori* (Hull *et al.*, 2010). It is suggested that once calcium enters the cell it binds to calmodulin to form a complex activating adenylate cyclase and/or phosphoprotein phosphatases (**Figure 14**). Calmodulin was characterized from pheromone glands of *B. mori* (Iwanaga *et al.*, 1998) and was shown to have an identical amino acid sequence to *Drosophila* calmodulin (Smith *et al.*, 1987). In *B. mori*, it is suggested that the Ca^{2+}/calmodulin complex directly or indirectly activates a phosphoprotein phosphatase (Matsumoto *et al.*, 1995). This phosphatase will then activate an acyl-CoA reductase in the biosynthetic pathway. Two genes encoding calcineurin heterosubunits were identified from the pheromone gland of *B. mori* and were found to be homologous to the catalytic subunit and regulatory subunits of other animal calcineurins (Yoshiga *et al.*,

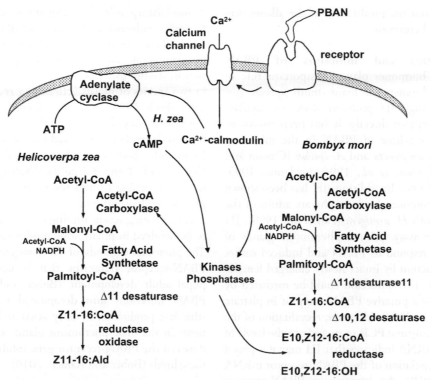

Figure 14 Proposed signal transduction mechanisms that stimulate the pheromone biosynthetic pathway in *Helicoverpa zea* and other heliothines as compared with that in *Bombyx mori*. It is proposed that PBAN binds to a receptor present in the cell membrane. Binding to the receptor somehow induces a receptor-activated calcium channel to open causing an influx of extracellular calcium. This calcium binds to calmodulin and in the case of *B. mori* will directly stimulate a phosphatase that will dephosphorylate and activate a reductase in the biosynthetic pathway. This activated reductase will then produce the pheromone bombykol. In *H. zea* and other heliothines like *Helicoverpa armigera*, the calcium-calmodulin will activate adenylate cyclase to produce cAMP that will then act through kinases and/or phosphatases to stimulate acetyl-CoA carboxylase in the biosynthetic pathway.

2002). The calcineurin complex will apparently dephosphorylate an acyl-CoA reductase, which catalyzes the formation of bombykol in *B. mori*.

12.5.2.4. Enzymes affected in the pheromone biosynthetic pathway

PBAN has been shown to stimulate the reductase that converts an acyl-CoA to an alcohol precursor (**Figure 14**) in several moths including *B. mori* (Arima *et al.*, 1991; Ozawa *et al.*, 1993), *Thaumetopoea pityocampa* (Gosalbo *et al.*, 1994), *S. littoralis* (Fabriàs *et al.*, 1994; Martinez *et al.*, 1990), and *M. sexta* (Fang *et al.*, 1995b; Tumlinson *et al.*, 1997). In *A. velutinana* (Tang *et al.*, 1989), *H. zea* (Jurenka *et al.*, 1991b), *Cadra cautella*, *S. exigua* (Jurenka, 1997), and *M. brassicae* (Jacquin *et al.*, 1994), it was demonstrated that PBAN controls pheromone biosynthesis by regulating a step during or prior to fatty acid biosynthesis (**Figure 14**). Circumstantial evidence in *A. segetum* (Zhu *et al.*, 1995) and *H. armigera* (Rafaeli *et al.*, 1990) also points to the regulation of fatty acid synthesis by PBAN. In one study using the moth *Sesamia nonagrioides* it was shown that the acetyltransferase enzyme might be regulated by PBAN (Mas *et al.*, 2000). There appears to be no particular pattern

as to which enzyme within the pheromone biosynthetic pathway will be regulated by PBAN. However, in the majority of moths studied it is either the reductase or fatty acid synthesis that is stimulated.

Several families of moths utilize hydrocarbons and/or their epoxides as sex pheromones (Millar, 2000, 2010). It is thought that PBAN does not regulate the production of hydrocarbon sex pheromones as demonstrated in *Scoliopteryx libatrix* (Subchev and Jurenka, 2001), *A. s. cretacea* (Wei *et al.*, 2004), *Utetheisa ornatrix* (Choi *et al.*, 2007b), and *Amyelois transitella* (Wang *et al.*, 2010b). However, PBAN is probably regulating the production of epoxide sex pheromones. This was demonstrated in *A. s. cretacea* where decapitation resulted in pheromone decline that could be restored by injecting PBAN (Miyamoto *et al.*, 1999). Decapitation also decreases the epoxide pheromone titer in the gypsy moth, *L. dispar*, and injection of PBAN restored pheromone production (Thyagaraja and Raina, 1994). However, decapitation did not decrease the levels of the hydrocarbon precursor in the gypsy moth (Jurenka, unpublished). These findings indicate that PBAN may regulate the epoxidation step in those moths that utilize epoxide pheromones but not the production of the alkene precursor or alkene pheromones.

12.5.2.5. Mediators and inhibitors of PBAN action

Juvenile hormones play an important role in reproductive development of many moth species (see Chapter 8). Although JH probably does not regulate pheromone biosynthesis directly, it has been shown to be involved in the release of PBAN in the migratory moths *Pseudaletia unipuncta* and *A. ipsilon* (Cusson and McNeil, 1989; Cusson *et al.*, 1994; Gadenne, 1993; Picimbon *et al.*, 1995). In addition, JH has been shown to prime the pheromone glands in pharate adults of the non-migratory moth *H. armigera* (Fan *et al.*, 1999). JH II, in an *in vitro* assay, primed pheromone glands of pharate adults to respond to PBAN and induced earlier pheromone production by intact newly emerged females (Fan *et al.*, 1999). This induction could be mediated by JH upregulation of a putative PBAN-receptor in pharate adults (Rafaeli *et al.*, 2003). However, reevaluation of the role of JH using real-time PCR to quantitate the levels of PBAN-receptor mRNA indicates that JH may not play a direct role in upregulation of the PBAN-receptor mRNA (Bober *et al.*, 2010). An increase in PBAN-receptor transcripts occurs in a short time frame — from 3 to 6 h post adult eclosion — and this increase was not affected by removal of the source of JH, but addition of JH decreased the transcript levels (Bober *et al.*, 2010).

The corpus bursae have been implicated in the mediation of PBAN stimulation of pheromone biosynthesis in some tortricids (Jurenka *et al.*, 1991c). A peptide was partially purified from corpus bursae that could stimulate pheromone production (Fabriàs *et al.*, 1992). A bursal factor has only been demonstrated in *A. velutinana* and the related tortricids *C. fumiferana* and *C. rosaceana* (Delisle *et al.*, 1999).

The role of the nervous system in pheromone biosynthesis in moths is not clearly understood. In several moths including *L. dispar* (Tang *et al.*, 1987; Thyagaraja and Raina, 1994), *H. virescens* (Christensen *et al.*, 1991), *S. littoralis* (Marco *et al.*, 1996), and *M. brassicae* (Iglesias *et al.*, 1998) an intact VNC was reported as necessary for pheromone biosynthesis. Christensen and co-workers (1991, 1992, 1994, 1995) proposed that the neurotransmitter octopamine might be involved as an intermediate messenger during the stimulation of sex pheromone production in *H. virescens*. These workers suggested that octopamine was involved in the regulation of pheromone production and that PBAN's role lies in the stimulation of octopamine release at nerve endings. However, contradicting results concerning VNC transection and octopamine-stimulated pheromone production were reported in the same species as well as in other moth species (Delisle *et al.*, 1999; Jurenka *et al.*, 1991c; Park and Ramaswamy, 1998; Rafaeli and Gileadi, 1996; Ramaswamy *et al.*, 1995).

A modulatory role for octopamine was suggested by research conducted on *H. armigera* (Rafaeli and Gileadi, 1995; Rafaeli *et al.*, 1997b). Octopamine and several octopaminergic analogs inhibited pheromone production in studies using both *in vitro* and *in vivo* bioassays in two species of moths (Hirashima *et al.*, 2001; Rafaeli and Gileadi, 1995, 1996; Rafaeli *et al.*, 1997b). The role of the VNC in pheromone production still requires clarification. It has been suggested that in some moth species (*S. littoralis*) both humoral and neural regulation occurs (Marco *et al.*, 1996). Given the diversity of moths it may not be surprising to find several mechanisms regulating pheromone biosynthesis.

A fascinating recent finding is that PBAN also appears to be involved in pheromone production in male moths. The pheromone glands of *H. armigera* males contain a PBAN-receptor transcript that is upregulated during pupal–adult development (Bober and Rafaeli, 2010). PBAN injections into decapitated males significantly stimulate production of fatty acids and alcohol components in the male pheromone gland, and RNAi knockdown of the PBAN-receptor gene inhibits production of these lipids (Bober and Rafaeli, 2010).

12.5.3. Juvenile Hormone Regulation in Beetles

12.5.3.1. General

In the Coleoptera, pheromone production or release is controlled by JH III (Tillman *et al.*, 1999). JH, or JH analogs, stimulate pheromone production in *T. molitor* (Menon, 1970), *Tr. castaneum* (both Tenebrionidae), some members of the Cucujidae (Plarre and Vanderwel, 1999), *A. grandis* (Curculionidae; Wiygul *et al.*, 1990), and various Scolytidae (see next section). Various factors may work upstream as cues to stimulate JH biosynthesis. Feeding stimulates pheromone production or release in many species (Vanderwel and Oehlschlager, 1987). Sexual maturity and population density may also be important, such as for some Scolytidae (Byers, 1983) and Cucujidae (Plarre and Vanderwel, 1999). The effect of population density may be mediated by sensitivity to pheromone concentrations. Antennectomy can raise pheromone biosynthetic rates in *A. grandis* (Dickens *et al.*, 1988), *Leptinotarsa decemlineata* (Dickens *et al.*, 2002), and *I. pini* (Ginzel *et al.*, 2007), suggesting that detection of pheromone components may inhibit their production. Various combinations of physiological and environmental factors therefore regulate JH titers, which in turn stimulate pheromone biosynthesis and/or release.

12.5.3.2. Endocrine regulation in bark beetles

Aggregation pheromone biosynthesis in bark beetles generally begins shortly after the beetle arrives at a new host tree (Wood, 1982). Unfed beetles can be artificially stimulated to produce pheromones by treatment with

JH III. For example, males are the pioneers of *Ips* spp., and starved male *I. paraconfusus* and *I. pini* can be induced to synthesize the pheromone components ipsenol and ipsdienol if JH III or a JH analog is applied topically (Borden, 1969; Chen *et al.*, 1988; Ivarsson and Birgersson, 1995; Tillman *et al.*, 1998). Similarly, starved *Dendroctonus* spp. can be induced to synthesize pheromone components following treatment with JH (Conn *et al.*, 1984; Bridges, 1982). Feeding apparently stimulates the synthesis of JH III in the CA (Tillman *et al.*, 1998), resulting in elevated JH titers that trigger pheromone biosynthesis in midgut cells (Hall *et al.*, 2002a,b).

Pheromone biosynthesis requires a shift in metabolic priorities, particularly for those beetles that produce large quantities of pheromone. JH must actuate this shift, and since pheromone components are synthesized *de novo* (Seybold *et al.*, 1995b), induction of the pheromone biosynthetic pathway involves elevated expression of at least some related genes. In the case of male *I. pini*, which can produce very large amounts of ipsdienol, there is coordinate induction of mevalonate pathway and pheromone-biosynthetic genes, with both effectively opening the entries and widening the pathway down which carbon must flow (Keeling *et al.*, 2004, 2006; Sandstrom *et al.*, 2006; Gilg *et al.*, 2005). Similar induction of known pheromone-biosynthetic genes has been observed in *I. paraconfusus* (Ivarsson *et al.*, 1997; Tittiger *et al.*, 1999), *I. confusus* (Bearfield *et al.*, 2009; Tillman *et al.*, 2004), *D. jeffreyi* (Tittiger *et al.*, 1999, 2003), and *D. ponderosae* (Barkawi *et al.*, 2003).

In addition to the known role for JH III to stimulate pheromone production, other regulatory schemes are clearly present. Basal expression levels of *I. pini* pheromone-biosynthetic genes are significantly higher in males (the pheromone-producing sex) than females, and *GPPS/MS* is not induced by JH III in females as it is in males. Both observations indicate developmental and sex-specific influences (Keeling *et al.*, 2006). Similarly, teneral male *D. jeffreyi* HMGR mRNA levels are not induced by JH III to the same extent as those in fully mature insects (Barkawi, 2002), suggesting developmental regulation. An antennally mediated negative-feedback regulation in *I. pini* suggests possible neural influences (Ginzel *et al.*, 2007) similar to that observed in *L. decemlineata* (Dickens *et al.*, 2002). Mevalonate pathway genes show a pulse of mRNA induction in fed male *I. pini*, whereas unfed, JH-III-treated males show a steady increase in mRNA levels over time (**Figure 11**; Keeling *et al.*, 2004, 2006). Perhaps most striking is the separation of regulation into transcriptional and post-transcriptional steps as observed in *I. confusus* and *I. paraconfusus*. In these insects, JH III treatment of unfed males elevates mRNA of some mevalonate pathway genes but does not stimulate enzymatic activity of their corresponding products or pheromone

biosynthesis (Bearfield, 2004; Tillman *et al.*, 2004). Instead, a secondary factor, hypothesized to be a peptide hormone, produced upon feeding, appears necessary for enzyme activity and concomitant pheromone production (Bearfield *et al.*, 2009). Finally, as noted earlier, some pheromone components are not regulated by JH III, and not all are produced in the midgut (e.g., *exo*-brevicomin). These examples illustrate that regulation of pheromone production is complex and diverse, even within a given genus.

12.5.4. 20-Hydroxyecdysone Regulation of Pheromone Production in the Housefly

12.5.4.1. Ecdysteroid regulation Newly emerged female houseflies do not have detectable amounts of any C23 sex pheromone components. Sex pheromone production correlates with ovarian development and vitellogenesis. The C23 sex pheromone components first appear when ovaries mature to the early vitellogenic stages and increase in amount until stages 9 and 10 (mature egg; Dillwith *et al.*, 1983; Mpuru *et al.*, 2001). Females ovariectomized within 6 h of adult emergence do not produce any of the C23 sex pheromone components, whereas control and allatectomized females produced abundant amounts of (*Z*)-9-tricosene. Ovariectomized insects that received ovary implants produced sex pheromone components in direct proportion to ovarian maturation. These data demonstrate that a hormone from the maturing ovary induced sex pheromone production (Blomquist *et al.*, 1993, 1998).

Juvenile hormone regulates both vitellogenesis and pheromone production in some insect species (Tillman *et al.*, 1999). In some Diptera, including the housefly, ovarian-produced ecdysteroids are involved in regulating vitellogenesis (Hagedorn, 1985; Adams *et al.*, 1997) at the transcriptional level (Martin *et al.*, 2001). Therefore, since ovariectomy abolished sex pheromone production and allatectomy (which abolishes JH production) had no effect on pheromone production (Blomquist *et al.*, 1992), it was hypothesized that an ecdysteroid, and not JH, regulated sex pheromone production in the housefly. Injection of 20E at doses as low as 0.5 ng every 6 h induced sex pheromone production in ovariectomized houseflies in a time- and dose-dependent manner (Adams *et al.*, 1984a, b, 1995). Multiple injections of 20E into ovariectomized insects over several days resulted in as much 23:1 produced as in intact control females. Application of JH or JH analogs, alone or in combination with ecdysteroids, had no effect on pheromone production (Blomquist *et al.*, 1992).

12.5.4.2. Ecdysteroids affect fatty acyl-CoA elongation enzymes There are two likely possibilities to account for the change in the chain length of the alkenes synthesized by the female housefly in the production of

(Z)-9 tricosene. They are (1) the chain length specificity of the reductive conversion of acyl CoAs to alkenes is altered such that 24:1-CoA becomes an efficient substrate, or (2) there is a change in the chain length specificity of the fatty acyl-CoA elongation enzymes such that 24:1-CoA is not efficiently elongated, resulting in an accumulation of 24:1-CoA. To determine which enzyme activities are affected by 20E to regulate the chain length of the alkenes, experiments were performed to examine the chain length specificity of the fatty acyl-CoA reductive conversion of acyl-CoAs to alkenes and elongation. Microsomal preparations from both males and females of all ages examined readily converted 24:1-CoA and the 24:1 aldehyde to (Z)-9-tricosene, indicating that 20-HE was not acting on this activity (Tillman-Wall et al., 1992; Reed et al., 1995). In contrast, microsomes from day 4 females (high ecdysteroid titer and production of (Z)-9 tricosene) did not elongate either 18:1-CoA or 24:1-CoA beyond 24 carbons, while microsomes from day 4 males or day 1 females (both of which produce alkenes of 27:1 and longer) readily elongated both 18:1-CoA and 24:1-CoA to 28:1-CoA (Tillman-Wall et al., 1992; Blomquist et al., 1995). Thus, 20-HE appears to regulate the fatty acyl-CoA elongases and not the enzymatic steps in the conversion of acyl-CoA to hydrocarbon.

12.5.4.3. Transport of pheromone

The role of hemolymph in transporting hydrocarbons and hydrocarbon pheromones has only recently become fully appreciated. Older models of hydrocarbon formation showed epidermal related cells (oenocytes) synthesizing and transporting hydrocarbons directly to the surface of the insect (Hadley, 1984). In the housefly, the role of hemolymph is most clearly seen when (Z)-9-tricosene production is initiated. (Z)-9-Tricosene first accumulates in the hemolymph, and then after a number of hours, is observed on the surface of the insect. Modeling of the process (Mpuru et al., 2001) showed that the delay is surprisingly long, more than 24 h are necessary for transport from site of synthesis to deposition on the surface of the insect.

In sexually mature females, (Z)-9-tricosene comprised a relatively large fraction of the hydrocarbon of the epicuticle and the hemolymph, but much smaller percentages of the hydrocarbons in other tissues, including the ovaries. It appears that certain hydrocarbons were selectively partitioned to certain tissues such as the ovaries, from which pheromone was relatively excluded (Schal et al., 2001). Both KBr gradient ultracentrifugation and specific immunoprecipitation showed that over 90% of the hemolymph hydrocarbon was associated with a high-density lipophorin. Lipophorin was composed of two aproproteins under denaturing conditions: apolipophorin I ($\approx$240 kDa) and apolipophorin II ($\approx$85 kDa) (Schal et al., 2001). These data suggest that lipophorin may play an important role

in an active mechanism that selectively delivers specific hydrocarbons to specific sites. A similar mechanism has been proposed in other dipterans, namely D. melanogaster, in which radio-tracing showed that as the internal pool of hydrocarbons (including diene pheromones) decreased, more labeled hydrocarbons appeared on the cuticle (Pho et al., 1996).

12.5.5. Regulation of Pheromone Production in Cockroaches

12.5.5.1. Development and cellular plasticity of pheromone glands In cockroaches, in striking contrast with many moths, pheromone glands acquire functional competence during an imaginal maturation period, and developmental regulation involves factors that also control adult reproductive readiness. Also, because reproduction in cockroaches is interrupted by periods of sexual inactivity (i.e., gestation), developmental regulation of the sex pheromone gland can result in alternating cycles of acquisition and subsequent waning of competence through maturation and retrogression, respectively, of cellular machinery. Consequently in female cockroaches, pheromone production is controlled by cyclic maturational changes in the gland in relation to the ovarian cycle.

Best exemplifying this phenomenon are the tergal and sternal glands of N. cinerea and B. germanica males. Both species possess class 3 glandular units, composed of two cells — a secretory cell and a duct cell (Quennedey, 1998). But after apolysis and before the imaginal molt the immature gland contains four concentric cells, including in addition to the two adult cells an enveloping cell and a ciliary cell (Sreng, 1998; Sreng and Quennedey, 1976). During several days after the adult molt the gland matures by undergoing apoptosis (programmed cell death). The ciliary cell gives rise to a part of the microvillar end apparatus, then dies, whereas the enveloping cell forms an upper portion of the duct, then it too dies (Sreng, 1998). In concert, before day 5, the immature sternal glands of N. cinerea males produce little pheromone, but after day 5 their pheromone content increases significantly (Sreng et al., 1999). Decapitation or allatectomy of N. cinerea males completely blocked the apoptotic process, while JH III treatment restored apoptosis (Sreng et al., 1999). Brain extracts or synthetic moth PBAN failed to restore gland differentiation or stimulate pheromone production.

Female B. germanica employ similar class 3 glands to produce blattellaquinone, a volatile sex pheromone (Nojima et al., 2005). Ultrastructural, behavioral, and electrophysiological studies have shown that, as in males, pheromone gland cells mature as the female sexually matures (Abed et al., 1993b; Liang and Schal, 1993; Tokro et al., 1993; Schal et al., 1996). The secretory cells of newly formed glands in the imaginal female are small and they contain little pheromone (determined with behavioral and EAG assays). As the female sexually matures, the size of pheromone-secreting cells increases, as does its pheromone content (Liang and Schal, 1993). The mature pheromone gland then undergoes cycles of cellular hypertrophy and retrogression in relation to the JH III titer in successive reproductive cycles. The gland becomes atrophied and its pheromone content declines during gestation, but as a new vitellogenic cycle begins after the egg case is deposited, the pheromone gland undergoes rapid re-growth and proliferation of cellular organelles and an increase in its pheromone content. Although this pattern corresponds well with the JH III titer in the hemolymph (Liang and Schal, 1993; Schal et al., 1996), no experimental manipulations of hormone titers have been conducted to verify the hypothesis that JH III controls the cellular plasticity of the pheromone gland.

12.5.5.2. Pheromone production regulated by juvenile hormone

Barth and Lester (1973) and Schal and Smith (1990) reviewed the early literature on hormone involvement in pheromone production in cockroaches. With the exception of *Nauphoeta* (detailed in the previous section), no studies on the regulation of volatile pheromone production are available that use analytical or biochemical approaches. The most detailed studies, with *B. germanica* and *S. longipalpa*, have employed behavioral and EAG responses of males to estimate the relative amount of pheromone in females or their pheromone glands. In both species, virgin females initiate pheromone production four days after the imaginal molt, in relation to increasing titers of JH III (Smith and Schal, 1990a; Liang and Schal, 1993). Ablation of the CA of newly emerged adult females prevents pheromone production in both species, and pheromone production is restored after reimplantation of active CA or by treatment with JH III or JH analogs. Interestingly, although growth of the vitellogenic oocytes is controlled by and highly correlated with JH III titers, direct or even intermediary involvement of the ovaries in regulating pheromone production and calling behavior in both species was excluded by ovariectomies (Smith and Schal, 1990a).

It is not known whether JH exerts its pheromonotropic effects directly on mature secretory cells of the pheromone gland, or if it acts indirectly by stimulating the synthesis and/or release of pheromonotropic neuropeptides. Although cockroach brain extracts induce PBAN-like pheromonotropic activity in moth pheromone glands (Raina et al., 1989), they fail to do so in allatectomized *Nauphoeta* males (Sreng et al., 1999) or *Supella* females (Schal, unpublished results). Moreover, lack of pheromone production in mated females that periodically produce large amounts of JH III suggests that JH plays a "permissive" role (Smith and Schal, 1990b; Schal et al., 1996), that is, its presence is *required* for pheromone to

be produced. Even when the JH III titer is high, pheromone production can be suppressed by neural or humoral pheromonostatic factors.

Blattella germanica has served as a useful model for delineating endocrine regulation of non-volatile cuticular pheromones, namely (3S,11S)-dimethylnonacosan-2-one. Both the amount of pheromone on the cuticular surface and *in vivo* incorporation of radiolabel from [1-^{14}C] propionate into the sex pheromone coincide with active stages of vitellogenesis, suggesting the involvement of JH (Schal et al., 1990a, 1994; Sevala et al., 1999). Indeed, females treated to reduce their JH III titer (allatectomy, anti-allatal drugs such as precocene, starvation, or implantation of an artificial egg case into the genital vestibulum, which inhibits JH biosynthesis) produce less pheromone (Schal et al., 1990a, 1994; Chase et al., 1992). Furthermore, pheromone production is greatly stimulated by treatments with JH III or with JH analogs. Because only the hydroxylation of 3,11-dimethylnonacosane to 3,11-dimethylnonacosan-2-ol is regulated in a sex-specific manner, it appears that this step is under JH III control (**Figure 12**; Chase et al., 1992).

Normally, adult male cockroaches have a much lower titer of JH III in the hemolymph (Piulachs et al., 1992; Wyatt and Davey, 1996). Because males also produce 3,11-dimethylnonacosane, metabolism of the alkane to contact pheromone may be contingent upon high JH titers in the adult female. In response to exposure to the JH analog hydroprene, female pheromone increased sixfold in treated males (Schal, 1988), showing some capacity to express the putative female-specific polysubstrate monooxygenase. The parallels are striking with estrogen induction of vitellogenin synthesis in the male liver of oviparous vertebrates, JH induction of vitellogenin synthesis in male cockroaches (e.g., Mundall et al., 1983), and ecdysteroid induction of female pheromone production in houseflies (see Section 12.5.4.1.; Blomquist et al., 1984b, 1987).

Contact pheromone production in *B. germanica* is also regulated through the regulated production of its precursor, 3,11-dimethylnonacosane. Biosynthesis of this alkane drops dramatically when food intake declines at the end of each vitellogenic phase (Schal et al., 1994, 1996), suggesting that hydrocarbon biosynthesis is linked to food intake, as in nymphs (Young et al., 1999), and not directly to either the ecdysteroid or JH titers. Dietary intake also stimulates the production of JH III (Schal et al., 1993; Osorio et al., 1998), which in turn stimulates the conversion of the alkane to contact pheromone. In allatectomized females large amounts of hydrocarbons accumulate in the hemolymph because food intake is not suppressed (i.e., no gestation) and hydrocarbons are not provisioned into oocytes (i.e., no vitellogenesis; Schal et al., 1994; Fan et al., 2002). As hydrocarbons accumulate in the hemolymph, the amount of cuticular pheromone also increases, suggesting that excess 3,11-dimethylnonaco-sane is metabolized to pheromone. These patterns suggest that under normal conditions, feeding in adult females is modulated in a stage-specific manner, regulating the amount of 3,11-dimethylnonacosane that is available for JH-mediated metabolism of 3,11-dimethylnonacosane to 3,11-dimethylnonacosan-2-one.

12.5.5.3. Transport and emission of pheromones

Little is known of the cellular processes that deliver volatile pheromones from secretory cells to the cuticular surface, even in the intensively researched Lepidoptera. In cockroaches, electron microscopy studies often show accumulation of secretion in the end apparatus, ducts, and around the cuticular pores of class 3 exocrine glands in both males and females. Studies in *L. maderae* have identified and sequenced an epicuticular protein, Lma-p54, which is expressed specifically in the tergites and sternites of adult males and females, but not in nymphs (Cornette *et al.*, 2002). The sequence of this protein is closely related to aspartic proteases, but because it appears to be enzymatically inactive the authors speculate that it serves as a ligand binding protein. Cornette *et al.* (2002) further hypothesize that Lma-p54, alone or together with a ligand, serves in sexual recognition. Other ligand-binding proteins, namely the lipocalins Lma-p22 and Lma-p18, have been isolated only from male tergal secretions of *L. maderae* (Cornette *et al.*, 2001). These exciting findings suggest that carrier proteins might be involved in transport of volatile pheromones to the cuticle, but functional studies will be needed to verify this hypothesis.

Attractant sex pheromones are usually emitted while the female or male cockroach performs a species-specific calling behavior (Gemeno and Schal, 2004; Gemeno *et al.*, 2003), as in most lepidopterans. JH III regulates calling behavior in both *B. germanica* and *S. longipalpa*. In both species, transection of the nerves connecting the CA to the brain, an operation that significantly accelerates the rate of JH III biosynthesis by the CA (Schal *et al.*, 1993), also hastens the age when calling first occurs (Smith and Schal, 1990a; Liang and Schal, 1994). In *B. germanica*, the central role of JH has been confirmed by ablation of the CA and with rescue experiments with a JH analog (Liang and Schal, 1994). In this species, JH is also required for females to become sexually receptive and accept courting males (Schal and Chiang, 1995).

In *B. germanica*, the transfer of contact pheromone components to the epicuticular surface is mediated by lipophorin, a high-density hemolymph lipoprotein (Schal *et al.*, 1998b, 2003). As in *M. domestica* (see Section 12.5.4.3.), newly biosynthesized pheromone appears first in the hemolymph before it turns up on the epicuticle (Gu *et al.*, 1995). That hemolymph is required to transport the pheromone to the cuticular surface was demonstrated by severing the veins that enter the forewings; the amount of

hydrocarbons and pheromone that appeared on the wings was significantly lower than on the intact forewings of the same insects.

In the cockroaches *P. americana* and *B. germanica* virtually all newly synthesized hydrocarbons that enter the hemolymph are bound to lipophorin (Chino, 1985; Gu *et al.*, 1995). Moreover, in both species newly synthesized hydrocarbons can only be transferred from the integument to an incubation medium if lipophorin is present, and other hemolymph lipoproteins, such as vitellogenin, cannot mediate this transfer (Katase and Chino, 1982, 1984; Fan *et al.*, 2002). The mechanisms by which hydrocarbons and pheromones are taken up by lipophorin are poorly understood. Takeuchi and Chino (1993) clearly demonstrated in the American cockroach that a very high density lipid transfer particle (LTP) catalyzes the transfer hydrocarbons between lipophorin particles. However, *in vitro* experiments with purified lipophorin of *P. americana* and *B. germanica* showed that lipophorin accepts hydrocarbons from oenocytes, apparently without involvement of LTP (Katase and Chino, 1982, 1984; Fan *et al.*, 2002). It remains to be determined whether this is because sufficient LTP remains bound to dissected tissues, if LTP is produced by the dissected tissues, or whether it plays no significant role in hydrocarbon and pheromone uptake by lipophorin.

Interestingly, while the uptake of hydrocarbons and pheromones by lipophorin *in vitro* appears to lack molecular specificity (Katase and Chino, 1984; Schal, unpublished results), their delivery to pheromone gland cells is highly specific (Schal *et al.*, 1998a,b; Matsuoka *et al.*, 2006). Uptake of lipophorin and its ligands might involve receptor-mediated endocytosis, as demonstrated in mosquito oocytes (Cheon *et al.*, 2001). This might explain why 3,11-dimethylnonacosan-2-one has been isolated from mature ovaries of the German cockroach (Gu *et al.*, 1995) and (Z)-9-tricosene is found in housefly ovaries (Schal *et al.*, 2001). However, an endocytic process would fail to discriminate various ligands, suggesting that alternative mechanisms need to be investigated.

Recent developments in functional genomic approaches hold promise for delineating lipophorin-hydrocarbon interactions. Lipophorin genes and lipophorin receptors have been described in a number of insects, and their interactions with dietary polar lipids have been extensively investigated. Different isoforms of lipophorin receptors have been described from the fat body and the ovary (e.g., in *B. germanica*; Ciudad *et al.*, 2007), but epidermal cells associated with pheromone glands or the cuticle have not yet been examined. Ciudad *et al.* (2007) showed that RNAi treatments that lower lipophorin receptor expression also lower lipophorin uptake by *B. germanica* oocytes. Since lipophorin delivers hydrocarbons to the oocytes and cuticle in this cockroach (Gu *et al.*, 1995; Fan *et al.*, 2002, 2008), it will be exciting to know whether RNAi treatment interferes with hydrocarbon deposition as well.

12.6. Concluding Remarks and Future Directions

The increase in our understanding of the biochemistry and regulation of pheromone production over the last three decades is nothing short of phenomenal. A 1983 review of the biochemistry and endocrine regulation of insect pheromone production (Blomquist and Dillwith, 1983) was 16 pages long. It was limited to early work on pheromone biosynthesis in moths and the housefly and recognized that JH and ecdysteroids may play a role in the regulation of pheromone production. Since that time the discovery of PBAN and its role in the regulation of lepidopteran pheromone production has been elucidated, along with determining which enzymes are affected by JH and ecdysteroids to regulate pheromone production in model cockroaches/beetles and flies, respectively. The work in Lepidoptera has moved from simply demonstrating that pheromone components were synthesized *de novo* to the molecular characterization of unique Δ11 desaturase and other desaturases that are involved in many female moths and their interplay with specific chain-shortening steps. While it is still true that in no system do we have a complete understanding of both the biochemical pathways and their endocrine regulation, we do have a much better understanding of how pheromones are made and in some systems are developing an understanding of their regulation at the molecular level. The continued application of the powerful tools of molecular biology along with studies using genomics and proteomics will only increase the rate at which we increase our understanding of pheromone production. Ultimately, just as behavioral chemicals have been extended into pest control, research on pheromone production will be directed toward practical applications in insect control.

References

Abed, D., Cheviet, P., Farine, J. P., Bonnard, O., Le Quéré, J. L., & Brossut, R. (1993a). Calling behaviour of female *Periplaneta americana*: behavioural analysis and identification of the pheromone source. *Journal of Insect Physiology, 39,* 709–720.

Abed, D., Tokro, P., Farine, J.-P., & Brossut, R. (1993b). Pheromones in *Blattella germanica* and *Blaberus craniifer* (Blaberoidea): Glandular source, morphology and analyses of pheromonally released behaviours. *Chemoecology, 4,* 46–54.

Adams, T. S., Holt, G. G., & Blomquist, G. J. (1984a). Endocrine control of pheromone biosynthesis and mating behavior in the housefly, *Musca domestica*. In W. Engels (Ed.), *Advances in Invertebrate Reproduction 3* (pp. 441–456). Amsterdam: Elsevier.

Adams, T. S., Dillwith, J. W., & Blomquist, G. J. (1984b). The role of 20-hydroxyecdysone in housefly sex pheromone biosynthesis. *Journal of Insect Physiology, 30,* 287–294.

Adams, T. S., Nelson, D. R., & Blomquist, G. J. (1995). Effect of endocrine organs and hormones on (*Z*)-9-tricosene levels in the internal and external lipids of female house flies, *Musca domestica*. *Journal of Insect Physiology, 41,* 609–615.

Adams, T. S., Gerst, J. W., & Masler, E. P. (1997). Regulation of ovarian ecdysteroid production in the housefly, *Musca domestica*. *Archives of Insect Biochemistry and Physiology, 35,* 135–148.

Ahmad, S., Kirkland, K. E., & Blomquist, G. J. (1987). Evidence for a sex pheromone metabolizing cytochrome P-450 monooxygenase in the housefly. *Archives of Insect Biochemistry and Physiology, 6,* 121–140.

Ando, T., Hase, T., Arima, R., & Uchiyama, M. (1988). Biosynthetic pathway of bombykol, the sex pheromone of the female silkworm moth. *Agricultural Biological Chemistry, 52,* 473–478.

Ando, T., Inomata, S., & Yamamoto, M. (2004). Lepidopteran Sex Pheromones. In S. Schulz (Ed.), *Topics in Current Chemistry Vol. 239.* (pp. 51–96). Heidelberg: Springer-Verlag.

Antony, C., & Jallon J.-M. (1982). The chemical basis for sex recognition in *Drosophila melanogaster*. *Journal of Insect Physiology, 28,* 873–880.

Antony, B., Fujii, T., Moto, K., Matsumoto, S., Fukuzawa, M., Nakano, R., Tatsuki, S., & Ishikawa, Y. (2009). Pheromone-gland-specific fatty-acyl reductase in the adzuki bean borer, *Ostrinia scapulalis* (Lepidoptera: Crambidae). *Insect Biochemistry and Molecular Biology, 39,* 90–95.

Arima, R., Takahara, K., Kadoshima, T., Numazaki, F., Ando, T., Uchiyama, N., Kitamusa, A., & Suzuki, A. (1991). Hormonal regulation of phermone biosynthesis in the silkwom moth, *Bombyx Mori* (Lepidoptera, Bombycidae). *Applied Entomology and Zoology, 26,* 137–147.

Arsequell, G., Fabriàs, G., & Camps, F. (1990). Sex pheromone biosynthesis in the processionary moth *Thaumetopoea pityocampa* by delta-13 desaturation. *Archives Insect Biochemistry Physiology, 14,* 47–56.

Aw, T., Schlauch, K., Keeling, C. I., Young, S., Bearfield, J. C., Blomquist, G. J., & Tittiger, C. (2010). Functional genomics of the mountain pine beetle (*Dendroctonus ponderosae*) midguts and fat bodies. *BMC Genomics, 11*(215).

Baker, T. C. (2002). Mechanism for saltational shifts in pheromone communication systems. *Proceedings of the National Academy of Sciences USA, 99,* 13368–13370.

Baltensweiler, W., & Priesner, E. (1988). A study of pheromone polymorphism in *Zeiraphera diniana* Gn. (Lep., Tortricidae) 3. Specificity of attraction to synthetic pheromone sources by different male response types from two host races. *Journal of Applied Entomology, 106,* 217–231.

Barak, A. V., & Burkholder, W. E. (1977). Behavior and pheromone studies with *Attagenus elongatulus* Casey (Coleoptera: Dermestidae). *Journal of Chemical Ecology, 3,* 219–237.

Barbier, J., & Lederer, E. (1960). Structure chemique de la substance royale de la reine d'abeille (*Apis mellifica* L. *Comptes Rendus de l'Academie des Sciences Paris, 251,* 1131–1135.

Barkawi, L. S. (2002). *Biochemical and molecular studies of aggreagation pheromones of bark beetles in the genus Dendroctonus (Coleoptera: Scolytidae) with special reference to the Jeffrey pine beetle, Dendroctonus jeffreyi Hopkins. Ph. D. Dissertation. Biochemistry.* Reno: University of Nevada, 220 pp.

Barkawi, L. S., Francke, W., Blomquist, G. J., & Seybold, S. J. (2003). Frontalin: de novo synthesis of an aggregation pheromone component by *Dendroctonus* spp. bark beetles (Coleoptera: Scolytidae). *Insect Biochemistry and Molecular Biology, 33,* 773–788.

Bartelt, R. J. (1999). Sap beetles. In J. Hardie, & A. K. Minks (Eds.), *Pheromones of Non-Lepidopteran Insects Associated with Agricultural Plants* (pp. 69–89). Oxon, UK: CABI Publishing.

Bartelt, R. J., Weisleder, D., Dowd, P. F., & Plattner, R. D. (1992). Male-specific tetraene and triene hydrocarbons of *Carpophilus hemipterus*: Structure and pheromonal activity. *Journal of Chemical Ecology, 18,* 379–402.

Bartelt, R. J. (2010). Volatile hydrocarbon pheromones from beetles. In G. J. Blomquist, & A.-G. Bagneres (Eds.), *Insect Hydrocarbons: Biology, Biochemistry and Chemical Ecology* (pp. 448–476). Cambridge: Cambridge University Press.

Barth, R. H., Jr. (1965). Insect mating behavior: Endocrine control of a chemical communication system. *Science, 149,* 882–883.

Barth, R. H., Jr., & Lester, L. J. (1973). Neuro-hormonal control of sexual behavior in insects. *Annual Review of Entomology, 18,* 445–472.

Bearfield, J. C. (2004). *Understanding juvenile hormone's mode of action in the pine engraver beetle, Ips pini (Coleoptera: Scolytidae). Ph. D. dissertation. Biochemistry and Molecular Biology.* Reno: University of Nevada, p. 230.

Bearfield, J. C., Henry, A. G., Tittiger, C., Blomquist, G. J., & Ginzel, M. D. (2009). Two regulatory mechanisms of monoterpenoid pheromone production in *Ips* spp. of bark beetles. *Journal of Chemical Ecology, 35,* 689–697.

Berger, R. S. (1966). Isolation, identification, and synthesis of the sex attractant of the Cabbage Looper, *Trichoplusia ni. Annals of the Entomological Society of America, 59,* 767–771.

Bestmann, H. J., Herrig, M., & Attygalle, A. B. (1987). Terminal acetylation in pheromone biosynthesis by *Mamestra brassicae* L. (Lepidoptera: Noctuidae). *Experientia, 43,* 1033–1034.

Billen, J., & Morgan, E. D. (1998). Pheromone communication in social insects: sources and secretions. In R. K. Vander Meer, M. D. Breed, M. L. Winston, & K. E. Espelie (Eds.), *Pheromone communication in social insects* (pp. 3–33). Boulder, CO: Westview Press.

Billeter, J. C., Atallah, J., Krupp, J. J., Millar, J. G., & Levine, J. D. (2009). Specialized cells tag sexual and species identity in *Drosophila melanogaster. Nature, 461,* 987–992.

Birgersson, G., Byers, J. A., Bergstrom, G., & Lofqvist, J. (1990). Production of pheromone components, chalcogran and methyl *(E, Z)*-2,4-decadienoate, in the spruce engraver *Pityogenes chalcographus. Journal of Insect Physiology, 36,* 391–395.

Bjostad, L. B., & Roelofs, W. L. (1981). Sex pheromone biosynthesis from radiolabeled fatty acids in the redbanded leafroller moth. *Journal of Biological Chemistry, 256,* 7936–7940.

Bjostad, L. B., & Roelofs, W. L. (1983). Sex pheromone biosynthesis in *Trichoplusia ni*: Key steps involve delta-11 desaturation and chain-shortening. *Science, 220,* 1387–1389.

Bjostad, L. B., & Roelofs, W. L. (1984). Biosynthesis of sex pheromone components and glycerolipid precursors from sodium [1-14C]acetate in redbanded leafroller moth. *Journal of Chemical Ecology, 10,* 681–691.

Bjostad, L. B., Linn, C. E., Du, J. -W., & Roelofs, W. L. (1984). Identification of new sex pheromone components in *Trichoplusia ni*, predicted from biosynthetic precursors. *Journal of Chemical Ecology, 10,* 1309–1323.

Bjostad, L. B., Wolf, W. A., & Roelofs, W. L. (1987). Pheromone biosynthesis in lepidopterans: Desaturation and chain shortening. In G. D. Prestwich, & G. J. Blomquist (Eds.), *Pheromone Biochemistry* (pp. 77–120). New York: Academic Press.

Blomquist, G. J., & Dillwith, J. W. (1983). Pheromones: Biochemistry and physiology. In R. G.H. Downer, & H. Laufer (Eds.), *Endocrinology of Insects* (pp. 527–542). New York: Alan R. Liss, Inc.

Blomquist, G. J., Dillwith, J. W., & Pomonis, J. G. (1984a). Sex pheromone of the housefly: Metabolism of *(Z)*-9-tricosene to *(Z)*-9,10-epoxytricosane and *(Z)*-14-tricosen-10-one. *Insect Biochemistry, 14,* 279–284.

Blomquist, G. J., Adams, T. S., & Dillwith, J. W. (1984b). Induction of female sex pheromone production in male houseflies by ovarian implants or 20-hydroxyecdysone. *Journal of Insect Physiology, 30,* 295–302.

Blomquist, G. J., Dillwith, J. W., & Adams, T. S. (1987). Biosynthesis and endocrine regulation of sex pheromone production in Diptera. In G. D. Prestwich, & G. J. Blomquist (Eds.), *Pheromone Biochemistry* (pp. 217–250). New York: Academic Press.

Blomquist, G. J., Adams, T. S., Halarnkar, P. P., Gu, P., Mackay, M. E., & Brown, L. (1992). Ecdysteroid induction of sex pheromone biosynthesis in the housefly, *Musca domestica-* Are other factors involved?. *Journal of Insect Physiology, 38,* 309–318.

Blomquist, G. J., Tillman-Wall, J. A., Guo, L., Quilici, D., Gu, P., & Schal, C. (1993). Hydrocarbon and hydrocarbon derived sex pheromones in insects: biochemistry and endocrine regulation. In D. W. Stanley-Samuelson, & D. R. Nelson (Eds.), *Insect Lipids: Chemistry and Biology* (pp. 317–351). Lincoln: University of Nebraska Press.

Blomquist, G. J., Guo, L., Gu, P., Blomquist, C., Reitz, R. C., & Reed, J. R. (1994). Methyl-branched fatty acids and their biosynthesis in the housefly, *Musca domestica* L. (Diptera: Muscidae). *Insect Biochemistry and Molecular Biology, 24,* 803–810.

Blomquist, G. J., Tillman, J. A., Reed, J. R., Gu, P., Vanderwel, D., Choi, S., & Reitz, R. C. (1995). Regulation of enzymatic activity involved in sex pheromone production in the housefly, *Musca domestica. Insect Biochemistry and Molecular Biology, 25,* 751–757.

Blomquist, G. J., Tillman, J. A., Mpuru, S., & Seybold, S. J. (1998). The cuticle and cuticular hydrocarbons of insects: structure, function and biochemistry. In R. K. Vander Meer, M. Breed, M. Winston, & C. Espelie (Eds.), *Pheromone Communication in Social Insects* (pp. 34–54). Boulder CO: Westview Press.

Blomquist, G. J. (2003). Biosynthesis and ecdysteroid regulation of housefly sex pheromone production. In G. J. Blomquist, & R. G. Vogt (Eds.), *Insect Pheromone Biochemistry and Molecular Biology* (pp. 231–252). London: Elsevier Academic Press.

Blomquist, G. J., & Howard, R. W. (2003). Pheromone biosynthesis in social insects. In G. J. Blomquist, & R. G. Vogt (Eds.), *Insect Pheromone Biochemistry and Molecular Biology* (pp. 323–230) London: Elsevier Academic Press.

Blomquist, G. J., Jurenka, R. A., Schal, C., & Tittiger, C. (2005). Biochemistry and molecular biology of pheromone production. In L. I. Gilbert, K. Iatrou, & S. S. Gill (Eds.), *Comprehensive Molecular Insect Sciences Vol 3*. (pp. 705–752). London: Elsevier.

Blomquist, G. J. (2010). Biosynthesis of cuticular hydrocarbons. In G. J. Blomquist, & A.-G. Bagneres (Eds.), *Insect Hydrocarbons: Biology, Biochemistry and Chemical Ecology* (pp. 35–52). Cambridge: Cambridge Press.

Blomquist, G. J., Figueroa-Teran, R., Aw, M., Song, M., Gorzalski, A., Abbott, N., Chang, E., & Tittiger, C. (2010). Pheromone production in bark beetles. *Insect Biochemistry and Molecular Biology, 40*, 699–712.

Bober, R., & Rafaeli, A. (2010). Gene-silencing reveals the functional significance of pheromone biosynthesis activating neuropeptide receptor (PBAN-R) in a male moth. *Proceedings of the National Academy of Sciences USA, 107*, 16858–16862.

Bober, R., Azrielli, A., & Rafaeli, A. (2010). Developmental regulation of the pheromone biosynthesis activating neuropeptide-receptor (PBAN-R): re-evaluating the role of juvenile hormone. *Insect Molecular Biology, 19*, 77–86.

Borden, J. H. (1985). Aggregation pheromones. In G. A. Kerkut, & L. I. Gilbert (Eds.), *Comprehensive Insect Physiology, Biochemistry, and Pharmacology Vol. 9*. (pp. 257–285). Oxford: Pergamon Press.

Borden, J. H., Nair, K. K., & Slater, C. E. (1969). Synthetic juvenile hormone: Induction of sex pheromone production in *Ips confusus*. *Science, 166*, 1626–1627.

Bridges, J. R. (1982). Effects of juvenile hormone on pheromone synthesis in *Dendroctonus frontalis*. *Environmental Entomology, 11*, 417–420.

Brossut, R., & Roth, L. M. (1977). Tergal modifications associated with abdominal glandular cells in the Blattaria. *Journal of Morphology, 151*, 259–297.

Butenandt, A., Beckmann, R., Stamm, D., & Hecker, E. (1959). Über dem sexual-lockstoff des seidenspinners *Bombyx mori*. Reindarstellung und konstitution. *Zeitschrift für Naturforschung A, 14*, 283–284.

Byers, J. A. (1983). Influence of sex, maturity and host substances on pheromones in the guts of the bark beetles, *Ips paraconfusus* and *Dendroctonus brevicomis*. *Journal of Insect Physiology, 29*, 5–13.

Byrne, K. J., Swigar, A. A., Silverstein, R. M., Borden, J. H., & Stokkink, E. (1974). Sulcatol: Population aggregation pheromone in the scolytid beetle, *Gnathotrichus sulcatus*. *Journal of Insect Physiology, 20*, 1895–1900.

Callow, R. K., & Johnston, N. C. (1960). The chemical constitution and synthesis of queen substance of honeybees (*Apis mellifera* L.). *Bee World, 41*, 152–153.

Chase, J., Jurenka, R. J., Schal, C., Halarnkar, P. P., & Blomquist, G. J. (1990). Biosynthesis of methyl-branched hydrocarbons in the German cockroach *Blattella germanica* (L.) (Orthoptera, Blattellidae). *Insect Biochemistry, 20*, 149–156.

Chase, J., Touhara, K., Prestwich, G. D., Schal, C., & Blomquist, G. J. (1992). Biosynthesis and endocrine control of the production of the German cockroach sex pheromone, 3,11-dimethylnonacosan-2-one. *Proceedings of the National Academy of Sciences USA, 89*, 6050–6054.

Cheesbrough, T. M., & Kolattukudy, P. E. (1984). Alkane biosynthesis by decarbonylation of aldehydes catalyzed by a particulate preparation from *Pisum sativum*. *Proceedings of the National Academy of Sciences USA, 81*, 6613–6617.

Cheesbrough, T. M., & Kolattukudy, P. E. (1988). Microsomal preparations from animal tissue catalyzes release of carbon monoxide from a fatty aldehyde to generate an alkane. *Journal of Biological Chemistry, 263*, 2738–2743.

Chen, N. M., Bordon, J. H., & Pierce, H. D., Jr. (1988). Effect of juvenile hormone analog fenoxycarb, on pheromone production by *Ips paraconfusus* (Coleoptera: Scolytidae). *Journal of Chemical Ecology, 14*, 1087–1098.

Cheon, H. M., Seo, S. J., Sun, J., Sappington, T. W., & Raikhel, A. S. (2001). Molecular characterization of the VLDL receptor homologue mediating binding of lipophorin in oocyte of the mosquito *Aedes aegypti*. *Insect Biochemistry and Molecular Biology, 31*, 753–760.

Chino, H. (1985). Lipid transport: biochemistry of hemolymph lipophorin. In G. A. Kerkut, & L. I. Gilbert (Eds.), *Comprehensive Insect Physiology, Biochemistry and Pharmacology Vol. 10*. (pp. 115–135). Oxford: Pergamon.

Choi, M.-Y., Tanaka, M., Kataoka, H., Boo, K. S., & Tatsuki, S. (1998). Isolation and identification of the cDNA encoding the pheromone biosynthesis activating neuropeptide and additional neuropeptides in the oriental tobacco budworm, *Helicoverpa assulta* (Lepidoptera: Noctuidae). *Insect Biochemistry and Molecular Biology, 28*, 759–766.

Choi, M.-Y., Han, K. S., Boo, K. S., & Jurenka, R. (2002). Pheromone biosynthetic pathway in *Helicoverpa zea* and *Helicoverpa assulta*. *Insect Biochemistry and Molecular Biology, 32*, 1353–1359.

Choi, M.-Y., Fuerst, E.-J., Rafaeli, A., & Jurenka, R. (2003). Identification of a G protein-coupled receptor for pheromone biosynthesis activating neuropeptide from pheromone glands of the moth, *Helicoverpa zea*. *Proceedings of the National Academy of Science USA, 100*, 9721–9726.

Choi, M.-Y., Fuerst, E.-J., Rafaeli, A., & Jurenka, R. (2007). Role of extracellular domains in PBAN/Pyrokinin GPCRs from insects using chimera receptors. *Insect Biochemistry and Molecular Biology, 37*, 296–306.

Choi, M.-Y., Lim, H., Park, K.-C., Adlof, R., Wang, S., Zhang, A., & Jurenka, R. (2007b). Identification and biosynthetic studies of the hydrocarbon sex pheromone in *Utetheisa ornatrix*. *Journal of Chemical Ecology, 33*, 1336–1345.

Christensen, T. A., & Hildebrand, J. G. (1995). Neural regulation of sex-pheromone glands in Lepidoptera. *Invertebrate Neuroscience, 1*, 97–103.

Christensen, T. A., Itagaki, H., Teal, P. E.A., Jasensky, R. D., Tumlinson, J. H., & Hildebrand, J. G. (1991). Innervation and neural regulation of the sex pheromone gland in female *Heliothis* moths. *Proceedings of the National Academy of Science USA, 88*, 4971–4975.

Christensen, T. A., Lehman, H. K., Teal, P. E.A., Itagaki, H., Tumlinson, J. H., & Hildebrand, J. G. (1992). Diel changes in the presence and physiological actions of octopamine in the female sex-pheromone glands of heliothine moths. *Insect Biochemistry and Molecular Biology, 22*, 841–849.

Christensen, T. A., Lashbrook, J. M., & Hildebrand, J. G. (1994). Neural activation of the sex-pheromone gland in the moth *Manduca sexta*: real-time measurement of pheromone release. *Physiological Entomology, 19*, 265–270.

Chu, A. J., & Blomquist, G. J. (1980). Biosynthesis of hydrocarbons in insects: Succinate is a precursor of the methyl branched alkanes. *Archives of Biochemistry and Biophysics, 201*, 304–312.

Ciudad, L., Bellés, X., & Piulachs M.-D. (2007). Structural and RNAi characterization of the German cockroach lipophorin receptor, and the evolutionary relationships of lipoprotein receptors. *BMC Molecular Biology, 8*, 53.

Conn, J. E., Bordon, J. H., Hunt, D. W.A., Holman, J., Whitney, H. S., Spanier, O. J., Pierce, H. D., & Oehlschlager, A. C. (1984). Pheromone production by axwnically reared *Dendroctonus ponderosae* and *Ips paraconfusus* (Coleoptera: Scolytidae). *Journal of Chemical Ecology, 10*, 281–290.

Cornette, R., Farine, J.-P., Quennedey, B., & Brossut, R. (2001). Molecular characterization of a new adult male putative calycin specific to tergal aphrodisiac secretion in the cockroach *Leucophaea maderae*. *FEBS, 507*, 313–317.

Cornette, R., Farine, J.-P., Quennedey, B., Riviere, S., & Brossut, R. (2002). Molecular characterization of Lma-p54, a new epicuticular surface protein in the cockroach *Leucophaea maderae* (Dictyoptera, Oxyhaloinae). *Insect Biochemistry and Molecular Biology, 32*, 1635–1642.

Cusson, M., & McNeil, J. N. (1989). Involvement of juvenile hormone in the regulation of pheromone release activities in a moth. *Science, 243*, 210–212.

Cusson, M., Tobe, S. S., & McNeil, J. N. (1994). Juvenile hormones - their role in the regulation of the pheromonal communication system of the armyworm moth, *Pseudaletia unipuncta*. *Archives Insect Biochemistry and Physiology, 25*, 329–345.

Davis, M. B., Vakharia, V. N., Henry, J., Kempe, T. G., & Raina, A. K. (1992). Molecular cloning of the pheromone biosynthesis-activating neuropeptide in *Helicoverpa zea*. *Proceedings of the National Academy of Science USA, 89*, 142–146.

Davis, N. T., Homberg, U., Teal, P. E.A., Altstein, M., Agricola, H.-J., & Hildebrand, J. G. (1996). Neuroanatomy and immunocytochemistry of the median neuroendocrine cells of the subesophageal ganglion of the tobacco hawkmoth, *Manduca sexta*: immunoreactivities to PBAN and other neuropeptides. *Microscopy Research Techniques, 35*, 201–229.

Delisle, J., Picimbon, J.-F., & Simard, J. (1999). Physiological control of pheromone production in *Choristoneura fummiferana* and *C. rosaceana*. *Archives of Insect Biochemistry and Physiology, 42*, 253–265.

de Marzo, L., & Vit, S. (1983). Contribution to the knowledge of Palearctic Batrisinae (Colopetera: Pselaphidae). Antennal male glands of *Batrisus* Aubé and *Batrisodes* Reitter: Morponogy, histology and taxanomical implications. *Entomolgicia, 18*, 77–110.

Dennis, M. W., & Kolattukudy, P. E. (1991). Alkane biosynthesis by decarbonylation of aldehyde catalyzed by a microsomal preparation from *Botryococcus brauni*. *Archives of Biochemistry and Biophysics, 287*, 268–275.

Dennis, M. W., & Kolattukudy, P. E. (1992). A cobalt-porphyrin enzyme converts a fatty aldehyde to a hydrocarbon and CO. *Proceedings of the National Academy of Sciences USA, 89*, 5306–5310.

de Renobales, M., & Blomquist, G. J. (1983). A developmental study of the composition and biosynthesis of the cuticular hydrocarbons of *Trichoplusia ni*. *Insect Biochemistry, 13*, 493–502.

de Renobales, M., Nelson, D. R., Zamboni, A. C., Mackay, M. E., Dwyer, L. A., Theisen, M. O., & Blomquist, G. J. (1989). Biosynthesis of very long-chain methyl branched alcohols during pupal development in the cabbage looper, *Trichoplusia ni*. *Insect Biochemistry, 19*, 209–214.

Dickens, J. C., McGovern, W. L., & Wiygul, G. (1988). Effects of antennectomy and a juvenile hormone analog on pheromone production in the boll weevil (Coleoptera: Curculionidae). *Journal of Entomological Science, 23*, 52–58.

Dickens, J. C., Oliver, J. E., Hollister, B., Davis, J. C., & Klun, J. A. (2002). Breaking a paradigm: male-produced aggregation pheromone for the Colorado potato beetle. *Journal of Experimental Biology, 205*, 1925–1933.

Dillwith, J. W., & Blomquist, G. J. (1982). Site of sex pheromone biosynthesis in the female housefly, *Musca domestica*. *Experientia, 38*, 471–473.

Dillwith, J. W., Blomquist, G. J., & Nelson, D. R. (1981). Biosynthesis of the hydrocarbon components of the sex pheromone of the housefly, *Musca domestica* L. *Insect Biochemistry, 11*, 247–253.

Dillwith, J. W., Nelson, J. H., Pomonis, J. G., Nelson, D. R., & Blomquist, G. J. (1982). A 13C-NMR study of methylbranched hydrocarbon biosynthesis in the housefly. *Journal of Biological Chemistry, 257*, 11305–11314.

Dillwith, J. W., Adams, T. S., & Blomquist, G. J. (1983). Correlation of housefly sex pheromone production with ovarian development. *Journal of Insect Physiology, 29*, 377–386.

Domingue, M., Starmer, W., & Teale, S. (2006). Genetic control of the enantiomeric composition of ipsdienol in the pine engraver, *Ips pini*. *Journal of Chemical Ecology, 32*, 1005–1026.

Domingue, M., & Teale, S. (2008). The genetic architecture of pheromone production between populations distant from the hybride zone of the pine engraver. *Ips pini*. *Chemoecology, 17*, 255–262.

Duportets, L., Gadenne, C., & Couillaud, F. (1999). A cDNA, from *Agrotis ipsilon*, that encodes the pheromone biosynthesis activating neuropeptide (PBAN) and other FXPRL peptides. *Peptides, 20*, 899–905.

Dwyer, L. A., de Renobales, M., & Blomquist, G. (1981). Biosynthesis of (Z,Z)-6,9-heptacosadiene in the American cockroach. *Lipids, 16*, 810–814.

Eigenheer, A. L., Keeling, C. I., Young, S., & Tittiger, C. (2003). Comparison of gene representation in midguts from two phytophagous insects, *Ips pini* and *Bombyx mori*, using expressed sequence tags. *Gene, 316*, 127–136.

Eisner, T., & Meinwald, J. (2003). Alkalid-derived pheromones and sexual selection in Lepidoptera. In G. J. Blomquist, & R. G. Vogt (Eds.), *Insect Pheromone Biochemistry and Molecular Biology* (pp. 341–370). Amsterdam: Elsevier Academic Press.

Eliyahu, D., Applebaum, S., & Rafaeli, A. (2003). Moth sex pheromone biosynthesis is inhibited by the herbicide diclofop. *Pesticide Biochemistry and Physiology, 77*, 75–81.

Eliyahu, D., Nojima, S., Mori, K., & Schal, C. (2008). New contact sex pheromone components of the German cockroach, *Blattella germanica*, predicted from the proposed biosynthetic pathway. *Journal of Chemical Ecology, 34*, 229–237.

Fabriès, G., Jurenka, R. A., & Roelofs, W. L. (1992). Stimulation of sex pheromone production by proteinaceous extracts of the bursa copulatrix in the redbanded leafroller moth. *Archives of Insect Biochemistry and Physiology, 20*, 75–86.

Fabriès, G., Marco, M.-P., & Camps, F. (1994). Effect of the pheromone biosynthesis activating neuropeptide on sex pheromone biosynthesis in *Spodoptera littoralis* isolated glands. *Archives of Insect Biochemistry and Physiology, 27*, 77–87.

Fan, Y., Rafaeli, A., Gileadi, C., & Applebaum, S. W. (1999). Juvenile hormone induction of pheromone gland PBAN-responsiveness in *Helicoverpa armigera* females. *Insect Biochemistry and Molecular Biology, 29*, 635–641.

Fan, Y., Chase, J., Sevala, V. L., & Schal, C. (2002). Lipophorin-facilitated hydrocarbon uptake by oocytes in the German cockroach, *Blattella germanica* (L). *Journal of Experimental Biology, 205*, 781–790.

Fan, Y., Zurek, L., Dykstra, M. J., & Schal, C. (2003). Hydrocarbon synthesis by enzymatically dissociated oenocytes of the abdominal integument of the German Cockroach. *Blattella germanica. Naturwissenschaften, 90*, 121–126.

Fan, Y., Eliyahu, D., & Schal, C. (2008). Cuticular hydrocarbons as maternal provisions in embryos and nymphs of the cockroach *Blattella germanica. Journal of Experimental Biology, 211*, 548–554.

Fang, N., Teal, P. E.A., Doolittle, R. E., & Tumlinson, J. H. (1995a). Biosynthesis of conjugated olefinic systems in the sex pheromone gland of female tobacco hornworm moths, *Manduca sexta* (L). *Insect Biochemistry and Molecular Biology, 25*, 39–48.

Fang, N., Teal, P. E.A., & Tumlinson, J. H. (1995b). Characterization of oxidase(s) associated with the sex pheromone gland in *Manduca sexta* (L.) females. *Archives of Insect Biochemistry and Physiology, 29*, 243–257.

Faustini, D. L., Post, D. C., & Burkholder, W. E. (1982). Histology of aggregation pheromone gland in the red flour beetle. *Annals of the Entomological Society of America, 75*, 187–190.

Ferveur, J. F., & Sureau, G. (1996). Simultaneous influence on male courtship of stimulatory and inhibitory pheromones produced by life sex mosaic *Drosophila melangaster. Proceedings of the Royal Society B, 263*, 967–973.

Ferveur, J. F., Savarit, F., O'Kane, C. J., Jureau, G., Greenspan, R. J., & Jallon, J. M. (1997). Genetic feminization of pheromones and its behavioral consequences in *Drosophila* males. *Science, 276*, 1555–1558.

Fónagy, A., Matsumoto, S., Uchhiumi, K., & Mitsui, T. (1992). Role of calcium ion and cyclic nucleotides in pheromone production in *Bombyx mori. Journal of Pesticide Science, 17*, 115–121.

Fónagy, A., Yokoyama, N., Ozawa, R., Okano, K., Tatsuki, S., Maeda, S., & Matsumoto, S. (1999). Involvement of calcineurin in the signal transduction of PBAN in the silkworm, *Bombyx mori* (Lepidoptera). *Comparative Biochemistry and Physiology B. Biochemistry and Molecular Biology, 124*, 51–60.

Foster, S. P. (2009). Sugar feeding via trehalose haemolymph concentration affects sex pheromone production in mated *Heliothis virescens* moths. *Journal of Experimental Biology, 212*, 2789–2794.

Foster, S. P., & Roelofs, W. L. (1988). Sex pheromone biosynthesis in the leafroller moth *Planotortrix excessana* by Δ10 desaturation. *Archives of Insect Biochemistry Physiology, 8*, 1–9.

Foster, S. P., & Roelofs, W. L. (1990). Biosynthesis of a monoene and a conjugated diene sex pheromone component of the light brown apple moth by E11-desaturation. *Experientia, 46*, 269–273.

Foster, S. P., & Roelofs, W. L. (1996). Sex pheromone biosynthesis in the tortricid moth, *Ctenopseustis herana* (Felder & Rogenhofer). *Archives of Insect Biochemistry and Physiology, 33*, 135–147.

Fukui, H., Matsumura, F., Barak, A. V., & Burkholder, W. E. (1977). Isolation and identification of a major sex-attracting component of *Attagenus elongatus* (Casey) (Coleoptera: Dermestidae). *Journal of Chemical Ecology, 3*, 539–548.

Gadenne, C. (1993). Effects of fenoxycarb, juvenile hormone mimetic, on female sexual behaviour of the black cutworm, *Agrotis ipsilon* (Lepidoptera: Noctuidae). *Journal of Insect Physiology, 39*, 25–29.

Gemeno, C., & Schal, C. (2004). Sex Pheromones of Cockroaches. In R. T. Cardé, & J. Millar (Eds.), *Advances in Insect Chemical Ecology* (pp. 179–247). New York: Cambridge University Press.

Gemeno, C., Snook, K., Benda, N., & Schal, C. (2003). Behavioral and electrophysiological evidence for volatile sex pheromones in *Parcoblatta* wood cockroaches. *Journal of Chemical Ecology, 29*, 37–54.

Gilg, A., Bearfield, J. C., Tittiger, C., Welch, W. H., & Blomquist, G. J. (2005). Isolation, and functional expression of the first animal geranyl diphosphate synthase and its role in bark beetle pheromone biosynthesis. *Proceedings of the National Academy of Science USA, 102*, 9760–9765.

Gilg, A., Tittiger, C., & Blomquist, G. J. (2009). Unique animal prenyltransferase with monoterpene synthase activity. *Naturwissenschaften, 96*, 731–735.

Ginzel, M. D., Bearfield, J. C., Keeling, C. I., McCormack, C. C., Blomquist, G. J., & Tittiger, C. (2007). Antennally-mediated negative-feedback regulation of pheromone production in the pine engraver beetle, *Ips pini. Naturwissenschaften, 94*, 61–64.

Goller, S., Szöcs, G., Francke, W., & Schulz, S. (2007). Biosynthesis of (3Z,6Z,9Z)- 3,6,9-octadecatriene the main component of the pheromone blend of *Erannis bajaria. Journal of Chemical Ecology, 33*, 1505–1509.

Gosalbo, L., Fabriès, G., & Camps, F. (1994). Inhibitory effect of 10,11-methylenetetradec-10-enoic acid on a Z9-desaturase in the sex pheromone biosynthesis of *Spodoptera littoralis. Archives of Insect Biochemistry and Physiology, 26*, 279–286.

Gu, P., Welch, W. H., & Blomquist, G. J. (1993). Methyl-branched fatty acid biosynthesis in the German cockroach, *Blattella germanica:* kinetic studies comparing a microsomal and soluble fatty acid synthetases. *Insect Biochemistry and Molecular Biology, 23*, 263–271.

Gu, X., Quilici, D., Juarez, P., Blomquist, G. J., & Schal, C. (1995). Biosynthesis of hydrocarbons and contact sex pheromone and their transport by lipophorin in females of the German cockroach *Blattella germanica. Journal of Insect Physiology, 41*, 257–267.

Gu, P., Welch, W. H., Guo, L., Schegg, K. M., & Blomquist, G. J. (1997). Characterization of a novel microsomal fatty acid synthetase (FAS) compared to a cytosolic FAS in the housefly, *Musca domestica. Comparative Biochemistry and Physiology, 118B*, 447–456.

Guerin, P. M., Baltensweiler, W., Arn, H., & Buser H.-R. (1984). Host race pheromone polymorphism in the larch budmoth. *Experientia, 40*, 892–894.

Guo, L., Latli, B., Prestwich, G. D., & Blomquist, G. J. (1991). Metabolically-blocked analogs of the housefly sex pheromone II. Metabolism studies. *Journal of Chemical Ecology, 17*, 1769–1782.

Hadley, N. F. (1984). Cuticle: biochemistry. In J. Bereiter-Hahn, A. G. Matoltsy, & K. S. Richards (Eds.), *Biology of the Integument* (pp. 685–702). Berlin: Springer-Verlag.

Hagedorn, H. H. (1985). In G. A. Kerkut, & L. I. Gilbert (Eds.), The role of ecdysteroids in reproduction. *Comprehensive insect physiology, biochemistry and pharmacology Vol. 8.* (pp. 205–262). Oxford: Pergamon.

Halarnkar, P. P., Nelson, J. H., Heisler, C. R., & Blomquist, G. J. (1985). Metabolism of propionate to acetate in the cockroach *Periplaneta americana. Archives of Biochemistry and Biophysics, 36*, 526–534.

Halarnkar, P. P., Heisler, C. R., & Blomquist, G. J. (1986). Propionate catabolism in the housefly *Musca domestica* and the termite *Zootermopsis nevadensis. Insect Biochemistry, 16*, 455–461.

Hall, G. M., Tittiger, C., Andrews, G., Mastick, G., Kuenzli, M., Luo, X., Seybold, S. J., & Blomquist, G. J. (2002a). Male pine engraver Beetles, *Ips pini*, synthesize the monoterpenoid pheromone ipsdienol de novo in midgut tissue. *Naturwissenschaften, 89*, 79–83.

Hall, G. M., Tittiger, C., Blomquist, G. J., Andrews, G. L., Mastick, G. S., Barkawi, L. S., Bengoa, C., & Seybold, S. J. (2002b). Male jeffrey pine beetle, *Dendroctonus jeffreyi*, synthesizes the pheromone component frontalin in anterior midgut tissue. *Insect Biochemistry and Molecular Biology, 32*, 1525–1532.

Hallberg, E., & Subchev, M. (1997). Unusual location and structure of female pheromone glands in *Theresimima* (= *Ino*) *ampelophaga* Bayle-Berelle (Lepidoptera:Zygaenidae). *International Journal of Insect Morphology and Embryology, 25*, 381–389.

Hao, G., Liu, W., O'Connor, M., & Roelofs, W. L. (2002). Acyl-CoA Z9- and Z10-desaturase genes from a New Zealand leafroller moth species, *Planotortrix octo. Insect Biochemistry and Molecular Biology, 32*, 961–966.

Hashimoto, T. (1996). Peroxisomal beta-oxidation: enzymology and molecular biology. *Annals New York Academy of Sciences, 804*, 86–98.

Haynes, K. F., & Hunt, R. E. (1990). A mutation in pheromonal communication system of cabbage looper moth, *Trichoplusia ni. Journal of Chemical Ecology, 16*, 1249–1257.

Hewes, R. S., & Taghert, P. H. (2001). Neuropeptides and neuropeptide receptors in the *Drosophila melanogaster* genome. *Genome Research, 11*, 1126–1142.

Hirashima, A., Eiraku, T., Watanabe, Y., Kuwano, E., Taniguchi, E., & Eto, M. (2001). Identification of novel inhibitors of calling and in vitro [C-14]acetate incorporation by pheromone glands of *Plodia interpunctella. Pest Management Science, 57*, 713–720.

Holman, G. M., Cook, B. J., & Nachman, R. J. (1986). Isolation, primary structure and synthesis of a blocked neuropeptide isolated from the cockroach, *Leucophaea maderae. Comparative Biochemistry and Physiology, 85C*, 219–224.

Homma, T., Watanabe, K., Tsurumaru, S., Kataoka, H., Imai, K., Kamba, M., Niimi, T., Yamashita, O., & Yaginuma, T. (2006). G protein-coupled receptor for diapause hormone, an inducer of *Bombyx* embryonic diapause. *Biochemical and Biophysical Research Communications, 344*, 386–393.

Hull, J. J., Ohnishi, A., Moto, K., Kawasaki, Y., Kurata, R., Suzuki, M. G., & Matsumoto, S. (2004). Cloning and characterization of the pheromone biosynthesis activating neuropeptide receptor from the Silkmoth, *Bombyx mori*: significance of the carboxyl terminus in receptor internalization. *Journal of Biological Chemistry, 279*, 51500–51507.

Hull, J. J., Kajigaya, R., Imai, K., & Matsumoto, S. (2007). Sex pheromone production in the silkworm, *Bombyx mori*, is mediated by store-operated Ca2+ channels. *Bioscience, Biotechnology, and Biochemistry, 71*, 1993–2001.

Hull, J. J., Lee, J. M., Kajigaya, R., & Matsumoto, S. (2009). *Bombyx mori* homologues of STIM1 and Orai1 are essential components of the signal transduction cascade that regulates sex pheromone production. *Journal of Biological Chemistry, 284*, 31200–31213.

Hull, J. J., Lee, J. M., & Matsumoto, S. (2010). Gqα-linked phospholipase Cβ1 and phospholipase CΥ are essential components of the pheromone biosynthesis activating neuropeptide (PBAN) signal transduction cascade. *Insect Molecular Biology, 19*, 553–566.

Hunt, D. W. A., Borden, J. H., Lindgren, B. S., & Gries, G. (1989). The role at autoxidation of alpha-pinene in the production of pheromones of Dendroctonus ponderoase (Coleoptera; Scellytidae). *Canadian Journal of Forest Research, 19*, 1275–1282.

Iglesias, F., Marco, M. P., Jacquin-Joly, E., Camps, F., & Fabriàs, G. (1998). Regulation of sex pheromone biosynthesis in two noctuid species, *S. littoralis* and *M. brassicae*, may involve both PBAN and the ventral nerve cord. *Archives of Insect Biochemistry and Physiology, 37*, 295–304.

Iglesias, F., Marco, P., François, M.-C., Camps, F., Fabriàs, G., & Jacquin-Joly, E. (2002). A new member of the PBAN family in *Spodoptera littoralis*: molecular cloning and immunovisualisation in scotophase hemolymph. *Insect Biochemistry and Molecular Biology, 32*, 901–908.

Imai, K., Konno, T., Nakazawa, Y., Komiya, T., Isobe, M., Koga, K., Goto, T., Yaginuma, T., Sakakibara, K., Hasegawa, K., & Yamashita, O. (1991). Isolation and structure of diapause hormone of the silkworm, *Bombyx mori. Proceedings of the Japan Academy, 67*(B), 98–101.

Islam, N., Bacala, R., Moore, A., & Vanderwel, D. (1999). Biosynthesis of 4-methyl-1- nonanol: Female-produced sex pheromone of the yellow mealworm beetle, *Tenebrio molitor* (Coleoptera: Tenebrionidae). *Insect Biochemistry and Molecular Biology, 29*, 201–208.

Ivarsson, P., & Birgirsson, G. (1995). Regulation and biosynthesis of pheromone components in the double spined bark beetles, *Ips duplicatus* (Coleoptera: Scolytidae). *Journal of Insect Physiology, 41*, 843–849.

Ivarsson, P., Blomquist, G. J., & Seybold, S. J. (1997). In vitro production of the pheromone intermediates ipsdienone and ipsenone by the bark beetles *Ips pini* (Say) and *I. paraconfusus* Lanier (Coleoptera: Scolytidae). *Naturwissenschaften, 84*, 454–457.

Jacquin, E., Jurenka, R. A., Ljungberg, H., Nagnan, P., Löfstedt, C., Descoins, C., & Roelofs, W. L. (1994). Control of sex pheromone biosynthesis in the moth *Mamestra brassicae* by the pheromone biosynthesis activating neuropeptide. *Insect Biochemistry and Molecular Biology, 24*, 203–211.

Jacquin-Joly, E., & Descoins, C. (1996). Identification of PBAN-like peptides in the brain-subesophageal ganglion complex of Lepidoptera using western-blotting. *Insect Biochemistry and Molecular Biology, 26*, 209–216.

Jacquin-Joly, E., Burnet, M., Francois, M., Ammar, D., Meillour, P., & Descoins, C. (1998). cDNA cloning and sequence determination of the pheromone biosynthesis activating neuropeptide of *Mamestra brassicae*: a new member of the PBAN family. *Insect Biochemistry Molecular Biology, 28*, 251–258.

Jallon, J.-M., & Wicker-Thomas, C. (2003). Genetic studies on pheromone production in *Drosophila*. In G. J. Blomquist, & R. G. Vogt (Eds.), *Insect Sex Pheromone Biochemistry and Molecular Biology* (pp. 253–282). London: Elsevier Academic Press.

Juarez, P., Chase, J., & Blomquist, G. J. (1992). A microsomal fatty acid synthetase from the integument of *Blattella germanica* synthesizes methyl-branched fatty acids, precursors to hydrocarbon and contact sex pheromone. *Archives of Biochemistry and Biophysics, 293*, 333–341.

Jurenka, R. A. (2003). Biochemistry of female moth sex pheromones. In G. J. Blomquist, & R. G. Vogt (Eds.), *Insect Pheromone Biochemistry and Molecular Biology* (pp. 53–80). Amsterdam: Elsevier Academic Press.

Jurenka, R. A., & Roelofs, W. L. (1989). Characterization of the acetyltransferase involved in pheromone biosynthesis in moths: Specificity for the Z isomer in Tortricidae. *Insect Biochemistry, 19*, 639–644.

Jurenka, R. A., Schal, C., Burns, E., Chase, J., & Blomquist, G. J. (1989). Structural correlation between cuticular hydrocarbons and female contact sex pheromone of German cockroach *Blattella germanica* (L.). *Journal of Chemical Ecology, 15*, 939–949.

Jurenka, R. A., Jacquin, E., & Roelofs, W. L. (1991a). Control of the pheromone biosynthetic pathway in *Helicoverpa zea* by the pheromone biosynthesis activating neuropeptide. *Archives of Insect Biochemistry and Physiology, 17*, 81–91.

Jurenka, R. A., Fabriàs, G., & Roelofs, W. L. (1991b). Hormonal control of female sex pheromone biosynthesis in the redbanded leafroller moth, *Argyrotaenia velutinana*. *Insect Biochemistry, 21*, 81–89.

Jurenka, R. A., Jacquin, E., & Roelofs, W. L. (1991c). Stimulation of sex pheromone biosynthesis in the moth *Helicoverpa zea*: Action of a brain hormone on pheromone glands involves Ca2+ and cAMP as second messengers. *Proceeding of the National Academy of Sciences, USA, 88*, 8621–8625.

Jurenka, R. A., Haynes, K. F., Adlof, R. O., Bengtsson, M., & Roelofs, W. L. (1994). Sex pheromone component ratio in the cabbage looper moth altered by a mutation affecting the fatty acid chain-shortening reactions in the pheromone biosynthetic pathway. *Insect Biochemistry Molecular Biology, 24*, 373–381.

Jurenka, R. A. (1997). Biosynthetic pathway for producing the sex pheromone component (*Z,E*)-9,12-tetradecadienyl acetate in moths involves a delta-12 desaturase. *Cellular and Molecular Life Sciences, 53*, 501–505.

Jurenka, R. A., & Subchev, M. (2000). Identification of cuticular hydrocarbons and the alkene precursor to the pheromone in hemolymph of the female gypsy moth, *Lymantria dispar*. *Archives of Insect Biochemistry and Physiology, 43*, 108–115.

Jurenka, R. A., Subchev, M., Abad, J.-L., Choi, M.-Y., & Fabriàs, G. (2003). Sex pheromone biosynthetic pathway for disparlure in the gypsy moth, *Lymantria dispar*. *Proceedings of the National Academy of Sciences, USA, 100*, 809–814.

Jurenka, R., & Nusawardani, T. (2011). The pyrokinin/pheromone biosynthesis-activating neuropeptide (PBAN) family of peptides and their receptors in Insecta evolutionary trace indicates potential receptor ligand-binding domains. *Insect Molecular Biology, 20*, 323–334.

Katase, H., & Chino, H. (1982). Transport of hydrocarbons by the lipophorin of insect hemolymph. *Biochimistry and Biophysics Acta, 710*, 341–348.

Katase, H., & Chino, H. (1984). Transport of hydrocarbons by haemolymph lipophorin in *Locusta migratoria*. *Insect Biochemistry, 14*, 1–6.

Kawano, T., Kataoka, H., Nagasawa, H., Isogai, A., & Suzuki, A. (1992). cDNA cloning and sequence determination of the pheromone biosynthesis activating neuropeptide of the silkworm, *Bombyx mori*. *Biochemical and Biophysical Research Communication, 189*, 221–226.

Keeling, C. I., Slessor, K. N., Higo, H. A., & Winston, M. L. (2003). New components of the honey bee (*Apis mellifera* L.) queen retinue pheromone. *Proceedings National Academy Sciences, USA, 100*, 4486–4491.

Keeling, C. I., Bearfield, J., Young, S., Blomquist, G., & Tittiger, C. (2006). Effects of juvenile hormone on gene expression in the pheromone-producing midgut of the pine engraver beetle, *Ips pini*. *Insect Molecular Biology, 15*, 207–216.

Keeling, C. I., Blomquist, G. J., & Tittiger, C. (2004). Coordinated gene expression for pheromone biosynthesis in the pine engraver beetle, *Ips pini* (Coleoptera). *Naturwissenschaften, 91*, 324–328.

Kim, Y.-J., Nachman, R. J., Aimanova, K., Gill, S., & Adams, M. E. (2008). The pheromone biosynthesis activating neuropeptide (PBAN) receptor of *Heliothis virescens*: identification, functional expression, and structure-activity relationships of ligand analogs. *Peptides, 29*, 268–275.

Kingan, T. G., Blackburn, M. B., & Raina, A. K. (1992). The distribution of PBAN immunoreactivity in the central nervous system of the corn earworm moth, *Helicoverpa zea*. *Cell and Tissue Research, 270*, 229–240.

Kitamura, A., Nagasawa, H., Kataoka, H., Inoue, T., Matsumoto, S., Ando, T., & Suzuki, A. (1989). Amino acid sequence of pheromone-biosynthesis-activating neuropeptide (PBAN) of the silkworm, *Bombyx mori*. *Biochemical and Biophysical Research Communication, 163*, 520–526.

Klun, J. A., & Maini, S. (1982). Genetic basis of an insect communication system: the European corn borer. *Environmental Entomology, 11*, 1084–1090.

Knipple, D. C., Rosenfield, C., Miller, S. J., Liu, W., Tang, J., Ma, P. W.K., & Roelofs, W. L. (1998). Cloning and functional expression of cDNA encoding a pheromone gland-specific acyl-CoA desaturase of the cabbage looper moth, *Trichoplusia ni. Proceedings National Academy of Sciences, USA, 95*, 15287–15292.

Knipple, D. C., Rosenfield, C.-L., Nielsen, R., You, K. M., & Jeong, S. E. (2002). Evolution of the integral membrane desaturase gene family in moths and flies. *Genetics, 162*, 1737–1752.

Knipple, D. C., & Roelofs, W. L. (2003). Molecular biological investigations of pheromone desaturases. In G. J. Blomquist, & R. G. Vogt (Eds.), *Insect Pheromone Biochemistry and Molecular Biology* (pp. 81–106). London: Elsevier Academic Press.

Kolattukudy, P. E. (1980). Cutin, suberin and waxes. In P. K. Stumpf, & E. V. Conn (Eds.), *The biochemistry of plants: a comprehensive treatise Vol. 4.* (pp. 571–645). New York: Academic Press.

Kolattukudy, P. E., Croteau, R., & Buckner, J. S. (1976). Biochemistry of plant waxes. In P. E. Kolattukudy (Ed.), *Chemistry and biochemistry of natural waxes* (pp. 289–347). Amsterdam: Elsevier.

Lanne, B. S., Ivarrson, P., Johnsson, P., Bergström, G., & Wassgren A.- B. (1989). Biosynthesis of 2-methyl-3-buten-2-ol, a pheromone component of *Ips typographus* (Coleoptera: Scolytidae). *Insect Biochemistry, 19*, 163–168.

Lassance, J.-M. (2010). Journey in the *Ostrinia*, world: from pest to model in chemical ecology. *Journal of Chemical Ecology, 36*, 1155–1169.

Lassance, J.-M., Groot, A. T., Lienard, M. A., Antony, B., Borgwardt, C., Andersson, F., Hedenstrom, E., Heckel, D. G., & Lofstedt, C. (2010). Allelic variation in a fatty-acyl reductase gene causes divergence in moth sex pheromones. *Nature, 466*, 486–489.

Leal, W. S., Zarbin, P. H.G., Wojtasek, H., & Ferreira, J. T. (1999). Biosynthesis of scarab beetle pheromones: Enantioselective 8-hydroxylation of fatty acids. *European Journal of Biochemistry, 259*, 175–180.

Lee, J. M., Choi, M. Y., Han, K. S., & Boo, K. S. (2001). Cloning of the cDNA encoding pheromone biosynthesis activating neuropeptide in Adoxophyes sp. (Lepidoptera: Tortricidae): a new member of the PBAN family. *GenBank Direct submission.*

Levinson, H. Z., Levinson, A. R., Ren, Z., & Mori, K. (1983). Occurrence of a pheromone-producing gland in female tobacco beetles. *Experientia, 39*, 1095–1097.

Liang, D., & Schal, C. (1993). Ultrastructure and maturation of a sex pheromone gland in the female German cockroach, *Blattella germanica. Tissue and Cell, 25*, 763–776.

Liang, D., & Schal, C. (1994). Neural and hormonal regulation of calling behavior in *Blattella germanica* females. *Journal of Insect Physiology, 40*, 251–258.

Liénard, M. A., Lassance, J.-M., Wang, H.-L., Zhao, C.-H., Piskur, J., Johansson, T., & Löfstedt, C. (2010). Elucidation of the sex-pheromone biosynthesis producing 5,7-dodecadienes in *Dendrolimus punctatus* (Lepidoptera: Lasiocampidae) reveals Δ11- and Δ9-desaturases with unusual catalytic properties. *Insect Biochemistry and Molecular Biology, 40*, 440–452.

Liu, W., Jiao, H., O'Connor, M., & Roelofs, W. L. (2002a). Moth desaturase characterized that produces both *Z* and *E* isomers of Δ11-tetradecenoic acids. *Insect Biochemistry and Molecular Biology, 32*, 1489–1495.

Liu, W., Jiao, H., Murray, N. C., O'Connor, M., & Roelofs, W. L. (2002b). Gene characterized for membrane desaturase that produces (*E*)-11 isomers of mono- and diunsaturated fatty acids. *Proceedings National Academy of Sciences USA, 99*, 620–624.

Liu, W., Rooney, A. P., Xue, B., & Roelofs, W. L. (2004). Desaturases from the spotted fireworm moth (*Choristoneura parallela*) shed light on the evolutionary origins of novel moth sex pheromone desaturases. *Gene, 342*, 303–311.

Löfstedt, C., & Bengtsson, M. (1988). Sex pheromone biosynthesis of (*E, E*)-8,10-dodecadienol in codling moth *Cydia pomonella* involves *E*9 desaturation. *Journal of Chemical Ecology, 14*, 903–915.

Lu, F. (1999). *Origin and endocrine regulation of pheromone biosynthesis in the pine bark beetles, Ips pini (Say) and Ips paraconfusus Lanier (Coleoptera: Scolytidae). Ph. D. Dissertation. Biochemistry.* Reno: University of Nevada, p. 152.

Luxova, A., & Svatos, A. (2006). Substrate specificity of membrane-bound alcohol oxidase from the tobacco hornworm moth (*Manduca sexta*) female pheromone glands. *Journal of Molecular Catalysis B: Enzymatic, 38*, 37–42.

Ma, P. W.K., Knipple, D. C., & Roelofs, W. L. (1994). Structural organization of the *Helicoverpa zea* gene encoding the precursor protein for pheromone biosynthesis-activating neuropeptide and other neuropeptides. In *Proceedings National Academy of Sciences, USA, 91*, 6506–6510.

Ma, P. W.K., & Roelofs, W. L. (1995a). Calcium involvement in the stimulation of sex pheromone production by PBAN in the European corn borer, *Ostrinia nubilalis* (Lepidoptera: Pyralidae). *Insect Biochemistry and Molecular Biology, 25*, 467–473.

Ma, P. W.K., & Roelofs, W. L. (1995a). Sites of synthesis and release of PBAN-like factor in female European corn borer, *Ostrinia nubilalis. Journal of Insect Physiology, 41*, 339–350.

Ma, P. W.K., & Roelofs, W. L. (1995b). Anatomy of the neurosecretory cells in the cerebral and subesophageal ganglia of the female European corn borer moth, *Ostrinia nubilalis* (Hubner)(Lepidoptera: Pyralidae). *International Journal of Insect Morphology and Embryology, 24*, 343–359.

Ma, P. W.K., Roelofs, W. L., & Jurenka, R. A. (1996). Characterization of PBAN and PBAN-encoding gene neuropeptides in the central nervous system of the corn earworm moth, *Helicoverpa zea. Journal of Insect Physiology, 42*, 257–266.

Ma, P. W.K., Garden, R. W., Niermann, J. T., O'Connor, M., Sweedler, J. V., & Roelofs, W. L. (2000). Characterizing the Hez-PBAN gene products in neuronal clusters with immunocytochemistry and MALDI MS. *Journal of Insect Physiology, 46*, 221–230.

Ma, P. W.K., & Ramaswamy, S. B. (2003). Biology and ultrastructure of sex pheromone producing tissue. In G. J. Blomquist, & R. G. Vogt (Eds.), *Insect Pheromone Biochemistry and Molecular Biology* (pp. 19–52). London: Elsevier Academic Press.

Marco, M. -P., Fabriàs, G., Lázaro, G., & Camps, F. (1996). Evidence for both humoral and neural regulation of sex pheromone biosynthesis in *Spodoptera littoralis. Archives of Insect Biochemistry and Physiology, 31*, 157–168.

Martin, D., Wang, S.-F., & Raikhel, A. S. (2001). The vitellogenin gene of the mosquito *Aedes aegypti* is a direct target of ecdysteroid receptor. *Molecular and Cellular Endocrinology*, *173*, 75–86.

Martin, D., Bohlmann, J., Gershenzon, J., Francke, W., & Seybold, S. J. (2003). A novel sex-specific and inducible monoterpene synthase activity associated with a pine bark beetle, the pine engraver, *Ips pini*. *Naturwissenschaften*, *90*, 173–179.

Martinez, T., Fabriàs, G., & Camps, F. (1990). Sex pheromone biosynthetic pathway in *Spodoptera littoralis* and its activation by a neurohormone. *Journal of Biological Chemistry*, *265*, 1381–1387.

Mas, E., Lloria, J., Quero, C., Camps, F., & Fabriàs, G. (2000). Control of the biosynthetic pathway of *Sesamia nonagrioides* sex pheromone by the pheromone biosynthesis activating neuropeptide. *Insect Biochemistry and Molecular Biology*, *30*, 455–459.

Masler, E. P., Raina, A. K., Wagner, R. M., & Kochansky, J. P. (1994). Isolation and identification of a pheromonotropic neuropeptide from the brain-suboesophageal ganglion complex of *Lymantria dispar*: A new member of the PBAN family. *Insect Biochemistry Molecular Biology*, *24*, 829–836.

Matsumoto, S. (2010). Molecular mechanisms underlying sex pheromone production in moths. *Bioscience Biotechnology and Biochemistry*, *74*, 223–231.

Matsumoto, S., Kitamura, A., Nagasawa, H., Kataoka, H., Orikasa, C., Mitsui, T., & Suzuki, A. (1990). Functional diversity of a neurohormone produced by the suboesophageal ganglion: Molecular identity of melanization and reddish colouration hormone and pheromone biosynthesis activating neuropeptide. *Journal of Insect Physiology*, *36*, 427–432.

Matsumoto, S., Hull, J. J., Ohnishi, A., Moto, K., & Fonagy, A. (2007). Molecular mechanisms underlying sex pheromone production in the silkmoth, *Bombyx mori*: Characterization of the molecular components involved in bombykol biosynthesis. *Journal of Insect Physiology*, *53*, 752–759.

Matsumoto, S., Ozawa, R. A., Nagamine, T., Kim, G., Uchiumi, K., Shono, T., & Mitsui, T. (1995). Intracellular transduction in the regulation of pheromone biosynthesis of the silkworm, *Bombyx mori*: suggested involvement of calmodulin and phosphoprotein phosphatase. *Bioscience Biotechnology and Biochemistry*, *59*, 560–562.

Matsuoka, K., Tabunoki, H., Kawai, T., Ishikawa, S., Yamamoto, M., Sato, R., & Ando, T. (2006). Transport of a hydrophobic biosynthetic precursor by lipophorin in the hemolymph of a geometrid female moth which secretes an epoxyalkenyl sex pheromone. *Insect Biochemistry and Molecular Biology*, *36*, 576–583.

Matsuoka, K., Yamamoto, M., Yamakawa, R., Muramatsu, M., Naka, H., Kondo, Y., & Ando, T. (2008). Identification of novel C_{20} and C_{22} trienoic acids from arctiid and geometrid female moths that produce polyenyl type II sex pheromone components. *Journal of Chemical Ecology*, *34*, 1437–1445.

Menon, M. (1970). Hormone-pheromone relationships in the beetle *Te. molitor*. *Journal of Insect Physiology*, *16*, 1123–1139.

Millar, J. G. (2000). Polyene hydrocarbons and epoxides: a second major class of lepidopteran sex attractant pheromones. *Annual Review of Entomology*, *45*, 575–604.

Millar, J. G. (2010). Polyene hydrocarbons, epoxides, and related compounds as components of lepidopteran pheromone blends. In G. J. Blomquist, & A.-G. Bagneres (Eds.), *Insect Hydrocarbons: Biology, Biochemistry and Chemical Ecology* (pp. 390–447). Cambridge: Cambridge University Press.

Miller, J. R., & Roelofs, W. L. (1980). Individual variation in sex pheromone component ratios in two populations of the redbanded leafroller moth, *Argyrotaenia velutinana*. *Environmental Entomology*, *9*, 359–363.

Mitlin, N., & Hedin, P. A. (1974). Biosynthesis of grandlure, the pheromone of the boll weevil, *Anthonomus grandis*, from acetate, mevalonate, and glucose. *Journal of Insect Physiology*, *20*, 1825–1831.

Miyamoto, T., Yamamoto, M., Ono, A., Ohtani, K., & Ando, T. (1999). Substrate specificity of the epoxidation reaction in sex pheromone biosynthesis of the Japanese giant looper (Lepidoptera: Geometridae). *Insect Biochemistry and Molecular Biology*, *29*, 63–69.

Mori, K. (2008). Synthesis of all the six components of the female-produced contact sex pheromone of the German cockroach, *Blattella germanica* (L.). *Tetrahedron*, *64*, 4060–4071.

Morse, D., & Meighen, E. A. (1987a). Biosynthesis of the acetate ester precursors of the spruce budworm sex pheromone by an acetyl CoA: fatty alcohol acetyltransferase. *Insect Biochemistry*, *17*, 53–59.

Morse, D., & Meighen, E. A. (1987b). Pheromone biosynthesis: Enzymatic studies in lepidoptera. In G. D. Prestwich, & G. J. Blomquist (Eds.), *Pheromone biochemistry* (pp. 121–158). London: Academic Press.

Moto, K., Yoshiga, T., Yamamoto, M., Takahashi, S., Okano, K., Ando, T., Nakata, T., & Matsumoto, S. (2003). Pheromone gland-specific fatty-acyl reductase of the silkmoth, *Bombyx mori*. *Proceedings of the National Academy of Sciences., USA*, *100*, 9156–9161.

Mpuru, S., Reed, J. R., Reitz, R. C., & Blomquist, G. J. (1996). Mechanism of hydrocarbon biosynthesis from aldehyde in selected insect species: requirement for O_2 and NADPH and carbonyl group released as CO_2. *Insect Biochemistry and Molecular Biology*, *26*, 203–208.

Mpuru, S., Blomquist, G. J., Schal, C., Kuenzli, M., Dusticier, G., Roux, M., & Bagneres A.-G. (2001). Effect of age and sex on the production of internal and external hydrocarbons and pheromones in the housefly, *Musca domestica*. *Insect Biochemistry and Molecular Biology*, *31*, 139–155.

Mundall, E. C., Szibbo, C. M., & Tobe, S. S. (1983). Vitellogenin induced in adult male *Diploptera punctata* by juvenile hormone and juvenile hormone analogue: identification and quantitative aspects. *Journal of Insect Physiology*, *29*, 201–207.

Nardi, J., Gilg-Young, A., Ujhelyi, E., Tittiger, C., Lehane, M., & Blomquist, G. (2002). Specialization of midgut cells for synthesis of male isoprenoid pheromone components in two scolytid beetles, *Dendroctonus jeffreyi* and *Ips pini*. *Tissue and Cell*, *226*, 1–11.

Neupert, S., Huetteroth, W., Schachtner, J., & Predel, R. (2009). Conservation of the function counts: homologous neurons express sequence-related neuropeptides that originate from different genes. *Journal of Neurochemistry, 111*, 757–765.

Nishida, R., & Fukami, H. (1983). Female sex pheromone of the German cockroach, *Blattella germanica. Memoirs of the College of Agriculture of the Kyoto University, 122*, 1–24.

Nishida, R., Fukami, H., & Ishii, S. (1974). Sex pheromone of the German cockroach (*Blattella germanica* L.) responsible for male wing-raising: 3,11-dimethyl-2-nonacosanone. *Experientia, 30*, 978–979.

Nojima, S., Schal, C., Webster, F. X., Santangelo, R. G., & Roelofs, W. L. (2005). Identification of the sex pheromone of the German cockroach, *Blattella germanica. Science, 307*, 1104–1106.

Nojima, S., Sakuma, M., Nishida, R., & Kuwahara, Y. (1999). A glandular gift in the German cockroach, *Blattella germanica* (L.) (Dictyoptera: Blattellidae): The courtship feeding of a female on secretions from male tergal glands. *Journal of Insect Behavior, 12*, 627–640.

Ohnishi, A., Hull, J. J., & Matsumoto, S. (2006). Targeted disruption of genes in the *Bombyx mori* sex pheromone biosynthetic pathway. *Proceedings of the National Academy of Sciences, USA, 103*, 4398–4403.

Ono, A., Imai, T., Inomata, S.-I., Watanabe, A., & Ando, T. (2002). Biosynthetic pathway for production of a conjugated dienyl sex pheromone of a plusiinae moth, *Thysanoplusia intermixta. Insect Biochemistry and Molecular Biology, 32*, 701–708.

Osorio, S., Piulachs, M. D., & Bellés, X. (1998). Feeding and activation of the corpora allata in the cockroach *Blattella germanica* (L.) (Dictyoptera, Blattellidae). *Journal of Insect Physiology, 44*, 31–38.

Ozawa, R. A., Ando, T., Nagasawa, H., Kataoka, H., & Suzuki, A. (1993). Reduction of the acyl group: the critical step in bombykol biosynthesis that is regulated in vitro by the neuropeptide hormone in the pheromone gland of Bombyx mori. *Bioscience Biotechnology and Biochemistry, 57*, 2144–2147.

Paine, T. D., Millar, J. G., Hanlon, C. C., & Hwang J.-S. (1999). Identification of semiochemicals associated with Jeffrey pine beetle, *Dendroctonus jeffreyi. Journal of Chemical Ecology, 25*, 433–453.

Park, Y. I., & Ramaswamy, S. B. (1998). Role of brain, ventral nerve cord and corpora cardiaca-corpora allata complex in the reproductive behavior of female tobacco budworm (Lepidoptera: Noctuidae). *Annals of the Entomological Society of America, 91*, 329–334.

Park, Y., Kim, Y.-J., & Adams, M. E. (2002). Identification of G protein-coupled receptors for *Drosophila* PRXamide peptides, CCAP, corazonin, and AKH supports a theory of ligand-receptor coevolution. *Proceedings of the National Academy of Sciences USA. 99*, 11423–11428.

Pennanec'h, M., Bricard, L., Kunesch, G., & Jallon, J. M. (1997). Incorporation of fatty acids into cuticular hydrocarbons of male and female *Drosophila melanogaster. Journal of Insect Physiology, 43*, 1111–1116.

Percy-Cunningham, J. E., & MacDonald, J. A. (1987). Biology and ultrastructure of sex pheromone-producing glands. In G. D. Prestwich, & G. J. Blomquist (Eds.), *Pheromone Biochemistry* (pp. 27–75). Orlando, Florida: Academic Press.

Perez, A. L., Gries, R., Gries, G., & Oehlschlager, A. C. (1996). Transformation of presumptive precursors to frontalin and exo-brevicomin by bark beetles and the West Indian sugarcane weevil (Coleoptera). *Bioorganic and Medicinal Chemistry, 4*, 445–450.

Petroski, R. J., Bartelt, R. J., & Weisleder, D. (1994). Biosynthesis of (2E,4E,6E)-5 ethyl-3-methyl-2,4,6-nonatriene: The aggregation pheromone of *Carpophilus freemani* (Coleoptera: Nitidulidae). *Insect Biochemistry and Molecular Biology, 24*, 69–78.

Pho, D. B., Pennanec'h, M., & Jallon, J. M. (1996). Purification of adult *Drosophila melanogaster* lipophorin and its role in hydrocarbon transport. *Archives of Insect Biochemistry and Physiology, 31*, 289–303.

Picimbon, J.-F., Becard, J.-M., Sreng, L., Clement, J.-L., & Gadenne, C. (1995). Juvenile hormone stimulates pheromonotropic brain factor release in the female black cutworm, *Agrotis ipsilon. Journal of Insect Physiology, 41*, 377–382.

Piulachs, M. D., Maestro, J. L., & Bellés, X. (1992). Juvenile hormone production and accessory gland development during sexual maturation of male *Blattella germanica* (L.) (Dictyoptera, Blattellidae). *Comparative Biochemistry and Physiology A, 102*, 477–480.

Plarre, R., & Vanderwel, D. C. (1999). Stored-product beetles. In J. Hardie, & A. K. Minks (Eds.), *Pheromones of non-lepidopteran insects associated with agricultural plants* (pp. 149–198). New York: CABI Publishing.

Plettner, E., Slessor, K. N., Winston, M. L., & Oliver, J. E. (1996). Caste-selective pheromone biosynthesis in honeybees. *Science, 271*, 1851–1853.

Plettner, E., Slessor, K. N., & Winston, M. L. (1998). Biosynthesis of mandibular acids in honeybees *(Apis mellifera):* de novo synthesis, route of fatty acid hydroxylation and caste selective β-oxidation. *Insect Biochemistry and Molecular Biology, 28*, 31–42.

Prestwich, G. D., & Blomquist, G. J. (1987). *Pheromone Biochemistry.* London: Academic Press, p. 565.

Quennedey, A. (1998). Insect epidermal gland cells: ultrastructure and morphogenesis. In F. W. Harrison, & M. Locke (Eds.), *Microscopic Anatomy of Invertebrates Vol 11A.,* (pp. 177–207). New York: Wiley-Liss, Inc.

Rafaeli, A. (1994). Pheromonotropic stimulation of moth pheromone gland cultures in vitro. *Archives of Insect Biochemistry and Physiolology, 25*, 287–299.

Rafaeli, A. (2009). Pheromone biosynthesis activating neuropeptide (PBAN): Regulatory role and mode of action. *General and Comparative Endocrinology, 162*, 69–78.

Rafaeli, A., & Gileadi, C. (1995). Modulation of the PBAN-stimulated pheromonotropic activity in *Helicoverpa armigera. Insect Biochemistry and Molecular Biology, 25*, 827–834.

Rafaeli, A., & Gileadi, C. (1996). Down regulation of pheromone biosynthesis: cellular mechanisms of pheromonostatic responses. *Insect Biochemistry and Molecular Biology, 26*, 797–808.

Rafaeli, A., Soroker, V., Kamensky, B., & Raina, A. K. (1990). Action of pheromone biosynthesis activating neuropeptide on in vitro pheromone glands of *Heliothis armigera* females. *Journal of Insect Physiology, 36*, 641–646.

Rafaeli, A., Hirsch, J., Soroker, V., Kamensky, B., & Raina, A. K. (1991). Spatial and temporal distribution of pheromone biosynthesis-activating neuropeptide in *Helicoverpa* (Heliothis) armigera using RIA and in vitro bioassay. *Archives of Insect Biochemistry and Physiology, 18,* 119–129.

Rafaeli, A., Soroker, V., Hirsch, J., Kamensky, B., & Raina, A. K. (1993). Influence of photoperiod and age on the competence of pheromone glands and on the distribution of immunoreactive PBAN in *Helicoverpa* spp. *Archives of Insect Biochemistry and Physiology, 22,* 169–180.

Rafaeli, A., Soroker, V., Kamensky, B., Gileadi, C., & Zisman, U. (1997a). Physiological and cellular mode of action of pheromone biosynthesis activating neuropeptide (PBAN) in the control of pheromonotropic activity of female moths. In R. T. Cardé, & A. K. Minks (Eds.), *Insect Pheromone Research: New Directions* (pp. 74–82). Chapman & Hall.

Rafaeli, A., Gileadi, C., & Cao, M. (1997b). Physiological mechanisms of pheromonostatic responses: effects of adrenergic agonists and antagonists on moth (*Helicoverpa armigera*) pheromone biosynthesis. *Journal of Insect Physiology, 43,* 261–269.

Rafaeli, A., & Gileadi, C. (1999). Synthesis and biological activity of a photoaffinity-biotinylated pheromone-biosynthesis activating neuropeptide (PBAN) analog. *Peptides, 20,* 787–794.

Rafaeli, A., Zakharova, T., Lapsker, Z., & Jurenka, R. A. (2003). The identification of an age- and female-specific putative PBAN membrane-receptor protein in pheromone glands of *Helicoverpa armigera*: possible up-regulation by juvenile hormone. *Insect Biochemistry and Molecular Biology, 33,* 371–380.

Rafaeli, A., & Jurenka, R. A. (2003). PBAN regulation of pheromone biosynthesis in female moths. In G. J. Blomquist, & R. G. Vogt (Eds.), *Insect Pheromone Biochemistry and Molecular Biology* (pp. 107–136). London: Elsevier Academic Press.

Rafaeli, A., Bober, R., Becker, L., Choi, M.-Y., Fuerst, E. J., & Jurenka, R. (2007). Spatial distribution and differential expression of the PBAN receptor in tissues of adult *Helicoverpa* spp. (Lepidoptera: Noctuidae). *Insect Molecular Biology, 16,* 287–293.

Raina, A. K., & Klun, J. A. (1984). Brain factor control of sex pheromone production in the female corn earworm moth. *Science, 225,* 531–533.

Raina, A. K., & Menn, J. J. (1987). Endocrine regulation of pheromone production in Lepidoptera. In G. D. Prestwich, & G. J. Blomquist (Eds.), *Pheromone Biochemistry* (pp. 159–174). Orlando, Florida: Academic Press.

Raina, A. K., Jaffe, H., Kempe, T. G., Keim, P., Blacher, R. W., Fales, H. M., Riley, C. T., Klun, J. A., Ridgeway, R. L., & Hayes, D. K. (1989). Identification of a neuropeptide hormone that regulates sex pheromone production in female moths. *Science, 244,* 796–798.

Ramaswamy, S. B., Jurenka, R. A., Linn, C. E., & Roelofs, W. L. (1995). Evidence for the presence of a pheromonotropic factor in hemolymph and regulation of sex pheromone production in *Helicoverpa zea. Journal of Insect Physiology, 41,* 501–508.

Reed, J. R., Vanderwel, D., Choi, S., Pomonis, J. G., Reitz, R. C., & Blomquist, G. J. (1994). Unusual mechanism of hydrocarbon formation in the housefly: cytochrome P450 converts aldehyde to the sex pheromone component (Z)-9-tricosene and CO_2. *Proceedings of the National Academy of Science. USA, 91,* 10,000–10,004.

Reed, J. R., Quilici, D. R., Blomquist, G. J., & Reitz, R. C. (1995). Proposed mechanism for the cytochrome P450 catalyzed conversion of aldehyde to hydrocarbon in the house fly, *Musca domestica. Biochemistry, 34,* 16,221–16,227.

Renwick, J. A.A., Hughes, P. R., & Krull, I. S. (1976). Selective production of cis- and trans-verbenol from (–)- and (+)-α-pinene by a bark beetle. *Science, 191,* 199–201.

Roelofs, W. L., Cardé, R. T., Benz, G., & von Salis, G. (1971). Sex attractant of the larch bud moth found by electroantennogram method. *Experientia, 27,* 1438.

Roelofs, W. L., Du, J.-W., Linn, C. E., Glover, T. J., & Bjostad, L. B. (1986). The potential for genetic manipulation of the redbanded leafroller moth sex pheromone blend. In M. D. Heuttel (Ed.), *Evolutionary Genetics of Invertebrate Behavior* (pp. 263–272). New York: Plenum Press.

Roelofs, W. L., & Jurenka, R. A. (1996). Biosynthetic enzymes regulating ratios of sex pheromone components in female redbanded leafroller moths. *Bioorganic and Medicinal Chemistry Letters, 4,* 461–466.

Roelofs, W. L., Liu, W., Hao, G., Jiao, H., Rooney, A. P., & Linn, C. E., Jr. (2002). Evolution of moth sex pheromones via ancestral genes. *Proceedings of the National Academy of Sciences. USA, 99,* 13621–13626.

Roelofs, W. L., & Wolf, W. A. (1988). Pheromone biosynthesis in Lepidoptera. *Journal of Chemical Ecology, 14,* 2019–2031.

Roelofs, W. L., & Rooney, A. P. (2003). Molecular genetics and evolution of pheromone biosynthesis in Lepidoptera. *Proceedings of the National Academy of Sciences. USA, 100,* 9179–9184.

Romer, F. (1991). The oenocytes of insects: differentiation, changes during molting, and their possible involvement in the secretion of moulting hormone. In A. P. Gupta (Ed.), *Morphogenetic Hormones of Arthropods Vol. 3.* (pp. 542–566). New Brunswick, New Jersey: Rutgers University Press.

Rosell, G., Hospital, S., Camps, F., & Guerrero, A. (1992). Inhibition of a chain shortening step in the biosynthesis of the sex pheromone of the Egyptian armyworm *Spodoptera littoralis. Insect Biochemistry and Molecular Biology, 22,* 679–685.

Rosenfield, C., You, K. M., Marsella-Herrick, P., Roelofs, W. L., & Knipple, D. C. (2001). Structural and functional conservation and divergence among acyl-CoA desaturases of two noctuid species, the corn earworm, *Helicoverpa zea,* and the cabbage looper, *Trichoplusia ni. Insect Biochemistry Molecular Biology, 31,* 949–964.

Rule, G. S., & Roelofs, W. L. (1989). Biosynthesis of sex pheromone components from linolenic acid in Arctiid moths. *Archives of Insect Biochemistry and Physiology, 12,* 89–97.

Sandstrom, P., Welch, W. H., Blomquist, G. J., & Tittiger, C. (2006). Functional expression of a bark beetle cytochrome P450 that hydroxylates myrcene to ipsdienol. *Insect Biochemistry Molecular Biology, 36,* 835–845.

Sandstrom, P., Ginzel, M. D., Bearfield, J. C., Welch, W. H., Blomquist, G. J., & Tittiger, C. (2008). Myrcene hydroxylases do not determine enantiomeric composition of pheromonal ipsdienol in *Ips* spp. *Journal of Chemical Ecology, 34,* 1584–1592.

Sato, Y., Oguchi, M., Menjo, N., Imai, K., Saito, H., Ikeda, M., Isobe, M., & Yamashita, O. (1993). Precursor polyprotein for multiple neuropeptides secreted from the suboesophageal ganglion of the silkworm *Bombyx mori*: Characterization of the cDNA encoding the diapause hormone precursor and identification of additional peptides. *Proceedings of the National Academy of Sciences. USA, 90,* 3251–3255.

Schal, C. (1988). Regulation of pheromone synthesis and release in cockroaches. In A. Zabza, F. Sehnal, & D. L. Denlinger (Eds.), *Endocrinological Frontiers in Physiological Insect Ecology* (pp. 695–700). Poland: Technical University of Wroclaw Press.

Schal, C., & Smith, A. F. (1990). Neuroendocrine regulation of pheromone production in cockoraches. In I. Huber, E. P. Masler, & B. R. Rao (Eds.), *Cockroaches as Models for Neurobiology: Applications in Biomedical Research* (pp. 179–200). Boca Raton: CRC Press.

Schal, C., & Chiang A.-S. (1995). Hormonal control of sexual receptivity in cockroaches. *Experientia, 51,* 994–998.

Schal, C., Burns, E. L., & Blomquist, G. J. (1990a). Endocrine regulation of female contact sex pheromone production in the German cockroach, *Blattella germanica. Physiological Entomology, 15,* 81–91.

Schal, C., Burns, E. L., Jurenka, R. A., & Blomquist, G. J. (1990b). A new component of the female sex pheromone of *Blattella germanica* (L.) (Dictyoptera: Blattellidae) and interaction with other pheromone components. *Journal of Chemical Ecology, 16,* 1997–2008.

Schal, C., Liang, D., Hazarika, L. K., Charlton, R. E., & Roelofs, W. L. (1992). Site of pheromone production in female *Supella longipalpa* (Dictyoptera: Blattellidae): Behavioral, electrophysiological, and morphological evidence. *Annals of the Entomological Society of America, 85,* 605–611.

Schal, C., Chiang, A.-S., Burns, E. L., Gadot, M., & Cooper, R. A. (1993). Role of the brain in juvenile hormone synthesis and oocyte development: Effects of dietary protein in the cockroach *Blattella germanica* (L.). *Journal of Insect Physiology, 39,* 303–313.

Schal, C., Gu, X., Burns, E. L., & Blomquist, G. J. (1994). Patterns of biosynthesis and accumulation of hydrocarbons and contact sex pheromone in the female German cockroach, *Blattella germanica. Archives of Insect Biochemistry and Physiology, 25,* 375–391.

Schal, C., Sevala, V., & Cardé, R. T. (1998a). Novel and highly specific transport of a volatile sex pheromone by hemolymph lipophorin in moths. *Naturwissenschaften, 85,* 339–342.

Schal, C., Sevala, V. L., Young, H. P., & Bachmann, J. A.S. (1998b). Synthesis and transport of hydrocarbons: cuticle and ovary as target tissues. *American Zoologist, 38,* 382–393.

Schal, C., Liang, D., & Blomquist, G. J. (1996). Neural and endocrine control of pheromone production and release in cockroaches. In R. T. Cardé, & A. K. Minks (Eds.), *Insect Pheromone Research: New Directions* (pp. 3–20). New York: Chapman and Hall.

Schal, C., Sevala, V., Capurro, M. d.L., Snyder, T. E., Blomquist, G. J., & Bagnères A.-G. (2001). Tissue distribution and lipophorin transport of hydrocarbon and sex pheromones in the house fly, *Musca domestica. Journal of Insect Sciences, 1,* 12.

Schal, C., Fan, Y., & Blomquist, G. J. (2003). Regulation of pheromone biosynthesis, transport and emission in cockroaches. In G. J. Blomquist, & R. G. Vogt (Eds.), *Insect Pheromone Biochemistry and Molecular Biology* (pp. 283–322). London: Elsevier Academic Press.

Schirmer, A., Rude, M. A., Xuezhi, L., Popova, E., & del Cardayre, S. B. (2010). Microbial biosynthesis of alkanes. *Science, 329,* 559–562.

Schlyter, F., & Birgersson, G. S. (1999). Forest beetles. In J. Hardie, & A. K. Minks (Eds.), *Pheromones of Non-Lepidopteran Insects Associated with Agricultural Plants* 113–148.

Sevala, V., Shu, S., Ramaswamy, S. B., & Schal, C. (1999). Lipophorin of female *Blattella germanica* (L.): Characterization and relation to hemolymph titers of juvenile hormone and hydrocarbons. *Journal of Insect Physiology, 45,* 431–441.

Seybold, S. J., Quilici, D. R., Tillman, J. A., Vanderwel, D., Wood, D. L., & Blomquist, G. J. (1995). De novo biosynthesis of the aggregation pheromone components ipsenol and ipsdienol by the pine bark beetles *Ips paraconfusus* Lanier and *Ips pini* (Say) (Coleoptera: Scolytidae). *Proceedings of the National Academy of Sciences. USA, 92,* 8393–8397.

Seybold, S. J., Bohlmann, J., & Raffa, K. F. (2000). Biosynthesis of coniferophagous bark beetle pheromones and conifer isoprenoids: Evolutionary perspective and synthesis. *Canadian Entomologist, 132,* 697–753.

Seybold, S. J., & Vanderwel, D. (2003). Biosynthesis and endocrine regulation of pheromone production in the Coleoptera. In G. J. Blomquist, & R. G. Vogt (Eds.), *Insect Pheromone Biochemistry and Molecular Biochemistry* (pp. 137–200). London: Elsevier Academic Press/Elsevier.

Shirangi, T. R., Dufour, H. D., Williams, T. M., & Carroll, S. B. (2009). Rapid evolution of sex pheromone-producing enzyme expression in *Drosophila. PLoS Biology, 7*(8), e1000168, doi:10.1371/journal.pbio.1000168.

Shorey, H. H. (1974). Environmental and physiological control of insext sex pheromone behavior. In M. C. Birch (Ed.), *Pheromones* (pp. 62–80). New York: American Elsevier.

Silverstein, R. M., Rodin, J. O., & Wood, D. L. (1966). Sex attractants in frass produced by male *Ips confusus* in ponderosa pine. *Science, 154,* 509–510.

Sirugue, D., Bonnard, O., Le Quere, J. L., Farine, J.-P., & Brossut, R. (1992). 2-Methylthiazolidine and 4-ethylguaiacol, male sex pheromone components of the cockroach *Nauphoeta cinerea* (Dictyoptera, Blaberidae): A reinvestigation. *Journal of Chemical Ecology, 18,* 2261–2276.

Smith, A. F., & Schal, C. (1990a). Corpus allatum control of sex pheromone production and calling in the female brown-banded cockroach, *Supella longipalpa* (F.) (Dictyoptera: Blattellidae). *Journal of Insect Physiology, 36,* 251–257.

Smith, A. F., & Schal, C. (1990b). The physiological basis for the termination of pheromone-releasing behaviour in the female brown-banded cockroach, *Supella longipalpa* (F.) (Dictyoptera: Blattellidae). *Journal of Insect Physiology, 36,* 369–373.

Smith, V. L., Doyle, K. E., Maune, J. F., Munjaal, R. P., & Beckingham, K. (1987). Structure and sequence of the *Drosophila melanogaster* calmodulin gene. *Journal of Molecular Biology, 196,* 471–485.

Soroker, V., & Rafaeli, A. (1989). In vitro hormonal stimulation of [14C]acetate incorporation by *Heliothis armigera* pheromone glands. *Insect Biochemistry, 19,* 1–5.

Sreng, I., Glover, T., & Roelofs, W. (1989). Canalization of the redbanded leafroller moth sex pheromone blend. *Archives of Insect Biochemistry and Physiology, 10,* 73–82.

Sreng, L. (1990). Seducin, male sex pheromone of the cockroach *Nauphoeta cinerea:* Isolation, identification, and bioassay. *Journal of Chemical Ecology, 16,* 2899–2912.

Sreng, L. (1998). Apostosis-inducing brain factors in maturation of an insect sex pheromone gland during differentiation. *Differentiation, 63,* 53–58.

Sreng, L., & Quennedey, A. (1976). Role of a temporary ciliary structure in the morphogenesis of insect glands. An electron microscope study of the tergal glands of male *Blattella germanica* L. (Dictyoptera, Blattellidae). *Journal of Ultrastructural Research, 56,* 78–95.

Sreng, L., Leoncini, I., & Clement, J. L. (1999). Regulation of sex pheromone production in the male *Nauphoeta cinerea* cockroach: role of brain extracts, corpora allata (CA), and juvenile hormone (JH). *Archives of Insect Biochemistry and Physiology, 40,* 165–172.

Stanley-Samuelson, D. W., Jurenka, R. A., Cripps, C., Blomquist, G. J., & deRenobales, M. (1988). Fatty acids in insects: Composition, metabolism and biological significance. *Archives of Insect Biochemistry and Physiology, 9,* 1–33.

Stern, P. S., Yu, L., Choi, M.-Y., Jurenka, R. A., Becker, L., & Rafaeli, A. (2007). Molecular modeling of the binding of pheromone biosynthesis activating neuropeptide to its receptor. *Journal of Insect Physiology, 53,* 803–818.

Subchev, M., & Jurenka, R. A. (2001). Identification of the pheromone in the hemolymph and cuticular hydrocarbons from the moth *Scoliopteryx libatrix* L. (Lepidoptera: Noctuidae). *Archives of Insect Biochemistry and Physiology, 47,* 35–43.

Taban, A. H., Fu, J., Blake, J., Awano, A., Tittiger, C., & Blomquist, G. J. (2006). Site of pheromone biosynthesis and isolation of HMG-CoA reductase in the cotton boll weevil, *Anthonomis grandis. Archives Insect Biochemistry and Physiology, 62,* 153–163.

Tada, S., & Leal, W. S. (1997). Localization and morphology of sex pheromone glands in scarab beetles. *Journal of Chemical Ecology, 23,* 903–915.

Tamaki, Y. (1985). Sex pheromones. In G. A. Kerkut, & L. I. Gilbert (Eds.), *Comprehensive Insect Physiology, Biochemistry and Pharmacology Vol. 9.* (pp. 145–191). Oxford: Pergamon Press.

Tang, J. D., Charlton, R. E., Cardé, R. T., & Yin C.-M. (1987). Effect of allatectomy and ventral nerve cord transection on calling, pheromone emission and pheromone production in *Lymantria dispar. Journal of Insect Physiology, 33,* 469–476.

Tang, J. D., Charlton, R. E., Jurenka, R. A., Wolf, W. A., Phelan, P. L., Sreng, L., & Roelofs, W. L. (1989). Regulation of pheromone biosynthesis by a brain hormone in two moth species. *Proceedings of the National Academy of Sciences. USA, 86,* 1806–1810.

Takeuchi, N., & Chino, H. (1993). Lipid transfer particle in the hemolymph of the American cockroach: Evidence for its capacity to transfer hydrocarbons between lipophorin particles. *Journal of Lipid Research, 34,* 543–551.

Teal, P. E. A., & Tumlinson, J. H. (1987). The role of alcohols in pheromone biosynthesis by two noctuid moths that use acetate pheromone components. *Archives of Insect Biochemistry and Physiology, 4,* 261–269.

Teal, P. E.A., & Tumlinson, J. H. (1988). Properties of cuticular oxidases used for sex pheromone biosynthesis by *Heliothis zea. Journal of Chemical Ecology, 14,* 2131–2145.

Thompson, A. C., & Mitlin, N. (1979). Biosynthesis of the sex pheromone of the male boll weevil from monoterpene precursors. *Insect Biochemistry, 9,* 293–294.

Thyagaraja, B. S., & Raina, A. K. (1994). Regulation of pheromone production in the gypsy moth, *Lymantria dispar,* and development of an in vitro bioassay. *Journal of Insect Physiology, 40,* 969–974.

Tillman, J.A., Holbrook, G. L., Dallara, P., Schal, C., Wood, D. L., Blomquist, G. J., & Seybold, S. J. (1998). Endocrine regulation of de novo aggregation pheromone biosynthesis in the pine engraver, *Ips pini* (Say) (Coleoptera: Scolytidae). *Insect Biochemistry and Molecular Biology, 28,* 705–715.

Tillman, J. A., Seybold, S. J., Jurenka, R. A., & Blomquist, G. J. (1999). Insect Pheromones - an overview of biosynthesis and endocrine regulation. *Insect Biochemistry and Molecular Biology, 29,* 481–514.

Tillman, J. A., Lu, F., Goodard, L. M., Donaldson, Z., Dwinell, S. C., Tittiger, C., Hall, G. M., Storer, A. J., Blomquist, G. J., & Seybold, S. J. (2004). Juvenile hormone regulates de novo isoprenoid aggregation pheromone biosynthesis in pine bark beetles, *Ips* spp. (Coleoptera: Scolytidae) through transcriptional control of HMG-CoA reductase. *Journal of Chemical Ecology, 30,* 2459–2494.

Tillman-Wall, J. A., Vanderwel, D., Kuenzli, M. E., Reitz, R. C., & Blomquist, G. J. (1992). Regulation of sex pheromone biosynthesis in the housefly, *Musca domestica:* relative contribution of the elongation and reductive step. *Archives of Biochemistry and Biophysics, 299,* 92–99.

Tittiger, C. (2003). Molecular biology of bark beetle pheromone production and endocrine regulation. In G. J. Blomquist, & R. G. Vogt (Eds.), *Insect Pheromone Biochemistry and Molecular Biology* (pp. 201–230). London: Elsevier Academic Press.

Tittiger, C., Blomquist, G. J., Ivarsson, P., Borgeson, C. E., & Seybold, S. J. (1999). Juvenile hormone regulation of HMG-R gene expression in the bark beetle *Ips paraconfusus* (Coleoptera: Scolytidae): implications for male aggregation pheromone biosynthesis. *Cellular and Molecular Life Sciences, 55,* 121–127.

Tokro, P. G., Brossut, R., & Sreng, L. (1993). Studies on the sex pheromone of female *Blattella germanica* L. *Insect Science and its Application, 14,* 115–126.

Torfs, P., Nieto, J., Cerstiaens, A., Boon, D., Baggerman, G., Poulos, C., Waelkens, E., Derua, R., Calderon, J., De Loof, A., & Schoofs, L. (2001). Pyrokinin neuropeptides in a crustacean - Isolation and identification in the white shrimp *Penaeus vannamei. European Journal of Biochemistry, 268,* 149–154.

Tsfadia, O., Azrielli, A., Falach, L., Zada, A., Roelofs, W., & Rafaeli, A. (2008). Pheromone biosynthetic pathways: PBAN-regulated rate-limiting steps and differential expression of desaturase genes in moth species. *Insect Biochemistry and Molecular Biology, 38,* 552–567.

Tumlinson, J. H., Fang, N., & Teal, P. E.A. (1997). The effect of PBAN on conversion of fatty acyls to pheromone aldehydes in female I. In R. T. Cardé, & A. K. Minks (Eds.), *Insect Pheromone Research: New Directions* (pp. 54–55). New York: Chapman & Hall.

Vanderwel, D. (1994). Factors affecting pheromone production in beetles. *Archives of Insect Biochemistry and Physiology, 25,* 347–362.

Vanderwel, D., & Oehlschlager, A. C. (1987). Biosynthesis of pheromones and endocrine regulation of pheromone production in Coleoptera. In G. D. Prestwich, & G. J. Blomquist (Eds.), *Pheromone Biochemistry* (pp. 175–215). Orlando: Academic Press.

Vaz, A. H., Jurenka, R. A., Blomquist, G. J., & Reitz, R. C. (1988). Tissue and chain length specificity of the fatty acyl-CoA elongation system in the American cockroach. *Archives of Biochemistry and Biophysics, 267,* 551–557.

Veelaert, D., Schoofs, L., Verhaert, P., & De Loof, A. (1997). Identification of two novel peptides from the central nervous system of the desert locust, *Schistocerca gregaria. Biochemical Biophysical Research Communications, 241,* 530–534.

Vogel, H., Heidel, A. J., Heckel, D. G., & Groot, A. T. (2010). Transcriptome analysis of the sex pheromone gland of the noctuid moth *Heliothis virescens. BMC Genomics, 11,* 29.

Wakayama, E. J., Dillwith, J. W., Howard, R. W., & Blomquist, G. J. (1984). Vitamin B12 levels in selected insects. *Insect Biochemistry, 14,* 175–179.

Wang, H.-L., Liénard, M. A., Zhao, C.-H., Wang, C.-Z., & Löfstedt, C. (2010a). Neofunctionalization in an ancestral insect desaturase lineage led to rare Δ6 pheromone signals in the Chinese tussah silkworm. *Insect Biochemistry and Molecular Biology, 40,* 742–751.

Wang, H.-L., Zhao, C.-H., Millar, J., Cardé, R., & Löfstedt, C. (2010b). Biosynthesis of unusual moth pheromone components involves two different pathways in the navel orangeworm, *Amyelois transitella. Journal of Chemical Ecology, 36,* 535–547.

Wei, W., Miyamoto, T., Endo, M., Murakawa, T., Pu, G.-Q., & Ando, T. (2003). Polyunsaturated hydrocarbons in the hemolymph: biosynthetic precursors of epoxy pheromones of geometrid and arctiid moths. *Insect Biochemistry and Molecular Biology, 33,* 397–405.

Wei, W., Yamamoto, M., Asatoa, T., Fujiia, T., Pub, G.-Q., & Ando, T. (2004). Selectivity and neuroendocrine regulation of the precursor uptake by pheromone glands from hemolymph in geometrid female moths, which secrete epoxyalkenyl sex pheromones. *Insect Biochemistry and Molecular Biology, 34,* 1215–1224.

Wicker, C., & Jallon, J. M. (1995a). Influence of ovary and ecdysteroids on pheromone biosynthesis in *Drosophila melanogaster* (Diptera, Drosophilidae). *European Journal of Entomology, 92,* 197–202.

Wicker, C., & Jallon J.-M. (1995b). Hormonal control of sex pheromone biosynthesis in *Drosophila melanogaster. Journal of Insect Physiology, 41,* 65–70.

Wicker-Thomas, C., Henriet, C., & Dallerac, R. (1997). Partial characterization of a fatty acid desaturase gene in *Drosophila melanogaster. Insect Biochemistry and Molecular Biology, 11,* 963–972.

Wicker-Thomas, C., Guenachi1, I., & Keita, Y. F. (2009). Contribution of oenocytes and pheromones to courtship behavior in *Drosophila. BMC Biochemistry, 10,* 21, doi:10.1186/1471-2091-10-2.

Wicker-Thomas, C., & Chertemps, T. (2010). Molecular biology and genetics of hydrocarbon production. In G. J. Blomquist, & A.-G. Bagneres (Eds.), *Insect Hydrocarbons: Biology, Biochemistry and Chemical Ecology* (pp. 53–74). Cambridge: Cambridge University Press.

Wiygul, G., Dickens, J. C., & Smith, J. W. (1990). Effect of juvenile hormone and beta-bisabolol on pheromone production in fat bodies of male boll weevils, *Anthonomus grandis* Boheman (Coleoptera; Curculionidae). *Comparative Biochemistry and Physiology, 95B,* 4898–4491.

Wolf, W. A., & Roelofs, W. L. (1987). Reinvestigation confirms action of Δ11-desaturase in spruce budworm moth sex pheromone biosynthesis. *Journal of Chemical Ecology, 13,* 1019–1027.

Wood, D. L. (1982). The role of pheromones, kairomones, and allomones in the host selection and colonization behavior of bark beetles Coleoptera. *Annual Review of Entomology, 27,* 411–446.

Wu, W. Q., Zhu, J. W., Millar, J., & Löfstedt, C. (1998). A comparative study of sex pheromone biosynthesis in two strains of the turnip moth, *Agrotis segetum,* producing different ratios of sex pheromone components. *Insect Biochemistry and Molecular Biology, 28,* 895–900.

Wyatt, G. R., & Davey, K. G. (1996). Cellular and molecular actions of juvenile hormone. II. Roles of juvenile hormone in adult insects. *Advances in Insect Physiology, 26,* 1–155.

Xue, B., Rooney, A. P., Kajikawa, M., Okada, N., & Roelofs, W. L. (2007). Novel sex pheromone desaturases in the genomes of corn borers generated through gene duplication and retroposon fusion. *Proceedings of the National Academy of Sciences. USA, 104,* 4467–4472.

Yang, H.-T., Chow, Y.-S., Peng, W.-K., & Hsu E.-L. (1998). Evidence for the site of female sex pheromone production in *Periplaneta americana. Journal of Chemical Ecology, 24,* 1831–1843.

Yoder, J. A., Denlinger, D. L., Dennis, M. W., & Kolattukudy, P. E. (1992). Enhancement of diapausing flesh fly puparia with additional hydrocarbons and evidence for alkane biosynthesis by a decarbonylation mechanism. *Insect Biochemistry and Molecular Biology, 22,* 237–243.

Yoshiga, T., Yokoyama, N., Imai, N., Ohnishi, A., Moto, K., & Matsumoto, S. (2002). cDNA cloning of calcineurin heterosubunits from the pheromone gland of the silkmoth, *Bombyx mori. Insect Biochemistry and Molecular Biology, 32,* 477–486.

Young, H. P., Bachmann, J. A.S., & Schal, C. (1999). Food intake in *Blattella germanica* (L.) nymphs affects hydrocarbon synthesis and its allocation in adults between epicuticle and reproduction. *Archives of Insect Biochemistry and Physiology, 41,* 214–224.

Zdarek, J., Nachman, R. J., & Hayes, T. K. (1998). Structure-activity relationships of insect neuropeptides of the pyrokinin/PBAN family and their selective action on pupariation in fleshfly (*Neobelleria bullata*) larvae (Diptera, Sarcophagidae). *European Journal of Entomology, 95,* 9–16.

Zhang, T., Zhang, L., Xu, W., & Shen, J. (2001). Cloning and characterization of the cDNA of diapause hormone-pheromone biosynthesis activating neuropeptide of *Helicoverpa armigera. GenBank Direct Submission.*

Zhao, C., Löfstedt, C., & Wang, X. (1990). Sex pheromone biosynthesis in the Asian corn borer *Ostrinia furnicalis* (II): biosynthesis of (*E*) and (*Z*)-12-tetradecenyl acetate involves Δ14 desaturation. *Archives of Insect Biochemistry and Physiology, 15*, 57–65.

Zhu, J., Millar, J., & Löfstedt, C. (1995). Hormonal regulation of sex pheromone biosynthesis in the turnip moth, *Agrotis segetum. Archives of Insect Biochemistry and Physiology, 30*, 41–59.

Zhu, J., Zhao, C. H., Lu, F., Bengtsson, M., & Löfstedt, C. (1996). Reductase specificity and the ratio regulation of E/Z isomers in pheromone biosynthesis of the European corn borer, *Ostrinia nubilalis* (Lepidoptera: Pyralidae). *Insect Biochemistry and Molecular Biology, 26*, 171–176.

Index

Note: Page numbers suffixed by *t* and *f* refer to tables and figures respectively.

Printed and bound by CPI Group (UK) Ltd, Croydon, CR0 4YY

03/10/2024

01040312-0018